## Fundamentals of Classical and Modern Error-Correcting Codes

Using easy-to-follow mathematics, this textbook provides comprehensive coverage of block codes and techniques for reliable communications and data storage. It covers major code designs and constructions from geometric, algebraic, and graph-theoretic points of view, decoding algorithms, error-control additive white Gaussian noise (AWGN) and erasure, and reliable data recovery. It simplifies a highly mathematical subject to a level that can be understood and applied with a minimum background in mathematics, provides step-by-step explanation of all covered topics, both fundamental and advanced, and includes plenty of practical illustrative examples to assist understanding. Numerous homework problems are included to strengthen student comprehension of new and abstract concepts, and a solution manual is available online for instructors. Modern developments, including polar codes, are also covered.

This is an essential textbook for senior undergraduates and graduates taking introductory coding courses, students taking advanced full-year graduate coding courses, and professionals working on coding for communications and data storage.

**Shu Lin** is Adjunct Professor in the Department of Electrical and Computer Engineering at the University of California, Davis, and an IEEE Life Fellow. He has authored and coauthored several books, including *LDPC Code Designs, Constructions, and Unification* (Cambridge University Press, 2016) and *Channel Codes: Classical and Modern* (Cambridge University Press, 2009).

**Juane Li** is Staff Systems Architect at Micron Technology Inc., San Jose, having previously completed her PhD at the University of California, Davis. She is also a coauthor of *LDPC Code Designs, Constructions, and Unification* (Cambridge University Press, 2016).

# Fundamentals of Classical and Modern Error-Correcting Codes

**SHU LIN**
*University of California, Davis*

**JUANE LI**
*Micron Technology Inc., San Jose*

# CAMBRIDGE
## UNIVERSITY PRESS

University Printing House, Cambridge CB2 8BS, United Kingdom

One Liberty Plaza, 20th Floor, New York, NY 10006, USA

477 Williamstown Road, Port Melbourne, VIC 3207, Australia

314–321, 3rd Floor, Plot 3, Splendor Forum, Jasola District Centre, New Delhi – 110025, India

103 Penang Road, #05–06/07, Visioncrest Commercial, Singapore 238467

Cambridge University Press is part of the University of Cambridge.

It furthers the University's mission by disseminating knowledge in the pursuit of
education, learning, and research at the highest international levels of excellence.

www.cambridge.org
Information on this title: www.cambridge.org/highereducation/isbn/9781316512623
DOI: 10.1017/9781009067928

First published 2022

Printed in the United Kingdom by TJ Books Limited, Padstow, Cornwall, 2022

*A catalogue record for this publication is available from the British Library.*

*Library of Congress Cataloging-in-Publication Data*
Names: Lin, Shu, 1937– author. | Li, Juane, author.
Title: Fundamentals of classical and modern error-correcting codes / Shu Lin, University of California,
   Davis, Juane Li, Micron Technology, San Jose.
Description: Cambridge, United Kingdom ; New York, NY, USA : Cambridge University Press, 2021. |
   Includes bibliographical references and index.
Identifiers: LCCN 2021025406 (print) | LCCN 2021025407 (ebook) |
   ISBN 9781316512623 (hardback) | ISBN 9781009067928 (epub)
Subjects: LCSH: Error-correcting codes (Information theory)
Classification: LCC TK5102.96 .L53 2021 (print) | LCC TK5102.96 (ebook) | DDC 005.7/2–dc23
LC record available at https://lccn.loc.gov/2021025406
LC ebook record available at https://lccn.loc.gov/2021025407

ISBN 978-1-316-51262-3 Hardback

Additional resources for this publication at www.cambridge.org/lin-li

# Contents

| | | |
|---|---|---|
| *List of Figures* | *page* **xiii** |
| *List of Tables* | **xxi** |
| *Preface* | **xxv** |
| *Acknowledgments* | **xxviii** |

**1 Coding for Reliable Digital Information Transmission and Storage** — **1**

| | | |
|---|---|---|
| 1.1 | Introduction | 2 |
| 1.2 | Categories of Error-Correcting Codes | 4 |
| 1.3 | Modulation and Demodulation | 6 |
| 1.4 | Hard-Decision and Soft-Decision Decodings | 8 |
| 1.5 | Maximum A Posteriori and Maximum Likelihood Decodings | 9 |
| 1.6 | Channel Capacity on Transmission Rate | 12 |
| 1.7 | Classification of Channel Errors | 13 |
| 1.8 | Error-Control Strategies | 14 |
| 1.9 | Measures of Performance | 16 |
| 1.10 | Contents of the Book | 18 |
| | References | 23 |

**2 Some Elements of Modern Algebra and Graphs** — **25**

| | | | |
|---|---|---|---|
| 2.1 | Groups | | 25 |
| | 2.1.1 | Basic Definitions and Concepts | 25 |
| | 2.1.2 | Finite Groups | 26 |
| | 2.1.3 | Subgroups and Cosets | 29 |
| 2.2 | Finite Fields | | 31 |
| | 2.2.1 | Basic Definitions and Concepts | 32 |
| | 2.2.2 | Prime Fields | 35 |
| | 2.2.3 | Finite Fields with Orders of Prime Powers | 36 |
| 2.3 | Polynomials over Galois Field GF(2) | | 39 |
| 2.4 | Construction of Galois Field GF($2^m$) | | 43 |
| 2.5 | Basic Properties and Structures of Galois Field GF($2^m$) | | 51 |
| 2.6 | Computations over Galois Field GF($2^m$) | | 58 |
| 2.7 | A General Construction of Finite Fields | | 59 |

| | | |
|---|---|---|
| 2.8 | Vector Spaces over Finite Fields | 60 |
| | 2.8.1 Basic Definitions and Concepts | 60 |
| | 2.8.2 Vector Spaces over Binary Field GF(2) | 62 |
| | 2.8.3 Vector Spaces over Nonbinary Field GF($q$) | 67 |
| 2.9 | Matrices over Finite Fields | 67 |
| | 2.9.1 Concepts of Matrices over GF(2) | 67 |
| | 2.9.2 Operations of Matrices over GF(2) | 69 |
| | 2.9.3 Matrices over Nonbinary Field GF($q$) | 73 |
| | 2.9.4 Determinants | 74 |
| 2.10 | Graphs | 78 |
| | 2.10.1 Basic Definitions and Concepts | 78 |
| | 2.10.2 Bipartite Graphs | 82 |
| | Problems | 83 |
| | References | 86 |

**3 Linear Block Codes** **89**
| | | |
|---|---|---|
| 3.1 | Definitions | 89 |
| 3.2 | Generator and Parity-Check Matrices | 90 |
| 3.3 | Systematic Linear Block Codes | 96 |
| 3.4 | Error Detection with Linear Block Codes | 99 |
| 3.5 | Syndrome and Error Patterns | 102 |
| 3.6 | Weight Distribution and Probability of Undetected Error | 103 |
| 3.7 | Minimum Distance of Linear Block Codes | 105 |
| 3.8 | Decoding of Linear Block Codes | 109 |
| 3.9 | Standard Array for Decoding Linear Block Codes | 110 |
| | 3.9.1 A Standard Array Decoding | 110 |
| | 3.9.2 Syndrome Decoding | 116 |
| 3.10 | Shortened and Extended Codes | 118 |
| 3.11 | Nonbinary Linear Block Codes | 120 |
| | Problems | 120 |
| | References | 123 |

**4 Binary Cyclic Codes** **125**
| | | |
|---|---|---|
| 4.1 | Characterization of Cyclic Codes | 125 |
| 4.2 | Structural Properties of Cyclic Codes | 127 |
| 4.3 | Existence of Cyclic Codes | 131 |
| 4.4 | Generator and Parity-Check Matrices of Cyclic Codes | 133 |
| 4.5 | Encoding of Cyclic Codes in Systematic Form | 136 |
| 4.6 | Syndrome Calculation and Error Detection | 142 |
| 4.7 | General Decoding of Cyclic Codes | 145 |
| 4.8 | Error-Trapping Decoding for Cyclic Codes | 150 |
| 4.9 | Shortened Cyclic Codes | 153 |
| 4.10 | Hamming Codes | 154 |
| 4.11 | Cyclic Redundancy Check Codes | 158 |
| 4.12 | Quadratic Residue Codes | 159 |

|  | 4.13 | Quasi-cyclic Codes | 161 |
|  |  | 4.13.1 Definitions and Fundamental Structures | 161 |
|  |  | 4.13.2 Generator and Parity-Check Matrices in Systematic Circulant Form | 163 |
|  |  | 4.13.3 Encoding of QC Codes | 164 |
|  |  | 4.13.4 Generator and Parity-Check Matrices in Semi-systematic Circulant Form | 168 |
|  |  | 4.13.5 Shortened QC Codes | 176 |
|  | 4.14 | Nonbinary Cyclic Codes | 176 |
|  | 4.15 | Remarks | 177 |
|  | Problems | | 177 |
|  | References | | 182 |

**5 BCH Codes** — **185**

|  | 5.1 | Primitive Binary BCH Codes | 185 |
|  | 5.2 | Structural Properties of BCH Codes | 190 |
|  | 5.3 | Minimum Distance of BCH Codes | 192 |
|  | 5.4 | Syndrome Computation and Error Detection | 196 |
|  | 5.5 | Syndromes and Error Patterns | 198 |
|  | 5.6 | Error-Location Polynomials of BCH Codes | 199 |
|  | 5.7 | A Procedure for Decoding BCH Codes | 200 |
|  | 5.8 | Berlekamp–Massey Iterative Algorithm | 201 |
|  | 5.9 | Simplification of Decoding Binary BCH Codes | 205 |
|  | 5.10 | Finding Error Locations and Error Correction | 211 |
|  | 5.11 | Nonprimitive Binary BCH Codes | 212 |
|  | 5.12 | Remarks | 216 |
|  | Problems | | 216 |
|  | References | | 218 |

**6 Nonbinary BCH Codes and Reed–Solomon Codes** — **220**

|  | 6.1 | Nonbinary Primitive BCH Codes | 221 |
|  | 6.2 | Decoding Steps of Nonbinary BCH Codes | 224 |
|  | 6.3 | Syndrome and Error Pattern of Nonbinary BCH Codes | 225 |
|  | 6.4 | Error-Location Polynomial of Nonbinary BCH Codes | 226 |
|  | 6.5 | Error-Value Evaluator | 231 |
|  | 6.6 | Decoding of Nonbinary BCH Codes | 232 |
|  | 6.7 | Key-Equation | 235 |
|  | 6.8 | Reed–Solomon Codes | 235 |
|  |  | 6.8.1 Primitive Reed–Solomon Codes | 236 |
|  |  | 6.8.2 Nonprimitive Reed–Solomon Codes | 237 |
|  | 6.9 | Decoding Reed–Solomon Codes with Berlekamp–Massey Iterative Algorithm | 238 |
|  | 6.10 | Euclidean Algorithm for Finding GCD of Two Polynomials | 243 |
|  | 6.11 | Solving the Key-Equation with Euclidean Algorithm | 247 |
|  | 6.12 | Weight Distribution and Probability of Undetected Error of Reed–Solomon Codes | 251 |

6.13 Remarks                                                        252
Problems                                                            252
References                                                          255

**7 Finite Geometries, Cyclic Finite-Geometry Codes, and
   Majority-Logic Decoding                                          258**
7.1 Fundamental Concepts of Finite Geometries                       259
7.2 Majority-Logic Decoding of Finite-Geometry Codes                261
7.3 Euclidean Geometries over Finite Fields                         266
    7.3.1 Basic Concepts and Properties                             266
    7.3.2 A Realization of Euclidean Geometries                     269
    7.3.3 Subgeometries of Euclidean Geometries                     275
7.4 Cyclic Codes Constructed Based on Euclidean Geometries          277
    7.4.1 Cyclic Codes on Two-Dimensional Euclidean Geometries      278
    7.4.2 Cyclic Codes on Multi-Dimensional Euclidean Geometries    284
7.5 Projective Geometries                                           288
7.6 Cyclic Codes Constructed Based on Projective Geometries         292
7.7 Remarks                                                         296
Problems                                                            297
References                                                          300

**8 Reed–Muller Codes                                               303**
8.1 A Review of Euclidean Geometries over GF(2)                     304
8.2 Constructing RM Codes from Euclidean Geometries over GF(2)      305
8.3 Encoding of RM Codes                                            311
8.4 Successive Retrieval of Information Symbols                     314
8.5 Majority-Logic Decoding through Successive Cancellations        319
8.6 Cyclic RM Codes                                                 324
8.7 Remarks                                                         325
Problems                                                            326
References                                                          327

**9 Some Coding Techniques                                          331**
9.1 Interleaving                                                    332
9.2 Direct Product                                                  334
9.3 Concatenation                                                   341
    9.3.1 Type-1 Serial Concatenation                               342
    9.3.2 Type-2 Serial Concatenation                               344
    9.3.3 Parallel Concatenation                                    346
9.4 $|\mathbf{u}|\mathbf{u} + \mathbf{v}|$-Construction             347
9.5 Kronecker Product                                               348
9.6 Automatic-Repeat-Request Schemes                                353
    9.6.1 Basic ARQ Schemes                                         353
    9.6.2 Mixed-Mode SR-ARQ Schemes                                 358
    9.6.3 Hybrid ARQ Schemes                                        359
Problems                                                            362
References                                                          363

**10 Correction of Error-Bursts and Erasures**     **367**
  10.1   Definitions and Structures of Burst-Error-Correcting Codes    368
  10.2   Decoding of Single Burst-Error-Correcting Cyclic Codes    370
  10.3   Fire Codes    373
  10.4   Short Optimal and Nearly Optimal Single
        Burst-Error-Correcting Cyclic Codes    376
  10.5   Interleaved Codes for Correcting Long Error-Bursts    377
  10.6   Product Codes for Correcting Error-Bursts    379
  10.7   Phased-Burst-Error-Correcting Codes    380
       10.7.1   Interleaved and Product Codes    380
       10.7.2   Codes Derived from RS Codes    380
       10.7.3   Burton Codes    381
  10.8   Characterization and Correction of Erasures    382
       10.8.1   Correction of Errors and Erasures over BSECs    383
       10.8.2   Correction of Erasures over BECs    385
       10.8.3   RM Codes for Correcting Random Erasures    388
  10.9   Correcting Erasure-Bursts over BBECs    390
       10.9.1   Cyclic Codes for Correcting Single Erasure-Burst    391
       10.9.2   Correction of Multiple Random Erasure-Bursts    394
  10.10 RS Codes for Correcting Random Errors and Erasures    394
  Problems    402
  References    403

**11 Introduction to Low-Density Parity-Check Codes**     **406**
  11.1   Definitions and Basic Concepts    407
  11.2   Graphical Representation of LDPC Codes    410
  11.3   Original Construction of LDPC Codes    414
       11.3.1   Gallager Codes    415
       11.3.2   MacKay Codes    416
  11.4   Decoding of LDPC Codes    416
       11.4.1   One-Step Majority-Logic Decoding    418
       11.4.2   Bit-Flipping Decoding    422
       11.4.3   Weighted One-Step Majority-Logic and Bit-Flipping
            Decodings    425
  11.5   Iterative Decoding Based on Belief-Propagation    427
       11.5.1   Message Passing    428
       11.5.2   Sum-Product Algorithm    429
       11.5.3   Min-Sum Algorithm    436
  11.6   Error Performance of LDPC Codes with Iterative Decoding    439
       11.6.1   Error-Floor    439
       11.6.2   Decoding Threshold    441
       11.6.3   Overall Performance and Its Determinating Factors    442
  11.7   Iterative Decoding of LDPC Codes over BECs    446
  11.8   Categories of LDPC Code Constructions    449
  11.9   Nonbinary LDPC Codes    450
  Problems    453
  References    456

**12 Cyclic and Quasi-cyclic LDPC Codes on Finite Geometries**      **464**
  12.1   Cyclic-FG-LDPC Codes                                           465
  12.2   A Complexity-Reduced Iterative Algorithm for Decoding
        Cyclic-FG-LDPC Codes                                            472
  12.3   QC-EG-LDPC Codes                                              480
  12.4   QC-PG-LDPC Codes                                              487
  12.5   Construction of QC-EG-LDPC Codes by CPM-Dispersion            489
  12.6   Masking Techniques                                            493
  12.7   Construction of QC-FG-LDPC Codes by
        Circulant-Decomposition                                         496
  12.8   A Complexity-Reduced Iterative Algorithm for Decoding
        QC-FG-LDPC Codes                                               503
  12.9   Remarks                                                       509
  Problems                                                            511
  References                                                          513

**13 Partial Geometries and Their Associated QC-LDPC Codes**       **518**
  13.1   CPM-Dispersions of Finite-Field Elements                     518
  13.2   Matrices with RC-Constrained Structure                        520
  13.3   Definitions and Structural Properties of Partial Geometries   522
  13.4   Partial Geometries Based on Prime-Order Cyclic Subgroups of
        Finite Fields and Their Associated QC-LDPC Codes               524
  13.5   Partial Geometries Based on Prime Fields and Their Associated
        QC-LDPC Codes                                                  531
  13.6   Partial Geometries Based on Balanced Incomplete Block Designs
        and Their Associated QC-LDPC Codes                             538
        13.6.1   BIBDs and Partial Geometries                          538
        13.6.2   Class-1 Bose $(N, M, t, r, 1)$-BIBDs                   541
        13.6.3   Class-2 Bose $(N, M, t, r, 1)$-BIBDs                   549
  13.7   Remarks                                                       556
  Problems                                                            556
  References                                                          559

**14 Quasi-cyclic LDPC Codes Based on Finite Fields**              **562**
  14.1   Construction of QC-LDPC Codes Based on CPM-Dispersion         563
  14.2   Construction of Type-I QC-LDPC Codes Based on Two Subsets
        of a Finite Field                                              564
  14.3   Construction of Type-II QC-LDPC Codes Based on Two Subsets
        of a Finite Field                                              576
  14.4   Masking-Matrix Design                                         580
        14.4.1   Type-1 Design                                         580
        14.4.2   Type-2 Design                                         582
        14.4.3   Type-3 Design                                         584
  14.5   A Search Algorithm for $2 \times 2/3 \times 3$ SM-Constrained Base
        Matrices for Constructing Rate-1/2 QC-LDPC Codes               586
  14.6   Designs of $2 \times 2$ SM-Constrained RS Base Matrices        590

14.7 Construction of Type-III QC-LDPC Codes Based on RS Codes 592
14.8 Construction of QC-RS-LDPC Codes with Girths at Least 8 598
14.9 A Special Class of QC-RS-LDPC Codes with Girth 8 603
14.10 Optimal Codes for Correcting Two Random CPM-Phased
        Erasure-Bursts 608
14.11 Globally Coupled LDPC Codes 612
14.12 Remarks 619
Problems 619
References 623

**15 Graph-Theoretic LDPC Codes** **628**
15.1 Protograph-Based LDPC Codes 629
15.2 A Matrix-Theoretic Method for Constructing Protograph-Based
        LDPC Codes 640
15.3 Masking Matrices as Protomatrices 655
15.4 LDPC Codes on Progressive Edge-Growth Algorithms 670
15.5 Remarks 676
Problems 677
References 679

**16 Collective Encoding and Soft-Decision Decoding of Cyclic
    Codes of Prime Lengths in Galois Fourier Transform
    Domain** **684**
16.1 Cyclic Codes of Prime Lengths and Their
        Hadamard Equivalents 685
16.2 Composing, Cascading, and Interleaving a Cyclic Code of
        Prime Length and Its Hadamard Equivalents 688
    16.2.1 Composing 688
    16.2.2 Cascading and Interleaving 689
16.3 Galois Fourier Transform of ICC Codes 692
16.4 Structural Properties of GFT-ICC Codes 694
16.5 Collective Encoding of GFT-ICC-LDPC Codes 696
16.6 Collective Iterative Soft-Decision Decoding of
        GFT-ICC-LDPC Codes 698
16.7 Analysis of the GFT-ISDD Scheme 700
    16.7.1 Performance Measurements 700
    16.7.2 Complexity 701
16.8 Joint Decoding of RS Codes with GFT-ISDD Scheme 701
16.9 Joint Decoding of BCH Codes with GFT-ISDD Scheme 706
16.10 Joint Decoding of QR Codes with GFT-ISDD Scheme 709
16.11 Code Shortening and Rate Reduction 710
    16.11.1 Shortened GFT-ICC Codes 710
    16.11.2 Reductions of Code Rate 713
16.12 Erasure Correction of GFT-ICC-RS-LDPC Codes 715
16.13 Remarks 717
Problems 717
References 719

**17 Polar Codes**                                                          **721**
   17.1  Kronecker Matrices and Their Structural Properties          721
   17.2  Kronecker Mappings and Their Logical Implementations         724
   17.3  Kronecker Vector Spaces and Codes                            735
   17.4  Definition and Polarized Encoding of Polar Codes             738
   17.5  Successive Information Retrieval from a Polarized Codeword    741
   17.6  Channel Polarization                                         744
       17.6.1  Some Elements of Information Theory                745
       17.6.2  Polarization Process                              747
       17.6.3  Channel Polarization Theorem                      760
   17.7  Construction of Polar Codes                                  766
   17.8  Successive Cancellation Decoding                             768
       17.8.1  SC Decoding of Polar Codes of Length $N = 2$      770
       17.8.2  SC Decoding of Polar Codes of Length $N = 4$      773
       17.8.3  SC Decoding of Polar Codes of Length $N = 2^{\ell}$   778
   17.9  Remarks                                                      779
   Problems                                                             780
   References                                                           781

**Appendix A  Factorization of $X^n + 1$ over $\mathbf{GF}(2)$**          **784**

**Appendix B  A $2 \times 2/3 \times 3$ SM-Constrained Masked Matrix
   Search Algorithm**                                                 **785**

**Appendix C  Proof of Theorem 14.4**                                  **786**

**Appendix D  The $2 \times 2$ CPM-Array Cycle Structure of the
   Tanner Graph of $C_{\mathrm{RS},n}(2, n)$**                       **791**

**Appendix E  Iterative Decoding Algorithm for Nonbinary
   LDPC Codes**                                                       **793**
   E.1  Introduction                                              793
   E.2  Algorithm Derivation                                      794
       E.2.1  VN Update                                        795
       E.2.2  CN Update: Complex Version                       795
       E.2.3  CN Update: Fast Hadamard Transform Version       796
   E.3  The Nonbinary LDPC Decoding Algorithm                     800

*Index*                                                                      **802**

# Figures

| 1.1 | Block diagram of a typical data-transmission (or data-storage) system | *page* 2 |
|---|---|---|
| 1.2 | A simplified model of a coded system | 4 |
| 1.3 | A binary convolutional encoder with $k = 1$, $n = 2$, and $m = 2$ | 6 |
| 1.4 | Transition probability diagrams: (a) BSC and (b) BI-DMC | 8 |
| 1.5 | A coded communication system with binary-input and $N$-ary-output discrete memoryless AWGN channel | 10 |
| 1.6 | The two-state Gilbert–Elliott model | 14 |
| 1.7 | Models for (a) BEC and (b) BSEC, where $p_e$ represents the erasure probability and $p_t$ represents the error probability | 15 |
| 1.8 | The BER performances of a coded communication using a $(127, 113)$ binary block code | 17 |
| 1.9 | Shannon limit $E_b/N_0$ (dB) as a function of code rate $R$ | 18 |
| 2.1 | A graph with seven nodes and eight edges | 79 |
| 2.2 | Simple graphs | 80 |
| 2.3 | A bipartite graph | 82 |
| 3.1 | Systematic format of codewords in an $(n, k)$ linear block code | 96 |
| 3.2 | Decoding regions for linear block codes | 110 |
| 4.1 | An encoding circuit for an $(n, k)$ cyclic code with generator polynomial $\mathbf{g}(X) = 1 + g_1 X + \cdots + g_{n-k-1} X^{n-k-1} + X^{n-k}$ | 139 |
| 4.2 | An encoding circuit for the $(7, 4)$ cyclic code given in Example 4.5 | 140 |
| 4.3 | Syndrome calculation circuit for an $(n, k)$ cyclic code | 143 |
| 4.4 | Syndrome calculation circuit for the $(7, 4)$ cyclic code in Example 4.7 | 143 |
| 4.5 | A general cyclic code decoder | 147 |
| 4.6 | A Meggitt decoder for the $(7, 4)$ cyclic code in Example 4.8 | 149 |
| 4.7 | An error-trapping decoder for cyclic codes | 151 |
| 4.8 | A Hamming code decoder | 156 |
| 4.9 | A CSRAA encoder circuit | 166 |
| 4.10 | A CSRAA-based QC code encoder | 167 |
| 5.1 | An error-location search and correction circuit | 212 |
| 6.1 | A decoding block diagram of a $q$-ary BCH decoder | 233 |
| 6.2 | Error performances of the $(255, 239)$ and $(255, 223)$ RS codes over $\mathrm{GF}(2^8)$ | 243 |

7.1   A five-point finite geometry   260

7.2   Error performances of the $(1057, 813)$ cyclic PG code
$C_{\mathrm{PG}}(2, 2^5)$ in Example 7.12 decoded with the OSMLD   296

7.3   A six-point finite geometry   297

9.1   A product codeword array $\mathbf{v}_{1 \times 2}$   335

9.2   Diagonal transmission of a codeword array in a cyclic
product code   340

9.3   A turbo codeword array   341

9.4   A type-1 serial concatenated coding system   342

9.5   A type-2 serial concatenated coding system   344

9.6   A parallel concatenated coding system   346

9.7   Stop-and-wait ARQ   354

9.8   Go-back-$N$ ARQ with $N = 4$   355

9.9   Selective-repeat ARQ   355

9.10  An SR/GBN-ARQ scheme with $\lambda = 1$ and $N = 4$   359

10.1  An error-trapping decoder for $l$-burst-error-correcting
cyclic codes   372

10.2  An error-trapping decoder for the $(5, 1, 1)$-Fire code
in Example 10.1   375

10.3  Mathematical models: (a) BEC and (b) BSEC   383

11.1  The Tanner graph of the $(10, 6)$ LDPC code given in Example 11.1   413

11.2  The Tanner graph of the $(15, 7)$ LDPC code given in Example 11.2   413

11.3  Message passing from VN $v_j$ to its neighbor (or adjacent) CNs   429

11.4  Message passing from CN $c_i$ to its neighbor (or adjacent) VNs   429

11.5  A VN decoder in an SPA decoder   430

11.6  A CN decoder in an SPA decoder   431

11.7  A Tanner graph with a cycle of length 4 and message passing   432

11.8  A plot of the $\phi(x)$ function together with its approximates
$2e^{-x}$ and $\log(x/2)$   435

11.9  The BER performances of the $(4095, 3367)$ LDPC code given
in Example 11.12 decoded with the OSML, BF, weighted BF,
MSA, and SPA decodings   438

11.10 The error-floor phenomenon   439

11.11 The BER and BLER performances of the $(3934, 3653)$ LDPC
code given in Example 11.13 decoded with SPA and MSA   440

11.12 (a) The Tanner graph of the $(10, 6)$ LDPC code given in
Example 11.1, (b) a $(4, 2)$ trapping set, and (c) a $(3, 2)$
elementary trapping set   443

11.13 The BER and BLER performances of the $(4095, 3367)$ LDPC
code given in Example 11.12 decoded with 5, 10, 50, and 100
iterations of the MSA   447

11.14 The UEBR and UEBLR performances of the $(4095, 3367)$
cyclic finite-geometry LDPC code given in Example 11.12
over a BEC   448

11.15 Two stopping sets of the $(10, 6)$ LDPC code given in Example 11.1   448

11.16 The Tanner graph of the 8-ary $(10,5)$ LDPC code given in Example 11.14     452

12.1 The BER and BLER performances of the $(1023,781)$ cyclic-EG-LDPC code $C_{\mathrm{EG,cyc}}(2,2^5)$ given in Example 12.1 decoded with 5, 10, and 50 iterations: (a) SPA and (b) MSA     467

12.2 The UEBR and UEBLR performances of the $(1023,781)$ cyclic-EG-LDPC code $C_{\mathrm{EG,cyc}}(2,2^5)$ given in Example 12.1 over the BEC     468

12.3 The error performances of the $(4095,3367)$ cyclic-EG-LDPC code given in Example 12.3 over: (a) AWGN channel and (b) BEC     470

12.4 The BER and BLER performances of the $(1057,813)$ cyclic-PG-LDPC code given in Example 12.4     471

12.5 The BER and BLER performances of the $(4095,3367)$ cyclic-EG-LDPC code in Example 12.5 using the RMSA with $\ell = 819$ and $f = 5$     478

12.6 The BER and BLER performances of the $(4095,3367)$ cyclic-EG-LDPC code decoded using the RMSA with $\ell = 1$ and $f = 16\,380$ given in Example 12.6     479

12.7 The BER and BLER performances of the $(1057,813)$ cyclic-PG-LDPC code given in Example 12.7 decoded with the RMSA using two grouping sizes: (a) $\ell = 151$, $f = 16$, and (b) $\ell = 244$, $f = 8$     481

12.8 The BER and BLER performances of the $(1023,909)$ QC-EG-LDPC code given in Example 12.9     486

12.9 The BER and BLER performances of the $(3780,3543)$ QC-EG-LDPC code given in Example 12.10     487

12.10 The BER and BLER performances of the $(906,662)$ QC-PG-LDPC code given in Example 12.11     488

12.11 The BER and BLER performances of the $(2016,1779)$ QC-EG-LDPC code given in Example 12.13     492

12.12 The performances of the $(16\,384,15\,363)$ QC-EG-LDPC code given in Example 12.14 over: (a) AWGN channel and (b) BEC     494

12.13 The BER and BLER performances of the unmasked $(2048,1027)$ and the masked $(2048,1024)$ QC-EG-LDPC codes given in Example 12.15     496

12.14 The BER and BLER performances of the $(8176,7156)$ QC-EG-LDPC code given in Example 12.16: (a) MSA and (b) hardware MSA decoder     501

12.15 The BER and BLER performances of the $(4088,2044)$ QC-EG-LDPC code given in Example 12.17     502

12.16 The BER and BLER performances of the $(4599,3068)$ QC-EG-LDPC code given in Example 12.18     503

12.17 The BER and BLER performances of the $(2016,1779)$ QC-EG-LDPC code $C_{\mathrm{EG,qc}}(4,32)$ given in Example 12.20 decoded with the CPM-RMSA of different sizes of decoding matrices: (a) $\ell = 1$ and (b) $\ell = 3$     510

13.1 The performances of the $(961, 840)$ QC-RS-PaG-LDPC code given in Example 13.4 over: (a) AWGN channel and (b) BEC 529

13.2 The BER and BLER performances of the $(7921, 7568)$ QC-RS-PaG-LDPC code $C_{\mathrm{RS,PaG},c}(4, 89)$ and the $(7921, 7566)$ PEG code $C_{\mathrm{peg}}$ given in Example 13.5 531

13.3 The BER and BLER performances of the two QC-RS-PaG-LDPC codes given in Example 13.6: (a) $(5696, 5343)$ and (b) $(2848, 2495)$ 532

13.4 (a) The BER and BLER performances of the $(11\,584, 10\,863)$ QC-PaG-LDPC code $C_{\mathrm{PaG},p}(4, 64)$ in Example 13.8 and (b) the BER performances of the four codes in Example 13.8 537

13.5 The performances of the $(1016, 508)$ QC-PaG-LDPC code given in Example 13.9 over: (a) AWGN channel and (b) BEC 539

13.6 The BER and BLER performances of the $(776, 680)$ QC-BIBD-PaG-LDPC code $C_{\mathrm{PaG,BIBD},1}(4, 32)$ given in Example 13.12 545

13.7 The performances of the $(44\,713, 41\,781)$ and $(23\,456, 20\,524)$ QC-BIBD-PaG-LDPC codes given in Example 13.13 over: (a) AWGN channel and (b) BEC 548

13.8 The BER and BLER performances of the four QC-BIBD-PaG-LDPC codes given in Example 13.14 549

13.9 The BER and BLER performances of the $(3934, 3653)$ QC-BIBD-PaG-LDPC code given in Example 13.15 552

13.10 The BER and BLER performances of the two QC-BIBD-PaG-LDPC codes given in Example 13.16 over the AWGN channel: (a) $(20\,512, 17\,951)$ and (b) $(5128, 2564)$ 554

13.11 The UEBR and UEBLR performances of the two QC-BIBD-PaG-LDPC codes given in Example 13.16 over the BEC: (a) $(20\,512, 17\,951)$ and (b) $(5128, 2564)$ 555

14.1 The BER and BLER performances of the $(180, 128)$ QC-LDPC code given in Example 14.2 567

14.2 The performances of the $(5080, 4589)$ QC-LDPC code $C_{\mathrm{s,qc}}(4, 40)$ given in Example 14.3 over: (a) AWGN channel and (b) BEC 569

14.3 The performances of the unmasked $(1016, 525)$ and masked $(1016, 508)$ QC-LDPC codes given in Example 14.3 over: (a) AWGN channel and (b) BEC 570

14.4 The BER performances of nine QC-LDPC codes in Example 14.4 572

14.5 The BER and BLER performances of the $(16\,120, 15\,345)$ QC-LDPC code given in Example 14.5 573

14.6 The BER and BLER performances of the three QC-LDPC codes given in Example 14.6 575

14.7 The BER and BLER performances of the $(4064, 3572)$ QC-LDPC code $C_{\mathrm{s,qc}}(4, 32)$ given in Example 14.7 577

14.8   The BER and BLER performances of the $(8192, 7171)$
       QC-LDPC code given in Example 14.8                                578

14.9   The BER and BLER performances of the $(3440, 2755)$ and
       $(6880, 6195)$ QC-LDPC codes given in Example 14.9               580

14.10  The BER performances of the nine QC-LDPC codes in
       Example 14.10                                                    583

14.11  The BER and BLER performances of the unmasked
       $(3960, 2643)$ and the masked $(3960, 2640)$ QC-LDPC codes in
       Example 14.11                                                    584

14.12  The BER and BLER performances of the unmasked
       $(5280, 3305)$ and the masked $(5280, 3302)$ QC-LDPC codes in
       Example 14.12                                                    586

14.13  The BER and BLER performances of the $(504, 252)$
       QC-LDPC code $C_{s,qc,mask}(4, 8)$ and two other $(504, 252)$
       LDPC codes given in Example 14.13                                590

14.14  The performances of the $(32\,704, 30\,153)$ QC-RS-LDPC code
       given in Example 14.14 over: (a) AWGN channel and (b) BEC        595

14.15  The BER and BLER performances of the two QC-RS-LDPC
       codes given in Example 14.15                                     596

14.16  The BER and BLER performances of the $(5696, 4985)$
       QC-RS-LDPC code given in Example 14.16                           598

14.17  The BER and BLER performances of the unmasked
       $(4088, 2047)$ and the masked $(4088, 2044)$ QC-RS-LDPC codes
       given in Example 14.17                                           601

14.18  The BER and BLER performances of the unmasked $(680, 343)$
       and the masked $(680, 340)$ QC-RS-LDPC codes given in
       Example 14.18                                                    604

14.19  The BER and BLER performances of the unmasked
       $(2040, 1025)$ and masked $(2040, 1020)$ QC-RS-LDPC codes
       given in Example 14.19                                           606

14.20  The BER and BLER performances of the four QC-RS-LDPC
       codes given in Example 14.20                                     608

14.21  The performances of the $(4672, 4383)$ QC-RS-LDPC code in
       Example 14.22 over: (a) AWGN channel and (b) BEC                 613

14.22  The performances of the $(15\,876, 14\,871)$ and $(15\,876, 13\,494)$
       CN-QC-GC-LDPC codes given in Examples 14.23 and 14.24,
       respectively, over: (a) AWGN channel and (b) BEC                 617

15.1   (a) The protograph $\mathcal{G}_{ptg}$, (b) three copies of the protograph
       $\mathcal{G}_{ptg}$ and the grouping of their VNs and CNs, and (c) the
       connected bipartite graph $\mathcal{G}_{ptg}(3, 3)$ given in Example 15.1   634

15.2   (a) The protograph $\mathcal{G}_{ptg}$, (b) the bipartite graph $\mathcal{G}_{ptg,2}$, (c)
       the performances of the $(680, 340)$ QC-PTG-LDPC code, and
       (d) the performances of the $(4088, 2044)$ QC-PTG-LDPC code
       given in Example 15.2                                            638

15.3 The UEBR and UEBLR performances of the $(680, 340)$ and $(4088, 2044)$ QC-PTG-LDPC codes given in Example 15.2 over BEC 639

15.4 (a) The protograph $\mathcal{G}_{\text{ptg}}$ and (b) the BER and BLER performances of the $(5792, 2896)$ QC-PTG-LDPC code given in Example 15.3 641

15.5 The performances of the $(2640, 1320)$ QC-PTG-LDPC code given in Example 15.6 over: (a) AWGN channel and (b) BEC 652

15.6 The protograph $\mathcal{G}_0$ specified by the protomatrix $\mathbf{B}_0$ given by (15.29) 653

15.7 (a) The protograph $\mathcal{G}_{\text{ptg}}$ and (b) the BER and BLER performances of the $(3060, 2040)$ QC-PTG-LDPC code given in Example 15.7 654

15.8 The BER and BLER performances of the $(8176, 7156)$ QC-PTG-LDPC code given in Example 15.8 657

15.9 The BER and BLER performances of the $(4080, 3060)$ QC-PTG-LDPC code given in Example 15.10 660

15.10 (a) The BER and BLER performances of the $(3969, 3213)$ QC-PTG-LDPC code $C_{\text{ptg,qc}}$ given in Example 15.11 and the $(3969, 3213)^*$ QC-LDPC code $C_{\text{qc}}^*$ in [33] and (b) the BER and BLER performances of the $(8001, 6477)$ QC-PTG-LDPC code given in Example 15.11 671

15.11 Tree representation of the neighborhood $N_{v_i}^{(l)}$ within depth $l$ of a VN $v_i$ 673

15.12 The BER and BLER performances of the $(4088, 2044)$ LDPC codes constructed by the PEG and ACE-PEG algorithms in Example 15.12 676

15.13 The Tanner graph of an SC-LDPC code 676

16.1 A collective encoding scheme for a GFT-ICC-LDPC code $C_{\text{LDPC}}$ 696

16.2 A collective iterative soft-decision decoding scheme for a GFT-ICC-LDPC code $C_{\text{LDPC}}$ 699

16.3 The FER and BLER performances of the $(31, 25)$ RS code given in Example 16.5 decoded by the GFT-ISDD/MSA and other decoding algorithms 704

16.4 (a) The FER and BLER performances of the $(127, 119)$ RS code given in Example 16.6 decoded by the GFT-ISDD/MSA and other decoding algorithms and (b) the average number of iterations required to decode the $(127, 119)$ RS code in Example 16.6 vs. $E_b/N_0$ (dB) 705

16.5 The BLER performances of the $(127, 113)$ BCH code given in Example 16.7 decoded by the GFT-ISDD/MSA, BM-HDDA, and MLD 707

16.6 The BLER performances of the $(127, 120)$ Hamming code given in Example 16.8 decoded by the GFT-ISDD/MSA, the BM-HDDA, and MLD 708

16.7 The BLER performances of the $(23, 12)$ Golay code given in
Example 16.9 decoded by the GFT-ISDD/MSA, HDDA, and MLD 710

16.8 (a) The BLER performances of the shortened $(64, 58)$ RS code
over $GF(2^7)$ and the $(127, 121)$ RS code over $GF(2^7)$ and (b)
the BLER performances of the shortened $(32, 26)$ RS code
over $GF(2^7)$ and the $(127, 121)$ RS code over $GF(2^7)$ given in
Example 16.10 decoded by the GFT-ISDD/MSA and the
BM-HDDA 714

16.9 The BLER performances of the $(16\,129, 11\,970)$ QC-LDPC
code $C_{\mathrm{BCH,LDPC}}(6, 6, \ldots, 6)$ in Example 16.11 decoded by the
GFT-ISDD/MSA 715

17.1 The 1-fold Kronecker mapping circuit 725

17.2 The 2-fold Kronecker mapping circuit 728

17.3 The 3-fold Kronecker mapping circuit 730

17.4 The $\ell$-fold Kronecker mapping circuit 733

17.5 (a) Two identical and independent channels $W$ and (b) a
combined 1-fold vector channel $W^2$ 747

17.6 (a) A combined vector channel with $W^+$ as the base channel
and (b) a combined vector channel with $W^-$ as the base channel 751

17.7 (a) A combined vector channel $W^4$ and (b) the combined
vector channel $W^4$ after rewire 752

17.8 A combined vector channel $W^8$ 755

17.9 An $\ell$-level channel polarization tree 759

17.10 The information transmission using the $(8, 4)$ polar code
$C_{\mathrm{p}}(4, 3)$ given in Example 17.8 over the BEC vector channel
$W^8$ with the base BEC channel $\mathrm{BEC}(0.5)$ 761

17.11 The bit-coordinate channel capacities for BEC channel
polarization with (a) $N = 16$ and (b) $N = 64$ for $\mathrm{BEC}(0.5)$ 762

17.12 The bit-coordinate channel capacities for BEC channel
polarization with (a) $N = 256$ and (b) $N = 1024$ for $\mathrm{BEC}(0.5)$ 763

17.13 The bit-coordinate channel capacities for BEC channel
polarization after sorting with (a) $N = 16$ and (b) $N = 64$ for
$\mathrm{BEC}(0.5)$ 764

17.14 The bit-coordinate channel capacities for BEC channel
polarization after sorting with (a) $N = 256$ and (b) $N = 1024$
for $\mathrm{BEC}(0.5)$ 765

17.15 Block diagram of a polar coded system 769

17.16 A block diagram of the SC decoding process for a polar code
of length $N = 2^\ell$ 770

17.17 An SC decoder for a polar code of length $N = 2$ 771

17.18 A tree structure of an SC decoder of size $N = 2$ 773

17.19 An SC decoder with size $N = 4$ 774

17.20 The SC decoding process of a polar code of length $N = 4$ 774

17.21 Message-passing and decision trees in decoding a polar code
of length 4 with SC decoding 776

17.22 The UEBR and UEBLR of the $(256, 128)$ polar code over
        BEC decoded with the SC decoding in Example 17.20            779
C.1    Location patterns of six configurations of a cycle-6 $C_6$      789
E.1    Diagram of implementation of $\mathbf{P} = \mathbf{p}\mathbf{H}_{16}$              799
E.2    Diagram of implementation of $\mathbf{P} = \mathbf{p}_0^7\mathbf{H}_8$              799
E.3    Diagram of the fast Hadamard transform                          800

# Tables

| | | |
|---|---|---|
| 1.1 | A binary block code with $k = 4$ and $n = 7$ | *page* 5 |
| 1.2 | Shannon limits, $E_b/N_0$ (dB), of a binary-input continuous-output AWGN channel with BPSK signaling for various code rates | 19 |
| 2.1 | The additive group $G = \{0, 1\}$ with modulo-2 addition | 27 |
| 2.2 | The additive group $G = \{0, 1, 2, 3, 4, 5, 6\}$ with modulo-7 addition | 27 |
| 2.3 | The multiplicative group $G = \{1, 2, 3, 4, 5, 6\}$ with modulo-7 multiplication | 28 |
| 2.4 | The additive group $G = \{0, 1, 2, 3, 4, 5, 6, 7\}$ under modulo-8 addition | 30 |
| 2.5 | A subgroup $S = \{0, 2, 4, 6\}$ of $G$ given in Table 2.4 under modulo-8 addition | 30 |
| 2.6 | The prime field GF(2) under modulo-2 addition and multiplication | 35 |
| 2.7 | The field GF(7) under modulo-7 addition and multiplication | 36 |
| 2.8 | A list of primitive polynomials over GF(2) | 43 |
| 2.9 | GF($2^4$) generated by $p(X) = 1 + X + X^4$ over GF(2) | 50 |
| 2.10 | GF($2^3$) generated by $p(X) = 1 + X + X^3$ over GF(2) | 56 |
| 2.11 | GF($2^6$) generated by $p_1(X) = 1 + X + X^6$ over GF(2) | 56 |
| 2.12 | The prime field GF(3) under modulo-3 addition and multiplication | 60 |
| 2.13 | GF($3^2$) generated by $p(X) = 2 + X + X^2$ over GF(3) | 60 |
| 2.14 | The vector space $\mathbf{V}_5$ over GF(2) given in Example 2.18 | 66 |
| 2.15 | A subspace $\mathbf{S}$ and its dual space $\mathbf{S}_d$ in $\mathbf{V}_5$ given in Example 2.18 | 66 |
| 3.1 | A codebook for a $(6, 3)$ linear block code over GF(2) | 91 |
| 3.2 | The dual code $C_d$ of the $(6, 3)$ linear block code $C$ given by Table 3.1 | 94 |
| 3.3 | The $(7, 4)$ linear block code generated by the matrix $\mathbf{G}$ given by (3.13) | 95 |
| 3.4 | The $(7, 3)$ linear block code generated by the matrix $\mathbf{H}$ (as a generator matrix) given by (3.14) | 95 |
| 3.5 | A standard array for an $(n, k)$ linear block code | 111 |
| 3.6 | A standard array for the $(6, 3)$ code given in Example 3.11 | 111 |
| 3.7 | A look-up table for syndrome decoding | 116 |
| 3.8 | A syndrome look-up decoding table for the $(6, 3)$ linear block code | 117 |
| 4.1 | A $(7, 4)$ cyclic code generated by $\mathbf{g}(X) = 1 + X + X^3$ | 133 |

xxi

4.2   The $(7, 4)$ systematic cyclic code generated by
      $\mathbf{g}(X) = 1 + X + X^3$ in Example 4.4                                      138
4.3   The register contents of syndrome calculation circuit of the
      $(7, 4)$ cyclic code with $\mathbf{r} = (0\ 0\ 0\ 1\ 0\ 1\ 1)$                   143
4.4   A syndrome look-up decoding table for the $(7, 4)$ cyclic code
      given in Example 4.8                                                             148
4.5   Decoding steps for the $(7, 4)$ cyclic code given in Example 4.8                 149
4.6   A list of standardized CRC codes                                                 158
4.7   A list of QR codes                                                               160
4.8   The eight codewords of the $(6, 3)$ QC code $C_{qc}$ given in
      Example 4.12                                                                     162
4.9   The eight codewords of the $(6, 3)$ QC code $C_{qc}$ given in
      Example 4.13                                                                     163
5.1   Binary primitive BCH codes of lengths less than $2^{10}$                         187
5.2   Weight distribution of the dual code of a
      double-error-correcting binary primitive BCH code of length
      $n = 2^m - 1$, where $m \geq 3$ and $m$ is odd                                   195
5.3   Weight distribution of the dual code of a
      double-error-correcting binary primitive BCH code of length
      $n = 2^m - 1$, where $m \geq 4$ and $m$ is even                                  195
5.4   Weight distribution of the dual code of a
      triple-error-correcting binary primitive BCH code of length
      $n = 2^m - 1$, where $m \geq 5$ and $m$ is odd                                   196
5.5   Weight distribution of the dual code of a
      triple-error-correcting binary primitive BCH code of length
      $n = 2^m - 1$, where $m \geq 6$ and $m$ is even                                  196
5.6   Berlekamp–Massey iterative procedure for finding the
      error-location polynomial of a BCH code                                         203
5.7   Steps for finding the error-location polynomial of
      $\mathbf{r}(X) = X^3 + X^5 + X^{12}$ for the $(15, 5)$ BCH code given in
      Example 5.3                                                                      206
5.8   Steps for finding the error-location polynomial of
      $\mathbf{r}(X) = X + X^3 + X^5 + X^7$ for the $(15, 5)$ BCH code given in
      Example 5.3                                                                      206
5.9   A simplified Berlekamp–Massey iterative procedure for finding
      the error-location polynomial of a binary BCH code                              207
5.10  Steps for finding the error-location polynomial of the binary
      $(15, 5)$ BCH code given in Example 5.4                                          207
5.11  GF($2^5$) generated by the primitive polynomial
      $\mathbf{p}(X) = 1 + X^2 + X^5$                                                  209
5.12  Steps for finding the error-location polynomial of
      $\mathbf{r}(X) = 1 + X^{12} + X^{20}$ for the binary $(31, 16)$ BCH code in
      Example 5.5                                                                      210
5.13  Steps for finding the error-location polynomial of
      $\mathbf{r}(X) = X + X^3 + X^5 + X^7$ for the binary $(31, 16)$ BCH code
      in Example 5.5                                                                   210

5.14    GF($2^6$) generated by $\mathbf{p}(X) = 1 + X + X^6$ over GF(2)    214

6.1    Berlekamp–Massey iterative procedure for finding the error-location polynomial of a nonbinary BCH code    230

6.2    Steps for finding the error-location polynomial of the received polynomial $\mathbf{r}(X)$ for the 4-ary $(15, 9)$ BCH code in Example 6.3    230

6.3    Steps for finding the error-location polynomial of the received polynomial $\mathbf{r}^*(X)$ for the 4-ary $(15, 9)$ BCH code in Example 6.3    231

6.4    GF($2^3$) generated by $\mathbf{p}(X) = 1 + X + X^3$ over GF(2)    237

6.5    Steps for finding the error-location polynomial of $\mathbf{r}(X)$ for the 16-ary $(15, 9)$ RS code over GF($2^4$) in Example 6.7    239

6.6    Steps for finding the error-location polynomial of $\mathbf{r}(X)$ for the 32-ary $(31, 25)$ RS code in Example 6.8    241

6.7    Euclidean algorithm for finding the GCD of two polynomials $\mathbf{a}(X)$ and $\mathbf{b}(X)$ over GF($q$)    246

6.8    Euclidean iterative algorithm for finding the GCD of two polynomials $\mathbf{a}(X)$ and $\mathbf{b}(X)$ over GF $(2^4)$ in Example 6.9    246

6.9    Euclidean algorithm for finding error-location polynomial $\boldsymbol{\sigma}(X)$ and error-value evaluator $\mathbf{Z}_0(X)$    248

6.10    Euclidean iterative algorithm for decoding the 16-ary $(15, 9)$ RS code given in Example 6.10    249

6.11    Euclidean iterative algorithm for decoding the 32-ary $(31, 25)$ RS code given in Example 6.11    250

7.1    GF($2^4$) generated by $\mathbf{p}(X) = 1 + X + X^4$ over GF(2)    272

7.2    GF($2^4$) as an extension field of GF($2^2$) $= \{0, 1, \beta, \beta^2\}$ with $\beta = \alpha^5$    272

7.3    GF($2^6$) as an extension field of GF($2^2$) $= \{0, 1, \beta, \beta^2\}$ with $\beta = \alpha^{21}$    273

7.4    Four parallel bundles of lines of EG($3, 2^2$) over GF($2^2$)    274

7.5    Two cyclic classes of lines of EG$^*$($3, 2^2$) over GF($2^2$)    276

7.6    A list of two-dimensional EG codes    281

7.7    Lines of the projective geometry PG($2, 2^2$) over GF($2^2$)    292

7.8    A list of two-dimensional PG codes    294

9.1    The 16 codewords of the $(7, 4)$ Hamming code    337

10.1    A list of Fire codes which have true burst-error-correcting capabilities larger than their designed values [27]    375

10.2    A list of optimal and nearly optimal cyclic and shortened cyclic $l$-burst-error-correcting codes [25]    376

10.3    Euclidean steps to find the solution $(\boldsymbol{\sigma}(X), \mathbf{Z}_0(X))$ of the key-equation for decoding the $(15, 9)$ RS code in Example 10.15    399

11.1    Decoding thresholds under IDBP for LDPC codes over AWGN channels    441

12.1    GF($2^4$) as an extension field of GF($2^2$) $= \{0, 1, \beta, \beta^2\}$ with $\beta = \alpha^5$    484

13.1    A list of $b$s for which $12b + 1$ is a prime and the field GF($12b + 1$) satisfies the condition given by (13.11)    542

13.2    A list of $b$s for which $20b + 1$ is a prime and the field GF($20b + 1$) satisfies the condition given by (13.19)    550

14.1    The nine QC-LDPC codes given in Example 14.4    571

14.2    The nine QC-LDPC codes constructed in Example 14.10    582

14.3 Numbers of $4 \times 8$ masked base matrices over various fields
$\mathrm{GF}(q)$ that give rate-1/2 QC-LDPC codes with girths 8 or 10
for $\eta = 1$   588

14.4 Numbers of $4 \times 8$ masked matrices over $\mathrm{GF}(331)$ that give
rate-1/2 QC-LDPC codes with girths 8 and 10 for different
choices of $\eta$s   589

15.1 The nonzero constituent matrices of the protomatrix $\mathbf{B}_{\mathrm{ptg},2}$
and the locations of their 1-entries with decomposition factor
$k = 85$ given in Example 15.5   648

15.2 The nonzero constituent matrices of the protomatrix $\mathbf{B}_{\mathrm{ptg},2}$
and the locations of their 1-entries with decomposition factor
$k = 511$ given in Example 15.5   649

15.3 The nonzero constituent matrices of the protomatrix $\mathbf{B}_{\mathrm{ptg},2}$
and the locations of their 1-entries with decomposition factor
$k = 330$ given in Example 15.6   650

15.4 The nonzero constituent matrices of the protomatrix $\mathbf{B}_{\mathrm{ptg},2}$
and the locations of their 1-entries with decomposition factor
$k = 255$ given in Example 15.7   653

15.5 The nonzero constituent matrices of the protomatrix $\mathbf{B}_{\mathrm{ptg}}$ and
the locations of their 1-entries with decomposition factor
$k = 511$ given in Example 15.8   656

15.6 The generators of the circulants in the array $\mathbf{H}_{\mathrm{ptg},\mathrm{qc}}(511,511)$
given in Example 15.8 and the locations of their 1-entries   656

15.7 The nonzero constituent matrices of the protomatrix $\mathbf{B}_{\mathrm{ptg}}$ and
the locations of their 1-entries with decomposition factor
$k = 330$ given in Example 15.9   658

15.8 The nonzero constituent matrices of the protomatrix $\mathbf{B}_{\mathrm{ptg}}$ and
the locations of their 1-entries with decomposition factor
$k = 255$ given in Example 15.10   659

15.9 Column and row weight distributions of the $12 \times 63$
protomatrix $\mathbf{B}_{\mathrm{ptg}}$ used in Example 15.11   661

15.10 The nonzero constituent matrices of the protomatrix $\mathbf{B}_{\mathrm{ptg}}$ and
the locations of their 1-entries with decomposition factor
$k = 63$ given in Example 15.11   662

15.11 The entries of $\mathbf{H}_{\mathrm{ptg},\mathrm{qc}}(63,63)$ given in Example 15.11   665

15.12 The nonzero constituent matrices of the protomatrix $\mathbf{B}_{\mathrm{ptg}}$ and
the locations of their 1-entries with decomposition factor
$k = 127$ given in Example 15.11   666

15.13 The entries of $\mathbf{H}_{\mathrm{ptg},\mathrm{qc}}(127,127)$ given in Example 15.11   669

16.1 $\mathrm{GF}(2^3)$ generated by $\mathbf{p}(X) = 1 + X + X^3$ over $\mathrm{GF}(2)$   688

17.1 The 2-fold Kronecker mappings of the 16 4-tuples over $\mathrm{GF}(2)$   727

A.1 Factorization of $X^n + 1$ over $\mathrm{GF}(2)$ with $1 \leq n \leq 31$   784

# Preface

One of the serious problems in a digital data communication or storage system is the occurrence of errors caused by noise and interference in communication channels or imperfections in storage mediums. A major concern to the communication or storage-system designers is the control of these errors such that reliable transmission or storage of data can be achieved. In 1948, Shannon demonstrated in a landmark paper that by proper encoding and decoding of the data, errors induced by a noisy channel or imperfect storage medium can be reduced to any desired level without sacrificing the rate of information transmission or storage, as long as the information rate is less than the capacity of the channel or the storage medium. Since Shannon's work, a tremendous amount of research effort has been expended on the problems of devising efficient encoding and decoding methods and techniques for error control on noisy channels or imperfect storage mediums. As a result of this research effort, various efficient encoding and decoding methods and techniques have been developed to achieve the reliability required by today's explosive high-speed and large-volume digital communication and storage systems.

Much of the work on error-correcting codes (or error-control codes) developed since 1948 is highly mathematical in nature, and a thorough understanding requires an extensive background in modern algebra, combinatorial mathematics, and graph theory. This requirement may impede senior and first-year graduate students in electrical and computer engineering who are interested in learning and pursuing research in coding theory, and practicing engineers in industry who are interested in applying error-control coding techniques to practical systems.

One of the objectives of this book is to bring this highly complex material down to a reasonably simple level such that it can be understood and applied with a minimum background in mathematics. To achieve this objective, we take a middle ground between mathematical rigor and heuristic reasoning as the first author did in his first book on the introduction to error-correcting codes published in 1970. Because of the extensive developments in error-correcting codes over the past 50 years, it is not possible to include certain categories of error-correcting codes in this book. The main coverage of this book is the fundamental and essential aspects of codes with block structure, called block codes, and their up-to-date developments in construction, encoding, and decoding techniques. The presentation of these subjects is intended to be comprehensive. In presenting every step of each topic in the book, illustrative examples are given to assist the readers to follow and fully understand the topic with a minimum barrier. Furthermore, derivations and long proofs that are not helpful in illustration of a topic are avoided. Long essential derivations or proofs of any topic are put in the appendices or referred to in published article(s).

In the following, a brief description of major coverage in each chapter is presented. Chapter 1 gives a brief overview of coding for error control in information transmission and data storage. Chapter 2 provides the readers with an elementary knowledge of modern algebra and graph theory that will aid in understanding the fundamental and essential aspects of error-correcting codes to be developed in the other chapters of the book. Chapter 3 gives an introduction to block codes with linear structure, called linear block codes, their structural properties, and general decoding methods. Chapter 4 introduces two special categories of linear block codes with cyclic and quasi-cyclic structures, called cyclic codes and quasi-cyclic codes, respectively. Also presented in this chapter are two small classes of cyclic codes, known as Hamming and quadratic-residue (QR) codes.

Chapters 5 and 6 present two well-known classes of cyclic codes constructed based on finite fields, called the Bose–Chaudhuri–Hocquenghem (BCH) and the Reed–Solomon (RS) codes. These two classes of cyclic codes have been widely used in digital-communication and data-storage systems. Major topics covered in these two chapters include code constructions, characterizations, and decoding algorithms. Presented in Chapter 7 are two classes of cyclic codes constructed based on two categories of finite geometries, named Euclidean and projective geometries. Finite-geometry codes in these two classes are low-density parity-check (LDPC) codes, which can be decoded with iterative soft-decision algorithms based on belief-propagation to achieve good error performance with practical implementation complexity.

Chapter 8 presents another well-known class of linear block codes, called Reed–Muller (RM) codes, for correcting multiple random errors. RM codes can be decoded with a simple majority-logic decoding algorithm using a successive-cancellation process. Chapter 9 presents several coding techniques that are commonly used in communication and storage systems for reliable information transmission and data storage. These coding techniques include: (1) interleaving; (2) direct product; (3) concatenation; (4) turbo coding; (5) $|\mathbf{u}|\mathbf{u} + \mathbf{v}|$-construction; (6) Kronecker (or tensor) product; and (7) automatic-request-retransmission (ARQ) schemes. Chapter 10 presents various types of codes and coding techniques for correcting bursts of errors, random erasures, bursts of erasures, and combinations of random errors and erasures. Also presented in this chapter is a simple successive-peeling algorithm for correcting random erasures.

Chapter 11 introduces LDPC codes which can achieve near-capacity (or close to Shannon-limit) performance with iterative soft-decision decoding based on belief-propagation over various communication and data-storage channels. Many LDPC codes have been adopted as standard codes for various current and next-generation communication systems. Major aspects covered in this chapter include: (1) basic concepts and characteristics; (2) matrix and graphical representations; (3) various iterative decoding algorithms based on belief-propagation; and (4) error performances over binary-input additive white Gaussian noise (AWGN) and erasure channels.

Chapters 12–14 present three classes of cyclic and quasi-cyclic LDPC codes that are constructed based on finite and partial geometries, finite fields, and experimental designs. These LDPC codes achieve good error performance on both binary-input AWGN and binary-erasure channels. Also included in these chapters are two reduced-complexity iterative decoding algorithms and a technique, called masking, for performance enhancement of LDPC codes. Chapter 15 presents two graphical methods for constructing LDPC codes, known as protograph and progressive-edge-growth methods. LDPC codes constructed based on protographs form a class of channel capacity approaching codes.

Chapter 16 presents a universal coding scheme for collective encoding and collective iterative soft-decision decoding of cyclic codes of prime lengths in the frequency domain. Collective encoding and decoding allows for reliability in information sharing among the received codewords during the decoding process. This collective decoding and information sharing can achieve a decoding gain over the maximum-likelihood decoding of individually received codewords. Collective encoding and decoding of BCH, RS, and QR codes are covered in this chapter.

Chapter 17 presents a class of channel-capacity-approaching codes, called polar codes. Essential aspects covered in this chapter include: (1) Kronecker matrices, mappings, and vector spaces; (2) polar codes from Kronecker mapping point of view; (3) multilevel encoding of polar codes; (4) construction of polar codes based on channel polarization; and (5) successive-cancellation decoding of polar codes.

Except for the first chapter, all other chapters contain a good number of problems. The problems are of various types, including those that require routine calculations, those that require computer solution or simulation, those that require derivations and proofs, and those that require designs and performance analysis. The problems are selected to strengthen students' or engineers' knowledge of the materials in each chapter.

This book can be used as a text for an introductory course on coding at the senior or beginning-graduate level or a more-comprehensive full-year graduate course. It can also be used as a self-guide for practicing engineers and computer scientists in industry who desire to learn the fundamentals and essentials of coding aspects and how they can be applied to the design of error-control systems. For a one-semester introductory course, the fundamentals presented in Chapters 1–6 and some selected topics in Chapters 7–11 can be used. For a two-semester sequence in coding theory, the first 10 chapters in fundamentals of coding can be used for the first semester and remaining chapters on advanced coding topics in the second semester. The book can also be used as a text for one-semester advanced-graduate course focused on LDPC and polar codes and collective encoding and decoding of cyclic codes of prime lengths. In this case, the instructor could use Chapters 11–17 or selected topics from Chapters 4–8. Furthermore, the materials covered in Chapters 1–4 can be used as supplementary subjects for an undergraduate course in information theory or digital-communication systems.

# Acknowledgments

The first author (Shu Lin) would like to take this opportunity to acknowledge the late Professor Paul E. Pfeiffer, the late Professor Wesley W. Peterson, the late Professor Tadao Kasami, and Professor Franklin F. Kuo, who motivated, inspired, and guided him into the wonderful, elegant, and practically useful domain of coding theory at the beginning of his research career. Professor Pfeiffer was his thesis advisor at Rice University. Professor Peterson and Professor Kasami were two pioneers in coding theory, and Professor Peterson wrote the first and most influential book on error-correcting codes. Professor Kou is one of the founders of the Aloha system. Next, the first author would like to thank Professor Ian F. Blake, Professor Dan J. Costello, Jr., Professor Khaled A. S. Abdel-Ghaffar, and Professor William E. Ryan who closely worked with him over many years and helped him to broaden his knowledge in coding theory and its applications in digital communication and data-storage systems. Professor Costello is also his coauthor of two other books.

The most important person behind the first author during the past 57 years, which include the final two years of his graduate study and his entire teaching and research career, is his wife, Ivy. Without her love, support, tolerance, and comfort, as well as raising their three children, his teaching and research career would not have reached as many students or lasted as long as it has. She even took a course in modern algebra to understand what the author was working on at the beginning of his teaching career and to improve their communication. Any success the first author has, he owes to her. She is the one who encouraged him to write this book with his coauthor. So, it is here the first author says to his wife, I love you. Also, the first author would like to give his love and special thanks to his children, their spouses, and grandchildren for their continuing love and affection through the years.

The second author (Juane Li) expresses her sincere gratitude to her advisors, Professor Shu Lin and Professor Khaled A. S. Abdel-Ghaffar, for their valuable guidance, support, and encouragement during her graduate study at the University of California at Davis and during days after her graduation. Professor Lin has rich knowledge, profound ideas, and endless enthusiasm in the error-correcting coding research area. Professor Abdel-Ghaffar is knowledgeable, mathematically rigorous, precise, and patient with students. These two professors' critical and constructive suggestions have pushed her work to a higher level. She also owes a special debt of gratitude to her family for their affection, encouragement, and support over the years.

Both authors are very grateful to Professor Ian F. Blake, Professor Harry Tan, Professor Shih-Chun Chang, Professor William E. Ryan, and Professor Khaled A. S. Abdel-Ghaffar who expended tremendous effort in reading through this book in detail and provided critical comments and numerous valuable suggestions. They would also like to express their appreciation to Professor Qin Huang and Professor Mona Nasseri, Dr. Keke Liu, and Dr. Xin Xiao for their contributions to several topics in this book. They also acknowledge the talented Dr. Zijian Wu who provided such a beautiful photograph of a lotus for the cover image of their book. Last but not least, the authors would like to thank Dr. Julie Lancashire at Cambridge University Press. Without her strong encouragement, warm support, and patience, we would not have been able to bring this book to fruition. Julie, you are a top and thoughtful editor and a good friend.

# 1

# Coding for Reliable Digital Information Transmission and Storage

This chapter presents a simple model of a digital communication (or storage) system and its key function units, which are relevant for reliable information transmission (or storage) over a transmission (or storage) medium subject to noise disturbance (or medium defect). The two function units that play the key role in protection of transmitted (or stored) information against noise (or medium defect) are *encoder* and *decoder*. The function of the encoder is to transform an information sequence into another sequence, called *a coded sequence*, which enables the detection and correction of transmission errors at the receiving end of the system. The transformation of an information sequence to a coded sequence is referred to as *encoding* and the coded sequence is called a *codeword*. At the receiving end of the system, the decoder reproduces the transmitted (or stored) information sequence reliably from the received (or stored) coded sequence. The information recovering process performed by the decoder is referred to as *decoding*. The combination of encoding and decoding for correcting transmission (or storage) errors is referred to as *error-control coding* or *error-correcting coding*.

The objective of this book is to present encoding and decoding methods that are effective in providing reliable information transmission (or storage) and can be practically implemented in a system. In this chapter, two categories of encoding, three decoding rules, and two error-control strategies are introduced. Their objective is to minimize the probability of decoding errors made at the decoder. Also, presented in this introduction chapter are two measurements of performances of a coded communication (or storage) system.

## 1.1   Introduction

In recent years, there has been an increasing demand for *efficient* and *reliable* digital data-transmission and data-storage systems. This demand has been accelerated by the emergence of large-scale and high-speed data networks for the exchange, processing, and storage of digital information. A major concern of the system designers is the control of transmission or storage errors caused by channel noise (or storage defects) so that *reliable reproduction* of transmitted (or stored) information can be obtained (retrieved).

In 1948, Shannon [1] demonstrated in a landmark paper that, by proper encoding of the information, errors induced by a noisy channel (or storage medium) can be reduced to *any desired level* without sacrificing the rate of information transmission (or storage). Since Shannon's work, a great deal of effort has been expended on the topic of devising efficient encoding and decoding methods for error control in a noisy environment. Recent developments have contributed toward achieving the reliability required by today's high-speed digital communication and high-density storage systems, and the use of error-control coding has become an integral part in the design of these systems.

The transmission and storage of digital information have much in common. They both transfer data from an information source to a destination through a channel.[1] A typical transmission (or storage) system may be represented by the block diagram shown in Fig. 1.1. The information source can be either a person or a machine (e.g., a computer). The source output, which is to be communicated to the destination, can be either a continuous waveform or a sequence of discrete symbols.

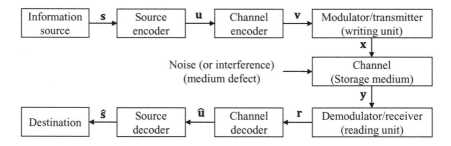

Figure 1.1 Block diagram of a typical data-transmission (or data-storage) system.

The source encoder transforms the source information **s** into a sequence **u** of binary digits (bits) called the *information sequence*. In the case of a continuous

---

[1]A channel is the physical medium through which the information is conveyed or by which it is stored. Microwave links, optical/coaxial cables, telephone circuits, and magnetic disks are examples of channels. Channels are subjected to various types of distortion, noise, and interference which cause the channel outputs to be different from its inputs. For our purposes, a channel is modeled as a probabilistic device and examples will be presented in later sections.

source, this involves *analog-to-digital* (A/D) conversion. The source encoder is ideally designed so that (1) the number of bits per unit time required to represent the source information **s** is minimized, and (2) the source information **s** can be reconstructed from the information sequence **u** without ambiguity. The subject of source coding is not discussed in this book. For a thorough treatment of this important topic, see References [2–6].

The *channel encoder* transforms the information sequence **u** into a discrete encoded sequence **v** called a *codeword*. The transformation of an information sequence **u** into a codeword **v** is called *encoding*. In most instances, **v** is also a binary sequence; however, in some applications, information sequences are encoded into nonbinary codewords. The design and implementation of channel encoders to combat the noisy environment in which codewords are transmitted (or stored) is one of the major topics of this book.

Discrete symbols at the output of the encoder are not suitable for transmission over a physical channel or recording on a digital storage medium. The *modulator* (or *writing unit*) transforms each output symbol of the channel encoder into a waveform of duration of $T$ seconds, which is suitable for transmission (or recording). This waveform enters the channel (or storage medium) and is corrupted by noise (or medium defect). The *demodulator* (or *reading unit*) processes each received waveform of duration of $T$ seconds and produces an output that may be discrete (quantized) or continuous (unquantized). The sequence **r** of demodulator outputs corresponding to the encoded sequence **v** is called the *received sequence*.

The *channel decoder* processes and decodes the received sequence **r** into a binary sequence **û** called the *estimated information sequence*. The decoding strategy is based on the *rules* of channel encoding and the noise characteristics of the channel (or storage medium). Ideally, the estimated information sequence **û** will be a replica of the transmitted information sequence **u**, although the channel noise (or storage medium defect) may cause some errors, i.e., **v** $\neq$ **r**. Another major topic of this book is the design and implementation of channel decoders that minimize the probability of decoding errors by detecting and correcting transmission errors in the received sequence **r**.

The source decoder transforms the estimated information sequence **û** into an estimate **ŝ** of the source output **s** and delivers this estimate to the destination. When the source is continuous, this involves *digital-to-analog* (D/A) conversion. In a well-designed system, the estimate will be a faithful reproduction of the source information **s** except the case when the channel (or storage medium) is very noisy.

To focus attention on the channel encoding and channel decoding, we simplify the system shown in Fig. 1.1 as follows: (1) the information source and source encoder are combined into a digital source with output **u**; (2) the modulator (or writing unit), the channel (or storage medium), and the demodulator (or reading unit) are combined into a coding channel with input **v** and output **r**; and (3) the source decoder and destination are combined into a digital sink with input **û**. The simplified model of a coded communication (or storage) system is shown in Fig. 1.2.

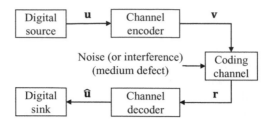

Figure 1.2 A simplified model of a coded system.

The major engineering problem addressed in this book is to design and implement the channel encoder–decoder pair such that (1) information can be transmitted (or recorded) in a noisy environment as fast as possible; (2) reliable reproduction (retrieval) of the information can be obtained at the output of the channel decoder; and (3) the cost of implementing the encoder and decoder falls within acceptable limits.

## 1.2   Categories of Error-Correcting Codes

The design of the channel encoder involves a mapping from an information sequence $\mathbf{u}$ to a coded sequence $\mathbf{v}$, which enables the error detection and/or correction at the channel decoder. The mapping can be achieved by so-called error-correcting codes or error-control codes.

Error-correcting codes can be classified into two categories, *block* and *convolutional* codes, which are structurally different. The encoder for a block code divides the information sequence into blocks, each consisting of $k$ information bits. A block of $k$ information bits is represented by a binary $k$-tuple $\mathbf{u} = (u_0, u_1, \ldots, u_{k-1})$ with $u_i = 0$ or 1, $0 \le i < k$, called a *message*. (In block coding, the symbol $\mathbf{u}$ is used to denote a $k$-bit message rather than the entire information sequence.) There is a total of $2^k$ different messages. The encoder encodes each message $\mathbf{u}$ independently into a codeword which is an $n$-tuple $\mathbf{v} = (v_0, v_1, \ldots, v_{n-1})$ of discrete symbols from a finite alphabet, where $n$ is called the *length* of the codeword. To avoid ambiguity, each message $\mathbf{u}$ is encoded into a *unique* codeword $\mathbf{v}$, i.e., the mapping between a message $\mathbf{u}$ and a codeword $\mathbf{v}$ is *one-to-one*. Because there are $2^k$ different messages, there are $2^k$ different codewords at the encoder output. This set of $2^k$ codewords of length $n$ is said to form an $(n, k)$ *block code*, where $n$ is called the *code length* and $k$ is called the *code dimension*. The ratio $R = k/n$ is called the *code rate* and can be interpreted as the number of information bits entering the channel per transmitted symbol. Because the $n$-symbol output codeword $\mathbf{v}$ of the encoder depends only on the corresponding $k$-bit input message $\mathbf{u}$, the encoder is *memoryless*, and thus can be implemented with a combinational logic circuit. (In block coding, the symbol $\mathbf{v}$ is used to denote an $n$-symbol block rather than the entire encoded sequence.)

In a binary code, each codeword $\mathbf{v}$ is also binary. In this case, we must have $k \leq n$ or $R \leq 1$. When $k < n$, $n - k$ bits are added to each message $\mathbf{u}$ to form a codeword $\mathbf{v}$. The $n - k$ bits added to each message $\mathbf{u}$ by the encoder are called *redundant bits*. These redundant bits carry no new information, and their major function is to provide the code with the capability of detecting and/or correcting errors introduced during the transmission of a codeword, i.e., combating the channel noise (or storage medium defect). How to form these redundant bits to achieve reliable transmission over a noisy channel is the major concern in designing the encoder (or the code). An example of a binary block code with $k = 4$ and $n = 7$ is shown in Table 1.1. This code is a $(7, 4)$ binary block code that is composed of 16 codewords of length 7 for 16 different messages. Each codeword in this code consists of four information bits and three redundant bits. In Chapter 3, it will be shown that this code is capable of correcting a single error, regardless of its location, caused by noise during the transmission of a codeword.

Table 1.1 A binary block code with $k = 4$ and $n = 7$.

| Messages $(u_0, u_1, u_2, u_3)$ | Codewords $(v_0, v_1, v_2, v_3, v_4, v_5, v_6)$ | Messages $(u_0, u_1, u_2, u_3)$ | Codewords $(v_0, v_1, v_2, v_3, v_4, v_5, v_6)$ |
|---|---|---|---|
| $(0\ 0\ 0\ 0)$ | $(0\ 0\ 0\ 0\ 0\ 0\ 0)$ | $(0\ 0\ 0\ 1)$ | $(1\ 0\ 1\ 0\ 0\ 0\ 1)$ |
| $(1\ 0\ 0\ 0)$ | $(1\ 1\ 0\ 1\ 0\ 0\ 0)$ | $(1\ 0\ 0\ 1)$ | $(0\ 1\ 1\ 1\ 0\ 0\ 1)$ |
| $(0\ 1\ 0\ 0)$ | $(0\ 1\ 1\ 0\ 1\ 0\ 0)$ | $(0\ 1\ 0\ 1)$ | $(1\ 1\ 0\ 0\ 1\ 0\ 1)$ |
| $(1\ 1\ 0\ 0)$ | $(1\ 0\ 1\ 1\ 1\ 0\ 0)$ | $(1\ 1\ 0\ 1)$ | $(0\ 0\ 0\ 1\ 1\ 0\ 1)$ |
| $(0\ 0\ 1\ 0)$ | $(1\ 1\ 1\ 0\ 0\ 1\ 0)$ | $(0\ 0\ 1\ 1)$ | $(0\ 1\ 0\ 0\ 0\ 1\ 1)$ |
| $(1\ 0\ 1\ 0)$ | $(0\ 0\ 1\ 1\ 0\ 1\ 0)$ | $(1\ 0\ 1\ 1)$ | $(1\ 0\ 0\ 1\ 0\ 1\ 1)$ |
| $(0\ 1\ 1\ 0)$ | $(1\ 0\ 0\ 0\ 1\ 1\ 0)$ | $(0\ 1\ 1\ 1)$ | $(0\ 0\ 1\ 0\ 1\ 1\ 1)$ |
| $(1\ 1\ 1\ 0)$ | $(0\ 1\ 0\ 1\ 1\ 1\ 0)$ | $(1\ 1\ 1\ 1)$ | $(1\ 1\ 1\ 1\ 1\ 1\ 1)$ |

The encoder for a convolutional code also accepts $k$-bit blocks of the information sequence $\mathbf{u}$ and produces an encoded sequence (codeword) $\mathbf{v}$ of $n$-symbol blocks. (In convolutional coding, the symbols $\mathbf{u}$ and $\mathbf{v}$ are used to denote sequences of blocks rather than a single block.) However, each encoded block depends not only on the corresponding $k$-bit message block at the same time unit, but also on the $m$ *previous message blocks*. Hence, the encoder has a *memory order of m*. The set of encoded sequences produced by a $k$-bit input and $n$-bit output encoder of memory order $m$ is called an $(n, k, m)$ *convolutional code*. The ratio $R = k/n$ is called the code rate. Because the encoder contains memory, it must be implemented with a sequential logic circuit. In a binary convolutional code, redundant bits for combating the channel noise are added to the information sequence when $k < n$ or $R < 1$. Typically, $k$ and $n$ are small integers and more redundancy is added by increasing the memory order $m$ of the code while holding $k$ and $n$, and hence the code rate $R$ is fixed. How to use the memory to achieve reliable transmission over a noisy channel is the major concern in designing the convolutional encoder.

An example of a binary convolutional encoder with $k = 1$, $n = 2$, and $m = 2$ is shown in Fig. 1.3. As an illustration of how codewords are generated, consider the information sequence $\mathbf{u} = (1\,1\,0\,1\,0\,0\,0\,\ldots)$, where the leftmost bit is assumed to enter the encoder first. Before any information bit enters the encoder, the contents of the two shift registers are initialized to zeros. Using the rules of EXCLUSIVE-OR (XOR) and assuming that the multiplexer takes the first encoded bit from the top output and the second bit from the bottom output, it is easy to see that the encoded sequence is $\mathbf{v} = (1\,1,\,1\,0,\,1\,0,\,0\,0,$ $0\,1,\,1\,1,\,0\,0,\,0\,0,\,0\,0,\,\ldots)$.

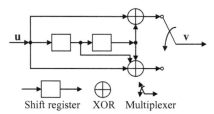

Figure 1.3 A binary convolutional encoder with $k = 1$, $n = 2$, and $m = 2$.

In this book, we only cover block codes, both *classical* and *modern*, because the most current and next generations of communication and storage systems are using and will use block codes in various forms. Readers who are interested in convolutional codes are referred to [7–10].

## 1.3    Modulation and Demodulation

The modulator in a communication system must select a waveform of duration of $T$ seconds, which is suitable for transmission, for each encoder output symbol. In the case of a binary code, the modulator shall generate one of two signals, $s_0(t)$ for an encoded "0" or $s_1(t)$ for an encoded "1" which are called *carriers*. For a wideband channel, the optimum choices of signals are

$$
\begin{aligned}
s_0(t) &= \sqrt{\tfrac{2E_s}{T}}\cos(2\pi f_0 t),\ 0 \le t \le T, \\
s_1(t) &= \sqrt{\tfrac{2E_s}{T}}\cos(2\pi f_0 t + \pi) = -\sqrt{\tfrac{2E_s}{T}}\cos(2\pi f_0 t),\ 0 \le t \le T,
\end{aligned}
\tag{1.1}
$$

where $f_0$ is a multiple of $1/T$ and $E_s$ is the energy of each signal. This type of modulation is called a *binary phase-shift-keyed* (BPSK) *modulation*, because the phase of the carrier $\cos(2\pi f_0 t)$ is shifted between $0$ and $\pi$, depending on the encoder output.

A common form of noise disturbance present in any communication system is *additive white Gaussian noise* (AWGN), which commonly includes the ambient heat in the modulator–demodulator hardware and the transmitted medium. A channel with such noise disturbance is called an *AWGN channel* and an AWGN channel with binary input is denoted by BI-AWGN channel. If the transmitted signal over an AWGN channel is $s(t)$ (either $s_0(t)$ or $s_1(t)$), then the received signal is

$$r(t) = s(t) + n(t), \tag{1.2}$$

where $n(t)$ is a *Gaussian random process* with *one-sided power spectral density* (PSD) $N_0$. Other forms of noise are also present in many systems. For example, in a communication system subject to multipath transmission, the received signal is observed to fade (lose strength) during certain time intervals. This fading can be modeled as a multiplicative noise component on the signal $s(t)$.

The demodulator must produce an output corresponding to the received signal in each $T$-second interval. This output may be a real number or one of a discrete set of preselected symbols, depending on the demodulator design. An optimum demodulator always includes a matched filter or correlation detector followed by a sampling switch. For BPSK modulation with coherent detection, the sampled output is a real number given by

$$y = \int_0^T r(t) \sqrt{\frac{2E_s}{T}} \cos(2\pi f_0 t) dt. \tag{1.3}$$

The sequence of unquantized demodulator outputs can be passed directly to the channel decoder for processing. In this case, the channel decoder must be capable of handling unquantized inputs, that is, it must be able to process real numbers. A much more common approach to decoding is to quantize the real-number detector output $y$ into one of a finite number $N$ of discrete output symbols. In this case, the channel decoder has discrete inputs and processes discrete values. Most coded communication (or storage) systems use some form of discrete processing.

If the detector output in a given interval depends only on the transmitted signal in that interval, and not on any previous transmission, the channel is said to be *memoryless*. In this case, the combination of an $M$-ary input modulator, the physical channel, and an $N$-ary output demodulator can be modeled as a *discrete memoryless channel* (DMC). A DMC is completely described by a set of transition probabilities, $P(j|i)$, $0 \le i < M$, $0 \le j < N$, where $i$ represents a modulator input symbol, $j$ represents a demodulator output symbol, and $P(j|i)$ is the probability of receiving $j$ given that $i$ was transmitted. The transition probability $P(j|i)$ can be calculated from a knowledge of the signals used, the probability distribution of the noise, and the output quantization threshold of the demodulator. If $M = 2$, the input of modular is binary, either 0 or 1. Such a DMC is called binary-input DMC (BI-DMC).

As an example, consider a communication system in which (1) binary modulation is used ($M = 2$), (2) the amplitude distribution of the noise is symmetric, and (3) the demodulator output is quantized to $N = 2$ levels. In this case, a particularly simple and practically important channel model, called the *binary symmetric channel* (BSC), results. The block diagram for a BSC is shown in Fig. 1.4(a). The transition probability $p$ is given by

$$p = Q\left(\sqrt{\frac{2E_s}{N_0}}\right), \tag{1.4}$$

where $Q(x)$ is the *complementary error function* of Gaussian statistics given as follows:

$$Q(x) \triangleq \frac{1}{\sqrt{2\pi}} \int_x^\infty e^{-y^2/2} dy, \ x > 0. \tag{1.5}$$

The transition probability $p$ is simply the BPSK bit-error probability for equally likely signals. An upper bound on $Q(x)$ that will be used later in evaluating the error performance of codes on a BSC is

$$Q(x) \leq \frac{e^{-x^2/2}}{2}, \ x > 0. \tag{1.6}$$

It is usually assumed that the power level of transmitter and the message symbol transmission rate are to remain constant. Applying various error-correcting codes, we introduce redundancy to the transmitted vector, which thus decreases the $E_s/N_0$ at the receiver side. With an error-correcting code of code rate $R$ applied to the system, (1.4) becomes

$$p = Q\left(\sqrt{\frac{2RE_s}{N_0}}\right). \tag{1.7}$$

Figure 1.4(b) shows a transition probability diagram of a BI-DMC.

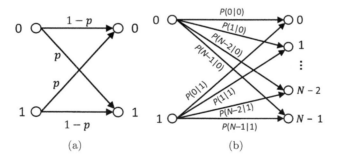

Figure 1.4 Transition probability diagrams: (a) BSC and (b) BI-DMC.

In this book, we focus only on error-correcting codes, their encoding and decoding for communication systems in which binary modulation is used, and the outputs of the demodulator are either quantized or unquantized.

## 1.4 Hard-Decision and Soft-Decision Decodings

Consider a binary block code of length $n$. Each codeword in the code is a sequence of 0s and 1s, denoted by $\mathbf{v} = (v_0, v_1, \ldots, v_{n-1})$. At the transmitter end, each bit $v_i$ of $\mathbf{v}$ is transformed into a specific signal waveform (BPSK signaling is the most widely used one). Therefore, each codeword $\mathbf{v}$ is mapped into a sequence of signal waveforms and transmitted over a noisy channel. At the receiver, the corrupted signal waveforms are demodulated and sampled into a sequence of real numbers $\mathbf{y} = (y_0, y_1, \ldots, y_{n-1})$, which is the output sequence of the demodulator.

Before entering the channel decoder, if each element of $\mathbf{y}$ is quantized into two levels (0 and 1), we say that a *hard-decision* is made. Then, the input to the channel decoder is a sequence $\mathbf{r} = (r_0, r_1, \ldots, r_{n-1})$ of $n$ bits. Based on this received sequence, the channel decoder performs error detection and/or error correction. Because decoding is based on the hard-decision output sequence of the binary demodulator, the decoding is called *hard-decision decoding*.

If each element of the output sequence $\mathbf{y} = (y_0, y_1, \ldots, y_{n-1})$ of the demodulator is quantized into *more than two* levels, then the input sequence $\mathbf{r} = (r_0, r_1, \ldots, r_{n-1})$ of the decoder is a sequence of discrete symbols from a finite alphabet set. In this case, the channel decoder has discrete inputs and it must process discrete values. If each element of the output sequence $\mathbf{y} = (y_0, y_1, \ldots, y_{n-1})$ of the demodulator is not quantized, the input sequence of the decoder is $\mathbf{y}$, i.e., $\mathbf{r} = \mathbf{y}$. In this case, the channel decoder must process real numbers. Decoding either a received sequence of discrete values or a received sequence of real numbers is referred to as *soft-decision decoding*.

Performing a hard-decision of a received signal at the demodulator, some useful information is lost. As a result, the decoding accuracy is decreased. However, the channel decoder needs only to process binary symbols, 0s and 1s, which results in a low decoding complexity and small decoding delay. Under the context of hard-decision decoding, long codes can be used. For communication systems that require low decoding complexity, low power consumption, and fast decoding process (small decoding delay), the hard-decision decoding is desired. Soft-decision decoding offers *significant* error-correcting improvement in performance compared to hard-decision decoding but requires higher decoding complexity, longer decoding delay, and larger power consumption. As the VLSI circuit design technology advances, the speed of decoding process can be increased significantly and higher decoding complexity can be handled with reasonable expense. Consequently, soft-decision decoding is used in more and more applications, such as 5G, satellite and optical communication and flash memories. Various soft-decision decoding algorithms have been developed, which offer effective tradeoffs between error-correcting performance and decoding complexity.

## 1.5 Maximum A Posteriori and Maximum Likelihood Decodings

Consider the block diagram of a block-coded communication system on an AWGN channel with finite output quantization as shown in Fig. 1.5. The source output $\mathbf{u}$ represents a $k$-bit message, the encoder output $\mathbf{v}$ represents an $n$-symbol codeword, the demodulator output $\mathbf{r}$ represents the corresponding

$N$-ary received $n$-tuple, and the decoder output $\hat{\mathbf{u}}$ represents the $k$-bit estimate of the source output message $\mathbf{u}$.

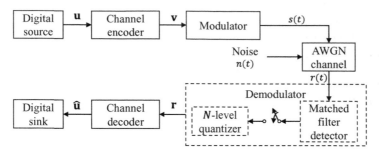

Figure 1.5 A coded communication system with binary-input and $N$-ary-output discrete memoryless AWGN channel.

The channel decoder must produce an estimate of the message $\mathbf{u}$ based on the received sequence $\mathbf{r}$. Equivalently, because there is a one-to-one correspondence between the message $\mathbf{u}$ and the codeword $\mathbf{v}$, the decoder can produce an estimate $\hat{\mathbf{v}}$ of the codeword $\mathbf{v}$. Clearly, $\hat{\mathbf{u}} = \mathbf{u}$ if and only if $\hat{\mathbf{v}} = \mathbf{v}$. A *decoding rule* is a strategy for choosing an estimate $\hat{\mathbf{v}}$ of the transmitted codeword $\mathbf{v}$ for each possible received sequence $\mathbf{r}$. A decoding error occurs if $\hat{\mathbf{v}} \neq \mathbf{v}$. Given that a sequence $\mathbf{r}$ is received, the error probability of the decoder is defined as

$$P(E|\mathbf{r}) \triangleq P(\hat{\mathbf{v}} \neq \mathbf{v}|\mathbf{r}), \tag{1.8}$$

where $E$ denotes the error event $\hat{\mathbf{v}} \neq \mathbf{v}$. The *error probability of the channel decoder* is then given by

$$P(E) = \sum_{\mathbf{r}} P(E|\mathbf{r})P(\mathbf{r}), \tag{1.9}$$

where $P(\mathbf{r})$ is the probability of receiving the sequence $\mathbf{r}$. The probability $P(\mathbf{r})$ is independent of the decoding rule, because $\mathbf{r}$ is produced prior to the decoding. Hence, *an optimum decoding rule*, that is, one that minimizes $P(E)$, must minimize $P(E|\mathbf{r})$ for all $\mathbf{r}$. As minimizing $P(\hat{\mathbf{v}} \neq \mathbf{v}|\mathbf{r})$ is equivalent to maximizing $P(\hat{\mathbf{v}} = \mathbf{v}|\mathbf{r})$, $P(\hat{\mathbf{v}} \neq \mathbf{v}|\mathbf{r})$ is minimized for a given $\mathbf{r}$ by choosing $\hat{\mathbf{v}}$ as the codeword $\mathbf{v}$ which maximizes

$$P(\mathbf{v}|\mathbf{r}) = P(\mathbf{r}|\mathbf{v})P(\mathbf{v})/P(\mathbf{r}), \tag{1.10}$$

that is, $\hat{\mathbf{v}}$ is chosen as the codeword which has the *largest a posteriori probability* (APP) $P(\hat{\mathbf{v}}|\mathbf{r})$ (or the *most likely codeword* given that $\mathbf{r}$ is received). This decoding is *optimal* and is called the *maximum a posteriori* (MAP) *decoding*. The decoding process is carried out as follows: (1) compute the APP $P(\mathbf{v}|\mathbf{r})$ for all codewords, $\mathbf{v}$, for a given received sequence $\mathbf{r}$, and (2) decode the received sequence $\mathbf{r}$ into the codeword $\hat{\mathbf{v}}$ which has the largest APP, i.e., $P(\hat{\mathbf{v}}|\mathbf{r}) \geq P(\mathbf{v}^*|\mathbf{r})$ for $\mathbf{v}^* \neq \hat{\mathbf{v}}$.

If all messages, and hence all codewords, are equally likely, that is, $P(\mathbf{v})$ is the same for all $\mathbf{v}$s. In this case, maximizing $P(\mathbf{v}|\mathbf{r})$ in (1.10) is equivalent to maximizing $P(\mathbf{r}|\mathbf{v})$. Hence, the estimate of the transmitted codeword $\mathbf{v}$ is chosen as the codeword that maximizes the conditional probability $P(\mathbf{r}|\mathbf{v})$. Such decoding rule is referred to as *maximum likelihood decoding* (MLD). For a DMC, we have

$$P(\mathbf{r}|\mathbf{v}) = \prod_{i=0}^{n-1} P(r_i|v_i), \tag{1.11}$$

as, for a memoryless channel, each received symbol depends only on the corresponding transmitted symbol. Because the function $\log(x)$ is a monotone increasing function of $x$, maximizing $P(\mathbf{r}|\mathbf{v})$ in (1.11) is equivalent to maximizing the following *log-likelihood function*

$$\log(P(\mathbf{r}|\mathbf{v})) = \sum_{i=0}^{n-1} \log(P(r_i|v_i)). \tag{1.12}$$

Hence, the MLD on a DMC is to choose the estimate of the transmitted codeword $\mathbf{v}$ that maximizes the sum in (1.12).

If the codewords are not equally likely, an MLD is not necessarily optimum, because the conditional probability $P(\mathbf{r}|\mathbf{v})$ must be weighted by the codeword probability $P(\mathbf{v})$ to determine the codeword which maximizes $P(\mathbf{v}|\mathbf{r})$. However, in many communication systems, the codeword probabilities are not known exactly at the receiver. In this case, making optimum decoding is impossible, and an MLD then becomes the *best feasible decoding*.

The probability $P(E)$ of decoding a received sequence $\mathbf{r}$ into a codeword that is not the transmitted codeword $\mathbf{v}$ is called the *block-error probability* or *block-error rate* (BLER) which is a measure of the decoder performance.

Now, we consider the application of the MLD decoding rule to the BSC with transition propability $p$. In this case, $\mathbf{r}$ is a binary sequence which may differ from the transmitted codeword $\mathbf{v}$ in some positions because of the channel noise. When $r_i \neq v_i$, $P(r_i|v_i) = p$, and when $r_i = v_i$, $P(r_i|v_i) = 1 - p$. Let $d(\mathbf{r}, \mathbf{v})$ be the number of positions in which $\mathbf{r}$ and $\mathbf{v}$ differ, called the *Hamming distance* [11] between $\mathbf{r}$ and $\mathbf{v}$. Then, for a block of length $n$, the conditional probability $P(\mathbf{r}|\mathbf{v})$ is given as follows:

$$P(\mathbf{r}|\mathbf{v}) = p^{d(\mathbf{r}, \mathbf{v})}(1 - p)^{n - d(\mathbf{r}, \mathbf{v})}, \tag{1.13}$$

and the log-likelihood function given by (1.12) becomes

$$\begin{aligned}
\log(P(\mathbf{r}|\mathbf{v})) &= d(\mathbf{r}, \mathbf{v})\log(p) + (n - d(\mathbf{r}, \mathbf{v}))\log(1 - p) \\
&= d(\mathbf{r}, \mathbf{v})\log(\tfrac{p}{1-p}) + n\log(1 - p).
\end{aligned} \tag{1.14}$$

Because $\log(p/(1 - p)) < 0$ for $p < 1/2$ and $n\log(1 - p)$ is a constant for all $\mathbf{v}$s, then the MLD decoding rule for a block code of length $n$ on the BSC is to choose the codeword $\mathbf{v}$ which minimizes the distance $d(\mathbf{r}, \mathbf{v})$ from $\mathbf{r}$, that is, it chooses the codeword that differs from the received sequence in the *fewest number of positions*. Hence, the MLD of a block code for the BSC is sometimes called a *minimum-distance decoding*.

The objective of the MAP or the MLD is to minimize the BLER at the output of the decoder. Another measure of the error performance of the decoder is the *bit-error probability* (or *bit-error rate* (BER)). The BER is defined as the probability that the estimate $\hat{u}_i$ of an information bit at the output of decoder is not the same as the information bit $u_i$ in the transmitted message $\mathbf{u} = (u_0, u_1, \ldots, u_{k-1})$ at the input of the encoder. Consider a coded communication system on a memoryless channel. Suppose, in decoding, we estimate the transmitted message $\mathbf{u}$ bit by bit. To carry out the decoding, we first compute the APPs $P(u_i = 0|\mathbf{r})$ and $P(u_i = 1|\mathbf{r})$, $0 \leq i < k$. Then, we choose the estimate $\hat{u}_i$ as follows:

$$\hat{u}_i = \begin{cases} 1, & \text{if } P(u_i = 1|\mathbf{r}) > P(u_i = 0|\mathbf{r}), \\ 0, & \text{otherwise.} \end{cases} \tag{1.15}$$

This symbol-by-symbol decoding rule is *optimal* and is called *symbol-by-symbol MAP decoding*. The APP $P(u|\mathbf{r})$ for an information symbol $u$ can be computed as follows:

$$P(u|\mathbf{r}) = P(\mathbf{r}|u)P(u)/P(\mathbf{r}). \tag{1.16}$$

If the two values, 0 and 1, of the information symbol $u$ are equally likely, then the decoding rule given by (1.15) can be reformulated as follows:

$$\hat{u}_i = \begin{cases} 1, & \text{if } P(\mathbf{r}|u_i = 1) > P(\mathbf{r}|u_i = 0), \\ 0, & \text{otherwise.} \end{cases} \tag{1.17}$$

If the two values of the information symbol $u$, 0 and 1, are not equally likely, the decoding rule given by (1.17) is not necessarily optimum. Although MLD and symbol-by-symbol MAP decoders use different criteria, they provide similar bit-error performance when applied to binary block codes.

## 1.6 Channel Capacity on Transmission Rate

The capability of a noisy channel to transmit information reliably was determined by Shannon [1] in his work published in 1948. The result, called the *noisy channel coding theorem*, states that every channel has a *channel capacity* $\boldsymbol{C}$, and that, for any rate $R < \boldsymbol{C}$, there exist codes of rate $R$ which, with MLD, have an arbitrarily small decoding error probability $P(E)$. In particular, for any $R < \boldsymbol{C}$, there exist block codes of length $n$ such that

$$P(E) \leq 2^{-nE(R)}, \tag{1.18}$$

where $E(R)$ is a positive function of $R$ for $R < \boldsymbol{C}$ and is completely determined by the channel characteristics. The bound of (1.18) implies that arbitrarily small error probabilities are achievable with block coding for any fixed $R < C$ by increasing the code length $n$ while holding the code rate $R = k/n$ constant.

The noisy channel coding theorem is based on an argument called *random coding*. The bound obtained is actually on the average error probability of the ensemble of all codes. Because some codes must perform better than the average, the noisy channel coding theorem guarantees the existence of codes satisfying (1.18), but does not indicate how to construct these codes. Furthermore, to achieve very low error probabilities for block codes of fixed rate $R < C$, long block lengths are needed. This requires that the number of codewords $2^k = 2^{nR}$ must be very large. Because an MLD decoder must compute $\log(P(\mathbf{r}|\mathbf{v}))$ for all the codewords and then choose the one that gives the maximum, the number of computations performed by an MLD decoder becomes excessively large. Hence, it is impractical to achieve very low error probabilities with MLD. Therefore, two major problems are encountered when designing a coded system to achieve low error probabilities: (1) to construct good long codes whose performances with MLD would satisfy (1.18), and (2) to find practically implementable methods of encoding and decoding these codes such that their actual performances are close to what could be achieved with MLD.

Since Shannon proved the theoretical achievable performance of codes in 1948, enormous research effort has been expended by coding theorists in designing codes and efficient decoding algorithms (practically implementable) to achieve a performance close to the capacity of a channel. The research effort has been very fruitful. Three families of channel capacity approaching codes, namely, *turbo*, *low-density parity-check* (LDPC), and *polar codes*, have been discovered or rediscovered and they can be decoded with various efficient decoding algorithms.

## 1.7 Classification of Channel Errors

On memoryless channels, the noise affects each transmitted symbol independently. As an example, consider the BSC whose transition diagram is shown in Fig. 1.4(a). Each transmitted bit has a probability $p$ of being received incorrectly and a probability $1 - p$ of being received correctly, independent of other transmitted bits. Hence, transmission errors occur *randomly* in the received sequence, and these memoryless channels are called *random-error channels*. The codes devised for correcting random errors are called *random-error-correcting codes*. Most of the codes presented in this book are random-error-correcting codes.

On channels with memory, the noise is not independent from transmission to transmission. A simple model of a channel with memory is shown in Fig. 1.6. This model, called the *two-state Gilbert–Elliott model*, contains two states, a *good state* $S_1$ in which transmission errors occur infrequently, i.e., $p_1 \approx 0$, and a *bad state* $S_2$ in which transmission errors are highly probable, i.e., $p_2 \approx 0.5$. The channel is in the good state $S_1$ most of the time, but on occasion it shifts to the bad state $S_2$ due to a change in the transmission characteristic of the channel such as *"deep fade"* caused by multipath transmission. As a consequence, transmission errors occur in clusters or bursts because of the high transition probability in the bad state $S_2$. Such channels with memory are called *burst-error channels*. Codes devised for correcting burst errors are called *burst-error-correcting codes*. Some channels contain a combination of both random and burst errors. They are called *compound channels*, and codes devised for correcting errors on these channels are called *burst-and-random-error-correcting codes*.

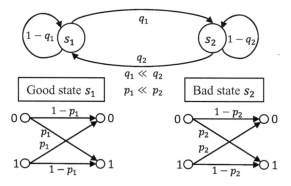

Figure 1.6 The two-state Gilbert–Elliott model.

In a communication or storage system, if the detector at the receiver side is able to detect the locations where the received symbols are unreliable, the receiver may choose to erase these received symbols rather than making hard-decisions of them, which are very likely to create errors. Erasing these unreliable received symbols results in *symbol losses at the known locations* of the received sequence. Transmitted symbols that have been erased are referred to as *erasures*. There are two basic types of erasures, *random* and *burst*. Erasures occur at the random locations, each with the same probability of occurrence, are referred to as *random erasures*. A binary-input channel with this pure random erasure characteristic is called a *binary erasure channel* (BEC) which is modeled in Fig. 1.7(a), where the symbol "?" represents an erasure. A binary-input channel with both random error and random erasure characteristics modeled in Fig. 1.7(b) is called a *binary symmetric erasure channel* (BSEC). A binary channel over which erasures cluster into bursts of locations is called a *binary burst erasure channel* (BBEC). Designs of codes and coding techniques for correcting random erasures on a BEC, random errors and random erasures on a BSEC, and bursts of erasures on a BBEC will be presented in Chapters 10, 14, and 16.

## 1.8   Error-Control Strategies

The block diagram shown in Fig. 1.1 represents a *one-way* communication (or storage) system. The transmission (or recording) is strictly in one direction, from transmitter to receiver. Error control for a one-way system must be

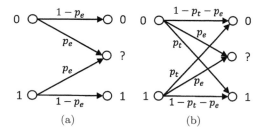

Figure 1.7 Models for (a) BEC and (b) BSEC, where $p_e$ represents the erasure probability and $p_t$ represents the error probability.

accomplished using *forward error correction* (FEC), that is, by employing error-correcting codes that automatically correct errors detected at the receiver. Most of the coded systems in use today apply some forms of FEC, even if the channel is not strictly one-way.

In some cases, a transmission system can be two-way, that is, information can be transmitted in both directions and the transmitter also acts as a receiver (a transceiver), and vice versa. Error control for a two-way system can be carried out in two modes: (1) error detection and/or correction, and (2) retransmission, called *automatic repeat request* (ARQ) [10, 12–15]. In an ARQ system, when errors are detected at the receiver, a request is sent to the transmitter to repeat the message, and this continues until the message is received correctly. In an ARQ system, a code with good error detection capability is needed.

There are two types of ARQ system: *stop-and-wait* ARQ and *continuous* ARQ. With stop-and-wait ARQ, the transmitter sends a codeword to the receiver and waits for a positive acknowledgment (ACK) or a negative acknowledgment (NAK) from the receiver. If an ACK is received (no errors detected), the transmitter sends the next codeword. If an NAK is received (errors detected), the transmitter resends the preceding codeword. When the noise is persistent, the same codeword may be retransmitted several times before it is correctly received and acknowledged.

With continuous ARQ, the transmitter sends codewords to the receiver continuously and receives acknowledgments continuously. When an NAK is received, the transmitter begins a retransmission. It may back up to the codeword being detected in error and resend that codeword plus the codewords that follow it, say $N-1$ of them. This is called a *go-back-N* ARQ. Alternatively, the transmitter may simply resend only those codewords that are acknowledged negatively. This is known as a *selective-repeat* ARQ (SR-ARQ). SR-ARQ is more efficient than go-back-$N$ ARQ, but requires more logic and buffering (memory to store the codewords which are not detected in error if messages must be delivered to the user(s) in order). Continuous ARQ is more efficient than stop-and-wait ARQ, but it is also more expensive.

The major advantage of ARQ over FEC is that error detection requires much simpler decoding equipment than error correction. Also, ARQ is *adaptive* in the sense that information is retransmitted only when errors occur. On the other hand, when the channel error rate is high, retransmissions must be sent frequently, and the system throughput, the rate at which newly generated messages are correctly received, is lowered by ARQ. In this situation, a combination of FEC for correcting the most frequent error patterns with error detection and retransmission for the less likely error patterns is more efficient than ARQ alone. Such a combination of FEC and ARQ is called a *hybrid-ARQ* (HARQ) [16, 17]. A well-designed HARQ not only provides highly reliable information transmission but also maintains high system throughput.

Various types of ARQ and HARQ schemes will be discussed in detail in Chapter 9.

## 1.9   Measures of Performance

The performance of a coded communication (or storage) system is, in general, measured by its probability of decoding error, called *error probability*, and its *coding gain* over an uncoded system that transmits information at the same rate. There are two types of error probabilities, *block-error probability* and *symbol-error probability*. As defined earlier, the block-error probability is the probability that the decoded codeword at the output of the decoder is different from the transmitted codeword at the output of the encoder. The block-error probability is commonly called *block-error rate* (BLER). The symbol-error probability, also called the *symbol-error rate* (SER), is defined as the probability that a decoded information symbol at the output of the decoder is in error. In binary case, the information symbols are bits and thus SER is referred to as *bit-error rate* (BER). A coded communication system should be designed to keep these two error probabilities as low as possible under certain system constraints, such as power, bandwidth, decoding complexity, and decoding delay.

The other performance measure of a coded communication system is *coding gain*. Coding gain is defined as a reduction in signal-to-noise ratio (SNR) required to achieve a specific error probability (BER or BLER) for a coded communication system compared with an uncoded system using the same modulation. Consider a binary coded communication system using BPSK modulation. Let $(E_b/N_0)_{\text{uncoded}}$ and $(E_b/N_0)_{\text{coded}}$ be the SNRs required by the uncoded and coded communication systems with BPSK modulation to achieve a specified error probability, respectively. Then, in decibels (dB), the coding gain of the coded system over the uncoded BPSK system is given by

$$\begin{aligned}
\zeta &= 10 \log_{10} \frac{(E_b/N_0)_{\text{uncoded}}}{(E_b/N_0)_{\text{coded}}} \\
&= 10 \log_{10}(E_b/N_0)_{\text{uncoded}} - 10 \log_{10}(E_b/N_0)_{\text{coded}}.
\end{aligned} \tag{1.19}$$

For example, if $(E_b/N_0)_{\text{uncoded}}/(E_b/N_0)_{\text{coded}} = 2$ at a BER of $10^{-8}$, then the coding gain is $\zeta = 10 \log_{10} 2 = 3$ dB. In this case, we say that the coded system achieves 3 dB coding gain over the uncoded system at the BER of $10^{-8}$.

Consider a coded communication system using a $(127, 113)$ binary block code for error control. Each codeword in the code consists of 113 information bits and 14 redundant bits. Therefore, the rate of the code is $113/127 = 0.8898$. The construction and decoding of this code will be explained in Chapter 5. Suppose BPSK modulation with coherent detection is used and the channel is an AWGN channel. The BER performances of the code decoded with both hard-decision decoding and unquantized soft-decision MLD decoding [10] versus SNR $E_b/N_0$ are shown in Fig. 1.8. (The hard-decision decoding is performed using the Berlekamp–Massey (BM) algorithm which will be presented in Chapter 5.) For comparison, the BER performance of an uncoded BPSK system is also included. We see that the coded system, with either hard-decision or soft-decision decoding, provides a much lower BER than the uncoded system at the same SNR. For hard-decision decoding, the coded system achieves 3.7 dB coding gain over the uncoded system at a BER of $10^{-8}$, and with unquantized MLD decoding, the coded system achieves 6.3 dB coding gain over the uncoded system at a BER of $10^{-8}$ (a large coding gain). At the BER of $10^{-8}$, the unquantized soft-decision MLD achieves around 2.6 dB coding gain over the hard-decision decoding. The coding gain of the soft-decision MLD over the hard-decision decoding of the coded system is achieved at the expense of much higher decoding complexity.

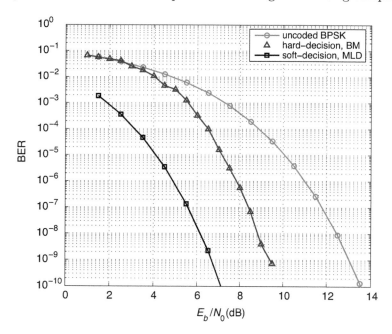

Figure 1.8 The BER performances of a coded communication using a $(127, 113)$ binary block code.

In designing a coded system, it is desired to minimize the SNR required to achieve a specific error rate. This is equivalent to maximizing the coding gain over an uncoded system. A theoretical limit on the minimum SNR required for a coded system with code rate $R$ to achieve error-free information transmission is called the *Shannon limit* [1]. This theoretical limit simply says that, for a coded system with rate $R$, error-free information transmission is achievable only if the SNR exceeds the limit. As long as the SNR exceeds the Shannon limit, Shannon's coding theorem guarantees the existence of a coded system, perhaps very complex, capable of achieving error-free information transmission.

Figure 1.9 gives the Shannon limit as a function of code rate for a binary input continuous output AWGN channel with BPSK signaling. The values of the Shannon limit for different code rates ranged from 0.01 to 0.99 are given in Table 1.2.

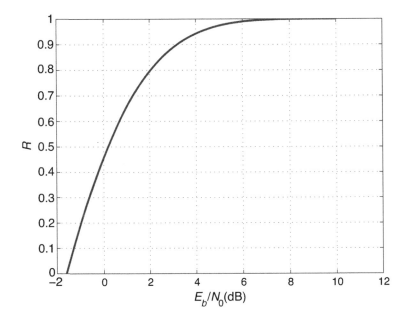

Figure 1.9 Shannon limit $E_b/N_0$ (dB) as a function of code rate $R$.

## 1.10    Contents of the Book

In the following, a synopsis of contents in each chapter of the book is given.

Chapter 2 presents some important elements of algebra and graphs that are necessary for understanding the aspects of error-correcting codes to be developed in the rest of the chapters of the book. The elements included are: groups, finite fields, polynomials, vector spaces over finite fields, matrices, and graphs. The treatments of these algebraic elements are mostly descriptive and make no attempt to be mathematically rigorous.

Chapter 3 gives an introduction to block codes with linear structure, called *linear block codes*. The coverage of this chapter includes: (1) fundamental concepts and structures of linear block codes; (2) generation of these codes in terms

Table 1.2 Shannon limits, $E_b/N_0$ (dB), of a binary-input continuous-output AWGN channel with BPSK signaling for various code rates.

| $R$ | $E_b/N_0$ | $R$ | $E_b/N_0$ | $R$ | $E_b/N_0$ | $R$ | $E_b/N_0$ |
|------|-----------|------|-----------|------|-----------|------|-----------|
| 0.01 | $-1.562$ | 0.26 | $-0.759$ | 0.51 | 0.233 | 0.76 | 1.704 |
| 0.02 | $-1.531$ | 0.27 | $-0.724$ | 0.52 | 0.279 | 0.77 | 1.784 |
| 0.03 | $-1.501$ | 0.28 | $-0.689$ | 0.53 | 0.326 | 0.78 | 1.866 |
| 0.04 | $-1.471$ | 0.29 | $-0.653$ | 0.54 | 0.374 | 0.79 | 1.952 |
| 0.05 | $-1.440$ | 0.30 | $-0.618$ | 0.55 | 0.432 | 0.80 | 2.040 |
| 0.06 | $-1.409$ | 0.31 | $-0.581$ | 0.56 | 0.472 | 0.81 | 2.132 |
| 0.07 | $-1.379$ | 0.32 | $-0.545$ | 0.57 | 0.522 | 0.82 | 2.228 |
| 0.08 | $-1.348$ | 0.33 | $-0.508$ | 0.58 | 0.574 | 0.83 | 2.328 |
| 0.09 | $-1.316$ | 0.34 | $-0.471$ | 0.59 | 0.626 | 0.84 | 2.432 |
| 0.10 | $-1.286$ | 0.35 | $-0.433$ | 0.60 | 0.679 | 0.85 | 2.543 |
| 0.11 | $-1.254$ | 0.36 | $-0.395$ | 0.61 | 0.733 | 0.86 | 2.685 |
| 0.12 | $-1.223$ | 0.37 | $-0.357$ | 0.62 | 0.788 | 0.87 | 2.781 |
| 0.13 | $-1.191$ | 0.38 | $-0.318$ | 0.63 | 0.844 | 0.88 | 2.911 |
| 0.14 | $-1.159$ | 0.39 | $-0.278$ | 0.64 | 0.901 | 0.89 | 3.049 |
| 0.15 | $-1.127$ | 0.40 | $-0.238$ | 0.65 | 0.959 | 0.90 | 3.198 |
| 0.16 | $-1.095$ | 0.41 | $-0.198$ | 0.66 | 1.019 | 0.91 | 3.359 |
| 0.17 | $-1.062$ | 0.42 | $-0.158$ | 0.67 | 1.080 | 0.92 | 3.533 |
| 0.18 | $-1.029$ | 0.43 | $-0.116$ | 0.68 | 1.143 | 0.93 | 3.728 |
| 0.19 | $-0.997$ | 0.44 | $-0.075$ | 0.69 | 1.206 | 0.94 | 3.943 |
| 0.20 | $-0.964$ | 0.45 | $-0.033$ | 0.70 | 1.272 | 0.95 | 4.190 |
| 0.21 | $-0.930$ | 0.46 | 0.010 | 0.71 | 1.339 | 0.96 | 4.481 |
| 0.22 | $-0.897$ | 0.47 | 0.054 | 0.72 | 1.408 | 0.97 | 4.835 |
| 0.23 | $-0.863$ | 0.48 | 0.098 | 0.73 | 1.479 | 0.98 | 5.301 |
| 0.24 | $-0.828$ | 0.49 | 0.142 | 0.74 | 1.552 | 0.99 | 6.013 |
| 0.25 | $-0.794$ | 0.50 | 0.187 | 0.75 | 1.626 | | |

of their generator and parity-check matrices; (3) decoding regions of a code; (4) minimum distance of a code; (5) error detection and correction capabilities of a code; and (6) general and minimum-distance decoding of a code.

Chapter 4 introduces a special category of linear block codes with cyclic structure, called *cyclic codes*. Cyclic codes are attractive for two reasons: first, encoding and error detection can be implemented easily by using simple shift-registers with feedback connections; and second, because they have considerable inherent algebraic structure, it is possible to devise various practical methods for hard-decision decoding. Several subclasses of linear block codes which have been widely used in digital communication and data storage systems are cyclic codes. Main coverage of this chapter includes: (1) basic concepts and structural properties of a cyclic code; (2) generation of a cyclic code in terms of its generator and parity-check polynomials; (3) encoding of a cyclic code; (4) general decoding of a cyclic code; (5) Hamming and quadratic-residue codes; (6) shortened cyclic and cyclic redundancy check (CRC) codes; and (7) a class of codes, called *quasi-cyclic* codes, which do not have a fully cyclic structure but do have a *partially* cyclic structure.

Chapter 5 presents a class of effective multiple random error-correcting codes, called *BCH codes*. BCH codes were discovered independently by Hoc-quenghem in 1959 and by Bose and Chaudhuri in 1960. In honor of them, these codes are named as BCH (Bose–Chaudhuri–Hocquenghem) codes. BCH codes are cyclic codes and constructed based on finite fields. An efficient hard-decision decoding algorithm for decoding BCH codes was devised by Berlekamp in 1965 and modified by Massey in the same year. This decoding algorithm is referred to as *Berlekamp–Massey iterative algorithm*. Since 1965, BCH codes have been widely used for error control in digital communication and storage systems. BCH codes presented in this chapter are binary codes. The major subjects covered in this chapter are: (1) generation of a BCH code by its generator polynomial; (2) structural properties of a BCH code; (3) error-correction capability of a BCH code; and (4) decoding a BCH code with the Berlekamp–Massey iterative decoding algorithm.

Chapter 6 presents a class of *nonbinary* BCH codes which are rich in structural properties. The code symbols of a nonbinary BCH code are elements of a finite field. Nonbinary BCH code are also cyclic and they are generalization of binary BCH codes. The most important and most widely used subclass of nonbinary BCH codes is the class of *Reed–Solomon* (RS) codes. Even though RS codes form a subclass of nonbinary BCH codes, they were dicovered independently by Reed and Solomon in 1960, the same year as the binary BCH codes discovered by Bose and Chaudhrui. The most effective algorithms for decoding RS codes are Belerkamp–Massey and Euclidean algorithms. The coverage of this chapter includes: (1) construction of nonbinary BCH codes; (2) structural properties of nonbinary BCH codes; (3) decoding of nonbinary BCH codes with Berlekamp–Massey iterative algorithm; (4) construction of RS codes; and (5) decoding RS codes with Berlekamp–Massey and Euclidean algorithms.

Chapter 7 presents another subclass of cyclic codes that are constructed based on *finite geometries*, called *finite-geometry codes*. These codes not only have attractive algebraic structures but also have distinctive geometric structures which allow them to be decoded with various hard-decision and soft-decision decoding methods and algorithms, ranging from low to high decoding complexity and from reasonably good to very good error performance. In this chapter, a very simple hard-decision decoding method, called *majority-logic decoding*, for decoding finite-geometry codes is presented. The coverage of this chapter includes: (1) basic concepts and fundamental structural properties of finite geometries; (2) construction of finite geometries based on finite fields; (3) construction of finite-geometry codes; and (4) majority-logic decoding of finite-geometry codes.

Chapter 8 presents another class of binary linear block codes for correcting multiple random errors. Codes in this class were first discovered by Muller in 1954, and in the same year Reed devised a simple method for decoding these codes. Codes in this class are called *Reed–Muller* (RM) codes. The decoding algorithm for RM codes devised by Reed is a majority-logic decoding algorithm which requires low-complexity logic circuit. RM codes have abundant algebraic structural properties. Besides the construction presented by Muller, there are several other methods for constructing these codes using algebraic, geometric, and logic approaches. Using a geometric approach, RM codes can be put in cyclic form and they form a special subclass of finite-geometry codes presented in Chapter 7. Coverage of this chapter includes: (1) construction of RM codes;

(2) investigation of their algebraic and geometric structures; and (3) majority-logic decoding of RM codes through *successive cancellation*.

Chapter 9 presents several coding techniques which are commonly used in communication and storage systems for reliable information transmission and data storage. These coding techniques include: (1) interleaving; (2) direct product; (3) concatenation; (4) turbo coding; (5) $|\mathbf{u}|\mathbf{u} + \mathbf{v}|$-construction; (6) Kronecker (or tensor) product; and (7) ARQ and HARQ schemes. The first six coding techniques allow us to construct long powerful codes from short codes for correcting multiple random errors with reduced decoding complexity. ARQ and HARQ are used to provide highly reliable information transmission in two-way communication systems.

Chapter 10 presents codes and coding techniques that are effective for correcting bursts of errors, random erasures, and bursts of erasures. The coverage of this chapter includes: (1) condition on correcting single burst of errors; (2) Fire codes for correcting single burst of errors; (3) short optimal and nearly optimal cyclic codes for correcting single burst of errors; (4) interleaved codes for correcting single long or multiple bursts of errors; (5) product codes for correcting multiple bursts of errors; (6) phased-burst-error-correcting codes; (7) interleaved and product codes for correcting multiple phased-bursts of errors; (8) correcting multiple phased-bursts of errors with RS codes; (9) Burton codes for correcting single phased-burst of errors; (10) characterization and correction of erasures; (11) correction of random errors and erasures over a BSEC; (12) correction of random erasures over a BEC; (13) RM codes for correcting random erasures; (14) correcting erasure-bursts over a BBEC; (15) cyclic codes for correcting single burst of erasures; (16) correction of multiple random bursts of erasures; and (17) RS codes for correcting nonbinary random symbol errors and erasures.

Chapter 11 introduces low-density parity-check (LDPC) codes which were discovered by Gallager in 1962 and rediscovered by MacKay, Luby, and others in middle of the 1990s. LDPC codes form a class of linear block codes which can achieve *near-capacity* (or *near Shannon limit*) performance on various communication and data-storage channels. Many LDPC codes have been adopted as standard codes for various current and next-generation communication systems, such as wireless (4G, 5G, etc.), optical, satellite, space, digital video broadcast (DVB), network communications, and others. This chapter covers the following aspects of LDPC codes: (1) basic concepts and definitions; (2) matrix and graph representations; (3) original construction methods for LDPC codes; (4) decoding algorithms, both hard-decision and soft-decision; (5) error performances and characteristics; and (6) categories of LDPC codes.

Chapter 12 presents a class of cyclic LDPC codes which are constructed based on finite geometries. This class of LDPC codes is the only class of LDPC codes with cyclic structure. Besides cyclic structure, they have other attractive algebraic and geometric structures. We call these codes *finite-geometry* (FG) *LDPC codes*. FG-LDPC codes have large minimum distances and their Tanner graphs have no trapping sets with size smaller than their minimum distances. Decoding of these codes with *iterative decoding based on belief-propagation*, such as the *sum-product* and *min-sum* algorithms, converges fast and can achieve a very low error rate without *error-floor*. In this chapter, the following subjects are covered: (1) construction of cyclic FG-LDPC codes; (2) characteristics and structural properties of the Tanner graphs of cyclic FG-LDPC codes; (3) complexity-reduced iterative decoding of cyclic FG-LDPC codes; (4) construction of *quasi-cyclic* (QC) LDPC codes; (5) masking for construction of QC-FG-LDPC codes; and (6) complexity-reduced iterative decoding of QC-LDPC codes.

Chapter 13 presents a class of QC-LDPC codes constructed based on *partial geometries* (PaGs) which have similar structural properties to the QC-FG-LDPC codes presented in Chapter 12. Codes in this class are called QC-PaG-LDPC codes. The construction of QC-PaG-LDPC codes is more flexible and gives a larger class of QC-LDPC codes than the construction of QC-FG-LDPC codes. The coverage of this chapter includes: (1) concepts and fundamental structural properties of PaGs; (2) construction of PaGs based on cyclic subgroups of prime order of finite fields and their associated QC-LDPC codes; (3) construction of PaGs based on prime fields and their associated QC-LDPC codes; and (4) construction of PaGs based on a special class of experimental designs, called *balanced incomplete block designs*, and their associated QC-LDPC codes.

Chapter 14 looks at the construction of QC-LDPC codes based on finite fields. Finite fields are not only effective in construction of algebraic codes, such as BCH and RS codes, but are also effective for constructing QC-LDPC codes. Decoding with the sum-product or min-sum algorithms, they can achieve good error performance in both waterfall and error-floor regions. The subjects covered in this chapter are: (1) two simple and flexible methods for constructing QC-LDPC codes based on two arbitrary subsets of a finite field; (2) construction of QC-LDPC codes based on finite fields in conjunction with masking; (3) design of masking matrices that produce QC-LDPC codes with good error performance in both waterfall and error-floor regions; (4) conditions on construction of finite-field-based QC-LDPC codes whose Tanner graphs have girth at least 8 and contain small numbers of short cycles; (5) construction of QC-LDPC codes based on RS codes; (6) construction of a subclass of QC-RS-LDPC codes whose Tanner graphs have girth 8; and (7) construction of globally coupled (GC) LDPC codes.

Chapter 15 presents LDPC codes constructed based on graphs. LDPC codes presented in Chapters 12 to 14 are constructed based on finite geometries and finite fields. In this chapter, two graphical methods to construct LDPC codes are presented. The constructed codes are called *graph-theoretic* LDPC codes. The best-known graphical methods for constructing this class of LDPC codes are the *protograph* (PTG)-based and the *progressive-edge-growth* (PEG) methods. The constructed LDPC codes are called PTG- and PEG-LDPC codes, respectively. Coverage of this chapter includes: (1) basic concepts and structures of graph-based LDPC codes; (2) construction of PTG-LDPC codes based on a chosen PTG with a copy-and-permute process; (3) a matrix-theoretic construction of QC-PTG-LDPC codes; (4) an algebraic construction of PTGs; and (5) construction of PEG-LDPC codes by building their Tanner graphs with progressive growing of their edges under certain constraints.

Chapter 16 presents a *universal coding scheme* for encoding and iterative soft-decision decoding of cyclic codes of prime lengths, *binary* and *nonbinary*. The key features of this coding scheme are *collective encoding* and *collective decoding*. The encoding of a cyclic code is performed on a *collection* of codewords which are mapped through *Hadamard permutations*, cascading, interleaving, and *Galois Fourier transform* (GFT) into a codeword in a *compound code* which is a nonbinary QC-LDPC code with a binary parity-check matrix. Using this matrix, *binary iterative decoding* is applied to jointly decode a *collection* of received vectors. The joint-decoding allows for *reliability information sharing* among the received vectors corresponding to the codewords in the collection during the iterative decoding process. This joint-decoding and information sharing can achieve a *joint-decoding gain* over the maximum likelihood decoding (MLD) of *individual* codewords. The binary iterative decoding can be performed efficiently and reduces the decoding complexity significantly. The coverage of this chapter includes: (1) a cyclic code of prime length and its Hadamard equivalents; (2) code composition, cascading, and interleaving of codewords in a cyclic code of prime length and its Hadamard equivalents; (3) GFT of an interleaved and cascaded composite code, called a *GFT interleaved cascaded composite* (GFT-ICC) code; (4) structural properties of a GFT-ICC code; (5) collective encoding of codewords in a cyclic code of prime length and its Hadamard equivalents in GFT domain; (6) collective iterative soft-decision decoding (ISDD) of codewords contained in a GFT-ICC code; (7) collective ISDD of RS-based GFT-ICC codes; (8) collective ISDD of BCH-based GFT-ICC codes; (9) collective ISDD of quadratic-residue (QR)-based GFT-ICC codes; (10) rate compatible GFT-ICC codes; and (11) erasure correction of a GFT-ICC-RS code.

Chapter 17 presents a class of channel capacity approaching codes, called *polar codes*. This chapter presents the construction, encoding, and decoding of polar codes following the Kronecker product in Section 9.5. Coverage of this chapter includes: (1) the Kronecker matrices and their structural properties; (2) Kronecker mappings and their logical implementations; (3) Kronecker vector spaces and Kronecker codes; (4) definition of polar codes and successive information retrieval; (5) channel polarization; (6) polar-code constructions based on channel polarization; and (7) successive cancellation decoding of polar codes.

# References

[1] C. Shannon, "A mathematical theory of communication," *Bell System Tech. J.*, **27**(3) (1948), 379–423 (Part I), 623–656 (Part II).

[2] R. Gallager, *Information Theory and Reliable Communication*, New York, Wiley, 1968.

[3] T. Berger, *Rate Distortion Theory*, Englewood Cliffs, NJ, Prentice-Hall, 1971.

[4] R. Gray, *Source Coding Theory*, Boston, MA, Kluwer Academic, 1990.

[5] J. Proakis and P. Massoud Salehi, *Digital Communications*, 5th ed., New York, McGraw-Hill, 2007.

[6] T. Cover and J. Thomas, *Elements of Information Theory*, 2nd ed., New York, Wiley, 2006.

[7] A. Dholakia, *Introduction to Convolutional Codes with Applications*, Boston, MA, Springer, 1994.

[8] C. Lee, *Convolutional Coding: Fundamentals and Applications*, Norwood, MA, Artech House, 1997.

[9] R. Johannesson and K. S. Zigangirov, *Fundamentals of Convolutional Coding*, Hoboken, NJ, Wiley-IEEE Press, 1999.

[10] S. Lin and D. J. Costello, Jr., *Error Control Coding: Fundamentals and Applications*, 2nd ed., Upper Saddle River, NJ, Prentice Hall, 2004.

[11] R. W. Hamming, "Error detecting and error correcting codes," *Bell Syst. Tech. J.*, **29** (1950), 147–160.

[12] J. J. Metzner and K. C. Morgan, "Coded feedback communication systems," *Proc. Nat. Electron. Conf.*, Chicago, Il, 1960, pp. 643–647.

[13] J. J. Metzner, "A study of an efficient retransmission strategy for data links," *NTC Conf. Rec.*, 1977, pp. 3B:1–1–3B:1–5.

[14] J. M. Morris, "Optimal block lengths for ARQ error control schemes," *IEEE Trans. Commun.*, **27**(2) (1979), 488–493.

[15] S. Lin, D. J. Costello, Jr., and M. J. Miller, "Automatic-repeat-request error control schemes," *IEEE Commun. Magazine*, **22**(12) (1984), 5–17.

[16] K. Brayer, "Error control techniques using binary symbol burst codes," *IEEE Trans. Commun.*, **16**(2) (1968), 199–214.

[17] H. O. Burton and D. D. Sullivan, "Errors and error control," *Proc. IEEE*, **10**(11) (1972), 1293–1310.

# 2

# Some Elements of Modern Algebra and Graphs

The objective of this chapter is to present some important elements of modern algebra and graphs that are pertinent to the understanding of the fundamental structural properties and constructions of some important classes of classical and modern error-correcting codes. The elements to be covered are groups, finite fields, polynomials, vector spaces, matrices, and graphs. The treatment of these elements is mostly descriptive and no attempt is made to be mathematically rigorous. There are many excellent textbooks on modern algebra, combinatorial mathematics, and graph theories. Some of these textbooks [1–21] are listed as references at the end of this chapter. The readers may refer to them for more advanced algebraic coding theories.

## 2.1  Groups

### 2.1.1  Basic Definitions and Concepts

Let $G$ be a set of elements. A binary operation $*$ on $G$ is a *rule* that assigns to each pair of elements $a$ and $b$ in $G$ a *uniquely* defined third element $c = a * b$ in $G$. When such a binary operation $*$ is defined on $G$, we say that $G$ is *closed* under $*$. A binary operation $*$ on $G$ is said to be *associative* if, for any $a$, $b$, and $c$ in $G$,

$$a * (b * c) = (a * b) * c.$$

A group is an *algebraic system* with an associative binary operation $*$ defined on it.

**Definition 2.1** A set $G$ on which a binary operation $*$ is defined is called a *group* if the following conditions (or axioms) are satisfied.
  (1) The binary operation $*$ is associative.

(2) $G$ contains a unique element $e$ such that for any element $a$ in $G$,

$$a * e = e * a = a.$$

The element $e$ is called the *identity element* of $G$.

(3) For any element $a$ in $G$, there exists a unique element $a'$ in $G$ such that

$$a * a' = a' * a = e.$$

The element $a'$ is called the *inverse* of $a$, and vice versa.

A group is said to be *commutative* (or *abelian*) if its binary operation $*$ also satisfies the following condition: for any elements $a$ and $b$ in $G$,

$$a * b = b * a.$$

The set $\mathbb{Z}$ of all integers is a commutative group under real addition "+." The integer 0 is the identity element of $\mathbb{Z}$ with respect to real addition "+," and the integer $-k$ is the (additive) inverse of the integer $k$ in $\mathbb{Z}$. The set $\mathbb{Q}$ of all rational numbers excluding zero is a commutative group under real multiplication "·." The integer 1 is the identity element of $\mathbb{Q}$ with respect to real multiplication "·," and the rational number $\frac{b}{a}$ is the (multiplicative) inverse of $\frac{a}{b}$ in $\mathbb{Q}$. However, the set of all integers does not form a group under real multiplication "·" because condition (3) cannot be satisfied.

**Definition 2.2** The number of elements in a group $G$ is called the *order* (or *cardinality*) of the group, denoted by $|G|$. A group with finite order is called a *finite group*; otherwise, it is called an *infinite group*.

The group $\mathbb{Z}$ of integers under real addition and the group $\mathbb{Q}$ of rational numbers under real multiplication are infinite groups. In Section 2.1.2, we will present the constructions of several types of finite groups.

## 2.1.2   Finite Groups

Finite groups are of practical importance in the area of coding theory and its applications. In this section, we will present two classes of finite groups under two binary operations which are similar to the real addition and multiplication, called *modulo-m addition* and *modulo-m multiplication*, where $m$ is a positive integer.

For any positive integer $m$, it is possible to construct a group of order $m$ under a binary operation that is very similar to the real addition. Define a binary operation, denoted by $\boxplus$, on $G$, where $G = \{0, 1, 2, \ldots, m-1\}$ is a set of $m$ nonnegative integers less than $m$. For any two integers $i$ and $j$ in $G$, define the following operation:

$$i \boxplus j = r, \tag{2.1}$$

where $r$ is the *remainder* resulting from dividing the real sum $i+j$ by $m$. By the Euclidean division algorithm, $r$ is a nonnegative integer between 0 and $m-1$ and is therefore an element in $G$. Thus, $G$ is closed under the binary operation $\boxplus$, which is called *modulo-m addition*.

The modulo-$m$ addition is associative and commutative, because it is derived from real addition "+," which is associative and commutative. $G$ is a group under modulo-$m$ addition where: (1) the integer 0 is the identity element; and (2) for any nonzero integer $i \in G$, $m - i$ is also in $G$ and is the inverse of $i$ with respect to $\boxplus$, called the *additive inverse* of $i$. Clearly, $i$ is the additive inverse of $m - i$. Note that the additive inverse of the element 0 in $G$ is itself. The above group $G$ under modulo-$m$ addition is called an *additive group*. The order of this group is $m$.

**Example 2.1** In this example, we construct an additive group with two elements. Let $m = 2$. Consider the set of two integers, $G = \{0, 1\}$. The modulo-2 addition is shown in Table 2.1, which is called a *Cayley table*.

Table 2.1 The additive group $G = \{0, 1\}$ with modulo-2 addition.

| $\boxplus$ | 0 | 1 |
|---|---|---|
| 0 | 0 | 1 |
| 1 | 1 | 0 |

From Table 2.1, we readily see that

$$0 \boxplus 0 = 0, 0 \boxplus 1 = 1, 1 \boxplus 0 = 1, 1 \boxplus 1 = 0.$$

From the above equalities, we verify that the set $G = \{0, 1\}$ is closed under modulo-2 addition $\boxplus$ and $\boxplus$ is commutative. The element 0 is the identity element. The additive inverse of 0 is itself. Because $1 \boxplus 1 = 0$, the additive inverse of 1 is itself. To prove that modulo-2 addition $\boxplus$ is associative, we simply need to show that for $a$, $b$, and $c$ in $G = \{0, 1\}$, the equality

$$(a \boxplus b) \boxplus c = a \boxplus (b \boxplus c)$$

holds for all the eight possible combinations of $a$, $b$, and $c$ using *perfect induction* (proof by exhaustion). Hence, the set $G = \{0, 1\}$ is an additive group under modulo-2 addition. It is also a commutative group. ▲▲

**Example 2.2** Let $m = 7$. Table 2.2 displays an additive group $G = \{0, 1, 2, 3, 4, 5, 6\}$ under modulo-7 addition.

Table 2.2 The additive group $G = \{0, 1, 2, 3, 4, 5, 6\}$ with modulo-7 addition.

| $\boxplus$ | 0 | 1 | 2 | 3 | 4 | 5 | 6 |
|---|---|---|---|---|---|---|---|
| 0 | 0 | 1 | 2 | 3 | 4 | 5 | 6 |
| 1 | 1 | 2 | 3 | 4 | 5 | 6 | 0 |
| 2 | 2 | 3 | 4 | 5 | 6 | 0 | 1 |
| 3 | 3 | 4 | 5 | 6 | 0 | 1 | 2 |
| 4 | 4 | 5 | 6 | 0 | 1 | 2 | 3 |
| 5 | 5 | 6 | 0 | 1 | 2 | 3 | 4 |
| 6 | 6 | 0 | 1 | 2 | 3 | 4 | 5 |

From Table 2.2, we see that 0 is the identity element and each element in $G$ has a unique additive inverse. The additive inverses of 0, 1, 2, 3, 4, 5, and 6 are 0, 6, 5, 4, 3, 2, and 1, respectively. Hence, $G$ is a commutative group of order 7 under modulo-7 addition. ▲▲

Next, we present a class of finite groups under a binary operation which is very similar to real multiplication. Let $p$ be a prime, say 2, 3, 5, 7, 11, 13, 17, 19, 23, 29, 31, ..., 127, .... Consider the set of nonzero integers less than $p$, $G = \{1, 2, \ldots, p-1\}$. Note that every integer in $G$ is relatively prime to $p$. Define a binary operation, denoted by $\boxdot$, on $G$, as follows: for any two integers $i$ and $j$ in $G$,

$$i \boxdot j = r, \tag{2.2}$$

where $r$ is the remainder resulting from dividing the real product $i \cdot j$ by $p$. Because both $i$ and $j$ are relatively prime to $p$, the real product $i \cdot j$ is not divisible by $p$. Hence, $r$ is a positive integer less than $p$. Therefore, $r$ is in $G$ and $G$ is closed under $\boxdot$. The binary operation $\boxdot$ is referred to as *modulo-p multiplication*. Because real multiplication is associative and commutative, the modulo-$p$ multiplication $\boxdot$ is also associative and commutative. The integer 1 is the identity element of $G$ with respect to $\boxdot$. It can be proved that every element $i$ in $G$ has a unique inverse with respect to $\boxdot$, which is called the *multiplicative inverse* of $i$. The multiplicative inverse of $i$ is the integer $j$ in $G$ for which the remainder of $i \cdot j$ divided by $p$ is 1. Such a group $G$ is called a *multiplicative group*.

**Example 2.3** Let $m = 7$. Table 2.3 displays the multiplicative group $G = \{1, 2, 3, 4, 5, 6\}$ under modulo-7 multiplication.

Table 2.3 The multiplicative group $G = \{1, 2, 3, 4, 5, 6\}$ with modulo-7 multiplication.

| $\boxdot$ | 1 | 2 | 3 | 4 | 5 | 6 |
|---|---|---|---|---|---|---|
| 1 | 1 | 2 | 3 | 4 | 5 | 6 |
| 2 | 2 | 4 | 6 | 1 | 3 | 5 |
| 3 | 3 | 6 | 2 | 5 | 1 | 4 |
| 4 | 4 | 1 | 5 | 2 | 6 | 3 |
| 5 | 5 | 3 | 1 | 6 | 4 | 2 |
| 6 | 6 | 5 | 4 | 3 | 2 | 1 |

From Table 2.3, we can easily check that every integer in $G$ has a unique multiplicative inverse, i.e., the multiplicative inverses of 1, 2, 3, 4, 5, 6 are 1, 4, 5, 2, 3, 6, respectively. ▲▲

Consider the multiplicative group $G = \{1, 2, \ldots, p-1\}$ under modulo-$p$ multiplication, where $p$ is a prime. Let $\alpha$ be an element in $G$. We define the powers of $\alpha$ as follows:

$$\alpha^1 = \alpha, \; \alpha^2 = \alpha \boxdot \alpha, \; \alpha^3 = \alpha \boxdot \alpha \boxdot \alpha, \dots, \alpha^i = \underbrace{\alpha \boxdot \alpha \boxdot \dots \boxdot \alpha}_{i \text{ elements}}, \dots.$$

Clearly, these powers of $\alpha$ are also elements of $G$.

**Definition 2.3** A multiplicative group $G$ is said to be *cyclic* if there exists an element $\alpha$ in $G$ such that, for any element $\beta$ in $G$, there is a nonnegative integer $i$ for which $\beta = \alpha^i$. Such an element $\alpha$ is called a *generator* of the cyclic group $G$ and we say that $G$ is generated by $\alpha$, denoted as $G = \langle \alpha \rangle$.

Consider the multiplicative group $G = \{1, 2, 3, 4, 5, 6\}$ under modulo-7 multiplication given by Table 2.3. The powers of 3 are:

$$3^1 = 3, \qquad 3^2 = 3 \boxdot 3 = 2, \qquad 3^3 = 3^2 \boxdot 3 = 2 \boxdot 3 = 6,$$
$$3^4 = 3^3 \boxdot 3 = 6 \boxdot 3 = 4, \; 3^5 = 3^4 \boxdot 3 = 4 \boxdot 3 = 5, \; 3^6 = 3^5 \boxdot 3 = 5 \boxdot 3 = 1.$$

The six powers of 3 give all the elements of the group $G = \{1, 2, 3, 4, 5, 6\}$ under modulo-7 multiplication. Hence, $G$ is a cyclic group with 3 as a generator. The six powers of 5 also give all the elements of $G$. Hence, 5 is also a generator of $G$. This says that a cyclic group may have more than one generator.

Consider the element 2 in $G = \{1, 2, 3, 4, 5, 6\}$. Its powers are:

$$2^1 = 2, 2^2 = 2 \boxdot 2 = 4, 2^3 = 2^2 \boxdot 2 = 4 \boxdot 2 = 1, 2^4 = 2^3 \boxdot 2 = 1 \boxdot 2 = 2.$$

It can be seen that 2 is not a generator of the group $G$ because its powers do not give all the elements in $G$.

### 2.1.3  Subgroups and Cosets

For a finite group $G$, there may exist a subset $S$ of $G$ which is itself a group under the binary operation on $G$. Such a subset $S$ is called a *subgroup* of $G$. In this section, we present the conditions under which a subset $S$ of $G$ is a subgroup and other aspects and concepts related to subgroups.

**Definition 2.4** A nonempty subset $S$ of a group $G$ with a binary operation $*$ defined on it is called a *subgroup* of $G$ if $S$ is itself a group with respect to the binary operation $*$ of $G$, i.e., $S$ satisfies all the axioms of a group under the same operation $*$ of $G$.

To determine whether a subset $S$ of a group $G$ with a binary operation $*$ is a subgroup, we do not need to verify all the axioms. The following two axioms are sufficient.

(**S1**) For any two elements $a$ and $b$ in $S$, $a * b$ is also an element in $S$.

(**S2**) For any element $a$ in $S$, its inverse is also an element in $S$.

The axiom **S1** implies that $S$ is closed under the operation $*$. Axioms **S1** and **S2** together imply that $S$ contains the identity element $e$ of $G$. $S$ satisfies the associative law by virtue that every element in $S$ is also in $G$ and $*$ is an operation on $G$.

**Example 2.4** Consider the additive group $G = \{0, 1, 2, 3, 4, 5, 6, 7\}$ under modulo-8 addition given by Table 2.4. The subset $S = \{0, 2, 4, 6\}$ forms a subgroup of $G$ under modulo-8 addition as shown in Table 2.5.

Table 2.4 The additive group $G = \{0, 1, 2, 3, 4, 5, 6, 7\}$ under modulo-8 addition.

| ⊞ | 0 | 1 | 2 | 3 | 4 | 5 | 6 | 7 |
|---|---|---|---|---|---|---|---|---|
| 0 | 0 | 1 | 2 | 3 | 4 | 5 | 6 | 7 |
| 1 | 1 | 2 | 3 | 4 | 5 | 6 | 7 | 0 |
| 2 | 2 | 3 | 4 | 5 | 6 | 7 | 0 | 1 |
| 3 | 3 | 4 | 5 | 6 | 7 | 0 | 1 | 2 |
| 4 | 4 | 5 | 6 | 7 | 0 | 1 | 2 | 3 |
| 5 | 5 | 6 | 7 | 0 | 1 | 2 | 3 | 4 |
| 6 | 6 | 7 | 0 | 1 | 2 | 3 | 4 | 5 |
| 7 | 7 | 0 | 1 | 2 | 3 | 4 | 5 | 6 |

Table 2.5 A subgroup $S = \{0, 2, 4, 6\}$ of $G$ given in Table 2.4 under modulo-8 addition.

| ⊞ | 0 | 2 | 4 | 6 |
|---|---|---|---|---|
| 0 | 0 | 2 | 4 | 6 |
| 2 | 2 | 4 | 6 | 0 |
| 4 | 4 | 6 | 0 | 2 |
| 6 | 6 | 0 | 2 | 4 |

Table 2.5 shows that $S$ is closed under modulo-8 addition and each element in $S$ has a unique inverse in $S$. Following the axioms **S1** and **S2**, $S$ is a subgroup of $G$. ▲▲

**Definition 2.5** Let $S$ be a subgroup of a group $G$ under binary operation $*$. Let $a$ be an element of $G$. Define two sets of elements in $G$: $a * S \triangleq \{a * s : s \in S\}$ and $S * a \triangleq \{s * a : s \in S\}$. The set $a * S$ is called a *left coset* of $S$ and the other set $S * a$ is called the *right coset* of $S$. The element $a$ is called a *coset leader* of $a * S$ (or $S * a$).

It is obvious that, if $G$ is commutative, every left coset $a * S$ is identical to its corresponding right coset $S * a$, i.e., $a * S = S * a$. If $a$ is the identity element in $G$, $a = e$, then $e * S = S * e = S$. Thus, $S$ is also considered as a coset of itself. If $a$ is an element in $S$, then $a * S = S$ and $S * a = S$, i.e., $a * S = S * a = S$. In the context of this book, we mainly deal with commutative groups. Therefore we do not differentiate between left cosets and right cosets, and just call them cosets. Additive and multiplicative groups both have cosets.

**Example 2.5** Consider the additive group $G = \{0, 1, 2, 3, 4, 5, 6, 7\}$ under modulo-8 addition given by Table 2.4 and its subgroup $S = \{0, 2, 4, 6\}$. Let $a = 3$. Then, the coset $3 \boxplus S$ is formed as below:

$$3 \boxplus S = \{3 \boxplus 0, 3 \boxplus 2, 3 \boxplus 4, 3 \boxplus 6\} = \{3, 5, 7, 1\}.$$

It can be seen that $S$ and its coset $3 \boxplus S$ are disjoint and they together form the group $G$, i.e., $G = S \cup (3 \boxplus S)$, where $\cup$ denotes the union operation.

If we choose 5 as the coset leader, the coset $5 \boxplus S$ is

$$5 \boxplus S = \{5 \boxplus 0, 5 \boxplus 2, 5 \boxplus 4, 5 \boxplus 6\} = \{5, 7, 1, 3\}.$$

We can see that $3 \boxplus S = 5 \boxplus S$, i.e., the two coset leaders 3 and 5 give the same coset of $S$. Similarly, we can verify that $3 \boxplus S = 5 \boxplus S = 7 \boxplus S = 1 \boxplus S$. That is, $S$ has only two cosets: $S = \{0, 2, 4, 6\}$ and $3 \boxplus S = \{3, 5, 7, 1\}$. ▲▲

In the following, we present several structural properties of a coset without proofs (see Problem 2.11).

(1) All elements in a coset of $S$ are distinct.
(2) Two different cosets of $S$ have no common element, i.e., they are disjoint.
(3) Every element in $G$ appears in one and only one coset of $S$.
(4) All the distinct cosets of $S$ together form the group $G$, i.e., $G$ is the union of all the distinct cosets of $S$.

Because two different cosets of the subgroup $S$ are disjoint and all the distinct cosets of $S$ form the group $G$, we say that all the distinct cosets of $S$ form a *partition* (or a *coset decomposition*) of $G$, denoted by $G/S$. In Example 2.5, the subgroup $S = \{0, 2, 4, 6\}$ and its coset $3 \boxplus S$ form a partition of the group $G = \{0, 1, 2, 3, 4, 5, 6, 7\}$, i.e., $G = S \cup (3 \boxplus S)$. A partition $G/S$ has a property given in Theorem 2.1.

**Theorem 2.1 (Lagrange's theorem)** *Let $G$ be a group of order $n$ and $S$ be a subgroup of $G$ with order $m$. Then, $m$ divides $n$, and the partition $G/S$ comprises $n/m$ cosets of $S$.*

In Example 2.5, we see that the subgroup $S$ has two cosets, $S$ and $3 \boxplus S$, $S \cap (3 \boxplus S) = \emptyset$ (where $\cap$ denotes the intersection operation and $\emptyset$ denotes the empty set), and $S \cup (3 \boxplus S) = G$. The order of $S$ ($m = 4$) divides the order of the group $G$ ($n = 8$). The partition $G/S$ is composed of $n/m = 2$ cosets of $S$, namely, $S$ and $3 \boxplus S$.

## 2.2 Finite Fields

In this section, we will extend the concept of groups to another algebraic system on which two binary operations are defined, called a *field*. A field is a set of elements in which we can perform addition, subtraction, multiplication, and division (only on nonzero elements) *without leaving the set*. Fields with finite number of elements, called *finite fields*, play an important role in developing algebraic coding theory and constructing error-correcting codes whose encoding and decoding can be efficiently implemented. Well-known and widely adopted error-correcting codes in communication and storage systems are *Reed–Solomon codes* (RS codes) [22] and *Bose–Chaudhuri–Hocquenghem codes* (BCH codes) [23, 24] which were discovered at the beginning of the 1960s. These two classes of codes are constructed based on finite fields. Most recently, finite fields have been successfully utilized to construct *low-density parity-check codes* (LDPC codes) [25] which perform close to the Shannon limit based on soft-decision iterative decoding. In this section, we will first introduce some basic concepts and fundamental properties of fields and then present a special class of finite fields.

## 2.2.1    Basic Definitions and Concepts

**Definition 2.6** Let $F$ be a set of elements on which two binary operations, called *addition* "+" and *multiplication* "·," are defined.[1] $F$ is a field under these two operations if the following conditions (or axioms) are satisfied.

    (1)  $F$ is a *commutative group* under addition "+." The identity element with respect to addition "+" is called the *zero element* (or *additive identity*) of $F$ and is denoted by 0.

    (2)  The set $F^* = F \backslash \{0\}$ of nonzero elements of $F$ forms a *commutative group* under multiplication "·." The identity element with respect to multiplication "·" is called the *unit element* (or *multiplicative identity*) of $F$ and is denoted by 1.

    (3)  For any three elements $a$, $b$, and $c$ in $F$, the following equality holds:

$$a \cdot (b + c) = a \cdot b + a \cdot c.$$

    The above equality is called the *distributive law*, i.e., multiplication is *distributive across addition*.

From the above definition, we see that a field consists of two groups with respect to its two binary operations, addition and multiplication. The group under addition is called the *additive group* of the field and the group under multiplication is called the *multiplicative group* of the field. Because each group must contain an identity element, a field must contain at least two elements, 0 and 1. In a field, the additive inverse of an element $a$ is denoted by "$-a$," and the multiplicative inverse of a *nonzero element* $b$ is denoted by "$b^{-1}$."

---

[1]Hereafter, we use "+" and "·" to denote both the real addition and multiplication and those defined on a field. In the context of a field, these two notations represent addition and multiplication defined on the field; otherwise, they represent real addition and multiplication.

Based on the addition and multiplication operations and additive and multiplicative inverses of elements of a field, two other operations, namely, *subtraction* "−" and *division* "÷," can be defined. Subtracting a field element $b$ from another field element $a$ is defined as adding the additive inverse $-b$ of $b$ to $a$, i.e.,

$$a - b \triangleq a + (-b).$$

It is clear that $a - a = 0$. Dividing a nonzero field element $a$ by a nonzero element $b$ is defined as multiplying $a$ by the multiplicative inverse $b^{-1}$ of $b$, i.e.,

$$a \div b \triangleq a \cdot (b^{-1}).$$

It is clear that $a \div a = 1$.

**Definition 2.7** The number of elements in a field is called the *order* (or *cardinality*) of the field. A field with finite order is called a *finite field*, otherwise called an *infinite field*.

**Definition 2.8** Let $F$ be a field. A subset $K$ of $F$ which is itself a field under the operations of $F$ is called a *subfield* of $F$. In this context, $F$ is called an *extension field* of $K$. If $K \neq F$, we say that $K$ is a *proper subfield* of $F$. A field containing no proper subfield is called a *prime field*.

The system of rational numbers with real addition and multiplication is a field, called *rational-number field* $\mathbb{Q}$. The system of real numbers is a field, called *real-number field* $\mathbb{R}$. The system of complex numbers is also a field, known as *complex-number field* $\mathbb{C}$. The fields $\mathbb{Q}$, $\mathbb{R}$, and $\mathbb{C}$ are examples of infinite fields. The complex-number field $\mathbb{C}$ is actually constructed from the real-number field $\mathbb{R}$ by requiring the symbol $i = \sqrt{-1}$ to be a root of the irreducible polynomial $X^2 + 1$ over the real-number field $\mathbb{R}$ (i.e., the polynomial $X^2 + 1$ has no root in $\mathbb{R}$), i.e., $\mathbb{C} = \{a + bi \mid a, b \in \mathbb{R}\}$. By setting $b = 0$, we obtain the real-number field $\mathbb{R}$. Thus, we say that the complex-number field $\mathbb{C}$ contains the real-number field $\mathbb{R}$ as a proper subfield and the complex-number field $\mathbb{C}$ is an extension field of the real-number field $\mathbb{R}$. The rational-number field $\mathbb{Q}$ is a prime field because it has no proper subfield.

It is noted that the complex-number field $\mathbb{C}$ is constructed based on a subfield (the real-number field $\mathbb{R}$) and an irreducible polynomial $X^2+1$ over that subfield. In later sections, we will show that other types of fields can be constructed in a similar way, i.e., choosing a subfield and an irreducible polynomial of a special type over the subfield.

A field $F$ has the following fundamental properties.
(1) For every element $a$ in $F$, $a \cdot 0 = 0 \cdot a = 0$.
(2) For any two nonzero elements $a$ and $b$ in $F$, $a \cdot b \neq 0$.
(3) For two elements $a$ and $b$ in $F$, $a \cdot b = 0$ implies that either $a = 0$ or $b = 0$.
(4) For any two elements $a$ and $b$ in $F$,

$$-(a \cdot b) = (-a) \cdot b = a \cdot (-b).$$

(5) For $a \neq 0$, $a \cdot b = a \cdot c$ implies that $b = c$ (called the *cancellation law*).

Because the unit element 1 is an element in $F$ and $F$ is an additive group under "+," the following sums:

$$1, 1 + 1, 1 + 1 + 1, \ldots, \underbrace{1 + 1 + \cdots + 1}_{\text{sum of } k \text{ unit element } 1}, \quad \ldots$$

are elements in $F$.

**Definition 2.9** Let $F$ be a field and 1 be its unit element (or multiplicative identity). The *characteristic* of $F$ is defined as the *smallest* positive integer $\lambda$ such that

$$\sum_{i=1}^{\lambda} 1 = \underbrace{1 + 1 + \cdots + 1}_{\lambda} = 0, \tag{2.3}$$

where the summation represents repeated applications of addition + of the unit element of $F$ $\lambda$ times. If no such $\lambda$ exists, $F$ is said to have *characteristic zero*, i.e., $\lambda = 0$, and $F$ is an infinite field in this case.

Because $\mathbb{Q}$, $\mathbb{R}$, and $\mathbb{C}$ are infinite fields, they have characteristic zero.

**Theorem 2.2** *The characteristic $\lambda$ of a finite field is a prime.*

*Proof* Suppose $\lambda$ is a not a prime. Then, $\lambda$ can be factored as a product of two smaller positive integers, say $k$ and $\ell$, with $1 < k, \ell < \lambda$. Then, $\lambda = k\ell$. It follows from the distributive law that

$$\sum_{i=1}^{\lambda} 1 = \left( \sum_{i=1}^{k} 1 \right) \cdot \left( \sum_{i=1}^{\ell} 1 \right) = 0.$$

The property (3) of a field implies that either $\sum_{i=1}^{k} 1 = 0$ or $\sum_{i=1}^{\ell} 1 = 0$. Because $1 < k, \ell < \lambda$, either $\sum_{i=1}^{k} 1 = 0$ or $\sum_{i=1}^{\ell} 1 = 0$ violates the minimality of the characteristic. Therefore, $\lambda$ must be a prime. ▲▲

Consider a finite field $F$ with characteristic $\lambda$. It follows from the definition of the characteristic of a finite field and Theorem 2.2 that the following $\lambda$ sums of the unit element of $F$:

$$1 = \sum_{i=1}^{1} 1, \sum_{i=1}^{2} 1, \sum_{i=1}^{3} 1, \ldots, \sum_{i=1}^{\lambda-1} 1, \sum_{i=1}^{\lambda} 1 = 0 \tag{2.4}$$

are $\lambda$ distinct elements in $F$ and they form a field, denoted by $\mathrm{GF}(\lambda)$, under the addition and multiplication operation of $F$ (see Problem 2.13). Because $\mathrm{GF}(\lambda)$

is a subset of $F$, it is a subfield of $F$ and is the smallest subfield of $F$. The field $GF(\lambda)$ is also a prime field because it contains no proper subfield.

It is possible to construct finite fields. Finite fields are also called *Galois fields* after their discovery by a French mathematician Evariste Galois. For any *prime* $p$ and any positive integer $m$, there exists a Galois field, denoted by $GF(q)$, with $q = p^m$ elements [26]. (In some of the literature, the Galois field is denoted by $\mathbb{F}_q$. In this book, we use the notation $GF(q)$.) In Section 2.2.2, we will start with finite fields constructed from prime numbers.

## 2.2.2 Prime Fields

Let $p$ be a prime. We have shown in Section 2.1 that the set $\{0, 1, \ldots, p-1\}$ forms a commutative group under modulo-$p$ addition and its subset $\{1, 2, \ldots, p-1\}$ forms a commutative group under modulo-$p$ multiplication. Based on the definitions of modulo-$p$ addition and multiplication and the fact that real addition and multiplication satisfy the distributive law, modulo-$p$ addition and multiplication also satisfy the distributive law. Therefore, the set

$$GF(p) = \{0, 1, ..., p-1\}$$

forms a finite field with $p$ elements under modulo-$p$ addition and multiplication, which is called a *prime field*. The zero element of the field $GF(p)$ is 0 and the unit element of the field $GF(p)$ is 1.

If $m$ is not a prime, the set $\{0, 1, 2, \ldots, m-1\}$ does not form a field under modulo-$m$ addition and multiplication (see Problem 2.15).

**Example 2.6** The smallest prime is $p = 2$. The set $\{0, 1\}$ forms a field $GF(2)$ with two elements under modulo-2 addition and multiplication as defined in Table 2.6.

Table 2.6 The prime field $GF(2)$ under modulo-2 addition and multiplication.

| ⊞ | 0 | 1 |
|---|---|---|
| 0 | 0 | 1 |
| 1 | 1 | 0 |

| ⊡ | 0 | 1 |
|---|---|---|
| 0 | 0 | 0 |
| 1 | 0 | 1 |

This field is also called a *binary field*, the smallest finite field. The binary field $GF(2)$ plays an extremely important role in coding theory and it is a commonly used alphabet of code symbols in error-correcting codes. ▲▲

**Example 2.7** Let $p = 7$. The set $GF(7) = \{0, 1, 2, 3, 4, 5, 6\}$ forms a prime field with 7 elements under modulo-7 addition and multiplication as defined in Tables 2.2 and 2.3, respectively. The field $GF(7)$ is given in Table 2.7.

Table 2.7 The field GF(7) under modulo-7 addition and multiplication.

| ⊞ | 0 1 2 3 4 5 6 |
|---|---|
| 0 | 0 1 2 3 4 5 6 |
| 1 | 1 2 3 4 5 6 0 |
| 2 | 2 3 4 5 6 0 1 |
| 3 | 3 4 5 6 0 1 2 |
| 4 | 4 5 6 0 1 2 3 |
| 5 | 5 6 0 1 2 3 4 |
| 6 | 6 0 1 2 3 4 5 |

| ⊡ | 0 1 2 3 4 5 6 |
|---|---|
| 0 | 0 0 0 0 0 0 0 |
| 1 | 0 1 2 3 4 5 6 |
| 2 | 0 2 4 6 1 3 5 |
| 3 | 0 3 6 2 5 1 4 |
| 4 | 0 4 1 5 2 6 3 |
| 5 | 0 5 3 1 6 4 2 |
| 6 | 0 6 5 4 3 2 1 |

Consider the calculation of $5 \boxdot (2 \boxplus 4)$. By using the distributive law, we have

$$5 \boxdot (2 \boxplus 4) = (5 \boxdot 2) \boxplus (5 \boxdot 4).$$

Using the multiplication table given in Table 2.7, we find $5 \boxdot 2 = 3$ and $5 \boxdot 4 = 6$. Then,

$$5 \boxdot (2 \boxplus 4) = 3 \boxplus 6.$$

Using the addition table in Table 2.7, we find $3 \boxplus 6 = 2$. Hence,

$$5 \boxdot (2 \boxplus 4) = 2.$$

If we first perform the addition $2 \boxplus 4$, we have $2 \boxplus 4 = 6$. Then,

$$5 \boxdot (2 \boxplus 4) = 5 \boxdot 6.$$

Using the multiplication table in Table 2.7, we have

$$5 \boxdot (2 \boxplus 4) = 5 \boxdot 6 = 2.$$

From above, we see that both ways of performing $5 \boxdot (2 \boxplus 4)$ give the same result. ▲▲

### 2.2.3   Finite Fields with Orders of Prime Powers

In Section 2.2.2, we showed that, for every prime number $p$, there exists a prime field GF($p$). For any positive integer $m$, it is possible to construct a Galois field GF($p^m$) with $p^m$ elements based on the prime field GF($p$). This will be shown in Section 2.4. The Galois field GF($p^m$) contains GF($p$) as a subfield and is an extension field of GF($p$). A very important feature of a finite field is that its order must be a power of a prime. That is to say that the order of any finite field is a power of a prime (see a possible proof in [27]). For simplicity, hereafter, we use GF($q$) to denote a Galois (or finite) field with $q$ elements, where $q$ is a power of a prime. Before we consider the construction of extension fields of prime fields, we continue to develop some important structural properties of a finite field that will be used in Chapters 12–16 for code construction.

Let $a$ be a nonzero element in $\mathrm{GF}(q)$. Because the set of nonzero elements of $\mathrm{GF}(q)$ is closed under multiplication, the following powers of $a$,

$$a^1 = a, a^2 = a \cdot a, a^3 = a \cdot a \cdot a, a^4 = a \cdot a \cdot a \cdot a, \ldots,$$

must also be nonzero elements in $\mathrm{GF}(q)$. Because $\mathrm{GF}(q)$ has only a finite number of elements, the powers of $a$ in the above sequence cannot all be distinct. Therefore, at some point in the sequence of powers of $a$, there must be a repetition; that is, there must exist two positive integers $k$ and $s$ such that $s > k$ and $a^k = a^s$. Let $a^{-1}$ be the multiplicative inverse of $a$. Then, $(a^{-1})^k = a^{-k}$ is the multiplicative inverse of $a^k$. Multiplying both sides of $a^k = a^s$ by $a^{-k}$, we obtain

$$1 = a^{s-k}.$$

The above equality implies that there must exist a *smallest positive integer* $n$ such that $a^n = 1$. The integer $n$ is called the *order* of the field element $a$. Therefore, the sequence, $a, a^2, a^3, \ldots,$ repeats itself after $a^n = 1$. Also, the powers $a, a^2, a^3, \ldots, a^{n-1}, a^n = 1$ are all distinct. In fact, they form a group under the multiplication of $\mathrm{GF}(q)$. First, we see that they contain the unit element 1. Consider two elements $a^i$ and $a^j$ in $\mathrm{GF}(q)$ and their product $a^i \cdot a^j$ with $1 \leq i, j \leq n$. If $i + j \leq n$,

$$a^i \cdot a^j = a^{i+j}.$$

If $i + j > n$, we have $i + j = n + r$, where $0 < r \leq n$. Hence,

$$a^i \cdot a^j = a^{i+j} = a^n \cdot a^r = a^r.$$

Therefore, the set of the powers $a, a^2, a^3, \ldots, a^{n-1}, a^n = 1$ is closed under the multiplication of $\mathrm{GF}(q)$. For $1 \leq i < n$, $a^{n-i}$ is the multiplicative inverse of $a^i$. Because the powers of $a$ are nonzero elements in $\mathrm{GF}(q)$, they satisfy the associative and commutative laws. Therefore, we conclude that $a^n = 1, a^1, a^2, \ldots, a^{n-1}$ form a commutative group under the multiplication of $\mathrm{GF}(q)$. This group is a *cyclic subgroup* of the multiplicative group of $\mathrm{GF}(q)$ with $a$ as its generator. The order of this cyclic subgroup is $n$.

Consider the element 2 of the prime field $\mathrm{GF}(7)$ given by Table 2.7. The powers of 2 are:

$$2, 2^2 = 4, 2^3 = 1.$$

Therefore, the smallest positive integer $n$ for which $2^n = 1$ is $n = 3$. Hence, the order of 2 is 3. The elements 1, 2, and 4 form a cyclic subgroup of the multiplicative group of the prime field $\mathrm{GF}(7)$. Suppose we consider the element 5 in $\mathrm{GF}(7)$. The power of 5 are $5, 5^2 = 4, 5^3 = 6, 5^4 = 2, 5^5 = 3$, and $5^6 = 1$. Hence, the order of 5 is 6. The 6 powers of 5 give all the 6 nonzero elements of $\mathrm{GF}(7)$ and they form the multiplicative group of $\mathrm{GF}(7)$.

**Theorem 2.3** *Let a be a nonzero element of a finite field GF(q). Then,* $a^{q-1} = 1$.

*Proof* Let $b_1, b_2, \ldots, b_{q-1}$ be the $q - 1$ nonzero elements of GF($q$) and $a$ be any nonzero element in GF($q$). Clearly, the $q - 1$ elements $a \cdot b_1, a \cdot b_2, \ldots, a \cdot b_{q-1}$ are nonzero, distinct, and form all the nonzero elements of GF($q$). Thus,

$$(a \cdot b_1) \cdot (a \cdot b_2) \cdots (a \cdot b_{q-1}) = b_1 \cdot b_2 \cdot b_3 \cdots b_{q-1}$$
$$a^{q-1} \cdot (b_1 \cdot b_2 \cdot b_3 \cdots b_{q-1}) = 1 \cdot (b_1 \cdot b_2 \cdot b_3 \cdots b_{q-1}).$$

Because $a \neq 0$ and $b_1 \cdot b_2 \cdot b_3 \cdots b_{q-1} \neq 0$, we must have $a^{q-1} = 1$ (using the cancellation law). ▲▲

**Theorem 2.4** *Let a be a nonzero element in a finite field GF(q). Let n be the order of a. Then, n divides q − 1.*

*Proof* Suppose $n$ does not divide $q - 1$. Dividing $q - 1$ by $n$, we obtain

$$q - 1 = kn + r,$$

where $k$ and $r$ are nonzero integers with $0 < r < n$. Then,

$$a^{q-1} = a^{kn+r} = a^{kn} \cdot a^r = (a^n)^k \cdot a^r.$$

Because $a^{q-1} = 1$ and $a^n = 1$, we must have $a^r = 1$. This is impossible, because $0 < r < n$ and $n$ is the smallest positive integer such that $a^n = 1$. Therefore, $n$ must divide $q - 1$. ▲▲

Let $a$ be an element of order $n$ in GF($q$). Let $b = a^i$ for $2 \leq i < n$. It can be verified that the order of the element $b = a^i$ is $n$ if and only if $i$ and $n$ are relatively prime, i.e., the greatest common divider (GCD) of $i$ and $n$ is GCD$\{i, n\} = 1$. If GCD$\{i, n\} \neq 1$, then the order of $a^i$ is $k \cdot n/i$ where $k$ is the smallest positive integer such that $k \cdot n/i$ is a positive integer greater than 1.

**Theorem 2.5** *The number of elements in a finite field GF(q) with order n is*

$$\phi(n) = n \prod_{t|n} \left(1 - \frac{1}{t}\right), \tag{2.5}$$

*where $t|n$ denotes that $t$ divides $n$, $t$ takes all distinct positive primes that divide $n$, and the function $\phi(n)$ is known as the Euler $\phi$ function or Euler totient function [17].*

**Definition 2.10** In a finite field GF($q$), a nonzero element $a$ is said to be *primitive* if its order is $q - 1$.

Therefore, the $q - 1$ powers of a primitive element $a$ generate all the nonzero elements of GF($q$). That is, all nonzero elements in GF($q$) can be represented as $q - 1$ consecutive powers of a primitive element $a$, i.e., GF($q$) = $\{0, a, a^2, \ldots, a^{q-2}, a^{q-1} = 1\}$. It can be shown that every finite field has at least one primitive element. It follows from Theorem 2.5 that there are exactly $\phi(q-1)$ primitive elements in GF($q$).

Consider the prime field GF(7) displayed by Table 2.7. If we take the powers of the element 3 in GF(7) using the multiplication table, we obtain

$$3^1 = 3, \qquad 3^2 = 3 \cdot 3 = 2, \ 3^3 = 3 \cdot 3^2 = 6,$$
$$3^4 = 3 \cdot 3^3 = 4, \ 3^5 = 3 \cdot 3^4 = 5, \ 3^6 = 3 \cdot 3^5 = 1.$$

Therefore, the order of the element 3 is 6, and it is a primitive element of GF(7). The field GF(7) can be represented as GF(7) $= \{0, 3, 3^2, 3^3, 3^4, 3^5, 3^6 = 1\}$. As shown earlier, the powers of 5 also give all the nonzero elements of GF(7). Hence, 5 is also a primitive element of GF(7). The powers of the element 4 in GF(7) are

$$4^1 = 4, 4^2 = 4 \cdot 4 = 2, 4^3 = 4 \cdot 4^2 = 1.$$

Clearly, the order of 4 is 3, which is a factor of 6. Hence, 4 is not a primitive element of GF(7). From Theorem 2.5, there are 2 primitive elements in GF(7).

## 2.3 Polynomials over Galois Field GF(2)

A polynomial $f(X) = f_0 + f_1 X + \cdots + f_n X^n$ in $X$ with coefficients $f_i$ from the binary field GF(2) is called *a polynomial over* GF(2) or a *binary polynomial*. If $f_n \neq 0$, $f(X)$ is said to have degree $n$; otherwise, it has degree less than $n$. For example, $1 + X + X^2$, $1 + X + X^4$, and $1 + X^3 + X^5$ are binary polynomials of degrees 2, 4, and 5, respectively. In general, there are $2^n$ polynomials over GF(2) with degree $n$.

Polynomials over GF(2) can be added (or subtracted), multiplied, and divided in the usual way. Let

$$g(X) = g_0 + g_1 X + \cdots + g_m X^m$$

be another polynomial over GF(2) of degree $m$. To add $f(X)$ and $g(X)$, we simply add the coefficients of the same power of $X$ in $f(X)$ and $g(X)$ as follows (assuming that $m \leq n$):

$$\begin{aligned} f(X) + g(X) = {} & (f_0 + g_0) + (f_1 + g_1)X + \cdots + (f_{m-1} + g_{m-1})X^{m-1} \\ & + (f_m + g_m)X^m + f_{m+1}X^{m+1} + \cdots + f_n X^n, \end{aligned} \tag{2.6}$$

where $f_i + g_i$ is carried out in modulo-2 addition. For example, adding $a(X) = 1 + X + X^3 + X^5$ and $b(X) = 1 + X^2 + X^3 + X^4 + X^7$, we obtain the following sum:

$$\begin{aligned} a(X) + b(X) &= (1+1) + X + X^2 + (1+1)X^3 + X^4 + X^5 + X^7 \\ &= X + X^2 + X^4 + X^5 + X^7. \end{aligned}$$

When we multiply $f(X)$ and $g(X)$, we obtain the following product:

$$f(X) \cdot g(X) = c(X) = c_0 + c_1 X + \cdots + c_{n+m-1} X^{n+m-1} + c_{n+m} X^{n+m},$$

where, for $0 \leq k \leq n + m$,

$$c_k = \sum_{i=0}^{n} \sum_{j=k-i}^{m} f_i \cdot g_j, \tag{2.7}$$

where the multiplication and addition of coefficients are modulo-2 operations. Multiplying $a(X)$ and $b(X)$ given above, we have

$$a(X) \cdot b(X) = 1 + X + X^2 + X^3 + X^5 + X^6 + X^7 + X^9 + X^{10} + X^{12}.$$

It is clear from (2.7) that, if $g(X) = 0$, then

$$f(X) \cdot 0 = 0.$$

We can readily verify that the polynomials over GF(2) satisfy the following conditions.

(1) Commutative:

$$a(X) + b(X) = b(X) + a(X).$$

(2) Associative:

$$a(X) + (b(X) + c(X)) = (a(X) + b(X)) + c(X),$$
$$a(X) \cdot (b(X) \cdot c(X)) = (a(X) \cdot b(X)) \cdot c(X).$$

(3) Distributive:

$$a(X) \cdot (b(X) + c(X)) = a(X) \cdot b(X) + a(X) \cdot c(X).$$

Suppose the degree of $g(X)$ is not zero. When $f(X)$ is divided by $g(X)$, we obtain a unique pair of polynomials over GF(2), $q(X)$ and $r(X)$, where $q(X)$ is called the *quotient* and $r(X)$ is called the *remainder*, such that

$$f(X) = q(X)g(X) + r(X),$$

and the degree of $r(X)$ is less than that of $g(X)$. This expression is known as *Euclid's division algorithm*. As an example, we divide $f(X) = 1 + X + X^4$ by $g(X) = 1 + X^2$. Using the long-division technique, we have

$$
\begin{array}{r}
X^2 \phantom{XX} + 1 \phantom{XXXXXXX} \\
\hline
X^2 + 1 ) X^4 \phantom{XXXX} + X + 1 \\
X^4 \phantom{XX} + X^2 \phantom{XXXXX} \\
\hline
X^2 + X + 1 \\
X^2 \phantom{XXX} + 1 \\
\hline
X \phantom{XXXXX}
\end{array}
$$

where the quotient is $q(X) = 1 + X^2$ and the remainder is $r(X) = X$. Then, $f(X)$ can be written as follows:

$$f(X) = 1 + X + X^4 = (1 + X^2)(1 + X^2) + X = (1 + X^2) \cdot g(X) + X.$$

When $f(X)$ is divided by $g(X)$, if the remainder $r(X)$ is identical to zero (i.e., $r(X) = 0$), we say that $f(X)$ is divisible by $g(X)$ and $g(X)$ is a factor of $f(X)$.

For real numbers, if $a$ is a *root* of a polynomial $f(X)$ (i.e., $f(a) = 0$), $f(X)$ is divisible by $X - a$. (This fact follows from Euclid's division algorithm.) This statement is still true for $f(X)$ over GF(2). For example, consider $f(X) = 1 + X^2 + X^3 + X^4$. Substituting $X = 1$ in $f(X)$, we obtain

$$f(1) = 1 + 1^2 + 1^3 + 1^4 = 1 + 1 + 1 + 1 = 0.$$

Thus, $f(X)$ has 1 as a root in GF(2) and it should be divisible by $X + 1$, as shown:

$$
\begin{array}{r}
X^3 + X\ \ + 1 \\
\hline
X + 1 \overline{)\, X^4 + X^3 + X^2 + \qquad 1} \\
X^4 + X^3 \\
\hline
X^2 + \qquad 1 \\
X^2 + X \\
\hline
X + 1 \\
X + 1 \\
\hline
0
\end{array}
$$

**Definition 2.11** A polynomial $p(X)$ over GF(2) of degree $m$ is said to be *irreducible* if it is not divisible by any polynomial over GF(2) of degree less than $m$ but greater than zero.

Among the four polynomials of degree 2, $X^2$, $1 + X^2$, and $X + X^2$ are not irreducible, because they are either divisible by $X$ or $1 + X$; however, $1 + X + X^2$ has neither 0 nor 1 as a root and hence is not divisible by any polynomial of degree 1. Therefore, $1 + X + X^2$ is an irreducible polynomial of degree 2 over GF(2). The polynomial $1 + X + X^3$ is an irreducible polynomial of degree 3 over GF(2). First, we note that $1 + X + X^3$ does not have either 0 or 1 as a root. Therefore, $1 + X + X^3$ is not divisible by $X$ or $1 + X$. Because the polynomial has degree 3 and is not divisible by any polynomial of degree 1, it cannot be divisible by a polynomial of degree 2. Consequently, $1 + X + X^3$ is irreducible over GF(2). We can verify that $1 + X + X^4$ is an irreducible polynomial over GF(2) of degree 4. It has been proved that for any $m \geq 1$, there exists an irreducible polynomial of degree $m$. An important theorem regarding irreducible polynomials over GF(2) is given in the following theorem without a proof (see a proof in [17]).

**Theorem 2.6** *Any irreducible polynomial over* GF(2) *of degree* $m$ *divides* $X^{2^m - 1} + 1$.

As an example of Theorem 2.6, we can check that $1 + X + X^3$ divides $X^{2^3-1} + 1 = X^7 + 1$:

$$
\begin{array}{r}
X^4 \quad\ + X^2 + X\ +1 \\ \hline
X^3 + X + 1 ) X^7 \qquad\qquad\qquad\qquad\qquad + 1 \\
\underline{X^7 \quad\ + X^5 + X^4} \\
X^5 + X^4 \qquad\qquad\quad + 1 \\
\underline{X^5 \qquad\quad + X^3 + X^2} \\
X^4 + X^3 + X^2 \qquad + 1 \\
\underline{X^4 \qquad\quad + X^2 + X} \\
X^3 \qquad + X + 1 \\
\underline{X^3 \qquad + X + 1} \\
0
\end{array}
$$

**Definition 2.12** An irreducible polynomial $p(X)$ of degree $m$ over GF(2) is said to be *primitive* if the *smallest* positive integer $n$ for which $p(X)$ divides $X^n + 1$ is $n = 2^m - 1$.

We may check that $p(X) = 1 + X + X^4$ over GF(2) divides $X^{15} + 1$ but does not divide any polynomial $X^n + 1$ for $1 \leq n < 15$. Hence, $p(X) = 1 + X + X^4$ is a primitive polynomial over GF(2). The polynomial $1 + X + X^2 + X^3 + X^4$ over GF(2) is irreducible but it is not primitive, because it divides $X^5 + 1$. It is not easy to recognize a primitive polynomial; however, there are tables of irreducible polynomials in which primitive ones are indicated. For any positive integer $m$, there may exist more than one primitive polynomial of degree $m$ over GF(2). Table 2.8 gives a list of primitive polynomials [28] of degrees less than 33 over GF(2). For each degree $m$, we list only a primitive polynomial with the smallest number of terms.

In the following, we derive another property of polynomials over GF(2) which will assist the analysis of some properties of extension fields of GF(2).

Let

$$f(X) = f_0 + f_1 X + \cdots + f_n X^n$$

be a polynomial over GF(2). Consider

$$
\begin{aligned}
f^2(X) &= (f_0 + f_1 X + \cdots + f_n X^n)^2 = (f_0 + (f_1 X + \cdots + f_n X^n))^2 \\
&= f_0^2 + f_0 \cdot (f_1 X + \cdots + f_n X^n) + f_0 \cdot (f_1 X + \cdots + f_n X^n) \\
&\quad + (f_1 X + \cdots + f_n X^n)^2 \\
&= f_0^2 + (f_1 X + \cdots + f_n X^n)^2.
\end{aligned}
$$

Table 2.8 A list of primitive polynomials over GF(2).

| Degree $m$ | Primitive polynomials | Degree $m$ | Primitive polynomials |
|---|---|---|---|
| 3 | $1 + X + X^3$ | 18 | $1 + X^7 + X^{18}$ |
| 4 | $1 + X + X^4$ | 19 | $1 + X + X^2 + X^5 + X^{19}$ |
| 5 | $1 + X^2 + X^5$ | 20 | $1 + X^3 + X^{20}$ |
| 6 | $1 + X + X^6$ | 21 | $1 + X^2 + X^{21}$ |
| 7 | $1 + X^3 + X^7$ | 22 | $1 + X + X^{22}$ |
| 8 | $1 + X^2 + X^3 + X^4 + X^8$ | 23 | $1 + X^5 + X^{23}$ |
| 9 | $1 + X^4 + X^9$ | 24 | $1 + X + X^2 + X^6 + X^{24}$ |
| 10 | $1 + X^3 + X^{10}$ | 25 | $1 + X^3 + X^{25}$ |
| 11 | $1 + X^2 + X^{11}$ | 26 | $1 + X + X^2 + X^6 + X^{26}$ |
| 12 | $1 + X + X^4 + X^6 + X^{12}$ | 27 | $1 + X + X^2 + X^5 + X^{27}$ |
| 13 | $1 + X + X^3 + X^4 + X^{13}$ | 28 | $1 + X^3 + X^{28}$ |
| 14 | $1 + X + X^6 + X^{10} + X^{14}$ | 29 | $1 + X^2 + X^{29}$ |
| 15 | $1 + X + X^{15}$ | 30 | $1 + X + X^2 + X^{23} + X^{30}$ |
| 16 | $1 + X + X^3 + X^{12} + X^{16}$ | 31 | $1 + X^3 + X^{31}$ |
| 17 | $1 + X^3 + X^{17}$ | 32 | $1 + X + X^2 + X^{22} + X^{32}$ |

Expanding the preceding equation repeatedly, we eventually obtain

$$f^2(X) = f_0^2 + (f_1 X)^2 + (f_2 X^2)^2 + \cdots + (f_n X^n)^2.$$

Because $f_i = 0$ or $1$, $f_i^2 = f_i$. Hence, we have

$$f^2(X) = f_0 + f_1 X^2 + f_2 (X^2)^2 + \cdots + f_n (X^2)^n$$
$$= f(X^2). \tag{2.8}$$

It follows from (2.8) that, for any $i \geq 0$,

$$[f(X)]^{2^i} = f(X^{2^i}). \tag{2.9}$$

Polynomials with coefficients from a finite field other than GF(2) have similar properties as those over GF(2). The general rules applied to polynomials over GF(2) can be applied to polynomials over any finite field. More on polynomials over a nonbinary field will be presented in later chapters.

## 2.4 Construction of Galois Field GF($2^m$)

Existing block codes are mostly constructed based on the binary field GF(2) given in Example 2.6 and its extension field GF($2^m$) with $2^m$ elements, where $m$ is a positive integer greater than one. In this section, we present the construction of an extension field of GF(2).

The construction of an extension field GF($2^m$) of GF(2) begins with a primitive polynomial $p(X) = p_0 + p_1 X + \cdots + p_{m-1} X^{m-1} + X^m$ of degree $m$ over

GF(2). Because $p(X)$ is irreducible over GF(2), it has no root in GF(2), i.e., neither 0 nor 1 is a root of $p(X)$. However, because $p(X)$ has degree $m$, it must have $m$ roots and these $m$ roots must be in a larger field which contains GF(2) as a subfield. Let $\alpha$ be such an element in a larger field which is a root of $p(X)$, i.e., $p(\alpha) = 0$.

Starting from the binary field GF(2) $= \{0, 1\}$ and $\alpha$, we define a multiplication "$\cdot$" to introduce a *sequence of powers* of $\alpha$ as follows.

$$
\begin{aligned}
0 \cdot 0 &= 0 \\
0 \cdot 1 &= 1 \cdot 0 = 0 \\
1 \cdot 1 &= 1 \\
0 \cdot \alpha &= \alpha \cdot 0 = 0 \\
1 \cdot \alpha &= \alpha \cdot 1 = \alpha \\
\alpha^2 &= \alpha \cdot \alpha \\
\alpha^3 &= \alpha \cdot \alpha \cdot \alpha \\
&\ \ \vdots \\
\alpha^j &= \alpha \cdot \alpha \cdot \alpha \cdots \alpha \ (j \text{ times}) \\
&\ \ \vdots
\end{aligned}
\tag{2.10}
$$

From the above definition of the multiplication "$\cdot$," we see that

$$
\begin{aligned}
0 \cdot a^j &= a^j \cdot 0 = 0, \\
1 \cdot \alpha^j &= \alpha^j \cdot 1 = \alpha^j, \\
\alpha^i \cdot \alpha^j &= \alpha^{i+j}.
\end{aligned}
\tag{2.11}
$$

Now, we have the following set of elements:

$$
F = \{0, 1, \alpha, \alpha^2, \ldots, \alpha^j, \ldots\},
\tag{2.12}
$$

which is closed under the multiplication "$\cdot$." Because $\alpha$ is a root of $p(X)$ and $p(X)$ divides $X^{2^m-1} + 1$, $\alpha$ must be a root of $X^{2^m-1} + 1$. Hence, $\alpha^{2^m-1} + 1 = 0$, which implies

$$
\alpha^{2^m-1} = 1.
\tag{2.13}
$$

As a result, $F$ comprises a finite number of elements as follows:

$$
F = \{0, 1, \alpha, \alpha^2, \ldots, \alpha^{2^m-2}\}.
\tag{2.14}
$$

Let $\alpha^0 = 1$ and $\alpha^{-\infty} = 0$. Then, the set $F$ can be expressed as the powers of $\alpha$, i.e., $F = \{\alpha^{-\infty}, \alpha^0, \alpha, \alpha^2, \ldots, \alpha^{2^m-2}\}$. The multiplication defined by (2.10) is carried out as follows using the equality of (2.13): for $0 \leq i, j < 2^m$,

$$
\alpha^i \cdot \alpha^j = \alpha^{i+j} = \alpha^r,
\tag{2.15}
$$

where $r$ is the remainder resulting from dividing $i + j$ by $2^m - 1$. Because $0 \le r < 2^m - 1$, $\alpha^r$ is an element in $F$ in the form of (2.14). Hence, the set of nonzero elements in $F$, $F \backslash \{0\}$, is closed under the multiplication "$\cdot$" defined by (2.10).

For $0 \le i < 2^m - 1$,

$$\alpha^i \cdot \alpha^{2^m - 1 - i} = \alpha^{2^m - 1} = 1.$$

Hence, $\alpha^{2^m - 1 - i}$ is the inverse of $\alpha^i$ (vice versa) with respect to the multiplication "$\cdot$" defined by (2.10). Using the fact $\alpha^{2^m - 1} = 1$, we can write

$$\alpha^{2^m - 1 - i} = \alpha^{2^m - 1} \cdot \alpha^{-i} = 1 \cdot \alpha^{-i} = \alpha^{-i}. \tag{2.16}$$

Hereafter, we use $\alpha^{-i}$ to denote the multiplicative inverse of $\alpha^i$.

The element "1" in $F$ is the *multiplicative identity* (or the *unit element*) of $F$. Based on the definition of the multiplication "$\cdot$" given by (2.10) (or (2.15)), we can readily prove that the multiplication "$\cdot$" is both associative and commutative. Hence, $F^* = F \backslash \{0\}$ is a group under the multiplication "$\cdot$" defined by (2.15). Note that $F^* = \{1, \alpha, \alpha^2, \ldots, \alpha^{2^m - 2}\}$ is also a cyclic group with $\alpha$ as its generator.

Next, we define an addition "$+$" on $F$ to make $F$ as a group under this addition operation. For $0 \le i < 2^m - 1$, dividing $X^i$ by $p(X)$ (using the polynomial division defined in Section 2.3), we obtain

$$X^i = q_i(X)p(X) + b_i(X), \tag{2.17}$$

where $q_i(X)$ and $b_i(X)$ are the quotient and remainder polynomials over GF(2), respectively. The remainder $b_i(X)$ is a polynomial over GF(2) with degree $m - 1$ or less and is of the following form:

$$b_i(X) = b_{i,0} + b_{i,1}X + \cdots + b_{i,m-1}X^{m-1} \tag{2.18}$$

with $b_{i,j} \in$ GF(2), $0 \le j < m$. Because $p(X)$ is an irreducible polynomial over GF(2) and the two polynomials $X$ and $p(X)$ are relatively prime, $X^i$ is not divisible by $p(X)$. Therefore, for $0 \le i < 2^m - 1$,

$$b_i(X) \neq 0.$$

For $0 \le i, j < 2^m - 1$ and $i \neq j$, we can readily prove that

$$b_i(X) \neq b_j(X). \tag{2.19}$$

Substituting $X$ in (2.17) by $\alpha$ (the root of $p(X)$), we have

$$\begin{aligned} \alpha^i &= q_i(\alpha)p(\alpha) + b_i(\alpha) \\ &= q_i(\alpha) \cdot 0 + b_i(\alpha) \\ &= b_{i,0} + b_{i,1}\alpha + \cdots + b_{i,m-1}\alpha^{m-1}. \end{aligned} \tag{2.20}$$

Because $b_i(X) \neq 0$, we must have $\alpha^i \neq 0$. Equation (2.20) shows that each nonzero element in $F$ can be expressed as a nonzero polynomial of $\alpha$ of degree $m-1$ or less with coefficients from GF(2). It follows from (2.19) and (2.20) that, for $i \neq j$,

$$\alpha^i \neq \alpha^j.$$

This says that all the nonzero elements in $F$ are *distinct*.

The 0-element, $\alpha^{-\infty} = 0$, in $F$ may be represented by the *zero polynomial*. Note that there are exactly $2^m$ polynomials of $\alpha$ over GF(2) with degree $m-1$ or less. Each of the $2^m$ elements in $F$ is represented by *one and only one* of these polynomials, and each of these polynomials represents *one and only one* element in $F$. The $m$ elements $\alpha^0 = 1, \alpha, \alpha^2, \ldots, \alpha^{m-2}, \alpha^{m-1}$ form a *polynomial basis* of $F$, i.e., every element in $F$ is a *unique* linear combination of these $m$ elements in $F$ as shown by (2.20). We call (2.20) the *polynomial representations* (or *forms*) of the elements in $F$.

Now, we define an addition "+" on $F$ using the polynomial representations of the elements in $F$. Consider two elements $\alpha^i$ and $\alpha^j$ of $F$ in polynomial form,

$$\alpha^i = b_{i,0} + b_{i,1}\alpha + \cdots + b_{i,m-1}\alpha^{m-1},$$
$$\alpha^j = b_{j,0} + b_{j,1}\alpha + \cdots + b_{j,m-1}\alpha^{m-1}.$$

The addition of the two elements $\alpha^i$ and $\alpha^j$ in $F$ is defined as the addition of their corresponding polynomial forms, i.e.,

$$\begin{aligned}
\alpha^i + \alpha^j &= (b_{i,0} + b_{i,1}\alpha + \cdots + b_{i,m-1}\alpha^{m-1}) + (b_{j,0} + b_{j,1}\alpha + \cdots + b_{j,m-1}\alpha^{m-1}) \\
&= (b_{i,0} + b_{j,0}) + (b_{i,1} + b_{j,1})\alpha + \cdots + (b_{i,m-1} + b_{j,m-1})\alpha^{m-1},
\end{aligned}$$
$$(2.21)$$

where $b_{i,k} + b_{j,k}$ is carried out over GF(2) (modulo-2 addition), $0 \leq k < m$.

The sum polynomial given by (2.21) is a polynomial of $\alpha$ with degree $m-1$ or less over GF(2) and hence represents an element in $F$, say $\alpha^\ell$ with $0 \leq \ell < 2^m - 1$ or $\ell = -\infty$. From (2.21), we readily see that

$$0 + \alpha^i = \alpha^i + 0. \tag{2.22}$$

Therefore, $F$ is close under the addition "+" defined by (2.21). Because modulo-2 addition is associative and commutative, we can readily verify that the addition defined by (2.21) is also associative and commutative. Note that

$$\begin{aligned}
\alpha^i + \alpha^i &= (b_{i,0} + b_{i,0}) + (b_{i,1} + b_{i,1})\alpha + \cdots + (b_{i,m-1} + b_{i,m-1})\alpha^{m-1} \\
&= 0 + 0 \cdot \alpha + \cdots + 0 \cdot \alpha^{m-1} = 0.
\end{aligned}$$
$$(2.23)$$

Hence, $\alpha^i$ is *its own inverse* with respect to the addition "+" defined by (2.21).

Based on the above developments, we conclude that $F$ is a commutative group under the addition defined by (2.21) with 0 as its additive identity. We can prove that the multiplication defined by (2.10) satisfies the distributive law across the addition defined by (2.21).

Up to this point, we have proved that: (1) $F$ is a commutative group under the addition "+" defined by (2.21); (2) $F\backslash\{0\}$ is a commutative group under the multiplication defined by (2.10); and (3) the multiplication and addition defined on $F$ satisfy the distributive law. Hence,

$$F = \{\alpha^{-\infty} = 0, \alpha^0 = 1, \alpha, \ldots, \alpha^{2^m-2}\} \qquad (2.24)$$

is a *field* of characteristic 2 with $2^m$ elements, denoted by GF($2^m$), and $\alpha$ is a *primitive element* of this field. Let $-\alpha^i$ denote the additive inverse of the element $\alpha^i$ in GF($2^m$). From (2.23), we see that

$$-\alpha^i = \alpha^i.$$

It is noted that the element $\alpha$ in GF($2^m$) is a root of the primitive polynomial $p(X) = p_0 + p_1 X + \cdots + p_{m-1} X^{m-1} + X^m$ of degree $m$ over GF(2). We have

$$p(\alpha) = p_0 + p_1 \alpha + \cdots + p_{m-1}\alpha^{m-1} + \alpha^m = 0. \qquad (2.25)$$

It follows that

$$\alpha^m = p_0 + p_1 \alpha + \cdots + p_{m-1}\alpha^{m-1}. \qquad (2.26)$$

Equation (2.26) indicates that every element $\alpha^i$ in GF($2^m$) can be expressed as a polynomial in $\alpha$ of degree $m-1$ or less. For example, for $0 \le i < m$, $\alpha^i = \alpha^i$, and $\alpha^{m+1} = \alpha \cdot \alpha^m = p_0\alpha + p_1\alpha^2 + \cdots + p_{m-2}\alpha^{m-1} + p_{m-1}\alpha^m = p_{m-1}p_0 + (p_{m-1}p_1 + p_0)\alpha + (p_{m-1}p_2 + p_1)\alpha^2 + \cdots + (p_{m-1} + p_{m-2})\alpha^{m-1}$. For $m+1 < i < 2^m - 1$, the polynomial representation of degree $m-1$ or less for $\alpha^i$ can be obtained following the same procedure. Because $\alpha$ is a primitive element in GF($2^m$), i.e., $\alpha$ has order $2^m - 1$, the $2^m - 1$ distinct powers of $\alpha$ (all the nonzero elements in GF($2^m$)) must have $2^m - 1$ distinct polynomials in the form of $b_{i,0} + b_{i,1}\alpha + \cdots + b_{i,m-1}\alpha^{m-1}$. For polynomials, $b_{i,0} + b_{i,1}\alpha + \cdots + b_{i,m-1}\alpha^{m-1}$, of degree $m-1$ or less in $\alpha$ with coefficients from GF(2), there are exactly $2^m$ of them in which $2^m - 1$ are nonzero. Thus, the correspondence

$$\alpha^i = b_{i,0} + b_{i,1}\alpha + \cdots + b_{i,m-1}\alpha^{m-1} \Longleftrightarrow (b_{i,0}, b_{i,1}, \ldots, b_{i,m-1}) \qquad (2.27)$$

is *one-to-one*. We call the $m$-tuple $\mathbf{b}_i = (b_{i,0}, b_{i,1}, \ldots, b_{i,m-1})$ over GF(2) the *vector representation* of the element $\alpha^i$ of GF($2^m$). With vector representation, the addition of two elements $\alpha^i$ and $\alpha^j$ over GF($2^m$) is carried out by the component-wise addition (modulo-2) of their corresponding $m$-tuples:

$$\begin{aligned}
\alpha^i + \alpha^j &\Longleftrightarrow (b_{i,0}, b_{i,1}, \ldots, b_{i,m-1}) + (b_{j,0}, b_{j,1}, \ldots, b_{j,m-1}) \\
&\Longleftrightarrow (b_{i,0} + b_{j,0}, b_{i,1} + b_{j,1}, \ldots, b_{i,m-1} + b_{j,m-1}).
\end{aligned} \qquad (2.28)$$

For every binary vector, we can represent it by an integer in $\mathbb{Z}$, i.e., a decimal representation. Let $\mathbf{b}_i = (b_{i,0}, b_{i,1}, \ldots, b_{i,m-1})$ be a binary vector of length $m$, the vector representation of $\alpha^i$. Then, its decimal form is calculated as

$$d_i = b_{i,0} + b_{i,1}2 + \cdots + b_{i,m-2}2^{m-2} + b_{i,m-1}2^{m-1}. \qquad (2.29)$$

Note that $d_i$ is an integer in the range between 0 and $2^m - 1$. The integer $d_i \in \mathbb{Z}$ is called the *decimal representation* of the field element $\alpha^i$.

Therefore, every element in $\mathrm{GF}(2^m)$ can be represented in four forms: (1) power form (2.24); (2) polynomial form (2.20); (3) vector form (2.27); and (4) decimal form (2.29). It is easier to carry out addition in either polynomial form or vector form. It is much easier to perform multiplication in power form. The decimal form provides a way to express the field in a similar way to a prime field $\mathrm{GF}(p) = \{0, 1, 2, \ldots, p-1\}$ where $p$ is a prime. It is much simpler in representing elements in a finite field.

Next, we show that the $m$ roots of the primitive polynomial $p(X)$ are elements in $\mathrm{GF}(2^m)$. Applying the equality given by (2.9) to $p(X)$, we have

$$p(X^{2^l}) = [p(X)]^{2^l}, \tag{2.30}$$

for any nonnegative integer $l$. Replacing $X$ by $\alpha$ in both sides of (2.30), we have

$$p(\alpha^{2^l}) = [p(\alpha)]^{2^l}. \tag{2.31}$$

Because $\alpha$ is a root of $p(X)$, $p(\alpha) = 0$. Then, from (2.31), we have

$$p(\alpha^{2^l}) = [p(\alpha)]^{2^l} = (0)^{2^l} = 0. \tag{2.32}$$

The equality of (2.32) says that the $m$ different elements, namely, $\alpha$, $\alpha^2$, $\alpha^{2^2}, \ldots, \alpha^{2^{m-1}}$, in $F$ are the $m$ roots of $p(X)$. (Note that $\alpha^{2^m} = \alpha$.) These roots are said to be *conjugate roots* of $p(X)$.

**Example 2.8** Let $m = 4$. In this example, we give the construction of the Galois field $\mathrm{GF}(2^4)$ with 16 elements. First, we choose a primitive polynomial of degree 4 over $\mathrm{GF}(2)$,

$$p(X) = 1 + X + X^4.$$

Let $\alpha$ be a root of $p(X)$. Then

$$p(\alpha) = 1 + \alpha + \alpha^4 = 0.$$

Using the fact that $\alpha^4 + \alpha^4 = 0$ and $\alpha^4 + 0 = \alpha^4$, and adding $\alpha^4$ to both sides of the above equation, we have

$$\alpha^4 = 1 + \alpha.$$

Consider the set

$$\{0, 1, \alpha, \alpha^2, \alpha^3, \alpha^4, \alpha^5, \alpha^6, \alpha^7, \alpha^8, \alpha^9, \alpha^{10}, \alpha^{11}, \alpha^{12}, \alpha^{13}, \alpha^{14}\}.$$

Note that $\alpha^{15} = 1$. Using the identity $\alpha^4 = 1 + \alpha$, every power $\alpha^i$, $0 \le i < 15$, can be expressed as a polynomial of $\alpha$ with degree 3 or less as follows:

$$0 \;=\; \alpha^{-\infty},$$
$$1 \;=\; \alpha^0,$$
$$\alpha \;=\; \alpha,$$
$$\alpha^2 \;=\; \alpha^2,$$
$$\alpha^3 \;=\; \alpha^3,$$
$$\alpha^4 \;=\; 1 + \alpha,$$
$$\alpha^5 \;=\; \alpha \cdot \alpha^4 = \alpha \cdot (\alpha + 1) = \alpha + \alpha^2,$$
$$\alpha^6 \;=\; \alpha \cdot \alpha^5 = \alpha \cdot (\alpha + \alpha^2) = \alpha^2 + \alpha^3,$$
$$\alpha^7 \;=\; \alpha \cdot \alpha^6 = \alpha \cdot (\alpha^2 + \alpha^3) = \alpha^3 + \alpha^4 = 1 + \alpha + \alpha^3,$$
$$\alpha^8 \;=\; \alpha \cdot \alpha^7 = \alpha \cdot (1 + \alpha + \alpha^3) = \alpha + \alpha^2 + \alpha^4 = 1 + \alpha^2,$$
$$\alpha^9 \;=\; \alpha \cdot \alpha^7 = \alpha \cdot (1 + \alpha^2) = \alpha + \alpha^3 ,$$
$$\alpha^{10} \;=\; \alpha \cdot \alpha^9 = \alpha \cdot (\alpha + \alpha^3) = \alpha^2 + \alpha^4 = 1 + \alpha + \alpha^2,$$
$$\alpha^{11} \;=\; \alpha \cdot \alpha^{10} = \alpha \cdot (1 + \alpha + \alpha^2) = \alpha + \alpha^2 + \alpha^3,$$
$$\alpha^{12} \;=\; \alpha \cdot \alpha^{11} = \alpha \cdot (\alpha + \alpha^2 + \alpha^3) = \alpha^2 + \alpha^3 + \alpha^4 = 1 + \alpha + \alpha^2 + \alpha^3,$$
$$\alpha^{13} \;=\; \alpha \cdot \alpha^{12} = \alpha \cdot (1 + \alpha + \alpha^2 + \alpha^3) = \alpha + \alpha^2 + \alpha^3 + \alpha^4 = 1 + \alpha^2 + \alpha^3,$$
$$\alpha^{14} \;=\; \alpha \cdot \alpha^{13} = \alpha \cdot (1 + \alpha^2 + \alpha^3) = \alpha + \alpha^3 + \alpha^4 = 1 + \alpha^3,$$
$$\alpha^{15} \;=\; \alpha \cdot \alpha^{14} = \alpha \cdot (1 + \alpha^3) = \alpha + \alpha^4 = 1.$$

The four representations, namely, power, polynomial, vector, and decimal forms, of the 16 elements in GF($2^4$) are given in Table 2.9.

The primitive polynomial $p(X) = 1 + X + X^4$ that generates GF($2^4$) has four roots, $\alpha$, $\alpha^2$, $\alpha^4$, and $\alpha^8$ in GF($2^4$). To verify this, we substitute $X$ of $p(X)$ with the four elements, $\alpha$, $\alpha^2$, $\alpha^4$, and $\alpha^8$, in turn as follows (using Table 2.9):

$$p(\alpha) \;=\; 1 + \alpha + \alpha^4 = 1 + \alpha + 1 + \alpha = 0,$$
$$p(\alpha^2) \;=\; 1 + \alpha^2 + \alpha^8 = 1 + \alpha^2 + 1 + \alpha^2 = 0,$$
$$p(\alpha^4) \;=\; 1 + \alpha^4 + \alpha^{16} = 1 + 1 + \alpha + \alpha = 0,$$
$$p(\alpha^8) \;=\; 1 + \alpha^8 + \alpha^{32} = 1 + 1 + \alpha^2 + \alpha^2 = 0.$$

The four roots $\alpha$, $\alpha^2$, $\alpha^4$, and $\alpha^8$ are the conjugate roots of $p(X)$. Following the above fact, we can write $p(X)$ as $p(X) = (X - \alpha)(X - \alpha^2)(X - \alpha^4)(X - \alpha^8) = (X + \alpha)(X + \alpha^2)(X + \alpha^4)(X + \alpha^8)$. Multiplying out the four terms and using Table 2.9, we find that

$$
\begin{aligned}
p(X) &= (X + \alpha)(X + \alpha^2)(X + \alpha^4)(X + \alpha^8) \\
&= (\alpha^3 + (\alpha^2 + \alpha)X + X^2)(\alpha^{12} + (\alpha^4 + \alpha^8)X + X^2) \\
&= (\alpha^3 + \alpha^5 X + X^2)(\alpha^{12} + \alpha^5 X + X^2) \\
&= \alpha^{15} + (\alpha^8 + \alpha^{17})X + (\alpha^{12} + \alpha^{10} + \alpha^3)X^2 + (\alpha^5 + \alpha^5)X^3 + X^4 \\
&= 1 + X + X^4.
\end{aligned}
$$

Note that the polynomial $p(X) = 1 + X + X^4$ has four roots $\alpha$, $\alpha^2$, $\alpha^4$, $\alpha^8$ in GF($2^4$), while it has no root in GF(2) because $p(0) = 1 + 0 + 0 = 1$ and $p(1) = 1 + 1 + 1 = 1$. ▲▲

Table 2.9 GF($2^4$) generated by $p(X) = 1 + X + X^4$ over GF(2).

| Power form | Polynomial form | | | | Vector form | Decimal form |
|---|---|---|---|---|---|---|
| 0 | 0 | | | | (0 0 0 0) | 0 |
| 1 | 1 | | | | (1 0 0 0) | 1 |
| $\alpha$ | | $\alpha$ | | | (0 1 0 0) | 2 |
| $\alpha^2$ | | | $\alpha^2$ | | (0 0 1 0) | 4 |
| $\alpha^3$ | | | | $\alpha^3$ | (0 0 0 1) | 8 |
| $\alpha^4$ | 1 | $+$ $\alpha$ | | | (1 1 0 0) | 3 |
| $\alpha^5$ | | $\alpha$ | $+$ $\alpha^2$ | | (0 1 1 0) | 6 |
| $\alpha^6$ | | | $\alpha^2$ | $+$ $\alpha^3$ | (0 0 1 1) | 12 |
| $\alpha^7$ | 1 | $+$ $\alpha$ | | $+$ $\alpha^3$ | (1 1 0 1) | 11 |
| $\alpha^8$ | 1 | | $+$ $\alpha^2$ | | (1 0 1 0) | 5 |
| $\alpha^9$ | | $\alpha$ | | $+$ $\alpha^3$ | (0 1 0 1) | 10 |
| $\alpha^{10}$ | 1 | $+$ $\alpha$ | $+$ $\alpha^2$ | | (1 1 1 0) | 7 |
| $\alpha^{11}$ | | $\alpha$ | $+$ $\alpha^2$ | $+$ $\alpha^3$ | (0 1 1 1) | 14 |
| $\alpha^{12}$ | 1 | $+$ $\alpha$ | $+$ $\alpha^2$ | $+$ $\alpha^3$ | (1 1 1 1) | 15 |
| $\alpha^{13}$ | 1 | | $+$ $\alpha^2$ | $+$ $\alpha^3$ | (1 0 1 1) | 13 |
| $\alpha^{14}$ | 1 | | | $+$ $\alpha^3$ | (1 0 0 1) | 9 |

The field GF($2^m$) is called an *extension field* of GF(2) and GF(2) is called the *ground field* of GF($2^m$). Every Galois field GF($2^m$) of $2^m$ elements is generated by a primitive polynomial of degree $m$ over GF(2). To construct an extension field GF($2^m$) over GF(2), we require a primitive polynomial $p(X)$ of degree $m$ over GF(2). If a different primitive polynomial $p^*(X)$ of degree $m$ over GF(2) is used, the construction results in an extension field GF$^*$($2^m$) over GF(2) that has the same set of elements as GF($2^m$). There is a one-to-one correspondence between GF($2^m$) and GF$^*$($2^m$) such that if $a \longleftrightarrow a^*$, $b \longleftrightarrow b^*$, and $c \longleftrightarrow c^*$, for $a$, $b$, $c \in$ GF($2^m$) and $a^*$, $b^*$, $c^* \in$ GF$^*$($2^m$), then

$$a + b \longleftrightarrow a^* + b^*$$
$$a \cdot b \longleftrightarrow a^* \cdot b^*$$
$$(a + b) + c \longleftrightarrow (a^* + b^*) + c^*$$
$$(a \cdot b) \cdot c \longleftrightarrow (a^* \cdot b^*) \cdot c^*$$
$$a \cdot (b + c) \longleftrightarrow a^* \cdot (b^* + c^*).$$

The two fields GF($2^m$) and GF$^*$($2^m$) are said to be *isomorphic*. That is, GF($2^m$) and GF$^*$($2^m$) are *structurally identical* up to the labeling of their elements. In this sense, we may say that any primitive polynomial $p(X)$ of degree $m$ over GF(2) gives the same extension field GF($2^m$). Reference [17] has a very nice development of this result.

## 2.5 Basic Properties and Structures of Galois Field GF($2^m$)

In this section, we present some fundamental properties of the extension field GF($2^m$) of the binary ground field GF(2). These properties are relevant in constructing and analyzing two important classes of error-correcting codes, namely, BCH and RS codes.

First, it follows from the equality of (2.9) (i.e., $[f(X)]^{2^i} = f(X^{2^i})$ for any $i \geq 0$) that we have the following theorem.

**Theorem 2.7** *Let $f(X)$ be a polynomial over* GF(2). *Let $\beta$ be a nonzero element of* GF($2^m$). *If $\beta$ is a root of $f(X)$ (i.e., $f(\beta) = 0$), then for any nonnegative integer $t$, $\beta^{2^t}$ is also a root of $f(X)$ (i.e., $f(\beta^{2^t}) = (f(\beta))^{2^t} = 0^{2^t} = 0$).*

The element $\beta^{2^t}$ is called a *conjugate* of $\beta$. It follows from Theorem 2.7 that the conjugates of $\beta$

$$\beta, \beta^2, \beta^{2^2}, \ldots, \beta^{2^t}, \ldots$$

are also roots of $f(X)$.

For a finite field, the number of conjugates of a nonzero element $\beta$ in GF($2^m$) must be finite. Let $e$ be the *smallest nonnegative integer* for which

$$\beta^{2^e} = \beta.$$

The integer $e$ is called the *exponent* of $\beta$. The powers of $\beta$, namely

$$\beta, \beta^2, \beta^{2^2}, \ldots, \beta^{2^{e-1}}$$

are distinct and form all the conjugates of $\beta$. Let $\Omega(\beta) = \{\beta, \beta^2, \beta^{2^2}, \ldots, \beta^{2^{e-1}}\}$ be a set which consists of all the conjugates of $\beta$. The set $\Omega(\beta)$ is called the *conjugate set* of $\beta$ which consists of $e$ elements in GF($2^m$). The integer $e$ is also called the order (or cardinality) of the conjugate set. It is clear that any element in a conjugate set can be used to generate all the elements in the set, i.e., for $0 \leq i < e$, $\Omega(\beta^{2^i}) = \Omega(\beta)$.

Let $\beta$ be a nonzero element of GF($2^m$) with order $n$ (i.e., $n$ is the smallest nonnegative integer such that $\beta^n = 1$). It follows from Theorem 2.4 that $n$ divides $2^m - 1$. Let $2^m - 1 = kn$. Then, $\beta^{2^m - 1} = \beta^{kn} = (\beta^n)^k$. Because $\beta^n = 1$, we have

$$\beta^{2^m - 1} = 1.$$

This implies that $\beta$ is a root of $X^{2^m - 1} + 1$. It follows from this fact that we have the following theorem.

**Theorem 2.8** *The $2^m - 1$ nonzero elements of* GF($2^m$), *namely, $\alpha^0 = 1, \alpha, \alpha^2, \ldots, \alpha^{2^m - 2}$, form all the roots of $X^{2^m - 1} + 1$.*

It follows from Theorem 2.8 that

$$X^{2^m-1} + 1 = (X + 1)(X + \alpha)(X + \alpha^2)\cdots(X + \alpha^{2^m-2}).$$

It is noted that $0$ is a root of the polynomial $X(X^{2^m-1} + 1)$. Based on this fact, we have the following corollary.

**Corollary 2.1** *The elements of* $\mathrm{GF}(2^m)$ *form all the roots of* $X^{2^m} + X$, *i.e.,*

$$X^{2^m} + X = X(X + 1)(X + \alpha)(X + \alpha^2)\cdots(X + \alpha^{2^m-2}).$$

**Definition 2.13** Let $\beta$ be an element of $\mathrm{GF}(2^m)$. Let $m(X)$ be the polynomial of the *smallest degree* over $\mathrm{GF}(2)$ that has $\beta$ as a root, i.e., $m(\beta) = 0$. The polynomial $m(X)$ is called the *minimal polynomial* of $\beta$.

**Theorem 2.9** *The minimal polynomial* $m(X)$ *of an element* $\beta$ *in* $\mathrm{GF}(2^m)$ *is irreducible over* $\mathrm{GF}(2)$.

*Proof* Suppose $m(X)$ is not irreducible, say $m(X) = m_1(X)m_2(X)$, where $m_1(X)$ and $m_2(X)$ are nontrivial, i.e., their degrees are greater than $0$. Both $m_1(X)$ and $m_2(X)$ have degrees lower than $m(X)$. Because $m(\beta) = m_1(\beta)m_2(\beta) = 0$, then either $m_1(\beta)$ or $m_2(\beta)$ must be zero. This contradicts the hypothesis that $m(X)$ is the polynomial of the smallest degree such that $m(\beta) = 0$. Therefore, $m(X)$ cannot have nontrivial factors and it must be irreducible. ▲▲

In the following, we give a number of theorems on the minimal polynomial of an element in $\mathrm{GF}(2^m)$ without proofs. The readers may prove them for practice.

**Theorem 2.10** *The minimal polynomial* $m(X)$ *of a field element* $\beta$ *is unique.*

**Theorem 2.11** *Let* $f(X)$ *be a polynomial over* $\mathrm{GF}(2)$. *Let* $m(X)$ *be the minimal polynomial of an element* $\beta$ *in* $\mathrm{GF}(2^m)$. *If* $\beta$ *is a root of* $f(X)$, *then* $f(X)$ *is divisible by* $m(X)$.

**Theorem 2.12** *Let* $p(X)$ *be an irreducible polynomial over* $\mathrm{GF}(2)$. *Let* $\beta$ *be an element of* $\mathrm{GF}(2^m)$ *and* $m(X)$ *be its minimal polynomial. If* $\beta$ *is a root of* $p(X)$, *then* $m(X) = p(X)$.

It follows from Theorems 2.8 and 2.11 that we have Theorem 2.13.

**Theorem 2.13** *Let* $m(X)$ *be the minimal polynomial of a nonzero element in* $\mathrm{GF}(2^m)$. *Then,* $m(X)$ *divides* $X^{2^m-1} + 1$.

It is noted that the minimal polynomial of the $0$-element of $\mathrm{GF}(2^m)$ is $X$. Then, combing this fact with Theorem 2.13, we conclude: if $m(X)$ is a minimal polynomial of an element of $\mathrm{GF}(2^m)$, then $m(X)$ divides $X^{2^m} + X$.

**Theorem 2.14** *Let* $m(X)$ *be the minimal polynomial of an element* $\beta$ *in* $\mathrm{GF}(2^m)$ *with exponent* $e$. *Then*

$$m(X) = \prod_{i=0}^{e-1}(X + \beta^{2^i}) = \prod_{\delta \in \Omega(\beta)} (X + \delta). \tag{2.33}$$

Following (2.33) and the definition of a conjugate set, it is clear that all the elements in a conjugate set have the same minimal polynomial.

It follows from Theorems 2.8, 2.10, and 2.14 that the minimal polynomials of the nonzero elements in the field GF($2^m$) provide a *complete factorization* of $X^{2^m-1} + 1$ over GF(2). Table A.1 in Appendix A gives a list of factorizations of $X^n + 1$ with $1 \leq n < 32$.

**Example 2.9** Consider the field GF($2^4$) given by Table 2.9 with a primitive element $\alpha$. Consider the element $\alpha^3 \in$ GF($2^4$). To find the minimal polynomial of $\alpha^3$, we need to find all its conjugates, which are:

$$\alpha^3, (\alpha^3)^2 = \alpha^6, (\alpha^3)^{2^2} = \alpha^{12}, (\alpha^3)^{2^3} = \alpha^{24} = \alpha^9.$$

Note that $(\alpha^3)^{2^4} = \alpha^{48} = \alpha^3$. Hence, the exponent of $\alpha^3$ is 4 and the minimal polynomial of $\alpha^3$ is

$$m_3(X) = (X + \alpha^3)(X + \alpha^6)(X + \alpha^9)(X + \alpha^{12}).$$

Expanding the right-hand side of the above equation using Table 2.9, we have

$$m_3(X) = 1 + X + X^2 + X^3 + X^4,$$

which is irreducible over GF(2).

Consider the element $\alpha^5$ of GF($2^4$). Its conjugates are $\alpha^5$ and $\alpha^{10}$. Because $(\alpha^5)^{2^2} = \alpha^{20} = \alpha^5$, the exponent of $\alpha^5$ is $e = 2$. The minimal polynomial of $\alpha^5$ is then

$$m_5(X) = (X + \alpha^5)(X + \alpha^{10}) = X^2 + (\alpha^5 + \alpha^{10})X + \alpha^{15} = 1 + X + X^2,$$

which is also irreducible over GF(2).

Similarly, we can find the minimal polynomials for the elements in the other conjugate sets: $\Omega(\alpha^0) = \{\alpha^0 = 1\}$, $\Omega(\alpha) = \{\alpha, \alpha^2, \alpha^4, \alpha^8\}$, $\Omega(\alpha^7) = \{\alpha^7, \alpha^{11}, \alpha^{13}, \alpha^{14}\}$. Their corresponding minimal polynomials are

$$m_0(X) = (X + 1) = 1 + X,$$
$$m_1(X) = (X + \alpha)(X + \alpha^2)(X + \alpha^4)(X + \alpha^8) = 1 + X^3 + X^4,$$
$$m_7(X) = (X + \alpha^7)(X + \alpha^{11})(X + \alpha^{13})(X + \alpha^{14}) = 1 + X + X^4.$$

Based on the minimal polynomials of all the nonzero elements in GF($2^4$), we obtain the following factorization of $X^{15} + 1$ over GF(2):

$$\begin{aligned} X^{15} + 1 &= m_0(X)m_1(X)m_3(X)m_5(X)m_7(X) \\ &= (1 + X)(1 + X^3 + X^4)(1 + X + X^2 + X^3 + X^4) \\ &\quad (1 + X + X^2)(1 + X + X^4). \end{aligned}$$

▲▲

As a field, GF($2^m$) consists of two groups, an additive group under the addition defined by (2.21) and a multiplicative group under the multiplication defined by (2.15). Each group of $GF(2^m)$ may contain proper subgroups. In the following, we present methods for constructing subgroups of the two groups of $GF(2^m)$.

Let $\alpha$ be a primitive element of $GF(2^m)$. In polynomial form, the additive group of $GF(2^m)$ is generated by $2^m$ linear sums over $GF(2)$ of the $m$ elements, $\alpha^0$, $\alpha$, $\alpha^2, \ldots, \alpha^{m-2}$, $\alpha^{m-1}$, which are linearly independent over $GF(2)$. For $0 \le k < m - 1$, let $j_0, j_1, \ldots, j_{k-1}$ be $k$ integers such that $0 \le j_0 < j_1 < \cdots < j_{k-1} < m - 1$. The $k$ elements $\alpha^{j_0}$, $\alpha^{j_1}, \ldots, \alpha^{j_{k-2}}$, $\alpha^{j_{k-1}}$ form a proper subset of $\{\alpha^0, \alpha, \alpha^2, \ldots, \alpha^{m-2}, \alpha^{m-1}\}$ and hence they are linearly independent. Then, the $2^k$ linear sums over $GF(2)$ of $\alpha^{j_0}$, $\alpha^{j_1}, \ldots, \alpha^{j_{k-2}}$, $\alpha^{j_{k-1}}$ give $2^k$ elements of $GF(2^m)$ which form a subgroup, denoted by $S_a^k$ (where the superscript "$k$" denotes the number of linearly independent elements in $S_a^k$ and the subscript "$a$" represents "additive subgroup"), of the additive group of $GF(2^m)$. Each element $\alpha^t$ in $S_a^k$ is of the following form:

$$\alpha^t = b_{t,0}\alpha^{j_0} + b_{t,1}\alpha^{j_1} + \cdots + b_{t,k-1}\alpha^{j_{k-1}},$$

where $b_{t,i} \in GF(2)$, $0 \le i < k$. The order of $S_a^k$ is $2^k$.

**Example 2.10** Consider the field $GF(2^4)$ constructed in Example 2.8 with $\alpha$ as its primitive element (a root of the primitive polynomial $p(X) = 1 + X + X^4$ over $GF(2)$). The elements $\alpha^0$, $\alpha$, $\alpha^2$, $\alpha^3$ form a polynomial basis of $GF(2^4)$ and their $2^4 = 16$ linear sums give the additive group of $GF(2^4)$.

Let $k = 2$ and $j_0 = 1$ and $j_1 = 2$. The four linear sums over $GF(2)$ of $\alpha^{j_0} = \alpha$ and $\alpha^{j_1} = \alpha^2$ in the following form:

$$\alpha^t = b_{t,0}\alpha + b_{t,1}\alpha^2,$$

where $b_{t,0}, b_{t,1} \in GF(2)$, give a subgroup $S_a^2$ of the additive group of $GF(2^4)$ with order 4. $S_a^2$ consists of 4 elements of $GF(2^4)$. Based on Table 2.9, we find that the 4 elements of $S_a^2$ are $0, \alpha, \alpha^2, \alpha^5$. Hence, $S_a^2 = \{0, \alpha, \alpha^2, \alpha^5\}$. ▲▲

Suppose $2^m - 1$ is not a prime. Then, $2^m - 1$ can be factored as a product of two integers greater than 1. Let $2^m - 1 = nk$ and $\alpha$ be a primitive element in $GF(2^m)$. Let $\beta = \alpha^k \in GF(2^m)$. Because $\beta^n = (\alpha^k)^n = \alpha^{2^m-1} = 1$, $\beta$ is an element of order $n$. The $n$ powers of $\beta$, namely, $\beta^0, \beta^1, \ldots, \beta^{n-1}$, form a cyclic subgroup of order $n$, denoted by $S_c^n$, of the multiplicative group of $GF(2^m)$, where the superscript "$n$" denotes the order of $S_c^n$ and the subscript "$c$" represents "cyclic subgroup." The element $\beta$ is the generator of the cyclic group $S_c^n$. The above implies that for any proper factor $n$ of $2^m - 1$, there exists a cyclic subgroup of the multiplicative group of $GF(2^m)$.

**Example 2.11** Consider the field $GF(2^4)$ constructed in Example 2.8 (see Table 2.9) with $\alpha$ as its primitive element. The number $2^4 - 1 = 15$ can be factored as the product of 3 and 5. Suppose we set $n = 3$ and $k = 5$. Let $\beta = \alpha^5$. Because $\beta^3 = (\alpha^5)^3 = 1$, the order of $\beta$ is 3. Hence, the set $S_c^3 = \{1, \alpha^5, \alpha^{10}\}$ generated by $\beta$ forms a cyclic subgroup of order 3 of the multiplicative group of $GF(2^4)$. ▲▲

The extension field GF($2^m$) of GF(2) generated by a primitive polynomial $p(X)$ of degree $m$ is the *smallest field* that contains the $m$ roots of $p(X)$. In the following, we present a method to construct an extension field which contains both GF(2) and GF($2^m$) as subfields and the roots of $p(X)$ as elements.

Let $c$ be a positive integer greater than 1. Choose a primitive polynomial $p_1(X)$ over GF(2) of degree $cm$. Then, $cm$ is the smallest integer for which $X^{2^{cm}-1} + 1$ is divisible by $p_1(X)$. The polynomial $X^{2^{cm}-1} + 1$ can be factored as follows:

$$X^{2^{cm}-1} + 1 = (X^{2^m-1} + 1)(X^{(2^{cm}-1)-(2^m-1)} + X^{(2^{cm}-1)-2(2^m-1)}$$
$$+ X^{(2^{cm}-1)-3(2^m-1)} + \cdots + X^{2(2^m-1)} + X^{2^m-1} + 1). \tag{2.34}$$

Because $p(X)$ divides $X^{2^m-1} + 1$, $p(X)$ divides $X^{2^{cm}-1} + 1$. Hence, the $m$ roots of $p(X)$ are the roots of $X^{2^{cm}-1} + 1$.

Now, we construct the extension field GF($2^{cm}$) of GF(2) using the primitive polynomial $p_1(X)$. Then, all $2^{cm} - 1$ nonzero elements of GF($2^{cm}$) are the roots of $X^{2^{cm}-1} + 1$. Because the $m$ roots of $p(X)$ are roots of $X^{2^{cm}-1} + 1$, the roots of $p(X)$ are elements in GF($2^{cm}$). Hence, all the nonzero elements of GF($2^m$) are elements in GF($2^{cm}$) and they form a subset of GF($2^{cm}$). This implies that GF($2^m$) is a subfield of GF($2^{cm}$). Therefore, GF($2^{cm}$) contains both GF(2) and GF($2^m$) as subfields and the roots of $p(X)$ as elements.

**Example 2.12** Let $m = 3$. From Table 2.8, we find that $p(X) = 1 + X + X^3$ is a primitive polynomial over GF(2). Then, $2^3 - 1 = 7$ is the smallest integer for which $X^7 + 1$ is divisible by $p(X)$. Let $\alpha$ be a root of $p(X)$. Then, $\alpha$ is also a root of $X^7 + 1$. Because $\alpha$ is a root of $p(X)$, we have

$$p(\alpha) = 1 + \alpha + \alpha^3 = 0,$$

and

$$\alpha^3 = 1 + \alpha.$$

Using $p(X)$ and $\alpha$, we construct the field GF($2^3$) = $\{0, 1, \alpha, \alpha^2, \alpha^3, \alpha^4, \alpha^5, \alpha^6\}$ which is shown in Table 2.10. Using (2.32), we find that the conjugate roots of $p(X)$ are $\alpha$, $\alpha^2$, and $\alpha^4$. The field GF($2^3$) is the smallest field that contains GF(2) as a subfield and the roots of $p(X)$.

Let $c = 2$. Then, $cm = 2 \cdot 3 = 6$. From Table 2.8, we find that $p_1(X) = 1 + X + X^6$ is a primitive polynomial over GF(2) of degree 6. Then, $cm = 6$ is the smallest integer for which $X^{2^{cm}-1} + 1 = X^{63} + 1$ is divisible by $p_1(X)$. It follows from (2.34) that $X^{63} + 1$ can be factored as follows:

$$X^{63} + 1 = (X^7 + 1)(X^{56} + X^{49} + X^{42} + X^{35} + X^{28} + X^{21} + X^{14} + X^7 + 1).$$

Note that $X^7 + 1$ is a factor of $X^{63} + 1$. Because the primitive polynomial $p(X) = 1 + X + X^3$ divides $X^7 + 1$, it divides $X^{63} + 1$. Hence, the three roots of $p(X)$ are also roots of $X^{63} + 1$.

Table 2.10 GF($2^3$) generated by $p(X) = 1 + X + X^3$ over GF(2).

| Power form | Polynomial form | Vector form | Decimal form |
|---|---|---|---|
| $0 = \alpha^{-\infty}$ | $0$ | (0 0 0) | 0 |
| $1 = \alpha^0$ | $1$ | (1 0 0) | 1 |
| $\alpha$ | $\alpha$ | (0 1 0) | 2 |
| $\alpha^2$ | $\alpha^2$ | (0 0 1) | 4 |
| $\alpha^3$ | $1 + \alpha$ | (1 1 0) | 3 |
| $\alpha^4$ | $\alpha + \alpha^2$ | (0 1 1) | 6 |
| $\alpha^5$ | $1 + \alpha + \alpha^2$ | (1 1 1) | 7 |
| $\alpha^6$ | $1 + \alpha^2$ | (1 0 1) | 5 |

Let $\beta$ be a root of $p_1(X)$. Then, $\beta$ is also a root of $X^{63} + 1$. Because $\beta$ is a root of $p_1(X)$, we have

$$p_1(X) = 1 + \beta + \beta^6 = 0,$$

and

$$\beta^6 = 1 + \beta.$$

Using $p_1(X) = 1 + X + X^6$ and $\beta$, we construct the field GF($2^6$) = {0, 1, $\beta$, $\beta^2, \ldots, \beta^{62}$} which is shown in Table 2.11. All 63 nonzero elements of GF($2^6$) are roots of $X^{63} + 1$. Because the roots of $p(X) = 1 + X + X^3$ are roots of $X^{63} + 1$, they are elements in GF($2^6$). Hence, the field GF($2^6$) contains both GF($2^3$) and GF(2) as subfields and the three roots of $p(X) = 1 + X + X^3$ as elements. Let $\alpha = \beta^9$. The order of $\alpha$ is 7. We can readily check that the roots of $p(X)$ in GF($2^6$) are the elements $\beta^{27}$, $\beta^{45}$, and $\beta^{54}$ in GF($2^6$). The seven nonzero elements of GF($2^3$) are $\alpha^0 = 1$, $\alpha = \beta^9$, $\alpha^2 = \beta^{18}$, $\alpha^3 = \beta^{27}$, $\alpha^4 = \beta^{36}$, $\alpha^5 = \beta^{45}$, and $\alpha^6 = \beta^{54}$. ▲▲

Table 2.11 GF($2^6$) generated by $p_1(X) = 1 + X + X^6$ over GF(2).

| Power form | Polynomial form | Vector form | Decimal form |
|---|---|---|---|
| $0 = \beta^{-\infty}$ | $0$ | (0 0 0 0 0 0) | 0 |
| $1 = \beta^0$ | $1$ | (1 0 0 0 0 0) | 1 |
| $\beta$ | $\beta$ | (0 1 0 0 0 0) | 2 |
| $\beta^2$ | $\beta^2$ | (0 0 1 0 0 0) | 4 |
| $\beta^3$ | $\beta^3$ | (0 0 0 1 0 0) | 8 |
| $\beta^4$ | $\beta^4$ | (0 0 0 0 1 0) | 16 |
| $\beta^5$ | $\beta^5$ | (0 0 0 0 0 1) | 32 |
| $\beta^6$ | $1 + \beta$ | (1 1 0 0 0 0) | 3 |
| $\beta^7$ | $\beta + \beta^2$ | (0 1 1 0 0 0) | 6 |
| $\beta^8$ | $\beta^2 + \beta^3$ | (0 0 1 1 0 0) | 12 |
| $\beta^9$ | $\beta^3 + \beta^4$ | (0 0 0 1 1 0) | 24 |

Table 2.11 (continued)

| | 1 | β | β² | β³ | β⁴ | β⁵ | | |
|---|---|---|---|---|---|---|---|---|
| $\beta^{10}$ | | | | | $\beta^4$ | $+\ \beta^5$ | (0 0 0 0 1 1) | 48 |
| $\beta^{11}$ | $1$ | $+\ \beta$ | | | | $+\ \beta^5$ | (1 1 0 0 0 1) | 35 |
| $\beta^{12}$ | $1$ | | $+\ \beta^2$ | | | | (1 0 1 0 0 0) | 5 |
| $\beta^{13}$ | | $\beta$ | | $+\ \beta^3$ | | | (0 1 0 1 0 0) | 10 |
| $\beta^{14}$ | | | $\beta^2$ | | $+\ \beta^4$ | | (0 0 1 0 1 0) | 20 |
| $\beta^{15}$ | | | | $\beta^3$ | | $+\ \beta^5$ | (0 0 0 1 0 1) | 40 |
| $\beta^{16}$ | $1$ | $+\ \beta$ | | | $+\ \beta^4$ | | (1 1 0 0 1 0) | 19 |
| $\beta^{17}$ | | $\beta$ | $+\ \beta^2$ | | | $+\ \beta^5$ | (0 1 1 0 0 1) | 38 |
| $\beta^{18}$ | $1$ | $+\ \beta$ | $+\ \beta^2$ | $+\ \beta^3$ | | | (1 1 1 1 0 0) | 15 |
| $\beta^{19}$ | | $\beta$ | $+\ \beta^2$ | $+\ \beta^3$ | $+\ \beta^4$ | | (0 1 1 1 1 0) | 30 |
| $\beta^{20}$ | | | $\beta^2$ | $+\ \beta^3$ | $+\ \beta^4$ | $+\ \beta^5$ | (0 0 1 1 1 1) | 60 |
| $\beta^{21}$ | $1$ | $+\ \beta$ | | $+\ \beta^3$ | $+\ \beta^4$ | $+\ \beta^5$ | (1 1 0 1 1 1) | 59 |
| $\beta^{22}$ | $1$ | | $+\ \beta^2$ | | $+\ \beta^4$ | $+\ \beta^5$ | (1 0 1 0 1 1) | 53 |
| $\beta^{23}$ | $1$ | | | $+\ \beta^3$ | | $+\ \beta^5$ | (1 0 0 1 0 1) | 41 |
| $\beta^{24}$ | $1$ | | | | $+\ \beta^4$ | | (1 0 0 0 1 0) | 17 |
| $\beta^{25}$ | | $\beta$ | | | | $+\ \beta^5$ | (0 1 0 0 0 1) | 34 |
| $\beta^{26}$ | $1$ | $+\ \beta$ | $+\ \beta^2$ | | | | (1 1 1 0 0 0) | 7 |
| $\beta^{27}$ | | $\beta$ | $+\ \beta^2$ | $+\ \beta^3$ | | | (0 1 1 1 0 0) | 14 |
| $\beta^{28}$ | | | $\beta^2$ | $+\ \beta^3$ | $+\ \beta^4$ | | (0 0 1 1 1 0) | 28 |
| $\beta^{29}$ | | | | $\beta^3$ | $+\ \beta^4$ | $+\ \beta^5$ | (0 0 0 1 1 1) | 56 |
| $\beta^{30}$ | $1$ | $+\ \beta$ | | | $+\ \beta^4$ | $+\ \beta^5$ | (1 1 0 0 1 1) | 51 |
| $\beta^{31}$ | $1$ | | $+\ \beta^2$ | | | $+\ \beta^5$ | (1 0 1 0 0 1) | 37 |
| $\beta^{32}$ | $1$ | | | $+\ \beta^3$ | | | (1 0 0 1 0 0) | 9 |
| $\beta^{33}$ | | $\beta$ | | | $+\ \beta^4$ | | (0 1 0 0 1 0) | 18 |
| $\beta^{34}$ | | | $\beta^2$ | | | $+\ \beta^5$ | (0 0 1 0 0 1) | 36 |
| $\beta^{35}$ | $1$ | $+\ \beta$ | | $+\ \beta^3$ | | | (1 1 0 1 0 0) | 11 |
| $\beta^{36}$ | | $\beta$ | $+\ \beta^2$ | | $+\ \beta^4$ | | (0 1 1 0 1 0) | 22 |
| $\beta^{37}$ | | | $\beta^2$ | $+\ \beta^3$ | | $+\ \beta^5$ | (0 0 1 1 0 1) | 44 |
| $\beta^{38}$ | $1$ | $+\ \beta$ | | $+\ \beta^3$ | $+\ \beta^4$ | | (1 1 0 1 1 0) | 27 |
| $\beta^{39}$ | | $\beta$ | $+\ \beta^2$ | | $+\ \beta^4$ | $+\ \beta^5$ | (0 1 1 0 1 1) | 54 |
| $\beta^{40}$ | $1$ | $+\ \beta$ | $+\ \beta^2$ | $+\ \beta^3$ | | $+\ \beta^5$ | (1 1 1 1 0 1) | 47 |
| $\beta^{41}$ | $1$ | | $+\ \beta^2$ | $+\ \beta^3$ | $+\ \beta^4$ | | (1 0 1 1 1 0) | 29 |
| $\beta^{42}$ | | $\beta$ | | $+\ \beta^3$ | $+\ \beta^4$ | $+\ \beta^5$ | (0 1 0 1 1 1) | 58 |
| $\beta^{43}$ | $1$ | $+\ \beta$ | $+\ \beta^2$ | | $+\ \beta^4$ | $+\ \beta^5$ | (1 1 1 0 1 1) | 55 |
| $\beta^{44}$ | $1$ | | $+\ \beta^2$ | $+\ \beta^3$ | | $+\ \beta^5$ | (1 0 1 1 0 1) | 45 |
| $\beta^{45}$ | $1$ | | | $+\ \beta^3$ | $+\ \beta^4$ | | (1 0 0 1 1 0) | 25 |
| $\beta^{46}$ | | $\beta$ | | | $+\ \beta^4$ | $+\ \beta^5$ | (0 1 0 0 1 1) | 50 |
| $\beta^{47}$ | $1$ | $+\ \beta$ | $+\ \beta^2$ | | | $+\ \beta^5$ | (1 1 1 0 0 1) | 39 |
| $\beta^{48}$ | $1$ | | $+\ \beta^2$ | $+\ \beta^3$ | | | (1 0 1 1 0 0) | 13 |
| $\beta^{49}$ | | $\beta$ | | $+\ \beta^3$ | $+\ \beta^4$ | | (0 1 0 1 1 0) | 26 |
| $\beta^{50}$ | | | $\beta^2$ | | $+\ \beta^4$ | $+\ \beta^5$ | (0 0 1 0 1 1) | 52 |
| $\beta^{51}$ | $1$ | $+\ \beta$ | | $+\ \beta^3$ | | $+\ \beta^5$ | (1 1 0 1 0 1) | 43 |
| $\beta^{52}$ | $1$ | | $+\ \beta^2$ | | $+\ \beta^4$ | | (1 0 1 0 1 0) | 21 |
| $\beta^{53}$ | | $\beta$ | | $+\ \beta^3$ | | $+\ \beta^5$ | (0 1 0 1 0 1) | 42 |
| $\beta^{54}$ | $1$ | $+\ \beta$ | $+\ \beta^2$ | | $+\ \beta^4$ | | (1 1 1 0 1 0) | 23 |
| $\beta^{55}$ | | $\beta$ | $+\ \beta^2$ | $+\ \beta^3$ | | $+\ \beta^5$ | (0 1 1 1 0 1) | 46 |
| $\beta^{56}$ | $1$ | $+\ \beta$ | $+\ \beta^2$ | $+\ \beta^3$ | $+\ \beta^4$ | | (1 1 1 1 1 0) | 31 |

Table 2.11 (*continued*)

| | | | |
|---|---|---|---|
| $\beta^{57}$ | | $\beta + \beta^2 + \beta^3 + \beta^4 + \beta^5$ (0 1 1 1 1 1) | 62 |
| $\beta^{58}$ | $1 + \beta +$ | $\beta^2 + \beta^3 + \beta^4 + \beta^5$ (1 1 1 1 1 1) | 63 |
| $\beta^{59}$ | $1$ | $+ \beta^2 + \beta^3 + \beta^4 + \beta^5$ (1 0 1 1 1 1) | 61 |
| $\beta^{60}$ | $1$ | $+ \beta^3 + \beta^4 + \beta^5$ (1 0 0 1 1 1) | 57 |
| $\beta^{61}$ | $1$ | $+ \beta^4 + \beta^5$ (1 0 0 0 1 1) | 49 |
| $\beta^{62}$ | $1$ | $+ \beta^5$ (1 0 0 0 0 1) | 33 |

## 2.6   Computations over Galois Field $\mathbf{GF}(2^m)$

Computations and solving linear equations over $GF(2^m)$ (or any finite field) are very similar to those over the real-number field $\mathbb{R}$. As an example, we consider the following two linear equations of two unknowns, $X$ and $Y$, with coefficients over $GF(2^4)$ (see Table 2.9):

$$\begin{aligned} X + \alpha^7 Y &= \alpha^2 \\ \alpha^{12}X + \alpha^8 Y &= \alpha^4. \end{aligned} \tag{2.35}$$

Suppose we determine $Y$ first and then $X$. Multiplying both sides of the second equation by $\alpha^3$ (the multiplicative inverse of $\alpha^{12}$) gives the following two equations:

$$\begin{aligned} X + \alpha^7 Y &= \alpha^2 \\ X + \alpha^{11}Y &= \alpha^7. \end{aligned}$$

By adding the above two equations and following Table 2.9, we get

$$\begin{aligned} (\alpha^7 + \alpha^{11})Y &= \alpha^2 + \alpha^7 \\ (1 + \alpha + \alpha^3 + \alpha + \alpha^2 + \alpha^3)Y &= \alpha^2 + 1 + \alpha + \alpha^3 \\ (1 + \alpha^2)Y &= 1 + \alpha + \alpha^2 + \alpha^3 \\ \alpha^8 Y &= \alpha^{12} \\ Y &= \alpha^4. \end{aligned}$$

Substituting $Y = \alpha^4$ into the first equation of (2.35), we obtain

$$\begin{aligned} X + \alpha^7 \cdot \alpha^4 &= \alpha^2 \\ X &= \alpha^2 + \alpha^{11} = \alpha^2 + (\alpha + \alpha^2 + \alpha^3) = \alpha + \alpha^3 = \alpha^9. \end{aligned}$$

Thus, the solution for (2.35) is $X = \alpha^9$ and $Y = \alpha^4$.

Alternately, (2.35) could be solved by using *Cramer's rule*:

$$X = \frac{\begin{vmatrix} \alpha^2 & \alpha^7 \\ \alpha^4 & \alpha^8 \end{vmatrix}}{\begin{vmatrix} 1 & \alpha^7 \\ \alpha^{12} & \alpha^8 \end{vmatrix}} = \frac{\alpha^{10} + \alpha^{11}}{\alpha^8 + \alpha^4} = \frac{\alpha^3 + 1}{\alpha^2 + \alpha} = \frac{\alpha^{14}}{\alpha^5} = \alpha^9,$$

$$Y = \frac{\begin{vmatrix} 1 & \alpha^2 \\ \alpha^{12} & \alpha^4 \end{vmatrix}}{\begin{vmatrix} 1 & \alpha^7 \\ \alpha^{12} & \alpha^8 \end{vmatrix}} = \frac{\alpha^4 + \alpha^{14}}{\alpha^8 + \alpha^4} = \frac{\alpha^3 + \alpha}{\alpha^2 + \alpha} = \frac{\alpha^9}{\alpha^5} = \alpha^4.$$

As one more example, suppose we want to solve the following equation over $\mathrm{GF}(2^4)$:

$$f(X) = X^2 + \alpha^7 X + \alpha = 0.$$

The ordinary method will not work for the above equation because it requires dividing by 2, and in the field $\mathrm{GF}(2^m)$, $2 = 0\,(\text{modulo-2})$. If $f(X) = 0$ has any solution in $\mathrm{GF}(2^4)$, the solutions can be found simply by substituting all the elements in $\mathrm{GF}(2^4)$ for $X$. By doing so, we find that $f(\alpha^6) = 0$ and $f(\alpha^{10}) = 0$, i.e.,

$$f(\alpha^6) = (\alpha^6)^2 + \alpha^7 \alpha^6 + \alpha = \alpha^{12} + \alpha^{13} + \alpha = 0,$$
$$f(\alpha^{10}) = (\alpha^{10})^2 + \alpha^7 \alpha^{10} + \alpha = \alpha^5 + \alpha^2 + \alpha = 0.$$

Thus, $\alpha^6$ and $\alpha^{10}$ are two roots of $f(X)$ in $\mathrm{GF}(2^4)$ and $f(X)$ can be factored as $f(X) = (X + \alpha^6)(X + \alpha^{10})$.

The above calculations are typical of those required for decoding BCH and RS codes and they can be programmed easily on a general-purpose computer.

## 2.7  A General Construction of Finite Fields

So far, we have considered only the construction of extension fields using the binary field $\mathrm{GF}(2)$ as the ground field. Any finite field can be used as the ground field for constructing extension fields. This is briefly described in this section.

Let $\mathrm{GF}(q)$ be a finite field with $q$ elements, where $q$ is a power of a prime $p$, say $q = p^s$, where $s$ is a positive integer. Let $p(X)$ be a primitive polynomial over $\mathrm{GF}(q)$ of degree $m$. Then, $n = q^m - 1$ is the smallest integer for which $X^n - 1$ is divisible by $p(X)$. Hence, all the roots of $p(X)$ are roots of $X^{q^m - 1} - 1$. Let $\alpha$ be a root of $p(X)$. Based on $p(X)$ and $\alpha$, we can construct an extension field $\mathrm{GF}(q^m)$ of $\mathrm{GF}(q)$ with $q^m$ elements. $\mathrm{GF}(q^m)$ is the smallest field which contains $\mathrm{GF}(q)$ as the subfield and all the roots of $p(X)$. The characteristic of $\mathrm{GF}(q^m)$ is $p$. Construction of $\mathrm{GF}(q^m)$ using $\mathrm{GF}(q)$ is similar to the construction of $\mathrm{GF}(2^m)$ using the binary field $\mathrm{GF}(2)$ as the ground field. Because 2 is a prime, if we set $p = 2$ and $s = 1$, the construction gives the extension field $\mathrm{GF}(2^m)$ of $\mathrm{GF}(2)$.

**Example 2.13** In this example, we give construction of an extension field of a ground field whose characteristic is not 2. Suppose we choose $p = 3$ and $s = 1$. Then, $q = 3$. Next, we construct the ground field $\mathrm{GF}(3) = \{0, 1, 2\}$ with modulo-3 addition and multiplication as shown in Table 2.12.

Set $m = 2$. To construct the extension field $\mathrm{GF}(3^2)$ of $\mathrm{GF}(3)$, we need a primitive polynomial over $\mathrm{GF}(3)$ of degree 2. The polynomial $p(X) = 2 + X + X^2$ is such a polynomial. Let $\alpha$ be a root of $p(X)$. Then, based on $p(X)$ and $\alpha$, we construct the extension field $GF(3^2)$ of $GF(3)$ as shown in Table 2.13. By substituting of the nonzero elements in $GF(3^2)$ in turn into $p(X)$, we find that $\alpha$ and $\alpha^3$ are the two roots of $p(X)$ in $GF(3^2)$.                    ▲▲

Table 2.12 The prime field $GF(3)$ under modulo-3 addition and multiplication.

| ⊞ | 0 1 2 |
|---|-------|
| 0 | 0 1 2 |
| 1 | 1 2 0 |
| 2 | 2 0 1 |

| ⊡ | 0 1 2 |
|---|-------|
| 0 | 0 0 0 |
| 1 | 0 1 2 |
| 2 | 0 2 1 |

Table 2.13 $GF(3^2)$ generated by $p(X) = 2 + X + X^2$ over $GF(3)$.

| Power form | Polynomial form | | | Vector form |
|------------|-----------------|---|---|-------------|
| $0 = \alpha^{-\infty}$ | 0 | | | (0 0) |
| $1 = \alpha^0$ | 1 | | | (1 0) |
| $\alpha$ | | | $\alpha$ | (0 1) |
| $\alpha^2$ | 1 | $+$ | $2\alpha$ | (1 2) |
| $\alpha^3$ | 2 | $+$ | $2\alpha$ | (2 2) |
| $\alpha^4$ | 2 | | | (2 0) |
| $\alpha^5$ | | | $2\alpha$ | (0 2) |
| $\alpha^6$ | 2 | $+$ | $\alpha$ | (2 1) |
| $\alpha^7$ | 1 | $+$ | $\alpha$ | (1 1) |
| $\alpha^8 = 1$ | | | | |

## 2.8    Vector Spaces over Finite Fields

In this section, we present another algebraic system called a *vector space* which is also a key ingredient for constructing error-correcting codes.

### 2.8.1    Basic Definitions and Concepts

Let $\mathbf{V}$ be a set of elements and $F$ be a field. On $\mathbf{V}$, a binary operation called addition, denoted by $+$, is defined. Between the elements in $F$ and elements in $\mathbf{V}$, a multiplication operation, denoted by $\cdot$, is defined.

**Definition 2.14** The set $\mathbf{V}$ is called a *vector space over the field $F$* if it satisfies the following axioms.
  (1) $\mathbf{V}$ is a commutative group under addition $+$ defined on $\mathbf{V}$.
  (2) For any element $a$ in $F$ and any element $\mathbf{v}$ in $\mathbf{V}$, $a \cdot \mathbf{v}$ is an element in $\mathbf{V}$.
  (3) For any two elements $a$ and $b$ in $F$ and any element $\mathbf{v}$ in $\mathbf{V}$, the following associative law holds:

$$(a \cdot b) \cdot \mathbf{v} = a \cdot (b \cdot \mathbf{v}).$$

(4) For any element $a$ in $F$ and any two elements $\mathbf{u}$ and $\mathbf{v}$ in $\mathbf{V}$, the following distributive law holds:

$$a \cdot (\mathbf{u} + \mathbf{v}) = a \cdot \mathbf{u} + a \cdot \mathbf{v}.$$

(5) For any two elements $a$ and $b$ in $F$ and any element $\mathbf{v}$ in $\mathbf{V}$, the following distributive law holds:

$$(a + b) \cdot \mathbf{v} = a \cdot \mathbf{v} + b \cdot \mathbf{v}.$$

(6) Let 1 be the unit element of $F$. Then, for any element $\mathbf{v}$ in $\mathbf{V}$,

$$1 \cdot \mathbf{v} = \mathbf{v}.$$

The elements of $\mathbf{V}$ are called *vectors* and denoted by boldface lower-case letters, $\mathbf{u}$, $\mathbf{v}$, $\mathbf{w}$, etc. The elements of $F$ are called *scalars* and denoted by italic lower-case letters, $a$, $b$, $c$, etc. The addition $+$ on $\mathbf{V}$ is called *vector addition*, the multiplication $\cdot$ that combines a scalar $a$ in $F$ and a vector $\mathbf{v}$ in $\mathbf{V}$ into a vector $a \cdot \mathbf{v}$ in $\mathbf{V}$ is referred to as *scalar multiplication*, and the vector $a \cdot \mathbf{v}$ is called the *product* of $a$ and $\mathbf{v}$. The additive identity of $\mathbf{V}$ is denoted by the boldface $\mathbf{0}$, called the *zero vector* of $\mathbf{V}$. Note that we use the notation $+$ for both additions on $\mathbf{V}$ and $F$. It should be clear that when we combine two vectors in $\mathbf{V}$, $\mathbf{u} + \mathbf{v}$, the addition $+$ means the vector addition, and when we combine two scalars in $F$, $a + b$, the addition $+$ means the addition defined on the field $F$. We also use the notation $\cdot$ for both scalar multiplication and multiplication defined on the field $F$. When a scalar in $F$ and a vector in $\mathbf{V}$ are multiplied, $a \cdot \mathbf{v}$, the multiplication $\cdot$ means scalar multiplication, and when two scalars in $F$ are multiplied, $a \cdot b$, the multiplication $\cdot$ means multiplication defined on $F$.

$F$ is commonly called the "scalar field" or "ground field" of the vector space $\mathbf{V}$.

Some basic properties of a vector space $\mathbf{V}$ over a field $F$ are given below.

(1) Let 0 be the zero element of the field $F$. For any vector $\mathbf{v}$ in $\mathbf{V}$,

$$0 \cdot \mathbf{v} = \mathbf{0}.$$

(2) For any element $a$ in $F$, $a \cdot \mathbf{0} = \mathbf{0}$.

(3) For any element $a$ in $F$ and any vector $\mathbf{v}$ in $\mathbf{V}$,

$$-a \cdot \mathbf{v} = a \cdot (-\mathbf{v}) = -(a \cdot \mathbf{v}),$$

i.e., either $-a \cdot \mathbf{v}$ or $a \cdot (-\mathbf{v})$ is the additive inverse of the vector $a \cdot \mathbf{v}$.

**Definition 2.15** Let $\mathbf{S}$ be a nonempty subset of a vector space $\mathbf{V}$ over a field $F$. $\mathbf{S}$ is called a *subspace* of $\mathbf{V}$ if it satisfies the axioms for a vector space given by Definition 2.14.

To determine whether a subset $\mathbf{S}$ of a vector space $\mathbf{V}$ over a field $F$ is a subspace, it is sufficient to prove the following.

**(S1)** For any two vectors $\mathbf{u}$ and $\mathbf{v}$ in $\mathbf{S}$, $\mathbf{u} + \mathbf{v}$ is also a vector in $\mathbf{S}$.

**(S2)** For any element $a$ in $F$ and any vector $\mathbf{u}$ in $\mathbf{S}$, $a \cdot \mathbf{u}$ is also in $\mathbf{S}$.

## 2.8.2 Vector Spaces over Binary Field GF(2)

The most commonly used vector space in error-correcting codes is the vector space of all the $n$-tuples over the binary field GF(2). This vector space consists of $2^n$ vectors (or $2^n$ $n$-tuples). In the following, we will focus on the construction of such a vector space over GF(2).

A binary $n$-tuple is an ordered sequence,

$$(a_0, a_1, \ldots, a_{n-1})$$

with components from GF(2), i.e., $a_i = 0$ or $1$ for $0 \le i < n$. There are $2^n$ distinct binary $n$-tuples. We denote this set of $2^n$ distinct binary $n$-tuples with $\mathbf{V}_n$. Let $\mathbf{a} = (a_0, a_1, \ldots, a_{n-1})$ and $\mathbf{b} = (b_0, b_1, \ldots, b_{n-1})$ be two $n$-tuples in $\mathbf{V}_n$. Define an *addition operation* on $\mathbf{V}_n$ as follows:

$$
\begin{aligned}
(a_0, a_1, &\ldots, a_{n-1}) + (b_0, b_1, \ldots, b_{n-1}) \\
&= (a_0 + b_0, a_1 + b_1, \ldots, a_{n-1} + b_{n-1}) \\
&= (c_0, c_1, \ldots, c_{n-1}),
\end{aligned}
\tag{2.36}
$$

where $a_i + b_i$ is carried out in addition defined on GF(2) (i.e., modulo-2 addition) and thus $c_i = a_i + b_i \in$ GF(2). The addition of two binary $n$-tuples in $\mathbf{V}_n$, $\mathbf{a}$ and $\mathbf{b}$, results in a third binary $n$-tuple in $\mathbf{V}_n$, $\mathbf{c} = (c_0, c_1, \ldots, c_{n-1})$. We refer to the addition defined by (2.36) as *component-wise addition*.

Define a *scalar multiplication* between an element $c$ in GF(2) and an $n$-tuple $(a_0, a_1, \ldots, a_{n-1})$ in $\mathbf{V}_n$ as follows:

$$
\begin{aligned}
c \cdot (a_0, a_1, \ldots, a_{n-1}) &= (c \cdot a_0, c \cdot a_1, \ldots, c \cdot a_{n-1}) \\
&= (d_0, d_1, \ldots, d_{n-1}),
\end{aligned}
\tag{2.37}
$$

where $d_i = c \cdot a_i$ is carried out in multiplication defined on GF(2) (i.e., modulo-2 multiplication) and thus $d_i \in$ GF(2). The $n$-tuple $\mathbf{d} = (d_0, d_1, \ldots, d_{n-1})$ is in $\mathbf{V}_n$. The scalar multiplication also results in a binary $n$-tuple in $\mathbf{V}_n$. Because the modulo-2 addition on GF(2) is associative and commutative, it follows that the addition defined on $\mathbf{V}_n$ is also associative and commutative:

$$
\begin{aligned}
(\mathbf{a} + \mathbf{b}) + \mathbf{c} &= ((a_0, a_1, \ldots, a_{n-1}) + (b_0, b_1, \ldots, b_{n-1})) + (c_0, c_1, \ldots, c_{n-1}) \\
&= (a_0 + b_0, a_1 + b_1, \ldots, a_{n-1} + b_{n-1}) + (c_0, c_1, \ldots, c_{n-1}) \\
&= (a_0 + b_0 + c_0, a_1 + b_1 + c_1, \ldots, a_{n-1} + b_{n-1} + c_{n-1}) \\
&= (a_0 + (b_0 + c_0), a_1 + (b_1 + c_1), \ldots, a_{n-1} + (b_{n-1} + c_{n-1})) \\
&= (a_0, a_1, \ldots, a_{n-1}) + (b_0 + c_0, b_1 + c_1, \ldots, b_{n-1} + c_{n-1}) \\
&= (a_0, a_1, \ldots, a_{n-1}) + ((b_0, b_1, \ldots, b_{n-1}) + (c_0, c_1, \ldots, c_{n-1})) \\
&= \mathbf{a} + (\mathbf{b} + \mathbf{c}),
\end{aligned}
$$

$$\begin{aligned}
\mathbf{a} + \mathbf{b} &= (a_0, a_1, \ldots, a_{n-1}) + (b_0, b_1, \ldots, b_{n-1}) \\
&= (a_0 + b_0, a_1 + b_1, \ldots, a_{n-1} + b_{n-1}) \\
&= (b_0 + a_0, b_1 + a_1, \ldots, b_{n-1} + a_{n-1}) \\
&= (b_0, b_1, \ldots, b_{n-1}) + (a_0, a_1, \ldots, a_{n-1}) \\
&= \mathbf{b} + \mathbf{a}.
\end{aligned}$$

Note that $\mathbf{V}_n$ contains the all-zero $n$-tuple, $\mathbf{0} = (0, 0, \ldots, 0)$. Based on the addition and multiplication defined on $\mathrm{GF}(2)$, we have $0 \cdot (a_0, a_1, \ldots, a_{n-1}) = \mathbf{0}$, and

$$(a_0, a_1, \ldots, a_{n-1}) + (a_0, a_1, \ldots, a_{n-1}) = (0, 0, \ldots, 0) = \mathbf{0}.$$

This shows that each $n$-tuple in $\mathbf{V}_n$ is its own additive inverse.

According to Definition 2.14, the set $\mathbf{V}_n$ together with the addition defined between two $n$-tuples in $\mathbf{V}_n$ and the scalar multiplication defined between an element in $\mathrm{GF}(2)$ and an $n$-tuple in $\mathbf{V}_n$ forms a *vector space* over $\mathrm{GF}(2)$. Vectors are represented by boldface letters, e.g., $\mathbf{a} = (a_0, a_1, \ldots, a_{n-1})$. The vector space $\mathbf{V}_n$ over the binary field $\mathrm{GF}(2)$ plays a central role in error-correcting coding.

**Example 2.14** Let $n = 4$. The vector space $\mathbf{V}_4$ over $\mathrm{GF}(2)$ consists of the following $2^4 = 16$ vectors:

$(0\ 0\ 0\ 0), (0\ 0\ 0\ 1), (0\ 0\ 1\ 0), (0\ 0\ 1\ 1), (0\ 1\ 0\ 0), (0\ 1\ 0\ 1), (0\ 1\ 1\ 0), (0\ 1\ 1\ 1),$
$(1\ 0\ 0\ 0), (1\ 0\ 0\ 1), (1\ 0\ 1\ 0), (1\ 0\ 1\ 1), (1\ 1\ 0\ 0), (1\ 1\ 0\ 1), (1\ 1\ 1\ 0), (1\ 1\ 1\ 1).$

Consider two vectors $\mathbf{a} = (0\ 1\ 0\ 1)$ and $\mathbf{b} = (1\ 1\ 1\ 0)$ in $\mathbf{V}_4$. According to the rule of vector addition defined by (2.36), we have

$$\begin{aligned}
\mathbf{a} + \mathbf{b} &= (0\ 1\ 0\ 1) + (1\ 1\ 1\ 0) = (0+1,\ 1+1,\ 0+1,\ 1+0) \\
&= (1\ 0\ 1\ 1),
\end{aligned}$$

which is a vector in $\mathbf{V}_4$. Let $c$ be an element (a scalar) in $\mathrm{GF}(2)$. The scalar multiplication of $c$ and $\mathbf{a} = (0\ 1\ 0\ 1)$ is

$$c \cdot \mathbf{a} = c \cdot (0\ 1\ 0\ 1) = (c \cdot 0,\ c \cdot 1,\ c \cdot 0,\ c \cdot 1).$$

For $c = 0$ and $c = 1$, we obtain the following two vectors:

$$\begin{aligned}
1 \cdot (0\ 1\ 0\ 1) &= (1 \cdot 0,\ 1 \cdot 1,\ 1 \cdot 0,\ 1 \cdot 1) = (0\ 1\ 0\ 1), \\
0 \cdot (0\ 1\ 0\ 1) &= (0 \cdot 0,\ 0 \cdot 1,\ 0 \cdot 0,\ 0 \cdot 1) = (0\ 0\ 0\ 0),
\end{aligned}$$

where the first vector is $\mathbf{a}$ itself and the second vector is the zero vector $\mathbf{0} = (0\ 0\ 0\ 0)$ in $\mathbf{V}_4$. ▲▲

**Example 2.15** Consider the following four vectors in the vector space $\mathbf{V}_4$ constructed in Example 2.14:

$$\mathbf{v}_0 = (0\ 0\ 0\ 0), \mathbf{v}_1 = (0\ 1\ 0\ 1), \mathbf{v}_2 = (1\ 0\ 1\ 0), \mathbf{v}_3 = (1\ 1\ 1\ 1).$$

The set $\mathbf{S} = \{\mathbf{v}_0, \mathbf{v}_1, \mathbf{v}_2, \mathbf{v}_3\}$ is a subset of $\mathbf{V}_4$ and contains the zero vector $\mathbf{v}_0 = (0\ 0\ 0\ 0)$ in $\mathbf{V}_4$. We can easily check that, for any two vectors in $\mathbf{S}$, their sum is also a vector in $\mathbf{S}$. For $c = 0$ and $0 \leq i < 4$, the scalar product $c \cdot \mathbf{v}_i$ gives the zero vector $\mathbf{0}$ in $\mathbf{V}_4$ and for $c = 1$, the scalar product $c \cdot \mathbf{v}_i$ gives the vector $\mathbf{v}_i$ in $\mathbf{S}$. Hence, $\mathbf{S}$ is a subspace of $\mathbf{V}_4$. ▲▲

Let $\mathbf{v}_0, \mathbf{v}_1, \ldots, \mathbf{v}_{k-1}$ be $k$ vectors of length $n$ in $\mathbf{V}_n$. A *linear combination* over GF(2) of $\mathbf{v}_0, \mathbf{v}_1, \ldots, \mathbf{v}_{k-1}$ results in a vector of the following form:

$$\mathbf{u} = c_0\mathbf{v}_0 + c_1\mathbf{v}_1 + \cdots + c_{k-1}\mathbf{v}_{k-1}, \tag{2.38}$$

where $c_i \in \mathrm{GF}(2)$ for $0 \leq i < k$ and $\mathbf{u} \in \mathbf{V}_n$. The scalars $c_0, c_1, \ldots, c_{k-1}$ are called the *coefficients* of the linear combination.

A set of vectors, $\mathbf{v}_0, \mathbf{v}_1, \ldots, \mathbf{v}_{k-1}$, in $\mathbf{V}_n$, is said to be *linearly independent* if

$$c_0\mathbf{v}_0 + c_1\mathbf{v}_1 + \cdots + c_{k-1}\mathbf{v}_{k-1} \neq \mathbf{0}$$

unless $c_0 = c_1 = \cdots = c_{k-1} = 0$, i.e., all the coefficients are the zero element of GF(2).

A set of vectors, $\mathbf{v}_0, \mathbf{v}_1, \ldots, \mathbf{v}_{k-1}$, in $\mathbf{V}_n$, is said to be *linearly dependent* if there exist $k$ scalars in GF(2) *not all zeros* such that

$$c_0\mathbf{v}_0 + c_1\mathbf{v}_1 + \cdots + c_{k-1}\mathbf{v}_{k-1} = \mathbf{0}.$$

The set given by the $2^k$ linear combinations of $k$ *linearly independent* vectors, $\mathbf{v}_0, \mathbf{v}_1, \ldots, \mathbf{v}_{k-1}$, in $\mathbf{V}_n$, is called a *$k$-dimensional subspace* $\mathbf{S}$ of $\mathbf{V}_n$. We say that $\mathbf{S}$ is spanned by these $k$ linearly independent vectors and has dimension $k$. The subspace $\mathbf{S}$ has $2^k$ $n$-tuples. These $k$ linearly independent vectors are said to form a *basis* $\mathcal{B}$ of $\mathbf{S}$. The vector space $\mathbf{V}_n$ is an $n$-dimensional vector space over GF(2) and is spanned by $n$ linearly independent vectors. For a given $k$-dimensional subspace $\mathbf{S}$ of the vector space $\mathbf{V}_n$ over GF(2), there may be more than one basis with $k$ linearly independent vectors spanning the subspace $\mathbf{S}$. Similarly, there is more than one basis for $\mathbf{V}_n$. One such basis is formed as follows. For $0 \leq i < n$, let $\mathbf{e}_i$ be a vector which has a single 1-component at the $i$th position and all the other $n - 1$ positions are zeros shown below:

$$\mathbf{e}_i = (0, 0, \ldots, 0, 1, 0, \ldots, 0).$$

It is straightforward to prove that the $n$ vectors $\mathbf{e}_0, \mathbf{e}_1, \ldots, \mathbf{e}_{n-1}$ are linearly independent. Any vector $\mathbf{a} = (a_0, a_1, \ldots, a_{n-1})$ in $\mathbf{V}_n$ can be expressed (or represented) as the following linear combination over GF(2) of $\mathbf{e}_0, \mathbf{e}_1, \ldots, \mathbf{e}_{n-1}$:

$$\mathbf{a} = a_0\mathbf{e}_0 + a_1\mathbf{e}_1 + \cdots + a_{n-1}\mathbf{e}_{n-1}, \tag{2.39}$$

and this expression (or representation) is *unique* (see Problem 2.36). Hence, the set $\mathcal{B} = \{\mathbf{e}_0, \mathbf{e}_1, \ldots, \mathbf{e}_{n-1}\}$ forms a basis for $\mathbf{V}_n$ which is called the *normal basis* of $\mathbf{V}_n$.

**Example 2.16** Consider the following three vectors:

$$\mathbf{v}_0 = (1\ 1\ 0\ 0), \mathbf{v}_1 = (1\ 0\ 1\ 0), \mathbf{v}_2 = (1\ 0\ 1\ 1)$$

in the vector space $\mathbf{V}_4$ over GF(2) given in Example 2.14. We can easily check that these three vectors are linearly independent. The eight linear combinations of these three vectors are:

$$(0\ 0\ 0\ 0), (0\ 1\ 1\ 0), (1\ 1\ 0\ 0), (0\ 1\ 1\ 1), (1\ 0\ 1\ 0), (0\ 0\ 0\ 1), (1\ 0\ 1\ 1), (1\ 1\ 0\ 1).$$

The above eight vectors form a three-dimensional subspace $\mathbf{S}$ of the vector space $\mathbf{V}_4$ over GF(2) and $\mathcal{B} = \{\mathbf{v}_0, \mathbf{v}_1, \mathbf{v}_2\}$ is a basis of the subspace $\mathbf{S}$. The set of the following four vectors in $\mathbf{V}_4$:

$$\{(1\ 0\ 0\ 0), (0\ 1\ 0\ 0), (0\ 0\ 1\ 0), (0\ 0\ 0\ 1)\}$$

forms a normal basis of $\mathbf{V}_4$. ▲▲

Let $\mathbf{a} = (a_0, a_1, \ldots, a_{n-1})$ and $\mathbf{b} = (b_0, b_1, \ldots, b_{n-1})$ be two vectors in $\mathbf{V}_n$ over GF(2). Define a product of $\mathbf{a}$ and $\mathbf{b}$ as follows:

$$\mathbf{a} \cdot \mathbf{b} = a_0 \cdot b_0 + a_1 \cdot b_1 + \cdots + a_{n-1} \cdot b_{n-1}, \tag{2.40}$$

where $a_i \cdot b_i$ and $a_i \cdot b_i + a_j \cdot b_j$ are carried out with the multiplication and addition defined on GF(2) (i.e., modulo-2 multiplication and addition), respectively. The result of the product defined by (2.40) is a scalar in GF(2) and is called the *inner product* of $\mathbf{a}$ and $\mathbf{b}$, which is also denoted by $\langle \mathbf{a}, \mathbf{b} \rangle$. Two vectors $\mathbf{a}$ and $\mathbf{b}$ in $\mathbf{V}_n$ are said to be *orthogonal* if their inner product equals 0, i.e., $\langle \mathbf{a}, \mathbf{b} \rangle = 0$.

**Example 2.17** The inner product of the two vectors $(1\ 1\ 0\ 1\ 1)$ and $(1\ 0\ 1\ 0\ 1)$ in the vector space $\mathbf{V}_5$ over GF(2) is

$$\begin{aligned}
\langle (1\ 1\ 0\ 1\ 1), (1\ 0\ 1\ 0\ 1) \rangle &= 1 \cdot 1 + 1 \cdot 0 + 0 \cdot 1 + 1 \cdot 0 + 1 \cdot 1 \\
&= 1 + 0 + 0 + 0 + 1 \\
&= 0.
\end{aligned}$$

Hence, the two vectors are orthogonal. ▲▲

Let $\mathbf{S}$ be a $k$-dimensional subspace of the vector space $\mathbf{V}_n$ of all the $n$-tuples over GF(2). Let $\mathbf{S}_d$ be the subset of vectors in $\mathbf{V}_n$ such that for any vector $\mathbf{a} \in \mathbf{S}$ and any vector $\mathbf{b} \in \mathbf{S}_d$,

$$\langle \mathbf{a}, \mathbf{b} \rangle = 0.$$

The set $\mathbf{S}_d$ contains at least the all-zero $n$-tuple $\mathbf{0}$ and hence it is nonempty. $\mathbf{S}_d$ is a subspace of $\mathbf{V}_n$ and is called the *dual* (or *null*) *space* of $\mathbf{S}$ and vice versa. The following theorem provides an important property between a space and its dual space. (The proof is omitted here. Readers are referred to [27] for a possible proof.)

**Theorem 2.15** *Let* **S** *be a k-dimensional subspace of the vector space* $\mathbf{V}_n$ *of all the n-tuples over* $\mathrm{GF}(2)$*. Then, the dual space* $\mathbf{S}_d$ *of* **S** *has dimension* $n - k$*.*

Therefore, $\mathbf{S}_d$ is an $(n - k)$-dimensional subspace of $\mathbf{V}_n$ and there must be $n - k$ linearly independent vectors $\mathbf{u}_0, \mathbf{u}_1, \dots, \mathbf{u}_{n-k-1}$ in $\mathbf{V}_n$ (also in $\mathbf{S}_d$) which span the subspace $\mathbf{S}_d$.

Note that the vector space **S** and its dual space $\mathbf{S}_d$ are not disjoint because they have at least the all-zero $n$-tuple **0** in common. In fact, for certain vector spaces, they are self-dual, i.e., $\mathbf{S} = \mathbf{S}_d$, as to be shown in Chapter 8. For some cases, it is far more easier to describe the vector spaces and analyze their properties from their dual spaces' point of view than based on the vector spaces themselves, e.g., the weight distribution of linear block codes to be presented in Chapter 3 and the low-density parity-check codes to be introduced in Chapter 11.

**Example 2.18** Consider the vector space $\mathbf{V}_5$ of all the 5-tuples over $\mathrm{GF}(2)$ shown in Table 2.14. It can be easily checked that the three-dimensional and two-dimensional subspaces **S** and $\mathbf{S}_d$ of $\mathbf{V}_5$ given in Table 2.15 are dual spaces of each other.

The three vectors (1 1 1 0 0), (0 1 0 1 0), (1 0 0 0 1) in **S** are linearly independent and they can be used as a basis to construct the subspace **S**. The two vectors (1 0 1 0 1) and (0 1 1 1 0) in $\mathbf{S}_d$ form a basis of $\mathbf{S}_d$ because they are linearly independent. From Table 2.15, it can be seen that **S** and its dual space $\mathbf{S}_d$ have two common vectors, (0 0 0 0 0) and (1 1 0 1 1), i.e., $\mathbf{S} \cap \mathbf{S}_d = \{(0\ 0\ 0\ 0\ 0), (1\ 1\ 0\ 1\ 1)\}$. ▲▲

Table 2.14 The vector space $\mathbf{V}_5$ over $\mathrm{GF}(2)$ given in Example 2.18.

| | | | |
|---|---|---|---|
| (0 0 0 0 0) | (0 0 1 0 1) | (1 1 1 1 1) | (1 1 1 1 0) |
| (1 0 0 0 0) | (1 0 1 1 0) | (1 1 0 1 1) | (0 1 1 1 1) |
| (0 1 0 0 0) | (0 1 0 1 1) | (1 1 0 0 1) | (1 0 0 1 1) |
| (0 0 1 0 0) | (1 0 0 0 1) | (1 1 0 0 0) | (1 1 1 0 1) |
| (0 0 0 1 0) | (1 1 1 0 0) | (0 1 1 0 0) | (1 1 0 1 0) |
| (0 0 0 0 1) | (0 1 1 1 0) | (0 0 1 1 0) | (0 1 1 0 1) |
| (1 0 1 0 0) | (0 0 1 1 1) | (0 0 0 1 1) | (1 0 0 1 0) |
| (0 1 0 1 0) | (1 0 1 1 1) | (1 0 1 0 1) | (0 1 0 0 1) |

Table 2.15 A subspace **S** and its dual space $\mathbf{S}_d$ in $\mathbf{V}_5$ given in Example 2.18.

| **S** | | $\mathbf{S}_d$ |
|---|---|---|
| (0 0 0 0 0) | (1 0 1 1 0) | (0 0 0 0 0) |
| (1 1 1 0 0) | (0 1 1 0 1) | (1 0 1 0 1) |
| (0 1 0 1 0) | (1 1 0 1 1) | (0 1 1 1 0) |
| (1 0 0 0 1) | (0 0 1 1 1) | (1 1 0 1 1) |

### 2.8.3   Vector Spaces over Nonbinary Field GF($q$)

The construction of vector space $\mathbf{V}_n$ of all $n$-tuples over a nonbinary field GF($q$) is similar to the construction of the vector space $\mathbf{V}_n$ of all the $n$-tuples over GF(2). In this case, the components of an $n$-tuple $\mathbf{a} = (a_0, a_1, \ldots, a_{n-1})$ are elements from the field GF($q$). There are $q^n$ $n$-tuples in $\mathbf{V}_n$ over GF($q$). The addition of two $n$-tuples $\mathbf{a} = (a_0, a_1, \ldots, a_{n-1})$ and $\mathbf{b} = (b_0, b_1, \ldots, b_{n-1})$ over GF($q$), $\mathbf{c} = \mathbf{a} + \mathbf{b} = (c_0, c_1, \ldots, c_{n-1})$, is performed by component-wise addition over GF($q$), i.e., $c_i = a_i + b_i$, $0 \le i < n$, where $a_i + b_i$ is carried out with the addition defined on GF($q$). The elements of GF($q$) are scalars. The scalar multiplication of a scalar $c \in$ GF($q$) and an $n$-tuple $\mathbf{a} = (a_0, a_1, \ldots, a_{n-1}) \in \mathbf{V}_n$ over GF($q$) is $c \cdot \mathbf{a} = (c \cdot a_0, c \cdot a_1, \ldots, c \cdot a_{n-1})$, where the product $c \cdot a_i$, $0 \le i < n$, is performed by using the multiplication operation defined on GF($q$). The most commonly used vector space over a nonbinary field is the one formed by all $n$-tuples over GF($2^m$), an extension field of GF(2).

More on vector spaces over finite fields, binary and nonbinary, will be developed in the remaining chapters in this book.

## 2.9   Matrices over Finite Fields

Matrices over finite fields are relevant in formulation, design, construction, encoding, and decoding of error-correcting codes which will be shown in the rest of this book. In this section, we present several aspects of matrices over binary and nonbinary fields, including basic concepts, characterization, operations, relationship with vector spaces over finite fields, and a special class of matrices.

### 2.9.1   Concepts of Matrices over GF(2)

Consider a $k \times n$ rectangular array of $k$ rows and $n$ columns as follows:

$$\mathbf{A} = \begin{bmatrix} a_{0,0} & a_{0,1} & \cdots & a_{0,n-1} \\ a_{1,0} & a_{1,1} & \cdots & a_{1,n-1} \\ \vdots & \vdots & \ddots & \vdots \\ a_{k-1,0} & a_{k-1,1} & \cdots & a_{k-1,n-1} \end{bmatrix} \tag{2.41}$$

where the element $a_{i,j}$ is either "0" or "1," i.e., $a_{i,j} \in$ GF(2), for $0 \le i < k$ and $0 \le j < n$. The array $\mathbf{A}$ is called a $k \times n$ *matrix over* GF(2) or a *binary matrix*. Each row of $\mathbf{A}$ is a binary $n$-tuple and each column of $\mathbf{A}$ is a binary $k$-tuple. If all the entries of $\mathbf{A}$ are zeros, $\mathbf{A}$ is called a zero matrix (ZM). A matrix is called a *square matrix* if the number of rows and the number of columns of the matrix are equal, i.e., $k = n$. Note that the rows and columns are labeled from 0 to $k-1$ and 0 to $n-1$, respectively. For simplicity, we often express the matrix $\mathbf{A}$ in the following compact form:

$$\mathbf{A} = [a_{i,j}]_{0 \le i < k, 0 \le j < n}. \tag{2.42}$$

For the $j$th column in $\mathbf{A}$, the number of nonzero entries in the column is called the *column weight*, denoted by $\omega_{c_j}$; and for the $i$th row in $\mathbf{A}$, the number of nonzero entries in the row is called the *row weight*, denoted by $\omega_{r_i}$. Apparently, $\mathbf{A}$ may have multiple column weights and multiple row weights. If $\mathbf{A}$ has a constant row weight and constant column weight, $\mathbf{A}$ is said to be *regular*; otherwise, it is *irregular*.

The *rank* of a matrix $\mathbf{A}$ over GF(2) is defined as the number of linearly independent rows (or columns) among the rows (or columns) of the matrix, denoted by rank($\mathbf{A}$). A matrix is said to be a *full row rank matrix* if all its rows are linearly independent, and called a *full column rank matrix* if all its columns are linearly independent. For a square matrix, these two concepts are equivalent and we say that the matrix is full rank if all its rows (columns) are linearly independent. A square matrix with full rank is said to be *nonsingular*; otherwise, it is *singular*. If not all the rows (columns) of a matrix are linearly independent, then there are some *redundant rows* (*columns*) in the matrix, i.e., some of the rows (columns) of the matrix are linear combinations of other rows (columns) of the matrix. For a nonsquare matrix, when we say that it is full rank, we mean that it is full column rank if the matrix has more rows than columns; otherwise, it is full row rank. Note that the row rank of a matrix equals its column rank.

A nonzero square matrix is called an *upper triangular matrix* if all the entries below its main (principal) diagonal are zero. It is called a *lower triangular matrix* if all entries above its main diagonal are zero. A square matrix is said to be *strictly triangular* if all the entries on and below (or above) its main diagonal are zero.

The transpose of a $k \times n$ matrix $\mathbf{A}$, denoted by $\mathbf{A}^T$, is an $n \times k$ matrix in which the rows and columns are the columns and rows of $\mathbf{A}$, respectively. The transpose of the $k \times n$ matrix $\mathbf{A}$ given by (2.41) is given as follows:

$$\mathbf{A}^T = \begin{bmatrix} a_{0,0} & a_{1,0} & \cdots & a_{k-1,0} \\ a_{0,1} & a_{1,1} & \cdots & a_{k-1,1} \\ \vdots & \vdots & \ddots & \vdots \\ a_{0,n-1} & a_{1,n-1} & \cdots & a_{k-1,n-1} \end{bmatrix}, \tag{2.43}$$

which is an $n \times k$ matrix over GF(2).

**Example 2.19** Consider the following $3 \times 6$ matrix over GF(2):

$$\mathbf{A} = \begin{bmatrix} 1 & 0 & 0 & 1 & 0 & 1 \\ 0 & 1 & 0 & 0 & 1 & 1 \\ 0 & 0 & 1 & 1 & 1 & 0 \end{bmatrix}.$$

The matrix $\mathbf{A}$ has a constant row weight 3 and two column weights, 1 and 2, with the leftmost three columns with weight 1 and the rightmost three columns with weight 2. Thus, $\mathbf{A}$ is an irregular matrix.

We can easily verify that the three rows of the matrix $\mathbf{A}$ are linearly independent. Hence, the row rank of $\mathbf{A}$ is 3 and $\mathbf{A}$ is a full row rank matrix. The transpose of $\mathbf{A}$ is a $6 \times 3$ matrix given as follows:

$$\mathbf{A}^T = \begin{bmatrix} 1 & 0 & 0 \\ 0 & 1 & 0 \\ 0 & 0 & 1 \\ 1 & 0 & 1 \\ 0 & 1 & 1 \\ 1 & 1 & 0 \end{bmatrix}.$$

The matrix $\mathbf{A}^T$ is not a full row rank matrix. It can be calculated that $\mathbf{A}^T$ has column rank 3 and its first three rows are linearly independent, i.e., there are three redundant rows in $\mathbf{A}^T$. ▲▲

### 2.9.2 Operations of Matrices over GF(2)

Two matrices over GF(2) can be added if they have the same number of rows and the same number of columns, i.e., they have the same size. Let $\mathbf{A} = [a_{i,j}]_{0 \le i < k, 0 \le j < n}$ and $\mathbf{B} = [b_{i,j}]_{0 \le i < k, 0 \le j < n}$ be two $k \times n$ matrices over GF(2). Adding these two matrices is defined as adding their corresponding entries using addition defined on GF(2) (modulo-2 addition). Hence, adding two matrices $\mathbf{A} = [a_{i,j}]_{0 \le i < k, 0 \le j < n}$ and $\mathbf{B} = [b_{i,j}]_{0 \le i < k, 0 \le j < n}$, we obtain the following $k \times n$ matrix over GF(2):

$$\mathbf{C} = \mathbf{A} + \mathbf{B}$$

$$= \begin{bmatrix} a_{0,0} + b_{0,0} & a_{0,1} + b_{0,1} & \cdots & a_{0,n-1} + b_{0,n-1} \\ a_{1,0} + b_{1,0} & a_{1,1} + b_{1,1} & \cdots & a_{1,n-1} + b_{1,n-1} \\ \vdots & \vdots & \ddots & \vdots \\ a_{k-1,0} + b_{k-1,0} & a_{k-1,1} + b_{k-1,1} & \cdots & a_{k-1,n-1} + b_{k-1,n-1} \end{bmatrix} \qquad (2.44)$$

$$= [a_{i,j} + b_{i,j}]_{0 \le i < k, 0 \le j < n},$$

where $a_{i,j} + b_{i,j}$ is performed over GF(2) for $0 \le i < k$, $0 \le j < n$. Recall that the addition of two vectors in the vector space $\mathbf{V}_n$ over GF(2) is defined as component-wise addition over GF(2) in Section 2.8.2. From (2.44), we see that the addition of two matrices is also a component-wise operation. From this point of view, we also call the addition of two matrices *component-wise addition* or *entry-wise addition*.

Two matrices can be multiplied provided that the number of columns of the first matrix is equal to the number of rows of the second matrix. Let $\mathbf{A} = [a_{i,j}]_{0 \le i < k, 0 \le j < n}$ and $\mathbf{B} = [b_{s,t}]_{0 \le s < n, 0 \le t < m}$ be a $k \times n$ matrix and an $n \times m$ matrix over GF(2), respectively. The matrix $\mathbf{A}$ has $n$ columns and the matrix $\mathbf{B}$ has $n$ rows. The product of $\mathbf{A}$ and $\mathbf{B}$, i.e., multiplying $\mathbf{A}$ by $\mathbf{B}$, is defined as the following $k \times m$ matrix over GF(2):

$$\mathbf{D} = \mathbf{A} \times \mathbf{B} = [d_{i,t}]_{0 \le i < k, 0 \le t < m}, \qquad (2.45)$$

where $d_{i,t}$ is the inner product of the $i$th row $\mathbf{a}_i$ of $\mathbf{A}$ and the $t$th column $\mathbf{b}_t$ of $\mathbf{B}$ over GF(2), i.e.,

$$d_{i,t} = \langle \mathbf{a}_i, \mathbf{b}_t \rangle = \sum_{j=0}^{n-1} a_{i,j} b_{j,t}. \qquad (2.46)$$

For a nonsingular matrix $\mathbf{A}$ of size $n \times n$, there exists a nonsingular matrix $\mathbf{B}$ of the same size such that $\mathbf{A} \times \mathbf{B} = \mathbf{I}_n$, where $\mathbf{I}_n$ is an *identity matrix* of size $n \times n$ (the subscript "$n$" of $\mathbf{I}_n$ represents the size of the identity matrix).[2] The matrix $\mathbf{B}$ is called the *inverse* of $\mathbf{A}$, denoted by $\mathbf{B} = \mathbf{A}^{-1}$.

**Definition 2.16** A square matrix is called a *permutation matrix* if each row (each column) contains a single 1-component (i.e., with column (row) weight 1). A square matrix is said to be a *circulant* if each row of the matrix is a cyclic-shift of the row above it *one place to the right* (or left) and the top row is the cyclic-shift of the last row one position to the right (or left). The top row of a circulant is called the *generator* of the circulant.

Based on the above definition, it is noted that a circulant has one column weight $\omega_c$ and one row weight $\omega_r$ and its column weight $\omega_c$ equals its row weight $\omega_r$. Thus, for a circulant, we do not distinguish its column weight and row weight and say that it has weight $\omega$. As an example, the following $7 \times 7$ matrix is a circulant with generator (1 0 1 1 0 0 0):

$$\begin{bmatrix} 1 & 0 & 1 & 1 & 0 & 0 & 0 \\ 0 & 1 & 0 & 1 & 1 & 0 & 0 \\ 0 & 0 & 1 & 0 & 1 & 1 & 0 \\ 0 & 0 & 0 & 1 & 0 & 1 & 1 \\ 1 & 0 & 0 & 0 & 1 & 0 & 1 \\ 1 & 1 & 0 & 0 & 0 & 1 & 0 \\ 0 & 1 & 1 & 0 & 0 & 0 & 1 \end{bmatrix},$$

which has weight $\omega_c = \omega_r = \omega = 3$.

Note that in a circulant, each column is the cyclic-shift of the column on its left (or right) one place downward (or upward). The leftmost (or rightmost) column is the cyclic-shift of the rightmost (or leftmost) column downward (or upward) one place.

A circulant whose generator has a single 1-component and all its other components are zeros is called a *circulant permutation matrix* (CPM). A $7 \times 7$ CPM is given below:

$$\begin{bmatrix} 0 & 0 & 0 & 1 & 0 & 0 & 0 \\ 0 & 0 & 0 & 0 & 1 & 0 & 0 \\ 0 & 0 & 0 & 0 & 0 & 1 & 0 \\ 0 & 0 & 0 & 0 & 0 & 0 & 1 \\ 1 & 0 & 0 & 0 & 0 & 0 & 0 \\ 0 & 1 & 0 & 0 & 0 & 0 & 0 \\ 0 & 0 & 1 & 0 & 0 & 0 & 0 \end{bmatrix}.$$

Obviously, a CPM of size $n \times n$ whose generator has the single 1-component in the zeroth position, i.e., (1 0 0 ... 0), is an *identity matrix* $\mathbf{I}_n$ of size $n \times n$. It is noted that any CPM of size $n \times n$ can be obtained by cyclically shifting all the rows of the identity matrix $\mathbf{I}_n$ to the right by certain positions. For example,

---

[2] An identity matrix is defined as a square matrix which has all 1s on its main diagonal and 0s elsewhere.

the above $7 \times 7$ CPM can be obtained by cyclically shifting all the rows of the $7 \times 7$ identity matrix $\mathbf{I}_7$ to the right by three positions.

Consider the following full-rank $k \times n$ matrix with $k \leq n$ (i.e., a full row rank matrix):

$$
\mathbf{G} = \begin{bmatrix} \mathbf{g}_0 \\ \mathbf{g}_1 \\ \vdots \\ \mathbf{g}_{k-1} \end{bmatrix} = \begin{bmatrix} g_{0,0} & g_{0,1} & \cdots & g_{0,n-1} \\ g_{1,0} & g_{1,1} & \cdots & g_{1,n-1} \\ \vdots & \vdots & \ddots & \vdots \\ g_{k-1,0} & g_{k-1,1} & \cdots & g_{k-1,n-1} \end{bmatrix}, \tag{2.47}
$$

whose $k$ rows $\mathbf{g}_0, \mathbf{g}_1, \ldots, \mathbf{g}_{k-1}$ (each as an $n$-tuple over $\mathrm{GF}(2)$) are linearly independent over $\mathrm{GF}(2)$. The $2^k$ linear combinations of these $k$ linearly independent rows give $2^k$ $n$-tuples over $\mathrm{GF}(2)$ which form a $k$-dimensional subspace $\mathbf{S}$ of the vector space $\mathbf{V}_n$ of all the $2^n$ $n$-tuples. The subspace $\mathbf{S}$ is called the *row space* of $\mathbf{G}$ because it is spanned by the rows of $\mathbf{G}$. The set of the $k$ rows in $\mathbf{G}$, $\mathcal{B} = \{\mathbf{g}_0, \mathbf{g}_1, \ldots, \mathbf{g}_{k-1}\}$, forms a basis of $\mathbf{S}$. Here, we connect a subspace $\mathbf{S}$ with a full-rank matrix $\mathbf{G}$. Any vector $\mathbf{v}$ in $\mathbf{S}$ is a linear combination of the $k$ rows in $\mathbf{G}$, i.e.,

$$
\mathbf{v} = f_0 \mathbf{g}_0 + f_1 \mathbf{g}_1 + \cdots + f_{k-1} \mathbf{g}_{k-1}, \tag{2.48}
$$

with $f_i \in \mathrm{GF}(2)$, $0 \leq i < k$.

**Example 2.20** Consider the following $4 \times 7$ matrix over $\mathrm{GF}(2)$

$$
\mathbf{G} = \begin{bmatrix} 1 & 1 & 0 & 1 & 0 & 0 & 0 \\ 0 & 1 & 1 & 0 & 1 & 0 & 0 \\ 1 & 1 & 1 & 0 & 0 & 1 & 0 \\ 1 & 0 & 1 & 0 & 0 & 0 & 1 \end{bmatrix}.
$$

It is easy to verify that the four rows of $\mathbf{G}$ are linearly independent. Thus, the row space of $\mathbf{G}$ is a four-dimensional subspace $\mathbf{S}$ of the vector space $\mathbf{V}_7$ of all the $2^7 = 128$ 7-tuples over $\mathrm{GF}(2)$. The 16 vectors in $\mathbf{S}$ are given below:

$$(0\ 0\ 0\ 0\ 0\ 0\ 0), (1\ 1\ 0\ 1\ 0\ 0\ 0), (0\ 1\ 1\ 0\ 1\ 0\ 0), (1\ 0\ 1\ 1\ 1\ 0\ 0),$$
$$(1\ 1\ 1\ 0\ 0\ 1\ 0), (0\ 0\ 1\ 1\ 0\ 1\ 0), (1\ 0\ 0\ 0\ 1\ 1\ 0), (0\ 1\ 0\ 1\ 1\ 1\ 0),$$
$$(1\ 0\ 1\ 0\ 0\ 0\ 1), (0\ 1\ 1\ 1\ 0\ 0\ 1), (1\ 1\ 0\ 0\ 1\ 0\ 1), (0\ 0\ 0\ 1\ 1\ 0\ 1),$$
$$(0\ 1\ 0\ 0\ 0\ 1\ 1), (1\ 0\ 0\ 1\ 0\ 1\ 1), (0\ 0\ 1\ 0\ 1\ 1\ 1), (1\ 1\ 1\ 1\ 1\ 1\ 1).$$

Consider the vector $\mathbf{v} = (0\ 0\ 1\ 1\ 0\ 1\ 0)$. It can be calculated by the following linear combination of the four rows of $\mathbf{G}$:

$$\mathbf{v} = 1 \cdot (1\ 1\ 0\ 1\ 0\ 0\ 0) + 0 \cdot (0\ 1\ 1\ 0\ 1\ 0\ 0) + 1 \cdot (1\ 1\ 1\ 0\ 0\ 1\ 0) + 0 \cdot (1\ 0\ 1\ 0\ 0\ 0\ 1),$$

i.e., $f_0 = 1$, $f_1 = 0$, $f_2 = 1$, and $f_3 = 0$.                                    ▲▲

For any $k \times n$ matrix $\mathbf{G}$ with $k$ linearly independent rows, there exists an $(n - k) \times n$ matrix $\mathbf{H}$ with $n - k$ linearly independent rows:

$$\mathbf{H} = \begin{bmatrix} \mathbf{h}_0 \\ \mathbf{h}_1 \\ \vdots \\ \mathbf{h}_{n-k-1} \end{bmatrix} = \begin{bmatrix} h_{0,0} & h_{0,1} & \cdots & h_{0,n-1} \\ h_{1,0} & h_{1,1} & \cdots & h_{1,n-1} \\ \vdots & \vdots & \ddots & \vdots \\ h_{n-k-1,0} & h_{n-k-1,1} & \cdots & h_{n-k-1,n-1} \end{bmatrix}, \quad (2.49)$$

such that any vector $\mathbf{v}$ in the row space $\mathbf{S}$ of $\mathbf{G}$ is orthogonal to all the rows of $\mathbf{H}$, i.e., for $0 \le j < n - k$,

$$\langle \mathbf{v}, \mathbf{h}_j \rangle = 0. \quad (2.50)$$

Because $\mathbf{g}_i$ is a vector in the row space of $\mathbf{G}$, we have the following inner product

$$\langle \mathbf{g}_i, \mathbf{h}_j \rangle = 0, \quad (2.51)$$

for $0 \le i < k$ and $0 \le j < n - k$, i.e., every row of $\mathbf{G}$ is orthogonal to every row of $\mathbf{H}$ and vice versa. The equality of (2.51) implies that

$$\mathbf{G} \times \mathbf{H}^T = \mathbf{O}, \quad (2.52)$$

where $\mathbf{O}$ is a $k \times (n - k)$ ZM.

Let $\mathbf{S}_d$ be the row space of $\mathbf{H}$ and $\mathbf{u}$ be a vector in $\mathbf{S}_d$. Then, $\mathbf{u}$ is a linear combination of rows of $\mathbf{H}$,

$$\mathbf{u} = c_0 \mathbf{h}_0 + c_1 \mathbf{h}_1 + \cdots + c_{n-k-1} \mathbf{h}_{n-k-1},$$

where $c_j = 0$ or $1$ for $0 \le j < n - k$. For any vector $\mathbf{v}$ in the row space $\mathbf{S}$ of $\mathbf{G}$, the inner product of $\mathbf{v}$ and $\mathbf{u}$ is

$$\begin{aligned} \langle \mathbf{v}, \mathbf{u} \rangle &= \mathbf{v} \cdot (c_0 \mathbf{h}_0 + c_1 \mathbf{h}_1 + \cdots + c_{n-k-1} \mathbf{h}_{n-k-1}) \\ &= c_0 \langle \mathbf{v}, \mathbf{h}_0 \rangle + c_1 \langle \mathbf{v}, \mathbf{h}_1 \rangle + \cdots + c_{n-k-1} \langle \mathbf{v}, \mathbf{h}_{n-k-1} \rangle. \end{aligned}$$

Because $\langle \mathbf{v}, \mathbf{h}_j \rangle = 0$ for $0 \le j < n - k$, the inner product $\langle \mathbf{v}, \mathbf{u} \rangle$ is 0. That is, any vector $\mathbf{v}$ in the row space $\mathbf{S}$ of $\mathbf{G}$ and any vector $\mathbf{u}$ in the row space $\mathbf{S}_d$ of $\mathbf{H}$ are orthogonal. Hence, the row space $\mathbf{S}$ of $\mathbf{G}$ is the *null space* (or *dual space*) of the row space $\mathbf{S}_d$ of $\mathbf{H}$ and vice versa. The dimensions of the subspace $\mathbf{S}$ and its dual space $\mathbf{S}_d$ are $k$ and $n - k$, respectively, which sum to $n$ (the dimension of the vector space $\mathbf{V}_n$).

**Example 2.21** Consider the following $3 \times 7$ matrix over $GF(2)$:

$$\mathbf{H} = \begin{bmatrix} 1 & 0 & 0 & 1 & 0 & 1 & 1 \\ 0 & 1 & 0 & 1 & 1 & 1 & 0 \\ 0 & 0 & 1 & 0 & 1 & 1 & 1 \end{bmatrix}.$$

We can easily check that the three rows of $\mathbf{H}$ are linearly independent. Furthermore, each row of $\mathbf{H}$ and each row of the $4 \times 7$ matrix $\mathbf{G}$ given in Example 2.20 are orthogonal. Hence, $\mathbf{G} \times \mathbf{H}^T = \mathbf{O}$. The row space of $\mathbf{H}$ is a three-dimensional subspace $\mathbf{S}_d$ of the vector space $\mathbf{V}_7$ over $GF(2)$ which is the dual space of the row space **S** of **G** given in Example 2.20. The row space $\mathbf{S}_d$ of **H** consists of the following eight vectors:

$$(0\ 0\ 0\ 0\ 0\ 0\ 0), (1\ 0\ 0\ 1\ 0\ 1\ 1), (0\ 1\ 0\ 1\ 1\ 1\ 0), (0\ 0\ 1\ 0\ 1\ 1\ 1),$$
$$(1\ 1\ 0\ 0\ 1\ 0\ 1), (1\ 0\ 1\ 1\ 1\ 0\ 0), (0\ 1\ 1\ 1\ 0\ 0\ 1), (1\ 1\ 1\ 0\ 0\ 1\ 0).$$

Consider the vector $\mathbf{u} = (0\ 1\ 1\ 1\ 0\ 0\ 1)$. It can be viewed as the following linear combination of the three rows in **H**:

$$\mathbf{u} = 0 \cdot (1\ 0\ 0\ 1\ 0\ 1\ 1) + 1 \cdot (0\ 1\ 0\ 1\ 1\ 1\ 0) + 1 \cdot (0\ 0\ 1\ 0\ 1\ 1\ 1),$$

i.e., $c_0 = 0$, $c_1 = 1$, and $c_2 = 1$. ▲▲

### 2.9.3 Matrices over Nonbinary Field GF($q$)

For the $k \times n$ matrix **A** in the form of (2.41), if $a_{i,j} \in \mathrm{GF}(q)$ where $q \neq 2$, we call **A** a *nonbinary matrix* or a *matrix over* $\mathrm{GF}(q)$. The addition and multiplication of two matrices over $\mathrm{GF}(q)$ are the same as those defined in Section 2.9.2 for matrices over $\mathrm{GF}(2)$ except that the addition $a_{i,j} + b_{i,j}$ and the multiplication $a_{i,j} \cdot b_{j,t}$ are carried out with the addition and multiplication defined on $\mathrm{GF}(q)$, respectively.

If the entries in the $k \times n$ matrix **G** given by (2.47) are elements from a nonbinary field $\mathrm{GF}(q)$, then **G** is a matrix over $\mathrm{GF}(q)$. The row space over $\mathrm{GF}(q)$ of **G** gives a $k$-dimensional subspace of the vector space $\mathbf{V}_n$ of all the $q^n$ $n$-tuples over $\mathrm{GF}(q)$. The dual space of the row space of **G** is the row space of an $(n-k) \times n$ matrix **H** over $\mathrm{GF}(q)$ of the form given by (2.49) with entries from $\mathrm{GF}(q)$.

Next, we introduce a type of nonbinary matrices with special structures. Let $a_1, a_2, \ldots, a_m$ be $m$ nonzero elements in $\mathrm{GF}(q)$. An $n \times m$ (or $m \times n$) matrix over $\mathrm{GF}(q)$ in one of the following two forms is called a *Van der Monde* (or *Vandermonde*) *matrix* over $\mathrm{GF}(q)$:

$$\text{Form 1} \rightarrow \quad \mathbf{V} = \begin{bmatrix} 1 & 1 & \cdots & 1 & 1 \\ a_1 & a_2 & \cdots & a_{m-1} & a_m \\ a_1^2 & a_2^2 & \cdots & a_{m-1}^2 & a_m^2 \\ \vdots & \vdots & \ddots & \vdots & \vdots \\ a_1^{n-2} & a_2^{n-2} & \cdots & a_{m-1}^{n-2} & a_m^{n-2} \\ a_1^{n-1} & a_2^{n-1} & \cdots & a_{m-1}^{n-1} & a_m^{n-1} \end{bmatrix}, \tag{2.53}$$

$$\text{Form 2} \rightarrow \quad \mathbf{V} = \begin{bmatrix} 1 & a_1 & a_1^2 & \cdots & a_1^{n-2} & a_1^{n-1} \\ 1 & a_2 & a_2^2 & \cdots & a_2^{n-2} & a_2^{n-1} \\ \vdots & \vdots & \ddots & \vdots & \vdots \\ 1 & a_{m-1} & a_{m-1}^2 & \cdots & a_{m-1}^{n-2} & a_{m-1}^{n-1} \\ 1 & a_m & a_m^2 & \cdots & a_m^{n-2} & a_m^{n-1} \end{bmatrix}. \tag{2.54}$$

Vandermonde matrices are instrumental in proving the minimum-distance properties of BCH and RS codes, as will be illustrated in Chapters 5 and 6. In the development of code design, analysis, and decoding algorithms based on the Galois Fourier transform [29–32], this type of matrices also plays a central role.

### 2.9.4   Determinants

A special characteristic of a square matrix over a finite field is its *determinant*. It can be used in the elementary study of the rank of a matrix, solution of simultaneous linear equations (as shown in Section 2.6), and analysis and design of some algebraic codes which will be shown in later chapters of the book. Before we define the determinant of a square matrix, we present some basic concepts and operations of permutations of the locations of its entries in terms of their indices.

Consider an $n \times n$ square matrix over $\mathrm{GF}(q)$:

$$\mathbf{A} = [a_{i,j}]_{0 \leq i < n, 0 \leq j < n} = \begin{bmatrix} a_{0,0} & a_{0,1} & \cdots & a_{0,n-1} \\ a_{1,0} & a_{1,1} & \cdots & a_{1,n-1} \\ \vdots & \vdots & \ddots & \vdots \\ a_{n-1,0} & a_{n-1,1} & \cdots & a_{n-1,n-1} \end{bmatrix}, \tag{2.55}$$

in which the rows and columns of $\mathbf{A}$ are labeled from 0 to $n - 1$. Both rows and columns are indexed in the order of $0, 1, 2, \ldots, n - 1$. For $0 \leq i, j < n$, the entry of $\mathbf{A}$ in the $i$th row and the $j$th column is $a_{i,j}$. Let $\mathbf{s} = (0, 1, \ldots, n - 1)$ denote the index sequence in an incremental order. The orders of the indices in $\mathbf{s} = (0, 1, \ldots, n - 1)$ can be permuted in $n! = n \cdot (n - 1) \cdots \cdots 2 \cdot 1$ distinct orders. Let $\Phi = \{\phi_0(\mathbf{s}), \phi_1(\mathbf{s}), \ldots, \phi_{n!-1}(\mathbf{s})\}$ denote the set of all the $n!$ distinct permutations of the orders of indices in $\mathbf{s}$. Permuting the orders of the indices in $\mathbf{s}$ with a permutation $\phi$ in $\Phi$, we obtain a permuted index sequence $\phi(\mathbf{s}) = (\phi(0), \phi(1), \ldots, \phi(n - 1))$. For clarity, we display the permutation, $\mathbf{s} \rightarrow \phi(\mathbf{s})$, as follows:

$$\phi(\mathbf{s}) = \begin{pmatrix} 0 & 1 & 2 & \cdots & n - 1 \\ \phi(0) & \phi(1) & \phi(2) & \cdots & \phi(n - 1) \end{pmatrix}. \tag{2.56}$$

**Example 2.22** Let $n = 3$. Consider the index sequence $\mathbf{s} = (0, 1, 2)$. There are $3! = 3 \cdot 2 \cdot 1 = 6$ distinct permutations of the orders of the three indices in $\mathbf{s} = (0, 1, 2)$ which are given below:

$$\phi_0(\mathbf{s}) = \begin{pmatrix} 0 & 1 & 2 \\ 0 & 1 & 2 \end{pmatrix}, \phi_1(\mathbf{s}) = \begin{pmatrix} 0 & 1 & 2 \\ 1 & 2 & 0 \end{pmatrix}, \phi_2(\mathbf{s}) = \begin{pmatrix} 0 & 1 & 2 \\ 2 & 0 & 1 \end{pmatrix},$$
$$\phi_3(\mathbf{s}) = \begin{pmatrix} 0 & 1 & 2 \\ 0 & 2 & 1 \end{pmatrix}, \phi_4(\mathbf{s}) = \begin{pmatrix} 0 & 1 & 2 \\ 1 & 0 & 2 \end{pmatrix}, \phi_5(\mathbf{s}) = \begin{pmatrix} 0 & 1 & 2 \\ 2 & 1 & 0 \end{pmatrix}. \tag{2.57}$$

▲▲

Let $\mathbf{s}_l = (i_0, i_1, \ldots, i_{l-1})$ be a subsequence of $\mathbf{s} = (0, 1, \ldots, n - 1)$ with $l$ indices, $0 \leq i_0 < i_1 < \cdots < i_{l-1} < n$. Let $\phi(\mathbf{s}_l) = (\phi(i_0), \phi(i_1), \ldots, \phi(i_{l-1}))$ be the $\phi$-permuted sequence of $\mathbf{s}_l = (i_0, i_1, \ldots, i_{l-1})$, a subsequence of $\phi(\mathbf{s})$.

A sequence $\mathbf{c}_l = (i_0, \phi(i_0), i_1, \phi(i_1), \ldots, i_{l-1}, \phi(i_{l-1}))$ of alternating indices in $\mathbf{s}_l$ and $\phi(\mathbf{s}_l)$ is said to form a *cycle* if the following conditions hold: (1) for $0 < k < l$, $\phi(i_{k-1}) = i_k$, and (2) $\phi(i_{l-1}) = i_0$. The length of this cycle is $2l$. The number $l$ is called the *order* of the cycle. This cycle can be represented as $i_0 \to i_1 \to \cdots \to i_{l-1} \to i_0$. It is clear that we obtain the same cycle if we trace the cycle $\mathbf{c}_l$ from any index in $(i_0, i_1, \ldots, i_{l-1})$ on it. A cycle of order $l = 1$ is called an identity permutation.

In the permutation $\phi(\mathbf{s})$, there may be more than one cycle. Two cycles are said to be *disjoint* if their index sets do not have any common index. For any permutation $\phi$ in $\Phi$, there exists a set of disjoint cycles in $\phi(\mathbf{s})$ such that the sum of their orders is equal to the number $n$ of indices in $\mathbf{s} = (0, 1, \ldots, n-1)$ [11, 12]. Let $\tau$ be the number of disjoint cycles contained in $(\mathbf{s}, \phi(\mathbf{s}))$ and let $l_0$, $l_1, \ldots, l_{\tau-1}$ be the orders of these cycles. The number $\lambda(\phi(\mathbf{s})) = (l_0 - 1) + (l_1 - 1) + \cdots + (l_{\tau-1} - 1)$ is called the *index of the permutation* $\phi(\mathbf{s})$. If $\lambda(\phi(\mathbf{s}))$ is even, then $\phi$ is said to be an *even permutation*; if $\lambda(\phi(\mathbf{s}))$ is odd, then $\phi$ is said to be an *odd permutation*.

**Example 2.23** Consider the first permutation $\phi_1(\mathbf{s})$ of $\mathbf{s} = (0, 1, 2)$ given in Example 2.22. It contains a single cycle, $c_3 = (0, 1, 1, 2, 2, 0)$ (i.e., $0 \to 1$, $1 \to 2$, $2 \to 0$), which is an order-3 cycle with length 6. Hence, $\tau = 1$ and the index $\lambda(\phi_1(\mathbf{s}))$ of the permutation $\phi_1(\mathbf{s})$ is 2. Hence, the permutation $\phi_1(\mathbf{s})$ is an even permutation. Next, consider the fifth permutation $\phi_5(\mathbf{s})$ of $\mathbf{s} = (0, 1, 2)$. It contains two disjoint cycles, $c_2 = (0, 2, 2, 0)$ (i.e., $0 \to 2$, $2 \to 0$) and $c_1 = (1, 1)$ (i.e., $1 \leftrightarrow 1$). The cycle $c_2$ has order 2 and length 4 and the cycle $c_1$ has order 1 and length 2 and is an identity permutation. The index $\lambda(\phi_5(\mathbf{s}))$ of the permutation $\phi_5(\mathbf{s})$ is $(2 - 1) + (1 - 1) = 1$. Hence, the permutation $\phi_5(\mathbf{s})$ is an odd permutation. The indices of the six permutations of $\mathbf{s} = (0, 1, 2)$ in Example 2.22 are 0, 2, 2, 1, 1, 1, respectively. Hence, there are three even permutations and three odd permutations of the index sequence $\mathbf{s} = (0, 1, 2)$. ▲▲

**Example 2.24** Let $n = 8$. Consider the index sequence $\mathbf{s} = (0, 1, 2, 3, 4, 5, 6, 7)$ with eight indices. There are $8! = 40\,320$ distinct permutations of $\mathbf{s} = (0, 1, 2, 3, 4, 5, 6, 7)$. Consider the following permutation of $\mathbf{s}$:

$$\phi_1(\mathbf{s}) = \begin{pmatrix} 0 & 1 & 2 & 3 & 4 & 5 & 6 & 7 \\ 7 & 6 & 1 & 5 & 4 & 3 & 0 & 2 \end{pmatrix}. \tag{2.58}$$

This permutation contains three disjoint cycles of orders 5, 2, and 1. These cycles are: (1) $c_5 = (0, 7, 7, 2, 2, 1, 1, 6, 6, 0)$ (i.e., $0 \to 7$, $7 \to 2$, $2 \to 1$, $1 \to 6$, $6 \to 0$); (2) $c_2 = (3, 5, 5, 3)$ (i.e., $3 \to 5$, $5 \to 3$); and (3) $c_1 = (4, 4)$ (i.e., $4 \leftrightarrow 4$). The index of the permutation $\phi_1(\mathbf{s})$ is $\lambda(\phi_1(\mathbf{s})) = (5-1) + (2-1) + (1-1) = 5$ and thus $\phi_1(\mathbf{s})$ is an odd permutation of $\mathbf{s}$.

Consider another permutation of $\mathbf{s} = (0, 1, 2, 3, 4, 5, 6, 7)$ given as follows:

$$\phi_2(\mathbf{s}) = \begin{pmatrix} 0 & 1 & 2 & 3 & 4 & 5 & 6 & 7 \\ 7 & 3 & 0 & 4 & 5 & 6 & 1 & 2 \end{pmatrix}. \tag{2.59}$$

This permeation $\phi_2(\mathbf{s})$ contains two disjoint cycles which are: (1) $c_3 = (0, 7, 7, 2, 2, 0)$ (i.e., $0 \to 7$, $7 \to 2$, $2 \to 0$); and (2) $c_5 = (1, 3, 3, 4, 4, 5, 5, 6, 6, 1)$ (i.e., $1 \to 3$, $3 \to 4$, $4 \to 5$, $5 \to 6$, $6 \to 1$). The orders of these two disjoint cycles are 3 and 5, respectively. The index of the permutation $\phi_2(\mathbf{s})$ is $\lambda(\phi_2(\mathbf{s})) = (3 - 1) + (5 - 1) = 6$. Hence, $\phi_2(\mathbf{s})$ is an even permutation of $\mathbf{s}$. ▲▲

The determinant of an $n \times n$ matrix $\mathbf{A}$ over $\mathrm{GF}(q)$ as given in (2.55) is a sum of products, $\pm a_{0,?} a_{1,?} \cdots a_{n-1,?}$, each consisting of $n$ entries in $\mathbf{A}$, where the positions with ?s are filled in by some permutation $\phi$ of the indices of $(0, 1, 2, \ldots, n-1)$. If the permutation $\phi$ carries $i$ into $\phi(i)$, then the product specified by the permutation $\phi$ is $\pm a_{0,\phi(0)} a_{1,\phi(1)} \cdots a_{n-1,\phi(n-1)}$, which contains exactly one entry from each row and one entry from each column, i.e., each entry appearing in a separate row and a separate column. If $\phi(0) = 0$, $\phi(1) = 1, \ldots, \phi(n-1) = n-1$, then the product specified by $\phi$ is $\pm a_{0,0} a_{1,1} \ldots a_{n-1,n-1}$, which is the product of $n$ entries lying on the main diagonal of $\mathbf{A}$. The sign $(\pm)$, called $\mathrm{sgn}(\phi)$ (for signum $\phi$), is determined by the characteristic of $\phi$. If $\phi$ is an even permutation, $\mathrm{sgn}(\phi)$ is $+1$; if $\phi$ is an odd permutation, $\mathrm{sgn}(\phi) = -1$. It is clear that the determinant of $\mathbf{A}$ is an element in $\mathrm{GF}(q)$. We denote the determinant of $\mathbf{A}$ by $\det(\mathbf{A})$ or $|\mathbf{A}|$.

Summarizing the above developments, we define the determinant of an $n \times n$ matrix over a finite field as follows.

**Definition 2.17** Let $\mathbf{A} = [a_{i,j}]_{0 \le i,j < n}$ be an $n \times n$ square matrix over a finite field $\mathrm{GF}(q)$ with rows and columns labeled from 0 to $n - 1$. Let $\mathbf{s} = (0, 1, 2, \ldots, n - 1)$ be the index sequence for the rows and columns of $\mathbf{A}$. Let $\Phi = \{\phi_0(\mathbf{s}), \phi_1(\mathbf{s}), \ldots, \phi_{n!-1}(\mathbf{s})\}$ be the set of all the $n!$ distinct permutations of the orders of indices in $\mathbf{s}$. Let $\phi(\mathbf{s}) = (\phi(0), \phi(1), \ldots, \phi(n-1))$ be the permuted sequence obtained by permuting the index sequence $\mathbf{s}$ using the permutation $\phi$ in $\Phi$. Then, the determinant $|\mathbf{A}|$ of $\mathbf{A}$ is defined as the following sum of products of the entries in $\mathbf{A}$:

$$\det(\mathbf{A}) = |\mathbf{A}| = \sum_{\phi \in \Phi} \mathrm{sgn}(\phi) a_{0,\phi(0)} a_{1,\phi(1)} \cdots a_{n-1,\phi(n-1)}. \tag{2.60}$$

Following the above developments and definition, the following theorems can be proved.

**Theorem 2.16** *Let $\mathbf{A}^T$ be the transpose of a square matrix $\mathbf{A}$ over $\mathrm{GF}(q)$. Then, $|\mathbf{A}^T| = |\mathbf{A}|$.*

**Theorem 2.17** *If two rows (or two columns) in a square matrix $\mathbf{A}$ over $\mathrm{GF}(q)$ are alike, then $|\mathbf{A}| = 0$.*

**Theorem 2.18** *A square matrix $\mathbf{A}$ over $\mathrm{GF}(q)$ is a full-rank matrix if its determinant is nonzero, i.e., $|\mathbf{A}| \neq 0$.*

**Theorem 2.19** *The determinant of a triangular square matrix $\mathbf{A}$ is the product of its diagonal entries.*

For a field $\mathrm{GF}(q)$ of characteristic 2, i.e., $q = 2^m$, because each element in $\mathrm{GF}(2^m)$ is its own additive inverse, the sign, $\mathrm{sgn}(\phi)$, of each product in the sum of (2.60) can be removed (i.e., replaced by $+$). In this case, we have

$$\det(\mathbf{A}) = |\mathbf{A}| = \sum_{\phi \in \Phi} a_{0,\phi(0)} a_{1,\phi(1)} \cdots a_{n-1,\phi(n-1)}. \tag{2.61}$$

It follows from Theorem 2.18 that a square matrix $\mathbf{A}$ over a finite field is nonsingular if its determinant is nonzero; otherwise, it is singular.

**Example 2.25** Consider the following $3 \times 3$ matrix over $\mathrm{GF}(2^3)$ given by Table 2.10 with $\alpha$ as a primitive element:

$$\mathbf{A} = \begin{bmatrix} \alpha & \alpha^2 & \alpha^5 \\ \alpha^5 & \alpha & \alpha^2 \\ \alpha^2 & \alpha^5 & \alpha \end{bmatrix}.$$

Note that the characteristic of $\mathrm{GF}(2^3)$ is 2. Using the six permutations given in Example 2.22 and (2.61), we calculate the determinant $|\mathbf{A}|$ of $\mathbf{A}$ as follows:

$$\begin{aligned}
|\mathbf{A}| &= \alpha\alpha\alpha + \alpha^2\alpha^2\alpha^2 + \alpha^5\alpha^5\alpha^5 + \alpha^5\alpha\alpha^2 + \alpha^2\alpha^5\alpha + \alpha\alpha^5\alpha^2 \\
&= \alpha^3 + \alpha^6 + \alpha^{15} + \alpha^8 + \alpha^8 + \alpha^8 \\
&= \alpha^3 + \alpha^6 + \alpha + \alpha + \alpha + \alpha \\
&= 1 + \alpha + 1 + \alpha^2 \\
&= \alpha^4.
\end{aligned}$$

Because $|\mathbf{A}|$ is nonzero, $\mathbf{A}$ is a full-rank matrix over $\mathrm{GF}(2^3)$ and nonsingular.

Next, we consider the following $3 \times 3$ matrix over $\mathrm{GF}(2^3)$:

$$\mathbf{A} = \begin{bmatrix} \alpha & \alpha^3 & \alpha^5 \\ \alpha^2 & \alpha & \alpha^3 \\ \alpha & \alpha^3 & \alpha^5 \end{bmatrix}.$$

Because $\mathbf{A}$ has two identical rows, it follows Theorem 2.17 that its determinant is zero. To show this, we calculate the determinant of $\mathbf{A}$ using the six permutations given in Example 2.22 and (2.61) as follows:

$$\begin{aligned}
|\mathbf{A}| &= \alpha\alpha\alpha^5 + \alpha^3\alpha^3\alpha + \alpha^5\alpha^3\alpha^2 + \alpha^5\alpha\alpha + \alpha^3\alpha^2\alpha^5 + \alpha\alpha^3\alpha^3 \\
&= \alpha^7 + \alpha^7 + \alpha^{10} + \alpha^7 + \alpha^{10} + \alpha^7 \\
&= 1 + 1 + \alpha^3 + 1 + \alpha^3 + 1 \\
&= 0.
\end{aligned}$$

Because $|\mathbf{A}| = 0$, $\mathbf{A}$ is a singular matrix over $\mathrm{GF}(2^3)$. ▲▲

If $m = n$, the Vandermonde matrix $\mathbf{V}$ given by either (2.53) or (2.54) is a square matrix with determinant expressed as

$$|\mathbf{V}| = \prod_{1 \le j < i \le n} (a_i - a_j). \tag{2.62}$$

This is called *Vandermonde determinant*. If all the elements $a_1, a_2, \ldots, a_n$ are distinct elements in $\mathrm{GF}(q)$, $|\mathbf{V}| \neq 0$, i.e., $\mathbf{V}$ is a nonsingular matrix. If there exist some $a_i$ and $a_j$ with $a_i = a_j$, we have $|\mathbf{V}| = 0$ because $\mathbf{V}$ has two identical rows (or columns) in this case, i.e., $\mathbf{V}$ is not full rank.

## 2.10 Graphs

Graph theory is an important part of combinatorial mathematics. A graph is basically a geometrical figure consisting of nodes (vertices) and edges (branches). In error-correcting codes, especially low-density parity-check codes, graphs also play an important role in code designs, decoding algorithms, and code performance analysis. In this section, we will give some basic concepts and important properties of graphs. Other structures and properties of graphs which are related to particular types of error-correcting codes will be covered in corresponding sections.

### 2.10.1 Basic Definitions and Concepts

**Definition 2.18** An undirected graph $\mathcal{G} = (\mathcal{W}, \mathcal{E})$ comprises a set of nodes, denoted by $\mathcal{W} = \{w_0, w_1, \ldots\}$, and a set of edges, denoted by $\mathcal{E} = \{e_0, e_1, \ldots\}$, where the edge $e_k$ is expressed as $e_k = \{w_i, w_j\}$ for $w_i, w_j \in \mathcal{W}$. Nodes $w_i$ and $w_j$ are said to be *adjacent* or be connected by edge $e_k$ (or be the two *end nodes* of edge $e_k$), edge $e_k$ is said to be *incident* with the two nodes $w_i$ and $w_j$, and node $w_i$ is said to be a *neighbor* of $w_j$ and vice versa. The sets $\mathcal{W}$ and $\mathcal{E}$ are called the *node* and *edge sets* of the graph $\mathcal{G}$, respectively.

In a graph $\mathcal{G}$, two edges $e_a$ and $e_b$ are said to be *adjacent* if they have a common end-node, i.e., there exist nodes $w_i, w_j, w_k$ in $\mathcal{W}$ such that $e_a = \{w_i, w_j\}$ and $e_b = \{w_j, w_k\}$. An edge $e_k = \{w_i, w_i\}$ is called a *self-loop*. There may be multiple edges connecting two nodes, which are said to be *parallel edges*. A node can be adjacent to multiple other nodes or it can have no neighbor.

**Definition 2.19** A graph $\mathcal{G}$ with a finite number of nodes and a finite number of edges is called a *finite graph*; otherwise, it is called an *infinite graph*.

**Definition 2.20** The number of edges that are incident with a node $w_i$ is called the degree of $w_i$, denoted by $d_{w_i}$. The set of degrees of all the nodes in $\mathcal{G}$, $\{d_{w_0}, d_{w_1}, d_{w_2}, \ldots\}$, is called *degree distribution* of the graph $\mathcal{G}$.

**Theorem 2.20** *For a finite graph $\mathcal{G} = (\mathcal{W}, \mathcal{E})$ with $n$ nodes, $w_0, w_1, \ldots, w_{n-1}$, and $m$ edges, $e_0, e_1, \ldots, e_{m-1}$, the sum of the degrees of all the nodes is equal to twice the number of edges, i.e.,*

$$\sum_{i=0}^{n-1} d_{w_i} = 2m.$$

**Example 2.26** Consider a graph $\mathcal{G}$ with a node set of seven nodes, $\mathcal{W} = \{w_0,$ $w_1, w_2, w_3, w_4, w_5, w_6\}$, and an edge set with eight edges, $\mathcal{E} = \{e_0, e_1, e_2, e_3,$ $e_4, e_5, e_6, e_7\}$, as shown in Fig. 2.1.

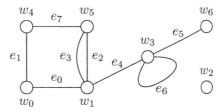

Figure 2.1 A graph with seven nodes and eight edges.

From the above figure, we can see that node $w_3$ has a self-loop $e_6$, there are two parallel edges $e_2$ and $e_3$ connecting nodes $w_1$ and $w_5$, and node $w_2$ has no adjacent node. The degrees of the nodes are $d_{w_0} = 2$, $d_{w_1} = 4$, $d_{w_2} = 0$, $d_{w_3} = 4$, $d_{w_4} = 2$, $d_{w_5} = 3$, $d_{w_6} = 1$. The sum of the degrees of all seven nodes is $\sum_{i=0}^{6} d_{w_i} = 16 = 2 \times 8$, which is twice the number of the edges in the graph. ▲▲

**Definition 2.21** A graph without self-loop and parallel edges is called a *simple graph*. A graph is said to be *regular* if all its nodes have the same degree; otherwise, it is an *irregular* graph.

**Definition 2.22** A simple graph is said to be *complete* if there exists an edge between every pair of nodes.

**Definition 2.23** Let $\mathcal{G} = (\mathcal{W}, \mathcal{E})$ and $\mathcal{G}_s = (\mathcal{W}_s, \mathcal{E}_s)$ be two graphs. The graph $\mathcal{G}_s$ is called a *subgraph* of $\mathcal{G}$ if $\mathcal{W}_s \subset \mathcal{W}$ and $\mathcal{E}_s \subset \mathcal{E}$.

The graph in Fig. 2.1 is not a simple graph because it has self-loop and parallel edges. The four graphs in Fig. 2.2 are simple graphs, where the graphs in Fig. 2.2(a) and Fig. 2.2(b) are subgraphs of the graph in Fig. 2.1. The graphs in Fig. 2.2(b) and Fig. 2.2(c) are regular, and the graph in Fig. 2.2(c) is a complete graph. The graph in Fig. 2.2(d) is an irregular graph.

**Definition 2.24** A *path* in a graph $\mathcal{G} = (\mathcal{W}, \mathcal{E})$ is defined as an alternating sequence of nodes and edges, such that each edge is incident with the nodes preceding and following it and no node appears more than once. The number of edges on a path is called the *length* of the path.

A path can be represented by listing its nodes in a consecutive order shown as below:

$$(w_{i_0}, w_{i_1}, \ldots, w_{i_k})$$

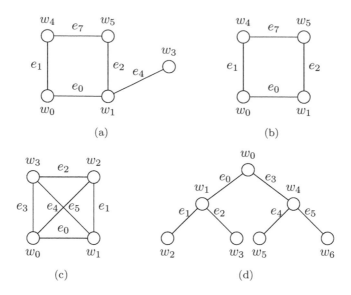

Figure 2.2 Simple graphs.

in which two consecutive nodes are connected by an edge and each edge on a path appears once and only once. We call the first node $w_{i_0}$ and the last node $w_{i_k}$ the *starting* and *ending* nodes of the path, respectively. Consider the graph shown in Fig. 2.2(a). The five nodes $w_0$, $w_4$, $w_5$, $w_1$, $w_3$ form a path of length 4 with $w_0$ and $w_3$ as its starting and ending nodes, respectively.

A graph is said to be *connected* if any two nodes in $\mathcal{G} = (\mathcal{W}, \mathcal{E})$ are connected by at least one path; otherwise, it is said to be disconnected. The graphs shown in Fig. 2.2(a), Fig. 2.2(b), Fig. 2.2(c), and Fig. 2.2(d) are simple and connected graphs. In a connected graph, the *distance* between two nodes $w_i$ and $w_j$, denoted by $d(w_i, w_j)$, is defined as the length of the *shortest* path between these two nodes. Consider the graph shown in Fig. 2.2(a). There are two paths connecting nodes $w_0$ and $w_3$: $(w_0, w_4, w_5, w_1, w_3)$ and $(w_0, w_1, w_3)$. The first path has length 4 and the second path has length 2. Thus, the shortest path connecting $w_0$ and $w_3$ is $(w_0, w_1, w_3)$ which has length 2, and hence the distance between $w_0$ and $w_3$ is $d(w_0, w_3) = 2$.

**Definition 2.25** In a graph $\mathcal{G} = (\mathcal{W}, \mathcal{E})$, a path is said to be *closed* if its starting and ending nodes are the same. A closed path is called a *cycle*. The number of edges on a cycle is called the *length* of the cycle.

**Definition 2.26** The *girth* of a graph is defined as the length of its shortest cycle. A graph with no cycle has an infinite girth. The girth of a graph is usually denoted by "*g*."

For example, in the graph shown in Fig. 2.2(c), the closed path $(w_0, w_1, w_2, w_3, w_0)$ is a cycle of length 4 and the closed path $(w_0, w_1, w_2, w_0)$ forms a cycle of length 3. The graph has girth 3 because it has no cycle of length less than 3.

**Definition 2.27** A graph with no cycle is said to be *acyclic*. A connected acyclic graph is called a *tree*. An edge in a tree is called a *branch* and a *pendant* in a tree is called an *end-node*.

For example, the graph in Fig. 2.2(d) is a tree. This tree has six branches and seven nodes among which nodes $w_2$, $w_3$, $w_5$, and $w_6$ are pendants.

A finite graph $\mathcal{G} = (\mathcal{W}, \mathcal{E})$ with node set $\mathcal{W} = \{w_0, w_1, \ldots, w_{n-1}\}$ and edge set $\mathcal{E} = \{e_0, e_1, \ldots, e_{m-1}\}$ can also be represented by two matrices, one called an *adjacency matrix* and the other called an *incidence matrix*. Let $\mathbf{A}(\mathcal{G}) = [a_{i,j}]_{0 \leq i,j < n}$ be an $n \times n$ matrix. The rows and columns of $\mathbf{A}(\mathcal{G})$ correspond to all the $n$ nodes in $\mathcal{G}$. For $0 \leq i, j < n$, the entry $a_{i,j}$ in $\mathbf{A}(\mathcal{G})$ is equal to $a_{i,j} = k$ if there are $k$ parallel edges in $\mathcal{E}$ connecting nodes $w_i$ and $w_j$, $a_{i,i} = 1$ if node $w_i$ is on a self-loop, and otherwise $a_{i,j} = 0$. The matrix $\mathbf{A}(\mathcal{G})$ is called the *adjacency matrix* of $\mathcal{G}$. If $\mathcal{G}$ is a simple graph, then $\mathbf{A}(\mathcal{G})$ is a binary matrix, i.e., either $a_{i,j} = 1$ or $a_{i,j} = 0$. Let $\mathbf{B}(\mathcal{G}) = [b_{i,j}]_{0 \leq i < m, 0 \leq j < n}$ be an $m \times n$ matrix. The $m$ rows of $\mathbf{B}(\mathcal{G})$ correspond to the $m$ edges in $\mathcal{G}$ and the $n$ columns of $\mathbf{B}(\mathcal{G})$ correspond to the $n$ nodes in $\mathcal{G}$. For $0 \leq i < m$ and $0 \leq j < n$, the entry $b_{i,j}$ in $\mathbf{B}(\mathcal{G})$ is equal to $b_{i,j} = 1$ if edge $e_i$ connects node $w_j$ and one other node in $\mathcal{W} \backslash \{w_j\}$, $b_{i,j} = 2$ if edge $e_i$ is the self-loop of the node $w_j$, and otherwise $b_{i,j} = 0$. The matrix $\mathbf{B}(\mathcal{G})$ is called the *incidence matrix* of $\mathcal{G}$. For a simple graph, $\mathbf{B}(\mathcal{G})$ is also a binary matrix. Based on the definition of the adjacency matrix $\mathbf{A}(\mathcal{G})$, the sum of the entries in the $j$th column (or $j$th row) is the number of neighbors of the $j$th node $w_j$ in $\mathcal{G}$. Based on the definitions of the incidence matrix $\mathbf{B}(\mathcal{G})$ and the degrees of nodes, the sum of entries in the $j$th column of $\mathbf{B}(\mathcal{G})$ is the degree of the $j$th node $w_j$ in $\mathcal{G}$.

Consider the graph in Fig. 2.1. Its adjacency and incidence matrices are given as follows:

$$
\mathbf{A}(\mathcal{G}) = \begin{array}{c} \\ w_0 \\ w_1 \\ w_2 \\ w_3 \\ w_4 \\ w_5 \\ w_6 \end{array} \begin{array}{c} \begin{array}{ccccccc} w_0 & w_1 & w_2 & w_3 & w_4 & w_5 & w_6 \end{array} \\ \left[ \begin{array}{ccccccc} 0 & 1 & 0 & 0 & 1 & 0 & 0 \\ 1 & 0 & 0 & 1 & 0 & 2 & 0 \\ 0 & 0 & 0 & 0 & 0 & 0 & 0 \\ 0 & 1 & 0 & 1 & 0 & 0 & 1 \\ 1 & 0 & 0 & 0 & 0 & 1 & 0 \\ 0 & 2 & 0 & 0 & 1 & 0 & 0 \\ 0 & 0 & 0 & 1 & 0 & 0 & 0 \end{array} \right] \end{array}, \quad \mathbf{B}(\mathcal{G}) = \begin{array}{c} \\ e_0 \\ e_1 \\ e_2 \\ e_3 \\ e_4 \\ e_5 \\ e_6 \\ e_7 \end{array} \begin{array}{c} \begin{array}{ccccccc} w_0 & w_1 & w_2 & w_3 & w_4 & w_5 & w_6 \end{array} \\ \left[ \begin{array}{ccccccc} 1 & 1 & 0 & 0 & 0 & 0 & 0 \\ 1 & 0 & 0 & 0 & 1 & 0 & 0 \\ 0 & 1 & 0 & 0 & 0 & 1 & 0 \\ 0 & 1 & 0 & 0 & 0 & 1 & 0 \\ 0 & 1 & 0 & 1 & 0 & 0 & 0 \\ 0 & 0 & 0 & 1 & 0 & 0 & 1 \\ 0 & 0 & 0 & 2 & 0 & 0 & 0 \\ 0 & 0 & 0 & 0 & 1 & 1 & 0 \end{array} \right] \end{array}.
$$

The following two matrices are the adjacency and incidence matrices of the simple graph in Fig. 2.2(c):

$$
\mathbf{A}(\mathcal{G}) = \begin{array}{c} \\ w_0 \\ w_1 \\ w_2 \\ w_3 \end{array} \begin{array}{c} \begin{array}{cccc} w_0 & w_1 & w_2 & w_3 \end{array} \\ \left[ \begin{array}{cccc} 0 & 1 & 1 & 1 \\ 1 & 0 & 1 & 1 \\ 1 & 1 & 0 & 1 \\ 1 & 1 & 1 & 0 \end{array} \right] \end{array}, \quad \mathbf{B}(\mathcal{G}) = \begin{array}{c} \\ e_0 \\ e_1 \\ e_2 \\ e_3 \\ e_4 \\ e_5 \end{array} \begin{array}{c} \begin{array}{cccc} w_0 & w_1 & w_2 & w_3 \end{array} \\ \left[ \begin{array}{cccc} 1 & 1 & 0 & 0 \\ 0 & 1 & 1 & 0 \\ 0 & 0 & 1 & 1 \\ 1 & 0 & 0 & 1 \\ 0 & 1 & 0 & 1 \\ 1 & 0 & 1 & 0 \end{array} \right] \end{array}.
$$

## 2.10.2 Bipartite Graphs

In this section, a special type of graphs is presented, called *bipartite graphs*, which is a particularly effective aid in the design, decoding, and performance analysis of a class of error-correcting codes.

**Definition 2.28** A graph $\mathcal{G} = (\mathcal{W}, \mathcal{E})$ is called a *bipartite graph* if its node set $\mathcal{W}$ can be partitioned into two disjoint subsets $\mathcal{V}$ and $\mathcal{C}$ such that every edge in $\mathcal{E}$ connects a node in $\mathcal{V}$ with a node in $\mathcal{C}$ and does not connect two nodes that are both either in $\mathcal{V}$ or in $\mathcal{C}$.

It is obvious that a bipartite graph has no self-loop. A cycle in a bipartite graph (if any) has even length. Thus, the girth of a bipartite graph is either infinite or an even number. Figure 2.3 shows a bipartite graph $\mathcal{G}$ with 15 nodes which are partitioned into two sets, $\mathcal{V} = \{v_0, v_1, v_2, \ldots, v_9\}$ and $\mathcal{C} = \{c_0, c_1, c_2, c_3, c_4\}$. There is a cycle of length 6 in the graph which is marked by boldface lines. It is easy to check that there is no cycle of length 2 or length 4 in the graph. Therefore, the girth of this graph is $g = 6$.

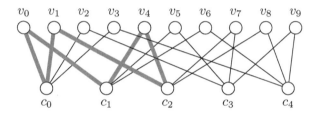

Figure 2.3 A bipartite graph.

A bipartite graph $\mathcal{G}$ can also be represented by the two matrices, namely, the adjacency matrix $\mathbf{A}(\mathcal{G})$ and incidence matrix $\mathbf{B}(\mathcal{G})$, as defined in Section 2.10.1. Considering the structure of bipartite graphs, we introduce another type of adjacency matrix which is slightly different from the one previously defined in Section 2.10.1 for general graphs. Let $\mathcal{G} = (\mathcal{W}, \mathcal{E})$ be a bipartite graph with $\mathcal{W} = \mathcal{V} \cup \mathcal{C}$. Suppose there are $n$ nodes in $\mathcal{V}$, $v_0, v_1, v_2, \ldots, v_{n-1}$, and $m$ nodes in $\mathcal{C}$, $c_0, c_1, \ldots, c_{m-1}$. The adjacency matrix $\mathbf{A}'(\mathcal{G}) = [a_{i,j}]_{0 \leq i < m, 0 \leq j < n}$ for a bipartite graph is an $m \times n$ matrix in which the $n$ columns correspond to the $n$ nodes in $\mathcal{V}$ and the $m$ rows correspond to the $m$ nodes in $\mathcal{C}$. For $0 \leq i < m$ and $0 \leq j < n$, the entry $a_{i,j}$ in $\mathbf{A}'(\mathcal{G})$ is equal to $a_{i,j} = k$ if there are $k$ parallel edges connecting nodes $c_i$ and $v_j$, otherwise, $a_{i,j} = 0$. Obviously, $\mathbf{A}'(\mathcal{G})$ is a submatrix of the adjacency matrix $\mathbf{A}(\mathcal{G})$ defined in Section 2.10.1; and $\mathbf{A}'(\mathcal{G})$ can be obtained from $\mathbf{A}(\mathcal{G})$ by removing columns corresponding to nodes in $\mathcal{C}$ and rows corresponding to nodes in $\mathcal{V}$ from $\mathbf{A}(\mathcal{G})$. The bipartite graph in Fig. 2.3 has the following adjacency matrix:

$$\mathbf{A}'(\mathcal{G}) = \begin{array}{c} \\ c_0 \\ c_1 \\ c_2 \\ c_3 \\ c_4 \end{array} \begin{array}{c} v_0\ v_1\ v_2\ v_3\ v_4\ v_5\ v_6\ v_7\ v_8\ v_9 \\ \left[\begin{array}{cccccccccc} 1 & 1 & 1 & 1 & 0 & 0 & 0 & 0 & 0 & 0 \\ 1 & 0 & 0 & 0 & 1 & 1 & 1 & 0 & 0 & 0 \\ 0 & 1 & 0 & 0 & 1 & 0 & 0 & 1 & 1 & 0 \\ 0 & 0 & 1 & 0 & 0 & 1 & 0 & 1 & 0 & 1 \\ 0 & 0 & 0 & 1 & 0 & 0 & 1 & 0 & 1 & 1 \end{array}\right] \end{array}.$$

Based on the above definition of the adjacency matrix $\mathbf{A}'(\mathcal{G})$ of a bipartite graph $\mathcal{G}$, it is clear that the sum of the entries in the $j$th column in $\mathbf{A}'(\mathcal{G})$ is the degree of node $v_j$ in the node set $\mathcal{V}$, the sum of the entries in the $i$th row in $\mathbf{A}'(\mathcal{G})$ is the degree of node $c_i$ in the node set $\mathcal{C}$, and the sum of all the entries in $\mathbf{A}'(\mathcal{G})$ is the number of edges in the graph $\mathcal{G}$. It is noted that there may be parallel edges in a bipartite graph.

For a simple bipartite graph $\mathcal{G}$, if there are no two nodes in $\mathcal{V}$ jointly connecting to two nodes in $\mathcal{C}$, then there is no cycle of length 4 in the graph. In term of the adjacency matrix, for this case, there are no four 1-entries at the four corners of a $2 \times 2$ submatrix in the adjacency matrix $\mathbf{A}'(\mathcal{G})$. Conversely, if there are no four 1-entries at the four corners of any $2 \times 2$ submatrix in $\mathbf{A}'(\mathcal{G})$, there is no cycle of length 4 in the corresponding bipartite graph $\mathcal{G}$. This provides a simple method to check whether a bipartite graph $\mathcal{G}$ has cycles of length 4 based on its adjacency matrix $\mathbf{A}'(\mathcal{G})$.

See more details about groups, fields, vector spaces, matrices, and graphs in References [1–21].

# Problems

**2.1** Construct a group $G$ under modulo-5 addition.

**2.2** Construct a group $G$ under modulo-10 addition.

**2.3** Let $G$ be a group of order $m$ under modulo-$m$ addition and $a \in G$. Prove that the additive inverse of $a$ is unique.

**2.4** Consider the additive group $G = \{0, 1, 2, 3, 4, 5, 6, 7\}$ under modulo-8 addition given by Table 2.4. Let $S = \{0, 1, 3, 5, 7\}$ be a subset of $G$. Is $S$ a subgroup of $G$? Justify your answer.

**2.5** Construct a group $G$ under modulo-5 multiplication and find a generator of the constructed group.

**2.6** Construct a group $G$ under modulo-11 multiplication and find a generator of the constructed group.

**2.7** Prove that the set $\{0, 1, 2, 3, 4, 5\}$ is not a group under modulo-6 multiplication. Find the zero divisors in this set. (A *zero divisor* is any nonzero element $a$ in the set $S$ for which there exists a nonzero element $b$ in $S$ such that $a \cdot b = 0$ modulo-$m$ (i.e., modulo-$m$ multiplication), where $m$ is the order of the set.)

**2.8** Let $G$ be a multiplicative group under modulo-$p$ multiplication and $a \in G$. Prove that the multiplicative inverse of $a$ is unique.

**2.9** Consider the additive group $G = \{0, 1, 2, 3, 4, 5, 6, 7\}$ under modulo-8 addition given by Table 2.4. Consider a subset $S = \{0, 4\}$ of $G$. Prove that $S = \{0, 4\}$ is a subgroup of $G$ and find all its cosets, i.e., find the partition $G/S$.

**2.10** Consider the multiplicative group $G = \{1, 2, 3, 4, 5, 6\}$ under modulo-7 multiplication given in Example 2.3. Find a subgroup $S$ with three elements of $G$ and find all the cosets of $S$, i.e., find the partition $G/S$.

**2.11** Prove the structural properties of a coset.

**2.12** Prove the structural properties of cosets of a proper subgroup of an additive group of order $m$ under modulo-$m$ addition.

**2.13** Prove that the $\lambda$ elements in $F$ given by the $\lambda$ sums in the form of (2.4) form a field.

**2.14** Construct the prime field GF(11) under modulo-11 addition and multiplication.

**2.15** Prove that the set $\{0, 1, 2, .., m-1\}$ is not a field under modulo-$m$ addition and multiplication if $m$ is not a prime.

**2.16** Prove the five fundamental properties of a finite field.

**2.17** Prove that the polynomial $X^m - 1$ divides $X^n - 1$ if and only if $m$ divides $n$.

**2.18** Show that $f(X) = 1 + X + X^3$ is an irreducible polynomial over GF(2).

**2.19** Consider the following polynomials over GF(5): $f_1(X) = 2 + 3X + 4X^2 + X^3$, $f_2(X) = 1 + X + X^2 + X^4$, $f_3(X) = 1 + X + X^5$. Identify the irreducible and reducible polynomials.

**2.20** The polynomial $p(X) = 1 + X^2 + X^5$ is a primitive polynomial over GF(2) of degree 5. Use this primitive polynomial to construct the extension field GF($2^5$) of GF(2) and find the roots of $p(X)$ in the constructed field GF($2^5$).

**2.21** Let $\alpha$ be a primitive element of GF($2^5$) constructed in Problem 2.20. Find the minimal polynomials of the two elements $\alpha^3$ and $\alpha^7$ in GF($2^5$).

**2.22** Are all the nonzero elements in GF($2^5$) constructed in Problem 2.20 primitive? Justify your answer.

**2.23** Consider the field GF($2^3$) given by Table 2.10 with a primitive element $\alpha$. Find the minimal polynomials of all the nonzero elements in GF($2^3$) and the factorization of $X^7 + 1$ over GF(2).

**2.24** Prove Theorem 2.10.

**2.25** Prove Theorem 2.11.

**2.26** Prove Theorem 2.12.

**2.27** Prove Theorem 2.13.

**2.28** Solve the following simultaneous equations for $X$, $Y$, $Z$, and $W$ with modulo-2 addition:

$$
\begin{aligned}
X + Y + Z + W &= 1 \\
X + Y + W &= 1 \\
X + Z + W &= 1 \\
Y + Z + W &= 1.
\end{aligned}
$$

**2.29** Use Table 2.9 to find the roots of $f(X) = X^3 + \alpha^6 X^2 + \alpha^9 X + \alpha^9 = 0$ in $\mathrm{GF}(2^4)$, where $\alpha$ is a primitive element in $\mathrm{GF}(2^4)$.

**2.30** Consider the field $\mathrm{GF}(2^4)$ constructed in Example 2.8. Construct an additive subgroup of $\mathrm{GF}(2^4)$ with a polynomial basis of $\{\alpha, \alpha^2, \alpha^3\}$.

**2.31** Consider the field $\mathrm{GF}(2^4)$ constructed in Example 2.8. Construct a cyclic subgroup of $\mathrm{GF}(2^4)$ of order 5.

**2.32** Consider the field $\mathrm{GF}(2^6)$ constructed in Example 2.12. Using Table 2.11, construct a cyclic subgroup of $\mathrm{GF}(2^6)$ of order 7.

**2.33** Construct the vector space $\mathbf{V}_5$ of all binary 5-tuples and find a basis for $\mathbf{V}_5$.

**2.34** Construct a three-dimensional subspace of $\mathbf{V}_5$ over $\mathrm{GF}(2)$ constructed in Problem 2.33.

**2.35** Construct the vector space $\mathbf{V}_2$ of all 2-tuples over $\mathrm{GF}(2^2)$.

**2.36** Prove that the expression given by (2.39) is unique.

**2.37** Given a matrix

$$
\mathbf{G} = \begin{bmatrix} 1 & 0 & 0 & 1 & 1 & 0 & 1 \\ 0 & 1 & 0 & 1 & 1 & 1 & 0 \\ 0 & 0 & 1 & 0 & 1 & 1 & 1 \end{bmatrix},
$$

show that the row space of $\mathbf{G}$ is the null space of the row space of the following matrix

$$
\mathbf{H} = \begin{bmatrix} 1 & 1 & 0 & 1 & 0 & 0 & 0 \\ 1 & 1 & 1 & 0 & 1 & 0 & 0 \\ 0 & 1 & 1 & 0 & 0 & 1 & 0 \\ 1 & 0 & 1 & 0 & 0 & 0 & 1 \end{bmatrix}.
$$

**2.38** Consider the following graph $\mathcal{G}$.

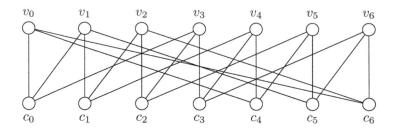

(a) Find the degree distributions of $\mathcal{G}$.
(b) Find the girth of $\mathcal{G}$ and one cycle with shortest length.
(c) Find the adjacency and incidence matrices of $\mathcal{G}$.

# References

[1] A. A. Albert, *Modern Higher Algebra*, Chicago, University of Chicago Press, 1937.

[2] R. D. Carmichael, *Introduction to the Theory of Groups of Finite Order*, Boston, MA, Ginn & Company, 1937.

[3] H. B. Mann, *Analysis and Design of Experiments*, New York, Dover, 1949.

[4] G. Birkhoff and S. Maclane, *A Survey of Morden Algebra*, New York, Macmillan, 1953.

[5] B. L. Van der Waerden, *Modern Algebra* (volumes 1 and 2), New York, Ungar Publishing Company, 1949.

[6] R. W. Marsh, *Table of Irreducible Polynomials over GF(2) through Degree 19*, Cambridge, MA, The MIT Press, 1957.

[7] W. W. Peterson, *Error-Correcting Codes*, Cambridge, MA, The MIT Press, 1961.

[8] N. Deo, *Graph Theory with Applications to Engineering and Computer Science*, Englewood Cliffs, NJ, Prentice-Hall, 1974.

[9] I. F. Blake and R. C. Mullin, *The Mathematical Theory of Coding*, New York, Academic Press, 1975.

[10] A. Clark, *Elements of Abstract Algebra*, New York, Dover, 1984.

[11] S. Lang, *Algebra*, 2nd ed., Reading, MA, Addison-Wesley, 1984.

[12] R. A. Horn and C. R. Johnson, *Matrix Analysis*, Cambridge, Cambridge University Press, 1985.

[13] R. A. Dean, *Classical Abstract Algebra*, New York, Harper & Row, 1990.

[14] J. E. Maxfield and M. W. Maxfield, *Abstract Algebra and Solution by Radicals*, New York, Dover, 1992.

[15] R. Lidl and H. Niederreiter, *Introduction to Finite Fields and Their Applications*, revised ed., Cambridge, Cambridge University Press, 1994.

[16] T. W. Hungerford, *Abstract Algebra: An Introduction*, 2nd ed., New York, Saunders College Publishing, 1997.

[17] R. J. McEliece, *Finite Fields for Computer Scientists and Engineers*, Boston, MA, Springer, 1987.

[18] R. E. Blahut, *Theory and Practice of Error Control Codes*, Reading, MA, Addison-Wesley, 1983.

[19] R. M. Roth, *Introduction to Coding Theory*, Cambridge, Cambridge University Press, 2006.

[20] D. B. West, *Introduction to Graph Theory*, 2nd ed., Upper Saddle River, NJ, Prentice Hall, 2001.

[21] R. J. Wilson, *Introduction to Graph Theory*, 5th ed., New York, Pearson, 2010.

[22] I. S. Reed and G. Solomon, "Polynomial codes over certain finite fields," *J. Soc. Indust. Appl. Math.*, **8** (1960), 300–304.

[23] A. Hocquenghem, "Codes correcteurs d'erreurs," *Chiffres*, **2** (1959), 147–156.

[24] R. C. Bose and D. K. Ray-Chaudhuri, "On a class of error correcting binary group codes," *Inf. Control*, **3** (1960), 68–79.

[25] R. G. Gallager, *Low-Density Parity-Check Codes*, Cambridge, MA, The MIT Press, 1963.

[26] E. R. Berlekamp, *Algebraic Coding Theory*, revised ed., Laguna Hills, CA, Aegean Park Press, 1984.

[27] S. B. Wicker, *Error Control Systems for Digital Communication and Storage*, Englewood Cliffs, NJ, Prentice-Hall, 1995.

[28] W. Stahnke, "Primitive binary polynomials," *Math. Comp.*, **27**(124) (1973), 977–980.

[29] Q. Diao, Q. Huang, S. Lin, and K. Abdel-Ghaffar, "A matrix-theoretic approach for analyzing quasi-cyclic low-density parity-check codes," *IEEE Trans. Inf. Theory*, **58**(6) (2012), 4030–4048.

[30] S. Lin, K. Abdel-Ghaffar, J. Li, and K. Liu, "A novel coding scheme for encoding and iterative soft-decision decoding of binary BCH codes of prime lengths," *Proc. Inf. Theory and Applications (ITA)*, San Diego, CA, February 11–16, 2018.

[31] S. Lin, K. Abdel-Ghaffar, J. Li, and K. Liu, "Collective encoding and iterative soft-decision decoding of cyclic codes of prime lengths in Galois–Fourier transform domain," *2018 10th International Symposium on Turbo Codes and Iterative Information Processing (ISTC)*, December 3–7, 2018, pp. 1–8.

[32] S. Lin, K. Abdel-Ghaffar, J. Li, and K. Liu, "A scheme for collective encoding and iterative soft-decision decoding of cyclic codes of prime lengths: applications to Reed–Solomon, BCH, and quadratic residue codes," *IEEE Trans. Inf. Theory*, **66**(9) (2020), 5358–5378.

# 3

# Linear Block Codes

As presented in Chapter 1, there are two categories of error-correcting codes, block codes and convolutional codes. This chapter gives an introduction to linear block codes, a subclass of block codes. The coverage of this chapter includes: (1) fundamental concepts and structures of linear block codes; (2) generation of these codes in terms of their generator and parity-check matrices; (3) their error detection and correction capabilities; and (4) general decoding of these codes. We will begin the introduction of linear block codes with symbols from the binary field GF(2). Linear block codes over nonbinary fields, which have similar structures and properties to those over the binary field GF(2), will be briefly discussed at the end of this chapter.

There are many excellent texts on the subject of error-correcting codes [1–22], which have extensive coverage of linear block codes. For an in-depth study of linear block codes, readers are referred to these texts.

## 3.1 Definitions

In block coding, an information sequence is segmented into blocks of fixed length, each consisting of $k$ information bits (binary digits). These blocks of information bits are called *messages*. Each message is represented by a $k$-tuple, $\mathbf{u} = (u_0, u_1, \ldots, u_{k-1})$, where $u_0, u_1, \ldots, u_{k-1} \in \mathrm{GF}(2)$ are the $k$ information bits. There are $2^k$ distinct messages. At the channel encoder, each input message $\mathbf{u} = (u_0, u_1, \ldots, u_{k-1})$ of $k$ information bits is encoded (or mapped) into a longer binary sequence $\mathbf{v} = (v_0, v_1, \ldots, v_{n-1})$ of $n$ bits with $n > k$, according to *certain encoding rules*. This longer binary sequence $\mathbf{v}$ is called the *codeword* for the message $\mathbf{u}$. The binary digits in a codeword are called *code bits*. Because there are $2^k$ distinct messages, there are $2^k$ codewords, one for each distinct message. This set of $2^k$ codewords is said to form *an $(n, k)$ block code*, where $n$ and $k$ are called the *length* and *dimension* of the code, respectively. The ratio $R = k/n$ is called the *code rate*, which is interpreted as the number of information bits carried by each code bit.

For a block code to be useful, the $2^k$ codewords for the $2^k$ distinct messages must be distinct, i.e., there is a *one-to-one correspondence* between a message and a codeword. The $n - k$ bits added to each input message by the channel encoder are called *redundant bits* or *parity-check bits*. The values and locations of the redundant bits depend on how the encoding is performed. They can be appended, prefixed, or interleaved with the $k$ messages bits to form an $n$-tuple codeword. These redundant bits carry no new information and their main function is to provide the code with capability of *detecting* and/or *correcting* transmission errors caused by the channel noise or interferences. How to form these redundant bits such that an $(n, k)$ block code has good error detection and/or correction capabilities is a major concern in designing the channel encoder.

For a block code of length $n$ with $2^k$ codewords, unless it has certain special structural properties, the encoding and decoding apparatus would be prohibitively complex for large $k$ because the encoder has to store $2^k$ codewords of length $n$ in a codebook and the decoder has to perform a table (with $2^n$ entries) look-up to determine the transmitted message. Therefore, we must consider block codes which can be implemented practically. A desirable structure for a block code is *linearity*.

**Definition 3.1** A binary block code of length $n$ with $2^k$ codewords is called an $(n, k)$ *linear block code* if and only if its $2^k$ codewords form a *k-dimensional subspace* $\mathbf{S}_k$ *of the vector space* $\mathbf{V}_n$ of all the $n$-tuples over GF(2). Such a linear block code is referred to as a code over GF(2) or a binary code.[1]

**Example 3.1** Consider an encoder which segments the information sequence from the source into messages, each consisting of $k = 3$ information bits, and transforms each message into a codeword of $n = 6$ bits as shown in Table 3.1. Because $k = 3$, there are $2^3 = 8$ possible distinct messages. Each message is transformed into a codeword with six code bits by the encoder. All codewords are distinct. Thus, from a codeword, one can tell which message is transmitted. We can easily check that the set of eight codewords forms a three-dimensional subspace $\mathbf{S}_3$ of the vector space $\mathbf{V}_6$ of all 6-tuples over GF(2). It contains the zero vector (zero codeword) and the sum of any two codewords is another codeword. Therefore, it is a $(6, 3)$ linear block code over GF(2). The rate of this code is $R = k/n = 1/2$. ▲▲

In this text, we mainly focus on linear block codes. Henceforth, the term "codes" shall refer to linear block codes unless otherwise stated.

## 3.2 Generator and Parity-Check Matrices

Because an $(n, k)$ linear block code $C$ is a $k$-dimensional subspace $\mathbf{S}_k$ of the vector space $\mathbf{V}_n$ of all the $n$-tuples over GF(2), there exist $k$ linearly independent

---

[1] In some literature, an $(n, k)$ linear block code is also denoted as $[n, k]$. To emphasize the minimum distance $d_{\min}$ (to be introduced in Section 3.7) of the code, the code is sometimes denoted as $(n, k, d_{\min})$ or $[n, k, d_{\min}]$.

Table 3.1 A codebook for a (6, 3) linear block code over GF(2).

| Messages $(u_0, u_1, u_2)$ | Codewords $(v_0, v_1, v_2, v_3, v_4, v_5)$ | Messages $(u_0, u_1, u_2)$ | Codewords $(v_0, v_1, v_2, v_3, v_4, v_5)$ |
|---|---|---|---|
| (0 0 0) | (0 0 0 0 0 0) | (1 0 0) | (0 1 1 1 0 0) |
| (0 0 1) | (1 1 0 0 0 1) | (1 0 1) | (1 0 1 1 0 1) |
| (0 1 0) | (1 0 1 0 1 0) | (1 1 0) | (1 1 0 1 1 0) |
| (0 1 1) | (0 1 1 0 1 1) | (1 1 1) | (0 0 0 1 1 1) |

codewords, $\mathbf{g}_0, \mathbf{g}_1, \ldots, \mathbf{g}_{k-1}$, where $\mathbf{g}_i = (g_{i,0}, g_{i,1}, \ldots, g_{i,n-1})$ with $g_{i,j} \in \mathrm{GF}(2)$ for $0 \le i < k$ and $0 \le j < n$, in $C$ such that every codeword $\mathbf{v}$ in $C$ is a linear combination of these $k$ linearly independent codewords.

These $k$ linearly independent codewords, $\mathbf{g}_0, \mathbf{g}_1, \ldots, \mathbf{g}_{k-1}$, in $C$ form a basis $\mathcal{B}_c$ of $C$. Using this basis, encoding can be accomplished as follows. Let $\mathbf{u} = (u_0, u_1, \ldots, u_{k-1})$ be the message to be encoded. The codeword $\mathbf{v} = (v_0, v_1, \ldots, v_{n-1})$ for this message is given by the following linear combination of $\mathbf{g}_0, \mathbf{g}_1, \ldots, \mathbf{g}_{k-1}$ with the $k$ message bits of $\mathbf{u}$ as the coefficients:

$$\mathbf{v} = u_0\mathbf{g}_0 + u_1\mathbf{g}_1 + \cdots + u_{k-1}\mathbf{g}_{k-1}, \tag{3.1}$$

where $u_i\mathbf{g}_i$ is the scalar multiplication and $u_i\mathbf{g}_i + u_j\mathbf{g}_j$ is the vector addition defined on the vector space $\mathbf{V}_n$ over $\mathrm{GF}(2)$. There are $2^k$ such linear combinations in the form given by (3.1) which correspond to the $2^k$ codewords in the code $C$.

We may arrange the $k$ linearly independent codewords, $\mathbf{g}_0, \mathbf{g}_1, \ldots, \mathbf{g}_{k-1}$, in $C$ as rows of a $k \times n$ matrix over $\mathrm{GF}(2)$ as follows:

$$\mathbf{G} = \begin{bmatrix} \mathbf{g}_0 \\ \mathbf{g}_1 \\ \vdots \\ \mathbf{g}_{k-1} \end{bmatrix} = \begin{bmatrix} g_{0,0} & g_{0,1} & \cdots & g_{0,n-1} \\ g_{1,0} & g_{1,1} & \cdots & g_{1,n-1} \\ \vdots & \vdots & \ddots & \vdots \\ g_{k-1,0} & g_{k-1,1} & \cdots & g_{k-1,n-1} \end{bmatrix}. \tag{3.2}$$

Then, the codeword $\mathbf{v} = (v_0, v_1, \ldots, v_{n-1})$ for the message $\mathbf{u} = (u_0, u_1, \ldots, u_{k-1})$ given by (3.1) can be expressed as the product of $\mathbf{u}$ and $\mathbf{G}$ as follows:

$$\mathbf{v} - \mathbf{uG}$$

$$= (u_0, u_1, \ldots, u_{k-1}) \begin{bmatrix} \mathbf{g}_0 \\ \mathbf{g}_1 \\ \vdots \\ \mathbf{g}_{k-1} \end{bmatrix} \tag{3.3}$$

$$= u_0\mathbf{g}_0 + u_1\mathbf{g}_1 + \cdots + u_{k-1}\mathbf{g}_{k-1}.$$

Therefore, the codeword $\mathbf{v}$ for a message $\mathbf{u}$ is simply a linear combination of the rows of the matrix $\mathbf{G}$ with the $k$ information bits in the message $\mathbf{u}$ as the coefficients. The matrix $\mathbf{G}$ given by (3.2) is called a *generator matrix* of the $(n, k)$ linear block code $C$. Because the code $C$ is spanned by the rows of $\mathbf{G}$, it is also called the *row space* of $\mathbf{G}$.

In general, an $(n, k)$ linear block code $C$ has more than one basis. Consequently, a generator matrix of a given $(n, k)$ linear block code is *not unique*. Any choice of a basis of $C$ gives a generator matrix of $C$. Obviously, the rank of a generator matrix of a linear block code $C$ equals the dimension of the code $C$.

**Example 3.2** Consider the $(6, 3)$ linear block code $C$ given in Example 3.1. The three codewords $\mathbf{g}_0 = (0\ 1\ 1\ 1\ 0\ 0)$, $\mathbf{g}_1 = (1\ 0\ 1\ 0\ 1\ 0)$, and $\mathbf{g}_2 = (1\ 1\ 0\ 0\ 0\ 1)$ given in Table 3.1 are linearly independent, which can be used as a basis of the three-dimensional subspace of the vector space $\mathbf{V}_6$ over GF(2), i.e., a basis for the code $C$. Using these three linearly independent codewords, we form the following $3 \times 6$ matrix:

$$\mathbf{G} = \begin{bmatrix} \mathbf{g}_0 \\ \mathbf{g}_1 \\ \mathbf{g}_2 \end{bmatrix} = \begin{bmatrix} 0\ 1\ 1\ 1\ 0\ 0 \\ 1\ 0\ 1\ 0\ 1\ 0 \\ 1\ 1\ 0\ 0\ 0\ 1 \end{bmatrix}. \tag{3.4}$$

The $2^3 = 8$ linear combinations of the rows of $\mathbf{G}$ give all the eight codewords of $C$ which are listed in Table 3.1. Hence, $\mathbf{G}$ is a generator matrix of the $(6, 3)$ linear block code over GF(2) given by Table 3.1. Suppose we have a message $\mathbf{u} = (0\ 0\ 1)$ of three information bits. The codeword $\mathbf{v}$ corresponding to the message $\mathbf{u}$ is

$$\mathbf{v} = \mathbf{u}\mathbf{G}$$

$$= (0\ 0\ 1) \cdot \begin{bmatrix} 0\ 1\ 1\ 1\ 0\ 0 \\ 1\ 0\ 1\ 0\ 1\ 0 \\ 1\ 1\ 0\ 0\ 0\ 1 \end{bmatrix}$$

$$= (1\ 1\ 0\ 0\ 0\ 1).$$

Consider another three codewords: $\mathbf{g}_0^* = (1\ 0\ 1\ 1\ 0\ 1)$, $\mathbf{g}_1^* = (1\ 1\ 0\ 1\ 1\ 0)$, and $\mathbf{g}_2^* = (0\ 0\ 0\ 1\ 1\ 1)$. It is easy to verify that these three codewords are also linearly independent. Thus, we can use them as a basis to form a generator matrix for the $(6, 3)$ linear block code $C$:

$$\mathbf{G}^* = \begin{bmatrix} \mathbf{g}_0^* \\ \mathbf{g}_1^* \\ \mathbf{g}_2^* \end{bmatrix} = \begin{bmatrix} 1\ 0\ 1\ 1\ 0\ 1 \\ 1\ 1\ 0\ 1\ 1\ 0 \\ 0\ 0\ 0\ 1\ 1\ 1 \end{bmatrix}. \tag{3.5}$$

With $\mathbf{G}^*$ as the generator matrix for the $(6, 3)$ linear block code $C$, the message $\mathbf{u} = (0\ 0\ 1)$ is mapped to the following codeword in $C$:

$$\mathbf{v} = \mathbf{u}\mathbf{G}^*$$

$$= (0\ 0\ 1) \cdot \begin{bmatrix} 1\ 0\ 1\ 1\ 0\ 1 \\ 1\ 1\ 0\ 1\ 1\ 0 \\ 0\ 0\ 0\ 1\ 1\ 1 \end{bmatrix}$$

$$= (0\ 0\ 0\ 1\ 1\ 1).$$

Note that with two different generator matrices, the same message $\mathbf{u} = (0\ 0\ 1)$ is mapped into two different codewords. ▲▲

Because a binary $(n, k)$ linear block code $C$ is a $k$-dimensional subspace of the vector space $\mathbf{V}_n$ of all the $n$-tuples over GF(2), its *null* (or *dual*) space, denoted by $C_d$, is an $(n - k)$-dimensional subspace of $\mathbf{V}_n$ that is given by the following set of $n$-tuples in $\mathbf{V}_n$:

$$C_d = \{\mathbf{w} \in \mathbf{V}_n : \langle \mathbf{w}, \mathbf{v} \rangle = 0 \ \text{ for all } \mathbf{v} \in C\}, \tag{3.6}$$

where $\langle \mathbf{w}, \mathbf{v} \rangle$ denotes the inner product of $\mathbf{w}$ and $\mathbf{v}$ (see Section 2.8.2). The set $C_d$ defined by (3.6) may be regarded as a binary $(n, n - k)$ linear block code and is called the *dual code* of $C$. The inner product of a codeword in $C$ and a codeword in $C_d$ is zero. Hence, every codeword in $C$ is *orthogonal* to every codeword in $C_d$ and vice versa.

Let $\mathcal{B}_d$ be a basis of $C_d$. Then, $\mathcal{B}_d$ consists of $n - k$ linearly independent codewords in $C_d$. Let $\mathbf{h}_0, \mathbf{h}_1, \ldots, \mathbf{h}_{n-k-1}$ be the $n - k$ linearly independent codewords in $C_d$, where for $0 \leq i < n - k$, $\mathbf{h}_i = (h_{i,0}, h_{i,1}, \ldots, h_{i,n-1})$ with $h_{i,j} \in$ GF(2) for $0 \leq j < n$. Then, every codeword in $C_d$ is a linear combination of these $n-k$ linearly independent codewords in $C_d$. Form the following $(n-k) \times n$ matrix over GF(2):

$$\mathbf{H} = \begin{bmatrix} \mathbf{h}_0 \\ \mathbf{h}_1 \\ \vdots \\ \mathbf{h}_{n-k-1} \end{bmatrix} = \begin{bmatrix} h_{0,0} & h_{0,1} & \cdots & h_{0,n-1} \\ h_{1,0} & h_{1,1} & \cdots & h_{1,n-1} \\ \vdots & \vdots & \ddots & \vdots \\ h_{n-k-1,0} & h_{n-k-1,1} & \cdots & h_{n-k-1,n-1} \end{bmatrix}. \tag{3.7}$$

Then, $\mathbf{H}$ is a generator matrix of the dual code $C_d$ of the $(n, k)$ linear block code $C$. It follows from (3.6) and the fact that every row in $\mathbf{G}$ is a codeword in $C$ and every row in $\mathbf{H}$ is a codeword in $C_d$ that

$$\mathbf{G} \cdot \mathbf{H}^T = \mathbf{O}, \tag{3.8}$$

where $\mathbf{O}$ is a $k \times (n - k)$ zero matrix. The equality of (3.8) says that the rows of $\mathbf{H}$ are orthogonal to the rows of $\mathbf{G}$ and vice versa. Furthermore, the code $C$ is also *completely* specified by the matrix $\mathbf{H}$ as follows: a binary $n$-tuple $\mathbf{v} \in \mathbf{V}_n$ is a codeword in $C$ if and only if

$$\mathbf{v}\mathbf{H}^T = \mathbf{0}, \tag{3.9}$$

where $\mathbf{0}$ is an all-zero $(n - k)$-tuple. Following the condition given by (3.9), the $(n, k)$ linear block code $C$ is specified by the matrix $\mathbf{H}$ as follows:

$$C = \{\mathbf{v} \in \mathbf{V}_n : \mathbf{v}\mathbf{H}^T = \mathbf{0}\}. \tag{3.10}$$

The $(n-k) \times n$ matrix $\mathbf{H}$ over GF(2) is called a *parity-check matrix* of $C$ and $C$ is said to be the *null space* of $\mathbf{H}$. *Therefore, a linear block code is completely specified by two matrices, a generator matrix $\mathbf{G}$ and a parity-check matrix $\mathbf{H}$.* A linear block code $C$ may have more than one parity-check matrix. From the above developments, we can see that a parity-check matrix of a linear block code $C$ is a generator matrix of its dual code $C_d$ and vice versa.

**Example 3.3** Again, we consider the $(6, 3)$ linear block code $C$ given by Table 3.1 in Example 3.1. Because $C$ is a three-dimensional subspace of the vector space $\mathbf{V}_6$ of all the 6-tuples over GF(2), its dual space $C_d$ is also a three-dimensional subspace of $\mathbf{V}_6$. Hence, $C_d$ should be a $(6, 3)$ linear block code. The three 6-tuples over GF(2), $\mathbf{h}_0 = (1\ 0\ 0\ 0\ 1\ 1)$, $\mathbf{h}_1 = (0\ 1\ 0\ 1\ 0\ 1)$, and $\mathbf{h}_2 = (0\ 0\ 1\ 1\ 1\ 0)$, are orthogonal to the rows of the generator matrix $\mathbf{G}$ of $C$ given in Example 3.2. They are three linearly independent codewords in $C_d$. Using $\mathbf{h}_0$, $\mathbf{h}_1$, and $\mathbf{h}_2$, we form the following $3 \times 6$ matrix over GF(2):

$$\mathbf{H} = \begin{bmatrix} \mathbf{h}_0 \\ \mathbf{h}_1 \\ \mathbf{h}_2 \end{bmatrix} = \begin{bmatrix} 1\ 0\ 0\ 0\ 1\ 1 \\ 0\ 1\ 0\ 1\ 0\ 1 \\ 0\ 0\ 1\ 1\ 1\ 0 \end{bmatrix}. \tag{3.11}$$

Because $\mathbf{G} \cdot \mathbf{H}^T = \mathbf{O}$ (this can be verified by taking the product of the matrix given by (3.4) and the transpose of the matrix $\mathbf{H}$ given by (3.11)), $\mathbf{H}$ is a parity-check matrix of the $(6, 3)$ linear block code $C$ given by Table 3.1 in Example 3.1. If we use $\mathbf{H}$ as a generator matrix, the row space of $\mathbf{H}$ gives a $(6, 3)$ linear block code $C_d$ which is the dual code of the code $C$. The codewords of $C_d$ are listed in Table 3.2.

Consider the following $3 \times 6$ matrix over GF(2):

$$\mathbf{H}^* = \begin{bmatrix} \mathbf{h}_0^* \\ \mathbf{h}_1^* \\ \mathbf{h}_2^* \end{bmatrix} = \begin{bmatrix} 1\ 0\ 1\ 1\ 0\ 1 \\ 0\ 1\ 1\ 0\ 1\ 1 \\ 0\ 1\ 0\ 1\ 0\ 1 \end{bmatrix}. \tag{3.12}$$

It can be easily checked that $\mathbf{G} \cdot (\mathbf{H}^*)^T = \mathbf{O}$. Hence, $\mathbf{H}^*$ is another parity-check matrix of the code $C$. This shows that a linear block code can have more than one parity-check matrix. ▲▲

Table 3.2 The dual code $C_d$ of the $(6, 3)$ linear block code $C$ given by Table 3.1.

| Messages $(u_0, u_1, u_2)$ | Codewords $(v_0, v_1, v_2, v_3, v_4, v_5)$ | Messages $(u_0, u_1, u_2)$ | Codewords $(v_0, v_1, v_2, v_3, v_4, v_5)$ |
|---|---|---|---|
| $(0\ 0\ 0)$ | $(0\ 0\ 0\ 0\ 0\ 0)$ | $(1\ 0\ 0)$ | $(1\ 0\ 0\ 0\ 1\ 1)$ |
| $(0\ 0\ 1)$ | $(0\ 0\ 1\ 1\ 1\ 0)$ | $(1\ 0\ 1)$ | $(1\ 0\ 1\ 1\ 0\ 1)$ |
| $(0\ 1\ 0)$ | $(0\ 1\ 0\ 1\ 0\ 1)$ | $(1\ 1\ 0)$ | $(1\ 1\ 0\ 1\ 1\ 0)$ |
| $(0\ 1\ 1)$ | $(0\ 1\ 1\ 0\ 1\ 0)$ | $(1\ 1\ 1)$ | $(1\ 1\ 1\ 0\ 0\ 0)$ |

**Example 3.4** Suppose we use the following $4 \times 7$ matrix as the generator matrix $\mathbf{G}$ to generate a $(7, 4)$ linear block code $C$ over GF(2),

$$\mathbf{G} = \begin{bmatrix} 1\ 1\ 0\ 1\ 0\ 0\ 0 \\ 0\ 1\ 1\ 0\ 1\ 0\ 0 \\ 1\ 1\ 1\ 0\ 0\ 1\ 0 \\ 1\ 0\ 1\ 0\ 0\ 0\ 1 \end{bmatrix}. \tag{3.13}$$

The 16 linear combinations of the four rows of $\mathbf{G}$ yield 16 codewords of the code $C$ as listed in Table 3.3. A parity-check matrix for this code is given as follows:

$$\mathbf{H} = \begin{bmatrix} 1 & 0 & 0 & 1 & 0 & 1 & 1 \\ 0 & 1 & 0 & 1 & 1 & 1 & 0 \\ 0 & 0 & 1 & 0 & 1 & 1 & 1 \end{bmatrix}. \tag{3.14}$$

It is noted that all the columns of $\mathbf{H}$ are distinct and they are all the nonzero 3-tuples over GF(2). Codes with this kind of parity-check matrices are *Hamming codes* [23]. Hamming codes will be covered in Chapter 4.

Suppose we use the parity-check matrix $\mathbf{H}$ given by (3.14) of the $(7,4)$ linear block code $C$ as a generator matrix $\mathbf{G}_d$, i.e., $\mathbf{G}_d = \mathbf{H}$. The code generated by $\mathbf{G}_d$ (i.e., the row space of $\mathbf{G}_d = \mathbf{H}$) is a $(7,3)$ linear block code $C_d$ which is the dual code of the $(7,4)$ code $C$. Table 3.4 lists all the eight codewords of the $(7,3)$ code $C_d$. It is interesting to observe that each of the nonzero codewords in $C_d$ has four 1-components. ▲▲

Table 3.3 The $(7,4)$ linear block code generated by the matrix $\mathbf{G}$ given by (3.13).

| Messages $(u_0, u_1, u_2, u_3)$ | Codewords $(v_0, v_1, v_2, v_3, v_4, v_5, v_6)$ | Messages $(u_0, u_1, u_2, u_3)$ | Codewords $(v_0, v_1, v_2, v_3, v_4, v_5, v_6)$ |
|---|---|---|---|
| $(0\ 0\ 0\ 0)$ | $(0\ 0\ 0\ 0\ 0\ 0\ 0)$ | $(1\ 0\ 0\ 0)$ | $(1\ 1\ 0\ 1\ 0\ 0\ 0)$ |
| $(0\ 1\ 0\ 0)$ | $(0\ 1\ 1\ 0\ 1\ 0\ 0)$ | $(1\ 1\ 0\ 0)$ | $(1\ 0\ 1\ 1\ 1\ 0\ 0)$ |
| $(0\ 0\ 1\ 0)$ | $(1\ 1\ 1\ 0\ 0\ 1\ 0)$ | $(1\ 0\ 1\ 0)$ | $(0\ 0\ 1\ 1\ 0\ 1\ 0)$ |
| $(0\ 1\ 1\ 0)$ | $(1\ 0\ 0\ 0\ 1\ 1\ 0)$ | $(1\ 1\ 1\ 0)$ | $(0\ 1\ 0\ 1\ 1\ 1\ 0)$ |
| $(0\ 0\ 0\ 1)$ | $(1\ 0\ 1\ 0\ 0\ 0\ 1)$ | $(1\ 0\ 0\ 1)$ | $(0\ 1\ 1\ 1\ 0\ 0\ 1)$ |
| $(0\ 1\ 0\ 1)$ | $(1\ 1\ 0\ 0\ 1\ 0\ 1)$ | $(1\ 1\ 0\ 1)$ | $(0\ 0\ 0\ 1\ 1\ 0\ 1)$ |
| $(0\ 0\ 1\ 1)$ | $(0\ 1\ 0\ 0\ 0\ 1\ 1)$ | $(1\ 0\ 1\ 1)$ | $(1\ 0\ 0\ 1\ 0\ 1\ 1)$ |
| $(0\ 1\ 1\ 1)$ | $(0\ 0\ 1\ 0\ 1\ 1\ 1)$ | $(1\ 1\ 1\ 1)$ | $(1\ 1\ 1\ 1\ 1\ 1\ 1)$ |

Table 3.4 The $(7,3)$ linear block code generated by the matrix $\mathbf{H}$ (as a generator matrix) given by (3.14).

| Messages $(u_0, u_1, u_2)$ | Codewords $(v_0, v_1, v_2, v_3, v_4, v_5, v_6)$ | Messages $(u_0, u_1, u_2)$ | Codewords $(v_0, v_1, v_2, v_3, v_4, v_5, v_6)$ |
|---|---|---|---|
| $(0\ 0\ 0)$ | $(0\ 0\ 0\ 0\ 0\ 0\ 0)$ | $(0\ 0\ 1)$ | $(0\ 0\ 1\ 0\ 1\ 1\ 1)$ |
| $(0\ 1\ 0)$ | $(0\ 1\ 0\ 1\ 1\ 1\ 0)$ | $(0\ 1\ 1)$ | $(0\ 1\ 1\ 1\ 0\ 0\ 1)$ |
| $(1\ 0\ 0)$ | $(1\ 0\ 0\ 1\ 0\ 1\ 1)$ | $(1\ 0\ 1)$ | $(1\ 0\ 1\ 1\ 1\ 0\ 0)$ |
| $(1\ 1\ 0)$ | $(1\ 1\ 0\ 0\ 1\ 0\ 1)$ | $(1\ 1\ 1)$ | $(1\ 1\ 1\ 0\ 0\ 1\ 0)$ |

In general, encoding of a linear block code is based on a generator matrix of the code and decoding is based on a parity-check matrix of the code (as will be shown in later chapters). Many classes of well-known linear block codes are constructed in terms of their parity-check matrices, as will be shown in later chapters.

## 3.3   Systematic Linear Block Codes

An $(n, k)$ linear block code is said to be *systematic* if it has the following structure: every codeword consists of two parts, a *message part* and a *parity-check part* (or redundant part). The message part consists of $k$ *unaltered* information bits and the parity-check part consists of $n - k$ *parity-check* (redundant) *bits*, as shown in Fig. 3.1.

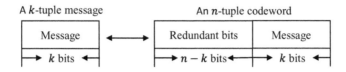

Figure 3.1 Systematic format of codewords in an $(n, k)$ linear block code.

The $(6, 3)$ and $(7, 4)$ linear block codes given in Examples 3.1 (Table 3.1) and 3.4 (Table 3.3), respectively, are in systematic form.

The generator matrix $\mathbf{G}_{\text{sys}}$ of an $(n, k)$ systematic linear block code is of the following form:

$$\mathbf{G}_{\text{sys}} = \begin{bmatrix} \mathbf{g}_0 \\ \mathbf{g}_1 \\ \vdots \\ \mathbf{g}_{k-1} \end{bmatrix} = \left[\underbrace{\begin{matrix} p_{0,0} & p_{0,1} & \cdots & p_{0,n-k-1} \\ p_{1,0} & p_{1,1} & \cdots & p_{1,n-k-1} \\ \vdots & \vdots & \ddots & \vdots \\ p_{k-1,0} & p_{k-1,1} & \cdots & p_{k-1,n-k-1} \end{matrix}}_{\mathbf{P}} \quad \underbrace{\begin{matrix} 1 & 0 & \cdots & 0 \\ 0 & 1 & \cdots & 0 \\ \vdots & \vdots & \ddots & \vdots \\ 0 & 0 & \cdots & 1 \end{matrix}}_{\mathbf{I}_k}\right], \qquad (3.15)$$

where $p_{i,j} = 0$ or $1$ for $0 \le i < k, 0 \le j < n - k$. The generator matrix $\mathbf{G}_{\text{sys}}$ consists of two submatrices, a $k \times (n - k)$ submatrix, denoted by $\mathbf{P}$, called the **P***-submatrix*, on the left with entries over GF(2), and a $k \times k$ identity matrix, denoted by $\mathbf{I}_k$, on the right. In terms of $\mathbf{P}$ and $\mathbf{I}_k$, the matrix $\mathbf{G}_{\text{sys}}$ can be expressed in the following form:

$$\mathbf{G}_{\text{sys}} = [\mathbf{P}\ \ \mathbf{I}_k\,]. \qquad (3.16)$$

Let $\mathbf{u} = (u_0, u_1, \dots, u_{k-1})$ be a message to be encoded. Performing a linear combination of the rows of $\mathbf{G}_{\text{sys}}$ given by (3.15) with the information bits in $\mathbf{u}$ as coefficients, we obtain the following corresponding codeword:

$$\mathbf{v} = (v_0, v_1, v_2, \dots, v_{n-1}) = (u_0, u_1, \dots, u_{k-1}) \cdot \mathbf{G}_{\text{sys}}$$

$$= (u_0, u_1, \dots, u_{k-1}) \cdot \begin{bmatrix} p_{0,0} & p_{0,1} & \cdots & p_{0,n-k-1} & 1 & 0 & \cdots & 0 \\ p_{1,0} & p_{1,1} & \cdots & p_{1,n-k-1} & 0 & 1 & \cdots & 0 \\ \vdots & \vdots & \ddots & \vdots & \vdots & \vdots & \ddots & \vdots \\ p_{k-1,0} & p_{k-1,1} & \cdots & p_{k-1,n-k-1} & 0 & 0 & \cdots & 1 \end{bmatrix}. \qquad (3.17)$$

Taking the matrix product of (3.17), we obtain the $n$ code bits of the code-word $\mathbf{v}$ as follows:

(1) for $0 \leq l < k$,

$$v_{n-k+l} = u_l, \tag{3.18}$$

and

(2) for $0 \leq j < n - k$,

$$v_j = u_0 p_{0,j} + u_1 p_{1,j} + \cdots + u_{k-1} p_{k-1,j}, \tag{3.19}$$

which is the inner product of $\mathbf{u}$ and the $j$th column of the $\mathbf{P}$-submatrix of $\mathbf{G}_{\text{sys}}$.

From (3.18) and (3.19), we see that the codeword $\mathbf{v}$ for the message $\mathbf{u} = (u_0, u_1, \ldots, u_{k-1})$ is

$$\mathbf{v} = (v_0, v_1 \ldots, v_{n-k-1}, u_0, u_1, \ldots, u_{k-1}),$$

which is in systematic form where the $k$ rightmost bits of $\mathbf{v}$ are identical to the message (unaltered information) bits $u_0, u_1, \ldots, u_{k-1}$ and the leftmost $n-k$ bits $v_0, v_1 \ldots, v_{n-k-1}$ are the parity-check bits. From (3.19), we see that each parity-check bit $v_j$, $0 \leq j < n - k$, of $\mathbf{v}$ is *a linear sum* of information bits. The $n - k$ parity-check bits of the codeword $\mathbf{v}$ are completely specified by the $n-k$ columns of the $\mathbf{P}$-submatrix of $\mathbf{G}_{\text{sys}}$ given by (3.16). The $n - k$ equations given by (3.19) are called the *parity-check equations (sums)* of the $(n, k)$ systematic linear block code $C$. These parity-check equations completely specify the code. Because the submatrix $\mathbf{P}$ of the generator matrix $\mathbf{G}_{\text{sys}}$ in systematic form uniquely specifies the $n-k$ parity-check bits of a codeword in an $(n, k)$ linear block code, we call $\mathbf{P}$ the *parity submatrix* of $\mathbf{G}_{\text{sys}}$. In code design, we need only to design the parity submatrix $\mathbf{P}$ of $\mathbf{G}_{\text{sys}}$.

**Example 3.5** Consider the $(7, 4)$ systematic linear code $C$ given in Example 3.4 whose generator $\mathbf{G}$ given by (3.13) is in systematic form. The $\mathbf{P}$-submatrix of $\mathbf{G}$ is a $4 \times 3$ matrix:

$$\mathbf{P} = \begin{bmatrix} 1 & 1 & 0 \\ 0 & 1 & 1 \\ 1 & 1 & 1 \\ 1 & 0 & 1 \end{bmatrix}. \tag{3.20}$$

Let $\mathbf{u} = (u_0, u_1, u_2, u_3)$ be a message to be encoded. In systematic form, the three parity-check bits of the codeword $\mathbf{v}$ for the message $\mathbf{u}$ are the inner product of $\mathbf{u} = (u_0, u_1, u_2, u_3)$ and the three columns of the parity submatrix $\mathbf{P}$ of the generator matrix $\mathbf{G}$, which are given as follows:

$$\begin{aligned} v_0 &= u_0 + u_2 + u_3, \\ v_1 &= u_0 + u_1 + u_2, \\ v_2 &= u_1 + u_2 + u_3. \end{aligned} \tag{3.21}$$

Hence, the codeword for $\mathbf{u} = (u_0, u_1, u_2, u_3)$ in systematic form is $\mathbf{v} = (u_0 + u_2 + u_3, u_0 + u_1 + u_2, u_1 + u_2 + u_3, u_0, u_1, u_2, u_3)$.

Suppose the message is $\mathbf{u} = (1\ 0\ 1\ 1)$. Then, by using (3.21), the three parity-check bits are calculated as $v_0 = 1$, $v_1 = 0$, and $v_2 = 0$. Encoding of $\mathbf{u} = (1\ 0\ 1\ 1)$ gives the codeword $\mathbf{v} = (1\ 0\ 0\ 1\ 0\ 1\ 1)$ which is the fourteenth codeword listed in Table 3.3.                                        ▲▲

Given a $k \times n$ generator matrix $\mathbf{G}$ of an $(n, k)$ linear block code $C$ not in systematic form, a generator matrix $\mathbf{G}_{\text{sys}}$ in systematic form of (3.15) can always be obtained by performing *elementary row operations* on the rows of $\mathbf{G}$ and then taking column permutation (if necessary). The $k \times n$ matrix $\mathbf{G}_{\text{sys}}$ is called a *combinatorially equivalent matrix* of $\mathbf{G}$. The systematic $(n, k)$ linear block code $C_{\text{sys}}$ generated by $\mathbf{G}_{\text{sys}}$ is called a *combinatorially equivalent code* of $C$.[2] The two codes, $C_{\text{sys}}$ and $C$, are only different in the arrangement (or order) of code bits in their codewords, i.e., a codeword in $C_{\text{sys}}$ can be obtained through performing a fixed permutation of the code bits in a codeword of $C$, and vice versa. Two combinatorially equivalent $(n, k)$ linear block codes give the same error performance.

If a generator matrix of an $(n, k)$ linear block code $C$ is given in systematic form of (3.15), its corresponding parity-check matrix in systematic form is given below:

$$
\mathbf{H}_{\text{sys}} = \begin{bmatrix} \mathbf{I}_{n-k} & \mathbf{P}^T \end{bmatrix} = \begin{bmatrix} \mathbf{p}_0 \\ \mathbf{p}_1 \\ \vdots \\ \mathbf{p}_{n-k-1} \end{bmatrix}
$$

$$
= \begin{bmatrix}
1 & 0 & 0 & 0 & \dots & 0 & p_{0,0} & p_{1,0} & \cdots & p_{k-1,0} \\
0 & 1 & 0 & 0 & \dots & 0 & p_{0,1} & p_{1,1} & \cdots & p_{k-1,1} \\
0 & 0 & 1 & 0 & \dots & 0 & p_{0,2} & p_{1,2} & \cdots & p_{k-1,2} \\
\vdots & \vdots & \vdots & \vdots & \ddots & \vdots & \vdots & \vdots & \ddots & \vdots \\
0 & 0 & 0 & 0 & \dots & 1 & p_{0,n-k-1} & p_{1,n-k-1} & \cdots & p_{k-1,n-k-1}
\end{bmatrix}, \tag{3.22}
$$

where $\mathbf{p}_0, \mathbf{p}_1, \dots, \mathbf{p}_{n-k-1}$ denote the $n - k$ rows of $\mathbf{H}_{\text{sys}}$. It can be proved that $\mathbf{G}_{\text{sys}} \cdot \mathbf{H}_{\text{sys}}^T = \mathbf{O}$ (see Problem 3.5).

The parity-check bits of a codeword in systematic form can be formed from $\mathbf{H}_{\text{sys}}$. Let $\mathbf{u} = (u_0, u_1, \dots, u_{k-1})$ be a message to be encoded. In systematic encoding, the codeword for $\mathbf{u} = (u_0, u_1, \dots, u_{k-1})$ must be in the form of $\mathbf{v} = (v_0, v_1, \dots, v_{n-k-1}, u_0, u_1, \dots, u_{k-1})$. Because $\mathbf{v}$ is a codeword, it is orthogonal to each row $\mathbf{p}_j = (0\ \dots\ 0\ 1\ 0\ \dots\ 0, p_{0,j}, p_{1,j}, \dots, p_{k-1,j})$, $0 \le j < n - k$, of $\mathbf{H}_{\text{sys}}$. Hence, the inner product of $\mathbf{v}$ and $\mathbf{p}_j$ is 0, i.e.,

$$
\langle \mathbf{v}, \mathbf{p}_j \rangle = v_j + u_0 p_{0,j} + u_1 p_{1,j} + \cdots + u_{k-1} p_{k-1,j} = 0.
$$

---

[2] We note here that being systematic is not a property of the linear block code $C$, instead is a property of the encoder of the code, i.e., the generator matrix $\mathbf{G}$.

From the above equality, we obtain the $j$th parity-check bit $v_j$, $0 \leq j < n - k$, as the following linear sum of message bits:

$$v_j = u_0 p_{0,j} + u_1 p_{1,j} + \cdots + u_{k-1} p_{k-1,j}$$

which is exactly the same as the parity-check equation given by (3.19).

The parity-check matrices given by (3.11) and (3.14) in Examples 3.3 and 3.4, respectively, are in systematic form.

**Example 3.6** Consider the following $4 \times 15$ parity-check matrix of a $(15, 11)$ systematic linear block code,

$$\mathbf{H}_{\mathrm{sys}} = \begin{bmatrix} 1 & 0 & 0 & 0 & 1 & 0 & 0 & 1 & 1 & 0 & 1 & 0 & 1 & 1 & 1 \\ 0 & 1 & 0 & 0 & 1 & 1 & 0 & 1 & 0 & 1 & 1 & 1 & 1 & 0 & 0 \\ 0 & 0 & 1 & 0 & 0 & 1 & 1 & 0 & 1 & 0 & 1 & 1 & 1 & 1 & 0 \\ 0 & 0 & 0 & 1 & 0 & 0 & 1 & 1 & 0 & 1 & 0 & 1 & 1 & 1 & 1 \end{bmatrix} = \begin{bmatrix} \mathbf{I}_4 & \mathbf{P}^T \end{bmatrix}. \tag{3.23}$$

Let $\mathbf{u} = (u_0, u_1, u_2, u_3, u_4, u_5, u_6, u_7, u_8, u_9, u_{10})$ be a message to be encoded. Using the above parity-check matrix, we find the four parity-check bits as follows:

$$\begin{aligned} v_0 &= u_0 + u_3 + u_4 + u_6 + u_8 + u_9 + u_{10}, \\ v_1 &= u_0 + u_1 + u_3 + u_5 + u_6 + u_7 + u_8, \\ v_2 &= u_1 + u_2 + u_4 + u_6 + u_7 + u_8 + u_9, \\ v_3 &= u_2 + u_3 + u_5 + u_7 + u_8 + u_9 + u_{10}. \end{aligned} \tag{3.24}$$

It follows from (3.16) and (3.23) that the generator matrix $\mathbf{G}_{\mathrm{sys}}$ of the $(15, 11)$ code in systematic form is given by

$$\mathbf{G}_{\mathrm{sys}} = \begin{bmatrix} \mathbf{P} & \mathbf{I}_{11} \end{bmatrix} = \begin{bmatrix} 1 & 1 & 0 & 0 \\ 0 & 1 & 1 & 0 \\ 0 & 0 & 1 & 1 \\ 1 & 1 & 0 & 1 \\ 1 & 0 & 1 & 0 \\ 0 & 1 & 0 & 1 \\ 1 & 1 & 1 & 0 \\ 0 & 1 & 1 & 1 \\ 1 & 1 & 1 & 1 \\ 1 & 0 & 1 & 1 \\ 1 & 0 & 0 & 1 \end{bmatrix} \mathbf{I}_{11} . \tag{3.25}$$

It is noted that all the columns of the parity-check matrix $\mathbf{H}_{\mathrm{sys}}$ are distinct and they are all the nonzero 4-tuples over $GF(2)$. As mentioned in Example 3.4, codes specified by these types of parity-check matrices are Hamming codes. ▲▲

## 3.4 Error Detection with Linear Block Codes

Consider an $(n, k)$ binary linear block code $C$ with an $(n - k) \times n$ parity-check matrix $\mathbf{H}$. Suppose a codeword $\mathbf{v} = (v_0, v_1, \ldots, v_{n-1})$ in $C$ is transmitted over a BSC. Let $\mathbf{r} = (r_0, r_1, \ldots, r_{n-1})$ be the corresponding hard-decision received vector, an $n$-tuple, at the input of the channel decoder. Because of the channel noise, the received vector $\mathbf{r}$ and the transmitted codeword $\mathbf{v}$ may differ in some positions.

Define the following vector subtraction (subtraction and addition operations are the same in GF(2)) of $\mathbf{r}$ and $\mathbf{v}$:

$$
\begin{aligned}
\mathbf{e} &\triangleq (e_0, e_1, \ldots, e_{n-1}) \\
&\triangleq \mathbf{r} - \mathbf{v} = \mathbf{r} + \mathbf{v} \\
&= (r_0, r_1, \ldots, r_{n-1}) + (v_0, v_1, \ldots, v_{n-1}) \\
&= (r_0 + v_0, r_1 + v_1, \ldots, r_{n-1} + v_{n-1}),
\end{aligned}
\tag{3.26}
$$

where $e_j = r_j + v_j$ for $0 \le j < n$ and the addition $+$ is the addition defined on GF(2) (i.e., modulo-2 addition). Note that $e_j = 1$ if $r_j \ne v_j$ and $e_j = 0$ if $r_j = v_j$. Therefore, the positions of the 1-components in the $n$-tuple $\mathbf{e}$ are the places where $\mathbf{r}$ and $\mathbf{v}$ differ. At these places, transmission errors have occurred. Because $\mathbf{e}$ displays the pattern of transmission errors in $\mathbf{r}$, we call $\mathbf{e}$ the *error pattern* (or vector) during the transmission of the codeword $\mathbf{v}$. The 1-components in $\mathbf{e}$ are called *transmission errors* caused by channel noise. A 1-component in $\mathbf{e}$ changes a transmitted code bit from either 1 to 0 or from 0 to 1.

For a BSC, an error can occur at any place with the same probability $p$ over a span of $n$ places (the length of the code). There are $2^n$ possible error patterns. Using (3.26), we can express the received vector $\mathbf{r}$ as the sum of the transmitted codeword $\mathbf{v}$ and the error pattern $\mathbf{e}$:

$$
\mathbf{r} = \mathbf{e} + \mathbf{v}.
\tag{3.27}
$$

At the receiver, neither the transmitted codeword $\mathbf{v}$ nor the error pattern $\mathbf{e}$ is known. Upon receiving $\mathbf{r}$, the decoder must first determine whether there are transmission errors in $\mathbf{r}$. If the presence of errors is detected, then the decoder must estimate the error pattern $\mathbf{e}$ based on the code $C$, the received vector $\mathbf{r}$, and the provided channel information. Let $\mathbf{e}^*$ denote the estimated error pattern. Then, the estimated transmitted codeword is given by

$$
\mathbf{v}^* = \mathbf{r} - \mathbf{e}^* = \mathbf{r} + \mathbf{e}^*.
$$

To check whether a received vector $\mathbf{r}$ contains transmission errors, we calculate the following $(n-k)$-tuple over GF(2):

$$
\begin{aligned}
\mathbf{s} &= (s_0, s_1, \ldots, s_{n-k-1}) \\
&= \mathbf{r} \cdot \mathbf{H}^T.
\end{aligned}
\tag{3.28}
$$

Note that $\mathbf{r}$ is an $n$-tuple in the vector space $\mathbf{V}_n$ of all the $n$-tuples over GF(2). Recall that an $n$-tuple $\mathbf{w}$ in $\mathbf{V}_n$ is a codeword in $C$ if and only if (see (3.9))

$$
\mathbf{w} \cdot \mathbf{H}^T = \mathbf{0}.
$$

Therefore, if $\mathbf{s} \ne \mathbf{0}$, $\mathbf{r}$ is not a codeword in $C$. In this case, the transmitter transmits a codeword but the receiver receives a vector which is not a codeword. Hence, the presence of transmission errors is detected. If $\mathbf{s} = \mathbf{0}$, then $\mathbf{r}$ is a codeword in $C$. In this case, the channel decoder assumes that $\mathbf{r}$ is *error free* and accepts $\mathbf{r}$ as the transmitted codeword. However, in the event that $\mathbf{r}$ is a codeword in $C$ but differs from the transmitted codeword $\mathbf{v}$, by accepting $\mathbf{r}$ as the transmitted codeword, the decoder commits a *decoding error*. This occurs when the error pattern $\mathbf{e}$ caused by the noise changes the transmitted codeword $\mathbf{v}$ into another codeword in $C$. That is when and only when the error pattern $\mathbf{e}$ is identical to a nonzero codeword in $C$, because the sum of two codewords in $C$ is another codeword in $C$. An error pattern of this type is called an *undetectable error pattern*. There are $2^k - 1$ such undetectable error patterns. Because the $(n - k)$-tuple $\mathbf{s} = (s_0, s_1, \ldots, s_{n-k-1})$ over GF(2) is used for detecting whether a received vector $\mathbf{r}$ contains transmission errors, it is called the *syndrome* of $\mathbf{r}$. From (3.28), we see that the $n - k$ syndrome bits are linear sums of the received bits in the received vector $\mathbf{r}$.

**Example 3.7** Consider the $(7, 4)$ systematic linear block code $C$ with parity-check matrix given by (3.14) in Example 3.4. Let $\mathbf{r} = (r_0, r_1, r_2, r_3, r_4, r_5, r_6)$ be the received vector. The syndrome of $\mathbf{r}$ is

$$\mathbf{s} = (s_0, s_1, s_2) = \mathbf{r} \cdot \mathbf{H}^T = (r_0, r_1, r_2, r_3, r_4, r_5, r_6) \cdot \begin{bmatrix} 1 & 0 & 0 \\ 0 & 1 & 0 \\ 0 & 0 & 1 \\ 1 & 1 & 0 \\ 0 & 1 & 1 \\ 1 & 1 & 1 \\ 1 & 0 & 1 \end{bmatrix}. \tag{3.29}$$

Then, the three syndrome bits are given by the following linear sums of received bits in $\mathbf{r}$:

$$\begin{aligned} s_0 &= r_0 + r_3 + r_5 + r_6, \\ s_1 &= r_1 + r_3 + r_4 + r_5, \\ s_2 &= r_2 + r_4 + r_5 + r_6. \end{aligned} \tag{3.30}$$

Suppose a codeword $\mathbf{v} = (0\ 1\ 0\ 0\ 0\ 1\ 1)$ in $C$ is transmitted. Let $\mathbf{r}_0 = (0\ 1\ 0\ 0\ 0\ 0\ 1)$ be the received vector. The syndrome of $\mathbf{r}$ is

$$\begin{aligned} \mathbf{s} = (s_0, s_1, s_2) &= \mathbf{r}_0 \cdot \mathbf{H}^T \\ &= (0\ 1\ 0\ 0\ 0\ 0\ 1) \cdot \begin{bmatrix} 1 & 0 & 0 \\ 0 & 1 & 0 \\ 0 & 0 & 1 \\ 1 & 1 & 0 \\ 0 & 1 & 1 \\ 1 & 1 & 1 \\ 1 & 0 & 1 \end{bmatrix} = (1\ 1\ 1). \end{aligned}$$

Alternatively, the three syndrome bits in $\mathbf{s}$ can be calculated using (3.30):

$$\begin{aligned} s_0 &= r_0 + r_3 + r_5 + r_6 = 0 + 0 + 0 + 1 = 1, \\ s_1 &= r_1 + r_3 + r_4 + r_5 = 1 + 0 + 0 + 0 = 1, \\ s_2 &= r_2 + r_4 + r_5 + r_6 = 0 + 0 + 0 + 1 = 1. \end{aligned}$$

Because $\mathbf{s} \neq \mathbf{0}$, $\mathbf{r}$ is not a codeword in $C$ and thus the presence of errors is detected. ▲▲

## 3.5   Syndrome and Error Patterns

Let $\mathbf{r} = \mathbf{v} + \mathbf{e}$ be the received vector, where $\mathbf{v}$ and $\mathbf{e}$ are the transmitted codeword and the error pattern, respectively. Then, the syndrome $\mathbf{s}$ of $\mathbf{r}$ is

$$\mathbf{s} = \mathbf{r} \cdot \mathbf{H}^T = (\mathbf{v} + \mathbf{e}) \cdot \mathbf{H}^T = \mathbf{v} \cdot \mathbf{H}^T + \mathbf{e} \cdot \mathbf{H}^T = \mathbf{e} \cdot \mathbf{H}^T, \qquad (3.31)$$

where $\mathbf{v} \cdot \mathbf{H}^T = \mathbf{0}$ is applied because $\mathbf{v}$ is a codeword. Equation (3.31) shows that the syndrome is the sum of the columns of $\mathbf{H}$ corresponding to the errors in $\mathbf{e}$. If there is only one bit error in $\mathbf{r}$, the syndrome $\mathbf{s}$ is identical to a certain column of the parity-check matrix corresponding to the error location. Equation (3.31) relates the unknown error pattern $\mathbf{e}$ to the computed syndrome $\mathbf{s}$.

Suppose the parity-check matrix $\mathbf{H}$ is in systematic form as shown in (3.22). Expanding $\mathbf{e} \cdot \mathbf{H}^T$, we obtain the following $n - k$ linear equations:

$$
\begin{aligned}
s_0 &= e_0 + e_{n-k}p_{0,0} + e_{n-k+1}p_{1,0} + \cdots + e_{n-1}p_{k-1,0}, \\
s_1 &= e_1 + e_{n-k}p_{0,1} + e_{n-k+1}p_{1,1} + \cdots + e_{n-1}p_{k-1,1}, \\
&\ \ \vdots
\end{aligned}
\qquad (3.32)
$$

$$s_{n-k-1} = e_{n-k-1} + e_{n-k}p_{0,n-k-1} + e_{n-k+1}p_{1,n-k-1} + \cdots + e_{n-1}p_{k-1,n-k-1}.$$

The $n - k$ equations given by (3.32) relate the error bits in the error pattern $\mathbf{e}$ to the computed syndrome bits. *Any method solving these $n - k$ equations for the error pattern $\mathbf{e}$ is a decoding method.* Because there are more unknowns ($n$ unknowns, $e_0, e_1, \ldots, e_{n-1}$) than the number of equations ($n - k$ equations), the equations given by (3.32) do not have a unique solution. In fact, there are $2^k$ possible solutions. Each solution gives an error pattern satisfying these $n - k$ equations. The true error pattern is just one of them. To minimize the probability of a decoding error, the *most probable* error pattern which satisfies the $n - k$ equations given by (3.32) is chosen as the *true error pattern*. If the channel is a BSC (assuming the transition probability $p$ less than $1/2$), the most probable error pattern is the one with the *least number* of nonzero bits.

**Example 3.8** Consider the $(7, 4)$ code given in Example 3.4 with the following parity-check matrix in systematic form:

$$\mathbf{H} = \begin{bmatrix} 1 & 0 & 0 & 1 & 0 & 1 & 1 \\ 0 & 1 & 0 & 1 & 1 & 1 & 0 \\ 0 & 0 & 1 & 0 & 1 & 1 & 1 \end{bmatrix}.$$

Suppose the codeword $\mathbf{v} = (1\ 0\ 0\ 1\ 0\ 1\ 1)$ is transmitted and $\mathbf{r} = (1\ 0\ 0\ 1\ 0\ 0\ 1)$ is received. The syndrome of $\mathbf{r}$ is $\mathbf{s} = (s_0, s_1, s_2) = \mathbf{r} \cdot \mathbf{H}^T = (1\ 1\ 1)$. Let $\mathbf{e} = (e_0, e_1, e_2, e_3, e_4, e_5, e_6)$ be the error pattern. Because $\mathbf{s} = \mathbf{e} \cdot \mathbf{H}^T$, we have the following three linear equations relating the errors to the computed syndrome bits:

$$
\begin{aligned}
s_0 &= 1 = e_0 + e_3 + e_5 + e_6, \\
s_1 &= 1 = e_1 + e_3 + e_4 + e_5, \\
s_2 &= 1 = e_2 + e_4 + e_5 + e_6.
\end{aligned}
$$

There are $2^4 = 16$ solutions for the above three linear equations, which are

$$(0\,0\,0\,0\,0\,1\,0), \ (1\,0\,1\,0\,0\,1\,1), \ (0\,1\,0\,1\,0\,1\,0), \ (0\,1\,1\,1\,0\,1\,1),$$
$$(0\,1\,1\,0\,1\,1\,0), \ (1\,1\,0\,0\,1\,1\,1), \ (1\,0\,1\,1\,1\,1\,0), \ (0\,0\,0\,1\,1\,1\,1),$$
$$(1\,1\,1\,0\,0\,0\,0), \ (0\,1\,0\,0\,0\,0\,1), \ (0\,0\,1\,1\,0\,0\,0), \ (1\,0\,0\,1\,0\,0\,1),$$
$$(1\,0\,0\,0\,1\,0\,0), \ (0\,0\,1\,0\,1\,0\,0), \ (0\,1\,0\,1\,1\,0\,0), \ (1\,1\,1\,1\,1\,0\,1).$$

For a BSC, the most probable error pattern among the above 16 solutions is $\mathbf{e}^* = (0\,0\,0\,0\,0\,1\,0)$ which contains a single error. Adding this error pattern to the received vector $\mathbf{r} = (1\,0\,0\,1\,0\,0\,1)$, we have

$$\mathbf{v}^* = \mathbf{r} + \mathbf{e}^* = (0\,0\,0\,0\,0\,1\,0) + (1\,0\,0\,1\,0\,0\,1) = (1\,0\,0\,1\,0\,1\,1),$$

which is the transmitted codeword $\mathbf{v}$. Hence, the decoding is correct.     ▲▲

Decoding of a linear block code with its parity-check matrix $\mathbf{H}$ can be carried out in three steps.
  (1) Compute the syndrome $\mathbf{s}$ of the received vector $\mathbf{r}$: $\mathbf{s} = \mathbf{r} \cdot \mathbf{H}^T$.
  (2) Identify the most probable error pattern $\mathbf{e}^*$ which satisfies the equality $\mathbf{s} = \mathbf{e}^* \cdot \mathbf{H}^T$, and take $\mathbf{e}^*$ as the estimated error pattern.
  (3) Add $\mathbf{e}^*$ to the received vector $\mathbf{r}$, and output $\mathbf{v}^* = \mathbf{r} + \mathbf{e}^*$ as the estimated transmitted codeword.

## 3.6 Weight Distribution and Probability of Undetected Error

Let $\mathbf{v} = (v_0, v_1, \ldots, v_{n-1})$ be an $n$-tuple over GF(2). The *Hamming weight* (or simply weight) of $\mathbf{v}$, denoted by $\omega(\mathbf{v})$, is defined as the number of nonzero components in $\mathbf{v}$. For example, if $\mathbf{v} = (1\,0\,0\,1\,0\,1\,1)$, the weight of $\mathbf{v}$ is $\omega(\mathbf{v}) = 4$ because $\mathbf{v}$ has four nonzero components. Consider an $(n, k)$ linear block code $C$ over GF(2). The smallest weight of a nonzero codeword in $C$, denoted by $\omega_{\min}(C)$, is called the *minimum weight* of $C$. Mathematically, the minimum weight of $C$ is given as follows:

$$\omega_{\min}(C) = \min\{\omega(\mathbf{v}) : \mathbf{v} \in C, \mathbf{v} \neq \mathbf{0}\}. \tag{3.33}$$

For $0 \leq i \leq n$, let $A_i$ be the number of codewords in $C$ with Hamming weight $i$. Then, the numbers $A_0, A_1, A_2, \ldots, A_n$ are referred to as the *weight distribution* (or *weight spectrum*) of $C$. It is clear that

$$A_0 + A_1 + \cdots + A_n = 2^k.$$

Because there is one and only one all-zero codeword in a linear block code, $A_0 = 1$.

Let $B_0, B_1, B_2, \ldots, B_n$ be the weight distribution of the dual code $C_d$ of $C$. It is also clear that

$$B_0 + B_1 + \cdots + B_n = 2^{n-k}.$$

Consider the following two polynomials with indeterminate $z$ and $A_0$, $A_1$, $A_2, \ldots, A_n$ and $B_0$, $B_1$, $B_2, \ldots, B_n$ as their coefficients:

$$\mathbf{A}(z) = A_0 + A_1 z + A_2 z^2 + \cdots + A_n z^n, \tag{3.34}$$

$$\mathbf{B}(z) = B_0 + B_1 z + B_2 z^2 + \cdots + B_n z^n. \tag{3.35}$$

The two polynomials $\mathbf{A}(z)$ and $\mathbf{B}(z)$ defined above are called the *weight enumerators* of the $(n, k)$ linear block code $C$ and its dual code $C_d$, respectively. These two polynomials are related to each other through the so-called *MacWilliams identity* [3] as follows:

$$\mathbf{A}(z) = 2^{-(n-k)}(1 + z)^n \mathbf{B}\left(\frac{1-z}{1+z}\right). \tag{3.36}$$

The MacWilliams identity provides more flexibility to compute the weight distribution of a linear block code. In some cases, it is impossible or very hard to compute the weight distribution of a linear block code; however, it is far easier to obtain the weight distribution of its dual code. Then, by using the MacWilliams identity in (3.36), we can compute the weight enumerator (or the weight distribution) of the linear block code.

Suppose $C$ is used for error control over a BSC with transition (or crossover) probability $p$. Recall that an undetectable error pattern is an error pattern identical to a nonzero codeword in $C$. When such an error pattern occurs, the decoder will not be able to detect the presence of transmission errors and hence will commit a decoding error.

For $0 \leq i \leq n$, the probability that $i$ transmission errors occur in specific $i$ positions of the received vector $\mathbf{r}$ during the transmission of $n$ code bits is $p^i(1-p)^{n-i}$. If an error pattern is a codeword in $C$, then it is undetectable to the decoder. There are $A_i$ such error patterns with weight $i$. Then, the probability that the decoder fails to detect the presence of transmission errors, called the *probability* (or *rate*) *of an undetected error*, denoted by $P_u(E)$, is equal to

$$P_u(E) = \sum_{i=1}^{n} A_i p^i (1-p)^{n-i} = \sum_{i=\omega_{\min}(C)}^{n} A_i p^i (1-p)^{n-i}. \tag{3.37}$$

Alternately, $P_u(E)$ can be computed based on the weight distribution of the dual code $C_d$ (see Problem 3.12):

$$P_u(E) = 2^{-(n-k)} \sum_{i=1}^{n} B_i (1-2p)^i - (1-p)^n. \tag{3.38}$$

Therefore, the weight distribution of a linear block code completely determines its probability of an undetected error. It has been proved that in the ensemble of $(n, k)$ linear block codes over GF(2), there exist codes with probability $P_u(E)$ of an undetected error, which is upper bounded by $2^{-(n-k)}$, i.e.,

$$P_u(E) \leq 2^{-(n-k)}. \tag{3.39}$$

In other words, there exist $(n,k)$ linear block codes with $P_u(E)$ which decreases *exponentially with the number of parity-check bits, $n - k$*. Even for a moderate $n-k$, these codes have a relatively small $P_u(E)$, i.e., their undetectable error rates are quite low. For example, if $n - k = 32$, $P_u(E) \approx 2.33 \times 10^{-10}$. Many linear block codes have been constructed over the years, but only a small portion of codes has been proved to have their $P_u(E)$ satisfying the upper bound $2^{-(n-k)}$. A code that satisfies the above upper bound is said to be a *good error detection code*. Reference [24] is an excellent source of error-detection codes.

**Example 3.9** Consider the $(7,4)$ systematic linear block code $C$ given in Example 3.4 whose codewords are listed in Table 3.3. We find the weight distribution of the code $C$: $A_0 = 1$, $A_1 = A_2 = 0$, $A_3 = 7$, $A_4 = 7$, $A_5 = A_6 = 0$, and $A_7 = 1$. The minimum weight of this code is 3, i.e., $\omega_{\min}(C) = 3$. Suppose we apply this code to a BSC with transition probability $p = 0.1$ for error detection. Then, the probability of undetected errors is

$$P_u(E) = 7 \cdot (0.1)^3 \cdot (0.9)^4 + 7 \cdot (0.1)^4 \cdot (0.9)^3 + (0.1)^7 = 0.005\,103,$$

which is much less than the upper bound $2^{-(n-k)} = 2^{-3} = 0.125$.  ▲▲

## 3.7  Minimum Distance of Linear Block Codes

Let $\mathbf{v}$ and $\mathbf{w}$ be two $n$-tuples over GF(2). The *Hamming distance* (or simply distance) between $\mathbf{v}$ and $\mathbf{w}$, denoted by $d(\mathbf{v},\mathbf{w})$, is defined as the number of places where $\mathbf{v}$ and $\mathbf{w}$ differ. The Hamming distance is a *metric function* that satisfies the *triangle inequality* (see Problem 3.14). Let $\mathbf{v}, \mathbf{w}$, and $\mathbf{x}$ be three $n$-tuples over GF(2). Then

$$d(\mathbf{v},\mathbf{w}) + d(\mathbf{w},\mathbf{x}) \geq d(\mathbf{v},\mathbf{x}). \tag{3.40}$$

It follows from the definitions of Hamming distance between two $n$-tuples and Hamming weight of an $n$-tuple that the Hamming distance between two vectors $\mathbf{v}$ and $\mathbf{w}$ is equal to the Hamming weight of the vector sum of $\mathbf{v}$ and $\mathbf{w}$, i.e.,

$$d(\mathbf{v},\mathbf{w}) = \omega(\mathbf{v} + \mathbf{w}). \tag{3.41}$$

The minimum distance of an $(n,k)$ linear block code $C$, denoted by $d_{\min}(C)$, is defined as the *smallest* Hamming distance between two different codewords in $C$, i.e.,

$$d_{\min}(C) = \min\{d(\mathbf{v},\mathbf{w}) : \mathbf{v},\mathbf{w} \in C, \mathbf{v} \neq \mathbf{w}\}. \tag{3.42}$$

Using the fact $d(\mathbf{v},\mathbf{w}) = \omega(\mathbf{v}+\mathbf{w})$, we can prove that the minimum distance $d_{\min}(C)$ of $C$ is equal to the minimum weight $\omega_{\min}(C)$ of $C$. Following from (3.41) and (3.42), we have

$$
\begin{aligned}
d_{\min}(C) &= \min\{d(\mathbf{v}, \mathbf{w}) : \mathbf{v}, \mathbf{w} \in C, \mathbf{v} \neq \mathbf{w}\} \\
&= \min\{\omega(\mathbf{v} + \mathbf{w}) : \mathbf{v}, \mathbf{w} \in C, \mathbf{v} \neq \mathbf{w}\} \\
&= \min\{\omega(\mathbf{x}) : \mathbf{x} \in C, \mathbf{x} \neq \mathbf{0}\} \\
&= \omega_{\min}(C).
\end{aligned}
\tag{3.43}
$$

Therefore, for a linear block code, determining its minimum distance is equivalent to determining its minimum weight. The weight structure (or weight distribution) of a linear block code $C$ can also be specified from its parity-check matrix $\mathbf{H}$ as described below.

Consider an $(n, k)$ linear block code $C$ with a parity-check matrix:

$$
\mathbf{H} = [\mathbf{h}_0 \ \mathbf{h}_1 \ \cdots \ \mathbf{h}_{n-1}],
\tag{3.44}
$$

where $\mathbf{h}_0, \mathbf{h}_1, \mathbf{h}_2, \ldots, \mathbf{h}_{n-1}$ are the $n$ columns of $\mathbf{H}$. Let $\mathbf{v} = (v_0, v_1, \ldots, v_{n-1})$ be a codeword of weight $l$ in $C$ and $v_{i_1}, v_{i_2}, \ldots, v_{i_l}, 0 \leq i_1 < i_2 < \cdots < i_l < n$, be the $l$ nonzero components of $\mathbf{v}$. Then, $v_{i_1} = v_{i_2} = \cdots = v_{i_l} = 1$. Because $\mathbf{v}$ is codeword in $C$, we have

$$
\mathbf{v} \cdot \mathbf{H}^T = (v_0, v_1, \ldots, v_{n-1}) \cdot
\begin{bmatrix}
\mathbf{h}_0^T \\
\mathbf{h}_1^T \\
\vdots \\
\mathbf{h}_{n-1}^T
\end{bmatrix}
= \mathbf{0}.
\tag{3.45}
$$

The matrix product of (3.45) results in the following equality:

$$
v_0 \mathbf{h}_0^T + v_1 \mathbf{h}_1^T + \cdots + v_{n-1} \mathbf{h}_{n-1}^T = \mathbf{0}.
\tag{3.46}
$$

Because $v_{i_1} = v_{i_2} = \ldots = v_{i_l} = 1$ and all the other components of $\mathbf{v}$ are zeros, the equality of (3.46) reduces to the following equality:

$$
\mathbf{h}_{i_1}^T + \mathbf{h}_{i_2}^T + \cdots + \mathbf{h}_{i_l}^T = \mathbf{0}.
\tag{3.47}
$$

Equality (3.47) says that for every codeword of weight $l$, there exist $l$ columns in $\mathbf{H}$ which sum to $\mathbf{0}$.

Next, we show that, for every $l$ columns in $\mathbf{H}$ which sum to $\mathbf{0}$, there exists a codeword with weight $l$ in $C$. Suppose $\mathbf{h}_{i_1}, \mathbf{h}_{i_2}, \ldots, \mathbf{h}_{i_l}$ are $l$ columns in $\mathbf{H}$ such that

$$
\mathbf{h}_{i_1}^T + \mathbf{h}_{i_2}^T + \cdots + \mathbf{h}_{i_l}^T = \mathbf{0}.
$$

We construct an $n$-tuple $\mathbf{v} = (v_0, v_1, \ldots, v_{n-1})$ whose nonzero components are $v_{i_1}, v_{i_2}, \ldots, v_{i_l}, 0 \leq i_1 < i_2 < \cdots < i_l < n$, i.e., $v_{i_1} = v_{i_2} = \cdots = v_{i_l} = 1$. Then,

$$
\begin{aligned}
\mathbf{v} \cdot \mathbf{H}^T &= v_0 \mathbf{h}_0^T + v_1 \mathbf{h}_1^T + \cdots + v_{n-1} \mathbf{h}_{n-1}^T \\
&= v_{i_1} \mathbf{h}_{i_1}^T + v_{i_2} \mathbf{h}_{i_2}^T + \cdots + v_{i_l} \mathbf{h}_{i_l}^T \\
&= \mathbf{h}_{i_1}^T + \mathbf{h}_{i_2}^T + \cdots + \mathbf{h}_{i_l}^T \\
&= \mathbf{0},
\end{aligned}
\tag{3.48}
$$

which implies that **v** is a codeword of weight $l$ in $C$. Summarizing the above results, we have the following theorem.

**Theorem 3.1** *Let $C$ be an $(n, k)$ linear block code over* GF(2) *with a parity-check matrix* **H**. *For each codeword in $C$ with weight $l$, there exist $l$ columns in* **H** *whose vector sum gives a zero vector. Conversely, if there are $l$ columns in* **H** *whose vector sum results in a zero vector, there is a codeword in $C$ with weight $l$.*

This theorem can be used to determine the weight structure of a linear block code, especially the minimum distance of the code, and can also be used for code construction. Two direct results of Theorem 3.1 are given in the following two corollaries.

**Corollary 3.1** *The minimum weight (or minimum distance) of an $(n, k)$ linear block code $C$ with a parity-check matrix* **H** *is equal to the smallest number of columns in* **H** *whose vector sum is a zero vector.*

**Corollary 3.2** *For an $(n, k)$ linear block code $C$ given by the null space of a parity-check matrix* **H**, *if there are no $d - 1$ or few columns in* **H** *that sum to a zero vector, the minimum distance (or weight) of $C$ is at least $d$.*

Corollary 3.2 gives a *lower bound* on the minimum distance (or weight) of a linear block code. In general, it is very hard to determine the exact minimum distance (or weight) of a linear block code. However, it is much easier to give a lower bound on its minimum distance (or weight).

**Example 3.10** Consider the $(7, 4)$ linear block code given in Example 3.4 whose parity-check matrix is

$$\mathbf{H} = \begin{bmatrix} 1 & 0 & 0 & 1 & 0 & 1 & 1 \\ 0 & 1 & 0 & 1 & 1 & 1 & 0 \\ 0 & 0 & 1 & 0 & 1 & 1 & 1 \end{bmatrix}.$$

Label the columns of **H** from 0 to 6. First, we notice that all seven columns of **H** are different. As a result, no two columns of **H** sum to **0**. It follows from Corollary 3.2 that the minimum distance of the code is at least 3. However, there are three columns in **H** which sum to **0**. For example, the zeroth, first, and third columns of **H** sum to **0**. It follows from Corollary 3.1 that the minimum distance of the code is 3.

Next, we consider the $(15, 11)$ linear block code given in Example 3.6. The parity-check matrix of this code is

$$\mathbf{H}_{\mathrm{sys}} = \begin{bmatrix} 1 & 0 & 0 & 0 & 1 & 0 & 0 & 1 & 1 & 0 & 1 & 0 & 1 & 1 & 1 \\ 0 & 1 & 0 & 0 & 1 & 1 & 0 & 1 & 0 & 1 & 1 & 1 & 1 & 0 & 0 \\ 0 & 0 & 1 & 0 & 0 & 1 & 1 & 0 & 1 & 0 & 1 & 1 & 1 & 1 & 0 \\ 0 & 0 & 0 & 1 & 0 & 0 & 1 & 1 & 0 & 1 & 0 & 1 & 1 & 1 & 1 \end{bmatrix}.$$

Label the columns of $\mathbf{H}_{\mathrm{sys}}$ from 0 to 14. We notice that all 15 columns of $\mathbf{H}_{\mathrm{sys}}$ are different. Hence, no two columns of $\mathbf{H}_{\mathrm{sys}}$ sum to **0**. It follows from Corollary 3.2 that the minimum distance of the code is at least 3. However, there are three columns in $\mathbf{H}_{\mathrm{sys}}$ which sum to **0**. For example, the zeroth, first, and fourth columns sum to **0**. Hence, the minimum distance of the code is 3. ▲▲

The $(7, 4)$ and $(15, 11)$ linear block codes given in Examples 3.4 and 3.6, respectively, belong to the same class of linear block codes with minimum distance 3. Codes in this class are called *Hamming codes* and they were discovered by Hamming in 1950 [23]. Hamming codes will be further discussed in Chapter 4.

The above developments for linear block codes over GF(2) apply to linear block codes over any nonbinary finite field GF($q$).

The following lemma gives an upper bound on the minimum distance of a linear block code.

**Lemma 3.1** *The minimum distance of an $(n, k)$ linear block code over* GF($q$) *is bounded by*

$$d_{\min}(C) \leq n - k + 1,$$

*which is called the* Singleton bound [25].

*Proof* Given an $(n, k)$ linear block code, there exists an $(n - k) \times n$ parity-check matrix $\mathbf{H}$, which has $n - k$ linearly independent rows (for example, the parity-check matrix $\mathbf{H}_{\mathrm{sys}}$ in systematic form). Hence, $\mathbf{H}$ has row rank $n - k$. Because $n > n - k$, the row rank of $\mathbf{H}$ is the same as its column rank, i.e., there are at most $n - k$ columns of $\mathbf{H}$ which are linearly independent. Thus, $d_{\min}(C) \leq n - k + 1$. ▲▲

**Definition 3.2** An $(n, k)$ linear block code $C$ with $d_{\min}(C) = n - k + 1$ is called a *maximum distance separable* (MDS) code.

If any $(n - k) \times (n - k)$ square submatrix of the $(n - k) \times n$ parity-check matrix of an $(n, k)$ linear block code $C$ over GF($q$) is nonsingular, then $C$ is an MSD code. A class of MDS codes will be presented in Chapter 6.

The weight distribution of a linear block code $C$ actually gives the *distance distribution* of the nonzero codewords with respect to the all-zero codeword $\mathbf{0}$. For $1 \leq i < n$, the number $A_i$ of codewords in $C$ with weight $i$ is simply equal to the number of codewords that are at a distance of $i$ from the all-zero codeword $\mathbf{0}$. Owing to the linear structure of $C$, $A_i$ also gives the number of codewords in $C$ that are at a distance of $i$ from any fixed codeword $\mathbf{v}$ in $C$. Therefore, the weight distribution of $C$ is also the distance distribution of $C$ with respect to any codeword in $C$.

The capabilities of a linear block code $C$ for detecting and correcting random errors over a BSC with hard-decision decoding are determined by its minimum distance $d_{\min}(C)$ and its error performance with soft-decision MLD is determined by its distance (or weight) distribution.

For an $(n, k)$ linear block code $C$ with minimum distance $d_{\min}(C)$, no error pattern with $d_{\min}(C) - 1$ or fewer errors can change a transmitted codeword into another codeword in $C$. Therefore, any error pattern with $d_{\min}(C) - 1$ or fewer errors will result in a received vector which is not a codeword in $C$ and hence its syndrome is not equal to zero. Therefore, all the error patterns with $d_{\min}(C) - 1$ or fewer errors are detectable by the channel decoder. However, if a codeword $\mathbf{v}$ is transmitted and an error pattern with $d_{\min}(C)$ or more errors occurs that happens to be a codeword in $C$ at a distance $d_{\min}(C)$ or greater from $\mathbf{v}$, then the received vector $\mathbf{r}$ is a codeword and its syndrome is zero. Such an error pattern is an *undetectable error pattern*. This is to say that all the error patterns with $d_{\min}(C) - 1$ or fewer errors are guaranteed to be detectable; however, detection is not guaranteed for error patterns with $d_{\min}(C)$ or more errors. For this reason, $d_{\min}(C) - 1$ is called the *error-detecting capability* of the code $C$.

The number of guaranteed detectable error patterns is equal to

$$\binom{n}{1} + \binom{n}{2} + \cdots + \binom{n}{d_{\min}(C) - 1}, \tag{3.49}$$

where $\binom{n}{i} = (n!)/(i!(n-i)!)$ is a binomial coefficient. For large $n$, the number of guaranteed detectable error patterns is only a small fraction of the $2^n - 2^k + 1$ detectable error patterns.

So far, only error detection of linear block codes has been discussed. Decoding and error-correcting capabilities of a linear block code will be discussed in the following sections.

## 3.8 Decoding of Linear Block Codes

Consider an $(n, k)$ linear block code $C$ with a parity-check matrix $\mathbf{H}$ and minimum distance $d_{\min}(C)$. Suppose a codeword in $C$ is transmitted and $\mathbf{r}$ is the received vector. For maximum-likelihood decoding (MLD) as described in Chapter 1, $\mathbf{r}$ is decoded into a codeword $\mathbf{v}$ such that the conditional probability $P(\mathbf{r}|\mathbf{v})$ is maximized. For a BSC, this is equivalent to decoding $\mathbf{r}$ into a codeword $\mathbf{v}$ such that the Hamming distance $d(\mathbf{r}, \mathbf{v})$ between $\mathbf{r}$ and $\mathbf{v}$ is minimized. This is called *minimum-distance* (or *nearest-neighbor*) *decoding*.

With minimum-distance decoding, the decoder has to compute the distance between $\mathbf{r}$ and every codeword in $C$ and then choose a codeword $\mathbf{v}$ (*not necessarily unique*) that is closest to $\mathbf{r}$ (i.e., $d(\mathbf{r}, \mathbf{v})$ is the smallest) as the decoded codeword. This decoding is called a *complete error-correction decoding* and requires a total of $2^k$ computations of distances between $\mathbf{r}$ and the $2^k$ codewords in $C$. For large $k$, implementation of the complete decoder is practically impossible. However, for many linear block codes, efficient algorithms have been developed for *incomplete error-correction decoding* to achieve good error performance with reduced decoding complexity.

No matter which codeword in $C$ is transmitted over a noisy channel, the received vector $\mathbf{r}$ is one of the $2^n$ $n$-tuples over GF(2). Let $\mathbf{v}_0 = \mathbf{0}, \mathbf{v}_1, \dots,$ $\mathbf{v}_{2^k-1}$ be the $2^k$ codewords in $C$. *Any decoding scheme used at the decoder is a rule to partition the $n$-dimensional space $\mathbf{V}_n$ of all the $n$-tuples into $2^k$ non-overlapped regions, $D(\mathbf{v}_0), D(\mathbf{v}_1), \dots, D(\mathbf{v}_{2^k-1})$, with each region containing one and only one codeword in $C$, as shown in Fig. 3.2.* Decoding is to find the region $D(\mathbf{v}_i)$ that contains the received vector $\mathbf{r}$, and then decode $\mathbf{r}$ into the codeword $\mathbf{v}_i$ which is the only codeword in that region. These regions, $D(\mathbf{v}_0)$, $D(\mathbf{v}_1), \dots, D(\mathbf{v}_{2^k-1})$, are called *decoding regions*.

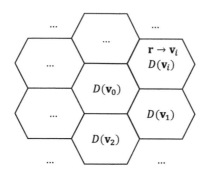

Figure 3.2 Decoding regions for linear block codes.

For MLD, the decoding regions for the $2^k$ codewords are: for $0 \leq i < 2^k$,

$$D(\mathbf{v}_i) = \{\mathbf{r} \in \mathbf{V}_n : P(\mathbf{r}|\mathbf{v}_i) \geq P(\mathbf{r}|\mathbf{v}_j), 0 \leq j < 2^k, j \neq i\}. \tag{3.50}$$

For minimum-distance decoding, the decoding regions for the $2^k$ codewords are: for $0 \leq i < 2^k$,

$$D(\mathbf{v}_i) = \{\mathbf{r} \in \mathbf{V}_n : d(\mathbf{r}, \mathbf{v}_i) \leq d(\mathbf{r}, \mathbf{v}_j), 0 \leq j < 2^k, j \neq i\}. \tag{3.51}$$

## 3.9   Standard Array for Decoding Linear Block Codes

### 3.9.1   A Standard Array Decoding

An algebraic method to partition the $2^n$ possible received vectors into $2^k$ decoding regions, called *standard array decoding*, is presented in this section.

First, we arrange the $2^k$ codewords in $C$ as the top row (or the zeroth row) of a $2^{n-k} \times 2^k$ array with the all-zero codeword $\mathbf{v}_0 = \mathbf{0}$ as the first (leftmost) entry as shown in Table 3.5. Then, we form the rest of the rows of the array *one row at a time*. Suppose we have formed the $(j-1)$th row of the array, $1 \leq j < 2^{n-k}$. To form the $j$th row of the array, we choose a vector $\mathbf{e}_j$ in $\mathbf{V}_n$ that *does not appear in the previous $j-1$ rows of the array*. Then, we form the $j$th row of the array by adding $\mathbf{e}_j$ to each codeword $\mathbf{v}_i$ in the top row of the array and placing the vector sum $\mathbf{e}_j + \mathbf{v}_i$ under $\mathbf{v}_i$. The array is completed when no vector can be chosen from $\mathbf{V}_n$. This array is called a *standard array* for the code $C$. Each row of the array is called a *coset* of $C$. The first element of each coset is called the *coset leader*. The coset leader of the zeroth row in the array is the all-zero codeword $\mathbf{e}_0 = \mathbf{v}_0 = (0, 0, \ldots, 0, 0)$.

Table 3.5 A standard array for an $(n, k)$ linear block code.

| $\mathbf{e}_0 = \mathbf{v}_0 = \mathbf{0}$ | $\mathbf{v}_1$ | $\cdots$ | $\mathbf{v}_i$ | $\cdots$ | $\mathbf{v}_{2^k-1}$ |
|---|---|---|---|---|---|
| $\mathbf{e}_1$ | $\mathbf{e}_1 + \mathbf{v}_1$ | $\cdots$ | $\mathbf{e}_1 + \mathbf{v}_i$ | $\cdots$ | $\mathbf{e}_1 + \mathbf{v}_{2^k-1}$ |
| $\mathbf{e}_2$ | $\mathbf{e}_2 + \mathbf{v}_1$ | $\cdots$ | $\mathbf{e}_2 + \mathbf{v}_i$ | $\cdots$ | $\mathbf{e}_2 + \mathbf{v}_{2^k-1}$ |
| $\vdots$ | $\vdots$ | $\cdots$ | $\vdots$ | $\cdots$ | $\vdots$ |
| $\mathbf{e}_\ell$ | $\mathbf{e}_\ell + \mathbf{v}_1$ | $\cdots$ | $\mathbf{e}_\ell + \mathbf{v}_i$ | $\cdots$ | $\mathbf{e}_\ell + \mathbf{v}_{2^k-1}$ |
| $\vdots$ | $\vdots$ | $\cdots$ | $\vdots$ | $\cdots$ | $\vdots$ |
| $\mathbf{e}_{2^{n-k}-1}$ | $\mathbf{e}_{2^{n-k}-1} + \mathbf{v}_1$ | $\cdots$ | $\mathbf{e}_{2^{n-k}-1} + \mathbf{v}_i$ | $\cdots$ | $\mathbf{e}_{2^{n-k}-1} + \mathbf{v}_{2^k-1}$ |

For $0 \leq j < 2^{n-k}$, the $j$th coset (or the $j$th row of the standard array) is given by

$$\mathbf{e}_j + C = \{\mathbf{e}_j + \mathbf{v}_i : \mathbf{v}_i \in C, 0 \leq i < 2^k\}, \tag{3.52}$$

where $\mathbf{e}_j$ is the coset leader.

**Example 3.11** Consider the $(6, 3)$ linear block code given in Example 3.1. A standard array for this code is shown in Table 3.6. It comprises eight cosets. The coset leaders consist of the all-zero 6-tuple, six 6-tuples of weight 1, and one 6-tuple with weight 2. ▲▲

Table 3.6 A standard array for the $(6, 3)$ code given in Example 3.11.

| Coset leader | | | | | | | |
|---|---|---|---|---|---|---|---|
| 000000 | 011100 | 101010 | 110001 | 110110 | 101101 | 011011 | 000111 |
| 100000 | 111100 | 001010 | 010001 | 010110 | 001101 | 111011 | 100111 |
| 010000 | 001100 | 111010 | 100001 | 100110 | 111101 | 001011 | 010111 |
| 001000 | 010100 | 100010 | 111001 | 111110 | 100101 | 010011 | 001111 |
| 000100 | 011000 | 101110 | 110101 | 110010 | 101001 | 011111 | 000011 |
| 000010 | 011110 | 101000 | 110011 | 110100 | 101111 | 011001 | 000101 |
| 000001 | 011101 | 101011 | 110000 | 110111 | 101100 | 011010 | 000110 |
| 100100 | 111000 | 001110 | 010101 | 010010 | 001001 | 111111 | 100011 |

Following from the construction of a coset, it is clear that the sum of two vectors in the same coset is a codeword in $C$. Because all the codewords in $C$ are distinct, no two vectors in the same coset are the same. In the following, several structural properties of a standard array for a linear block code are presented in the form of theorems.

**Theorem 3.2** *No two vectors in two different cosets of a standard array of a linear block code $C$ are the same.*

*Proof* Suppose two vectors $\mathbf{e}_i + \mathbf{v}_l$ and $\mathbf{e}_j + \mathbf{v}_m$ in the $i$th coset with coset leader $\mathbf{e}_i$ and the $j$th coset with coset leader $\mathbf{e}_j$ with $i < j$, respectively, of a standard array for a linear block code $C$ are the same. Then,

$$\mathbf{e}_i + \mathbf{v}_l = \mathbf{e}_j + \mathbf{v}_m,$$

which implies

$$\mathbf{e}_j = \mathbf{e}_i + \mathbf{v}_l + \mathbf{v}_m.$$

Because $\mathbf{v}_l + \mathbf{v}_m = \mathbf{v}_t$ is a codeword in $C$, $\mathbf{e}_j = \mathbf{e}_i + \mathbf{v}_t$ appears in the $i$th coset. This contradicts the construction of a standard array where every coset leader should not appear in any coset which has been already formed. Hence, no two vectors in two different cosets are the same. ▲▲

It follows from the construction of a standard array for a linear block code $C$ and Theorem 3.2 that every $n$-tuple in the vector space $\mathbf{V}_n$ of all the $n$-tuples over GF(2) appears *once and only once* in the array. Only the top row of the standard array consists of codewords in the linear block code $C$ and all the other rows contain no codeword in $C$.

**Theorem 3.3** *All the vectors in a coset of a standard array for a linear block code $C$ with parity-check matrix $\mathbf{H}$ have the same syndrome, which is equal to the syndrome of the coset leader.*

*Proof* Let $\mathbf{e}_i + \mathbf{v}_l$ be a vector in the $i$th coset of a standard array for a linear block code $C$. Then, the syndrome of $\mathbf{e}_i + \mathbf{v}_l$ is

$$(\mathbf{e}_i + \mathbf{v}_l) \cdot \mathbf{H}^T = \mathbf{e}_i \cdot \mathbf{H}^T + \mathbf{v}_l \cdot \mathbf{H}^T.$$

Because $\mathbf{v}_l$ is a codeword in $C$, we have $\mathbf{v}_l \cdot \mathbf{H}^T = \mathbf{0}$. Hence,

$$(\mathbf{e}_i + \mathbf{v}_l) \cdot \mathbf{H}^T = \mathbf{e}_i \cdot \mathbf{H}^T.$$

This says that every vector in the same coset of a standard array for the linear block code $C$ has the same syndrome which is equal to the syndrome of the coset leader. ▲▲

**Theorem 3.4** *Different coset leaders of a standard array for a linear block code $C$ have different syndromes.*

*Proof* For $i \neq j$ and $i < j$, let $\mathbf{e}_i$ and $\mathbf{e}_j$ be two different coset leaders of a standard array. Suppose the two coset leaders $\mathbf{e}_i$ and $\mathbf{e}_j$ have the same syndrome, i.e., $\mathbf{e}_i \cdot \mathbf{H}^T = \mathbf{e}_j \cdot \mathbf{H}^T$. Then,

$$(\mathbf{e}_i + \mathbf{e}_j) \cdot \mathbf{H}^T = \mathbf{0}.$$

This implies that $\mathbf{e}_i + \mathbf{e}_j$ is a codeword in $C$, say $\mathbf{e}_i + \mathbf{e}_j = \mathbf{v}_l$. Hence, $\mathbf{e}_j = \mathbf{e}_i + \mathbf{v}_l$, which indicates that $\mathbf{e}_j$ appears in the $i$th coset of a standard array for the code $C$. This contradicts to the construction of a standard array that each coset leader should be an $n$-tuple which does not appear before. Hence, two different coset leaders have different syndromes. ▲▲

Because there are $2^{n-k}$ different $(n-k)$-tuple syndromes with respect to the parity-check matrix $\mathbf{H}$ of an $(n,k)$ linear block code $C$ and there are $2^{n-k}$ cosets in a standard array for $C$, it follows from Theorems 3.3 and 3.4 that there is a *one-to-one correspondence* between a coset leader and an $(n-k)$-tuple syndrome.

A standard array for an $(n,k)$ linear block code $C$ consists of $2^k$ columns and each column contains *one and only one codeword* at the top of the column. For $0 \leq j < 2^k$, the $j$th column of the standard array consists of the following set of $2^{n-k}$ $n$-tuples:

$$D(\mathbf{v}_j) = \{\mathbf{e}_0 + \mathbf{v}_j, \mathbf{e}_1 + \mathbf{v}_j, \mathbf{e}_2 + \mathbf{v}_j, \ldots, \mathbf{e}_{2^{n-k}-1} + \mathbf{v}_j\}, \qquad (3.53)$$

in which each element is the vector sum of the $j$th codeword $\mathbf{v}_j$ and a coset leader $\mathbf{e}_i$ with $0 \leq i < 2^{n-k}$. We see that the $j$th codeword $\mathbf{v}_j$ is the only codeword in $D(\mathbf{v}_j)$. (Note that $\mathbf{e}_0 = (0, 0, \ldots, 0)$ is the all-zero $n$-tuple.)

The $2^k$ columns of a standard array for $C$ form a partition of the vector space $\mathbf{V}_n$ of all the $n$-tuples over GF(2). These $2^k$ columns of the standard array can be used as decoding regions for decoding $C$. If the received vector $\mathbf{r}$ is found in the $j$th column $D(\mathbf{v}_j)$, we decode $\mathbf{r}$ into $\mathbf{v}_j$. From the structure of a standard array for $C$, we can easily check that if the $j$th codeword $\mathbf{v}_j$ is transmitted and the error pattern caused by the channel noise is a coset leader $\mathbf{e}_i$, then the received vector $\mathbf{r} = \mathbf{v}_j + \mathbf{e}_i$ is in the column $D(\mathbf{v}_j)$ which contains $\mathbf{v}_j$. However, if $\mathbf{v}_j$ is transmitted but the error pattern is not a coset leader, then the received vector $\mathbf{r}$ is not in the column $D(\mathbf{v}_j)$ (see Problem 3.16). Therefore, by using the columns of a standard array of an $(n,k)$ linear block code $C$ as decoding regions, decoding is correct (i.e., $\mathbf{r}$ is decoded into the transmitted codeword) *if and only if* the error pattern caused by the channel noise is identical to a coset leader. This is to say that the $2^{n-k} - 1$ nonzero coset leaders of a standard array are all the correctable error patterns (i.e., they result in a correct decoding). To minimize the probability of a decoding error, the error patterns that are *most likely to occur* for a given channel should be chosen as the coset leaders.

For a BSC, an error pattern of smaller weight (corresponding to smaller number of errors) is *more probable* than an error pattern with larger weight (corresponding to larger number of errors). Hence, when we form a standard array for a linear block code $C$, each coset leader should be chosen to be an $n$-tuple of the least weight from the remaining $n$-tuples in $\mathbf{V}_n$. Hence, decoding based on this standard array for $C$ is the *minimum-distance decoding* (or MLD for a BSC). A standard array formed in this way is called an *optimal standard array*.

Let $\ell_i$ denote the number of coset leaders of weight $i$. Then, $\ell_0, \ell_1, \ldots, \ell_n$ are the weight distribution of the coset leaders in the optimal standard array. With minimum-distance decoding, a decoding error happens if and only if the error pattern is not a coset leader. Thus, the decoding error probability over a BSC with transition probability $p$ is given by

$$P(E) = 1 - \sum_{i=0}^{n} \ell_i p^i (1-p)^{n-i}.$$

The standard array of the $(6, 3)$ linear block code given in Example 3.1 as shown in Table 3.6 is an optimal standard array for the code. The weight distribution of the coset leaders is $\ell_0 = 1$, $\ell_1 = 6$, $\ell_2 = 1$, $\ell_3 = \ell_4 = \ell_5 = \ell_6 = 0$. Then, the decoding error probability with the optimal standard array over BSC is $P(E) = 1 - (1-p)^6 - 6 \cdot p(1-p)^5 - p^2(1-p)^4$. If the transition probability of the BSC is $p = 10^{-2}$, $P(E) \approx 1.37 \times 10^{-3}$.

The minimum distance $d_{\min}(C)$ of a linear block code $C$ is either odd or even. Let

$$t = \lfloor (d_{\min}(C) - 1)/2 \rfloor, \tag{3.54}$$

where $\lfloor x \rfloor$ denotes the integer part of $x$.[3] Then,

$$2t + 1 \le d_{\min}(C) \le 2t + 2. \tag{3.55}$$

In the following, we show that all the $n$-tuples over GF(2) of weight $t$ or less can be used as coset leaders in an optimal standard array for $C$. Hence, these coset leaders form correctable error patterns. We also show that there are $n$-tuples of weight $t + 1$ or larger that cannot be used as coset leaders. During transmission, if such an $n$-tuple occurs as an error pattern, the received vector $\mathbf{r}$ will be decoded incorrectly. Therefore, for a linear block code $C$ with minimum distance $d_{\min}(C)$, any error pattern with $t = \lfloor (d_{\min}(C) - 1)/2 \rfloor$ or fewer errors is guaranteed to be correctable (i.e., resulting in a correct decoding), but not all the error patterns with $t + 1$ or more errors.

**Theorem 3.5** *No two vectors with weight $t$ or less can be in the same coset.*

*Proof* Let $\mathbf{a}_1$ and $\mathbf{a}_2$ be two vectors for which $w(\mathbf{a}_1) \le t$ and $w(\mathbf{a}_2) \le t$. Suppose they are in the same coset. Then, $\mathbf{a}_1 + \mathbf{a}_2$ must be a codeword $\mathbf{v}$, i.e., $\mathbf{a}_1 + \mathbf{a}_2 = \mathbf{v}$. Then,

$$w(\mathbf{a}_1 + \mathbf{a}_2) = w(\mathbf{v}) \le w(\mathbf{a}_1) + w(\mathbf{a}_2) \le 2t < d_{\min}(C).$$

This contradicts the fact that the minimum weight of $C$ is $d_{\min}(C)$. Therefore, $\mathbf{a}_1$ and $\mathbf{a}_2$ cannot be in the same coset.                     ▲▲

A direct result of Theorem 3.5 is the following corollary.

**Corollary 3.3** *All the vectors of weight $t$ or less are in different cosets and hence can be used as coset leaders.*

**Theorem 3.6** *For a linear block code with $2t + 1 \le d_{\min}(C) \le 2t + 2$, there is at least one $n$-tuple with weight $t + 1$ which cannot be used as a coset leader.*

---

[3]The function $f(x) = \lfloor x \rfloor$ is called the *floor function*, which evaluates the largest integer less than or equal to $x$. There is another function as a counterpart of the floor function, $g(x) = \lceil x \rceil$, called the *ceiling function*, which evaluates the smallest integer greater than or equal to $x$.

*Proof* First, we consider the case for which $d_{\min}(C) = 2t+1$. Let $\mathbf{v}$ be a codeword with $\omega(\mathbf{v}) = 2t + 1$. Let $\mathbf{a}_1$ and $\mathbf{a}_2$ be two vectors such that: (1) $\mathbf{a}_1$ and $\mathbf{a}_2$ do not have any common nonzero components; (2) $\omega(\mathbf{a}_1) = t$ and $\omega(\mathbf{a}_2) = t + 1$; and (3) $\mathbf{a}_1 + \mathbf{a}_2 = \mathbf{v}$. Suppose $\mathbf{a}_1$ is used as a coset leader, then $\mathbf{a}_2 = \mathbf{a}_1 + \mathbf{v}$ is in the coset with $\mathbf{a}_1$ as the coset leader. Therefore, $\mathbf{a}_2$ cannot be used as a coset leader.

Now, we consider the case for which $d_{\min}(C) = 2t + 2$. Let $\mathbf{v}$ be a codeword with weight $2t+2$. We split $\mathbf{v}$ into two $n$-tuples $\mathbf{a}_1$ and $\mathbf{a}_2$ such that: (1) $\omega(\mathbf{a}_1) = \omega(\mathbf{a}_2) = t+1$; (2) $\mathbf{a}_1$ and $\mathbf{a}_2$ do not have any common nonzero components; and (3) $\mathbf{a}_1 + \mathbf{a}_2 = \mathbf{v}$. Then, $\mathbf{a}_2 = \mathbf{a}_1 + \mathbf{v}$. In this case, if $\mathbf{a}_1$ is used as a coset leader, then $\mathbf{a}_2$ is in the coset with $\mathbf{a}_1$ as the coset leader. Hence, $\mathbf{a}_2$ cannot be used as a coset leader. This proves the theorem. ▲▲

Theorem 3.6 can be put in a more general form. That is, for $l > t$, there is at least one $n$-tuple of weight $l$ which cannot be used as a coset leader.

It follows from Theorems 3.5 and 3.6 and Corollary 3.3 that all the $n$-tuples of weight $t = \lfloor (d_{\min}(C) - 1)/2 \rfloor$ or less can be used as coset leaders, but not all the $n$-tuples with weight $t + 1$ or larger. Consequently, if we use all the $n$-tuples of weight $t$ or less as coset leaders, they form correctable error patterns. These error patterns are *guaranteed correctable error patterns*. However, there are error patterns with $t + 1$ or more errors which are not correctable.

The parameter $t = \lfloor (d_{\min}(C) - 1)/2 \rfloor$ is referred to as the *error-correcting capability* of the code $C$. We say that $C$ is capable of correcting $t$ or fewer random errors and is called a *t-error-correcting code*. There are

$$N_t = \binom{n}{0} + \binom{n}{1} + \cdots + \binom{n}{t} \leq 2^{n-k} \qquad (3.56)$$

guaranteed correctable error patterns which form a small fraction of $2^{n-k}$ correctable error patterns for large $n - k$, except for a very few cases. The upper bound given by (3.56) is called the *Hamming bound* [23]. An $(n, k)$ linear block code is called a *perfect code* for error correction if

$$N_t = \binom{n}{0} + \binom{n}{1} + \cdots + \binom{n}{t} = 2^{n-k}. \qquad (3.57)$$

The $(7, 4)$ and $(15, 11)$ linear block codes[4] given in Examples 3.4 and 3.6, respectively, are perfect codes which can correct all error patterns, each containing a single error, but no others. Another well-known perfect code is a $(23, 12)$ linear block code, which has minimum distance 7 and is capable of correcting all error patterns with three or fewer errors but no others. The sum $\binom{23}{0} + \binom{23}{1} + \binom{23}{2} + \binom{23}{3}$ is equal to $2^{11}$. This code is known as *Golay code*

---

[4]It is noted that these two linear block codes are Hamming code which will be presented in Section 4.10. Hamming codes form a class of perfect codes. Repetition codes of odd lengths also form a class of perfect codes. Readers are referred to [3, 26] for more on perfect codes and their existence.

discovered by Golay in 1949 [27]. (The Golay code will be further discussed in Chapter 4.) A parity-check matrix of the $(23, 12)$ Golay code is given below:

$$\mathbf{H} = \left[ \; \mathbf{I}_{11} \; \left| \begin{array}{ccccccccccc} 1 & 0 & 0 & 1 & 1 & 1 & 0 & 0 & 0 & 1 & 1 & 1 \\ 1 & 0 & 1 & 0 & 1 & 1 & 0 & 1 & 1 & 0 & 0 & 1 \\ 1 & 0 & 1 & 1 & 0 & 1 & 1 & 0 & 1 & 0 & 1 & 0 \\ 1 & 0 & 1 & 1 & 1 & 0 & 1 & 1 & 0 & 1 & 0 & 0 \\ 1 & 1 & 0 & 0 & 1 & 1 & 1 & 0 & 1 & 1 & 0 & 0 \\ 1 & 1 & 0 & 1 & 0 & 1 & 1 & 1 & 0 & 0 & 0 & 1 \\ 1 & 1 & 0 & 1 & 1 & 0 & 0 & 1 & 1 & 0 & 1 & 0 \\ 1 & 1 & 1 & 0 & 0 & 1 & 0 & 1 & 0 & 1 & 1 & 0 \\ 1 & 1 & 1 & 0 & 1 & 0 & 1 & 0 & 0 & 0 & 1 & 1 \\ 1 & 1 & 1 & 1 & 0 & 0 & 0 & 0 & 1 & 1 & 0 & 1 \\ 0 & 1 & 1 & 1 & 1 & 1 & 1 & 1 & 1 & 1 & 1 & 1 \end{array} \right. \right] .$$

With all the above developments in this section, it is clear that if a linear block code $C$ is capable of correcting all the error patterns of $t$ or fewer errors, the minimum distance of the code is at least $2t + 1$, i.e., $d_{\min}(C) \geq 2t + 1$.

In practice, a code is often used for correcting $\lambda$ or fewer random errors and simultaneously detecting $\ell$ ($\ell > \lambda$) or fewer errors. That is, when $\lambda$ or fewer errors occur, the code is able to correct them; when more than $\lambda$ but fewer than $\ell + 1$ errors occur, the code is capable of detecting their existence without making a decoding error. For those purposes, the minimum distance of the code should be at least $\lambda + \ell + 1$, i.e., $d_{\min}(C) \geq \lambda + \ell + 1$ (see Problem 3.18).

### 3.9.2 Syndrome Decoding

Decoding of an $(n, k)$ linear block code $C$ based on an optimal standard array for the code requires a memory to store $2^n$ $n$-tuples, i.e., the size of the standard array. For large $n$, the size of the memory will be prohibitively large and the implementation of a standard array-based decoding becomes practically impossible. However, decoding can be significantly simplified by using the following facts: (1) the coset leaders form all the correctable error patterns; and (2) there is a one-to-one correspondence between an $(n - k)$-tuple syndrome and a coset leader. Based on these two facts, we form a table with only two columns which consists of the $2^{n-k}$ coset leaders (correctable error patterns) in one column and their corresponding syndromes in another column as shown in Table 3.7.

Table 3.7 A look-up table for syndrome decoding.

| Syndromes | Correctable error patterns |
|:---:|:---:|
| $\mathbf{0}$ | $\mathbf{e}_0 = \mathbf{0}$ |
| $\mathbf{s}_1$ | $\mathbf{e}_1$ |
| $\mathbf{s}_2$ | $\mathbf{e}_2$ |
| $\cdots$ | $\cdots$ |
| $\mathbf{s}_{2^{n-k}-1}$ | $\mathbf{e}_{2^{n-k}-1}$ |

With this table, decoding of a received vector $\mathbf{r}$ is carried out in the following three steps.

(1) Compute the syndrome of $\mathbf{r}$, $\mathbf{s} = \mathbf{r} \cdot \mathbf{H}^T$.

(2) Find the coset leader $\mathbf{e}$ in the table whose syndrome is equal to $\mathbf{s}$. Then, $\mathbf{e}$ is assumed to be the error pattern caused by the channel noise.

(3) Decode $\mathbf{r}$ into a codeword $\mathbf{v} = \mathbf{r} + \mathbf{e}$.

The above decoding is called *syndrome decoding* or *table look-up decoding*. With this decoding, the decoder complexity is drastically reduced compared to the standard array-based decoding.

**Example 3.12** Consider the standard array for the $(6, 3)$ linear block code given in Example 3.1 with parity-check matrix

$$\mathbf{H} = \begin{bmatrix} 1 & 0 & 0 & 0 & 1 & 1 \\ 0 & 1 & 0 & 1 & 0 & 1 \\ 0 & 0 & 1 & 1 & 1 & 0 \end{bmatrix}.$$

A standard array for this code is shown in Table 3.6. Using the coset leaders of the standard array and their corresponding syndromes, the syndrome look-up decoding table for the code is shown in Table 3.8.

Table 3.8 A syndrome look-up decoding table for the $(6, 3)$ linear block code.

| Syndromes $(s_0, s_1, s_2)$ | Correctable error patterns $(e_0, e_1, e_2, e_3, e_4, e_5)$ | Syndromes $(s_0, s_1, s_2)$ | Correctable error patterns $(e_0, e_1, e_2, e_3, e_4, e_5)$ |
|---|---|---|---|
| (0 0 0) | (0 0 0 0 0 0) | (0 1 1) | (0 0 0 1 0 0) |
| (1 0 0) | (1 0 0 0 0 0) | (1 0 1) | (0 0 0 0 1 0) |
| (0 1 0) | (0 1 0 0 0 0) | (1 1 0) | (0 0 0 0 0 1) |
| (0 0 1) | (0 0 1 0 0 0) | (1 1 1) | (1 0 0 1 0 0) |

Suppose a codeword $\mathbf{v} = (1\,0\,1\,0\,1\,0)$ is transmitted and $\mathbf{r} = (1\,0\,1\,1\,1\,0)$ is received. The true error vector is $\mathbf{e} = \mathbf{r} + \mathbf{v} = (0\,0\,0\,1\,0\,0)$. Using Table 3.8 for decoding, we first compute the syndrome of $\mathbf{r}$:

$$\mathbf{s} = (s_0, s_1, s_2) = \mathbf{r} \cdot \mathbf{H}^T = (1\,0\,1\,1\,1\,0) \cdot \begin{bmatrix} 0 & 1 & 1 \\ 1 & 0 & 1 \\ 1 & 1 & 0 \\ 1 & 0 & 0 \\ 0 & 1 & 0 \\ 0 & 0 & 1 \end{bmatrix} = (0\,1\,1).$$

From Table 3.8, we find that $(0\,1\,1)$ is the syndrome of the error pattern $\mathbf{e} = (0\,0\,0\,1\,0\,0)$, the fourth coset leader of Table 3.6. Adding the error pattern $\mathbf{e} = (0\,0\,0\,1\,0\,0)$ to the receive vector $\mathbf{r} = (1\,0\,1\,1\,1\,0)$, we obtain the decoded vector $\mathbf{r} + \mathbf{e} = (1\,0\,1\,1\,1\,0) + (0\,0\,0\,1\,0\,0) = (1\,0\,1\,0\,1\,0)$ which is the transmitted codeword $\mathbf{v} = (1\,0\,1\,0\,1\,0)$. Hence, the decoding is correct.

Suppose the same codeword $\mathbf{v} = (1\,0\,1\,0\,1\,0)$ is transmitted and the received vector is $\mathbf{r} = (1\,1\,1\,1\,1\,0)$. The true error vector is $\mathbf{e} = \mathbf{r} + \mathbf{v} = (0\,1\,0\,1\,0\,0)$ which contains two errors. Using Table 3.8 for decoding, we compute the syndrome of $\mathbf{r}$ which is (0 0 1). The error pattern in the table that has this syndrome is (0 0 1 0 0 0). Adding this error pattern to the received vector, we obtain the decoded vector $\mathbf{v}^* = (1\ 1\ 1\ 1\ 1\ 0) + (0\ 0\ 1\ 0\ 0\ 0) = (1\ 1\ 0\ 1\ 1\ 0)$ which is a codeword in the $(6,3)$ linear block code. However, it is not the transmitted codeword $\mathbf{v} = (1\ 0\ 1\ 0\ 1\ 0)$. Hence, decoding the received vector $\mathbf{r} = (1\ 1\ 1\ 1\ 1\ 0)$ into the codeword $\mathbf{v}^* = (1\ 1\ 0\ 1\ 1\ 0)$ commits a decoding error. This is because the error pattern (0 1 0 1 0 0) caused by the channel noise is not a coset leader in the standard array of the $(6,3)$ code given in Table 3.6. Hence, it is not correctable. Because all error patterns of weight one are coset leaders in the standard array shown by Table 3.6, this code can correct a single error at any location of a received vector. The code also can correct one error pattern with two errors, i.e., $\mathbf{e} = (1\ 0\ 0\ 1\ 0\ 0)$, but no other error patterns. Thus, it is a single-error-correcting code. ▲▲

Despite the significant decoding reduction compared with the standard array decoding, for a long code with large $n - k$, a complete syndrome table look-up decoder is still very complex, requiring a very large memory to store the table. If we limit this table to only correct the error patterns guaranteed to be correctable by the error-correcting capability $t = \lfloor (d_{\min}(C) - 1)/2 \rfloor$ of the code, then the size of the look-up table can be further reduced. The new table comprises only $N_t = \binom{n}{0} + \binom{n}{1} + \cdots + \binom{n}{t}$ correctable error patterns guaranteed by the minimum distance $d_{\min}(C)$ of the code.

Based on this new table, decoding of a received vector $\mathbf{r}$ consists of the following four steps.
(1) Compute the syndrome $\mathbf{s}$ of $\mathbf{r}$, $\mathbf{s} = \mathbf{r} \cdot \mathbf{H}^T$.
(2) If the syndrome $\mathbf{s}$ corresponds to a correctable error pattern $\mathbf{e}$ listed in the table, go to step (3); otherwise, go to step (4).
(3) Decode $\mathbf{r}$ into the codeword $\mathbf{v} = \mathbf{r} + \mathbf{e}$.
(4) Declare a *decoding failure*. In this case, the presence of errors is detected but the decoder fails to correct the errors.

With the above partial table look-up decoding, the number of errors to be corrected is bounded by the error-correcting capability of the code. Thus, it is called *bounded distance decoding*.

Many classes of linear block codes with good error-correcting capabilities have been constructed over the years. Efficient algorithms to carry out the bounded distance decoding of these classes of codes have been devised.

## 3.10   Shortened and Extended Codes

In system (communication or storage) design, if a code of suitable natural length or suitable number of information bits cannot be found, it may be necessary to shorten or extend an existing code to meet the design requirements.

Let $C$ be an $(n, k)$ linear block code over GF(2) in systematic form (see Fig. 3.1) with rate $k/n$ and minimum distance $d_{\min}(C)$. Let $\lambda$ be a nonnegative integer less than $k$, i.e., $0 \leq \lambda < k$. Consider the set of codewords in $C$ for which the first $\lambda$ (the rightmost) information bits are zero bits. There are $2^{k-\lambda}$ such codewords in $C$ and they form a *linear subcode* $C_0$ of $C$ with dimension $k - \lambda$. It is clear that the minimum distance of $C_0$ is at least $d_{\min}(C)$. If we delete the first $\lambda$ zero information bits from each codeword in the subcode $C_0$ of $C$, we obtain an $(n - \lambda, k - \lambda)$ linear code, denoted by $C_s(\lambda)$, with rate $(k - \lambda)/(n - \lambda)$ and minimum distance at least $d_{\min}(C)$, where the subscript "$s$" in $C_s(\lambda)$ stands for "shortened." The code $C_s(\lambda)$ is called a *shortened code* of $C$ and $C$ is called the *mother code*. This type of shortening reduces both the length and dimension of $C$ by $\lambda$. If $n$, $k$, and $\lambda$ are chosen appropriately, a shortened code may be constructed to meet the system requirements in length, rate, and error-correcting capability.

Shortening of a chosen linear block code can also be obtained by deleting a fixed number of parity-check bits. Consider the generator matrix $\mathbf{G}_{\text{sys}}$ of an $(n, k)$ linear systematic block code $C$ given by (3.15). Let $\gamma$ be a nonnegative integer such that $0 \leq \gamma < n-k$. If we delete $\gamma$ columns from the parity submatrix $\mathbf{P}$ of $\mathbf{G}_{\text{sys}}$, we obtain a $k \times (n - \gamma)$ matrix $\mathbf{G}_{\text{sys}}(\gamma)$ which is also in systematic form. The code generated by $\mathbf{G}_{\text{sys}}(\gamma)$ is an $(n - \gamma, k)$ linear systematic block code $C_s(\gamma)$ with rate $k/(n - \gamma)$. The minimum distance of $C_s(\gamma)$, in general, is smaller than the minimum distance $d_{\min}(C)$ of $C$. The above type of shortening is commonly referred to as *puncturing*; this reduces both the length and the minimum distance of the chosen mother code but increases the code rate.

Shortening a linear block code $C$ can also be achieved by deleting some chosen parity-check bits and some chosen information bits.

Next, we consider a simple way to extend a chosen linear block code. Suppose the chosen $(n, k)$ linear block code $C$ consists of codewords with both even and odd weights and has an *odd* minimum distance $d_{\min}(C) = 2t + 1$. Then, it can be proved that half of the codewords in $C$ have odd weights and the other half have even weights (see Problem 3.21). For each codeword $\mathbf{v} = (v_0, v_1, \ldots, v_{n-1})$ in $C$, we add an extra bit, denoted by $v_{-1}$, to its left. This results in an extended codeword $\mathbf{v}_{\text{ext}} = (v_{-1}, v_0, v_1, \ldots, v_{n-1})$. The added bit $v_{-1}$ is the modulo-2 sum of the $n$ code bits in $\mathbf{v}$, i.e., $v_{-1} = v_0 + v_1 + \cdots + v_{n-1}$. It is clear that $v_{-1} = 1$ if the weight of $\mathbf{v}$ is odd and $v_{-1} = 0$ if the weight of $\mathbf{v}$ is even. Hence, the weight of the extended codeword $\mathbf{v}_{\text{ext}}$ is even. The bit $v_{-1}$ is called an *overall parity-check bit*. Adding an overall parity-check bit to each codeword $\mathbf{v}$ in $C$, we obtain an $(n+1, k)$ linear block code $C_{\text{ext}}$ with rate $k/(n+1)$ and minimum distance $d_{\min}(C_{\text{ext}}) = d_{\min}(C) + 1 = 2t + 2$, where the subscript "ext" in $C_{\text{ext}}$ represents "extended." The code $C_{\text{ext}}$ is called an *extended code* of $C$ and all the codewords in $C_{\text{ext}}$ have even weights. Besides being capable of correcting all the error patterns with $t$ or fewer errors, the extended code $C_{\text{ext}}$ is capable of detecting all the error patterns with $2t + 1$ or fewer errors and all the error patterns with *odd number* of errors greater than $2t + 1$. This simple extension of a linear block code is quite frequently used in system design.

**Example 3.13** Consider the $(15, 11)$ Hamming code $C$ with minimum distance 3 given in Example 3.6. Adding an extra overall parity-check bit to each codeword in $C$, we obtain a $(16, 11)$ extended Hamming code with minimum distance 4. Besides being capable of correcting all the single error patterns, the extended code is capable of detecting all the error patterns with 3 or fewer errors and all the error patterns with odd number of errors greater than 3. ▲▲

A special case of the above extension is that $C$ is the entire vector space $\mathbf{V}_n$ of all the $n$-tuples over GF(2). Then, $C$ is an $(n, n)$ code with minimum distance 1. In this case, the extended code $C_{\text{ext}}$ of $C$ obtained by adding a single overall parity-check bit to each codeword in $C$ is an $(n+1, n)$ linear code with minimum distance 2. Because each codeword in $C_{\text{ext}}$ has a single parity-check bit, $C_{\text{ext}}$ is called an $(n + 1, n)$ *single parity-check* (SPC) code.

Other extensions of an $(n, k)$ linear block code $C$ by adding more than one parity-check bit will be discussed in Chapter 9.

## 3.11   Nonbinary Linear Block Codes

So far, only linear block codes with symbols from the binary field GF(2) have been presented. Linear block codes with symbols from nonbinary fields can be constructed in exactly the same manner as binary linear block codes. A $q$-ary $(n, k)$ linear block code of length $n$ with $q^k$ codewords is simply a $k$-dimensional subspace of the vector space of all the $q^n$ $n$-tuples over a nonbinary field GF($q$) with $q$ elements, where $q$ is a power of a prime.

All the fundamental concepts and structural properties developed for binary linear block codes apply to $q$-ary linear block codes. We simply replace the binary field GF(2) with a nonbinary field GF($q$). A $q$-ary $(n, k)$ linear block code is specified by either a $k \times n$ generator matrix $\mathbf{G}$ over GF($q$) or an $(n - k) \times n$ parity-check matrix $\mathbf{H}$ over GF($q$). Generator and parity-check matrices of $q$-ary $(n, k)$ linear block codes in systematic form are exactly the same as those given by (3.15) and (3.22) for binary codes, except that the entries of the matrices are from GF($q$). A message for a $q$-ary $(n, k)$ linear block code consists of $k$ information symbols from GF($q$). Encoding and decoding of $q$-ary linear block codes are the same as those for binary linear block codes, except that computation operations are performed over GF($q$).

In practical applications, $q$, in general, is a power of 2, say $q = 2^m$. In this case, GF($2^m$) is an extension field of the binary field GF(2) as presented in Chapter 2. A class of $2^m$-ary linear block codes which have been widely used in communication and storage systems will be presented in Chapter 6. These codes are known as *Reed–Solomon* (RS) codes, which were discovered by Reed and Solomon in 1960 [28]. RS codes are MDS codes.

## Problems

**3.1** Prove (3.8): $\mathbf{G} \cdot \mathbf{H}^T = \mathbf{O}$.

**3.2** Find the dual code $C_d$ of the $(7, 4)$ linear block code constructed in Example 3.4 (i.e., find all the codewords in $C_d$).

**3.3** Consider a systematic $(8, 4)$ linear block code with the four parity-check bits specified by

$$
\begin{aligned}
v_0 &= u_1 + u_2 + u_3, \\
v_1 &= u_0 + u_1 + u_2, \\
v_2 &= u_0 + u_1 + u_3, \\
v_3 &= u_0 + u_2 + u_3.
\end{aligned}
$$

Find its generator and parity-check matrices.

**3.4** Consider the following generator and parity-check matrices of a $(7, 4)$ linear block code:

$$
\mathbf{G} = \begin{bmatrix} 0\,1\,1\,1\,0\,0\,1 \\ 0\,1\,1\,0\,1\,0\,0 \\ 1\,0\,0\,0\,1\,1\,0 \\ 1\,0\,1\,0\,0\,0\,1 \end{bmatrix}, \quad
\mathbf{H} = \begin{bmatrix} 1\,0\,0\,1\,0\,1\,1 \\ 1\,1\,0\,0\,1\,0\,1 \\ 0\,0\,1\,0\,1\,1\,1 \end{bmatrix}.
$$

Put the above two matrices in systematic form.

**3.5** Prove $\mathbf{G}_{\mathrm{sys}} \cdot \mathbf{H}_{\mathrm{sys}}^T = \mathbf{O}$, where $\mathbf{G}_{\mathrm{sys}}$ is the systematic generator matrix given by (3.15) and $\mathbf{H}_{\mathrm{sys}}$ is the systematic parity-check matrix given by (3.22) of a linear block code.

**3.6** Consider an $(n, n-1)$ single parity-check code $C_{\mathrm{SPC}}$. Let $\mathbf{u} = (u_0, u_1, \ldots, u_{n-2})$ be an $(n-1)$-tuple message and $\mathbf{v} = (v_0, v_1, \ldots, v_{n-1})$ be the corresponding codeword. Based on the definition of single parity-check codes, we have

$$
\begin{aligned}
v_0 &= u_0 + u_1 + \cdots + u_{n-3} + u_{n-2}, \\
v_1 &= u_0, \\
v_2 &= u_1, \\
&\;\;\vdots \\
v_{n-2} &= u_{n-3}, \\
v_{n-1} &= u_{n-2}.
\end{aligned}
$$

Find the systematic generator and parity-check matrices of the code $C_{\mathrm{SPC}}$.

**3.7** Prove that there are $2^k - 1$ undetectable error patterns for an $(n, k)$ binary linear block code.

**3.8** Consider the $(7, 4)$ systematic linear block code $C$ with parity-check matrix given by (3.14) in Example 3.4. Suppose $\mathbf{r}_1 = (1\ 0\ 1\ 0\ 1\ 1\ 1)$ and $\mathbf{r}_2 = (0\ 0\ 1\ 0\ 1\ 1\ 1)$ are two received vectors. Calculate their corresponding syndromes and show whether there is any error in the received vectors.

**3.9** Prove that the equations given by (3.32) have $2^k$ possible solutions.

**3.10** Consider the $(7, 4)$ linear block code given in Example 3.8. Suppose the codeword $\mathbf{v} = (1\ 0\ 0\ 1\ 0\ 1\ 1)$ is transmitted and $\mathbf{r} = (0\ 0\ 0\ 1\ 0\ 1\ 1)$ is received. Find all possible error patterns and the most probable error pattern, and check whether the decoding is successful.

**3.11** In Problem 3.10, suppose the codeword $\mathbf{v} = (1\,0\,0\,1\,0\,1\,1)$ is transmitted and $\mathbf{r} = (0\,1\,0\,1\,0\,1\,1)$ is received. Find all possible error patterns and the most probable error pattern, and check whether the decoding is successful.

**3.12** Show the derivation of (3.38).

**3.13** Consider the $(6,3)$ linear block code given by Table 3.1 in Example 3.1. Find its weight distribution and calculate the probability of an undetectable error over a BSC with transition probability $p = 0.1$.

**3.14** Prove that the Hamming distance satisfies the triangle inequality shown by (3.40).

**3.15** Find a standard array for the $(7,4)$ linear block code given in Example 3.8.

**3.16** Prove that if a codeword $\mathbf{v}_j$ is transmitted but the error pattern is not a coset leader, then the received vector $\mathbf{r}$ is not in the column $D(\mathbf{v}_j)$ of a standard array.

**3.17** Construct the syndrome look-up decoding table for the $(7,4)$ linear block code given in Example 3.8.

**3.18** Prove that for a code $C$ to correct $\lambda$ or fewer random errors and simultaneously detect $\ell$ ($\ell > \lambda$) or fewer errors, the minimum distance of the code should be at least $\lambda + \ell + 1$, i.e., $d_{\min}(C) \geq \lambda + \ell + 1$.

**3.19** Consider the $(7,4)$ linear block code given in Example 3.8. Suppose $\mathbf{v} = (1\,0\,0\,1\,0\,1\,1)$ is the transmitted codeword and $\mathbf{r}_0 = (1\,0\,0\,1\,0\,1\,0)$ is a received vector. Use the table constructed in Problem 3.17 to decode $\mathbf{r}_0$. If $\mathbf{r}_1 = (1\,0\,0\,1\,0\,0\,0)$ is received, calculate the decoded vector based on the syndrome look-up decoding table and analyze the decoding result.

**3.20** Let $\mathbf{H}$ be the parity-check matrix for an $(n,k)$ linear block code $C$ with odd minimum distance $d$. Construct a new code $C^*$ whose parity-check matrix is

$$\mathbf{H}^* = \left[\begin{array}{c|cccccc} 0 & & & & & & \\ 0 & & & & & & \\ \vdots & & & \mathbf{H} & & & \\ 0 & & & & & & \\ \hline 1 & 1 & 1 & 1 & 1 & 1 \end{array}\right].$$

(a) Prove that $C^*$ is an $(n+1, k)$ linear block code.
(b) Prove that every codeword in $C^*$ has even weight.
(c) Prove that the minimum distance of $C^*$ is $d + 1$.

**3.21** Let $C$ be an $(n,k)$ linear block code which consists of codewords with both odd and even weights. Prove that half of the codewords in $C$ have odd weights and the other half have even weights.

**3.22** Consider an $(n, k)$ linear code $C$ over GF(2) whose generator matrix has no column with all 0s. Arrange the $2^k$ codewords in $C$ as rows of a $2^k \times n$ array $\mathbf{A}$.

(a) Prove that there is no column in the array $\mathbf{A}$ with all 0s.

(b) Prove that each column of $\mathbf{A}$ has $2^{k-1}$ 1s and $2^{k-1}$ 0s.

(c) Prove that the set of all codewords with zeros in a particular component position forms a subspace of $C$. Compute the dimension of such a subspace.

(d) Prove that the minimum distance of the code $C$ is at most $n2^{k-1}/(2^k - 1)$.

# References

[1] W. W. Peterson and E. J. Weldon, Jr., *Error-Correcting Codes*, 2nd ed., Cambridge, MA, The MIT Press, 1972.

[2] I. F. Blake and R.C. Mullin, *The Mathematical Theory of Coding*, New York, Academic Press, 1975.

[3] F. J. MacWilliams and N. J. A. Sloane, *The Theory of Error-Correcting Codes*, Amsterdam, North-Holland, 1977.

[4] R. J. McEliece, *The Theory of Information Theory and Coding*, Cambridge, Cambridge University Press, 2002.

[5] G. Clark and J. Cain, *Error-Correcting Codes for Digital Communications*, New York, Plenum Press, 1981.

[6] E. R. Berlekamp, *Algebraic Coding Theory*, revised ed., Laguna Hills, CA, Aegean Park Press, 1984.

[7] A. M. Michaelson and A. J. Levesque, *Error Control Coding Techniques for Digital Communications*, New York, John Wiley & Sons, 1985.

[8] S. A. Vanstone and P. C. van Oorshot, *An Introduction to Error Correcting Codes with Applications*, Boston, MA, Kluwer Academic, 1989.

[9] A. Poli and L. Huguet, *Error Correcting Codes: Theory and Applications*, Hemel Hempstead, Prentice-Hall, 1992.

[10] S. B. Wicker, *Error Control Systems for Digital Communication and Storage*, Englewood Cliffs, NJ, Prentice-Hall, 1995.

[11] V. Pless, *Introduction to the Theory of Error Correcting Codes*, 3rd ed., New York, Wiley, 1998.

[12] P. Sweeney, *Error Control Coding: From Theory to Practice*, Chichester, Wiley, 2002.

[13] W. Cary Huffman and Vera Pless, *Fundamentals of Error-Correcting Codes*, Cambridge, Cambridge University Press, 2003.

[14] R. E. Blahut, *Algebraic Codes for Data Transmission*, Cambridge, Cambridge University Press, 2003.

[15] S. Lin and D. J. Costello, Jr., *Error Control Coding: Fundamentals and Applications*, 2nd ed., Upper Saddle River, NJ, Prentice-Hall, 2004.

[16] T. K. Moon, *Error Correction Coding: Mathematical Methods and Algorithms*, New York, Wiley-Interscience, 2005.

[17] J. C. Moreira and P. G. Farrell, *Essentials of Error-Control Coding*, West Sussex, John Wiley & Sons, 2006.

[18] R. M. Roth, *Introduction to Coding Theory*, Cambridge, Cambridge University Press, 2006.

[19] T. Richardson and R. Urbanke, *Modern Coding Theory*, New York, Cambridge University Press, 2008.

[20] W. E. Ryan and S. Lin, *Channel Codes: Classical and Modern,* New York: Cambridge University Press, 2009.

[21] M. Tomlinson, C. J. Tjhai, M. A. Ambroze, M. Ahmed, and M. Jibril, *Error-Correction Coding and Decoding Bounds, Codes, Decoders, Analysis and Applications*, Cham, Switzerland, Springer International Publishing, 2017.

[22] E. Sanvicente, *Understanding Error Control Coding*, Cham, Switzerland, Springer, 2019.

[23] R. W. Hamming, "Error detecting and error correcting codes," *Bell Syst. Tech. J.*, **29** (1950), 147–160.

[24] T. Klove and V. I. Korzhik, *Error Detecting Codes*, Boston, MA, Kluwer Academic, 1995.

[25] R. C. Singleton, "Maximum distance $q$-nary codes," *IEEE Trans. Inf. Theory*, **IT-10** (1964), 116–118.

[26] A. Tietavainen, "On the nonexistence of perfect codes over finite fields," *SIAM J. Appl. Math.*, **24**(1) (1973), 88–96.

[27] M. J. E. Golay, "Notes on digital coding," *Proc. IEEE*, **37** (1949), 657.

[28] I. S. Reed and G. Solomon, "Polynomial codes over certain finite fields," *J. Soc. Indust. Appl. Math.*, **8** (1960), 300–304.

# 4

# Binary Cyclic Codes

Cyclic codes form an important subclass of linear block codes. These codes are attractive for two reasons: first, encoding and syndrome computation can be implemented easily by using simple shift-registers with linear feedback connections, namely, *linear feedback shift-registers* (LFSRs); and second, because they have considerable inherent algebraic structure, it is possible to devise various practical algorithms for decoding them. Cyclic codes have been widely used in communication and storage systems for error control. They are particularly efficient for error detection.

Cyclic codes were first studied by Eugene Prange in 1957 [1]. Many classes of cyclic codes have been constructed over the years. The most well-known classes of cyclic codes are *Bose–Chaudhuri–Hocquenghem* (BCH) *codes* [2, 3], *Reed–Solomon* (RS) *codes* [4], *quadratic residue* (QR) *codes* [5], and *finite–geometry low-density parity-check* (LDPC) *codes* [6], which will be covered in Section 4.12 or in Chapters 5, 6, 7, and 12. As known from Chapter 3, there are two categories of linear block codes, binary and nonbinary. Cyclic codes, as a subclass of linear block codes, also consist of these two categories of codes, i.e., binary cyclic codes and nonbinary cyclic codes. Binary and nonbinary cyclic codes have similar properties and constructions. Thus, in this chapter, we focus on binary cyclic codes. Nonbinary cyclic codes can be analyzed in a similar way.

In this chapter, we present characterization, fundamental structural properties, generation, encoding, and a general simple decoding method for cyclic codes. For demonstration, we also introduce two small, but well-known, classes of cyclic codes, called *Hamming codes* [7] and *quadratic-residue codes* [5]. References [8–18] give extensive (also in-depth) coverage of cyclic codes.

## 4.1 Characterization of Cyclic Codes

Let $\mathbf{v} = (v_0, v_1, v_2, \ldots, v_{n-2}, v_{n-1})$ be an $n$-tuple over GF(2). If we shift every component of $\mathbf{v}$ *cyclically one place to the right*, we obtain the following $n$-tuple:

$$\mathbf{v}^{(1)} = (v_{n-1}, v_0, v_1, \ldots, v_{n-2}), \tag{4.1}$$

which is called a *right cyclic-shift* (or simply cyclic-shift) of $\mathbf{v}$.

For $0 \leq i < n$, if we cyclically shift the $n$-tuple $\mathbf{v}$ to the right $i$ places, we obtain the following $n$-tuple:

$$\mathbf{v}^{(i)} = (v_{n-i}, v_{n-i+1}, \ldots, v_{n-1}, v_0, v_1, \ldots, v_{n-i-1}), \tag{4.2}$$

which is called the *$i$th cyclic-shift* of $\mathbf{v}$. It is clear that $\mathbf{v}^{(0)} = \mathbf{v}^{(n)} = \mathbf{v}$.

**Definition 4.1** An $(n, k)$ linear block code $C$ is said to be *cyclic* if the cyclic-shift of a codeword in $C$ is another codeword in $C$.

Cyclic codes are normally studied in terms of polynomials over GF(2) (or GF($q$) for nonbinary codes). There is a *one-to-one correspondence* between an $n$-tuple $\mathbf{v}$ and a polynomial $\mathbf{v}(X)$ of degree $n-1$ or less as follows:

$$\mathbf{v} = (v_0, v_1, \ldots, v_{n-1}) \Longleftrightarrow \mathbf{v}(X) = v_0 + v_1 X + \cdots + v_{n-1} X^{n-1}. \tag{4.3}$$

Thus, each $n$-tuple corresponds one-to-one to a polynomial of degree $n-1$ or less. If $v_{n-1} \neq 0$, the degree of $\mathbf{v}(X)$ is $n-1$; if $v_{n-1} = 0$, the degree of $\mathbf{v}(X)$ is less than $n-1$. We shall call $\mathbf{v}(X)$ the *code polynomial* of $\mathbf{v}$. Hereafter, we shall use the terms codeword and code polynomial interchangeably. Note that the weight of a codeword $\mathbf{v}$ is equal to the number of nonzero coefficients of its code polynomial $\mathbf{v}(X)$. Because the code is linear, the sum of two codewords is another codeword. In terms of code polynomials, the sum of two code polynomials is another code polynomial.

For $0 \leq i < n$, the code polynomial which corresponds to the codeword $\mathbf{v}^{(i)}$, the $i$th cyclic-shift of $\mathbf{v}$, is

$$\begin{aligned} \mathbf{v}^{(i)}(X) = v_{n-i} + v_{n-i+1}X + \cdots + v_{n-1}X^{i-1} + v_0 X^i \\ + v_1 X^{i+1} + \cdots + v_{n-i-1}X^{n-1}. \end{aligned} \tag{4.4}$$

The relation between the code polynomial $\mathbf{v}(X)$ and its shift $\mathbf{v}^{(i)}(X)$ is specified in the following theorem.

**Theorem 4.1** *The code polynomial $\mathbf{v}^{(i)}(X)$ is the remainder resulting from dividing $X^i \mathbf{v}(X)$ by $X^n + 1$ (or $X^n - 1$ because $X^n - 1 = X^n + 1$ in GF(2)).*

*Proof* We first express $X^i \mathbf{v}(X)$ as follows:

$$X^i \mathbf{v}(X) = v_0 X^i + v_1 X^{i+1} + \cdots + v_{n-i-1}X^{n-1} + v_{n-i}X^n + \cdots + v_{n-1}X^{n-1+i}.$$

Next, we add and subtract $v_{n-i} + v_{n-i+1}X + \cdots + v_{n-1}X^{i-1}$ to and from the above expression of $X^i \mathbf{v}(X)$. This results in the following expression of $X^i \mathbf{v}(X)$:

$$\begin{aligned} X^i \mathbf{v}(X) = v_{n-i} + v_{n-i+1}X + \cdots + v_{n-1}X^{i-1} + v_0 X^i + v_1 X^{i+1} + \cdots + v_{n-i-1} \\ X^{n-1} + v_{n-i}(X^n + 1) + v_{n-i+1}X(X^n+1) + \cdots + v_{n-1}X^{i-1}(X^n+1) \\ = \mathbf{q}(X)(X^n + 1) + \mathbf{v}^{(i)}(X), \end{aligned} \tag{4.5}$$

where

$$\mathbf{q}(X) = v_{n-i} + v_{n-i+1}X + \cdots + v_{n-1}X^{i-1},$$

and

$$\mathbf{v}^{(i)}(X) = v_{n-i} + v_{n-i+1}X + \cdots + v_{n-1}X^{i-1} + v_0 X^i + v_1 X^{i+1} + \cdots + v_{n-i-1}X^{n-1}.$$

The expression of $X^i \mathbf{v}(X)$ given by (4.5) shows that $\mathbf{v}^{(i)}(X)$ is the remainder resulting from dividing $X^i \mathbf{v}(X)$ by $X^n + 1$. This proves the theorem. ▲▲

With polynomial representation, it is possible to develop some important structural properties of a cyclic code which enable simple implementation of encoding, syndrome computation, and decoding of a cyclic code.

It is noted that, for an element $a \in GF(2)$, we have $a = -a$, which means that addition and subtraction operations are the same in $GF(2)$. Because we focus just on binary codes (codes over $GF(2)$), we use addition to represent both addition and subtraction when developing properties and proving theorems in this chapter.

## 4.2   Structural Properties of Cyclic Codes

Consider an $(n, k)$ cyclic code $C$ in which each codeword is represented by a code polynomial of degree $n - 1$ or less. Let $r$ be a positive integer less than $n$ and

$$\mathbf{g}(X) = g_0 + g_1 X + \cdots + g_{r-1}X^{r-1} + X^r$$

be a nonzero code polynomial in $C$ of *minimum degree* $r$, where $g_j \in GF(2)$ with $0 \le j < r$, corresponding to the codeword $\mathbf{g} = (g_0, g_1, \ldots, g_{r-1}, 1, 0, 0, \ldots, 0)$. For $0 \le i < n$, $\mathbf{g}^{(i)}(X)$ is also a code polynomial in $C$. In the following, we develop some fundamental properties of this minimum-degree code polynomial $\mathbf{g}(X)$ in a sequence of lemmas and show that this code polynomial completely specifies the cyclic code $C$.

**Lemma 4.1** *The minimum–degree code polynomial* $\mathbf{g}(X)$ *in an* $(n, k)$ *cyclic code* $C$ *is unique.*

*Proof* Suppose $\mathbf{g}(X)$ is not unique. Let

$$\mathbf{g}^*(X) = g_0^* + g_1^* X + \cdots + g_{r-1}^* X^{r-1} + X^r,$$

be another minimum-degree nonzero code polynomial. Because the code is linear, the sum

$$\mathbf{g}(X) + \mathbf{g}^*(X) = (g_0 + g_0^*) + (g_1 + g_1^*)X + \cdots + (g_{r-1} + g_{r-1}^*)X^{r-1}$$

is also a code polynomial. If $\mathbf{g}(X) \ne \mathbf{g}^*(X)$, then $\mathbf{g}(X) + \mathbf{g}^*(X)$ is a nonzero code polynomial with degree less than the minimum degree $r$. This is impossible because the minimum degree of a nonzero code polynomial in $C$ is $r$. Hence, $\mathbf{g}(X) = \mathbf{g}^*(X)$ and $\mathbf{g}(X)$ is unique. ▲▲

**Lemma 4.2** *The constant coefficient $g_0$ of* $\mathbf{g}(X)$ *is equal to "1."*

*Proof* Assume that $g_0 = 0$. Then,

$$\mathbf{g}(X) = g_1 X + \cdots + g_{r-1} X^{r-1} + X^r.$$

Cyclically shifting $\mathbf{g}(X)$ $n-1$ places to the right (or one place to the left), we obtain a code polynomial,

$$\mathbf{g}^{(n-1)}(X) = g_1 + g_2 X + \cdots + g_{r-1} X^{r-2} + X^{r-1},$$

which has degree one less than the minimum degree $r$. This is a contradiction to the fact that the minimum degree of the nonzero code polynomial in $C$ is $r$. Hence, $g_0$ must be nonzero, i.e., $g_0 = 1$. ▲▲
    It follows from Lemma 4.2 that $\mathbf{g}(X)$ must be of the following form:

$$\mathbf{g}(X) = 1 + g_1 X + \cdots + g_{r-1} X^{r-1} + X^r. \tag{4.6}$$

**Lemma 4.3** *Every code polynomial* $\mathbf{v}(X)$ *in an* $(n, k)$ *cyclic code $C$ is divisible by its unique minimum-degree nonzero code polynomial* $\mathbf{g}(X) = 1 + g_1 X + \cdots + g_{r-1} X^{r-1} + X^r$.

*Proof* Dividing $\mathbf{v}(X)$ by $\mathbf{g}(X)$, we have

$$\mathbf{v}(X) = \mathbf{a}(X)\mathbf{g}(X) + \mathbf{b}(X), \tag{4.7}$$

in which $\mathbf{a}(X)$ and $\mathbf{b}(X)$ are the quotient and remainder, respectively, with

$$\mathbf{a}(X) = a_0 + a_1 X + \cdots + a_{n-r-1} X^{n-r-1},$$

and

$$\mathbf{b}(X) = b_0 + b_1 X + \cdots + b_{r-1} X^{r-1},$$

where $a_i \in \mathrm{GF}(2)$, $0 \le i < n - r$, and $b_j \in \mathrm{GF}(2)$, $0 \le j < r$. Note that, for $0 \le i < n - r$, $X^i \mathbf{g}(X) = \mathbf{g}^{(i)}(X)$ (see Problem 4.1). Hence,

$$\mathbf{a}(X)\mathbf{g}(X) = a_0 \mathbf{g}(X) + a_1 X \mathbf{g}(X) + \cdots + a_{n-r-1} X^{n-r-1} \mathbf{g}(X)$$
$$= a_0 \mathbf{g}(X) + a_1 \mathbf{g}^{(1)}(X) + \cdots + a_{n-r-1} \mathbf{g}^{(n-r-1)}(X),$$

which is a linear combination of the code polynomials, $\mathbf{g}(X)$, $\mathbf{g}^{(1)}(X), \ldots,$ $\mathbf{g}^{(n-r-1)}(X)$. Hence, $\mathbf{a}(X)\mathbf{g}(X)$ is a code polynomial. Rearranging the expression of (4.7), we obtain

$$\mathbf{b}(X) = \mathbf{v}(X) + \mathbf{a}(X)\mathbf{g}(X).$$

Because $\mathbf{b}(X)$ is the sum of two code polynomials, it is also a code polynomial. If $\mathbf{b}(X) \ne 0$, then $\mathbf{b}(X)$ is a nonzero code polynomial with degree less than the degree $r$ of the minimum-degree code polynomial $\mathbf{g}(X)$. This is not possible. Hence, $\mathbf{b}(X)$ must be 0. As a result, $\mathbf{v}(X) = \mathbf{a}(X)\mathbf{g}(X)$. Therefore, $\mathbf{v}(X)$ is divisible by $\mathbf{g}(X)$. ▲▲

**Lemma 4.4** *A polynomial over* $\mathrm{GF}(2)$ *of degree* $n-1$ *or less is a code polynomial in an* $(n,k)$ *cyclic code* $C$ *if and only if it is divisible by the minimum-degree nonzero code polynomial* $\mathbf{g}(X) = 1 + g_1 X + \cdots + g_{r-1}X^{r-1} + X^r$ *of* $C$.

*Proof* Let $\mathbf{v}(X) = v_0 + v_1 X + \cdots + v_{n-2}X^{n-2} + v_{n-1}X^{n-1}$ be a polynomial over $\mathrm{GF}(2)$ of degree $n-1$ or less. Assume $\mathbf{v}(X) = \mathbf{a}(X)\mathbf{g}(X)$. Then

$$\mathbf{a}(X) = a_0 + a_1 X + \cdots + a_{n-r-1}X^{n-r-1}$$

is a polynomial of degree $n-r-1$ or less. Multiplying out the product $\mathbf{a}(X)\mathbf{g}(X)$, we have

$$\mathbf{v}(X) = a_0\mathbf{g}(X) + a_1 X\mathbf{g}(X) + \cdots + a_{n-r-1}X^{n-r-1}\mathbf{g}(X)$$
$$= a_0\mathbf{g}(X) + a_1\mathbf{g}^{(1)}(X) + \cdots + a_{n-r-1}\mathbf{g}^{(n-r-1)}(X).$$

Because $\mathbf{v}(X)$ is a linear combination of the code polynomials, $\mathbf{g}(X), \mathbf{g}^{(1)}(X), \ldots,$ $\mathbf{g}^{(n-r-1)}(X)$, it is a code polynomial in $C$.

On the other side, following from Lemma 4.3, it is known that a code polynomial is divisible by its unique minimum-degree nonzero code polynomial $\mathbf{g}(X)$. This proves the lemma. ▲▲

**Lemma 4.5** *The degree of the minimum-degree nonzero code polynomial* $\mathbf{g}(X)$ *of an* $(n,k)$ *cyclic code* $C$ *is* $n-k$.

*Proof* It follows from Lemma 4.4 that a code polynomial $\mathbf{v}(X)$ in $C$ can be expressed as

$$\mathbf{v}(X) = \mathbf{a}(X)\mathbf{g}(X),$$

where $\mathbf{a}(X) = a_0 + a_1 X + \cdots + a_{n-r-1}X^{n-r-1}$ with $a_i \in \mathrm{GF}(2)$, $0 \leq i < n-r$. There are $2^{n-r}$ such polynomials, $\mathbf{a}(X)$. Thus, there are $2^{n-r}$ such code polynomials, $\mathbf{v}(X)$. It follows Lemma 4.4 that these code polynomials form all the code polynomials of $C$. Because there are $2^k$ code polynomials in an $(n,k)$ linear block code (corresponding to the $2^k$ codewords in $C$), then $2^{n-r} = 2^k$. Hence, $r = n - k$. ▲▲

It follows from Lemmas 4.2 and 4.5 that the code polynomial $\mathbf{g}(X)$ of minimum degree $r$ in an $(n,k)$ cyclic code $C$ must be of the following form:

$$\mathbf{g}(X) = 1 + g_1 X + \cdots + g_{n-k-1}X^{n-k-1} + X^{n-k}. \tag{4.8}$$

The degree of $\mathbf{g}(X)$ is equal to the number $n-k$ of parity-check bits of the $(n,k)$ code $C$. It follows from Lemma 4.4 that every code polynomial $\mathbf{v}(X)$ in an $(n,k)$ cyclic code can be expressed in the following form:

$$\mathbf{v}(X) = \mathbf{u}(X)\mathbf{g}(X)$$
$$= (u_0 + u_1 X + u_2 X^2 + \cdots + u_{k-1}X^{k-1})\mathbf{g}(X), \tag{4.9}$$

where $u_i \in \mathrm{GF}(2)$, $0 \leq i < k$. If the coefficients of $\mathbf{u}(X)$, $u_0$, $u_1$, $u_2, \ldots,$ $u_{k-1}$, are viewed as the $k$ information bits to be encoded, then $\mathbf{v}(X)$ would be the corresponding code polynomial. *Thus, the encoding of a message* $\mathbf{u} = (u_0, u_1, \ldots, u_{k-1})$ *is equivalent to multiplying the message polynomial* $\mathbf{u}(X)$ *by the nonzero minimum-degree code polynomial* $\mathbf{g}(X)$. Therefore, an $(n, k)$ cyclic code is *completely specified* by the polynomial $\mathbf{g}(X)$ given by (4.8).

**Definition 4.2** The minimum-degree code polynomial

$$\mathbf{g}(X) = 1 + g_1 X + \cdots + g_{n-k-1} X^{n-k-1} + X^{n-k}$$

is called the *generator polynomial* of the $(n, k)$ cyclic code $C$.

**Lemma 4.6** *The generator polynomial* $\mathbf{g}(X)$ *of an* $(n, k)$ *cyclic code* $C$ *divides* $X^n + 1$, *i.e.,* $\mathbf{g}(X)$ *is a factor of* $X^n + 1$:

$$X^n + 1 = \mathbf{g}(X)\mathbf{h}(X). \tag{4.10}$$

*Proof* Multiplying the generator polynomial $\mathbf{g}(X)$ by $X^k$, we obtain the following polynomial of degree $n$:

$$X^k \mathbf{g}(X) = X^k + g_1 X^{k+1} + g_2 X^{k+2} + \cdots + g_{n-k-1} X^{n-1} + X^n.$$

Dividing $X^k \mathbf{g}(X)$ by $X^n + 1$, we obtain

$$X^k \mathbf{g}(X) = (X^n + 1) + 1 + X^k + g_1 X^{k+1} + g_2 X^{k+2} + \cdots + g_{n-k-1} X^{n-1}, \tag{4.11}$$

where the remainder $1 + X^k + g_1 X^{k+1} + g_2 X^{k+2} + \cdots + g_{n-k-1} X^{n-1}$ is simply the code polynomial $\mathbf{g}^{(k)}(X)$ obtained by shifting the generator polynomial $\mathbf{g}(X)$ cyclically $k$ places to the right. It follows from Lemma 4.3 that $\mathbf{g}^{(k)}(X)$ is divisible by $\mathbf{g}(X)$ and hence $\mathbf{g}^{(k)}(X) = \mathbf{q}(X)\mathbf{g}(X)$. Replacing the remainder in (4.11) by $\mathbf{g}^{(k)}(X) = \mathbf{q}(X)\mathbf{g}(X)$, we have

$$X^k \mathbf{g}(X) = (X^n + 1) + \mathbf{q}(X)\mathbf{g}(X).$$

The above equality can be put in the following form:

$$X^n + 1 = (X^k + \mathbf{q}(X))\mathbf{g}(X),$$

which shows that $\mathbf{g}(X)$ divides $X^n + 1$. This completes the proof. ▲▲

Summarizing the results given above, we have the following theorem.

**Theorem 4.2** *An* $(n, k)$ *cyclic code* $C$ *over* GF(2) *is completely specified by a unique nonzero code polynomial of degree* $n - k$ *over* GF(2),

$$\mathbf{g}(X) = 1 + g_1 X + \cdots + g_{n-k-1} X^{n-k-1} + X^{n-k},$$

*which is called the* generator polynomial *of* $C$. *We say that the cyclic code* $C$ *is generated by* $\mathbf{g}(X)$. *A polynomial* $\mathbf{v}(X) = v_0 + v_1 X + \cdots + v_{n-1} X^{n-1}$ *over* GF(2) *of degree* $n - 1$ *or less is a code polynomial in* $C$ *if and only if* $\mathbf{v}(X)$ *is divisible by the generator polynomial* $\mathbf{g}(X)$.

## 4.3  Existence of Cyclic Codes

In this section, we will show the existence of cyclic codes.

**Theorem 4.3** *If $\mathbf{g}(X)$ is a polynomial of degree $n-k$ and is a factor of $X^n+1$, then $\mathbf{g}(X)$ generates an $(n,k)$ cyclic code.*

*Proof* Because $\mathbf{g}(X)$ is a polynomial of degree $n-k$, it must be of the following form:

$$\mathbf{g}(X) = g_0 + g_1X + \cdots + g_{n-k-1}X^{n-k-1} + X^{n-k}.$$

Because $\mathbf{g}(X)$ is a factor of $X^n + 1$, we have

$$X^n + 1 = \mathbf{g}(X)\mathbf{h}(X),$$

where $\mathbf{h}(X)$ is a polynomial of degree $k$ in the form of $\mathbf{h}(X) = h_0 + h_1X + h_2X^2 + \cdots + h_{k-1}X^{k-1} + X^k$. Factoring out the right-hand side of the above equation gives $g_0h_0 = 1$, which implies $g_0 = h_0 = 1$. Thus, $\mathbf{g}(X)$ is in the following form:

$$\mathbf{g}(X) = 1 + g_1X + \cdots + g_{n-k-1}X^{n-k-1} + X^{n-k}.$$

Construct an $n$-tuple $\mathbf{g}$ over GF(2) using the $n-k+1$ coefficients of $\mathbf{g}(X)$ as the first $n-k+1$ components and zeros as the last $k-1$ components. The $n$-tuple $\mathbf{g}$ is of the following form:

$$\mathbf{g} = (1, g_1, g_2, \ldots, g_{n-k-1}, 1, 0, 0, \ldots, 0). \tag{4.12}$$

Then, the $k-1$ $n$-tuples corresponding to the $k-1$ polynomials, $X\mathbf{g}(X)$, $X^2\mathbf{g}(X), \ldots, X^{k-1}\mathbf{g}(X)$, are simply the cyclic-shifts of $\mathbf{g}$ to the right by 1, $2, \ldots, k-1$ positions, respectively. We denote these $k-1$ $n$-tuples by $\mathbf{g}^{(1)}$, $\mathbf{g}^{(2)}, \ldots, \mathbf{g}^{(k-1)}$. It is straightforward to show that the $k$ $n$-tuples $\mathbf{g}$, $\mathbf{g}^{(1)}$, $\mathbf{g}^{(2)}, \ldots, \mathbf{g}^{(k-1)}$ are linearly independent. Hence, $\mathbf{g}(X)$, $X\mathbf{g}(X) = \mathbf{g}^{(1)}(X)$, $X^2\mathbf{g}(X) = \mathbf{g}^{(2)}(X), \ldots, X^{k-1}\mathbf{g}(X) = \mathbf{g}^{(k-1)}(X)$ are $k$ linearly independent polynomials over GF(2), where for $1 \leq i < k$, $\mathbf{g}^{(i)}(X)$ is simply the $i$th cyclic-shift of $\mathbf{g}(X)$ (regarding $\mathbf{g}(X)$ as a polynomial of degree $n-1$ or less).

Any linear combination of the $k$ polynomials $\mathbf{g}(X)$, $X\mathbf{g}(X)$, $X^2\mathbf{g}(X), \ldots$, $X^{k-1}\mathbf{g}(X)$,

$$\begin{aligned}
\mathbf{v}(X) &= u_0\mathbf{g}(X) + u_1X\mathbf{g}(X) + \cdots + u_{k-1}X^{k-1}\mathbf{g}(X) \\
&= (u_0 + u_1X + \cdots + u_{k-1}X^{k-1})\mathbf{g}(X) \\
&= \mathbf{u}(X)\mathbf{g}(X),
\end{aligned}$$

is a polynomial of degree $n-1$ or less and is divisible by $\mathbf{g}(X)$, where $\mathbf{u}(X) = u_0 + u_1X + \cdots + u_{k-1}X^{k-1}$ is a polynomial of degree $k-1$ or less. There are $2^k$ such polynomials, $\mathbf{v}(X)$, and their corresponding $2^k$ $n$-tuples form an $(n,k)$ linear code $C$.

Next, we show that this code is cyclic. Let $\mathbf{v}(X)$ be a code polynomial in $C$. Dividing $X\mathbf{v}(X)$ by $X^n + 1$, we have

$$X\mathbf{v}(X) = v_0 X + v_1 X^2 + \cdots + v_{n-1}X^n$$
$$= v_{n-1}(X^n + 1) + \mathbf{v}^{(1)}(X).$$

Rearranging the above equation, we have

$$\mathbf{v}^{(1)}(X) = X\mathbf{v}(X) + v_{n-1}(X^n + 1).$$

It follows from Theorem 4.1 that $\mathbf{v}^{(1)}(X)$ is a cyclic-shift of $\mathbf{v}(X)$. Because $X^n + 1 = \mathbf{g}(X)\mathbf{h}(X)$ and $\mathbf{v}(X)$ is divisible by $\mathbf{g}(X)$, we have

$$\mathbf{v}^{(1)}(X) = X\mathbf{v}(X) + v_{n-1}(X^n + 1)$$
$$= X\mathbf{u}(X)\mathbf{g}(X) + v_{n-1}\mathbf{g}(X)\mathbf{h}(X)$$
$$= [X\mathbf{u}(X) + v_{n-1}\mathbf{h}(X)]\mathbf{g}(X)$$
$$= \mathbf{u}^*(X)\mathbf{g}(X),$$

where $\mathbf{u}^*(X) = u_0^* + u_1^* X + \cdots + u_{k-1}^* X^{k-1} + u_k^* X^k$ is a polynomial of degree $k$ or less. Hence, $\mathbf{v}^{(1)}(X)$ is divisible by $\mathbf{g}(X)$. Thus, $\mathbf{v}^{(1)}(X)$ is a code polynomial in $C$. Because $\mathbf{v}^{(1)}(X)$ is a cyclic-shift of $\mathbf{v}(X)$, it follows from Definition 4.1 that the linear code $C$ generated by taking all the $2^k$ linear combinations of $\mathbf{g}(X), X\mathbf{g}(X), X^2\mathbf{g}(X), \ldots, X^{k-1}\mathbf{g}(X)$ is an $(n, k)$ cyclic code.    ▲▲

Theorem 4.3 says that any factor of $X^n + 1$ with degree $n - k$ generates an $(n, k)$ cyclic code over GF(2). For large $n$, $X^n + 1$ may have many factors of degree $n - k$ (see Table A.1 in Appendix A for the factorizations of $X^n + 1$ with $1 \leq n \leq 31$). Some of these polynomials generate good codes with large minimum distances. How to select generator polynomials to produce good cyclic codes is a very challenging problem, and coding theorists have expended tremendous research effort in searching for good cyclic codes. Several classes of good cyclic codes have been discovered. These codes have distinct algebraic structures, and efficient and practically implementable hard-decision decoding algorithms have been devised. Some such classes of codes will be presented in Sections 4.10 and 4.12, and Chapters 5 and 6.

Let $\mathbf{g}(X)$ be the generator polynomial of an $(n, k)$ cyclic code $C$ and $\mathbf{u}(X) = u_0 + u_1 X + \cdots + u_{k-1} X^{k-1}$ be a message polynomial corresponding to a $k$-tuple message $\mathbf{u}$. From the development of Theorem 4.3, the polynomial

$$\mathbf{v}(X) = \mathbf{u}(X)\mathbf{g}(X)$$
$$= u_0\mathbf{g}(X) + u_1 X\mathbf{g}(X) + \cdots + u_{k-1}X^{k-1}\mathbf{g}(X) \tag{4.13}$$

is a code polynomial in $C$ generated by $\mathbf{g}(X)$. Equation (4.13) provides an encoding method for an $(n, k)$ cyclic code in terms of its generator polynomial, $\mathbf{g}(X)$, and a message polynomial, $\mathbf{u}(X)$.

**Example 4.1** Let $n = 7$. The polynomial $X^7 + 1$ can be factored as follows:

$$X^7 + 1 = (1 + X)(1 + X + X^3)(1 + X^2 + X^3).$$

The $(7, 4)$ cyclic code generated by $\mathbf{g}(X) = 1 + X + X^3$ has 16 code polynomials or codewords as shown in Table 4.1. Let $\mathbf{u}(X) = 1 + X^3$ be the message polynomial corresponding to the message $\mathbf{u} = (1\ 0\ 0\ 1)$. Using (4.13), the code polynomial for $\mathbf{u}(X)$ is $\mathbf{v}(X) = (1 + X^3)(1 + X + X^3) = 1 + X + X^4 + X^6$ as shown in Table 4.1. Notice that this code is not systematic. The minimum distance of this code is 3. It is a single-error-correcting code.

Consider another polynomial $\mathbf{g}^*(X)$ which is the product of two factors of $X^7 + 1$, $1 + X$ and $1 + X + X^3$, i.e., $\mathbf{g}^*(X) = (1 + X)(1 + X + X^3) = 1 + X^2 + X^3 + X^4$. Using $\mathbf{g}^*(X)$ as a generator polynomial, we obtain a $(7, 3)$ cyclic code with minimum distance 4. This code is capable of correcting a single error and simultaneously detecting double errors. ▲▲

Table 4.1 A $(7, 4)$ cyclic code generated by $\mathbf{g}(X) = 1 + X + X^3$.

| Messages | Code polynomials | Codewords |
|---|---|---|
| $(0\ 0\ 0\ 0)$ | $0 \cdot (1 + X + X^3) = 0$ | $(0\ 0\ 0\ 0\ 0\ 0\ 0)$ |
| $(1\ 0\ 0\ 0)$ | $1 \cdot (1 + X + X^3) = 1 + X + X^3$ | $(1\ 1\ 0\ 1\ 0\ 0\ 0)$ |
| $(0\ 1\ 0\ 0)$ | $X \cdot (1 + X + X^3) = X + X^2 + X^4$ | $(0\ 1\ 1\ 0\ 1\ 0\ 0)$ |
| $(1\ 1\ 0\ 0)$ | $(1 + X) \cdot (1 + X + X^3) = 1 + X^2 + X^3 + X^4$ | $(1\ 0\ 1\ 1\ 1\ 0\ 0)$ |
| $(0\ 0\ 1\ 0)$ | $X^2 \cdot (1 + X + X^3) = X^2 + X^3 + X^5$ | $(0\ 0\ 1\ 1\ 0\ 1\ 0)$ |
| $(1\ 0\ 1\ 0)$ | $(1 + X^2) \cdot (1 + X + X^3) = 1 + X + X^2 + X^5$ | $(1\ 1\ 1\ 0\ 0\ 1\ 0)$ |
| $(0\ 1\ 1\ 0)$ | $(X + X^2) \cdot (1 + X + X^3) = X + X^3 + X^4 + X^5$ | $(0\ 1\ 0\ 1\ 1\ 1\ 0)$ |
| $(1\ 1\ 1\ 0)$ | $(1 + X + X^2) \cdot (1 + X + X^3) = 1 + X^4 + X^5$ | $(1\ 0\ 0\ 0\ 1\ 1\ 0)$ |
| $(0\ 0\ 0\ 1)$ | $X^3 \cdot (1 + X + X^3) = X^3 + X^4 + X^6$ | $(0\ 0\ 0\ 1\ 1\ 0\ 1)$ |
| $(1\ 0\ 0\ 1)$ | $(1 + X^3) \cdot (1 + X + X^3) = 1 + X + X^4 + X^6$ | $(1\ 1\ 0\ 0\ 1\ 0\ 1)$ |
| $(0\ 1\ 0\ 1)$ | $(X + X^3) \cdot (1 + X + X^3) = X + X^2 + X^3 + X^6$ | $(0\ 1\ 1\ 1\ 0\ 0\ 1)$ |
| $(1\ 1\ 0\ 1)$ | $(1 + X + X^3) \cdot (1 + X + X^3) = 1 + X^2 + X^6$ | $(1\ 0\ 1\ 0\ 0\ 0\ 1)$ |
| $(0\ 0\ 1\ 1)$ | $(X^2 + X^3) \cdot (1 + X + X^3) = X^2 + X^4 + X^5 + X^6$ | $(0\ 0\ 1\ 0\ 1\ 1\ 1)$ |
| $(1\ 0\ 1\ 1)$ | $(1 + X^2 + X^3) \cdot (1 + X + X^3) = 1 + X + X^2 + X^3 + X^4 + X^5 + X^6$ | $(1\ 1\ 1\ 1\ 1\ 1\ 1)$ |
| $(0\ 1\ 1\ 1)$ | $(X + X^2 + X^3) \cdot (1 + X + X^3) = X + X^5 + X^6$ | $(0\ 1\ 0\ 0\ 0\ 1\ 1)$ |
| $(1\ 1\ 1\ 1)$ | $(1 + X + X^2 + X^3) \cdot (1 + X + X^3) = 1 + X^3 + X^5 + X^6$ | $(1\ 0\ 0\ 1\ 0\ 1\ 1)$ |

## 4.4 Generator and Parity-Check Matrices of Cyclic Codes

Consider an $(n, k)$ cyclic code $C$ over GF(2) with generator polynomial

$$\mathbf{g}(X) = 1 + g_1 X + g_2 X^2 + \cdots + g_{n-k-1} X^{n-k-1} + X^{n-k}.$$

The codeword corresponding to $\mathbf{g}(X)$ is

$$\mathbf{g} = (1, g_1, g_2, \ldots, g_{n-k-1}, 1, 0, 0, \ldots, 0), \tag{4.14}$$

which is called the *generator codeword* of $C$.

In the proof of Theorem 4.3, we showed that the generator codeword $\mathbf{g}$ and its $k - 1$ cyclic-shifts $\mathbf{g}^{(1)}, \mathbf{g}^{(2)}, \ldots, \mathbf{g}^{(k-1)}$ are linearly independent. Thus, these $k$ $n$-tuples, $\mathbf{g}, \mathbf{g}^{(1)}, \mathbf{g}^{(2)}, \ldots, \mathbf{g}^{(k-1)}$, form a basis of a $k$-dimensional subspace of the vector space $\mathbf{V}_n$ of all the $n$-tuples over GF(2). Recall from Chapter 3 that an $(n, k)$ linear block code is defined as a $k$-dimensional subspace of $\mathbf{V}_n$. Therefore, the $k$ $n$-tuples ($k$ codewords in $C$) $\mathbf{g}, \mathbf{g}^{(1)}, \mathbf{g}^{(2)}, \ldots, \mathbf{g}^{(k-1)}$ form a *basis* of the $(n, k)$ cyclic code $C$. Any codeword in $C$ is a linear combination of these $k$ codewords.

Using the $k$ codewords $\mathbf{g}, \mathbf{g}^{(1)}, \mathbf{g}^{(2)}, \ldots, \mathbf{g}^{(k-1)}$ in $C$, we form a $k \times n$ matrix over GF(2) as follows:

$$\mathbf{G} = \begin{bmatrix} \mathbf{g} \\ \mathbf{g}^{(1)} \\ \mathbf{g}^{(2)} \\ \vdots \\ \mathbf{g}^{(k-1)} \end{bmatrix} = \begin{bmatrix} 1 \, g_1 \, g_2 \, \cdot & \cdot \cdot \cdot \, g_{n-k-1} & 1 & 0 & 0 \, 0 \cdots & 0 & 0 \\ 0 \, 1 \, g_1 \, g_2 \, \cdot & \cdot \cdot & \cdot & g_{n-k-1} & 1 & 0 \, 0 \cdots & 0 & 0 \\ 0 \, 0 \, 1 \, g_1 \, g_2 \cdot \cdot & \cdot & \cdot & g_{n-k-1} \, 1 \, 0 \cdots & 0 & 0 \\ \vdots \, \vdots \, \vdots \, \vdots \, \vdots \, \vdots \, \vdots & \vdots & \vdots & \vdots & \vdots \, \vdots \, \vdots & \vdots & \vdots \\ 0 \, 0 & \cdot \cdot \cdot \, 0 \, 1 & g_1 & g_2 & \cdot & \cdot \cdot \cdot \, g_{n-k-1} \, 1 \end{bmatrix}. \quad (4.15)$$

The row space of $\mathbf{G}$ gives the $(n, k)$ cyclic code $C$. Hence, $\mathbf{G}$ is a *generator matrix* of the cyclic code $C$, but not in systematic form.

**Example 4.2** A generator matrix in the form of (4.15) for the $(7, 4)$ cyclic code generated by $\mathbf{g}(X) = 1 + X + X^3$ in Example 4.1 is

$$\mathbf{G} = \begin{bmatrix} 1 \, 1 \, 0 \, 1 \, 0 \, 0 \, 0 \\ 0 \, 1 \, 1 \, 0 \, 1 \, 0 \, 0 \\ 0 \, 0 \, 1 \, 1 \, 0 \, 1 \, 0 \\ 0 \, 0 \, 0 \, 1 \, 1 \, 0 \, 1 \end{bmatrix},$$

which is not in systematic form. If we perform elementary row operations on $\mathbf{G}$, i.e., adding the zeroth row to the second row and adding the sum of the first two rows to the third row, we obtain the following generator matrix in systematic form:

$$\mathbf{G}_{\text{sys}} = \begin{bmatrix} 1 \, 1 \, 0 \, 1 \, 0 \, 0 \, 0 \\ 0 \, 1 \, 1 \, 0 \, 1 \, 0 \, 0 \\ 1 \, 1 \, 1 \, 0 \, 0 \, 1 \, 0 \\ 1 \, 0 \, 1 \, 0 \, 0 \, 0 \, 1 \end{bmatrix}.$$

This matrix generates the same code as $\mathbf{G}$. ▲▲

Recall that the generator polynomial $\mathbf{g}(X)$ of an $(n, k)$ cyclic code $C$ is a factor of $X^n + 1$, say

$$X^n + 1 = \mathbf{g}(X)\mathbf{h}(X), \quad (4.16)$$

where the polynomial $\mathbf{h}(X)$ has degree $k$ and is of the following form:

$$\mathbf{h}(X) = h_0 + h_1 X + h_2 X^2 + \cdots + h_k X^k \quad (4.17)$$

with $h_0 = h_k = 1$. Next, we show that a parity-check matrix of $C$ can be obtained from $\mathbf{h}(X)$.

Let $\mathbf{v} = (v_0, v_1, \ldots, v_{n-1})$ be a codeword in $C$ and $\mathbf{v}(X)$ be the corresponding code polynomial. Then, $\mathbf{v}(X) = \mathbf{u}(X)\mathbf{g}(X)$ for some polynomial $\mathbf{u}(X)$ of degree $k - 1$ or less. Multiplying $\mathbf{v}(X)$ by $\mathbf{h}(X)$, we obtain

$$\begin{aligned}
\mathbf{v}(X)\mathbf{h}(X) &= \mathbf{u}(X)\mathbf{g}(X)\mathbf{h}(X) \\
&= \mathbf{u}(X)(X^n + 1) \\
&= \mathbf{u}(X) + \mathbf{u}(X)X^n.
\end{aligned}$$

Because the degree of $\mathbf{u}(X)$ is $k - 1$ or less, the power terms $X^k$, $X^{k+1}, \ldots,$ $X^{n-1}$ do not appear in $\mathbf{u}(X) + \mathbf{u}(X)X^n$. If we expand the product $\mathbf{v}(X)\mathbf{h}(X)$ on the left-hand side of the above equation, the coefficients of $X^k$, $X^{k+1}, \ldots,$ $X^{n-1}$ must be equal to zero. Therefore, we obtain the following $n - k$ equalities:

$$\sum_{i=0}^{k} h_{k-i} v_{i+j} = 0, \quad \text{for } 0 \leq j \leq n - k. \tag{4.18}$$

Now, we take the *reciprocal* of $\mathbf{h}(X)$, which is defined as follows:

$$\mathbf{h}^*(X) = X^k \mathbf{h}(X^{-1}) \triangleq h_k + h_{k-1}X + h_{k-2}X^2 + \cdots + h_0 X^k, \tag{4.19}$$

i.e., the reciprocal polynomial $\mathbf{h}^*(X)$ just has the coefficients of $\mathbf{h}(X)$ in reversed order. It is easy to show that the reciprocal polynomial $\mathbf{h}^*(X)$ of $\mathbf{h}(X)$ is also a factor of $X^n + 1$. The polynomial $\mathbf{h}^*(X) = X^k \mathbf{h}(X^{-1})$ generates an $(n, n - k)$ cyclic code with the following $(n - k) \times n$ matrix as a generator matrix:

$$\mathbf{H} = \begin{bmatrix}
h_k & h_{k-1} & h_{k-2} & . & & . & . & . & . & h_0 & 0 & 0 & 0 & \cdots & 0 \\
0 & h_k & h_{k-1} & h_{k-2} & . & . & . & & . & & h_0 & 0 & 0 & \cdots & 0 \\
0 & 0 & h_k & h_{k-1} & h_{k-2} & . & . & & . & & . & h_0 & 0 & \cdots & 0 \\
\vdots & \vdots & \vdots & \vdots & \vdots & \vdots & & \vdots & & \vdots & \vdots & \vdots & \vdots & & \vdots \\
0 & 0 & . & & . & 0 & h_k & h_{k-1} & h_{k-2} & . & . & . & . & \cdots & h_0
\end{bmatrix}. \tag{4.20}$$

It follows from the $n - k$ equalities given by (4.18) that any codeword $\mathbf{v}$ in $C$ is orthogonal to every row of $\mathbf{H}$. Therefore, $\mathbf{H}$ is a parity-check matrix of the cyclic code $C$, and the row space of $\mathbf{H}$ is the dual code $C_d$ of $C$. Because the parity-check matrix $\mathbf{H}$ is obtained from the polynomial $\mathbf{h}(X)$, we call $\mathbf{h}(X)$ the *parity-check polynomial* of $C$. Hence, a cyclic code is also completely specified by its parity-check polynomial. The dual code $C_d$ of $C$ is an $(n, n - k)$ cyclic code with generator polynomial $\mathbf{h}^*(X)$ given by (4.19). Note that the parity-check matrix $\mathbf{H}$ given by (4.20) is not in systematic form. Elementary row operations can be performed to put it in systematic form.

Because $h_0 = h_k = 1$, we see that each row of the parity-check matrix $\mathbf{H}$ contains $n - k - 1$ consecutive zeros confined between two 1-entries (including the *end-around case*). These $n - k - 1$ consecutive zeros are said to form a *zero-span*. The zero-span structure of the parity-check matrix of an $(n, k)$ cyclic code in the form of (4.20) allows us to correct *erasures* clustered in a span of $n - k$ bit positions in a received vector. Codes for erasure correction will be discussed in Chapter 10.

**Example 4.3** Consider the $(7,4)$ cyclic code $C$ generated by $\mathbf{g}(X) = 1+X+X^3$ in Example 4.1. By long division, we find

$$X^7 + 1 = \mathbf{g}(X)\mathbf{h}(X) = (1 + X + X^3)(1 + X + X^2 + X^4).$$

Hence, the parity-check polynomial of $C$ is $\mathbf{h}(X) = 1 + X + X^2 + X^4$. The reciprocal of $\mathbf{h}(X)$ is $\mathbf{h}^*(X) = 1 + X^2 + X^3 + X^4$. Then, the parity-check matrix $\mathbf{H}$ in the form of $(4.20)$ of the $(7,4)$ cyclic code is

$$\mathbf{H} = \begin{bmatrix} 1 & 0 & 1 & 1 & 1 & 0 & 0 \\ 0 & 1 & 0 & 1 & 1 & 1 & 0 \\ 0 & 0 & 1 & 0 & 1 & 1 & 1 \end{bmatrix}. \tag{4.21}$$

It can be easily verified that $\mathbf{G} \cdot \mathbf{H}^T = \mathbf{O}$, where $\mathbf{G}$ is the generator matrix of the code obtained in Example 4.2. Note that the parity-check matrix $\mathbf{H}$ is not in systematic form. Adding the last row to the zeroth row of $\mathbf{H}$, we obtain a systematic parity-check matrix for the code $C$:

$$\mathbf{H}_{\text{sys}} = \begin{bmatrix} 1 & 0 & 0 & 1 & 0 & 1 & 1 \\ 0 & 1 & 0 & 1 & 1 & 1 & 0 \\ 0 & 0 & 1 & 0 & 1 & 1 & 1 \end{bmatrix}.$$

The dual code $C_d$ of the $(7,4)$ cyclic code $C$ is a $(7,3)$ cyclic code with generator polynomial $\mathbf{h}^*(X) = 1 + X^2 + X^3 + X^4$. ▲▲

## 4.5 Encoding of Cyclic Codes in Systematic Form

The encoding of cyclic codes, as a subclass of linear block codes, can be done by multiplying a message vector $\mathbf{u}$ and its generator matrix, either in non-systematic form ($\mathbf{u}\mathbf{G}$) or in systematic form ($\mathbf{u}\mathbf{G}_{\text{sys}}$). The previous section presented a way to find a generator matrix $\mathbf{G}$ for cyclic codes. Elementary operations can be performed to convert $\mathbf{G}$ into systematic form. In this section, we will first present the systematic encoding of cyclic codes from polynomial point of view, and then construct directly systematic generator matrices for cyclic codes without any requirement of elementary operations.

Consider an $(n,k)$ cyclic code with generator polynomial,

$$\mathbf{g}(X) = 1 + g_1 X + g_2 X^2 + \cdots + g_{n-k-1} X^{n-k-1} + X^{n-k}.$$

Let $\mathbf{u} = (u_0, u_1, \ldots, u_{k-1})$ be a message to be encoded. Represent $\mathbf{u}$ with a polynomial of degree $k - 1$ or less,

$$\mathbf{u}(X) = u_0 + u_1 X + \cdots + u_{k-1} X^{k-1}. \tag{4.22}$$

In general, the message $\mathbf{u}$ can be encoded to a codeword $\mathbf{v}$ in $C$ by multiplying the message polynomial $\mathbf{u}(X)$ with the generator polynomial $\mathbf{g}(X)$, i.e.,

$\mathbf{v}(X) = \mathbf{u}(X)\mathbf{g}(X)$ (also see (4.13)). The polynomial $\mathbf{v}(X)$ is the code polynomial corresponding to the message polynomial $\mathbf{u}(X)$. However, this kind of encoding does not give a cyclic code in systematic form.

In the following, an encoding method is presented to obtain a cyclic code in systematic form. To encode a message polynomial $\mathbf{u}(X)$, we first multiply it by $X^{n-k}$. This results in the following polynomial:

$$X^{n-k}\mathbf{u}(X) = u_0 X^{n-k} + u_1 X^{n-k+1} + \cdots + u_{k-1} X^{n-1}. \qquad (4.23)$$

Dividing $X^{n-k}\mathbf{u}(X)$ by $\mathbf{g}(X)$, we have

$$X^{n-k}\mathbf{u}(X) = \mathbf{a}(X)\mathbf{g}(X) + \mathbf{b}(X), \qquad (4.24)$$

where $\mathbf{a}(X)$ and $\mathbf{b}(X)$ are the quotient and remainder polynomials, respectively. Because the degree of the generator polynomial $\mathbf{g}(X)$ is $n - k$, the degree of the remainder $\mathbf{b}(X)$ must be $n - k - 1$ or less. That is,

$$\mathbf{b}(X) = b_0 + b_1 X + \cdots + b_{n-k-1} X^{n-k-1}. \qquad (4.25)$$

Rearrange the expression of (4.24) as follows:

$$\mathbf{b}(X) + X^{n-k}\mathbf{u}(X) = \mathbf{a}(X)\mathbf{g}(X). \qquad (4.26)$$

The expression of (4.26) shows that the polynomial $\mathbf{b}(X) + X^{n-k}\mathbf{u}(X)$ is divisible by $\mathbf{g}(X)$ and has degree $n - 1$ or less. Therefore, $\mathbf{b}(X) + X^{n-k}\mathbf{u}(X)$ is a code polynomial of the cyclic code $C$ generated by $\mathbf{g}(X)$.

Plugging (4.23) and (4.25) into $\mathbf{b}(X) + X^{n-k}\mathbf{u}(X)$, we obtain the following polynomial of degree $n - 1$ or less:

$$\begin{aligned} \mathbf{b}(X) + X^{n-k}\mathbf{u}(X) = {} & b_0 + b_1 X + \cdots + b_{n-k-1} X^{n-k-1} \\ & + u_0 X^{n-k} + u_1 X^{n-k+1} + \cdots + u_{k-1} X^{n-1}, \end{aligned} \qquad (4.27)$$

which corresponds to the following codeword

$$(\underbrace{b_0, b_1, \ldots, b_{n-k-1}}_{\text{Parity-check bits}}, \underbrace{u_0, u_1, \ldots, u_{k-1}}_{\text{Information bits}}), \qquad (4.28)$$

for the message $\mathbf{u} = (u_0, u_1, \ldots, u_{k-1})$. So, the codeword (or code polynomial) given by (4.28) (or (4.27)) is in systematic form as shown in Fig. 3.1.

**Example 4.4** Consider the $(7, 4)$ cyclic code generated by $\mathbf{g}(X) = 1 + X + X^3$ in Example 4.1, where the code in nonsystematic form is generated. In this example, we show the encoding of this code in systematic form. Let $\mathbf{u} = (1\,0\,1\,1)$

be the message to be encoded. Then, the message polynomial is $\mathbf{u}(X) = 1 + X^2 + X^3$. Dividing $X^3\mathbf{u}(X) = X^3 + X^5 + X^6$ by $\mathbf{g}(X)$, we obtain the remainder $\mathbf{b}(X) = 1$ as shown below.

$$
\begin{array}{r}
X^3 +X^2 +X +1 \\
X^3 + X + 1 \overline{)\; X^6 +X^5 \qquad\;\; +X^3} \\
\underline{X^6 \qquad\quad +X^4 +X^3} \\
X^5 +X^4 \\
\underline{X^5 \qquad +X^3 +X^2} \\
X^4 +X^3 +X^2 \\
\underline{X^4 \qquad\quad +X^2 +X} \\
X^3 \qquad\quad +X \\
\underline{X^3 \qquad\quad +X +1} \\
1
\end{array}
$$

Thus, the code polynomial for the message $\mathbf{u}(X) = 1 + X^2 + X^3$ is

$$\mathbf{v}(X) = \mathbf{b}(X) + X^3\mathbf{u}(X) = 1 + X^3 + X^5 + X^6$$

and the corresponding codeword is (1 0 0 1 0 1 1), where the rightmost four bits, $1, 0, 1, 1$, are the message (information) bits and the leftmost three bits, $1, 0, 0$, are the parity-check bits. The 16 codewords of the $(7, 4)$ cyclic code in systematic form are listed in Table 4.2 which are identical to these in Table 3.3.　　　　　　　　　　　　　　　　　　　　▲▲

Table 4.2 The $(7, 4)$ systematic cyclic code generated by $\mathbf{g}(X) = 1 + X + X^3$ in Example 4.4.

| Messages | Codewords | Messages | Codewords |
|---|---|---|---|
| (0 0 0 0) | (0 0 0 0 0 0 0) | (0 0 0 1) | (1 0 1 0 0 0 1) |
| (1 0 0 0) | (1 1 0 1 0 0 0) | (1 0 0 1) | (0 1 1 1 0 0 1) |
| (0 1 0 0) | (0 1 1 0 1 0 0) | (0 1 0 1) | (1 1 0 0 1 0 1) |
| (1 1 0 0) | (1 0 1 1 1 0 0) | (1 1 0 1) | (0 0 0 1 1 0 1) |
| (0 0 1 0) | (1 1 1 0 0 1 0) | (0 0 1 1) | (0 1 0 0 0 1 1) |
| (1 0 1 0) | (0 0 1 1 0 1 0) | (1 0 1 1) | (1 0 0 1 0 1 1) |
| (0 1 1 0) | (1 0 0 0 1 1 0) | (0 1 1 1) | (0 0 1 0 1 1 1) |
| (1 1 1 0) | (0 1 0 1 1 1 0) | (1 1 1 1) | (1 1 1 1 1 1 1) |

The encoding of an $(n,\ k)$ cyclic code $C$ in systematic form can be achieved with a division circuit which divides the message polynomial $X^{n-k}\mathbf{u}(X)$ by the generator polynomial $\mathbf{g}(X)$ and takes the remainder as the parity-check part of the codeword. This division circuit can be implemented with an $(n-k)$-stage shift-register with linear feedback connections (called *linear feedback shift-registers* (LFSRs)) [19] based on the coefficients of the generator polynomial $\mathbf{g}(X)$ as shown in Fig. 4.1. The premultiplication of the message polynomial $\mathbf{u}(X)$ by $X^{n-k}$ is accomplished by shifting the message polynomial from the right-hand end of the encoding feedback shift-register as shown in Fig. 4.1. There are several components in the encoding circuit shown by Fig. 4.1, where (1) the symbol $\rightarrow\boxed{b_i}\rightarrow$ denotes a single binary shift-register stage (for instance, a flip-flop) which is shifted by an external synchronous clock so that its input at a particular time appears at its output one unit of time later; (2) the symbol $\rightarrow\bigoplus\rightarrow$ denotes an Exclusive-OR (XOR) gate (or modulo-2 adder), whose output is defined as $\mathrm{XOR}(a,b) = a + b$ modulo-2 with $a, b \in \mathrm{GF}(2)$ as its inputs; and (3) the symbol $\widehat{g_i}$ represents a connection, i.e., there is a connection if $g_i = 1$ and no connection if $g_i = 0$.

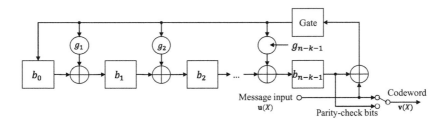

Figure 4.1 An encoding circuit for an $(n, k)$ cyclic code with generator polynomial $\mathbf{g}(X) = 1 + g_1 X + \cdots + g_{n-k-1}X^{n-k-1} + X^{n-k}$.

The encoding process of a message of $k$ information bits consists of the following three steps.

(1) With the input Gate turned on, the $k$ information bits $u_0, u_1, \ldots, u_{k-1}$ are shifted into the register in the order of $u_{k-1}, u_{k-2}, \ldots, u_1, u_0$ and simultaneously into a transmitted channel. As soon as the $k$ information bits have completely entered the LFSR, the $n-k$ bits in the shift-register are the parity-check bits $b_0, b_1, \ldots, b_{n-k-1}$.

(2) Break the feedback connection by turning off the Gate (i.e., the output of the gate is now "0").

(3) Shift the contents of the shift-register out and send them into the channel. These $n-k$ parity-check bits $b_0, b_1, \ldots, b_{n-k-1}$ together with the $k$ information bits $u_0, u_1, \ldots, u_{k-1}$ form a complete codeword $\mathbf{v} = (b_0, b_1, \ldots, b_{n-k-1}, u_0, u_1, \ldots, u_{k-1})$, where $u_{k-1}$ is the first bit sent to the channel and $b_0$ is the last bit sent to the channel.

The encoding circuit consists of: (1) $n-k$ shift-register stages (say $n-k$ flip-flops); (2) approximately $(n-k)/2$ XOR gates; and (3) a counter to control the output switch and the gate in the feedback connection. Let $m$ be the smallest integer such that $n < 2^m$. Then, the counter requires $m$ shift-register stages ($m$ flip-flops).

**Example 4.5** Consider the $(7, 4)$ cyclic code given in Example 4.1 with generator polynomial $\mathbf{g}(X) = 1 + X + X^3$ which divides $X^7 + 1$. The encoding circuit for this code is shown in Fig. 4.2.

Suppose a 4-bit message $\mathbf{u} = (1\ 0\ 1\ 1)$ $(\mathbf{u}(X) = 1 + X^2 + X^3)$ is to be encoded. As the information bits shift into the register sequentially, the contents in the register are shown in the following table.

| Input | Register contents $(b_0, b_1, b_2)$ | |
|---|---|---|
| | 0 0 0 | (initial state) |
| $u_3 = 1$ | 1 1 0 | (first shift) |
| $u_2 = 1$ | 1 0 1 | (second shift) |
| $u_1 = 0$ | 1 0 0 | (third shift) |
| $u_0 = 1$ | 1 0 0 | (fourth shift) |

After four shifts, Gate is turned off and the contents of the register are $(1\ 0\ 0)$. Thus, the complete codeword is $(1\ 0\ 0\ 1\ 0\ 1\ 1)$ and the code polynomial is $\mathbf{v}(X) = 1 + X^3 + X^5 + X^6$. ▲▲

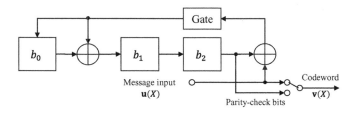

Figure 4.2 An encoding circuit for the $(7, 4)$ cyclic code given in Example 4.5.

Based on the above method for systematic encoding of a message, we can construct the systematic generator and parity-check matrices of an $(n, k)$ cyclic code $C$.

For $0 \le i < k$, let $\mathbf{u}_i(X) = X^i$ be the message polynomial corresponding to the message $\mathbf{u}_i$ with a single nonzero information bit at the $i$th position to be encoded. Dividing $X^{n-k}\mathbf{u}_i(X) = X^{n-k+i}$ by $\mathbf{g}(X)$, we obtain

$$X^{n-k+i} = \mathbf{a}_i(X)\mathbf{g}(X) + \mathbf{b}_i(X), \qquad (4.29)$$

where the remainder $\mathbf{b}_i(X)$ is of the following form:

$$\mathbf{b}_i(X) = b_{i,0} + b_{i,1}X + \cdots + b_{i,n-k-1}X^{n-k-1}. \qquad (4.30)$$

Let $\mathbf{v}_i(X) = \mathbf{b}_i(X) + X^{n-k+i}$. Equation (4.29) implies that $\mathbf{v}_i(X)$ is divisible by $\mathbf{g}(X)$; thus, it is a code polynomial in $C$. Obviously, the $k$ code polynomials, $\mathbf{v}_0(X), \mathbf{v}_1(X), \ldots, \mathbf{v}_{k-1}(X)$, are linearly independent. Hence, their $n$-tuple representations, $\mathbf{v}_0, \mathbf{v}_1, \ldots, \mathbf{v}_{k-1}$, can be used as a basis for the $(n, k)$ cyclic code. Arranging the $n$-tuple representations, $\mathbf{v}_0, \mathbf{v}_1, \ldots, \mathbf{v}_{k-1}$, of the $k$ code polynomials, $\mathbf{v}_0(X)$, $\mathbf{v}_1(X), \ldots,$ $\mathbf{v}_{k-1}(X)$, respectively, as the rows of a $k \times n$ matrix over $\mathrm{GF}(2)$, we obtain the following matrix:

$$
\mathbf{G}_{\mathrm{sys}} = \begin{bmatrix} \mathbf{v}_0 \\ \mathbf{v}_1 \\ \vdots \\ \mathbf{v}_{k-1} \end{bmatrix} = \begin{bmatrix} b_{0,0} & b_{0,1} & \cdots & b_{0,n-k-1} & 1 & 0 & 0 & \cdots & 0 \\ b_{1,0} & b_{1,1} & \cdots & b_{1,n-k-1} & 0 & 1 & 0 & \cdots & 0 \\ \vdots & \vdots & \ddots & \vdots & \vdots & \vdots & \vdots & \ddots & \vdots \\ b_{k-1,0} & b_{k-1,1} & \cdots & b_{k-1,n-k-1} & 0 & 0 & 0 & \cdots & 1 \end{bmatrix} \tag{4.31}
$$
$$
= [\mathbf{P} \ \mathbf{I}_k].
$$

Based on the development of linear block codes, we know that the row space of $\mathbf{G}_{\mathrm{sys}}$ gives an $(n, k)$ linear block code. Thus, $\mathbf{G}_{\mathrm{sys}}$ is the *generator matrix* of the $(n, k)$ cyclic code $C$ in systematic form. The top row of $\mathbf{G}_{\mathrm{sys}}$ corresponds to the coefficients of the first $n - k$ terms of the generator polynomial $\mathbf{g}(X)$ of the code (see Problem 4.11). If the $k$-tuple $\mathbf{u} = (u_0, u_1, \ldots, u_{k-1})$ is the message to be encoded, then the corresponding codeword is given by

$$
\mathbf{v} = \mathbf{u}\mathbf{G}_{\mathrm{sys}} = (u_0, u_1, \ldots, u_{k-1}) \begin{bmatrix} \mathbf{v}_0 \\ \mathbf{v}_1 \\ \vdots \\ \mathbf{v}_{k-1} \end{bmatrix} \tag{4.32}
$$
$$
= u_0 \mathbf{v}_0 + u_1 \mathbf{v}_1 + \cdots + u_{k-1} \mathbf{v}_{k-1}.
$$

In polynomial form, we have

$$
\begin{aligned}
\mathbf{v}(X) &= u_0 \mathbf{v}_0(X) + u_1 \mathbf{v}_1(X) + \cdots + u_{k-1} \mathbf{v}_{k-1}(X). \\
&= (u_0 \mathbf{b}_0(X) + u_1 \mathbf{b}_1(X) + \cdots + u_{k-1} \mathbf{b}_{k-1}(X)) \\
&\quad + u_0 X^{n-k} + u_1 X^{n-k+1} + \cdots + u_{n-k} X^{n-1} \\
&= \mathbf{b}(X) + X^{n-k} \mathbf{u}(X),
\end{aligned} \tag{4.33}
$$

where $\mathbf{u}(X) = u_0 + u_1 X + \cdots + u_{k-1} X^{k-1}$ and $\mathbf{b}(X) = u_0 \mathbf{b}_0(X) + u_1 \mathbf{b}_1(X) + \cdots + u_{k-1} \mathbf{b}_{k-1}(X)$. Thus, $\mathbf{v}(X)$ is in exactly the same form as in (4.27).

From (4.31), we obtain the systematic parity-check matrix of the cyclic code $C$:

$$
\mathbf{H}_{\mathrm{sys}} = \begin{bmatrix} \mathbf{I}_{n-k} & \mathbf{P}^T \end{bmatrix} = \begin{bmatrix} 1 & 0 & 0 & \cdots & 0 & b_{0,0} & b_{1,0} & \cdots & b_{k-1,0} \\ 0 & 1 & 0 & \cdots & 0 & b_{0,1} & b_{1,1} & \cdots & b_{k-1,1} \\ \vdots & \vdots & \vdots & \ddots & \vdots & \vdots & \vdots & \ddots & \vdots \\ 0 & 0 & 0 & \cdots & 1 & b_{0,n-k-1} & b_{1,n-k-1} & \cdots & b_{k-1,n-k-1} \end{bmatrix}. \tag{4.34}
$$

**Example 4.6** Again, we consider the $(7, 4)$ cyclic code $C$ generated by $\mathbf{g}(X) = 1 + X + X^3$ in Example 4.1. Dividing $X^3 \mathbf{u}_0(X) = X^3$, $X^3 \mathbf{u}_1(X) = X^4$, $X^3 \mathbf{u}_2(X) = X^5$, $X^3 \mathbf{u}_3(X) = X^6$ by $\mathbf{g}(X)$ result in four remainders, $\mathbf{b}_0(X) = 1 + X$, $\mathbf{b}_1(X) = X + X^2$, $\mathbf{b}_2(X) = 1 + X + X^2$, and $\mathbf{b}_3(X) = 1 + X^2$, respectively. Based on these four remainders, we form the following generator and parity-check matrices of $C$ in systematic form:

$$
\mathbf{G}_{\mathrm{sys}} = \begin{bmatrix} 1 & 1 & 0 & 1 & 0 & 0 & 0 \\ 0 & 1 & 1 & 0 & 1 & 0 & 0 \\ 1 & 1 & 1 & 0 & 0 & 1 & 0 \\ 1 & 0 & 1 & 0 & 0 & 0 & 1 \end{bmatrix}, \quad \mathbf{H}_{\mathrm{sys}} = \begin{bmatrix} 1 & 0 & 0 & 1 & 0 & 1 & 1 \\ 0 & 1 & 0 & 1 & 1 & 1 & 0 \\ 0 & 0 & 1 & 0 & 1 & 1 & 1 \end{bmatrix},
$$

which are the same as those given in Examples 4.2 and 4.3, respectively.  ▲▲

## 4.6   Syndrome Calculation and Error Detection

Consider an $(n, k)$ cyclic code with generator polynomial $\mathbf{g}(X)$. Suppose a code polynomial

$$\mathbf{v}(X) = v_0 + v_1 X + \cdots + v_{n-1} X^{n-1}$$

is transmitted, and

$$\mathbf{r}(X) = r_0 + r_1 X + \cdots + r_{n-1} X^{n-1}$$

is the received polynomial. Then,

$$\mathbf{r}(X) = \mathbf{v}(X) + \mathbf{e}(X), \tag{4.35}$$

where $\mathbf{e}(X)$ is the *error polynomial* in the form of $\mathbf{e}(X) = e_0 + e_1 X + \cdots + e_{n-1} X^{n-1}$. In terms of vectors, we have $\mathbf{r} = \mathbf{v} + \mathbf{e}$.

Both $\mathbf{v}(X)$ and $\mathbf{e}(X)$ are unknown to the receiver. Before decoding the received polynomial $\mathbf{r}(X)$, we first compute its syndrome to check whether $\mathbf{r}(X)$ is a code polynomial, i.e., whether $\mathbf{r}(X)$ is divisible by the generator polynomial $\mathbf{g}(X)$ of the code.

Dividing $\mathbf{r}(X)$ by $\mathbf{g}(X)$, we have

$$\mathbf{r}(X) = \mathbf{a}(X)\mathbf{g}(X) + \mathbf{s}(X), \tag{4.36}$$

where

$$\mathbf{s}(X) = s_0 + s_1 X + \cdots + s_{n-k-1} X^{n-k-1} \tag{4.37}$$

is the remainder. If $\mathbf{s}(X) = 0$, $\mathbf{r}(X)$ is a code polynomial. In this case, $\mathbf{r}(X)$ is accepted as the transmitted code polynomial and delivered to the user. If $\mathbf{s}(X) \neq 0$, then $\mathbf{r}(X)$ is not a code polynomial. In this case, the presence of errors in $\mathbf{r}(X)$ is detected and then the decoder performs error correction based on a designed decoding algorithm for the code. Because the remainder $\mathbf{s}(X)$ gives the information whether the received polynomial $\mathbf{r}(X)$ is contagious with errors, it is an error syndrome and we call it *syndrome polynomial* of the received polynomial $\mathbf{r}(X)$.

The syndrome calculation is accomplished by a division circuit shown in Fig. 4.3, which is identical to the encoding circuit at the transmitter except that the received polynomial $\mathbf{r}(X)$ is shifted into the register from the left-hand end. The received bits are shifted into the circuit in the order of $r_{n-1}, r_{n-2}, \ldots, r_1, r_0$. Gate 1 is turned on when the received bits are shifted in. After all the received bits are shifted into the circuit, Gate 1 is turned off and the contents in the registers are the syndrome bits. The syndrome can be shifted out sequentially in the order of $s_{n-k-1}, s_{n-k-2}, \ldots, s_1, s_0$. After all syndrome bits are shifted out, the circuit is ready for next received vector.

**Example 4.7** Consider the (7, 4) cyclic code given in Example 4.1 with generator polynomial $\mathbf{g}(X) = 1 + X + X^3$. The encoding circuit for this code is given

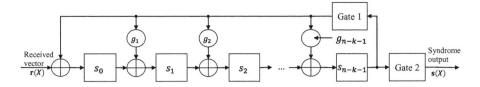

Figure 4.3 Syndrome calculation circuit for an $(n, k)$ cyclic code.

in Example 4.5 and shown in Fig. 4.2. Let $\mathbf{r} = (0\ 0\ 0\ 1\ 0\ 1\ 1)$ be the received vector. The corresponding received polynomial is $\mathbf{r}(X) = X^3 + X^5 + X^6$. By long division, we find that $\mathbf{r}(X) = (1 + X + X^2 + X^3)\mathbf{g}(X) + 1$. Thus, $\mathbf{s}(X) = 1$, i.e., $\mathbf{s} = (1\ 0\ 0)$.

The syndrome calculation circuit of the code is shown in Fig. 4.4. Table 4.3 shows the contents of registers in each stage. After all the received bits are shifted into the register (at stage 7), Gate 1 is turned off and the register contents give the syndrome which is $\mathbf{s} = (s_0, s_1, s_2) = (1\ 0\ 0)$. ▲▲

Table 4.3 The register contents of syndrome calculation circuit of the $(7, 4)$ cyclic code with $\mathbf{r} = (0\ 0\ 0\ 1\ 0\ 1\ 1)$.

| Stage | Input | Register contents $(s_0, s_1, s_2)$ |
|---|---|---|
| 0 | — | 0 0 0 (initial) |
| 1 | $r_6 = 1$ | 1 0 0 |
| 2 | $r_5 = 1$ | 1 1 0 |
| 3 | $r_4 = 0$ | 0 1 1 |
| 4 | $r_3 = 1$ | 0 1 1 |
| 5 | $r_2 = 0$ | 1 1 1 |
| 6 | $r_1 = 0$ | 1 0 1 |
| 7 | $r_0 = 0$ | 1 0 0 (syndrome $\mathbf{s}$) |
| 8 | — | 0 1 0 (syndrome $\mathbf{s}^{(1)}$) |
| 9 | — | 0 0 1 (syndrome $\mathbf{s}^{(2)}$) |
| 10 | — | 1 1 0 (syndrome $\mathbf{s}^{(3)}$) |

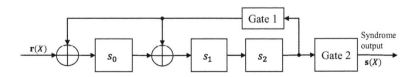

Figure 4.4 Syndrome calculation circuit for the $(7, 4)$ cyclic code in Example 4.7.

Because $\mathbf{v}(X) = \mathbf{u}(X)\mathbf{g}(X)$, it follows from (4.35) and (4.36) that

$$\mathbf{e}(X) = [\mathbf{a}(X) + \mathbf{u}(X)]\mathbf{g}(X) + \mathbf{s}(X). \tag{4.38}$$

This gives a relationship between the syndrome $\mathbf{s}(X)$ and the error polynomial $\mathbf{e}(X)$. Equation (4.38) has $2^k$ solutions. The real error polynomial $\mathbf{e}(X)$ is simply one of them.

Over the years since the discovery of cyclic codes, many methods (or algorithms) have been devised for decoding these codes. Because there are many different types of cyclic codes, their decoding methods were devised based on their specific constructions and characteristics. Several types of cyclic codes and their decoding methods will be discussed in Chapters 5–7. In the following, we develop some properties of the syndrome polynomial $\mathbf{s}(X)$ which will be used in the decoding of a cyclic code.

Let $\mathbf{r}^{(1)}(X)$ be the first cyclic-shift of $\mathbf{r}(X)$. Then, $\mathbf{r}^{(1)}(X)$ is the remainder resulting from dividing $X\mathbf{r}(X)$ by $X^n + 1$. That is,

$$X\mathbf{r}(X) = r_{n-1}(X^n + 1) + \mathbf{r}^{(1)}(X) = r_{n-1}\mathbf{g}(X)\mathbf{h}(X) + \mathbf{r}^{(1)}(X), \qquad (4.39)$$

where $X^n + 1 = \mathbf{g}(X)\mathbf{h}(X)$ is based on the fact that the generator polynomial $\mathbf{g}(X)$ of an $(n, k)$ cyclic code is a factor of $X^n + 1$. Let $\mathbf{s}(X)$ and $\mathbf{s}^{(1)}(X)$ be the syndromes of $\mathbf{r}(X)$ and $\mathbf{r}^{(1)}(X)$, respectively. That is, $\mathbf{r}(X) = \mathbf{a}(X)\mathbf{g}(X) + \mathbf{s}(X)$ and $\mathbf{r}^{(1)}(X) = \mathbf{a}_1(X)\mathbf{g}(X) + \mathbf{s}^{(1)}(X)$. Substituting $\mathbf{r}(X)$ and $\mathbf{r}^{(1)}(X)$ in (4.39) with $\mathbf{a}(X)\mathbf{g}(X) + \mathbf{s}(X)$ and $\mathbf{a}_1(X)\mathbf{g}(X) + \mathbf{s}^{(1)}(X)$, respectively, we have

$$X\{\mathbf{a}(X)\mathbf{g}(X) + \mathbf{s}(X)\} = r_{n-1}\mathbf{g}(X)\mathbf{h}(X) + \mathbf{a}_1(X)\mathbf{g}(X) + \mathbf{s}^{(1)}(X). \qquad (4.40)$$

Rearranging the terms in (4.40), we have

$$X\mathbf{s}(X) = (r_{n-1}\mathbf{h}(X) + X\mathbf{a}(X) + \mathbf{a}_1(X))\mathbf{g}(X) + \mathbf{s}^{(1)}(X). \qquad (4.41)$$

Equation (4.41) says that the syndrome $\mathbf{s}^{(1)}(X)$ of the first cyclic-shift $\mathbf{r}^{(1)}(X)$ of a received polynomial $\mathbf{r}(X)$ is the *remainder* resulting from dividing $X\mathbf{s}(X)$ by $\mathbf{g}(X)$. A direct consequence of this result is the following theorem.

**Theorem 4.4** *For $1 \leq i < n$, let $\mathbf{s}^{(i-1)}(X)$ and $\mathbf{s}^{(i)}(X)$ be the syndromes of the $(i-1)$th and the $i$th cyclic-shifts, $\mathbf{r}^{(i-1)}(X)$ and $\mathbf{r}^{(i)}(X)$, of the received polynomial $\mathbf{r}(X)$, respectively.[1] Then, $\mathbf{s}^{(i)}(X)$ is the remainder resulting from dividing $X\mathbf{s}^{(i-1)}(X)$ by $\mathbf{g}(X)$.*

Theorem 4.4 says that the syndrome sequence $\mathbf{s}^{(1)}(X), \mathbf{s}^{(2)}(X), \ldots, \mathbf{s}^{(n-1)}(X)$ can be generated iteratively from the syndrome $\mathbf{s}(X)$ of the received polynomial $\mathbf{r}(X)$. This syndrome sequence will be used to decode a cyclic code in general. The iterative calculation of the syndrome sequence simplifies the decoding, and this will be shown in Section 4.7.

First, we shall show how to generate $\mathbf{s}^{(1)}(X)$ from $\mathbf{s}(X)$ using the syndrome circuit shown in Fig. 4.3. Because the degree of $X\mathbf{s}(X)$ is $n - k$ or less, the quotient $r_{n-1}\mathbf{h}(X) + X\mathbf{a}(X) + \mathbf{a}_1(X)$ in (4.41) is a constant, which is equal to $s_{n-k-1}$. Consequently, the expression of (4.41) reduces to the following form:

$$X\mathbf{s}(X) = s_{n-k-1}\mathbf{g}(X) + \mathbf{s}^{(1)}(X). \qquad (4.42)$$

---

[1] For $i = 1$, it is defined that $\mathbf{s}^{(i-1)}(X) = \mathbf{s}^{(0)}(X) = \mathbf{s}(X)$ and $\mathbf{r}^{(i-1)}(X) = \mathbf{r}^{(0)}(X) = \mathbf{r}(X)$.

Equation (4.42) can be rearranged in the following form:

$$\mathbf{s}^{(1)}(X) = X\mathbf{s}(X) + s_{n-k-1}\mathbf{g}(X). \tag{4.43}$$

From (4.43), we find

$$\begin{aligned}
\mathbf{s}^{(1)}(X) &= s_{n-k-1}g_0 + (s_0 + s_{n-k-1}g_1)X + (s_1 + s_{n-k-1}g_2)X^2 \\
&\quad + \cdots + (s_{n-k-2} + s_{n-k-1}g_{n-k-1})X^{n-k-1}.
\end{aligned} \tag{4.44}$$

From (4.44), we see that $\mathbf{s}^{(1)}(X)$ is obtained by shifting (clocking) the syndrome circuit shown in Fig. 4.3 once with Gate 1 turned on. It follows from Theorem 4.4 that the syndrome sequence, $\mathbf{s}^{(1)}(X)$, $\mathbf{s}^{(2)}(X), \ldots, \mathbf{s}^{(n-1)}(X)$, is obtained by shifting the syndrome circuit $n-1$ times. Consider Example 4.7 with the syndrome calculation circuit shown in Fig. 4.4, where the syndrome $\mathbf{s}(X)$ is calculated at stage 7. At stage 8 (with Gate 1 turned on), clocking the circuit once with no input produces $\mathbf{s}^{(1)}(X)$ as shown in Table 4.3. Continuing this process, we can obtain the syndrome sequence, $\mathbf{s}^{(1)}(X)$, $\mathbf{s}^{(2)}(X), \ldots,$ $\mathbf{s}^{(6)}(X)$.

## 4.7 General Decoding of Cyclic Codes

In this section, we will present a general algorithm to decode cyclic codes. The syndrome sequence $\mathbf{s}(X)$, $\mathbf{s}^{(1)}(X)$, $\mathbf{s}^{(2)}(X), \ldots,$ $\mathbf{s}^{(n-1)}(X)$ developed in Section 4.6 allows us to decode a received polynomial $\mathbf{r}(X) = r_0 + r_1 X + \cdots + r_{n-1}X^{n-1}$ in a serial manner, one received bit at a time in the order of $r_{n-1}, r_{n-2}, \ldots, r_2, r_1, r_0$.

Decoding of $r_{n-1}$ is based on the syndrome $\mathbf{s}(X)$ of $\mathbf{r}(X)$. After decoding $r_{n-1}$, we cyclically shift $\mathbf{r}(X)$ once to generate the first cyclic-shift of $\mathbf{r}(X)$,

$$\mathbf{r}^{(1)}(X) = r_{n-1} + r_0 X + r_1 X^2 + \cdots + r_{n-2}X^{n-1}.$$

Then, we decode $r_{n-2}$ based on the syndrome $\mathbf{s}^{(1)}(X)$ of $\mathbf{r}^{(1)}(X)$. After decoding $r_{n-2}$, we cyclically shift $\mathbf{r}^{(1)}(X)$ once to generate the second cyclic-shift of $\mathbf{r}(X)$,

$$\mathbf{r}^{(2)}(X) = r_{n-2} + r_{n-1}X + r_0 X^2 + r_1 X^3 + \cdots + r_{n-3}X^{n-1}.$$

Then, we decode $r_{n-3}$ based on the syndrome $\mathbf{s}^{(2)}(X)$ of $\mathbf{r}^{(2)}(X)$. Repeat the above decoding process until $r_0$ is decoded.

To decode $r_{n-1}$, the decoder checks whether $\mathbf{s}(X)$ corresponds to a correctable error pattern, $\mathbf{e}(X) = e_0 + e_1 X + \cdots + e_{n-1}X^{n-1}$, with $e_{n-1} = 1$ (i.e., with an error at the highest-order position $X^{n-1}$).

(1) If there is no such error pattern, i.e., $e_{n-1} = 0$, no correction for $r_{n-1}$ is needed. Then, we cyclically shift $\mathbf{r}(X)$ to generate $\mathbf{r}^{(1)}(X)$. At the same time, we shift the syndrome register once to form $\mathbf{s}^{(1)}(X)$. The same circuit checks whether $\mathbf{s}^{(1)}(X)$ corresponds to a correctable error pattern with an error at the highest-order position $X^{n-1}$ of $\mathbf{r}^{(1)}(X)$.

(2) If there exists such an error pattern, $e_{n-1} = 1$, $r_{n-1}$ is an erroneous bit. Correction is performed by adding $e_{n-1}$ to $r_{n-1}$, i.e., $v_{n-1} = r_{n-1} + e_{n-1}$. This correction results in a modified received polynomial,

$$
\begin{aligned}
\mathbf{r}_{n-1}(X) &= r_0 + r_1 X + \cdots + r_{n-2} X^{n-2} + (r_{n-1} + e_{n-1}) X^{n-1} \\
&= e_{n-1} X^{n-1} + \mathbf{r}(X).
\end{aligned} \tag{4.45}
$$

After correcting $r_{n-1}$, the effect of $e_{n-1}$ on the syndrome $\mathbf{s}(X)$ can be removed. From the expression of $\mathbf{r}_{n-1}(X)$ given by (4.45), we see that the syndrome of $\mathbf{r}_{n-1}(X)$ is the sum of $\mathbf{s}(X)$ and the syndrome of $e_{n-1} X^{n-1}$. To decode $r_{n-2}$, we cyclically shift $\mathbf{r}_{n-1}(X)$ to obtain $\mathbf{r}_{n-1}^{(1)}(X)$ as follows:

$$
\begin{aligned}
\mathbf{r}_{n-1}^{(1)}(X) &= (e_{n-1} + r_{n-1}) + r_0 X + \cdots + r_{n-2} X^{n-1} \\
&= e_{n-1} + \mathbf{r}^{(1)}(X).
\end{aligned} \tag{4.46}
$$

From (4.46), it is obvious that the syndrome of $\mathbf{r}_{n-1}^{(1)}(X)$ is

$$
\mathbf{s}_{n-1}^{(1)}(X) = e_{n-1} + \mathbf{s}^{(1)}(X). \tag{4.47}
$$

Next, we decode $r_{n-2}$ based on $\mathbf{r}_{n-1}^{(1)}(X)$ and the modified syndrome $\mathbf{s}_{n-1}^{(1)}(X)$. (Note that the coefficient of the highest-order $X^{n-1}$ of $\mathbf{r}_{n-1}^{(1)}(X)$ is now $r_{n-2}$.) The process of decoding $r_{n-2}$ is the same as the decoding of $r_{n-1}$. The corrected polynomial is

$$
\begin{aligned}
\mathbf{r}_{n-2}(X) &= (e_{n-1} + r_{n-1}) + r_0 X + \cdots + (r_{n-2} + e_{n-2}) X^{n-1} \\
&= e_{n-2} X^{n-1} + \mathbf{r}_{n-1}^{(1)}(X).
\end{aligned} \tag{4.48}
$$

After decoding of $r_{n-2}$, we cyclically shift the corrected polynomial $\mathbf{r}_{n-2}(X)$ to obtain the next shifted and partially corrected received polynomial as follows:

$$
\begin{aligned}
\mathbf{r}_{n-2}^{(1)}(X) &= (r_{n-2} + e_{n-2}) + (r_{n-1} + e_{n-1}) X + r_0 X^2 + r_1 X^3 + \cdots + r_{n-3} X^{n-1} \\
&= e_{n-2} + \mathbf{r}_{n-1}^{(2)}(X).
\end{aligned} \tag{4.49}
$$

At the same time, we modify the syndrome $\mathbf{s}_{n-1}^{(2)}(X)$ by adding $e_{n-2}$. This gives the syndrome $\mathbf{s}_{n-2}^{(1)}(X) = e_{n-2} + \mathbf{s}_{n-1}^{(2)}(X)$ of $\mathbf{r}_{n-2}^{(1)}(X)$ for decoding the received bit $r_{n-3}$. (Note that the coefficient of the highest-order $X^{n-1}$ of $\mathbf{r}_{n-2}^{(1)}(X)$ is now $r_{n-3}$.)

This shift and error-correction process continues until the last received bit $r_0$ gets corrected. At the end of the decoding process, if the syndrome register contains only zeros, decoding is *successful* and the decoded vector is a codeword. If the content of the syndrome register is nonzero, an uncorrectable error pattern is detected and we declare a decoding failure. The above process for decoding a cyclic code is called the *Meggitt decoding process* [20]. A general cyclic decoder

[15, 20] is shown in Fig. 4.5. There are three major modules in the decoding diagram: (1) a buffer register to store the received polynomial; (2) a syndrome calculation circuit; and (3) an error pattern detection circuit which can detect an error pattern based on the calculated syndrome. The decoding procedure is described below.

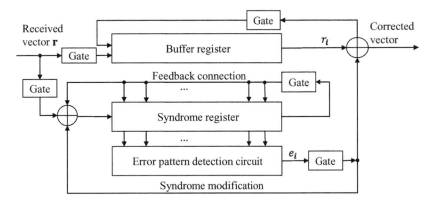

Figure 4.5 A general cyclic code decoder.

**Step 1**. The syndrome is formed by shifting the entire received vector into the syndrome register. At the same time, the received vector is stored in the buffer register.

**Step 2**. The syndrome is read into the error pattern detection circuit to find the corresponding error pattern. The error pattern detection circuit is a combination logic circuit which is designed in such a way that its output is "1" if and only if the syndrome in the syndrome register corresponds to a correctable error pattern with an error at the highest-order position $X^{n-1}$. Thus, if the error pattern detection circuit outputs "1," the rightmost bit (at position $n-1$) of the received vector in the buffer is assumed to be erroneous and must be corrected; if the error pattern detection circuit outputs "0," the rightmost bit (at position $n-1$) of the received vector in the buffer is assumed to be correct and no correction is required. That is, the output of the error pattern detection circuit is the estimated error value for the bit to be shifted out from the buffer.

**Step 3**. The buffer register is shifted once. The first received bit is read out from the buffer. At the same time, the syndrome register is shifted once. If the first received bit is detected to be erroneous, it is corrected by the output of the detection circuit. The output of the detection circuit is also fed into the syndrome register to modify the syndrome (i.e., remove the error effect from the syndrome). This results in a new syndrome, which corresponds to another vector obtained by shifting the corrected received vector one place to the right.

**Step 4**. The new syndrome formed in Step 3 is used to detect whether the second received bit which is now the rightmost bit in the buffer is an erroneous bit. The detection circuit repeats Steps 2 and 3. The second received bit is corrected in exactly the same way as that for the first received bit.

**Step 5**. The decoder decodes the received vector bit by bit following Steps 2–4 until the entire received vector is read out of the buffer.

From the above development of the decoder, it is obvious that the error pattern detection circuit needs only to detect the syndrome which corresponds to an error pattern whose highest-order bit $e_{n-1}$ is "1." As we know from Chapter 3, a linear block code can be decoded based on the syndrome look-up decoding table in which a syndrome is mapped to a correctable error pattern. Thus, the syndrome look-up decoding table provides a possible way to construct an error pattern detection circuit.

**Example 4.8** Consider the $(7, 4)$ cyclic code $C$ generated by $\mathbf{g}(X) = 1 + X + X^3$ in Example 4.1. Table 4.4 shows a syndrome look-up decoding table of the code $C$.

Table 4.4 A syndrome look-up decoding table for the $(7, 4)$ cyclic code given in Example 4.8.

| Syndromes $(s_0, s_1, s_2)$ | Correctable error patterns $(e_0, e_1, e_2, e_3, e_4, e_5, e_6)$ | Syndromes $(s_0, s_1, s_2)$ | Correctable error patterns $(e_0, e_1, e_2, e_3, e_4, e_5, e_6)$ |
|---|---|---|---|
| (1 0 0) | (1 0 0 0 0 0 0) | (0 1 1) | (0 0 0 0 1 0 0) |
| (0 1 0) | (0 1 0 0 0 0 0) | (1 1 1) | (0 0 0 0 0 1 0) |
| (0 0 1) | (0 0 1 0 0 0 0) | (1 0 1) | (0 0 0 0 0 0 1) |
| (1 1 0) | (0 0 0 1 0 0 0) | | |

From the table, the syndrome $\mathbf{s} = (1\ 0\ 1)$ corresponds to a correctable error pattern $\mathbf{e} = (0\ 0\ 0\ 0\ 0\ 0\ 1)$ with an error bit at the highest order. Thus, the error pattern detection circuit can be a combination logic circuit which outputs "1" if the input syndrome is $\mathbf{s} = (1\ 0\ 1)$; otherwise, it outputs "0." The error pattern detection circuit can be achieved by an AND-gate with $s_0$, $\bar{s}_1$, and $s_2$ as inputs where $\bar{s}_1$ is the complement of $s_1$. The syndrome calculation circuit of the code $C$ is shown in Fig. 4.4. Then, we can build a decoder for the $(7, 4)$ cyclic code as shown in Fig. 4.6.

Suppose the codeword $\mathbf{v} = (1\ 0\ 0\ 1\ 0\ 1\ 1)$ is transmitted and $\mathbf{r} = (1\ 0\ 0\ 1\ 0\ 0\ 1)$ is received. In the decoding, before all the received bits are shifted into the buffer, the multiplexer in Fig. 4.6 chooses $\mathbf{r}$ as its output; after all the received bits are shifted into the buffer, the multiplexer chooses the output from the XOR gate as its output. After all the received bits are shifted into the buffer, Gate is turned off. At the initial stage (shift = 0), the syndrome register is initialized to 0. After seven shifts (shift = 7), the received vector $\mathbf{r}$ is shifted into the buffer

($r_6$ is the first received bit) and the syndrome register contains the syndrome $\mathbf{s} = (1\ 1\ 1)$ corresponding to the received vector. Then, Gate is turned off. The output of the error pattern detection circuit is 0, which indicates that $r_6$ is correct and no error correction is needed. The buffer is shifted once to the right with the bit $r_6$ shifted to the leftmost of the buffer. At the same time, the syndrome register is also shifted once to the right. Then, the syndrome register becomes $\mathbf{s} = (1\ 0\ 1)$. Now, the output of the error pattern detection circuit is 1 which indicates that the rightmost bit in the buffer is erroneous. The rightmost bit $r_5$ in the buffer is read out and corrected by the output of the error pattern detection circuit. The corrected bit is shifted to the leftmost of the buffer and the buffer is shifted once to the right. The output of the error pattern detection circuit is also fed into the syndrome register. The decoding repeats this process until all the received bits are read out from the buffer. At the end of decoding, the contents in the buffer are the decoded vector. The decoding process is shown in Table 4.5. At shift = 14, the contents in the syndrome register are 0s which indicates the decoding is successful. The decoded vector in the buffer is (1 0 0 1 0 1 1), which is identical to the transmitted codeword. Thus, the decoding is successful. ▲▲

Table 4.5 Decoding steps for the $(7, 4)$ cyclic code given in Example 4.8.

| Shift | Buffer contents | Syndrome $(s_0, s_1, s_2)$ | Output of detection circuit |
|---|---|---|---|
| 0 | — | (0 0 0) | 0 (initial) |
| 7 | (1 0 0 1 0 0 1) | (1 1 1) | 0 ($\mathbf{r}$ shifted into the buffer) |
| 8 | (1 1 0 0 1 0 0) | (1 0 1) | 1 |
| 9 | (1 1 1 0 0 1 0) | (0 0 0) | 0 |
| 10 | (0 1 1 1 0 0 1) | (0 0 0) | 0 |
| 11 | (1 0 1 1 1 0 0) | (0 0 0) | 0 |
| 12 | (0 1 0 1 1 1 0) | (0 0 0) | 0 |
| 13 | (0 0 1 0 1 1 1) | (0 0 0) | 0 |
| 14 | (1 0 0 1 0 1 1) | (0 0 0) | 0 (end of the decoding) |

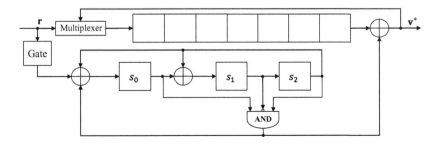

Figure 4.6 A Meggitt decoder for the $(7, 4)$ cyclic code in Example 4.8.

The above decoding method applies in principle to any cyclic code. But whether or not this decoder is practical depends entirely on its combination logic circuit, i.e., the error pattern detection circuit. There are cases in which the logic circuit is simple. In Example 4.8, the error pattern detection circuit is built based on the syndrome look-up decoding table. In Section 4.8, another simple circuit will be presented.

The decoding process presented in this section is basically an *error peeling procedure*, which peels off errors in an erroneously received vector one at a time. This peeling procedure is very effective in correcting erasures (or recovering erased symbols) over a BEC, as will be discussed in Chapter 10.

## 4.8 Error-Trapping Decoding for Cyclic Codes

In principle, the general Meggitt decoding method applies to any cyclic code. As pointed out earlier, there are *many different types* of cyclic codes. Each type of cyclic codes has its own special structural properties. Based on their special structural properties, efficient decoding methods or algorithms for each type of cyclic codes have been devised. Several important types of cyclic codes and their decoding methods will be presented in Chapters 5–7 and 16. Before we leave this chapter, we present a practical variation of Meggitt decoding method, called *error-trapping decoding* [21]. This decoding method is quite effective for correcting errors which are confined to $n - k$ consecutive positions (*including the end-around case*). A decoder based on this decoding technique employs a very simple combinational logic circuit for error detection and correction.

Suppose a $t$-error-correcting $(n, k)$ cyclic code is used for error-control purposes. Let $\mathbf{v}(X)$ be the transmitted code polynomial and $\mathbf{r}(X)$ be the received polynomial. Then, the error pattern caused by the channel noise is $\mathbf{e}(X) = \mathbf{v}(X) + \mathbf{r}(X)$. As shown in (4.38), the syndrome $\mathbf{s}(X)$ of $\mathbf{r}(X)$ is equal to the remainder resulting from dividing the error pattern $\mathbf{e}(X)$ by the code generator polynomial $\mathbf{g}(X)$, i.e.,

$$\mathbf{e}(X) = \mathbf{q}(X)\mathbf{g}(X) + \mathbf{s}(X). \tag{4.50}$$

If the errors are confined to the $n - k$ parity-check positions $X^0$, $X^1$, $X^2$, ..., $X^{n-k-1}$ of $\mathbf{r}(X)$, then $\mathbf{e}(X)$ is a polynomial of degree $n - k - 1$ or less. Because the degree of $\mathbf{g}(X)$ is $n - k$, it follows from Equation (4.50) that $\mathbf{q}(X) = 0$ and $\mathbf{e}(X) = \mathbf{s}(X)$. That is, if the errors in $\mathbf{r}(X)$ are confined to the $n - k$ parity-check positions, then the syndrome of $\mathbf{r}(X)$ is identical to the error pattern, i.e., $\mathbf{s}(X) = \mathbf{e}(X)$. Thus, correction can be accomplished simply by adding (modulo-2) the syndrome to the $n - k$ received parity-check bits.

Suppose the errors are not confined to the $n - k$ parity-check positions of $\mathbf{r}(X)$ but are confined to $n - k$ consecutive positions (*including the end-around case*), say $X^i$, $X^{i+1}$, ..., $X^{(n-k)+i-1}$. After $n - i$ cyclic-shifts of $\mathbf{r}(X)$, the errors will be shifted to the $n - k$ parity-check positions of the cyclically shifted received polynomial $\mathbf{r}^{(n-i)}(X)$. Then the syndrome of $\mathbf{r}^{(n-i)}(X)$ is identical to the errors confined to the positions $X^i$, $X^{i+1}$, ..., $X^{(n-k)+i-1}$ of $\mathbf{r}(X)$. As a result, the errors can be corrected. Based on the above facts, *an error-trapping decoding* is devised [21] as shown in Fig. 4.7.

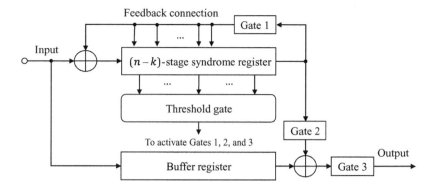

Figure 4.7 An error-trapping decoder for cyclic codes.

The operation of the error-trapping decoder is described in the following steps.

**Step 1.** Gate 1 is turned on and Gates 2 and 3 are turned off. The received polynomial $\mathbf{r}(X)$ is read into the syndrome register and simultaneously into the buffer register. Because we are interested only in the recovery of the $k$ transmitted information bits, the buffer register has only to store the $k$ received information bits. As soon as the entire received vector has been shifted into the syndrome register, the contents of the register are the syndrome of the received polynomial $\mathbf{r}(X)$.

**Step 2.** The weight of the syndrome is tested by an $(n-k)$-input *Threshold gate*. The output of this gate is "1" when $t$ (the error-correcting capability of the cyclic code) or fewer of its inputs are "1"; otherwise, the output is "0."

**Step 3. (a)** If the weight of the syndrome is $t$ or less, the errors are confined to the $(n-k)$ parity-check positions $X^0, X^1, X^2, \ldots, X^{n-k-1}$ of the received polynomial. Thus, the $k$ received information bits in the buffer register are *error free*. Gate 3 is then turned on and the information bits are sent to the data sink with Gate 2 turned off. The decoding is completed. Return to Step 1 to decode the next received vector. **(b)** If the weight of the syndrome calculated in the first step is greater than $t$, the syndrome register is then shifted once with Gate 1 turned on and Gates 2 and 3 turned off. Go to Step 4.

**Step 4.** The weight of the new contents of the syndrome register is tested. **(a)** If the weight is $t$ or less, the errors are confined to the positions $X^{n-1}, X^0, X^1, \ldots, X^{n-k-2}$ of the received polynomial. The leftmost digit in the syndrome register matches the error at the position $X^{n-1}$ of the received vector; the other $n-k-1$ bits in the syndrome register match the errors at the parity-check positions $X^0, X^1, \ldots, X^{n-k-2}$ of the received polynomial. The output of the threshold gate turns Gate 1 off and sets a clock to count from 1. The syndrome register is then shifted (in step with the clock) with Gate 1 turned off. As soon as the clock has counted to $n - k$, the contents of the syndrome register will be $(0, 0, \ldots, 0, 1)$. The rightmost digit matches the error at the position $X^{n-1}$ of the received vector. The $k$ received information bits are then read out of the buffer. The first received information bit is corrected by the "1" coming out from the syndrome register. The decoding is thus completed. Return to Step 1 to process the next received vector. **(b)** If the weight of the contents of the syndrome register calculated in Step 3(b) is greater than $t$, the syndrome register is shifted once again with Gate 1 turned on and Gates 2 and 3 turned off. Go to Step 5.

**Step 5**. Repeat Step 4(b) until the weight of the contents of the syndrome register goes down to $t$ or less. If the weight of the syndrome register never goes down to $t$ or less after the $(n - k)$th shift, go to Step 6. If the weight goes down to $t$ or less after the $i$th shift, for $1 \leq i \leq n - k$, the clock starts to count from $i$. At the same time, the syndrome register is shifted with Gate 1 turned off. As soon as the clock has counted to $n - k$, the rightmost $i$ bits in the syndrome register match the errors in the first $i$ received information bits in the buffer register. The other information bits are error free. Then, Gates 2 and 3 are turned on. The received information bits are read out of the buffer for correction. Return to Step 1.

**Step 6**. If the weight of the contents of the syndrome register never goes down to $t$ or less by the time when the syndrome register has been shifted $n - k$ times with Gate 1 turned on, Gate 3 is then turned on and the received information bits are read out of the buffer one at a time. As soon as the weight of the contents of the syndrome register goes down to $t$ or less, the contents match the errors in the next $n - k$ bits to come out of the buffer. Gate 2 is then turned on and the erroneous information bits are corrected by the bits coming out from the syndrome register with Gate 1 turned off. Gate 3 is turned off as soon as $k$ information bits have been read out of the buffer.

If the weight of the contents of the syndrome register never goes down to $t$ or less by the time the $k$ received information bits have been read out of the buffer, then either an uncorrectable error pattern has occurred or a correctable error pattern with errors not confined to $n - k$ consecutive positions has occurred. In this case, we declare a decoding failure.

Error-trapping decoding is most effective for correcting a single random error and errors which are clustered together within a certain span of consecutive positions called *burst of errors*. Application of error-trapping decoding for correcting a single burst of errors will be discussed in Chapter 10. Several modifications or refinements of this simple decoding method have been devised to decode multiple-error-correcting codes [22–27]. For more on decoding cyclic codes, readers are referred to [9, 13, 15].

## 4.9   Shortened Cyclic Codes

As discussed in Section 3.10, in system design, if a code of suitable natural length or suitable number of information bits cannot be found, we may need to shorten the length and/or reduce the dimension of an existing code. This can be achieved with the shortening techniques presented in Section 3.10. In the following, we present a commonly used technique to shorten a cyclic code.

Let $C$ be an $(n, k)$ cyclic code over GF(2) with generator polynomial $\mathbf{g}(X)$. Suppose we want to shorten $C$ to obtain an $(n-\lambda, k-\lambda)$ code $C_0$ with $0 \leq \lambda < k$. To achieve this, the information sequence at the output of the source encoder is first segmented into a sequence of messages, each consisting of $k - \lambda$ information bits. Before encoding a message $\mathbf{u}_0 = (u_0, u_1, \ldots, u_{k-\lambda-1})$ of $k - \lambda$ information bits, we append $\lambda$ zeros to the front (rightmost side) of the message $\mathbf{u}_0$ to obtain the following augmented message:

$$\mathbf{u} = (u_0, u_1, \ldots, u_{k-\lambda-1}, 0, 0, \ldots, 0).$$

Then, encode the augmented message $\mathbf{u}$ into the following codeword in $C$ (in systematic form):

$$\begin{aligned}
\mathbf{v} &= (v_0, v_1, \ldots, v_{n-k-1}, v_{n-k}, \ldots, v_{n-\lambda-1}, 0, 0, \ldots, 0) \\
&= (v_0, v_1, \ldots, v_{n-k-1}, u_0, u_1, \ldots, u_{k-\lambda-1}, 0, 0, \ldots, 0),
\end{aligned}$$

whose first $\lambda$ code bits are zeros. For transmission, the first $\lambda$ zero code bits of $\mathbf{v}$ are deleted. This results in a shortened codeword

$$\mathbf{v}_0 = (v_0, v_1, \ldots, v_{n-k-1}, u_0, u_1, \ldots, u_{k-\lambda-1}).$$

Then, the shortened codeword $\mathbf{v}_0$ is transmitted.

The above shortening gives an $(n - \lambda, k - \lambda)$ shortened cyclic code $C_0$ which is no longer cyclic. However, the encoding and syndrome computation for $C_0$ can be accomplished by the same circuit as used for the original cyclic code $C$. This is because the deletion of the $\lambda$ leading zero code bits does not affect the parity-check bit and syndrome computations.

The decoder for the original cyclic code can also be used for decoding the shortened cyclic code $C_0$ by (virtually) prefixing each received vector

$$\mathbf{r}_0 = (r_0, r_1, \ldots, r_{n-k-1}, r_{n-k}, \ldots, r_{n-\lambda-1}),$$

with $\lambda$ zeros. This zero-prefixing results in an extended received vector

$$\mathbf{r} = (r_0, r_1, \ldots, r_{n-k-1}, r_{n-k}, \ldots, r_{n-\lambda-1}, 0, 0, \ldots, 0).$$

Prior to the error-correction process, the syndrome register is first cyclically shifted $\lambda$ times to generate the proper syndrome for decoding the first bit $r_{n-\lambda-1}$ of the received vector $\mathbf{r}_0$ (the first information bit $u_{k-\lambda-1}$ of the message $\mathbf{u}_0$).

## 4.10 Hamming Codes

Hamming codes were discovered by Richard Hamming in 1950 [7]. These codes form the first class of error-correcting codes constructed algebraically. Hamming codes in their original form were not cyclic. When cyclic codes were discovered by Eugene Prange in 1957 [1], it was found that Hamming codes can be put in cyclic form.

Binary cyclic Hamming codes are generated by primitive polynomials over $GF(2)$. Let $\mathbf{p}(X)$ be a primitive polynomial of degree $m$ over $GF(2)$. As shown in Chapter 2, the smallest positive integer $n$ such that $X^n + 1$ is divisible by $\mathbf{p}(X)$ is $n = 2^m - 1$. Hence, $\mathbf{p}(X)$ is a factor of $X^{2^m-1} + 1$. Using $\mathbf{p}(X)$ as the generator polynomial, we obtain a cyclic Hamming code $C_{\text{Ham}}$ with the following parameters:

$$
\begin{aligned}
&\text{code length} && n = 2^m - 1, \\
&\text{number of parity-check bits } && n - k = m, \\
&\text{number of information bits } && k = 2^m - m - 1, \\
&\text{error-correcting capability } && t = 1 \ (d_{\min} = 3).
\end{aligned}
$$

Based on the parameters of a Hamming code, its parity-check matrix $\mathbf{H}$ is an $m \times (2^m - 1)$ matrix over $GF(2)$. Hence, $\mathbf{H}$ must contain all the nonzero $m$-tuples as columns. Because all the columns of $\mathbf{H}$ are distinct, the Hamming code has minimum distance at least 3. Because the columns of $\mathbf{H}$ are all the nonzero $m$-tuples, we can always find 3 $m$-tuples which sum to the zero $m$-tuple, e.g., (1 0 0 ... 0 0), (0 1 0 ... 0 0), and (1 1 0 ... 0 0). Then, it follows from Corollary 3.1 that the minimum distance of a Hamming code is exactly 3. The $(7, 4)$ and $(15, 11)$ linear block codes given in Examples 3.4 and 3.6, respectively, are Hamming codes. A parity-check matrix associated with the $(7, 4)$ Hamming code is

$$
\mathbf{H} = \begin{bmatrix} 1 & 0 & 0 & 1 & 0 & 1 & 1 \\ 0 & 1 & 0 & 1 & 1 & 1 & 0 \\ 0 & 0 & 1 & 0 & 1 & 1 & 1 \end{bmatrix}.
$$

In the following, we shall show that for any positive integer $m$, a Hamming code is a single-error-correcting code and also a perfect code [15].

Let $\mathbf{e}_i(X) = X^i$ be an error pattern with a single error at position $X^i$ and let $\mathbf{e}_j(X) = X^j$ be another error pattern with a single error at position $X^j$, where $i \neq j$ and $0 \leq i, j < n$. The syndrome $\mathbf{s}_i(X)$ which corresponds to $\mathbf{e}_i(X)$ is equal to the remainder resulting from dividing $\mathbf{e}_i(X)$ by the code generator polynomial $\mathbf{p}(X)$, i.e.,

$$
\mathbf{e}_i(X) = \mathbf{q}_i(X)\mathbf{p}(X) + \mathbf{s}_i(X). \tag{4.51}
$$

Similarly, the syndrome $\mathbf{s}_j(X)$ which corresponds to $\mathbf{e}_j(X)$ is equal to the remainder resulting from dividing $\mathbf{e}_j(X)$ by $\mathbf{p}(X)$, i.e.,

$$
\mathbf{e}_j(X) = \mathbf{q}_j(X)\mathbf{p}(X) + \mathbf{s}_j(X). \tag{4.52}
$$

Next, we show $\mathbf{s}_i(X) \neq \mathbf{s}_j(X)$. Suppose $\mathbf{s}_i(X) = \mathbf{s}_j(X)$. Then, combining (4.51) and (4.52), we obtain

$$\mathbf{e}_i(X) + \mathbf{e}_j(X) = (\mathbf{q}_i(X) + \mathbf{q}_j(X))\mathbf{p}(X). \tag{4.53}$$

Assuming $i < j$, we have

$$X^i(X^{j-i} + 1) = (\mathbf{q}_i(X) + \mathbf{q}_j(X))\mathbf{p}(X). \tag{4.54}$$

Equation (4.54) implies that $\mathbf{p}(X)$ divides $X^{j-i}+1$, which is impossible because $j - i < 2^m - 1$ and $\mathbf{p}(X)$ is a primitive polynomial of degree $m$. Therefore, $\mathbf{s}_i(X) \neq \mathbf{s}_j(X)$ for $i \neq j$. That is, different error patterns with a single error have *different syndromes*. From another point of view, (4.53) indicates $\mathbf{e}_i(X) + \mathbf{e}_j(X)$ is a code polynomial in $C_{\text{Ham}}$. This contradicts the fact that $C_{\text{Ham}}$ has minimum distance (weight) 3. Therefore, the two syndromes $\mathbf{s}_i(X)$ and $\mathbf{s}_j(X)$ cannot be the same.

There are $n = 2^m - 1$ error patterns with a single error. By Theorem 3.6, it is possible to form a standard array with all the $2^m - 1$ error patterns of single error as coset leaders because they have different syndromes. Thus, they are correctable error patterns. Because there are $2^{2^m - 1}$ $(2^m - 1)$-tuples and $2^{2^m - m - 1}$ codewords, there are $2^{2^m - 1}/2^{2^m - m - 1} = 2^m$ cosets. Therefore, a standard array can be formed with the zero codeword and all the $2^m - 1$ error patterns of single error as all the coset leaders. This proves that a Hamming code can correct all the error patterns of single error and *no others*. Hence, it is a perfect code.

The encoding circuit in Fig. 4.1 can be used to encode a Hamming code, which consists of an $m$-stage LFSR with feedback connections according to the generator polynomial $\mathbf{p}(X)$:

$$\mathbf{p}(X) = 1 + p_1 X + p_2 X^2 + \cdots + p_{m-1}X^{m-1} + X^m,$$

where $p_i = 0$ or $1$, $1 \leq i < m$. This encoding circuit consists of: (1) $m$ flip-flops; (2) approximately $m/2$ XOR gates; and (3) a counter that consists of $m$ flip-flops.

Hamming codes can be decoded using error-trapping decoding in a simple manner. A decoder as shown in Fig. 4.8 consists of a syndrome register, a buffer, and an $m$-input AND gate. The syndrome register is identical to that of the encoder. The detection of the occurrence of a single error is based on the following fact. Suppose a single error has occurred at the position $X^{n-1}$ in the received polynomial $\mathbf{r}(X)$. Because $\mathbf{r}(X)$ is read into the syndrome calculator from the rightmost stage, *which is equivalent to preshifting $\mathbf{r}(X)$ $m$ times cyclically*, the error is then shifted to the position $X^{m-1}$. Therefore, the syndrome register contains the syndrome corresponding to $\mathbf{e}_{m-1}(X) = X^{m-1}$, which is just $\mathbf{s}_{m-1}(X) = X^{m-1}$, or $\mathbf{s}_{m-1} = (0, 0, \ldots, 0, 1)$. That is, the syndrome register contains $(0, 0, \ldots, 0, 1)$ when a single error occurs at the position $X^{n-1}$ in the received polynomial $\mathbf{r}(X)$.

The decoding procedure can be described in the following steps.

**Step 1.** The syndrome is obtained by shifting the entire received vector into the syndrome register. At the same time, the received vector is stored

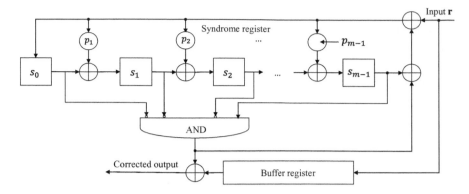

Figure 4.8 A Hamming code decoder.

into the buffer. If the syndrome is zero, the decoder assumes that no error has occurred and just reads out the received vector from the buffer register and goes to Step 4. If the syndrome is not zero, the decoder assumes that a single error has occurred and goes to Step 2.

**Step 2**. The received vector is read out of the buffer bit by bit. As each bit is read out of the buffer, the syndrome register is shifted cyclically once. As soon as the syndrome in the register is $(0, 0, \ldots, 0, 1)$, the next bit to come out of the buffer is the erroneous bit, and the output of the $m$-input AND gate is "1."

**Step 3**. The erroneous bit is read out of the buffer and is corrected by the output of the $m$-input AND gate. The correction is accomplished by an XOR gate.

**Step 4**. The syndrome register is reset to zero after the entire received vector is read out of the buffer. Go to Step 1 to decode the next received vector.

The above decoder is very simple. An $m$-stage shift-register is needed for syndrome computation. This requires $m$ flip-flops and approximately $m/2$ XOR gates. A $(2^m - 1)$-stage buffer register is needed for storing the received vector. This requires $2^m - 1$ flip-flops. An $m$-input AND gate and an XOR gate are also needed.

The weight distribution of a Hamming code $C$ of length $n = 2^m - 1$ is completely known [15]. For $0 \leq i < 2^m$, the number of codewords of weight $i$, $A_i$, is simply the coefficient of $z^i$ in the expansion of the following polynomial in $z$:

$$A(z) = \frac{1}{n+1}\{(1+z)^n + n(1-z)(1-z^2)^{(n-1)/2}\}. \qquad (4.55)$$

The polynomial $A(z)$ is referred to as the *weight enumerator* of the Hamming code $C$.

**Example 4.9** Let $m = 4$. Consider the $(15, 11)$ Hamming code $C_{\text{Ham}}$ generated by the primitive polynomial $\mathbf{p}(X) = 1 + X + X^4$ over GF(2). It follows from (4.55) that the weight enumerator of the $(15, 11)$ Hamming code is

$$A(z) = 1/16 \cdot \{(1 + z)^{15} + 15(1 - z)(1 - z^2)^7\}$$

$$= 1 + 35z^3 + 105z^4 + 168z^5 + 280z^6 + 435z^7 + 435z^8$$

$$+ 280z^9 + 168z^{10} + 105z^{11} + 35z^{12} + z^{15}.$$

Hence, the weight distribution of the $(15, 11)$ Hamming code is

$$A_0 = A_{15} = 1, \qquad A_3 = A_{12} = 35, \qquad A_4 = A_{11} = 105,$$
$$A_5 = A_{10} = 168, \qquad A_6 = A_9 = 280, \qquad A_7 = A_8 = 435.$$

From the weight distribution, we see that 1024 (half) of the codewords in $C_{\text{Ham}}$ have odd weights and 1024 (the other half) of codewords in $C_{\text{Ham}}$ have even weights. ▲▲

If a $(2^m - 1, 2^m - 1 - m)$ Hamming code is used for error detection on a BSC with transition probability $p$, its probability $P_u(E)$ of an undetected error can be computed from (3.37) and (4.55). It has been proved in [28] that $P_u(E)$ is upper bounded as follows:

$$P_u(E) \leq 2^{-m}$$

(the upper bound given by (3.37)). Hence, Hamming codes are good for error detection.

The dual code of a $(2^m - 1, 2^m - 1 - m)$ Hamming code $C_{\text{Ham}}$ is a $(2^m - 1, m)$ code with minimum distance $2^{m-1}$ which is called a *simplex code* [29] or a *maximum-length feedback shift register code* or a *maximum-length sequence code*. Its generator polynomial is $\mathbf{h}^*(X) = X^{2^m - m - 1} \mathbf{h}(X^{-1})$, where $\mathbf{h}(X) = (X^{2^m - 1} + 1)/\mathbf{p}(X)$ and $\mathbf{p}(X)$ is the generator polynomial of the Hamming code $C_{\text{Ham}}$.

The dual code of the $(15, 11)$ Hamming code $C_{\text{Ham}}$ given in Example 4.9 is a $(15, 4)$ code $C_{\text{Ham},d}$ which is a simplex code with generator polynomial $\mathbf{h}^*(X) = 1 + X^3 + X^4 + X^6 + X^8 + X^9 + X^{10} + X^{11}$. The code $C_{\text{Ham},d}$ has the following 16 codewords:

$$
\begin{array}{ll}
(0\,0\,0\,0\,0\,0\,0\,0\,0\,0\,0\,0\,0\,0\,0), & (1\,0\,1\,0\,1\,0\,1\,0\,1\,0\,1\,0\,1\,0\,1), \\
(0\,1\,1\,0\,0\,1\,1\,0\,0\,1\,1\,0\,0\,1\,1), & (1\,1\,0\,0\,1\,1\,0\,0\,1\,1\,0\,0\,1\,1\,0), \\
(0\,0\,0\,1\,1\,1\,1\,0\,0\,0\,0\,1\,1\,1\,1), & (1\,0\,1\,1\,0\,1\,0\,0\,1\,0\,1\,1\,0\,1\,0), \\
(0\,1\,1\,1\,1\,0\,0\,0\,0\,1\,1\,1\,1\,0\,0), & (1\,1\,0\,1\,0\,0\,1\,0\,1\,1\,0\,1\,0\,0\,1), \\
(0\,0\,0\,0\,0\,0\,0\,1\,1\,1\,1\,1\,1\,1\,1), & (1\,0\,1\,0\,1\,0\,1\,1\,0\,1\,0\,1\,0\,1\,0), \\
(0\,1\,1\,0\,0\,1\,1\,1\,1\,0\,0\,1\,1\,0\,0), & (1\,1\,0\,0\,1\,1\,0\,1\,0\,0\,1\,1\,0\,0\,1), \\
(0\,0\,0\,1\,1\,1\,1\,1\,1\,1\,0\,0\,0\,0), & (1\,0\,1\,1\,0\,1\,0\,1\,0\,1\,0\,0\,1\,0\,1), \\
(0\,1\,1\,1\,1\,0\,0\,1\,1\,0\,0\,0\,0\,1\,1), & (1\,1\,0\,1\,0\,0\,1\,1\,0\,0\,1\,0\,1\,1\,0).
\end{array}
$$

Each of the nonzero codewords in $C_{\text{Ham},d}$ has weight 8 and thus the minimum distance of the code is $d_{\min} = 8$. In general, all of the nonzero codewords of a $(2^m - 1, m)$ simplex code have weight $2^{m-1}$.

# 4.11 Cyclic Redundancy Check Codes

Shortened cyclic codes are widely used for error detection in communication systems, particularly in computer communications, where they are usually called *cyclic redundancy check codes* (CRC codes). CRC codes are basically derived from cyclic codes by shortening. As pointed out in Section 4.9, shortened cyclic codes are not cyclic in general. Hence, CRC codes are not cyclic. However, they take the advantages of their cyclic mother codes in encoding and error detection (syndrome computation) implementations.

A CRC code, denoted by $C_{\mathrm{CRC}}$, is generated by a polynomial $\mathbf{g}(X)$ over GF(2) of degree $m$. Hence, the number of parity-check bits of a CRC code $C_{\mathrm{CRC}}$ is $m$, but its length and rate can be varied depending on the degree of shortening the length $n$ of the original cyclic code $C$ generated by $\mathbf{g}(X)$. If $\mathbf{g}(X) = \mathbf{p}(X)$ is a primitive polynomial, then the CRC codes generated by $\mathbf{g}(X)$ are shortened codes of the Hamming code of length $n = 2^m - 1$.

In many cases, the generator polynomials $\mathbf{g}(X)$ of a CRC code contain $1+X$ as a factor, i.e., $\mathbf{g}(X) = (1 + X)\mathbf{f}(X)$. The factor $1 + X$ ensures that a CRC code is capable of detecting all the error patterns with odd number of errors. If $\mathbf{f}(X) = \mathbf{p}(X)$ is a primitive polynomial, the cyclic code generated by $\mathbf{g}(X) = (1+X)\mathbf{p}(X)$ is a $(2^m - 1, 2^m - m - 2)$ cyclic code $C_{\mathrm{Ham},4}$ with minimum distance 4 and $m + 1$ parity-check bits, which is a subcode of the $(2^m - 1, 2^m - m - 1)$ Hamming code $C_{\mathrm{Ham}}$ generated by $\mathbf{p}(X)$. The subcode $C_{\mathrm{Ham},4}$ consists of all the even-weight codewords in the Hamming code $C_{\mathrm{Ham}}$ and is called a *distance*-4 *Hamming code* (see more in Chapter 5). On a BSC, the probability $P_u(E)$ of an undetectable error of the distance-4 Hamming code $C_{\mathrm{Ham},4}$ is upper bounded by $2^{-(m+1)}$. The probability of an undetectable error of a CRC code derived from $C_{\mathrm{Ham},4}$ is also upper bounded by $2^{-(m+1)}$.

References [30–33] present detailed examinations of different generator polynomials used for CRC codes. Reference [16] analyzes the error detection capabilities of CRC codes. Table 4.6 gives a list of generator polynomials of various degrees for generating CRC codes [34]. A generator polynomial with an even number of terms has $1 + X$ as a factor. The CRC codes generated by the generator polynomials listed in Table 4.6 have good error detection performance.

Table 4.6 A list of standardized CRC codes.

| Codes | Applications | $\mathbf{g}(X)$ |
|---|---|---|
| CRC-1 | Single parity-check code | $1 + X$ |
| CRC-5-EPC | Gen 2 RFID | $1 + X^3 + X^5$ |
| CRC-8-WCDMA | Mobile networks | $1 + X + X^3 + X^4 + X^7 + X^8$ |
| CRC-16-CCITT | X.25, V.41, HDLC FCS, XMODEM, Bluetooth, etc. | $1 + X^5 + X^{12} + X^{16}$ |
| CRC-32 | ISO 3309, FED-STD-1003, IEEE 802-3 (Ethernet), etc. | $1 + X + X^2 + X^4 + X^5 + X^7 + X^8 + X^{10} + X^{11} + X^{12} + X^{16} + X^{22} + X^{23} + X^{26} + X^{32}$ |
| CRC-64-ISO | ISO 3309 (HDLC), etc. | $1 + X + X^3 + X^4 + X^{64}$ |

## 4.12 Quadratic Residue Codes

Another well-known class of cyclic codes is the class of binary *quadratic residue* (QR) codes, which have prime lengths. QR codes were first introduced by Gleason [5] and later were investigated by many mathematicians and coding theorists. A good introduction and coverage of QR codes can be found in [8, 9, 11, 14, 17]. QR codes are good in terms of their minimum distances, i.e., for a given length $n$ and dimension $k$, an $(n, k)$ QR code may have the largest possible minimum distance compared with other block codes. Many short binary QR codes with good (even the best) minimum distances have been found. Among them, the most well-known QR code is the $(23, 12)$ *Golay code* with minimum distance 7. QR codes are good but, in general, they are hard to decode algebraically up to their error-correcting capabilities guaranteed by their minimum distances.

In this section, we give a brief introduction to these codes. Let $p$ be an odd prime. Then, there exists a prime field, denoted by $\mathrm{GF}(p)$, with integer elements, 0, 1, 2, ..., $p-1$, under modulo-$p$ addition and multiplication. There are exactly $(p-1)/2$ (half) of the nonzero elements that have square roots in $\mathrm{GF}(p)$ [8, 9, 11, 14, 17], i.e., they are even powers of a primitive element in $\mathrm{GF}(p)$. These elements are referred to as *quadratic residues* [8, 9, 11, 14, 17].[2] Let $\alpha$ be a primitive element of $\mathrm{GF}(p)$ and $m = (p-1)/2$. Then, $m$ even powers of $\alpha$, namely, $\alpha^0, \alpha^2, \alpha^4, \ldots, \alpha^{2(m-1)}$, modulo-$p$, give the $(p-1)/2$ quadratic residues in $\mathrm{GF}(p)$. Notice that after the power $\alpha^{2(m-1)}$, the set of quadratic residues will repeat. As an example, consider the prime field $\mathrm{GF}(7) = \{0, 1, 2, 3, 4, 5, 6\}$ under modulo-7 addition and multiplication given in Example 2.7. The integer 5 is a primitive element of $\mathrm{GF}(7)$ and $m = (7-1)/2 = 3$. If we take the powers, $5^0, 5^2$, and $5^{2\times2}$, modulo-7, we obtain the quadratic residues, 1, 4, and 2, in $\mathrm{GF}(7)$, respectively. Notice that $5^{2\times3}$ modulo-7 is 1. From here, the elements in the set $\{1, 4, 2\}$ of quadratic residues will repeat. Another way to find the $m = (p-1)/2$ quadratic residues in $\mathrm{GF}(p)$ is to take the *powers of two* of the nonzero elements, $1, 2, 3, \ldots, m$, i.e., $1^2, 2^2, 3^2, \ldots, m^2$ modulo-$p$ [11]. Consider the prime field $\mathrm{GF}(7)$ given above. The powers of two of the elements, 1, 2, and 3 (i.e., $1^2$, $2^2$, and $3^2$ modulo-7) give the three quadratic residues 1, 4, and 2 in $\mathrm{GF}(7)$ which are the same as those computed above.

In the following, we define a class of binary QR codes. Let $p$ be an odd prime of the form $p = 8\delta \pm 1$ where $\delta$ is a positive integer and let $s$ be the smallest positive integer such that $p$ is a factor of $2^s - 1$. Let $\Omega_p$ be the set of $m = (p-1)/2$ quadratic residues in the prime field $\mathrm{GF}(p)$. Let $\beta$ be an element of order $p$ in the field $\mathrm{GF}(2^s)$. Define the following polynomial:

$$\mathbf{g}_{\mathrm{QR}}(X) = \prod_{\ell \in \Omega_p} (X - \beta^\ell), \tag{4.56}$$

---

[2]A more general definition of quadratic residue is given as follows. If there is an integer $x$, $0 < x < p$, such that $x^2 = y$ modulo-$p$, then $y$ is said to be a quadratic residue (modulo-$p$).

i.e., $\mathbf{g}_{\mathrm{QR}}(X)$ is a polynomial which has $\beta^\ell$, $\ell \in \Omega_p$, as roots. The polynomial $\mathbf{g}_{\mathrm{QR}}(X)$ defined by (4.56) is a polynomial of degree $m$ over $\mathrm{GF}(2)$ which divides $X^p - 1$. The cyclic code with $\mathbf{g}_{\mathrm{QR}}(X)$ as the generator polynomial is a binary $(p, p-m)$ QR code, denoted by $C_{\mathrm{QR}}$, with rate $(p+1)/2p$ which is slightly above or below $1/2$. Many short binary QR codes of this type have been constructed and they have very good (or even the best) minimum distances $d_{\min}$ [35]. Table 4.7 gives a list of QR codes.

Table 4.7 A list of QR codes.

| $\delta$ | Length | Dimension | $d_{\min}$ | $\delta$ | Length | Dimension | $d_{\min}$ |
|---|---|---|---|---|---|---|---|
| 1 | 7 | 4 | 3 | 9 | 71 | 36 | 11 |
| 2 | 17 | 9 | 5 | 9 | 73 | 37 | 13 |
| 3 | 23 | 12 | 7 | 10 | 79 | 40 | 15 |
| 4 | 31 | 16 | 7 | 11 | 89 | 45 | 17 |
| 5 | 41 | 21 | 9 | 12 | 97 | 49 | 15 |
| 6 | 47 | 24 | 11 | 13 | 103 | 52 | 19 |

**Example 4.10** Consider the field $\mathrm{GF}(2^8)$ constructed based on the primitive polynomial $\mathbf{p}(X) = 1 + X^2 + X^3 + X^4 + X^8$ over $\mathrm{GF}(2)$. The integer $2^8 - 1 = 255$ can be factored as the product of 15 and 17. The odd prime factor 17 of $2^8 - 1$ is in the form of $17 = 8 \times 2 + 1$ and 8 is the smallest positive integer such that 17 divides $2^8 - 1$, i.e., $p = 17$, $\delta = 2$, and $s = 8$. The quadratic residue set in the prime field $\mathrm{GF}(17)$ is $\Omega_{17} = \{1, 2, 4, 8, 9, 13, 15, 16\}$. Let $\alpha$ be a primitive element of $\mathrm{GF}(2^8)$ and $\beta = \alpha^{15}$. Then, $\beta$ is an element of order 17 in $\mathrm{GF}(2^8)$. The polynomial over $\mathrm{GF}(2)$ which has $\beta, \beta^2, \beta^4, \beta^8, \beta^9, \beta^{13}, \beta^{15}$, and $\beta^{16}$ in $\mathrm{GF}(2^8)$ as roots is $\mathbf{g}_{\mathrm{QR}}(X) = 1 + X^3 + X^4 + X^5 + X^8$, which is an irreducible polynomial over $\mathrm{GF}(2)$. The cyclic code generated by $\mathbf{g}_{\mathrm{QR}}(X)$ is a binary $(17, 9)$ QR code $C_{\mathrm{QR}}$ with minimum distance 5, which is the best linear block code in terms of minimum distance for codes with length 17 and dimension 9. ▲▲

**Example 4.11** The most well-known QR code is the $(23, 12)$ Golay code $C_{\mathrm{Golay}}$ [36] with minimum distance 7, which is a perfect code. The length 23 of the code is in the form of $23 = 8 \times 3 - 1$. The set of 11 quadratic residues in the prime field $\mathrm{GF}(23)$ is $\Omega_{23} = \{1, 2, 3, 4, 6, 8, 9, 12, 13, 16, 18\}$. The smallest positive integer $s$ for which 23 divides $2^s - 1$ is $s = 11$. The integer $2^{11} - 1 = 2047$ can be factored as the product of two prime factors 23 and 89. Let $\alpha$ be a primitive element of $\mathrm{GF}(2^{11})$ and $\beta = \alpha^{89}$. Then, the order of $\beta$ is 23. The generator polynomial of the $(23, 12)$ Golay code $C_{\mathrm{Golay}}$ is

$$\mathbf{g}_{\mathrm{Golay}}(X) = 1 + X + X^5 + X^6 + X^7 + X^9 + X^{11},$$

which has $\beta, \beta^2, \beta^3, \beta^4, \beta^6, \beta^8, \beta^9, \beta^{12}, \beta^{13}, \beta^{16}$, and $\beta^{18}$ over $\mathrm{GF}(2^{11})$ as roots.

Note that 89 is also a prime factor of $2^{11} - 1$. Using the quadratic residues of the prime field $\mathrm{GF}(89)$, an $(89, 45)$ QR code with minimum distance 17 and $\delta = 11$ can be constructed (see Problem 4.29). ▲▲

# 4.13   Quasi-cyclic Codes

Cyclic codes possess a fully cyclic structure, i.e., cyclically shifting a codeword by any number of positions to the right or left results in another codeword. With this fully cyclic structure, we can implement their encoding and decoding with shift-registers and simple logic circuits. However, most of the linear block codes do not have this fully cyclic structure. There is another type of linear block codes which do not have the fully cyclic structure but a *partially* cyclic structure, called *quasi-cyclic* (QC) *structure*. Linear block codes with QC structure are called *QC codes*. The QC structure of these codes can also facilitate their encoding and decoding implementations [37–39]. This section will present the basic structural properties and encoding of QC codes. Several classes of QC codes and their decoding will be explored in Chapters 12–15.

## 4.13.1   Definitions and Fundamental Structures

**Definition 4.3 (Classic definition [14, 40])** An $(n, k)$ linear block code with length $n = cn_0$, $n_0 \geq 1$, is called a *QC code*, denoted by $C_{qc}$, if cyclically shifting a codeword in $C_{qc}$ by $n_0$ code symbol positions, either to the right or left, yields another codeword in $C_{qc}$. This cyclic-shift structure is called the *QC structure*. The parameter $n_0$ is called the *shifting constraint*.

From the above definition, we can see that QC codes are a generalization of cyclic codes, i.e., cyclic codes are QC codes with shifting constraint $n_0 = 1$.

The generator matrix of a QC code possesses a QC structure. We use the following example to show this structure. More details about the generator matrices of QC codes can be found in [13, 14].

**Example 4.12** Consider the following generator matrix $\mathbf{G}_{qc}$ of a $(6, 3)$ linear block code $C_{qc}$:

$$\mathbf{G}_{qc} = \begin{bmatrix} 1 & 1 & 0 & 1 & 0 & 0 \\ 0 & 0 & 1 & 1 & 0 & 1 \\ 0 & 1 & 0 & 0 & 1 & 1 \end{bmatrix}. \tag{4.57}$$

The eight codewords of $C_{qc}$ are listed in Table 4.8. From this table, we can see that the code $C_{qc}$ is a QC code with shifting constraint $n_0 = 2$, i.e., cyclically shifting any codeword in $C_{qc}$ to the right (or left) by 2 positions results in another codeword. The generator matrix $\mathbf{G}_{qc}$ has a QC structure: each row of $\mathbf{G}_{qc}$ is the cyclic-shift to the right by $n_0 = 2$ positions of the row above it and the top row is the cyclic-shift to the right by $n_0 = 2$ positions of the last row. ▲▲

Next, we present another definition of QC codes which are most commonly used nowadays. Let $t$ and $b$ be two positive integers. Consider a *tb*-tuple over GF(2):

Table 4.8 The eight codewords of the $(6,3)$ QC code $C_{qc}$ given in Example 4.12.

| Messages $(u_0, u_1, u_2)$ | Codewords $(v_0, v_1, v_2, v_3, v_4, v_5)$ | Messages $(u_0, u_1, u_2)$ | Codewords $(v_0, v_1, v_2, v_3, v_4, v_5)$ |
|:---:|:---:|:---:|:---:|
| (0 0 0) | (0 0 0 0 0 0) | (1 0 0) | (1 1 0 1 0 0) |
| (0 0 1) | (0 1 0 0 1 1) | (1 0 1) | (1 0 0 1 1 1) |
| (0 1 0) | (0 0 1 1 0 1) | (1 1 0) | (1 1 1 0 0 1) |
| (0 1 1) | (0 1 1 1 1 0) | (1 1 1) | (1 0 1 0 1 0) |

$$\mathbf{v} = (\underbrace{v_0, v_1, \ldots, v_{b-1}}_{\mathbf{v}_0}, \underbrace{v_b, v_{b+1}, \ldots, v_{2b-1}}_{\mathbf{v}_1}, \ldots, \underbrace{v_{(t-1)b}, v_{(t-1)b+1}, \ldots, v_{tb-1}}_{\mathbf{v}_{t-1}}) \tag{4.58}$$
$$= (\mathbf{v}_0, \mathbf{v}_1, \ldots, \mathbf{v}_{t-1}),$$

which comprises $t$ sections, $\mathbf{v}_0, \mathbf{v}_1, \ldots, \mathbf{v}_{t-1}$, each section consisting of $b$ bits, i.e., a $b$-tuple. For $0 \le j < t$, denote the $b$ bits in the $j$th section $\mathbf{v}_j$ of $\mathbf{v}$ as follows:

$$\mathbf{v}_j = (v_{j,0}, v_{j,1}, \ldots, v_{j,b-1}). \tag{4.59}$$

Let $\mathbf{v}_j^{(1)} = (v_{j,b-1}, v_{j,0}, v_{j,1}, \ldots, v_{j,b-2})$ be the $b$-tuple over GF(2) obtained by cyclically shifting each component of $\mathbf{v}_j$ one place to the right. We call $\mathbf{v}_j^{(1)}$ the (right) cyclic-shift of $\mathbf{v}_j$. Consider the cyclic-shifts of all the $t$ sections, $\mathbf{v}_0^{(1)}, \mathbf{v}_1^{(1)}, \ldots, \mathbf{v}_{t-1}^{(1)}$, of $\mathbf{v}$ and form the following $tb$-tuple:

$$\mathbf{v}^{(1)} = (\mathbf{v}_0^{(1)}, \mathbf{v}_1^{(1)}, \ldots, \mathbf{v}_{t-1}^{(1)}). \tag{4.60}$$

The $tb$-tuple $\mathbf{v}^{(1)}$ comprises $t$ sections, each consisting of $b$ bits. We call $\mathbf{v}^{(1)}$ the *section-wise cyclic-shift* of $\mathbf{v}$ and the parameter $b$ is called the *section size*.

**Definition 4.4** Let $b$, $k$, and $t$ be positive integers such that $k < tb$. A $(tb, k)$ linear block code $C_{qc}$ over GF(2) is called a QC code if it satisfies the following two conditions: (1) each codeword in $C_{qc}$ comprises $t$ sections, each consisting of $b$ bits; and (2) section-wise cyclic-shifting a codeword in $C_{qc}$ $l$ places to the right for $0 \le l < b$ yields another codeword in $C_{qc}$.

Under the above definition, if $t = 1$, the QC code $C_{qc}$ is a cyclic code. Therefore, cyclic codes form a subclass of QC codes.

**Example 4.13** Consider the following generator matrix $\mathbf{G}_{qc}$ of a $(6,3)$ linear block code $C_{qc}$:

$$\mathbf{G}_{qc} = \begin{bmatrix} 1 & 0 & 0 & 1 & 1 & 0 \\ 0 & 1 & 0 & 0 & 1 & 1 \\ 0 & 0 & 1 & 1 & 0 & 1 \end{bmatrix}. \tag{4.61}$$

The eight codewords of $C_{qc}$ are listed in Table 4.9. From this table, we can see that the code $C_{qc}$ is a QC code with $t = 2$ and $b = 3$, i.e., each codeword in $C_{qc}$

consists of two sections of three bits each and section-wise cyclically shifting any codeword in $C_{qc}$ results in another codeword. Consider the the second codeword $\mathbf{v} = (\mathbf{v}_0, \mathbf{v}_1) = (0\ 0\ 1, 1\ 0\ 1)$ of $C_{qc}$ in Table 4.9. The codeword $\mathbf{v}$ consists of two sections, $\mathbf{v}_0 = (0\ 0\ 1)$ and $\mathbf{v}_1 = (1\ 0\ 1)$. Section-wise cyclically shifting $\mathbf{v}$ one position to the right results in a 6-tuple $\mathbf{v}^{(1)} = (1\ 0\ 0, 1\ 1\ 0)$, which is the fifth codeword of $C_{qc}$ in Table 4.9. Section-wise cyclically shifting $\mathbf{v}^{(1)}$ to the right by one position yields a 6-tuple, $\mathbf{v}^{(2)} = (0\ 1\ 0, 0\ 1\ 1)$, which is the third codeword of $C_{qc}$ listed in Table 4.9. Section-wise cyclically shifting $\mathbf{v}^{(2)}$ one position to the right, we obtain the codeword $\mathbf{v}$ in $C_{qc}$.

Note that the generator matrix $\mathbf{G}_{qc}$ for this QC code is a $1 \times 2$ array of circulants of size $3 \times 3$. ▲▲

Table 4.9 The eight codewords of the $(6, 3)$ QC code $C_{qc}$ given in Example 4.13.

| Messages | Codewords | Messages | Codewords |
|----------|-----------|----------|-----------|
| $(u_0, u_1, u_2)$ | $(v_0, v_1, v_2, v_3, v_4, v_5)$ | $(u_0, u_1, u_2)$ | $(v_0, v_1, v_2, v_3, v_4, v_5)$ |
| $(0\ 0\ 0)$ | $(0\ 0\ 0\ 0\ 0\ 0)$ | $(1\ 0\ 0)$ | $(1\ 0\ 0\ 1\ 1\ 0)$ |
| $(0\ 0\ 1)$ | $(0\ 0\ 1\ 1\ 0\ 1)$ | $(1\ 0\ 1)$ | $(1\ 0\ 1\ 0\ 1\ 1)$ |
| $(0\ 1\ 0)$ | $(0\ 1\ 0\ 0\ 1\ 1)$ | $(1\ 1\ 0)$ | $(1\ 1\ 0\ 1\ 0\ 1)$ |
| $(0\ 1\ 1)$ | $(0\ 1\ 1\ 1\ 1\ 0)$ | $(1\ 1\ 1)$ | $(1\ 1\ 1\ 0\ 0\ 0)$ |

It can be shown that Definition 4.3 and Definition 4.4 are equivalent, i.e., a QC code $C_{qc,0}$ under Definition 4.3 is a QC code $C_{qc,1}$ under Definition 4.4 through a certain coordinate (or symbol) permutation of the codewords in $C_{qc,0}$ and vice versa (see Problem 4.33).

## 4.13.2 Generator and Parity-Check Matrices in Systematic Circulant Form

For a $(tb, k)$ QC code $C_{qc}$, if the dimension $k$ of the code is also a multiple of the section size $b$, i.e., $k = cb$ for some integer $c$ ($1 \leq c < t$), the generator matrix $\mathbf{G}_{qc}$ of a $(tb, cb)$ QC code can be put as a $c \times t$ array of $b \times b$ circulants. For example, the generator matrix $\mathbf{G}_{qc}$ of the $(6, 3)$ QC code given in Example 4.13 is a $1 \times 2$ array of $3 \times 3$ circulants.

The generator matrix of a $(tb, cb)$ QC code in systematic form is a $cb \times tb$ matrix over $\mathrm{GF}(2)$ given as follows:

$$
\mathbf{G}_{qc,sys} = \begin{bmatrix} \mathbf{G}_0 \\ \mathbf{G}_1 \\ \vdots \\ \mathbf{G}_{c-1} \end{bmatrix} = \left[ \underbrace{\begin{matrix} \mathbf{G}_{0,0} & \mathbf{G}_{0,1} & \cdots & \mathbf{G}_{0,t-c-1} \\ \mathbf{G}_{1,0} & \mathbf{G}_{1,1} & \cdots & \mathbf{G}_{1,t-c-1} \\ \vdots & \vdots & \ddots & \vdots \\ \mathbf{G}_{c-1,0} & \mathbf{G}_{c-1,1} & \cdots & \mathbf{G}_{c-1,t-c-1} \end{matrix}}_{\mathbf{P}} \underbrace{\begin{matrix} \mathbf{I} & \mathbf{O} & \cdots & \mathbf{O} \\ \mathbf{O} & \mathbf{I} & \cdots & \mathbf{O} \\ \vdots & \vdots & \ddots & \vdots \\ \mathbf{O} & \mathbf{O} & \cdots & \mathbf{I} \end{matrix}}_{\mathbf{I}_{cb}} \right], \qquad (4.62)
$$

where $\mathbf{I}$ is a $b \times b$ identity matrix, $\mathbf{O}$ is a $b \times b$ zero matrix (ZM), and $\mathbf{G}_{i,j}$ is a $b \times b$ circulant with $0 \le i < c$ and $0 \le j < t - c$. The generator matrix of the form given by (4.62) is said to be in *systematic circulant form*. $\mathbf{G}_{\text{qc,sys}}$ consists of $c$ row-blocks, $\mathbf{G}_0, \mathbf{G}_1, \ldots, \mathbf{G}_{c-1}$, each being a $1 \times t$ array of $b \times b$ circulants and ZMs, and $t$ column-blocks, each being a $c \times 1$ array of $b \times b$ circulants and ZMs. The $\mathbf{P}$-submatrix on the left-hand side of $\mathbf{G}_{\text{qc,sys}}$ is a $c \times (t - c)$ array of $b \times b$ circulants. Then, the parity-check matrix $\mathbf{H}_{\text{qc,sys}}$ of the $(tb, cb)$ QC code in systematic circulant form is given by:

$$
\mathbf{H}_{\text{qc,sys}} = \underbrace{\begin{bmatrix} \mathbf{I} & \mathbf{O} & \cdots & \mathbf{O} \\ \mathbf{O} & \mathbf{I} & \cdots & \mathbf{O} \\ \vdots & \vdots & \ddots & \vdots \\ \mathbf{O} & \mathbf{O} & \cdots & \mathbf{I} \end{bmatrix}}_{\mathbf{I}_{(t-c)b}} \underbrace{\begin{matrix} \mathbf{G}_{0,0}^T & \mathbf{G}_{1,0}^T & \cdots & \mathbf{G}_{c-1,0}^T \\ \mathbf{G}_{0,1}^T & \mathbf{G}_{1,1}^T & \cdots & \mathbf{G}_{c-1,1}^T \\ \vdots & \vdots & \ddots & \vdots \\ \mathbf{G}_{0,t-c-1}^T & \mathbf{G}_{1,t-c-1}^T & \cdots & \mathbf{G}_{c-1,t-c-1}^T \end{matrix}}_{\mathbf{P}^T}, \tag{4.63}
$$

which is a $(t - c) \times t$ array of $b \times b$ circulants and ZMs.

For example, consider the following generator matrix $\mathbf{G}_{\text{qc,sys}}$ for a $(12, 6)$ QC code $C_{\text{qc}}$:

$$
\mathbf{G}_{\text{qc,sys}} = \begin{bmatrix} 0\,0\,1 & 0\,1\,0 & 1\,0\,0 & 0\,0\,0 \\ 1\,0\,0 & 0\,0\,1 & 0\,1\,0 & 0\,0\,0 \\ 0\,1\,0 & 1\,0\,0 & 0\,0\,1 & 0\,0\,0 \\ 0\,1\,0 & 0\,0\,1 & 0\,0\,0 & 1\,0\,0 \\ 0\,0\,1 & 1\,0\,0 & 0\,0\,0 & 0\,1\,0 \\ 1\,0\,0 & 0\,1\,0 & 0\,0\,0 & 0\,0\,1 \end{bmatrix},
$$

which is a $2 \times 4$ array of circulants of size $3 \times 3$ in the form of (4.62). Then, the parity-check matrix $\mathbf{H}_{\text{qc,sys}}$ of the $(12, 6)$ QC code $C_{\text{qc}}$ in systematic circulant form of (4.63) is given as follows:

$$
\mathbf{H}_{\text{qc,sys}} = \begin{bmatrix} 1\,0\,0 & 0\,0\,0 & 0\,1\,0 & 0\,0\,1 \\ 0\,1\,0 & 0\,0\,0 & 0\,0\,1 & 1\,0\,0 \\ 0\,0\,1 & 0\,0\,0 & 1\,0\,0 & 0\,1\,0 \\ 0\,0\,0 & 1\,0\,0 & 0\,0\,1 & 0\,1\,0 \\ 0\,0\,0 & 0\,1\,0 & 1\,0\,0 & 0\,0\,1 \\ 0\,0\,0 & 0\,0\,1 & 0\,1\,0 & 1\,0\,0 \end{bmatrix}.
$$

### 4.13.3 Encoding of QC Codes

Let

$$
\mathbf{u} = (\underbrace{u_0, u_1, \ldots, u_{b-1}}_{\mathbf{u}_0}, \underbrace{u_b, u_{b+1}, \ldots, u_{2b-1}}_{\mathbf{u}_1}, \ldots, \underbrace{u_{(c-1)b}, u_{(c-1)b+1}, \ldots, u_{cb-1}}_{\mathbf{u}_{c-1}})
$$

$$
= (\mathbf{u}_0, \mathbf{u}_1, \ldots, \mathbf{u}_{c-1}) \tag{4.64}
$$

be a message $\mathbf{u}$ (a $cb$-tuple) over GF(2) to be encoded. The message $\mathbf{u}$ consists of $c$ sections, $\mathbf{u}_0, \mathbf{u}_1, \ldots, \mathbf{u}_{c-1}$, each composed of $b$ bits. The codeword $\mathbf{v}$ corresponding to $\mathbf{u}$ in systematic form is given by:

$$
\begin{aligned}
\mathbf{v} &= \mathbf{u} \cdot \mathbf{G}_{\text{qc,sys}} \\
&= \mathbf{u}_0 \cdot \mathbf{G}_0 + \mathbf{u}_1 \cdot \mathbf{G}_1 + \cdots + \mathbf{u}_{c-1} \cdot \mathbf{G}_{c-1}.
\end{aligned}
\tag{4.65}
$$

Encoding of a QC code in systematic circulant form can be implemented using simple shift-registers with linear complexity [37]. For $0 \le i < c$ and $0 \le j < t - c$, let $\mathbf{g}_{i,j}$ be the generator (the top row) of the circulant $\mathbf{G}_{i,j}$ in the $\mathbf{P}$-submatrix of the generator matrix $\mathbf{G}_{\text{qc,sys}}$ given by (4.62). For $0 \le l < b$, let $\mathbf{g}_{i,j}^{(l)}$ be the $b$-tuple obtained by cyclically shifting every component of $\mathbf{g}_{i,j}$ $l$ places to the right. The $b$-tuple $\mathbf{g}_{i,j}^{(l)}$ is called the *$l$th right cyclic-shift* of $\mathbf{g}_{i,j}$. It is clear that $\mathbf{g}_{i,j}^{(0)} = \mathbf{g}_{i,j}^{(b)} = \mathbf{g}_{i,j}$. Let $\mathbf{u} = (u_0, u_1, \ldots, u_{cb-1}) = (\mathbf{u}_0, \mathbf{u}_1, \ldots, \mathbf{u}_{c-1})$ in the form of (4.64) be the information sequence of $cb$ bits to be encoded. Then, the codeword for $\mathbf{u}$ is $\mathbf{v} = \mathbf{u}\mathbf{G}_{\text{qc,sys}}$, which has the following systematic form:

$$
\mathbf{v} = (\mathbf{p}_0, \mathbf{p}_1, \ldots, \mathbf{p}_{t-c}, \mathbf{u}_0, \mathbf{u}_1, \ldots, \mathbf{u}_{c-1}),
\tag{4.66}
$$

where, for $0 \le j < t - c$, $\mathbf{p}_j = (p_{j,0}, p_{j,1}, \ldots, p_{j,b-1})$ is a section of $b$ parity-check bits. It follows from $\mathbf{v} = \mathbf{u}\mathbf{G}_{\text{qc,sys}}$ that, for $0 \le j < t - c$,

$$
\mathbf{p}_j = \mathbf{u}_0 \mathbf{G}_{0,j} + \mathbf{u}_1 \mathbf{G}_{1,j} + \cdots + \mathbf{u}_{c-1} \mathbf{G}_{c-1,j},
\tag{4.67}
$$

where, for $0 \le i < c$,

$$
\mathbf{u}_i \mathbf{G}_{i,j} = u_{ib} \mathbf{g}_{i,j}^{(0)} + u_{ib+1} \mathbf{g}_{i,j}^{(1)} + \cdots + u_{(i+1)b-1} \mathbf{g}_{i,j}^{(b-1)}.
\tag{4.68}
$$

It follows from (4.67) and (4.68) that the $j$th parity-check section $\mathbf{p}_j$ can be computed, step by step, as the information sequence $\mathbf{u}$ is shifted into the encoder. The information sequence $\mathbf{u} = (\mathbf{u}_0, \mathbf{u}_1, \ldots, \mathbf{u}_{c-1})$ is shifted into the encoder in the order from $\mathbf{u}_{c-1}$ to $\mathbf{u}_0$, i.e., the section $\mathbf{u}_{c-1}$ is shifted into the encoder first and $\mathbf{u}_0$ last. For $1 \le l \le c$, at the $l$th step, the accumulated sum

$$
\mathbf{s}_{l,j} = \mathbf{u}_{c-1} \mathbf{G}_{c-1,j} + \mathbf{u}_{c-2} \mathbf{G}_{c-2,j} + \cdots + \mathbf{u}_{c-l} \mathbf{G}_{c-l,j}
\tag{4.69}
$$

is formed and stored in an accumulator. At the $(l+1)$th step, the partial sum $\mathbf{u}_{c-l-1} \mathbf{G}_{c-l-1,j}$ is computed from (4.68) and added to $\mathbf{s}_{l,j}$ to form the accumulated sum $\mathbf{s}_{l+1,j}$. At the $c$th step, the accumulated sum $\mathbf{s}_{c,j}$ gives the $j$th parity-check section $\mathbf{p}_j$ as shown by (4.67).

By application of the above encoding process and (4.68), the $j$th parity-check section $\mathbf{p}_j$ can be formed with a *cyclic shift-register-adder-accumulator* (CSRAA) circuit as shown in Fig. 4.9 [37]. At the beginning of the first step, $\mathbf{g}_{c-1,j}^{(b-1)}$, the $(b-1)$th right cyclic-shift of the generator $\mathbf{g}_{c-1,j}$ of the circulant $\mathbf{G}_{c-1,j}$ is stored in the feedback shift-register B, and the content of register A is set to zero. When the information bit $u_{cb-1}$ is shifted into the encoder and the channel, the product $u_{cb-1} \mathbf{g}_{c-1,j}^{(b-1)}$ is formed at the outputs of AND gates,

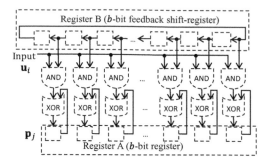

Figure 4.9 A CSRAA encoder circuit.

and is added (through XOR gates) to the content stored in register A (zero at this time). The sum is then stored back into register A. The feedback shift-register B is shifted once to the left. The new content in B is $\mathbf{g}_{c-1,j}^{(b-2)}$. When the next information bit $u_{cb-2}$ is shifted into the encoder and the channel, the product $u_{cb-2}\mathbf{g}_{c-1,j}^{(b-2)}$ is formed at the outputs of the AND gates. This product is then added to the sum $u_{cb-1}\mathbf{g}_{c-1,j}^{(b-1)}$ in the accumulator register A. The sum $u_{cb-2}\mathbf{g}_{c-1,j}^{(b-2)} + u_{cb-1}\mathbf{g}_{c-1,j}^{(b-1)}$ is then stored back into A. The above *shift-add-store* process continues. When the last information bit $u_{(c-1)b}$ of the information section $\mathbf{u}_{c-1}$ has been shifted into the encoder, the register A stores the partial sum $\mathbf{u}_{c-1}\mathbf{G}_{c-1,j}$, which is the contribution to the parity-check section $\mathbf{p}_j$ from the information section $\mathbf{u}_{c-1}$. At this time, $\mathbf{g}_{c-2,j}^{(b-1)}$, the $(b-1)$th right cyclic-shift of the generator $\mathbf{g}_{c-2,j}$ of circulant $\mathbf{G}_{c-2,j}$ is loaded into register B. The shift-add-store process repeats. When the information section $\mathbf{u}_{c-2}$ has been completely shifted into the encoder, the register A stores the accumulated partial sum $\mathbf{u}_{c-2}\mathbf{G}_{c-2,j} + \mathbf{u}_{c-1}\mathbf{G}_{c-1,j}$, which is the contribution to the parity-check section $\mathbf{p}_j$ from the information sections $\mathbf{u}_{c-1}$ and $\mathbf{u}_{c-2}$. The above process repeats until the last information section $\mathbf{u}_0$ has been shifted into the encoder. At this time, the accumulator register A contains the $j$th parity-check section $\mathbf{p}_j$.

To form the $t-c$ parity-check sections, $t-c$ CSRAA circuits are needed, one for computing each parity-check section. A block diagram for the entire encoder is shown in Fig. 4.10. All the parity-check sections are formed at the same time in parallel, and they are then shifted into the channel serially. The encoding circuit consists of $t-c$ CSRAA circuits with a total of $2(t-c)b$ flip-flops, $(t-c)b$ AND gates, and $(t-c)b$ two-input XOR gates (modulo-2 adders). The encoding is accomplished in linear time with complexity linearly proportional to the number of parity-check bits $(t-c)b$.

For most methods of constructing QC codes (which will be shown in Chapters 12–15), the code construction starts with a parity-check matrix $\mathbf{H}_{qc}$ which is an array of circulants. In general, $\mathbf{H}_{qc}$ is not in systematic circulant form. In the following, we present a method for finding a generator matrix in systematic circulant form based on the parity-check array $\mathbf{H}_{qc}$ of a QC code.

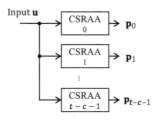

Figure 4.10 A CSRAA-based QC code encoder.

Consider the following parity-check matrix:

$$\mathbf{H}_{qc} = \begin{bmatrix} \mathbf{A}_{0,0} & \mathbf{A}_{0,1} & \cdots & \mathbf{A}_{0,t-1} \\ \mathbf{A}_{1,0} & \mathbf{A}_{1,1} & \cdots & \mathbf{A}_{1,t-1} \\ \vdots & \vdots & \ddots & \vdots \\ \mathbf{A}_{t-c-1,0} & \mathbf{A}_{t-c-1,1} & \cdots & \mathbf{A}_{t-c-1,t-1} \end{bmatrix}, \tag{4.70}$$

which is a $(t-c) \times t$ array of $b \times b$ circulants over GF(2). It is a $(t-c)b \times tb$ matrix over GF(2). The null space of $\mathbf{H}_{qc}$ gives a QC code $C_{qc}$ of length $tb$. If $\mathbf{H}_{qc}$ is a full-rank matrix, i.e., rank$(\mathbf{H}_{qc}) = (t - c)b$, $C_{qc}$ is a $(tb, cb)$ QC code. If $\mathbf{H}_{qc}$ is not a full-rank matrix, i.e., rank$(\mathbf{H}_{qc}) < (t - c)b$, then $C_{qc}$ has dimension larger than $cb$. Given the parity-check matrix $\mathbf{H}_{qc}$ of a QC code $C_{qc}$ in the form of (4.70), we need to find its generator matrix in the systematic circulant form of (4.62) for efficient encoding. There are two cases to be considered: (1) $\mathbf{H}_{qc}$ is a full-rank matrix; and (2) $\mathbf{H}_{qc}$ is not a full-rank matrix.

We first consider the case that the parity-check matrix $\mathbf{H}_{qc}$ of a QC code is a full-rank matrix. In this case, we assume that the columns of circulants of $\mathbf{H}_{qc}$ given by (4.70) are arranged in such a way that the following $(t - c) \times (t - c)$ subarray $\mathbf{D}$ of $\mathbf{H}_{qc}$ (the leftmost $t - c$ column-blocks of $\mathbf{H}_{qc}$) has the same rank as $\mathbf{H}_{qc}$, i.e., rank$(\mathbf{D}) = $ rank$(\mathbf{H}_{qc}) = (t - c)b$,

$$\mathbf{D} = \begin{bmatrix} \mathbf{A}_{0,0} & \mathbf{A}_{0,1} & \cdots & \mathbf{A}_{0,t-c-1} \\ \mathbf{A}_{1,0} & \mathbf{A}_{1,1} & \cdots & \mathbf{A}_{1,t-c-1} \\ \vdots & \vdots & \ddots & \vdots \\ \mathbf{A}_{t-c-1,0} & \mathbf{A}_{t-c-1,1} & \cdots & \mathbf{A}_{t-c-1,t-c-1} \end{bmatrix}. \tag{4.71}$$

The matrix in the form of (4.62) is a generator matrix of the QC code $C_{qc}$ given by the null space of $\mathbf{H}_{qc}$ if and only if

$$\mathbf{H}_{qc}\mathbf{G}_{qc,sys}^{T} = \mathbf{O}, \tag{4.72}$$

where $\mathbf{O}$ is a $(t - c)b \times cb$ ZM.

For $0 \le i < c$ and $0 \le j < t-c$, let $\mathbf{g}_{i,j}$ be the generator of the circulant $\mathbf{G}_{i,j}$ in the $\mathbf{P}$-submatrix of $\mathbf{G}_{qc,sys}$. Once we have found the generators, $\mathbf{g}_{i,j}$, from (4.72), we can form all the circulants $\mathbf{G}_{i,j}$ of $\mathbf{G}_{qc,sys}$. Let $\mathbf{x} = (1, 0, \ldots, 0)$ be a $b$-tuple with a "1" in the first position (which can be viewed as the generator of the $b \times b$ identity matrix $\mathbf{I}$), and $\mathbf{0} = (0, 0, \ldots, 0)$ be the all-zero $b$-tuple (which can be viewed as the generator of the $b \times b$ ZM $\mathbf{O}$). For $0 \le i < c$, the first row of the $i$th row-block $\mathbf{G}_i$ of $\mathbf{G}_{\text{qc,sys}}$ given by (4.62) in terms of the generators of its constituent circulants is:

$$\mathbf{g}_i = (\mathbf{g}_{i,0} \; \mathbf{g}_{i,1} \cdots \mathbf{g}_{i,t-c-1} \; \mathbf{0} \cdots \mathbf{0} \; \mathbf{x} \; \mathbf{0} \cdots \mathbf{0}), \tag{4.73}$$

where the $b$-tuple $\mathbf{x} = (1, 0, \ldots, 0)$ is at the $(t - c + i)$th position of $\mathbf{g}_i$ (section-wise). The row $\mathbf{g}_i$ comprises $t$ sections, each consisting of $b$ bits. The first $t - c$ sections are simply the generators of the $t - c$ circulants $\mathbf{G}_{i,0}, \mathbf{G}_{i,1}, \ldots, \mathbf{G}_{i,t-c-1}$ of the $i$th row-block $\mathbf{G}_i$ of $\mathbf{G}_{\text{qc,sys}}$ given by (4.62).

It follows from (4.72) that we have

$$\mathbf{H}_{\text{qc}} \mathbf{g}_i^T = \mathbf{0}, \tag{4.74}$$

where $\mathbf{0}$ is the all-zero $(t - c)b$-tuple for $0 \le i < c$.

For $0 \le i < c$, let $\mathbf{z}_i$ be the first $t - c$ sections of $\mathbf{g}_i$, i.e.,

$$\mathbf{z}_i = (\mathbf{g}_{i,0} \; \mathbf{g}_{i,1} \; \cdots \; \mathbf{g}_{i,t-c-1}) \tag{4.75}$$

and $\mathbf{M}$ be the $(t - c + i)$th column-block of $\mathbf{H}_{\text{qc}}$, i.e.,

$$\mathbf{M}_{t-c+i} = \begin{bmatrix} \mathbf{A}_{0,t-c+i} \\ \mathbf{A}_{1,t-c+i} \\ \vdots \\ \mathbf{A}_{t-c-1,t-c+i} \end{bmatrix}. \tag{4.76}$$

Carrying out the product $\mathbf{H}_{\text{qc}} \mathbf{g}_i^T = \mathbf{0}$ given by (4.74), we have

$$\mathbf{D} \mathbf{z}_i^T + \mathbf{M}_{t-c+i} \mathbf{x}^T = \mathbf{0}. \tag{4.77}$$

Because $\mathbf{D}$ is a $(t - c)b \times (t - c)b$ square matrix and has full rank $(t - c)b$, it is nonsingular and has an inverse $\mathbf{D}^{-1}$ (over GF(2)). Because all the matrices and vectors are over GF(2), it follows from (4.77) that

$$\mathbf{z}_i^T = \mathbf{D}^{-1} \mathbf{M}_{t-c+i} \mathbf{x}^T. \tag{4.78}$$

Solving (4.78) for $0 \le i < c$, we obtain $\mathbf{z}_0, \mathbf{z}_1, \ldots, \mathbf{z}_{c-1}$. From $\mathbf{z}_0, \mathbf{z}_1, \ldots, \mathbf{z}_{c-1}$ and (4.75), we obtain all the generators, $\mathbf{g}_{i,j}$, of the circulants, $\mathbf{G}_{i,j}$, of $\mathbf{G}_{\text{qc,sys}}$. Then, $\mathbf{G}_{\text{qc,sys}}$ can be constructed readily.

## 4.13.4 Generator and Parity-Check Matrices in Semi-systematic Circulant Form

The above development is based on the assumption that there is $(t - c) \times (t - c)$ subarray $\mathbf{D}$ in $\mathbf{H}_{\text{qc}}$ with the same rank as $\mathbf{H}_{\text{qc}}$. However, there may not exist such a subarray in the parity-check matrix for some cases. In the following, we consider this case and the case in which the parity-check matrix $\mathbf{H}_{\text{qc}}$ is not full rank. In either case, the generator matrix of the QC code $C_{qc}$ given by the null space of $\mathbf{H}_{qc}$ cannot be put exactly into the systematic circulant form given by (4.62); however, it can be put into a *semi-systematic circulant form* that allows encoding with simple shift-registers as shown in Fig. 4.9.

To construct the generator matrix in semi-systematic circulant form, we first find the *least number* of column-blocks in $\mathbf{H}_{qc}$, say $l$ with $t - c \leq l \leq t$, such that these $l$ column-blocks form a $(t - c) \times l$ subarray $\mathbf{D}^*$, whose rank is equal to the rank $r$ of $\mathbf{H}_{qc}$. Next, we permute the column-blocks of $\mathbf{H}_{qc}$ to form a new $(t - c) \times t$ array $\mathbf{H}^*_{qc}$ of circulants,[3] such that the first (or the leftmost) $l$ column-blocks of $\mathbf{H}^*_{qc}$ form the subarray $\mathbf{D}^*$. Then, we have

$$
\mathbf{D}^* = \begin{bmatrix} \mathbf{A}_{0,0} & \mathbf{A}_{0,1} & \cdots & \mathbf{A}_{0,l-1} \\ \mathbf{A}_{1,0} & \mathbf{A}_{1,1} & \cdots & \mathbf{A}_{1,l-1} \\ \vdots & \vdots & \ddots & \vdots \\ \mathbf{A}_{t-c,0} & \mathbf{A}_{t-c,1} & \cdots & \mathbf{A}_{t-c,l-1} \end{bmatrix}, \tag{4.79}
$$

which has rank $r$. Thus, there are $(t - c)b - r$ linearly dependent rows in $\mathbf{D}^*$ and $lb - r$ linearly dependent columns in $\mathbf{D}^*$.

The generator matrix of the QC code $C_{qc}$ given by the null space of $\mathbf{H}^*_{qc}$ can be put in the following *semi-systematic circulant form*:

$$
\mathbf{G}_{qc,\text{semi-sys}} = \begin{bmatrix} \mathbf{Q} \\ \mathbf{G}^*_{qc,\text{sys}} \end{bmatrix}, \tag{4.80}
$$

which consists of two submatrices $\mathbf{Q}$ and $\mathbf{G}^*_{qc,\text{sys}}$. The submatrix $\mathbf{Q}$ is an $(lb - r) \times tb$ matrix. The submatrix $\mathbf{G}^*_{qc,\text{sys}}$ is a $(t - l)b \times tb$ matrix in systematic circulant form,

$$
\mathbf{G}^*_{qc,\text{sys}} = \begin{bmatrix} \mathbf{G}_{0,0} & \mathbf{G}_{0,1} & \cdots & \mathbf{G}_{0,l-1} & \mathbf{I} & \mathbf{O} & \cdots & \mathbf{O} \\ \mathbf{G}_{1,0} & \mathbf{G}_{1,1} & \cdots & \mathbf{G}_{1,l-1} & \mathbf{O} & \mathbf{I} & \cdots & \mathbf{O} \\ \vdots & \vdots & \ddots & \vdots & \vdots & \vdots & \ddots & \vdots \\ \mathbf{G}_{t-l-1,0} & \mathbf{G}_{t-l-1,1} & \cdots & \mathbf{G}_{t-l-1,l-1} & \mathbf{O} & \mathbf{O} & \cdots & \mathbf{I} \end{bmatrix}, \tag{4.81}
$$

where each submatrix $\mathbf{G}_{i,j}$ is a $b \times b$ circulant, $\mathbf{I}$ is a $b \times b$ identity matrix, and $\mathbf{O}$ is a $b \times b$ ZM. Following (4.72), we have

$$
\mathbf{H}^*_{qc} \begin{bmatrix} \mathbf{Q} \\ \mathbf{G}^*_{qc,\text{sys}} \end{bmatrix}^T = \mathbf{O}, \tag{4.82}
$$

where $\mathbf{O}$ is a $(t - c)b \times cb$ ZM. Carrying out the product on the left side of the above equation, we have

$$
\mathbf{H}^*_{qc} \mathbf{Q}^T = \mathbf{O}, \tag{4.83}
$$

---

[3]Note that the null space of $\mathbf{H}^*_{qc}$ gives a code that is combinatorially equivalent to the code given by the null space of $\mathbf{H}_{qc}$, i.e., they give the same error performance with the same decoding.

where $\mathbf{O}$ is a $(t-c)b \times (lb-r)$ ZM, and

$$\mathbf{H}_{qc}^* \mathbf{G}_{qc,sys}^{* \ T} = \mathbf{O}, \tag{4.84}$$

where $\mathbf{O}$ is a $(t-c)b \times (t-l)b$ ZM.

Following (4.84), the generator $\mathbf{g}_{i,j}$ of the $b \times b$ circulant $\mathbf{G}_{i,j}$ in $\mathbf{G}_{qc,sys}^*$ with $0 \le i < t-l$ and $0 \le j < l$ can be obtained by solving (4.77) with $\mathbf{D}$ replaced by $\mathbf{D}^*$ and setting the $lb-r$ elements in $\mathbf{z}_i = (\mathbf{g}_{i,0}\mathbf{g}_{i,1} \cdots \mathbf{g}_{i,l-1})$ whose positions correspond to the $lb-r$ linearly dependent columns of $\mathbf{D}^*$ to zeros.

The submatrix $\mathbf{Q}$ of $\mathbf{G}_{qc,semi-sys}$ specified by (4.83) is an $(lb-r) \times tb$ matrix whose rows are linearly independent. To obtain $\mathbf{Q}$, let $d_0, d_1, \ldots, d_{l-1}$ be the number of linearly dependent columns in the zeroth, first, $\ldots, (l-1)$th column-blocks of $\mathbf{D}^*$, respectively, such that

$$d_0 + d_1 + \cdots + d_{l-1} = lb - r. \tag{4.85}$$

For a $b \times b$ circulant, if it is full rank, all the columns (or rows) are linearly independent. If it is not full rank, then its rank $\lambda$ is less than $b$ and any $\lambda$ consecutive columns (or rows) of the circulant are linearly independent and the other $b - \lambda$ columns (or rows) are linearly dependent. With this structure, we take the last $b - d_j$ columns of the $j$th column-block of $\mathbf{D}^*$ as the linearly independent columns. Then, the first $d_0, d_1, \ldots, d_{l-1}$ columns of the zeroth, first, $\ldots, (l-1)$th column-blocks of $\mathbf{D}^*$ are linearly dependent columns. The matrix $\mathbf{Q}$ can be put into the following partial circulant form:

$$\mathbf{Q} = \begin{bmatrix} \mathbf{Q}_0 \\ \mathbf{Q}_1 \\ \vdots \\ \mathbf{Q}_{l-1} \end{bmatrix} = \begin{bmatrix} \mathbf{Q}_{0,0} & \mathbf{Q}_{0,1} & \cdots & \mathbf{Q}_{0,l-1} & \mathbf{O}_{0,0} & \mathbf{O}_{0,1} & \cdots & \mathbf{O}_{0,t-l-1} \\ \mathbf{Q}_{1,0} & \mathbf{Q}_{1,1} & \cdots & \mathbf{Q}_{1,l-1} & \mathbf{O}_{1,0} & \mathbf{O}_{1,1} & \cdots & \mathbf{O}_{1,t-l-1} \\ \vdots & \vdots & \ddots & \vdots & \vdots & \vdots & \ddots & \vdots \\ \mathbf{Q}_{l-1,0} & \mathbf{Q}_{l-1,1} & \cdots & \mathbf{Q}_{l-1,l-1} & \mathbf{O}_{l-1,0} & \mathbf{O}_{l-1,1} & \cdots & \mathbf{O}_{l-1,t-l-1} \end{bmatrix}, \tag{4.86}$$

where (1) $\mathbf{O}_{i,j}$ is a $d_i \times b$ zero matrix with $0 \le i < l$ and $0 \le j < t-l$; and (2) $\mathbf{Q}_{i,j}$ is a $d_i \times b$ *partial circulant* obtained by cyclically shifting its first row $d_i - 1$ times to form the other $d_i - 1$ rows with $0 \le i, j < l$.

For $0 \le i < l$, let

$$\begin{aligned} \mathbf{q}_i &= (\mathbf{w}_i, \mathbf{0}) \\ &= (q_{i,0}, q_{i,1}, \ldots, q_{i,lb-1}, 0, 0, \ldots, 0) \end{aligned} \tag{4.87}$$

be the first row of the $i$th row-block $\mathbf{Q}_i$ of $\mathbf{Q}$, which consists of two parts, the right part $\mathbf{0}$ and left part $\mathbf{w}_i$. The right part $\mathbf{0} = (0, 0, \ldots, 0)$ is the all-zero $(t-l)b$-tuple. The left part $\mathbf{w}_i = (q_{i,0}, q_{i,1}, \ldots, q_{i,lb-1})$ is an $lb$-tuple, which correspond to the $lb$ columns of $\mathbf{D}^*$. The $lb - r$ bits of $\mathbf{w}_i$ that correspond to the linearly dependent columns of $\mathbf{D}^*$, called *dependent bits*, have the following form:

$$(\mathbf{0}_0, \ldots, \mathbf{0}_{i-1}, \mathbf{x}_i, \mathbf{0}_{i+1}, \ldots, \mathbf{0}_{l-1}), \tag{4.88}$$

where, for $e \neq i$, $\mathbf{0}_e$ is a zero $d_e$-tuple and $\mathbf{x}_i = (1, 0, \ldots, 0)$ is a $d_i$-tuple with "1" in its first position. From the structure of $\mathbf{w}_i$, the number of unknown components in $\mathbf{w}_i$ is $r$, the rank of $\mathbf{D}^*$ (or $\mathbf{H}^*_{qc}$).

The condition $\mathbf{H}^*_{qc} \mathbf{Q}^T = \mathbf{O}$ in (4.83) gives the following equation for $0 \leq i < l$:

$$\mathbf{D}^* \mathbf{w}_i^T = \begin{bmatrix} \mathbf{A}_{0,0} & \mathbf{A}_{0,1} & \cdots & \mathbf{A}_{0,l-1} \\ \mathbf{A}_{1,0} & \mathbf{A}_{1,1} & \cdots & \mathbf{A}_{1,l-1} \\ \vdots & \vdots & \ddots & \vdots \\ \mathbf{A}_{t-c-1,0} & \mathbf{A}_{t-c-1,1} & \cdots & \mathbf{A}_{t-c-1,l-1} \end{bmatrix} \begin{bmatrix} q_{i,0} \\ q_{i,1} \\ \vdots \\ q_{i,lb-1} \end{bmatrix} = \mathbf{0}. \quad (4.89)$$

Solving (4.89), we find $\mathbf{w}_i = (q_{i,0}, q_{i,1}, \ldots, q_{i,lb-1})$ for $0 \leq i < l$. Divide $\mathbf{w}_i$ into $l$ sections, denoted by $\mathbf{w}_{i,0}, \mathbf{w}_{i,1}, \ldots, \mathbf{w}_{i,l-1}$, each consisting of $b$ consecutive components of $\mathbf{w}_i$. For $0 \leq i, j < l$, the partial circulant $\mathbf{Q}_{i,j}$ of $\mathbf{Q}$ is obtained by using $\mathbf{w}_{i,j}$ as the first row, and then cyclically shifting it $d_i - 1$ times to form the other $d_i - 1$ rows. From the partial circulants $\mathbf{Q}_{i,j}$s with $0 \leq i, j < l$, we form the $\mathbf{Q}$ submatrix of $\mathbf{G}_{qc,semi-sys}$. By combining $\mathbf{Q}$ and $\mathbf{G}^*_{qc,sys}$ into the form of (4.80), we obtain the generator matrix $\mathbf{G}_{qc,semi-sys}$ of the QC code $\mathcal{C}_{qc}$ given by the null space of a nonfull-rank array $\mathbf{H}_{qc}$ of circulants.

With $\mathbf{G}_{qc,sem-sys}$ in the form of (4.80) as a generator matrix, an encoder for the QC code $\mathcal{C}_{qc}$ with $l$ CSRAA circuits of the form given by Fig. 4.10 can be implemented. Encoding consists of two phases. An information sequence $\mathbf{a} = (a_0, a_1, \ldots, a_{tb-r-1})$ of $tb - r$ bits is divided into two parts, $\mathbf{a}^{(1)} = (a_{lb-r}, a_{lb-r+1}, \ldots, a_{tb-r-1})$ and $\mathbf{a}^{(2)} = (a_0, a_1, \ldots, a_{lb-r-1})$. Then, $\mathbf{a} = (\mathbf{a}^{(2)}, \mathbf{a}^{(1)})$. The information bits are shifted into the encoder serially in the order of bit $a_{tb-r-1}$ first and bit $a_0$ last. The first part $\mathbf{a}^{(1)}$ of $\mathbf{a}$, consisting of $(t-l)b$ bits, is first shifted into the encoder and is encoded into a codeword $\mathbf{v}^{(1)}$ in the subcode $\mathcal{C}_{qc}^{(1)}$ generated by $\mathbf{G}^*_{qc,sys}$. The $l$ parity sections are generated at the same time when $\mathbf{a}^{(1)}$ has been completely shifted into the encoder, as described in Section 4.13.3. After encoding of $\mathbf{a}^{(1)}$, the second part $\mathbf{a}^{(2)}$ of $\mathbf{a}$ is then shifted into the encoder and is encoded into a codeword $\mathbf{v}^{(2)}$ in the subcode $\mathcal{C}_{qc}^{(2)}$ generated by $\mathbf{Q}$. By adding $\mathbf{v}^{(1)}$ and $\mathbf{v}^{(2)}$, we obtain the codeword $\mathbf{v} = \mathbf{v}^{(1)} + \mathbf{v}^{(2)}$ for the information sequence $\mathbf{a} = (\mathbf{a}^{(2)}, \mathbf{a}^{(1)})$.

To encode $\mathbf{a}^{(2)}$ using $\mathbf{Q}$, we divide $\mathbf{a}^{(2)}$ into $l$ sections, $\mathbf{a}_0^{(2)}, \mathbf{a}_1^{(2)}, \ldots, \mathbf{a}_{l-1}^{(2)}$, with $d_0, d_1, \ldots, d_{l-1}$ bits, respectively. Then, the codeword for $\mathbf{a}^{(2)}$ is of the form

$$\mathbf{v}^{(2)} = (\mathbf{v}_0^{(2)}, \mathbf{v}_1^{(2)}, \ldots, \mathbf{v}_{l-1}^{(2)}, \mathbf{0}, \mathbf{0}, \ldots, \mathbf{0}), \quad (4.90)$$

which consists of $t - l$ zero sections and $l$ nonzero sections, $\mathbf{v}_0^{(2)}, \mathbf{v}_1^{(2)}, \ldots, \mathbf{v}_{l-1}^{(2)}$, each section, zero or nonzero, consisting of $b$ bits. For $0 \leq j < l$,

$$\mathbf{v}_j^{(2)} = \mathbf{a}_0^{(2)} \mathbf{Q}_{0,j} + \mathbf{a}_1^{(2)} \mathbf{Q}_{1,j} + \cdots + \mathbf{a}_{l-1}^{(2)} \mathbf{Q}_{l-1,j}. \quad (4.91)$$

Because each $\mathbf{Q}_{i,j}$ in $\mathbf{Q}$ with $0 \le i, j < l$ is a partial circulant with $d_i$ rows, encoding of $\mathbf{a}^{(2)}$ can be accomplished with the same $l$ CSRAA circuits as used for encoding $\mathbf{a}^{(1)}$. For $0 \le j < l$, at the end of encoding $\mathbf{a}^{(1)}$, the accumulator A of the $j$th CSRAA circuit stores the $j$th parity-check section $\mathbf{v}_j^{(1)}$ of $\mathbf{v}^{(1)}$. In the second phase of encoding, $\mathbf{a}_0^{(2)}, \mathbf{a}_1^{(2)}, \ldots, \mathbf{a}_{l-1}^{(2)}$ are shifted into the encoder one at a time and the generators, $\mathbf{w}_{0,j}, \mathbf{w}_{1,j}, \ldots, \mathbf{w}_{l-1,j}$, of the partial circulants $\mathbf{Q}_{0,j}, \mathbf{Q}_{1,j}, \ldots, \mathbf{Q}_{l-1,j}$ are stored in register B of the $j$th CSRAA circuit in turn. Then, cyclically shifts the register B, $d_0, d_1, \ldots, d_{l-1}$ times, in turn. At the end of $d_0 + d_1 + \cdots + d_{l-1} = lb - r$ shifts, the $j$th parity-check section, $\mathbf{v}_j^{(1)} + \mathbf{v}_j^{(2)}$, is stored in the accumulator register A of the $j$th CSRAA circuit.

Note that in the above encoding, the bits in the first part $\mathbf{a}^{(1)}$ are shifted simultaneously into the CSRAA circuits and the channel; however, the bits in the second part $\mathbf{a}^{(2)}$ are shifted only into the CSRAA circuits. Therefore, the codeword $\mathbf{v} = (v_0, v_1, \ldots, v_{tb-1})$ for the information sequence $\mathbf{a} = (\mathbf{a}^{(2)}, \mathbf{a}^{(1)})$ is not completely systematic. Only the rightmost $(t - l)b$ bits of $\mathbf{v}$ are identical to the information bits in $\mathbf{a}^{(1)}$, i.e., $(v_{lb}, v_{lb+1}, \ldots, v_{tb-1}) = \mathbf{a}^{(1)}$. The next $lb - r$ bits, $v_r, v_{r+1}, \ldots, v_{lb-1}$, of $\mathbf{v}$ are not identical to the information bits in $\mathbf{a}^{(2)}$. The leftmost $r$ bits $v_0, v_1, \ldots, v_{r-1}$ of $\mathbf{v}$ are the parity-check bits.

**Example 4.14** Consider the following parity-check matrix $\mathbf{H}_{\mathrm{qc}}$ of size $16 \times 20$:

$$
\mathbf{H}_{\mathrm{qc}} = \left[\begin{array}{cccc|cccc|cccc|cccc|cccc}
0&0&0&0&1&0&0&0&0&1&0&0&0&0&1&0&0&0&0&1\\
0&0&0&0&0&1&0&0&0&0&1&0&0&0&0&1&1&0&0&0\\
0&0&0&0&0&0&1&0&0&0&0&1&1&0&0&0&0&1&0&0\\
0&0&0&0&0&0&0&1&1&0&0&0&0&1&0&0&0&0&1&0\\
\hline
1&0&0&0&0&0&0&1&0&0&1&0&0&0&0&0&0&1&0&0\\
0&1&0&0&1&0&0&0&0&0&0&1&0&0&0&0&0&0&1&0\\
0&0&1&0&0&1&0&0&1&0&0&0&0&0&0&0&0&0&0&1\\
0&0&0&1&0&0&1&0&0&1&0&0&0&0&0&0&1&0&0&0\\
\hline
0&1&0&0&0&0&1&0&1&0&0&0&0&0&0&1&0&0&0&0\\
0&0&1&0&0&0&0&1&0&1&0&0&1&0&0&0&0&0&0&0\\
0&0&0&1&1&0&0&0&0&0&1&0&0&1&0&0&0&0&0&0\\
1&0&0&0&0&1&0&0&0&0&0&1&0&0&1&0&0&0&0&0\\
\hline
0&0&1&0&0&0&0&0&0&0&0&1&0&1&0&0&1&0&0&0\\
0&0&0&1&0&0&0&0&1&0&0&0&0&0&1&0&0&1&0&0\\
1&0&0&0&0&0&0&0&0&1&0&0&0&0&0&1&0&0&1&0\\
0&1&0&0&0&0&0&0&0&0&1&0&1&0&0&0&0&0&0&1
\end{array}\right],
$$

which is a $4 \times 5$ array of $4 \times 4$ circulants. $\mathbf{H}_{\mathrm{qc}}$ is a full-rank matrix with rank$(\mathbf{H}_{\mathrm{qc}}) = 16$. The null space of $\mathbf{H}_{\mathrm{qc}}$ gives a $(20, 4)$ QC code $C_{\mathrm{qc}}$. The left four column-blocks of $\mathbf{H}_{\mathrm{qc}}$ form a $16 \times 16$ matrix with rank 16, i.e., there is a $4 \times 4$ subarray of $\mathbf{H}_{\mathrm{qc}}$:

$$D = \begin{bmatrix} 0\ 0\ 0\ 0 & 1\ 0\ 0\ 0 & 0\ 1\ 0\ 0 & 0\ 0\ 1\ 0 \\ 0\ 0\ 0\ 0 & 0\ 1\ 0\ 0 & 0\ 0\ 1\ 0 & 0\ 0\ 0\ 1 \\ 0\ 0\ 0\ 0 & 0\ 0\ 1\ 0 & 0\ 0\ 0\ 1 & 1\ 0\ 0\ 0 \\ 0\ 0\ 0\ 0 & 0\ 0\ 0\ 1 & 1\ 0\ 0\ 0 & 0\ 1\ 0\ 0 \\ 1\ 0\ 0\ 0 & 0\ 0\ 0\ 1 & 0\ 0\ 1\ 0 & 0\ 0\ 0\ 0 \\ 0\ 1\ 0\ 0 & 1\ 0\ 0\ 0 & 0\ 0\ 0\ 1 & 0\ 0\ 0\ 0 \\ 0\ 0\ 1\ 0 & 0\ 1\ 0\ 0 & 1\ 0\ 0\ 0 & 0\ 0\ 0\ 0 \\ 0\ 0\ 0\ 1 & 0\ 0\ 1\ 0 & 0\ 1\ 0\ 0 & 0\ 0\ 0\ 0 \\ 0\ 1\ 0\ 0 & 0\ 0\ 1\ 0 & 1\ 0\ 0\ 0 & 0\ 0\ 0\ 1 \\ 0\ 0\ 1\ 0 & 0\ 0\ 0\ 1 & 0\ 1\ 0\ 0 & 1\ 0\ 0\ 0 \\ 0\ 0\ 0\ 1 & 1\ 0\ 0\ 0 & 0\ 0\ 1\ 0 & 0\ 1\ 0\ 0 \\ 1\ 0\ 0\ 0 & 0\ 1\ 0\ 0 & 0\ 0\ 0\ 1 & 0\ 0\ 1\ 0 \\ 0\ 0\ 1\ 0 & 0\ 0\ 0\ 0 & 0\ 0\ 0\ 1 & 0\ 1\ 0\ 0 \\ 0\ 0\ 0\ 1 & 0\ 0\ 0\ 0 & 1\ 0\ 0\ 0 & 0\ 0\ 1\ 0 \\ 1\ 0\ 0\ 0 & 0\ 0\ 0\ 0 & 0\ 1\ 0\ 0 & 0\ 0\ 0\ 1 \\ 0\ 1\ 0\ 0 & 0\ 0\ 0\ 0 & 0\ 0\ 1\ 0 & 1\ 0\ 0\ 0 \end{bmatrix},$$

which has the same rank as $\mathbf{H}_{qc}$. Then, it follows from (4.62) that the generator matrix of the code $C_{qc}$ in the systematic circulant form is given below:

$$\mathbf{G}_{qc,sys} = \begin{bmatrix} \mathbf{G}_0 \end{bmatrix} = \begin{bmatrix} \mathbf{G}_{0,0}\ \mathbf{G}_{0,1}\ \mathbf{G}_{0,2}\ \mathbf{G}_{0,3}\ \mathbf{I} \end{bmatrix}.$$

Following (4.78), we can find the generators of the four circulants, $\mathbf{G}_{0,0}$, $\mathbf{G}_{0,1}$, $\mathbf{G}_{0,2}$, $\mathbf{G}_{0,3}$, in the generator matrix $\mathbf{G}_{qc,sys}$:

$$\begin{aligned} \mathbf{z}_0^T &= \mathbf{D}^{-1}\mathbf{M}_4\mathbf{x}^T \\ &= (1\ 0\ 0\ 0,\ 1\ 0\ 0\ 0,\ 0\ 1\ 1\ 1,\ 1\ 0\ 0\ 0)^T, \end{aligned}$$

where $\mathbf{M}_4$ is the last column-block of $\mathbf{H}_{qc}$ and $\mathbf{x} = (1\ 0\ 0\ 0)$. Then, the generators of the four circulants $\mathbf{G}_{0,0}$, $\mathbf{G}_{0,1}$, $\mathbf{G}_{0,2}$, $\mathbf{G}_{0,3}$ are

$$\mathbf{g}_{0,0} = (1\ 0\ 0\ 0), \mathbf{g}_{0,1} = (1\ 0\ 0\ 0), \mathbf{g}_{0,2} = (0\ 1\ 1\ 1), \mathbf{g}_{0,3} = (1\ 0\ 0\ 0).$$

Then, the generator matrix $\mathbf{G}_{qc,sys}$ of the $(20, 4)$ QC code $C_{qc}$ in systematic circulant form is

$$\mathbf{G}_{qc,sys} = \begin{bmatrix} 1\ 0\ 0\ 0 & 1\ 0\ 0\ 0 & 0\ 1\ 1\ 1 & 1\ 0\ 0\ 0 & 1\ 0\ 0\ 0 \\ 0\ 1\ 0\ 0 & 0\ 1\ 0\ 0 & 1\ 0\ 1\ 1 & 0\ 1\ 0\ 0 & 0\ 1\ 0\ 0 \\ 0\ 0\ 1\ 0 & 0\ 0\ 1\ 0 & 1\ 1\ 0\ 1 & 0\ 0\ 1\ 0 & 0\ 0\ 1\ 0 \\ 0\ 0\ 0\ 1 & 0\ 0\ 0\ 1 & 1\ 1\ 1\ 0 & 0\ 0\ 0\ 1 & 0\ 0\ 0\ 1 \end{bmatrix}.$$

It can be verified that $\mathbf{H}_{qc}\mathbf{G}_{qc,sys}^T = \mathbf{O}$.  ▲▲

In the next example, we consider a parity-check matrix which is not full rank.

**Example 4.15** Consider the following parity-check matrix $\mathbf{H}_{qc}$ of size $9 \times 12$:

$$\mathbf{H}_{qc} = \begin{bmatrix} 0 & 0 & 0 & 1 & 0 & 0 & 0 & 1 & 0 & 0 & 0 & 1 \\ 0 & 0 & 0 & 0 & 1 & 0 & 0 & 0 & 1 & 1 & 0 & 0 \\ 0 & 0 & 0 & 0 & 0 & 1 & 1 & 0 & 0 & 0 & 1 & 0 \\ 1 & 0 & 0 & 0 & 0 & 0 & 0 & 0 & 1 & 0 & 1 & 0 \\ 0 & 1 & 0 & 0 & 0 & 0 & 1 & 0 & 0 & 0 & 0 & 1 \\ 0 & 0 & 1 & 0 & 0 & 0 & 0 & 1 & 0 & 1 & 0 & 0 \\ 0 & 1 & 0 & 0 & 0 & 1 & 0 & 0 & 0 & 1 & 0 & 0 \\ 0 & 0 & 1 & 1 & 0 & 0 & 0 & 0 & 0 & 0 & 1 & 0 \\ 1 & 0 & 0 & 0 & 1 & 0 & 0 & 0 & 0 & 0 & 0 & 1 \end{bmatrix},$$

which is a $3 \times 4$ array of circulants of size $3 \times 3$, i.e., $t = 4$, $c = 1$. The rank of $\mathbf{H}_{qc}$ is 7, i.e., $\mathbf{H}_{qc}$ is not a full-rank matrix and has two redundant rows. The null space of $\mathbf{H}_{qc}$ gives a $(12, 5)$ QC code $C_{qc}$. Through calculation, we find that the rightmost nine columns of $\mathbf{H}_{qc}$, as a $3 \times 3$ subarray $\mathbf{D}^*$ of $3 \times 3$ circulants, have rank 7, which is the same as $\mathbf{H}_{qc}$. Then, we have

$$\mathbf{D}^* = \begin{bmatrix} 1 & 0 & 0 & 0 & 1 & 0 & 0 & 0 & 1 \\ 0 & 1 & 0 & 0 & 0 & 1 & 1 & 0 & 0 \\ 0 & 0 & 1 & 1 & 0 & 0 & 0 & 1 & 0 \\ 0 & 0 & 0 & 0 & 0 & 1 & 0 & 1 & 0 \\ 0 & 0 & 0 & 1 & 0 & 0 & 0 & 0 & 1 \\ 0 & 0 & 0 & 0 & 1 & 0 & 1 & 0 & 0 \\ 0 & 0 & 1 & 0 & 0 & 0 & 1 & 0 & 0 \\ 1 & 0 & 0 & 0 & 0 & 0 & 0 & 1 & 0 \\ 0 & 1 & 0 & 0 & 0 & 0 & 0 & 0 & 1 \end{bmatrix}.$$

We permute the column-blocks of $\mathbf{H}_{qc}$ in such a way that the first $l = 3$ column-blocks form the subarray $\mathbf{D}^*$. The permuted array $\mathbf{H}_{qc}^*$ is

$$\mathbf{H}_{qc}^* = \begin{bmatrix} 1 & 0 & 0 & 0 & 1 & 0 & 0 & 0 & 1 & 0 & 0 & 0 \\ 0 & 1 & 0 & 0 & 0 & 1 & 1 & 0 & 0 & 0 & 0 & 0 \\ 0 & 0 & 1 & 1 & 0 & 0 & 0 & 1 & 0 & 0 & 0 & 0 \\ 0 & 0 & 0 & 0 & 0 & 1 & 0 & 1 & 0 & 1 & 0 & 0 \\ 0 & 0 & 0 & 1 & 0 & 0 & 0 & 0 & 1 & 0 & 1 & 0 \\ 0 & 0 & 0 & 0 & 1 & 0 & 1 & 0 & 0 & 0 & 0 & 1 \\ 0 & 0 & 1 & 0 & 0 & 0 & 1 & 0 & 0 & 0 & 1 & 0 \\ 1 & 0 & 0 & 0 & 0 & 0 & 0 & 1 & 0 & 0 & 0 & 1 \\ 0 & 1 & 0 & 0 & 0 & 0 & 0 & 0 & 1 & 1 & 0 & 0 \end{bmatrix}.$$

Because $\mathbf{H}_{qc}^*$ is not a full-rank matrix, the corresponding generator matrix $\mathbf{G}_{qc,\text{semi-sys}}$ of $C_{qc}$ has the semi-systematic circulant form given by (4.80), which has two subarrays $\mathbf{G}_{qc,\text{sys}}^*$ and $\mathbf{Q}$. In the following, we will calculate the arrays $\mathbf{G}_{qc,\text{sys}}^*$ in the form of (4.81) and $\mathbf{Q}$ in the form of (4.86).

The $3 \times 3$ array $\mathbf{D}^*$ has rank 7, i.e., it has two redundant columns. Through calculation, we find that the first two columns of $\mathbf{D}^*$ are the two linearly dependent columns. Then, we have $d_0 = 2$, $d_1 = 0$, $d_2 = 0$, i.e., two linearly dependent columns in the zeroth column-block, no linearly dependent columns in the first and second column-blocks of $\mathbf{D}^*$.

Following the development above, we have the vector $\mathbf{z}_i$ (we have $i = 0$ in this example) given by (4.75) with the following form:

$$\begin{aligned} \mathbf{z}_0 &= (\mathbf{g}_{0,0}, \mathbf{g}_{0,1}, \mathbf{g}_{0,2}) \\ &= (0, 0, g_{0,0,2}, g_{0,1,0}, g_{0,1,1}, g_{0,1,2}, g_{0,2,0}, g_{0,2,1}, g_{0,2,2}), \end{aligned}$$

i.e., set the first two entries in $\mathbf{g}_{0,0}$ to zeros which correspond to the first two linearly dependent columns in the zeroth column-block of $\mathbf{D}^*$.

Solving (4.77), we obtain

$$\mathbf{z}_0 = (0, 0, 1, 1, 1, 1, 1, 0, 1),$$

and then have $\mathbf{g}_{0,0} = (0, 0, 1)$, $\mathbf{g}_{0,1} = (1, 1, 1)$, and $\mathbf{g}_{0,2} = (1, 0, 1)$. The subarray $\mathbf{G}^*_{\mathrm{qc,sys}}$ is

$$\mathbf{G}^*_{\mathrm{qc,sys}} = \begin{bmatrix} 0 & 0 & 1 & 1 & 1 & 1 & 1 & 0 & 1 & 1 & 0 & 0 \\ 1 & 0 & 0 & 1 & 1 & 1 & 1 & 1 & 1 & 0 & 0 & 1 & 0 \\ 0 & 1 & 0 & 1 & 1 & 1 & 0 & 1 & 1 & 0 & 0 & 1 \end{bmatrix}.$$

Next, we compute the subarray $\mathbf{Q}$. Considering the linearity of the columns in the matrix $\mathbf{D}^*$, the vector $\mathbf{w}_i$ in (4.87) has the following format:

$$\mathbf{w}_0 = (1, 0, q_{0,2}, q_{0,3}, \ldots, q_{0,8}).$$

Solving (4.89), we obtain

$$\mathbf{w}_0 = (1, 0, 1, 0, 1, 1, 1, 1, 0)$$

and then have $\mathbf{w}_{0,0} = (1, 0, 1)$, $\mathbf{w}_{0,1} = (0, 1, 1)$, and $\mathbf{w}_{0,2} = (1, 1, 0)$. Cyclically shifting the three vectors, $\mathbf{w}_{0,0}$, $\mathbf{w}_{0,1}$, and $\mathbf{w}_{0,2}$, one place to the right, we obtain the three $2 \times 3$ submatrices, $\mathbf{Q}_{0,0}$, $\mathbf{Q}_{0,1}$, and $\mathbf{Q}_{0,2}$, in the subarray $\mathbf{Q}$ of $\mathbf{G}_{\mathrm{qc,semi-sys}}$:

$$\mathbf{Q} = [\mathbf{Q}_{0,0}\ \mathbf{Q}_{0,1}\ \mathbf{Q}_{0,2}] = \begin{bmatrix} 1 & 0 & 1 & 0 & 1 & 1 & 1 & 1 & 0 & 0 & 0 & 0 \\ 1 & 1 & 0 & 1 & 0 & 1 & 0 & 1 & 1 & 0 & 0 & 0 \end{bmatrix}.$$

Combining $\mathbf{G}^*_{\mathrm{qc,sys}}$ and $\mathbf{Q}$ to the form of (4.80), we obtain the generator matrix $\mathbf{G}_{\mathrm{qc,semi-sys}}$ of $C_{\mathrm{qc}}$ in semi-systematic circulant form:

$$\mathbf{G}_{\mathrm{qc,semi-sys}} = \begin{bmatrix} 1 & 0 & 1 & 0 & 1 & 1 & 1 & 1 & 0 & 0 & 0 & 0 \\ 1 & 1 & 0 & 1 & 0 & 1 & 0 & 1 & 1 & 0 & 0 & 0 \\ 0 & 0 & 1 & 1 & 1 & 1 & 1 & 0 & 1 & 1 & 0 & 0 \\ 1 & 0 & 0 & 1 & 1 & 1 & 1 & 1 & 1 & 0 & 0 & 1 & 0 \\ 0 & 1 & 0 & 1 & 1 & 1 & 0 & 1 & 1 & 0 & 0 & 1 \end{bmatrix}.$$

It can be verified that $\mathbf{H}^*_{\mathrm{qc}} \mathbf{G}^T_{\mathrm{qc,semi-sys}} = \mathbf{O}$. ▲▲

### 4.13.5   Shortened QC Codes

Given a QC code $C_{\mathrm{qc}}$, various shortenings can be performed to obtain shortened QC codes of lower or higher rates. Consider a $(tb, cb)$ systematic section-wise QC code $C_{\mathrm{qc,sys}}$ of length $tb$ and rate $c/t$ with generator matrix $\mathbf{G}_{\mathrm{qc,sys}}$ in the form given by (4.62). Let $\delta$ be a nonnegative integer such that $0 \le \delta < c$. Suppose we delete the $\delta$ rightmost column-blocks and the bottom $\delta$ row-blocks from $\mathbf{G}_{\mathrm{qc,sys}}$. We obtain a $(c - \delta)b \times (t - \delta)b$ matrix $\mathbf{G}_{\mathrm{qc,sys},1}(\delta)$, a $(c - \delta) \times (t - \delta)$ array of circulants of size $b \times b$ in the form of (4.62). The code $C_{\mathrm{qc,sys},1}(\delta)$ generated by $\mathbf{G}_{\mathrm{qc,sys},1}(\delta)$ is a $((t - \delta)b, (c - \delta)b)$ systematic section-wise QC code of length $(t - \delta)b$ and rate $(c - \delta)/(t - \delta)$. Let $\tau$ be a nonnegative integer such that $0 \le \tau < t - c$. Deleting any $\tau$ column-blocks from the subarray $\mathbf{P}$ of $\mathbf{G}_{\mathrm{qc,sys}}$, we obtain a $c \times (t - \tau)$ array $\mathbf{G}_{\mathrm{qc,sys},2}(\tau)$ of circulants of size $b \times b$ in the form of (4.62). The row space of $\mathbf{G}_{\mathrm{qc,sys},2}(\tau)$ gives a $((t - \tau)b, cb)$ systematic section-wise QC code $C_{\mathrm{qc,sys},2}(\tau)$ of length $(t - \tau)b$ and rate $c/(t - \tau)$ which is higher than the rate of the mother code $C_{\mathrm{qc,sys}}$. Suppose we delete $\delta$ rightmost column-blocks of $\mathbf{G}_{\mathrm{qc,sys}}$ and any $\tau$ column-blocks from the subarray $\mathbf{P}$ of $\mathbf{G}_{\mathrm{qc,sys}}$, and then remove the bottom $\delta$ row-blocks from the resultant array. We obtain a $(c - \delta) \times (t - \delta - \tau)$ array $\mathbf{G}_{\mathrm{qc,sys},3}(\delta, \tau)$ of circulants of size $b \times b$. The row space of $\mathbf{G}_{\mathrm{qc,sys},3}(\delta, \tau)$ gives a $((t - \delta - \tau)b, (c - \delta)b)$ systematic section-wise QC code of length $(t - \delta - \tau)b$ and rate $(c - \delta)/(t - \delta - \tau)$.

Similarly, we can shorten a QC code $C_{\mathrm{qc,semi\text{-}sys}}$ with generator matrix $\mathbf{G}_{\mathrm{qc,semi\text{-}sys}}$ in the form of (4.80) to obtain shortened versions of $C_{\mathrm{qc,semi\text{-}sys}}$ with various lengths and rates in systematic or semi-systematic form.

The above developments in shortening show that section-wise QC codes are very flexible in shortening. Note that shortening a section-wise QC code can also be carried out by deleting column-blocks and/or row-blocks from its parity-check matrix in systematic or nonsystematic form.

## 4.14   Nonbinary Cyclic Codes

So far, this chapter has focused only on binary cyclic codes. As mentioned at the beginning of the chapter, nonbinary cyclic codes can be constructed and their properties can be analyzed in a similar way as binary codes. In this section, we briefly describe this category of cyclic codes.

An $(n, k)$ $q$-ary cyclic code $C$ is generated by a *monic* polynomial of degree $n - k$ over $\mathrm{GF}(q)$,

$$\mathbf{g}(X) = g_0 + g_1 X + g_2 X^2 + \cdots + g_{n-k-1}X^{n-k-1} + X^{n-k}, \tag{4.92}$$

where $g_0 \ne 0$ and $g_i \in \mathrm{GF}(q)$ for $0 \le i < n - k$.

The generator polynomial $\mathbf{g}(X)$ is a factor of $X^n - 1$. Let $m$ be the smallest positive integer for which $n$ divides $q^m - 1$. Then, the roots of $\mathbf{g}(X)$ are elements of the extension field $\mathrm{GF}(q^m)$ of $\mathrm{GF}(q)$. Suppose the $n - k$ roots of $\mathbf{g}(X)$ belong to $l$ different conjugate sets.[4] For $1 \leq i \leq l$, let $m_i(X)$ be the (monic) minimal polynomial associated with the $i$th conjugate set. Each of these minimal polynomials is an irreducible polynomial over GF($q$). The generator polynomial $\mathbf{g}(X)$ of the $(n, k)$ $q$-ary cyclic code $C$ is the product of these $l$ minimal polynomials, i.e.,

$$\mathbf{g}(X) = \prod_{i=1}^{l} m_i(X). \tag{4.93}$$

A polynomial $\mathbf{v}(X)$ of degree $n - 1$ or less over GF($q$) is a code polynomial in $C$ if and only if $\mathbf{v}(X)$ is divisible by the generator polynomial $\mathbf{g}(X)$.

The generator and parity-check matrices of nonbinary cyclic codes are in the form of (4.15) and (4.20), respectively, where the entries in these matrices are from GF($q$). More on nonbinary cyclic codes will be discussed in Chapters 6 and 16.

## 4.15   Remarks

The Meggitt decoding presented in Section 4.7 is a general scheme for decoding a cyclic code in a serial manner, symbol by symbol. The key part of this decoding scheme is how to determine whether a received symbol is erroneous or not when it is ready to be shifted out from the decoder to the user. This part of decoding function dominates the decoder performance and complexity. Hence, an efficient, good-performance achieving, and practical implementable algorithm to perform this part of decoding function is needed. We call this a *symbol-error-location* (SEL) algorithm. Devising an SEL-algorithm for a cyclic code very much depends on the specific structure(s), algebraic and/or geometric, of the code. Efficient SEL-algorithms for decoding several special classes of well-known cyclic codes, such as BCH, RS, and finite-geometry codes, have been devised, which will be presented in the next three chapters.

## Problems

**4.1** Let $\mathbf{g}(X)$ be the code polynomial given by (4.6) and $\mathbf{g}^{(i)}(X)$ be the $i$th cyclic-shift of $\mathbf{g}(X)$. Prove that for $0 \leq i < n - r$, $X^i \mathbf{g}(X) = \mathbf{g}^{(i)}(X)$.

**4.2** Show that $X + 1$ is a factor of $X^n + 1$ over GF(2) for any positive integer $n$.

**4.3** Consider an $(n, k)$ cyclic code $C$ generated by $\mathbf{g}(X)$. Show that, if $\mathbf{g}(X)$ has $1 + X$ as a factor, all the codewords in the code $C$ have even weight.

---

[4]Note that the conjugates of the $n - k$ roots of $\mathbf{g}(X)$ are with respect to the ground field GF($q$). The conjugate set of an element $\beta \in$ GF($q^m$) of exponent $e$ with respect to the ground field GF($q$) is $\{\beta, \beta^q, \beta^{q^2}, \ldots, \beta^{q^{e-1}}\}$. If $q = 2$, we have the definition for conjugate roots as given in Chapter 2.

**4.4** Let $n$ be the smallest integer such that the polynomial $\mathbf{g}(X)$ divides $X^n+1$. Consider the cyclic code of length $n$ generated by $\mathbf{g}(X)$. Prove that the minimum distance of this cyclic code is at least 3.

**4.5** Consider an $(n,k)$ cyclic code generated by $\mathbf{g}(X)$. Suppose $n$ is odd and $1+X$ is not a factor of $\mathbf{g}(X)$. Show that the cyclic code contains an $n$-tuple of all ones as a codeword.

**4.6** Consider the $(7,3)$ cyclic code generated by $\mathbf{g}(X) = 1+X^2+X^3+X^4$ given in Example 4.1.
  (a) Find all the codewords and code polynomials of the code.
  (b) Find the minimum distance of the code.
  (c) Find the generator matrix $\mathbf{G}$ of the code and the generator matrix $\mathbf{G}_{\mathrm{sys}}$ in systematic form.
  (d) Find the parity-check polynomial $\mathbf{h}(X)$ and the corresponding parity-check matrix $\mathbf{H}$, and put $\mathbf{H}$ in systematic form.

**4.7** Construct a cyclic code of length 7 with the following generator polynomial:
$$\mathbf{g}(X) = 1 + X + X^2 + X^4.$$
  (a) Find the dimension of the code.
  (b) Find all the codewords and code polynomials of the code.
  (c) Find the minimum distance of the code.
  (d) Find the generator matrix $\mathbf{G}$ of the code and the generator matrix $\mathbf{G}_{\mathrm{sys}}$ in systematic form.
  (e) Find the parity-check polynomial $\mathbf{h}(X)$ and the corresponding parity-check matrix $\mathbf{H}$, and put $\mathbf{H}$ in systematic form.
  (f) Show that the constructed cyclic code is a dual code of the $(7,4)$ cyclic code constructed in Example 4.3.

**4.8** Consider the $(7,4)$ cyclic code generated by $\mathbf{g}(X) = 1 + X + X^3$ given in Example 4.1. Let $\mathbf{u}_0(X) = 1$, $\mathbf{u}_1(X) = X$, $\mathbf{u}_2(X) = X^2$, $\mathbf{u}_3(X) = X^3$ be four message polynomials.
  (a) Find the code polynomials in systematic form of the four message polynomials.
  (b) Use the encoding circuit shown in Fig. 4.2 to encode the four messages and show the contents of the shift-register in each step.

**4.9** Consider the factorization of $X^7 + 1$: $X^7 + 1 = (1 + X)(1 + X + X^3)(1 + X^2 + X^3)$.
  (a) Based on the factorization of $X^7 + 1$, how many cyclic codes of length 7 can be constructed?
  (b) Find the generator and parity-check polynomials of all the cyclic codes of length 7 and find the code dimensions and rates.
  (c) Find the generator matrices of all the cyclic codes of length 7.

**4.10** Find the encoding circuit for the cyclic code constructed in Problem 4.6, and use the circuit to encode a message $\mathbf{u} = (1\ 1\ 1)$.

**4.11** Consider the systematic generator matrix $\mathbf{G}_{\text{sys}}$ of an $(n, k)$ cyclic code $C$ in the form of (4.31). Let $\mathbf{g}(X) = 1 + g_1 X + g_2 X^2 + \cdots + g_{n-k-1} X^{n-k-1} + X^{n-k}$ be the generator polynomial of the code $C$. Show the following relation between the first row entries of $\mathbf{G}_{\text{sys}}$ and the coefficients of the generator polynomial $\mathbf{g}(X)$:

$$b_{0,i} = g_i, \text{ for } 0 \leq i < n - k, \text{ where } b_{0,0} = g_0 = 1.$$

**4.12** For a cyclic code, if an error pattern $\mathbf{e}(X)$ is detectable, show that its $i$th cyclic-shift $\mathbf{e}^{(i)}(X)$ is also detectable.

**4.13** Consider the $(15, 7)$ cyclic code generated by $\mathbf{g}(X) = 1 + X^4 + X^6 + X^7 + X^8$. Is $\mathbf{r}(X) = 1 + X + X^5 + X^{14}$ a code polynomial? If not, do the following.
(a) Find the syndrome $\mathbf{s}(X)$ of $\mathbf{r}(X)$ using polynomial long division.
(b) Construct the syndrome calculation circuit and use it to compute the syndrome $\mathbf{s}$.
(c) Find the systematic parity-check matrix $\mathbf{H}_{\text{sys}}$ of the code and use it to calculate the syndrome $\mathbf{s}$ $(\mathbf{s} = \mathbf{r} \cdot \mathbf{H}_{\text{sys}}^T)$.

**4.14** Prove (4.44).

**4.15** Consider the decoding of the received vector $\mathbf{r}$ in Problem 4.13. Calculate the first four components, $\mathbf{s}(X)$, $\mathbf{s}^{(1)}(X)$, $\mathbf{s}^{(2)}(X)$, $\mathbf{s}^{(3)}(X)$, in the syndrome sequence.

**4.16** Consider the $(7, 4)$ cyclic code $C$ generated by $\mathbf{g}(X) = 1 + X + X^3$ in Example 4.1. Suppose the codeword $\mathbf{v} = (1\ 0\ 0\ 1\ 0\ 1\ 1)$ is transmitted and the two vectors are received, $\mathbf{r}_0 = (1\ 0\ 0\ 0\ 0\ 1\ 1)$ and $\mathbf{r}_1 = (1\ 1\ 0\ 0\ 0\ 1\ 1)$. Use the decoding circuit in Fig. 4.6 to decode the two received vectors $\mathbf{r}_0$ and $\mathbf{r}_1$.

**4.17** Let $d_{\min}$ be the minimum distance of a cyclic code $C$. Prove that every error polynomial with Hamming weight less than $\frac{1}{2}d_{\min}$ has a unique syndrome polynomial.

**4.18** Find the weight distribution of the $(7, 4)$ Hamming code by using its weight enumerator $A(z)$.

**4.19** Consider the $(2^m - 1, 2^m - m - 2)$ cyclic Hamming code $C$ generated by $\mathbf{g}(X) = (1 + X)\mathbf{p}(X)$, where $\mathbf{p}(X)$ is a primitive polynomial of degree $m$. An error pattern of the form

$$\mathbf{e}(X) = X^i + X^{i+1}$$

is called a *double-adjacent-error pattern*. Show that no two double-adjacent-error patterns can be in the same coset of a standard array for $C$. Therefore, the code is capable of correcting single-error patterns and all the double-adjacent-error patterns.

**4.20** Consider the $(7,4)$ cyclic code generated by $\mathbf{g}(X) = 1 + X + X^3$ given in Example 4.1. A shortened $(5,2)$ code can be constructed by shortening the $(7,4)$ cyclic code.

(a) Find all the codewords of the shortened $(5,2)$ code.

(b) Find the generator and parity-check matrices of the $(5,2)$ code.

(c) Use the encoding circuit in Fig. 4.2 for the $(7,4)$ cyclic code to encode a message $\mathbf{u} = (1\ 0)$ for the $(5,2)$ code.

(d) Use the decoding circuit in Fig. 4.6 for the $(7,4)$ cyclic code to decode a received vector $\mathbf{r} = (1\ 0\ 0\ 0\ 0)$ for the $(5,2)$ code.

**4.21** Consider a $(2^m - 1, 2^m - 1 - m)$ Hamming code $C$ of length $n = 2^m - 1$ with an $m \times n$ parity-check matrix $\mathbf{H}$.

(a) Show that the sum (component-wise) of any two distinct columns of $\mathbf{H}$ is equal to a unique column of $\mathbf{H}$.

(b) Show that the code $C$ has an all-one codeword.

(c) Based on the result obtained in (a), show that the number of codewords in $C$ with minimum weight is $n(n-1)/6$.

**4.22** Derive the weight enumerator given by $(4.55)$ of a $(2^m - 1, 2^m - 1 - m)$ Hamming code based on the weight distribution of its dual code. (*Hint*: MacWilliams identity given by $(3.36)$.)

**4.23** Consider the $(15,11)$ Hamming code generated by $\mathbf{g}(X) = 1 + X + X^4$.

(a) Find the dual code $C_d$ of the Hamming code and its generator polynomial.

(b) Construct the generator matrix of $C_d$ in systematic form.

(c) Find all the codewords of $C_d$ and determine its minimum distance.

**4.24** Consider the $(15,11)$ Hamming code generated by $\mathbf{g}(X) = 1 + X + X^4$.

(a) Construct the generator matrix $\mathbf{G}_{\text{sys}}$ of the code in systematic form.

(b) Using the constructed generator matrix $\mathbf{G}_{\text{sys}}$, construct a $(11,8)$ shortened Hamming code with minimum distance at least 3 by deleting three information bits and one parity-check bit.

**4.25** Show that $\mathbf{g}(X) = 1 + X^2 + X^4 + X^6 + X^7 + X^{10}$ generates a $(21,11)$ cyclic code. Devise a syndrome computation circuit for this code. Let $\mathbf{r}(X) = 1 + X^5 + X^{17}$ be a received polynomial. Compute the syndrome of $\mathbf{r}(X)$. Display the contents of the syndrome register after each bit of $\mathbf{r}$ has been shifted into the syndrome computation circuit.

**4.26** Shorten the $(15,11)$ Hamming code by deleting the seven leading high-order information bits. The resultant code is an $(8,4)$ shortened cyclic code. Devise a decoder for the shortened code.

**4.27** Shorten the $(31,26)$ cyclic Hamming code by deleting the 11 leading high-order information bits. The resultant code is a $(20,15)$ shortened cyclic code. Devise a decoding circuit for the shortened code.

**4.28** Consider a $(2^m - 1, 2^m - 1 - m)$ Hamming code $C$ with a parity-check matrix $\mathbf{H}$. Because the minimum distance of $C$ is 3, $C$ is able to correct a single bit error but not able to simultaneously detect any two bit errors. Let $\mathbf{v} = (v_0, v_1, v_2, \ldots, v_{2^m-2})$ be a codeword in $C$. Calculate the following overall parity-check bit:

$$\mathbf{v}_\infty = \sum_{i=0}^{2^m-2} v_i,$$

and form a $2^m$-tuple, $\mathbf{v}^* = (\mathbf{v}_\infty, v_0, v_1, v_2, \ldots, v_{2^m-2})$.

(a) Prove that the set of the $2^m$-tuples, $\mathbf{v}^*$, forms a $(2^m, 2^m - m)$ linear block code $C_{\text{ext}}$ which can correct a single bit error and simultaneously detect any two bit errors, i.e., the code $C_{\text{ext}}$ has minimum distance 4. (The code $C_{\text{ext}}$ is called an *extended Hamming code*.)

(b) Find a parity-check matrix for the constructed code $C_{\text{ext}}$ based on the parity-check matrix of the Hamming code $C$.

**4.29** Construct the $(89, 45)$ QR code mentioned in Example 4.11, i.e., find the generator polynomial of the code.

**4.30** Construct a $(31, 16)$ QR code, i.e., find the generator polynomial of the code.

**4.31** Consider a binary $(n, k)$ cyclic code $C$ generated by $\mathbf{g}(X)$. Let

$$\mathbf{g}^*(X) = X^{n-k}\mathbf{g}(X^{-1})$$

be the reciprocal polynomial of $\mathbf{g}(X)$.

(a) Show that $\mathbf{g}^*(X)$ also generates an $(n, k)$ cyclic code.

(b) Let $C^*$ denote the cyclic code generated by $\mathbf{g}^*(X)$. Show that $C$ and $C^*$ have the same weight distribution.

(*Hint*: Show that

$$\mathbf{v}(X) = v_0 + v_1 X + \cdots + v_{n-2} X^{n-2} + v_{n-1} X^{n-1}$$

is a code polynomial in $C$ if and only if

$$X^{n-1}\mathbf{v}(X^{-1}) = v_{n-1} + v_{n-2} X + \cdots + v_1 X^{n-2} + v_0 X^{n-1}$$

is a code polynomial in $C^*$.)

**4.32** Let $\mathbf{Q}_0, \mathbf{Q}_1, \ldots, \mathbf{Q}_{m-1}$ be $m$ binary circulants of size $n \times n$.

(a) If we use $\mathbf{Q}_0$ as a parity-check matrix, $\mathbf{H}_1 = \mathbf{Q}_0$, show that the code given by the null space over $\text{GF}(2)$ of $\mathbf{H}_1$ is a cyclic code of length $n$.

(b) Consider the following matrix constructed by arranging the $m$ circulants in a column,

$$\mathbf{H}_2 = \begin{bmatrix} \mathbf{Q}_0 \\ \mathbf{Q}_1 \\ \vdots \\ \mathbf{Q}_{m-1} \end{bmatrix}.$$

Show that the code given by the null space over $\text{GF}(2)$ of $\mathbf{H}_2$ is also a cyclic code of length $n$.

**4.33** Prove that Definition 4.3 and Definition 4.4 are equivalent. Verify this by considering the two QC codes given in Examples 4.12 and 4.13.

**4.34** Find the generator matrix in systematic circulant form from the following $5 \times 5$ parity-check array $\mathbf{H}_{qc}$ of $4 \times 4$ circulants:

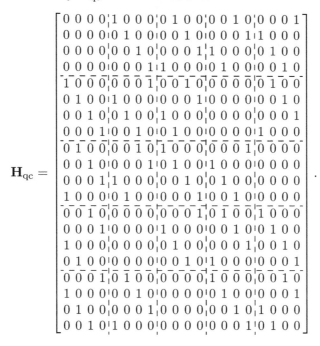

$$
\mathbf{H}_{qc} =
\begin{bmatrix}
0\,0\,0\,0 & 1\,0\,0\,0 & 0\,1\,0\,0 & 0\,0\,1\,0 & 0\,0\,0\,1 \\
0\,0\,0\,0 & 0\,1\,0\,0 & 0\,0\,1\,0 & 0\,0\,0\,1 & 1\,0\,0\,0 \\
0\,0\,0\,0 & 0\,0\,1\,0 & 0\,0\,0\,1 & 1\,0\,0\,0 & 0\,1\,0\,0 \\
0\,0\,0\,0 & 0\,0\,0\,1 & 1\,0\,0\,0 & 0\,1\,0\,0 & 0\,0\,1\,0 \\
1\,0\,0\,0 & 0\,0\,0\,1 & 0\,0\,1\,0 & 0\,0\,0\,0 & 0\,1\,0\,0 \\
0\,1\,0\,0 & 1\,0\,0\,0 & 0\,0\,0\,1 & 0\,0\,0\,0 & 0\,0\,1\,0 \\
0\,0\,1\,0 & 0\,1\,0\,0 & 1\,0\,0\,0 & 0\,0\,0\,0 & 0\,0\,0\,1 \\
0\,0\,0\,1 & 0\,0\,1\,0 & 0\,1\,0\,0 & 0\,0\,0\,0 & 1\,0\,0\,0 \\
0\,1\,0\,0 & 0\,0\,1\,0 & 1\,0\,0\,0 & 0\,0\,1\,0 & 0\,0\,0\,0 \\
0\,0\,1\,0 & 0\,0\,0\,1 & 0\,1\,0\,0 & 1\,0\,0\,0 & 0\,0\,0\,0 \\
0\,0\,0\,1 & 1\,0\,0\,0 & 0\,0\,1\,0 & 0\,1\,0\,0 & 0\,0\,0\,0 \\
1\,0\,0\,0 & 0\,1\,0\,0 & 0\,0\,0\,1 & 0\,0\,1\,0 & 0\,0\,0\,0 \\
0\,0\,1\,0 & 0\,0\,0\,0 & 0\,0\,0\,1 & 0\,1\,0\,0 & 1\,0\,0\,0 \\
0\,0\,0\,1 & 0\,0\,0\,1 & 0\,0\,0\,0 & 0\,0\,1\,0 & 0\,1\,0\,0 \\
1\,0\,0\,0 & 0\,0\,0\,0 & 0\,1\,0\,0 & 0\,0\,0\,1 & 0\,0\,1\,0 \\
0\,1\,0\,0 & 0\,0\,0\,0 & 0\,0\,1\,0 & 1\,0\,0\,0 & 0\,0\,0\,1 \\
0\,0\,0\,1 & 0\,1\,0\,0 & 0\,0\,0\,0 & 1\,0\,0\,0 & 0\,0\,1\,0 \\
1\,0\,0\,0 & 0\,0\,1\,0 & 0\,0\,0\,0 & 0\,1\,0\,0 & 0\,0\,0\,1 \\
0\,1\,0\,0 & 0\,0\,0\,1 & 0\,0\,0\,0 & 0\,0\,1\,0 & 1\,0\,0\,0 \\
0\,0\,1\,0 & 1\,0\,0\,0 & 0\,0\,0\,0 & 0\,0\,0\,1 & 0\,1\,0\,0
\end{bmatrix}.
$$

# References

[1] Eugene Prange, *Cyclic Error-Correcting Codes in Two Symbols*, AFCRC-TN-57, 103, Air Force Cambridge Research Center, Cambridge, MA, September 1957.

[2] A. Hocquenghem, "Codes correcteurs d'erreurs," *Chiffres*, **2** (1959), 147–156.

[3] R. C. Bose and D. K. Ray-Chaudhuri, "On a class of error correcting binary group codes," *Inf. Control*, **3** (1960), 68–79.

[4] I. S. Reed and G. Solomon, "Polynomial codes over certain finite fields," *J. Soc. Indust. Appl. Math.*, **8** (1960), 300–304.

[5] A. M. Gleason, "Weight polynomials of self dual and the MacWilliams identities," *Actes Congr. Int. Math.*, **3**, 211–215 (Paris: Gauthier-Villars, 1970).

[6] Y. Kou, S. Lin, and M. Fossorier, "Low density parity check codes based on finite geometries: a rediscovery," *Proc. IEEE Int. Symp. Inf. Theory*, Sorrento, June 25–30, 2000, p. 200.

[7] R. W. Hamming, "Error detecting and error correcting codes," *Bell Syst. Tech. J.*, **29** (1950), 147–160.

[8] R. E. Blahut, *Theory and Practice of Error Control Codes*, Reading, MA, Addison-Wesley, 1983.

[9] E. R. Berlekamp, *Algebraic Coding Theory*, revised ed., Laguna Hills, CA, Aegean Park Press, 1984.

[10] R. E. Blahut, *Algebraic Codes for Data Transmission*, Cambridge, Cambridge University Press, 2003.

[11] I. F. Blake and R. C. Mullin, *The Mathematical Theory of Coding*, New York, Academic Press, 1975.

[12] G. Clark and J. Cain, *Error-Correcting Codes for Digital Communications*, New York, Plenum Press, 1981.

[13] S. Lin and D. J. Costello, Jr., *Error Control Coding: Fundamentals and Applications*, 2nd ed., Upper Saddle River, NJ, Prentice-Hall, 2004.

[14] F. J. MacWilliams and N. J. A. Sloane, *The Theory of Error-Correcting Codes*, Amsterdam, North-Holland, 1977.

[15] W. W. Peterson and E. J. Weldon, Jr., *Error-Correcting Codes*, 2nd ed., Cambridge, MA, The MIT Press, 1972.

[16] S. B. Wicker, *Error Control Systems for Digital Communication and Storage*, Englewood Cliffs, NJ, Prentice-Hall, 1995.

[17] V. Pless, *Introduction to the Theory of Error-Correcting Codes*, 3rd ed., New York, Wiley, 1998.

[18] R. M. Roth, *Introduction to Coding Theory*, Cambridge, Cambridge University Press, 2006.

[19] S. Golomb, *Shift Register Sequences*, Laguna Hills, CA, Aegean Park Press, 1982.

[20] J. E. Meggitt, "Error correcting codes and their implementation," *IRE Trans. Inf. Theory*, **IT-7** (1961), 232–244.

[21] M. E. Mitchell, "Error-trap decoding of cyclic codes," *G. E. Report No. 62MCD3*, General Electric Military Communication Dept., Oklahoma City, December 1962.

[22] T. Kasami, "A decoding method for multiple-error-correcting cyclic codes by using threshold logic," *Conf. Rec. Inform. Process. Soc. Jap.*, Tokyo, November 1961, pp. 134–138.

[23] E. Prange, "The use of information sets in decoding cyclic codes," *IEEE Trans. Inf. Theory*, **IT-8** (1962), 80–85.

[24] L. Rudolph, "Easily implementation error-correction encoding-decoding," *G. E. Report No. 62MCD2*, General Electric Corporation, Oklahoma City, December 1962.

[25] T. Kasami, "A decoding procedure for multiple-error-correction cyclic codes," *IEEE Trans. Inf. Theory*, **IT-10** (1964), 134–139.

[26] F. J. MacWilliams, "Permutation decoding of systematic codes," *Bell Syst. Tech. J.*, **43** (1964), 485–505.

[27] L. Rudolph and M. E. Mitchell, "Implementation of decoders for cyclic codes," *IEEE Trans. Inf. Theory*, **IT-10** (1964), 259–260.

[28] T. Klove and V. I. Korzhik, *Error Detecting Codes*, Boston, MA, Kluwer Academic, 1995.

[29] B. Dunbridge, "Asymmetric signal design for coherent Gaussian channel," *IEEE Trans. Inf. Theory*, **IT-13** (1967), 422–431.

[30] G. Castagnoli, J. Ganz, and P. Graber, "Optimum cyclic redundancy-check codes with 16-bit redundancy," *IEEE Trans. Commun.*, **38**(1) (1990), 111–114.

[31] P. Merkey and E. C. Posner, "Optimum cyclic redundancy check codes for noisy channels," *IEEE Trans. Inf. Theory*, **IT-30**(6) (1984), 865–867.

[32] K. A. Witzke and C. Leung, "A comparison of some error detecting CRC code standards," *IEEE Trans. Commun.*, **33**(9) (1985), 996–998.

[33] P. Koopman, "Best CRC polynomials," Online: https://users.ece.cmu .edu/~koopman/crc/, accessed 2020.

[34] Online: https://en.wikipedia.org/wiki/Cyclic_redundancy_check, accessed 2020.

[35] Y. Li, Q. Chen, H. Liu, and T.-K. Truong, "Performance and analysis of quadratic residue codes of lengths less than 100," 2014 [Online]. Available: arXiv:1408.5674v1 [cs.IT].

[36] M. J. E. Golay, "Notes on digital coding," *Proc. IRE*, **37** (1949), 657.

[37] Z. Li, L. Chen, L. Zeng, S. Lin, and W. Fong, "Efficient encoding of quasi-cyclic low-density parity-check codes," *IEEE Trans. Commun.*, **54**(1) (2006), 71–81.

[38] Y. Chen and K. Parhi, "Overlapped message passing for quasi-cyclic low-density parity-check codes," *IEEE Trans. Circuits and Systems I*, **51**(6) (2004), 1106–1113.

[39] W. E. Ryan and S. Lin, *Channel Codes: Classical and Modern*, New York, Cambridge University Press, 2009.

[40] R. L. Townsend and E. J. Weldon, Jr., "Self-orthogonal quasi-cyclic codes," *IEEE Trans. Inf. Theory*, **IT-13**(2) (1967), 183–195.

# 5

# BCH Codes

Bose–Chaudhuri–Hocquenghem (BCH) codes form a large class of cyclic codes for correcting multiple random errors. This class of codes was first discovered by Hocquenghem in 1959 [1] and independently by Bose and Chaudhuri in 1960 [2]. The first algorithm for decoding binary BCH codes was devised by Peterson in 1960 [3]. Since then, Peterson's decoding algorithm has been improved and generalized by others [4–12]. The most efficient algorithm for decoding BCH codes is the Berlekamp–Massey iterative algorithm developed first by Berkekamp [6] and then improved by Massey [11]. The Berlekamp–Massey algorithm has been reformulated and improved to facilitate high-throughput and low-complexity hardware implementations of BCH decoders in [12–17]. In this chapter, we focus on binary BCH codes. For a more detailed description of BCH codes, their algebraic structures, and decoding algorithms, readers are referred to [6, 11, 18–23]. Nonbinary BCH codes will be presented in Chapter 6.

## 5.1 Primitive Binary BCH Codes

For any positive integers $m \geq 3$ and $t < 2^{m-1}$, there exists an $(n, k)$ binary cyclic BCH code with the following parameters:

$$
\begin{aligned}
&\text{code length} && n = 2^m - 1, \\
&\text{number of parity-check bits} && n - k \leq mt, \\
&\text{minimum distance} && d_{\min} \geq 2t + 1.
\end{aligned}
$$

This code is capable of correcting $t$ or fewer random errors over a span of $n = 2^m - 1$ bit positions and hence is called a *t-error-correcting BCH code*, denoted by $C$. The parameter $t$ is called the *designed error-correcting capability* and the parameter $2t + 1$ is called the *designed minimum distance* of the code.

For example, for $m = 6$ and $t = 3$, there exists a triple-error-correcting BCH code with $n = 2^6 - 1 = 63$, $n - k = 6 \times 3 = 18$, and $d_{\min} = 2 \times 3 + 1 = 7$.

Let $\alpha$ be a primitive element of the field $\mathrm{GF}(2^m)$ generated by a primitive polynomial $\mathbf{p}(X)$ of degree $m$ over $\mathrm{GF}(2)$, which is an extension field of $\mathrm{GF}(2)$. The order of $\alpha$ is $2^m - 1$ and its minimal polynomial is the primitive polynomial $\mathbf{p}(X)$.

The generator polynomial $\mathbf{g}(X)$ of a $t$-error-correcting BCH code of length $n = 2^m - 1$ is a polynomial over $\mathrm{GF}(2)$ of the *least degree* which has the following $2t$ *consecutive powers* of $\alpha$,

$$\alpha, \alpha^2, \ldots, \alpha^{2t}, \tag{5.1}$$

as roots. It follows from Theorem 2.7 in Chapter 2 that $\mathbf{g}(X)$ also has all the conjugates of $\alpha, \alpha^2, \ldots, \alpha^{2t}$ as roots.

For $1 \le i \le 2t$, let $m_i(X)$ be the minimal polynomial of $\alpha^i$. Then, it follows from the definition of $\mathbf{g}(X)$ that $\mathbf{g}(X)$ is the *least common multiple* (LCM) of $m_1(X), m_2(X), \ldots, m_{2t}(X)$, i.e.,

$$\mathbf{g}(X) = \mathrm{LCM}\{m_1(X), m_2(X), \ldots, m_{2t}(X)\}. \tag{5.2}$$

Suppose $i$ is an even positive integer. Then, $i$ can be expressed as a product of an odd integer $i'$ and a power of 2, say $2^l$, as follows:

$$i = i' 2^l. \tag{5.3}$$

Consider

$$\alpha^i = \alpha^{i' 2^l} = (\alpha^{i'})^{2^l}. \tag{5.4}$$

This says that every even power $\alpha^i$ of $\alpha$ in the sequence of (5.1) is a conjugate of some preceding odd power $\alpha^{i'}$ of $\alpha$ in (5.1). Therefore, $\alpha^i$ and $\alpha^{i'}$ are in the same conjugate set and thus have the same minimal polynomial, i.e.,

$$m_i(X) = m_{i'}(X). \tag{5.5}$$

Consequently, we can remove all the minimal polynomials with even subscripts from the expression (5.2). This results in the following expression of $\mathbf{g}(X)$:

$$\mathbf{g}(X) = \mathrm{LCM}\{m_1(X), m_3(X), \ldots, m_{2t-1}(X)\}. \tag{5.6}$$

Note that every minimal polynomial $m_i(X)$ in (5.6) has degree $m$ or less and there are at most $t$ different minimal polynomials in (5.6). Thus, the degree of $\mathbf{g}(X)$ is at most $mt$. Because every minimal polynomial $m_i(X)$ in (5.6) divides $X^{2^m-1} + 1$, $\mathbf{g}(X)$ divides $X^{2^m-1} + 1$. Because the minimal polynomial $m_1(X)$ of $\alpha$ (a primitive element) is a primitive polynomial of degree $m$, $2^m - 1$ is the smallest positive integer such that $m_1(X)$ divides $X^{2^m-1} + 1$. Therefore, $2^m - 1$ is the smallest positive integer for which $\mathbf{g}(X)$ divides $X^{2^m-1} + 1$.

The binary BCH code generated by $\mathbf{g}(X)$ is a cyclic code $C$ of length $2^m - 1$ with no more than $mt$ parity-check bits, i.e., its dimension $k$ is at least $2^m - mt - 1$. There is no simple formula to calculate the number $n - k$ of parity-check bits of the code $C$; but for small $t$ or $2^m - 1$ being a prime, $n - k$ is equal to $mt$ [6, 24, 25]. The commonly used method to find the number of parity-check bits is to compute the conjugate sets of all the roots of the generator polynomial $\mathbf{g}(X)$; then, the total number of elements in these conjugate sets gives the number of parity-check bits. It will be proved that the minimum distance $d_{\min}$ of $C$ is at least $2t + 1$. The field $\mathrm{GF}(2^m)$ is called the *code construction field* (or *root field*). The BCH code generated by $\mathbf{g}(X)$ in the form of (5.6) is referred to as a *primitive BCH code*. The parameters for all binary primitive BCH codes of length $2^m - 1$ with $m \le 10$ are given in Table 5.1.

Table 5.1 Binary primitive BCH codes of lengths less than $2^{10}$.

| $n$ | $k$ | $t$ | $n$ | $k$ | $t$ | $n$ | $k$ | $t$ | $n$ | $k$ | $t$ |
|---|---|---|---|---|---|---|---|---|---|---|---|
| 7 | 4 | 1 | 255 | 87 | 26 | 511 | 94 | 62 | 1023 | 553 | 52 |
| 15 | 11 | 1 | 255 | 79 | 27 | 511 | 85 | 63 | 1023 | 543 | 53 |
| 15 | 7 | 2 | 255 | 71 | 29 | 511 | 76 | 85 | 1023 | 533 | 54 |
| 15 | 5 | 3 | 255 | 63 | 30 | 511 | 67 | 87 | 1023 | 523 | 55 |
| 31 | 26 | 1 | 255 | 55 | 31 | 511 | 58 | 91 | 1023 | 513 | 57 |
| 31 | 21 | 2 | 255 | 47 | 42 | 511 | 49 | 93 | 1023 | 503 | 58 |
| 31 | 16 | 3 | 255 | 45 | 43 | 511 | 40 | 95 | 1023 | 493 | 59 |
| 31 | 11 | 5 | 255 | 37 | 45 | 511 | 31 | 109 | 1023 | 483 | 60 |
| 31 | 6 | 7 | 255 | 29 | 47 | 511 | 28 | 111 | 1023 | 473 | 61 |
| 63 | 57 | 1 | 255 | 21 | 55 | 511 | 19 | 119 | 1023 | 463 | 62 |
| 63 | 51 | 2 | 255 | 13 | 59 | 511 | 10 | 121 | 1023 | 453 | 63 |
| 63 | 45 | 3 | 255 | 9 | 63 | 1023 | 1013 | 1 | 1023 | 443 | 73 |
| 63 | 39 | 4 | 511 | 502 | 1 | 1023 | 1003 | 2 | 1023 | 433 | 74 |
| 63 | 36 | 5 | 511 | 493 | 2 | 1023 | 993 | 3 | 1023 | 423 | 75 |
| 63 | 30 | 6 | 511 | 484 | 3 | 1023 | 983 | 4 | 1023 | 413 | 77 |
| 63 | 24 | 7 | 511 | 475 | 4 | 1023 | 973 | 5 | 1023 | 403 | 78 |
| 63 | 18 | 10 | 511 | 466 | 5 | 1023 | 963 | 6 | 1023 | 393 | 79 |
| 63 | 16 | 11 | 511 | 457 | 6 | 1023 | 953 | 7 | 1023 | 383 | 82 |
| 63 | 10 | 13 | 511 | 448 | 7 | 1023 | 943 | 8 | 1023 | 378 | 83 |
| 63 | 7 | 15 | 511 | 439 | 8 | 1023 | 933 | 9 | 1023 | 368 | 85 |
| 127 | 120 | 1 | 511 | 430 | 9 | 1023 | 923 | 10 | 1023 | 358 | 86 |
| 127 | 113 | 2 | 511 | 421 | 10 | 1023 | 913 | 11 | 1023 | 348 | 87 |
| 127 | 106 | 3 | 511 | 412 | 11 | 1023 | 903 | 12 | 1023 | 338 | 89 |
| 127 | 99 | 4 | 511 | 403 | 12 | 1023 | 893 | 13 | 1023 | 328 | 90 |
| 127 | 92 | 5 | 511 | 394 | 13 | 1023 | 883 | 14 | 1023 | 318 | 91 |
| 127 | 85 | 6 | 511 | 385 | 14 | 1023 | 873 | 15 | 1023 | 308 | 93 |
| 127 | 78 | 7 | 511 | 376 | 15 | 1023 | 863 | 16 | 1023 | 298 | 94 |
| 127 | 71 | 9 | 511 | 367 | 16 | 1023 | 858 | 17 | 1023 | 288 | 95 |
| 127 | 64 | 10 | 511 | 358 | 18 | 1023 | 848 | 18 | 1023 | 278 | 102 |
| 127 | 57 | 11 | 511 | 349 | 19 | 1023 | 838 | 19 | 1023 | 268 | 103 |

Table 5.1 (*continued*)

| 127 | 50  | 13 | 511 | 340 | 20 | 1023 | 828 | 20 | 1023 | 258 | 106 |
|-----|-----|----|-----|-----|----|------|-----|----|------|-----|-----|
| 127 | 43  | 14 | 511 | 331 | 21 | 1023 | 818 | 21 | 1023 | 249 | 107 |
| 127 | 36  | 14 | 511 | 322 | 22 | 1023 | 808 | 22 | 1023 | 238 | 109 |
| 127 | 29  | 21 | 511 | 313 | 23 | 1023 | 798 | 23 | 1023 | 228 | 110 |
| 127 | 22  | 23 | 511 | 304 | 25 | 1023 | 788 | 24 | 1023 | 218 | 111 |
| 127 | 15  | 27 | 511 | 295 | 26 | 1023 | 778 | 25 | 1023 | 208 | 115 |
| 127 | 8   | 31 | 511 | 286 | 27 | 1023 | 768 | 26 | 1023 | 203 | 117 |
| 255 | 247 | 1  | 511 | 277 | 28 | 1023 | 758 | 27 | 1023 | 193 | 118 |
| 255 | 239 | 2  | 511 | 268 | 29 | 1023 | 748 | 28 | 1023 | 183 | 119 |
| 255 | 231 | 3  | 511 | 259 | 30 | 1023 | 738 | 29 | 1023 | 173 | 122 |
| 255 | 223 | 4  | 511 | 250 | 31 | 1023 | 728 | 30 | 1023 | 163 | 123 |
| 255 | 215 | 5  | 511 | 241 | 36 | 1023 | 718 | 31 | 1023 | 153 | 125 |
| 255 | 207 | 6  | 511 | 238 | 37 | 1023 | 708 | 34 | 1023 | 143 | 126 |
| 255 | 199 | 7  | 511 | 229 | 38 | 1023 | 698 | 35 | 1023 | 133 | 127 |
| 255 | 191 | 8  | 511 | 220 | 39 | 1023 | 688 | 36 | 1023 | 123 | 170 |
| 255 | 187 | 9  | 511 | 211 | 41 | 1023 | 678 | 37 | 1023 | 121 | 171 |
| 255 | 179 | 10 | 511 | 202 | 42 | 1023 | 668 | 38 | 1023 | 111 | 173 |
| 255 | 171 | 11 | 511 | 193 | 43 | 1023 | 658 | 39 | 1023 | 101 | 175 |
| 255 | 163 | 12 | 511 | 184 | 45 | 1023 | 648 | 41 | 1023 | 91  | 181 |
| 255 | 155 | 13 | 511 | 175 | 46 | 1023 | 638 | 42 | 1023 | 86  | 183 |
| 255 | 147 | 14 | 511 | 166 | 47 | 1023 | 628 | 43 | 1023 | 76  | 187 |
| 255 | 139 | 18 | 511 | 157 | 51 | 1023 | 618 | 44 | 1023 | 66  | 189 |
| 255 | 131 | 19 | 511 | 148 | 53 | 1023 | 608 | 45 | 1023 | 56  | 191 |
| 255 | 123 | 21 | 511 | 139 | 54 | 1023 | 598 | 46 | 1023 | 46  | 219 |
| 255 | 115 | 22 | 511 | 130 | 55 | 1023 | 588 | 47 | 1023 | 36  | 223 |
| 255 | 107 | 23 | 511 | 121 | 58 | 1023 | 578 | 49 | 1023 | 26  | 239 |
| 255 | 99  | 24 | 511 | 112 | 59 | 1023 | 573 | 50 | 1023 | 16  | 147 |
| 255 | 91  | 25 | 511 | 103 | 61 | 1023 | 563 | 51 | 1023 | 11  | 255 |

For $t = 1$,

$$\mathbf{g}(X) = m_1(X).$$

Because $m_1(X)$ is a primitive polynomial of degree $m$, $\mathbf{g}(X)$ generates a single-error-correcting BCH code $C$ with $n = 2^m - 1$, $n - k = m$, and $d_{\min} = 3$. This single-error-correcting BCH code is simply a *cyclic Hamming code* presented in Section 4.10. Thus, Hamming codes form a subclass of the (primitive) BCH codes.

**Example 5.1** In this example, we consider $m = 4$ and $t = 2$. Let $\alpha$ be a primitive element of $\mathrm{GF}(2^4)$ that is constructed based on the primitive polynomial $\mathbf{p}(X) = 1 + X + X^4$ over $\mathrm{GF}(2)$ (see Table 2.9), and let $m_1(X)$ and $m_3(X)$ be the minimal polynomials of $\alpha$ and $\alpha^3$, respectively. We shall use the technique described in Chapter 2 to find $m_1(X)$ and $m_3(X)$. To find $m_1(X)$, we first form the following sequence:

$$\alpha, \alpha^2, \alpha^{2^2} = \alpha^4, \alpha^{2^3} = \alpha^8, \alpha^{2^4} = \alpha^{16} = \alpha, \alpha^{2^5} = \alpha^{32} = \alpha^2, \ldots.$$

In this sequence, there are only four distinct elements, $\alpha$, $\alpha^2$, $\alpha^4$, and $\alpha^8$, which form a conjugate set. Thus, $m_1(X)$ has $\alpha$, $\alpha^2$, $\alpha^4$, and $\alpha^8$ over $\mathrm{GF}(2^4)$ as all its roots, and

$$m_1(X) = (X + \alpha)(X + \alpha^2)(X + \alpha^4)(X + \alpha^8).$$

Expanding the above polynomial $m_1(X)$ with the aid of Table 2.9, we obtain

$$m_1(X) = 1 + X + X^4.$$

In the same manner, we can find

$$m_3(X) = 1 + X + X^2 + X^3 + X^4.$$

According to (5.6), the double-error-correcting BCH code $C$ of length $n = 2^4 - 1 = 15$ is generated by

$$\mathbf{g}(X) = \mathrm{LCM}\{m_1(X), m_3(X)\}.$$

Because $m_1(X)$ and $m_3(X)$ are two distinct irreducible polynomials over $\mathrm{GF}(2)$, we have

$$\begin{aligned}
\mathbf{g}(X) &= m_1(X)m_3(X) \\
&= (1 + X + X^4)(1 + X + X^2 + X^3 + X^4) \\
&= 1 + X^4 + X^6 + X^7 + X^8.
\end{aligned}$$

Thus, the BCH code generated by $\mathbf{g}(X)$ is a $(15, 7)$ cyclic code with minimum distance $d_{\min} \geq 5$. Because the generator polynomial $\mathbf{g}(X)$ is a code polynomial of weight 5, the minimum distance of this code is exactly 5.

If we set $t = 3$, the generator polynomial of the triple-error-correcting BCH code of length 15 is given by

$$\mathbf{g}(X) = \mathrm{LCM}\{m_1(X), m_3(X), m_5(X)\},$$

where $m_5(X)$ is the minimal polynomial of $\alpha^5$. Through calculation, we find $m_5(X) = 1 + X + X^2$. Because $m_1(X)$, $m_3(X)$, and $m_5(X)$ are distinct and irreducible over $\mathrm{GF}(2)$, we have

$$\begin{aligned}
\mathbf{g}(X) &= \mathrm{LCM}\{m_1(X), m_3(X), m_5(X)\} \\
&= m_1(X)m_3(X)m_5(X) \\
&= (1 + X + X^4)(1 + X + X^2 + X^3 + X^4)(1 + X + X^2) \\
&= 1 + X + X^2 + X^4 + X^5 + X^8 + X^{10}.
\end{aligned}$$

The triple-error-correcting BCH code $C$ generated by the above polynomial $\mathbf{g}(X)$ is a $(15, 5)$ cyclic code with minimum distance $d_{\min} \geq 7$. Because $\mathbf{g}(X)$ is a code polynomial and has weight 7, the minimum distance of the $(15, 5)$ BCH code is exactly 7.

The single-error-correcting BCH code generated by

$$\mathbf{g}(X) = m_1(X) = 1 + X + X^4$$

is a $(15, 11)$ cyclic code which is a Hamming code with $d_{\min} = 3$. ▲▲

It is noted that the $2t$ roots of the generator polynomial $\mathbf{g}(X)$ are not necessarily $\alpha, \alpha^2, \ldots, \alpha^{2t}$; they can be any $2t$ consecutive powers of $\alpha$. In general, the $2t$ roots of $\mathbf{g}(X)$ can be $\alpha^{m_0}, \alpha^{m_0+1}, \alpha^{m_0+2}, \ldots, \alpha^{m_0+2t-2}, \alpha^{m_0+2t-1}$ for some integer $m_0 \geq 0$. In this book, we always choose $m_0 = 1$, in which case the constructed codes are known as *narrow-sense BCH codes*.

## 5.2 Structural Properties of BCH Codes

Consider a $t$-error-correcting primitive BCH code $C$ of length $n = 2^m - 1$ with generator polynomial $\mathbf{g}(X)$ which has $\alpha, \alpha^2, \ldots, \alpha^{2t}$ as roots, i.e.,

$$\mathbf{g}(\alpha^i) = 0, \tag{5.7}$$

for $1 \leq i \leq 2t$. Because a code polynomial

$$\mathbf{v}(X) = v_0 + v_1 X + v_2 X^2 + \cdots + v_{n-1} X^{n-1} \tag{5.8}$$

in $C$ is a multiple of its generator polynomial $\mathbf{g}(X)$, $\mathbf{v}(X)$ also has $\alpha, \alpha^2, \ldots, \alpha^{2t}$ as roots, i.e.,

$$\mathbf{v}(\alpha^i) = 0, \tag{5.9}$$

for $1 \leq i \leq 2t$. Conversely, if a polynomial $\mathbf{v}(X) = v_0 + v_1 X + \cdots + v_{n-1} X^{n-1}$ over GF(2) with degree $n - 1$ or less has $\alpha, \alpha^2, \ldots, \alpha^{2t}$ as roots, then $\mathbf{v}(X)$ must be divisible by $\mathbf{g}(X)$ and hence is a code polynomial in $C$.

Summarizing the above results, we have the following theorem.

**Theorem 5.1** *Let $\alpha$ be a primitive element of* GF($2^m$). *A polynomial $\mathbf{v}(X)$ of degree less than or equal to $2^m - 2$ over* GF(2) *is a code polynomial in a binary $t$-error-correcting primitive BCH code $C$ of length $n = 2^m - 1$ if and only if it has $\alpha, \alpha^2, \ldots, \alpha^{2t}$ as roots.*

It follows from (5.8) and (5.9) that, for $1 \leq i \leq 2t$,

$$\mathbf{v}(\alpha^i) = v_0 + v_1 \alpha^i + v_2 \alpha^{2i} + \cdots + v_{n-1} \alpha^{(n-1)i} = 0. \tag{5.10}$$

The equality of (5.10) can be expressed as the following matrix product form:

$$[v_0, v_1, \ldots, v_{n-1}] \cdot \begin{bmatrix} 1 \\ \alpha^i \\ \alpha^{2i} \\ \vdots \\ \alpha^{(n-1)i} \end{bmatrix} = 0, \tag{5.11}$$

for $1 \le i \le 2t$. This simply says that the inner product of a codeword $\mathbf{v} = (v_0, v_1, \ldots, v_{n-1})$ in $C$ and the $n$-tuple $(1, \alpha^i, \alpha^{2i}, \ldots, \alpha^{(n-1)i})$ over $\mathrm{GF}(2^m)$, $1 \le i \le 2t$, is equal to 0.

Form the following $2t \times n$ Vandermonde matrix over $\mathrm{GF}(2^m)$:

$$\mathbf{H} = \begin{bmatrix} 1 & \alpha & \alpha^2 & \ldots & \alpha^{n-1} \\ 1 & \alpha^2 & (\alpha^2)^2 & \ldots & (\alpha^2)^{n-1} \\ 1 & \alpha^3 & (\alpha^3)^2 & \ldots & (\alpha^3)^{n-1} \\ \vdots & \vdots & \vdots & & \vdots \\ 1 & \alpha^{2t} & (\alpha^{2t})^2 & \ldots & (\alpha^{2t})^{n-1} \end{bmatrix}. \tag{5.12}$$

It follows from (5.11) that for every codeword $\mathbf{v} = (v_0, v_1, \ldots, v_{n-1})$ in the binary $t$-error-correcting primitive BCH code $C$, the following condition holds:

$$\mathbf{v} \cdot \mathbf{H}^T = \mathbf{0}, \tag{5.13}$$

where $\mathbf{0} = (0, 0, \ldots, 0)$ is a zero $2t$-tuple. On the other hand, if an $n$-tuple $\mathbf{v} = (v_0, v_1, \ldots, v_{n-1})$ over $\mathrm{GF}(2)$ satisfies the condition given by (5.13), then it follows from (5.10) that its corresponding polynomial $\mathbf{v}(X) = v_0 + v_1 X + \cdots + v_{n-1} X^{n-1}$ has $\alpha, \alpha^2, \ldots, \alpha^{2t}$ as roots. Consequently, $\mathbf{v}(X)$ is divisible by the generator polynomial $\mathbf{g}(X)$ of $C$ and hence is a code polynomial.

Hence, the $t$-error-correcting BCH code $C$ generated by $\mathbf{g}(X)$ in the form of (5.6) is the null space over $\mathrm{GF}(2)$ of $\mathbf{H}$ given by (5.12) and $\mathbf{H}$ is a nonbinary parity-check matrix of the code. Consequently, we have the following theorem.

**Theorem 5.2** *Let* $n = 2^m - 1$. *An* $n$-*tuple* $\mathbf{v}$ *over* $\mathrm{GF}(2)$ *is a codeword in a* $t$-*error-correcting BCH code generated by* $\mathbf{g}(X)$ *in the form of* (5.6) *if and only if*

$$\mathbf{v} \cdot \mathbf{H}^T = \mathbf{0}, \tag{5.14}$$

*where the matrix* $\mathbf{H}$ *is given by* (5.12).

The matrix $\mathbf{H}$ is a parity-check matrix of $C$ over $\mathrm{GF}(2^m)$. If each entry of $\mathbf{H}$ is represented by an $m$-tuple over $\mathrm{GF}(2)$ in column form, we obtain a binary parity-check matrix $\mathbf{H}_b$ of $C$. For decoding the BCH code $C$, the parity-check matrix $\mathbf{H}$ over $\mathrm{GF}(2^m)$ given by (5.12) is used.

Consider the $(15, 7)$ BCH code $C$ given in Example 5.1 with $t = 2$ and $\mathbf{g}(X) = 1 + X^4 + X^6 + X^7 + X^8$. The generator polynomial $\mathbf{g}(X)$ has $\alpha$, $\alpha^2$,

$\alpha^3$, and $\alpha^4$ as roots, where $\alpha$ is a primitive element in $\mathrm{GF}(2^4)$. The nonbinary parity-check matrix $\mathbf{H}$ of this code in the form of (5.12) is

$$\mathbf{H} = \begin{bmatrix} 1 & \alpha & \alpha^2 & \cdots & \alpha^{13} & \alpha^{14} \\ 1 & \alpha^2 & (\alpha^2)^2 & \cdots & (\alpha^2)^{13} & (\alpha^2)^{14} \\ 1 & \alpha^3 & (\alpha^3)^2 & \cdots & (\alpha^3)^{13} & (\alpha^3)^{14} \\ 1 & \alpha^4 & (\alpha^4)^2 & \cdots & (\alpha^4)^{13} & (\alpha^4)^{14} \end{bmatrix} = \begin{bmatrix} 1 & \alpha & \alpha^2 & \alpha^3 & \cdots & \alpha^{12} & \alpha^{13} & \alpha^{14} \\ 1 & \alpha^2 & \alpha^4 & \alpha^6 & \cdots & \alpha^9 & \alpha^{11} & \alpha^{13} \\ 1 & \alpha^3 & \alpha^6 & \alpha^9 & \cdots & \alpha^6 & \alpha^9 & \alpha^{12} \\ 1 & \alpha^4 & \alpha^8 & \alpha^{12} & \cdots & \alpha^3 & \alpha^{12} & \alpha^{11} \end{bmatrix} .$$

The null space over $\mathrm{GF}(2)$ of $\mathbf{H}$ gives the binary $(15, 7)$ BCH code $C$. Replacing the entries in the above parity-check matrix $\mathbf{H}$ over $\mathrm{GF}(2^4)$ by their corresponding 4-tuples (see Table 2.9), we obtain a binary parity-check matrix $\mathbf{H}_b$ for the $(15, 7)$ BCH code $C$:

$$\mathbf{H}_b = \left[ \begin{array}{ccccccccccccccc} 1 & 0 & 0 & 0 & 1 & 0 & 0 & 1 & 1 & 0 & 1 & 0 & 1 & 1 & 1 \\ 0 & 1 & 0 & 0 & 1 & 1 & 0 & 1 & 0 & 1 & 1 & 1 & 1 & 0 & 0 \\ 0 & 0 & 1 & 0 & 0 & 1 & 1 & 0 & 1 & 0 & 1 & 1 & 1 & 1 & 0 \\ 0 & 0 & 0 & 1 & 0 & 0 & 1 & 1 & 0 & 1 & 0 & 1 & 1 & 1 & 1 \\ \hline 1 & 0 & 1 & 0 & 1 & 1 & 1 & 0 & 0 & 0 & 1 & 0 & 0 & 1 \\ 0 & 0 & 1 & 0 & 0 & 1 & 1 & 0 & 1 & 0 & 1 & 1 & 1 & 1 & 0 \\ 0 & 1 & 0 & 1 & 1 & 1 & 1 & 0 & 0 & 0 & 1 & 0 & 0 & 1 & 1 \\ 0 & 0 & 0 & 1 & 0 & 0 & 1 & 1 & 0 & 1 & 0 & 1 & 1 & 1 & 1 \\ \hline 1 & 0 & 0 & 0 & 1 & 1 & 0 & 0 & 0 & 1 & 1 & 0 & 0 & 0 & 1 \\ 0 & 0 & 0 & 1 & 1 & 0 & 0 & 0 & 1 & 1 & 0 & 0 & 0 & 1 & 1 \\ 0 & 0 & 1 & 0 & 1 & 0 & 0 & 1 & 0 & 1 & 0 & 0 & 1 & 0 & 1 \\ 0 & 1 & 1 & 1 & 1 & 0 & 1 & 1 & 1 & 1 & 0 & 1 & 1 & 1 & 1 \\ \hline 1 & 1 & 1 & 1 & 0 & 0 & 0 & 1 & 0 & 0 & 1 & 1 & 0 & 1 & 0 \\ 0 & 1 & 0 & 1 & 1 & 1 & 1 & 0 & 0 & 0 & 1 & 0 & 0 & 1 & 1 \\ 0 & 0 & 1 & 1 & 0 & 1 & 0 & 1 & 1 & 1 & 1 & 0 & 0 & 0 & 1 \\ 0 & 0 & 0 & 1 & 0 & 0 & 1 & 1 & 0 & 1 & 0 & 1 & 1 & 1 & 1 \end{array} \right] .$$

The null space over $\mathrm{GF}(2)$ of $\mathbf{H}_b$ gives the $(15, 7)$ BCH code $C$. Note that $\mathbf{H}_b$ is a $16 \times 15$ matrix over $\mathrm{GF}(2)$. Thus, the parity-check matrix $\mathbf{H}_b$ has eight redundant rows.

At this point, we see that binary primitive BCH codes are simply a special type of cyclic codes whose parity-check matrices are specified by the roots of their generator polynomials.

## 5.3   Minimum Distance of BCH Codes

Now, we are ready to prove that the minimum distance of the BCH code $C$ generated by $\mathbf{g}(X)$ in the form of (5.6) is at least $2t + 1$. All we need to do is to show that no nonzero codeword of $C$ has weight less than $2t + 1$. Let $\mathbf{v} = (v_0, v_1, \ldots, v_{n-1})$ be a nonzero codeword in $C$ whose nonzero components are $v_{j_1}, v_{j_2}, \ldots, v_{j_\delta}$, $0 \le j_1 < j_2 < \cdots < j_\delta < n$, i.e.,

$$v_{j_1} = v_{j_2} = \cdots = v_{j_\delta} = 1.$$

Hence, the Hamming weight of $\mathbf{v}$ is $\delta$. Suppose $\delta \leq 2t$. If we can prove that this hypothesis is invalid, then the weight of any nonzero codeword in $C$ is at least $2t + 1$. This implies that the minimum distance $d_{\min}$ of the code is at least $2t + 1$. It follows from (5.12) and (5.13) that

$$
\mathbf{0} = \mathbf{v} \cdot \mathbf{H}^T
$$

$$
= (v_0, v_1, \ldots, v_{n-1}) \cdot
\begin{bmatrix}
1 & 1 & \cdots & 1 \\
\alpha & \alpha^2 & \cdots & \alpha^{2t} \\
\alpha^2 & (\alpha^2)^2 & \cdots & (\alpha^{2t})^2 \\
\alpha^3 & (\alpha^2)^3 & \cdots & (\alpha^{2t})^3 \\
\vdots & \vdots & & \vdots \\
\alpha^{n-1} & (\alpha^2)^{n-1} & \cdots & (\alpha^{2t})^{n-1}
\end{bmatrix}
$$

$$
= (v_{j_1}, v_{j_2}, \ldots, v_{j_\delta}) \cdot
\begin{bmatrix}
\alpha^{j_1} & (\alpha^2)^{j_1} & \cdots & (\alpha^{2t})^{j_1} \\
\alpha^{j_2} & (\alpha^2)^{j_2} & \cdots & (\alpha^{2t})^{j_2} \\
\vdots & \vdots & & \vdots \\
\alpha^{j_\delta} & (\alpha^2)^{j_\delta} & \cdots & (\alpha^{2t})^{j_\delta}
\end{bmatrix}
$$

$$
= (v_{j_1}, v_{j_2}, \ldots, v_{j_\delta}) \cdot
\begin{bmatrix}
\alpha^{j_1} & (\alpha^{j_1})^2 & \cdots & (\alpha^{j_1})^{2t} \\
\alpha^{j_2} & (\alpha^{j_2})^2 & \cdots & (\alpha^{j_2})^{2t} \\
\vdots & \vdots & & \vdots \\
\alpha^{j_\delta} & (\alpha^{j_\delta})^2 & \cdots & (\alpha^{j_\delta})^{2t}
\end{bmatrix}. \tag{5.15}
$$

Equation (5.15) implies that

$$
\mathbf{0} = (v_{j_1}, v_{j_2}, \ldots, v_{j_\delta}) \cdot
\begin{bmatrix}
\alpha^{j_1} & (\alpha^{j_1})^2 & \cdots & (\alpha^{j_1})^\delta \\
\alpha^{j_2} & (\alpha^{j_2})^2 & \cdots & (\alpha^{j_2})^\delta \\
\vdots & \vdots & & \vdots \\
\alpha^{j_\delta} & (\alpha^{j_\delta})^2 & \cdots & (\alpha^{j_\delta})^\delta
\end{bmatrix}. \tag{5.16}
$$

Note that all the components of the $\delta$-tuple $(v_{j_1}, v_{j_2}, \ldots, v_{j_\delta})$ are nonzero, in fact equal to 1. For the equality of (5.16) to hold, the determinant of the $\delta \times \delta$ matrix in (5.16) must be zero, i.e.,

$$
\begin{vmatrix}
\alpha^{j_1} & (\alpha^{j_1})^2 & \cdots & (\alpha^{j_1})^\delta \\
\alpha^{j_2} & (\alpha^{j_2})^2 & \cdots & (\alpha^{j_2})^\delta \\
\vdots & \vdots & & \vdots \\
\alpha^{j_\delta} & (\alpha^{j_\delta})^2 & \cdots & (\alpha^{j_\delta})^\delta
\end{vmatrix} = 0.
$$

The above determinant can be simplified as follows:

$$
\alpha^{j_1 + j_2 + \cdots + j_\delta}
\begin{vmatrix}
1 & \alpha^{j_1} & \cdots & (\alpha^{j_1})^{\delta-1} \\
1 & \alpha^{j_2} & \cdots & (\alpha^{j_2})^{\delta-1} \\
\vdots & \vdots & & \vdots \\
1 & \alpha^{j_\delta} & \cdots & (\alpha^{j_\delta})^{\delta-1}
\end{vmatrix} = 0, \tag{5.17}
$$

which implies

$$\Delta = \begin{vmatrix} 1 & \alpha^{j_1} & \dots & (\alpha^{j_1})^{\delta-1} \\ 1 & \alpha^{j_2} & \dots & (\alpha^{j_2})^{\delta-1} \\ \vdots & \vdots & & \vdots \\ 1 & \alpha^{j_\delta} & \dots & (\alpha^{j_\delta})^{\delta-1} \end{vmatrix} = 0.$$

Note that $\Delta$ is a *Vandermonde deteminant* and

$$\Delta = \prod_{1 \leq k < i \leq \delta} (\alpha^{j_i} - \alpha^{j_k}). \tag{5.18}$$

Because $\alpha^{j_1}, \alpha^{j_2}, \dots, \alpha^{j_\delta}$ are distinct elements in $\mathrm{GF}(2^m)$, $\Delta \neq 0$. Hence, $\mathbf{v} \cdot \mathbf{H}^T \neq \mathbf{0}$. This is a contradiction to Theorem 5.2 which declares that a codeword $\mathbf{v}$ in $C$ satisfies the condition, $\mathbf{v} \cdot \mathbf{H}^T = \mathbf{0}$. Therefore, the hypothesis that $\delta \leq 2t$ is invalid. As a result, we must have

$$d_{\min} \geq 2t + 1. \tag{5.19}$$

Summarizing the above results, we conclude that the primitive BCH code $C$ generated by $\mathbf{g}(X)$ in the form of (5.6) has minimum distance $d_{\min}$ *at least* $2t + 1$. The number $2t + 1$ is a *lower bound* on the minimum distance of a $t$-error-correcting BCH code. This bound is referred to as the *BCH bound*, which is actually based on the fact that the generator polynomial $\mathbf{g}(X)$ has $2t$ consecutive powers of a primitive element $\alpha$ in $\mathrm{GF}(2^m)$ as roots. In general, when constructing an arbitrary cyclic code (see Chapter 4), we cannot guarantee the minimum distance of the constructed code. To find the minimum distance of such cyclic code, we may need to search through all the codewords (computer aids are often required). However, in constructing a BCH code, by applying the BCH bound to put a constraint on the generator polynomial $\mathbf{g}(X)$ (i.e., $\mathbf{g}(X)$ has $2t$ consecutive powers of a primitive element $\alpha$ in $\mathrm{GF}(2^m)$ as roots), we can ensure a designed minimum distance of the constructed BCH code as shown by (5.19).

Note that the polynomial $X^{2^m-1} + 1$ can be factored as the product of $X + 1$ and $X^{2^m-2} + X^{2^m-3} + \cdots + X + 1$. Because all the $2^m - 1$ nonzero elements in $\mathrm{GF}(2^m)$ are the roots of $X^{2^m-1} + 1$ and the element 1 is the root of $X + 1$, the polynomial $X^{2^m-2} + X^{2^m-3} + \cdots + X + 1$ must have the elements $\alpha, \alpha^2, \dots,$ $\alpha^{2^m-2}$ in $\mathrm{GF}(2^m)$ as roots. Hence, the polynomial $X^{2^m-2} + X^{2^m-3} + \cdots + X + 1$ is a code polynomial in the primitive BCH code $C$ generated by the polynomial $\mathbf{g}(X)$ defined by (5.6). The weight of this code polynomial is $2^m - 1$ which is odd and its corresponding codeword is the all-one $(2^m - 1)$-tuple, $\mathbf{v} = (1, 1, \dots, 1)$.

Because a linear block code cannot have all its codewords with only odd weights, the primitive BCH code $C$ generated by the polynomial $\mathbf{g}(X)$ defined by (5.6) must have half of its codewords with even weights and the other half with odd weights. Because the all-one $(2^m - 1)$-tuple $\mathbf{v} = (1, 1, \dots, 1)$ is a

Table 5.2 Weight distribution of the dual code of a double-error-correcting binary primitive BCH code of length $n = 2^m - 1$, where $m \geq 3$ and $m$ is odd.

| Weight $i$ | Number of codewords with weight $i$, $B_i$ |
|---|---|
| 0 | 1 |
| $2^{m-1} - 2^{(m+1)/2-1}$ | $\left(2^{m-2} + 2^{(m-1)/2-1}\right)\left(2^m - 1\right)$ |
| $2^{m-1}$ | $\left(2^m - 2^{m-1} + 1\right)\left(2^m - 1\right)$ |
| $2^{m-1} + 2^{(m+1)/2-1}$ | $\left(2^{m-2} - 2^{(m-1)/2-1}\right)\left(2^m - 1\right)$ |

Table 5.3 Weight distribution of the dual code of a double-error-correcting binary primitive BCH code of length $n = 2^m - 1$, where $m \geq 4$ and $m$ is even.

| Weight $i$ | Number of codewords with weight $i$, $B_i$ |
|---|---|
| 0 | 1 |
| $2^{m-1} - 2^{(m+2)/2-1}$ | $2^{(m-2)/2-1}\left(2^{(m-2)/2} + 1\right)\left(2^m - 1\right)/3$ |
| $2^{m-1} - 2^{m/2-1}$ | $2^{(m+2)/2-1}\left(2^{m/2} + 1\right)\left(2^m - 1\right)/3$ |
| $2^{m-1}$ | $\left(2^{m-2} + 1\right)\left(2^m - 1\right)$ |
| $2^{m-1} + 2^{m/2-1}$ | $2^{(m+2)/2-1}\left(2^{m/2} - 1\right)\left(2^m - 1\right)/3$ |
| $2^{m-1} + 2^{(m+2)/2-1}$ | $2^{(m-2)/2-1}\left(2^{(m-2)/2} - 1\right)\left(2^m - 1\right)/3$ |

codeword, we readily see that, for each codeword $\mathbf{w}$ with even weight $\omega$, there is a codeword $\mathbf{v} + \mathbf{w}$ with odd weight $2^m - 1 - \omega$ and vice versa. All the even-weight codewords in $C$ form a subcode $C_e$ of $C$ with minimum distance at least $2t + 2$. The even-weight subcode $C_e$ of $C$ is cyclic and its generator polynomial is $(1 + X)\mathbf{g}(X)$. We call the code $C_e$ *even-weight BCH code*.

A special case is that $\mathbf{g}(X)$ is a primitive polynomial $\mathbf{p}(X)$ with degree $m$ over GF(2). In this case, the cyclic code generated by $(1 + X)\mathbf{p}(X)$ is a $(2^m - 1, 2^m - m - 2)$ code with minimum distance exactly 4, which is commonly known as a *distance-4 Hamming code* as defined in Section 4.10. Some CRC codes for error detection are derived from distance-4 Hamming codes, as pointed out in Section 4.11.

The weight distributions of general BCH codes are still unknown. However, the weight distributions for double-error-correcting and triple-error-correcting BCH codes have been completely determined [23, 26]. Reference [23] provides tables to show the weight distributions of dual codes of the double- and triple-error-correcting BCH codes; these are presented here in Tables 5.2–5.5. The weight distributions of the double- and triple-error-correcting BCH codes can be calculated by using the MacWilliams identity given by (3.36) and the weight distributions given in these four tables.

Table 5.4 Weight distribution of the dual code of a triple-error-correcting binary primitive BCH code of length $n = 2^m - 1$, where $m \geq 5$ and $m$ is odd.

| Weight $i$ | Number of codewords with weight $i$, $B_i$ |
|---|---|
| 0 | 1 |
| $2^{m-1} - 2^{(m+1)/2}$ | $2^{(m-5)/2} \left(2^{(m-3)/2} + 1\right) \left(2^{m-1} - 1\right) \left(2^m - 1\right) / 3$ |
| $2^{m-1} - 2^{(m-1)/2}$ | $2^{(m-3)/2} \left(2^{(m-1)/2} + 1\right) \left(5 \cdot 2^{m-1} + 4\right) \left(2^m - 1\right) / 3$ |
| $2^{m-1}$ | $\left(9 \cdot 2^{2m-4} + 3 \cdot 2^{m-3} + 1\right) \left(2^m - 1\right)$ |
| $2^{m-1} + 2^{(m-1)/2}$ | $2^{(m-3)/2} \left(2^{(m-1)/2} - 1\right) \left(5 \cdot 2^{m-1} + 4\right) \left(2^m - 1\right) / 3$ |
| $2^{m-1} + 2^{(m+1)/2}$ | $2^{(m-5)/2} \left(2^{(m-3)/2} - 1\right) \left(2^{m-1} - 1\right) \left(2^m - 1\right) / 3$ |

Table 5.5 Weight distribution of the dual code of a triple-error-correcting binary primitive BCH code of length $n = 2^m - 1$, where $m \geq 6$ and $m$ is even.

| Weight $i$ | Number of codewords with weight $i$, $B_i$ |
|---|---|
| 0 | 1 |
| $2^{m-1} - 2^{(m+4)/2-1}$ | $\left(2^{m-1} + 2^{(m+4)/2-1}\right) \left(2^m - 4\right) \left(2^m - 1\right) / 960$ |
| $2^{m-1} - 2^{(m+2)/2-1}$ | $7 \cdot \left(2^{m-1} + 2^{(m+2)/2-1}\right) 2^m \left(2^m - 1\right) / 48$ |
| $2^{m-1} - 2^{m/2-1}$ | $2 \cdot \left(2^{m-1} + 2^{m/2-1}\right) \left(3 \cdot 2^m + 8\right) \left(2^m - 1\right) / 15$ |
| $2^{m-1}$ | $\left(29 \cdot 2^{2m} - 4 \cdot 2^m + 64\right) \left(2^m - 1\right) / 64$ |
| $2^{m-1} + 2^{m/2-1}$ | $2 \cdot \left(2^{m-1} - 2^{m/2-1}\right) \left(3 \cdot 2^m + 8\right) \left(2^m - 1\right) / 15$ |
| $2^{m-1} + 2^{(m+2)/2-1}$ | $7 \cdot \left(2^{m-1} - 2^{(m+2)/2-1}\right) 2^m \left(2^m - 1\right) / 48$ |
| $2^{m-1} + 2^{(m+4)/2-1}$ | $\left(2^{m-1} - 2^{(m+4)/2-1}\right) \left(2^m - 4\right) \left(2^m - 1\right) / 960$ |

## 5.4 Syndrome Computation and Error Detection

Consider a $t$-error-correcting primitive BCH code $C$ of length $n = 2^m - 1$ with generator polynomial $\mathbf{g}(X)$ given by (5.6). Suppose a code polynomial

$$\mathbf{v}(X) = v_0 + v_1 X + v_2 X^2 + \cdots + v_{n-1} X^{n-1}$$

is transmitted. Let

$$\mathbf{r}(X) = r_0 + r_1 X + r_2 X^2 + \cdots + r_{n-1} X^{n-1}$$

be the received polynomial. Then

$$\mathbf{r}(X) = \mathbf{v}(X) + \mathbf{e}(X), \tag{5.20}$$

where

$$\mathbf{e}(X) = e_0 + e_1 X + e_2 X^2 + \cdots + e_{n-1} X^{n-1}$$

is the error pattern caused by the channel noise and/or interferences.

Error detection is to check whether the received polynomial $\mathbf{r}(X)$ is a code polynomial. To accomplish this, we simply need to check whether $\mathbf{r}(X)$ has $\alpha, \alpha^2, \ldots, \alpha^{2t}$ as roots according to Theorem 5.1. Therefore, error detection can be performed by computing

$$
\begin{aligned}
S_i &= \mathbf{r}(\alpha^i) \\
&= r_0 + r_1 \alpha^i + r_2 \alpha^{2i} + \cdots + r_{n-1} \alpha^{(n-1)i},
\end{aligned} \tag{5.21}
$$

for $1 \le i \le 2t$. The $2t$-tuple over $\mathrm{GF}(2^m)$

$$
\mathbf{S} = (S_1, S_2, \ldots, S_{2t}) \tag{5.22}
$$

is defined as the *syndrome* of $\mathbf{r}(X)$ and $S_1, S_2, \ldots, S_{2t}$ are the $2t$ syndrome components.

From (5.21), we see that if $\mathbf{S} = \mathbf{0} = (0, 0, \ldots, 0)$, the received polynomial $\mathbf{r}(X)$ has all the $2t$ elements $\alpha, \alpha^2, \ldots, \alpha^{2t}$ as roots and hence it is a code polynomial. In this case, we assume that $\mathbf{r}(X)$ is error free and deliver it to the user. If $\mathbf{S} \ne \mathbf{0}$, the received polynomial $\mathbf{r}(X)$ does not have all the $2t$ elements $\alpha, \alpha^2, \ldots, \alpha^{2t}$ as roots and hence it is *not* a code polynomial. In this case, $\mathbf{r}(X)$ is detected in error and error correction must be initiated.

In Chapter 2, we showed that for a polynomial $\mathbf{f}(X)$ over $\mathrm{GF}(2)$, $(\mathbf{f}(X))^2 = \mathbf{f}(X^2)$ (see (2.9)). Because the received polynomial $\mathbf{r}(X)$ is a polynomial over $\mathrm{GF}(2)$, we have $(\mathbf{r}(X))^2 = \mathbf{r}(X^2)$. Substituting $X$ with $\alpha^i$ in $\mathbf{r}(X)$, we have

$$
(\mathbf{r}(\alpha^i))^2 = \mathbf{r}(\alpha^{2i}).
$$

From the above equality and (5.21), we obtain

$$
S_{2i} = S_i^2. \tag{5.23}
$$

Equality (5.23) says that to compute the syndrome $\mathbf{S} = (S_1, S_2, \ldots, S_{2t})$, we need only to compute the syndrome components with odd subscripts, i.e., $S_1$, $S_3, \ldots, S_{2t-1}$.

**Example 5.2** Consider the $(15, 5)$ triple-error-correcting BCH code $C$ (i.e., $t = 3$) constructed in Example 5.1, whose generator polynomial $\mathbf{g}(X)$ has $\alpha, \alpha^2, \alpha^3, \alpha^4, \alpha^5$, and $\alpha^6$ as roots. Suppose the zero code polynomial $\mathbf{v}(X) = 0$ is transmitted and

$$
\mathbf{r}(X) = X^3 + X^5 + X^{12}
$$

is the received polynomial. The syndrome of $\mathbf{r}(X)$ is a 6-tuple $\mathbf{S} = (S_1, S_2, S_3, S_4, S_5, S_6)$ over $\mathrm{GF}(2^4)$. Using the field $\mathrm{GF}(2^4)$ given by Table 2.9, we calculate the syndrome components with odd subscripts,

$$
\begin{aligned}
S_1 &= \mathbf{r}(\alpha) = \alpha^3 + \alpha^5 + \alpha^{12} = 1, \\
S_3 &= \mathbf{r}(\alpha^3) = \alpha^9 + \alpha^{15} + \alpha^{36} = \alpha^9 + 1 + \alpha^6 = \alpha^{10}, \\
S_5 &= \mathbf{r}(\alpha^5) = \alpha^{15} + \alpha^{25} + \alpha^{60} = 1 + \alpha^{10} + 1 = \alpha^{10}.
\end{aligned}
$$

Using (5.23), we can easily obtain the syndrome components with even subscripts based on those with odd subscripts,

$$S_2 = S_1^2 = 1, S_4 = S_2^2 = 1, S_6 = S_3^2 = \alpha^{20} = \alpha^5.$$

Hence, the syndrome of the received polynomial $\mathbf{r}(X)$ is $\mathbf{S} = (1, 1, \alpha^{10}, 1, \alpha^{10}, \alpha^5)$. Because $\mathbf{S} \neq \mathbf{0}$, the presence of errors in $\mathbf{r}(X)$ is detected.

Suppose the decoder receives the following polynomial $\mathbf{r}^*(X) = 1 + X + X^2 + X^4 + X^5 + X^8 + X^{10}$. The syndrome of this received polynomial is $\mathbf{S}^* = (0, 0, 0, 0, 0, 0)$. In this case, the decoder assumes that the received polynomial $\mathbf{r}^*(X)$ is error free and is the transmitted code polynomial. However, the received polynomial $\mathbf{r}^*(X)$ is not identical to the transmitted code polynomial $\mathbf{v}(X) = 0$. In this case, the decoder fails to detect the errors in $\mathbf{r}^*(X)$, i.e., an undetectable error occurs. ▲▲

## 5.5   Syndromes and Error Patterns

Because $\mathbf{r}(X) = \mathbf{v}(X) + \mathbf{e}(X)$, then

$$S_i = \mathbf{r}(\alpha^i) = \mathbf{v}(\alpha^i) + \mathbf{e}(\alpha^i) = \mathbf{e}(\alpha^i), \tag{5.24}$$

for $1 \leq i \leq 2t$. Equality (5.24) gives a relationship between the syndrome $\mathbf{S} = (S_1, S_2, \ldots, S_{2t})$ and the error pattern $\mathbf{e}(X)$. Suppose $\mathbf{e}(X)$ has $\nu$ errors at the locations $X^{j_1}, X^{j_2}, \ldots, X^{j_\nu}$. Then

$$\mathbf{e}(X) = X^{j_1} + X^{j_2} + \cdots + X^{j_\nu}, \tag{5.25}$$

where $0 \leq j_1 < j_2 < \cdots < j_\nu < n$. From (5.24) and (5.25), we obtain the following $2t$ equations which relate the error locations to the computed syndrome components:

$$
\begin{aligned}
S_1 &= \mathbf{e}(\alpha) &&= \alpha^{j_1} + \alpha^{j_2} + \cdots + \alpha^{j_\nu}, \\
S_2 &= \mathbf{e}(\alpha^2) &&= (\alpha^{j_1})^2 + (\alpha^{j_2})^2 + \cdots + (\alpha^{j_\nu})^2, \\
&\;\;\vdots \\
S_{2t} &= \mathbf{e}(\alpha^{2t}) &&= (\alpha^{j_1})^{2t} + (\alpha^{j_2})^{2t} + \cdots + (\alpha^{j_\nu})^{2t}.
\end{aligned}
\tag{5.26}
$$

If we can solve the above $2t$ equations, we can determine $\alpha^{j_1}, \alpha^{j_2}, \ldots, \alpha^{j_\nu}$ whose exponents $j_1, j_2, \ldots, j_\nu$ give the error locations in the error pattern $\mathbf{e}(X)$. Because the elements $\alpha^{j_1}, \alpha^{j_2}, \ldots, \alpha^{j_\nu}$ give the locations of errors, they are called *error-location numbers*. The above $2t$ equations have $S_1, S_2, \ldots, S_{2t}$ as their $2t$ knowns and $\alpha^{j_1}, \alpha^{j_2}, \ldots, \alpha^{j_\nu}$ as their $\nu$ unknowns. Any method for solving the equations in (5.26) is a decoding algorithm for BCH codes.

To simplify the notations in (5.26), we define

$$\beta_l = \alpha^{j_l}, \tag{5.27}$$

with $1 \le l \le \nu$. Then, the $2t$ equations can be simplified as follows:

$$
\begin{aligned}
S_1 &= \beta_1 + \beta_2 + \cdots + \beta_\nu, \\
S_2 &= \beta_1^2 + \beta_2^2 + \cdots + \beta_\nu^2, \\
&\vdots \\
S_{2t} &= \beta_1^{2t} + \beta_2^{2t} + \cdots + \beta_\nu^{2t}.
\end{aligned} \tag{5.28}
$$

The equations in (5.28) are known as the *power-sum symmetric functions*, which are nonlinear equations.

## 5.6 Error-Location Polynomials of BCH Codes

Define the following polynomial over $\mathrm{GF}(2^m)$:

$$
\begin{aligned}
\boldsymbol{\sigma}(X) &= (1 + \beta_1 X)(1 + \beta_2 X)\cdots(1 + \beta_\nu X) \\
&= \sigma_0 + \sigma_1 X + \sigma_2 X^2 + \cdots + \sigma_\nu X^\nu,
\end{aligned} \tag{5.29}
$$

where $\sigma_0 = 1$. From (5.29), we see that the polynomial $\boldsymbol{\sigma}(X)$ has $\beta_1^{-1}$, $\beta_2^{-1}, \ldots, \beta_\nu^{-1}$ (the reciprocals (or inverses) of the error-location numbers) as roots. The polynomial $\boldsymbol{\sigma}(X)$ is called the *error-location polynomial*. If we can determine $\boldsymbol{\sigma}(X)$ from the syndrome $\mathbf{S} = (S_1, S_2, \ldots, S_{2t})$, the inverses of the roots of $\boldsymbol{\sigma}(X)$ give the error-location numbers, and then the error pattern $\mathbf{e}(X)$ can be determined.

Expanding the product of (5.29), we find the following $\nu$ equalities which relate the coefficients of the error-location polynomial $\boldsymbol{\sigma}(X)$ and the $\nu$ error-location numbers:

$$
\begin{aligned}
\sigma_0 &= 1, \\
\sigma_1 &= \beta_1 + \beta_2 + \cdots + \beta_\nu, \\
\sigma_2 &= \beta_1\beta_2 + \beta_1\beta_3 + \cdots + \beta_{\nu-1}\beta_\nu, \\
\sigma_3 &= \beta_1\beta_2\beta_3 + \beta_1\beta_2\beta_4 + \cdots + \beta_{\nu-2}\beta_{\nu-1}\beta_\nu, \\
&\vdots \\
\sigma_\nu &= \beta_1\beta_2\cdots\beta_\nu.
\end{aligned} \tag{5.30}
$$

These equalities are called the *elementary-symmetric functions*. From (5.28) and (5.30), we can derive the following equations that relate the coefficients of the error-location polynomial $\boldsymbol{\sigma}(X)$ and the computed syndrome components:

$$S_1 + \sigma_1 = 0,$$
$$S_2 + \sigma_1 S_1 + 2\sigma_2 = 0,$$
$$S_3 + \sigma_1 S_2 + \sigma_2 S_1 + 3\sigma_3 = 0,$$
$$\vdots$$
$$S_{\nu-1} + \sigma_1 S_{\nu-2} + \sigma_2 S_{\nu-3} + \cdots + \sigma_{\nu-2} S_1 + (\nu-1)\sigma_{\nu-1} = 0,$$
$$S_\nu + \sigma_1 S_{\nu-1} + \sigma_2 S_{\nu-2} + \cdots + \sigma_{\nu-1} S_1 + \nu\sigma_\nu = 0,$$
$$(5.31)$$
$$S_{\nu+1} + \sigma_1 S_\nu + \sigma_2 S_{\nu-1} + \cdots + \sigma_{\nu-1} S_2 + \sigma_\nu S_1 = 0,$$
$$S_{\nu+2} + \sigma_1 S_{\nu+1} + \sigma_2 S_\nu + \cdots + \sigma_{\nu-1} S_3 + \sigma_\nu S_2 = 0,$$
$$\vdots$$
$$S_{2t} + \sigma_1 S_{2t-1} + \sigma_2 S_{2t-2} + \cdots + \sigma_{\nu-1} S_{2t-\nu+1} + \sigma_\nu S_{2t-\nu} = 0.$$

Note that $1 + 1 = 0$ in GF(2). Then

$$i\sigma_i = \begin{cases} \sigma_i, & \text{for odd } i; \\ 0, & \text{for even } i. \end{cases}$$

The equations of (5.31) are referred to as *Newton's identities* [27], which are linear equations. If we can determine the coefficients $\sigma_1, \sigma_2, \ldots, \sigma_\nu$ of the error-location polynomial $\boldsymbol{\sigma}(X)$ from Newton's identities, then we can determine the error-location numbers, $\beta_1, \beta_2, \ldots, \beta_\nu$, of the error pattern $\mathbf{e}(X)$ by finding the roots of $\boldsymbol{\sigma}(X)$.

## 5.7   A Procedure for Decoding BCH Codes

Based on the developments in Sections 5.5 and 5.6, a procedure for decoding a binary $t$-error-correcting BCH code can be formulated into the following steps.
  (1) Compute the syndrome $\mathbf{S} = (S_1, S_2, \ldots, S_{2t})$ from the received polynomial $\mathbf{r}(X)$.
  (2) Determine the error-location polynomial $\boldsymbol{\sigma}(X)$ from Newton's identities.
  (3) Find the roots, $\beta_1^{-1}, \beta_2^{-1}, \ldots, \beta_\nu^{-1}$, of $\boldsymbol{\sigma}(X)$ in GF($2^m$). Take the inverses of these roots to obtain the error-location numbers, $\beta_1 = \alpha^{j_1}, \beta_2 = \alpha^{j_2}, \ldots, \beta_\nu = \alpha^{j_\nu}$. Then, the error pattern is $\mathbf{e}(X) = X^{j_1} + X^{j_2} + \cdots + X^{j_\nu}$.
  (4) Perform the error correction by subtracting $\mathbf{e}(X)$ from $\mathbf{r}(X)$ (or adding $\mathbf{e}(X)$ to $\mathbf{r}(X)$). This gives the decoded vector, $\mathbf{v}^*(X) = \mathbf{r}(X) + \mathbf{e}(X)$.

Steps (1), (3), and (4) can be carried out easily; however, Step (2) involves solving Newton's identities. There is, in general, more than one error pattern for which the coefficients of its error-location polynomial satisfy Newton's identities. To minimize the probability of a decoding error, we need to find the *most probable* error pattern. For a BSC, finding the most probable error pattern is to determine the error-location polynomial of the *minimum degree* whose coefficients satisfy Newton's identities.

The first algorithm to carry out this procedure to decode BCH codes was devised by Peterson [3, 18]. Since then, many algorithms have been proposed for decoding BCH codes. The most efficient algorithm for decoding BCH codes is the one developed first by Berlekamp [6] and then improved by Massey [11]. This algorithm is called the *Berlekamp–Massey iterative algorithm*.

## 5.8 Berlekamp–Massey Iterative Algorithm

In this section, we present the Berlekamp–Massey iterative algorithm to find the error-location polynomial $\boldsymbol{\sigma}(X)$ without any proof. For more details of the proof, the readers are referred to [6, 18, 19]. The algorithm is described as follows.

The error-location polynomial $\boldsymbol{\sigma}(X)$ can be computed *iteratively* in $2t$ steps. At the $\mu$th step, we determine a *minimum-degree* polynomial

$$\boldsymbol{\sigma}^{(\mu)}(X) = 1 + \sigma_1^{(\mu)} X + \sigma_2^{(\mu)} X^2 + \cdots + \sigma_{l_\mu}^{(\mu)} X^{l_\mu}, \tag{5.32}$$

whose coefficients, $\sigma_1^{(\mu)}, \sigma_2^{(\mu)}, \dots, \sigma_{l_\mu}^{(\mu)} \in \mathrm{GF}(2^m)$, satisfy the first $\mu$ Newton's identities. The superscript "$\mu$" stands for "the $\mu$th step" in the iterative process and $l_\mu$ is the degree of the polynomial $\boldsymbol{\sigma}^{(\mu)}(X)$.

At the $(\mu + 1)$th step, we derive the next minimum-degree polynomial $\boldsymbol{\sigma}^{(\mu+1)}(X)$ based on $\boldsymbol{\sigma}^{(\mu)}(X)$, whose coefficients satisfy the first $\mu + 1$ Newton's identities. First, we check whether the coefficients of $\boldsymbol{\sigma}^{(\mu)}(X)$ also satisfy the $(\mu + 1)$th Newton's identity. If it does, then $\boldsymbol{\sigma}^{(\mu+1)}(X) = \boldsymbol{\sigma}^{(\mu)}(X)$ is the minimum-degree polynomial whose coefficients satisfy the first $\mu + 1$ Newton's identities. If it does not, a *correction term* is added to $\boldsymbol{\sigma}^{(\mu)}(X)$ to form the next solution $\boldsymbol{\sigma}^{(\mu+1)}(X)$ whose coefficients satisfy the first $\mu + 1$ Newton's identities. To test whether the coefficients of $\boldsymbol{\sigma}^{(\mu)}(X)$ satisfy the $(\mu + 1)$th Newton's identity, we compute

$$d_\mu = S_{\mu+1} + \sigma_1^{(\mu)} S_\mu + \sigma_2^{(\mu)} S_{\mu-1} + \cdots + \sigma_{l_u}^{(\mu)} S_{\mu+1-l_\mu}. \tag{5.33}$$

The above quantity $d_\mu$ is called the $\mu$th *discrepancy*. The sum on the right-hand side of (5.33) is actually the left-hand side of the $(\mu + 1)$th Newton's identity.

If $d_\mu = 0$, then the coefficients of $\boldsymbol{\sigma}^{(\mu)}(X)$ satisfy the $(\mu + 1)$th Newton's identity. In this case, we set

$$\boldsymbol{\sigma}^{(\mu+1)}(X) = \boldsymbol{\sigma}^{(\mu)}(X),$$

i.e., the current solution $\boldsymbol{\sigma}^{(\mu)}(X)$ is also the next solution $\boldsymbol{\sigma}^{(\mu+1)}(X)$. If $d_\mu \neq 0$, then $\boldsymbol{\sigma}^{(\mu)}(X)$ needs to be adjusted to obtain a new minimum-degree polynomial $\boldsymbol{\sigma}^{(\mu+1)}(X)$ whose coefficients satisfy the first $\mu + 1$ Newton's identities.

The correction is made as described below. Go back to the steps *prior to* the $\mu$th step and determine a step $\rho$ at which the partial solution is $\boldsymbol{\sigma}^{(\rho)}(X)$ such that $d_\rho \neq 0$ and $\mu - \rho + l_\rho$ has the *smallest value*, where $l_\rho$ is the degree of $\boldsymbol{\sigma}^{(\rho)}(X)$. Then, the solution at the $(\mu+1)$th step of the iteration process for finding the error-location polynomial $\boldsymbol{\sigma}(X)$ is given by

$$\boldsymbol{\sigma}^{(\mu+1)}(X) = \boldsymbol{\sigma}^{(\mu)}(X) + d_\mu d_\rho^{-1} X^{\mu-\rho} \boldsymbol{\sigma}^{(\rho)}(X), \qquad (5.34)$$

where

$$d_\mu d_\rho^{-1} X^{\mu-\rho} \boldsymbol{\sigma}^{(\rho)}(X) \qquad (5.35)$$

is the correction term with degree $\mu - \rho + l_\rho$. Because $\rho$ is chosen to minimize $\mu - \rho + l_\rho$, this choice of $\rho$ minimizes the degree $\mu - \rho + l_\rho$ of the correction term.

The polynomial $\boldsymbol{\sigma}^{(\mu+1)}(X)$ given by (5.34) is the minimum-degree polynomial whose coefficients satisfy the first $\mu + 1$ Newton's identities in (5.31). (The proof is omitted here because of its complexity. See the proof in [6, 18, 19].) The degree of $\boldsymbol{\sigma}^{(\mu+1)}(X)$ is

$$l_{\mu+1} = \max\{l_\mu, \mu - \rho + l_\rho\}. \qquad (5.36)$$

In either case whether $d_\mu = 0$ or $d_\mu \neq 0$, the discrepancy $d_{\mu+1}$ at the $(\mu+1)$th iteration step is

$$d_{\mu+1} = S_{\mu+2} + \sigma_1^{(\mu+1)} S_{\mu+1} + \cdots + \sigma_{l_{u+1}}^{(\mu+1)} S_{\mu+2-l_{\mu+1}}, \qquad (5.37)$$

where $\sigma_1^{(\mu+1)}, \ldots, \sigma_{l_{u+1}}^{(\mu+1)}$ are the coefficients of $\boldsymbol{\sigma}^{(\mu+1)}(X)$.

Repeat the above testing and correction process until we reach the $2t$th step. Then,

$$\boldsymbol{\sigma}(X) = \boldsymbol{\sigma}^{(2t)}(X) \qquad (5.38)$$

is the computed error-location polynomial. If the number of errors in the received polynomial $\mathbf{r}(X)$ is $t$ or less, the degree of $\boldsymbol{\sigma}^{(2t)}(X)$ is $t$ or less. In this case, the roots of $\boldsymbol{\sigma}^{(2t)}(X)$ are elements in $\mathrm{GF}(2^m)$ and whose inverses give the error-location numbers. If the number of roots of the error-location polynomial $\boldsymbol{\sigma}^{(2t)}(X)$ in $\mathrm{GF}(2^m)$ is less than its degree (e.g., $\boldsymbol{\sigma}^{(2t)}(X)$ may have no root in $\mathrm{GF}(2^m)$), this indicates that the number of errors in the received polynomial is larger than the error-correcting capability of the BCH code. In this case, it is in general not possible to locate the errors.

From the first Newton's identity, we readily see that

$$\boldsymbol{\sigma}^{(1)}(X) = 1 + S_1 X. \qquad (5.39)$$

To carry out the iteration process to find the error-location polynomial $\boldsymbol{\sigma}(X)$, we begin with Table 5.6 and proceed to fill it out. The rows corresponding to $\mu = -1$ and $\mu = 0$ in Table 5.6 give the initial conditions for the iteration process to find $\boldsymbol{\sigma}(X)$.

Table 5.6 Berlekamp–Massey iterative procedure for finding the error-location polynomial of a BCH code.

| Step | Partial solution | Discrepancy | Degree | Step/degree difference |
|------|------------------|-------------|--------|------------------------|
| $\mu$ | $\sigma^{(\mu)}(X)$ | $d_\mu$ | $l_\mu$ | $\mu - l_\mu$ |
| $-1$ | $1$ | $1$ | $0$ | $-1$ |
| $0$ | $1$ | $S_1$ | $0$ | $0$ |
| $1$ | $1 + S_1 X$ | | | |
| $2$ | | | | |
| $\vdots$ | | | | |
| $2t$ | | | | |

The above iterative decoding algorithm to decode BCH codes is known as the *Berlekamp–Massey iterative decoding algorithm*. It applies to both binary and nonbinary BCH codes. For more detailed proofs of the algorithm, readers are referred to References [6, 11, 18, 19].

**Example 5.3** Consider the $(15, 5)$ triple-error-correcting BCH code (i.e., $t = 3$) given in Example 5.1 whose generator polynomial $\mathbf{g}(X)$ has $\alpha, \alpha^2, \alpha^3, \alpha^4, \alpha^5$, and $\alpha^6$ as roots, where $\alpha$ is a primitive element of $\mathrm{GF}(2^4)$ (see Table 2.9). Suppose the all-zero codeword

$$\mathbf{v} = (0\,0\,0\,0\,0\,0\,0\,0\,0\,0\,0\,0\,0\,0\,0)$$

is transmitted and

$$\mathbf{r} = (0\,0\,0\,1\,0\,1\,0\,0\,0\,0\,0\,0\,1\,0\,0)$$

is the received vector. Then, the received polynomial is $\mathbf{r}(X) = X^3 + X^5 + X^{12}$. From the computations given in Example 5.2, we find that the syndrome of the received polynomial $\mathbf{r}(X)$ is

$$\mathbf{S} = (S_1, S_2, S_3, S_4, S_5, S_6),$$

with $S_1 = 1, S_2 = 1, S_3 = \alpha^{10}, S_4 = 1, S_5 = \alpha^{10}$ and $S_6 = \alpha^5$, i.e., $\mathbf{S} = (1, 1, \alpha^{10}, 1, \alpha^{10}, \alpha^5)$.

Because $t = 3$, we need to find an error-location polynomial $\boldsymbol{\sigma}(X) = \sigma_0 + \sigma_1 X + \sigma_2 X^2 + \sigma_3 X^3$ of degree 3 or less based on the syndrome $\mathbf{S} = (1, 1, \alpha^{10}, 1, \alpha^{10}, \alpha^5)$ whose coefficients satisfy the following six Newton's identities:

$$
\begin{aligned}
1 + \sigma_1 &= 0, \\
1 + \sigma_1 + 2\sigma_2 &= 0, \\
\alpha^{10} + \sigma_1 + \sigma_2 + 3\sigma_3 &= 0, \\
1 + \alpha^{10}\sigma_1 + \sigma_2 + \sigma_3 &= 0, \\
\alpha^{10} + \sigma_1 + \alpha^{10}\sigma_2 + \sigma_3 &= 0, \\
\alpha^5 + \alpha^{10}\sigma_1 + \sigma_2 + \alpha^{10}\sigma_3 &= 0.
\end{aligned}
\tag{5.40}
$$

Carrying out the Berlekamp–Massey iterative algorithm, we obtain Table 5.7, which gives partial solutions of the error-location polynomial at each iteration step up to the sixth step.

For illustration, we consider iteration step $\mu = 4$. At the completion of this step, we find $\sigma^{(4)}(X) = 1 + X + \alpha^5 X^2$ whose coefficients satisfy the first four Newton's identities given by (5.40). We then proceed to find the next partial solution $\sigma^{(5)}(X)$. First, we check whether the coefficients of $\sigma^{(4)}(X)$ also satisfy the fifth Newton's identity. Through calculation, we find that the coefficients of $\sigma^{(4)}(X)$ do not satisfy the fifth Newton's identity and the discrepancy is

$$d_4 = \alpha^{10} + 1 + \alpha^5 \alpha^{10} = \alpha^{10}.$$

In this case, we go back to steps prior to the fourth step and determine a step $\rho$ at which the partial solution is $\sigma^{(\rho)}(X)$ such that $d_\rho \neq 0$ and $\mu - \rho + l_\rho$ has the smallest value. By search, we find that $\rho = 2$ is such step. At this step, the partial solution is $\sigma^{(2)}(X) = 1 + X$, $d_2 = \alpha^5$, and $\mu - \rho + l_\rho = 4 - 2 + 1 = 3$. Next, we update the partial solution $\sigma^{(4)}(X)$ at the fourth step by adding the following correction term:

$$d_4 d_2^{-1} X^{4-2} \sigma^{(2)}(X) = \alpha^{10} \alpha^{-5} X^2 (1 + X) = \alpha^5 X^2 + \alpha^5 X^3.$$

The correction results in the following partial solution:

$$\sigma^{(5)}(X) = \sigma^{(4)}(X) + \alpha^5 X^2 + \alpha^5 X^3 = 1 + X + \alpha^5 X^3,$$

whose coefficients satisfy the first five Newton's identities. Through calculation, we find $d_5 = 0$. This indicates that the coefficients of $\sigma^{(5)}(X)$ also satisfy the sixth Newton's identity. Hence, the solution at the sixth step is

$$\sigma^{(6)}(X) = \sigma^{(5)}(X) = 1 + X + \alpha^5 X^3,$$

which has degree 3. By checking, we find that the coefficients of $\sigma^{(6)}(X)$ satisfy the six Newton's identities given by (5.40). Hence, $\sigma^{(6)}(X)$ is the error-location polynomial, i.e.,

$$\sigma(X) = \sigma^{(6)}(X) = 1 + X + \alpha^5 X^3.$$

Substituting the variable $X$ of $\sigma(X)$ with the elements, $\alpha^0, \alpha, \ldots, \alpha^{14}$ of $GF(2^4)$ in turn, we find that

$$\sigma(\alpha^3) = \sigma(\alpha^{10}) = \sigma(\alpha^{12}) = 0.$$

Hence, $\alpha^3$, $\alpha^{10}$, and $\alpha^{12}$ are the roots of $\sigma(X)$. The inverses of these three roots of $\sigma(X)$ are: $\alpha^{-3} = \alpha^{12}, \alpha^{-10} = \alpha^5$, and $\alpha^{-12} = \alpha^3$, which give the error-location numbers. The powers of these three error-location numbers are 12, 5, and 3. Consequently, the error pattern is

$$\mathbf{e}(X) = X^3 + X^5 + X^{12},$$

which indicates that the received vector has three errors. The number of errors is within the error-correcting capability of the code, $t = 3$. Subtracting $\mathbf{e}(X)$ from the received polynomial $\mathbf{r}(X)$, we obtain the decoded code polynomial:

$$\mathbf{v}(X) = \mathbf{r}(X) - \mathbf{e}(X) = \mathbf{r}(X) + \mathbf{e}(X) = 0 \text{ (zero polynomial)},$$

which is identical to the transmitted code polynomial. Hence, the decoding is correct.

Consider another case in which the received polynomial is $\mathbf{r}(X) = X + X^3 + X^5 + X^7$. The syndrome $\mathbf{S}$ of this received polynomial is $\mathbf{S} = (S_1, S_2, S_3, S_4, S_5, S_6) = (\alpha^{10}, \alpha^5, \alpha^{12}, \alpha^{10}, \alpha^5, \alpha^9)$. The six Newton's identities are

$$\begin{aligned}
\alpha^{10} + \sigma_1 &= 0, \\
\alpha^5 + \alpha^{10}\sigma_1 + 2\sigma_2 &= 0, \\
\alpha^{12} + \alpha^5\sigma_1 + \alpha^{10}\sigma_2 + 3\sigma_3 &= 0, \\
\alpha^{10} + \alpha^{12}\sigma_1 + \alpha^5\sigma_2 + \alpha^{10}\sigma_3 &= 0, \\
\alpha^5 + \alpha^{10}\sigma_1 + \alpha^{12}\sigma_2 + \alpha^5\sigma_3 &= 0, \\
\alpha^9 + \alpha^5\sigma_1 + \alpha^{10}\sigma_2 + \alpha^{12}\sigma_3 &= 0.
\end{aligned}$$

Following the Berlekamp–Massey iterative decoding algorithm to fill out the decoding table as shown in Table 5.6, we obtain Table 5.8 for decoding the received polynomial $\mathbf{r}(X) = X + X^3 + X^5 + X^7$.

We find that the computed error-location polynomial is $\boldsymbol{\sigma}(X) = 1 + \alpha^{10}X + \alpha^5 X^2 + \alpha^{12}X^3$. By searching, we find that $\boldsymbol{\sigma}(X)$ has no root in GF($2^4$). Thus, the decoder is not able to locate the errors. Then, the decoder declares a decoding failure. Because the received polynomial $\mathbf{r}(X)$ has 4 errors which are beyond the error-correcting capability $t = 3$ of the $(15, 5)$ BCH code, the Berlekamp-Massey iterative decoding algorithm is not able to decode this received polynomial.▲▲

## 5.9 Simplification of Decoding Binary BCH Codes

For decoding a binary BCH code, it can be proved that if the first, third, ..., $(2t-1)$th Newton's identities hold, then the second, fourth, ..., $2t$th Newton's identities also hold [6, 11, 18]. This implies that, with the iterative algorithm for finding the error-location polynomial $\boldsymbol{\sigma}(X)$, the solution $\boldsymbol{\sigma}^{(2\mu-1)}(X)$ at the $(2\mu - 1)$th step of iteration is also the solution $\boldsymbol{\sigma}^{(2\mu)}(X)$ at the $2\mu$th step of iteration, i.e.,

$$\boldsymbol{\sigma}^{(2\mu)}(X) = \boldsymbol{\sigma}^{(2\mu-1)}(X), \tag{5.41}$$

for $1 \leq \mu \leq t$. This fact is also demonstrated in Tables 5.7 and 5.8.

This suggests that the $(2\mu - 1)$th and $2\mu$th steps can be combined into one step. As a result, the forgoing algorithm for finding the error-location polynomial $\boldsymbol{\sigma}(X)$ can be reduced to $t$ steps. This simplification applies only to the decoding of binary BCH codes, but not to the decoding of nonbinary codes.

Table 5.7 Steps for finding the error-location polynomial of $\mathbf{r}(X) = X^3 + X^5 + X^{12}$ for the $(15,5)$ BCH code given in Example 5.3.

| $\mu$ | $\boldsymbol{\sigma}^{(\mu)}(X)$ | $d_\mu$ | $l_\mu$ | $\mu - l_\mu$ |
|-------|------------------|---------|---------|---------------|
| $-1$ | $1$ | $1$ | $0$ | $-1$ |
| $0$ | $1$ | $1$ | $0$ | $0$ |
| $1$ | $1 + X$ | $0$ | $1$ | $0$ (take $\rho = -1$) |
| $2$ | $1 + X$ | $\alpha^5$ | $1$ | $1$ |
| $3$ | $1 + X + \alpha^5 X^2$ | $0$ | $2$ | $1$ (take $\rho = 0$) |
| $4$ | $1 + X + \alpha^5 X^2$ | $\alpha^{10}$ | $2$ | $2$ |
| $5$ | $1 + X + \alpha^5 X^3$ | $0$ | $3$ | $2$ (take $\rho = 2$) |
| $6$ | $1 + X + \alpha^5 X^3$ | $-$ | $-$ | $-$ |

Table 5.8 Steps for finding the error-location polynomial of $\mathbf{r}(X) = X + X^3 + X^5 + X^7$ for the $(15,5)$ BCH code given in Example 5.3.

| $\mu$ | $\boldsymbol{\sigma}^{(\mu)}(X)$ | $d_\mu$ | $l_\mu$ | $\mu - l_\mu$ |
|-------|------------------|---------|---------|---------------|
| $-1$ | $1$ | $1$ | $0$ | $-1$ |
| $0$ | $1$ | $\alpha^{10}$ | $0$ | $0$ |
| $1$ | $1 + \alpha^{10} X$ | $0$ | $1$ | $0$ (take $\rho = -1$) |
| $2$ | $1 + \alpha^{10} X$ | $\alpha^{11}$ | $1$ | $1$ |
| $3$ | $1 + \alpha^{10} X + \alpha X^2$ | $0$ | $2$ | $1$ (take $\rho = 0$) |
| $4$ | $1 + \alpha^{10} X + \alpha X^2$ | $\alpha^{13}$ | $2$ | $2$ |
| $5$ | $1 + \alpha^{10} X + \alpha^5 X^2 + \alpha^{12} X^3$ | $0$ | $3$ | $2$ (take $\rho = 2$) |
| $6$ | $1 + \alpha^{10} X + \alpha^5 X^2 + \alpha^{12} X^3$ | $-$ | $-$ | $-$ |

The simplified algorithm for finding the error-location polynomial in decoding of a binary BCH code can be carried out by filling out a table with only $t$ steps, as shown in Table 5.9.

Suppose we have filled out all the rows up to and including the $\mu$th row. We fill out the $(\mu + 1)$th row as follows.

(1) If $d_\mu = 0$, then we set

$$\boldsymbol{\sigma}^{(\mu+1)}(X) = \boldsymbol{\sigma}^{(\mu)}(X).$$

(2) If $d_\mu \neq 0$, we find a row prior to the $\mu$th row, say the $\rho$th row, with partial solution $\boldsymbol{\sigma}^{(\rho)}(X)$ such that $d_\rho \neq 0$ and $2(\mu - \rho) + l_\rho$ is the smallest value. Then,

$$\boldsymbol{\sigma}^{(\mu+1)}(X) = \boldsymbol{\sigma}^{(\mu)}(X) + d_\mu d_\rho^{-1} X^{2(\mu-\rho)} \boldsymbol{\sigma}^{(\rho)}(X). \qquad (5.42)$$

The degree of $\boldsymbol{\sigma}^{(\mu+1)}(X)$ is $l_{\mu+1} = \max\{l_\mu, 2(\mu - \rho) + l_\rho\}$.

Table 5.9 A simplified Berlekamp–Massey iterative procedure for finding the error-location polynomial of a binary BCH code.

| Step $\mu$ | Partial solution $\boldsymbol{\sigma}^{(\mu)}(X)$ | Discrepancy $d_\mu$ | Degree $l_\mu$ | Step/degree difference $2\mu - l_\mu$ |
|---|---|---|---|---|
| $-\frac{1}{2}$ | 1 | 1 | 0 | $-1$ |
| 0 | 1 | $S_1$ | 0 | 0 |
| 1 | | | | |
| 2 | | | | |
| $\vdots$ | | | | |
| $t$ | | | | |

(3) Compute the discrepancy

$$d_{\mu+1} = S_{2\mu+3} + \sigma_1^{(\mu+1)} S_{2\mu+2} + \sigma_2^{(\mu+1)} S_{2\mu+1} + \cdots + \sigma_{l_{\mu+1}}^{(\mu+1)} S_{2\mu+3-l_{\mu+1}}. \tag{5.43}$$

Note that with the simplified algorithm, the computation required to find the error-location polynomial $\boldsymbol{\sigma}(X)$ is *half* of that required by the general algorithm for decoding binary BCH codes.

**Example 5.4** Using the simplified algorithm for finding the error-location polynomial, Table 5.7 given in Example 5.3 is reduced to Table 5.10. We find that the error-location polynomial is $\boldsymbol{\sigma}(X) = 1 + X + \alpha^5 X^3$ which is the same as the one found in Example 5.3. ▲▲

Table 5.10 Steps for finding the error-location polynomial of the binary $(15, 5)$ BCH code given in Example 5.4.

| $\mu$ | $\boldsymbol{\sigma}^{(\mu)}(X)$ | $d_\mu$ | $l_\mu$ | $2\mu - l_\mu$ |
|---|---|---|---|---|
| $-\frac{1}{2}$ | 1 | 1 | 0 | $-1$ |
| 0 | 1 | $S_1 = 1$ | 0 | 0 |
| 1 | $1 + X$ | $\alpha^5$ | 1 | 1 (take $\rho = -\frac{1}{2}$) |
| 2 | $1 + X + \alpha^5 X^2$ | $\alpha^{10}$ | 2 | 2 (take $\rho = 0$) |
| 3 | $1 + X + \alpha^5 X^3$ | $-$ | 3 | 3 (take $\rho = 1$) |

**Example 5.5** Let $\alpha$ be a primitive element of the field $\mathrm{GF}(2^5)$ constructed by using the primitive polynomial $\mathbf{p}(X) = 1 + X^2 + X^5$ over $\mathrm{GF}(2)$ shown in Table 5.11. Set $t = 3$. Consider the triple-error-correcting BCH code of length 31 whose generator polynomial $\mathbf{g}(X)$ has $\alpha, \alpha^2, \alpha^3, \alpha^4, \alpha^5, \alpha^6$ and their conjugates as roots. The roots of $\mathbf{g}(X)$ are contained in the following three conjugate sets:

$$\Omega(\alpha) \;= \{\alpha, \alpha^2, \alpha^4, \alpha^8, \alpha^{16}\},$$
$$\Omega(\alpha^3) = \{\alpha^3, \alpha^6, \alpha^{12}, \alpha^{17}, \alpha^{24}\},$$
$$\Omega(\alpha^5) = \{\alpha^5, \alpha^9, \alpha^{10}, \alpha^{18}, \alpha^{20}\}.$$

Next, we compute the minimal polynomials of elements in the above conjugate sets by using Table 5.11:

$$m_1(X) = (X + \alpha)(X + \alpha^2)(X + \alpha^4)(X + \alpha^8)(X + \alpha^{16})$$
$$= 1 + X^2 + X^5,$$
$$m_3(X) = (X + \alpha^3)(X + \alpha^6)(X + \alpha^{12})(X + \alpha^{17})(X + \alpha^{24})$$
$$= 1 + X^2 + X^3 + X^4 + X^5,$$
$$m_5(X) = (X + \alpha^5)(X + \alpha^9)(X + \alpha^{10})(X + \alpha^{18})(X + \alpha^{20})$$
$$= 1 + X + X^2 + X^4 + X^5.$$

Hence, the generator polynomial is

$$\mathbf{g}(X) = \mathrm{LCM}\{m_1(X), m_3(X), m_5(X)\}$$
$$= m_1(X)m_3(X)m_5(X)$$
$$= 1 + X + X^2 + X^3 + X^5 + X^7 + X^8 + X^9 + X^{10} + X^{11} + X^{15}.$$

The BCH code generated by $\mathbf{g}(X)$ is a $(31, 16)$ cyclic code $C$ with minimum distance at least 7 which is capable of correcting three or fewer errors with the simplified Berlekamp–Massey iterative algorithm.

Suppose the zero code polynomial $\mathbf{v}(X)$ in $C$ is transmitted and the received polynomial is

$$\mathbf{r}(X) = 1 + X^{12} + X^{20}.$$

To decode this received polynomial, we first compute its corresponding syndrome. Substituting $X$ of $\mathbf{r}(X)$ by $\alpha$, $\alpha^2$, $\alpha^3$, $\alpha^4$, $\alpha^5$, $\alpha^6$ in turn, we obtain the following six syndrome components:

$$S_1 = \mathbf{r}(\alpha) = 1 + \alpha^{12} + \alpha^{20} = \alpha^{18}, \quad S_2 = \mathbf{r}(\alpha^2) = 1 + \alpha^{24} + \alpha^{40} = \alpha^5,$$
$$S_3 = \mathbf{r}(\alpha^3) = 1 + \alpha^{36} + \alpha^{60} = \alpha^8, \quad S_4 = \mathbf{r}(\alpha^4) = 1 + \alpha^{48} + \alpha^{80} = \alpha^{10},$$
$$S_5 = \mathbf{r}(\alpha^5) = 1 + \alpha^{60} + \alpha^{100} = \alpha^{13}, \ S_6 = \mathbf{r}(\alpha^6) = 1 + \alpha^{72} + \alpha^{120} = \alpha^{16}.$$

Hence, the syndrome of the received polynomial $\mathbf{r}(X)$ is $\mathbf{S} = (\alpha^{18}, \alpha^5, \alpha^8, \alpha^{10}, \alpha^{13}, \alpha^{16})$. By using the simplified Berlekamp–Massey algorithm and the computed syndrome, we fill out Table 5.12 to find the error-location polynomial $\boldsymbol{\sigma}(X)$. In filling out the table, we find the partial solution for the error-location polynomial at each step.

From Table 5.12, we see that the partial solution at step 2 ($\mu = 2$) is $\sigma^{(2)}(X) = 1 + \alpha^{18}X + \alpha^{14}X^2$. Because the discrepancy at this step is $d_2 = \alpha^{15}$ which is not zero, to obtain the next solution, we need to make adjustment of

Table 5.11 $\mathrm{GF}(2^5)$ generated by the primitive polynomial $\mathbf{p}(X) = 1 + X^2 + X^5$.

| Power form | Polynomial form | | | | Vector form | Decimal form |
|---|---|---|---|---|---|---|
| 0 | 0 | | | | $(0\,0\,0\,0\,0)$ | 0 |
| 1 | 1 | | | | $(1\,0\,0\,0\,0)$ | 1 |
| $\alpha$ | $\alpha$ | | | | $(0\,1\,0\,0\,0)$ | 2 |
| $\alpha^2$ | | $\alpha^2$ | | | $(0\,0\,1\,0\,0)$ | 4 |
| $\alpha^3$ | | | $\alpha^3$ | | $(0\,0\,0\,1\,0)$ | 8 |
| $\alpha^4$ | | | | $\alpha^4$ | $(0\,0\,0\,0\,1)$ | 16 |
| $\alpha^5$ | $1 +$ | $\alpha^2$ | | | $(1\,0\,1\,0\,0)$ | 5 |
| $\alpha^6$ | $\alpha +$ | | $\alpha^3$ | | $(0\,1\,0\,1\,0)$ | 10 |
| $\alpha^7$ | | $\alpha^2$ | | $\alpha^4$ | $(0\,0\,1\,0\,1)$ | 20 |
| $\alpha^8$ | $1 +$ | $\alpha^2 +$ | $\alpha^3$ | | $(1\,0\,1\,1\,0)$ | 13 |
| $\alpha^9$ | $\alpha +$ | | $\alpha^3 +$ | $\alpha^4$ | $(0\,1\,0\,1\,1)$ | 26 |
| $\alpha^{10}$ | $1 +$ | | | $\alpha^4$ | $(1\,0\,0\,0\,1)$ | 17 |
| $\alpha^{11}$ | $1 + \alpha + \alpha^2$ | | | | $(1\,1\,1\,0\,0)$ | 7 |
| $\alpha^{12}$ | $\alpha + \alpha^2 + \alpha^3$ | | | | $(0\,1\,1\,1\,0)$ | 14 |
| $\alpha^{13}$ | | $\alpha^2 +$ | $\alpha^3 +$ | $\alpha^4$ | $(0\,0\,1\,1\,1)$ | 28 |
| $\alpha^{14}$ | $1 +$ | $\alpha^2 +$ | $\alpha^3 +$ | $\alpha^4$ | $(1\,0\,1\,1\,1)$ | 29 |
| $\alpha^{15}$ | $1 + \alpha + \alpha^2 +$ | | $\alpha^3 +$ | $\alpha^4$ | $(1\,1\,1\,1\,1)$ | 31 |
| $\alpha^{16}$ | $1 + \alpha +$ | | $\alpha^3 +$ | $\alpha^4$ | $(1\,1\,0\,1\,1)$ | 27 |
| $\alpha^{17}$ | $1 + \alpha +$ | | | $\alpha^4$ | $(1\,1\,0\,0\,1)$ | 19 |
| $\alpha^{18}$ | $1 + \alpha$ | | | | $(1\,1\,0\,0\,0)$ | 3 |
| $\alpha^{19}$ | $\alpha + \alpha^2$ | | | | $(0\,1\,1\,0\,0)$ | 6 |
| $\alpha^{20}$ | | $\alpha^2 +$ | $\alpha^3$ | | $(0\,0\,1\,1\,0)$ | 12 |
| $\alpha^{21}$ | | | $\alpha^3 +$ | $\alpha^4$ | $(0\,0\,0\,1\,1)$ | 24 |
| $\alpha^{22}$ | $1 +$ | $\alpha^2 +$ | | $\alpha^4$ | $(1\,0\,1\,0\,1)$ | 21 |
| $\alpha^{23}$ | $1 + \alpha + \alpha^2 + \alpha^3$ | | | | $(1\,1\,1\,1\,0)$ | 15 |
| $\alpha^{24}$ | $\alpha + \alpha^2 + \alpha^3 +$ | | | $\alpha^4$ | $(0\,1\,1\,1\,1)$ | 30 |
| $\alpha^{25}$ | $1 +$ | | $\alpha^3 +$ | $\alpha^4$ | $(1\,0\,0\,1\,1)$ | 25 |
| $\alpha^{26}$ | $1 + \alpha + \alpha^2 +$ | | | $\alpha^4$ | $(1\,1\,1\,0\,1)$ | 23 |
| $\alpha^{27}$ | $1 + \alpha +$ | | $\alpha^3$ | | $(1\,1\,0\,1\,0)$ | 11 |
| $\alpha^{28}$ | $\alpha + \alpha^2 +$ | | | $\alpha^4$ | $(0\,1\,1\,0\,1)$ | 22 |
| $\alpha^{29}$ | $1 +$ | | $\alpha^3$ | | $(1\,0\,0\,1\,0)$ | 9 |
| $\alpha^{30}$ | $\alpha +$ | | | $\alpha^4$ | $(0\,1\,0\,0\,1)$ | 18 |

$\boldsymbol{\sigma}^{(2)}(X)$. We find $\rho = 1$ for which $d_\rho \neq 0$ and $2(\mu - \rho) + l_\rho$ is the smallest. The correction term is:

$$d_2 d_1^{-1} X^2 \boldsymbol{\sigma}^{(1)}(X) = \alpha^{15} \alpha^{-1} X^2 (1 + \alpha^{18} X)$$
$$= \alpha^{14} X^2 + \alpha X^3.$$

Adding the correction term to $\boldsymbol{\sigma}^{(2)}(X)$, we obtain the next solution,

$$\boldsymbol{\sigma}^{(3)}(X) = \boldsymbol{\sigma}^{(2)}(X) + \alpha^{14} X^2 + \alpha X^3$$
$$= 1 + \alpha^{18} X + \alpha X^3,$$

Table 5.12 Steps for finding the error-location polynomial of $\mathbf{r}(X) = 1 + X^{12} + X^{20}$ for the binary $(31, 16)$ BCH code in Example 5.5.

| $\mu$ | $\sigma^{(\mu)}(X)$ | $d_\mu$ | $l_\mu$ | $2\mu - l_\mu$ | |
|---|---|---|---|---|---|
| $-\frac{1}{2}$ | $1$ | $1$ | $0$ | $-1$ | |
| $0$ | $1$ | $S_1 = \alpha^{18}$ | $0$ | $0$ | |
| $1$ | $1 + \alpha^{18}X$ | $\alpha$ | $1$ | $1$ | (take $\rho = -\frac{1}{2}$) |
| $2$ | $1 + \alpha^{18}X + \alpha^{14}X^2$ | $\alpha^{15}$ | $2$ | $2$ | (take $\rho = 0$) |
| $3$ | $1 + \alpha^{18}X + \alpha X^3$ | $-$ | $3$ | $3$ | (take $\rho = 1$) |

which is the final solution. Hence, $\boldsymbol{\sigma}(X) = \boldsymbol{\sigma}^{(3)}(X) = 1 + \alpha^{18}X + \alpha X^3$.

The coefficients of $\boldsymbol{\sigma}(X)$ satisfy the six Newton's identities as shown below:

$$\alpha^{18} + \alpha^{18} = 0,$$
$$\alpha^5 + \alpha^{18}\alpha^{18} + 2\alpha = \alpha^5 + \alpha^5 = 0,$$
$$\alpha^8 + \alpha^{18}\alpha^8 + 0\alpha^{18} + 3\alpha = \alpha^8 + \alpha^{23} + \alpha = 0,$$
$$\alpha^{10} + \alpha^{18}\alpha^8 + 0\alpha^5 + \alpha\alpha^{18} = \alpha^{10} + \alpha^{26} + \alpha^{19} = 0,$$
$$\alpha^{13} + \alpha^{18}\alpha^{10} + 0\alpha^8 + \alpha\alpha^5 = \alpha^{13} + \alpha^{28} + \alpha^6 = 0,$$
$$\alpha^{16} + \alpha^{18}\alpha^{13} + 0\alpha^{10} + \alpha\alpha^8 = \alpha^{16} + 1 + \alpha^9 = 0.$$

The roots of $\boldsymbol{\sigma}(X) = 1 + \alpha^{18}X + \alpha X^3$ in $\mathrm{GF}(2^5)$ are $1$, $\alpha^{11}$, and $\alpha^{19}$. The inverses of $1$, $\alpha^{11}$, and $\alpha^{19}$ are $1$, $\alpha^{20}$, and $\alpha^{12}$, respectively, which are the error-location numbers of the error polynomial $\mathbf{e}(X)$. Hence, the estimated error polynomial is $\mathbf{e}(X) = 1 + X^{12} + X^{20}$. Adding the estimated error polynomial to the received polynomial $\mathbf{r}(X) = 1 + X^{12} + X^{20}$, we obtain the decoded code polynomial which is the zero code polynomial. Thus, the decoding is correct.

Suppose the received polynomial is $\mathbf{r}(X) = X + X^3 + X^5 + X^7$. The syndrome of this received polynomial is $\mathbf{S} = (S_1, S_2, S_3, S_4, S_5, S_6) = (\alpha^6, \alpha^{12}, \alpha^{29}, \alpha^{24}, \alpha^3, \alpha^{27})$. Following the decoding steps of the simplified Berlekamp–Massey algorithm, we obtain Table 5.13.

Table 5.13 Steps for finding the error-location polynomial of $\mathbf{r}(X) = X + X^3 + X^5 + X^7$ for the binary $(31, 16)$ BCH code in Example 5.5.

| $\mu$ | $\sigma^{(\mu)}(X)$ | $d_\mu$ | $l_\mu$ | $2\mu - l_\mu$ | |
|---|---|---|---|---|---|
| $-\frac{1}{2}$ | $1$ | $1$ | $0$ | $-1$ | |
| $0$ | $1$ | $S_1 = \alpha^6$ | $0$ | $0$ | |
| $1$ | $1 + \alpha^6 X$ | $\alpha^6$ | $1$ | $1$ | (take $\rho = -\frac{1}{2}$) |
| $2$ | $1 + \alpha^6 X + X^2$ | $\alpha^{17}$ | $2$ | $2$ | (take $\rho = 0$) |
| $3$ | $1 + \alpha^6 X + \alpha^{19}X^2 + \alpha^{17}X^3$ | $-$ | $3$ | $3$ | (take $\rho = 1$) |

The error-location polynomial of the received polynomial is $\boldsymbol{\sigma}(X) = 1 + \alpha^6 X + \alpha^{19} X^2 + \alpha^{17} X^3$. By calculation (substituting $X$ in $\boldsymbol{\sigma}(X)$ with the nonzero elements in $GF(2^5)$ in turn), we find that $\boldsymbol{\sigma}(X)$ only has one root in $GF(2^5)$, $\alpha^{14}$. As we know, the degree of $\boldsymbol{\sigma}(X)$ indicates that the received polynomial has three errors. However, $\boldsymbol{\sigma}(X)$ only has one root in $GF(2^5)$. This indicates that the decoding has failed. Because the received polynomial $\mathbf{r}(X)$ has four errors which are beyond the error-correcting capability $t = 3$ of the $(31, 16)$ BCH code, the Berlekamp–Massey iterative decoding algorithm is not able to decode this received polynomial. ▲▲

## 5.10 Finding Error Locations and Error Correction

The last step in decoding a BCH code is to find the error-location numbers which are reciprocals of the roots of the error-location polynomial $\boldsymbol{\sigma}(X) = \sigma_0 + \sigma_1 X + \sigma_2 X^2 + \cdots + \sigma_\nu X^\nu$. Let $n = 2^m - 1$. The roots of $\boldsymbol{\sigma}(X)$ can be found simply by substituting the variable $X$ with the nonzero elements of $GF(2^m)$, $\alpha^0$, $\alpha$, $\alpha^2, \ldots, \alpha^{n-1}$, in turn. For $1 \leq i \leq n$, if $\alpha^i$ is a root of $\boldsymbol{\sigma}(X)$, then $\alpha^{n-i}$ (the inverse of $\alpha^i$) is an error-location number. In this case, the received bit $r_{n-i}$ at the location $n - i$ is an erroneously received bit and it must be corrected.

Determining the error locations and making error corrections can be carried out simultaneously using the *Chien search procedure* [7]. The received polynomial $\mathbf{r}(X) = r_0 + r_1 X + r_2 X^2 + \cdots + r_{n-1} X^{n-1}$ is decoded on a *bit-by-bit basis* in the order of $r_{n-1}, r_{n-2}, r_{n-3}, \ldots, r_1, r_0$. The highest-order bit $r_{n-1}$ of $\mathbf{r}(X)$ is decoded first and the lowest-order bit $r_0$ is decoded last.

To decode $r_{n-1}$, the decoder tests whether $\alpha^{n-1}$ is an error-location number. This is equivalent to testing whether $\alpha$ is a root of the error-location polynomial $\boldsymbol{\sigma}(X)$. If $\alpha$ is a root of $\boldsymbol{\sigma}(X)$, we have

$$\sigma_0 + \sigma_1 \alpha + \sigma_2 \alpha^2 + \cdots + \sigma_\nu \alpha^\nu = 0.$$

Because $\sigma_0 = 1$,

$$\sigma_1 \alpha + \sigma_2 \alpha^2 + \cdots + \sigma_\nu \alpha^\nu = 1.$$

Therefore, to decode $r_{n-1}$, the decoder first forms the products $\sigma_1 \alpha$, $\sigma_2 \alpha^2, \ldots, \sigma_\nu \alpha^\nu$ and then takes the sum $\sigma_1 \alpha + \sigma_2 \alpha^2 + \cdots + \sigma_\nu \alpha^\nu$. If the sum is equal to 1, then $\alpha^{n-1}$ is an error-location number and $r_{n-1}$ is an erroneous bit which must be corrected; otherwise, $r_{n-1}$ is a correct bit.

For $1 \leq i \leq n$, to decode $r_{n-i}$, the decoder forms the products $\sigma_1 \alpha^i$, $\sigma_2 \alpha^{2i}, \ldots, \sigma_\nu \alpha^{\nu i}$ and tests the sum

$$\sigma_1 \alpha^i + \sigma_2 \alpha^{2i} + \cdots + \sigma_\nu \alpha^{\nu i}.$$

If the sum is equal to 1, then $\alpha^i$ is a root of $\boldsymbol{\sigma}(X)$, $\alpha^{n-i}$ is an error-location number, and $r_{n-i}$ is an erroneous bit which must be corrected; otherwise, $r_{n-i}$ is a correctly received bit.

The testing procedure for error locations described above can be implemented in a straightforward manner by a circuit shown in Fig. 5.1. There are $t$ $\sigma$-registers, each of $m$-bits, and $t$ multipliers $\otimes$ which perform the multiplication of two elements in GF($2^m$). Block A is a logic circuit which computes the sum (over GF($2^m$)) of $t$ inputs from the $t$ $\sigma$-registers. The output of A is 1 if the sum equals 1; otherwise, it outputs 0. The buffer register stores the received $n$ bits. The $t$ $\sigma$-registers are initially loaded with $\sigma_1, \sigma_2, \ldots, \sigma_t$ computed in Step (2) of the decoding. If $\nu < t$, set $\sigma_{\nu+1} = \sigma_{\nu+2} = \cdots = \sigma_t = 0$.

Immediately before $r_{n-1}$ is read out of the buffer, the $t$ $\sigma$-registers are pulsed once and the $t$ multipliers $\otimes$ operate once. The multiplications are performed and the results $\sigma_1\alpha$, $\sigma_2\alpha^2$, $\cdots$, $\sigma_\nu\alpha^\nu$ are stored in the $\sigma$-registers. The output of the logic circuit A is 1 if and only if the sum $\sigma_1\alpha + \sigma_2\alpha^2 + \cdots + \sigma_\nu\alpha^\nu$ is equal to 1; otherwise, the output of A is 0. The bit $r_{n-1}$ is read out of the buffer and corrected by the output of A with an XOR gate. Having decoded $r_{n-1}$, the $t$ multipliers are pulsed again. Now, $\sigma_1\alpha^2$, $\sigma_2\alpha^4, \ldots, \sigma_\nu\alpha^{2\nu}$ are stored in the $\sigma$-registers. The sum

$$\sigma_1\alpha^2 + \sigma_2\alpha^4 + \cdots + \sigma_\nu\alpha^{2\nu}$$

is tested for a 1. The bit $r_{n-2}$ is read out of the buffer and corrected by the output of A in the same manner as the correction of $r_{n-1}$. The sum test and error-correction process continues until the entire received vector $\mathbf{r}$ is read out of the buffer. If the number of errors contained in $\mathbf{r}$ is $t$ or less, the output of the circuit given in Fig. 5.1 is the transmitted codeword.

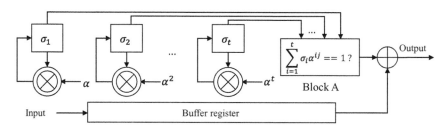

Figure 5.1 An error-location search and correction circuit.

The Chien search algorithm requires multiplication and addition operations over GF($2^m$). It is noted that the syndrome calculation and the Berlekamp–Massey iterative algorithm to find the error-location polynomial also need these operations. More details on how to implement multiplication and addition operations over GF($2^m$) in hardware can be found in [6, 18, 23].

## 5.11 Nonprimitive Binary BCH Codes

So far, we have considered an important subclass of BCH codes, known as the primitive BCH codes of length $2^m - 1$, where $m$ is a positive integer greater than 2. A more general class of binary BCH codes is presented in this section, called *nonprimitive BCH codes*.

Consider the field $GF(2^m)$. Let $\alpha$ be a primitive element in $GF(2^m)$ and let $n$ be a factor of $2^m - 1$, say $2^m - 1 = n\ell$. Let $\beta = \alpha^\ell$. Then, $\beta$ is an element of order $n$ in $GF(2^m)$, i.e., $\beta^n = 1$. As shown in Chapter 2, the set $\{1, \beta, \beta^2, \ldots, \beta^{n-1}\}$ forms a cyclic subgroup of the multiplicative group of $GF(2^m)$. Let $t$ be a positive integer such that $2t < n$. The cyclic code $C$ over $GF(2)$ of length $n$ generated by the polynomial $\mathbf{g}(X)$ over $GF(2)$ of the *least degree* which has $\beta$, $\beta^2, \ldots, \beta^{2t}$ and their conjugates as roots is a BCH code with minimum distance at least $2t + 1$. Such a BCH code is called a *nonprimitive BCH code*. The polynomial $\mathbf{g}(X)$ is the generator polynomial of $C$, which is simply the LCM of the minimal polynomials of $\beta$, $\beta^3, \ldots, \beta^{2t-1}$ and $\mathbf{g}(X)$ divides $X^n + 1$. The proof that the minimum distance of the nonprimitive BCH code $C$ is at least $2t + 1$ is exactly the same as the proof of the minimum distance of a $t$-error-correcting primitive BCH code. This nonprimitive BCH code is capable of correcting $t$ or fewer random errors using the Berlekamp–Massey decoding algorithm. A nonprimitive BCH code has the same structural properties as a primitive BCH code. Note that, if $2^m - 1$ is a prime, then $n = 2^m - 1$. In this case, $\beta$ is a primitive element of $GF(2^m)$ and the BCH code generated by $\mathbf{g}(X)$ is a primitive BCH code.

**Example 5.6** Consider the field $GF(2^6)$ given by Table 5.14 constructed based on the primitive polynomial $\mathbf{p}(X) = 1 + X + X^6$ over $GF(2)$. Note that $2^6 - 1$ can be factored as the product $3 \times 21$. Let $\alpha$ be a primitive element of $GF(2^6)$ and $\beta = \alpha^3$. The order of $\beta$ is 21. Set $n = 21$ and $t = 2$. Suppose we construct a nonprimitive BCH code of length 21 whose generator polynomial $\mathbf{g}(X)$ has $\beta$, $\beta^2$, $\beta^3$, $\beta^4$ and their conjugates as roots. We find that $\beta$, $\beta^2$, and $\beta^4$ have the same minimal polynomial which is

$$m_1(X) = (X + \alpha^3)(X + \alpha^6)(X + \alpha^{12})(X + \alpha^{24})(X + \alpha^{33})(X + \alpha^{48})$$
$$= 1 + X + X^2 + X^4 + X^6.$$

The minimal polynomial of $\beta^3$ is

$$m_3(X) = (X + \alpha^9)(X + \alpha^{18})(X + \alpha^{36})$$
$$= 1 + X^2 + X^3.$$

Hence,

$$\mathbf{g}(X) = \mathrm{LCM}\{m_1(X), m_3(X)\}$$
$$= m_1(X)m_3(X)$$
$$= 1 + X + X^4 + X^5 + X^7 + X^8 + X^9.$$

The nonprimitive BCH code generated by $\mathbf{g}(X)$ is a $(21, 12)$ cyclic code which has error-correcting capability $t = 2$. ▲▲

Table 5.14 GF($2^6$) generated by $\mathbf{p}(X) = 1 + X + X^6$ over GF(2).

| Power form | Polynomial form | Vector form | Decimal form |
|---|---|---|---|
| $0 = \alpha^{-\infty}$ | $0$ | $(0\,0\,0\,0\,0\,0)$ | 0 |
| $1 = \alpha^0$ | $1$ | $(1\,0\,0\,0\,0\,0)$ | 1 |
| $\alpha$ | $\alpha$ | $(0\,1\,0\,0\,0\,0)$ | 2 |
| $\alpha^2$ | $\alpha^2$ | $(0\,0\,1\,0\,0\,0)$ | 4 |
| $\alpha^3$ | $\alpha^3$ | $(0\,0\,0\,1\,0\,0)$ | 8 |
| $\alpha^4$ | $\alpha^4$ | $(0\,0\,0\,0\,1\,0)$ | 16 |
| $\alpha^5$ | $\alpha^5$ | $(0\,0\,0\,0\,0\,1)$ | 32 |
| $\alpha^6$ | $1 + \alpha$ | $(1\,1\,0\,0\,0\,0)$ | 3 |
| $\alpha^7$ | $\alpha + \alpha^2$ | $(0\,1\,1\,0\,0\,0)$ | 6 |
| $\alpha^8$ | $\alpha^2 + \alpha^3$ | $(0\,0\,1\,1\,0\,0)$ | 12 |
| $\alpha^9$ | $\alpha^3 + \alpha^4$ | $(0\,0\,0\,1\,1\,0)$ | 24 |
| $\alpha^{10}$ | $\alpha^4 + \alpha^5$ | $(0\,0\,0\,0\,1\,1)$ | 48 |
| $\alpha^{11}$ | $1 + \alpha + \alpha^5$ | $(1\,1\,0\,0\,0\,1)$ | 35 |
| $\alpha^{12}$ | $1 + \alpha^2$ | $(1\,0\,1\,0\,0\,0)$ | 5 |
| $\alpha^{13}$ | $\alpha + \alpha^3$ | $(0\,1\,0\,1\,0\,0)$ | 10 |
| $\alpha^{14}$ | $\alpha^2 + \alpha^4$ | $(0\,0\,1\,0\,1\,0)$ | 20 |
| $\alpha^{15}$ | $\alpha^3 + \alpha^5$ | $(0\,0\,0\,1\,0\,1)$ | 40 |
| $\alpha^{16}$ | $1 + \alpha + \alpha^4$ | $(1\,1\,0\,0\,1\,0)$ | 19 |
| $\alpha^{17}$ | $\alpha + \alpha^2 + \alpha^5$ | $(0\,1\,1\,0\,0\,1)$ | 38 |
| $\alpha^{18}$ | $1 + \alpha + \alpha^2 + \alpha^3$ | $(1\,1\,1\,1\,0\,0)$ | 15 |
| $\alpha^{19}$ | $\alpha + \alpha^2 + \alpha^3 + \alpha^4$ | $(0\,1\,1\,1\,1\,0)$ | 30 |
| $\alpha^{20}$ | $\alpha^2 + \alpha^3 + \alpha^4 + \alpha^5$ | $(0\,0\,1\,1\,1\,1)$ | 60 |
| $\alpha^{21}$ | $1 + \alpha + \alpha^3 + \alpha^4 + \alpha^5$ | $(1\,1\,0\,1\,1\,1)$ | 59 |
| $\alpha^{22}$ | $1 + \alpha^2 + \alpha^4 + \alpha^5$ | $(1\,0\,1\,0\,1\,1)$ | 53 |
| $\alpha^{23}$ | $1 + \alpha^3 + \alpha^5$ | $(1\,0\,0\,1\,0\,1)$ | 41 |
| $\alpha^{24}$ | $1 + \alpha^4$ | $(1\,0\,0\,0\,1\,0)$ | 17 |
| $\alpha^{25}$ | $\alpha + \alpha^5$ | $(0\,1\,0\,0\,0\,1)$ | 34 |
| $\alpha^{26}$ | $1 + \alpha + \alpha^2$ | $(1\,1\,1\,0\,0\,0)$ | 7 |
| $\alpha^{27}$ | $\alpha + \alpha^2 + \alpha^3$ | $(0\,1\,1\,1\,0\,0)$ | 14 |
| $\alpha^{28}$ | $\alpha^2 + \alpha^3 + \alpha^4$ | $(0\,0\,1\,1\,1\,0)$ | 28 |
| $\alpha^{29}$ | $\alpha^3 + \alpha^4 + \alpha^5$ | $(0\,0\,0\,1\,1\,1)$ | 56 |
| $\alpha^{30}$ | $1 + \alpha + \alpha^4 + \alpha^5$ | $(1\,1\,0\,0\,1\,1)$ | 51 |
| $\alpha^{31}$ | $1 + \alpha^2 + \alpha^5$ | $(1\,0\,1\,0\,0\,1)$ | 37 |
| $\alpha^{32}$ | $1 + \alpha^3$ | $(1\,0\,0\,1\,0\,0)$ | 9 |
| $\alpha^{33}$ | $\alpha + \alpha^4$ | $(0\,1\,0\,0\,1\,0)$ | 18 |
| $\alpha^{34}$ | $\alpha^2 + \alpha^5$ | $(0\,0\,1\,0\,0\,1)$ | 36 |
| $\alpha^{35}$ | $1 + \alpha + \alpha^3$ | $(1\,1\,0\,1\,0\,0)$ | 11 |
| $\alpha^{36}$ | $\alpha + \alpha^2 + \alpha^4$ | $(0\,1\,1\,0\,1\,0)$ | 22 |
| $\alpha^{37}$ | $\alpha^2 + \alpha^3 + \alpha^5$ | $(0\,0\,1\,1\,0\,1)$ | 44 |
| $\alpha^{38}$ | $1 + \alpha + \alpha^3 + \alpha^4$ | $(1\,1\,0\,1\,1\,0)$ | 27 |
| $\alpha^{39}$ | $\alpha + \alpha^2 + \alpha^4 + \alpha^5$ | $(0\,1\,1\,0\,1\,1)$ | 54 |
| $\alpha^{40}$ | $1 + \alpha + \alpha^2 + \alpha^3 + \alpha^5$ | $(1\,1\,1\,1\,0\,1)$ | 47 |
| $\alpha^{41}$ | $1 + \alpha^2 + \alpha^3 + \alpha^4$ | $(1\,0\,1\,1\,1\,0)$ | 29 |

Table 5.14 (*continued*)

| | | | |
|---|---|---|---|
| $\alpha^{42}$ | $\alpha \qquad + \alpha^3 + \alpha^4 + \alpha^5$ | (0 1 0 1 1 1) | 58 |
| $\alpha^{43}$ | $1 + \alpha + \alpha^2 \qquad + \alpha^4 + \alpha^5$ | (1 1 1 0 1 1) | 55 |
| $\alpha^{44}$ | $1 \qquad + \alpha^2 + \alpha^3 \qquad + \alpha^5$ | (1 0 1 1 0 1) | 45 |
| $\alpha^{45}$ | $1 \qquad + \alpha^3 + \alpha^4$ | (1 0 0 1 1 0) | 25 |
| $\alpha^{46}$ | $\alpha \qquad + \alpha^4 + \alpha^5$ | (0 1 0 0 1 1) | 50 |
| $\alpha^{47}$ | $1 + \alpha + \alpha^2 \qquad + \alpha^5$ | (1 1 1 0 0 1) | 39 |
| $\alpha^{48}$ | $1 \qquad + \alpha^2 + \alpha^3$ | (1 0 1 1 0 0) | 13 |
| $\alpha^{49}$ | $\alpha \qquad + \alpha^3 + \alpha^4$ | (0 1 0 1 1 0) | 26 |
| $\alpha^{50}$ | $\alpha^2 \qquad + \alpha^4 + \alpha^5$ | (0 0 1 0 1 1) | 52 |
| $\alpha^{51}$ | $1 + \alpha \qquad + \alpha^3 \qquad + \alpha^5$ | (1 1 0 1 0 1) | 43 |
| $\alpha^{52}$ | $1 \qquad + \alpha^2 \qquad + \alpha^4$ | (1 0 1 0 1 0) | 21 |
| $\alpha^{53}$ | $\alpha \qquad + \alpha^3 \qquad + \alpha^5$ | (0 1 0 1 0 1) | 42 |
| $\alpha^{54}$ | $1 + \alpha + \alpha^2 \qquad + \alpha^4$ | (1 1 1 0 1 0) | 23 |
| $\alpha^{55}$ | $\alpha + \alpha^2 + \alpha^3 \qquad + \alpha^5$ | (0 1 1 1 0 1) | 46 |
| $\alpha^{56}$ | $1 + \alpha + \alpha^2 + \alpha^3 + \alpha^4$ | (1 1 1 1 1 0) | 31 |
| $\alpha^{57}$ | $\alpha + \alpha^2 + \alpha^3 + \alpha^4 + \alpha^5$ | (0 1 1 1 1 1) | 62 |
| $\alpha^{58}$ | $1 + \alpha + \alpha^2 + \alpha^3 + \alpha^4 + \alpha^5$ | (1 1 1 1 1 1) | 63 |
| $\alpha^{59}$ | $1 \qquad + \alpha^2 + \alpha^3 + \alpha^4 + \alpha^5$ | (1 0 1 1 1 1) | 61 |
| $\alpha^{60}$ | $1 \qquad + \alpha^3 + \alpha^4 + \alpha^5$ | (1 0 0 1 1 1) | 57 |
| $\alpha^{61}$ | $1 \qquad + \alpha^4 + \alpha^5$ | (1 0 0 0 1 1) | 49 |
| $\alpha^{62}$ | $1 \qquad + \alpha^5$ | (1 0 0 0 0 1) | 33 |

**Example 5.7** Consider the field $GF(2^{11})$ constructed based on the primitive polynomial $\mathbf{p}(X) = 1 + X^2 + X^{11}$ over $GF(2)$. The number $2^{11} - 1 = 2047$ can be factored as the product of two primes 23 and 89, i.e., $2047 = 23 \times 89$. Let $\alpha$ be a primitive element of $GF(2^{11})$ and $\beta = \alpha^{89}$. The order of $\beta$ is 23. Set $n = 23$ and $t = 2$. Suppose we construct a double-error-correcting nonprimitive BCH code of length 23 whose generator polynomial $\mathbf{g}(X)$ has $\beta$, $\beta^2$, $\beta^3$, $\beta^4$ and their conjugates $\beta^6$, $\beta^8$, $\beta^9$, $\beta^{12}$, $\beta^{13}$, $\beta^{16}$, $\beta^{18}$ as roots. With computer aid, we find that

$$\mathbf{g}(X) = 1 + X + X^5 + X^6 + X^7 + X^9 + X^{11}.$$

The nonprimitive BCH code generated by $\mathbf{g}(X)$ is a $(23, 12)$ cyclic code $C$ with designed error-correcting capability $t = 2$. Based on the BCH bound given by (5.19), the minimum distance $d_{\min}$ of $C$ is lower bounded by $2t + 1 = 5$, i.e., $d_{\min} \geq 5$. However, the true minimum distance of this code was found to be 7 [28]. Hence, the true error-correcting capability of $C$ is 3. By using the Berlekamp–Massey algorithm for decoding this code, only two or fewer errors can be corrected.

The generator polynomial of the $(23, 12)$ nonprimitive BCH code constructed above is identical to the generator polynomial of the $(23, 12)$ Golay code constructed in Example 4.11. Hence, the $(23, 12)$ nonprimitive BCH code is the $(23, 12)$ Golay code and has minimum distance 7, which is two greater than its designed value.

Because the $(23, 12)$ nonprimitive BCH code is short and the number of parity-check bits is only 11, the simplest and fastest way to decode the code is to use the table look-up decoding presented in Section 3.9; this requires a memory to store $2^{11} = 1024$ error patterns with 0, 1, 2, and 3 errors and their corresponding syndromes. ▲▲

## 5.12 Remarks

Binary BCH codes using the Berlekamp–Massey iterative decoding algorithm can be used to correct combinations of random errors and erasures over a binary symmetric erasure channel (BSEC) as shown in Fig. 1.7(b). They can also be used to correct random bursts of erasures over a binary burst erasure channel (BBEC) by using the Berlekamp–Massey iterative decoding algorithm in conjunction with a coding technique known as interleaving. These topics will be discussed in Chapter 10. A soft-decision decoding algorithm for binary BCH codes will be presented in Chapter 16.

Decoding of BCH codes presented in this chapter is actually the Meggitt decoding using the Berlekamp–Massey iterative algorithm as the SEL-algorithm for error detection and correction.

## Problems

**5.1** Let $C_0$ and $C_1$ be two BCH codes of the same length $n = 2^m - 1$ with designed error-correcting capabilities $t_0$ and $t_1$. Assume $t_1 \geq t_0$. Show that $C_1$ is a subcode of $C_0$, i.e., all the codewords in $C_1$ are codewords in $C_0$.

**5.2** Consider the Galois field $GF(2^4)$ given by Table 2.9. The element $\beta = \alpha^7$ is also a primitive element of $GF(2^4)$. Let $\mathbf{g}_0(X)$ be the lowest-degree polynomial over $GF(2)$ that has $\beta$, $\beta^2$, $\beta^3$, and $\beta^4$ as its roots. This polynomial also generates a double-error-correcting primitive BCH code $C$ of length 15.
  (a) Determine $\mathbf{g}_0(X)$.
  (b) Find the parity-check matrix for the BCH code $C$.
  (c) Find the weight distribution of the BCH code $C$.
  (d) Show that $\mathbf{g}_0(X)$ is the reciprocal polynomial of the generator polynomial $\mathbf{g}(X)$ of the $(15, 7)$ double-error-correcting BCH code given in Example 5.1.

**5.3** Let $\alpha$ be a primitive element of the field $GF(2^m)$ and $m_0$ and $t$ be two nonnegative integers. Let $\mathbf{g}(X)$ be the polynomial over $GF(2)$ with $\alpha^{m_0}$, $\alpha^{m_0+1}$, $\alpha^{l_0+2}, \ldots,$ $\alpha^{m_0+2t-1}$ and their conjugates as roots. Prove that the cyclic code generated by $\mathbf{g}(X)$ has minimum distance at least $2t + 1$. (This code is a BCH code in general sense.)

**5.4** Prove that the parity-check matrix given by (5.12) of a BCH code can be reduced to

$$\mathbf{H}^* = \begin{bmatrix} 1 & \alpha & \alpha^2 & \cdots & \alpha^{n-1} \\ 1 & \alpha^3 & (\alpha^3)^2 & \cdots & (\alpha^3)^{n-1} \\ 1 & \alpha^5 & (\alpha^5)^2 & \cdots & (\alpha^5)^{n-1} \\ \vdots & \vdots & \vdots & & \vdots \\ 1 & \alpha^{2t-1} & (\alpha^{2t-1})^2 & \cdots & (\alpha^{2t-1})^{n-1} \end{bmatrix},$$

i.e., the parity-check matrix $\mathbf{H}^*$ of a BCH code can be constructed based on the roots with odd powers, $\alpha, \alpha^3, \alpha^5, \ldots, \alpha^{2t-3}, \alpha^{2t-1}$, of its generator polynomial $\mathbf{g}(X)$.

**5.5** Construct two binary BCH codes $C_1$ and $C_2$ with the following parameters:
  (a) $C_1$: $m = 5$, $t = 2$,
  (b) $C_2$: $m = 5$, $t = 3$.
Find their lengths, dimensions, minimum distances, and weight distributions. (Assume that the field $\mathrm{GF}(2^5)$ is generated by the primitive polynomial $\mathbf{p}(X) = 1 + X^2 + X^5$ over $\mathrm{GF}(2)$ (see Table 5.11).)

**5.6** Consider the $(15, 5)$ BCH code $C$ given in Example 5.1 with $t = 3$.
  (a) Let $\mathbf{u}(X) = 1 + X^4$ be a message polynomial. Find the corresponding code polynomials in both nonsystematic and systematic forms.
  (b) Find the generator and parity-check matrices, $\mathbf{G}_{\mathrm{sys}}$ and $\mathbf{H}_{\mathrm{sys}}$ in systematic form, of the code $C$.
  (c) Find the parity-check matrix over $\mathrm{GF}(2^4)$ of the code $C$.
  (d) Find the binary parity-check matrix $\mathbf{H}_b$ of the code $C$ based on the parity-check matrix $\mathbf{H}$ over $\mathrm{GF}(2^4)$ of the code $C$.
  (e) Find the simplified nonbinary parity-check matrix $\mathbf{H}^*$ (see Problem 5.4) and the corresponding binary parity-check matrix $\mathbf{H}_b^*$ of the code $C$.

**5.7** Consider the $(15, 5)$ BCH code $C$ given in Example 5.1 with $t = 3$. Calculate the syndromes of the following two received polynomials and check that any error is detected in these two received polynomials:
  (a) $\mathbf{r}_1(X) = X + X^2 + X^8$,
  (b) $\mathbf{r}_2(X) = 1 + X + X^2 + X^8$.

**5.8** Based on the syndromes calculated in Problem 5.7, decode the two received polynomials $\mathbf{r}_1(X)$ and $\mathbf{r}_2(X)$ using Berlekamp–Massey iterative decoding algorithm.

**5.9** Consider the field $\mathrm{GF}(2^6)$ given by Table 5.14 constructed based on the primitive polynomial $\mathbf{p}(X) = 1 + X + X^6$ over $\mathrm{GF}(2)$. Construct a nonprimitive BCH code with length $n = 9$ and error-correcting capability $t = 1$.

**5.10** Let $\mathbf{g}(X)$ be the generator polynomial of the $(15, 7)$ BCH code given in Example 5.1. Construct a cyclic code with $\mathbf{g}_e(X) = (1 + X)\mathbf{g}(X)$ as the generator polynomial. Determine its minimum distance.

**5.11** Consider a $t$-error-correcting primitive binary BCH code of length $n = 2^m - 1$. If $2t + 1$ is a factor of $n$, prove that the minimum distance $d_{\min}$ of the code is exactly $2t + 1$. (*Hint*: Let $n = \ell(2t + 1)$. Show that $(X^n + 1)/(X^\ell + 1)$ is a code polynomial of weight $2t + 1$.)

**5.12** Show that the equations given by (5.26) have $2^k$ solutions.

# References

[1] A. Hocquenghem, "Codes correcteurs d'erreurs," *Chiffres*, **2** (1959), 147–156.

[2] R. C. Bose and D. K. Ray-Chaudhuri, "On a class of error correcting binary group codes," *Inf. Control*, **3**(1) (1960), 68–79.

[3] W. W. Peterson, "Encoding and error-correction procedures for the Bose–Chaudhuri codes," *IRE Trans. Inf. Theory*, **6**(5) (1960), 459–470.

[4] E. R. Berlekamp, "On decoding binary Bose-Chaudhuri-Hocquenghem codes," *IEEE Trans. Inf. Theory*, **IT-11**(4) (1965), 577–580.

[5] E. R. Berlekamp, "Nonbinary BCH decoding," *Proc. IEEE Int. Symp. Inf. Theory*, San Remo, Italy, 1967.

[6] E. R. Berlekamp, *Algebraic Coding Theory*, revised ed., Laguna Hills, CA, Aegean Park Press, 1984.

[7] R. T. Chien, "Cyclic decoding procedure for the Bose–Chaudhuri–Hocquenghem codes," *IEEE Trans. Inf. Theory*, **IT-10**(4) (1964), 357–363.

[8] G. D. Forney, "On decoding BCH codes," *IEEE Trans. Inf. Theory*, **IT-11**(4) (1965), 549–557.

[9] D. Gorenstein and N. Zierler, "A class of cyclic linear error-correcting codes in $p^m$ symbols," *J. Soc. Indust. Appl. Math.*, **9** (1961), 107–214.

[10] J. L. Massey, "Step-by-step decoding of the Bose–Chaudhuri–Hocquenghem codes," *IEEE Trans. Inf. Theory*, **11**(4) (1965), 580–585.

[11] J. L. Massey, "Shift-register synthesis and BCH decoding," *IEEE Trans. Inf. Theory*, **15**(1) (1969), 122–127.

[12] H. O. Burton, "Inversionless decoding of binary BCH decoding," *IEEE Trans. Inf. Theory*, **17**(4) (1971), 464–466.

[13] D. V. Sarwate and N. R. Shanbhag, "High-speed architecture for Reed–Solomon decoders," *IEEE Trans. Very Large Scale Integration (VLSI) Systems*, **9**(5) (2001), 641–655.

[14] W. Liu, J. Rho, and W. Sung, "Lower-power high-throughput BCH error correction VLSI design for multi-level cell NAND flash memories," *IEEE Workshop on Signal Processing Systems Design and Implementation*, Banff, Alta, Canada, October 2–4, 2006, pp. 303–308.

[15] H. Choi, W. Liu, and W. Sung, "VLSI implementation of BCH error correction for multilevel cell NAND flash memory," *IEEE Trans. Very Large Scale Integration (VLSI) Systems*, **18**(5) (2010), 843–847.

[16] M. Yin, M. Xie, and B. Yi, "Optimized algorithm for binary BCH codes," *IEEE International Symposium on Circuits and Systems*, Beijing China, May 19–23, 2013, pp. 1552–1555.

[17] B. Park, S. An, J. Park, and Y. Lee, "Novel folded-KES architecture for high-speed and area-efficient BCH decoders," *IEEE Trans. Circuits and Systems II: Express Briefs*, **64**(5) (2017), 535–539.

[18] W. W. Peterson and E. J. Weldon, Jr., *Error-Correcting Codes*, 2nd ed., Cambridge, MA, The MIT Press, 1972.

[19] F. J. MacWilliams and N. J. A. Sloane, *The Theory of Error-Correcting Codes*, Amsterdam, North-Holland, 1977.

[20] G. Clark and J. Cain, *Error-Correcting Codes for Digital Communications*, New York, Plenum Press, 1981.

[21] S. B. Wicker, *Error Control Systems for Digital Communication and Storage*, Englewood Cliffs, NJ, Prentice-Hall, 1995.

[22] R. E. Blahut, *Algebraic Codes for Data Transmission*, Cambridge, Cambridge University Press, 2003.

[23] S. Lin and D. J. Costello, Jr., *Error Control Coding: Fundamentals and Applications*, 2nd ed., Upper Saddle River, NJ, Prentice-Hall, 2004.

[24] E. R. Berlekamp, "The enumeration of information symbols in BCH codes," *Bell Systems Tech. J.*, **46**(8) (1967), 1861–1880.

[25] H. B. Mann, "On the number of information symbols in Bose–Chaudhuri codes," *Inf. Control*, **5**(2) (1962), 153–162.

[26] T. Kasami, S. Lin, and W. W. Peterson, "Some results on weight distributions of BCH codes," *IEEE Trans. Inf. Theory*, **12**(2) (1966), 247.

[27] J. Riordan, *An Introduction to Combinatorial Analysis*, New York, John Wiley & Sons, 1958.

[28] M. J. E. Golay, "Notes on digital coding," *Proc. IRE*, **37** (1949), 657.

# 6

# Nonbinary BCH Codes and Reed–Solomon Codes

Right after the discovery of binary BCH codes by Hocquenghem in 1959 [1] and by Bose and Chaudhuri in 1960 [2] independently, Gorenstein and Zierler extended this class of codes to the nonbinary case in 1961 [3]. The code symbols of a nonbinary BCH code are from a nonbinary field $GF(q)$, where $q$ is a power of a prime $p$ and the roots of its generator polynomial are elements in an extension field $GF(q^m)$, $m \geq 1$, of $GF(q)$. The most important and widely used subclass of nonbinary BCH codes is the class of *Reed–Solomon* (RS) codes. Even though RS codes form a subclass of nonbinary BCH codes, they were constructed independently using a totally different approach by Reed and Solomon in 1960 [4], the same year as binary BCH codes were discovered by Bose and Chaudhuri. The relationship between RS codes and nonbinary BCH codes was proved by Gorenstein and Zierler in 1961 [3]. RS codes are widely used in many applications, such as communication systems (internet, DSL, WiMAX, satellite communication, DVB, etc.), data-storage systems (CDs, DVDs, hard drives, etc.), barcodes, quick response codes, and many other areas.

The first decoding algorithm for nonbinary BCH codes or RS codes was devised by Gorenstein and Zierler [3] and then improved by Chien [5] and Forney [6]. The most effective algorithm for decoding both binary and nonbinary BCH codes is the Berlekamp–Massey iterative decoding algorithm [7–11] presented in the previous chapter. Sugiyama, Kasahara, Hirasawa, and Namekawa proposed an algorithm to decode BCH and RS codes using the Euclidean algorithm to find the greatest common divisor of two polynomials in 1975 [12]. BCH and RS codes can also be decoded in the frequency domain [13–15]. Algorithms for decoding RS codes have been reformulated and improved to facilitate high-throughput and low-complexity hardware implementation of RS decoders in [16–18]. Algebraic soft-decision decoding algorithms and efficient hardware architectures for decoding BCH and RS codes to improve their error performance over the hard-decision decoding algorithms have also been devised and implemented [18–30].

In this chapter, we present the construction of nonbinary BCH codes and RS codes, and two of their commonly used decoding algorithms, namely, the Berlekamp–Massey iterative algorithm and the one based on the Euclidean algorithm. Additional materials on nonbinary BCH codes and RS codes and their decoding algorithms can be found in [9, 15, 31–37].

## 6.1 Nonbinary Primitive BCH Codes

Let $\mathrm{GF}(q^m)$ be an extension field of $\mathrm{GF}(q)$ and $\alpha$ be a primitive element of $\mathrm{GF}(q^m)$. A $q$-ary $t$-symbol-error-correcting primitive BCH code $C$ of length $n = q^m - 1$ is a cyclic code generated by the smallest-degree polynomial $\mathbf{g}(X)$ over $\mathrm{GF}(q)$ that has $\alpha, \alpha^2, \ldots, \alpha^{2t}$ and their conjugates with respect to $\mathrm{GF}(q)$ as roots.[1] For $1 \leq i \leq 2t$, let $m_i(X)$ be the (monic) minimal polynomial of $\alpha^i$ over $\mathrm{GF}(q)$. Then,

$$\mathbf{g}(X) = \mathrm{LCM}\{m_1(X), m_2(X), \ldots, m_{2t}(X)\}, \tag{6.1}$$

where $m_1(X)$ is a primitive polynomial over $\mathrm{GF}(q)$. Because the degree of the minimal polynomial of an element in $\mathrm{GF}(q^m)$ is at most $m$, the degree of $\mathbf{g}(X)$ is at most $2mt$ and $\mathbf{g}(X)$ divides $X^{q^m-1} - 1$. Hence, the $q$-ary $t$-symbol-error-correcting primitive BCH code $C$ has length $n = q^m - 1$ with dimension at least $q^m - 2mt - 1$.

A $q$-ary $t$-symbol-error-correcting BCH code can be characterized by a theorem similar to Theorem 5.1 that characterizes a binary $t$-error-correcting BCH code.

**Theorem 6.1** *Let $n = q^m - 1$ and $\alpha$ be a primitive element of $\mathrm{GF}(q^m)$. A polynomial*

$$\mathbf{v}(X) = v_0 + v_1 X + v_2 X^2 + \cdots + v_{n-1} X^{n-1}$$

*over $\mathrm{GF}(q)$ is a code polynomial in the $q$-ary $t$-symbol-error-correcting BCH code $C$ if and only if it has $\alpha, \alpha^2, \ldots, \alpha^{2t}$ as roots.*

The parity-check matrix $\mathbf{H}$ of a $q$-ary BCH code in terms of its roots is exactly the same as the one given by (5.12) for a binary BCH code,

$$\mathbf{H} = \begin{bmatrix} 1 & \alpha & \alpha^2 & \cdots & \alpha^{n-1} \\ 1 & \alpha^2 & (\alpha^2)^2 & \cdots & (\alpha^2)^{n-1} \\ 1 & \alpha^3 & (\alpha^3)^2 & \cdots & (\alpha^3)^{n-1} \\ \vdots & & & & \vdots \\ 1 & \alpha^{2t} & (\alpha^{2t})^2 & \cdots & (\alpha^{2t})^{n-1} \end{bmatrix}. \tag{6.2}$$

In the same manner, we can prove that no $2t$ or fewer columns of $\mathbf{H}$ can be summed to a zero column vector. Hence, the code has minimum distance at

---

[1]The conjugates of an element $\beta \in \mathrm{GF}(q^m)$ of exponent $e$ with respect to the ground field $\mathrm{GF}(q)$ are $\beta, \beta^q, \beta^{q^2}, \ldots, \beta^{q^{e-1}}$. If $q = 2$, we have the definition for conjugate roots as given in Chapter 2.

least $2t + 1$ (BCH bound) and is capable of correcting $t$ or fewer random symbol errors over a span of $n = q^m - 1$ symbol positions. We can characterize a $q$-ary $t$-symbol-error-correcting BCH code $C$ in terms of the parity-check matrix given by (6.2) as shown in the following theorem.

**Theorem 6.2** *Let $n = q^m - 1$. An $n$-tuple $\mathbf{v} = (v_0, v_1, \ldots, v_{n-1})$ over GF($q$) is a codeword in the $q$-ary $t$-symbol-error-correcting BCH code $C$ if and only if*

$$\mathbf{v} \cdot \mathbf{H}^T = \mathbf{0}, \tag{6.3}$$

*where $\mathbf{0} = (0, 0, \ldots, 0)$ is a zero $2t$-tuple.*

For a given field GF($q^m$), a family of $q$-ary BCH codes can be constructed. For practical applications, $q$ is commonly a power of 2, say $q = 2^s$ where $s$ is a nonnegative integer. In the following, we illustrate the construction of $q$-ary BCH codes by two examples.

**Example 6.1** Consider a 4-ary 2-symbol-error-correcting BCH code $C$ of length 15. Then, we have $s = 2$, $q = 2^2$, and $m = 2$. The construction field is GF($2^4$). Let $\alpha$ be a primitive element in GF($2^4$) and $\beta = \alpha^5$. The order of $\beta$ is 3, i.e., $\beta^3 = (\alpha^5)^3 = \alpha^{15} = 1$. The subfield GF($2^2$) of GF($2^4$) can be formed as GF($2^2$) = $\{0, 1, \beta = \alpha^5, \beta^2 = \alpha^{10}\}$.

Next, we calculate the generator polynomial $\mathbf{g}(X)$ of the 4-ary BCH code with $t = 2$. Based on the definition of a $q$-ary BCH code, $\mathbf{g}(X)$ is a polynomial over GF($2^2$) which has $2t = 4$ consecutive powers of $\alpha$, namely, $\alpha$, $\alpha^2$, $\alpha^3$, $\alpha^4$, and their conjugates with respect to GF($2^2$) as roots. The conjugates of $\alpha$ over GF($2^4$) with respect to GF($2^2$) are

$$\alpha^4, \alpha^{4^2} = \alpha^{16} = \alpha.$$

Following the same calculation, we find the following three conjugate sets which contain $\alpha$, $\alpha^2$, $\alpha^3$, $\alpha^4$:

$$\Omega(\alpha) = \{\alpha, \alpha^4\}, \quad \Omega(\alpha^2) = \{\alpha^2, \alpha^8\}, \quad \Omega(\alpha^3) = \{\alpha^3, \alpha^{12}\}.$$

The minimal polynomials over GF($2^2$) of $\alpha$, $\alpha^2$, and $\alpha^3$ are (using Table 2.9)

$$m_1(X) = (X + \alpha)(X + \alpha^4)$$
$$= \alpha^5 + (\alpha + \alpha^4)X + X^2 = \alpha^5 + X + X^2 = \beta + X + X^2,$$
$$m_2(X) = (X + \alpha^2)(X + \alpha^8)$$
$$= \alpha^{10} + (\alpha^2 + \alpha^8)X + X^2 = \alpha^{10} + X + X^2 = \beta^2 + X + X^2,$$
$$m_3(X) = (X + \alpha^3)(X + \alpha^{12})$$
$$= \alpha^{15} + (\alpha^3 + \alpha^{12})X + X^2 = 1 + \alpha^{10}X + X^2 = 1 + \beta^2 X + X^2.$$

Then, we have

$$\mathbf{g}(X) = \text{LCM}\{m_1(X), m_2(X), m_3(X)\}$$
$$= m_1(X)m_2(X)m_3(X)$$
$$= 1 + \beta X + \beta X^2 + X^3 + X^4 + \beta^2 X^5 + X^6.$$

Thus, the code $C$ is a 4-ary 2-symbol-error-correcting BCH code of length 15 and dimension 9. Let $\mathbf{u} = (1, \beta, 1, 0, 0, 0, 0, 0, 0)$ be a 9-tuple over $\mathrm{GF}(2^2)$ that is a message to be encoded. The message polynomial is $\mathbf{u}(X) = 1 + \beta X + X^2$. Following the method of systematic encoding for cyclic codes (see Chapter 4), we have

$$X^{n-k}\mathbf{u}(X) = X^6\mathbf{u}(X) = \mathbf{a}(X)\mathbf{g}(X) + \mathbf{b}(X).$$

By performing long division $(X^6\mathbf{u}(X)/\mathbf{g}(X))$, we obtain

$$\mathbf{b}(X) = \beta^2 + \beta X^2 + \beta^2 X^3 + \beta X^5,$$

whose coefficients are the four parity-check symbols, $\beta^2$, $\beta$, $\beta^2$, and $\beta$. Then, the code polynomial $\mathbf{v}(X)$ in systematic form of the message polynomial $\mathbf{u}(X)$ is

$$\begin{aligned}
\mathbf{v}(X) &= \mathbf{b}(X) + X^6\mathbf{u}(X) \\
&= \beta^2 + \beta X^2 + \beta^2 X^3 + \beta X^5 + X^6 + \beta X^7 + X^8,
\end{aligned}$$

and the corresponding 4-ary codeword is $\mathbf{v} = (\beta^2, 0, \beta, \beta^2, 0, \beta, 1, \beta, 1, 0, 0, 0, 0, 0, 0)$. ▲▲

**Example 6.2** In this example, we consider the construction of a longer nonbinary BCH code. Let $s = 2$, $q = 2^2$, $m = 3$, and $t = 2$. Based on these parameters and the construction field $\mathrm{GF}(2^6)$ (see Table 5.14), we can construct a 4-ary BCH code $C$ of length 63 which is capable of correcting two or fewer symbol errors. Let $\alpha$ be a primitive element of $\mathrm{GF}(2^6)$ and $\beta = \alpha^{21}$. Then, the order of $\beta$ is 3 and the subfield $\mathrm{GF}(2^2)$ of $\mathrm{GF}(2^6)$ can be obtained as $\mathrm{GF}(2^2) = \{0, 1, \beta = \alpha^{21}, \beta^2 = \alpha^{42}\}$.

The generator polynomial $\mathbf{g}(X)$ of the BCH code $C$ is a polynomial over $\mathrm{GF}(2^2)$ which has $\alpha$, $\alpha^2$, $\alpha^3$, $\alpha^4$ and their conjugates over $\mathrm{GF}(2^6)$ with respect to $\mathrm{GF}(2^2)$ as roots. Through calculation, we find the following three conjugate sets of the roots of $\mathbf{g}(X)$ in $\mathrm{GF}(2^6)$:

$$\Omega(\alpha) = \{\alpha, \alpha^4, \alpha^{16}\}, \Omega(\alpha^2) = \{\alpha^2, \alpha^8, \alpha^{32}\}, \Omega(\alpha^3) = \{\alpha^3, \alpha^{12}, \alpha^{48}\}.$$

Their corresponding minimal polynomials over $\mathrm{GF}(2^2)$ are found as follows:

$$\begin{aligned}
m_1(X) &= (X + \alpha)(X + \alpha^4)(X + \alpha^{16}) = \beta + \beta^2 X + X^2 + X^3, \\
m_2(X) &= (X + \alpha^2)(X + \alpha^8)(X + \alpha^{32}) = \beta^2 + \beta X + X^2 + X^3, \\
m_3(X) &= (X + \alpha^3)(X + \alpha^{12})(X + \alpha^{48}) = 1 + \beta^2 X + X^3.
\end{aligned}$$

Then, the generator polynomial $\mathbf{g}(X)$ of the code $C$ is

$$\begin{aligned}
\mathbf{g}(X) &= \mathrm{LCM}\{m_1(X), m_2(X), m_3(X)\} \\
&= m_1(X)m_2(X)m_3(X) \\
&= 1 + \beta X + \beta^2 X^2 + X^3 + X^4 + X^6 + \beta^2 X^7 + X^9.
\end{aligned}$$

Thus, the code $C$ is a 4-ary $(63, 54)$ 2-symbol-error-correcting BCH code. Suppose the message $\mathbf{u}$ to be encoded is

$$\mathbf{u} = (0, \beta, 1, \underbrace{0, 0, \ldots, 0}_{51\ 0\text{'s}})$$

and the message polynomial over $\mathrm{GF}(2^2)$ is $\mathbf{u}(X) = \beta X + X^2$. Following the systematic encoding and the long division, we compute the code polynomial $\mathbf{v}(X)$ of $\mathbf{u}(X)$:

$$\begin{aligned}
\mathbf{v}(X) &= \mathbf{b}(X) + X^9 \mathbf{u}(X) \\
&= \beta^2 + \beta^2 X + \beta X^4 + \beta^2 X^5 + \beta X^6 + \beta X^{10} + X^{11},
\end{aligned}$$

where $\mathbf{b}(X) = \beta^2 + \beta^2 X + \beta X^4 + \beta^2 X^5 + \beta X^6$. Thus, the corresponding 4-ary codeword of $\mathbf{u}$ is

$$\mathbf{v} = (\beta^2, \beta^2, 0, 0, \beta, \beta^2, \beta, 0, 0, 0, \beta, 1, \underbrace{0, 0, \ldots, 0}_{51\ 0\text{'s}}).$$

▲▲

## 6.2  Decoding Steps of Nonbinary BCH Codes

Let $n = q^m - 1$. Suppose a code polynomial

$$\mathbf{v}(X) = v_0 + v_1 X + v_2 X^2 + \cdots + v_{n-1} X^{n-1}$$

in a $q$-ary BCH code $C$ of length $n$ is transmitted and

$$\mathbf{r}(X) = r_0 + r_1 X + r_2 X^2 + \cdots + r_{n-1} X^{n-1}$$

is the corresponding received polynomial. Both $\mathbf{v}(X)$ and $\mathbf{r}(X)$ are polynomials over $\mathrm{GF}(q)$. The difference between them

$$\begin{aligned}
\mathbf{e}(X) &= \mathbf{r}(X) - \mathbf{v}(X) \\
&= e_0 + e_1 X + e_2 X^2 + \cdots + e_{n-1} X^{n-1}
\end{aligned}$$

is defined as the *error polynomial*, where the coefficients of $\mathbf{e}(X)$, $e_i = r_i - v_i$, $0 \leq i < n$, are elements in $\mathrm{GF}(q)$.

Decoding a $q$-ary $t$-symbol-error-correcting BCH code can be accomplished in a manner similar to the decoding of a binary $t$-error-correcting BCH code. However, an additional step is needed to determine the values of errors at the error locations. This is because, for a binary BCH code, the error values are always 1 at the error locations; however, for a $q$-ary BCH code, the error values at the error locations can be any nonzero element in $\mathrm{GF}(q)$. The decoding of $q$-ary BCH codes consists of the following steps.

(1) Compute the syndrome of the received polynomial $\mathbf{r}(X)$.
(2) Find the error-location polynomial $\boldsymbol{\sigma}(X)$.
(3) Determine the error locations.
(4) Compute the values of errors at the error locations.
(5) Perform error correction.

The Berlekamp–Massey iterative algorithm presented in Section 5.8 can be used to find the error-location polynomial $\boldsymbol{\sigma}(X)$, but $2t$ steps are required.

## 6.3 Syndrome and Error Pattern of Nonbinary BCH Codes

For a $q$-ary $t$-symbol-error-correcting BCH code, the syndrome of a received polynomial $\mathbf{r}(X) = r_0 + r_1 X + r_2 X^2 + \cdots + r_{n-1} X^{n-1}$ is given by a $2t$-tuple over $\mathrm{GF}(q^m)$,

$$\mathbf{S} = (S_1, S_2, \ldots, S_{2t})$$

with

$$S_i = \mathbf{r}(\alpha^i) = r_0 + r_1 \alpha^i + \cdots + r_{n-1} \alpha^{(n-1)i}, \tag{6.4}$$

for $1 \leq i \leq 2t$, where the additions and multiplications are carried out in $\mathrm{GF}(q^m)$.

Suppose the error polynomial $\mathbf{e}(X)$ contains $\nu$ errors at the locations $X^{j_1}$, $X^{j_2}, \ldots, X^{j_\nu}$ with $0 \leq j_1 < j_2 \cdots < j_{\nu-1} < j_\nu < n$. Then,

$$\mathbf{e}(X) = e_{j_1} X^{j_1} + e_{j_2} X^{j_2} + \cdots + e_{j_\nu} X^{j_\nu}, \tag{6.5}$$

where $e_{j_1}, e_{j_2}, \ldots, e_{j_\nu} \in \mathrm{GF}(q)$ are the values of errors at the error locations $X^{j_1}, X^{j_2}, \ldots, X^{j_\nu}$, respectively. Because $\mathbf{r}(X) = \mathbf{v}(X) + \mathbf{e}(X)$, then for $1 \leq i \leq 2t$, the $i$th component $S_i$ of the syndrome $\mathbf{S} = (S_1, S_2, \ldots, S_{2t})$ of the received polynomial $\mathbf{r}(X)$ is related to the error pattern as follows:

$$S_i = \mathbf{r}(\alpha^i) = \mathbf{v}(\alpha^i) + \mathbf{e}(\alpha^i) = \mathbf{e}(\alpha^i). \tag{6.6}$$

From (6.5) and (6.6), we obtain the following $2t$ equalities that relate the error locations and values to the computed syndrome components:

$$
\begin{aligned}
S_1 &= e_{j_1} \alpha^{j_1} + e_{j_2} \alpha^{j_2} + \cdots + e_{j_\nu} \alpha^{j_\nu}, \\
S_2 &= e_{j_1} \alpha^{2j_1} + e_{j_2} \alpha^{2j_2} + \cdots + e_{j_\nu} \alpha^{2j_\nu}, \\
&\vdots \\
S_{2t} &= e_{j_1} \alpha^{2tj_1} + e_{j_2} \alpha^{2tj_2} + \cdots + e_{j_\nu} \alpha^{2tj_\nu}.
\end{aligned}
\tag{6.7}
$$

For $1 \leq i \leq \nu$, let

$$\beta_i \triangleq \alpha^{j_i}, \qquad \delta_i \triangleq e_{j_i}. \tag{6.8}$$

The elements $\beta_1, \beta_2, \ldots, \beta_\nu$ and $\delta_1, \delta_2, \ldots, \delta_\nu$ are called the *error-location num-bers* and *error values*, respectively. With the above definitions of $\beta_i$s and $\delta_i$s, the equalities of (6.7) can be simplified as follows:

$$
\begin{aligned}
S_1 &= \delta_1\beta_1 + \delta_2\beta_2 + \cdots + \delta_\nu\beta_\nu, \\
S_2 &= \delta_1\beta_1^2 + \delta_2\beta_2^2 + \cdots + \delta_\nu\beta_\nu^2, \\
&\;\;\vdots \\
S_{2t} &= \delta_1\beta_1^{2t} + \delta_2\beta_2^{2t} + \cdots + \delta_\nu\beta_\nu^{2t}.
\end{aligned}
\tag{6.9}
$$

## 6.4  Error-Location Polynomial of Nonbinary BCH Codes

The error-location polynomial of a $q$-ary BCH code whose generator polynomial $\mathbf{g}(X)$ has $\alpha, \alpha^2, \ldots, \alpha^{2t}$ and their conjugates with respect to GF$(q)$ as roots is defined as follows:

$$
\begin{aligned}
\boldsymbol{\sigma}(X) &= (1 - \beta_1 X)(1 - \beta_2 X) \cdots (1 - \beta_\nu X) \\
&= \sigma_0 + \sigma_1 X + \sigma_2 X^2 + \cdots + \sigma_\nu X^\nu,
\end{aligned}
\tag{6.10}
$$

where

$$
\begin{aligned}
\sigma_0 &= 1, \\
\sigma_1 &= -(\beta_1 + \beta_2 + \cdots + \beta_\nu), \\
\sigma_2 &= (-1)^2(\beta_1\beta_2 + \beta_1\beta_3 + \cdots + \beta_{\nu-1}\beta_\nu), \\
&\;\;\vdots \\
\sigma_\nu &= (-1)^\nu \beta_1\beta_2 \cdots \beta_\nu.
\end{aligned}
\tag{6.11}
$$

We readily see that $\beta_1^{-1}, \beta_2^{-1}, \ldots, \beta_\nu^{-1}$ are the roots of the error-location polynomial $\boldsymbol{\sigma}(X)$ and their inverses, $\beta_1, \beta_2, \ldots, \beta_\nu$, are the error-location numbers.

From (6.9) and (6.11), it is possible to obtain the relation between the coefficients of the error-location polynomial $\boldsymbol{\sigma}(X)$ and the computed syndrome components. Define a syndrome power series as follows:

$$
\begin{aligned}
\mathbf{S}(X) &= S_1 + S_2 X + \cdots + S_{2t} X^{2t-1} + S_{2t+1} X^{2t} + \cdots \\
&= \sum_{j=1}^{\infty} S_j X^{j-1}.
\end{aligned}
\tag{6.12}
$$

Note that the coefficients of only the first $2t$ terms of $\mathbf{S}(X)$ are known.

Recall (from (6.9)) that

$$
S_j = \sum_{l=1}^{\nu} \delta_l \beta_l^j.
\tag{6.13}
$$

Substituting $S_j$ in (6.12) with the expression of (6.13), we have

$$\mathbf{S}(X) = \sum_{j=1}^{\infty} X^{j-1} \sum_{l=1}^{\nu} \delta_l \beta_l^j$$

$$= \sum_{l=1}^{\nu} \delta_l \beta_l \sum_{j=1}^{\infty} (\beta_l X)^{j-1}. \tag{6.14}$$

Note that

$$\frac{1}{(1 - \beta_l X)} = \sum_{j=1}^{\infty} (\beta_l X)^{j-1}. \tag{6.15}$$

Combining (6.14) and (6.15), we obtain

$$\mathbf{S}(X) = \sum_{l=1}^{\nu} \frac{\delta_l \beta_l}{1 - \beta_l X}. \tag{6.16}$$

Recall that the expression of the error-location polynomial $\boldsymbol{\sigma}(X)$ given in (6.10) is

$$\boldsymbol{\sigma}(X) = \prod_{i=1}^{\nu} (1 - \beta_i X) = 1 + \sigma_1 X + \cdots + \sigma_\nu X^\nu. \tag{6.17}$$

Taking the product of $\boldsymbol{\sigma}(X)$ given by (6.17) and $\mathbf{S}(X)$ given by (6.12), we have

$$\boldsymbol{\sigma}(X)\mathbf{S}(X) = (1 + \sigma_1 X + \cdots + \sigma_\nu X^\nu) \cdot (S_1 + S_2 X + S_3 X^2 + \cdots)$$

$$= S_1 + (S_2 + \sigma_1 S_1)X + (S_3 + \sigma_1 S_2 + \sigma_2 S_1)X^2 + \cdots + \tag{6.18}$$

$$(S_{2t} + \sigma_1 S_{2t-1} + \cdots + \sigma_\nu S_{2t-\nu})X^{2t-1} + \cdots.$$

If we use the sum expression of $\mathbf{S}(X)$ given by (6.16) and the product expression of $\boldsymbol{\sigma}(X)$ given by (6.17), the product $\boldsymbol{\sigma}(X)\mathbf{S}(X)$ can be expressed as follows:

$$\boldsymbol{\sigma}(X)\mathbf{S}(X) = \left\{ \prod_{i=1}^{\nu} (1 - \beta_i X) \right\} \cdot \left\{ \sum_{l=1}^{\nu} \frac{\delta_l \beta_l}{1 - \beta_l X} \right\}$$

$$= \sum_{l=1}^{\nu} \frac{\delta_l \beta_l}{1 - \beta_l X} \cdot \prod_{i=1}^{\nu} (1 - \beta_i X) \tag{6.19}$$

$$= \sum_{l=1}^{\nu} \delta_l \beta_l \prod_{i=1, i \neq l}^{\nu} (1 - \beta_i X).$$

From the above expression, we see that $\boldsymbol{\sigma}(X)\mathbf{S}(X)$ is a polynomial of degree at most $\nu - 1$.

Equating the two expressions of $\boldsymbol{\sigma}(X)\mathbf{S}(X)$ given by (6.18) and (6.19), we find that the coefficients of $X^\nu$ to $X^{2t-1}$ in the right-hand side of (6.18) must be equal to zero, i.e.,

$$
\begin{aligned}
S_{\nu+1} + \sigma_1 S_\nu + \sigma_2 S_{\nu-1} + \cdots + \sigma_\nu S_1 &= 0, \\
S_{\nu+2} + \sigma_1 S_{\nu+1} + \sigma_2 S_\nu + \cdots + \sigma_\nu S_2 &= 0, \\
&\ \ \vdots \\
S_{2t} + \sigma_1 S_{2t-1} + \sigma_2 S_{2t-2} + \cdots + \sigma_\nu S_{2t-\nu} &= 0.
\end{aligned}
\tag{6.20}
$$

The above $2t - \nu$ equalities are called the *generalized Newton's identities*, which relate the known syndrome $\mathbf{S} = (S_1, S_2, \ldots, S_{2t})$ to the unknown coefficients of the error-location polynomial $\boldsymbol{\sigma}(X)$.

Our objective is to find the minimum-degree polynomial $\boldsymbol{\sigma}(X)$ whose coefficients satisfy the $2t - \nu$ generalized Newton's identities given by (6.20). This can be accomplished with the Berlekamp–Massey iterative algorithm with $2t$ steps presented in Section 5.8 as described below.

At the $\mu$th step, find a polynomial of minimum-degree

$$
\boldsymbol{\sigma}^{(\mu)}(X) = \sigma_0^{(\mu)} + \sigma_1^{(\mu)} X + \cdots + \sigma_{l_\mu}^{(\mu)} X^{l_\mu}
\tag{6.21}
$$

whose coefficients satisfy the following $\mu - l_\mu$ generalized Newton's identities:

$$
\begin{aligned}
S_{l_\mu+1} + \sigma_1^{(\mu)} S_{l_\mu} + \cdots + \sigma_{l_\mu}^{(\mu)} S_1 &= 0, \\
S_{l_\mu+2} + \sigma_1^{(\mu)} S_{l_\mu+1} + \cdots + \sigma_{l_\mu}^{(\mu)} S_2 &= 0, \\
&\ \ \vdots \\
S_\mu + \sigma_1^{(\mu)} S_{\mu-1} + \cdots + \sigma_{l_\mu}^{(\mu)} S_{\mu-l_\mu} &= 0.
\end{aligned}
\tag{6.22}
$$

To find the solution $\boldsymbol{\sigma}^{(\mu+1)}(X)$ at the $(\mu+1)$th step, we check whether the coefficients of $\boldsymbol{\sigma}^{(\mu)}(X)$ satisfy the next generalized Newton's identity. To do this, we compute the discrepancy

$$
d_\mu = S_{\mu+1} + \sigma_1^{(\mu)} S_\mu + \cdots + \sigma_{l_\mu}^{(\mu)} S_{\mu+1-l_\mu}.
\tag{6.23}
$$

If $d_\mu = 0$, the coefficients of the current solution $\boldsymbol{\sigma}^{(\mu)}(X)$ satisfy the $(\mu+1)$th identity, i.e.,

$$
S_{\mu+1} + \sigma_1^{(\mu)} S_\mu + \cdots + \sigma_{l_\mu}^{(\mu)} S_{\mu+1-l_\mu} = 0.
$$

In this case, we set

$$
\boldsymbol{\sigma}^{(\mu+1)}(X) = \boldsymbol{\sigma}^{(\mu)}(X).
$$

If $d_\mu \neq 0$, a correction term with minimum degree is added to $\boldsymbol{\sigma}^{(\mu)}(X)$ to obtain the solution for the $(\mu+1)$th step,

$$
\boldsymbol{\sigma}^{(\mu+1)}(X) = \sigma_0^{(\mu+1)} + \sigma_1^{(\mu+1)} X + \cdots + \sigma_{l_{\mu+1}}^{(\mu+1)} X^{l_{\mu+1}}
\tag{6.24}
$$

whose coefficients satisfy the following $\mu + 1 - l_\mu$ generalized Newton's identities of (6.20),

$$
\begin{aligned}
S_{l_{\mu+1}+1} + \sigma_1^{(\mu+1)} S_{l_{\mu+1}} + \cdots + \sigma_{l_{\mu+1}}^{(\mu+1)} S_1 &= 0, \\
S_{l_{\mu+1}+2} + \sigma_1^{(\mu+1)} S_{l_{\mu+1}+1} + \cdots + \sigma_{l_{\mu+1}}^{(\mu+1)} S_2 &= 0, \\
&\vdots \\
S_{\mu+1} + \sigma_1^{(\mu+1)} S_\mu + \cdots + \sigma_{l_{\mu+1}}^{(\mu+1)} S_{\mu+1-l_{\mu+1}} &= 0.
\end{aligned}
\tag{6.25}
$$

The correction is computed as follows. Go back to the steps prior to the $\mu$th step and determine a polynomial $\boldsymbol{\sigma}^{(\rho)}(X)$ such that $d_\rho \neq 0$ and $\mu - \rho + l_\rho$ has the smallest value, where $l_\rho$ is the degree of $\boldsymbol{\sigma}^{(\rho)}(X)$. Then,

$$
\boldsymbol{\sigma}^{(\mu+1)}(X) = \boldsymbol{\sigma}^{(\mu)}(X) + d_\mu d_\rho^{-1} X^{\mu-\rho} \boldsymbol{\sigma}^{(\rho)}(X)
\tag{6.26}
$$

is the solution at the $(\mu+1)$th step of the iteration process. Continue the above iterative process until $2t$ steps have been completed. At the $2t$th step, we have

$$
\boldsymbol{\sigma}(X) = \boldsymbol{\sigma}^{(2t)}(X),
\tag{6.27}
$$

which is the minimum-degree polynomial whose coefficients satisfy the $2t - \nu$ generalized Newton's identities given by (6.20). If $\nu \leq t$ (the designed error-correcting capability), $\boldsymbol{\sigma}^{(2t)}(X)$ is unique and is the decoded error-location polynomial with all its roots in $\mathrm{GF}(q^m)$. If $\nu > t$, the Berlekamp–Massey iterative algorithm is, in general, not able to locate the errors because the number of errors is beyond the designed error-correcting capability of the code.

The roots of the error-location polynomial $\boldsymbol{\sigma}(X)$ over $\mathrm{GF}(q^m)$ can be found by applying the Chien search algorithm. If $\alpha^i \in \mathrm{GF}(q^m)$ is a root of $\boldsymbol{\sigma}(X)$, i.e., $\boldsymbol{\sigma}(\alpha^i) = 0$, then the inverse of $\alpha^i$,

$$
\alpha^{-i} = \alpha^{q^m-1-i}
$$

is an error-location number which indicates that the received symbol at the location $q^m - 1 - i$ is erroneous.

To execute the above algorithm for finding the error-location polynomial, we set up and fill out Table 6.1.

**Example 6.3** Consider the 4-ary 2-symbol-error-correcting $(15, 9)$ BCH code $C$ constructed in Example 6.1. Here, we shall adopt all the definitions and notations in Example 6.1. Suppose the transmitted codeword is $\mathbf{v} = (\beta^2, 0, \beta, \beta^2, 0, \beta, 1, \beta, 1, 0, 0, 0, 0, 0, 0)$ and the corresponding code polynomial is $\mathbf{v}(X) = \beta^2 + \beta X^2 + \beta^2 X^3 + \beta X^5 + X^6 + \beta X^7 + X^8$.

Owing to the channel noise or interference, the received vector is $\mathbf{r} = (\beta^2, 0, \beta, \beta^2, 0, \beta, 1, \beta, 1, 0, 0, 0, 0, \beta, 1)$. The corresponding received polynomial is $\mathbf{r}(X) = \beta^2 + \beta X^2 + \beta^2 X^3 + \beta X^5 + X^6 + \beta X^7 + X^8 + \beta X^{13} + X^{14}$. Two symbol errors are introduced by the channel.

The syndrome $\mathbf{S}$ of the received vector $\mathbf{r}$ is

$$
S_1 = \mathbf{r}(\alpha) = 1, \ S_2 = \mathbf{r}(\alpha^2) = \alpha^{12}, \ S_3 = \mathbf{r}(\alpha^3) = \alpha^5, \ S_4 = \mathbf{r}(\alpha^4) = 1.
$$

Table 6.1 Berlekamp–Massey iterative procedure for finding the error-location polynomial of a nonbinary BCH code.

| Step $\mu$ | Partial solution $\sigma^{(\mu)}(X)$ | Discrepancy $d_\mu$ | Degree $l_\mu$ | Step/degree difference $\mu - l_\mu$ |
|---|---|---|---|---|
| $-1$ | $1$ | $1$ | $0$ | $-1$ |
| $0$ | $1$ | $S_1$ | $0$ | $0$ |
| $1$ | $1+S_1X$ | | | |
| $2$ | | | | |
| $\vdots$ | | | | |
| $2t$ | | | | |

Using the Berlekamp–Massey iterative algorithm to decode the received vector, we fill out Table 6.1 based on the computed syndrome. The result is shown in Table 6.2.

Table 6.2 Steps for finding the error-location polynomial of the received polynomial $\mathbf{r}(X)$ for the 4-ary $(15,9)$ BCH code in Example 6.3.

| Step $\mu$ | Partial solution $\sigma^{(\mu)}(X)$ | Discrepancy $d_\mu$ | Degree $l_\mu$ | Step/degree difference $\mu - l_\mu$ |
|---|---|---|---|---|
| $-1$ | $1$ | $1$ | $0$ | $-1$ |
| $0$ | $1$ | $1$ | $0$ | $0$ |
| $1$ | $1 + X$ | $\alpha^{11}$ | $1$ | $0$ (take $\rho = -1$) |
| $2$ | $1 + \alpha^{12}X$ | $\alpha^6$ | $1$ | $1$ (take $\rho = 0$) |
| $3$ | $1 + \alpha^3 X + \alpha^{10}X^2$ | $\alpha^{12}$ | $2$ | $1$ (take $\rho = 1$) |
| $4$ | $1 + \alpha^2 X + \alpha^{12}X^2$ | $0$ | $2$ | $2$ (take $\rho = 2$) |

The error-location polynomial for the received polynomial $\mathbf{r}(X)$ is

$$\sigma(X) = 1 + \alpha^2 X + \alpha^{12}X^2.$$

By applying the Chien search algorithm, we find that the roots over $GF(2^4)$ of $\sigma(X)$ are $\alpha$ and $\alpha^2$. Thus, the error-location numbers are $\alpha^{-1} = \alpha^{15-1} = \alpha^{14}$ and $\alpha^{-2} = \alpha^{15-2} = \alpha^{13}$, which are the true error locations.

Consider another case in which $\mathbf{v}(X) = \beta^2 + \beta X^2 + \beta^2 X^3 + \beta X^5 + X^6 + \beta X^7 + X^8$ is transmitted and the received polynomial is $\mathbf{r}^*(X) = \beta^2 + \beta X^2 + \beta^2 X^3 + \beta X^5 + X^6 + \beta X^7 + X^8 + \beta^2 X^{12} + \beta X^{13} + X^{14}$. The syndrome $\mathbf{S}$ of $\mathbf{r}^*(X)$ is

$$S_1 = \mathbf{r}(\alpha) = \alpha^9, S_2 = \mathbf{r}(\alpha^2) = \alpha^6, S_3 = \mathbf{r}(\alpha^3) = \alpha^2, S_4 = \mathbf{r}(\alpha^4) = \alpha^6.$$

To decode this received polynomial, we fill out Table 6.1 and the result is shown in Table 6.3.

Table 6.3 Steps for finding the error-location polynomial of the received polynomial $\mathbf{r}^*(X)$ for the 4-ary $(15, 9)$ BCH code in Example 6.3.

| Step $\mu$ | Partial solution $\sigma^{(\mu)}(X)$ | Discrepancy $d_\mu$ | Degree $l_\mu$ | Step/degree difference $\mu - l_\mu$ |
|---|---|---|---|---|
| $-1$ | $1$ | $1$ | $0$ | $-1$ |
| $0$ | $1$ | $\alpha^9$ | $0$ | $0$ |
| $1$ | $1 + \alpha^9 X$ | $\alpha^2$ | $1$ | $0$ (take $\rho = -1$) |
| $2$ | $1 + \alpha^{12} X$ | $\alpha^6$ | $1$ | $1$ (take $\rho = 0$) |
| $3$ | $1 + \alpha^6 X + \alpha^{13} X^2$ | $\alpha^9$ | $2$ | $1$ (take $\rho = 1$) |
| $4$ | $1 + \alpha^2 X + \alpha^6 X^2$ | $0$ | $2$ | $2$ (take $\rho = 2$) |

The error-location polynomial for the received polynomial $\mathbf{r}^*(X)$ is

$$\sigma^*(X) = 1 + \alpha^2 X + \alpha^6 X^2.$$

By applying a search, we find that the roots of $\sigma^*(X)$ are $\alpha^{10}$ and $\alpha^{14}$. Then, the error-location numbers are $\alpha^{-10} = \alpha^{15-10} = \alpha^5$ and $\alpha^{-14} = \alpha^{15-14} = \alpha$. Note that the received polynomial $\mathbf{r}^*(X)$ actually has three symbol errors which are located at positions 12, 13, and 14. However, the Berlekamp–Massey iterative algorithm gives error locations at positions 1 and 5, i.e., it fails to locate the errors. So, the error pattern is an uncorrectable error pattern and the decoding would be incorrect. ▲▲

## 6.5  Error-Value Evaluator

Once the error-location polynomial $\sigma(X) = \sigma_0 + \sigma_1 X + \cdots + \sigma_\nu X^\nu$ is found, we determine its roots $\beta_1^{-1}, \beta_2^{-1}, \ldots, \beta_\nu^{-1}$ to obtain the error-location numbers $\beta_1$, $\beta_2, \ldots, \beta_\nu$. Next, we need to calculate the error values at these error locations.

Define a polynomial of degree $\nu - 1$ or less from the syndrome components and the coefficients of the error-location polynomial as follows (by using the first $\nu$ terms in the product $\sigma(X)\mathbf{S}(X)$ given by (6.18)):

$$\mathbf{Z}_0(X) = S_1 + (S_2 + \sigma_1 S_1)X + (S_3 + \sigma_1 S_2 + \sigma_2 S_1)X^2$$
$$+ \cdots + (S_\nu + \sigma_1 S_{\nu-1} + \cdots + \sigma_{\nu-1} S_1)X^{\nu-1}, \tag{6.28}$$

which is called the *error-value evaluator*. From (6.19), we obtain

$$\mathbf{Z}_0(X) = \sum_{l=1}^{\nu} \delta_l \beta_l \prod_{i=1, i \neq l}^{\nu} (1 - \beta_i X). \tag{6.29}$$

Substituting the variable $X$ in $\mathbf{Z}_0(X)$ with $\beta_k^{-1}$ for $1 \le k \le \nu$, we have

$$
\begin{aligned}
\mathbf{Z}_0(\beta_k^{-1}) &= \sum_{l=1}^{\nu} \delta_l \beta_l \prod_{i=1, i \ne l}^{\nu} (1 - \beta_i \beta_k^{-1}) \\
&= \delta_k \beta_k \prod_{i=1, i \ne k}^{\nu} (1 - \beta_i \beta_k^{-1}).
\end{aligned}
\tag{6.30}
$$

Taking the derivative of the error-location polynomial $\boldsymbol{\sigma}(X)$ given by (6.17), we have

$$
\begin{aligned}
\boldsymbol{\sigma}'(X) &= \frac{d}{dX} \prod_{i=1}^{\nu} (1 - \beta_i X) \\
&= -\sum_{l=1}^{\nu} \beta_l \prod_{i=1, i \ne l}^{\nu} (1 - \beta_i X).
\end{aligned}
\tag{6.31}
$$

Substituting $X$ in $\boldsymbol{\sigma}'(X)$ with $\beta_k^{-1}$, we have

$$
\boldsymbol{\sigma}'(\beta_k^{-1}) = -\beta_k \prod_{i=1, i \ne k}^{\nu} (1 - \beta_i \beta_k^{-1}).
\tag{6.32}
$$

From (6.30) and (6.32), we obtain the error value $\delta_k$ at the error location $\beta_k$ [6],

$$
\delta_k = \frac{-\mathbf{Z}_0(\beta_k^{-1})}{\boldsymbol{\sigma}'(\beta_k^{-1})}.
\tag{6.33}
$$

## 6.6  Decoding of Nonbinary BCH Codes

Summarizing the results developed in previous sections, the decoding procedure of a nonbinary BCH code can be described as follows.

(1) Compute the syndrome $\mathbf{S} = (S_1, S_2, \ldots, S_{2t})$ of the received polynomial $\mathbf{r}(X)$.

(2) Determine the error-location polynomial $\boldsymbol{\sigma}(X)$.

(3) Determine the error-value evaluator $\mathbf{Z}_0(X)$.

(4) Determine the error-location numbers and evaluate the error values at the error locations.

(5) Compute the error pattern $\mathbf{e}(X)$.

(6) Perform error correction by subtracting the error pattern $\mathbf{e}(X)$ from the received polynomial $\mathbf{r}(X)$.

Once the error-location polynomial is found, the roots of $\boldsymbol{\sigma}(X)$ in $\mathrm{GF}(q^m)$ can be determined by substituting the variable $X$ in $\boldsymbol{\sigma}(X)$ with the elements $\alpha^0, \alpha, \ldots, \alpha^{q^m-2}$ of $\mathrm{GF}(q^m)$ in turn, i.e., applying the Chien search algorithm. If $\boldsymbol{\sigma}(\alpha^i) = 0$, then $\alpha^i$ is a root of $\boldsymbol{\sigma}(X)$ and

$$
\alpha^{-i} = \alpha^{q^m-1-i}
$$

is an error-location number. Then, the decoded symbol at the error location $q^m - 1 - i$ is

$$v_{q^m - 1 - i} = r_{q^m - 1 - i} - e_{q^m - 1 - i}.$$

A general organization block diagram of a $q$-ary BCH decoder is shown in Fig. 6.1.

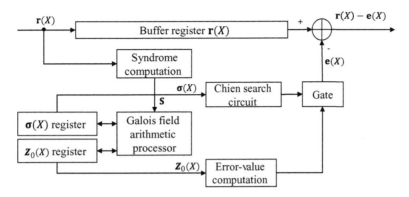

Figure 6.1 A decoding block diagram of a $q$-ary BCH decoder.

**Example 6.4** Consider the 4-ary 2-symbol-error-correcting $(15, 9)$ BCH code $C$ constructed in Example 6.1. In Example 6.3, the code polynomial $\mathbf{v}(X) = \beta^2 + \beta X^2 + \beta^2 X^3 + \beta X^5 + X^6 + \beta X^7 + X^8$ was transmitted and we computed the syndrome and error-location polynomials of the received polynomial $\mathbf{r}(X) = \beta^2 + \beta X^2 + \beta^2 X^3 + \beta X^5 + X^6 + \beta X^7 + X^8 + \beta X^{13} + X^{14}$:

$$\mathbf{S}(X) = 1 + \alpha^{12}X + \alpha^5 X^2 + X^3$$

and

$$\boldsymbol{\sigma}(X) = 1 + \alpha^2 X + \alpha^{12} X^2.$$

(All notations follow those used in Example 6.3.) The roots of $\boldsymbol{\sigma}(X)$ are $\alpha$ and $\alpha^2$ and the error-location numbers are $\alpha^{13}$ and $\alpha^{14}$. In this example, we compute the error values in these two error locations.

First, we compute the error-evaluator polynomial $\mathbf{Z}_0(X)$ (see (6.28)):

$$\mathbf{Z}_0(X) = S_1 + (S_2 + \sigma_1 S_1)X = 1 + (\alpha^{12} + \alpha^2)X = 1 + \alpha^7 X,$$

and the derivative of the error-location polynomial

$$\boldsymbol{\sigma}'(X) = \alpha^2.$$

Following (6.33), we compute the error values at the error locations 13 and 14:

$$e_{13} = \frac{-\mathbf{Z}_0(\alpha^{-13})}{\sigma'(\alpha^{-13})} = \frac{1 + \alpha^2 \alpha^{-13}}{\alpha^2} = \alpha^5 = \beta,$$

$$e_{14} = \frac{-\mathbf{Z}_0(\alpha^{-14})}{\sigma'(\alpha^{-14})} = \frac{1 + \alpha^2 \alpha^{-14}}{\alpha^2} = 1.$$

Then, the error polynomial is

$$\mathbf{e}(X) = \beta X^{13} + X^{14},$$

and the decoded polynomial (i.e., the estimated transmitted code polynomial) is

$$\begin{aligned}
\mathbf{v}^*(X) &= \mathbf{r}(X) - \mathbf{e}(X) \\
&= \beta^2 + \beta X^2 + \beta^2 X^3 + \beta X^5 + X^6 + \beta X^7 + X^8 \\
&\quad + \beta X^{13} + X^{14} - (\beta X^{13} + X^{14}) \\
&= \beta^2 + \beta X^2 + \beta^2 X^3 + \beta X^5 + X^6 + \beta X^7 + X^8.
\end{aligned}$$

We can see $\mathbf{v}^*(X) = \mathbf{v}(X)$. Hence, the decoding is successful and correct.

Consider the other case given in Example 6.3. In this case, the received polynomial is $\mathbf{r}^*(X) = \beta^2 + \beta X^2 + \beta^2 X^3 + \beta X^5 + X^6 + \beta X^7 + X^8 + \beta^2 X^{12} + \beta X^{13} + X^{14}$, the syndrome is $\mathbf{S} = (\alpha^9, \alpha^6, \alpha^{12}, \alpha^6)$, the error-location polynomial is $\sigma^*(X) = 1 + \alpha^2 X + \alpha^6 X^2$ with roots $\alpha^{10}$ and $\alpha^{14}$, and the error-location numbers are $\alpha^5$ and $\alpha$. Based on these polynomials, we compute the error-evaluator

$$\mathbf{Z}_0^*(X) = S_1 + (S_2 + \sigma_1 S_1)X = \alpha^9 + (\alpha^6 + \alpha^2 \alpha^9)X = \alpha^9 + \alpha X,$$

and the derivative of the error-location polynomial

$$\boldsymbol{\sigma}^{*'}(X) = \alpha^2.$$

The error values at the error locations 1 and 5 are computed as follows:

$$e_1 = \frac{-\mathbf{Z}_0^*(\alpha^{-1})}{\sigma^{*'}(\alpha^{-1})} = \frac{\alpha^9 + \alpha \alpha^{-1}}{\alpha^2} = \alpha^5 = \beta,$$

$$e_5 = \frac{-\mathbf{Z}_0^*(\alpha^{-5})}{\sigma^{*'}(\alpha^{-5})} = \frac{\alpha^9 + \alpha \alpha^{-5}}{\alpha^2} = 1.$$

Then, the estimated error polynomial is

$$\mathbf{e}^*(X) = \beta X + X^5$$

and the decoded polynomial is

$$\begin{aligned}
\mathbf{v}^{**}(X) &= \mathbf{r}^*(X) - \mathbf{e}^*(X) \\
&= \beta^2 + \beta X^2 + \beta^2 X^3 + \beta X^5 + X^6 + \beta X^7 + X^8 \\
&\quad + \beta^2 X^{12} + \beta X^{13} + X^{14} - (\beta X + X^5) \\
&= \beta^2 + \beta X + \beta X^2 + \beta^2 X^3 + \beta^2 X^5 + X^6 + \beta X^7 \\
&\quad + X^8 + \beta^2 X^{12} + \beta X^{13} + X^{14}.
\end{aligned}$$

It can be verified that $\mathbf{v}^{**}$ is a codeword of the 4-ary $(15, 9)$ BCH code (i.e., $\mathbf{v}^{**} \cdot \mathbf{H}^T = \mathbf{0}$). However, $\mathbf{v}^{**} \neq \mathbf{v}$. The decoder commits a decoding error. ▲▲

## 6.7   Key-Equation

From the expression of $\boldsymbol{\sigma}(X)\mathbf{S}(X)$, the generalized Newton's identities, and the error-evaluator polynomial $\mathbf{Z}_0(X)$ given by (6.18), (6.20), and (6.28), respectively, we find that: (1) the first $\nu$ terms of $\boldsymbol{\sigma}(X)\mathbf{S}(X)$ give the error-value evaluator $\mathbf{Z}_0(X)$; and (2) the next $2t - \nu$ terms of $\boldsymbol{\sigma}(X)\mathbf{S}(X)$ give the generalized Newton's identities. Mathematically, the three polynomials $\boldsymbol{\sigma}(X)$, $\mathbf{S}(X)$, and $\mathbf{Z}_0(X)$ are connected by the following equation:

$$\boldsymbol{\sigma}(X)\mathbf{S}(X) \equiv \mathbf{Z}_0(X) \quad \text{modulo } X^{2t}, \tag{6.34}$$

which means that the error-value evaluator $\mathbf{Z}_0(X)$ is the remainder resulting from dividing $\boldsymbol{\sigma}(X)\mathbf{S}(X)$ by $X^{2t}$, which amounts to erasing all the terms of degrees equal or greater than $2t$ in $\boldsymbol{\sigma}(X)\mathbf{S}(X)$.

Any method of solving (6.34) to find $\boldsymbol{\sigma}(X)$ and $\mathbf{Z}_0(X)$ is a decoding method for BCH and RS codes. The Berlekamp–Massey iterative algorithm presented in Sections 6.2–6.6 is such a method. The equation given by (6.34) is known as the *key-equation* [9] for decoding BCH and RS codes. Mathematically, the key-equation given by (6.34) is read as follows: $\boldsymbol{\sigma}(X)\mathbf{S}(X)$ is *congruent* to $\mathbf{Z}_0(X)$ modulo $X^{2t}$.

Note that the degree of $\mathbf{Z}_0(X)$ is at least one less than that of $\boldsymbol{\sigma}(X)$. If the number of errors in an error pattern $\mathbf{e}(X)$ is less than or equal to the error-correcting capability $t$ of the code, then the key-equation has a *unique pair of solution*, $(\boldsymbol{\sigma}(X), \mathbf{Z}_0(X))$, with

$$\deg(\mathbf{Z}_0(X)) < \deg(\boldsymbol{\sigma}(X)) \leq t.$$

In Section 6.11, we will present another method to solve the key-equation.

## 6.8   Reed–Solomon Codes

In the previous sections, we have considered nonbinary BCH codes with code symbols from $\mathrm{GF}(q)$ and the roots of their generator polynomials from an extension field $\mathrm{GF}(q^m)$, $m \geq 1$, of $\mathrm{GF}(q)$. In the case of $m = 1$, we obtain a special subclass of nonbinary BCH codes. For a code in this subclass, *the code symbols and the roots of its generator polynomial are from the same field* $\mathrm{GF}(q)$, i.e., the code symbol field is the same as the code construction field. Such a $q$-ary BCH code is called a *Reed–Solomon (RS) code* over $\mathrm{GF}(q)$ discovered by Reed and Solomon in 1960 [4]. RS codes have been proved to be good error detection codes [38] and their weight distributions have been completely determined [39–41]. RS codes are effective for combating mixed types of noise and interferences. They have been widely used in communication and storage systems over the years since the early 1970s. In this section, we present the construction of RS codes and their structural properties, including primitive and nonprimitive RS codes.

## 6.8.1 Primitive Reed–Solomon Codes

Let $\alpha$ be a primitive element of $\mathrm{GF}(q)$. For $1 \leq t < q$, the generator polynomial $\mathbf{g}(X)$ over $\mathrm{GF}(q)$ of a *q-ary t-symbol-error-correcting RS code $C$* has

$$\alpha, \alpha^2, \alpha^3, \ldots, \alpha^{2t}$$

over $\mathrm{GF}(q)$ as roots. Because $\alpha, \alpha^2, \ldots, \alpha^{2t}$ are elements in $\mathrm{GF}(q)$, their minimal polynomials over $\mathrm{GF}(q)$ are simply $X - \alpha$, $X - \alpha^2, \ldots, X - \alpha^{2t}$, respectively. Hence,

$$\begin{aligned}
\mathbf{g}(X) &= (X - \alpha)(X - \alpha^2) \cdots (X - \alpha^{2t}) \\
&= g_0 + g_1 X + \cdots + g_{2t-1} X^{2t-1} + g_{2t} X^{2t},
\end{aligned} \tag{6.35}$$

with $g_i \in \mathrm{GF}(q)$ for $0 \leq i \leq 2t$, $g_0 \neq 0$, and $g_{2t} = 1$.

Because $\alpha, \alpha^2, \ldots, \alpha^{2t}$ are also roots of $X^{q-1} - 1$, $\mathbf{g}(X)$ divides $X^{q-1} - 1$. The RS code $C$ over $\mathrm{GF}(q)$ generated by $\mathbf{g}(X)$ has the following parameters:

| | |
|---|---|
| length | $n = q - 1$, |
| dimension | $k = q - 2t - 1$, |
| number of parity-check symbols | $n - k = 2t$, |
| minimum distance | $d_{\min} = 2t + 1$, |

which is called a *primitive RS code*. Based on the development of nonbinary BCH codes, it is known that the minimum distance of such an RS code (as a subclass of BCH codes) is at least $2t + 1$. To prove that the minimum distance of the above RS code $C$ is exactly $2t + 1$, we need only to show that there is a nonzero code polynomial in $C$ with weight $2t + 1$. This is straightforward to verify because the generator polynomial $\mathbf{g}(X)$ given by (6.35) is such a code polynomial. From (6.35), we see that $\mathbf{g}(X)$ has $2t + 1$ terms. None of the coefficients of these $2t + 1$ terms can be zero; otherwise, the codeword corresponding to the generator polynomial $\mathbf{g}(X)$ would have weight less than the BCH bound $2t + 1$ on the minimum weight of a BCH code whose generator polynomial has $2t$ consecutive powers of $\alpha$ as roots. Therefore, the RS code generated by the generator polynomial $\mathbf{g}(X)$ given by (6.35) is a $(q - 1, q - 2t - 1)$ code over $\mathrm{GF}(q)$ with minimum distance exactly $2t + 1$ and its generator polynomial $\mathbf{g}(X)$ is a minimum-weight code polynomial.

Another characteristic of a $q$-ary $(q-1, q-2t-1)$ RS code is that its minimum distance $2t + 1$ is one greater than the number $2t$ of parity-check symbols of the code, i.e., it meets the Singleton bound, $n - k + 1 = 2t + 1$. Hence, RS codes form a class of maximum-distance separable (MDS) codes. The parity-check matrices of RS codes are in the same form as those of nonbinary BCH codes.

In practical applications of RS codes in digital communication and data-storage systems, $q$ is commonly chosen as a power of 2, say $q = 2^s$, and the code symbols are from $\mathrm{GF}(2^s)$. If each code symbol is represented by an $s$-tuple over $\mathrm{GF}(2)$, then an RS code can be transmitted by using binary signaling, such as

BPSK. In decoding, every $s$ received bits are grouped into a received symbol in GF$(2^s)$. This results in a received vector of $n = 2^s - 1$ symbols in GF$(2^s)$. Then, decoding is performed on the received symbol vector. In the rest of this chapter, we only consider RS codes over GF$(2^s)$ with $s > 1$.

**Example 6.5** Let $q = 2^3$ and $t = 2$. Then, an 8-ary double-symbol-error-correcting $(7, 3)$ RS code $C$ can be constructed. The construction field is GF$(2^3)$; this is constructed by using a primitive polynomial $\mathbf{p}(X) = 1 + X + X^3$ over GF$(2)$, as shown in Table 6.4.

Let $\alpha$ be a primitive element of GF$(2^3)$. The generator polynomial $\mathbf{g}(X)$ of the RS code $C$ has $\alpha$, $\alpha^2$, $\alpha^3$, and $\alpha^4$ as roots:

$$\mathbf{g}(X) = (X + \alpha)(X + \alpha^2)(X + \alpha^3)(X + \alpha^4)$$
$$= \alpha^3 + \alpha X + X^2 + \alpha^3 X^3 + X^4.$$

Let $\mathbf{u} = (1, 1, \alpha)$ be the message to be encoded which is a 3-tuple over GF$(2^3)$. The corresponding message polynomial is $\mathbf{u}(X) = 1 + X + \alpha X^2$. For systematic encoding, dividing $X^4 \mathbf{u}(X)$ by $\mathbf{g}(X)$ gives

$$X^4 \mathbf{u}(X) = (1 + \alpha^5 X + \alpha X^2)\mathbf{g}(X) + (\alpha^3 + \alpha X^2).$$

Then, the systematic code polynomial is

$$\mathbf{v}_{\text{sys}}(X) = \alpha^3 + \alpha X^2 + X^4 \mathbf{u}(X) = \alpha^3 + \alpha X^2 + X^4 + X^5 + \alpha X^6,$$

and the corresponding codeword is $\mathbf{v}_{\text{sys}} = (\alpha^3, 0, \alpha, 0, 1, 1, \alpha)$.

For binary transmission, the codeword $\mathbf{v}_{\text{sys}}$ is mapped into a 21-tuple over GF$(2)$: $(1, 1, 0, 0, 0, 0, 0, 1, 0, 0, 0, 0, 1, 0, 0, 1, 0, 0, 0, 1, 0)$. ▲▲

Table 6.4 GF$(2^3)$ generated by $\mathbf{p}(X) = 1 + X + X^3$ over GF$(2)$.

| Power form | Polynomial form | | | | Vector form | Decimal form |
|---|---|---|---|---|---|---|
| $0 = \alpha^{-\infty}$ | 0 | | | | (0 0 0) | 0 |
| $1 = \alpha^0$ | 1 | | | | (1 0 0) | 1 |
| $\alpha$ | | $\alpha$ | | | (0 1 0) | 2 |
| $\alpha^2$ | | | | $\alpha^2$ | (0 0 1) | 4 |
| $\alpha^3$ | 1 | + | $\alpha$ | | (1 1 0) | 3 |
| $\alpha^4$ | | | $\alpha$ + | $\alpha^2$ | (0 1 1) | 6 |
| $\alpha^5$ | 1 | + | $\alpha$ + | $\alpha^2$ | (1 1 1) | 7 |
| $\alpha^6$ | 1 | | + | $\alpha^2$ | (1 0 1) | 5 |

## 6.8.2 Nonprimitive Reed–Solomon Codes

So far, we have considered an important subclass of RS codes, known as primitive RS codes of length $q - 1$. Similar to nonprimitve BCH codes, there is a subclass of RS codes, known as nonprimitive RS codes.

Consider the field $GF(q)$. Let $\alpha$ be a primitive element in $GF(q)$ and $n$ be a factor of $q - 1$, say $q - 1 = n\ell$. Let $\beta = \alpha^{\ell}$. Then, $\beta$ is an element of order $n$ in $GF(q)$, i.e., $\beta^n = 1$. As shown in Chapter 2, the set $\{1, \beta, \beta^2, \ldots, \beta^{n-1}\}$ forms a cyclic subgroup of the multiplicative group of $GF(q)$. Let $t$ be a positive integer such that $2t < n$. The cyclic code $C$ over $GF(q)$ of length $n$ generated by the polynomial $\mathbf{g}(X)$ over $GF(q)$ of the *least degree* which has $\beta$, $\beta^2, \ldots, \beta^{2t}$ over $GF(q)$ as roots is an RS code with minimum distance $2t + 1$. Such an RS code is called a *nonprimitive RS code*. The polynomial $\mathbf{g}(X)$ is the generator polynomial of $C$, which is simply the LCM of the minimal polynomials of $\beta$, $\beta^2, \ldots, \beta^{2t}$ and divides $X^n - 1$. Because $\beta$, $\beta^2, \ldots, \beta^{2t}$ are elements in $GF(q)$, their minimal polynomials over $GF(q)$ are simply $X - \beta$, $X - \beta^2, \ldots, X - \beta^{2t}$, respectively. Hence,

$$
\begin{aligned}
\mathbf{g}(X) &= (X - \beta)(X - \beta^2) \cdots (X - \beta^{2t}) \\
&= g_0 + g_1 X + \cdots + g_{2t-1} X^{2t-1} + g_{2t} X^{2t},
\end{aligned}
\tag{6.36}
$$

with $g_i \in GF(q)$ for $0 \leq i \leq 2t$, $g_0 \neq 0$, and $g_{2t} = 1$.

This nonprimitive RS code is capable of correcting $t$ or fewer random symbol errors using the Berlekamp–Massey decoding algorithm. A nonprimitive RS code has the same structural properties as a primitive RS code. Note that if $q - 1$ is a prime, then $n = q - 1$. In this case, $\beta$ is a primitive element of $GF(q)$ and the RS code generated by $\mathbf{g}(X)$ is a primitive RS code.

**Example 6.6** Consider the field $GF(2^6)$ given by Table 5.14 constructed based on the primitive polynomial $\mathbf{p}(X) = 1 + X + X^6$ over $GF(2)$. Note that $2^6 - 1$ can be factored as the product $3 \times 21$. Let $\alpha$ be a primitive element of $GF(2^6)$ and $\beta = \alpha^3$. The order of $\beta$ is 21. Set $n = 21$ and $t = 2$. Suppose we construct a nonprimitive RS code of length 21 whose generator polynomial $\mathbf{g}(X)$ has $\beta$, $\beta^2$, $\beta^3$, $\beta^4$ as roots. Following (6.36), we have

$$
\begin{aligned}
\mathbf{g}(X) &= (X - \beta)(X - \beta^2)(X - \beta^3)(X - \beta^4) \\
&= (X + \alpha^3)(X + \alpha^6)(X + \alpha^9)(X + \alpha^{12}) \\
&= \alpha^{30} + \alpha^{51} X + \alpha^{23} X^2 + \alpha^{36} X^3 + X^4.
\end{aligned}
$$

The nonprimitive RS code generated by $\mathbf{g}(X)$ is a 64-ary $(21, 17)$ cyclic code which has symbol-error-correcting capability $t = 2$.                    ▲▲

## 6.9 Decoding Reed–Solomon Codes with Berlekamp–Massey Iterative Algorithm

Because RS codes are special $q$-ary BCH codes, they can be decoded with the Berlekamp–Massey iterative algorithm. The major step of decoding is to find the error-location polynomial $\boldsymbol{\sigma}(X)$ based on the syndrome of a received polynomial. To find the error-location polynomial $\boldsymbol{\sigma}(X)$, we fill out Table 6.1. Once $\boldsymbol{\sigma}(X)$ is found, we determine the locations and values of the errors in the received polynomial. Then, we correct the errors.

**Example 6.7** Let $q = 2^4$ and $GF(2^4)$ (see Table 2.9) be the field for code construction. Let $\alpha$ be a primitive element of $GF(2^4)$. Suppose we want to construct a triple-symbol-error-correcting RS code $C$ of length $n = 2^4 - 1 = 15$ over $GF(2^4)$. The generator polynomial $\mathbf{g}(X)$ of this code is

$$\mathbf{g}(X) = (X + \alpha)(X + \alpha^2)(X + \alpha^3)(X + \alpha^4)(X + \alpha^5)(X + \alpha^6)$$
$$= \alpha^6 + \alpha^9 X + \alpha^6 X^2 + \alpha^4 X^3 + \alpha^{14} X^4 + \alpha^{10} X^5 + X^6.$$

The RS code generated by $\mathbf{g}(X)$ is a 16-ary $(15, 9)$ cyclic code with minimum distance 7.

Suppose the zero code polynomial is transmitted and

$$\mathbf{r}(X) = \alpha^7 X^3 + \alpha^3 X^6 + \alpha^4 X^{12}$$

is the received polynomial. The syndrome of $\mathbf{r}(X)$ is

$$\mathbf{S} = (S_1, S_2, S_3, S_4, S_5, S_6),$$

where

$$S_1 = \mathbf{r}(\alpha) = \alpha^{10} + \alpha^9 + \alpha = \alpha^{12}, \quad S_2 = \mathbf{r}(\alpha^2) = \alpha^{13} + 1 + \alpha^{13} = 1,$$
$$S_3 = \mathbf{r}(\alpha^3) = \alpha + \alpha^6 + \alpha^{10} = \alpha^{14}, \quad S_4 = \mathbf{r}(\alpha^4) = \alpha^4 + \alpha^{12} + \alpha^7 = \alpha^{10},$$
$$S_5 = \mathbf{r}(\alpha^5) = \alpha^7 + \alpha^3 + \alpha^4 = 0, \quad S_6 = \mathbf{r}(\alpha^6) = \alpha^{10} + \alpha^9 + \alpha = \alpha^{12}.$$

Using the syndrome $\mathbf{S} = (\alpha^{12}, 1, \alpha^{14}, \alpha^{10}, 0, \alpha^{12})$ and the Berlekamp–Massey algorithm, we fill out Table 6.1 to find the error-location polynomial $\boldsymbol{\sigma}(X)$. The result is shown in Table 6.5.

Table 6.5 Steps for finding the error-location polynomial of $\mathbf{r}(X)$ for the 16-ary $(15, 9)$ RS code over $GF(2^4)$ in Example 6.7.

| Step | Partial solution | Discrepancy | Degree | Step/degree difference |
|------|------------------|-------------|--------|------------------------|
| $\mu$ | $\sigma^{(\mu)}(X)$ | $d_\mu$ | $l_\mu$ | $\mu - l_\mu$ |
| $-1$ | $1$ | $1$ | $0$ | $-1$ |
| $0$ | $1$ | $\alpha^{12}$ | $0$ | $0$ |
| $1$ | $1 + \alpha^{12} X$ | $\alpha^7$ | $1$ | $0$ (take $\rho = -1$) |
| $2$ | $1 + \alpha^3 X$ | $1$ | $1$ | $1$ (take $\rho = 0$) |
| $3$ | $1 + \alpha^3 X + \alpha^3 X^2$ | $\alpha^7$ | $2$ | $1$ (take $\rho = 0$) |
| $4$ | $1 + \alpha^4 X + \alpha^{12} X^2$ | $\alpha^{10}$ | $2$ | $2$ (take $\rho = 2$) |
| $5$ | $1 + \alpha^7 X + \alpha^4 X^2 + \alpha^6 X^3$ | $0$ | $3$ | $2$ (take $\rho = 3$) |
| $6$ | $1 + \alpha^7 X + \alpha^4 X^2 + \alpha^6 X^3$ | $-$ | $-$ | $-$ |

For illustration of filling out Table 6.5, we consider step 3 ($\mu = 3$). At this step, the partial solution for the error-location polynomial is

$$\sigma^{(3)}(X) = 1 + \alpha^3 X + \alpha^3 X^2.$$

Because the discrepancy is $d_3 = \alpha^7$ which is not zero, a correction term must be added to $\boldsymbol{\sigma}^{(3)}(X)$ to obtain the next partial solution for the error-location polynomial. Prior to step 3, we find step $\rho = 2$, for which $d_\rho \neq 0$ and $\mu - \rho + l_\rho$ is the smallest. Hence, the correction term is

$$d_3 d_2^{-1} X^{3-2} \boldsymbol{\sigma}^{(2)}(X) = \alpha^7 X(1 + \alpha^3 X) = \alpha^7 X + \alpha^{10} X^2.$$

Adding the above correction term to $\boldsymbol{\sigma}^{(3)}(X)$, we obtain the partial solution for the error-location polynomial at step 4,

$$\begin{aligned}
\boldsymbol{\sigma}^{(4)}(X) &= \boldsymbol{\sigma}^{(3)}(X) + \alpha^7 X + \alpha^{10} X^2 \\
&= 1 + \alpha^3 X + \alpha^3 X^2 + \alpha^7 X + \alpha^{10} X^2 \\
&= 1 + \alpha^4 X + \alpha^{12} X^2.
\end{aligned}$$

From Table 6.5, we find the error-location polynomial

$$\boldsymbol{\sigma}(X) = 1 + \alpha^7 X + \alpha^4 X^2 + \alpha^6 X^3.$$

By substituting $X$ in $\boldsymbol{\sigma}(X)$ with the 15 nonzero elements $1, \alpha, \alpha^2, \ldots, \alpha^{14}$ of $GF(2^4)$ in turn, we find that $\alpha^3$, $\alpha^9$, and $\alpha^{12}$ are the roots of $\boldsymbol{\sigma}(X)$. The inverses of these roots give the error-location numbers, $\alpha^{12}$, $\alpha^6$, and $\alpha^3$, of the received polynomial $\mathbf{r}(X) = \alpha^7 X^3 + \alpha^3 X^6 + \alpha^4 X^{12}$. Hence, the errors in $\mathbf{r}(X)$ occur at locations $X^3$, $X^6$, and $X^{12}$.

Using the coefficients of $\boldsymbol{\sigma}(X)$ and the syndrome components $S_1, S_2, S_3$ and following from (6.28), we find the error-value evaluator

$$\begin{aligned}
\mathbf{Z}_0(X) &= S_1 + (S_2 + \sigma_1 S_1)X + (S_3 + \sigma_1 S_2 + \sigma_2 S_1)X^2 \\
&= \alpha^{12} + (1 + \alpha^7 \alpha^{12})X + (\alpha^{14} + \alpha^7 + \alpha^4 \alpha^{12})X^2 \\
&= \alpha^{12} + \alpha X.
\end{aligned}$$

The derivative of $\boldsymbol{\sigma}(X)$ is

$$\boldsymbol{\sigma}'(X) = \alpha^7 + \alpha^6 X^2.$$

Using (6.32) and (6.33), we find the error values at the locations $X^3$, $X^6$, and $X^{12}$,

$$e_3 = \frac{\mathbf{Z}_0(\alpha^{-3})}{\boldsymbol{\sigma}'(\alpha^{-3})} = \alpha^7, e_6 = \frac{\mathbf{Z}_0(\alpha^{-6})}{\boldsymbol{\sigma}'(\alpha^{-6})} = \alpha^3, e_{12} = \frac{\mathbf{Z}_0(\alpha^{-12})}{\boldsymbol{\sigma}'(\alpha^{-12})} = \alpha^4.$$

So, the estimated error polynomial is

$$\mathbf{e}(X) = \alpha^7 X^3 + \alpha^3 X^6 + \alpha^4 X^{12}.$$

Subtracting $\mathbf{e}(X)$ from the received polynomial $\mathbf{r}(X) = \alpha^7 X^3 + \alpha^3 X^6 + \alpha^4 X^{12}$, we obtain the transmitted zero code polynomial. Therefore, the decoding is correct.                                                               ▲▲

In the next example, we consider the construction and decoding of a longer RS code with a larger code symbol field.

**Example 6.8** Let $\alpha$ be a primitive element of the field $GF(2^5)$ constructed by using the primitive polynomial $\mathbf{p}(X) = 1 + X^2 + X^5$ over $GF(2)$ as shown in Table 5.11. The triple-symbol-error-correcting RS code ($t = 3$) whose generator polynomial has $\alpha, \alpha^2, \alpha^3, \alpha^4, \alpha^5,$ and $\alpha^6$ over $GF(2^5)$ as roots is a 32-ary $(31, 25)$ cyclic code. The generator polynomial of this code is

$$\mathbf{g}(X) = (X + \alpha)(X + \alpha^2)(X + \alpha^3)(X + \alpha^4)(X + \alpha^5)(X + \alpha^6)$$
$$= \alpha^{21} + \alpha^{24}X + \alpha^{16}X^2 + \alpha^{24}X^3 + \alpha^9 X^4 + \alpha^{10}X^5 + X^6.$$

Suppose the zero code polynomial $\mathbf{v}(X) = 0$ is transmitted and the received polynomial is

$$\mathbf{r}(X) = \alpha^2 + \alpha^{21}X^{12} + \alpha^7 X^{20}.$$

To decode the received polynomial, we first compute its syndrome $\mathbf{S}$. Substituting $X$ in $\mathbf{r}(X)$ by $\alpha, \alpha^2, \alpha^3, \alpha^4, \alpha^5,$ and $\alpha^6$, we obtain the following six syndrome components of $\mathbf{S}$:

$$S_1 = \mathbf{r}(\alpha) = \alpha^2 + \alpha^{21}\alpha^{12} + \alpha^7\alpha^{20} = \alpha^{27},$$
$$S_2 = \mathbf{r}(\alpha^2) = \alpha^2 + \alpha^{21}(\alpha^2)^{12} + \alpha^7(\alpha^2)^{20} = \alpha,$$
$$S_3 = \mathbf{r}(\alpha^3) = \alpha^2 + \alpha^{21}(\alpha^3)^{12} + \alpha^7(\alpha^3)^{20} = \alpha^{28},$$
$$S_4 = \mathbf{r}(\alpha^4) = \alpha^2 + \alpha^{21}(\alpha^4)^{12} + \alpha^7(\alpha^4)^{20} = \alpha^{29},$$
$$S_5 = \mathbf{r}(\alpha^5) = \alpha^2 + \alpha^{21}(\alpha^5)^{12} + \alpha^7(\alpha^5)^{20} = \alpha^{15},$$
$$S_6 = \mathbf{r}(\alpha^6) = \alpha^2 + \alpha^{21}(\alpha^6)^{12} + \alpha^7(\alpha^6)^{20} = \alpha^8.$$

Based on the above syndrome components, we need to find an error-location polynomial $\boldsymbol{\sigma}(X)$ of degree $\nu$ less than or equal to $t$ whose coefficients satisfy the $6 - \nu$ generalized Newton's identities given by (6.20). To do so, we fill out Table 6.1. The completely filled table is Table 6.6.

Table 6.6 Steps for finding the error-location polynomial of $\mathbf{r}(X)$ for the 32-ary $(31, 25)$ RS code in Example 6.8.

| Step $\mu$ | Partial solution $\boldsymbol{\sigma}^{(\mu)}(X)$ | Discrepancy $d_\mu$ | Degree $l_\mu$ | Step/degree difference $\mu - l_\mu$ |
|---|---|---|---|---|
| $-1$ | $1$ | $1$ | $0$ | $-1$ |
| $0$ | $1$ | $\alpha^{27}$ | $0$ | $0$ |
| $1$ | $1 + \alpha^{27}X$ | $\alpha^8$ | $1$ | 1 (take $\rho = -1$) |
| $2$ | $1 + \alpha^5 X$ | $\alpha^{13}$ | $1$ | 1 (take $\rho = 0$) |
| $3$ | $1 + \alpha X^2$ | $\alpha^8$ | $2$ | 1 (take $\rho = 1$) |
| $4$ | $1 + \alpha^{26}X + \alpha^{18}X^2$ | $\alpha^{24}$ | $2$ | 2 (take $\rho = 2$) |
| $5$ | $1 + \alpha^{20}X + \alpha^{18}X^2 + \alpha^{17}X^3$ | $\alpha^{16}$ | $3$ | 2 (take $\rho = 3$) |
| $6$ | $1 + \alpha^{18}X + \alpha X^3$ | $0$ | $3$ | 3 (take $\rho = 4$) |

For illustration of filling out Table 6.6, we consider step 4 ($\mu = 4$). The partial solution for the error-location polynomial is $\boldsymbol{\sigma}^{(4)}(X) = 1 + \alpha^{26}X + \alpha^{18}X^2$. Because the discrepancy at this step is $d_4 = \alpha^{24}$, which is not zero, the coefficients of $\boldsymbol{\sigma}^{(4)}(X)$ do not satisfy the next generalized Newton's identity. Hence, correction must be made. Prior to step 4, the step $\rho$ for which $d_\rho \neq 0$ and $\mu - \rho + l_\rho$ is the smallest is step 3, i.e., $\rho = 3$. In this case, the correction term to be added to $\boldsymbol{\sigma}^{(4)}(X)$ is

$$d_4 d_3^{-1} X^{4-3} \boldsymbol{\sigma}^{(3)}(X) = \alpha^{24} \alpha^{-8} X(1 + \alpha X^2)$$
$$= \alpha^{16} X + \alpha^{17} X^3.$$

Adding the above correction term to $\boldsymbol{\sigma}^{(4)}(X)$, we obtain the next partial solution for the error-location polynomial,

$$\boldsymbol{\sigma}^{(5)}(X) = \boldsymbol{\sigma}^{(4)}(X) + \alpha^{16} X + \alpha^{17} X^3$$
$$= 1 + \alpha^{26} X + \alpha^{18} X^2 + \alpha^{16} X + \alpha^{17} X^3$$
$$= 1 + (\alpha^{26} + \alpha^{16})X + \alpha^{18} X^2 + \alpha^{17} X^3$$
$$= 1 + \alpha^{20} X + \alpha^{18} X^2 + \alpha^{17} X^3.$$

The solution $\boldsymbol{\sigma}^{(6)}(X)$ at step 6 (the last step) of Table 6.6 gives the error-location polynomial,

$$\boldsymbol{\sigma}(X) = \boldsymbol{\sigma}^{(6)}(X) = 1 + \alpha^{18} X + \alpha X^3,$$

whose derivative is $\boldsymbol{\sigma}'(X) = \alpha^{18} + \alpha X^2$. Using the coefficients of $\boldsymbol{\sigma}(X)$ and (6.28), we compute the error-value evaluator

$$\mathbf{Z}_0(X) = S_1 + (S_2 + \sigma_1 S_1)X + (S_3 + \sigma_1 S_2 + \sigma_2 S_1)X^2$$
$$= \alpha^{27} + (1 + \alpha^{18}\alpha^{27})X + (\alpha^{28} + \alpha^{18}\alpha + 0\alpha^{27})X^2$$
$$= \alpha^{27} + \alpha^{15} X + \alpha^4 X^2.$$

The degree of $\mathbf{Z}_0(X)$ is one less than that of $\boldsymbol{\sigma}(X)$.

The roots over $\mathrm{GF}(2^5)$ of $\boldsymbol{\sigma}(X)$ are 1, $\alpha^{11}$, and $\alpha^{19}$. The inverses of these roots are $\alpha^0 = 1$, $\alpha^{20}$, and $\alpha^{12}$, which are the error-location numbers. Hence, errors in the received polynomial $\mathbf{r}(X)$ are at the locations $X^0$, $X^{12}$, and $X^{20}$. Using (6.32) and (6.33), we find the error values at the positions $X^0$, $X^{12}$, and $X^{20}$, which are:

$$e_0 = \frac{\mathbf{Z}_0(\alpha^0)}{\sigma'(\alpha^0)} = \frac{\alpha^{27} + \alpha^{15} + \alpha^4}{\alpha^{18} + \alpha(\alpha^0)^2} = \alpha^2,$$

$$e_{12} = \frac{\mathbf{Z}_0(\alpha^{-12})}{\sigma'(\alpha^{-12})} = \frac{\alpha^{27} + \alpha^{15}\alpha^{-12} + \alpha^4(\alpha^{-12})^2}{\alpha^{18} + \alpha(\alpha^{-12})^2} = \alpha^{21},$$

$$e_{20} = \frac{\mathbf{Z}_0(\alpha^{-20})}{\sigma'(\alpha^{-20})} = \frac{\alpha^{27} + \alpha^{15}\alpha^{-20} + \alpha^4(\alpha^{-20})^2}{\alpha^{18} + \alpha(\alpha^{-20})^2} = \alpha^7.$$

Hence, the estimated error pattern is

$$\mathbf{e}(X) = \alpha^2 + \alpha^{21} X^{12} + \alpha^7 X^{20}.$$

Subtracting $\mathbf{e}(X)$ from the received polynomial $\mathbf{r}(X) = \alpha^2 + \alpha^{21}X^{12} + \alpha^7 X^{20}$, we obtain the decoded code polynomial which is the transmitted zero code polynomial. The decoding is correct. ▲▲

Two widely used RS codes are the (255, 239) and (255, 223) RS codes over $GF(2^8)$ with minimum distances 17 and 33, respectively. The first code is capable of correcting any combination of eight or fewer random symbol errors. It was a standard code for optical communications, data storage, and hard-disk drives for many years. The second code is the NASA (National Aeronautics and Space Administration) standard code for deep space, satellite communications, and other missions. This code is capable of correcting 16 and fewer random symbol errors. The symbol and block-error performances, i.e., symbol error rate (SER) and block-error rate (BLER), of these two codes over the AWGN channel are shown in Fig. 6.2.

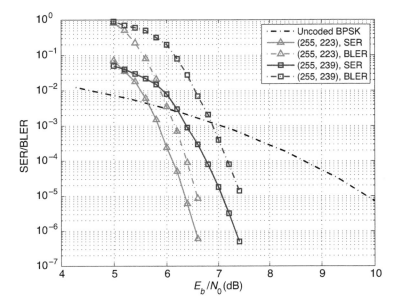

Figure 6.2 Error performances of the (255, 239) and (255, 223) RS codes over $GF(2^8)$.

## 6.10 Euclidean Algorithm for Finding GCD of Two Polynomials

Besides the Berlekamp–Massey algorithm for solving the key-equation given by (6.34) to find the error-location polynomial and the error-value evaluator in decoding nonbinary BCH codes, there is another effective algorithm for solving the key-equation which is based on the Euclidean algorithm for finding the *greatest common divisor* (GCD) of two polynomials over the same field [12]. Before we present this decoding algorithm, we briefly describe the Euclidean algorithm for finding the GCD of two polynomials over the same field.

Consider two polynomials, $\mathbf{a}(X)$ and $\mathbf{b}(X)$, over $GF(q)$. Assume that

$$\deg(\mathbf{a}(X)) \geq \deg(\mathbf{b}(X)).$$

Let $\text{GCD}\{\mathbf{a}(X), \mathbf{b}(X)\}$ denote the GCD of the two polynomials $\mathbf{a}(X)$ and $\mathbf{b}(X)$. Then, $\text{GCD}\{\mathbf{a}(X),\ \mathbf{b}(X)\}$ can be found by applying the Euclidean iterative division algorithm as described below.

Set $\mathbf{a}(X)$ as the dividend and $\mathbf{b}(X)$ as the divisor. Divide $\mathbf{a}(X)$ by $\mathbf{b}(X)$. Let $\mathbf{q}_1(X)$ and $\mathbf{r}_1(X)$ be the quotient and remainder resulting from the division, respectively. Then,

$$\mathbf{a}(X) = \mathbf{q}_1(X)\mathbf{b}(X) + \mathbf{r}_1(X).$$

In the next step, we use $\mathbf{b}(X)$ as the dividend and $\mathbf{r}_1(X)$ as the divisor. Dividing $\mathbf{b}(X)$ by $\mathbf{r}_1(X)$, we obtain

$$\mathbf{b}(X) = \mathbf{q}_2(X)\mathbf{r}_1(X) + \mathbf{r}_2(X),$$

where $\mathbf{q}_2(X)$ and $\mathbf{r}_2(X)$ are the quotient and remainder, respectively. At the third step, we set $\mathbf{r}_1(X)$ and $\mathbf{r}_2(X)$ as the dividend and divisor, respectively. The division results in the following equation:

$$\mathbf{r}_1(X) = \mathbf{q}_3(X)\mathbf{r}_2(X) + \mathbf{r}_3(X),$$

where $\mathbf{q}_3(X)$ and $\mathbf{r}_3(X)$ are the quotient and remainder, respectively. The above division process simply uses the current remainder as the divisor to divide the remainder immediately before it to obtain the next quotient and remainder. At step $i$, $i \geq 3$, we use remainder $\mathbf{r}_{i-2}(X)$ at the $(i-2)$th step as the dividend and the remainder $\mathbf{r}_{i-1}(X)$ at the $(i-1)$th step as the divisor. Dividing $\mathbf{r}_{i-2}(X)$ by $\mathbf{r}_{i-1}(X)$, we obtain the following equality:

$$\mathbf{r}_{i-2}(X) = \mathbf{q}_i(X)\mathbf{r}_{i-1}(X) + \mathbf{r}_i(X),$$

where $\mathbf{q}_i(X)$ and $\mathbf{r}_i(X)$ are the quotient and remainder, respectively, at the $i$th step.

The alternating division process continues until the step, say the $n$th step, is reached where the remainder $\mathbf{r}_n(X)$ at the $(n-1)$th step divides the remainder $\mathbf{r}_{n-1}(X)$ at the $(n-2)$th step. Then, it can be shown that

$$\mathbf{r}_n(X) = \lambda \text{GCD}\{\mathbf{a}(X), \mathbf{b}(X)\}, \qquad (6.37)$$

where $\lambda \in \text{GF}(q)$ is a scalar.

If we regard $\mathbf{a}(X)$ and $\mathbf{b}(X)$ as $\mathbf{r}_{-1}(X)$ and $\mathbf{r}_0(X)$, respectively, i.e., $\mathbf{a}(X) = \mathbf{r}_{-1}(X)$ and $\mathbf{b}(X) = \mathbf{r}_0(X)$, the above alternating Euclidean division process can be formulated as follows: $1 \leq i \leq n$

$$
\begin{aligned}
\mathbf{r}_{-1}(X) &= \mathbf{q}_1(X)\mathbf{r}_0(X) + \mathbf{r}_1(X), \\
\mathbf{r}_0(X) &= \mathbf{q}_2(X)\mathbf{r}_1(X) + \mathbf{r}_2(X), \\
\mathbf{r}_1(X) &= \mathbf{q}_3(X)\mathbf{r}_2(X) + \mathbf{r}_3(X), \\
&\vdots \\
\mathbf{r}_{i-2}(X) &= \mathbf{q}_i(X)\mathbf{r}_{i-1}(X) + \mathbf{r}_i(X), \\
&\vdots \\
\mathbf{r}_{n-2}(X) &= \mathbf{q}_n(X)\mathbf{r}_{n-1}(X) + \mathbf{r}_n(X), \\
\mathbf{r}_{n-1}(X) &= \mathbf{q}_{n+1}(X)\mathbf{r}_n(X) \quad \text{(Stop)}.
\end{aligned}
\qquad (6.38)
$$

From (6.38) and through mathematical manipulations, it is possible to express the remainder at each division step as follows:

$$\mathbf{r}_1(X) = \mathbf{f}_1(X)\mathbf{a}(X) + \mathbf{g}_1(X)\mathbf{b}(X),$$
$$\mathbf{r}_2(X) = \mathbf{f}_2(X)\mathbf{a}(X) + \mathbf{g}_2(X)\mathbf{b}(X),$$
$$\vdots$$
$$\mathbf{r}_i(X) = \mathbf{f}_i(X)\mathbf{a}(X) + \mathbf{g}_i(X)\mathbf{b}(X), \qquad (6.39)$$
$$\vdots$$
$$\mathbf{r}_n(X) = \mathbf{f}_n(X)\mathbf{a}(X) + \mathbf{g}_n(X)\mathbf{b}(X).$$

It follows from (6.37) and the last equation of (6.39) that we can express $\text{GCD}\{\mathbf{a}(X), \mathbf{b}(X)\}$ as follows:

$$\text{GCD}\{\mathbf{a}(X), \mathbf{b}(X)\} = \lambda^{-1}\mathbf{f}_n(X)\mathbf{a}(X) + \lambda^{-1}\mathbf{g}_n(X)\mathbf{b}(X)$$
$$= \mathbf{f}(X)\mathbf{a}(X) + \mathbf{g}(X)\mathbf{b}(X). \qquad (6.40)$$

From (6.38) and (6.39), we have the following iteration equations for finding $\mathbf{r}_i(X)$, $\mathbf{f}_i(X)$, and $\mathbf{g}_i(X)$:

$$\mathbf{r}_i(X) = \mathbf{r}_{i-2}(X) - \mathbf{q}_i(X)\mathbf{r}_{i-1}(X)$$
$$\mathbf{f}_i(X) = \mathbf{f}_{i-2}(X) - \mathbf{q}_i(X)\mathbf{f}_{i-1}(X) \qquad (6.41)$$
$$\mathbf{g}_i(X) = \mathbf{g}_{i-2}(X) - \mathbf{q}_i(X)\mathbf{g}_{i-1}(X)$$

for $1 \le i \le n$. The initial conditions for this recursion process are:

$$\mathbf{r}_{-1}(X) = \mathbf{a}(X), \qquad \mathbf{r}_0(X) = \mathbf{b}(X),$$
$$\mathbf{f}_{-1}(X) = \mathbf{g}_0(X) = 1, \ \mathbf{f}_0(X) = \mathbf{g}_{-1}(X) = 0. \qquad (6.42)$$

An important property of the Euclidean algorithm is that, for $i \ge 1$,

$$\deg(\mathbf{a}(X)) = \deg(\mathbf{g}_i(X)) + \deg(\mathbf{r}_{i-1}(X)). \qquad (6.43)$$

From (6.43), we see that as $i$ increases, the degree of $\mathbf{r}_{i-1}(X)$ decreases and the degree of $\mathbf{g}_i(X)$ increases. This fact will be used for solving the key-equation.

To find $\text{GCD}\{\mathbf{a}(X), \mathbf{b}(X)\}$, we fill out Table 6.7 iteratively with $\mathbf{r}_i(X)$, $\mathbf{q}_i(X)$, $\mathbf{f}_i(X)$, and $\mathbf{g}_i(X)$ as entries. When $\mathbf{r}_{n+1}(X) = 0$, we stop the iterative division process. Then, the last nonzero remainder $\mathbf{r}_n(X)$ gives $\text{GCD}\{\mathbf{a}(X), \mathbf{b}(X)\}$ or $\lambda\text{GCD}\{\mathbf{a}(X), \mathbf{b}(X)\}$ where $\lambda$ is a nonzero element in $\text{GF}(q)$.

**Example 6.9** Let

$$\mathbf{a}(X) = \alpha^6 + \alpha^9 X + \alpha^6 X^2 + \alpha^4 X^3 + \alpha^{14} X^4 + \alpha^{10} X^5 + X^6$$

and

$$\mathbf{b}(X) = \alpha + \alpha^3 X + \alpha^{10} X^2 + \alpha^{10} X^3 + \alpha^4 X^4$$

be two polynomials over $\text{GF}(2^4)$, where $\alpha$ is a primitive element of $\text{GF}(2^4)$.

Table 6.7 Euclidean algorithm for finding the GCD of two polynomials $\mathbf{a}(X)$ and $\mathbf{b}(X)$ over GF$(q)$.

| $i$ | $\mathbf{r}_i(X)$ | $\mathbf{q}_i(X)$ | $\mathbf{f}_i(X)$ | $\mathbf{g}_i(X)$ |
|---|---|---|---|---|
| $-1$ | $\mathbf{r}_{-1}(X) = \mathbf{a}(X)$ | $-$ | $1$ | $0$ |
| $0$ | $\mathbf{r}_0(X) = \mathbf{b}(X)$ | $-$ | $0$ | $1$ |
| $1$ | $\mathbf{r}_1(X)$ | $\mathbf{q}_1(X)$ | $\mathbf{f}_1(X)$ | $\mathbf{g}_1(X)$ |
| $2$ | $\mathbf{r}_2(X)$ | $\mathbf{q}_2(X)$ | $\mathbf{f}_2(X)$ | $\mathbf{g}_2(X)$ |
| $\vdots$ | | | | |
| $n$ | $\mathbf{r}_n(X)$ | $\mathbf{q}_n(X)$ | $\mathbf{f}_n(X)$ | $\mathbf{g}_n(X)$ |
| $n+1$ | $0$ | (Stop) | | |

Because the degree of $\mathbf{a}(X)$ is greater than that of $\mathbf{b}(X)$, we use $\mathbf{a}(X)$ as the dividend and $\mathbf{b}(X)$ as the divisor. To find the GCD of $\mathbf{a}(X)$ and $\mathbf{b}(X)$, we fill Table 6.8 iteratively, using the iterative equations given by (6.41) and the initial conditions given by (6.42). The result is shown in Table 6.8. From the table, we find

$$\lambda\text{GCD}\{\mathbf{a}(X), \mathbf{b}(X)\} = \alpha^{12} + \alpha^{10}X = \alpha^{10}(\alpha^2 + X).$$

Hence, $\alpha^2 + X$ is the GCD of $\mathbf{a}(X)$ and $\mathbf{b}(X)$. ▲▲

Table 6.8 Euclidean iterative algorithm for finding the GCD of two polynomials $\mathbf{a}(X)$ and $\mathbf{b}(X)$ over GF $(2^4)$ in Example 6.9.

| $i$ | $\mathbf{r}_i(X)$ | $\mathbf{q}_i(X)$ | $\mathbf{f}_i(X)$ | $\mathbf{g}_i(X)$ |
|---|---|---|---|---|
| $-1$ | $\alpha^6 + \alpha^9 X + \alpha^6 X^2 + \alpha^4 X^3 + \alpha^{14} X^4 + \alpha^{10} X^5 + X^6$ | $-$ | $1$ | $0$ |
| $0$ | $\alpha + \alpha^3 X + \alpha^{10} X^2 + \alpha^{10} X^3 + \alpha^4 X^4$ | $-$ | $0$ | $1$ |
| $1$ | $\alpha^{13} + \alpha^{13} X + \alpha^8 X^2 + \alpha^{13} X^3$ | $\alpha^{14} + \alpha^3 X + \alpha^{11} X^2$ | $1$ | $\alpha^{14} + \alpha^3 X + \alpha^{11} X^2$ |
| $2$ | $\alpha^6 + \alpha^8 X + \alpha^3 X^2$ | $\alpha^{13} + \alpha^6 X$ | $\alpha^{13} + \alpha^6 X$ | $\alpha^{11} + \alpha^2 X + \alpha^2 X^2$ |
| $3$ | $\alpha^{12} + \alpha^{10} X$ | $\alpha^{10} + \alpha^{10} X$ | $\alpha^2 + \alpha^{10} X + \alpha X^2$ | $\alpha^8 + \alpha^7 X + X^2 + \alpha^{12} X^3 + \alpha^{12} X^4$ |
| $4$ | $0$ | $\alpha^9 + \alpha^8 X$ | $\alpha^4 + \alpha^3 X + \alpha^{12} X^2 + \alpha^9 X^3$ | $\alpha^9 + \alpha^2 X + \alpha^7 X^2 + \alpha^{13} X^3 + \alpha^9 X^4 + \alpha^5 X^5$ |

# 6.11 Solving the Key-Equation with Euclidean Algorithm

The key-equation given by (6.34) can be expressed in the following form:

$$\boldsymbol{\sigma}(X)\mathbf{S}(X) = \mathbf{Q}(X)X^{2t} + \mathbf{Z}_0(X). \tag{6.44}$$

Rearranging (6.44), we have

$$\mathbf{Z}_0(X) = -\mathbf{Q}(X)X^{2t} + \boldsymbol{\sigma}(X)\mathbf{S}(X). \tag{6.45}$$

Setting $\mathbf{a}(X) = X^{2t}$ and $\mathbf{b}(X) = \mathbf{S}(X)$, we see that (6.45) is in exactly the same form as the one given by (6.40). This suggests that $\boldsymbol{\sigma}(X)$ and $\mathbf{Z}_0(X)$ can be found by the Euclidean iterative division algorithm for the two polynomials:

$$\mathbf{a}(X) = X^{2t}, \quad \mathbf{b}(X) = \mathbf{S}(X). \tag{6.46}$$

For $1 \leq i \leq n$, let

$$
\begin{aligned}
\mathbf{Z}_0^{(i)}(X) &= \mathbf{r}_i(X), \\
\boldsymbol{\sigma}^{(i)}(X) &= \mathbf{g}_i(X), \\
\boldsymbol{\gamma}^{(i)}(X) &= \mathbf{f}_i(X).
\end{aligned}
\tag{6.47}
$$

Then, it follows from (6.46) and (6.47) that (6.39) and (6.41) with initial condition of (6.42) can be put in the following forms, respectively, for $i \geq 1$:

$$\mathbf{Z}_0^{(i)}(X) = \boldsymbol{\gamma}^{(i)}(X)X^{2t} + \boldsymbol{\sigma}^{(i)}(X)\mathbf{S}(X), \tag{6.48}$$

and

$$
\begin{aligned}
\mathbf{Z}_0^{(i)}(X) &= \mathbf{Z}_0^{(i-2)}(X) - \mathbf{q}_i(X)\mathbf{Z}_0^{(i-1)}(X), \\
\boldsymbol{\sigma}^{(i)}(X) &= \boldsymbol{\sigma}^{(i-2)}(X) - \mathbf{q}_i(X)\boldsymbol{\sigma}^{(i-1)}(X), \\
\boldsymbol{\gamma}^{(i)}(X) &= \boldsymbol{\gamma}^{(i-2)}(X) - \mathbf{q}_i(X)\boldsymbol{\gamma}^{(i-1)}(X),
\end{aligned}
\tag{6.49}
$$

with

$$
\begin{aligned}
\mathbf{Z}_0^{(-1)}(X) &= X^{2t}, & \mathbf{Z}_0^{(0)}(X) &= \mathbf{S}(X), \\
\boldsymbol{\gamma}^{(-1)}(X) = \boldsymbol{\sigma}^{(0)}(X) &= 1, & \boldsymbol{\gamma}^{(0)}(X) = \boldsymbol{\sigma}^{(-1)}(X) &= 0.
\end{aligned}
\tag{6.50}
$$

To find $\boldsymbol{\sigma}(X)$ and $\mathbf{Z}_0(X)$, we carry out the iterative process given by (6.49) as follows.

At the $i$th step, $i \geq 1$,

(1) divide $\mathbf{Z}_0^{(i-2)}(X)$ by $\mathbf{Z}_0^{(i-1)}(X)$ to obtain the quotient $\mathbf{q}_i(X)$ and remainder $\mathbf{Z}_0^{(i)}(X)$;
(2) find $\boldsymbol{\sigma}^{(i)}(X)$ from

$$\boldsymbol{\sigma}^{(i)}(X) = \boldsymbol{\sigma}^{(i-2)}(X) - \mathbf{q}_i(X)\boldsymbol{\sigma}^{(i-1)}(X). \tag{6.51}$$

Iteration stops when a step $\rho$ is reached for which

$$\deg(\mathbf{Z}_0^{(\rho)}(X)) < \deg(\boldsymbol{\sigma}^{(\rho)}(X)) \le t. \tag{6.52}$$

Then,

$$\begin{aligned} \mathbf{Z}_0(X) &= \mathbf{Z}_0^{(\rho)}(X), \\ \boldsymbol{\sigma}(X) &= \boldsymbol{\sigma}^{(\rho)}(X). \end{aligned} \tag{6.53}$$

If the number of errors is $t$ or less, there always exists a step $\rho$ for which the condition given by (6.52) holds. We can see that

$$\rho \le 2t.$$

The iteration process for finding $\boldsymbol{\sigma}(X)$ and $\mathbf{Z}_0(X)$ can be carried out by filling out Table 6.9 iteratively.

Table 6.9 Euclidean algorithm for finding error-location polynomial $\boldsymbol{\sigma}(X)$ and error-value evaluator $\mathbf{Z}_0(X)$.

| $i$ | $\mathbf{Z}_0^{(i)}(X)$ | $\mathbf{q}_i(X)$ | $\boldsymbol{\sigma}^{(i)}(X)$ |
|---|---|---|---|
| $-1$ | $X^{2t}$ | $-$ | $0$ |
| $0$ | $\mathbf{S}(X)$ | $-$ | $1$ |
| $1$ | | | |
| $\vdots$ | | | |
| $i$ | | | |
| $\vdots$ | | | |

**Example 6.10** Consider the triple-symbol-error-correcting RS code of length $n = 15$ over $\mathrm{GF}(2^4)$ given in Example 6.7 whose generator polynomial has $\alpha$, $\alpha^2$, $\alpha^3$, $\alpha^4$, $\alpha^5$, and $\alpha^6$ as roots, i.e.,

$$\begin{aligned} \mathbf{g}(X) &= (X + \alpha)(X + \alpha^2)(X + \alpha^3)(X + \alpha^4)(X + \alpha^5)(X + \alpha^6) \\ &= \alpha^6 + \alpha^9 X + \alpha^6 X^2 + \alpha^4 X^3 + \alpha^{14} X^4 + \alpha^{10} X^5 + X^6. \end{aligned}$$

Suppose the received polynomial is

$$\mathbf{r}(X) = \alpha^7 X^3 + \alpha^{11} X^{10}.$$

The syndrome components of the syndrome $\mathbf{S}$ of $\mathbf{r}(X)$ are

$$\begin{aligned} S_1 &= \mathbf{r}(\alpha) = \alpha^{10} + \alpha^{21} = \alpha^7, \quad S_2 = \mathbf{r}(\alpha^2) = \alpha^{13} + \alpha^{31} = \alpha^{12}, \\ S_3 &= \mathbf{r}(\alpha^3) = \alpha^{16} + \alpha^{41} = \alpha^6, \quad S_4 = \mathbf{r}(\alpha^4) = \alpha^{19} + \alpha^{51} = \alpha^{12}, \\ S_5 &= \mathbf{r}(\alpha^5) = \alpha^7 + \alpha = \alpha^{14}, \quad S_6 = \mathbf{r}(\alpha^6) = \alpha^{10} + \alpha^{11} = \alpha^{14}. \end{aligned}$$

Hence, the syndrome polynomial is

$$\mathbf{S}(X) = \alpha^7 + \alpha^{12}X + \alpha^6 X^2 + \alpha^{12}X^3 + \alpha^{14}X^4 + \alpha^{14}X^5.$$

Using the Euclidean algorithm, we fill out Table 6.9. The resultant table is given by Table 6.10.

Table 6.10 Euclidean iterative algorithm for decoding the 16-ary $(15, 9)$ RS code given in Example 6.10.

| $i$ | $\mathbf{Z}_0^{(i)}(X)$ | $\mathbf{q}_i(X)$ | $\boldsymbol{\sigma}^{(i)}(X)$ |
|---|---|---|---|
| $-1$ | $X^6$ | $-$ | $0$ |
| $0$ | $\mathbf{S}(X)$ | $-$ | $1$ |
| $1$ | $\alpha^8 + \alpha^3 X + \alpha^5 X^2 + \alpha^5 X^3 + \alpha^6 X^4$ | $\alpha + \alpha X$ | $\alpha + \alpha X$ |
| $2$ | $\alpha^3 + \alpha^2 X$ | $\alpha^{11} + \alpha^8 X$ | $\alpha^{11} + \alpha^8 X + \alpha^9 X^2$ |

We find that, at iteration step 2, the condition (6.52) holds. Hence,

$$\begin{aligned}\boldsymbol{\sigma}(X) &= \boldsymbol{\sigma}^{(2)}(X) = \alpha^{11} + \alpha^8 X + \alpha^9 X^2, \\ \mathbf{Z}_0(X) &= \mathbf{Z}_0^{(2)}(X) = \alpha^3 + \alpha^2 X.\end{aligned}$$

The derivative of $\boldsymbol{\sigma}(X)$ is $\boldsymbol{\sigma}'(X) = \alpha^8$. The roots over $\mathrm{GF}(2^4)$ of $\boldsymbol{\sigma}(X)$ are $\alpha^5$ and $\alpha^{12}$. Their reciprocals are $\alpha^{10}$ and $\alpha^3$ and hence there are two errors in the estimated error pattern $\mathbf{e}(X)$ at the locations $X^3$ and $X^{10}$. The error values at these two error locations are

$$e_3 = \frac{-\mathbf{Z}_0(\alpha^{-3})}{\boldsymbol{\sigma}'(\alpha^{-3})} = \frac{\alpha^3 + \alpha^2 \alpha^{-3}}{\alpha^8} = \frac{1}{\alpha^8} = \alpha^7,$$

$$e_{10} = \frac{-\mathbf{Z}_0(\alpha^{-10})}{\boldsymbol{\sigma}'(\alpha^{-10})} = \frac{\alpha^3 + \alpha^2 \alpha^{-10}}{\alpha^8} = \frac{\alpha^4}{\alpha^8} = \alpha^{11}.$$

Therefore, the estimated error pattern is $\mathbf{e}(X) = \alpha^7 X^3 + \alpha^{11}X^{10}$ and the decoded code polynomial $\mathbf{v}^*(X) = \mathbf{r}(X) - \mathbf{e}(X)$ is the zero code polynomial. ▲▲

**Example 6.11** In this example, we use the Euclidean algorithm to decode the $(31, 25)$ RS code with $t = 3$ over $\mathrm{GF}(2^5)$ given in Example 6.8. The generator polynomial of this code has $\alpha$, $\alpha^2$, $\alpha^3$, $\alpha^4$, $\alpha^5$, and $\alpha^6$ as roots. The received polynomial to be decoded is

$$\mathbf{r}(X) = \alpha^2 + \alpha^{21}X^{12} + \alpha^7 X^{20}.$$

Using the syndrome $\mathbf{S} = (\alpha^{27}, \alpha, \alpha^{28}, \alpha^{29}, \alpha^{15}, \alpha^8)$ of $\mathbf{r}(X)$ computed in Example 6.8, we form the following syndrome polynomial:

$$\mathbf{S}(X) = \alpha^{27} + \alpha X + \alpha^{28}X^2 + \alpha^{29}X^3 + \alpha^{15}X^4 + \alpha^8 X^5.$$

Using the Euclidean algorithm to find the error-location polynomial $\boldsymbol{\sigma}(X)$ and the error-value evaluator $\mathbf{Z}_0(X)$, we fill out Table 6.9 iteratively, with the initial conditions: $\mathbf{Z}_0^{(-1)}(X) = X^6$, $\mathbf{Z}_0^{(0)}(X) = \mathbf{S}(X)$, $\boldsymbol{\sigma}^{(-1)}(X) = 0$, and $\boldsymbol{\sigma}^{(0)}(X) = 1$. The iterative filling process results in Table 6.11.

Table 6.11 Euclidean iterative algorithm for decoding the 32-ary $(31, 25)$ RS code given in Example 6.11.

| $i$ | $\mathbf{Z}_0^{(i)}(X)$ | $\mathbf{q}_i(X)$ | $\boldsymbol{\sigma}^{(i)}(X)$ |
|---|---|---|---|
| $-1$ | $X^6$ | $-$ | $0$ |
| $0$ | $\mathbf{S}(X)$ | $-$ | $1$ |
| $1$ | $\alpha^{26} + \alpha^{11}X + \alpha^{22}X^2 + \alpha^9 X^3 + \alpha^5 X^4$ | $\alpha^{30} + \alpha^{23}X$ | $\alpha^{30} + \alpha^{23}X$ |
| $2$ | $\alpha^6 + \alpha^4 X + \alpha^8 X^3$ | $\alpha^5 + \alpha^3 X$ | $\alpha^{10} + \alpha^{30}X + \alpha^{26}X^2$ |
| $3$ | $\alpha^{18} + \alpha^6 X + \alpha^{26}X^2$ | $\alpha + \alpha^{28}X$ | $\alpha^{22} + \alpha^9 X + \alpha^{23}X^3$ |

In order to illustrate the filling out of Table 6.11, we carry out the computations at each step. At step 1, we divide $X^6$ by $\mathbf{S}(X)$. The remainder and quotient resulting from the division are:

$$\mathbf{Z}_0^{(1)}(X) = \alpha^{26} + \alpha^{11}X + \alpha^{22}X^2 + \alpha^9 X^3 + \alpha^5 X^4$$

and

$$\mathbf{q}_1(X) = \alpha^{30} + \alpha^{23}X,$$

respectively. Using (6.51), we compute $\boldsymbol{\sigma}^{(1)}(X)$. Because the computations are performed on the extension field $\mathrm{GF}(2^5)$ of the binary field $\mathrm{GF}(2)$, we can change the subtraction operation "$-$" to the addition operation "$+$." Hence,

$$\boldsymbol{\sigma}^{(1)}(X) = \boldsymbol{\sigma}^{(-1)}(X) + \mathbf{q}_1(X)\boldsymbol{\sigma}^{(0)}(X) = \alpha^{30} + \alpha^{23}X.$$

At step 2, we divide $\mathbf{S}(X)$ by $\mathbf{Z}_0^{(1)}(X)$. The division results in the following remainder and quotient, respectively:

$$\mathbf{Z}_0^{(2)}(X) = \alpha^6 + \alpha^4 X + \alpha^8 X^3$$

and

$$\mathbf{q}_2(X) = \alpha^5 + \alpha^3 X.$$

Then,

$$\begin{aligned}
\boldsymbol{\sigma}^{(2)}(X) &= \boldsymbol{\sigma}^{(0)}(X) + \mathbf{q}_2(X)\boldsymbol{\sigma}^{(1)}(X) \\
&= 1 + (\alpha^5 + \alpha^3 X)(\alpha^{30} + \alpha^{23}X) \\
&= \alpha^{10} + \alpha^{30}X + \alpha^{26}X^2.
\end{aligned}$$

At step 3, we divide $\mathbf{Z}_0^{(1)}(X)$ by $\mathbf{Z}_0^{(2)}(X)$. The division results in the following remainder and quotient, respectively:

$$\mathbf{Z}_0^{(3)}(X) = \alpha^{18} + \alpha^6 X + \alpha^{26} X^2$$

and

$$\mathbf{q}_3(X) = \alpha + \alpha^{28} X.$$

Then,

$$
\begin{aligned}
\boldsymbol{\sigma}^{(3)}(X) &= \boldsymbol{\sigma}^{(1)}(X) + \mathbf{q}_3(X)\boldsymbol{\sigma}^{(2)}(X) \\
&= \alpha^{30} + \alpha^{23} X + (\alpha + \alpha^{28} X)(\alpha^{10} + \alpha^{30} X + \alpha^{26} X^2) \\
&= \alpha^{22} + \alpha^9 X + \alpha^{23} X^3.
\end{aligned}
$$

At step 3, we find that the condition given by (6.52) holds, i.e.,

$$\deg(\mathbf{Z}_0^{(3)}(X)) < \deg(\boldsymbol{\sigma}^{(3)}(X)) = 3 \leq t = 3.$$

Hence, $\boldsymbol{\sigma}^{(3)}(X)$ and $\mathbf{Z}_0^{(3)}(X)$ are the error-location polynomial $\boldsymbol{\sigma}(X)$ and error-value evaluator $\mathbf{Z}_0(X)$, respectively, i.e.,

$$
\begin{aligned}
\boldsymbol{\sigma}(X) &= \boldsymbol{\sigma}^{(3)}(X) = \alpha^{22} + \alpha^9 X + \alpha^{23} X^3 = \alpha^{22}(1 + \alpha^{18} X + \alpha X^3), \\
\mathbf{Z}_0(X) &= \mathbf{Z}_0^{(3)}(X) = \alpha^{18} + \alpha^6 X + \alpha^{26} X^2 = \alpha^{22}(\alpha^{27} + \alpha^{15} X + \alpha^4 X^2),
\end{aligned}
$$

which are the scaled versions ($\alpha^{22}$ times) of the error-location polynomial $\boldsymbol{\sigma}(X)$ and error-value evaluator $\mathbf{Z}_0(X)$ found by the Berlekamp–Massey algorithm in Example 6.8. Based on $\boldsymbol{\sigma}(X)$ and $\mathbf{Z}_0(X)$, we correct the errors in the received polynomial in exactly the same way as in Example 6.8. ▲▲

## 6.12 Weight Distribution and Probability of Undetected Error of Reed–Solomon Codes

The weight distribution of an RS code has been completely determined [39–42]. For a $t$-symbol-error-correcting RS code $C$ of length $q - 1$ with symbols from $GF(q)$, the number of codewords of weight $i$ in $C$ is given by

$$A_i = \binom{q - 1}{i} \left( \sum_{j=0}^{i-2t-1} (-1)^j \binom{i}{j} (q^{i-2t-j} - 1) \right), \tag{6.54}$$

for $2t + 1 \leq i \leq q - 1$.

Suppose a $q$-ary RS code is used for error detection on a discrete memoryless channel with $q$ inputs and $q$ outputs. Let $(1 - \varepsilon)$ be the probability that a transmitted symbol is received correctly and let $\varepsilon/(q-1)$ be the probability that a transmitted symbol is changed into each of the other $q-1$ symbols. Using the weight distribution given by (6.54), it can be shown that the probability of an undetected error for an RS code is [38]

$$P_u(E) = q^{-2t}\left\{1 + \sum_{i=0}^{2t-1}\binom{q-1}{i}(q^{2t} - q^i)\left(\frac{\varepsilon}{q-1}\right)^i \right.$$
$$\left. \times\left(1 - \frac{q\varepsilon}{q-1}\right)^{q-1-i} - q^{2t}(1-\varepsilon)^{q-1}\right\}. \tag{6.55}$$

It has been also shown in [38] that $P_u(E) < q^{-2t}$ and $P_u(E)$ decreases monotonically as $\varepsilon$ decreases from $(q-1)/q$ to 0. Hence, RS codes are good for error detection.

## 6.13   Remarks

Besides being capable of correcting random errors, RS codes are very effective for correcting combinations of random symbol errors and erasures or purely random symbol erasures over erasure channels by using either the Berlekamp–Massey iterative algorithm or the Euclidean algorithm. Section 10.10 will present details about the erasure-correcting capabilities of RS codes.

## Problems

**6.1** Consider the 4-ary $(15, 9)$ BCH code in Example 6.1.
  (a) Verify that the 15-tuples $\mathbf{v}$ and $\mathbf{v}^*$ over $GF(2^2)$ are valid codewords of the code. (*Hint*: check $\mathbf{v} \cdot \mathbf{H}^T$ or check that $\mathbf{v}(X)$ has $\alpha, \alpha^2, \alpha^3, \alpha^4$ as roots.)
  (b) Find the parity-check and generator matrices, $\mathbf{H}$ and $\mathbf{G}$, of the code.
  (c) Find the systematic parity-check and generator matrices, $\mathbf{H}_{sys}$ and $\mathbf{G}_{sys}$, of the code.
  (d) Let $\mathbf{u} = (1, \beta, 1, 0, 0, 0, 0, 1, 1)$ be a 9-tuple over $GF(2^2)$ which is a message to be encoded. Compute the codeword $\mathbf{v}$ for $\mathbf{u}$.

**6.2** Verify that the 63-tuple $\mathbf{v}$ over $GF(2^2)$ obtained in Example 6.2 is a valid codeword in the 4-ary $(63, 54)$ BCH code. (*Hint*: check $\mathbf{v} \cdot \mathbf{H}^T$ or check that $\mathbf{v}(X)$ has $\alpha, \alpha^2, \alpha^3, \alpha^4$ as roots.)

**6.3** Construct a 4-ary single-symbol-error-correcting BCH code of length 63. Find the codeword $\mathbf{v}$ of the message $\mathbf{u}$ which has 1 at its first three symbols and 0 at the other symbols.

**6.4** In Example 6.4, assume that the received polynomial is $\mathbf{r}(X) = \beta^2 + \beta X^2 + \beta^2 X^3 + \beta X^5 + X^6 + \beta X^7 + X^8 + \beta X^{13} + \beta X^{14}$. Use the Berlekamp–Massey iterative algorithm to decode this received polynomial.

**6.5** Suppose the codeword $\mathbf{v}$ obtained in Problem 6.3 is transmitted. During the transmission, two symbol errors are introduced to locations 0 and 1, i.e.,

$r_0 = v_0 + 1$, $r_1 = v_1 + 1$, and $r_i = v_i$ for $2 \leq i < 63$. Use the Berlekamp–Massey iterative algorithm to decode this received vector.

**6.6** Consider the triple-symbol-error-correcting RS code given in Example 6.7.
(a) Find the weight distribution of this RS code.
(b) Find the code polynomial for the message

$$\mathbf{u}(X) = \alpha + \alpha^5 X + \alpha X^4 + \alpha^7 X^8.$$

**6.7** Construct a double-symbol-error-correcting RS code $C$ of length 15.
(a) Find the weight distribution of the constructed RS code.
(b) Encode a message with the first three symbols as 1 and all other symbols as 0.

**6.8** Consider the $(15, 9)$ RS code over $\mathrm{GF}(2^4)$ in Example 6.7. Suppose $\mathbf{r}(X) = \alpha^7 X^3 + \alpha^3 X^6$ is the received polynomial. Use the Berlekamp–Massey iterative algorithm to decode $\mathbf{r}(X)$.

**6.9** Consider the $(15, 9)$ RS code over $\mathrm{GF}(2^4)$ in Example 6.7. Suppose $\mathbf{r}(X) = \alpha^7 X^3 + \alpha^3 X^6$ is the received polynomial. Use the Euclidean algorithm to decode $\mathbf{r}(X)$.

**6.10** Using the Galois field $\mathrm{GF}(2^5)$ given in Table 5.11, find the generator polynomial of the 32-ary double-symbol-error-correcting RS code of length 31.

**6.11** Consider the double-symbol-error-correcting RS code of length 31 constructed in Problem 6.10. Decode the following received polynomial with the Berlelamp–Massey iterative and Euclidean algorithms

$$\mathbf{r}(X) = \alpha^2 + \alpha^{21} X^{12}.$$

**6.12** Using the Galois field $\mathrm{GF}(2^6)$ given in Table 5.14, find the generator polynomials of the double- and triple-symbol-error-correcting RS codes of length 63.

**6.13** Prove that the dual code of an RS code is also an RS code and thus an MSD code.

**6.14** Consider a $2^m$-ary $(2^m - 1, 2^m - 2t - 1)$ $t$-symbol-error-correcting RS code $C$. Let $\mathbf{v} = (v_0, v_1, \ldots, v_{2^m - 2})$ be a codeword in $C$ with $2^m - 1$ coordinates. Show that any combination of $2^m - 2t - 1$ coordinates of the $2^m - 1$ coordinates in a codeword of $C$ can be used as information (message) coordinates, i.e., any $(2^m - 2t - 1) \times (2^m - 2t - 1)$ submatrix of the generator matrix of the RS code is nonsingular. (*Hint*: use the result from Problem 6.13.)

**6.15** Prove that the $2^m$-ary $(2^m - 1, 2^m - 2t - 1)$ $t$-symbol-error-correcting RS code contains the binary $t$-error-correcting primitive BCH code of length $2^m - 1$ as a subcode. This subcode is called a *subfield subcode*.

**6.16** Let $\alpha$ be a primitive element of $GF(2^m)$. Consider the $2^m$-ary $(2^m - 1, 2^m - 2t - 1)$ $t$-symbol-error-correcting RS code $C$ with generator polynomial

$$\mathbf{g}(X) = (X - \alpha)(X - \alpha^2) \cdots (X - \alpha^{2t}).$$

For each codeword in $C$, $\mathbf{v} = (v_0, v_1, \ldots, v_{2^m-2})$, compute the following symbol $v_\infty \in GF(2^m)$ (an overall parity-check symbol)

$$v_\infty = - \sum_{i=0}^{2^m-2} v_i$$

and form a $2^m$-tuple $\mathbf{v}^* = (v_\infty, v_0, v_1, \ldots, v_{2^m-2})$.
  (a) Prove that the set of $2^m$-tuples, $\mathbf{v}^*$, is a $2^m$-ary $(2^m, 2^m - 2t - 1)$ code, denoted by $C_{RS,ext,1}$, with minimum distance $2t + 2$, i.e., an MDS code.
  (b) Find a parity-check matrix for the code $C_{RS,ext,1}$.

**6.17** Consider the $2^m$-ary $(2^m, 2^m - 2t - 1)$ MDS code $C_{RS,ext,1}$ with minimum distance $2t + 2$ constructed in Problem 6.16. Let $\alpha$ be a primitive element of $GF(2^m)$. For each codeword $\mathbf{v} = (v_0, v_1, \ldots, v_{2^m-2}, v_{2^m-1})$ in $C_{RS,ext,1}$, we prefix it with the following symbol:

$$v_\infty = - \sum_{i=0}^{2^m-1} v_i \cdot \alpha^{(2t+1)i},$$

i.e., form a $(2^m + 1)$-tuple $\mathbf{v}' = (v_\infty, v_0, v_1, \ldots, v_{2^m-2}, v_{2^m-1})$.
  (a) Prove that the set of $(2^m + 1)$-tuples, $\mathbf{v}'$, is a $2^m$-ary $(2^m + 1, 2^m - 2t - 1)$ code, denoted by $C_{RS,ext,2}$, with minimum distance $2t + 3$, i.e., an MDS code.
  (b) Find a parity-check matrix for the code $C_{RS,ext,2}$.

**6.18** Consider a $t$-symbol-error-correcting RS code over $GF(2^m)$ with the following parity-check matrix

$$\mathbf{H} = \begin{bmatrix} 1 & \alpha & \alpha^2 & \cdots & \alpha^{n-1} \\ 1 & \alpha^2 & (\alpha^2)^2 & \cdots & (\alpha^2)^{n-1} \\ 1 & \alpha^3 & (\alpha^3)^2 & \cdots & (\alpha^3)^{n-1} \\ \vdots & & & & \vdots \\ 1 & \alpha^{2t} & (\alpha^{2t})^2 & \cdots & (\alpha^{2t})^{n-1} \end{bmatrix},$$

where $n = 2^m - 1$ and $\alpha$ is a primitive element of $GF(2^m)$. Consider the following matrix obtained by adding two columns to $\mathbf{H}$:

$$\mathbf{H}_1 = \begin{bmatrix} 0 & 1 & & \\ 0 & 0 & & \\ \vdots & \vdots & \mathbf{H} & \\ 0 & 0 & & \\ 1 & 0 & & \end{bmatrix}.$$

Prove that the extended code given by the null space over $GF(2^m)$ of $\mathbf{H}_1$ also has minimum distance $2t + 1$.

**6.19** Let $\alpha$ be a primitive element in $GF(2^m)$ and $n$ be a positive integer with $3 < n < 2^m$. Consider the following $3 \times n$ matrix $\mathbf{H}$ over $GF(2^m)$:

$$\mathbf{H} = \begin{bmatrix} 1 & 0 & 0 & 1 & 1 & \ldots & 1 & 1 \\ 0 & 1 & 0 & \alpha & \alpha^2 & \ldots & \alpha^{n-4} & \alpha^{n-3} \\ 0 & 0 & 1 & \alpha^2 & \alpha^4 & \ldots & \alpha^{2(n-4)} & \alpha^{2(n-3)} \end{bmatrix}.$$

(a) Let $C$ be the $2^m$-ary code given by the null space over $\mathrm{GF}(2^m)$ of $\mathbf{H}$. Compute the code length, dimension, and the minimum distance of the code.

(b) Show that $C$ is able to correct one symbol error and simultaneously detect two symbol errors.

(c) Device a simple decoding algorithm based on $\mathbf{H}$ for the code $C$.

# References

[1] A. Hocquenghem, "Codes correcteurs d'erreurs," *Chiffres*, **2** (1959), 147–156.

[2] R. C. Bose and D. K. Ray-Chaudhuri, "On a class of error correcting binary group codes," *Inf. Control*, **3**(1) (1960), 68–79.

[3] D. Gorenstein and N. Zierler, "A class of cyclic linear error-correcting codes in $p^m$ symbols," *J. Soc. Indust. Appl. Math.*, **9** (1961), 107–214.

[4] I. S. Reed and G. Solomon, "Polynomial codes over certain finite fields," *J. Soc. Indust. Appl. Math.*, **8**(8) (1960), 300–304.

[5] R. T. Chien, "Cyclic decoding procedure for the Bose–Chaudhuri–Hocquenghem codes," *IEEE Trans. Inf. Theory*, **IT-10**(4) (1964), 357–363.

[6] G. D. Forney, "On decoding BCH codes," *IEEE Trans. Inf. Theory*, **IT-11**(4) (1965), 549–557.

[7] E. R. Berlekamp, "On decoding binary Bose–Chaudhuri–Hocquenghem codes," *IEEE Trans. Inf. Theory*, **IT-11**(4) (1965), 577–580.

[8] E. R. Berlekamp, "Nonbinary BCH decoding," *Proc. IEEE Int. Symp. Inf. Theory*, San Remo, Italy, 1967.

[9] E. R. Berlekamp, *Algebraic Coding Theory*, revised ed., Laguna Hills, CA, Aegean Park Press, 1984.

[10] J. L. Massey, "Step-by-step decoding of the Bose–Chaudhuri–Hocquenghem codes," *IEEE Trans. Inf. Theory*, **11**(4) (1965), 580–585.

[11] J. L. Massey, "Shift-register synthesis and BCH decoding," *IEEE Trans. Inf. Theory*, **15**(1) (1969), 122–127.

[12] Y. Sugiyama, M. Kasahara, S. Hirasawa, and T. Namekawa, "A method for solving key equation for decoding Gappa codes," *Inf. Control*, **27**(1) (1975), 87–99.

[13] W. C. Gore, "Transmitting binary symbols with Reed–Solomon codes," *Proc. Conf. Infor. Sci. and Syst.*, Princeton, NJ, 1973, pp. 495–497.

[14] R. E. Blahut, "Transform techniques for error-control codes," *IBM J. Res. Dev.*, **23**(3) (1979), 299–315.

[15] S. Lin and D. J. Costello, Jr., *Error Control Coding: Fundamentals and Applications*, 2nd ed., Upper Saddle River, NJ, Prentice-Hall, 2004.

[16] D. V. Sarwate and N. R. Shanbhag, "High-speed architecture for Reed–Solomon decoders," *IEEE Trans. on Very Large Scale Integration (VLSI) Systems,* **9**(5) (2001), 641–655.

[17] K. Seth, K. N. Viswajith, S. Srinivasan, and V. Kamakoti, "Ultra folded high-speed architectures for Reed Solomon decoders," *Proc. IEEE Intl. Conf. VLSI Design*, Hyderabad, India, 2006, pp. 517–520.

[18] X. Zhang, *VLSI Architectures for Modern Error-Correcting Codes,* Boca Raton, FL, CRC Press, 2017.

[19] R. Koetter and A. Vardy, "Algebraic soft-decision decoding of Reed–Solomon codes," *IEEE Trans. Inf. Theory*, **49**(11) (2003), 2809–2825.

[20] F. Parvaresh and A. Vardy, "Multiplicity assignments for algebraic soft decoding of Reed–Solomon codes," *Proc. IEEE Intl. Symp. Info. Theory*, Yokohama, Japan, July 2003, p. 205.

[21] M. El-Khamy and R. J. McEliece, "Interpolation multiplicity assignment algorithms for algebraic soft-decision decoding of Reed–Solomon codes," *AMS-DIMACS volume on Algebraic Coding Theory and Info. Theory*, vol. 68, 2005.

[22] N. Ratnakar and R. Koetter, "Exponential error bounds for algebraic soft decision decoding of Reed–Solomon codes," *IEEE Trans. Inf. Theory*, **51**(11) (2005), 3899-3917.

[23] J. Bellorado and A. Kavcic, "Low-complexity soft-decoding algorithms for Reed–Solomon codes – part I: an algebraic soft-in hard-out Chase decoder," *IEEE Trans. Inf. Theory*, **56**(3) (2010), 945–959.

[24] J. Jiang and K. Narayanan, "Algebraic soft-decision decoding of Reed–Solomon codes using bit-level soft information," *IEEE Trans. Inf. Theory*, **54**(9) (2008), 3907–3928.

[25] J. Zhu, X. Zhang, and Z. Wang, "Backward interpolation for algebraic soft-decision Reed-Solomon decoding," *IEEE Trans. VLSI Syst.*, **17**(11) (2009), 1602–1615.

[26] X. Zhang, J. Zhu, and W. Zhang, "Efficient re-encoder architectures for algebraic soft-decision Reed–Solomon decoding," *IEEE Trans. Circuits and Syst.-II*, **59**(3) (2012), 163–167.

[27] R. Koetter, *On algebraic decoding of algebraic-geometric and cyclic codes*, Ph.D. thesis, Department of Electrical Engineering, Linköping University, Linköping, Sweden, 1996.

[28] Z. Wang and J. Ma, "High-speed interpolation architecture for soft decision decoding of Reed–Solomon codes," *IEEE Trans. VLSI Syst.*, **14**(9) (2006), 937–950.

[29] N. Kamiya, "An algebraic soft-decision decoding algorithms for BCH codes," *IEEE Trans. Inf. Theory*, **47**(1) (2001), 45–58.

[30] X. Zhang, J. Zhu, and Y. Wu, "Efficient one-pass Chase soft-decision BCH decoder for multi-level cell NAND flash memory," *Proc. IEEE Intl. Midwest Symp. Circuits and Syst.*, Seoul, South Korea, August 2011, pp. 1–4.

[31] W. W. Peterson and E. J. Weldon, Jr., *Error-Correcting Codes*, 2nd ed., Cambridge, MA, The MIT Press, 1972.

[32] F. J. MacWilliams and N. J. A. Sloane, *The Theory of Error-Correcting Codes*, Amsterdam, North-Holland, 1977.

[33] G. Clark and J. Cain, *Error-Correcting Codes for Digital Communications*, New York, Plenum Press, 1981.

[34] A. M. Michelson and A. H. Levesque, *Error-Control Techniques for Digital Communications*, New York, John Wiley, 1985.

[35] S. B. Wicker and V. K. Bhargava, *Reed–Solomon Codes and Their Applications*, New York, IEEE Press, 1994.

[36] S. B. Wicker, *Error Control Systems for Digital Communication and Storage*, Englewood Cliffs, NJ, Prentice-Hall, 1995.

[37] R. E. Blahut, *Algebraic Codes for Data Transmission*, Cambridge, Cambridge University Press, 2003.

[38] T. Kasami and S. Lin, "On the probability of undetected error for the maximum distance separable codes," *IEEE Trans. Commun.*, **32**(9) (1984), 998–1006.

[39] E. F. Assmus, Jr., H. F. Mattson, Jr., and R. J. Turyn, *Cyclic Codes, Scientific Report No. AFCRL-65-332*, Air Force Cambridge Research Center, Cambridge, MA, April 1965.

[40] T. Kasami, S. Lin, and W. W. Peterson, "Some results on weight distributions of BCH codes," *IEEE Trans. Inf. Theory*, **12**(2) (1966), 247.

[41] G. D. Forney, Jr., *Concatenated Codes*, Cambridge, MA, The MIT Press, 1966.

[42] M. F. Ezerman, M. Grassl, and P. Sole, "The weights in MDS codes," *IEEE Trans. Inf. Theory*, **57**(1) (2010), 392–396.

# 7

# Finite Geometries, Cyclic Finite-Geometry Codes, and Majority-Logic Decoding

In Chapters 5 and 6, two types of cyclic codes, namely, BCH and RS codes, are constructed based on finite fields, and they not only have distinctive algebraic structures but are also powerful in correcting errors. In this chapter, we show that cyclic codes can be constructed based on *finite geometries*. These codes are referred to as *finite-geometry codes* and they have not only attractive algebraic structures but also distinctive geometric structures which allow for decoding them with various hard-decision and soft-decision decoding methods and algorithms, ranging from low to high complexity and from reasonably good to very good error performance. Finite-geometry codes were discovered and extensively investigated in the late 1960s and early 1970s [1–3]. In the year 2000, finite-geometry codes were found to form a class of powerful *low-density parity-check* (LDPC) *codes* [4–7]. As LDPC codes, they have very distinct structural properties which allow them to achieve *very low error rates without visible error-floors under iterative decoding based on belief-propagation* [8–15]. Two well-known and well-developed categories of finite geometries are *Euclidean* and *projective geometries* [16, 17]. In this chapter, we present these two categories of finite geometries and cyclic codes constructed from them. Also presented in this chapter is a simple method for decoding this class of cyclic codes, known as *one-step majority-logic decoding*. Finite geometries can also be used to construct general linear block codes.

Finite-geometry LDPC codes and their associated iterative algorithms, both hard-decision and soft-decision, will be presented in Chapters 11 and 12.

# 7.1  Fundamental Concepts of Finite Geometries

A finite geometry, denoted by FG, is a mathematical system composed of a set of $n$ points and a set of $M$ lines. Each line $L$ consists of (or passes through) a *fixed number* of $\rho$ points with $\rho \geq 2$. Two points are on (or connected by) *at most* one line. Two lines either do not have any point in common or they have *one and only one* point in common. If two lines have a common point $\mathbf{a}$, we say that they intersect at the point $\mathbf{a}$. Every point is intersected by a *fixed number* of $\gamma$ lines with $\gamma \geq 2$, i.e., every point is on $\gamma$ lines. The $\gamma$ lines that intersect at a point are said to form an *intersecting bundle*. Two lines are said to be *parallel* if they have no point in common, i.e., they do not intersect. It is clear that $n\gamma = M\rho$.

Denote the points and lines in FG by $\mathbf{a}_0, \mathbf{a}_1, \ldots, \mathbf{a}_{n-1}$ and $L_0, L_1, \ldots, L_{M-1}$, respectively, with the points labeled from 0 to $n-1$ and lines labeled from 0 to $M-1$. For the $i$th line $L_i$, $0 \leq i < M$, in FG, we define an $n$-tuple over GF(2) as follows:

$$\mathbf{h}_i = (h_{i,0}, h_{i,1}, \ldots, h_{i,n-1}),$$

whose components correspond to the $n$ points in FG, where $h_{i,j} = 1$ *if and only if* the $j$th point $\mathbf{a}_j$ of FG is on the $i$th line $L_i$; otherwise, $h_{i,j} = 0$, $0 \leq j < n$. The $n$-tuple $\mathbf{h}_i$ is called the *incidence vector* of the $i$th line $L_i$ in FG. Because each line in FG consists of $\rho$ points, the weight (i.e., the number of nonzero components) of the incidence vector $\mathbf{h}_i$ is $\rho$.

Next, we form the following $M \times n$ matrix over GF(2) with the incidence vectors of the $M$ lines in FG as rows:

$$\mathbf{H}_{\mathrm{FG}} = \begin{bmatrix} \mathbf{h}_0 \\ \mathbf{h}_1 \\ \vdots \\ \mathbf{h}_{M-1} \end{bmatrix} = \begin{bmatrix} h_{0,0} & h_{0,1} & \cdots & h_{0,n-1} \\ h_{1,0} & h_{1,1} & \cdots & h_{1,n-1} \\ \vdots & \vdots & \ddots & \vdots \\ h_{M-1,0} & h_{M-1,1} & \cdots & h_{M-1,n-1} \end{bmatrix}. \tag{7.1}$$

The matrix $\mathbf{H}_{\mathrm{FG}}$ is called the *line-point incidence* (or *line-point adjacency*) *matrix* of FG whose rows and columns correspond to the lines and points in FG, respectively. A row in $\mathbf{H}_{\mathrm{FG}}$ simply displays the points on a specific line of FG and has weight $\rho$. A column in $\mathbf{H}_{\mathrm{FG}}$ simply displays the lines that intersect at a specific point in FG and has weight $\gamma$. Because two points in FG are on *one and only one line*, every two columns have *exactly one place* where they both have 1-components. Because two lines in FG are either *parallel* or they intersect at *one and only one point*, two rows of $\mathbf{H}_{\mathrm{FG}}$ have *at most one place* where they both have 1-components. Because each point in FG is on (or is intersected by) $\gamma$ lines in FG, we can easily see that $\mathbf{H}_{\mathrm{FG}}$ has the following structure: for each column position $j$, $0 \leq j < n$, of $\mathbf{H}_{\mathrm{FG}}$, there is a set $\Omega_j$ of $\gamma$ rows in $\mathbf{H}_{\mathrm{FG}}$ for which each row has a 1-component in its $j$th position and for $\ell \neq j$, there is *at most one row* in $\Omega_j$ which has 1-component in its $\ell$th position. The $\gamma$ rows in $\Omega_j$ are said to be *orthogonal* on the $j$th column position of $\mathbf{H}_{\mathrm{FG}}$. The above structure of the matrix $\mathbf{H}_{\mathrm{FG}}$ is referred to as an *orthogonal structure*.

**Example 7.1** Consider the geometry given in Fig. 7.1, which comprises $n = 5$ points:

$$\mathbf{a}_0, \mathbf{a}_1, \mathbf{a}_2, \mathbf{a}_3, \mathbf{a}_4$$

and $M = 10$ lines, each consisting of $\rho = 2$ points:

$$L_0 = \{\mathbf{a}_0, \mathbf{a}_1\}, \quad L_1 = \{\mathbf{a}_0, \mathbf{a}_2\}, \quad L_2 = \{\mathbf{a}_0, \mathbf{a}_3\}, \quad L_3 = \{\mathbf{a}_0, \mathbf{a}_4\}, \quad L_4 = \{\mathbf{a}_1, \mathbf{a}_2\},$$
$$L_5 = \{\mathbf{a}_1, \mathbf{a}_3\}, \quad L_6 = \{\mathbf{a}_1, \mathbf{a}_4\}, \quad L_7 = \{\mathbf{a}_2, \mathbf{a}_3\}, \quad L_8 = \{\mathbf{a}_2, \mathbf{a}_4\}, \quad L_9 = \{\mathbf{a}_3, \mathbf{a}_4\}.$$

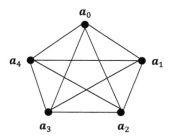

Figure 7.1 A five-point finite geometry.

Each point is intersected by $\gamma = 4$ lines. For example, the four lines $L_0$, $L_1$, $L_2$, and $L_3$ intersect at the point $\mathbf{a}_0$ and they form an intersecting bundle. We can easily check that any two points in this geometry are on at most one line, and any two lines either have no point in common or they have one and only one point in common. Thus, this geometry is a finite geometry.

The two lines $L_0$ and $L_7$ are parallel because they have no point in common. For each line, there are three lines parallel to it. For example, the lines $L_7$, $L_8$, and $L_9$ are parallel to the line $L_0$.

The incidence vectors for the 10 lines are

$$\mathbf{h}_0 = (1\ 1\ 0\ 0\ 0), \quad \mathbf{h}_1 = (1\ 0\ 1\ 0\ 0), \quad \mathbf{h}_2 = (1\ 0\ 0\ 1\ 0), \quad \mathbf{h}_3 = (1\ 0\ 0\ 0\ 1),$$
$$\mathbf{h}_4 = (0\ 1\ 1\ 0\ 0), \quad \mathbf{h}_5 = (0\ 1\ 0\ 1\ 0), \quad \mathbf{h}_6 = (0\ 1\ 0\ 0\ 1), \quad \mathbf{h}_7 = (0\ 0\ 1\ 1\ 0),$$
$$\mathbf{h}_8 = (0\ 0\ 1\ 0\ 1), \quad \mathbf{h}_9 = (0\ 0\ 0\ 1\ 1).$$

Then, the incidence matrix $\mathbf{H}_{\mathrm{FG}}$ of this finite geometry is

$$\mathbf{H}_{\mathrm{FG}} = \begin{bmatrix} \mathbf{h}_0 \\ \mathbf{h}_1 \\ \mathbf{h}_2 \\ \mathbf{h}_3 \\ \mathbf{h}_4 \\ \mathbf{h}_5 \\ \mathbf{h}_6 \\ \mathbf{h}_7 \\ \mathbf{h}_8 \\ \mathbf{h}_9 \end{bmatrix} = \begin{bmatrix} 1\ 1\ 0\ 0\ 0 \\ 1\ 0\ 1\ 0\ 0 \\ 1\ 0\ 0\ 1\ 0 \\ 1\ 0\ 0\ 0\ 1 \\ 0\ 1\ 1\ 0\ 0 \\ 0\ 1\ 0\ 1\ 0 \\ 0\ 1\ 0\ 0\ 1 \\ 0\ 0\ 1\ 1\ 0 \\ 0\ 0\ 1\ 0\ 1 \\ 0\ 0\ 0\ 1\ 1 \end{bmatrix}. \tag{7.2}$$

The matrix $\mathbf{H}_{\text{FG}}$ is a $10 \times 5$ matrix with column weight $\gamma = 4$ and row weight $\rho = 2$. Consider the first four rows in $\mathbf{H}_{\text{FG}}$, i.e., $\Omega_0 = \{\mathbf{h}_0, \mathbf{h}_1, \mathbf{h}_2, \mathbf{h}_3\}$. All the four rows in $\Omega_0$ have a 1-entry in its zeroth position and for the other positions (the first, second, third, and fourth), there is one and only one line which has 1-entry at those positions. The four rows in $\Omega_0$ are orthogonal at the zeroth column position of $\mathbf{H}_{\text{FG}}$. It is easy to find rows that are orthogonal to each of the other four column positions of $\mathbf{H}_{\text{FG}}$. ▲▲

## 7.2 Majority-Logic Decoding of Finite-Geometry Codes

If we use the line-point incidence matrix $\mathbf{H}_{\text{FG}}$ of a finite-geometry FG given by (7.1) as the parity-check matrix of a linear block code, the null space over $\text{GF}(2)$ of $\mathbf{H}_{\text{FG}}$ gives a linear block code $C_{\text{FG}}$ of length $n$ which is referred to as a *finite-geometry code*. The incidence matrix $\mathbf{H}_{\text{FG}}$ of FG may not be a full-rank matrix, i.e., the rows of $\mathbf{H}_{\text{FG}}$ may not be linearly independent. Let $m$ be the rank of $\mathbf{H}_{\text{FG}}$, $1 \leq m \leq M$. Then, the FG code $C_{\text{FG}}$ is a binary $(n, n-m)$ linear block code. Consider the finite geometry in Example 7.1 whose incidence matrix $\mathbf{H}_{\text{FG}}$ given by (7.2) is a $10 \times 5$ matrix. The rank of $\mathbf{H}_{\text{FG}}$ is 4. Using $\mathbf{H}_{\text{FG}}$ as a parity-check matrix, we obtain a $(5, 1)$ finite-geometry code.

Based on the orthogonal structure of $\mathbf{H}_{\text{FG}}$, a simple decoding method for $C_{\text{FG}}$ can be devised to correct $\lfloor \gamma/2 \rfloor$ or fewer random errors over a BSC. Let $\mathbf{v} = (v_0, v_1, \dots, v_{n-1})$ be a codeword in $C_{\text{FG}}$. Because $\mathbf{v}$ is in the null space of $\mathbf{H}_{\text{FG}}$, $\mathbf{v} \cdot \mathbf{H}_{\text{FG}}^T = \mathbf{0}$. For $0 \leq j < n$, the $j$th code symbol $v_j$ of $\mathbf{v}$ corresponds to the $j$th column of $\mathbf{H}_{\text{FG}}$. Because $\mathbf{H}_{\text{FG}}$ has $M$ rows, the product $\mathbf{v} \cdot \mathbf{H}_{\text{FG}}^T$ gives $M$ *parity-check sums* of the code symbols of $\mathbf{v}$, where each is a sum (with modulo-2 addition) of $\rho$ code symbols. Suppose $\mathbf{v}$ is transmitted over a BSC. Let $\mathbf{r} = (r_0, r_1, \dots, r_{n-1})$ be the hard-decision received vector and let $\mathbf{e} = (e_0, e_1, \dots, e_{n-1})$ be the error vector caused by the channel noise. Then, $\mathbf{r} = \mathbf{v} + \mathbf{e}$. The syndrome of $\mathbf{r}$ is an $M$-tuple over $\text{GF}(2)$ calculated as follows:

$$\mathbf{S} = (S_0, S_1, \dots, S_{M-1})$$
$$= \mathbf{r} \cdot \mathbf{H}_{\text{FG}}^T$$
$$= (\mathbf{v} + \mathbf{e}) \cdot \mathbf{H}_{\text{FG}}^T$$
$$= \mathbf{v} \cdot \mathbf{H}_{\text{FG}}^T + \mathbf{e} \cdot \mathbf{H}_{\text{FG}}^T.$$

Because $\mathbf{v} \cdot \mathbf{H}_{\text{FG}}^T = \mathbf{0}$, we have

$$\mathbf{S} = \mathbf{e} \cdot \mathbf{H}_{\text{FG}}^T. \tag{7.3}$$

The product $\mathbf{e} \cdot \mathbf{H}_{\text{FG}}^T$ gives $M$ linear equations which relate the $n$ unknown error bits of the error vector $\mathbf{e}$ to the $M$ syndrome bits of $\mathbf{S}$ computed from the received vector $\mathbf{r}$. Expanding (7.3), we have

$$
\begin{aligned}
S_0 &= \langle \mathbf{e}, \mathbf{h}_0 \rangle &&= e_0 h_{0,0} + e_1 h_{0,1} + \cdots + e_{n-1} h_{0,n-1}, \\
S_1 &= \langle \mathbf{e}, \mathbf{h}_1 \rangle &&= e_0 h_{1,0} + e_1 h_{1,1} + \cdots + e_{n-1} h_{1,n-1}, \\
&\;\;\vdots \\
S_{M-1} &= \langle \mathbf{e}, \mathbf{h}_{M-1} \rangle &&= e_0 h_{M-1,0} + e_1 h_{M-1,1} + \cdots + e_{n-1} h_{M-1,n-1}.
\end{aligned}
\tag{7.4}
$$

Because the weight of each row in $\mathbf{H}_{FG}$ is $\rho$, each syndrome bit $S_i$, $0 \le i < M$, is a sum (with modulo-2 addition) of $\rho$ error bits in the error vector $\mathbf{e}$, called a *syndrome check-sum of error bits*. For $0 \le j < n$, let

$$
\Omega_j = \{ \mathbf{h}_\ell^{(j)} = (h_{\ell,0}^{(j)}, h_{\ell,1}^{(j)}, \ldots, h_{\ell,n-1}^{(j)}) : h_{\ell,j}^{(j)} = 1, 0 \le \ell < \gamma \}
\tag{7.5}
$$

be the set of $\gamma$ rows in $\mathbf{H}_{FG}$ that are orthogonal on the $j$th column position of $\mathbf{H}_{FG}$, where $\mathbf{h}_\ell^{(j)}$ represents a certain row $\mathbf{h}_i$ of $\mathbf{H}_{FG}$ with the entry $h_{i,j} = 1$.

Because the $j$th code symbol $v_j$ of the codeword $\mathbf{v}$ corresponds to the $j$th column of $\mathbf{H}_{FG}$, the $\gamma$ rows in $\Omega_j$ are orthogonal on the code symbol $v_j$. Let

$$
\Lambda_j = \{ S_\ell^{(j)} = \langle \mathbf{r}, \mathbf{h}_\ell^{(j)} \rangle : \mathbf{h}_\ell^{(j)} \in \Omega_j, 0 \le \ell < \gamma \}
\tag{7.6}
$$

be a set of the $\gamma$ syndrome bits computed from $\mathbf{r}$ and the $\gamma$ rows of $\mathbf{H}_{FG}$ in $\Omega_j$, where $\langle \mathbf{r}, \mathbf{h}_\ell^{(j)} \rangle$ represents the inner product of $\mathbf{r}$ and the row $\mathbf{h}_\ell^{(j)}$ of $\mathbf{H}_{FG}$. The $\gamma$ syndrome check-sums in $\Lambda_j$ are said to be *orthogonal* on the $j$th received bit $r_j$.

It follows from (7.4) to (7.6) that we obtain the following relationship between the $\gamma$ syndrome check-sums in $\Lambda_j$ and the error bits $e_0, e_1, \ldots, e_{n-1}$ in the error vector $\mathbf{e}$: for $0 \le \ell < \gamma$,

$$
S_\ell^{(j)} = e_j + \sum_{0 \le k < n, k \ne j} e_k h_{\ell,k}^{(j)}.
\tag{7.7}
$$

From (7.7), we see that all the $\gamma$ syndrome check-sums in $\Lambda_j$ contain (or check on) the $j$th error bit $e_j$. Owing to the orthogonal structure of the rows in $\Omega_j$, any error bit *other than* $e_j$ is checked by (or appears in) *at most one* of the $\gamma$ syndrome check-sums in $\Lambda_j$. These syndrome check-sums are said to be orthogonal on the $j$th error bit $e_j$ and are called *orthogonal syndrome check-sums* on the $j$th error bit $e_j$ in the error vector $\mathbf{e}$.

Next, we demonstrate that the $\gamma$ orthogonal syndrome check-sums given by (7.7) can be used to decode the error vector $\mathbf{e}$. First, we see that if all the error bits in the sum $\sum_{0 \le k < n, k \ne j} e_k h_{\ell,k}^{(j)}$ of (7.7) are zero, then $e_j = S_\ell^{(j)}$, $0 \le \ell < \gamma$. Suppose there are $\lfloor \gamma/2 \rfloor$ or fewer errors in the error vector $\mathbf{e}$. If $e_j = 1$, the other nonzero error bits can distribute among *at most* $\lfloor \gamma/2 \rfloor - 1$ syndrome check-sums orthogonal on $e_j$. Hence, *at least* $\gamma - (\lfloor \gamma/2 \rfloor - 1) = \gamma - \lfloor \gamma/2 \rfloor + 1$ (*more than half*) of the syndrome check-sums orthogonal on $e_j$ are equal to $e_j = 1$. However, if $e_j = 0$, the nonzero error bits can distribute among *at most* $\lfloor \gamma/2 \rfloor$ syndrome check-sums orthogonal on $e_j$. Hence, *at least* $\gamma - \lfloor \gamma/2 \rfloor$ (*half or more than half*) of the syndrome check-sums orthogonal on $e_j$ are equal to $e_j = 0$. Thus, the value of the error bit $e_j$ is equal to the value assumed by a *clear majority* of the syndrome check-sums orthogonal on $e_j$; if no value is assumed by a clear majority (i.e., there is a *tie*) of the syndrome check-sums orthogonal on $e_j$, then the error bit $e_j$ is equal to zero. Based on the above facts, an algorithm for decoding $e_j$, $0 \leq j < n$, can be formulated as follows.

> *The error bit $e_j$ is decoded as "1" if a clear majority of the $\gamma$ syndrome check-sums orthogonal on $e_j$ is "1," otherwise, $e_j$ is decoded as "0."*

Correct decoding of $e_j$ is guaranteed if the error vector $\mathbf{e}$ contains $\lfloor \gamma/2 \rfloor$ or fewer errors. The decoding algorithm given above is called *one-step majority-logic decoding* (OSMLD). The first majority-logic decoding algorithm was devised by Reed [18] to decode RM codes (to be presented in Chapter 8). Reed's algorithm was later extended and generalized by many coding theorists. The first unified formulation of majority-logic decoding was due to Massey [19]. The parameter $\lfloor \gamma/2 \rfloor$ is referred to as the *majority-logic error-correcting capability* of the FG code $C_{\mathrm{FG}}$. Because the code is capable of correcting $\lfloor \gamma/2 \rfloor$ or fewer errors, its minimum distance is *at least* $\gamma + 1$. If the number of errors in the error vector $\mathbf{e}$ is greater than $\lfloor \gamma/2 \rfloor$, but the distribution of errors results in less than half of syndrome check-sums orthogonal on an error bit that contains the error bit alone, the error bit can be decoded correctly. That is to say that OSMLD can correct error patterns with more than $\lfloor \gamma/2 \rfloor$ errors. Hence, it is not a *bounded distance decoding*.

A code that can be decoded with OSMLD is said to be *OSMLD decodable*. The decoding complexity of OSMLD is extremely low because it requires only modulo-2 computations and counting to determine the majority value of the orthogonal syndrome check-sums.

**Example 7.2** Consider the following $20 \times 16$ matrix $\mathbf{H}_{\mathrm{FG}}$ constructed based on the lines and points of a finite geometry:

$$
\mathbf{H}_{\mathrm{FG}} = \begin{bmatrix} \mathbf{h}_0 \\ \mathbf{h}_1 \\ \vdots \\ \mathbf{h}_{18} \\ \mathbf{h}_{19} \end{bmatrix} = \begin{bmatrix}
1 & 1 & 0 & 0 & 0 & 0 & 1 & 0 & 0 & 0 & 0 & 1 & 0 & 0 & 0 & 0 \\
0 & 0 & 1 & 1 & 0 & 1 & 0 & 0 & 0 & 1 & 0 & 0 & 0 & 0 & 0 & 0 \\
0 & 0 & 0 & 0 & 1 & 0 & 0 & 0 & 0 & 0 & 0 & 0 & 1 & 1 & 0 & 1 \\
0 & 0 & 0 & 0 & 0 & 0 & 0 & 1 & 1 & 0 & 1 & 0 & 0 & 0 & 1 & 0 \\
1 & 0 & 1 & 0 & 0 & 0 & 0 & 1 & 0 & 0 & 0 & 0 & 1 & 0 & 0 & 0 \\
0 & 1 & 0 & 0 & 0 & 1 & 0 & 0 & 0 & 0 & 0 & 0 & 0 & 1 & 1 & 0 \\
0 & 0 & 0 & 1 & 1 & 0 & 1 & 0 & 0 & 0 & 1 & 0 & 0 & 0 & 0 & 0 \\
0 & 0 & 0 & 0 & 0 & 0 & 0 & 0 & 1 & 1 & 0 & 1 & 0 & 0 & 0 & 1 \\
1 & 0 & 0 & 1 & 0 & 0 & 0 & 0 & 1 & 0 & 0 & 0 & 0 & 1 & 0 & 0 \\
0 & 1 & 0 & 0 & 0 & 0 & 0 & 0 & 0 & 1 & 1 & 0 & 1 & 0 & 0 & 0 \\
0 & 0 & 1 & 0 & 0 & 0 & 1 & 0 & 0 & 0 & 0 & 0 & 0 & 0 & 1 & 1 \\
0 & 0 & 0 & 0 & 1 & 1 & 0 & 1 & 0 & 0 & 0 & 1 & 0 & 0 & 0 & 0 \\
1 & 0 & 0 & 0 & 1 & 0 & 0 & 0 & 0 & 1 & 0 & 0 & 0 & 0 & 1 & 0 \\
0 & 1 & 0 & 1 & 0 & 0 & 0 & 1 & 0 & 0 & 0 & 0 & 0 & 0 & 0 & 1 \\
0 & 0 & 1 & 0 & 0 & 0 & 0 & 0 & 0 & 1 & 1 & 0 & 1 & 0 & 0 & 0 \\
0 & 0 & 0 & 0 & 0 & 1 & 1 & 0 & 1 & 0 & 0 & 0 & 1 & 0 & 0 & 0 \\
1 & 0 & 0 & 0 & 0 & 1 & 0 & 0 & 0 & 0 & 1 & 0 & 0 & 0 & 0 & 1 \\
0 & 1 & 1 & 0 & 1 & 0 & 0 & 0 & 1 & 0 & 0 & 0 & 0 & 0 & 0 & 0 \\
0 & 0 & 0 & 1 & 0 & 0 & 0 & 0 & 0 & 0 & 0 & 1 & 1 & 0 & 1 & 0 \\
0 & 0 & 0 & 0 & 0 & 0 & 1 & 1 & 0 & 1 & 0 & 0 & 0 & 1 & 0 & 0
\end{bmatrix}.
$$

From the matrix $\mathbf{H}_{\text{FG}}$, we see that there are 16 points and 20 lines in this finite geometry, where each line consists of $\rho = 4$ points and each point is intersected by $\gamma = 5$ lines. The null space over GF(2) of $\mathbf{H}_{\text{FG}}$ gives a $(16, 7)$ FG code $C_{\text{FG}}$. Because, for each code symbol location, five orthogonal syndrome check-sums can be formed, the code is capable of correcting $\lfloor 5/2 \rfloor = 2$ or fewer errors with the OSMLD.

Assume that the all-zero codeword, $\mathbf{v} = (0\ 0\ 0\ 0\ 0\ 0\ 0\ 0\ 0\ 0\ 0\ 0\ 0\ 0\ 0\ 0)$, is transmitted over the channel. Let $\mathbf{r} = (1\ 0\ 0\ 0\ 0\ 0\ 0\ 0\ 0\ 0\ 0\ 0\ 0\ 0\ 0\ 0)$ be the received hard-decision vector which contains a single error at the zeroth position. The decoder first calculates the syndrome to check whether there are errors in the received vector $\mathbf{r}$:

$$\mathbf{S} = (S_0, S_1, \ldots, S_{18}, S_{19})$$

$$= \mathbf{r} \cdot \mathbf{H}_{\text{FG}}^T$$

$$= (1\ 0\ 0\ 0\ 1\ 0\ 0\ 0\ 1\ 0\ 0\ 0\ 1\ 0\ 0\ 0\ 1\ 0\ 0\ 0).$$

Because $\mathbf{S} \neq \mathbf{0}$, the presence of error is detected and the decoder proceeds to decode the received vector $\mathbf{r}$. Without any information about the error vector $\mathbf{e}$, the decoder will start with the decoding of the first error bit $e_0$. Form the set $\Omega_0$ of rows in $\mathbf{H}_{\text{FG}}$ which have 1-component in their zeroth position (i.e., rows corresponding to the lines which intersect at the zeroth point):

$$\Omega_0 = \{\mathbf{h}_0, \mathbf{h}_4, \mathbf{h}_8, \mathbf{h}_{12}, \mathbf{h}_{16}\}.$$

The following matrix $\mathbf{H}_{\text{FG}}^*$ is formed by taking all the elements in $\Omega_0$ as its rows, which is simply a submatrix of $\mathbf{H}_{\text{FG}}$:

$$\mathbf{H}_{\text{FG}}^* = \begin{bmatrix} \mathbf{h}_0 \\ \mathbf{h}_4 \\ \mathbf{h}_8 \\ \mathbf{h}_{12} \\ \mathbf{h}_{16} \end{bmatrix} = \begin{bmatrix} 1 & 1 & 0 & 0 & 0 & 0 & 1 & 0 & 0 & 0 & 0 & 1 & 0 & 0 & 0 & 0 \\ 1 & 0 & 1 & 0 & 0 & 0 & 0 & 1 & 0 & 0 & 0 & 0 & 1 & 0 & 0 & 0 \\ 1 & 0 & 0 & 1 & 0 & 0 & 0 & 0 & 1 & 0 & 0 & 0 & 0 & 1 & 0 & 0 \\ 1 & 0 & 0 & 0 & 1 & 0 & 0 & 0 & 0 & 1 & 0 & 0 & 0 & 0 & 1 & 0 \\ 1 & 0 & 0 & 0 & 0 & 1 & 0 & 0 & 0 & 0 & 1 & 0 & 0 & 0 & 0 & 1 \end{bmatrix}.$$

The above matrix $\mathbf{H}_{\text{FG}}^*$ shows the orthogonal structure of the set $\Omega_0$: all the rows in $\Omega_0$ have 1-component at their zeroth position (i.e., the zeroth column in $\mathbf{H}_{\text{FG}}^*$ contains all 1s), and for the $\ell$th position with $\ell \neq 0$, there is one and only one row in $\Omega_0$ with a 1-component at that position (i.e., the $\ell$th column in $\mathbf{H}_{\text{FG}}^*$, $\ell \neq 0$, has one 1-component). The syndrome check-sums formed by the rows in $\Omega_0$ are orthogonal on the zeroth bit of the received vector $\mathbf{r}$.

Next, based on $\Omega_0$, we form the syndrome check-sum set $\Lambda_0$:

$$\Lambda_0 = \{S_\ell = \langle \mathbf{r}, \mathbf{h}_\ell \rangle : \ell \in \{0, 4, 8, 12, 16\}\},$$

where the five syndrome check-sums are $S_0 = 1$, $S_4 = 1$, $S_8 = 1$, $S_{12} = 1$, and $S_{16} = 1$. In terms of error bits, we have

$$S_0 = 1 = e_0 + e_1 + e_6 + e_{11},$$
$$S_4 = 1 = e_0 + e_2 + e_7 + e_{12},$$
$$S_8 = 1 = e_0 + e_3 + e_8 + e_{13},$$
$$S_{12} = 1 = e_0 + e_4 + e_9 + e_{14},$$
$$S_{16} = 1 = e_0 + e_5 + e_{10} + e_{15}.$$

From above, we see that there are five syndrome check-sums in $\Lambda_0$ which are equal to 1 (more than half, $\lfloor \gamma/2 \rfloor = \lfloor 5/2 \rfloor = 2$). Therefore, the error bit $e_0$ in the error vector $\mathbf{e}$ is decoded as "1." Correction is then made to the received bit $r_0$ in $\mathbf{r}$ which results in a decoded vector $\mathbf{v}^* = (r_0 + e_0, r_1, r_2, \ldots, r_{n-1}) = (0\ 0\ 0\ 0\ 0\ 0\ 0\ 0\ 0\ 0\ 0\ 0\ 0\ 0\ 0\ 0)$. Next, the decoder checks whether the decoded vector $\mathbf{v}^*$ is a valid codeword by calculating its syndrome $\mathbf{S}^*$:

$$\mathbf{S}^* = \mathbf{v}^* \cdot \mathbf{H}_{\mathrm{FG}}^T$$
$$= (0\ 0\ 0\ 0\ 0\ 0\ 0\ 0\ 0\ 0\ 0\ 0\ 0\ 0\ 0\ 0) \cdot \mathbf{H}_{\mathrm{FG}}^T$$
$$= (0\ 0\ 0\ 0\ 0\ 0\ 0\ 0\ 0\ 0\ 0\ 0\ 0\ 0\ 0\ 0).$$

Because $\mathbf{S}^* = \mathbf{0}$, the decoder outputs $\mathbf{v}^*$ as a valid codeword, i.e., it assumes $\mathbf{v}^* = \mathbf{v}$, declares a successful decoding, and stops the decoding.

Consider another received vector $\mathbf{r}_1 = (1\ 1\ 0\ 0\ 0\ 0\ 0\ 0\ 0\ 0\ 0\ 0\ 0\ 0\ 0\ 0)$ which contains two bit errors at the first two positions. The syndrome of $\mathbf{r}_1$ is:

$$\mathbf{S} = (S_0, S_1, \ldots, S_{18}, S_{19})$$
$$= \mathbf{r}_1 \cdot \mathbf{H}_{\mathrm{FG}}^T$$
$$= (0\ 0\ 0\ 0\ 1\ 1\ 0\ 0\ 1\ 1\ 0\ 0\ 1\ 1\ 0\ 0\ 1\ 1\ 0\ 0).$$

Because $\mathbf{S} \neq \mathbf{0}$, the decoder detects the existence of errors in the received vector $\mathbf{r}_1$ and proceeds to correct the errors. Following the calculation given above, the five syndrome check-sums orthogonal on the zeroth bit are:

$$S_0 = 0, S_4 = 1, S_8 = 1, S_{12} = 1, S_{16} = 1,$$

among which there are four syndrome check-sums equaling 1 (more than half). Thus, the error bit $e_0$ is decoded as "1." Correction is then made to the received vector $\mathbf{r}_1$ to produce a decoded vector $\mathbf{v}^* = (r_0 + e_0, r_1, r_2, \ldots, r_{n-1}) = (0\ 1\ 0\ 0\ 0\ 0\ 0\ 0\ 0\ 0\ 0\ 0\ 0\ 0\ 0\ 0)$. To check whether the decoded vector $\mathbf{v}^*$ is a valid codeword, the decoder calculates its syndrome:

$$\mathbf{S}^* = (S_0, S_1, \ldots, S_{18}, S_{19})$$
$$= \mathbf{v}^* \cdot \mathbf{H}_{\mathrm{FG}}^T$$
$$= (1\ 0\ 0\ 0\ 0\ 1\ 0\ 0\ 0\ 1\ 0\ 0\ 0\ 1\ 0\ 0\ 0\ 1\ 0\ 0).$$

Because $\mathbf{S}^* \neq \mathbf{0}$, the decoder continues to make corrections on the first received bit $r_1$ in $\mathbf{r}$. The five syndrome check-sums orthogonal on the first bit $r_1$ are

$$S_1 = 0, S_5 = 1, S_9 = 1, S_{13} = 1, S_{17} = 1.$$

Because four syndrome check-sums (more than half) orthogonal on the first bit $r_1$ are equal to 1, the first error bit $e_1$ is decoded as "1." The decoded vector is $\mathbf{v}^* = (r_0 + e_0, r_1 + e_1, r_2, \ldots, r_{n-1}) = (0\ 0\ 0\ 0\ 0\ 0\ 0\ 0\ 0\ 0\ 0\ 0\ 0\ 0\ 0\ 0)$ and its syndrome $\mathbf{S}^* = \mathbf{v}^* \cdot \mathbf{H}^T = \mathbf{0}$. Because $\mathbf{S}^* = \mathbf{0}$, the decoder stops, outputs $\mathbf{v}^*$ as the decoded codeword, and declares a decoding success.

Consider another case where the received vector is $\mathbf{r}_2 = (1\ 1\ 1\ 0\ 0\ 0\ 0\ 0$ $0\ 0\ 0\ 0\ 0\ 0\ 0\ 0)$. Following the steps in the OSMLD, we find that at the end of the decoding process, the decoded vector is $\mathbf{v}^* = (0\ 0\ 0\ 1\ 0\ 1\ 0\ 0\ 0\ 1\ 1\ 0$ $0\ 1\ 1\ 1)$. However, the syndrome of the decoded vector $\mathbf{v}^*$ is $\mathbf{S}^* = \mathbf{v}^* \cdot \mathbf{H}^T =$ $(0\ 1\ 0\ 0\ 0\ 1\ 0\ 0\ 0\ 0\ 0\ 1\ 0\ 0\ 0\ 1\ 1\ 0\ 0\ )$, which is not a zero vector. Hence, the decoding fails. In this case, because the number of errors in the received vector is larger than the correcting capability ($\lfloor \gamma/2 \rfloor = \lfloor 5/2 \rfloor = 2$) of the OSMLD, the OSMLD is not guaranteed to correct these errors.

Suppose $\mathbf{r}_3 = (1\ 1\ 1\ 1\ 0\ 0\ 0\ 0\ 0\ 0\ 0\ 0\ 0\ 0\ 0\ 0)$ is received. Using the OSMLD, the decoded vector $\mathbf{v}^*$ is $\mathbf{v}^* = (1\ 1\ 1\ 1\ 0\ 0\ 0\ 0\ 0\ 0\ 1\ 0\ 0\ 0\ 1\ 0)$ which is a valid codeword because $\mathbf{v}^* \cdot \mathbf{H}^T = \mathbf{0}$. The decoder assumes that the decoded vector is the transmitted codeword and accepts it. However, $\mathbf{v}^* \neq \mathbf{v}$. In this case, the decoder commits a decoding error.                                                  ▲▲

Codes constructed based on finite geometry are OSMLD decodable. Because this class of codes has a large minimum distance (i.e., $\gamma$ is large), decoding FG codes with OSMLD provides excellent error-correcting performance considering its low decoding complexity. In the following sections, two specific families of finite geometries and their associated codes will be presented.

## 7.3 Euclidean Geometries over Finite Fields

In this section, the first category of finite geometries, called *Euclidean geometries* (also known as *affine geometries*), is presented. The section starts with the concepts of points, lines, and flats of Euclidean geometries. A realization of Euclidean geometries from finite fields is demonstrated. At the end of this section, subgeometries with *cyclic structures* of Euclidean geometries are introduced.

### 7.3.1 Basic Concepts and Properties

Let $\mathrm{GF}(q)$ be a Galois field with $q$ elements, where $q$ is a power of a prime $p$, say $q = p^m$ where $m$ is a nonnegative integer. Denote an $m$-tuple over $\mathrm{GF}(q)$ by $\mathbf{a} = (a_0, a_1, \ldots, a_{m-1})$, where $a_i$, $0 \leq i < m$, is an element in $\mathrm{GF}(q)$. There are $q^m$ such $m$-tuples and they form a *vector space* $\mathbf{V}_m$ of dimension $m$ over $\mathrm{GF}(q)$ (see Section 2.8). Let $\mathbf{a}$ and $\mathbf{b}$ be two $m$-tuples in $\mathbf{V}_m$ and $\beta$ be an element in $\mathrm{GF}(q)$. Recall that the vector addition, scalar multiplication, and inner product of vectors are defined as follows (see Section 2.8):

$$(a_0, a_1, \ldots, a_{m-1}) + (b_0, b_1, \ldots, b_{m-1}) = (a_0 + b_0, a_1 + b_1, \ldots, a_{m-1} + b_{m-1}),$$
$$\beta \cdot (a_0, a_1, \ldots, a_{m-1}) = (\beta \cdot a_0, \beta \cdot a_1, \ldots, \beta \cdot a_{m-1}),$$
$$\langle (a_0, a_1, \ldots, a_{m-1}), (b_0, b_1, \ldots, b_{m-1}) \rangle = \sum_{i=0}^{m-1} a_i b_i,$$

where the addition $a_i + b_i$ and multiplication $\beta \cdot a_i$ and $a_i b_i$ are carried out in $\mathrm{GF}(q)$ (i.e., addition and multiplication defined on $\mathrm{GF}(q)$). If the inner product of $\mathbf{a}$ and $\mathbf{b}$ is $\langle \mathbf{a}, \mathbf{b} \rangle = 0$, the two vectors $\mathbf{a}$ and $\mathbf{b}$ are said to be *orthogonal*.

In combinatorial mathematics, the $q^m$ $m$-tuples over $\mathrm{GF}(q)$ are known to form an $m$-dimensional *Euclidean geometry* over $\mathrm{GF}(q)$, denoted by $\mathrm{EG}(m, q)$. Each $m$-tuple $\mathbf{a} = (a_0, a_1, \ldots, a_{m-1})$ is regarded as a *point* and the all-zero $m$-tuple $\mathbf{0} = (0, 0, \ldots, 0)$ is regarded as the *origin* of the geometry. Hence, the total number of points in $\mathrm{EG}(m, q)$ is

$$n = q^m. \tag{7.8}$$

Let $\mathbf{a}$ be a *nonorigin point* in $\mathrm{EG}(m, q)$. Then, the following collection of $q$ points:

$$\{\beta \mathbf{a}\} \triangleq \{\beta \mathbf{a} : \beta \in \mathrm{GF}(q)\} \tag{7.9}$$

forms a *line* in $\mathrm{EG}(m, q)$. Because for $\beta = 0$, $0 \cdot \mathbf{a} = \mathbf{0}$, the line $\{\beta \mathbf{a}\}$ contains the origin $\mathbf{0}$ as a point. We say that the line $\{\beta \mathbf{a}\}$ passes through the origin. The $q$ points on the line $\{\beta \mathbf{a}\}$ actually form a *one-dimensional subspace* $\mathbf{W}_1$ of the vector space $\mathbf{V}_m$ of all the $m$-tuples over $\mathrm{GF}(q)$.

Let $\mathbf{a}_0$ be a nonzero point in $\mathrm{EG}(m, q)$ that is linearly independent of $\mathbf{a}$, i.e., $\mathbf{a}_0 \neq \beta \mathbf{a}$ for any $\beta$ in $\mathrm{GF}(q)$. Then, the following collection of $q$ points:

$$\{\mathbf{a}_0 + \beta \mathbf{a}\} \triangleq \{\mathbf{a}_0 + \beta \mathbf{a} : \beta \in \mathrm{GF}(q)\} \tag{7.10}$$

forms a *line* in $\mathrm{EG}(m, q)$ that passes through the point $\mathbf{a}_0$. The $q$ points in $\{\mathbf{a}_0 + \beta \mathbf{a}\}$ actually form a *coset* of the one-dimensional subspace $\mathbf{W}_1$ of the vector space $\mathbf{V}_m$. There are $q^{m-1}$ cosets of $\mathbf{W}_1$ (including $\mathbf{W}_1$ itself). Therefore, a line in $\mathrm{EG}(m, q)$ is either a one-dimensional subspace $\mathbf{W}_1$ of the vector space $\mathbf{V}_m$ of all the $m$-tuples over $\mathrm{GF}(q)$ or a coset of $\mathbf{W}_1$. The two lines $\{\beta \mathbf{a}\}$ and $\{\mathbf{a}_0 + \beta \mathbf{a}\}$ have no common point and hence they are parallel lines.

Because the line $\{\beta \mathbf{a}\}$ corresponds to a one-dimensional subspace of the $m$-dimensional vector space $\mathbf{V}_m$ which has $q^{m-1}$ cosets including itself (see Lagrange's theorem given by Theorem 2.1), the line $\{\beta \mathbf{a}\}$ has $q^{m-1} - 1$ lines in $\mathrm{EG}(m, q)$ parallel to it. The line $\{\beta \mathbf{a}\}$ and the lines parallel to it form *a bundle of parallel lines* in $\mathrm{EG}(m, q)$, called a *parallel bundle*. Therefore, a parallel bundle in $\mathrm{EG}(m, q)$ consists of

$$q^{m-1} \tag{7.11}$$

parallel lines among which one line passes through the origin. Because each line in $\mathrm{EG}(m, q)$ consists of $q$ points, a parallel bundle of $q^{m-1}$ lines contains all the $q^m$ points of $\mathrm{EG}(m, q)$, and each point appears on *one and only one* line.

Let $\mathbf{a}_0$, $\mathbf{a}_1$, and $\mathbf{a}_2$ be three linearly independent points in $\mathrm{EG}(m, q)$. The two lines $\{\mathbf{a}_0 + \beta \mathbf{a}_1\}$ and $\{\mathbf{a}_0 + \beta \mathbf{a}_2\}$ have only the point $\mathbf{a}_0$ in common. Hence, these two lines intersect at the point $\mathbf{a}_0$. The two lines $\{\beta \mathbf{a}_1\}$ and $\{\beta \mathbf{a}_2\}$ intersect at the origin $\mathbf{0}$. Given a point $\mathbf{a}_0$ in $\mathrm{EG}(m, q)$, there are

$$\frac{q^m - 1}{q - 1} \tag{7.12}$$

lines in $\mathrm{EG}(m,q)$ intersecting at $\mathbf{a}_0$ (including the line $\{\beta\mathbf{a}_0\}$ that passes through the origin of $\mathrm{EG}(m,q)$). These $(q^m-1)/(q-1)$ lines form a bundle of lines which intersect at the point $\mathbf{a}_0$, called an *intersecting bundle*. An intersecting bundle of lines contains $q^m$ points of $\mathrm{EG}(m,q)$ and each point except the intersecting point is on *one and only one* line in the intersecting bundle.

If the point $\mathbf{a}_0$ is the origin of $\mathrm{EG}(m,q)$, then the lines in $\mathrm{EG}(m,q)$ passing through the origin correspond to all different one-dimensional subspaces of the vector space $\mathbf{V}_m$ of all the $m$-tuples over $\mathrm{GF}(q)$. Hence, there are $(q^m-1)/(q-1)$ different one-dimensional subspaces in the $m$-dimensional vector space $\mathbf{V}_m$. Because each one-dimensional subspace of $\mathbf{V}_m$ has $q^{m-1}$ cosets that form a parallel bundle of $q^{m-1}$ lines in $\mathrm{EG}(m,q)$, the total number of lines in $\mathrm{EG}(m,q)$ is

$$M = \frac{q^{m-1}(q^m - 1)}{q - 1} \tag{7.13}$$

which is the *product* of the number of lines in $\mathrm{EG}(m,q)$ passing through the origin (or any point) and the number of lines in a parallel bundle of $\mathrm{EG}(m,q)$. It follows from (7.12) and (7.13) that the lines in $\mathrm{EG}(m,q)$ can be partitioned into $(q^m - 1)/(q - 1)$ parallel bundles.

Let $\mathbf{a}_1$ and $\mathbf{a}_2$ be two points in $\mathrm{EG}(m,q)$. If $\mathbf{a}_1$ and $\mathbf{a}_2$ are linearly dependent, then $\mathbf{a}_2 = \beta\mathbf{a}_1$ for some element $\beta$ in $\mathrm{GF}(q)$. In this case, the two points $\mathbf{a}_1$ and $\mathbf{a}_2$ are connected by the line $\{\beta\mathbf{a}_1\}$. Suppose $\mathbf{a}_1$ and $\mathbf{a}_2$ are linearly independent. Then, $\mathbf{a}_1$ and $\mathbf{a}_2$ are connected by the line $\{\mathbf{a}_1 + \beta(\mathbf{a}_2 - \mathbf{a}_1)\}$, where $-\mathbf{a}_1$ is the additive inverse $m$-tuple of $\mathbf{a}_1$. Therefore, any two points in $\mathrm{EG}(m,q)$ are connected by a line in $\mathrm{EG}(m,q)$.

The above developments show that $\mathrm{EG}(m,q)$ has all the fundamental properties of a finite geometry. More structural properties of an Euclidean geometry will be developed in the following.

For $1 \leq \mu \leq m$, let $\mathbf{a}_1, \mathbf{a}_2, \ldots, \mathbf{a}_\mu$ be $\mu$ linearly independent vectors in the $m$-dimensional vector space $\mathbf{V}_m$ of all the $m$-tuples over $\mathrm{GF}(q)$. The set

$$\{\beta_1\mathbf{a}_1 + \beta_2\mathbf{a}_2 + \cdots + \beta_\mu\mathbf{a}_\mu\} \triangleq \{\beta_1\mathbf{a}_1 + \beta_2\mathbf{a}_2 + \cdots + \beta_\mu\mathbf{a}_\mu : \beta_i \in \mathrm{GF}(q), 1 \leq i \leq \mu\}$$

of $q^\mu$ linear combinations of the vectors $\mathbf{a}_1, \mathbf{a}_2, \ldots, \mathbf{a}_\mu$ forms a *$\mu$-dimensional subspace* $\mathbf{W}_\mu$ of $\mathbf{V}_m$. If $\mathbf{a}_0$ is not an $m$-tuple in $\{\beta_1\mathbf{a}_1 + \beta_2\mathbf{a}_2 + \cdots + \beta_\mu\mathbf{a}_\mu\}$, then the set

$$\{\mathbf{a}_0 + \beta_1\mathbf{a}_1 + \cdots + \beta_\mu\mathbf{a}_\mu\} \triangleq \{\mathbf{a}_0 + \beta_1\mathbf{a}_1 + \cdots + \beta_\mu\mathbf{a}_\mu : \beta_i \in \mathrm{GF}(q), 1 \leq i \leq \mu\}$$

of $q^\mu$ $m$-tuples forms a coset of $\{\beta_1\mathbf{a}_1 + \beta_2\mathbf{a}_2 + \cdots + \beta_\mu\mathbf{a}_\mu\}$. Regarding an $m$-tuple in $\mathbf{V}_m$ as a point in $\mathrm{EG}(m,q)$, the set $\{\beta_1\mathbf{a}_1 + \beta_2\mathbf{a}_2 + \cdots + \beta_\mu\mathbf{a}_\mu\}$ of $q^\mu$ points in $\mathrm{EG}(m,q)$ forms a $\mu$-dimensional *hyperplane*, commonly called a *$\mu$-flat*, which contains the origin of $\mathrm{EG}(m,q)$. The set $\{\mathbf{a}_0 + \beta_1\mathbf{a}_1 + \beta_2\mathbf{a}_2 + \cdots + \beta_\mu\mathbf{a}_\mu\}$ of $q^\mu$ points in $\mathrm{EG}(m,q)$ also forms a $\mu$-flat in $\mathrm{EG}(m,q)$. Because the two $\mu$-flats $\{\beta_1\mathbf{a}_1 + \beta_2\mathbf{a}_2 + \cdots + \beta_\mu\mathbf{a}_\mu\}$ and $\{\mathbf{a}_0 + \beta_1\mathbf{a}_1 + \beta_2\mathbf{a}_2 + \cdots + \beta_\mu\mathbf{a}_\mu\}$ do not have any point in common, they are parallel.

It follows from the definition given above that a $\mu$-flat in $\mathrm{EG}(m, q)$ is either a $\mu$-dimensional subspace of $\mathbf{V}_m$ or a coset of a $\mu$-dimensional subspace of $\mathbf{V}_m$. A 1-flat is a line in $\mathrm{EG}(m, q)$ and the $m$-flat is the entire geometry $\mathrm{EG}(m, q)$. For each $\mu$-flat in $\mathrm{EG}(m, q)$, there are $q^{m-u} - 1$ $\mu$-flats parallel to it and they together form a parallel bundle of $\mu$-flats in $\mathrm{EG}(m, q)$. A $\mu$-flat contains

$$\frac{q^{\mu-1}(q^{\mu} - 1)}{q - 1} \tag{7.14}$$

lines.

For $1 \leq \mu < m$, let $\mathbf{a}_{\mu+1}$ be a point not on the $\mu$-flat $\{\mathbf{a}_0 + \beta_1\mathbf{a}_1 + \beta_2\mathbf{a}_2 + \cdots + \beta_\mu\mathbf{a}_\mu\}$ and $\beta_{\mu+1}$ be an element in $\mathrm{GF}(q)$. Then, the $(\mu + 1)$-flat $\{\mathbf{a}_0 + \beta_1\mathbf{a}_1 + \beta_2\mathbf{a}_2 + \cdots + \beta_\mu\mathbf{a}_\mu + \beta_{\mu+1}\mathbf{a}_{\mu+1}\}$ contains the $\mu$-flat $\{\mathbf{a}_0 + \beta_1\mathbf{a}_1 + \beta_2\mathbf{a}_2 + \cdots + \beta_\mu\mathbf{a}_\mu\}$. Let $\mathbf{b}_{\mu+1}$ be a point in $\mathrm{EG}(m, q)$ but not on the $(\mu+1)$-flat $\{\mathbf{a}_0 + \beta_1\mathbf{a}_1 + \beta_2\mathbf{a}_2 + \cdots + \beta_\mu\mathbf{a}_\mu + \beta_{\mu+1}\mathbf{a}_{\mu+1}\}$. Then, the two $(\mu + 1)$-flats $\{\mathbf{a}_0 + \beta_1\mathbf{a}_1 + \beta_2\mathbf{a}_2 + \cdots + \beta_\mu\mathbf{a}_\mu + \beta_{\mu+1}\mathbf{a}_{\mu+1}\}$ and $\{\mathbf{a}_0 + \beta_1\mathbf{a}_1 + \beta_2\mathbf{a}_2 + \cdots + \beta_\mu\mathbf{a}_\mu + \beta_{\mu+1}\mathbf{b}_{\mu+1}\}$ intersect on the $\mu$-flat $\{\mathbf{a}_0 + \beta_1\mathbf{a}_1 + \beta_2\mathbf{a}_2 + \cdots + \beta_\mu\mathbf{a}_\mu\}$, i.e., they have the points on the $\mu$-flat $\{\mathbf{a}_0 + \beta_1\mathbf{a}_1 + \beta_2\mathbf{a}_2 + \cdots + \beta_\mu\mathbf{a}_\mu\}$ as the common points. Given a $\mu$-flat in $\mathrm{EG}(m, q)$, there are

$$\frac{q^{m-\mu} - 1}{q - 1} \tag{7.15}$$

$(\mu + 1)$-flats in $\mathrm{EG}(m, q)$ intersecting on the $\mu$-flat. It can be shown that the total number of $\mu$-flats in $\mathrm{EG}(m, q)$ is [20, 21]

$$q^{m-\mu} \cdot \prod_{i=1}^{\mu} \frac{q^{m-i+1} - 1}{q^{\mu-i+1} - 1}. \tag{7.16}$$

Note that for $\mu = 0$, the 0-flat is a point.

## 7.3.2 A Realization of Euclidean Geometries

Consider the extension field $\mathrm{GF}(q^m)$ of $\mathrm{GF}(q)$. Next, we show that $\mathrm{GF}(q^m)$ is a realization of the $m$-dimensional Euclidean geometry $\mathrm{EG}(m, q)$ over $\mathrm{GF}(q)$ with its elements as the points of $\mathrm{EG}(m, q)$. Let $\alpha$ be a primitive element of $\mathrm{GF}(q^m)$. Then, the powers of $\alpha$, namely, $\alpha^{-\infty} = 0$, $\alpha^0 = 1$, $\alpha, \alpha^2, \ldots, \alpha^{q^m-2}$, give the $q^m$ elements of $\mathrm{GF}(q^m)$. Every element $\alpha^j$ in $\mathrm{GF}(q^m)$ can be expressed as a polynomial in $\alpha$ of degree less than $m$ as follows (see Section 2.4):

$$\alpha^j = a_{j,0} + a_{j,1}\alpha + a_{j,2}\alpha^2 + \cdots + a_{j,m-2}\alpha^{m-2} + a_{j,m-1}\alpha^{m-1}, \tag{7.17}$$

where $a_{j,k} \in \mathrm{GF}(q)$ for $0 \leq k < m$. There is a *one-to-one correspondence* between the element $\alpha^j$ in $\mathrm{GF}(q^m)$ and the $m$-tuple $(a_{j,0}, a_{j,1}, \ldots, a_{j,m-2}, a_{j,m-1})$ over $\mathrm{GF}(q)$. Therefore, $\mathrm{GF}(q^m)$ may be regarded as a vector space $\mathbf{V}_m$ of all the $m$-tuples over $\mathrm{GF}(q)$ and hence as an $m$-dimensional Euclidean geometry $\mathrm{EG}(m, q)$ over $\mathrm{GF}(q)$. In this case, the $q^m$ elements of $\mathrm{GF}(q^m)$, namely,

$\alpha^{-\infty} = 0, \alpha^0 = 1, \alpha, \alpha^2, \ldots, \alpha^{q^m-2}$, represent the $q^m$ points of $\mathrm{EG}(m,q)$. The element $\alpha^j$ in $\mathrm{GF}(q^m)$ represents the $j$th point of $\mathrm{EG}(m,q)$. The origin of $\mathrm{EG}(m,q)$ is represented by $\alpha^{-\infty} = 0$ in $\mathrm{GF}(q^m)$. Hereafter, we label the points of $\mathrm{EG}(m,q)$ by the powers of $\alpha$ in the order of $-\infty, 0, 1, 2, \ldots, q^m - 2$.

Let $\alpha^{j_1}$ be a nonorigin point in $\mathrm{EG}(m,q)$. Then, the set

$$\{\beta\alpha^{j_1} : \beta \in \mathrm{GF}(q)\} \tag{7.18}$$

of $q$ points forms a line passing through the origin. From a vector point of view, $\{\beta\alpha^{j_1} : \beta \in \mathrm{GF}(q)\}$ is a one-dimensional subspace of the vector space $\mathbf{V}_m$ over $\mathrm{GF}(q)$. Let $\alpha^{j_0}$ be a point linearly independent of $\alpha^{j_1}$, Then, the set

$$\{\alpha^{j_0} + \beta\alpha^{j_1} : \beta \in \mathrm{GF}(q)\} \tag{7.19}$$

of $q$ points forms a line passing through the point $\alpha^{j_0}$. From a vector point of view, $\{\alpha^{j_0} + \beta\alpha^{j_1} : \beta \in \mathrm{GF}(q)\}$ is a coset of $\{\beta\alpha^{j_1} : \beta \in \mathrm{GF}(q)\}$. The line $\{\alpha^{j_0} + \beta\alpha^{j_1} : \beta \in \mathrm{GF}(q)\}$ is parallel to the line $\{\beta\alpha^{j_1} : \beta \in \mathrm{GF}(q)\}$, because they have no point in common.

Let $\alpha^{j_0}, \alpha^{j_1}$, and $\alpha^{j_2}$ be three linearly independent points of $\mathrm{EG}(m,q)$. Then, the two lines $\{\alpha^{j_0} + \beta\alpha^{j_1} : \beta \in \mathrm{GF}(q)\}$ and $\{\alpha^{j_0} + \beta\alpha^{j_2} : \beta \in \mathrm{GF}(q)\}$ intersect at the point $\alpha^{j_0}$. For $0 \leq \mu < m$, let $\alpha^{j_0}, \alpha^{j_1}, \ldots, \alpha^{j_\mu}$ be $\mu+1$ linearly independent points in $\mathrm{EG}(m,q)$. Then, the set

$$\{\alpha^{j_0} + \beta_{j_1}\alpha^{j_1} + \beta_{j_2}\alpha^{j_2} + \cdots + \beta_{j_\mu}\alpha^{j_\mu} : \beta_{j_i} \in \mathrm{GF}(q), 1 \leq i \leq \mu\} \tag{7.20}$$

of $q^\mu$ points forms a $\mu$-flat of $\mathrm{EG}(m,q)$.

**Example 7.3** Consider the two-dimensional Euclidean geometry $\mathrm{EG}(2,2^2)$ over $\mathrm{GF}(2^2)$. This geometry consists of $2^{2\times 2} = 16$ points. Each point in $\mathrm{EG}(2,2^2)$ is a 2-tuple over $\mathrm{GF}(2^2)$ and each line in $\mathrm{EG}(2,2^2)$ consists of four points. Consider the Galois field $\mathrm{GF}(2^4)$ generated by the primitive polynomial $\mathbf{p}(X) = 1 + X + X^4$ over $\mathrm{GF}(2)$ given by Table 7.1. Let $\alpha$ be a primitive element of $\mathrm{GF}(2^4)$ and $\beta = \alpha^5$. The order of $\beta$ is 3. The set

$$\mathrm{GF}(2^2) = \{0, \beta^0 = 1, \beta = \alpha^5, \beta^2 = \alpha^{10}\}$$

forms the subfield $\mathrm{GF}(2^2)$ of $\mathrm{GF}(2^4)$. Every element $\alpha^i$ in $\mathrm{GF}(2^4)$ can be expressed as the following polynomial form in $\alpha$ of degree less than 2:

$$\alpha^i = \beta_0\alpha^0 + \beta_1\alpha = \beta_0 + \beta_1\alpha,$$

with $\beta_0, \beta_1 \in \mathrm{GF}(2^2)$. In vector form, $\alpha^i$ is represented by a 2-tuple $(\beta_0, \beta_1)$ over $\mathrm{GF}(2^2)$. The field $\mathrm{GF}(2^4)$ as an extension field of $\mathrm{GF}(2^2)$ is given by Table 7.2. Regard $\mathrm{GF}(2^4)$ as a realization of the two-dimensional Euclidean geometry $\mathrm{EG}(2,2^2)$ over $\mathrm{GF}(2^2)$. Then, the set

$$\begin{aligned} L_0 &= \{\beta_1\alpha\} = \{\beta_1\alpha : \beta_1 \in \mathrm{GF}(2^2)\} \\ &= \{0 \cdot \alpha, \beta^0 \cdot \alpha, \beta^1 \cdot \alpha, \beta^2 \cdot \alpha\} \\ &= \{0, \alpha, \alpha^6, \alpha^{11}\} \end{aligned}$$

of four points forms a line passing through the origin of $\text{EG}(2, 2^2)$. The three lines parallel to $L_0$ are (using Table 7.1 or Table 7.2)

$$
\begin{aligned}
L_1 &= \{1 + \beta_1 \alpha\} &&= \{1, \alpha^4, \alpha^{12}, \alpha^{13}\}, \\
L_2 &= \{\alpha^5 + \beta_1 \alpha\} &&= \{\alpha^2, \alpha^3, \alpha^5, \alpha^9\}, \\
L_3 &= \{\alpha^{10} + \beta_1 \alpha\} &&= \{\alpha^8, \alpha^7, \alpha^{10}, \alpha^{14}\}.
\end{aligned}
$$

The four lines $L_0$, $L_1$, $L_2$, and $L_3$ together form a parallel bundle of lines in $\text{EG}(2, 2^2)$, which contains all the points in $\text{EG}(2, 2^2)$, i.e., all the elements in $\text{GF}(2^4)$.

For every point in $\text{EG}(2, 2^2)$, there are $(2^{2 \times 2} - 1)/(2^2 - 1) = 5$ lines intersecting at it. For example, consider the point $\alpha$ in $\text{EG}(2, 2^2)$. The lines that intersect at $\alpha$ are

$$
\begin{aligned}
L_0' &= \{\beta_1 \alpha\} &&= \{0, \alpha, \alpha^6, \alpha^{11}\}, \\
L_1' &= \{\alpha + \beta_1 \alpha^0\} &&= \{\alpha, \alpha^2, \alpha^4, \alpha^8\}, \\
L_2' &= \{\alpha + \beta_1 \alpha^2\} &&= \{\alpha, \alpha^5, \alpha^{13}, \alpha^{14}\}, \\
L_3' &= \{\alpha + \beta_1 \alpha^3\} &&= \{\alpha, \alpha^9, \alpha^{10}, \alpha^{12}\}, \\
L_4' &= \{\alpha + \beta_1 \alpha^4\} &&= \{1, \alpha, \alpha^3, \alpha^7\}.
\end{aligned}
$$

The five lines, $L_0'$, $L_1'$, $L_2'$, $L_3'$, and $L_4'$, form an intersecting bundle of $\text{EG}(2, 2^2)$, which contains all the points in $\text{EG}(2, 2^2)$. In this intersecting bundle, except the intersected point $\alpha$, each point in $\text{EG}(2, 2^2)$ is on one and only one line. ▲▲

**Example 7.4** In this example, a three-dimensional Euclidean geometry $\text{EG}(3, 2^2)$ over $\text{GF}(2^2)$ is considered. There are $2^{2 \times 3} = 64$ points in $\text{EG}(3, 2^2)$ where each point is a 3-tuple over $\text{GF}(2^2)$. There are $4^{3-1}(4^3 - 1)/(4 - 1) = 336$ lines in $\text{EG}(3, 2^2)$, each consisting of $q = 4$ points. From (7.16), $\text{EG}(3, 2^2)$ has 84 2-flats in which each 2-flat contains 16 points and 20 lines. Based on the previous development, the finite field $\text{GF}(2^6)$ is a realization of $\text{EG}(3, 2^2)$. Let $\mathbf{p}(X) = 1 + X + X^6$ be the primitive polynomial over $\text{GF}(2)$ which generates $\text{GF}(2^6)$. Let $\alpha$ be a primitive element of $\text{GF}(2^6)$ and $\beta = \alpha^{21}$. Then, $\beta$ has order 3 and its powers, namely, $\beta^{-\infty} = 0$, $\beta^0 = 1$, $\beta = \alpha^{21}$, $\beta^2 = \alpha^{42}$, form the subfield $\text{GF}(2^2)$ of $\text{GF}(2^6)$, i.e., $\text{GF}(2^2) = \{\beta^{-\infty} = 0, \beta^0 = 1, \beta = \alpha^{21}, \beta^2 = \alpha^{42}\}$. Every element $\alpha^i$ in $\text{GF}(2^6)$ can be represented by a polynomial of $\alpha$ with degree less than 3:

$$
\alpha^i = \beta_0 + \beta_1 \alpha + \beta_2 \alpha^2,
$$

where $\beta_0, \beta_1, \beta_2 \in \text{GF}(2^2)$. In vector form, we use a 3-tuple $(\beta_0, \beta_1, \beta_2)$ over $\text{GF}(2^2)$ to represent $\alpha^i$. The field $\text{GF}(2^6)$ as an extension field of $\text{GF}(2)$ is given in Table 5.14 and as an extension field of $\text{GF}(2^2)$ is provided in Table 7.3.

With $\text{GF}(2^6)$ as the realization of $\text{EG}(3, 2^2)$, the set

$$
L_0 = \{\beta_1 \alpha^0 : \beta_1 \in \text{GF}(2^2)\} = \{0, 1, \alpha^{21}, \alpha^{42}\}
$$

Table 7.1 GF($2^4$) generated by $\mathbf{p}(X) = 1 + X + X^4$ over GF(2).

| Power form | Polynomial form | | | | Vector form | Decimal form |
|---|---|---|---|---|---|---|
| $0 = \alpha^{-\infty}$ | 0 | | | | $(0\ 0\ 0\ 0)$ | 0 |
| $1 = \alpha^0$ | 1 | | | | $(1\ 0\ 0\ 0)$ | 1 |
| $\alpha$ | | $\alpha$ | | | $(0\ 1\ 0\ 0)$ | 2 |
| $\alpha^2$ | | | $\alpha^2$ | | $(0\ 0\ 1\ 0)$ | 4 |
| $\alpha^3$ | | | | $\alpha^3$ | $(0\ 0\ 0\ 1)$ | 8 |
| $\alpha^4$ | 1 | $+\ \alpha$ | | | $(1\ 1\ 0\ 0)$ | 3 |
| $\alpha^5$ | | $\alpha$ | $+\ \alpha^2$ | | $(0\ 1\ 1\ 0)$ | 6 |
| $\alpha^6$ | | | $\alpha^2$ | $+\ \alpha^3$ | $(0\ 0\ 1\ 1)$ | 12 |
| $\alpha^7$ | 1 | $+\ \alpha$ | | $+\ \alpha^3$ | $(1\ 1\ 0\ 1)$ | 11 |
| $\alpha^8$ | 1 | | $+\ \alpha^2$ | | $(1\ 0\ 1\ 0)$ | 5 |
| $\alpha^9$ | | $\alpha$ | | $+\ \alpha^3$ | $(0\ 1\ 0\ 1)$ | 10 |
| $\alpha^{10}$ | 1 | $+\ \alpha$ | $+\ \alpha^2$ | | $(1\ 1\ 1\ 0)$ | 7 |
| $\alpha^{11}$ | | $\alpha$ | $+\ \alpha^2$ | $+\ \alpha^3$ | $(0\ 1\ 1\ 1)$ | 14 |
| $\alpha^{12}$ | 1 | $+\ \alpha$ | $+\ \alpha^2$ | $+\ \alpha^3$ | $(1\ 1\ 1\ 1)$ | 15 |
| $\alpha^{13}$ | 1 | | $+\ \alpha^2$ | $+\ \alpha^3$ | $(1\ 0\ 1\ 1)$ | 13 |
| $\alpha^{14}$ | 1 | | | $+\ \alpha^3$ | $(1\ 0\ 0\ 1)$ | 9 |

Table 7.2 GF($2^4$) as an extension field of GF($2^2$) $= \{0, 1, \beta, \beta^2\}$ with $\beta = \alpha^5$.

| Power form | Polynomial form | | Vector form | Power form | Polynomial form | | Vector form |
|---|---|---|---|---|---|---|---|
| 0 | 0 | | $(0,0)$ | $\alpha^7$ | $\beta^2 +$ | $\beta\alpha$ | $(\beta^2, \beta)$ |
| 1 | 1 | | $(1,0)$ | $\alpha^8$ | $\beta^2 +$ | $\alpha$ | $(\beta^2, 1)$ |
| $\alpha$ | | $\alpha$ | $(1,0)$ | $\alpha^9$ | $\beta +$ | $\beta\alpha$ | $(\beta, \beta)$ |
| $\alpha^2$ | $\beta +$ | $\alpha$ | $(\beta, 1)$ | $\alpha^{10}$ | $\beta^2$ | | $(\beta^2, 0)$ |
| $\alpha^3$ | $\beta +$ | $\beta^2\alpha$ | $(\beta, \beta^2)$ | $\alpha^{11}$ | | $\beta^2\alpha$ | $(0, \beta^2)$ |
| $\alpha^4$ | $1 +$ | $\alpha$ | $(1,1)$ | $\alpha^{12}$ | $1 +$ | $\beta^2\alpha$ | $(1, \beta^2)$ |
| $\alpha^5$ | $\beta$ | | $(\beta, 0)$ | $\alpha^{13}$ | $1 +$ | $\beta\alpha$ | $(1, \beta)$ |
| $\alpha^6$ | | $\beta\alpha$ | $(0, \beta)$ | $\alpha^{14}$ | $\beta^2 +$ | $\beta^2\alpha$ | $(\beta^2, \beta^2)$ |

forms a line in EG($3, 2^2$) which passes through the origin of EG($3, 2^2$). Choosing another point $\alpha$ which is linearly independent of $\alpha^0$ (i.e., $\alpha$ is not on the line $L_0$), form the following line (using Table 5.14):

$$L_1 = \{\alpha + \beta_1\alpha^0\} = \{\alpha, \alpha^6, \alpha^{29}, \alpha^{60}\}.$$

Because $L_1$ and $L_0$ have no point in common, they are parallel. Choosing another point $\alpha^2$ that is not on $L_1$ nor $L_0$, we can form the following line which is parallel to $L_0$ and $L_1$:

$$L_2 = \{\alpha^2 + \beta_1\alpha^0\} = \{\alpha^2, \alpha^{12}, \alpha^{57}, \alpha^{58}\}.$$

Table 7.3 $GF(2^6)$ as an extension field of $GF(2^2) = \{0, 1, \beta, \beta^2\}$ with $\beta = \alpha^{21}$.

| Power form | Polynomial form | Vector form | Power form | Polynomial form | Vector form |
|---|---|---|---|---|---|
| $0 = \alpha^{-\infty}$ | $0$ | $(0,0,0)$ | $\alpha^{31}$ | $\beta + \beta^2\alpha + \beta\alpha^2$ | $(\beta,\beta^2,\beta)$ |
| $1 = \alpha^0$ | $1$ | $(1,0,0)$ | $\alpha^{32}$ | $\beta^2 + \beta^2\alpha + \alpha^2$ | $(\beta^2,\beta^2,1)$ |
| $\alpha$ | $\alpha$ | $(0,1,0)$ | $\alpha^{33}$ | $\beta + \beta\alpha^2$ | $(\beta,0,\beta)$ |
| $\alpha^2$ | $\alpha^2$ | $(0,0,1)$ | $\alpha^{34}$ | $\beta^2 + \beta^2\alpha + \beta\alpha^2$ | $(\beta^2,\beta^2,\beta)$ |
| $\alpha^3$ | $\beta + \beta^2\alpha + \alpha^2$ | $(\beta,\beta^2,1)$ | $\alpha^{35}$ | $\beta^2 + \beta\alpha + \alpha^2$ | $(\beta^2,\beta,1)$ |
| $\alpha^4$ | $\beta + \alpha + \beta\alpha^2$ | $(\beta,1,\beta)$ | $\alpha^{36}$ | $\beta + \beta^2\alpha^2$ | $(\beta,0,\beta^2)$ |
| $\alpha^5$ | $\beta^2 + \beta^2\alpha + \beta^2\alpha^2$ | $(\beta^2,\beta^2,\beta^2)$ | $\alpha^{37}$ | $1 + \beta^2\alpha^2$ | $(1,0,\beta^2)$ |
| $\alpha^6$ | $1 + \alpha$ | $(1,1,0)$ | $\alpha^{38}$ | $1 + \beta^2\alpha + \beta^2\alpha^2$ | $(1,\beta^2,\beta^2)$ |
| $\alpha^7$ | $\alpha + \alpha^2$ | $(0,1,1)$ | $\alpha^{39}$ | $1 + \beta^2\alpha$ | $(1,\beta^2,0)$ |
| $\alpha^8$ | $\beta + \beta^2\alpha$ | $(\beta,\beta^2,0)$ | $\alpha^{40}$ | $\alpha + \beta^2\alpha^2$ | $(0,1,\beta^2)$ |
| $\alpha^9$ | $\beta\alpha + \beta^2\alpha^2$ | $(0,\beta,\beta^2)$ | $\alpha^{41}$ | $1 + \beta\alpha + \beta\alpha^2$ | $(1,\beta,\beta)$ |
| $\alpha^{10}$ | $1 + \beta\alpha + \alpha^2$ | $(1,\beta,1)$ | $\alpha^{42}$ | $\beta^2$ | $(\beta^2,0,0)$ |
| $\alpha^{11}$ | $\beta + \beta\alpha + \beta^2\alpha^2$ | $(\beta,\beta,\beta^2)$ | $\alpha^{43}$ | $\beta^2\alpha$ | $(0,\beta^2,0)$ |
| $\alpha^{12}$ | $1 + \alpha^2$ | $(1,0,1)$ | $\alpha^{44}$ | $\beta^2\alpha^2$ | $(0,0,\beta^2)$ |
| $\alpha^{13}$ | $\beta + \beta\alpha + \alpha^2$ | $(\beta,\beta,1)$ | $\alpha^{45}$ | $1 + \beta\alpha + \beta^2\alpha^2$ | $(1,\beta,\beta^2)$ |
| $\alpha^{14}$ | $\beta + \alpha + \beta^2\alpha^2$ | $(\beta,1,\beta^2)$ | $\alpha^{46}$ | $1 + \beta^2\alpha + \alpha^2$ | $(1,\beta^2,1)$ |
| $\alpha^{15}$ | $1 + \beta\alpha^2$ | $(1,0,\beta)$ | $\alpha^{47}$ | $\beta + \beta\alpha + \beta\alpha^2$ | $(\beta,\beta,\beta)$ |
| $\alpha^{16}$ | $\beta^2 + \beta\alpha^2$ | $(\beta^2,0,\beta)$ | $\alpha^{48}$ | $\beta^2 + \beta^2\alpha$ | $(\beta^2,\beta^2,0)$ |
| $\alpha^{17}$ | $\beta^2 + \beta\alpha + \beta\alpha^2$ | $(\beta^2,\beta,\beta)$ | $\alpha^{49}$ | $\beta^2\alpha + \beta^2\alpha^2$ | $(0,\beta^2,\beta^2)$ |
| $\alpha^{18}$ | $\beta^2 + \beta\alpha$ | $(\beta^2,\beta,0)$ | $\alpha^{50}$ | $1 + \beta\alpha$ | $(1,\beta,0)$ |
| $\alpha^{19}$ | $\beta^2\alpha + \beta\alpha^2$ | $(0,\beta^2,\beta)$ | $\alpha^{51}$ | $\alpha + \beta\alpha^2$ | $(0,1,\beta)$ |
| $\alpha^{20}$ | $\beta^2 + \alpha + \alpha^2$ | $(\beta^2,1,1)$ | $\alpha^{52}$ | $\beta^2 + \alpha + \beta^2\alpha^2$ | $(\beta^2,1,\beta^2)$ |
| $\alpha^{21}$ | $\beta$ | $(\beta,0,0)$ | $\alpha^{53}$ | $1 + \alpha + \beta\alpha^2$ | $(1,1,\beta)$ |
| $\alpha^{22}$ | $\beta\alpha$ | $(0,\beta,0)$ | $\alpha^{54}$ | $\beta^2 + \beta^2\alpha^2$ | $(\beta^2,0,\beta^2)$ |
| $\alpha^{23}$ | $\beta\alpha^2$ | $(0,0,\beta)$ | $\alpha^{55}$ | $1 + \alpha + \beta^2\alpha^2$ | $(1,1,\beta^2)$ |
| $\alpha^{24}$ | $\beta^2 + \alpha + \beta\alpha^2$ | $(\beta^2,1,\beta)$ | $\alpha^{56}$ | $1 + \beta^2\alpha + \beta\alpha^2$ | $(1,\beta^2,\beta)$ |
| $\alpha^{25}$ | $\beta^2 + \beta\alpha + \beta^2\alpha^2$ | $(\beta^2,\beta,\beta^2)$ | $\alpha^{57}$ | $\beta^2 + \alpha^2$ | $(\beta^2,0,1)$ |
| $\alpha^{26}$ | $1 + \alpha + \alpha^2$ | $(1,1,1)$ | $\alpha^{58}$ | $\beta + \alpha^2$ | $(\beta,0,1)$ |
| $\alpha^{27}$ | $\beta + \beta\alpha$ | $(\beta,\beta,0)$ | $\alpha^{59}$ | $\beta + \alpha + \alpha^2$ | $(\beta,1,1)$ |
| $\alpha^{28}$ | $\beta\alpha + \beta\alpha^2$ | $(0,\beta,\beta)$ | $\alpha^{60}$ | $\beta + \alpha$ | $(\beta,1,0)$ |
| $\alpha^{29}$ | $\beta^2 + \alpha$ | $(\beta^2,1,0)$ | $\alpha^{61}$ | $\beta\alpha + \alpha^2$ | $(0,\beta,1)$ |
| $\alpha^{30}$ | $\beta^2\alpha + \alpha^2$ | $(0,\beta^2,1)$ | $\alpha^{62}$ | $\beta + \beta^2\alpha + \beta^2\alpha^2$ | $(\beta,\beta^2,\beta^2)$ |

By continuing this process, we can find all the $16 = 4^{3-1}$ lines in a parallel bundle:

$L_0 = \{0, \alpha^0, \alpha^{21}, \alpha^{42}\}$, $\quad L_1 = \{\alpha^1, \alpha^6, \alpha^{29}, \alpha^{60}\}$, $\quad L_2 = \{\alpha^2, \alpha^{12}, \alpha^{57}, \alpha^{58}\}$,

$L_3 = \{\alpha^3, \alpha^{30}, \alpha^{32}, \alpha^{46}\}$, $\quad L_4 = \{\alpha^4, \alpha^{24}, \alpha^{51}, \alpha^{53}\}$, $\quad L_5 = \{\alpha^5, \alpha^{38}, \alpha^{49}, \alpha^{62}\}$,

$L_6 = \{\alpha^7, \alpha^{20}, \alpha^{26}, \alpha^{59}\}$, $\quad L_7 = \{\alpha^8, \alpha^{39}, \alpha^{43}, \alpha^{48}\}$, $\quad L_8 = \{\alpha^9, \alpha^{11}, \alpha^{25}, \alpha^{45}\}$,

$L_9 = \{\alpha^{10}, \alpha^{13}, \alpha^{35}, \alpha^{61}\}$, $L_{10} = \{\alpha^{14}, \alpha^{40}, \alpha^{52}, \alpha^{55}\}$, $L_{11} = \{\alpha^{15}, \alpha^{16}, \alpha^{23}, \alpha^{33}\}$,

$L_{12} = \{\alpha^{17}, \alpha^{28}, \alpha^{41}, \alpha^{47}\}$, $L_{13} = \{\alpha^{18}, \alpha^{22}, \alpha^{27}, \alpha^{50}\}$, $L_{14} = \{\alpha^{19}, \alpha^{31}, \alpha^{34}, \alpha^{56}\}$,

$L_{15} = \{\alpha^{36}, \alpha^{37}, \alpha^{44}, \alpha^{54}\}$.

Table 7.4 Four parallel bundles of lines of $EG(3, 2^2)$ over $GF(2^2)$.

**Lines**

| | | | |
|---|---|---|---|
| $\{0, \alpha^0, \alpha^{21}, \alpha^{42}\}$ | $\{\alpha^1, \alpha^6, \alpha^{29}, \alpha^{60}\}$ | $\{\alpha^2, \alpha^{12}, \alpha^{57}, \alpha^{58}\}$ | $\{\alpha^3, \alpha^{30}, \alpha^{32}, \alpha^{46}\}$ |
| $\{\alpha^4, \alpha^{24}, \alpha^{51}, \alpha^{53}\}$ | $\{\alpha^5, \alpha^{38}, \alpha^{49}, \alpha^{62}\}$ | $\{\alpha^7, \alpha^{20}, \alpha^{26}, \alpha^{59}\}$ | $\{\alpha^8, \alpha^{39}, \alpha^{43}, \alpha^{48}\}$ |
| $\{\alpha^9, \alpha^{11}, \alpha^{25}, \alpha^{45}\}$ | $\{\alpha^{10}, \alpha^{13}, \alpha^{35}, \alpha^{61}\}$ | $\{\alpha^{14}, \alpha^{40}, \alpha^{52}, \alpha^{55}\}$ | $\{\alpha^{15}, \alpha^{16}, \alpha^{23}, \alpha^{33}\}$ |
| $\{\alpha^{17}, \alpha^{28}, \alpha^{41}, \alpha^{47}\}$ | $\{\alpha^{18}, \alpha^{22}, \alpha^{27}, \alpha^{50}\}$ | $\{\alpha^{19}, \alpha^{31}, \alpha^{34}, \alpha^{56}\}$ | $\{\alpha^{36}, \alpha^{37}, \alpha^{44}, \alpha^{54}\}$ |
| $\{0, \alpha^1, \alpha^{22}, \alpha^{43}\}$ | $\{\alpha^0, \alpha^6, \alpha^{39}, \alpha^{50}\}$ | $\{\alpha^2, \alpha^7, \alpha^{30}, \alpha^{61}\}$ | $\{\alpha^3, \alpha^{13}, \alpha^{58}, \alpha^{59}\}$ |
| $\{\alpha^4, \alpha^{31}, \alpha^{33}, \alpha^{47}\}$ | $\{\alpha^5, \alpha^{25}, \alpha^{52}, \alpha^{54}\}$ | $\{\alpha^8, \alpha^{21}, \alpha^{27}, \alpha^{60}\}$ | $\{\alpha^9, \alpha^{40}, \alpha^{44}, \alpha^{49}\}$ |
| $\{\alpha^{10}, \alpha^{12}, \alpha^{26}, \alpha^{46}\}$ | $\{\alpha^{11}, \alpha^{14}, \alpha^{36}, \alpha^{62}\}$ | $\{\alpha^{15}, \alpha^{41}, \alpha^{53}, \alpha^{56}\}$ | $\{\alpha^{16}, \alpha^{17}, \alpha^{24}, \alpha^{34}\}$ |
| $\{\alpha^{18}, \alpha^{29}, \alpha^{42}, \alpha^{48}\}$ | $\{\alpha^{19}, \alpha^{23}, \alpha^{28}, \alpha^{51}\}$ | $\{\alpha^{20}, \alpha^{32}, \alpha^{35}, \alpha^{57}\}$ | $\{\alpha^{37}, \alpha^{38}, \alpha^{45}, \alpha^{55}\}$ |
| $\{0, \alpha^2, \alpha^{23}, \alpha^{44}\}$ | $\{\alpha^0, \alpha^{12}, \alpha^{15}, \alpha^{37}\}$ | $\{\alpha^1, \alpha^7, \alpha^{40}, \alpha^{51}\}$ | $\{\alpha^3, \alpha^8, \alpha^{31}, \alpha^{62}\}$ |
| $\{\alpha^4, \alpha^{14}, \alpha^{59}, \alpha^{60}\}$ | $\{\alpha^5, \alpha^{32}, \alpha^{34}, \alpha^{48}\}$ | $\{\alpha^6, \alpha^{26}, \alpha^{53}, \alpha^{55}\}$ | $\{\alpha^9, \alpha^{22}, \alpha^{28}, \alpha^{61}\}$ |
| $\{\alpha^{10}, \alpha^{41}, \alpha^{45}, \alpha^{50}\}$ | $\{\alpha^{11}, \alpha^{13}, \alpha^{27}, \alpha^{47}\}$ | $\{\alpha^{16}, \alpha^{42}, \alpha^{54}, \alpha^{57}\}$ | $\{\alpha^{17}, \alpha^{18}, \alpha^{25}, \alpha^{35}\}$ |
| $\{\alpha^{19}, \alpha^{30}, \alpha^{43}, \alpha^{49}\}$ | $\{\alpha^{20}, \alpha^{24}, \alpha^{29}, \alpha^{52}\}$ | $\{\alpha^{21}, \alpha^{33}, \alpha^{36}, \alpha^{58}\}$ | $\{\alpha^{38}, \alpha^{39}, \alpha^{46}, \alpha^{56}\}$ |
| $\{0, \alpha^3, \alpha^{24}, \alpha^{45}\}$ | $\{\alpha^0, \alpha^4, \alpha^9, \alpha^{32}\}$ | $\{\alpha^1, \alpha^{13}, \alpha^{16}, \alpha^{38}\}$ | $\{\alpha^2, \alpha^8, \alpha^{41}, \alpha^{52}\}$ |
| $\{\alpha^5, \alpha^{15}, \alpha^{60}, \alpha^{61}\}$ | $\{\alpha^6, \alpha^{33}, \alpha^{35}, \alpha^{49}\}$ | $\{\alpha^7, \alpha^{27}, \alpha^{54}, \alpha^{56}\}$ | $\{\alpha^{10}, \alpha^{23}, \alpha^{29}, \alpha^{62}\}$ |
| $\{\alpha^{11}, \alpha^{42}, \alpha^{46}, \alpha^{51}\}$ | $\{\alpha^{12}, \alpha^{14}, \alpha^{28}, \alpha^{48}\}$ | $\{\alpha^{17}, \alpha^{43}, \alpha^{55}, \alpha^{58}\}$ | $\{\alpha^{18}, \alpha^{19}, \alpha^{26}, \alpha^{36}\}$ |
| $\{\alpha^{20}, \alpha^{31}, \alpha^{44}, \alpha^{50}\}$ | $\{\alpha^{21}, \alpha^{25}, \alpha^{30}, \alpha^{53}\}$ | $\{\alpha^{22}, \alpha^{34}, \alpha^{37}, \alpha^{59}\}$ | $\{\alpha^{39}, \alpha^{40}, \alpha^{47}, \alpha^{57}\}$ |

Similarly, we can find all the parallel bundles. Table 7.4 shows four parallel bundles in $EG(3, 2^2)$. The 16 lines in four rows of this table separated by horizontal lines form a parallel bundle. As seen from this table, the first line in each parallel bundle is a line passing through the origin.

Because $\alpha^0 = 1$ and $\alpha$ are two linearly independent points in $EG(3, 2^2)$, then the following set

$$F_0 = \{\beta_0 \alpha^0 + \beta_1 \alpha : \beta_0, \beta_1 \in GF(2^2)\}$$
$$= \{0, 1, \alpha, \alpha^6, \alpha^8, \alpha^{18}, \alpha^{21}, \alpha^{22}, \alpha^{27}, \alpha^{29}, \alpha^{39}, \alpha^{42}, \alpha^{43}, \alpha^{48}, \alpha^{50}, \alpha^{60}\}$$

forms a 2-flat of $EG(3, 2^2)$ which passes through the origin and consists of 16 points of $EG(3, 2^2)$. Choose another point $\alpha^2$ which is not on the flat $F_0$ (i.e., $\alpha^2$ is linearly independent of $\alpha^0 = 1$ and $\alpha$) and form the following set

$$F_1 = \{\alpha^2 + \beta_0 \alpha^0 + \beta_1 \alpha : \beta_0, \beta_1 \in GF(2^2)\}$$
$$= \{\alpha^2, \alpha^3, \alpha^7, \alpha^{10}, \alpha^{12}, \alpha^{13}, \alpha^{20}, \alpha^{26}, \alpha^{30}, \alpha^{32}, \alpha^{35}, \alpha^{46}, \alpha^{57}, \alpha^{58}, \alpha^{59}, \alpha^{61}\}.$$

Because $F_0$ and $F_1$ have no point in common, they are parallel 2-flats. Similarly, we can construct another two 2-flats which are parallel to $F_0$:

$$F_2 = \{\alpha^4, \alpha^{15}, \alpha^{16}, \alpha^{17}, \alpha^{19}, \alpha^{23}, \alpha^{24}, \alpha^{28}, \alpha^{31}, \alpha^{33}, \alpha^{34}, \alpha^{41}, \alpha^{47}, \alpha^{51}, \alpha^{53}, \alpha^{56}\},$$
$$F_3 = \{\alpha^5, \alpha^9, \alpha^{11}, \alpha^{14}, \alpha^{25}, \alpha^{36}, \alpha^{37}, \alpha^{38}, \alpha^{40}, \alpha^{44}, \alpha^{45}, \alpha^{49}, \alpha^{52}, \alpha^{54}, \alpha^{55}, \alpha^{62}\}.$$

The four 2-flats, $F_0$, $F_1$, $F_2$, and $F_3$, form a parallel bundle of 2-flats in $\mathrm{EG}(3, 2^2)$. ▲▲

### 7.3.3 Subgeometries of Euclidean Geometries

By removing the origin and all the lines passing through the origin from the $m$-dimensional Euclidean geometry $\mathrm{EG}(m, q)$ over $\mathrm{GF}(q)$, we obtain a *subgeometry* of $\mathrm{EG}(m, q)$. The subgeometry possesses a special structural property, called *cyclic structure*. In this section, we demonstrate this special property of the subgeometry.

Let $\mathrm{EG}^*(m, q)$ be the *subgeometry* of $\mathrm{EG}(m, q)$ obtained by removing the origin and all the lines passing through the origin. Then, $\mathrm{EG}^*(m, q)$ has $q^m - 1$ points and

$$\frac{(q^{m-1} - 1)(q^m - 1)}{q - 1} \tag{7.21}$$

lines not passing through the origin. The lines in $\mathrm{EG}^*(m, q)$ can be partitioned into $(q^m - 1)/(q - 1)$ parallel bundles of lines, each consisting of $q^{m-1} - 1$ parallel lines not passing through the origin of $\mathrm{EG}(m, q)$. For each point in $\mathrm{EG}^*(m, q)$, there are

$$\frac{q^m - 1}{q - 1} - 1 \tag{7.22}$$

lines intersecting at it.

Let $\alpha^{j_0}$ and $\alpha^{j_1}$ be two linearly independent points in $\mathrm{EG}^*(m, q)$ and let $L = \{\alpha^{j_0} + \beta\alpha^{j_1} : \beta \in \mathrm{GF}(q)\}$ be a line in $\mathrm{EG}^*(m, q)$ which passes through the point $\alpha^{j_0}$. For $0 \leq i < q^m - 1$, we construct the following set of $q$ points:

$$\alpha^i L \triangleq \{\alpha^{j_0 + i} + \beta\alpha^{j_1 + i} : \beta \in \mathrm{GF}(q)\}. \tag{7.23}$$

Because $\alpha^{j_0}$ and $\alpha^{j_1}$ are two linearly independent points, $\alpha^{j_0 + i}$ and $\alpha^{j_1 + i}$ are also linearly independent. Hence, the set $\alpha^i L$ of $q$ points given by (7.23) forms a line in $\mathrm{EG}^*(m, q)$ which passes through the point $\alpha^{j_0 + i}$. It can be proved that $L, \alpha L, \alpha^2 L, \ldots, \alpha^{q^m - 2} L$ form $q^m - 1$ *different lines* in $\mathrm{EG}^*(m, q)$ (see Problem 7.9). Note that $\alpha^{q^m - 1} = 1$. Then, we have $\alpha^{q^m - 1} L = L$. Because a point in $\alpha^i L$ is obtained by *cyclically shifting* (or *increasing*) the power of each point on $L$ by $i$, the line $\alpha^i L$ is referred to as the $i$th *cyclic-shift* of the line $L$. The set

$$Q = \{L, \alpha L, \alpha^2 L, \ldots, \alpha^{q^m - 2} L\} \tag{7.24}$$

of $q^m - 1$ lines in $\mathrm{EG}^*(m, q)$ is said to form a *cyclic class* of lines in $\mathrm{EG}^*(m, q)$. The line $L$ is called the *generator* of the cyclic class $Q$. The $(q^{m-1} - 1)$ $(q^m - 1)/(q - 1)$ lines in $\mathrm{EG}^*(m, q)$ can be partitioned into

$$\frac{q^{m-1} - 1}{q - 1} \tag{7.25}$$

cyclic classes, each consisting of $q^m - 1$ lines.

Based on the cyclic structure of lines in $\mathrm{EG}^*(m, q)$, we can construct all the lines in $\mathrm{EG}(m, q)$ not passing through the origin in the following manner. Take a line $L_0$ in $\mathrm{EG}(m, q)$ not passing through the origin as the generator and cyclically shift it $q^m - 2$ times. This results in a cyclic class $Q_0$ of $q^m - 1$ lines in $\mathrm{EG}(m, q)$ not passing through the origin. Next, we take another line $L_1$ in $\mathrm{EG}^*(m, q)$ but not in $Q_0$ and cyclically shift it $q^m - 2$ times. This results in another cyclic class $Q_1$ of lines. We continue this cyclic-shift process until we construct all the $(q^{m-1} - 1)/(q - 1)$ cyclic classes of lines in $\mathrm{EG}^*(m, q)$.

The flats in the subgeometry $\mathrm{EG}^*(m, q)$ of $\mathrm{EG}(m, q)$ also have a cyclic structure.

**Example 7.5** Example 7.4 shows the line and 2-flat structures of $\mathrm{EG}(3, 2^2)$. In this example, we consider the subgeometry $\mathrm{EG}^*(3, 2^2)$ of $\mathrm{EG}(3, 2^2)$ obtained by removing the origin and all the lines passing through the origin. Choose a line not passing through the origin of $\mathrm{EG}(3, 2^2)$ from Table 7.4:

$$L_0 = \{\alpha, \alpha^6, \alpha^{29}, \alpha^{60}\}.$$

Then, the first and second cyclic-shifts of the line $L_0$ are

$$\begin{aligned} \alpha L_0 &= \{\alpha^2, \alpha^7, \alpha^{30}, \alpha^{61}\}, \\ \alpha^2 L_0 &= \{\alpha^3, \alpha^8, \alpha^{31}, \alpha^{62}\}. \end{aligned}$$

Continuing this process, we can obtain the 63 lines in a cyclic class:

$$Q_0 = \{L_0, \alpha L_0, \alpha^2 L_0, \ldots, \alpha^{62} L_0\}.$$

Choose a line in $\mathrm{EG}^*(3, 2^2)$ but not in the cyclic class $Q_0$: $L_1 = \{\alpha^2, \alpha^{12}, \alpha^{57}, \alpha^{58}\}$. Similarly, we can find the cyclic class $Q_1$ by using $L_1$ as its generator. Continuing this process, we can find all the cyclic classes in $\mathrm{EG}^*(3, 2^2)$. Table 7.5 shows two cyclic classes of lines in $\mathrm{EG}^*(3, 2^2)$. ▲▲

Table 7.5 Two cyclic classes of lines of $\mathrm{EG}^*(3, 2^2)$ over $\mathrm{GF}(2^2)$.

| Lines | Lines | lines | lines |
|---|---|---|---|
| $\{\alpha^1, \alpha^6, \alpha^{29}, \alpha^{60}\}$ | $\{\alpha^2, \alpha^7, \alpha^{30}, \alpha^{61}\}$ | $\{\alpha^3, \alpha^8, \alpha^{31}, \alpha^{62}\}$ | $\{\alpha^0, \alpha^4, \alpha^9, \alpha^{32}\}$ |
| $\{\alpha^1, \alpha^5, \alpha^{10}, \alpha^{33}\}$ | $\{\alpha^2, \alpha^6, \alpha^{11}, \alpha^{34}\}$ | $\{\alpha^3, \alpha^7, \alpha^{12}, \alpha^{35}\}$ | $\{\alpha^4, \alpha^8, \alpha^{13}, \alpha^{36}\}$ |
| $\{\alpha^5, \alpha^9, \alpha^{14}, \alpha^{37}\}$ | $\{\alpha^6, \alpha^{10}, \alpha^{15}, \alpha^{38}\}$ | $\{\alpha^7, \alpha^{11}, \alpha^{16}, \alpha^{39}\}$ | $\{\alpha^8, \alpha^{12}, \alpha^{17}, \alpha^{40}\}$ |
| $\{\alpha^9, \alpha^{13}, \alpha^{18}, \alpha^{41}\}$ | $\{\alpha^{10}, \alpha^{14}, \alpha^{19}, \alpha^{42}\}$ | $\{\alpha^{11}, \alpha^{15}, \alpha^{20}, \alpha^{43}\}$ | $\{\alpha^{12}, \alpha^{16}, \alpha^{21}, \alpha^{44}\}$ |

Table 7.5 (*continued*)

$\{\alpha^{13},\alpha^{17},\alpha^{22},\alpha^{45}\}$ $\{\alpha^{14},\alpha^{18},\alpha^{23},\alpha^{46}\}$ $\{\alpha^{15},\alpha^{19},\alpha^{24},\alpha^{47}\}$ $\{\alpha^{16},\alpha^{20},\alpha^{25},\alpha^{48}\}$
$\{\alpha^{17},\alpha^{21},\alpha^{26},\alpha^{49}\}$ $\{\alpha^{18},\alpha^{22},\alpha^{27},\alpha^{50}\}$ $\{\alpha^{19},\alpha^{23},\alpha^{28},\alpha^{51}\}$ $\{\alpha^{20},\alpha^{24},\alpha^{29},\alpha^{52}\}$
$\{\alpha^{21},\alpha^{25},\alpha^{30},\alpha^{53}\}$ $\{\alpha^{22},\alpha^{26},\alpha^{31},\alpha^{54}\}$ $\{\alpha^{23},\alpha^{27},\alpha^{32},\alpha^{55}\}$ $\{\alpha^{24},\alpha^{28},\alpha^{33},\alpha^{56}\}$
$\{\alpha^{25},\alpha^{29},\alpha^{34},\alpha^{57}\}$ $\{\alpha^{26},\alpha^{30},\alpha^{35},\alpha^{58}\}$ $\{\alpha^{27},\alpha^{31},\alpha^{36},\alpha^{59}\}$ $\{\alpha^{28},\alpha^{32},\alpha^{37},\alpha^{60}\}$
$\{\alpha^{29},\alpha^{33},\alpha^{38},\alpha^{61}\}$ $\{\alpha^{30},\alpha^{34},\alpha^{39},\alpha^{62}\}$ $\{\alpha^{0},\alpha^{31},\alpha^{35},\alpha^{40}\}$ $\{\alpha^{1},\alpha^{32},\alpha^{36},\alpha^{41}\}$
$\{\alpha^{2},\alpha^{33},\alpha^{37},\alpha^{42}\}$ $\{\alpha^{3},\alpha^{34},\alpha^{38},\alpha^{43}\}$ $\{\alpha^{4},\alpha^{35},\alpha^{39},\alpha^{44}\}$ $\{\alpha^{5},\alpha^{36},\alpha^{40},\alpha^{45}\}$
$\{\alpha^{6},\alpha^{37},\alpha^{41},\alpha^{46}\}$ $\{\alpha^{7},\alpha^{38},\alpha^{42},\alpha^{47}\}$ $\{\alpha^{8},\alpha^{39},\alpha^{43},\alpha^{48}\}$ $\{\alpha^{9},\alpha^{40},\alpha^{44},\alpha^{49}\}$
$\{\alpha^{10},\alpha^{41},\alpha^{45},\alpha^{50}\}$ $\{\alpha^{11},\alpha^{42},\alpha^{46},\alpha^{51}\}$ $\{\alpha^{12},\alpha^{43},\alpha^{47},\alpha^{52}\}$ $\{\alpha^{13},\alpha^{44},\alpha^{48},\alpha^{53}\}$
$\{\alpha^{14},\alpha^{45},\alpha^{49},\alpha^{54}\}$ $\{\alpha^{15},\alpha^{46},\alpha^{50},\alpha^{55}\}$ $\{\alpha^{16},\alpha^{47},\alpha^{51},\alpha^{56}\}$ $\{\alpha^{17},\alpha^{48},\alpha^{52},\alpha^{57}\}$
$\{\alpha^{18},\alpha^{49},\alpha^{53},\alpha^{58}\}$ $\{\alpha^{19},\alpha^{50},\alpha^{54},\alpha^{59}\}$ $\{\alpha^{20},\alpha^{51},\alpha^{55},\alpha^{60}\}$ $\{\alpha^{21},\alpha^{52},\alpha^{56},\alpha^{61}\}$
$\{\alpha^{22},\alpha^{53},\alpha^{57},\alpha^{62}\}$ $\{\alpha^{0},\alpha^{23},\alpha^{54},\alpha^{58}\}$ $\{\alpha^{1},\alpha^{24},\alpha^{55},\alpha^{59}\}$ $\{\alpha^{2},\alpha^{25},\alpha^{56},\alpha^{60}\}$
$\{\alpha^{3},\alpha^{26},\alpha^{57},\alpha^{61}\}$ $\{\alpha^{4},\alpha^{27},\alpha^{58},\alpha^{62}\}$ $\{\alpha^{0},\alpha^{5},\alpha^{28},\alpha^{59}\}$

$\{\alpha^{2},\alpha^{12},\alpha^{57},\alpha^{58}\}$ $\{\alpha^{3},\alpha^{13},\alpha^{58},\alpha^{59}\}$ $\{\alpha^{4},\alpha^{14},\alpha^{59},\alpha^{60}\}$ $\{\alpha^{5},\alpha^{15},\alpha^{60},\alpha^{61}\}$
$\{\alpha^{6},\alpha^{16},\alpha^{61},\alpha^{62}\}$ $\{\alpha^{0},\alpha^{7},\alpha^{17},\alpha^{62}\}$ $\{\alpha^{0},\alpha^{1},\alpha^{8},\alpha^{18}\}$ $\{\alpha^{1},\alpha^{2},\alpha^{9},\alpha^{19}\}$
$\{\alpha^{2},\alpha^{3},\alpha^{10},\alpha^{20}\}$ $\{\alpha^{3},\alpha^{4},\alpha^{11},\alpha^{21}\}$ $\{\alpha^{4},\alpha^{5},\alpha^{12},\alpha^{22}\}$ $\{\alpha^{5},\alpha^{6},\alpha^{13},\alpha^{23}\}$
$\{\alpha^{6},\alpha^{7},\alpha^{14},\alpha^{24}\}$ $\{\alpha^{7},\alpha^{8},\alpha^{15},\alpha^{25}\}$ $\{\alpha^{8},\alpha^{9},\alpha^{16},\alpha^{26}\}$ $\{\alpha^{9},\alpha^{10},\alpha^{17},\alpha^{27}\}$
$\{\alpha^{10},\alpha^{11},\alpha^{18},\alpha^{28}\}$ $\{\alpha^{11},\alpha^{12},\alpha^{19},\alpha^{29}\}$ $\{\alpha^{12},\alpha^{13},\alpha^{20},\alpha^{30}\}$ $\{\alpha^{13},\alpha^{14},\alpha^{21},\alpha^{31}\}$
$\{\alpha^{14},\alpha^{15},\alpha^{22},\alpha^{32}\}$ $\{\alpha^{15},\alpha^{16},\alpha^{23},\alpha^{33}\}$ $\{\alpha^{16},\alpha^{17},\alpha^{24},\alpha^{34}\}$ $\{\alpha^{17},\alpha^{18},\alpha^{25},\alpha^{35}\}$
$\{\alpha^{18},\alpha^{19},\alpha^{26},\alpha^{36}\}$ $\{\alpha^{19},\alpha^{20},\alpha^{27},\alpha^{37}\}$ $\{\alpha^{20},\alpha^{21},\alpha^{28},\alpha^{38}\}$ $\{\alpha^{21},\alpha^{22},\alpha^{29},\alpha^{39}\}$
$\{\alpha^{22},\alpha^{23},\alpha^{30},\alpha^{40}\}$ $\{\alpha^{23},\alpha^{24},\alpha^{31},\alpha^{41}\}$ $\{\alpha^{24},\alpha^{25},\alpha^{32},\alpha^{42}\}$ $\{\alpha^{25},\alpha^{26},\alpha^{33},\alpha^{43}\}$
$\{\alpha^{26},\alpha^{27},\alpha^{34},\alpha^{44}\}$ $\{\alpha^{27},\alpha^{28},\alpha^{35},\alpha^{45}\}$ $\{\alpha^{28},\alpha^{29},\alpha^{36},\alpha^{46}\}$ $\{\alpha^{29},\alpha^{30},\alpha^{37},\alpha^{47}\}$
$\{\alpha^{30},\alpha^{31},\alpha^{38},\alpha^{48}\}$ $\{\alpha^{31},\alpha^{32},\alpha^{39},\alpha^{49}\}$ $\{\alpha^{32},\alpha^{33},\alpha^{40},\alpha^{50}\}$ $\{\alpha^{33},\alpha^{34},\alpha^{41},\alpha^{51}\}$
$\{\alpha^{34},\alpha^{35},\alpha^{42},\alpha^{52}\}$ $\{\alpha^{35},\alpha^{36},\alpha^{43},\alpha^{53}\}$ $\{\alpha^{36},\alpha^{37},\alpha^{44},\alpha^{54}\}$ $\{\alpha^{37},\alpha^{38},\alpha^{45},\alpha^{55}\}$
$\{\alpha^{38},\alpha^{39},\alpha^{46},\alpha^{56}\}$ $\{\alpha^{39},\alpha^{40},\alpha^{47},\alpha^{57}\}$ $\{\alpha^{40},\alpha^{41},\alpha^{48},\alpha^{58}\}$ $\{\alpha^{41},\alpha^{42},\alpha^{49},\alpha^{59}\}$
$\{\alpha^{42},\alpha^{43},\alpha^{50},\alpha^{60}\}$ $\{\alpha^{43},\alpha^{44},\alpha^{51},\alpha^{61}\}$ $\{\alpha^{44},\alpha^{45},\alpha^{52},\alpha^{62}\}$ $\{\alpha^{0},\alpha^{45},\alpha^{46},\alpha^{53}\}$
$\{\alpha^{1},\alpha^{46},\alpha^{47},\alpha^{54}\}$ $\{\alpha^{2},\alpha^{47},\alpha^{48},\alpha^{55}\}$ $\{\alpha^{3},\alpha^{48},\alpha^{49},\alpha^{56}\}$ $\{\alpha^{4},\alpha^{49},\alpha^{50},\alpha^{57}\}$
$\{\alpha^{5},\alpha^{50},\alpha^{51},\alpha^{58}\}$ $\{\alpha^{6},\alpha^{51},\alpha^{52},\alpha^{59}\}$ $\{\alpha^{7},\alpha^{52},\alpha^{53},\alpha^{60}\}$ $\{\alpha^{8},\alpha^{53},\alpha^{54},\alpha^{61}\}$
$\{\alpha^{9},\alpha^{54},\alpha^{55},\alpha^{62}\}$ $\{\alpha^{0},\alpha^{10},\alpha^{55},\alpha^{56}\}$ $\{\alpha^{1},\alpha^{11},\alpha^{56},\alpha^{57}\}$

# 7.4 Cyclic Codes Constructed Based on Euclidean Geometries

Cyclic codes can be constructed using the lines of an Euclidean geometry not passing through its origin. These codes are called *cyclic Euclidean geometry (EG) codes*. In this section, we demonstrate the constructions of two classes of cyclic EG codes, one based on two-dimensional Euclidean geometries over fields of characteristic 2 and the other one based on multi-dimensional Euclidean geometries over fields of characteristic 2.

## 7.4.1 Cyclic Codes on Two-Dimensional Euclidean Geometries

In this section, we focus on the construction of cyclic EG codes based on the two-dimensional Euclidean geometries over fields of characteristic 2, i.e., $\mathrm{GF}(2^s)$ with $s \geq 1$. Construction of cyclic EG codes based on Euclidean geometries with dimensions greater than 2 will be presented in Section 7.4.2. General construction of cyclic EG codes using lines and flats of Euclidean geometries over fields of characteristics other than 2 can be found in [4–6, 20–31].

Consider the two-dimensional ($m = 2$) Euclidean geometry $\mathrm{EG}(2, 2^s)$ over $\mathrm{GF}(2^s)$ ($q = 2^s$) which can be realized by the field $\mathrm{GF}(2^{2s})$, an extension field of $\mathrm{GF}(2^s)$. Let $\alpha$ be a primitive element in $\mathrm{GF}(2^{2s})$. Then, the powers of $\alpha$, namely, $\alpha^{-\infty} = 0$, $\alpha^0 = 1$, $\alpha$, $\alpha^2, \ldots, \alpha^{2^{2s}-2}$, give all the points of $\mathrm{EG}(2, 2^s)$. The geometry $\mathrm{EG}(2, 2^s)$ has the following properties.

(1) It consists of $2^{2s}$ points and $2^s(2^s + 1)$ lines.
(2) Each line in $\mathrm{EG}(2, 2^s)$ consists of $2^s$ points.
(3) Each point in $\mathrm{EG}(2, 2^s)$ is intersected by $2^s + 1$ lines.
(4) Each line in $\mathrm{EG}(2, 2^s)$ has $2^s - 1$ lines parallel to it.

Let $\mathrm{EG}^*(2, 2^s)$ be the subgeometry of $\mathrm{EG}(2, 2^s)$ obtained by removing the origin and the $2^s + 1$ lines in $\mathrm{EG}(2, 2^s)$ passing through the origin. From the development in Section 7.3, the subgeometry $\mathrm{EG}^*(2, 2^s)$ has the following properties.

(1) It has $2^{2s} - 1$ points and $2^{2s} - 1$ lines.
(2) Each line consists of $2^s$ points.
(3) Each point in $\mathrm{EG}^*(2, 2^s)$ is intersected by $2^s$ lines.
(4) Each line in $\mathrm{EG}^*(2, 2^s)$ has $2^s - 2$ lines parallel to it.
(5) The $2^{2s} - 1$ lines in $\mathrm{EG}^*(2, 2^s)$ can be partitioned into $2^s + 1$ parallel bundles, each consisting of $2^s - 1$ parallel lines.

The lines in $\mathrm{EG}^*(2, 2^s)$ form a *single cyclic class* according to (7.25). Hence, all the lines in $\mathrm{EG}^*(2, 2^s)$ can be generated by taking any line in $\mathrm{EG}^*(2, 2^s)$ as a generator and cyclically shifting it $2^{2s} - 2$ times. Let $L_0$ be a line in $\mathrm{EG}^*(2, 2^s)$. It follows from the cyclic structure of $\mathrm{EG}^*(2, 2^s)$ that the set

$$Q = \{L_0 = \alpha^0 L_0, L_1 = \alpha L_0, L_2 = \alpha^2 L_0, \ldots, L_{2^{2s}-2} = \alpha^{2^{2s}-2} L_0\} \qquad (7.26)$$

forms the single cyclic class of $\mathrm{EG}^*(2, 2^s)$ which consists of all the $2^{2s} - 1$ lines of $\mathrm{EG}^*(2, 2^s)$. Note that any line in $\mathrm{EG}^*(2, 2^s)$ can be used to generate all the lines in $\mathrm{EG}^*(2, 2^s)$, i.e., each line in $\mathrm{EG}^*(2, 2^s)$ can be used as a generator of the single cyclic class $Q$ of $\mathrm{EG}^*(2, 2^s)$.

Label all the points in $\mathrm{EG}^*(2, 2^s)$ by the powers of $\alpha$, i.e., $\alpha^0$, $\alpha$, $\alpha^2, \ldots,$ $\alpha^{2s-2}$. For $0 \le i < 2^{2s} - 1$, we form the following $(2^{2s} - 1)$-tuple over $\mathrm{GF}(2)$:

$$\mathbf{h}_i = (h_{i,0}, h_{i,1}, \ldots, h_{i,2^{2s}-2}), \tag{7.27}$$

where $h_{i,j} = 1$ if and only if the $j$th point $\alpha^j$ of $\mathrm{EG}^*(2, 2^s)$ is on the $i$th line $L_i$ of $\mathrm{EG}^*(2, 2^s)$ (or in $Q$); otherwise, $h_{i,j} = 0$, $0 \le j < 2^{2s} - 1$. The $(2^{2s} - 1)$-tuple $\mathbf{h}_i$ over $\mathrm{GF}(2)$ is the incidence vector of the line $L_i$. Because each line in $\mathrm{EG}^*(2, 2^s)$ has $2^s$ points, the weight of $\mathbf{h}_i$ is $2^s$. It follows from the cyclic structure of $\mathrm{EG}^*(2, 2^s)$ that the incidence vector $\mathbf{h}_i$ of the $i$th line $L_i$ in $\mathrm{EG}^*(2, 2^s)$ is the $i$th cyclic-shift of the incidence vector $\mathbf{h}_0$ of the line $L_0$.

Form the following $(2^{2s} - 1) \times (2^{2s} - 1)$ matrix over $\mathrm{GF}(2)$ with the incidence vectors $\mathbf{h}_0, \mathbf{h}_1, \ldots, \mathbf{h}_{2^{2s}-2}$ of the lines $L_0, L_1, \ldots, L_{2^{2s}-2}$ in $\mathrm{EG}^*(2, 2^s)$ as rows:

$$\mathbf{H}_{\mathrm{EG}}(2, 2^s) = \begin{bmatrix} \mathbf{h}_0 \\ \mathbf{h}_1 \\ \vdots \\ \mathbf{h}_{2^{2s}-2} \end{bmatrix} = \begin{bmatrix} h_{0,0} & h_{0,1} & \cdots & h_{0,2^{2s}-2} \\ h_{1,0} & h_{1,1} & \cdots & h_{1,2^{2s}-2} \\ \vdots & \vdots & \ddots & \vdots \\ h_{2^{2s}-2,0} & h_{2^{2s}-2,1} & \cdots & h_{2^{2s}-2,2^{2s}-2} \end{bmatrix}, \tag{7.28}$$

where the notations 2 and $2^s$ used in $\mathbf{H}_{\mathrm{EG}}(2, 2^s)$ indicate that the matrix is constructed based on the points and lines of $\mathrm{EG}^*(2, 2^s)$. The matrix $\mathbf{H}_{\mathrm{EG}}(2, 2^s)$ is the line-point incidence matrix of $\mathrm{EG}^*(2, 2^s)$ whose columns correspond to the points $\alpha^0, \alpha, \alpha^2, \ldots, \alpha^{2^{2s}-2}$ of $\mathrm{EG}^*(2, 2^s)$ and rows correspond to the lines $L_0 = \alpha^0 L_0, L_1 = \alpha L_0, L_2 = \alpha^2 L_0, \ldots, L_{2^{2s}-2} = \alpha^{2^{2s}-2} L_0$ of $\mathrm{EG}^*(2, 2^s)$. From the cyclic structure of the lines in $\mathrm{EG}^*(2, 2^s)$, it can be seen that the matrix $\mathbf{H}_{\mathrm{EG}}(2, 2^s)$ is a $(2^{2s} - 1) \times (2^{2s} - 1)$ *circulant* with weight $2^s$. As shown in Section 7.1, the matrix $\mathbf{H}_{\mathrm{EG}}(2, 2^s)$ has the *orthogonal* structure. For each column position $j$, $0 \le j < 2^{2s} - 1$, of $\mathbf{H}_{\mathrm{EG}}(2, 2^s)$, there are $2^s$ rows in $\mathbf{H}_{\mathrm{EG}}(2, 2^s)$ orthogonal to it. It can be proved that the rank of $\mathbf{H}_{\mathrm{EG}}(2, 2^s)$ [16, 17, 32, 33] is

$$\mathrm{rank}(\mathbf{H}_{\mathrm{EG}}(2, 2^s)) = 3^s - 1. \tag{7.29}$$

Hence, the matrix $\mathbf{H}_{\mathrm{EG}}(2, 2^s)$ has $2^{2s} - 3^s$ *redundant* (linearly dependent) rows. If we use $\mathbf{H}_{\mathrm{EG}}(2, 2^s)$ as the parity-check matrix of a linear block code, the null space over $\mathrm{GF}(2)$ of $\mathbf{H}_{\mathrm{EG}}(2, 2^s)$ gives a binary $(2^{2s} - 1, 2^{2s} - 3^s)$ cyclic code $C_{\mathrm{EG}}(2, 2^s)$ (see Problem 4.32) of length $2^{2s} - 1$ and dimension $2^{2s} - 3^s$. This code is called a *two-dimensional cyclic EG code*.

Because $C_{\mathrm{EG}}(2, 2^s)$ is a binary cyclic code, it is uniquely specified by a generator polynomial over $\mathrm{GF}(2)$. In the following, a theorem which specifies how to find the generator polynomial of a cyclic EG code is described without proof.

Let $k$ be a nonnegative integer less than $2^{2s}$. Then, we can express $k$ in radix-$2^s$ form as follows:

$$k = k_0 + k_1 2^s, \tag{7.30}$$

where $k_0$ and $k_1$ are two nonnegative integers with $0 \leq k_0, k_1 < 2^s$. We define the $2^s$-weight of $k$, denoted by $W_{2^s}(k)$, as the following sum

$$W_{2^s}(k) = k_0 + k_1. \tag{7.31}$$

For a nonnegative integer $\ell$ with $0 \leq \ell < s$, let $k^{(\ell)}$ be the remainder resulting from dividing $k2^\ell$ by $2^{2s} - 1$, i.e., $k^{(\ell)} = \left(k2^\ell \bmod (2^{2s} - 1)\right)$. Then, $0 \leq k^{(\ell)} < 2^{2s} - 1$. Express $k^{(\ell)}$ in radix-$2^s$ form:

$$k^{(\ell)} = k_0^{(\ell)} + k_1^{(\ell)} 2^s, \tag{7.32}$$

where $k_0^{(\ell)}$ and $k_1^{(\ell)}$ are two nonnegative integers with $0 \leq k_0^{(\ell)}, k_1^{(\ell)} < 2^s$. Then, the $2^s$-weight of $k^{(\ell)}$ is

$$W_{2^s}(k^{(\ell)}) = k_0^{(\ell)} + k_1^{(\ell)}. \tag{7.33}$$

The following theorem characterizes the roots of the generator polynomial of the cyclic EG code $C_{\text{EG}}(2, 2^s)$. The proof of the theorem can be found in [32–34] and is omitted here.

**Theorem 7.1** *Let $\alpha$ be a primitive element of the field $\text{GF}(2^{2s})$ and $k$ be a nonnegative integer less than $2^{2s}$. The generator polynomial $\mathbf{g}_{\text{EG}}(X)$ of the cyclic EG code $C_{\text{EG}}(2, 2^s)$ has $\alpha^k$ as a root if and only if*

$$0 < \max_{0 \leq \ell < s} \left(W_{2^s}(k^{(\ell)})\right) \leq 2^s - 1. \tag{7.34}$$

It can be easily checked that the $2^s$ integers equal or less than $2^s$, namely, $1, 2, \ldots, 2^s$, have $2^s$-weights which satisfy the condition given by (7.34). Hence, the generator polynomial $\mathbf{g}_{\text{EG}}(X)$ of the cyclic EG code $C_{\text{EG}}(2, 2^s)$ has $\alpha$, $\alpha^2, \ldots, \alpha^{2^s}$ ($2^s$ consecutive powers of $\alpha$) as roots. Then, it follows from the BCH bound given by (5.19) that the minimum distance of $C_{\text{EG}}(2, 2^s)$ is *at least* $2^s + 1$ which is the column weight $2^s$ of the parity-check matrix $\mathbf{H}_{\text{EG}}(2, 2^s)$ of $C_{\text{EG}}(2, 2^s)$ plus one.

**Theorem 7.2** *The minimum distance of a cyclic EG code $C_{\text{EG}}(2, 2^s)$ constructed based on $\text{EG}^*(2, 2^s)$ is $2^s + 1$.*

*Proof* Because the length of the code is $n = 2^{2s} - 1$, the generator polynomial $\mathbf{g}_{\text{EG}}(X)$ of $C_{\text{EG}}(2, 2^s)$ must be a factor of $X^{2^{2s}-1} + 1$. Factor $X^{2^{2s}-1} + 1$ as the following product:

$$X^{2^{2s}-1} + 1 = (X^{2^s-1} + 1)(X^{2^s(2^s-1)}$$
$$+ X^{(2^s-1)(2^s-1)} + X^{(2^s-2)(2^s-1)} + \cdots + X^{2^s-1} + 1).$$

The first factor $X^{2^s-1} + 1$ of $X^{2^{2s}-1} + 1$ has $\alpha^0 = 1$, $\alpha^{2^s+1}$, $\alpha^{2(2^s+1)}$, $\alpha^{3(2^s+1)}, \ldots, \alpha^{(2^s-2)(2^s+1)}$ as all its roots. The powers of these roots do not satisfy the condition given by (7.34) and hence the roots of $X^{2^s-1} + 1$ are not the roots of the generator polynomial $\mathbf{g}_{\mathrm{EG}}(X)$ of $C_{\mathrm{EG}}(2, 2^s)$. Therefore, the second factor $X^{2^s(2^s-1)} + X^{(2^s-1)(2^s-1)} + X^{(2^s-2)(2^s-1)} + \cdots + X^{2^s-1} + 1$ of $X^{2^{2s}-1} + 1$ has all the roots of $\mathbf{g}_{\mathrm{EG}}(X)$ as roots. This means that the polynomial $X^{2^s(2^s-1)} + X^{(2^s-1)(2^s-1)} + X^{(2^s-2)(2^s-1)} + \cdots + X^{2^s-1} + 1$ is a code polynomial of the EG code $C_{\mathrm{EG}}(2, 2^s)$. This code polynomial has weight $2^s + 1$ which meets the BCH bound $2^s + 1$ on the minimum distance of $C_{\mathrm{EG}}(2, 2^s)$. Hence, the minimum distance of the EG code $C_{\mathrm{EG}}(2, 2^s)$ is exactly $2^s + 1$. ▲▲

It can be proved that the total number of roots of the generator polynomial $\mathbf{g}_{\mathrm{EG}}(X)$ of the cyclic EG code $C_{\mathrm{EG}}(2, 2^s)$ is $3^s - 1$ [32, 33]. Hence, the number of parity-check symbols of the code $C_{\mathrm{EG}}(2, 2^s)$ is

$$n - k = 3^s - 1.$$

Hence, the two-dimensional EG code $C_{\mathrm{EG}}(2, 2^s)$ has the following parameters:

| | |
|---|---|
| length | $n = 2^{2s} - 1,$ |
| dimension | $k = 2^{2s} - 3^s,$ |
| number of parity-check symbols | $n - k = 3^s - 1,$ |
| minimum distance | $d_{\min} = 2^s + 1.$ |

Because $\mathbf{H}_{\mathrm{EG}}(2, 2^s)$ has an orthogonal structure and has column weight $2^s$, the code $C_{\mathrm{EG}}(2, 2^s)$ can be decoded with the OSMLD which is capable of correcting $2^{s-1}$ or fewer errors. A list of two-dimensional EG codes is given in Table 7.6.

Table 7.6 A list of two-dimensional EG codes.

| $s$ | $n$ | $k$ | rate | $d_{\min}$ | $\gamma$ | $\rho$ |
|---|---|---|---|---|---|---|
| 2 | 15 | 7 | 0.467 | 5 | 4 | 4 |
| 3 | 63 | 37 | 0.587 | 9 | 8 | 8 |
| 4 | 255 | 175 | 0.686 | 17 | 16 | 16 |
| 5 | 1 023 | 781 | 0.763 | 33 | 32 | 32 |
| 6 | 4 095 | 3 367 | 0.822 | 65 | 64 | 64 |
| 7 | 16 383 | 14 197 | 0.867 | 129 | 128 | 128 |
| 8 | 65 535 | 58 975 | 0.900 | 257 | 256 | 256 |

**Example 7.6** Let $m = 2$ and $s = 2$. Consider the two-dimensional Euclidean geometry $\mathrm{EG}(2, 2^2)$ over the field $\mathrm{GF}(2^2)$ which can be realized by the field $\mathrm{GF}(2^4)$. The subgeometry $\mathrm{EG}^*(2, 2^2)$ is constructed by removing the origin and all the lines passing through the origin from $\mathrm{EG}(2, 2^2)$. Based on this subgeometry, a two-dimensional EG code $C_{\mathrm{EG}}(2, 2^2)$ with length $n = 2^{2 \times 2} - 1 = 15$ and dimension $k = 2^{2 \times 2} - 3^2 = 7$ can be constructed. Let $\alpha$ be a primitive element in $\mathrm{GF}(2^4)$ (see Table 7.1). Let $k$ be a nonnegative integer less than 15.

It follows from Theorem 7.1 that the generator polynomial $\mathbf{g}_{EG}(X)$ of the EG code $C_{EG}(2, 2^2)$ has $\alpha^k$ as a root if and only if

$$0 < \max_{0 \le \ell < 2} \left( W_{2^2}(k^{(\ell)}) \right) \le 3.$$

By calculating, it is found that there are eight integers less than 15 satisfying the above condition, which are 1, 2, 3, 4, 6, 8, 9, and 12. Thus, the generator polynomial $\mathbf{g}_{EG}(X)$ has $\alpha$, $\alpha^2$, $\alpha^3$, $\alpha^4$, $\alpha^6$, $\alpha^8$, $\alpha^9$, and $\alpha^{12}$ over $GF(2^4)$ as roots. The elements $\alpha$, $\alpha^2$, $\alpha^4$, and $\alpha^8$ are in the same conjugate set and thus have the same minimal polynomial, $m_1(X) = 1 + X + X^4$; and the elements $\alpha^3$, $\alpha^6$, $\alpha^{12}$, and $\alpha^{24} = \alpha^9$ are in the same conjugate set and thus have the same minimal polynomial, $m_3(X) = 1 + X + X^2 + X^3 + X^4$. Therefore,

$$\begin{aligned} \mathbf{g}_{EG}(X) &= \text{LCM}\{m_1(X), m_3(X)\} = m_1(X)m_3(X) \\ &= (1 + X + X^4)(1 + X + X^2 + X^3 + X^4) \\ &= 1 + X^4 + X^6 + X^7 + X^8. \end{aligned}$$

Thus, the code $C_{EG}(2, 2^2)$ is a $(15, 7)$ cyclic code. The minimum distance $d_{\min}$ of $C_{EG}(2, 2^2)$ is $2^2 + 1 = 5$ which can also be verified by the generator polynomial $\mathbf{g}_{EG}(X)$ because $\mathbf{g}_{EG}(X)$ is a valid code polynomial of weight 5.

The cyclic code $C_{EG}(2, 2^2)$ can also be specified by a parity-check matrix in the form of (7.28) as developed above. Consider the following line not passing through the origin of $EG(2, 2^2)$

$$L_0 = \{1, \alpha^4, \alpha^{12}, \alpha^{13}\}.$$

The incidence vector of the line $L_0$ is

$$\mathbf{h}_0 = (1\ 0\ 0\ 0\ 1\ 0\ 0\ 0\ 0\ 0\ 0\ 0\ 1\ 1\ 0).$$

Using $\mathbf{h}_0$ and its 14 cyclic-shifts, we construct the following $15 \times 15$ matrix (a circulant) in the form of (7.28):

$$\mathbf{H}_{EG}(2, 2^2) = \begin{bmatrix} 1&0&0&0&1&0&0&0&0&0&0&0&1&1&0 \\ 0&1&0&0&0&1&0&0&0&0&0&0&0&1&1 \\ 1&0&1&0&0&0&1&0&0&0&0&0&0&0&1 \\ 1&1&0&1&0&0&0&1&0&0&0&0&0&0&0 \\ 0&1&1&0&1&0&0&0&1&0&0&0&0&0&0 \\ 0&0&1&1&0&1&0&0&0&1&0&0&0&0&0 \\ 0&0&0&1&1&0&1&0&0&0&1&0&0&0&0 \\ 0&0&0&0&1&1&0&1&0&0&0&1&0&0&0 \\ 0&0&0&0&0&1&1&0&1&0&0&0&1&0&0 \\ 0&0&0&0&0&0&1&1&0&1&0&0&0&1&0 \\ 0&0&0&0&0&0&0&1&1&0&1&0&0&0&1 \\ 1&0&0&0&0&0&0&0&1&1&0&1&0&0&0 \\ 0&1&0&0&0&0&0&0&0&1&1&0&1&0&0 \\ 0&0&1&0&0&0&0&0&0&0&1&1&0&1&0 \\ 0&0&0&1&0&0&0&0&0&0&0&1&1&0&1 \end{bmatrix}.$$

The rank of the parity-check matrix $\mathbf{H}_{EG}(2, 2^2)$ is 8. For each column position in $\mathbf{H}_{EG}(2, 2^2)$, there are four rows orthogonal to it. Hence, the code $C_{EG}(2, 2^2)$ is capable of correcting two or fewer random errors with OSML decoding. ▲▲

**Example 7.7** In this example, we consider the construction of the $(63, 37)$ cyclic EG code $C_{EG}(2, 2^3)$ listed in Table 7.6. The code is constructed based on the two-dimensional Euclidean geometry $EG(2, 2^3)$ over the field $GF(2^3)$, i.e., $m = 2$ and $s = 3$. The extension field $GF(2^6)$ (see Table 5.14) of $GF(2^3)$ is a realization of $EG(2, 2^3)$. Let $\alpha$ be a primitive element of $GF(2^6)$. To find the generator polynomial $\mathbf{g}_{EG}(X)$ of the EG code $C_{EG}(2, 2^3)$, we need to find all its roots, $\alpha^k$, where the nonnegative integer $k$ less than 63 satisfies the following condition:

$$0 < \max_{0 \le \ell < 2} (W_{2^3}(k^{(\ell)})) \le 7.$$

There are 26 such nonnegative integers: 1, 2, 3, 4, 5, 6, 7, 8, 10, 12, 14, 16, 17, 20, 21, 24, 28, 32, 33, 34, 35, 40, 42, 48, 49, 56. Thus, $\mathbf{g}_{EG}(X)$ has 26 roots, $\alpha^1$, $\alpha^2$, $\alpha^3$, $\alpha^4$, $\alpha^5$, $\alpha^6$, $\alpha^7$, $\alpha^8$, $\alpha^{10}$, $\alpha^{12}$, $\alpha^{14}$, $\alpha^{16}$, $\alpha^{17}$, $\alpha^{20}$, $\alpha^{21}$, $\alpha^{24}$, $\alpha^{28}$, $\alpha^{32}$, $\alpha^{33}$, $\alpha^{34}$, $\alpha^{35}$, $\alpha^{40}$, $\alpha^{42}$, $\alpha^{48}$, $\alpha^{49}$, $\alpha^{56}$, which can be partitioned into the following five conjugate sets:

$$\Phi_1 = \{\alpha^1, \alpha^2, \alpha^4, \alpha^8, \alpha^{16}, \alpha^{32}\},$$
$$\Phi_3 = \{\alpha^3, \alpha^6, \alpha^{12}, \alpha^{24}, \alpha^{48}, \alpha^{96} = \alpha^{33}\},$$
$$\Phi_5 = \{\alpha^5, \alpha^{10}, \alpha^{20}, \alpha^{40}, \alpha^{80} = \alpha^{17}, \alpha^{34}\},$$
$$\Phi_7 = \{\alpha^7, \alpha^{14}, \alpha^{28}, \alpha^{56}, \alpha^{112} = \alpha^{49}, \alpha^{98} = \alpha^{35}\},$$
$$\Phi_{21} = \{\alpha^{21}, \alpha^{42}\}.$$

Elements in a conjugate set have the same minimal polynomial. The minimal polynomials corresponding to the elements in the five conjugate sets are

$$m_1(X) = 1 + X + X^6, \qquad\qquad m_3(X) = 1 + X + X^2 + X^4 + X^6,$$
$$m_5(X) = 1 + X + X^2 + X^5 + X^6, \quad m_7(X) = 1 + X^3 + X^6,$$
$$m_{21}(X) = 1 + X + X^2.$$

Then, the generator polynomial $\mathbf{g}_{EG}(X)$ of the $(63, 37)$ EG code $C_{EG}(2, 2^3)$ is calculated as follows:

$$
\begin{aligned}
\mathbf{g}_{EG}(X) &= \text{LCM}\{m_1(X), m_3(X), m_5(X), m_7(X), m_{21}(X)\} \\
&= m_1(X)m_3(X)m_5(X)m_7(X)m_{21}(X) \\
&= (1 + X + X^6)(1 + X + X^2 + X^4 + X^6)(1 + X + X^2 + X^5 + X^6) \\
&\quad (1 + X^3 + X^6)(1 + X + X^2) \\
&= 1 + X^2 + X^6 + X^{10} + X^{12} + X^{13} + X^{14} + X^{15} + X^{16} + X^{17} \\
&\quad + X^{18} + X^{19} + X^{20} + X^{21} + X^{22} + X^{23} + X^{25} + X^{26}.
\end{aligned}
$$

A parity-check matrix (a $63 \times 63$ circulant) in the form of (7.28) of the $(63, 37)$ EG code $C_{EG}(2, 2^3)$ can also be constructed by using a line in $EG^*(2, 2^3)$ (see Problem 7.12). ▲▲

## 7.4.2 Cyclic Codes on Multi-Dimensional Euclidean Geometries

In Section 7.4.1, we presented a subclass of cyclic EG codes which is constructed based on the lines not passing through the origin of the class of two-dimensional Euclidean geometries over fields of characteristic 2. In this section, we present another subclass of cyclic EG codes which are constructed based on the lines not passing through the origin of the class of Euclidean geometries over fields of characteristic 2 with dimensions larger than 2.

For $m > 2$, consider the $m$-dimensional Euclidean geometry $\mathrm{EG}(m, 2^s)$ over $\mathrm{GF}(2^s)$. Let $\mathrm{EG}^*(m, 2^s)$ be the subgeometry of $\mathrm{EG}(m, 2^s)$ obtained by removing the origin of $\mathrm{EG}(m, 2^s)$ and the lines passing through the origin. In Section 7.3, we showed that the lines in $\mathrm{EG}^*(m, 2^s)$ can be partitioned into

$$K = \frac{2^{(m-1)s} - 1}{2^s - 1} \tag{7.35}$$

cyclic classes (see (7.25)), denoted by $Q_0, Q_1, \ldots, Q_{K-1}$, each consisting of $2^{ms} - 1$ lines not passing through the origin of $\mathrm{EG}(m, 2^s)$.

For each cyclic class $Q_i$, $0 \le i < K$, we form a $(2^{ms} - 1) \times (2^{ms} - 1)$ circulant $\mathbf{H}_i$ over $\mathrm{GF}(2)$ with the incidence vectors of lines in $Q_i$ as rows arranged in such a way that each row of $\mathbf{H}_i$ is a cyclic-shift of the row above it and the first row is the cyclic-shift of the last row. Each circulant $\mathbf{H}_i$ has weight $2^s$. Next, we form the following $K(2^{ms} - 1) \times (2^{ms} - 1)$ matrix with $\mathbf{H}_0, \mathbf{H}_1, \ldots, \mathbf{H}_{K-1}$ as submatrices arranged in a column:

$$\mathbf{H}_{\mathrm{EG}}(m, 2^s) = \begin{bmatrix} \mathbf{H}_0 \\ \mathbf{H}_1 \\ \vdots \\ \mathbf{H}_{K-1} \end{bmatrix}, \tag{7.36}$$

which has row and column weights $2^s$ and $K2^s = 2^s(2^{(m-1)s} - 1)/(2^s - 1)$, respectively. The matrix $\mathbf{H}_{\mathrm{EG}}(m, 2^s)$ is the incidence matrix of the subgeometry $\mathrm{EG}^*(m, 2^s)$. The columns and rows of $\mathbf{H}_{\mathrm{EG}}(m, 2^s)$ correspond to the points and lines in $\mathrm{EG}^*(m, 2^s)$, respectively. The matrix $\mathbf{H}_{\mathrm{EG}}(m, 2^s)$ has an orthogonal structure. Label the rows and columns of $\mathbf{H}_{\mathrm{EG}}(m, 2^s)$ from 0 to $K(2^{ms} - 1) - 1$ and 0 to $2^{ms} - 2$, respectively. For $0 \le j < 2^{ms} - 1$, there are $K2^s$ rows in $\mathbf{H}_{\mathrm{EG}}(m, 2^s)$ that are orthogonal on the $j$th column of $\mathbf{H}_{\mathrm{EG}}(m, 2^s)$. Suppose we use $\mathbf{H}_{\mathrm{EG}}(m, 2^s)$ as the parity-check matrix of a linear block code. The null space over $\mathrm{GF}(2)$ of $\mathbf{H}_{\mathrm{EG}}(m, 2^s)$ gives an EG code $C_{\mathrm{EG}}(m, 2^s)$ of length $2^{ms} - 1$. Based on the structure of the parity-check matrix $\mathbf{H}_{\mathrm{EG}}(m, 2^s)$, if $\mathbf{v}$ is a codeword in $C_{\mathrm{EG}}(m, 2^s)$, then a cyclic-shift of $\mathbf{v}$ is also a codeword in $C_{\mathrm{EG}}(m, 2^s)$. Thus, $C_{\mathrm{EG}}(m, 2^s)$ is a cyclic EG code (see Problem 4.32). Next, we describe how to find the generator polynomial $\mathbf{g}_{\mathrm{EG}}(X)$ of the cyclic code $C_{\mathrm{EG}}(m, 2^s)$.

Let $k$ be a nonnegative integer less than $2^{ms}$. Express $k$ in radix-$2^s$ form as follows:

$$k = k_0 + k_1 2^s + k_2 2^{2s} + \cdots + k_{m-1} 2^{(m-1)s}, \tag{7.37}$$

where $0 \leq k_i < 2^s$ for $0 \leq i < m$. The $2^s$-weight $W_{2^s}(k)$ of $k$ is

$$W_{2^s}(k) = k_0 + k_1 + k_2 + \cdots + k_{m-1}. \tag{7.38}$$

For a nonnegative integer $\ell$ with $0 \leq \ell < s$, let $k^{(\ell)}$ be the remainder resulting from dividing $k2^\ell$ by $2^{ms} - 1$. Then, $0 \leq k^{(\ell)} < 2^{ms} - 1$. Express $k^{(\ell)}$ in radix-$2^s$ form,

$$k^{(\ell)} = k_0^{(\ell)} + k_1^{(\ell)} 2^s + k_2^{(\ell)} 2^{2s} + \cdots + k_{m-1}^{(\ell)} 2^{(m-1)s}. \tag{7.39}$$

The $2^s$-weight of $k^{(\ell)}$ is

$$W_{2^s}(k^{(\ell)}) = k_0^{(\ell)} + k_1^{(\ell)} + k_2^{(\ell)} + \cdots + k_{m-1}^{(\ell)}. \tag{7.40}$$

The following theorem characterizes the roots of the generator polynomial $\mathbf{g}_{\mathrm{EG}}(X)$ of the cyclic EG code $C_{\mathrm{EG}}(m, 2^s)$ (see the proofs in [32–34]).

**Theorem 7.3** *Let $\alpha$ be a primitive element of* $\mathrm{GF}(2^{ms})$. *Then, the generator polynomial $\mathbf{g}_{\mathrm{EG}}(X)$ of the cyclic EG code $C_{\mathrm{EG}}(m, 2^s)$ has $\alpha^k$ as a root if and only if the following condition holds*

$$0 < \max_{0 \leq \ell < s} \left( W_{2^s}(k^{(\ell)}) \right) \leq (m-1)(2^s - 1). \tag{7.41}$$

For $1 \leq k \leq 2^s K$ where $K = (2^{(m-1)s} - 1)/(2^s - 1)$, it can be proved that the $2^s$-weight of $k^{(\ell)}$, $0 \leq \ell < s$, satisfies the condition given by (7.41). Hence, the generator polynomial $\mathbf{g}_{\mathrm{EG}}(X)$ of the cyclic EG code $C_{\mathrm{EG}}(m, 2^s)$ has $\alpha, \alpha^2, \ldots, \alpha^{2^s K}$ ($2^s K$ consecutive powers of $\alpha$) as roots. It follows from the BCH bound that the minimum distance of $C_{\mathrm{EG}}(m, 2^s)$ is at least $2^s K + 1 = (2^{(m-1)s} - 1)2^s/(2^s - 1) + 1$ which equals the column weight of the parity-check matrix $\mathbf{H}_{\mathrm{EG}}(m, 2^s)$ of $C_{\mathrm{EG}}(m, 2^s)$ plus one.

Because the parity-check matrix $\mathbf{H}_{\mathrm{EG}}(m, 2^s)$ of the EG code $C_{\mathrm{EG}}(m, 2^s)$ has an orthogonal structure, $C_{\mathrm{EG}}(m, 2^s)$ can be decoded with the OSMLD. For each column position $j$, $0 \leq j < 2^{ms} - 1$, there are $(2^{(m-1)s} - 1)2^s/(2^s - 1)$ rows in $\mathbf{H}_{\mathrm{EG}}(m, 2^s)$ which are orthogonal on it. Hence, with the OSMLD, $(2^{(m-1)s} - 1)2^s/(2^s - 1)$ syndrome check-sums orthogonal on each error bit $e_j$ in an error vector $\mathbf{e}$ can be formed. Based on these orthogonal syndrome check-sums, the code is capable of correcting

$$K2^s/2 = \frac{(2^{(m-1)s} - 1)2^{s-1}}{2^s - 1} \tag{7.42}$$

or fewer errors with the OSMLD.

**Example 7.8** Let $m = 3$ and $s = 2$. Consider the three-dimensional Euclidean geometry $\mathrm{EG}(3, 2^2)$ over $\mathrm{GF}(2^2)$. By removing the origin and all the lines passing through the origin, we obtain a subgeometry $\mathrm{EG}^*(3, 2^2)$ which comprises $K = (2^{(3-1)\times 2} - 1)/(2^2 - 1) = 5$ cyclic classes, each consisting of $2^{ms} - 1 = 2^{3\times 2} - 1 = 63$ lines. Based on these cyclic classes of lines of $\mathrm{EG}^*(3, 2^2)$, we can construct a three-dimensional cyclic EG code $C_{\mathrm{EG}}(3, 2^2)$. The generator polynomial $\mathbf{g}_{\mathrm{EG}}(X)$ of this EG code can be found based on Theorem 7.3 which is calculated as follows.

Let $\alpha$ be a primitive element of the field $\mathrm{GF}(2^6)$ which is a realization of $\mathrm{EG}(3, 2^2)$. It follows from Theorem 7.3 that $\alpha^k$ is a root of $\mathbf{g}_{\mathrm{EG}}(X)$ if and only if its power $k$ satisfies the following condition:

$$0 < \max_{0 \le \ell < 2} \left( W_{2^2}(k^{(\ell)}) \right) \le 6.$$

Through calculation, we find that there are 50 integers satisfying the above condition: 1, 2, 3, 4, 5, 6, 7, 8, 9, 10, 11, 12, 13, 14, 15, 16, 17, 18, 19, 20, 21, 22, 24, 25, 26, 27, 28, 30, 32, 33, 34, 35, 36, 37, 38, 39, 40, 41, 42, 44, 45, 48, 49, 50, 51, 52, 54, 56, 57, 60. Thus, $\mathbf{g}_{\mathrm{EG}}(X)$ has 50 roots over $\mathrm{GF}(2^6)$: $\alpha, \alpha^2, \alpha^3, \alpha^4, \alpha^5, \alpha^6, \alpha^7, \alpha^8, \alpha^9, \alpha^{10}, \alpha^{11}, \alpha^{12}, \alpha^{13}, \alpha^{14}, \alpha^{15}, \alpha^{16}, \alpha^{17}, \alpha^{18}, \alpha^{19}, \alpha^{20}, \alpha^{21}, \alpha^{22}, \alpha^{24}, \alpha^{25}, \alpha^{26}, \alpha^{27}, \alpha^{28}, \alpha^{30}, \alpha^{32}, \alpha^{33}, \alpha^{34}, \alpha^{35}, \alpha^{36}, \alpha^{37}, \alpha^{38}, \alpha^{39}, \alpha^{40}, \alpha^{41}, \alpha^{42}, \alpha^{44}, \alpha^{45}, \alpha^{48}, \alpha^{49}, \alpha^{50}, \alpha^{51}, \alpha^{52}, \alpha^{54}, \alpha^{56}, \alpha^{57}, \alpha^{60}$, which can be partitioned into the following 10 conjugate sets:

$$\Phi_1 = \{\alpha, \alpha^2, \alpha^4, \alpha^8, \alpha^{16}, \alpha^{32}\}, \quad \Phi_3 = \{\alpha^3, \alpha^6, \alpha^{12}, \alpha^{24}, \alpha^{48}, \alpha^{33}\},$$

$$\Phi_5 = \{\alpha^5, \alpha^{10}, \alpha^{20}, \alpha^{40}, \alpha^{17}, \alpha^{34}\}, \quad \Phi_7 = \{\alpha^7, \alpha^{14}, \alpha^{28}, \alpha^{56}, \alpha^{49}, \alpha^{35}\},$$

$$\Phi_9 = \{\alpha^9, \alpha^{18}, \alpha^{36}\}, \quad \Phi_{11} = \{\alpha^{11}, \alpha^{22}, \alpha^{44}, \alpha^{25}, \alpha^{50}, \alpha^{37}\},$$

$$\Phi_{13} = \{\alpha^{13}, \alpha^{26}, \alpha^{52}, \alpha^{41}, \alpha^{19}, \alpha^{38}\}, \quad \Phi_{15} = \{\alpha^{15}, \alpha^{30}, \alpha^{60}, \alpha^{57}, \alpha^{51}, \alpha^{39}\},$$

$$\Phi_{21} = \{\alpha^{21}, \alpha^{42}\}, \quad \Phi_{27} = \{\alpha^{27}, \alpha^{54}, \alpha^{45}\}.$$

The minimal polynomials of the elements in the above 10 conjugate sets are

$$m_1(X) = 1 + X + X^6, \quad m_3(X) = 1 + X + X^2 + X^4 + X^6,$$

$$m_5(X) = 1 + X + X^2 + X^5 + X^6, \quad m_7(X) = 1 + X^3 + X^6,$$

$$m_9(X) = 1 + X^2 + X^3, \quad m_{11}(X) = 1 + X^2 + X^3 + X^5 + X^6,$$

$$m_{13}(X) = 1 + X + X^3 + X^4 + X^6, \quad m_{15}(X) = 1 + X^2 + X^4 + X^5 + X^6,$$

$$m_{21}(X) = 1 + X + X^2, \quad m_{27}(X) = 1 + X + X^3.$$

Then,

$$\mathbf{g}_{\mathrm{EG}}(X) = \mathrm{LCM}\{m_1(X)m_3(X)m_5(X)m_7(X)m_9(X)m_{11}(X)m_{13}(X)m_{15}(X)m_{21}(X)m_{27}(X)\}$$

$$= m_1(X)m_3(X)m_5(X)m_7(X)m_9(X)m_{11}(X)m_{13}(X)m_{15}(X)m_{21}(X)m_{27}(X)$$

$$= 1 + X^2 + X^5 + X^{10} + X^{12} + X^{15} + X^{17} + X^{18} + X^{19} + X^{23} + X^{25} + X^{30}$$

$$+ X^{32} + X^{34} + X^{35} + X^{37} + X^{41} + X^{43} + X^{44} + X^{45} + X^{48} + X^{49} + X^{50}.$$

Therefore, the three-dimensional EG code $C_{\mathrm{EG}}(3, 2^2)$ is a $(63, 13)$ cyclic code, which is capable of correcting 10 or fewer random errors with the OSMLD. It is noted that the generator polynomial $\mathbf{g}_{\mathrm{EG}}(X)$ of $C_{\mathrm{EG}}(3, 2^2)$ has 22 consecutive powers of $\alpha$ as roots. According to the BCH bound, $C_{\mathrm{EG}}(3, 2^2)$ has minimum distance $d_{\min} \ge 23$. Because its generator polynomial has weight 23, the minimum distance of $C_{\mathrm{EG}}(3, 2^2)$ is exactly 23.

A parity-check matrix of the code in the form of (7.36) can be constructed by using the generators of the five cyclic classes of lines in $\mathrm{EG}^*(3, 2^2)$.     ▲▲

For $1 \leq \kappa \leq K$, we can form a cyclic code by using any $\kappa$ cyclic classes of lines in $\mathrm{EG}^*(m, 2^s)$. Let $Q_0, Q_1, \ldots, Q_{\kappa-1}$ be the $\kappa$ chosen cyclic classes of lines in $\mathrm{EG}^*(m, 2^s)$. Using these $\kappa$ cyclic classes of lines in $\mathrm{EG}^*(m, 2^s)$, we form $\kappa$ circulants, $\mathbf{H}_0, \mathbf{H}_1, \ldots, \mathbf{H}_{\kappa-1}$, each of size $(2^{ms} - 1) \times (2^{ms} - 1)$, and arrange them in a column as follows:

$$\mathbf{H}_{\mathrm{EG},\kappa}(m, 2^s) = \begin{bmatrix} \mathbf{H}_0 \\ \mathbf{H}_1 \\ \vdots \\ \mathbf{H}_{\kappa-1} \end{bmatrix}. \qquad (7.43)$$

The matrix $\mathbf{H}_{\mathrm{EG},\kappa}(m, 2^s)$ is a $\kappa(2^{ms} - 1) \times (2^{ms} - 1)$ matrix over $\mathrm{GF}(2)$ with row and column weights $2^s$ and $\kappa 2^s$, respectively. Using $\mathbf{H}_{\mathrm{EG},\kappa}(m, 2^s)$ as the parity-check matrix of a linear block code, the null space over $\mathrm{GF}(2)$ of $\mathbf{H}_{\mathrm{EG},\kappa}(m, 2^s)$ gives a cyclic EG code $C_{\mathrm{EG},\kappa}(m, 2^s)$ of length $2^{ms} - 1$ and minimum distance at least $\kappa 2^s + 1$ which is OSML decodable and is capable of correcting $\kappa 2^{s-1}$ or fewer errors.

With various choices of $\kappa$, we can construct a family of OSML decodable EG codes based on an $m$-dimensional Euclidean geometry $\mathrm{EG}(m, 2^s)$.

The cyclic code generated by the parity-check matrix $\mathbf{H}_{\mathrm{EG},\kappa}(m, 2^s)$ is, in general, a low-rate code even though it has a large minimum distance. To construct a high-rate code using the $\kappa$ circulants in $\mathbf{H}_{\mathrm{EG},\kappa}(m, 2^s)$, we can arrange them in a row as follows:

$$\mathbf{H}_{\mathrm{EG},\kappa}^*(m, 2^s) = [\mathbf{H}_0, \mathbf{H}_1, \ldots, \mathbf{H}_{\kappa-1}]. \qquad (7.44)$$

The null space over $\mathrm{GF}(2)$ of $\mathbf{H}_{\mathrm{EG},\kappa}^*(m, 2^s)$ gives a linear block code of length $\kappa(2^{ms} - 1)$ with rate at least $(\kappa - 1)/\kappa$ and minimum distance at least $2^s + 1$. Such a code is no longer cyclic but is quasi-cyclic (QC), as shown in Section 4.13. It is OSML decodable and capable of correcting $2^{s-1}$ or fewer random errors. These QC finite-geometry codes, as a class of LDPC codes, will be discussed further in Chapter 12.

**Example 7.9** Using the five cyclic classes in the subgeometry $\mathrm{EG}^*(3, 2^2)$ of $\mathrm{EG}(3, 2^2)$, we construct five circulants $\mathbf{H}_0, \mathbf{H}_1, \mathbf{H}_2, \mathbf{H}_3, \mathbf{H}_4$, each of size $63 \times 63$ with weight 4. Choosing two circulants, say $\mathbf{H}_0$ and $\mathbf{H}_1$, we form a $126 \times 63$ matrix $\mathbf{H}_{\mathrm{EG},2}(3, 2^2)$ in the form of (7.43) with row and column weights 4 and 8, respectively. The null space over $\mathrm{GF}(2)$ of $\mathbf{H}_{\mathrm{EG},2}(3, 2^2)$ gives a $(63, 13)$ cyclic EG code $C_{\mathrm{EG},2}(3, 2^2)$ with minimum distance at least 9. With the OSMLD, the code $C_{\mathrm{EG},2}(3, 2^2)$ is capable of correcting four or fewer errors.

If we use the five circulants constructed above and arrange them in a row (i.e., in the form of (7.44)), we obtain a $63 \times 315$ matrix $\mathbf{H}_{\mathrm{EG},5}^*(3, 2^2) = [\mathbf{H}_0, \mathbf{H}_1, \mathbf{H}_2, \mathbf{H}_3, \mathbf{H}_4]$ with column and row weights 4 and 20, respectively. The null space over $\mathrm{GF}(2)$ of $\mathbf{H}_{\mathrm{EG},5}^*(3, 2^2)$ gives a $(315, 265)$ QC-EG code with rate 0.84 and minimum distance at least 5. The code is capable of correcting up to two random errors with the OSMLD. ▲▲

## 7.5 Projective Geometries

In this section, we introduce the other category of finite geometries, named *projective geometries*. This category of finite geometries can also be constructed based on finite fields.

Consider the Galois field $GF(q^{m+1})$ which is an extension of the ground field $GF(q)$. Let $\alpha$ be a primitive element of $GF(q^{m+1})$. Then, $\alpha^0, \alpha, \alpha^2, \ldots, \alpha^{q^{m+1}-2}$ form all the nonzero elements of $GF(q^{m+1})$. Let

$$n = \frac{q^{m+1} - 1}{q - 1} \tag{7.45}$$

and $\beta = \alpha^n$. Then, the order of $\beta$ is $q - 1$ and $0, 1, \beta, \beta^2, \ldots, \beta^{q-2}$ give the $q$ elements of $GF(q)$.

Consider the set of the first $n$ powers of $\alpha$:

$$\Gamma = \{\alpha^0, \alpha, \alpha^2, \ldots, \alpha^{n-1}\}. \tag{7.46}$$

No element $\alpha^i \in \Gamma$ can be a product of an element $\beta^\ell \in GF(q)$ and another element $\alpha^j \in \Gamma$, i.e., $\alpha^i \neq \beta^\ell \alpha^j$ for $0 \leq i, j < n$, $i \neq j$, and $0 \leq \ell < q - 1$.

Partition the nonzero elements of $GF(q^{m+1})$ into $n$ disjoint subsets as follows:

$$\begin{aligned}
&\{\alpha^0, \beta\alpha^0, \beta^2\alpha^0, \ldots, \beta^{q-2}\alpha^0\}, \\
&\{\alpha^1, \beta\alpha^1, \beta^2\alpha^1, \ldots, \beta^{q-2}\alpha^1\}, \\
&\{\alpha^2, \beta\alpha^2, \beta^2\alpha^2, \ldots, \beta^{q-2}\alpha^2\}, \\
&\quad\vdots \\
&\{\alpha^{n-1}, \beta\alpha^{n-1}, \beta^2\alpha^{n-1}, \ldots, \beta^{q-2}\alpha^{n-1}\}.
\end{aligned} \tag{7.47}$$

Each subset consists of $q - 1$ nonzero elements of $GF(q^{m+1})$ and each element in a subset is a *multiple* of the first element in the subset. Represent each subset by its first element as follows: for $0 \leq i < n$,

$$(\alpha^i) = \{\alpha^i, \beta\alpha^i, \ldots, \beta^{q-2}\alpha^i\}. \tag{7.48}$$

For any element $\alpha^j \in GF(q^{m+1})$, if $\alpha^j = \beta^\ell \alpha^i$ with $0 \leq i < n$ and $0 \leq \ell < q-1$, then $\alpha^j$ is represented by $\alpha^i$, the first element in $(\alpha^i)$.

If each element in $GF(q^{m+1})$ is represented by an $(m+1)$-tuple over $GF(q)$, then $(\alpha^i)$ consists of $q - 1$ nonzero $(m+1)$-tuples over $GF(q)$, in which each $(m+1)$-tuple is a multiple of the $(m+1)$-tuple representation of $\alpha^i$. Suppose we regard all the $q - 1$ $(m+1)$-tuples in $(\alpha^i)$ as a single point in a finite geometry over $GF(q)$ and represent the point by the first $(m+1)$-tuple in $(\alpha^i)$. Then, the points

$$(\alpha^0), (\alpha^1), \ldots, (\alpha^{n-1}) \tag{7.49}$$

are said to form an $m$-dimensional *projective geometry* (PG) over $GF(q)$, denoted by $PG(m, q)$.

Note that, if the 0-element of $\mathrm{GF}(q^{m+1})$ is included in the set $(\alpha^i)$, we obtain a set $\{0, \alpha^i, \beta\alpha^i, \ldots, \beta^{q-2}\alpha^i\}$ of $q$ elements. This set of $q$ elements, viewed as $(m+1)$-tuples over $\mathrm{GF}(q)$, is simply a one-dimensional subspace of the vector space $\mathbf{V}_{m+1}$ of all the $(m+1)$-tuples over $\mathrm{GF}(q)$ and hence it is a line in the $(m+1)$-dimensional Euclidean geometry $\mathrm{EG}(m+1, q)$ over $\mathrm{GF}(q)$, which passes through the origin of $\mathrm{EG}(m+1, q)$. Therefore, we may regard a point $(\alpha^i)$ of $\mathrm{PG}(m, q)$ as a *projection* of a line of $\mathrm{EG}(m+1, q)$ passing through the origin of $\mathrm{EG}(m+1, q)$. From this point of view, we call $\mathrm{PG}(m, q)$ a projective geometry. The all-zero $(m+1)$-tuple is not a point in $\mathrm{PG}(m, q)$. Hence, $\mathrm{PG}(m, q)$ does not have an origin.

Let $(\alpha^i)$ and $(\alpha^j)$ be two distinct points in $\mathrm{PG}(m, q)$. Then, the line passing through (or connecting) $(\alpha^i)$ and $(\alpha^j)$ consists of points of the following form:

$$(\delta_1\alpha^i + \delta_2\alpha^j), \tag{7.50}$$

where $\delta_1$ and $\delta_2$ are elements from $\mathrm{GF}(q)$ and are *not both equal to zero*. We denote this line by

$$\{(\delta_1\alpha^i + \delta_2\alpha^j)\} \triangleq \{(\delta_1\alpha^i + \delta_2\alpha^j) : \delta_1, \delta_2 \in \mathrm{GF}(q) \text{ and not both } \delta_1 \text{ and } \delta_2 \text{ are zero}\}. \tag{7.51}$$

There are $q^2 - 1$ choices of $\delta_1$ and $\delta_2$ from $\mathrm{GF}(q)$ (excluding $\delta_1 = \delta_2 = 0$). Because $(\delta_1\alpha^i + \delta_2\alpha^j)$ and $(\beta^\ell\delta_1\alpha^i + \beta^\ell\delta_2\alpha^j)$ are the same point in $\mathrm{PG}(m, q)$ where $\beta^\ell \in \mathrm{GF}(q)$ with $0 \leq \ell < q - 1$ (i.e., a choice of $(\delta_1, \delta_2)$ and its $q - 2$ multiples, $(\beta^\ell\delta_1, \beta^\ell\delta_2)$, $0 < \ell < q - 1$, result in the same point), the line connecting points $(\alpha^i)$ and $(\alpha^j)$ consists of

$$\frac{q^2 - 1}{q - 1} = q + 1 \tag{7.52}$$

points. For any two points in $\mathrm{PG}(m, q)$, there is *one and only one line* connecting them. From the number of points ($n = (q^{m+1} - 1)/(q - 1)$ points in $\mathrm{PG}(m, q)$) and the definition of a line in $\mathrm{PG}(m, q)$ (each line consists of $q + 1$ points), we find that the number of lines in $\mathrm{PG}(m, q)$ is

$$M = \frac{(q^{m+1} - 1)(q^m - 1)}{(q^2 - 1)(q - 1)}. \tag{7.53}$$

Let $(\alpha^k)$ be a point not on the line $\{(\delta_1\alpha^i + \delta_2\alpha^j)\}$. Then, the line $L_1 = \{(\delta_1\alpha^i + \delta_2\alpha^j)\}$ and the line $L_2 = \{(\delta_1\alpha^i + \delta_2\alpha^k)\}$ have $(\alpha^i)$ as the common point (with $\delta_1 = 1$ and $\delta_2 = 0$), which is the only common point. Hence, the two lines $L_1$ and $L_2$ intersect at the point $(\alpha^i)$. The number of lines in $\mathrm{PG}(m, q)$ that intersect at a point in $\mathrm{PG}(m, q)$ is equal to

$$\frac{q^m - 1}{q - 1}. \tag{7.54}$$

The lines that intersect at a point $(\alpha^i)$ of $\mathrm{PG}(m, q)$ are said to form an *intersecting bundle* of lines at the point $(\alpha^i)$. Similar to a Euclidean geometry, two lines in a projective geometry are either disjoint or they intersect at one and only one point. However, a projective geometry does not have the parallel bundle structure that a Euclidean geometry has.

Similar to a Euclidean geometry, the lines in a projective geometry also have cyclic structure. Let $L_0 = \{(\delta_1 \alpha^{j_0} + \delta_2 \alpha^{j_1})\}$ be a line in $\mathrm{PG}(m, q)$. For $0 \leq i < n = (q^{m+1} - 1)/(q - 1)$,

$$\alpha^i L_0 = (\delta_0 \alpha^{j_0 + i} + \delta_1 \alpha^{j_1 + i}), \tag{7.55}$$

is also a line in $\mathrm{PG}(m, q)$, which is called the $i$th cyclic-shift of $L_0$. If $m$ is even, $m + 1$ is odd. In this case, the lines $L_0, \alpha L_0, \ldots, \alpha^{n-1} L_0$ are all different and they form a cyclic class. Then, the lines of $\mathrm{PG}(m, q)$ can be partitioned into

$$\frac{q^m - 1}{q^2 - 1} \tag{7.56}$$

cyclic classes, each consisting of $(q^{m+1} - 1)/(q - 1)$ lines. For even $m$, each line in $\mathrm{PG}(m, q)$ can be used to generate a cyclic class by cyclically shifting it $(q^{m+1} - 1)/(q - 1) - 1$ times. Such a line is said to be *primitive* [21]. However, for odd $m$ with $m \geq 3$, there are $(q^{m+1} - 1)/(q^2 - 1)$ lines in $\mathrm{PG}(m, q)$ which are not primitive. These nonprimitive lines form a cyclic class of size $(q^{m+1} - 1)/(q^2 - 1)$. All the other lines in $\mathrm{PG}(m, q)$ are primitive and they can be partitioned into

$$\frac{q(q^{m-1} - 1)}{q^2 - 1} \tag{7.57}$$

cyclic classes, each consisting of $(q^{m+1} - 1)/(q - 1)$ lines.

Similar to an $m$-dimensional Euclidean geometry $\mathrm{EG}(m, q)$ over $\mathrm{GF}(q)$ with $m \geq 2$, an $m$-dimensional projective geometry $\mathrm{PG}(m, q)$ over $\mathrm{GF}(q)$ contains flats. For $0 \leq \mu < m$, let $(\alpha^{j_0}), (\alpha^{j_1}), \ldots, (\alpha^{j_\mu})$ be $\mu + 1$ linearly independent points. Then, a $\mu$-flat in $\mathrm{PG}(m, q)$ consists of points of the following form:

$$(\delta_0 \alpha^{j_0} + \delta_1 \alpha^{j_1} + \cdots + \delta_\mu \alpha^{j_\mu}), \tag{7.58}$$

where $\delta_k \in \mathrm{GF}(q)$ with $0 \leq k \leq \mu$, and not all of $\delta_0, \delta_1, \ldots, \delta_\mu$ are zero. We denote this $\mu$-flat by

$$\{(\delta_0 \alpha^{j_0} + \delta_1 \alpha^{j_1} + \cdots + \delta_\mu \alpha^{j_\mu})\} \triangleq$$

$$\{(\delta_0 \alpha^{j_0} + \delta_1 \alpha^{j_1} + \cdots + \delta_\mu \alpha^{j_\mu}) : \delta_k \in \mathrm{GF}(q), 0 \leq k \leq \mu, \text{ not all } \delta_k \text{s are zero}\}.$$

The number of points on a $\mu$-flat is

$$\frac{q^{\mu+1} - 1}{q - 1} \tag{7.59}$$

and there are

$$\prod_{i=0}^{\mu} \frac{q^{m-i+1} - 1}{q^{\mu-i+1} - 1} \tag{7.60}$$

$\mu$-flats in $\mathrm{PG}(m, q)$.

**Example 7.10** Let $m = 2$ and $s = 2$. Consider the projective geometry $PG(2, 2^2)$ over $GF(2^2)$. The geometry $PG(2, 2^2)$ can be constructed from the extension field $GF(2^6)$ of $GF(2^2)$. The field $GF(2^6)$ is given in Table 5.14 with $GF(2)$ as the ground field and in Table 7.3 with $GF(2^2)$ as the ground field. Let $\alpha$ be a primitive element of $GF(2^6)$, $n = (4^{2+1} - 1)/(4 - 1) = 21$, and $\beta = \alpha^n = \alpha^{21}$. Then, $GF(2^2) = \{0, 1, \beta = \alpha^{21}, \beta^2 = \alpha^{42}\}$. The geometry $PG(2, 2^2)$ consists of the following 21 lines:

$$(\alpha^0), (\alpha^1), (\alpha^2), (\alpha^3), (\alpha^4), (\alpha^5), (\alpha^6), (\alpha^7), (\alpha^8), (\alpha^9), (\alpha^{10}),$$
$$(\alpha^{11}), (\alpha^{12}), (\alpha^{13}), (\alpha^{14}), (\alpha^{15}), (\alpha^{16}), (\alpha^{17}), (\alpha^{18}), (\alpha^{19}), (\alpha^{20}).$$

Consider a line passing through the two points, $(\alpha^0)$ and $(\alpha^1)$:

$$L = \{(\delta_1 \alpha^0 + \delta_2 \alpha) : \delta_1, \delta_2 \in GF(2^2)\},$$

where $\delta_1$ and $\delta_2$ are not all zero. The points on this line are calculated as follows:

$$\delta_1 = 0, \quad \delta_2 = 1 \quad : (\alpha),$$
$$\delta_1 = 0, \quad \delta_2 = \beta \quad : (\beta\alpha) = (\alpha),$$
$$\delta_1 = 0, \quad \delta_2 = \beta^2 \quad : (\beta^2\alpha) = (\alpha),$$
$$\delta_1 = 1, \quad \delta_2 = 0 \quad : (1 \cdot \alpha^0 + 0 \cdot \alpha) = (\alpha^0),$$
$$\delta_1 = 1, \quad \delta_2 = 1 \quad : (1 \cdot \alpha^0 + 1 \cdot \alpha) = (\alpha^6),$$
$$\delta_1 = 1, \quad \delta_2 = \beta \quad : (1 \cdot \alpha^0 + \beta \cdot \alpha) = (\alpha^{50}) = (\beta^2\alpha^8) = (\alpha^8),$$
$$\delta_1 = 1, \quad \delta_2 = \beta^2 : (1 \cdot \alpha^0 + \beta^2 \cdot \alpha) = (\alpha^{39}) = (\beta\alpha^{18}) = (\alpha^{18}),$$
$$\delta_1 = \beta, \quad \delta_2 = 0 \quad : (\beta \cdot \alpha^0 + 0 \cdot \alpha) = (\alpha^0),$$
$$\delta_1 = \beta, \quad \delta_2 = 1 \quad : (\beta \cdot \alpha^0 + 1 \cdot \alpha) = (\alpha^{60}) = (\beta^2\alpha^{18}) = (\alpha^{18}),$$
$$\delta_1 = \beta, \quad \delta_2 = \beta \quad : (\beta \cdot \alpha^0 + \beta \cdot \alpha) = (\alpha^{27}) = (\beta\alpha^6) = (\alpha^6),$$
$$\delta_1 = \beta, \quad \delta_2 = \beta^2 : (\beta \cdot \alpha^0 + \beta^2 \cdot \alpha) = (\alpha^8),$$
$$\delta_1 = \beta^2, \delta_2 = 0 \quad : (\beta^2 \cdot \alpha^0 + 0 \cdot \alpha) = (\beta^2\alpha^0) = (\alpha^0),$$
$$\delta_1 = \beta^2, \delta_2 = 1 \quad : (\beta^2 \cdot \alpha^0 + 1 \cdot \alpha) = (\alpha^{29}) = (\beta\alpha^8) = (\alpha^8),$$
$$\delta_1 = \beta^2, \delta_2 = \beta \quad : (\beta^2 \cdot \alpha^0 + \beta \cdot \alpha) = (\alpha^{18}),$$
$$\delta_1 = \beta^2, \delta_2 = \beta^2 : (\beta^2 \cdot \alpha^0 + \beta^2 \cdot \alpha) = (\alpha^{48}) = (\beta^2\alpha^6) = (\alpha^6).$$

Based on the above calculation, it can be seen that there are five distinct points on the line $L$: $(\alpha^0)$, $(\alpha^1)$, $(\alpha^6)$, $(\alpha^8)$, and $(\alpha^{18})$, i.e.,

$$L = \{(\alpha^0), (\alpha^1), (\alpha^6), (\alpha^8), (\alpha^{18})\}.$$

Choosing a point $(\alpha^0)$ on the line $L$ and another point $(\alpha^2)$ which is not on $L$, we can obtain another line

$$L^* = \{(\delta_1 \alpha^0 + \delta_2 \alpha^2) : \delta_1, \delta_2 \in GF(2^2)\}$$
$$= \{(\alpha^0), (\alpha^2), (\alpha^{12}), (\alpha^{15}), (\alpha^{16})\}.$$

The two lines $L$ and $L^*$ have one common point $(\alpha^0)$, i.e., they intersect at point $(\alpha^0)$. Following this process, we can find all the 21 lines in $\mathrm{PG}(2, 2^2)$ which are listed in Table 7.7. ▲▲

Table 7.7 Lines of the projective geometry $\mathrm{PG}(2, 2^2)$ over $\mathrm{GF}(2^2)$.

| Lines | |
|---|---|
| $\{(\alpha^0), (\alpha^1), (\alpha^6), (\alpha^8), (\alpha^{18})\}$ | $\{(\alpha^0), (\alpha^2), (\alpha^{12}), (\alpha^{15}), (\alpha^{16})\}$ |
| $\{(\alpha^0), (\alpha^3), (\alpha^4), (\alpha^9), (\alpha^{11})\}$ | $\{(\alpha^0), (\alpha^5), (\alpha^7), (\alpha^{17}), (\alpha^{20})\}$ |
| $\{(\alpha^0), (\alpha^{10}), (\alpha^{13}), (\alpha^{14}), (\alpha^{19})\}$ | $\{(\alpha^1), (\alpha^2), (\alpha^7), (\alpha^9), (\alpha^{19})\}$ |
| $\{(\alpha^1), (\alpha^3), (\alpha^{13}), (\alpha^{16}), (\alpha^{17})\}$ | $\{(\alpha^1), (\alpha^4), (\alpha^5), (\alpha^{10}), (\alpha^{12})\}$ |
| $\{(\alpha^1), (\alpha^{11}), (\alpha^{14}), (\alpha^{15}), (\alpha^{20})\}$ | $\{(\alpha^2), (\alpha^3), (\alpha^8), (\alpha^{10}), (\alpha^{20})\}$ |
| $\{(\alpha^2), (\alpha^4), (\alpha^{14}), (\alpha^{17}), (\alpha^{18})\}$ | $\{(\alpha^2), (\alpha^5), (\alpha^6), (\alpha^{11}), (\alpha^{13})\}$ |
| $\{(\alpha^3), (\alpha^5), (\alpha^{15}), (\alpha^{18}), (\alpha^{19})\}$ | $\{(\alpha^3), (\alpha^6), (\alpha^7), (\alpha^{12}), (\alpha^{14})\}$ |
| $\{(\alpha^4), (\alpha^6), (\alpha^{16}), (\alpha^{19}), (\alpha^{20})\}$ | $\{(\alpha^4), (\alpha^7), (\alpha^8), (\alpha^{13}), (\alpha^{15})\}$ |
| $\{(\alpha^5), (\alpha^8), (\alpha^9), (\alpha^{14}), (\alpha^{16})\}$ | $\{(\alpha^6), (\alpha^9), (\alpha^{10}), (\alpha^{15}), (\alpha^{17})\}$ |
| $\{(\alpha^7), (\alpha^{10}), (\alpha^{11}), (\alpha^{16}), (\alpha^{18})\}$ | $\{(\alpha^8), (\alpha^{11}), (\alpha^{12}), (\alpha^{17}), (\alpha^{19})\}$ |
| $\{(\alpha^9), (\alpha^{12}), (\alpha^{13}), (\alpha^{18}), (\alpha^{20})\}$ | |

## 7.6 Cyclic Codes Constructed Based on Projective Geometries

Using projective geometries, cyclic codes can be constructed based on the cyclic structure of lines in these geometries. These cyclic codes are called *projective-geometry* (PG) *codes* and are OSML decodable. In this section, we focus on the construction of codes based on PGs over fields of characteristic 2, i.e., $q = 2^s$.

Consider an $m$-dimensional projective geometry $\mathrm{PG}(m, 2^s)$ over $\mathrm{GF}(2^s)$ with $m \geq 2$. This geometry consists of (see (7.53))

$$n = (2^{s(m+1)} - 1)/(2^s - 1) \tag{7.61}$$

points and

$$M = \frac{(2^{ms} - 1)(2^{s(m+1)} - 1)}{(2^{2s} - 1)(2^s - 1)} \tag{7.62}$$

lines.

The incidence vector of each line is an $n$-tuple $\mathbf{v} = (v_0, v_1, v_2, \ldots, v_{n-1})$ over $\mathrm{GF}(2)$. Form an $M \times n$ matrix $\mathbf{H}_{\mathrm{PG}}(m, 2^s)$ whose rows are the incidence vectors of the lines in $\mathrm{PG}(m, 2^s)$ and columns correspond to the points of $\mathrm{PG}(m, 2^s)$.

The columns of $\mathbf{H}_{\mathrm{PG}}(m, 2^s)$ are arranged (or labeled) in the order of $(\alpha^0)$, $(\alpha^1)$, $(\alpha^2), \ldots, (\alpha^{n-1})$. Because each line in $\mathrm{PG}(m, 2^s)$ consists of $2^s + 1$ points, each row of $\mathbf{H}_{\mathrm{PG}}(m, 2^s)$ has weight $2^s + 1$. Because each point is intersected by $(2^{ms} - 1)/(2^s - 1)$ lines, each column of $\mathbf{H}_{\mathrm{PG}}(m, 2^s)$ has weight $(2^{ms} - 1)/(2^s - 1)$. The matrix $\mathbf{H}_{\mathrm{PG}}(m, 2^s)$ has orthogonal structure. For $0 \le j < n$, there are $(2^{ms} - 1)/(2^s - 1)$ rows in $\mathbf{H}_{\mathrm{PG}}(m, 2^s)$ which are orthogonal on the $j$th column position $(\alpha^j)$. Because the lines in $\mathrm{PG}(m, 2^s)$ have a cyclic structure and they can be grouped in cyclic classes, $\mathbf{H}_{\mathrm{PG}}(m, 2^s)$ can be arranged as columns of circulants. For even $m$, each circulant in $\mathbf{H}_{\mathrm{PG}}(m, 2^s)$ is of size $n \times n$ with $n = (2^{(m+1)s} - 1)/(2^s - 1)$. For odd $m$, every circulant in $\mathbf{H}_{\mathrm{PG}}(m, 2^s)$ is of size $n \times n$ except for one which is an $n/(2^s + 1) \times n$ *nonsquare* circulant.

Using $\mathbf{H}_{\mathrm{PG}}(m, 2^s)$ as the parity-check matrix of a linear block code, the null space over $\mathrm{GF}(2)$ of $\mathbf{H}_{\mathrm{PG}}(m, 2^s)$ gives a cyclic PG code $C_{\mathrm{PG}}(m, 2^s)$ of length $n = (2^{(m+1)s} - 1)/(2^s - 1)$. Based on the orthogonal structure of its parity-check matrix $\mathbf{H}_{\mathrm{PG}}(m, 2^s)$, the code $C_{\mathrm{PG}}(m, 2^s)$ is OSML decodable and is capable of correcting $\lfloor \gamma/2 \rfloor = \lfloor (2^{ms} - 1)/2(2^s - 1) \rfloor$ or fewer errors. Hence, the PG code $C_{\mathrm{PG}}(m, 2^s)$ has minimum distance at least $\gamma + 1 = (2^{ms} - 1)/(2^s - 1) + 1$.

Let $k$ be a nonnegative integer less than $2^{(m+1)s} - 1$. For a nonnegative integer $\ell$ with $0 \le \ell < s$, let $k^{(\ell)}$ be the remainder resulting from dividing $k2^\ell$ by $2^{(m+1)s} - 1$, i.e., $k^{(\ell)} = \left( k2^\ell \text{ modulo } (2^{(m+1)s} - 1) \right)$. Express $k^{(\ell)}$ in radix-$2^s$ form,

$$k^{(\ell)} = k_0^{(\ell)} + k_1^{(\ell)} 2^s + k_2^{(\ell)} 2^{2s} + \cdots + k_{m-1}^{(\ell)} 2^{(m-1)s} + k_m^{(\ell)} 2^{ms}, \qquad (7.63)$$

where $0 \le k_i^{(\ell)} < 2^s$ and $0 \le i \le m$. The $2^s$-weight of $k^{(\ell)}$ is

$$W_{2^s}(k^{(\ell)}) = k_0^{(\ell)} + k_1^{(\ell)} + k_2^{(\ell)} + \cdots + k_{m-1}^{(\ell)} + k_m^{(\ell)}. \qquad (7.64)$$

The generator polynomial of the cyclic PG code $C_{\mathrm{PG}}(m, 2^s)$ is characterized by the following theorem [32, 35, 36].

**Theorem 7.4** *Let $\alpha$ be a primitive element of $\mathrm{GF}(2^{(m+1)s})$ and $k$ be a nonnegative integer less than $2^{(m+1)s} - 1$. Then, the generator polynomial $\mathbf{g}_{\mathrm{PG}}(X)$ of the PG code $C_{\mathrm{PG}}(m, 2^s)$ constructed based on the lines of $\mathrm{PG}(m, 2^s)$ has $\alpha^k$ as a root if and only if $k$ is divisible by $2^s - 1$ and*

$$0 < \max_{0 \le \ell < s} \left( W_{2^s}(k^{(\ell)}) \right) = t(2^s - 1), \qquad (7.65)$$

*with $0 \le t \le m$.*

Let $\zeta = \alpha^{2^s - 1} \in \mathrm{GF}(2^{(m+1)s})$. Then, the order of $\zeta$ is $n = (2^{(m+1)s} - 1)/(2^s - 1)$. From the characterization of the roots of $\mathbf{g}_{\mathrm{PG}}(X)$ specified by (7.65), it can be shown that $\mathbf{g}_{\mathrm{PG}}(X)$ has the following $(2^{ms} - 1)/(2^s - 1)$ consecutive powers of $\zeta$,

$$\zeta^0 = 1, \zeta, \zeta^2, \ldots, \zeta^{(2^{ms}-1)/(2^s-1)-1} \qquad (7.66)$$

as roots. Therefore, it follows from the BCH bound that the minimum distance of the $m$-dimensional cyclic PG code $C_{\mathrm{PG}}(m, 2^s)$ is at least $(2^{ms}-1)/(2^s-1)+1$. The number of parity-check symbols of the $m$-dimensional PG code $C_{\mathrm{PG}}(m, 2^s)$ for a given $s$ can be enumerated by determining the total number of roots of its generator polynomial. A combinatorial expression for this number can be found in [33].

A special subclass of PG codes is the class of codes constructed based on two-dimensional projective geometries $\mathrm{PG}(2, 2^s)$ over $\mathrm{GF}(2^s)$ for various $s \geq 2$. The projective geometry $\mathrm{PG}(2, 2^s)$ over $\mathrm{GF}(2^s)$ consists of $n = 2^{2s} + 2^s + 1$ points and $M = 2^{2s} + 2^s + 1$ lines. Each line consists of $2^s + 1$ points and each point is intersected by $2^s + 1$ lines. The lines in $\mathrm{PG}(2, 2^s)$ form a single cyclic class.

Form a $(2^{2s} + 2^s + 1) \times (2^{2s} + 2^s + 1)$ matrix $\mathbf{H}_{\mathrm{PG}}(2, 2^s)$ using the incidence vectors of the lines in $\mathrm{PG}(2, 2^s)$ as rows. Because the lines in $\mathrm{PG}(2, 2^s)$ have cyclic structure, $\mathbf{H}_{\mathrm{PG}}(2, 2^s)$ is a circulant with weight $2^s + 1$. The null space over $\mathrm{GF}(2)$ of $\mathbf{H}_{\mathrm{PG}}(2, 2^s)$ gives a two-dimensional cyclic PG code $C_{\mathrm{PG}}(2, 2^s)$. Let $\alpha$ be a primitive element of $\mathrm{GF}(2^{3s})$ and $\zeta = \alpha^{2^s-1}$. It follows from Theorem 7.4 and (7.66) that the generator polynomial $\mathbf{g}_{\mathrm{PG}}(X)$ of $C_{\mathrm{PG}}(2, 2^s)$ has $\zeta^0 = 1$, $\zeta, \zeta^2, \ldots, \zeta^{2^s}$ (i.e., $2^s + 1$ consecutive powers of $\zeta$) as roots. Following from the BCH bound, the code has minimum distance at least $2^s + 2$. It can be proved that number of roots of $\mathbf{g}_{\mathrm{PG}}(X)$ is $3^s + 1$, i.e., the degree of $\mathbf{g}_{\mathrm{PG}}(X)$ is $3^s + 1$. Hence, the two-dimensional cyclic PG code $C_{\mathrm{PG}}(2, 2^s)$ has the following parameters [16, 17, 32, 33]:

length $\qquad\qquad\qquad\qquad\qquad\qquad n = 2^{2s} + 2^s + 1,$

dimension $\qquad\qquad\qquad\qquad\qquad k = 2^{2s} + 2^s - 3^s,$

number of parity-check symbols $\quad n - k = 3^s + 1,$

minimum distance $\qquad\qquad\qquad d_{\min} \geq 2^s + 2.$

A list of two-dimensional PG codes is given in Table 7.8, where the column labeled with $d_{\min}$ shows the lower bounds of the minimum distances of the corresponding codes.

Table 7.8 A list of two-dimensional PG codes.

| $s$ | $n$ | $k$ | rate | $d_{\min}$ | $\gamma$ | $\rho$ |
|---|---|---|---|---|---|---|
| 2 | 21 | 11 | 0.524 | 6 | 5 | 5 |
| 3 | 73 | 45 | 0.616 | 10 | 9 | 9 |
| 4 | 273 | 191 | 0.700 | 18 | 17 | 17 |
| 5 | 1 057 | 813 | 0.769 | 34 | 33 | 33 |
| 6 | 4 161 | 3 431 | 0.825 | 66 | 65 | 65 |
| 7 | 16 513 | 14 325 | 0.867 | 130 | 129 | 129 |
| 8 | 65 793 | 59 231 | 0.900 | 258 | 257 | 257 |

**Example 7.11** Let $m = 2$ and $s = 3$. Consider the PG code $C_{\mathrm{PG}}(2, 2^3)$ constructed based on the projective geometry $\mathrm{PG}(2, 2^3)$ with length

$$n = 2^{2 \times 3} + 2^3 + 1 = 73,$$

and minimum distance $d_{\min} \geq 2^s + 2 = 10$.

Let $\alpha$ be a primitive element of $\mathrm{GF}(2^9)$. Let $k$ be a nonnegative integer less than $2^{3 \times 3} - 1 = 511$. It follows from Theorem 7.4 that $\alpha^k$ is a root of the generator polynomial $\mathbf{g}_{\mathrm{PG}}(X)$ of the PG code $C_{\mathrm{PG}}(2, 2^3)$ if and only if its power $k$ is divisible by $2^3 - 1 = 7$ and satisfies the following equation:

$$0 < \max_{0 \leq \ell < 3} \left( W_{2^3}(k^{(\ell)}) \right) = 7t,$$

with $0 \leq t \leq 2$. Such integers, $k$, are 0, 7, 14, 21, 28, 35, 42, 49, 56, 70, 84, 98, 112, 133, 140, 161, 168, 196, 224, 259, 266, 273, 280, 322, 336, 385, 392, and 448. Thus, $\mathbf{g}_{\mathrm{PG}}(X)$ has $\alpha^0$, $\alpha^7$, $\alpha^{14}$, $\alpha^{21}$, $\alpha^{28}$, $\alpha^{35}$, $\alpha^{42}$, $\alpha^{49}$, $\alpha^{56}$, $\alpha^{70}$, $\alpha^{84}$, $\alpha^{98}$, $\alpha^{112}$, $\alpha^{133}$, $\alpha^{140}$, $\alpha^{161}$, $\alpha^{168}$, $\alpha^{196}$, $\alpha^{224}$, $\alpha^{259}$, $\alpha^{266}$, $\alpha^{273}$, $\alpha^{280}$, $\alpha^{322}$, $\alpha^{336}$, $\alpha^{385}$, $\alpha^{392}$, and $\alpha^{448}$ over $\mathrm{GF}(2^9)$ as roots. These $3^3 + 1 = 28$ roots of $\mathbf{g}_{\mathrm{PG}}(X)$ can be partitioned into the following four conjugate sets:

$$\begin{aligned}
\Phi_0 &= \{\alpha^0\}, \\
\Phi_7 &= \{\alpha^7, \alpha^{14}, \alpha^{28}, \alpha^{56}, \alpha^{112}, \alpha^{224}, \alpha^{448}, \alpha^{385}, \alpha^{259}\}, \\
\Phi_{21} &= \{\alpha^{21}, \alpha^{42}, \alpha^{84}, \alpha^{168}, \alpha^{336}, \alpha^{161}, \alpha^{322}, \alpha^{133}, \alpha^{266}\}, \\
\Phi_{35} &= \{\alpha^{35}, \alpha^{70}, \alpha^{140}, \alpha^{280}, \alpha^{49}, \alpha^{98}, \alpha^{196}, \alpha^{392}, \alpha^{273}\}.
\end{aligned}$$

Elements in the same conjugate set have the same minimal polynomial. The four minimal polynomials corresponding to the elements of the above four conjugate sets are

$$\begin{aligned}
m_0(X) &= 1 + X, & m_7(X) &= 1 + X^3 + X^4 + X^7 + X^9, \\
m_{21}(X) &= 1 + X + X^2 + X^4 + X^9, & m_{35}(X) &= 1 + X^8 + X^9.
\end{aligned}$$

Thus,

$$\begin{aligned}
\mathbf{g}_{\mathrm{PG}}(X) &= \mathrm{LCM}\{m_0(X), m_7(X), m_{21}(X), m_{35}(X)\} \\
&= m_0(X) m_7(X) m_{21}(X) m_{35}(X) \\
&= 1 + X^5 + X^6 + X^7 + X^8 + X^{11} + X^{12} + X^{22} + X^{24} + X^{28}.
\end{aligned}$$

Hence, the PG code $C_{\mathrm{PG}}(2, 2^3)$ is a $(73, 45)$ cyclic code. The minimum distance $d_{\min}$ of this code is equal to 10 because its generator polynomial $\mathbf{g}_{\mathrm{PG}}(X)$ is a code polynomial with 10 nonzero coefficients.

A parity-check matrix $\mathbf{H}_{\mathrm{PG}}(2, 2^3)$ of size $73 \times 73$ can be constructed by using the lines in the cyclic class of $\mathrm{PG}(2, 2^3)$ for the PG code $C_{\mathrm{PG}}(2, 2^3)$. Because the column weight of $\mathbf{H}_{\mathrm{PG}}(2, 2^3)$ is 9, the code $C_{\mathrm{PG}}(2, 2^3)$ is capable of correcting four or fewer random errors with the OSMLD. ▲▲

**Example 7.12** Let $m = 2$ and $s = 5$. Consider the two-dimensional projective geometry $PG(2, 2^5)$ which consists of $n = 2^{2 \times 5} + 2^5 + 1 = 1057$ points and $M = 2^{2 \times 5} + 2^5 + 1 = 1057$ lines. The 1057 lines in $PG(2, 2^5)$ form a single cycle class of lines. Using the incidence vectors of these 1057 lines, we construct a $1057 \times 1057$ circulant $\mathbf{H}_{PG}(2, 2^5)$ of weight 33. The null space over $GF(2)$ of $\mathbf{H}_{PG}(2, 2^5)$ gives a $(1057, 813)$ cyclic PG code $C_{PG}(2, 2^5)$ with minimum distance at least 34.

Suppose the PG code $C_{PG}(2, 2^5)$ is applied for error correcting over the AWGN channel with BPSK singling. The error performances (bit-error rate (BER) and block-error rate (BLER)) of the code $C_{PG}(2, 2^5)$ decoded with the OSMLD are shown in Fig. 7.2. The code can achieve a BER of $10^{-9}$ at SNR = 6.8 dB. ▲▲

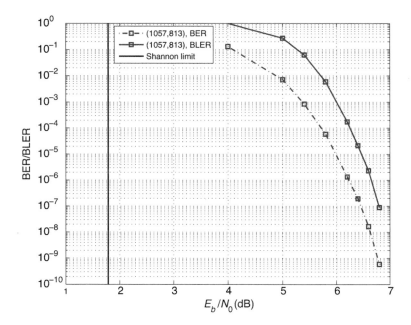

Figure 7.2 Error performances of the $(1057, 813)$ cyclic PG code $C_{PG}(2, 2^5)$ in Example 7.12 decoded with the OSMLD.

## 7.7 Remarks

In this chapter, we focused on two categories of finite geometries, namely, Euclidean and projective geometries, and the constructions of cyclic codes based on the points and lines of these two categories of finite geometries that have cyclic structures. If we remove the requirement for lines with cyclic structure, we can construct a much larger class of Euclidean and projective-geometry codes, including codes with *quasi-cyclic structures*. Constructions of Euclidean and projective-geometry codes, especially on constructions of Euclidean and projective-geometry LDPC codes, will be discussed further in Chapter 12. Besides Euclidean and projective geometries, there is another category of finite geometries, called *partial geometries*. The lines of a partial geometry, in general, do not have cyclic structure but they have a quasi-cyclic structure. Partial geometries are particularly effective in construction of *quasi-cyclic LDPC codes*. Quasi-cyclic LDPC codes are commonly used in current and next generations of communication and data-storage systems. Partial geometries and their adoption for constructing LDPC codes will be discussed in Chapter 13. For more general constructions of Euclidean and projective-geometry codes, readers are referred to [1, 4–6, 20–25, 27–31].

In this chapter, we also presented a simple hard-decision decoding method, known as one-step majority-logic decoding, for Euclidean and projective-geometry cyclic codes. Another simple hard-decision decoding method for these codes, with an iterative nature, will be presented in Chapter 11. Here, we would like to point out that the one-step majority-logic decoding can also be applied to linear block codes other than Euclidean and projective-geometry codes as long as their parity-check matrices have orthogonal structure. For more on the majority-logic decoding, readers are referred to [2, 3, 35, 37–39].

Note that one-step majority-logic decoding of a cyclic EG or PG code in serial one bit at a time is the Meggitt decoding with the majority-logic decoding rule as the SEL-algorithm for error location and correction.

## Problems

**7.1** Find the rows orthogonal at the first column position of the matrix $\mathbf{H}_{\mathrm{FG}}$ given in Example 7.1.

**7.2** Consider the finite geometry shown in Fig. 7.3, which consists of $M = 4$ lines where each line contains $\rho = 3$ points.

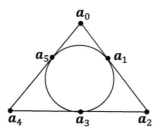

Figure 7.3 A six-point finite geometry.

(a) Find all the points on the four lines, $L_0$, $L_1$, $L_2$, and $L_3$.
(b) Find the lines intersecting at the point $\mathbf{a}_5$.

(c) Find the incidence vectors of all the lines and the incidence matrix $\mathbf{H}_{\mathrm{FG}}$ of this finite geometry.

(d) Show the orthogonal structure of the incidence matrix $\mathbf{H}_{\mathrm{FG}}$.

**7.3** Suppose the FG code in Example 7.2 is applied for error correcting. Suppose the decoder receives two hard-decision vectors $\mathbf{r}_0 = (0\ 0\ 0\ 0\ 0\ 1\ 0\ 0\ 0\ 0\ 0\ 0\ 0\ 0\ 0\ 0)$ and $\mathbf{r}_1 = (1\ 1\ 0\ 0\ 0\ 0\ 0\ 0\ 0\ 0\ 0\ 0\ 0\ 0\ 0\ 0)$. Find the error vectors $\mathbf{e}_0$ and $\mathbf{e}_1$ for the two received vectors by using OSMLD and verify the decoding results.

**7.4** Consider the two lines $\{\beta\mathbf{a}\}$ and $\{\mathbf{a}_0 + \beta\mathbf{a}\}$ in EG$(m,q)$ specified by (7.9) and (7.10), respectively, where $\mathbf{a}$ and $\mathbf{a}_0$ are two linearly independent points in EG$(m,q)$. Show that the two lines $\{\beta\mathbf{a}\}$ and $\{\mathbf{a}_0 + \beta\mathbf{a}\}$ have no common point.

**7.5** Consider the one-dimensional subspace $\mathbf{W}_1$ of the vector space $\mathbf{V}_m$ of all the $m$-tuples over GF$(q)$ formed by the $q$ points on the line $\{\beta\mathbf{a}\}$ defined by (7.9) in EG$(m,q)$. Show that there are $q^{m-1}$ cosets of $\mathbf{W}_1$ (including $\mathbf{W}_1$ itself).

**7.6** Show that there are $(q^m - 1)/(q - 1)$ lines in EG$(m,q)$ intersecting at a point $\mathbf{a}_0$ in EG$(m,q)$.

**7.7** In Example 7.3, one parallel bundle of four lines and the five lines passing through the point $\alpha$ are calculated. Find a different parallel bundle and the five lines passing through the point $\alpha^2$.

**7.8** Consider the three-dimensional Euclidean geometry EG$(3, 2^2)$ over GF$(2^2)$ in Example 7.4.

(a) Find all the one-dimensional subspaces.

(b) Find all the lines intersecting at the point $\alpha^2$.

(c) Find all the lines in the 2-flat $F_0$.

(d) Consider the two linearly independent points $\alpha$ and $\alpha^2$ in EG$(3, 2^2)$. Use these two points to construct a 2-flat $F_0^*$ and find all the lines in this 2-flat.

**7.9** Consider a subgeometry EG$^*(m,q)$ of EG$(m,q)$ obtained by removing the origin and the lines passing through the origin. Let $L$ be a line in EG$^*(m,q)$. Prove that the cyclic-shifts of $L$, namely, $L$, $\alpha L$, $\alpha^2 L, \ldots,$ and $\alpha^{q^m-2}L$, form $q^m - 1$ different lines in EG$^*(m,q)$.

**7.10** Consider the subgeometry EG$^*(3, 2^2)$ of EG$(3, 2^2)$ in Example 7.5.

(a) Find the first five lines in the cyclic class $Q_1$ of lines.

(b) Consider the 2-flat $F_0 = \{\alpha^2,\ \alpha^3,\ \alpha^7,\ \alpha^{10},\ \alpha^{12},\ \alpha^{13},\ \alpha^{20},\ \alpha^{26},\ \alpha^{30},\ \alpha^{32},$ $\alpha^{35},\ \alpha^{46},\ \alpha^{57},\ \alpha^{58},\ \alpha^{59},\ \alpha^{61}\}$ in EG$^*(3, 2^2)$. Find the first four 2-flats in the cyclic class of 2-flats generated by $F_0$.

**7.11** Consider the cyclic code $C_{\mathrm{EG}}(2, 2^2)$ constructed based on the lines in subgeometry EG$^*(2, 2^2)$ of EG$(2, 2^2)$ over the field GF$(2^2)$ in Example 7.6.

(a) The generator polynomial of $C_{\mathrm{EG}}(2, 2^2)$ is $\mathbf{g}_{\mathrm{EG}}(X) = 1 + X^4 + X^6 + X^7 + X^8$. Based on $\mathbf{g}_{\mathrm{EG}}(X)$, find the generator and parity-check matrices of the cyclic code $C_{\mathrm{EG}}(2, 2^2)$ (systematic or nonsystematic form).

(b) A parity-check matrix for the cyclic code $C_{\mathrm{EG}}(2, 2^2)$ is given in the form of (7.28). Suppose a line $L = \{\alpha^2, \alpha^3, \alpha^5, \alpha^9\}$ is chosen as a generator. Calculate the parity-check matrix $\mathbf{H}_{\mathrm{EG}}(2, 2^2)$ in the form of (7.28) by using the incidence vector of the line $L$.

(c) Show the orthogonal structure of the parity-check matrix $\mathbf{H}_{\mathrm{EG}}(2, 2^2)$.

**7.12** Consider the cyclic code $C_{\mathrm{EG}}(2, 2^3)$ constructed based on the lines in the subgeometry $\mathrm{EG}^*(2, 2^3)$ of $\mathrm{EG}(2, 2^3)$ over the field $\mathrm{GF}(2^3)$ in Example 7.7.

(a) Based on the generator polynomial $\mathbf{g}_{\mathrm{EG}}(X)$ of the cyclic code, find the generator and parity-check matrices of the cyclic code (systematic or nonsystematic form).

(b) A parity-check matrix for the cyclic code $C_{\mathrm{EG}}(2, 2^3)$ is given in the form of (7.28). Choose a line $L$ not passing through the origin of $\mathrm{EG}(2, 2^3)$ and use it as a generator. Calculate the parity-check matrix $\mathbf{H}_{\mathrm{EG}}(2, 2^3)$ in the form of (7.28) by using the incidence vector of the line $L$.

(c) Show the orthogonal structure of the parity-check matrix $\mathbf{H}_{\mathrm{EG}}(2, 2^3)$.

**7.13** Suppose $k$ is a nonnegative integer less than $2^{ms} - 1$. Prove: for $1 \le k \le 2^s K$ where $K = (2^{(m-1)s} - 1)/(2^s - 1)$, the $2^s$-weight of $k^{(\ell)}$, $0 \le \ell < s$, satisfies the condition given by (7.41) in Theorem 7.3.

**7.14** Consider the $(63, 13)$ EG code $C_{\mathrm{EG}}(3, 2^2)$ constructed in Example 7.8.

(a) Based on the generator polynomial found in the example, compute the generator and parity-check matrices for $C_{\mathrm{EG}}(3, 2^2)$.

(b) Find the generators (five lines) of the five cyclic classes of lines in $\mathrm{EG}^*(3, 2^2)$.

(c) Use the five incidence vectors of the five generators to construct the parity-check matrix $\mathbf{H}_{\mathrm{EG}}(3, 2^2)$ in the form of (7.36) of the code $C_{\mathrm{EG}}(3, 2^2)$.

(d) Show the orthogonal structure of the parity-check matrix $\mathbf{H}_{\mathrm{EG}}(3, 2^2)$.

**7.15** Let $m = 3$ and $s = 3$. Consider the three-dimensional Euclidean geometry $\mathrm{EG}(3, 2^3)$ over $\mathrm{GF}(2^3)$ and its subgeometry $\mathrm{EG}^*(3, 2^3)$ obtained by removing the origin and all the lines passing through the origin.

(a) Compute the number of cyclic classes of lines in $\mathrm{EG}^*(3, 2^3)$.

(b) Construct a cyclic code based on all the cyclic classes of lines in $\mathrm{EG}^*(3, 2^3)$, i.e., find the generator polynomial.

(c) Compute the parity-check matrix of the code in the form of (7.36).

**7.16** Consider the projective geometry $\mathrm{PG}(2, 2^2)$ constructed in Example 7.10.

(a) Find the other points on the line $L$ passing through $(\alpha)$ and $(\alpha^2)$.

(b) Find the lines that intersect at the point $(\alpha)$.

(c) Put the lines of $\mathrm{PG}(2, 2^2)$ given in Table 7.7 in a cyclic order.

**7.17** Let $m = 2$ and $s = 3$. Construct the projective geometry $\mathrm{PG}(2, 2^3)$ by finding all its points and lines.

**7.18** Consider the $(73, 45)$ PG code $C_{\mathrm{PG}}(2, 2^3)$ constructed in Example 7.11 based on the projective geometry $\mathrm{PG}(2, 2^3)$. Based on the lines found in Problem 7.17, compute the parity-check matrix $\mathbf{H}_{\mathrm{PG}}(2, 2^3)$ of the code.

**7.19** Let $m = 2$ and $s = 2$.

    (a) Construct a cyclic code based on projective geometry $\text{PG}(m, 2^s)$, i.e., compute the code length, dimension, and its generator polynomial.

    (b) Compute the parity-check matrix $\mathbf{H}_{\text{PG}}(2, 2^2)$ of the constructed PG code.

# References

[1] E. J. Weldon, Jr., "Euclidean geometry cyclic codes," *Proc. Symp. Combinatorial Mathematics*, University of North Carolina, Chapel Hill, NC, April 1967.

[2] K. J. C. Smith, "Majority decodable codes derived from finite geometries," *Institute of Statistics Memo Series*, no. 561, University of North Carolina, Chapel Hill, NC, 1967.

[3] L. D. Rudolph, *Geometric configuration and majority logic decodable codes*, M.E.E. thesis, University of Oklahoma, Norman, OK, 1964.

[4] Y. Kou, S. Lin, and M. Fossorier, "Low density parity check codes based on finite geometries: a rediscovery," *Proc. IEEE Int. Symp. Inf. Theory*, Sorrento, June 25–30, 2000, p. 200.

[5] Y. Kou, S. Lin, and M. Fossorier, "Construction of low density parity check codes: a geometric approach," *Proc. 2nd Int. Symp. Turbo Codes and Related Topics*, Brest, September 2000, pp. 137–140.

[6] Y. Kou, S. Lin, and M. Fossorier, "Low density parity-check codes: construction based on finite geometries," *Proc. IEEE Globecom*, San Francisco, CA, November 27–December 1, 2000, pp. 825–829.

[7] R. G. Gallager, "Low-density parity-check codes," *IRE Trans. Inf. Theory*, **IT-8**(1) (1962), 21–28.

[8] T. Richardson, A. Shokrollahi, and R. Urbanke, "Design of capacity approaching irregular codes," *IEEE Trans. Inf. Theory*, **47**(2) (2001), 619–637.

[9] L. Chen, J. Xu, I. Djurdjevic, and S. Lin, "Near Shannon limit quasi-cyclic low-density parity-check codes," *IEEE Trans. Commun.*, **52**(7) (2004), 1038–1042.

[10] J. Chen and M. P. C. Fossorier, "Near optimum universal belief propagation based decoding of low-density parity check codes," *IEEE Trans. Commun.*, **50**(3) (2002), 406–414.

[11] M. Jiang, C. Zhao, Z. Shi, and Y. Chen, "An improvement on the modified weighted bit-flipping decoding algorithm for LDPC codes," *IEEE Commun. Lett.*, **9**(7) (2005), 814–816.

[12] Z. Liu and D. A. Pados, "A decoding algorithm for finite-geometry LDPC codes," *IEEE Trans. Commun.*, **53**(3) (2005), 415–421.

[13] M. Shan, C. Zhao, and M. Jian, "Improved weighted bit-flipping algorithm for decoding LDPC codes," *IEE Proc. Commun.*, **152**(6) (2005), 919–922.

[14] N. Miladinovic and M. P. Fossorier, "Improved bit-flipping decoding of low-density parity-check codes," *IEEE Trans. Inf. Theory*, **51**(4) (2005), 1594–1606.

[15] T. J. Richardson, "Error floors of LDPC codes," *Proc. 41st Allerton Conf. on Communications, Control and Computing*, Monticello, IL, October 2003, pp. 1426–1435.

[16] W. E. Ryan and S. Lin, *Channel Codes: Classical and Modern*, New York, Cambridge University Press, 2009.

[17] S. Lin and D. J. Costello, Jr., *Error Control Coding: Fundamentals and Applications*, 2nd ed., Upper Saddle River, NJ, Prentice-Hall, 2004.

[18] I. S. Reed, "A class of multiple-error-correcting codes and the decoding scheme," *IRE Trans.*, **IT-4**(4) (1954), 38–49.

[19] J. L. Massey, *Threshold Decoding*, Cambridge, MA, The MIT Press, 1963.

[20] Y. Kou, S. Lin, and M. Fossorier, "Low-density parity-check codes based on finite geometries: a rediscovery and new results," *IEEE Trans. Inf. Theory*, **47**(7) (2001), 2711–2736.

[21] H. Tang, J. Xu, S. Lin, and K. Abdel-Ghaffar, "Codes on finite geometries," *IEEE Trans. Inf. Theory*, **51**(2) (2005), 572–596.

[22] Y. Kou, J. Xu, H. Tang, S. Lin, and K. Abdel-Ghaffar, "On circulant low density parity check codes," *Proc. IEEE Int. Symp. Inf. Theory*, Lausanne, June 30–July 5, 2002, p. 200.

[23] S. Lin, H. Tang, and Y. Kou, "On a class of finite geometry low density parity check codes," *Proc. IEEE Int. Symp. Inf. Theory*, Washington, DC, June 29, 2001, p. 24.

[24] S. Lin, H. Tang, Y. Kou, and K. Abdel-Ghaffar, "Codes on finite geometries," *Proc. IEEE Inf. Theory Workshop*, Cairns, Australia, September 2–7, 2001, pp. 14–16.

[25] H. Tang, *Codes on finite geometries*, Ph.D. dissertation, University of California, Davis, CA, 2002.

[26] H. Tang, Y. Kou, J. Xu, S. Lin, and K. Abdel-Ghaffar, "Codes on finite geometries: old, new, majority-logic and iterative decoding," *Proc. 6th Int. Symp. Communication Theory and Applications*, Ambleside, UK, July 2001, pp. 381–386.

[27] H. Tang, J. Xu, Y. Kou, S. Lin, and K. Abdel-Ghaffar, "On algebraic construction of Gallager low density parity-check codes," *Proc. IEEE Int. Symp. Inf. Theory*, Lausanne, June 30–July 5, 2002, p. 482.

[28] H. Tang, J. Xu, Y. Kou, S. Lin, and K. Abdel-Ghaffar, "On algebraic construction of Gallager and circulant low density parity check codes," *IEEE Trans. Inf. Theory*, **50**(6) (2004), 1269–1279.

[29] J. Xu, L. Chen, I. Djurdjevic, S. Lin, and K. Abdel-Ghaffar, "Construction of regular and irregular LDPC codes: geometry decomposition and masking," *IEEE Trans. Inf. Theory*, **53**(1) (2007), 121–134.

[30] L.-Q. Zeng, L. Lan, Y. Y. Tai, B. Zhou, S. Lin, and K. Abdel-Ghaffar, "Construction of nonbinary quasi-cyclic LDPC codes: a finite geometry approach," *IEEE Trans. Commun.*, **56**(3) (2008), 378–387.

[31] B. Zhou, J.-Y. Kang, Y. Y. Tai, S. Lin, and Z. Ding, "High performance nonbinary quasi-cyclic LDPC codes on Euclidean geometries," *IEEE Trans. Commun.*, **57**(5) (2009), 1298–1311.

[32] T. Kasami, S. Lin, and W. W. Peterson, "Polynomial codes," *IEEE Trans. Inf. Theory*, **14**(6) (1968), 807–814.

[33] S. Lin, "On the number of information symbols in polynomial codes," *IEEE Trans. Inf. Theory*, **18**(6) (1972), 785–794.

[34] J. E. Maxfield and M. W. Maxfield, *Abstract Algebra and Solution by Radicals*, New York, Dover, 1992.

[35] J. M. Goethals and P. Delsarte, "On a class of majority-logic decodable codes," *IEEE Trans. Inf. Theory*, **14**(2) (1968), 182–189.

[36] E. J. Weldon, Jr., "New generations of the Reed–Muller codes, part II: nonprimitive codes," *IEEE Trans. Inf. Theory*, **IT-14** (1968), 199–205.

[37] T. Kasami and S. Lin, "On majority-logic decoding for duals of primitive polynomial codes," *IEEE Trans. Inf. Theory*, **17**(3) (1971), 322–331.

[38] E. J. Weldon, Jr., "Some results on majority-logic decoding," chapter 8 in H. Mann, ed., *Error-Correcting Codes*, New York, John Wiley, 1968.

[39] L. D. Rudolph, "A class of majority logic decodable codes," *IEEE Trans. Inf. Theory*, **13**(3) (1967), 305–307.

# 8

# Reed–Muller Codes

Reed–Muller (RM) codes form a class of multiple error-correcting codes. These codes were first discovered by Muller in 1954 [1]. In the same year, Reed devised a simple method for decoding these codes [2]. Owing to their simple construction and ease of decoding, these codes were intensively investigated in the 1960s and 1970s. Many algebraic and geometric structural properties of these codes were developed in [3–11]. They were also generalized and found to be finite-geometry codes. Cyclic version of these codes was also developed [6]. In the 1980s and 1990s, these codes were found to have beautiful trellis structure [12–17]. Based on their trellis structure, some efficient trellis-based soft-decision decoding algorithms were devised for these codes [17–21]. In 2009, a class of channel-capacity-approaching codes closely related to RM codes, called *polar codes*, was discovered [22]. Lately, RM codes have also been found to be effective for correcting erasures [23, 24].

In this chapter, we give an introduction to binary RM codes. There are several algebraic and geometric approaches to the construction of RM codes. We will take a geometric approach to the construction of RM codes and show that these codes are finite-geometry codes which are constructed based on multi-dimensional Euclidean geometries over GF(2). The development of this chapter is as follows. In Section 8.1, we give a brief review of Euclidean geometries over GF(2) and present some structural properties of these geometries. In Section 8.2, we present a geometric construction of RM codes whose generator matrices are not systematic. In Section 8.3, a specific encoding of an RM code is presented. In encoding, a sequence of information symbols, called *a message*, is divided into *groups*, where each group of information symbols is encoded into *a constituent codeword*. Then, the constituent codewords are combined into a codeword for the entire message. In Section 8.4, we present a method to retrieve information symbols of a message in groups *successively* from the symbols of its corresponding codeword, one group at a time. In Section 8.5, we present a *multi-level majority-logic decoding* for RM codes. When a vector is received, the information symbols contained in the received vector are decoded in groups. In each decoding level, a group of information symbols is decoded. At the completion of a decoding level, the effect of decoded information symbols is removed (or canceled) from the received vector. This results in a modified received vector. Then, we decode the next group of information symbols based on the modified received vector. In each decoding level, majority-logic decoding is performed. Decoding continues level by level until all the information symbols contained in the transmitted codeword are decoded. This successive multilevel decoding was first presented in [21, chapter 4]. Finally, a cyclic version of RM codes is presented in Section 8.6.

More on construction, structures, and erasure correction of RM codes will be presented in Chapters 9, 10, and 17.

## 8.1 A Review of Euclidean Geometries over GF(2)

Let $m$ be a positive integer greater than 2, i.e., $m \geq 2$. Consider the $m$-dimensional Euclidean geometry $\mathrm{EG}(m, 2)$ over $\mathrm{GF}(2)$. It follows from (7.8) and (7.13) that $\mathrm{EG}(m, 2)$ consists of $2^m$ points and $2^{m-1}(2^m - 1)$ lines. For $1 \leq \mu \leq m$, it follows from (7.16) that $\mathrm{EG}(m, 2)$ contains

$$2^{m-\mu} \prod_{i=1}^{\mu} \frac{2^{m-i+1} - 1}{2^{\mu-i+1} - 1} \tag{8.1}$$

$\mu$-flats. Each $\mu$-flat in $\mathrm{EG}(m, 2)$ contains $2^\mu$ points. A 1-flat in $\mathrm{EG}(m, 2)$ is a line which consists of two points and a point is regarded as a 0-flat.

Two $\mu$-flats in $\mathrm{EG}(m, 2)$ are either parallel or they intersect on a flat of smaller dimension. The largest flat that two $\mu$-flats can intersect on is a $(\mu - 1)$-flat. The smallest flat that two $\mu$-flats can intersect on is a point. The number of $\mu$-flats in $\mathrm{EG}(m, 2)$ that intersect on a given $(\mu - 1)$-flat is

$$2^{m-\mu+1} - 1. \tag{8.2}$$

A point in $\mathrm{EG}(m, 2)$ is intersected by $2^m - 1$ lines. The incidence vector $\mathbf{v} = (v_0, v_1, \ldots, v_{2^m-1})$ of a $\mu$-flat $F$ is a $2^m$-tuple over $\mathrm{GF}(2)$ with $2^\mu$ 1-components at the locations corresponding to the $2^\mu$ points on the flat $F$. Thus, the incidence vector $\mathbf{v}$ has weight $2^\mu$.

Let $\mathbf{a} = (a_0, a_1, \ldots, a_{n-1})$ and $\mathbf{b} = (b_0, b_1, \ldots, b_{n-1})$ be two $n$-tuples over $\mathrm{GF}(2)$. Define the following component-wise product of $\mathbf{a}$ and $\mathbf{b}$:

$$\mathbf{a} \cdot \mathbf{b} \triangleq (a_0 \cdot b_0, a_1 \cdot b_1, \ldots, a_{n-1} \cdot b_{n-1}), \tag{8.3}$$

where $a_j \cdot b_j$, $0 \leq j < n$, is the modulo-2 multiplication of $a_j$ and $b_j$. Hence, $\mathbf{a} \cdot \mathbf{b}$ is an $n$-tuple over $\mathrm{GF}(2)$. The weight $w(\mathbf{a} \cdot \mathbf{b})$ of the $n$-tuple $\mathbf{a} \cdot \mathbf{b}$ is equal to the number of places where both $\mathbf{a}$ and $\mathbf{b}$ have 1-components. We call $\mathbf{a} \cdot \mathbf{b}$ the *vector product* (or simply product) of $\mathbf{a}$ and $\mathbf{b}$.

Let $\mathbf{v}_1$ and $\mathbf{v}_2$ be the incidence vectors of two $\mu$-flats $F_1$ and $F_2$, respectively. If $F_1$ and $F_2$ are parallel, then the product $\mathbf{v}_1 \cdot \mathbf{v}_2$ of their incidence vectors $\mathbf{v}_1$

and $\mathbf{v}_2$ is the zero $2^m$-tuple $(0, 0, \ldots, 0)$, i.e., $\mathbf{v}_1 \cdot \mathbf{v}_2 = (0, 0, \ldots, 0)$. If $F_1$ and $F_2$ intersect on an $l$-flat with $0 \le l < \mu$, then the weight of $\mathbf{v}_1 \cdot \mathbf{v}_2$ is $2^l$, which is the number of points on the intersected $l$-flat. The product concept of the incidence vectors of flats in the Euclidean geometry $\mathrm{EG}(m, 2)$ will be used in constructing RM codes in the next section.

Hereafter, we use $\mathbf{ab}$ instead of $\mathbf{a} \cdot \mathbf{b}$ to represent the vector product for simplicity.

## 8.2 Constructing RM Codes from Euclidean Geometries over GF(2)

Let $m$ and $r$ be two nonnegative integers such that $m \ge 3$ and $0 \le r \le m$. There exists an RM code, denoted by $C_{\mathrm{RM}}(r, m)$, of order $r$, with the following parameters:

$$
\begin{array}{lll}
\text{code length} & n = 2^m, \\
\text{code dimension} & k(r, m) = 1 + \binom{m}{1} + \binom{m}{2} + \cdots + \binom{m}{r}, & (8.4) \\
\text{minimum distance} & d_{\min} = 2^{m-r},
\end{array}
$$

where $\binom{m}{l} = m!/((m - l)! \times l!)$, $0 \le l \le m$, is the binomial coefficient.

There are several approaches for constructing RM codes [1, 5, 8, 11, 21, 25, 26]. In this section, we present a geometric approach to the construction of RM codes based on multi-dimensional Euclidean geometries over GF(2). This geometric approach will reveal many structural properties of RM codes which are useful for decoding, weight distribution analysis, and applications of RM codes.

It was shown in Chapter 7 that the field $\mathrm{GF}(2^m)$ (or the vector space of all the $2^m$ $m$-tuples over GF(2)) is a *realization* of the $m$-dimensional Euclidean geometry $\mathrm{EG}(m, 2)$ over GF(2). Let $\alpha$ be a primitive element of the field $\mathrm{GF}(2^m)$ constructed based on a primitive polynomial over GF(2) (see Chapter 2). The $2^m$ elements of $\mathrm{GF}(2^m)$, $\alpha^{-\infty} = 0$, $\alpha^0 = 1$, $\alpha, \alpha^2, \ldots, \alpha^{2^m-2}$, represent the $2^m$ points in $\mathrm{EG}(m, 2)$. Let $\{-\infty, 0, 1, \ldots, 2^m - 2\}$ be the set of powers of the $2^m$ points in $\mathrm{EG}(m, 2)$. For any $j \in \{-\infty, 0, 1, \ldots, 2^m - 2\}$, the element $\alpha^j$ can be expressed as a linear sum of the $m$ elements $\alpha^0 = 1, \alpha, \alpha^2, \ldots, \alpha^{m-1}$ in $\mathrm{GF}(2^m)$ as follows:

$$
\alpha^j = c_{0,j} + c_{1,j}\alpha + c_{2,j}\alpha^2 + \cdots + c_{m-1,j}\alpha^{m-1}, \quad (8.5)
$$

where $c_{l,j}$ is an element in GF(2) for $0 \le l < m$. The $m$-tuple $(c_{0,j}, c_{1,j}, \ldots, c_{m-1,j})$ is a vector representation of the element $\alpha^j$ in $\mathrm{GF}(2^m)$. Hence, $(c_{0,j}, c_{1,j}, \ldots, c_{m-1,j})$ is a vector representation of the point $\alpha^j$ in $\mathrm{EG}(m, 2)$, and the $2^m$ $m$-tuples, $(c_{0,j}, c_{1,j}, \ldots, c_{m-1,j})$ with $j \in \{-\infty, 0, 1, \ldots, 2^m - 2\}$, over GF(2) represent the $2^m$ points in $\mathrm{EG}(m, 2)$. For $j = -\infty$, the point $\alpha^{-\infty} = 0$ represents the zero $m$-tuple $(0, 0, \ldots, 0)$, which is the origin of $\mathrm{EG}(m, 2)$.

Form the following $m \times 2^m$ matrix over GF(2):

$$
\mathbf{B} = 
\begin{bmatrix}
\overset{\alpha^{-\infty}}{c_{0,-\infty}} & \overset{\alpha^{0}}{c_{0,0}} & \overset{\alpha}{c_{0,1}} & \cdots & \overset{\alpha^{2^m-2}}{c_{0,2^m-2}} \\
c_{1,-\infty} & c_{1,0} & c_{1,1} & \cdots & c_{1,2^m-2} \\
\vdots & \vdots & \ddots & \vdots & \vdots \\
c_{m-1,-\infty} & c_{m-1,0} & c_{m-1,1} & \cdots & c_{m-1,2^m-2}
\end{bmatrix},
\tag{8.6}
$$

where the $j$th column, $(c_{0,j}, c_{1,j}, \ldots, c_{m-1,j})^T$, is the transpose of the vector representation of the point $\alpha^j$ in EG$(m, 2)$. Hence, the columns of the matrix $\mathbf{B}$ correspond to all the $2^m$ points in EG$(m, 2)$. Because the columns of $\mathbf{B}$ form all the $2^m$ $m$-tuples over GF(2), each row of $\mathbf{B}$ has weight $2^{m-1}$. Let $\alpha^t$ be the point whose $m$-tuple representation is the all-one $m$-tuple $(1, 1, \ldots, 1)$. Then, each row of $\mathbf{B}$ can be viewed as the incidence vector of an $(m-1)$-flat in EG$(m, 2)$ that contains (or passes through) the point $\alpha^t$. It is noted that all the rows in $\mathbf{B}$ are linearly independent.

Next, we permute the columns of $\mathbf{B}$ to obtain the following $m \times 2^m$ matrix:

$$
\mathbf{G}_1 = 
\begin{bmatrix}
\mathbf{v}_m \\
\mathbf{v}_{m-1} \\
\vdots \\
\mathbf{v}_1
\end{bmatrix},
\tag{8.7}
$$

where, for $1 \leq i \leq m$, the $i$th row is a $2^m$-tuple over GF(2) of the following form:

$$
\mathbf{v}_i = (\underbrace{0 \ldots 0}_{2^{i-1}}, \underbrace{1 \ldots 1}_{2^{i-1}}, \underbrace{0 \ldots 0}_{2^{i-1}}, \ldots, \underbrace{1 \ldots 1}_{2^{i-1}})
\tag{8.8}
$$

which consists of $2^{m-i+1}$ alternate *all-zero* and *all-one sections*, where each section is a $2^{i-1}$-tuple over GF(2). Note that each row $\mathbf{v}_i$ in $\mathbf{G}_1$ is still the incidence vector of an $(m-1)$-flat containing the point $\alpha^t$ except with reordering of the points in EG$(m, 2)$. From (8.8), we see that all the $m$ entries in the *rightmost column* (the last column) of $\mathbf{G}_1$ are 1-entries, i.e., the point $\alpha^t$ is reordered to the location corresponding to the last (rightmost) column of the matrix $\mathbf{G}_1$. With the above reordering of points in EG$(m, 2)$ (permutation of columns of $\mathbf{B}$), the rows of $\mathbf{G}_1$ are the incidence vectors in the form of (8.8) of the $(m-1)$-flats in EG$(m, 2)$ passing through the point represented by the all-one $m$-tuple $(1, 1, \ldots, 1)$.

For $1 < l \leq m$, we can divide each $2^m$-tuple $\mathbf{v}_l$ into $2^{m-l+1}$ sections, each consisting of $2^{l-1}$ components of $\mathbf{v}_l$. From (8.8), we see that, the modulo-2 sum of the components in each section is zero. For $l = 1$, we divide $\mathbf{v}_1$ into $2^m$ sections, each consisting of either a zero or a one component. Except for $\mathbf{v}_1$, the modulo-2 sum of the components in each section of $\mathbf{v}_l$ is zero. We call this property of $\mathbf{G}_1$ a *section-regularity structure*.

The rows of the matrix $\mathbf{G}_1$ given by (8.8) will be used to construct RM codes of length $2^m$ with various dimensions and rates. The section-regularity structure of the rows in $\mathbf{G}_1$ is helpful for demonstrating a majority-logic decoding scheme for RM codes which will be presented in Section 8.5.

**Example 8.1** Let $m = 4$. Consider the field $\mathrm{GF}(2^4)$ given in Table 2.9 which is constructed based on the primitive polynomial $\mathbf{p}(X) = 1 + X + X^4$ over $\mathrm{GF}(2)$. The field $\mathrm{GF}(2^4)$ is a realization of the four-dimensional Euclidean geometry $\mathrm{EG}(4, 2)$ over $\mathrm{GF}(2)$ which consists of 16 points, 120 lines, 140 2-flats, 30 3-flats, and 1 4-flat. Let $\alpha$ be a primitive element in $\mathrm{GF}(2^4)$. The elements 0, 1, $\alpha$, $\alpha^2$, ..., $\alpha^{13}$, $\alpha^{14}$ of $\mathrm{GF}(2^4)$ give the 16 points in $\mathrm{EG}(4, 2)$, each represented by a 4-tuple over $\mathrm{GF}(2)$ as shown in Table 2.9.

Using the 4-tuple representations of the 16 points in $\mathrm{EG}(4, 2)$, we construct the following $4 \times 16$ matrix in the form of (8.6):

$$
\mathbf{B} = \begin{array}{c}
\begin{array}{cccccccccccccccc}
0 & 1 & \alpha & \alpha^2 & \alpha^3 & \alpha^4 & \alpha^5 & \alpha^6 & \alpha^7 & \alpha^8 & \alpha^9 & \alpha^{10} & \alpha^{11} & \alpha^{12} & \alpha^{13} & \alpha^{14}
\end{array} \\
\left[\begin{array}{cccccccccccccccc}
0 & 1 & 0 & 0 & 0 & 1 & 0 & 0 & 1 & 1 & 0 & 1 & 0 & 1 & 1 & 1 \\
0 & 0 & 1 & 0 & 0 & 1 & 1 & 0 & 1 & 0 & 1 & 1 & 1 & 1 & 0 & 0 \\
0 & 0 & 0 & 1 & 0 & 0 & 1 & 1 & 0 & 1 & 0 & 1 & 1 & 1 & 1 & 0 \\
0 & 0 & 0 & 0 & 1 & 0 & 0 & 1 & 1 & 0 & 1 & 0 & 1 & 1 & 1 & 1
\end{array}\right].
\end{array} \quad (8.9)
$$

The four rows of $\mathbf{B}$ are the incidence vectors of four 3-flats in $\mathrm{EG}(4, 2)$, each containing the point $\alpha^{12}$ represented by the all-one 4-tuple $(1\ 1\ 1\ 1)$. To put the rows of the matrix $\mathbf{B}$ given by (8.9) in the form of (8.8), we permute the columns of $\mathbf{B}$ to obtain

$$
\mathbf{G}_1 = \begin{bmatrix} \mathbf{v}_4 \\ \mathbf{v}_3 \\ \mathbf{v}_2 \\ \mathbf{v}_1 \end{bmatrix}
$$

$$
\begin{array}{c}
\begin{array}{cccccccccccccccc}
0 & \alpha^3 & \alpha^2 & \alpha^6 & \alpha & \alpha^9 & \alpha^5 & \alpha^{11} & 1 & \alpha^{14} & \alpha^8 & \alpha^{13} & \alpha^4 & \alpha^7 & \alpha^{10} & \alpha^{12}
\end{array} \\
= \left[\begin{array}{cccccccccccccccc}
0 & 0 & 0 & 0 & 0 & 0 & 0 & 0 & 1 & 1 & 1 & 1 & 1 & 1 & 1 & 1 \\
0 & 0 & 0 & 0 & 1 & 1 & 1 & 1 & 0 & 0 & 0 & 0 & 1 & 1 & 1 & 1 \\
0 & 0 & 1 & 1 & 0 & 0 & 1 & 1 & 0 & 0 & 1 & 1 & 0 & 0 & 1 & 1 \\
0 & 1 & 0 & 1 & 0 & 1 & 0 & 1 & 0 & 1 & 0 & 1 & 0 & 1 & 0 & 1
\end{array}\right].
\end{array} \quad (8.10)
$$

We see that each row of $\mathbf{G}_1$ is indeed in the form of (8.8). The column of $\mathbf{B}$ corresponding to the point $\alpha^{12}$ is permuted to the last column of $\mathbf{G}_1$. From (8.10), we see that the row $\mathbf{v}_4$ of $\mathbf{G}_1$ consists of two sections of length 8, one consisting of eight consecutive zeros and the other consisting of eight consecutive ones. The row $\mathbf{v}_3$ of $\mathbf{G}_1$ consists of four sections of length 4, two 0-sections and two 1-sections. The row $\mathbf{v}_2$ of $\mathbf{G}_1$ consists of eight sections of length 2, four 0-sections and four 1-sections. The row $\mathbf{v}_1$ of $\mathbf{G}_1$ consists of 16 sections of length 1, each consisting of a single 0-entry or a single 1-entry. The rightmost column of $\mathbf{G}_1$ is a column with four 1-entries. ▲▲

For $l$ different positive integers $i_1, i_2, \ldots, i_l$ with $1 \le i_1 < i_2 < \cdots < i_l \le m$, we take the vector product

$$
\mathbf{v}_{i_1} \mathbf{v}_{i_2} \cdots \mathbf{v}_{i_l} \quad (8.11)
$$

of the $l$ rows $\mathbf{v}_{i_1}, \mathbf{v}_{i_2}, \ldots, \mathbf{v}_{i_l}$ in $\mathbf{G}_1$. The vector product $\mathbf{v}_{i_1}\mathbf{v}_{i_2}\ldots\mathbf{v}_{i_l}$ is said to have *degree* $l$ (or $l$ is the product degree of the vector $\mathbf{v}_{i_1}\mathbf{v}_{i_2}\ldots\mathbf{v}_{i_l}$). Following the structural properties of flats in EG$(m,2)$ and the structure of the rows of $\mathbf{G}_1$ given by (8.8), the vector product $\mathbf{v}_{i_1}\mathbf{v}_{i_2}\ldots\mathbf{v}_{i_l}$ is the incidence vector of an $(m-l)$-flat in EG$(m,2)$ which contains the point $\alpha^t$ and has weight $2^{m-l}$ (see Problem 8.4). The last (rightmost) component of $\mathbf{v}_{i_1}\mathbf{v}_{i_2}\ldots\mathbf{v}_{i_l}$ is 1. There are $\binom{m}{l}$ vector products of $l$ rows in $\mathbf{G}_1$. Each of these vector products of degree $l$ is the incidence vector of an $(m-l)$-flat in EG$(m,2)$ with weight $2^{m-l}$ and the last component being 1.

For a positive integer $l$ such that $1 \le l \le m$, let

$$\Omega_l = \{\mathbf{v}_{i_1}\mathbf{v}_{i_2}\ldots\mathbf{v}_{i_l} : i_j \in \{1,2,3,\ldots,m\}, 1 \le j \le l, 1 \le i_1 < i_2 < \cdots < i_l \le m\} \tag{8.12}$$

be the set of $\binom{m}{l}$ vector products of $l$ rows among the $m$ rows $\mathbf{v}_1, \mathbf{v}_2, \ldots, \mathbf{v}_m$ in $\mathbf{G}_1$.

Let $\mathbf{G}_l$ be a $\binom{m}{l} \times 2^m$ matrix over GF$(2)$ with the vector products in $\Omega_l$ as rows. Let $\mathbf{v}_0 = (1\ 1\ \cdots\ 1)$ be the all-one $2^m$-tuple over GF$(2)$. The vector $\mathbf{v}_0$ is the incidence vector of the $m$-flat in EG$(m,2)$ which consists of all the $2^m$ points in EG$(m,2)$. For $0 \le r \le m$, we form the following $k(m,r) \times 2^m$ matrix over GF$(2)$:[1]

$$\mathbf{G}_{\mathrm{RM}}(r,m) = \begin{bmatrix} \mathbf{v}_0 \\ \mathbf{G}_1 \\ \mathbf{G}_2 \\ \cdots \\ \mathbf{G}_r \end{bmatrix} \tag{8.13}$$

where

$$k(r,m) = 1 + \binom{m}{1} + \binom{m}{2} + \cdots + \binom{m}{r}. \tag{8.14}$$

The rows of $\mathbf{G}_{\mathrm{RM}}(r,m)$ are the incidence vectors of the $m$-flat, the $(m-1)$-flats, $\ldots$, the $(m-r)$-flats in EG$(m,2)$ which contain the point $\alpha^t$ represented by the all-one $m$-tuple $(1\ 1\ \ldots\ 1)$ over GF$(2)$. From (8.8) and (8.11), we see that all the entries of the rightmost column of $\mathbf{G}_{\mathrm{RM}}(r,m)$ are 1-entries. All the rows of $\mathbf{G}_{\mathrm{RM}}(r,m)$ are linearly independent. The weights of the rows in $\mathbf{G}_{\mathrm{RM}}(r,m)$ are $2^m, 2^{m-1}, \ldots, 2^{m-r+1}, 2^{m-r}$, i.e., powers of 2.

The vector space spanned by the rows of $\mathbf{G}_{\mathrm{RM}}(r,m)$ (i.e., using $\mathbf{G}_{\mathrm{RM}}(r,m)$ as a generator matrix) gives a $(2^m, k(r,m))$ RM code $C_{\mathrm{RM}}(r,m)$ of length $2^m$, dimension $k(r,m)$, and minimum distance $2^{m-r}$ (the smallest row weight of $\mathbf{G}_{\mathrm{RM}}(r,m)$) as given in (8.4) [5, 7, 21]. Such an RM code is called an *$r$th-order RM code*. For $r = 0$, $\mathbf{G}_{\mathrm{RM}}(0,m)$ consists of a single row which is $\mathbf{v}_0$, the all-one $2^m$-tuple. The zeroth-order RM code $C_{\mathrm{RM}}(0,m)$ generated by $\mathbf{G}_{\mathrm{RM}}(0,m)$ consists of two codewords, an all-zero and an all-one codewords. It is a *repetition*

---

[1] In some context, we also define $\mathbf{G}_0 = \mathbf{v}_0$. If $r = 0$, the matrix $\mathbf{G}_{\mathrm{RM}}(r,m)$ is defined as $\mathbf{G}_{\mathrm{RM}}(r,m) = \mathbf{v}_0 = \mathbf{G}_0$.

*code.* For $r = m-1$, $\mathbf{G}_{\mathrm{RM}}(m-1, m)$ generates a $(2^m, 2^m - 1)$ code with minimum distance 2, a *single parity-check* (SPC) *code.*

From (8.13), we see that $\mathbf{G}_{\mathrm{RM}}(r-1, m)$ is a submatrix of $\mathbf{G}_{\mathrm{RM}}(r, m)$. Hence, for $1 \leq r \leq m$, the $(r-1)$th-order RM code $C_{\mathrm{RM}}(r-1, m)$ is a subcode of the $r$th-order RM code $C_{\mathrm{RM}}(r, m)$. Consequently, RM codes of length $2^m$ with $m \geq 3$ form the following *inclusion chain*:

$$C_{\mathrm{RM}}(0, m) \subset C_{\mathrm{RM}}(1, m) \subset \cdots \subset C_{\mathrm{RM}}(r, m). \tag{8.15}$$

For $r < m$, because all the rows in $\mathbf{G}_{\mathrm{RM}}(r, m)$ have even weights, all the codewords in the $r$th-order RM code $C_{\mathrm{RM}}(r, m)$ have even weights. As a result, the inner product of any two codewords in $C_{\mathrm{RM}}(r, m)$ is zero. In fact, for any two nonnegative integers $l$ and $r$ less than $m$, the inner product of a row in $\mathbf{G}_{\mathrm{RM}}(l, m)$ and a row in $\mathbf{G}_{\mathrm{RM}}(r, m)$ is zero. We refer to this property as *a mutually orthogonal structure*.

Consider the $(m-r-1)$th-order RM code $C_{\mathrm{RM}}(m-r-1, m)$ whose dimension is

$$k(m - r - 1, m) = 1 + \binom{m}{1} + \binom{m}{2} + \cdots + \binom{m}{m - r - 1}. \tag{8.16}$$

Because $\binom{m}{l} = \binom{m}{m-l}$, $k(m - r - 1, m)$ can be expressed as the following sum:

$$k(m - r - 1, m) = 1 + \binom{m}{m - 1} + \binom{m}{m - 2} + \cdots + \binom{m}{r + 1}. \tag{8.17}$$

It follows from (8.14) and (8.17) that we have

$$
\begin{aligned}
k(r, m) &+ k(m - r - 1, m) \\
&= 1 + \binom{m}{1} + \binom{m}{2} + \cdots + \binom{m}{r} + \binom{m}{r+1} + \binom{m}{m-1} + 1 \\
&= 2^m.
\end{aligned} \tag{8.18}
$$

The expression given in (8.18) shows that the sum of dimensions of the $r$th-order RM code $C_{\mathrm{RM}}(r, m)$ and the $(m-r-1)$th-order RM code $C_{\mathrm{RM}}(m-r-1, m)$ is equal to the length $2^m$ of both codes. Because the inner product of any row in the generator matrix $\mathbf{G}_{\mathrm{RM}}(r, m)$ of the $r$th-order RM code $C_{\mathrm{RM}}(r, m)$ and any row in the generator matrix $\mathbf{G}_{\mathrm{RM}}(m - r - 1, m)$ of the $(m - r - 1)$th-order RM code $C_{\mathrm{RM}}(m - r - 1, m)$ is zero, the following equality holds

$$\mathbf{G}_{\mathrm{RM}}(r, m)\mathbf{G}_{\mathrm{RM}}^{T}(m - r - 1, m) = \mathbf{O}, \tag{8.19}$$

where $\mathbf{O}$ is a $k(r, m) \times k(m-r-1, m)$ zero matrix, i.e., the product of $\mathbf{G}_{\mathrm{RM}}(r, m)$ and the transpose of $\mathbf{G}_{\mathrm{RM}}(m - r - 1, m)$ gives a $k(r, m) \times k(m - r - 1, m)$ zero matrix. It follows from (8.18) and (8.19) that the $(m - r - 1)$th-order RM code $C_{\mathrm{RM}}(m - r - 1, m)$ is the *dual code* of the $r$th-order RM code $C_{\mathrm{RM}}(r, m)$ and vice versa. Hence, the generator matrix $\mathbf{G}_{\mathrm{RM}}(m-r-1, m)$ of the $(m-r-1)$th-order RM code $C_{\mathrm{RM}}(m - r - 1, m)$ is a parity-check matrix of the $r$th-order

RM code $C_{\mathrm{RM}}(r, m)$ and vice versa. For odd $m$, let $r = (m-1)/2$. Then, $C_{\mathrm{RM}}(m-r-1, m) = C_{\mathrm{RM}}(r, m) = C_{\mathrm{RM}}((m-1)/2, m)$. In this case, $C_{\mathrm{RM}}(r, m)$ is the dual code of itself. We say that $C_{\mathrm{RM}}(r, m)$ is *self-dual*. As mentioned earlier, the $r$th-order RM code $C_{\mathrm{RM}}(r, m)$ is the vector space given by the incidence vectors of the $m$-flat, the $(m-1)$-flats, $\ldots$, the $(m-r)$-flats in $\mathrm{EG}(m, 2)$ which contain the point $\alpha^t$ represented by the all-one $m$-tuple $(1\ 1\ \ldots\ 1)$ over $\mathrm{GF}(2)$. From the dual-code point of view, the $r$th-order RM code $C_{\mathrm{RM}}(r, m)$ is given by the null space of the incidence vectors of the $(r+1)$-flat, the $(r+2)$-flats, $\ldots$, the $(m-1)$-flats, and the $m$-flat in $\mathrm{EG}(m, 2)$ which contain the point $\alpha^t$ represented by the all-one $m$-tuple $(1\ 1\ \ldots\ 1)$ over $\mathrm{GF}(2)$.

It was proved in [3, 7, 27] that the minimum-weight codewords in the $r$th-order RM code $C_{\mathrm{RM}}(r, m)$ are the incidence vectors of all the $(m-r)$-flats in $\mathrm{EG}(m, 2)$. As a result, the number of minimum-weight codewords in $C_{\mathrm{RM}}(r, m)$ is given by the number of $(m-r)$-flats in $\mathrm{EG}(m, 2)$ which is

$$2^r \prod_{i=1}^{m-r} \frac{2^{m-i+1} - 1}{2^{m-r-i+1} - 1}. \tag{8.20}$$

For more on the weight distributions and structural properties of RM codes, readers are referred to [25, 28–31]. In particular, the first-order RM code $C_{\mathrm{RM}}(1, m)$ has only three weights, $0$, $2^{m-1}$, and $2^m$. The weight distribution of $C_{\mathrm{RM}}(1, m)$ is

$$A_0 = 1, A_{2^{m-1}} = 2^{m+1} - 2, A_{2^m} = 1. \tag{8.21}$$

The second-order RM code $C_{\mathrm{RM}}(2, m)$ has the following weight distribution:

$$
\begin{aligned}
A_0 &= 1, \\
A_{2^{m-1} \pm 2^{m-1-l}} &= 2^{l(l+1)} \frac{\prod_{i=m-2l+1}^{m}(2^i - 1)}{\prod_{i=1}^{l}(2^{2i} - 1)}, \text{ for } 1 \leq l \leq \lfloor \tfrac{m}{2} \rfloor, \\
A_{2^{m-1}} &= 2^{(m^2+m+2)/2} - 2 - 2\sum_{l=1}^{\lfloor \frac{m}{2} \rfloor} 2^{l(l+1)} \frac{\prod_{i=m-2l+1}^{m}(2^i - 1)}{\prod_{i=1}^{l}(2^{2i} - 1)}, \\
A_{2^m} &= 1.
\end{aligned}
\tag{8.22}
$$

Following from (8.19), (8.21), (8.22), and the MacWilliams identity given by (3.36), the weight distributions of the $(m-2)$th-order RM code $C_{\mathrm{RM}}(m-2, m)$ and the $(m-3)$th-order RM code $C_{\mathrm{RM}}(m-3, m)$ can be derived (see Problem 8.8).

**Example 8.2** Consider the $4 \times 16$ matrix $\mathbf{G}_1$ constructed based on the four-dimensional Euclidean geometry $\mathrm{EG}(4, 2)$ given by (8.10) in Example 8.1. For $r = 1$, the generator matrix $\mathbf{G}_{\mathrm{RM}}(1, 4)$ of the first-order RM code $C_{\mathrm{RM}}(1, 4)$ is given below:

$$
\mathbf{G}_{\mathrm{RM}}(1, 4) = \begin{bmatrix} \mathbf{v}_0 \\ \mathbf{G}_1 \end{bmatrix} = \begin{bmatrix} \mathbf{v}_0 \\ \mathbf{v}_4 \\ \mathbf{v}_3 \\ \mathbf{v}_2 \\ \mathbf{v}_1 \end{bmatrix} = \begin{bmatrix} 1\,1\,1\,1\,1\,1\,1\,1\,1\,1\,1\,1\,1\,1\,1\,1 \\ 0\,0\,0\,0\,0\,0\,0\,0\,1\,1\,1\,1\,1\,1\,1\,1 \\ 0\,0\,0\,0\,1\,1\,1\,1\,0\,0\,0\,0\,1\,1\,1\,1 \\ 0\,0\,1\,1\,0\,0\,1\,1\,0\,0\,1\,1\,0\,0\,1\,1 \\ 0\,1\,0\,1\,0\,1\,0\,1\,0\,1\,0\,1\,0\,1\,0\,1 \end{bmatrix}. \tag{8.23}
$$

The first-order RM code $C_{\mathrm{RM}}(1,4)$ generated by $\mathbf{G}_{\mathrm{RM}}(1,4)$ is a $(16,5)$ code with minimum distance 8. The code $C_{\mathrm{RM}}(1,4)$ has 30 codewords of weight 8, one codeword of weight 16, and one zero codeword. The 30 minimum-weight codewords are the incidence vectors of the 30 3-flats in $\mathrm{EG}(4,2)$. The first-order RM code $C_{\mathrm{RM}}(1,4)$ is spanned by the incidence vectors of four 3-flats and the single 4-flat in $\mathrm{EG}(4,2)$.

For $r = 2$, the generator matrix $\mathbf{G}_{\mathrm{RM}}(2,4)$ of the second-order RM code $C_{\mathrm{RM}}(2,4)$ is given below:

$$\mathbf{G}_{\mathrm{RM}}(2,4) = \begin{bmatrix} \mathbf{v}_0 \\ \mathbf{G}_1 \\ \mathbf{G}_2 \end{bmatrix} = \begin{bmatrix} \mathbf{v}_0 \\ \mathbf{v}_4 \\ \mathbf{v}_3 \\ \mathbf{v}_2 \\ \mathbf{v}_1 \\ \mathbf{v}_3\mathbf{v}_4 \\ \mathbf{v}_2\mathbf{v}_4 \\ \mathbf{v}_1\mathbf{v}_4 \\ \mathbf{v}_2\mathbf{v}_3 \\ \mathbf{v}_1\mathbf{v}_3 \\ \mathbf{v}_1\mathbf{v}_2 \end{bmatrix} = \begin{bmatrix} 1\,1\,1\,1\,1\,1\,1\,1\,1\,1\,1\,1\,1\,1\,1\,1 \\ 0\,0\,0\,0\,0\,0\,0\,0\,1\,1\,1\,1\,1\,1\,1\,1 \\ 0\,0\,0\,0\,1\,1\,1\,1\,0\,0\,0\,0\,1\,1\,1\,1 \\ 0\,0\,1\,1\,0\,0\,1\,1\,0\,0\,1\,1\,0\,0\,1\,1 \\ 0\,1\,0\,1\,0\,1\,0\,1\,0\,1\,0\,1\,0\,1\,0\,1 \\ 0\,0\,0\,0\,0\,0\,0\,0\,0\,0\,0\,0\,1\,1\,1\,1 \\ 0\,0\,0\,0\,0\,0\,0\,0\,0\,0\,1\,1\,0\,0\,1\,1 \\ 0\,0\,0\,0\,0\,0\,0\,0\,1\,0\,1\,0\,1\,0\,1 \\ 0\,0\,0\,0\,0\,0\,1\,1\,0\,0\,0\,0\,0\,0\,1\,1 \\ 0\,0\,0\,0\,0\,1\,0\,1\,0\,0\,0\,0\,0\,1\,0\,1 \\ 0\,0\,0\,1\,0\,0\,0\,1\,0\,0\,0\,1\,0\,0\,0\,1 \end{bmatrix}. \qquad (8.24)$$

The second-order RM code $C_{\mathrm{RM}}(2,4)$ generated by $\mathbf{G}_{\mathrm{RM}}(2,4)$ is a $(16,11)$ code with minimum distance 4. Using (8.20), we find that the $(16,11)$ RM code has 140 codewords of weight 4 that are the incidence vectors of the 140 2-flats in $\mathrm{EG}(4,2)$. The second-order RM code $C_{\mathrm{RM}}(2,4)$ is spanned by the incidence vectors of the 2-flats, the 3-flats, and the single 4-flat in $\mathrm{EG}(4,2)$. The RM codes $C_{\mathrm{RM}}(1,4)$ and $C_{\mathrm{RM}}(2,4)$ are dual to each other. The generator matrix $\mathbf{G}_{\mathrm{RM}}(1,4)$ of $C_{\mathrm{RM}}(1,4)$ is a parity-check matrix of $C_{\mathrm{RM}}(2,4)$ and vice versa. Besides, the first-order RM code $C_{\mathrm{RM}}(1,4)$ is a subcode of the second-order RM code $C_{\mathrm{RM}}(2,4)$. ▲▲

**Example 8.3** Set $m = 5$ and $r = 2$. The second-order RM code $C_{\mathrm{RM}}(2,5)$ constructed based on the 3-flats, the 4-flats, and the single 5-flat in the five-dimensional Euclidean geometry $\mathrm{EG}(5,2)$ over $\mathrm{GF}(2)$ containing the point $(1\ 1\ 1\ 1\ 1)$ is a $(32,16)$ code with minimum distance 8. This code is self-dual.

For $m = 7$ and $r = 3$, the third-order RM code $C_{\mathrm{RM}}(3,7)$ constructed based on the 4-flats, the 5-flats, the 6-flats, and the single 7-flat in the seven-dimensional Euclidean geometry $\mathrm{EG}(7,2)$ over $\mathrm{GF}(2)$ containing the point $(1\ 1\ 1\ 1\ 1\ 1\ 1)$ is a $(128,64)$ code with minimum distance 16 and is also self-dual. ▲▲

## 8.3 Encoding of RM Codes

The generator matrix of an RM code with rows in the form of (8.13) is not systematic. It can be put in systematic form by elementary row operations and/or column permutations. However, the generator matrix $\mathbf{G}_{\mathrm{RM}}(r,m)$ in the form of (8.13) can be used to decode the $r$th-order RM code $C_{\mathrm{RM}}(r,m)$ in a simple way through majority-logic decoding with $r+1$ steps.

In the following, we consider a specific encoding of the $r$th-order RM code $C_{\mathrm{RM}}(r,m)$ with the generator matrix $\mathbf{G}_{\mathrm{RM}}(r,m)$ in the form of (8.13). A message $\mathbf{a}$ consisting of $k(r,m)$ information symbols (bits) is encoded into a codeword $\mathbf{b} = (b_0, b_1, b_2, \ldots, b_{2^m-1})$ of $2^m$ code symbols (bits) in $C_{\mathrm{RM}}(r,m)$ through

$$\mathbf{b} = \mathbf{a}\mathbf{G}_{\mathrm{RM}}(r,m) = (b_0, b_1, b_2, \ldots, b_{2^m-1}). \qquad (8.25)$$

For decoding purpose, we divide the $k(r,m)$ information symbols of the message $\mathbf{a}$ into $r+1$ sections, denoted by $\mathbf{a}_0, \mathbf{a}_1, \ldots, \mathbf{a}_r$. Then, $\mathbf{a} = (\mathbf{a}_0, \mathbf{a}_1, \ldots, \mathbf{a}_r)$. For $1 \leq l \leq r$, the $l$th section $\mathbf{a}_l$ of $\mathbf{a}$ consists of $\binom{m}{l}$ consecutive information symbols which correspond to the $\binom{m}{l}$ product rows of degree $l$ in the submatrix $\mathbf{G}_l$ of $\mathbf{G}_{\mathrm{RM}}(r,m)$. We denote the information symbol corresponding to the product row $\mathbf{v}_{i_1} \mathbf{v}_{i_2} \ldots \mathbf{v}_{i_l}$ by $a_{i_1, i_2, \ldots, i_l}$. For $1 \leq l \leq r$, let $\mathcal{I}_l = \{(i_1, i_2, \ldots, i_l) : i_t \in \{1, 2, \ldots, m\}, 1 \leq t \leq l\}$ be a set of $l$ indices of the rows $\mathbf{v}_{i_1}, \mathbf{v}_{i_2}, \ldots, \mathbf{v}_{i_l}$ in the product $\mathbf{v}_{i_1} \mathbf{v}_{i_2} \ldots \mathbf{v}_{i_l}$. Then, the $l$th section $\mathbf{a}_l$ of the message $\mathbf{a}$ consists of the following set of $\binom{m}{l}$ information symbols:

$$\mathbf{A}_l = \{a_{i_1, i_2, \ldots, i_l} : a_{i_1, i_2, \ldots, i_l} \in \mathrm{GF}(2), (i_1, i_2, \ldots, i_l) \in \mathcal{I}_l\}. \qquad (8.26)$$

For $l = 0$, the zeroth section $\mathbf{a}_0$ of $\mathbf{a}$ consists of the first information symbol $a_0$ of $\mathbf{a}$, i.e., $\mathbf{A}_0 = \{a_0\}$ and $\mathbf{a}_0 = (a_0)$. With the above grouping of information symbols, the message $\mathbf{a}$ can be put in the following form:

$$\mathbf{a} = (\underbrace{a_0}_{\mathbf{a}_0}, \underbrace{a_m, a_{m-1}, \ldots, a_2, a_1}_{\mathbf{a}_1}, \underbrace{a_{m-1,m}, a_{m-2,m}, \ldots, a_{1,2}}_{\mathbf{a}_2}, \ldots,$$
$$\underbrace{a_{m-r+1,m-r+2,\ldots,m}, \ldots, a_{1,2,\ldots,r}}_{\mathbf{a}_r}), \qquad (8.27)$$

where the $l$th section, $0 \leq l \leq r$, is a tuple of length $\binom{m}{l}$:

$$\mathbf{a}_l = (a_{m-l+1,m-l+2,\ldots,m}, \ldots, a_{1,2,\ldots,l}). \qquad (8.28)$$

For $1 \leq l \leq r$, define the following $2^m$-tuple:

$$\mathbf{b}_l = \mathbf{a}_l \mathbf{G}_l$$
$$= \sum_{(i_1,i_2,\ldots,i_l) \in \mathcal{I}_l} a_{i_1,i_2,\ldots,i_l} \mathbf{v}_{i_1} \mathbf{v}_{i_2} \ldots \mathbf{v}_{i_l} \qquad (8.29)$$

which is a linear combination of the rows in $\mathbf{G}_l$ with the information symbols in $\mathbf{A}_l$ as coefficients. For $l = 0$, define a $2^m$-tuple $\mathbf{b}_0$ as $\mathbf{b}_0 = a_0 \mathbf{v}_0$. Then, the codeword $\mathbf{b}$ for the message $\mathbf{a}$ is

$$\mathbf{b} = \mathbf{a}\mathbf{G}_{\mathrm{RM}}(r,m) = (\mathbf{a}_0, \mathbf{a}_1, \ldots, \mathbf{a}_r) \begin{bmatrix} \mathbf{v}_0 \\ \mathbf{G}_1 \\ \mathbf{G}_2 \\ \cdots \\ \mathbf{G}_r \end{bmatrix} = \mathbf{b}_0 + \mathbf{b}_1 + \cdots + \mathbf{b}_r. \qquad (8.30)$$

The vectors ($2^m$-tuples) $\mathbf{b}_0, \mathbf{b}_1, \ldots, \mathbf{b}_r$ are called the *constituent codewords* of $\mathbf{b}$.

**Example 8.4** Consider the encoding of the $(16, 11)$ second-order RM code $C_{RM}(2, 4)$ given in Example 8.2. Let

$$\mathbf{a} = (a_0, a_4, a_3, a_2, a_1, a_{3,4}, a_{2,4}, a_{1,4}, a_{2,3}, a_{1,3}, a_{1,2})$$

be the message of length 11 to be encoded. The message $\mathbf{a}$ is divided into $r + 1 = 2 + 1 = 3$ sections as follows:

$$\begin{aligned}
\mathbf{a}_0 &= (a_0), \\
\mathbf{a}_1 &= (a_4, a_3, a_2, a_1), \\
\mathbf{a}_2 &= (a_{3,4}, a_{2,4}, a_{1,4}, a_{2,3}, a_{1,3}, a_{1,2}).
\end{aligned} \tag{8.31}$$

The constituent codewords for the three sections $\mathbf{a}_0$, $\mathbf{a}_1$, $\mathbf{a}_2$ of the message $\mathbf{a}$ are:

$$\begin{aligned}
\mathbf{b}_0 &= a_0 \mathbf{v}_0, \\
\mathbf{b}_1 &= a_4 \mathbf{v}_4 + a_3 \mathbf{v}_3 + a_2 \mathbf{v}_2 + a_1 \mathbf{v}_1, \\
\mathbf{b}_2 &= a_{3,4} \mathbf{v}_3 \mathbf{v}_4 + a_{2,4} \mathbf{v}_2 \mathbf{v}_4 + a_{1,4} \mathbf{v}_1 \mathbf{v}_4 + a_{2,3} \mathbf{v}_2 \mathbf{v}_3 + a_{1,3} \mathbf{v}_1 \mathbf{v}_3 + a_{1,2} \mathbf{v}_1 \mathbf{v}_2.
\end{aligned} \tag{8.32}$$

Then, the codeword $\mathbf{b}$ for the entire message $\mathbf{a}$ is

$$\begin{aligned}
\mathbf{b} &= (b_0, b_1, b_2, b_3, b_4, b_5, b_6, b_7, b_8, b_9, b_{10}, b_{11}, b_{12}, b_{13}, b_{14}, b_{15}) \\
&= \mathbf{b}_0 + \mathbf{b}_1 + \mathbf{b}_2 \\
&= a_0 \mathbf{v}_0 + a_4 \mathbf{v}_4 + a_3 \mathbf{v}_3 + a_2 \mathbf{v}_2 + a_1 \mathbf{v}_1 + a_{3,4} \mathbf{v}_3 \mathbf{v}_4 + \\
&\quad a_{2,4} \mathbf{v}_2 \mathbf{v}_4 + a_{1,4} \mathbf{v}_1 \mathbf{v}_4 + a_{2,3} \mathbf{v}_2 \mathbf{v}_3 + a_{1,3} \mathbf{v}_1 \mathbf{v}_3 + a_{1,2} \mathbf{v}_1 \mathbf{v}_2.
\end{aligned} \tag{8.33}$$

Using (8.24) and (8.33), we find the code symbols in $\mathbf{b}$ which are given as follows:

$$\begin{aligned}
b_0 &= a_0, \\
b_1 &= a_0 + a_1, \\
b_2 &= a_0 + a_2, \\
b_3 &= a_0 + a_1 + a_2 + a_{1,2}, \\
b_4 &= a_0 + a_3, \\
b_5 &= a_0 + a_1 + a_3 + a_{1,3}, \\
b_6 &= a_0 + a_2 + a_3 + a_{2,3}, \\
b_7 &= a_0 + a_1 + a_2 + a_3 + a_{2,3} + a_{1,3} + a_{1,2}, \\
b_8 &= a_0 + a_4, \\
b_9 &= a_0 + a_1 + a_4 + a_{1,4}, \\
b_{10} &= a_0 + a_2 + a_4 + a_{2,4}, \\
b_{11} &= a_0 + a_1 + a_2 + a_4 + a_{2,4} + a_{1,4} + a_{1,2}, \\
b_{12} &= a_0 + a_3 + a_4 + a_{3,4}, \\
b_{13} &= a_0 + a_1 + a_3 + a_4 + a_{3,4} + a_{1,4} + a_{1,3}, \\
b_{14} &= a_0 + a_2 + a_3 + a_4 + a_{3,4} + a_{2,4} + a_{2,3}, \\
b_{15} &= a_0 + a_1 + a_2 + a_3 + a_4 + a_{3,4} + a_{2,4} + a_{1,4} + a_{2,3} + a_{1,3} + a_{1,2}.
\end{aligned} \tag{8.34}$$

▲▲

The above encoding scheme is referred to as *multilevel encoding*. Based on this multilevel encoding of messages in groups, we will present a multilevel decoding scheme in Section 8.5. In the scheme, the transmitted information symbols in a message $\mathbf{a} = (\mathbf{a}_0, \mathbf{a}_1, \ldots, \mathbf{a}_r)$ are decoded in groups from a received vector, one group at a time. The decoding is carried out by levels in the order of $\mathbf{a}_r, \mathbf{a}_{r-1}, \ldots, \mathbf{a}_1, \mathbf{a}_0$.

## 8.4 Successive Retrieval of Information Symbols

With the encoding of the information symbols of a message $\mathbf{a}$ in groups, information symbols can be retrieved from the code symbols of its corresponding codeword $\mathbf{b}$ in groups *successively through $r + 1$ levels of cancellation*, called *successive cancellation*. In the zeroth level, the information symbols in $\mathbf{A}_r = \{a_{i_1, i_2, \ldots, i_r} : (i_1, i_2, \ldots, i_r) \in \mathcal{I}_r\}$ are retrieved from $\mathbf{b}$ (denote $\mathbf{b}^{(0)} = \mathbf{b}$ to simplify the notation in the following description) based on the rows of $\mathbf{G}_r$. After the recovery of the information symbols in $\mathbf{A}_r$, we remove (or cancel) the $r$th constituent codeword $\mathbf{b}_r$ of $\mathbf{b}$ from $\mathbf{b}$. This cancellation results in a descendant codeword, $\mathbf{b}^{(1)} = \mathbf{b} - \mathbf{b}_r = \mathbf{b}^{(0)} - \mathbf{b}_r$, of $\mathbf{b}$. In the first level of information retrieval, we retrieve the information symbols in $\mathbf{A}_{r-1} = \{a_{i_1, i_2, \ldots, i_{r-1}} : (i_1, i_2, \ldots, i_{r-1}) \in \mathcal{I}_{r-1}\}$ from $\mathbf{b}^{(1)}$ based on the rows of $\mathbf{G}_{r-1}$. After recovery of the information symbols in $\mathbf{A}_{r-1}$, we remove the $(r-1)$th constituent codeword $\mathbf{b}_{r-1}$ from $\mathbf{b}^{(1)}$ to obtain a descendant codeword $\mathbf{b}^{(2)} = \mathbf{b}^{(1)} - \mathbf{b}_{r-1} = \mathbf{b} - \mathbf{b}_r - \mathbf{b}_{r-1}$ of $\mathbf{b}$. Then, we retrieve the information symbols in $\mathbf{A}_{r-2}$ from $\mathbf{b}^{(2)}$ based on the rows of $\mathbf{G}_{r-2}$. We continue this information retrieval process level by level until the information symbol $a_0$ is retrieved from $\mathbf{b}^{(r)} = \mathbf{b}^{(r-1)} - \sum_{(i_1) \in \mathcal{I}_1} a_{i_1} \mathbf{v}_{i_1} = \mathbf{b} - \mathbf{b}_r - \mathbf{b}_{r-1} - \cdots - \mathbf{b}_1 = a_0 \mathbf{v}_0$.

To retrieve the information symbols in $\mathbf{A}_r = \{a_{i_1, i_2, \ldots, i_r} : (i_1, i_2, \ldots, i_r) \in \mathcal{I}_r\}$ from $\mathbf{b}$, we need to know the code symbols in $\mathbf{b}$ (i.e., the code symbol indices in the codeword $\mathbf{b}$) whose calculations involve the information symbols in $\mathbf{A}_r$. To find these code symbol indices, we first form the following set of integers in radix-2 form:

$$\mathbf{S}_r \triangleq \{c_{i_1-1} 2^{i_1-1} + c_{i_2-1} 2^{i_2-1} + \cdots + c_{i_r-1} 2^{i_r-1} : c_{i_j-1} \in \{0, 1\}, 1 \le j \le r\}. \tag{8.35}$$

The set $\mathbf{S}_r$ consists of $2^r$ nonnegative integers less than $2^m$. Let $\{i_1 - 1, i_2 - 1, \ldots, i_r - 1\}$ be the set of the exponents of the radix-2 expansions of the integers in $\mathbf{S}_r$ which is a subset of $\{0, 1, 2, \ldots, m-1\}$, i.e., $\{i_1 - 1, i_2 - 1, \ldots, i_r - 1\} \subseteq \{0, 1, 2, \ldots, m-1\}$. Define the following set of integers:

$$\begin{aligned} \mathbf{Q}_r &\triangleq \{0, 1, 2, \ldots, m-1\} \backslash \{i_1 - 1, i_2 - 1, \ldots, i_r - 1\} \\ &= \{j_1, j_2, \ldots, j_{m-r}\}, \end{aligned} \tag{8.36}$$

where $0 \le j_1 < j_2 < \cdots < j_{m-r} < m$ and the notation "$\backslash$" stands for removing the elements in the set $\{i_1 - 1, i_2 - 1, \ldots, i_r - 1\}$ from the set $\{0, 1, 2, \ldots, m-1\}$. Next, we form the following set of integers:

$$\mathbf{S}_r^c \triangleq \{d_{j_1} 2^{j_1} + d_{j_2} 2^{j_2} + \cdots + d_{j_{m-r}} 2^{j_{m-r}} : d_{j_t} \in \{0, 1\}, 1 \le t \le m-r\} \tag{8.37}$$

which consists of $2^{m-r}$ nonnegative integers in radix-2 form. It follows from (8.35) and (8.37) that $\mathbf{S}_r \cap \mathbf{S}_r^c = \{0\}$.

For each nonnegative integer $t_e \in \mathbf{S}_r^c$, $1 \leq e \leq 2^{m-r}$, we form the following set of $2^r$ integers:

$$\mathbf{B}_{r,e}(t_e) \triangleq t_e + \mathbf{S}_r = \{t_e + s : s \in \mathbf{S}_r\}. \tag{8.38}$$

Note that for $t_{e'} \neq t_e$, $\mathbf{B}_{r,e'}(t_{e'})$ and $\mathbf{B}_{r,e}(t_e)$ are disjoint, i.e., $\mathbf{B}_{r,e'}(t_{e'}) \cap \mathbf{B}_{r,e}(t_e) = \emptyset$. Because there are $2^{m-r}$ integers in $\mathbf{S}_r^c$, there are $2^{m-r}$ sets of integers in the form of (8.38), denoted by $\mathbf{B}_{r,1}(t_1), \mathbf{B}_{r,2}(t_2), \ldots, \mathbf{B}_{r,2^{m-r}}(t_{2^{m-r}})$, each consisting of $2^r$ nonnegative integers. For $1 \leq e \leq 2^{m-r}$, the $2^r$ integers in $\mathbf{B}_{r,e}(t_e)$ are the *indices* of the $2^r$ code symbols in the codeword $\mathbf{b}$ whose sum (modulo-2 addition) gives the information symbol $a_{i_1,i_2,\ldots,i_r}$, i.e.,

$$a_{i_1,i_2,\ldots,i_r} = \sum_{k \in \mathbf{B}_{r,e}(t_e)} b_k. \tag{8.39}$$

There are $2^{m-r}$ sums in the form of (8.39), each giving the information symbol $a_{i_1,i_2,\ldots,i_r}$. Hence, there are $2^{m-r}$ *independent determinations* of the information symbol $a_{i_1,i_2,\ldots,i_r}$. So, for each index sequence $(i_1, i_2, \ldots, i_r) \in \mathcal{I}_r$, we can retrieve the information symbol $a_{i_1,i_2,\ldots,i_r}$ in $\mathbf{A}_r = \{a_{i_1,i_2,\ldots,i_r} : (i_1, i_2, \ldots, i_r) \in \mathcal{I}_r\}$ from the code symbols in $\mathbf{b}$. This is the zeroth level of information symbol retrieval. The sets $\mathbf{B}_{r,1}(t_1), \mathbf{B}_{r,2}(t_2), \ldots, \mathbf{B}_{r,2^{m-r}}(t_{2^{m-r}})$ are called the *index sets* for the retrieval of the information symbol $a_{i_1,i_2,\ldots,i_r}$ in $\mathbf{A}_r$.

**Example 8.5** Consider the $(16, 11)$ second-order RM code $C_{\mathrm{RM}}(2, 4)$ constructed in Example 8.2 whose encoding was given in Example 8.4. Suppose we want to retrieve the information symbol $a_{1,2}$ in $\mathbf{A}_2 = \{a_{1,2}, a_{1,3}, a_{1,4}, a_{2,3}, a_{2,4}, a_{3,4}\}$ from the code symbols in the codeword

$$\mathbf{b} = (b_0, b_1, b_2, b_3, b_4, b_5, b_6, b_7, b_8, b_9, b_{10}, b_{11}, b_{12}, b_{13}, b_{14}, b_{15}).$$

Because $m = 4$, $r = 2$, $i_1 = 1$, and $i_2 = 2$, it follows from (8.35), (8.36), and (8.37) that we obtain the following set of integers:

$$\begin{aligned}
\mathbf{S}_2 &= \{c_0 + c_1 2 : c_0, c_1 \in \{0, 1\}\} = \{0, 1, 2, 3\}, \\
\mathbf{Q}_2 &= \{0, 1, 2, 3\} \backslash \{0, 1\} = \{2, 3\}, \\
\mathbf{S}_2^c &= \{d_2 2^2 + d_3 2^3 : d_2, d_3 \in \{0, 1\}\} = \{0, 4, 8, 12\}.
\end{aligned} \tag{8.40}$$

Based on $\mathbf{S}_2$, $\mathbf{S}_2^c$, and (8.38), we form the following index sets:

$$\begin{aligned}
\mathbf{B}_{2,1}(0) &= \{0, 1, 2, 3\}, & \mathbf{B}_{2,2}(4) &= \{4, 5, 6, 7\}, \\
\mathbf{B}_{2,3}(8) &= \{8, 9, 10, 11\}, & \mathbf{B}_{2,4}(12) &= \{12, 13, 14, 15\}.
\end{aligned} \tag{8.41}$$

Using (8.38), (8.39), and (8.41) with $r = 2$, we form the following four independent determinations of the information symbol $a_{1,2}$:

$$\begin{aligned}
a_{1,2} &= b_0 + b_1 + b_2 + b_3, & a_{1,2} &= b_4 + b_5 + b_6 + b_7, \\
a_{1,2} &= b_8 + b_9 + b_{10} + b_{11}, & a_{1,2} &= b_{12} + b_{13} + b_{14} + b_{15}.
\end{aligned} \tag{8.42}$$

To retrieve the information symbol $a_{3,4}$ with $i_1 = 3$ and $i_2 = 4$, we construct the following three sets:

$$\mathbf{S}_2 = \{c_2 2^2 + c_3 2^3 : c_2, c_3 \in \{0,1\}\} = \{0,4,8,12\},$$
$$\mathbf{Q}_2 = \{0,1,2,3\}\backslash\{2,3\} = \{0,1\}, \qquad (8.43)$$
$$\mathbf{S}_2^c = \{d_0 + d_1 2 : d_0, d_1 \in \{0,1\}\} = \{0,1,2,3\}.$$

Based on $\mathbf{S}_2$, $\mathbf{S}_2^c$, and (8.38), we form the following index sets:

$$\mathbf{B}_{2,1}(0) = \{0,4,8,12\}, \qquad \mathbf{B}_{2,2}(1) = \{1,5,9,13\},$$
$$\mathbf{B}_{2,3}(2) = \{2,6,10,14\}, \qquad \mathbf{B}_{2,4}(3) = \{3,7,11,15\}. \qquad (8.44)$$

Using (8.38), (8.39), and (8.44) with $r = 2$, we form the following four independent determinations of the information symbol $a_{3,4}$:

$$a_{3,4} = b_0 + b_4 + b_8 + b_{12}, \qquad a_{3,4} = b_1 + b_5 + b_9 + b_{13},$$
$$a_{3,4} = b_2 + b_6 + b_{10} + b_{14}, \qquad a_{3,4} = b_3 + b_7 + b_{11} + b_{15}. \qquad (8.45)$$

Similarly, we can retrieve the other four information symbols in the set $\mathbf{A}_2$ from the symbols of the codeword $\mathbf{b}$ as follows:

$$a_{1,3} = b_0 + b_1 + b_4 + b_5, \qquad a_{1,3} = b_2 + b_3 + b_6 + b_7,$$
$$a_{1,3} = b_8 + b_9 + b_{12} + b_{13}, \qquad a_{1,3} = b_{10} + b_{11} + b_{14} + b_{15},$$

$$a_{2,3} = b_0 + b_2 + b_4 + b_6, \qquad a_{2,3} = b_1 + b_3 + b_5 + b_7,$$
$$a_{2,3} = b_8 + b_{10} + b_{12} + b_{14}, \qquad a_{2,3} = b_9 + b_{11} + b_{13} + b_{15},$$

$$a_{1,4} = b_0 + b_1 + b_8 + b_9, \qquad a_{1,4} = b_2 + b_3 + b_{10} + b_{11}, \qquad (8.46)$$
$$a_{1,4} = b_4 + b_5 + b_{12} + b_{13}, \qquad a_{1,4} = b_6 + b_7 + b_{14} + b_{15},$$

$$a_{2,4} = b_0 + b_2 + b_8 + b_{10}, \qquad a_{2,4} = b_1 + b_3 + b_9 + b_{11},$$
$$a_{2,4} = b_4 + b_6 + b_{12} + b_{14}, \qquad a_{2,4} = b_5 + b_7 + b_{13} + b_{15}.$$

▲▲

After the completion of the zeroth level retrieval of the information symbols in $\mathbf{A}_r = \{a_{i_1,i_2,\ldots,i_r} : (i_1, i_2, \ldots, i_r) \in \mathcal{I}_r\}$, we form the $r$th constituent codeword $\mathbf{b}_r$ of $\mathbf{b}$ as follows:

$$\mathbf{b}_r = \sum_{(i_1,i_2,\ldots,i_r)\in\mathcal{I}_r} a_{i_1,i_2,\ldots,i_r} \mathbf{v}_{i_1} \mathbf{v}_{i_2} \cdots \mathbf{v}_{i_r}. \qquad (8.47)$$

Next, we remove (or cancel) $\mathbf{b}_r$ from $\mathbf{b}$. This results in a modified codeword (also a codeword in the $(r-1)$th-order RM code $C_{\mathrm{RM}}(r-1,m)$),

$$\mathbf{b}^{(1)} = \mathbf{b} - \mathbf{b}_r. \qquad (8.48)$$

Then, we retrieve the information symbols in $\mathbf{A}_{r-1} = \{a_{i_1,i_2,\ldots,i_{r-1}} : (i_1, i_2, \ldots, i_{r-1}) \in \mathcal{I}_{r-1}\}$ from the code symbols in $\mathbf{b}^{(1)}$.

The retrieval of the information symbols in $\mathbf{A}_{r-1}$ from the code symbols in $\mathbf{b}^{(1)}$ is exactly the same as those in $\mathbf{A}_r$ from the code symbols in $\mathbf{b}$. This is accomplished by replacing $r$ by $r-1$ in forming the sets given by (8.35), (8.36), (8.37), (8.38), and the information retrieval sums given by (8.39). For each information symbol $a_{i_1,i_2,\ldots,i_{r-1}}$ in $\mathbf{A}_{r-1}$, we can form $2^{m-r+1}$ independent sums of code symbols in $\mathbf{b}^{(1)}$ which are equal to $a_{i_1,i_2,\ldots,i_{r-1}}$.

For $0 \le l \le r$, we retrieve the information symbols in $\mathbf{A}_{r-l} = \{a_{i_1,i_2,\ldots,i_{r-l}} : (i_1,i_2,\ldots,i_{r-l}) \in \mathcal{I}_{r-l}\}$ from the code symbols in the codeword

$$\begin{aligned}
\mathbf{b}^{(l)} &= \mathbf{b} - \mathbf{b}_r - \mathbf{b}_{r-1} - \cdots - \mathbf{b}_{r-l+1} \\
&= (b_0^{(l)}, b_1^{(l)}, \ldots, b_{2^m-2}^{(l)}, b_{2^m-1}^{(l)}).
\end{aligned} \tag{8.49}$$

Note that $\mathbf{b}^{(l)}$ is also a codeword in the $(r-l)$th-order RM code $C_{\mathrm{RM}}(r-l,m)$. To retrieve the information symbols in $\mathbf{A}_{r-l}$, we form the following three sets:

$$\mathbf{S}_l \triangleq \{c_{i_1-1}2^{i_1-1}+c_{i_2-1}2^{i_2-1}+\cdots+c_{i_{r-l}-1}2^{i_{r-l}-1} : c_{i_j-1} \in \{0,1\}, 1 \le j \le r-l\}, \tag{8.50}$$

$$\mathbf{Q}_l \triangleq \{0,1,2,\ldots,m-1\}\setminus\{i_1-1,i_2-1,\ldots,i_{r-l}-1\} = \{j_1,j_2,\ldots,j_{m-r+l}\}, \tag{8.51}$$

$$\mathbf{S}_l^c \triangleq \{d_{j_1}2^{j_1} + d_{j_2}2^{j_2} + \cdots + d_{j_{m-r+l}}2^{j_{m-r+l}} : d_{j_t} \in \{0,1\}, 1 \le t \le m-r+l\}. \tag{8.52}$$

For each integer $t_e$ in $\mathbf{S}_l^c$, we form the following set of indices

$$\mathbf{B}_{l,e}(t_e) \triangleq t_e + S_l = \{t_e + s : s \in S_l\}, \tag{8.53}$$

with $1 \le e \le 2^{m-r+l}$. Each set $\mathbf{B}_{l,e}(t_e)$ contains $2^{r-l}$ indices. There are $2^{m-r+l}$ index sets in the form of (8.53). Based on each of these index sets, $\mathbf{B}_{l,e}(t_e)$ with $1 \le e \le 2^{m-r+l}$, we can retrieve the information symbol $a_{i_1,i_2,\ldots,i_{r-l}}$ from the symbols in the modified codeword $\mathbf{b}^{(l)}$ given by (8.49). The information symbol $a_{i_1,i_2,\ldots,i_{r-l}}$ is retrieved from the following sum:

$$a_{i_1,i_2,\ldots,i_{r-l}} = \sum_{k \in \mathbf{B}_{l,e}(t_e)} b_k^{(l)}. \tag{8.54}$$

There are $2^{m-r+l}$ independent determinations for the information symbols $a_{i_1,i_2,\ldots,i_{r-l}}$ in $\mathbf{A}_{r-l}$. For $l = 0$, the retrievals of information symbols in $\mathbf{A}_r = \{a_{i_1,i_2,\ldots,i_r} : (i_1,i_2,\ldots,i_r) \in \mathcal{I}_r\}$ are based on the symbols in the codeword $\mathbf{b}$ as given by (8.39).

The above process of retrieving information symbols in a message of the $r$th-order RM code $C_{\mathrm{RM}}(r,m)$ from its corresponding codeword is carried out successively in $r+1$ levels, one level at a time. At the completion of each level, we cancel the constituent codeword from the descendant codeword formed in the previous level. From this cancellation point of view, we refer to this process of retrieval of information symbols as *successive cancellation information retrieval* (SCIR).

Note that, for $0 \leq l \leq r$, each of the $2^{m-r+l}$ sums given by (8.54) is formed by an $(r-l)$-flat in a parallel bundle $P$ which consists of $2^{m-r+l}$ parallel $(r-l)$-flats in $\mathrm{EG}(m,2)$. Each of these sums is the inner product of the incidence vector of an $(r-l)$-flat in $P$ and the descendant codeword $\mathbf{b}^{(l)}$ given by (8.49) (see Problem 8.13). So, the construction, encoding, information retrieval, and majority-logic decoding of RM codes presented in this chapter are based on EGs over $\mathrm{GF}(2)$.

**Example 8.6** In this example, we continue Example 8.5 to retrieve information symbols in $\mathbf{A}_1 = \{a_1, a_2, a_3, a_4\}$. First, we construct the second constituent codeword of $\mathbf{b}$ by using (8.32) and the information symbols in $\mathbf{A}_2$ retrieved in Example 8.5,

$$\mathbf{b}_2 = a_{3,4}\mathbf{v}_3\mathbf{v}_4 + a_{2,4}\mathbf{v}_2\mathbf{v}_4 + a_{1,4}\mathbf{v}_1\mathbf{v}_4 + a_{2,3}\mathbf{v}_2\mathbf{v}_3 + a_{1,3}\mathbf{v}_1\mathbf{v}_3 + a_{1,2}\mathbf{v}_1\mathbf{v}_2. \tag{8.55}$$

Canceling $\mathbf{b}_2$ from $\mathbf{b}$, we obtain

$$\mathbf{b}^{(1)} = \mathbf{b} - \mathbf{b}_2$$
$$= a_0\mathbf{v}_0 + a_1\mathbf{v}_1 + a_2\mathbf{v}_2 + a_3\mathbf{v}_3 + a_4\mathbf{v}_4 \tag{8.56}$$
$$= (b_0^{(1)}, b_1^{(1)}, b_2^{(1)}, b_3^{(1)}, b_4^{(1)}, b_5^{(1)}, b_6^{(1)}, b_7^{(1)}, b_8^{(1)}, b_9^{(1)}, b_{10}^{(1)}, b_{11}^{(1)}, b_{12}^{(1)}, b_{13}^{(1)}, b_{14}^{(1)}, b_{15}^{(1)}).$$

Next, we retrieve the information symbols $a_1, a_2, a_3, a_4$ from the code symbols in $\mathbf{b}^{(1)}$. To retrieve $a_1$, we set $i_1 = 1$ and form the following three sets:

$$\mathbf{S}_1 = \{0, 1\},$$
$$\mathbf{Q}_1 = \{0, 1, 2, 3\}\backslash\{0\} = \{1, 2, 3\}, \tag{8.57}$$
$$\mathbf{S}_1^c = \{d_1 2 + d_2 2^2 + d_3 2^3 : d_1, d_2, d_3 \in \{0, 1\}\} = \{0, 2, 4, 6, 8, 10, 12, 14\}.$$

Based on $\mathbf{S}_1$, $\mathbf{S}_1^c$, and (8.38), we form the following index sets:

$$\mathbf{B}_{1,1}(0) = \{0, 1\}, \mathbf{B}_{1,2}(2) = \{2, 3\}, \quad \mathbf{B}_{1,3}(4) = \{4, 5\}, \quad \mathbf{B}_{1,4}(6) = \{6, 7\},$$
$$\mathbf{B}_{1,5}(8) = \{8, 9\}, \mathbf{B}_{1,6}(10) = \{10, 11\}, \mathbf{B}_{1,7}(12) = \{12, 13\}, \mathbf{B}_{1,8}(14) = \{14, 15\}. \tag{8.58}$$

Using (8.54) and (8.58) with $r = 2$ and $l = 1$, we form the following eight independent determinations of the information symbol $a_1$:

$$a_1 = b_0^{(1)} + b_1^{(1)}, \, a_1 = b_2^{(1)} + b_3^{(1)}, \, a_1 = b_4^{(1)} + b_5^{(1)}, \, a_1 = b_6^{(1)} + b_7^{(1)},$$
$$a_1 = b_8^{(1)} + b_9^{(1)}, \, a_1 = b_{10}^{(1)} + b_{11}^{(1)}, \, a_1 = b_{12}^{(1)} + b_{13}^{(1)}, \, a_1 = b_{14}^{(1)} + b_{15}^{(1)}. \tag{8.59}$$

Similarly, we can form the eight independent determinations for each of the other three information symbols in $\mathbf{A}_1$ as follows:

$$a_2 = b_0^{(1)} + b_2^{(1)}, \ a_2 = b_1^{(1)} + b_3^{(1)}, \ a_2 = b_4^{(1)} + b_6^{(1)}, \ a_2 = b_5^{(1)} + b_7^{(1)},$$
$$a_2 = b_8^{(1)} + b_{10}^{(1)}, \ a_2 = b_9^{(1)} + b_{11}^{(1)}, \ a_2 = b_{12}^{(1)} + b_{14}^{(1)}, \ a_2 = b_{13}^{(1)} + b_{15}^{(1)},$$
$$a_3 = b_0^{(1)} + b_4^{(1)}, \ a_3 = b_1^{(1)} + b_5^{(1)}, \ a_3 = b_2^{(1)} + b_6^{(1)}, \ a_3 = b_3^{(1)} + b_7^{(1)},$$
$$a_3 = b_8^{(1)} + b_{12}^{(1)}, \ a_3 = b_9^{(1)} + b_{13}^{(1)}, \ a_3 = b_{10}^{(1)} + b_{14}^{(1)}, \ a_3 = b_{11}^{(1)} + b_{15}^{(1)},$$
$$a_4 = b_0^{(1)} + b_8^{(1)}, \ a_4 = b_1^{(1)} + b_9^{(1)}, \ a_4 = b_2^{(1)} + b_{10}^{(1)}, \ a_4 = b_3^{(1)} + b_{11}^{(1)},$$
$$a_4 = b_4^{(1)} + b_{12}^{(1)}, \ a_4 = b_5^{(1)} + b_{13}^{(1)}, \ a_4 = b_6^{(1)} + b_{14}^{(1)}, \ a_4 = b_7^{(1)} + b_{15}^{(1)}.$$

$$(8.60)$$

From (8.59) and (8.60), we retrieve the information symbols $a_1, a_2, a_3$, and $a_4$.

Next, we form the first constituent codeword $\mathbf{b}_1$ of $\mathbf{b}$ by using the retrieved symbols in $\mathbf{A}_1$,

$$\mathbf{b}_1 = a_4\mathbf{v}_4 + a_3\mathbf{v}_3 + a_2\mathbf{v}_2 + a_1\mathbf{v}_1. \tag{8.61}$$

Removing $\mathbf{b}_1$ from $\mathbf{b}^{(1)}$, we obtain

$$\begin{aligned}
\mathbf{b}^{(2)} &= \mathbf{b}^{(1)} - \mathbf{b}_1 = \mathbf{b} - \mathbf{b}_2 - \mathbf{b}_1 \\
&= \mathbf{b}_0 = a_0\mathbf{v}_0 \\
&= (a_0, a_0, a_0, a_0, a_0, a_0, a_0, a_0, a_0, a_0, a_0, a_0, a_0, a_0, a_0, a_0),
\end{aligned} \tag{8.62}$$

and

$$a_0 = b_i^{(2)}, \ \ 0 \le i < 16, \tag{8.63}$$

i.e., every code symbol in $\mathbf{b}^{(2)}$ gives the information symbol $a_0$. At this point, we have completed the retrieval of all the 11 information symbols of the message $\mathbf{a}$ for the $(16, 11)$ second-order RM code $C_{\mathrm{RM}}(2, 4)$ given in Example 8.2. ▲▲

## 8.5 Majority-Logic Decoding through Successive Cancellations

Based on the multilevel retrieval of information symbols in groups, one group at a time, from the code symbols of an RM codeword presented in Section 8.4, a *multilevel* (or *multistep*) *majority-logic decoding scheme* for RM codes can be devised, i.e., the algorithm proposed by Reed [2] to decode RM codes. At each level of decoding, a group of information symbols is decoded based on the symbols of a received vector or its modified versions. At the end of each decoding level, an *estimated* constituent codeword is removed from the modified received vector formed in the previous level of decoding to form a new modified received vector. Then, the new modified received vector is used to recover information symbols in the next group at the next level of decoding. The decoding continues level by level until all the transmitted information symbols are recovered or an *incorrect decoding* is detected.

Consider the decoding of the $r$th-order RM code $C_{\mathrm{RM}}(r, m)$ of length $2^m$. Let $\mathbf{b} = (b_0, b_1, \ldots, b_{2^m-1})$ be the codeword for the message

$$\mathbf{a} = (\underbrace{a_0}_{\mathbf{a}_0}, \underbrace{a_m, a_{m-1}, \ldots, a_2, a_1}_{\mathbf{a}_1}, \underbrace{a_{m-1,m}, a_{m-2,m}, \ldots, a_{1,2}, \ldots,}_{\mathbf{a}_2}$$
$$\underbrace{a_{m-r+1,m-r+2,\ldots,m}, \cdots, a_{1,2,\ldots,r}}_{\mathbf{a}_r}),$$

which consists of $r + 1$ groups of information symbols. Then, the codeword $\mathbf{b}$ for $\mathbf{a}$ is

$$\mathbf{b} = a_0 \mathbf{v}_0 + \sum_{1 \le i_1 \le m} a_{i_1} \mathbf{v}_{i_1} + \sum_{1 \le i_1 < i_2 \le m} a_{i_1,i_2} \mathbf{v}_{i_1} \mathbf{v}_{i_2} + \cdots$$
$$+ \sum_{1 \le i_1 < i_2 < \cdots < i_r \le m} a_{i_1,i_2,\ldots,i_r} \mathbf{v}_{i_1} \mathbf{v}_{i_2} \cdots \mathbf{v}_{i_r}. \tag{8.64}$$

Suppose the codeword $\mathbf{b}$ is transmitted. Let $\mathbf{r} = (r_0, r_1, r_2, \ldots, r_{2^m-1})$ be the received vector. The decoding of the information symbols in $\mathbf{a}$ based on $\mathbf{r}^{(0)} = \mathbf{r}$ is similar to the retrieval of the information symbols in $\mathbf{a}$ from $\mathbf{b}^{(0)} = \mathbf{b}$ presented in Section 8.4. At the zeroth level of decoding, the information symbols in the group $\mathbf{A}_r = \{a_{i_1,i_2,\ldots,i_r} : (i_1, i_2, \ldots, i_r) \in \mathcal{I}_r\}$ are decoded. To decode each information symbol $a_{i_1,i_2,\ldots,i_r}$ in $\mathbf{A}_r$, we form $2^{m-r}$ *independent estimates* of $a_{i_1,i_2,\ldots,i_r}$ using the received code symbols in $\mathbf{r}$ as follows:

$$a^*_{i_1,i_2,\ldots,i_r} \triangleq E_e(a_{i_1,i_2,\ldots,i_r}) = \sum_{k \in \mathbf{B}_{r,e}(t_e)} r_k \tag{8.65}$$

with $1 \le e \le 2^{m-r}$, where the index set $\mathbf{B}_{r,e}(t_e)$ is defined by (8.38). The sum given by (8.65) is derived from (8.39) with the received symbol $r_k$ replacing the transmitted code symbol $b_k$ for $k \in \mathbf{B}_{r,e}(t_e)$. If the received code symbols in the sum are error free, then $a^*_{i_1,i_2,\ldots,i_r} = a_{i_1,i_2,\ldots,i_r}$. If there are $2^{m-r-1} - 1$ or fewer errors in the received vector $\mathbf{r}$, then more than half of the estimates formed by (8.65) give the true value of $a_{i_1,i_2,\ldots,i_r}$ and hence the decoding of $a_{i_1,i_2,\ldots,i_r}$ using the majority-decision rule is correct, i.e., by taking the value assumed by the majority of the estimates computed using (8.65). However, if $\mathbf{r}$ contains $2^{m-r-1}$ or more errors, there is no guarantee that a majority of the estimates computed from (8.65) will assume the true value of $a_{i_1,i_2,\ldots,i_r}$. In this case, the majority-logic decision may result in an incorrect decoding of $a_{i_1,i_2,\ldots,i_r}$.

Once all the information symbols in $\mathbf{A}_r = \{a_{i_1,i_2,\ldots,i_r} : (i_1, i_2, \ldots, i_r) \in \mathcal{I}_r\}$ have been decoded, we form the $r$th *estimated constituent codeword* of $\mathbf{b}$,

$$\mathbf{b}^*_r = \sum_{(i_1,i_2,\ldots,i_r) \in \mathcal{I}_r} a^*_{i_1,i_2,\ldots,i_r} \mathbf{v}_{i_1} \mathbf{v}_{i_2} \cdots \mathbf{v}_{i_r}. \tag{8.66}$$

Next, we remove $\mathbf{b}_r^*$ from $\mathbf{r}$ to obtain the following modified received vector,

$$\mathbf{r}^{(1)} = \mathbf{r} - \mathbf{b}_r^* = (r_0^{(1)}, r_1^{(1)}, \ldots, r_{2^m-1}^{(1)}), \qquad (8.67)$$

which is called a *descendant* of $\mathbf{r}$.

In the first level of decoding, the information symbols in $\mathbf{A}_{r-1} = \{a_{i_1,i_2,\ldots,i_{r-1}} : (i_1, i_2, \ldots, i_{r-1}) \in \mathcal{I}_{r-1}\}$ are decoded using the symbols in $\mathbf{r}^{(1)}$. There are $2^{m-r+1}$ independent estimates of $a_{i_1,i_2,\ldots,i_{r-1}}$ that can be formed. Each estimate of $a_{i_1,i_2,\ldots,i_{r-1}}$ is given below:

$$a_{i_1,i_2,\ldots,i_{r-1}}^* = E_e(a_{i_1,i_2,\ldots,i_{r-1}}) = \sum_{k \in \mathbf{B}_{r-1,e}(t_e)} r_k^{(1)}, \qquad (8.68)$$

which is obtained by setting $l = 1$ and replacing $b_k^{(l)}$ in the sum of (8.54) by $r_k^{(1)}$. In decoding of $a_{i_1,i_2,\ldots,i_{r-1}}$, we take the value assumed by the majority of the $2^{m-r+1}$ independent estimates of $a_{i_1,i_2,\ldots,i_{r-1}}$ as an estimate of $a_{i_1,i_2,\ldots,i_{r-1}}$.

At the completion of decoding the information symbols in $\mathbf{A}_{r-1}$, we form the estimated $(r-1)$th estimated constituent codeword $\mathbf{b}_{r-1}^*$ as follows:

$$\mathbf{b}_{r-1}^* = \sum_{(i_1,i_2,\ldots,i_{r-1}) \in \mathcal{I}_{r-1}} a_{i_1,i_2,\ldots,i_{r-1}}^* \mathbf{v}_{i_1} \mathbf{v}_{i_2} \cdots \mathbf{v}_{i_{r-1}}. \qquad (8.69)$$

Removing $\mathbf{b}_{r-1}^*$ from $\mathbf{r}^{(1)}$, we obtain the following descendant of $\mathbf{r}$:

$$\mathbf{r}^{(2)} = \mathbf{r}^{(1)} - \mathbf{b}_{r-1}^* = \mathbf{r} - \mathbf{b}_r^* - \mathbf{b}_{r-1}^* = (r_0^{(2)}, r_1^{(2)}, \ldots, r_{2^m-1}^{(2)}). \qquad (8.70)$$

The second-level decoding is to decode the information symbols in $\mathbf{A}_{r-2} = \{a_{i_1,i_2,\ldots,i_{r-2}} : (i_1, i_2, \ldots, i_{r-2}) \in \mathcal{I}_{r-2}\}$ based on the symbols in the modified received vector $\mathbf{r}^{(2)}$. The second-level decoding is exactly the same as the zeroth and first-level decodings. For each information symbol $a_{i_1,i_2,\ldots,i_{r-2}}$ in $\mathbf{A}_{r-2}$, we can form $2^{m-r+2}$ independent estimations from the symbols in $\mathbf{r}^{(2)}$.

In the $l$th level of decoding, $0 \le l \le r$, the information symbols in $\mathbf{A}_{r-l} = \{a_{i_1,i_2,\ldots,i_{r-l}} : (i_1, i_2, \ldots, i_{r-l}) \in \mathcal{I}_{r-l}\}$ are decoded. For each information symbol $a_{i_1,i_2,\ldots,i_{r-l}}$ in $\mathbf{A}_{r-l}$, we can form $2^{m-r+l}$ independent estimations of $a_{i_1,i_2,\ldots,i_{r-l}}$ based on the symbols in the modified received vector $\mathbf{r}^{(l)} = \mathbf{r} - \mathbf{b}_r^* - \mathbf{b}_{r-1}^* - \ldots - \mathbf{b}_{r-l+1}^*$ which is obtained at the completion of the $(l-1)$th level of decoding. Decoding continues level by level based on the majority-logic decision until the information symbol $a_0$ is decoded at the $r$th level of decoding.

Note that the number of estimates of each information symbol is doubled at each subsequent level of decoding. If the number of errors in the received vector $\mathbf{r}$ is $2^{m-r-1} - 1$ or less, the majority-logic decision based on the estimates computed at each level results in a correct decoding. Hardware implementation of majority-logic decoding is simple. With the successive majority-logic decoding, incorrect decoding of some information symbols at a decoding level can result in *error propagation* and cause the subsequent levels of decoding to be incorrect.

Basically, the above decoding of the $r$th-order RM code $C_{\mathrm{RM}}(r, m)$ is to perform decodings of its subcodes $C_{\mathrm{RM}}(r, m)$, $C_{\mathrm{RM}}(r - 1, m), \ldots, C_{\mathrm{RM}}(1, m)$, $C_{\mathrm{RM}}(0, m)$ in this order. We may consider the successive cancellation decoding method as a *successive peeling-off decoding procedure*.

**Example 8.7** Consider the $(16, 11)$ second-order RM code $C_{\mathrm{RM}}(2, 4)$ with minimum distance 4 constructed in Example 8.2. Suppose the codeword $\mathbf{b} = (b_0,$ $b_1, b_2, b_3, b_4, b_5, b_6, b_7, b_8, b_9, b_{10}, b_{11}, b_{12}, b_{13}, b_{14}, b_{15})$ for the message $\mathbf{a} = (a_0, a_4, a_3, a_2, a_1, a_{3,4}, a_{2,4}, a_{1,4}, a_{2,3}, a_{1,3}, a_{1,2})$ is transmitted. Let

$$\mathbf{r} = (r_0, r_1, r_2, r_3, r_4, r_5, r_6, r_7, r_8, r_9, r_{10}, r_{11}, r_{12}, r_{13}, r_{14}, r_{15})$$

be the received vector. The zeroth level decoding of $\mathbf{r}$ is to decode the information symbols $a_{1,2}$, $a_{1,3}$, $a_{1,4}$, $a_{2,3}$, $a_{2,4}$, $a_{3,4}$. To decode these six information symbols, we form the following 24 estimates of the six information symbols $a_{1,2}$, $a_{1,3}$, $a_{1,4}$, $a_{2,3}$, $a_{2,4}$, $a_{3,4}$, four for each information symbol (similar to the derivations of (8.42), (8.45), and (8.46)):

$$
\begin{aligned}
a_{1,2}^* &= r_0 + r_1 + r_2 + r_3, & a_{1,2}^* &= r_4 + r_5 + r_6 + r_7, \\
a_{1,2}^* &= r_8 + r_9 + r_{10} + r_{11}, & a_{1,2}^* &= r_{12} + r_{13} + r_{14} + r_{15}, \\
a_{1,3}^* &= r_0 + r_1 + r_4 + r_5, & a_{1,3}^* &= r_2 + r_3 + r_6 + r_7, \\
a_{1,3}^* &= r_8 + r_9 + r_{12} + r_{13}, & a_{1,3}^* &= r_{10} + r_{11} + r_{14} + r_{15}, \\
a_{1,4}^* &= r_0 + r_1 + r_8 + r_9, & a_{1,4}^* &= r_2 + r_3 + r_{10} + r_{11}, \\
a_{1,4}^* &= r_4 + r_5 + r_{12} + r_{13}, & a_{1,4}^* &= r_6 + r_7 + r_{14} + r_{15}, \\
a_{2,3}^* &= r_0 + r_2 + r_4 + r_6, & a_{2,3}^* &= r_1 + r_3 + r_5 + r_7, \\
a_{2,3}^* &= r_8 + r_{10} + r_{12} + r_{14}, & a_{2,3}^* &= r_9 + r_{11} + r_{13} + r_{15}, \\
a_{2,4}^* &= r_0 + r_2 + r_8 + r_{10}, & a_{2,4}^* &= r_1 + r_3 + r_9 + r_{11}, \\
a_{2,4}^* &= r_4 + r_6 + r_{12} + r_{14}, & a_{2,4}^* &= r_5 + r_7 + r_{13} + r_{15}, \\
a_{3,4}^* &= r_0 + r_4 + r_8 + r_{12}, & a_{3,4}^* &= r_1 + r_5 + r_9 + r_{13}, \\
a_{3,4}^* &= r_2 + r_6 + r_{10} + r_{14}, & a_{3,4}^* &= r_3 + r_7 + r_{11} + r_{15}.
\end{aligned}
\tag{8.71}
$$

From the above estimates, we find that if there is a single error in $\mathbf{r}$, three estimates, *a clear majority*, for each information symbol in $\mathbf{A}_2 = \{a_{1,2}, a_{1,3}, a_{2,3}, a_{2,3}, a_{1,4}, a_{2,4}, a_{3,4}\}$ give the true value of the information symbol. Hence, in the zeroth level of decoding, each information symbol is decoded correctly. However, if there are two errors in $\mathbf{r}$, there is no guarantee that for each information symbol in $\mathbf{A}_2$ there is a clear majority of estimates to give the true value of each information symbol in $\mathbf{A}_2$. For example, if the received symbols $r_1$ and $r_7$ are erroneous, there is no majority of the estimates to give the true value of any information symbol in $\mathbf{A}_2$. Choosing one of the two values of the four estimates for each information symbol in $\mathbf{A}_2$ at random, an incorrect decoding may occur. If there are three errors in $\mathbf{r}$, say in the positions 1, 3, and 8, then three estimates for the information symbols $a_{1,3}$ will give an incorrect value of $a_{1,3}$. Decoding of $a_{1,3}$ by the majority-decision will then be incorrect.

After the decoding of the information symbols $a_{1,2}$, $a_{1,3}$, $a_{1,4}$, $a_{2,3}$, $a_{2,4}$, $a_{3,4}$, we form the estimated second constituent codeword of $\mathbf{b}$,

$$\mathbf{b}_2^* = a_{3,4}^* \mathbf{v}_3 \mathbf{v}_4 + a_{2,4}^* \mathbf{v}_2 \mathbf{v}_4 + a_{1,4}^* \mathbf{v}_1 \mathbf{v}_4 + a_{2,3}^* \mathbf{v}_2 \mathbf{v}_3 + a_{1,3}^* \mathbf{v}_1 \mathbf{v}_3 + + a_{1,2}^* \mathbf{v}_1 \mathbf{v}_2.$$

Canceling $\mathbf{b}_2^*$ from $\mathbf{r}$, we obtain the following modified received vector:

$$\mathbf{r}^{(1)} = \mathbf{r} - \mathbf{b}_2^* = (r_0^{(1)}, r_1^{(1)}, \ldots, r_{14}^{(1)}, r_{15}^{(1)}).$$

Using (8.60) given in Example 8.6 with the symbols in $\mathbf{r}^{(1)}$ (replacing symbol $b_i^{(1)}$ by $r_i^{(1)}$, $0 \le i < 2^m$), we can form 32 estimates for the information symbols in $\mathbf{A}_1$, namely, $a_1$, $a_2$, $a_3$, $a_4$, eight for each. From these estimates, we decode these four information symbols. They will be decoded correctly if the received vector $\mathbf{r}$ contains no more than one error.

At the completion of the first level of decoding, we form the estimated first constituent codeword of $\mathbf{b}$,

$$\mathbf{b}_1^* = a_4^* \mathbf{v}_4 + a_3^* \mathbf{v}_3 + a_2^* \mathbf{v}_2 + a_1^* \mathbf{v}_1.$$

Canceling $\mathbf{b}_1^*$ from $\mathbf{r}^{(1)}$, we obtain the modified received vector $\mathbf{r}^{(2)} = \mathbf{r} - \mathbf{b}_2^* - \mathbf{b}_1^*$ to decode the information symbol $a_0$ in $\mathbf{A}_0$ of the message $\mathbf{a}$. In the second level of decoding, we form the following 16 estimates of $a_0$:

$$a_0^* = r_i^{(2)}, \quad 0 \le i < 16.$$

The true value of $a_0$ is given by the majority of these 16 estimates if the received vector $\mathbf{r}$ contains no more than one error. This completes the decoding of $\mathbf{a}$. ▲▲

The inclusion chain structure of RM codes described in Section 8.2 (see (8.15)) in conjunction with the multilevel majority-logic decoding presented above makes RM codes quite suitable for adaptive communications. Suppose we design a high-order (high-rate) RM code for a communication system in its normal low-level channel noise situation. When the noise level is detected to be becoming high, we lower the order (rate) of the chosen RM code, say from the $r$th order to $(r - l)$th order with $0 < l \le r$, for information transmission. This increases the error-correcting capability of the code from $2^{m-r-1} - 1$ to $2^{m-r+l-1} - 1$. In this case, we can use the same encoder and the same decoder of the chosen RM code for encoding and decoding. In encoding, we set the last $l$ sections of the information sequence $\mathbf{a}$ (see (8.27)) to zeros (virtual). In decoding, we carry out the last $r - l + 1$ levels of the majority-logic decoding, i.e., decoding the $(r - l)$th-order RM code.

## 8.6 Cyclic RM Codes

If we delete the leftmost column of the matrix $\mathbf{B}$ given by (8.6), we obtain the following $m \times (2^m - 1)$ matrix over GF(2):

$$\mathbf{B}_{\text{cyclic}} = \begin{bmatrix} \mathbf{w}_1 \\ \mathbf{w}_2 \\ \vdots \\ \mathbf{w}_m \end{bmatrix} = \begin{bmatrix} \overset{\alpha^0}{c_{0,0}} & \overset{\alpha}{c_{0,1}} & \cdots & \overset{\alpha^{2^m-2}}{c_{0,2^m-2}} \\ c_{1,0} & c_{1,1} & \cdots & c_{1,2^m-2} \\ \vdots & & \ddots & \vdots \\ c_{m-1,0} & c_{m-1,1} & \cdot & c_{m-1,2^m-2} \end{bmatrix}, \tag{8.72}$$

where $\mathbf{w}_1, \mathbf{w}_2, \ldots, \mathbf{w}_m$ denote the $m$ rows in $\mathbf{B}_{\text{cyclic}}$. Adding the $(2^m - 1)$-tuple $\mathbf{w}_0$ of all-one to the top of $\mathbf{B}_{\text{cyclic}}$, we obtain the following $(m + 1) \times (2^m - 1)$ matrix:

$$\mathbf{G}_{1,\text{cyclic}} = \begin{bmatrix} \mathbf{w}_0 \\ \mathbf{w}_1 \\ \mathbf{w}_2 \\ \cdots \\ \mathbf{w}_m \end{bmatrix} = \begin{bmatrix} 1 & 1 & \cdots & 1 \\ c_{0,0} & c_{0,1} & \cdots & c_{0,2^m-2} \\ c_{1,0} & c_{0,1} & \cdots & c_{1,2^m-2} \\ \vdots & & \ddots & \vdots \\ c_{m-1,0} & c_{m-1,1} & \cdot & c_{m-1,2^m-2} \end{bmatrix}. \tag{8.73}$$

Form a $k(r, m) \times (2^m - 1)$ matrix $\mathbf{G}_{\text{RM,cyclic}}(r, m)$ as follows:

$$\mathbf{G}_{\text{RM,cyclic}}(r, m) = \begin{bmatrix} \mathbf{G}_{1,\text{cyclic}} \\ \mathbf{G}_{2,\text{cyclic}} \\ \cdots \\ \mathbf{G}_{r,\text{cyclic}} \end{bmatrix}. \tag{8.74}$$

The matrix $\mathbf{G}_{\text{RM,cyclic}}(r, m)$ given by (8.74) consists of the $m + 1$ rows, $\mathbf{w}_0, \mathbf{w}_1, \mathbf{w}_2, \ldots, \mathbf{w}_m$, and the vector products of $\mathbf{w}_1, \mathbf{w}_2, \ldots, \mathbf{w}_m$ of degree 2 to degree $r$ as rows in the form of (8.13) where $k(r, m)$ is given by (8.14).

It was proved in [4, 6, 21, 32] that the space spanned by the rows of the matrix $\mathbf{G}_{\text{RM,cyclic}}(r, m)$ gives a *cyclic RM code* $C_{\text{RM,cyclic}}(r, m)$ of length $2^m - 1$, dimension $k(r, m)$, and minimum distance $2^{m-r} - 1$. The successive multilevel majority decoding presented in Section 8.5 can be applied to decode the code $C_{\text{RM,cyclic}}(r, m)$ based on its generator matrix $\mathbf{G}_{\text{RM,cyclic}}(r, m)$.

From the geometric point of view, the cyclic $r$th-order RM code $C_{\text{RM,cyclic}}(r, m)$ is a finite-geometry code which is the null space spanned by the incidence vectors of the $(r + 1)$-flats, the $(r + 2)$-flats, $\ldots$, and the $m$-flat in the Euclidean geometry EG$(m, 2)$ over GF(2) not containing the origin. Let $h$ be a nonnegative integer less than $2^m$. Express $h$ in radix-2 form as follows:

$$h = h_0 + h_1 2 + h_2 2^2 + \cdots + h_{m-1} 2^{m-1}, \tag{8.75}$$

where $h_j = 0$ or 1 for $0 \leq j < m$. The real sum of $h_0, h_1, \ldots, h_{m-1}$ given below:

$$W(h) = h_0 + h_1 + h_2 + \cdots + h_{m-1} \tag{8.76}$$

is the radix-2 weight of $h$. Let $\mathbf{g}_{\mathrm{RM}}(X)$ be the generator polynomial of the code $C_{\mathrm{RM,cyclic}}(r, m)$ and $\alpha$ be a primitive element of $\mathrm{GF}(2^m)$. Then, $\alpha^h \in \mathrm{GF}(2^m)$ is a root of $\mathbf{g}_{\mathrm{RM}}(X)$ if and only if

$$0 < W(h) \leq m - r - 1. \tag{8.77}$$

**Example 8.8** For $m = 4$ and $r = 1$, the first-order cyclic RM code $C_{\mathrm{RM,cyclic}}$ $(1, 4)$ is a $(15, 5)$ code with minimum distance 7. Let $\alpha$ be a primitive element of $\mathrm{GF}(2^4)$. The nonnegative integers less than 16 whose radix-2 weights satisfy the condition $0 < W(h) \leq 2$ are 1, 2, 3, 4, 5, 6, 8, 9, 10, and 12. Hence, the elements $\alpha$, $\alpha^2$, $\alpha^3$, $\alpha^4$, $\alpha^5$, $\alpha^6$, $\alpha^8$, $\alpha^9$, $\alpha^{10}$, and $\alpha^{12}$ over $\mathrm{GF}(2^4)$ are the roots of the generator polynomial $\mathbf{g}_{\mathrm{RM}}(X)$ of the $(15, 5)$ cyclic RM code. From these roots, we find

$$\mathbf{g}_{\mathrm{RM}}(X) = 1 + X + X^2 + X^4 + X^5 + X^8 + X^{10}.$$

This $(15, 5)$ cyclic RM code is also a BCH code (see Problem 8.15).    ▲▲

**Example 8.9** For $m = 6$ and $r = 3$, the third-order cyclic RM code $C_{\mathrm{RM,cyclic}}$ $(3, 6)$ is a $(63, 42)$ code with minimum distance 7. Let $\alpha$ be a primitive element of $\mathrm{GF}(2^6)$. The nonnegative integers less than 64 whose radix-2 weights satisfy the condition $0 < W(h) \leq 2$ are 1, 2, 3, 4, 5, 6, 8, 9, 10, 12, 16, 17, 18, 20, 24, 32, 33, 34, 36, 40, and 48. Hence, the elements $\alpha$, $\alpha^2$, $\alpha^3$, $\alpha^4$, $\alpha^5$, $\alpha^6$, $\alpha^8$, $\alpha^9$, $\alpha^{10}$, $\alpha^{12}$, $\alpha^{16}$, $\alpha^{17}$, $\alpha^{18}$, $\alpha^{20}$, $\alpha^{24}$, $\alpha^{32}$, $\alpha^{33}$, $\alpha^{34}$, $\alpha^{36}$, $\alpha^{40}$, and $\alpha^{48}$ over $\mathrm{GF}(2^6)$ are the roots of the generator polynomial $\mathbf{g}_{\mathrm{RM}}(X)$ of the $(63, 42)$ cyclic RM code. From these roots, we find

$$\mathbf{g}_{\mathrm{RM}}(X) = 1 + X + X^3 + X^7 + X^8 + X^9 + X^{10} + X^{11} + X^{12} + X^{15} + X^{16} + X^{18} + X^{21}.$$

This $(63, 42)$ cyclic RM code is a subcode (see Problem 8.16) of the $(63, 45)$ primitive BCH code given in Table 5.1.    ▲▲

## 8.7  Remarks

In Section 8.2, we presented a construction of RM codes using a geometric approach. As mentioned in the introduction, besides the geometric construction, there are several other approaches for constructing RM codes, such as $|\mathbf{u}|\mathbf{u} + \mathbf{v}|$-construction (to be presented in Section 9.4), Kronecker product (to be presented in Section 9.5), squaring construction, Boolean algebra, and polynomial approach [12, 21, 25, 26]. Each construction approach reveals certain unique figures of RM codes which allow them for certain specific applications and/or to be decoded with certain decoding methods or algorithms. For example, squaring construction shows that RM codes have elegant trellis structure which allows RM codes to be decoded with trellis-based MLD, MAP (*maximum a posteriori probability*), and other simplified suboptimal trellis-based decoding algorithms. Good coverage of these subjects can be found in [17, 21]. Constructions of RM codes using Kronecker product and $|\mathbf{u}|\mathbf{u} + \mathbf{v}|$-construction will be presented in Chapters 9 and 17. In Chapter 17, we will show that RM codes and polar codes are two subclasses of a special class of linear block codes, called *Kronecker codes*. Recently, RM codes have been shown to be effective in correcting erasures [24]. Correcting random erasures with RM codes will be presented in Chapter 10. Furthermore, some quantum codes are constructed based on RM codes [33–35], and some experimental designs (balanced incomplete block designs) can be constructed based on the minimum-weight codewords of RM codes [32]. The purpose of the remarks is just to show that RM codes are rich in geometric, algebraic, graphical, and combinatorial structural properties.

## Problems

**8.1** Show that each row of the matrix $\mathbf{B}$ given by (8.6) has weight $2^{m-1}$.

**8.2** Show that rows of the matrix $\mathbf{B}$ given by (8.6) are linearly independent.

**8.3** Set $m = 3$ and consider the field $\mathrm{GF}(2^3)$. Construct the matrices $\mathbf{B}$ and $\mathbf{G}_1$ in the forms of (8.6) and (8.7), respectively.

**8.4** Show that the product vector $\mathbf{v}_{i_1}\mathbf{v}_{i_2}\cdots\mathbf{v}_{i_l}$ given by (8.11) is the incidence vector of an $(m - l)$-flat in $\mathrm{EG}(m, 2)$ which contains the point $\alpha^t$ represented by the all-one $m$-tuple and has weight $2^{m-l}$.

**8.5** In Example 8.2, the first-order and second-order RM codes $C_{\mathrm{RM}}(1, 4)$ and $C_{\mathrm{RM}}(2, 4)$ of length 16 are constructed.
(a) Show that the two codes $C_{\mathrm{RM}}(1, 4)$ and $C_{\mathrm{RM}}(2, 4)$ are dual, i.e.,

$$\mathbf{G}_{\mathrm{RM}}(1, 4)\mathbf{G}_{\mathrm{RM}}^T(2, 4) = \mathbf{O}.$$

(b) Find the weight distributions of the two codes $C_{\mathrm{RM}}(1, 4)$ and $C_{\mathrm{RM}}(2, 4)$.
(c) Construct the third-order RM code $C_{\mathrm{RM}}(3, 4)$ and its dual code.

**8.6** Compute the generator matrix $\mathbf{G}_{\mathrm{RM}}(2, 5)$ of the second-order RM code $C_{\mathrm{RM}}(2, 5)$ in Example 8.3 and show that the code is self-dual, i.e., $\mathbf{G}_{\mathrm{RM}}(2, 5) \cdot \mathbf{G}_{\mathrm{RM}}^T(2, 5) = \mathbf{O}$. Find the weight distribution of $C_{\mathrm{RM}}(2, 5)$.

**8.7** Let $r = 3$ and $m = 5$. Construct the third-order RM code $C_{\mathrm{RM}}(3, 5)$ and find its weight distribution.

**8.8** Find the weight distributions of the $(m-2)$th-order RM code $C_{\mathrm{RM}}(m-2, m)$ and the $(m - 3)$th-order RM code $C_{\mathrm{RM}}(m - 3, m)$.

**8.9** Consider the first-order RM code $C_{\mathrm{RM}}(1, 4)$ constructed in Example 8.2 whose generator matrix $\mathbf{G}_{\mathrm{RM}}(1, 4)$ given by (8.23) is not systematic. Find a systematic generator matrix for the code $C_{\mathrm{RM}}(1, 4)$.

**8.10** Consider the $(16, 11)$ second-order RM code $C_{\mathrm{RM}}(2, 4)$ constructed in Example 8.2. Find the codeword **b** for the following message **a** by using the multilevel encoding,

$$\mathbf{a} = (1, 0, 1, 1, 0, 1, 0, 1, 1, 0, 0).$$

**8.11** Consider the third-order RM code $C_{\mathrm{RM}}(3, 4)$ constructed in Problem 8.5. Find the codeword **b** for the following message **a** by using the multilevel encoding,

$$\mathbf{a} = (1, 0, 1, 1, 0, 1, 0, 1, 1, 0, 0, 1, 1, 0, 1).$$

**8.12** Consider the third-order RM code $C_{\mathrm{RM}}(3, 4)$ constructed in Problem 8.5. Derive the independent determinations of the information symbols $a_{1,2,3}$, $a_{1,2}$, and $a_1$ for a message **a**.

**8.13** Prove the sum given by (8.54) is the inner product of the incidence vector of an $(r - l)$-flat in a parallel bundle $P$ and the descendant codeword $\mathbf{b}^{(l)}$ given by (8.49), where $P$ consists of $2^{m-r+l}$ parallel $(r - l)$-flats in $\mathrm{EG}(m, 2)$.

**8.14** Suppose the codeword **b** computed in Problem 8.10 is transmitted. Decode the received vectors, $\mathbf{r} = \mathbf{b} + \mathbf{e}_i$, for the following error vectors, $0 \le i < 3$,

$$\mathbf{e}_0 = (1, 0, 0, 0, 0, 0, 0, 0, 0, 0, 0, 0, 0, 0, 0, 0),$$
$$\mathbf{e}_1 = (1, 0, 0, 1, 0, 0, 0, 0, 0, 0, 0, 0, 0, 0, 0, 0),$$
$$\mathbf{e}_2 = (1, 0, 0, 0, 1, 1, 0, 0, 0, 0, 0, 0, 0, 0, 0, 0).$$

**8.15** Prove that the $(15, 5)$ cyclic RM code given in Example 8.8 is a BCH code.

**8.16** Prove that the $(63, 42)$ cyclic RM code given in Example 8.9 is a subcode of the $(63, 45)$ primitive BCH code.

**8.17** Let $m = 4$ and $r = 2$. Find the generator polynomial of the second-order cyclic RM code $C_{\mathrm{RM}}(2, 4)$.

**8.18** Let $m = 5$ and $r = 2$. Find the generator polynomial of the second-order cyclic RM code $C_{\mathrm{RM}}(2, 5)$ and show the code is a primitive BCH code.

# References

[1] D. E. Muller, "Applications of boolean algebra to switching circuits design to error detection," *IRE Trans.*, **EC-3**(3) (1954), 6–12.

[2] I. S. Reed, "A class of multiple-error-correcting codes and the decoding scheme," *IRE Trans.*, **IT-4**(4) (1954), 38–49.

[3] L. D. Rudolph, *Geometric configuration and majority logic decodable codes*, M. E. E. thesis, University of Oklahoma, Norman, 1964.

[4] T. Kasami, S. Lin, and W. W. Peterson, "Linear codes which are invariant under the affine group and some results on minimum weights in BCH codes," *Electron. Commun. Jap.*, **50**(10) (1967), 100–106.

[5] T. Kasami, S. Lin, and W. W. Peterson, "New generalizations of the Reed–Muller codes, Pt. I: primitive codes," *IEEE Trans. Inf. Theory*, **IT-14**(2) (1968), 189–199.

[6] V. D. Kolesnik and E. T. Mironchikov, "Cyclic Reed–Muller codes and their decoding," *Probl. Inf. Trans.*, **4**(4) (1968), 15–19.

[7] C. R. P. Hartmann, J. B. Ducey, and L. D. Rudolph, "On the structure of generalized finite geometry codes," *IEEE Trans. Inf. Theory*, **IT-20**(2) (1974), 240–252.

[8] D. K. Chow, "A generalized approach to coding theory with applications to information retrieval," *CSL Report No. R-368*, University of Illinois, Urbana, 1967.

[9] T. Kasami, S. Lin, and W. W. Peterson, "Polynomial codes," *IEEE Trans. Inf. Theory*, **14** (1968), 807–814.

[10] P. Delsarte, "A geometric approach to a class of cyclic codes," *J. Combinatorial Theory*, **6**(4) (1969), 340–358.

[11] P. Delsarte, J. M. Goethals, and J. MacWilliams, "On GRM and related codes," *Inf. Control*, **16** (1970), 403–442.

[12] G. D. Forney, Jr., "Coset codes II: binary lattices and related codes," *IEEE Trans. Inf. Theory*, **34**(5) (1988), 1152–1187.

[13] T. Kasami, T. Takata, T. Fujiwara, and S. Lin, "On the optimum bit orders with respect to the state complexity of trellis diagrams for binary linear codes," *IEEE Trans. Inf. Theory*, **39**(1) (1993), 242–243.

[14] F. R. Kschischang and V. Sorokine, "On the trellis structure of block codes," *IEEE Trans. Inf. Theory*, **41**(6) (1995), 1924–1937.

[15] A. Lafourcade and A. Vardy, "Lower bounds on trellis complexity of block codes," *IEEE Trans. Inf. Theory*, **41**(6) (1995), 1938–1954.

[16] C. C. Lu and S. H. Huang, "On bit-level trellis complexity of Reed–Muller codes," *IEEE Trans. Inf. Theory*, **41**(6) (1995), 2061–2064.

[17] S. Lin, T. Kasami, T. Fujiwara, and M. Fossorier, *Trellis and Trellis-Based Decoding Algorithms for Linear Block Codes*, Boston, MA, Kluwer Academic, 1998.

[18] R. J. McEliece, "On the BCJR trellis for linear block codes," *IEEE Trans. Inf. Theory*, **42**(4) (1996), 1072–1092.

[19] H. Moorthy, S. Lin, and T. Kasami, "Soft-decision decoding of binary linear block code based on an iterative search algorithm," *IEEE Trans. Inf. Theory*, **43**(3) (1997), 1030–1040.

[20] B. Honary and G. Markarian, *Trellis Decoding of Block Codes: A Practical Approach*, Boston, MA, Kluwer Academic, 1997.

[21] S. Lin and D. J. Costello, Jr., *Error Control Coding: Fundamentals and Applications*, 2nd ed., Upper Saddle River, NJ, Prentice-Hall, 2004.

[22] E. Arikan, "Channel polarization: a method for constructing capacity-achieving codes for symmetric binary-input memoryless channels," *IEEE Trans. Inf. Theory*, **55**(7) (2009), 3051–3073.

[23] E. Abbe, A. Shpilka, and A. Wigderson, "Reed-Muller codes for random erasures and errors," *IEEE Trans. Inf. Theory*, **61**(10) (2015), 5229–5252.

[24] S. Kudekar, S. Kumar, M. Mondelli, *et al.*, "Reed-Muller codes achieve capacity on erasure channels," *IEEE Trans. Inf. Theory*, **63**(7) (2017), 4298–4316.

[25] F. J. MacWilliams and N. J. A. Sloane, *The Theory of Error-Correcting Codes*, Amsterdam, North-Holland, 1977.

[26] M. Plotkin, "Binary codes with specific minimum distance," *IEEE Trans. Inf. Theory*, **IT-6**(4) (1960), 445–450.

[27] E. J. Weldon, Jr., "Euclidean geometry cyclic codes," *Proc. Symp. Combinatorial Math.*, University of North Carolina, Chapel Hill, April 1967.

[28] T. Kasami, "The weight enumerators for several classes of subcodes of the second-order binary Reed–Muller codes," *Inf. Control*, **18**(4) (1971), 369–394.

[29] M. Sugino, Y. Ienaga, N. Tokura, and T. Kasami, "Weight distribution of $(128, 64)$ Reed–Muller code," *IEEE Trans. Inf. Theory*, **17**(5) (1971), 627–628.

[30] T. Kasami, N. Tokura, and S. Azumi, "On the weight enumerator of weight less than $2.5d$ of Reed–Muller codes," *Inf. Control*, **30**(4) (1976), 380–395.

[31] T. Sugita, T. Kasami, and T. Fujiwara, "The weight distributions of the third-order Reed–Muller code of length 512," *IEEE Trans. Inf. Theory*, **42**(5) (1996), 1622–1625.

[32] S. Lin, "Some codes which are invariant under a transitive permutation group and their connection with balanced incomplete block designs," in R. C. Bose, ed., *Proc. Combinational Mathematics and Its Applications*, University of Carolina Press, Chapel Hill, pp. 388–401, April 10–14, 1967.

[33] A. R. Calderbank and P. W. Shor, "Good quantum error-correcting codes exist," *Phys. Rev. A*, **54**(2) (1996), 1098–1105.

[34] A. R. Calderbank, E. M. Rains, N. J. A. Sloane, and P. W. Shor, "Quantum error correction and orthogonal geometry," *Phys. Rev. Lett.*, **78**(3) (1997), 405–409.

[35] D. J. C. MacKay, G. Mitchison, and P. L. McFadden, "Sparse-graph codes for quantum error correction," *IEEE Trans. Inf. Theory*, **50**(10) (2004), 2315–2330.

# 9

# Some Coding Techniques

Apart from the construction of BCH, RS, finite-geometry, and RM codes based on finite fields and finite geometries, there are other methods (or techniques) for constructing *long powerful codes from good short codes*. Among these methods, the well-known ones are interleaving, direct product, concatenation, $|\mathbf{u}|\mathbf{u} + \mathbf{v}|$-construction, and Kronecker product. Codes constructed based on these coding methods are capable of correcting both random errors and bursts of errors (or erasures). These coding methods and their combinations are commonly used in various communication and data-storage systems to achieve good error performance with a significant reduction in decoding complexity. In this chapter, we present these well-known coding methods for constructing codes and various ways of encoding and decoding of the codes constructed. In this presentation, we focus only on correcting random errors. Correction of bursts of errors and/or erasures will be discussed separately in Chapter 10.

As pointed out in Chapter 1, there are two general types of error-control coding schemes, namely, the forward error control (FEC) scheme and the automatic-repeat-request (ARQ) scheme. In a FEC system, an error-correcting code is used. When the receiver detects the existence of errors in a received vector, its decoder attempts to correct the errors. If decoding is successful, the receiver retrieves the message from the decoded codeword and delivers it to the user(s). If decoding fails, the receiver declares a decoding failure, discards the received vector, and continues to decode the next received vector. In an ARQ system, when a vector is received, the decoder in the receiver computes its syndrome. If the syndrome is zero, the received vector is assumed to be error free. Then, the message contained in the received vector is retrieved and delivered to the user(s). At the same time, the receiver notifies the transmitter, *via a return channel*, that the transmitted codeword has been successfully received. If the syndrome is not zero, errors are detected in the received vector. Then, the transmitter is requested, through the return channel, to retransmit the same codeword. Retransmission continues under *a certain retransmission protocol* until the codeword is successfully received. If a code in an ARQ system is designed to correct a certain number of errors besides error detection, the decoder in the receiver first attempts to correct the errors when the syndrome of a received vector is not zero. If the decoder fails to correct the errors, the transmitter is requested to retransmit the same codeword (or simply the parity-check symbols of the codeword) under a certain retransmission protocol. Retransmission and error correction continue until the codeword is either received or decoded successfully. Such an ARQ scheme is referred to as a *hybrid ARQ* (HARQ) *scheme*. In this chapter, some basic ARQ and HARQ schemes will be presented.

## 9.1 Interleaving

This section presents the straightforward but important concept of interleaving. Let $C$ be an $(n, k)$ linear block code over $\mathrm{GF}(q)$ with rate $k/n$. Let $\lambda$ be a positive integer and $\mathbf{v}_0, \mathbf{v}_1, \ldots, \mathbf{v}_{\lambda-1}$ be $\lambda$ codewords in $C$ where, for $0 \leq i < \lambda$,

$$\mathbf{v}_i = (v_{i,0}, v_{i,1}, \ldots, v_{i,n-1}).$$

Using the $\lambda$ codewords $\mathbf{v}_0, \mathbf{v}_1, \ldots, \mathbf{v}_{\lambda-1}$ in $C$, we form a $\lambda n$-tuple over $\mathrm{GF}(q)$,

$$\mathbf{w} = (\mathbf{w}_0, \mathbf{w}_1, \ldots, \mathbf{w}_{n-1}), \tag{9.1}$$

which consists of $n$ sections, $\mathbf{w}_0, \mathbf{w}_1, \ldots, \mathbf{w}_{n-1}$, each composed of $\lambda$ code symbols. For $0 \leq j < n$, the $j$th section $\mathbf{w}_j$ of $\mathbf{w}$ consists of the $j$th code symbols in the $\lambda$ codewords $\mathbf{v}_0, \mathbf{v}_1, \ldots, \mathbf{v}_{\lambda-1}$, i.e.,

$$\mathbf{w}_j = (v_{0,j}, v_{1,j}, \ldots, v_{\lambda-2,j}, v_{\lambda-1,j}), \tag{9.2}$$

where $v_{i,j}$ is the $j$th code symbol of the $i$th codeword $\mathbf{v}_i$ in $C$, $0 \leq i < \lambda$ and $0 \leq j < n$.

This arrangement of the $\lambda$ codewords, $\mathbf{v}_0, \mathbf{v}_1, \ldots, \mathbf{v}_{\lambda-1}$, in $C$ to form a $\lambda n$-tuple $\mathbf{w}$ defined by (9.1) and (9.2) is referred to as *interleaving*. The parameter $\lambda$ is called the *interleaving degree*. The vector $\mathbf{w}$ is called an *interleaved vector*. The $\lambda$ codewords, $\mathbf{v}_0, \mathbf{v}_1, \ldots, \mathbf{v}_{\lambda-1}$, are called the *constituent codewords* of $\mathbf{w}$. The vector $\mathbf{w}$ is obtained by simply interleaving the code symbols of the $\lambda$ codewords, $\mathbf{v}_0, \mathbf{v}_1, \ldots, \mathbf{v}_{\lambda-1}$, in $C$.

There are $q^{\lambda k}$ distinct collections of $\lambda$ codewords in $C$. Based on these distinct collections of codewords in $C$, we can construct $q^{\lambda k}$ interleaved vectors, each of length $\lambda n$. These $q^{\lambda k}$ interleaved vectors form a $(\lambda n, \lambda k)$ linear block code over $\mathrm{GF}(q)$ of rate $k/n$, denoted by $C_{\mathrm{int}}^{(\lambda)}$, which is called an *interleaved code* of $C$ with interleaving degree $\lambda$. The $\lambda n$-tuple $\mathbf{w}$ is called an *interleaved codeword* of $C_{\mathrm{int}}^{(\lambda)}$. If the minimum distance of $C$ is $d_{\min}$, the minimum distance of the interleaved code $C_{\mathrm{int}}^{(\lambda)}$ is also $d_{\min}$. The $(n, k)$ code $C$ is called the *base code* of the interleaved code $C_{\mathrm{int}}^{(\lambda)}$. Interleaving increases the length of a code without changing its rate and minimum distance.

A simple way to perform the interleaving of the $(n, k)$ linear block code $C$ by a degree of $\lambda$ is to arrange the $\lambda$ codewords, $\mathbf{v}_0, \mathbf{v}_1, \ldots, \mathbf{v}_{\lambda-1}$, in a collection as rows of a $\lambda \times n$ array as follows:

$$\mathbf{W} = \begin{bmatrix} v_{0,0} & v_{0,1} & \cdots & v_{0,j} & \cdots & v_{0,n-1} \\ v_{1,0} & v_{1,1} & \cdots & v_{1,j} & \cdots & v_{1,n-1} \\ \vdots & \vdots & \vdots & \vdots & \vdots & \vdots \\ v_{\lambda-1,0} & v_{\lambda-1,1} & \cdots & v_{\lambda-1,j} & \cdots & v_{\lambda-1,n-1} \end{bmatrix}. \tag{9.3}$$

From (9.2) and (9.3), we see that the $j$th column of $\mathbf{W}$ is the $j$th section $\mathbf{w}_j = (v_{0,j}, v_{1,j}, \ldots, v_{\lambda-2,j}, v_{\lambda-1,j})$ of the interleaved codeword $\mathbf{w}$ in $C_{\text{int}}^{(\lambda)}$. In transmission, we transmit the array $\mathbf{W}$ column by column from right to left, which gives the interleaved codeword $\mathbf{w}$.

Let

$$\mathbf{H} = \begin{bmatrix} \mathbf{h}_0 & \mathbf{h}_1 & \cdots & \mathbf{h}_{n-1} \end{bmatrix} \tag{9.4}$$

be the $(n-k) \times n$ parity-check matrix of the base code $C$ where, for $0 \le j < n$, $\mathbf{h}_j$ is the $j$th column of $\mathbf{H}$. For $0 \le j < n$, we form the following $\lambda \times \lambda$ diagonal array:

$$\mathbf{H}_j = \begin{bmatrix} \mathbf{h}_j & & & \\ & \mathbf{h}_j & & \\ & & \ddots & \\ & & & \mathbf{h}_j \end{bmatrix} \tag{9.5}$$

with $\lambda$ copies of the $j$th column $\mathbf{h}_j$ of $\mathbf{H}$ lying on its main diagonal and zeros elsewhere. Then, the parity-check matrix of the interleaved code $C_{\text{int}}^{(\lambda)}$ is the following $\lambda(n-k) \times \lambda n$ matrix:

$$\mathbf{H}_{\text{int}}^{(\lambda)} = \begin{bmatrix} \mathbf{H}_0 & \mathbf{H}_1 & \cdots & \mathbf{H}_{n-1} \end{bmatrix}. \tag{9.6}$$

The matrix $\mathbf{H}_{\text{int}}^{(\lambda)}$ is simply obtained by taking $\lambda$ copies of $\mathbf{H}$ and interleaving their columns. We called $\mathbf{H}_{\text{int}}^{(\lambda)}$ the *interleaved matrix* of $\mathbf{H}$ with degree $\lambda$.

In encoding, the $\lambda$ constituent codewords of an interleaved codeword $\mathbf{w}$ in $C_{\text{int}}^{(\lambda)}$ can be generated in serial, one at a time, using a single encoder for the base code $C$, or they can be generated in parallel by $\lambda$ identical encoders. Suppose an interleaved codeword $\mathbf{w} = (\mathbf{w}_0, \mathbf{w}_1, \ldots, \mathbf{w}_{n-1})$ in $C_{\text{int}}^{(\lambda)}$ is transmitted. Let $\mathbf{w}^* = (\mathbf{w}_0^*, \mathbf{w}_1^*, \ldots, \mathbf{w}_{n-1}^*)$ be the received vector. In decoding, the received vector $\mathbf{w}^*$ is first deinterleaved into $\lambda$ vectors $\mathbf{v}_0^*, \mathbf{v}_1^*, \ldots, \mathbf{v}_{\lambda-1}^*$ which are the received versions of the $\lambda$ transmitted constituent codewords $\mathbf{v}_0, \mathbf{v}_1, \ldots, \mathbf{v}_{\lambda-1}$ in $\mathbf{w}$, respectively. After deinterleaving $\mathbf{w}^*$, the $\lambda$ received constituent vectors, $\mathbf{v}_0^*, \mathbf{v}_1^*, \ldots, \mathbf{v}_{\lambda-1}^*$, of $\mathbf{w}^*$ can be decoded by a single decoder for $C$ in serial, one at a time, or decoded in parallel by $\lambda$ identical decoders to reduce decoding latency. An error pattern contained in $\mathbf{w}^*$ is correctable if and only if the error pattern contained in each received constituent vector $\mathbf{v}_j^*$ of $\mathbf{w}^*$ is a correctable error pattern for the base code $C$.

The interleaving technique is very effective for constructing long and powerful codes for correcting errors (or erasures) that cluster in bursts. This topic will be discussed in Chapter 10.

If the base code $C$ is an $(n, k)$ cyclic code over GF($q$) with generator polynomial

$$\mathbf{g}(X) = g_0 + g_1 X + g_2 X^2 + \cdots g_{n-k-1} X^{n-k-1} + X^{n-k}, \tag{9.7}$$

then the interleaved code $C_{\text{int}}^{(\lambda)}$ of $C$ is a $(\lambda n, \lambda k)$ cyclic code over GF($q$) with generator polynomial (see Problem 9.3)

$$\mathbf{g}_{\text{int}}^{(\lambda)}(X) = g_0 + g_1 X^{\lambda} + g_2 X^{2\lambda} + \cdots g_{n-k-1} X^{(n-k-1)\lambda} + X^{(n-k)\lambda}. \tag{9.8}$$

In cyclic form, the encoding and error correction of an interleaved code can be accomplished by using shift registers as shown in Chapter 4. The decoder can be derived from the decoder of the base cyclic code $C$ simply by replacing each register stage of the base decoder by $\lambda$ stages without changing the other connections. This essentially allows the decoder circuitry to look at successive rows of the codeword array $\mathbf{W}$ on successive decoder cycles. Therefore, if the decoder of the base code is simple (such as Hamming codes), so is the decoder for the interleaved code.

**Example 9.1** Consider the $(31, 16)$ BCH code $C$ given in Example 5.5 whose generator polynomial is

$$\mathbf{g}(X) = 1 + X + X^2 + X^3 + X^5 + X^7 + X^8 + X^9 + X^{10} + X^{11} + X^{15}.$$

The code $C$ is capable of correcting any combination of three or fewer errors. Suppose we interleave this code with degree $\lambda = 10$. Interleaving $C$ with degree 10 gives a $(310, 160)$ cyclic interleaved code $C_{\text{int}}^{(10)}$ whose generator polynomial is

$$\mathbf{g}_{\text{int}}^{(10)}(X) = 1 + X^{10} + X^{20} + X^{30} + X^{50} + X^{70} + X^{80} + X^{90} + X^{100} + X^{110} + X^{150}.$$

To decode $C_{\text{int}}^{(10)}$, we first deinterleave the received vector into 10 received constituent vectors and decode each received constituent vector with the Berlekamp–Massey iterative algorithm presented in Chapter 5. ▲▲

Let $C_0, C_1, \ldots, C_{\lambda-1}$ be $\lambda$ linear block codes over GF($q$) of the same length $n$ and different dimensions $k_0, k_1, \ldots, k_{\lambda-1}$, respectively. If we interleave the $\lambda$ codes $C_0, C_1, \ldots, C_{\lambda-1}$, we obtain an interleaved code, denoted by $C_{\text{int}}^{(\lambda)} = C_0 * C_1 * \cdots * C_{\lambda-1}$. Each interleaved codeword $\mathbf{w}$ in $C_{\text{int}}^{(\lambda)}$ is composed of $\lambda$ constituent codewords, each from one of the $\lambda$ codes $C_0, C_1, \ldots, C_{\lambda-1}$. If we arrange a codeword $\mathbf{w}$ in $C_{\text{int}}^{(\lambda)}$ into a $\lambda \times n$ array $\mathbf{W}$ in the form of (9.3), the $i$th row of $\mathbf{W}$ is a codeword from the $i$th base code $C_i$, $0 \leq i < \lambda$. The interleaved code $C_{\text{int}}^{(\lambda)}$ is a $(\lambda n, k_0 + k_1 + \cdots + k_{\lambda-1})$ linear block code over GF($q$).

## 9.2 Direct Product

Let $C_1$ be an $(n_1, k_1)$ linear block code over GF($q$) with minimum distance $d_1$, and let $C_2$ be an $(n_2, k_2)$ linear block code over GF($q$) with minimum distance

$d_2$. Using $C_1$ and $C_2$ as component codes, we can construct an $(n_1 n_2, k_1 k_2)$ linear block code over $\mathrm{GF}(q)$, denoted by $C_1 \times C_2$, with minimum distance $d_1 d_2$. Each codeword in $C_1 \times C_2$ is a rectangular array, denoted by $\mathbf{v}_{1 \times 2}$, of $n_2$ rows and $n_1$ columns, as shown in Fig. 9.1, in which every row is a codeword in $C_1$ and every column is a codeword in $C_2$. The array $\mathbf{v}_{1 \times 2}$ is called a *codeword array* in $C_1 \times C_2$.

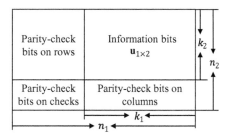

Figure 9.1 A product codeword array $\mathbf{v}_{1 \times 2}$.

Suppose both $C_1$ and $C_2$ are in systematic form. To form such a codeword array in $C_1 \times C_2$, we first arrange a sequence of $k_1 k_2$ information symbols into a $k_2 \times k_1$ information array, denoted by $\mathbf{u}_{1 \times 2}$, as shown in the upper right subarray of $\mathbf{v}_{1 \times 2}$, which consists of $k_2$ rows, each composed of $k_1$ information symbols, and $k_1$ columns, each composed of $k_2$ information symbols. Next, we encode the information array $\mathbf{u}_{1 \times 2}$ in two steps. In the first step, we encode each row of $\mathbf{u}_{1 \times 2}$ into a codeword in $C_1$. This row encoding results in an array of $k_2$ rows and $n_1$ columns which is the upper subarray of $\mathbf{v}_{1 \times 2}$ as shown in Fig. 9.1. In the second step of encoding, each column of the $k_2 \times n_1$ array formed in the first step of encoding is encoded into a codeword in $C_2$. The column encoding of the $k_2 \times n_1$ array formed in the first step of encoding results in an $n_2 \times n_1$ array $\mathbf{v}_{1 \times 2}$ as shown in Fig. 9.1. Because each row of the lower subarray of $\mathbf{v}_{1 \times 2}$ formed in the second step of encoding is a linear combination of the $k_2$ rows in the upper subarray of $\mathbf{v}_{1 \times 2}$ formed in the first step of encoding which are codewords in $C_1$, each row of the lower subarray of $\mathbf{v}_{1 \times 2}$ is a codeword in $C_1$. Hence, each row of the array $\mathbf{v}_{1 \times 2}$ is a codeword in $C_1$ and each column of $\mathbf{v}_{1 \times 2}$ is a codeword in $C_2$. Because there are $q^{k_1 k_2}$ distinct information sequences of length $k_1 k_2$, with the above two-step encoding, we can construct $q^{k_1 k_2}$ distinct codeword arrays. Because both $C_1$ and $C_2$ are linear block codes, a linear sum of two codeword arrays is a codeword array. Hence, the $q^{k_1 k_2}$ codeword arrays form an $(n_1 n_2, k_1 k_2)$ linear block code $C_1 \times C_2$ over $\mathrm{GF}(q)$. The code $C_1 \times C_2$ is called the *direct product code* of $C_1$ and $C_2$ which is a *two-dimensional linear block code* [1].[1] In forming a product code, the two component codes $C_1$ and $C_2$

---

[1]It is noted that if the two component codes $C_1$ and $C_2$ of product codes are both cyclic, this class of product codes forms a class of *two-dimensional cyclic codes*, also called *bicyclic codes* [2, 3]. A bicyclic code is generally defined as a linear two-dimensional code in which every cyclic-shift of a codeword array in the row direction is a codeword array and every cyclic-shift of a codeword array in the column direction is a codeword array.

can be the same code. It is noted that if we choose $k_2 = n_2$ (i.e., the code $C_2$ is the entire vector space $\mathbf{V}_{n_2}$ of all the $n_2$-tuples over $\mathrm{GF}(q)$), the codeword array in Fig. 9.1 is reduced to the interleaved codeword $\mathbf{w}$ given by (9.3) of an interleaved code. Thus, an interleaved code can be viewed as a degenerate form of a product code.

As shown in Fig. 9.1, a codeword array $\mathbf{v}_{1\times 2}$ in the product code $C_1 \times C_2$ consists of four subarrays, the subarray at the upper right corner, the subarray at the upper left corner, the subarray at the lower right corner, and the subarray at the lower left corner. The subarray at the upper right corner of $\mathbf{v}_{1\times 2}$ is the information array which consists of $k_1 k_2$ information symbols. The subarray at the upper left corner of $\mathbf{v}_{1\times 2}$ consists of $k_2(n_1 - k_1)$ parity-check symbols which are formed based on the $(n_1, k_1)$ code $C_1$. The subarray at the lower right corner of $\mathbf{v}_{1\times 2}$ consists of $k_1(n_2 - k_2)$ parity-check symbols which are formed based on the $(n_2, k_2)$ code $C_2$. The subarray at the lower left corner of $\mathbf{v}_{1\times 2}$ consists of $(n_1 - k_1)(n_2 - k_2)$ *overall parity-check symbols* which check both the parity-check symbols at the upper left corner of $\mathbf{v}_{1\times 2}$ and the parity-check symbols at the lower right corner of $\mathbf{v}_{1\times 2}$. We call the overall parity-check symbols *checks on checks*. Furthermore, the $n_2 - k_2$ rows of the lower part of the codeword array $\mathbf{v}_{1\times 2}$ are linear combinations of the $k_2$ rows in the upper part of $\mathbf{v}_{1\times 2}$. Because the $k_2$ rows in the upper part of $\mathbf{v}_{1\times 2}$ are codewords in $C_1$, the $n_2 - k_2$ rows in the lower part of $\mathbf{v}_{1\times 2}$ are also codewords in $C_1$ which may be considered as parity-check codewords for the $k_2$ codewords of $C_1$ in the upper part of $\mathbf{v}_{1\times 2}$. The $n_1 - k_1$ columns in the left part of the codeword array $\mathbf{v}_{1\times 2}$ are linear combinations of the $k_1$ columns in the right part of $\mathbf{v}_{1\times 2}$ which are codewords in $C_2$. Hence, the $n_1 - k_1$ columns of the left part of $\mathbf{v}_{1\times 2}$ are also codewords in $C_2$ which may be considered as parity-check codewords for the $k_1$ codeword of $C_2$ in the right part of $\mathbf{v}_{1\times 2}$.

Let $\mathbf{v}_{1\times 2}$ be a nonzero codeword array in the product code $C_1 \times C_2$. Because the minimum distances of $C_1$ and $C_2$ are $d_1$ and $d_2$, respectively, each nonzero row in $\mathbf{v}_{1\times 2}$ (a nonzero codeword in $C_1$) has weight at least $d_1$ and each nonzero column in $\mathbf{v}_{1\times 2}$ (a nonzero codeword in $C_2$) has weight at least $d_2$. For each nonzero row $\mathbf{b}$ in $\mathbf{v}_{1\times 2}$, there are at least $d_1$ columns in $\mathbf{v}_{1\times 2}$ which intersect with $\mathbf{b}$ at its nonzero positions. Because each of these columns has weight at least $d_2$, the number of nonzero components in $\mathbf{v}_{1\times 2}$ is at least $d_1 d_2$. This implies that the weight of $\mathbf{v}_{1\times 2}$ is at least $d_1 d_2$. Therefore, the minimum weight of $C_1 \times C_2$ is at least $d_1 d_2$. To show that the minimum weight of $C_1 \times C_2$ is exactly $d_1 d_2$, we simply construct a codeword array in $C_1 \times C_2$ with weight $d_1 d_2$. A codeword array in $C_1 \times C_2$ with weight $d_1 d_2$ can be constructed as follows: (1) choose a minimum-weight codeword $\mathbf{v}_1$ in $C_1$ and a minimum-weight codeword $\mathbf{v}_2$ in $C_2$; and (2) form a codeword array $\mathbf{v}_{1\times 2}$ in which all columns corresponding to the zero components in $\mathbf{v}_1$ are zero columns and all the columns corresponding to the nonzero components of $\mathbf{v}_1$ are the minimum-weight codeword $\mathbf{v}_2$ chosen from $C_2$. Then, the weight of the constructed codeword array $\mathbf{v}_{1\times 2}$ is exactly $d_1 d_2$ which is the minimum weight of $C_1 \times C_2$. Hence, the minimum distance of $C_1 \times C_2$ is $d_1 d_2$.

**Example 9.2** Let $C_1$ and $C_2$ both be the $(7, 4)$ Hamming code with minimum distance 3 given in Example 3.4. The 16 codewords of $C_1$ in systematic form are given in Table 9.1.

Table 9.1 The 16 codewords of the $(7, 4)$ Hamming code.

| Messages $(u_0, u_1, u_2, u_3)$ | Codewords $(v_0, v_1, v_2, v_3, v_4, v_5, v_6)$ | Messages $(u_0, u_1, u_2, u_3)$ | Codewords $(v_0, v_1, v_2, v_3, v_4, v_5, v_6)$ |
|---|---|---|---|
| (0 0 0 0) | (0 0 0 0 0 0 0) | (0 0 0 1) | (1 0 1 0 0 0 1) |
| (1 0 0 0) | (1 1 0 1 0 0 0) | (1 0 0 1) | (0 1 1 1 0 0 1) |
| (0 1 0 0) | (0 1 1 0 1 0 0) | (0 1 0 1) | (1 1 0 0 1 0 1) |
| (1 1 0 0) | (1 0 1 1 1 0 0) | (1 1 0 1) | (0 0 0 1 1 0 1) |
| (0 0 1 0) | (1 1 1 0 0 1 0) | (0 0 1 1) | (0 1 0 0 0 1 1) |
| (1 0 1 0) | (0 0 1 1 0 1 0) | (1 0 1 1) | (1 0 0 1 0 1 1) |
| (0 1 1 0) | (1 0 0 0 1 1 0) | (0 1 1 1) | (0 0 1 0 1 1 1) |
| (1 1 1 0) | (0 1 0 1 1 1 0) | (1 1 1 1) | (1 1 1 1 1 1 1) |

The direct product $C_1 \times C_2$ of $C_1$ and $C_2$ is a $(49, 16)$ linear code with minimum distance 9. A codeword array in $C_1 \times C_2$ is

$$
\mathbf{v}_{1 \times 2} = 
\begin{bmatrix}
1 & 0 & 0 & 0 & 1^* & 1^* & 0 \\
0 & 1 & 0 & 1 & 1^* & 1^* & 0 \\
1 & 0 & 1 & 0 & 0 & 0 & 1 \\
0 & 1 & 0 & 0 & 0 & 1 & 1 \\
0 & 1 & 1 & 1 & 0 & 0 & 1 \\
1 & 0 & 1 & 1 & 1 & 0 & 0 \\
1 & 0 & 0 & 1 & 0 & 1 & 1
\end{bmatrix}.
\tag{9.9}
$$

Label the rows and columns of $\mathbf{v}_{1 \times 2}$ from 0 to 6. Each row and each column of the codeword array $\mathbf{v}_{1 \times 2}$ given by (9.9) are codewords in the $(7, 4)$ Hamming code. The top row (i.e., the fourth row) of the lower part of $\mathbf{v}_{1 \times 2}$ is the vector-sum of the top three rows of the upper part of $\mathbf{v}_{1 \times 2}$. The first row (i.e., the fifth row) of the lower part of $\mathbf{v}_{1 \times 2}$ is the vector-sum of the last three rows of the upper part of $\mathbf{v}_{1 \times 2}$. The last row of the lower part of $\mathbf{v}_{1 \times 2}$ is the vector-sum of the zeroth, first, and third rows of the upper part of $\mathbf{v}_{1 \times 2}$. Similarly, the three columns of the left part of $\mathbf{v}_{1 \times 2}$ are linear sums of the four columns of the right part of $\mathbf{v}_{1 \times 2}$.

Choose two minimum-weight codewords, $\mathbf{v}_1 = (1\ 1\ 0\ 1\ 0\ 0\ 0)$ and $\mathbf{v}_2 = (1\ 0\ 0\ 0\ 1\ 1\ 0)$, from the $(7, 4)$ Hamming code. Following the steps to construct a minimum-weight codeword array $\mathbf{v}_{1 \times 2}$ of $C_1 \times C_2$, we obtain the following codeword array:

$$\mathbf{v}_{1\times2} = \begin{bmatrix} 1 & 1 & 0 & 1 & 0 & 0 & 0 \\ 0 & 0 & 0 & 0 & 0 & 0 & 0 \\ 0 & 0 & 0 & 0 & 0 & 0 & 0 \\ 0 & 0 & 0 & 0 & 0 & 0 & 0 \\ 1 & 1 & 0 & 1 & 0 & 0 & 0 \\ 1 & 1 & 0 & 1 & 0 & 0 & 0 \\ 0 & 0 & 0 & 0 & 0 & 0 & 0 \end{bmatrix}. \tag{9.10}$$

There are nine nonzero components in $\mathbf{v}_{1\times2}$, i.e., $\mathbf{v}_{1\times2}$ has weight 9 which equals the minimum weight of the code $C_1 \times C_2$. ▲▲

In transmission, a codeword array $\mathbf{v}_{1\times2}$ in $C_1 \times C_2$ can be transmitted column by column from right to left (or transmitted row by row from top to bottom). This ordered transmission of a codeword array in $C_1 \times C_2$ interleaves the code-words in $C_1$ by a degree of $n_2$ (or the codewords in $C_2$ by a degree of $n_1$). In the encoding of a $k_2 \times k_1$ information array $\mathbf{u}_{1\times2}$, we can perform the column encoding first and then the row encoding. The resultant codeword array is the same as the codeword array obtained by performing the row encoding first and then the column encoding (see Problem 9.4).

Decoding of a received codeword array can be performed in two steps itera-tively to reduce decoding complexity. Suppose a codeword array $\mathbf{v}_{1\times2}$ is trans-mitted row by row. When a sequence of $n_1 n_2$ symbols is received, it is arranged back into an $n_2 \times n_1$ array $\mathbf{v}_{1\times2}^*$ which corresponds to the transmitted codeword array $\mathbf{v}_{1\times2}$. The array $\mathbf{v}_{1\times2}^*$ is referred to as the *received array*. With a two-step decoding, the rows of the received array are first decoded based on $C_1$, one at a time, from top to bottom. The row decoding of the received array $\mathbf{v}_{1\times2}^*$ results in a decoded array. Then, we check whether the syndromes of all the rows and all the columns of the decoded array are zero. If they are all zeros, we stop the decoding and deliver the decoded information array to the user(s). If not all the syndromes of the rows and columns of the decoded array are zeros, we initiate the second step of decoding and decode the decoded array column by column based on $C_2$. At the end of the second step of decoding, we obtain a new decoded array. Again, we check the syndromes of the rows and columns of the newly decoded array. If they are all zeros, we stop the decoding; other-wise, we start the next two-step row–column decoding process. We continue the row–column decoding of the received array iteratively until either the decoded array at the end of a decoding iteration is a codeword array in the product code $C_1 \times C_2$ or a preset maximum number of iterations is reached. The two-step row–column iterative decoding of a product code was first introduced by Elias [1].

With the two-step iterative row–column decoding of a received array $\mathbf{v}_{1\times2}^*$, an error pattern is correctable if and only if the uncorrectable error patterns on rows of $\mathbf{v}_{1\times2}^*$ after row decoding are reduced to correctable error patterns in the column decoding and vice versa. The product code $C_1 \times C_2$ can correct any combination of $\lfloor (d_1 d_2 - 1)/2 \rfloor$ errors; however, the two-step iterative row–column decoding will not achieve this. For example, consider the product of two Hamming single-error-correcting codes given in Example 9.1. The minimum distance of the product code is 9. However, with the two-step iterative row–column decoding, there are error patterns of four errors which are not correctable. For example, consider the error pattern of four errors occurring at the four corners of a $2 \times 2$ rectangle in a received array $\mathbf{v}_{1 \times 2}^*$ as marked by "$*$" in (9.9). This error pattern gives two errors in each of the two rows and two errors in each of the two columns and hence is not correctable by the two-step iterative row–column iterative decoding no matter how many iterations are performed.

The row decoding and the column decoding of a received array can be both hard-decision, both soft-decision, or one hard-decision and the other soft-decision. The choice of the type of decoding depends on the performance and/or decoding requirements.

Let $\mathbf{H}_1 = [\mathbf{h}_{1,0} \ \mathbf{h}_{1,1} \ \cdots \ \mathbf{h}_{1,n_1-1}]$ and $\mathbf{H}_2 = [\mathbf{h}_{2,0} \ \mathbf{h}_{2,1} \ \cdots \ \mathbf{h}_{2,n_2-1}]$ be the parity-check matrices of $C_1$ and $C_2$, respectively. The matrix $\mathbf{H}_1$ is an $(n_1 - k_1) \times n_1$ matrix consisting of $n_1$ columns, $\mathbf{h}_{1,0}, \mathbf{h}_{1,1}, \ldots, \mathbf{h}_{1,n_1-1}$; and $\mathbf{H}_2$ is an $(n_2 - k_2) \times n_2$ matrix consisting of $n_2$ columns, $\mathbf{h}_{2,0}, \mathbf{h}_{2,1}, \ldots, \mathbf{h}_{2,n_2-1}$. Form the following $(n_2(n_1 - k_1) + n_1(n_2 - k_2)) \times n_1 n_2$ matrix:

$$\mathbf{H}_{1 \times 2} = \begin{bmatrix} \mathbf{H}_1 & & & & \\ & \mathbf{H}_1 & & & \\ & & \ddots & & \\ & & & & \mathbf{H}_1 \\ \mathbf{H}_{2,0} & \mathbf{H}_{2,1} & \cdots & \mathbf{H}_{2,n_2-1} \end{bmatrix} \tag{9.11}$$

where, for $0 \leq j < n_2$,

$$\mathbf{H}_{2,j} = \begin{bmatrix} \mathbf{h}_{2,j} & & & \\ & \mathbf{h}_{2,j} & & \\ & & \ddots & \\ & & & \mathbf{h}_{2,j} \end{bmatrix} \tag{9.12}$$

is an $n_1 \times n_1$ diagonal array with $n_1$ copies of the $j$th column $\mathbf{h}_{2,j}$ of $\mathbf{H}_2$ lying on the main diagonal, which is an $n_1(n_2 - k_2) \times n_1$ matrix in the same form as given in (9.5).

The matrix $\mathbf{H}_{1 \times 2}$ is the parity-check matrix of the product code $C_1 \times C_2$ with the $n_2$ rows of each codeword array arranged as a sequence (a sequence of $n_2$ codewords in $C_1$) (see Problem 9.6). Note that $\mathbf{H}_{1 \times 2}$ consists of two submatrices, the upper one and the lower one. The upper submatrix, denoted by $\mathbf{H}_{1 \times 2, \text{upper}} = \text{diag}(\mathbf{H}_1, \mathbf{H}_1, \ldots, \mathbf{H}_1)$, of $\mathbf{H}_{1 \times 2}$ is a diagonal array with $n_2$ copies of the parity-check matrix $\mathbf{H}_1$ of $C_1$ lying on its main diagonal. The null space of $\mathbf{H}_{1 \times 2, \text{upper}}$ gives an $(n_1 n_2, k_1 n_2)$ linear code obtained by cascading $n_2$ copies of $C_1$, called a *cascaded code* of $C_1$ with a *cascading degree* of $n_2$ and denoted by $C_{1, \text{casc}}^{(n_2)}$. The lower submatrix, denoted by $\mathbf{H}_{1 \times 2, \text{lower}} = [\mathbf{H}_{2,0} \ \mathbf{H}_{2,1} \ \cdots \ \mathbf{H}_{2,n_2-1}]$, of $\mathbf{H}_{1 \times 2}$ is an $n_1(n_2 - k_2) \times n_1 n_2$ matrix obtained by interleaving the parity-check matrix $\mathbf{H}_2$ of $C_2$ with a degree of $n_1$. Hence, the null space of $\mathbf{H}_{1 \times 2, \text{lower}}$ gives an $(n_1 n_2, k_2 n_1)$ interleaved code $C_{2, \text{int}}^{(n_1)}$ of $C_2$ with an interleaving degree of $n_1$. The product code $C_1 \times C_2$ is simply the intersection of the cascaded code $C_{1, \text{casc}}^{(n_2)}$ and the interleaved code $C_{2, \text{int}}^{(n_1)}$.

One way for the product code $C_1 \times C_2$ to correct errors up to its guaranteed error-correcting capability $\lfloor (d_1 d_2 - 1)/2 \rfloor$ is to devise a decoding algorithm (or method) based on its parity-check matrix $\mathbf{H}_{1\times 2}$ given by (9.11). If both $C_1$ and $C_2$ are one-step majority-logic decodable, such as finite-geometry codes presented in Chapter 7, it can be proved that $\mathbf{H}_{1\times 2}$ has orthogonal structure (see Problem 9.7) and the product code $C_1 \times C_2$ is also one-step majority-logic decodable based on $\mathbf{H}_{1\times 2}$. Let $\sigma_1$ and $\sigma_2$ be the numbers of orthogonal check-sums that can be formed based on the parity-check matrices, $\mathbf{H}_1$ and $\mathbf{H}_2$, of $C_1$ and $C_2$, respectively. Then, $\sigma_1 + \sigma_2$ orthogonal check-sums can be formed based on $\mathbf{H}_{1\times 2}$ and hence the product code $C_1 \times C_2$ can correct up to $\lfloor (\sigma_1 + \sigma_2)/2 \rfloor$ errors with one-step majority-logic decoding [4]. More on decoding product codes will be discussed in Chapter 10.

If $C_1$ and $C_2$ are cyclic codes and their lengths $n_1$ and $n_2$ are relatively prime, then their product $C_1 \times C_2$ is cyclic (i.e., $C_1 \times C_2$ is an $(n_1 n_2, k_1 k_2)$ cyclic code) if the code symbols of a codeword array are transmitted in *a diagonal order*. We start the transmission of a codeword array $\mathbf{v}_{1\times 2}$ from its upper right corner and move down to the left on a 45° diagonal as shown in Fig. 9.2. When we reach the end of a column, we move to the top of the next column on its left. Then, we move down to the left on a 45° diagonal. When we reach the end of a row, we move to the rightmost code symbol of the next row as shown in Fig. 9.2 and then move down to the left on a 45° diagonal. The diagonal transmission continues until all the code symbols of the codeword array are transmitted. Because $n_1$ and $n_2$ are relatively prime, no column and no row of the codeword array $\mathbf{v}_{1\times 2}$ will be repeated with the 45° downward transmission.

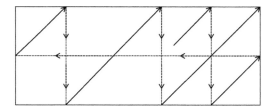

Figure 9.2 Diagonal transmission of a codeword array in a cyclic product code.

Let $\mathbf{g}_1(X)$ and $\mathbf{h}_1(X)$ be the generator and parity-check polynomials of $C_1$, respectively, and let $\mathbf{g}_2(X)$ and $\mathbf{h}_2(X)$ be the generator and parity-check polynomials of $C_2$, respectively. Because $n_1$ and $n_2$ are relatively prime, there exists a pair of integers, $a$ and $b$, such that

$$an_1 + bn_2 = 1.$$

It was proved in [5, 6] that the generator polynomial $\mathbf{g}(X)$ of the cyclic product code $C_1 \times C_2$ of $C_1$ and $C_2$ is the greatest common divisor (GCD) of $X^{n_1 n_2} - 1$ and $\mathbf{g}_1(X^{bn_2})\mathbf{g}_2(X^{an_1})$, i.e.,

$$\mathbf{g}(X) = \mathrm{GCD}\{X^{n_1 n_2} - 1, \mathbf{g}_1(X^{bn_2})\mathbf{g}_2(X^{an_1})\}. \tag{9.13}$$

The parity-check polynomial $\mathbf{h}(X)$ of $C_1 \times C_2$ is the GCD of $\mathbf{h}_1(X^{bn_2})$ and $\mathbf{h}_2(X^{an_1})$, i.e.,

$$\mathbf{h}(X) = \text{GCD}\{\mathbf{h}_1(X^{bn_2}), \mathbf{h}_2(X^{an_1})\}. \tag{9.14}$$

If both $C_1$ and $C_2$ are BCH codes, their product $C_1 \times C_2$ can be decoded in two steps using the Berlekamp–Massey algorithm for both row and column decodings of a codeword array in $C_1 \times C_2$.

In encoding a $k_1 \times k_2$ information array $\mathbf{u}_{1\times2}$, if we only encode the rows of $\mathbf{u}_{1\times2}$ with the $(n_1, k_1)$ code $C_1$ and the columns of $\mathbf{u}_{1\times2}$ with the $(n_2, k_2)$ code $C_2$ without forming the $(n_1 - k_1)(n_2 - k_2)$ checks on checks, we obtain an incomplete codeword array as shown in Fig. 9.3. This incomplete product of $C_1$ and $C_2$ results in a $(k_2 n_1 + k_1 n_2 - k_1 k_2, k_1 k_2)$ linear code, denoted by $C_{1\times2,\text{turbo}}$, with minimum distance $d_1 + d_2 - 1$ (see Problem 9.8). This incomplete product code is referred to as a *turbo product code* which has a shorter length and a higher rate but smaller minimum distance than the complete product code $C_1 \times C_2$. With two-step iterative row–column decoding, only $k_2$ rows and $k_1$ columns are decoded in each iteration. The reliability information of decoded parity-check symbols of one code is not passed to decode the other code. Decoding is performed in a *turbo manner* [7–11].

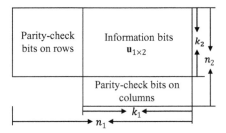

Figure 9.3 A turbo codeword array.

## 9.3 Concatenation

Concatenation is another effective technique for constructing long powerful codes from short codes to achieve reliable information transmission and data storage. This coding technique was introduced by Forney in 1965 [12] and ever since it has been widely used for error control in a wide range of communication and data-storage systems.

A typical concatenated coding system consists of two linear component codes, one for correcting *local errors* and the other for correcting *global errors*. Encoding of the two component codes can be performed either in serial or in parallel. Decoding of the two component codes can also be performed either in serial or in parallel. A concatenated coding system in which both encoding and decoding are performed in serial is referred to as a *serial concatenated coding system*. A concatenated coding system in which both the encoding and decoding are performed in parallel is referred to as a *parallel concatenated coding system*.

In this section, we present two types of serial concatenations and one type of parallel concatenation.

## 9.3.1  Type-1 Serial Concatenation

The setup of a type-1 serial concatenated coding system is shown in Fig. 9.4.

Figure 9.4 A type-1 serial concatenated coding system.

The two component codes, $C_1$ and $C_2$, used in the concatenation are an $(n_1, k_1)$ linear code and an $(n_2, k_2)$ linear code with $n_1 < n_2$ and $k_2 = mn_1$, where $m$ is a positive integer. Both codes, $C_1$ and $C_2$, are over the same field $\mathrm{GF}(q)$, with $q = 2$ or $2^s$, and are in systematic form. Let $\mathbf{u}$ be an information sequence of $k = mk_1$ information symbols. Divide this information sequence into $m$ sections, $\mathbf{u}_0, \mathbf{u}_1, \ldots, \mathbf{u}_{m-1}$, each consisting of $k_1$ information symbols, called messages. In the first stage of encoding, each message $\mathbf{u}_j$, $0 \le j < m$, is encoded into a codeword $\mathbf{v}_j$ of $n_1$ code symbols in code $C_1$. This stage of encoding is referred to as the *outer encoding* and $C_1$ is called the *outer code*. Hence, in the first stage of encoding, the information sequence $\mathbf{u}$ is encoded into a sequence of $m$ outer codewords, denoted by $\mathbf{v} = (\mathbf{v}_0, \mathbf{v}_1, \ldots, \mathbf{v}_{m-1})$. In the second stage of concatenated encoding, the sequence $\mathbf{v}$ of $m$ outer codewords in $C_1$ is treated as a message of $k_2 = mn_1$ information symbols and is encoded into a codeword $\mathbf{y} = (y_0, y_1, \ldots, y_{n_2-1})$ of $n_2$ code symbols in the code $C_2$. This stage of concatenated encoding is referred to as the *inner encoding* and $C_2$ is called the *inner code*. The two encoders in the system are called the *outer* and *inner encoders*, respectively.

With the above two-stage concatenated encoding, an information sequence $\mathbf{u}$ of $k = mk_1$ information symbols is encoded into a codeword of $n_2$ code symbols in the inner code $C_2$. There are $q^{mk_1}$ distinct information sequences of length $mk_1$ with symbols from $\mathrm{GF}(q)$. They are encoded into $q^{mk_1}$ codewords in $C_2$ which form an $(n_2, mk_1)$ linear code, denoted by $C_{1 \times 2, \mathrm{conc}}^{(1)}$. The code $C_{1 \times 2, \mathrm{conc}}^{(1)}$ is referred to as the *type-1 serial concatenated code of $C_1$ and $C_2$*. The rate of $C_{1 \times 2, \mathrm{conc}}^{(1)}$ is $R = mk_1/n_2$. Because $k_2 = mn_1$, we find $R = k_1 k_2 / n_1 n_2$, the product of the rates of the two component codes $C_1$ and $C_2$. Note that the concatenated code $C_{1 \times 2, \mathrm{conc}}^{(1)}$ is a subcode of the inner code $C_2$. If the minimum distance of $C_2$ is $d_2$, the minimum distance of $C_{1 \times 2, \mathrm{conc}}^{(1)}$ is at least $d_2$ (see Problem 9.9). Each codeword $\mathbf{y} = (y_0, y_1, \ldots, y_{n_2-1})$ in $C_{1 \times 2, \mathrm{conc}}^{(1)}$ consists of $m$ outer codewords, $\mathbf{v}_0, \mathbf{v}_1, \ldots, \mathbf{v}_{m-1}$, and a section $\mathbf{z}$ of $n_2 - k_2$ parity-check symbols as follows:

$$\mathbf{y} = (\mathbf{z}, \mathbf{v}_0, \mathbf{v}_1, \ldots, \mathbf{v}_{m-1}), \tag{9.15}$$

where the $n_2 - k_2$ parity-check symbols in the section $\mathbf{z}$ are computed based on the inner code $C_2$. We call $\mathbf{z}$ the *parity-check section* of $\mathbf{y}$.

Decoding of the concatenated code $C_{1 \times 2, \text{conc}}^{(1)}$ also consists of two stages. Let $\mathbf{y} = (\mathbf{z}, \mathbf{v}_0, \mathbf{v}_1, \dots, \mathbf{v}_{m-1})$ and $\mathbf{y}^* = (\mathbf{z}^*, \mathbf{v}_0^*, \mathbf{v}_1^*, \dots, \mathbf{v}_{m-1}^*)$ be the transmitted codeword and received vector, respectively. As soon as a received outer vector $\mathbf{v}_j^*$, $0 \leq j < m$, is received, it is decoded based on the outer code $C_1$. At the completion of decoding the $m$ received outer vectors, if the syndromes of the $m$ decoded outer vectors are all zeros, the decoded vectors are codewords in the outer code $C_1$. In this case, we remove the $n_1 - k_1$ parity-check symbols from each decoded outer codeword. The removal of parity-check symbols from the $m$ decoded codewords results in a decoded information sequence of $k = mk_1$ information symbols. Then, the decoded information sequence is delivered to the user(s).

If not all the received vectors in $\mathbf{y}^* = (\mathbf{z}^*, \mathbf{v}_0^*, \mathbf{v}_1^*, \dots, \mathbf{v}_{m-1}^*)$ are successfully decoded, we replace the successfully decoded vectors in $\mathbf{y}^*$ by their decoded versions, $\mathbf{v}_0^{**}, \mathbf{v}_1^{**}, \dots, \mathbf{v}_{m-1}^{**}$. This results in a modified received sequence $\mathbf{y}^{**} = (\mathbf{z}^*, \mathbf{v}_0^{**}, \mathbf{v}_1^{**}, \dots, \mathbf{v}_{m-1}^{**})$ which has the same received parity-check section $\mathbf{z}^*$ as $\mathbf{y}^*$. Then, the inner decoder decodes $\mathbf{y}^{**}$ based on the inner code $C_2$. The replacement of the successfully decoded vectors in $\mathbf{y}^*$ by their decoded versions reduces the number of errors in $\mathbf{y}^{**}$ if the successfully decoded codewords are identical to the transmitted outer codewords in $\mathbf{y} = (\mathbf{z}, \mathbf{v}_0, \mathbf{v}_1, \dots, \mathbf{v}_{m-1})$. In this case, $\mathbf{y}^{**}$ is more likely to be decoded successfully. If the inner decoder fails to decode $\mathbf{y}^{**}$, we declare a decoding failure.

In the type-1 concatenated coding scheme, the outer decoding is first performed followed by the inner decoding. The errors in a received outer vector are referred to as *local errors*. Errors in received outer vectors that cannot be corrected by the outer decoder are referred to as *global errors*. When global errors are detected, the inner decoder is activated to correct them. If the number of global errors is within the error-correcting capability of the inner code with a specific decoding algorithm, they will be corrected.

The outer decoding and inner decoding can be both hard-decision, both soft-decision, or one hard-decision and the other soft-decision. Decoding both outer and inner codes with soft-decision decoding algorithms gives the best error performance. However, the decoding complexity may be too high for practical applications if the inner code is a relatively long code. Because the outer code is a shorter code, using soft-decision decoding for the outer code and hard-decision decoding for the inner code may offer an effective tradeoff between error performance and decoding complexity. Decoding of both outer and inner codes with hard-decision decoding algorithms gives the lowest decoding complexity among the four combinations of outer and inner decoding.

If $C_1$ and $C_2$ are *low-density parity-check* (LDPC) codes (to be presented in later chapters), both codes can be decoded effectively with a soft-decision decoding algorithm based on belief-propagation (to be presented in Chapter 11). The outer decoding and inner decoding can be performed *iteratively*.

Let $\mathbf{H}_1$ and $\mathbf{H}_2$ be the parity-check matrices of the outer and inner codes, respectively, where $\mathbf{H}_1$ is an $(n_1 - k_1) \times n_1$ matrix and $\mathbf{H}_2$ is an $(n_2 - k_2) \times n_2$ matrix. Divide the parity-check matrix $\mathbf{H}_2$ into two submatrices in the following form:

$$\mathbf{H}_2 = [\mathbf{H}_{2,\text{info}} \quad \mathbf{H}_{2,\text{p}}], \tag{9.16}$$

where $\mathbf{H}_{2,\text{info}}$ and $\mathbf{H}_{2,\text{p}}$ are the $(n_2 - k_2) \times (n_2 - k_2)$ submatrix and the $(n_2 - k_2) \times k_2$ submatrix of $\mathbf{H}_2$, respectively. The submatrices $\mathbf{H}_{2,\text{info}}$ and $\mathbf{H}_{2,\text{p}}$ of $\mathbf{H}_2$ are called the *information* and *parity-check submatrices* of $\mathbf{H}_2$, respectively. Based on the construction of the type-1 concatenated code $C_{1 \times 2,\text{conc}}^{(1)}$, we readily see that the parity-check matrix of $C_{1 \times 2,\text{conc}}^{(1)}$ has the following form:

$$\mathbf{H}_{1 \times 2,\text{conc}}^{(1)} = \left[ \begin{array}{c|c} \mathbf{O} & \text{diag}(\mathbf{H}_1, \mathbf{H}_1, \ldots, \mathbf{H}_1) \\ \hline \mathbf{H}_{2,\text{info}} & \mathbf{H}_{2,\text{p}} \end{array} \right] \tag{9.17}$$

which is an $(n_2 - mk_1) \times n_2$ matrix. The parity-check matrix $\mathbf{H}_{1 \times 2,\text{conc}}^{(1)}$ consists of two submatrices, the upper one and the lower one. The upper submatrix of $\mathbf{H}_{1 \times 2,\text{conc}}^{(1)}$ consists of two parts. The first part $\text{diag}(\mathbf{H}_1, \mathbf{H}_1, \ldots, \mathbf{H}_1)$ is an $m \times m$ diagonal array with $m$ copies of the parity-check matrix $\mathbf{H}_1$ of the outer code $C_1$ lying on its main diagonal which is an $m(n_1 - k_1) \times mn_1$ matrix. The second part $\mathbf{O}$ of $\mathbf{H}_{1 \times 2,\text{conc}}^{(1)}$ is an $m(n_1 - k_1) \times (n_2 - k_2)$ zero matrix. The lower submatrix of $\mathbf{H}_{1 \times 2,\text{conc}}^{(1)}$ is the parity-check matrix $\mathbf{H}_2$ of the inner code $C_2$. Note that the matrix $\mathbf{H}_{1 \times 2,\text{conc}}^{(1)}$ consists of $m$ copies of the parity-check matrix $\mathbf{H}_1$ of the outer code $C_1$ which are connected by the parity-check matrix $\mathbf{H}_2$ of the inner code $C_2$.

### 9.3.2 Type-2 Serial Concatenation

The setup of a type-2 serial concatenated coding system is shown in Fig. 9.5.

Figure 9.5 A type-2 serial concatenated coding system.

The outer and inner codes, $C_1$ and $C_2$, are $(n_1, k_1)$ and $(n_2, k_2)$ linear codes, respectively, with $n_1 = mk_2$ where $m$ is a positive integer. Again, both codes are over the same field $\text{GF}(q)$ and are in systematic form. The outer code is longer than the inner code (opposite to the type-1 serial concatenation). Let $\mathbf{u}$ be an information sequence of $k_1$ information symbols from $\text{GF}(q)$. In the outer stage of encoding, the information sequence $\mathbf{u}$ is encoded into a codeword $\mathbf{v}$ of $n_1$ code symbols in $C_1$. In the inner stage of encoding, the codeword $\mathbf{v}$ is first divided into $m$ sections, denoted by $\mathbf{v}_0, \mathbf{v}_1, \ldots, \mathbf{v}_{m-1}$, each consisting of $k_2$ code symbols. Then, each section $\mathbf{v}_j$, $0 \leq j < m$, of the codeword $\mathbf{v}$ is treated as a message and is encoded into a codeword $\mathbf{y}_j$ of $n_2$ code symbols in $C_2$.

With the above two stages of outer/inner encoding, the information sequence $\mathbf{u}$ is encoded into a sequence $\mathbf{y} = (\mathbf{y}_0, \mathbf{y}_1, \ldots, \mathbf{y}_{m-1})$ of $m$ codewords in the inner code $C_2$ which is a sequence of $mn_2$ code symbols. There are $q^{k_1}$ distinct information sequences of length $k_1$ over GF($q$). Corresponding to these $q^{k_1}$ information sequences, there are $q^{k_1}$ distinct sequences of inner codewords through the outer/inner encoding. These $q^{k_1}$ sequences of inner codewords form an $(mn_2, k_1)$ linear code which is called a *type-2 serial concatenated code*, denoted by $C_{1 \times 2, \text{conc}}^{(2)}$, of $C_1$ and $C_2$. Because $m = n_1/k_2$, the rate of the concatenated code $C_{1 \times 2, \text{conc}}^{(2)}$ is $R = k_1/mn_2 = k_1 k_2/n_1 n_2$, the product of the rates of the two component codes $C_1$ and $C_2$.

The order of decoding the concatenated code $C_{1 \times 2, \text{conc}}^{(2)}$ is inner/outer decoding (opposite to the outer/inner decoding of a type-1 concatenated code). Let $\mathbf{y} = (\mathbf{y}_0, \mathbf{y}_1, \ldots, \mathbf{y}_{m-1})$ and $\mathbf{y}^* = (\mathbf{y}_0^*, \mathbf{y}_1^*, \ldots, \mathbf{y}_{m-1}^*)$ be the transmitted and received sequences, respectively. In the first stage of decoding, the inner decoder decodes each received inner vector $\mathbf{y}_j^*$, $0 \leq j < m$, as soon as it is received. At the completion of decoding the $m$ received inner vectors, if the syndromes of the $m$ decoded vectors are all zeros, the decoded vectors are codewords in the inner code $C_2$. In this case, we remove the parity-check symbols from each decoded inner codeword. The removal of parity-check symbols from the $m$ decoded inner codewords results in $m$ decoded sections, $\mathbf{v}_0^*, \mathbf{v}_1^*, \ldots, \mathbf{v}_{m-1}^*$, corresponding to the codeword $\mathbf{v}$ at the output of the outer encoder. Form a vector $\mathbf{v}^* = (\mathbf{v}_0^*, \mathbf{v}_1^*, \ldots, \mathbf{v}_{m-1}^*)$ and compute its syndrome based on the outer code $C_1$. If the syndrome of $\mathbf{v}^*$ is zero, then $\mathbf{v}^*$ is a codeword in $C_1$. In this case, we stop the decoding process and deliver the information sequence contained in $\mathbf{v}^*$ to the user(s).

If the syndrome of $\mathbf{v}^*$ is not zero, $\mathbf{v}^*$ is not a codeword in the outer code $C_1$. In this case, the outer code decoder is activated to decode the vector $\mathbf{v}^*$ based on $C_1$. In the outer decoding, we replace those sections in $\mathbf{v}^* = (\mathbf{v}_0^*, \mathbf{v}_1^*, \ldots, \mathbf{v}_{m-1}^*)$ by their successfully decoded versions. The replacement results in a modified vector $\mathbf{v}^{**} = (\mathbf{v}_0^{**}, \mathbf{v}_1^{**}, \ldots, \mathbf{v}_{m-1}^{**})$. Then, $\mathbf{v}^{**}$ is decoded by the outer decoder based on $C_2$. The replacement of the sections in $\mathbf{v}^*$ by their successfully decoded versions reduces the number of errors in $\mathbf{v}^{**}$ if these decoded sections are identical to the transmitted sections in $\mathbf{v} = (\mathbf{v}_0, \mathbf{v}_1, \ldots, \mathbf{v}_{m-1})$. In this case, $\mathbf{v}^{**}$ is more likely to be decoded successfully into the transmitted codeword $\mathbf{v}$. If the decoding of $\mathbf{v}^{**}$ is successful, we stop the decoding process and deliver the information sequence $\mathbf{u}$ retrieved from the decoded codeword to the user(s); otherwise, we declare a decoding failure.

Note that if all the inner received vectors are successfully decoded at the first stage of decoding, we could choose to stop the entire decoding process.

Let $\mathbf{H}_1$ and $\mathbf{H}_2$ be the parity-check matrices of the outer and inner codes, respectively, where $\mathbf{H}_1$ is an $(n_1 - k_1) \times n_1$ matrix and $\mathbf{H}_2$ is an $(n_2 - k_2) \times n_2$ matrix. Divide the parity-check matrix $\mathbf{H}_1$ into $m$ submatrices, $\mathbf{H}_{1,0}, \mathbf{H}_{1,1}, \ldots, \mathbf{H}_{1,m-1}$, each consisting of $k_2$ consecutive columns of $\mathbf{H}_1$. Let $\mathbf{O}$ be an

$(n_1 - k_1) \times (n_2 - k_2)$ zero matrix. Next, we form the following $(n_1 - k_1) \times mn_2$ matrix:

$$\mathbf{H}' = [\mathbf{O}\ \mathbf{H}_{1,0}\ \mathbf{O}\ \mathbf{H}_{1,1}\ \cdots\ \mathbf{O}\ \mathbf{H}_{1,m-1}]. \tag{9.18}$$

Then, it follows from the construction of a type-2 serial concatenated code $C^{(2)}_{1\times2,\text{conc}}$ that the parity-check matrix of $C^{(2)}_{1\times2,\text{conc}}$ is the following $(mn_2 - k_1) \times mn_2$ matrix:

$$\mathbf{H}^{(2)}_{1\times2,\text{conc}} = \left[ \begin{array}{c} \mathbf{O}\ \mathbf{H}_{1,0}\ \mathbf{O}\ \mathbf{H}_{1,1}\ \cdots\ \mathbf{O}\ \mathbf{H}_{1,m-1} \\ \hline \text{diag}(\mathbf{H}_2, \mathbf{H}_2, \ldots, \mathbf{H}_2) \end{array} \right], \tag{9.19}$$

where the submatrix $\text{diag}(\mathbf{H}_2, \mathbf{H}_2, \ldots, \mathbf{H}_2)$ is an $m \times m$ diagonal array with $m$ copies of the parity-check $\mathbf{H}_2$ of the inner code $C_2$ lying on its main diagonal which is an $(mn_2 - n_1) \times mn_2$ matrix. If we want to decode the concatenated code $C^{(2)}_{1\times2,\text{conc}}$ in one step, we may use $\mathbf{H}^{(2)}_{1\times2,\text{conc}}$ for decoding.

### 9.3.3 Parallel Concatenation

The two types of concatenations presented in Sections 9.3.1 and 9.3.2 are in serial form; however, concatenation can be also achieved in *parallel*. One such setup is shown in Fig. 9.6 in which both codes, $C_1$ and $C_2$, have the same dimension, i.e., $k_1 = k_2 = k$. Commonly, both codes are binary and are the same code. Here, we use an $(n, k)$ binary code $C$ for both $C_1$ and $C_2$ in the following parallel concatenated coding system.

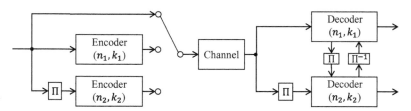

Figure 9.6 A parallel concatenated coding system.

An information sequence $\mathbf{u}$ of $k$ binary information symbols is encoded by two encoders independently using a *pseudorandom interleaver*. Let $\Pi$ denote the permutation induced by the interleaver. The interleaver permutes the information sequence $\mathbf{u}$ into another sequence $\mathbf{u}_{\text{int}} = \Pi(\mathbf{u})$ which consists of the same set of information symbols as $\mathbf{u}$. The information sequence $\mathbf{u}$ and its permuted version $\mathbf{u}_{\text{int}}$ are then encoded in parallel by the encoders of $C_1$ and $C_2$, independently, as shown in Fig. 9.6. Encoding generates two independent sets, $\mathbf{z}$ and $\mathbf{z}_{\text{int}}$, of parity-check symbols, one for $\mathbf{u}$ and the other for $\mathbf{u}_{\text{int}}$, respectively. Both $\mathbf{z}$ and $\mathbf{z}_{\text{int}}$ have the same number of parity-check symbols. Then, $\mathbf{u}$, $\mathbf{z}$, and $\mathbf{z}_{\text{int}}$ are multiplexed to form a code sequence $\mathbf{y} = (\mathbf{z}_{\text{int}}, \mathbf{z}, \mathbf{u})$ of $2n - k$ code symbols for transmission. Corresponding to the $2^k$ information sequences, there are $2^k$

code sequences in the form of $\mathbf{y} = (\mathbf{z}_{\text{int}}, \mathbf{z}, \mathbf{u})$. These $2^k$ code sequences form a $(2n - k, k)$ linear code, called a *turbo concatenated code* (or *parallel concatenated code*). Note that both $(\mathbf{z}, \mathbf{u})$ and $(\mathbf{z}_{\text{int}}, \mathbf{u}_{\text{int}})$ are codewords in $C$.

A turbo concatenated code is commonly decoded with soft-decision decoding iteratively, as shown in Fig. 9.6 where $\Pi^{-1}$ denotes the inverse of the permutation $\Pi$. Let $\mathbf{y} = (\mathbf{z}_{\text{int}}, \mathbf{z}, \mathbf{u})$ and $\mathbf{y}^* = (\mathbf{z}_{\text{int}}^*, \mathbf{z}^*, \mathbf{u}^*)$ be the transmitted and received code sequences, respectively. When $\mathbf{y}^*$ is received, the received information sequence $\mathbf{u}^*$ in $\mathbf{y}^*$ is permuted into an information sequence $\mathbf{u}_{\text{int}}^* = \Pi(\mathbf{u}^*)$. Then, $(\mathbf{z}^*, \mathbf{u}^*)$ and $(\mathbf{z}_{\text{int}}^*, \mathbf{u}_{\text{int}}^*)$ are decoded by two identical decoders in parallel. The decoded information of one decoder is used as the input to the other decoder in the next decoding iteration. Decoding iteration continues until a stopping criterion is met or a preset maximum number of iterations is reached. The above iterative decoding of a parallel concatenated code is referred to as *parallel turbo decoding*.

For more on turbo concatenated codes and their decoding, readers are referred to [9, 11, 12].

## 9.4 |**u**|**u** + **v**|-**Construction**

Let $C_1$ and $C_2$ be an $(n, k_1)$ linear block code and an $(n, k_2)$ linear block code over $\text{GF}(q)$ with minimum distances $d_1$ and $d_2$, respectively. Let $\mathbf{u} = (u_0, u_1, \ldots, u_{n-1})$ and $\mathbf{v} = (v_0, v_1, \ldots, v_{n-1})$ be a codeword in $C_1$ and a codeword in $C_2$, respectively. Form the following $2n$-tuple over $\text{GF}(q)$:

$$|\mathbf{u}|\mathbf{u} + \mathbf{v}| = (u_0, u_1, \ldots, u_{n-1}, u_0 + v_0, u_1 + v_1, \ldots, u_{n-1} + v_{n-1}). \quad (9.20)$$

The first half of the tuple $|\mathbf{u}|\mathbf{u} + \mathbf{v}|$ is the codeword $\mathbf{u}$ in $C_1$ and its second half is the vector-sum over $\text{GF}(q)$ of the codewords $\mathbf{u}$ and $\mathbf{v}$. All the combinations of codewords in $C_1$ and codewords in $C_2$ in the form of (9.20) give $q^{k_1 + k_2}$ $2n$-tuples in the form of (9.20) over $\text{GF}(q)$ which form a $(2n, k_1 + k_2)$ linear code, denoted by $C = |C_1|C_1 + C_2|$, over $\text{GF}(q)$, i.e.,

$$C = |C_1|C_1 + C_2| = \{|\mathbf{u}|\mathbf{u} + \mathbf{v}| : \mathbf{u} \in C_1, \mathbf{v} \in C_2\}. \quad (9.21)$$

The construction of $C = |C_1|C_1 + C_2|$ is called $|\mathbf{u}|\mathbf{u} + \mathbf{v}|$-*construction* [13, 14] with component codes $C_1$ and $C_2$. The minimum distance $d_{\min}$ of $C = |C_1|C_1 + C_2|$ (see Problem 9.12) is

$$d_{\min} \geq \min\{2d_1, d_2\}. \quad (9.22)$$

For $i = 1, 2$, let $\mathbf{G}_i$ and $\mathbf{H}_i$ be the generator and parity-check matrices of $C_i$, respectively. It follows from (9.21) that the generator and parity-check matrices of $C = |C_1|C_1 + C_2|$ are:

$$\mathbf{G} = \begin{bmatrix} \mathbf{G}_1 & \mathbf{G}_1 \\ \mathbf{O} & \mathbf{G}_2 \end{bmatrix}, \quad \mathbf{H} = \begin{bmatrix} \mathbf{H}_1 & \mathbf{O} \\ \mathbf{H}_2 & \mathbf{H}_2 \end{bmatrix}. \quad (9.23)$$

Decoding of $C = |C_1|C_1 + C_2|$ can be carried out based on the parity-check matrix $\mathbf{H}$ given by (9.23) in one stage or in two stages. Let $|\mathbf{u}|\mathbf{u} + \mathbf{v}| = (u_0, u_1, \ldots, u_{n-1}, u_0 + v_0, u_1 + v_1, \ldots, u_{n-1} + v_{n-1})$ and $|\mathbf{u}^*|\mathbf{u}^* + \mathbf{v}^*| = (u_0^*, u_1^*, \ldots, u_{n-1}^*, u_0^* + v_0^*, u_1^* + v_1^*, \ldots, u_{n-1}^* + v_{n-1}^*)$ be the transmitted and received sequences, respectively. In the two-stage decoding of the received vector, we first decode the first half $(u_0^*, u_1^*, \ldots, u_{n-1}^*)$ of $|\mathbf{u}^*|\mathbf{u}^* + \mathbf{v}^*|$ using $\mathbf{H}_1$. Let $(u_0^{**}, u_1^{**}, \ldots, u_{n-1}^{**})$ be the vector obtained by decoding $(u_0^*, u_1^*, \ldots, u_{n-1}^*)$. If $(u_0^{**}, u_1^{**}, \ldots, u_{n-1}^{**})$ is a codeword in $C_1$, we subtract $(u_0^{**}, u_1^{**}, \ldots, u_{n-1}^{**})$ from the second half $(u_0^* + v_0^*, u_1^* + v_1^*, \ldots, u_{n-1}^* + v_{n-1}^*)$ of $|\mathbf{u}^*|\mathbf{u}^* + \mathbf{v}^*|$. This results in a vector

$$(v_0^{**}, v_1^{**}, \ldots, v_{n-1}^{**}) = ((u_0^* + u_0^{**}) + v_0^*, (u_1^* + u_1^{**}) + v_1^*, \ldots, (u_{n-1}^* + u_{n-1}^{**}) + v_{n-1}^*).$$

If $(v_0^{**}, v_1^{**}, \ldots, v_{n-1}^{**})$ is a codeword in $C_2$, we stop the decoding process; otherwise, we decode $(v_0^{**}, v_1^{**}, \ldots, v_{n-1}^{**})$ using $\mathbf{H}_2$. If the decoded vector is a codeword in $C_2$, we stop the decoding process. Then, we retrieve the information symbols from the two successfully decoded codewords and deliver them to the user(s). With the above two-stage decoding, if the first stage of decoding fails, we stop the entire decoding process and declare a decoding failure.

The two-stage decoding of $C = |C_1|C_1 + C_2|$ reduces decoding complexity but may result in some loss of error performance of the code. The decoding of $C_1$ and $C_2$ can be both hard-decision, both soft-decision, or one hard-decision and the other soft-decision.

**Example 9.3** Let $\alpha$ be a primitive element of the field $\mathrm{GF}(2^5)$. Let $C_1$ be the $(31, 25)$ BCH code whose generator polynomial has $1$, $\alpha$, and their conjugates as roots and let $C_2$ be the $(31, 15)$ BCH code whose generator polynomial has $1$, $\alpha$, $\alpha^3$, $\alpha^5$, and their conjugates as roots. The minimum distances of $C_1$ and $C_2$ are $d_1 = 4$ and $d_2 = 8$, respectively. Using $C_1$ and $C_2$ as component codes, the $|\mathbf{u}|\mathbf{u} + \mathbf{v}|$-construction gives a $(62, 40)$ linear code with minimum distance $d_{\min} = 8$. ▲▲

## 9.5 Kronecker Product

In this section, we will show that an RM code can be constructed using shorter RM codes by a technique called the *Kronecker product*.

**Definition 9.1** Let $\mathbf{A} = [a_{i,j}]_{0 \leq i < m, 0 \leq j < n}$ and $\mathbf{B} = [b_{i,j}]_{0 \leq i < s, 0 \leq j < t}$ be an $m \times n$ matrix and an $s \times t$ matrix over $\mathrm{GF}(q)$, respectively. Define the following $ms \times nt$ matrix:

$$\mathbf{A} \otimes \mathbf{B} = [a_{i,j}\mathbf{B}]_{0 \leq i < m, 0 \leq j < n} = \begin{bmatrix} a_{0,0}\mathbf{B} & a_{0,1}\mathbf{B} & \cdots & a_{0,n-1}\mathbf{B} \\ a_{1,0}\mathbf{B} & a_{1,1}\mathbf{B} & \cdots & a_{1,n-1}\mathbf{B} \\ \vdots & \vdots & \vdots & \vdots \\ a_{m-1,0}\mathbf{B} & a_{m-1,1}\mathbf{B} & \cdots & a_{m-1,n-1}\mathbf{B} \end{bmatrix}. \quad (9.24)$$

The matrix $\mathbf{A} \otimes \mathbf{B}$ is called the *Kronecker* (or *tensor*) *product* of matrices of $\mathbf{A}$ and $\mathbf{B}$ which is simply obtained by replacing the entry $a_{i,j}$ in $\mathbf{A}$ with the matrix $a_{i,j}\mathbf{B}$, $0 \le i < m, 0 \le j < n$. If $a_{i,j} = 0$, $a_{i,j}\mathbf{B}$ is an $s \times t$ zero matrix (ZM); otherwise, $a_{i,j}\mathbf{B}$ is an $s \times t$ matrix obtained by multiplying each entry in $\mathbf{B}$ by $a_{i,j}$. The matrix $\mathbf{A} \otimes \mathbf{B}$ is an $m \times n$ array containing copies of the matrix $\mathbf{B}$.

In the following, we present a special class of Kronecker product matrices which are used for constructing codes, such RM codes and polar codes [15] (to be presented in Chapter 17). Let

$$\mathbf{G}_{2\times 2} = \begin{bmatrix} 1 & 1 \\ 0 & 1 \end{bmatrix} \tag{9.25}$$

be a $2 \times 2$ matrix over GF(2). The Kronecker product of $\mathbf{G}_{2\times 2}$ with itself is a $2 \times 2$ array as follows:

$$\mathbf{G}_{2^2 \times 2^2} = \mathbf{G}_{2\times 2} \otimes \mathbf{G}_{2\times 2} = \begin{bmatrix} \mathbf{G}_{2\times 2} & \mathbf{G}_{2\times 2} \\ \mathbf{O} & \mathbf{G}_{2\times 2} \end{bmatrix}. \tag{9.26}$$

The array $\mathbf{G}_{2^2 \times 2^2}$ contains three copies of the matrix $\mathbf{G}_{2\times 2}$ lying on and above its main diagonal. The array $\mathbf{G}_{2^2 \times 2^2}$ is called the *2-fold Kronecker product* of $\mathbf{G}_{2\times 2}$. It is a $4 \times 4$ matrix over GF(2) as follows:

$$\mathbf{G}_{2^2 \times 2^2} = \begin{bmatrix} 1 & 1 & 1 & 1 \\ 0 & 1 & 0 & 1 \\ 0 & 0 & 1 & 1 \\ 0 & 0 & 0 & 1 \end{bmatrix}. \tag{9.27}$$

From (9.27), we see that all the 1-entries of $\mathbf{G}_{2^2 \times 2^2}$ lie on and above its main diagonal and $\mathbf{G}_{2^2 \times 2^2}$ has three different row weights: $\binom{2}{0} = 1$ row with weight $2^0 = 1$, $\binom{2}{1} = 2$ rows with weight $2^1 = 2$, and $\binom{2}{2} = 1$ row with weight $2^2 = 4$.

The 3-fold Kronecker product of $\mathbf{G}_{2\times 2}$ is defined as the Kronecker product of $\mathbf{G}_{2\times 2}$ and $\mathbf{G}_{2^2 \times 2^2}$ given below:

$$\mathbf{G}_{2^3 \times 2^3} = \mathbf{G}_{2\times 2} \otimes \mathbf{G}_{2^2 \times 2^2}$$

$$= \begin{bmatrix} \mathbf{G}_{2^2 \times 2^2} & \mathbf{G}_{2^2 \times 2^2} \\ \mathbf{O} & \mathbf{G}_{2^2 \times 2^2} \end{bmatrix}$$

$$= \begin{bmatrix} \mathbf{G}_{2\times 2} & \mathbf{G}_{2\times 2} & \mathbf{G}_{2\times 2} & \mathbf{G}_{2\times 2} \\ \mathbf{O} & \mathbf{G}_{2\times 2} & \mathbf{O} & \mathbf{G}_{2\times 2} \\ \mathbf{O} & \mathbf{O} & \mathbf{G}_{2\times 2} & \mathbf{G}_{2\times 2} \\ \mathbf{O} & \mathbf{O} & \mathbf{O} & \mathbf{G}_{2\times 2} \end{bmatrix}. \tag{9.28}$$

From (9.28), we see that $\mathbf{G}_{2^3 \times 2^3}$ is a $2^2 \times 2^2$ array with $3^2 = 9$ copies of $\mathbf{G}_{2\times 2}$ lying on and above its main diagonal and $2 \times 2$ ZMs elsewhere. The array $\mathbf{G}_{2^3 \times 2^3}$ is an $8 \times 8$ matrix as follows:

$$\mathbf{G}_{2^3 \times 2^3} = \begin{bmatrix} 1 & 1 & 1 & 1 & 1 & 1 & 1 & 1 \\ 0 & 1 & 0 & 1 & 0 & 1 & 0 & 1 \\ 0 & 0 & 1 & 1 & 0 & 0 & 1 & 1 \\ 0 & 0 & 0 & 1 & 0 & 0 & 0 & 1 \\ 0 & 0 & 0 & 0 & 1 & 1 & 1 & 1 \\ 0 & 0 & 0 & 0 & 0 & 1 & 0 & 1 \\ 0 & 0 & 0 & 0 & 0 & 0 & 1 & 1 \\ 0 & 0 & 0 & 0 & 0 & 0 & 0 & 1 \end{bmatrix}. \tag{9.29}$$

From (9.29), we see that all the 1-entries in $\mathbf{G}_{2^3 \times 2^3}$ lie on and above its main diagonal and $\mathbf{G}_{2^3 \times 2^3}$ has four different row weights, $\binom{3}{0} = 1$ row with weight $2^0 = 1$, $\binom{3}{1} = 3$ rows with weight $2^1 = 2$, $\binom{3}{2} = 3$ rows with weight $2^2 = 4$, and $\binom{3}{3} = 1$ row with weight $2^3 = 8$.

The 4-fold Kronecker product of $\mathbf{G}_{2 \times 2}$ is a $16 \times 16$ matrix given below:

$$\mathbf{G}_{2^4 \times 2^4} = \begin{bmatrix} 1&1&1&1&1&1&1&1&1&1&1&1&1&1&1&1 \\ 0&1&0&1&0&1&0&1&0&1&0&1&0&1&0&1 \\ 0&0&1&1&0&0&1&1&0&0&1&1&0&0&1&1 \\ 0&0&0&1&0&0&0&1&0&0&0&1&0&0&0&1 \\ 0&0&0&0&1&1&1&1&0&0&0&0&1&1&1&1 \\ 0&0&0&0&0&1&0&1&0&0&0&0&0&1&0&1 \\ 0&0&0&0&0&0&1&1&0&0&0&0&0&0&1&1 \\ 0&0&0&0&0&0&0&1&0&0&0&0&0&0&0&1 \\ 0&0&0&0&0&0&0&0&1&1&1&1&1&1&1&1 \\ 0&0&0&0&0&0&0&0&0&1&0&1&0&1&0&1 \\ 0&0&0&0&0&0&0&0&0&0&1&1&0&0&1&1 \\ 0&0&0&0&0&0&0&0&0&0&0&1&0&0&0&1 \\ 0&0&0&0&0&0&0&0&0&0&0&0&1&1&1&1 \\ 0&0&0&0&0&0&0&0&0&0&0&0&0&1&0&1 \\ 0&0&0&0&0&0&0&0&0&0&0&0&0&0&1&1 \\ 0&0&0&0&0&0&0&0&0&0&0&0&0&0&0&1 \end{bmatrix}, \tag{9.30}$$

which has five different row weights: $\binom{4}{0} = 1$ rows with weight $2^0 = 1$, $\binom{4}{1} = 4$ rows with weight $2^1 = 2$, $\binom{4}{2} = 6$ rows with weight $2^2 = 4$, $\binom{4}{3} = 4$ rows with weight $2^3 = 8$, and $\binom{4}{4} = 1$ rows with weight $2^4 = 16$.

Let $m$ be a positive integer. The $m$-fold Kronecker product $\mathbf{G}_{2^m \times 2^m}$ of $\mathbf{G}_{2 \times 2}$ is defined as the Kronecker product of $\mathbf{G}_{2 \times 2}$ and the $(m-1)$-fold Kronecker product, $\mathbf{G}_{2^{m-1} \times 2^{m-1}}$, of $\mathbf{G}_{2 \times 2}$ as follows:

$$\mathbf{G}_{2^m \times 2^m} = \mathbf{G}_{2 \times 2} \otimes \mathbf{G}_{2^{m-1} \times 2^{m-1}} = \begin{bmatrix} \mathbf{G}_{2^{m-1} \times 2^{m-1}} & \mathbf{G}_{2^{m-1} \times 2^{m-1}} \\ \mathbf{O} & \mathbf{G}_{2^{m-1} \times 2^{m-1}} \end{bmatrix}. \tag{9.31}$$

It follows from the recursive construction of multifold Kronecker products of $\mathbf{G}_{2 \times 2}$ that $\mathbf{G}_{2^m \times 2^m}$ is a $2^{m-1} \times 2^{m-1}$ array with $3^{m-1}$ copies of $\mathbf{G}_{2 \times 2}$ lying on and above its main diagonal. The array $\mathbf{G}_{2^m \times 2^m}$ is a $2^m \times 2^m$ square matrix with all the 1-entries lying on and above its main diagonal. Hence, as a matrix, $\mathbf{G}_{2^m \times 2^m}$ consists of an *upper triangular submatrix* which contains all the 1-entries of $\mathbf{G}_{2^m \times 2^m}$ and a lower triangular submatrix which contains only 0-entries. The matrix $\mathbf{G}_{2^m \times 2^m}$ has $m + 1$ different row weights which are $2^0 = 1$, $2^1 = 2$, $2^2 = 4, \ldots, 2^{m-1}$, and $2^m$. The number of rows in $\mathbf{G}_{2^m \times 2^m}$ with weight $2^l$, $0 \le l \le m$, is $\binom{m}{l}$ (binomial coefficient). Except for one row (the bottom row of $\mathbf{G}_{2^m \times 2^m}$), every row of $\mathbf{G}_{2^m \times 2^m}$ has even weight. The entries of the top row and the last (rightmost) column of $\mathbf{G}_{2^m \times 2^m}$ are all 1-entries. The rows of $\mathbf{G}_{2^m \times 2^m}$ are *linearly independent* and they span the vector space of all the $2^{2^m}$ $2^m$-tuples over GF(2).

Suppose we set $\mathbf{G}_{2 \times 2}$ as the following $2 \times 2$ matrix:

$$\mathbf{G}^*_{2 \times 2} = \begin{bmatrix} 1 & 0 \\ 1 & 1 \end{bmatrix}. \tag{9.32}$$

Then, the $m$-fold Kronecker product $\mathbf{G}^*_{2^m \times 2^m}$ of $\mathbf{G}^*_{2 \times 2}$ is a $2^{m-1} \times 2^{m-1}$ array with $3^{m-1}$ copies of $\mathbf{G}^*_{2 \times 2}$ lying on and below its main diagonal. The array $\mathbf{G}^*_{2^m \times 2^m}$ is a $2^m \times 2^m$ square matrix with all the 1-entries lying on and below its main diagonal. In this case, $\mathbf{G}^*_{2^m \times 2^m}$ consists of a *lower triangular submatrix* which contains all the 1-entries of $\mathbf{G}^*_{2^m \times 2^m}$ and an *upper triangular submatrix* which contains only 0-entries. All the entries in the leftmost column and the bottom row of $\mathbf{G}^*_{2^m \times 2^m}$ are all 1-entries. For example, for $m = 4$, the 4-fold Kronecker product of $\mathbf{G}^*_{2 \times 2}$ is the following $16 \times 16$ matrix:

$$
\mathbf{G}^*_{2^4 \times 2^4} =
\begin{bmatrix}
1&0&0&0&0&0&0&0&0&0&0&0&0&0&0&0\\
1&1&0&0&0&0&0&0&0&0&0&0&0&0&0&0\\
1&0&1&0&0&0&0&0&0&0&0&0&0&0&0&0\\
1&1&1&1&0&0&0&0&0&0&0&0&0&0&0&0\\
1&0&0&0&1&0&0&0&0&0&0&0&0&0&0&0\\
1&1&0&0&1&1&0&0&0&0&0&0&0&0&0&0\\
1&0&1&0&1&0&1&0&0&0&0&0&0&0&0&0\\
1&1&1&1&1&1&1&1&0&0&0&0&0&0&0&0\\
1&0&0&0&0&0&0&0&1&0&0&0&0&0&0&0\\
1&1&0&0&0&0&0&0&1&1&0&0&0&0&0&0\\
1&0&1&0&0&0&0&0&1&0&1&0&0&0&0&0\\
1&1&1&1&0&0&0&0&1&1&1&1&0&0&0&0\\
1&0&0&0&1&0&0&0&1&0&0&0&1&0&0&0\\
1&1&0&0&1&1&0&0&1&1&0&0&1&1&0&0\\
1&0&1&0&1&0&1&0&1&0&1&0&1&0&1&0\\
1&1&1&1&1&1&1&1&1&1&1&1&1&1&1&1
\end{bmatrix}. \tag{9.33}
$$

We see that the 4-fold Kronecker product $\mathbf{G}^*_{2^4 \times 2^4}$ of the $2 \times 2$ matrix $\mathbf{G}^*_{2 \times 2}$ given by (9.32) has a lower triangular submatrix which contains the 1-entries of $\mathbf{G}^*_{2^4 \times 2^4}$ and an upper triangular submatrix which contains only 0-entries.

The RM codes presented in Chapter 8 can be constructed from multifold Kronecker products of the $2 \times 2$ matrix $\mathbf{G}_{2 \times 2}$ given in the form of (9.25) (or $\mathbf{G}^*_{2 \times 2}$ in the form of (9.32)). For $0 \leq r < m$, the generator matrix $\mathbf{G}_{\mathrm{RM}}(r, m)$ of the $r$th order RM code $C_{\mathrm{RM}}(r, m)$ of length $2^m$ simply consists of the rows of the $m$-fold Kronecker product $\mathbf{G}_{2^m \times 2^m}$ of $\mathbf{G}_{2 \times 2}$ (or $\mathbf{G}^*_{2^m \times 2^m}$ of $\mathbf{G}^*_{2 \times 2}$) with weights $2^{m-r}, 2^{m-r+1}, \ldots, 2^m$. Permuting these rows, we can put $\mathbf{G}_{\mathrm{RM}}(r, m)$ in the normal form given by (8.7). For example, consider the 4-fold Kronecker product of the $2 \times 2$ matrix $\mathbf{G}_{2 \times 2}$ in the form of (9.25). If we take the 11 rows of weights 4, 8, and 16 from $\mathbf{G}_{2^4 \times 2^4}$ given by (9.30), we obtain the following $11 \times 16$ matrix:

$$
\begin{bmatrix}
1&1&1&1&1&1&1&1&1&1&1&1&1&1&1&1\\
0&1&0&1&0&1&0&1&0&1&0&1&0&1&0&1\\
0&0&1&1&0&0&1&1&0&0&1&1&0&0&1&1\\
0&0&0&1&0&0&0&1&0&0&0&1&0&0&0&1\\
0&0&0&0&1&1&1&1&0&0&0&0&1&1&1&1\\
0&0&0&0&0&1&0&1&0&0&0&0&0&1&0&1\\
0&0&0&0&0&0&1&1&0&0&0&0&0&0&1&1\\
0&0&0&0&0&0&0&0&1&1&1&1&1&1&1&1\\
0&0&0&0&0&0&0&0&0&1&0&1&0&1&0&1\\
0&0&0&0&0&0&0&0&0&0&1&1&0&0&1&1\\
0&0&0&0&0&0&0&0&0&0&0&0&0&1&1&1
\end{bmatrix}.
$$

Permuting the rows of the above matrix, we obtain the following $11 \times 16$ matrix:

$$
\mathbf{G}_{\mathrm{RM}}(2,4) = 
\begin{bmatrix}
1\,1\,1\,1\,1\,1\,1\,1\,1\,1\,1\,1\,1\,1\,1\,1\\
0\,0\,0\,0\,0\,0\,0\,0\,1\,1\,1\,1\,1\,1\,1\,1\\
0\,0\,0\,0\,1\,1\,1\,1\,0\,0\,0\,0\,1\,1\,1\,1\\
0\,0\,1\,1\,0\,0\,1\,1\,0\,0\,1\,1\,0\,0\,1\,1\\
0\,1\,0\,1\,0\,1\,0\,1\,0\,1\,0\,1\,0\,1\,0\,1\\
0\,0\,0\,0\,0\,0\,0\,0\,0\,0\,0\,0\,1\,1\,1\,1\\
0\,0\,0\,0\,0\,0\,0\,0\,0\,0\,1\,1\,0\,0\,1\,1\\
0\,0\,0\,0\,0\,0\,0\,0\,1\,0\,1\,0\,1\,0\,1\\
0\,0\,0\,0\,0\,0\,1\,1\,0\,0\,0\,0\,0\,0\,1\,1\\
0\,0\,0\,0\,0\,1\,0\,1\,0\,0\,0\,0\,0\,1\,0\,1\\
0\,0\,0\,1\,0\,0\,0\,1\,0\,0\,0\,1\,0\,0\,0\,1
\end{bmatrix},
\tag{9.34}
$$

which is the generator matrix of the $(16, 11)$ second-order RM code $C_{\mathrm{RM}}(2,4)$ in the normal form given in Example 8.2.

From (9.31), we readily see that the generator matrix $\mathbf{G}_{\mathrm{RM}}(r, m)$ of the $r$th-order RM code $C_{\mathrm{RM}}(r, m)$ of length $2^m$ can be expressed in terms of the generator matrices $\mathbf{G}_{\mathrm{RM}}(r, m-1)$ and $\mathbf{G}_{\mathrm{RM}}(r-1, m-1)$ of the $r$th order and the $(r-1)$th-order RM codes, $C_{\mathrm{RM}}(r, m-1)$ and $C_{\mathrm{RM}}(r-1, m-1)$, of length $2^{m-1}$ in the $|\mathbf{u}|\mathbf{u} + \mathbf{v}|$-construction as follows:

$$
\mathbf{G}_{\mathrm{RM}}(r, m) = 
\begin{bmatrix}
\mathbf{G}_{\mathrm{RM}}(r, m-1) & \mathbf{G}_{\mathrm{RM}}(r, m-1)\\
\mathbf{O} & \mathbf{G}_{\mathrm{RM}}(r-1, m-1)
\end{bmatrix}.
\tag{9.35}
$$

Hence, the RM code $C_{\mathrm{RM}}(r, m)$ can be constructed from the RM codes, $C_{\mathrm{RM}}(r, m-1)$ and $C_{\mathrm{RM}}(r-1, m-1)$, using the $|\mathbf{u}|\mathbf{u} + \mathbf{v}|$-construction. In this case, we can decode the $r$th-order RM code $C_{\mathrm{RM}}(r, m)$ of length $2^m$ in two stages (or two layers) as presented in Section 9.4. In the first layer, we decode the RM code $C_{\mathrm{RM}}(r, m-1)$ of length $2^{m-1}$ and then decode the RM code $C_{\mathrm{RM}}(r-1, m-1)$ of length $2^{m-1}$ in the second layer.

The parity-check matrix of the RM code $C_{\mathrm{RM}}(r, m)$ corresponding to its generator matrix in the form of (9.35) is given below:

$$
\mathbf{G}_{\mathrm{RM}}(m - r - 1, m) = 
\begin{bmatrix}
\mathbf{G}_{\mathrm{RM}}(m - r - 1, m-1) & \mathbf{G}_{\mathrm{RM}}(m - r - 1, m-1)\\
\mathbf{O} & \mathbf{G}_{\mathrm{RM}}(m - r - 2, m-1)
\end{bmatrix},
\tag{9.36}
$$

which is the generator matrix of the $(m - r - 1)$th-order RM code $C_{\mathrm{RM}}(m - r - 1, m)$.

From (9.35), we see that an RM code can be constructed *recursively* from shorter RM codes using the $|\mathbf{u}|\mathbf{u} + \mathbf{v}|$-construction. From a different point of view, Equation (9.35) demonstrates that an RM code can be decomposed into a chain of RM codes of shorter lengths and smaller dimensions. Several multilayer decoding schemes (or algorithms) for an RM code based on its multilayer structure have been devised [11, 16–19].

**Example 9.4** Consider the generator matrix $\mathbf{G}_{\mathrm{RM}}(2,4)$ of the RM code $C_{\mathrm{RM}}(2,4)$ in Kronecker-product form given by (9.34). It follows from (9.35) that $\mathbf{G}_{\mathrm{RM}}(2,4)$ can be put in $|\mathbf{u}|\mathbf{u}+\mathbf{v}|$-construction form as follows:

$$\mathbf{G}_{\mathrm{RM}}(2,4) = \begin{bmatrix} 1 & 1 & 1 & 1 & 1 & 1 & 1 & 1 & 1 & 1 & 1 & 1 & 1 & 1 & 1 & 1 \\ 0 & 1 & 0 & 1 & 0 & 1 & 0 & 1 & 0 & 1 & 0 & 1 & 0 & 1 & 0 & 1 \\ 0 & 0 & 1 & 1 & 0 & 0 & 1 & 1 & 0 & 0 & 1 & 1 & 0 & 0 & 1 & 1 \\ 0 & 0 & 0 & 1 & 0 & 0 & 0 & 1 & 0 & 0 & 0 & 1 & 0 & 0 & 0 & 1 \\ 0 & 0 & 0 & 0 & 1 & 1 & 1 & 1 & 0 & 0 & 0 & 0 & 1 & 1 & 1 & 1 \\ 0 & 0 & 0 & 0 & 0 & 1 & 0 & 1 & 0 & 0 & 0 & 0 & 0 & 1 & 0 & 1 \\ 0 & 0 & 0 & 0 & 0 & 0 & 1 & 1 & 0 & 0 & 0 & 0 & 0 & 0 & 1 & 1 \\ 0 & 0 & 0 & 0 & 0 & 0 & 0 & 0 & 1 & 1 & 1 & 1 & 1 & 1 & 1 & 1 \\ 0 & 0 & 0 & 0 & 0 & 0 & 0 & 0 & 0 & 1 & 0 & 1 & 0 & 1 & 0 & 1 \\ 0 & 0 & 0 & 0 & 0 & 0 & 0 & 0 & 0 & 0 & 1 & 1 & 0 & 0 & 1 & 1 \\ 0 & 0 & 0 & 0 & 0 & 0 & 0 & 0 & 0 & 0 & 0 & 0 & 1 & 1 & 1 & 1 \end{bmatrix}.$$

The top seven rows of $\mathbf{G}_{\mathrm{RM}}(2,4)$ are two copies of the generator matrix $\mathbf{G}_{\mathrm{RM}}(2,3)$ of the $(8,7)$ second-order RM code $C_{\mathrm{RM}}(2,3)$. The eight rightmost columns of the last 4 rows of $\mathbf{G}_{\mathrm{RM}}(2,4)$ form the generator matrix $\mathbf{G}_{\mathrm{RM}}(1,3)$ of the first-order $(8,4)$ RM code $C_{\mathrm{RM}}(1,3)$. Hence, $\mathbf{G}_{\mathrm{RM}}(2,4)$ is in the $|\mathbf{u}|\mathbf{u}+\mathbf{v}|$-construction form given by (9.23). ▲▲

Multifold Kronecker products of the $2 \times 2$ matrix $\mathbf{G}^*_{2\times2}$ given in the form of (9.32) are used to construct polar codes [15] which will be presented in Chapter 17. In polar-code construction, the matrix $\mathbf{G}^*_{2\times2}$ is referred to as the *kernel*.

From geometric point of view, the rows of the $m$-fold Kronecker matrix $\mathbf{G}_{2^m\times2^m}$ (or $\mathbf{G}^*_{2^m\times2^m}$) are simply the incidence vectors of the 0-flats, 1-flats, 2-flats, ..., $(m-1)$-flats, and $m$-flat in the $m$-dimensional Euclidean geometry $\mathrm{EG}(m,2)$ that contains a fixed point in $\mathrm{EG}(m,2)$. Hence, $\mathbf{G}_{2^m\times2^m}$ (or $\mathbf{G}^*_{2^m\times2^m}$) specifies $\mathrm{EG}(m,2)$.

## 9.6 Automatic-Repeat-Request Schemes

In this section, some basic automatic-repeat-request (ARQ) and hybrid ARQ (HARQ) schemes will be presented. For more on ARQ and HARQ, readers are referred to References [11, 20–39].

### 9.6.1 Basic ARQ Schemes

Based on retransmission strategies, there are three basic types of ARQ schemes: *stop-and-wait* (SW) ARQ, *go-back-N* (GBN) ARQ, and *selective-repeat* (ST) ARQ [11, 20–24]. In an SW-ARQ system, the transmitter sends a codeword to the receiver and waits for an acknowledgment, as shown in Fig. 9.7. A positive acknowledgment (ACK) from the receiver indicates that the transmitted codeword has been successfully received, and the transmitter can send the next codeword. A negative acknowledgment (NAK) from the receiver indicates that the transmitted codeword has been detected in error, and the transmitter then resends the codeword and again waits for acknowledgment. Retransmissions continue until the transmitter receives an ACK. This ARQ scheme is simple but inherently inefficient because of the time waiting for an acknowledgment of each transmitted codeword.

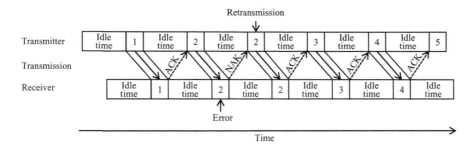

Figure 9.7 Stop-and-wait ARQ.

In a GBN-ARQ system, the transmitter continuously transmits codewords in order and stores them pending receipt of an ACK/NAK for each, as shown in Fig. 9.8. The acknowledgment for a codeword arrives after a *round-trip delay*, which is the time interval between the transmission of a codeword and the receipt of an acknowledgment for that codeword. During this interval, $N - 1$ successive codewords are also transmitted. Whenever the transmitter receives an NAK indicating that a specific codeword, say codeword $l$, was received in error, it stops transmitting new codewords. Then, it goes back to codeword $l$ and proceeds to retransmit the codeword $l$ and the $N - 1$ succeeding codewords which were transmitted during the one round-trip delay. At the receiving end, the receiver discards the erroneously received codeword $l$ and all $N - 1$ subsequently received codewords regardless of whether they are error free or not. Retransmission continues until the codeword $l$ is positively acknowledged. In each retransmission for the codeword $l$, the transmitter resends the same sequence of $N$ codewords. As soon as the codeword $l$ is positively acknowledged, the transmitter proceeds to transmit new codewords. The main drawback of the GBN-ARQ scheme is that, whenever a received vector is detected in error, the receiver also rejects the next $N - 1$ received vectors, even though many of them (or all of them) may be error free. As a result, they must be retransmitted. This represents a waste of transmission time, which can result in a severe deterioration of the throughput performance if the round-trip delay is large. The GBN-ARQ scheme is not effective for communication systems with high data rates and large round-trip delays.

In an SR-ARQ system, codewords are also transmitted continuously. However, the transmitter only resends those codewords which are negatively acknowledged. After resending the negatively acknowledged codewords, the transmitter continues transmitting new codewords in the transmitter buffer, as shown in Fig. 9.9. If the codewords must be delivered to the user(s) in a correct

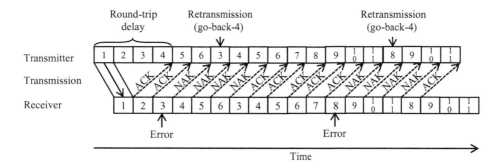

Figure 9.8 Go-back-$N$ ARQ with $N = 4$.

order, a buffer is also needed at the receiver to store the error-free codewords following a received vector detected in error. The SR-ARQ scheme overcomes the weaknesses of both the SW-ARQ and GBN-ARQ schemes, but at the expense of higher receiver complexity. In high-speed communication systems, an SR-ARQ scheme is needed.

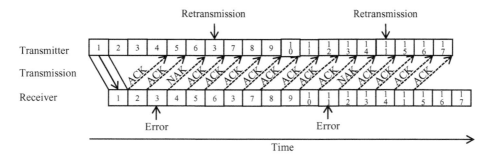

Figure 9.9 Selective-repeat ARQ.

The performance of an ARQ error-control system is normally measured in terms of its *reliability* and *throughput efficiency*. In an ARQ system, the receiver commits a decoding error whenever it accepts a received vector with undetected errors. Such an event is called an *error event*. Let $P(E)$ denote the probability of an error event. Clearly, for an ARQ system to be reliable, $P(E)$ should be very small. The throughput efficiency of an ARQ system is defined as the ratio of the average number of information bits successfully accepted by the receiver per unit time to the total number of bits that could be transmitted per unit time [21, 40]. All three basic ARQ schemes using the same code for error detection achieve the same system reliability, but they provide different throughput efficiencies.

Suppose an $(n, k)$ linear code $C$ is used for error detection in an ARQ system. Define the following three probabilities.

$P_c$: probability that a received vector contains no error;
$P_d$: probability that a received vector contains a detectable error pattern; and

$P_e$: probability that a received vector contains an undetectable error pattern.

The sum of these three probabilities is equal to 1, i.e., $P_c + P_d + P_e = 1$. The probability $P_c$ depends on the channel error statistics, and the probabilities $P_d$ and $P_e$ depend on both the channel statistics and the error-detecting capability of the chosen $(n, k)$ code $C$. The probability $P_e$ is normally called the *probability of undetected error* of the code $C$.

A received vector is accepted by the receiver only if it either contains no errors or contains an undetectable error pattern. The probability that a received vector will be accepted by the receiver is

$$P = P_e + P_c. \tag{9.37}$$

If the received vector contains a detectable error pattern, it is not accepted by the receiver and a retransmission is requested. Because an undetectable error pattern can occur in the initial transmission of a codeword or in any subsequent retransmission, the error event probability $P(E)$ that the received vector in an ARQ system is accepted and a decoding error is commit is given by

$$\begin{aligned} P(E) &= P_e + P_d P_e + P_d^2 P_e + P_d^3 P_e + \cdots \\ &= P_e(1 + P_d + P_d^2 + P_d^3 + \cdots) \\ &= P_e/(1 - P_d) = P_e/(P_c + P_e). \end{aligned} \tag{9.38}$$

If the code $C$ is properly chosen, $P_e$ can be very small relative to $P_c$ ($P_e \ll P_c$), and hence $P(E)$ can be very small and $P \approx P_c$. For a BSC with transition probability $p$ (bit-error rate), we have

$$P_c = (1 - p)^n. \tag{9.39}$$

It has been proved that linear block codes for the BSC exist with the probability $P_e$ of undetected error satisfying the following upper bound [11, 41, 42]:

$$P_e \leq (1 - (1 - p)^n)2^{-(n-k)}. \tag{9.40}$$

Well-known codes that satisfy this bound are cyclic Hamming codes, double-error-correcting primitive BCH codes, and triple-error-correcting BCH codes of length $2^m - 1$ with $m$ odd and $m \geq 5$ [11, 42]. If a code satisfying the bound given by (9.40) is used for error detection and if the number of parity-check bits, $n - k$, is sufficiently large, $P_e$ can be very small relative to $P_c$ and hence $P(E) \ll 1$. For example, let $C$ be the $(2047, 2014)$ triple-error-correcting primitive BCH code of length $2^{11} - 1$ ($m = 11$) and rate 0.935. This code satisfies the bound given by (9.40). Suppose this code is used for error detection in an ARQ system. Let $p = 10^{-3}$, a very high bit-error rate. Then, $P_c \approx 1.25 \times 10^{-1}$ and $P_e \leq 10^{-10}$. From (9.38), we have $P(E) \leq 8 \times 10^{-10}$. From this example, we see that high system reliability can be achieved by an ARQ scheme using very little parity overhead.

Even though the three basic ARQ schemes presented above have the same reliability for information transmission, they have quite different throughput efficiencies. For simplicity, we assume that the forward channel is a random-error channel with bit-error rate $p$ and the feedback (or return) channel is noiseless. First, we consider the SW-ARQ scheme. Let $\tau$ be the idle time of the transmitter between two successive transmissions. Let $\rho$ be the bit rate of the transmitter. Even though the transmitter does not transmit anything during the idle period, the effect of the idle period on the throughput must be taken into consideration. In one round-trip delay time, the transmitter could transmit $n + \rho\tau$ bits if it does not remain idle. For a codeword to be received correctly, the average number of bits that the transmitter could have transmitted is given by

$$T_{\text{SW}} = \frac{n + \rho\tau}{P} \tag{9.41}$$

(see Problem 9.15). Therefore, the throughput efficiency of an SW-ARQ scheme is given by

$$\xi_{\text{SW}} = \frac{k}{T_{\text{SW}}} = \frac{P}{1 + \rho\tau/n} \cdot \frac{k}{n}, \tag{9.42}$$

where $k/n$ is the rate of the code used by the system. For communication systems with high data rates and large round-trip delays, $\rho\tau/n$ becomes very large unless the length $n$ of the code is large. In this case, the throughput performance of the SW-ARQ scheme becomes unacceptable.

In a GBN-ARQ system, retransmission of a negatively acknowledged codeword always involves resending $N$ codewords. Consequently, for a codeword to be successfully received, the average number of transmissions is given by

$$T_{\text{GBN}} = 1 + N(1 - P)/P \tag{9.43}$$

and the throughput of a GBN-ARQ system is given by

$$\xi_{\text{GBN}} = \frac{1}{T_{\text{GBN}}} \cdot \frac{k}{n} = \frac{P}{P + (1 - P)N} \cdot \frac{k}{n}. \tag{9.44}$$

From (9.44), we see that the throughput of a GBN-ARQ scheme depends on both the channel error rate $1 - P$ and the round-trip delay ($N$). The term $(1 - P)N$ represents both the effect of the channel error rate and the round-trip delay. For communication systems where the channel error rate is low, the effect of the round-trip delay, $(1 - P)N$ is insignificant and then the GBN-ARQ scheme provides a high-throughput performance. However, for a communication system with a high channel error rate, $(1 - P)N$ becomes significantly large, and this would make the throughput performance of the GBN-ARQ scheme inadequate.

Now, we consider an SR-ARQ system in which the receiver has an infinite buffer to store the error-free codewords when a received vector is detected in error. We call such a system *an ideal SR-ARQ* system. For a codeword to be successfully accepted by the receiver, the average number of transmissions needed is given by

$$T_{\text{SR}} = 1/P. \tag{9.45}$$

Hence, the throughput of an ideal SR-ARQ system is given by

$$\xi_{SR} = \frac{1}{T_{SR}} \cdot \frac{k}{n} = P \cdot \frac{k}{n}. \qquad (9.46)$$

We see that the throughput does not depend on the round-trip delay. As a result, the SR-ARQ scheme offers the best throughput performance among the three basic ARQ schemes.

The high-throughput performance of an ideal SR-ARQ is achieved at the expense of extensive buffering (theoretically infinite buffering) and more complex logic at both the transmitter and the receiver. If a finite buffer is used at the receiver (as in the case for practical systems), *buffer overflow* may occur which degrades the throughput performance of the system [11]. However, if a sufficiently long buffer is used and if buffer overflow is properly handled, even with a reduction in throughput, the SR-ARQ scheme still significantly outperforms the other two ARQ schemes, especially for systems where the channel error rate is high and the round-trip delay is large.

### 9.6.2 Mixed-Mode SR-ARQ Schemes

The buffer overflow in an SR-ARQ system can be prevented by using a *mixed-mode retransmission strategy*. In the following, we present three mixed-mode SR-ARQ schemes. The first one is the SR-ARQ used in conjunction with the GBN retransmission [43], denoted by *SR/GBN-ARQ*. The second one is the SR-ARQ used in conjunction with a stuttering (ST) retransmission [43, 44], denoted by *SR/ST-ARQ*. The third one is the SR-ARQ used in conjunction with a combination of the ST and the GBN retransmission modes, denoted by *SR/ST/GBN-ARQ*.

In the SR/GBN-ARQ scheme, when the transmitter first receives an NAK for a given codeword (say codeword $l$), it retransmits that codeword and then continues transmitting other new codewords, as in the basic SR mode. If another NAK is received for the codeword $l$, indicating that its second transmission attempt was unsuccessful, the transmitter switches to the GBN retransmission mode. That is, it sends no new codewords but backs up to the codeword $l$ and resends that codeword and the $N - 1$ succeeding codewords that were transmitted after the previous transmission attempt of the codeword $l$ as shown in Fig. 9.10. The transmitter stays in the GBN retransmission mode until the codeword $l$ is positively acknowledged. At the receiver, when the second transmission attempt of the codeword $l$ is detected in error, the subsequent $N - 1$ received vectors are discarded, regardless of whether they were received error free or not. This scheme achieves a superior throughput performance compared to the GBN-ARQ scheme, because of the benefits gained from the use of the SR mode for the first retransmission attempt. The use of the secondary mode (GBN) guarantees that buffer overflows cannot occur at the receiver if a buffer is provided for storing $N$ codewords. This scheme can have an even higher throughput if larger buffer storage is provided at the receiver, and the transmitter is designed to permit more than one retransmission attempt for a given negatively acknowledged codeword in the SR mode before switching to the GBN mode. If $\lambda$ retransmissions in the SR mode are allowed before the transmitter switches to the GBN mode, the receiver buffer must be able to store $\lambda(N-1)+1$ codewords to prevent buffer overflow. As $\lambda$ increases, the throughput performance of the SR/GBN-ARQ approaches that of the ideal SR-ARQ. The throughput efficiency of the SR/GBN-ARQ with $\lambda$ retransmissions in the SR mode is given as follows [43]:

$$\xi_{\text{SR/GBN}} = \frac{P}{1 - (N-1)(1-P)^{\lambda+1}} \cdot \frac{k}{n}. \tag{9.47}$$

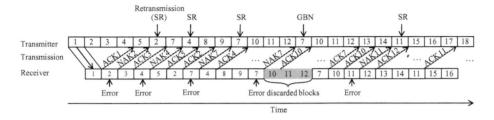

Figure 9.10 An SR/GBN-ARQ scheme with $\lambda = 1$ and $N = 4$.

In the SR/ST-ARQ scheme, when the transmitter first receives an NAK for a given codeword (say codeword $l$), it switches to the ST mode in which it repeatedly retransmits, referred to as *stuttering*, that codeword until it receives an ACK. The SR/ST-ARQ scheme is simpler than the SR/GBN-ARQ scheme but less efficient.

In the SR/ST/GBN-ARQ scheme, when the transmitter is in the SR mode following the receipt of an NAK, it switches to ST mode and retransmits the negatively acknowledged codeword $t$ times. Then, it proceeds to transmit other codewords waiting in the transmitter buffer in sequence from which it left off. The number $t$ can be chosen to provide maximum throughput for a given error rate and delay. If all the $t$ retransmissions of a codeword are received with detectable errors, then the transmitter reverts to the GBN mode. This mixed-mode ARQ scheme provides higher throughput than the SR/GBN-ARQ scheme for high bit-error rates. However, it is more complex.

### 9.6.3 Hybrid ARQ Schemes

A hybrid ARQ (HARQ) system consists of a FEC subsystem in an ARQ system [21, 25]. The function of the FEC subsystem is to reduce the frequency of retransmission by correcting the error patterns which occur most frequently. This increases the system throughput performance. When a less-frequent error pattern occurs and is detected, the receiver requests a retransmission rather than passing the unreliably decoded message to the user(s). This also increases the system reliability. As a result, a proper combination of FEC and ARQ provides higher reliability than a FEC system alone and higher throughput than a system with ARQ alone.

A straightforward HARQ scheme uses a code which is designed for simultaneous error detection and correction. When a received vector is detected in error, the receiver first attempts to correct the errors. If the number of errors is within the designed error-correcting capability of the code, the errors will be corrected and the decoded message will be delivered to the user(s) or saved in the buffer until it is ready to be passed to the user(s). If an uncorrectable error pattern is detected, the receiver rejects the received vector and requests a retransmission. The retransmission is the same codeword. When the retransmitted codeword is received, the receiver again attempts to correct the errors (if any). If the decoding is not successful, the receiver again rejects the received vector and requests another retransmission. This continues until the codeword is either successfully received or successfully decoded. Error correction may be included in any type of ARQ schemes. Because a code is used for both error correction and detection in an HARQ system, it requires more parity-check bits than a code used only for error detection in a pure ARQ system. As a result, the overhead for each transmission is increased. The above HARQ scheme is referred to as a *type-I HARQ*.

When the channel error rate is low, a type-I HARQ system has lower throughput than its corresponding ARQ system. However, when the channel error rate is high, a type-I HARQ system provides higher throughput than its corresponding ARQ system, because its error-correcting capability reduces the retransmission frequency.

Type-I HARQ schemes are best suited for communication systems in which a relatively constant level of noise and interference is anticipated on the channel. In this case, enough error correction can be designed into the system to correct the vast majority of received vectors, thereby greatly reducing the number of retransmissions and enhancing the system performance. However, for a nonstationary channel where the bit-error rate changes, a type-I HARQ scheme has some drawbacks. When the channel bit-error rate is low, the transmission is smooth and no (or little) error correction is needed. As a result, the extra parity-check bits for error correction included in each transmission represent a waste. When the channel is very noisy, the designed error-correcting capability may become inadequate. As a result, the frequency of retransmission increases and hence the system throughput is reduced. Several type-I HARQ schemes have been proposed and analyzed in [25–32].

For a channel with a nonstationary bit-error rate, one would like to design an adaptive HARQ system. When the channel is quiet, the system behaves just like a pure ARQ system, with only parity-check bits for error detection included in each transmission. Therefore, the throughput performance is the same as that of a pure ARQ system. However, when the channel becomes noisy, extra parity-check bits are needed. This concept forms the basis of an adaptive HARQ scheme. A message in its first transmission is coded with parity-check bits for error detection only, as in a pure ARQ scheme. When the receiver detects the presence of errors in a received vector, it saves the erroneous vector in a buffer and at the same time requests a retransmission. The retransmission is not the original codeword but a block of parity-check bits which is formed based on the original message and an error-correcting code. When this block of parity-check bits is received, it is used to correct the errors in the erroneous vector stored in the receiver buffer. If error correction is not successful, the receiver requests a second retransmission of the negatively acknowledged codeword. The second retransmission may be either a repetition of the original codeword or another block of parity-check bits. This depends on the retransmission strategy and the type of error-correcting code. An adaptive HARQ is referred to as a *type-II HARQ*. Several type-II HARQ schemes have been proposed and analyzed in [31, 33–39].

The concatenated codes, type-1 or type-2, are quite adaptive for a HARQ system. For illustration, we consider a type-1 concatenated code. Suppose a concatenated codeword $\mathbf{y} = (\mathbf{z}, \mathbf{v}_0, \mathbf{v}_1, \ldots, \mathbf{v}_{m-1})$ at the output of the inner encoder is transmitted. Let $\mathbf{v}_{i_0}$, $\mathbf{v}_{i_1}, \ldots, \mathbf{v}_{i_{m-l}}$ be the $m - l$ outer codewords which the decoder fails to decode. The receiver sends a message to the transmitter requesting retransmission of these outer codewords. Before the retransmission of these $m - l$ codewords, the transmitter cascades these $m - l$ outer codewords and $l$ zero codewords to form a cascaded codeword $\mathbf{v}_{\text{retran}}$ in the transmission order. The $l$ zero codewords in $\mathbf{v}_{\text{retran}}$ correspond to the $l$ successfully decoded outer codewords at the receiver. Then, $\mathbf{v}_{\text{retran}}$ is encoded into a codeword $\mathbf{y}_{\text{retran}} = (\mathbf{z}_{\text{retran}}, \mathbf{v}_{\text{retran}})$ in the inner code for retransmission. The insertion of the $l$ zero codewords allows us to use the same inner encoder. In retransmission, the $l$ zero codewords in $\mathbf{y}_{\text{retran}}$ are not transmitted. Only the outer codewords $\mathbf{v}_{i_0}, \mathbf{v}_{i_1}, \ldots, \mathbf{v}_{i_{m-l}}$ in $\mathbf{v}_{\text{retran}}$ and the party-check vector $\mathbf{z}_{\text{retran}}$ in $\mathbf{y}_{\text{retran}}$ are transmitted.

Let $\mathbf{v}_{i_0}^*$, $\mathbf{v}_{i_1}^*, \ldots, \mathbf{v}_{i_{m-l}}^*$ and $\mathbf{z}_{\text{retran}}^*$ be the received vectors and the received parity-check vector corresponding to the transmitted outer codewords $\mathbf{v}_{i_0}$, $\mathbf{v}_{i_1}, \ldots, \mathbf{v}_{i_{m-l}}$ and the parity-check vector $\mathbf{z}_{\text{retran}}$, respectively. Before starting the inner–outer decoding, we cascade $\mathbf{v}_{i_0}^*$, $\mathbf{v}_{i_1}^*, \ldots, \mathbf{v}_{i_{m-l}}^*$, $\mathbf{z}_{\text{retran}}^*$, and $l$ zero vectors in order to form a received vector $\mathbf{y}_{\text{retran}}^*$. Then, we start the inner–outer iterative decoding of $\mathbf{y}_{\text{retran}}^*$. If the inner–outer iterative decoder successfully decodes all the $m - l$ retransmitted outer codewords, then all the parity-check bits are removed from all the $m$ successfully decoded outer codewords and the $mk_1$ decoded information bits are delivered to the user(s). If the inner–outer iterative decoder fails to decode all the retransmitted outer codewords, another retransmission is requested. Retransmission continues until all the transmitted outer codewords are successfully decoded or a preset maximum number of retransmissions is reached. During retransmission, the successfully decoded codewords are saved in the received buffer if the transmitted information bits must be delivered to the user(s) in order.

We refer to the above HARQ scheme as a *concatenated HARQ scheme*, a type-I concatenated HARQ scheme. Similarly, we can use the type-2 concatenation coding scheme for a concatenated HARQ. For more about HARQ, the readers are referred to References [11, 45–47].

# Problems

**9.1** Show that the minimum distance of the interleaved code $C_{\text{int}}^{(\lambda)}$ is $d_{\min}$ which is the minimum distance of the $(n, k)$ base code $C$.

**9.2** Prove that the matrix $\mathbf{H}_{\text{int}}^{(\lambda)}$ given by (9.6) is the parity-check matrix of the interleaved code $C_{\text{int}}^{(\lambda)}$.

**9.3** Prove that the generator polynomial $\mathbf{g}_{\text{int}}^{(\lambda)}(X)$ of an interleaved code $C_{\text{int}}^{(\lambda)}$ of a cyclic code $C$ with interleaved degree $\lambda$ is in the form of (9.8).

**9.4** For the encoding of a product code, prove that the codeword array obtained by performing the row encoding first and then the column encoding is the same as the one obtained by performing the column encoding first and then the row encoding.

**9.5** Consider the $(49, 16)$ product code $C_1 \times C_2$ in Example 9.2.
    (a) Construct two different codeword arrays in $C_1 \times C_2$ and find the distance of these two codeword arrays.
    (b) Find a minimum-weight codeword array in $C_1 \times C_2$.

**9.6** Prove that the matrix $\mathbf{H}_{1 \times 2}$ given by (9.11) is the parity-check matrix of the product code $C_1 \times C_2$.

**9.7** For product codes, prove that if both the component codes $C_1$ and $C_2$ are one-step majority-logic decodable, the matrix $\mathbf{H}_{1 \times 2}$ given by (9.11) has an orthogonal structure.

**9.8** Prove that the minimum distance of a turbo product code is $d_1 + d_2 - 1$, where $d_1$ and $d_2$ are the minimum distances of its two component codes $C_1$ and $C_2$, respectively.

**9.9** Prove that the minimum distance of a type-1 serial concatenated code $C_{1 \times 2, \text{conc}}^{(1)}$ is at least $d_2$, where $d_2$ is the minimum distance of its inner code $C_2$.

**9.10** Prove that the matrix $\mathbf{H}_{1 \times 2, \text{conc}}^{(1)}$ in the form of (9.17) is a parity-check matrix of a type-1 concatenated code $C_{1 \times 2, \text{conc}}^{(1)}$.

**9.11** Prove that the matrix $\mathbf{H}_{1 \times 2, \text{conc}}^{(2)}$ in the form of (9.19) is a parity-check matrix of a type-2 concatenated code $C_{1 \times 2, \text{conc}}^{(2)}$.

**9.12** Prove that the minimum distance of the code $C = |C_1|C_1 + C_2|$ constructed by the $|\mathbf{u}|\mathbf{u} + \mathbf{v}|$-construction is in the form of (9.22).

**9.13** Show that the matrices $\mathbf{G}$ and $\mathbf{H}$ in the form of (9.23) are the generator and parity-check matrices of the code $C = |C_1|C_1 + C_2|$ constructed by using the $|\mathbf{u}|\mathbf{u} + \mathbf{v}|$-construction.

**9.14** Consider the following two matrices $\mathbf{A}$ and $\mathbf{B}$. Compute the Kronecker product $\mathbf{A} \otimes \mathbf{B}$.

$$\mathbf{A} = \begin{bmatrix} 1 & 0 \\ 1 & 1 \end{bmatrix}, \ \mathbf{B} = \begin{bmatrix} 1 & 0 & 1 & 0 \\ 0 & 1 & 0 & 1 \\ 1 & 1 & 1 & 0 \end{bmatrix}.$$

**9.15** Prove (9.41).

**9.16** Prove (9.43).

**9.17** Prove (9.45).

**9.18** The throughput of a GBN-ARQ system depends on the probability $P_c = (1 - p)^n$ that a received vector contains no error where $p$ is the channel (BSC) transition probability and $n$ is the block length, as shown by (9.44). Let $\tau$ be the data rate in bits per second. Let $T$ be the round-trip delay time in seconds. Then, $N = \tau \cdot T / n$. Suppose $p$ and the code rate $k/n$ is fixed. Determine the code length $n_0$ that maximizes the throughput $\xi_{\text{GBN}}$ of a GBN-ARQ system. The code length $n_0$ is called the *optical code length*. Optimal code lengths for the three basic ARQ schemes were investigated by Morris [23].

# References

[1] P. Elias, "Error-free coding," *IRE Trans. Inf. Theory*, **4**(4) (1954), 29–37.

[2] H. Imai, "A theory of two-dimensional cyclic codes," *Inf. Control*, **43**(1) (1977), 1–21.

[3] R. E. Blahut, *Algebraic Codes on Lines, Planes and Curves*, Cambridge, Cambridge University Press, 2008.

[4] J. L. Massey, *Threshold Decoding*, Cambridge, MA, The MIT Press, 1963.

[5] H. O. Burton and E. J. Weldon, Jr., "Cyclic product codes," *IEEE Trans. Inf. Theory*, **11**(3) (1965), 433–440.

[6] S. Lin and E. J. Weldon, Jr., "Further results on cyclic product codes," *IEEE Trans. Inf. Theory*, **16**(4) (1970), 452–459.

[7] C. Berrou and A. Glavieux, "Near optimum error correcting coding and decoding: turbo-codes," *IEEE Trans. Commun.*, **44**(10) (1996), 1261–1271.

[8] R. Pyndiah, "Near-optimum decoding of product codes: block turbo codes," *IEEE Trans. Commun.*, **46**(8) (1998), 1003–1010.

[9] C. Heegard and S. B. Wicker, *Turbo Coding*, Boston, MA, Kluwer Academic, 1999.

[10] B. Vucetic and J. Yuan, *Turbo Codes*, Boston, MA, Kluwer Academic, 2000.

[11] S. Lin and D. J. Costello, Jr., *Error Control Coding: Fundamentals and Applications*, 2nd ed., Upper Saddle River, NJ, Prentice-Hall, 2004.

[12] G. D. Forney, Jr., *Concatenated Codes*, Cambridge, MA, The MIT Press, 1966.

[13] F. J. MacWilliams and N. J. A. Sloane, *The Theory of Error-Correcting Codes*, Amsterdam, North-Holland, 1977.

[14] M. Plotkin, "Binary codes with specific minimum distance," *IEEE Trans. Inf. Theory*, **6**(4) (1960), 445–450.

[15] E. Arikan, "Channel polarization: a method for constructing capacity-achieving codes for symmetric binary-input memoryless channels," *IEEE Trans. Inf. Theory*, **55**(7) (2009), 3051–3073.

[16] A. R. Calderbank, "Multilevel codes and multistage decoding," *IEEE Trans. Commun.*, **37**(3) (1989), 222–229.

[17] U. Dettmar, J. Portugheis, and H. Hentsch, "New multistage decoding algorithms," *Electronics Lett.*, **28**(7) (1992), 635–636.

[18] G. Schnabl and M. Bossert, "Soft-decision decoding of Reed–Muller codes as generalized multiple concatenated codes," *IEEE Trans. Inf. Theory*, **41**(1) (1995), 304–308, January 1995.

[19] D. Stojanovic, M. P. C. Fossorier, and S. Lin, "Iterative multistage maximum likelihood decoding of multilevel codes," *Proc. Coding and Cryptography*, Paris, France, January 11–14, 1999, pp. 91–101.

[20] J. J. Metzner and K. C. Morgan, "Coded feedback communication systems," *Proc. Nat. Electron. Conf.*, Chicago, II, January 1960, pp. 643–647.

[21] H. O. Burton and D. D. Sullivan, "Errors and error control," *Proc. IEEE*, **10**(11) (1972), 1293–1310.

[22] J. J. Metzner, "A study of an efficient retransmission strategy for data links," *NTC Conf. Rec.*, Los Angeles, CA, December 5–7, 1977, pp. 3B:1–1–3B:1–5.

[23] J. M. Morris, "Optimal block lengths for ARQ error control schemes," *IEEE Trans. Commun.*, **27**(2) (1979), 488–493.

[24] S. Lin, D. J. Costello, Jr., and M. J. Miller, "Automatic-repeat-request error control schemes," *IEEE Commun. Magazine*, **22**(12) (1984), 5–17.

[25] K. Brayer, "Error control techniques using binary symbol burst codes," *IEEE Trans. Commun.*, **16**(2) (1968), 199–214.

[26] A. R. K. Sastry, "Performance of hybrid error control scheme on satellite channels," *IEEE Trans. Commun.*, **23**(7) (1975), 689–694.

[27] H. Yamamoto and K. Itoh, "Viterbi decoding algorithm for convolutional codes with repeat requests," *IEEE Trans. Inf. Theory*, **26**(5) (1980), 540–547.

[28] C. S. K. Leung and A. Lam, "Forward error correcting for an ARQ scheme," *IEEE Trans. Commun.*, **29**(10) (1981), 1514–1519.

[29] A. Drukarev and D. J. Costello, Jr., "A comparison of block and convolutional codes in ARQ error control schemes," *IEEE Trans. Commun.*, **30**(11) (1982), 2449–2455.

[30] A. Drukarev and D. J. Costello, Jr., "Hybrid ARQ error control using sequential decoding," *IEEE Trans. Inf. Theory*, **29**(4) (1983), 521–535.

[31] S. B. Wicker and M. J. Bartz, "The design and implementation of type-I and type-II hybrid-ARQ protocols based on first-order Reed–Muller codes," *IEEE Trans. Commun.*, **42**(234) (1994), 979–987.

[32] L. K. Rasmussen and S. B. Wicker, "Trellis-coded type-I hybrid ARQ protocols based on CRC error-detecting codes," *IEEE Trans. Commun.*, **43**(10) (1995), 2569–2575.

[33] D. M. Mandelbaum, "Adaptive-feedback coding scheme using incremental redundancy," *IEEE Trans. Inf. Theory*, **20**(3) (1974), 388–389.

[34] S. Lin and J. S. Ma, "A hybrid ARQ system with parity retransmission for error correction," *IBM Res. Rep.*, 7478 (#32232), January 1979.

[35] T. C. Ancheta, "Convolutional parity-check automatic repeat request," *Proc. IEEE Intl. Symp. Inform. Theory*, Grignano, Italy, June 25–29, 1979.

[36] S. Lin and P. S. Yu, "A hybrid ARQ scheme with parity retransmission for error control of satellite channels," *IEEE Trans. Commun.*, **30**(7) (1982), 1701–1719.

[37] S. Kallel and D. Haccoun, "Generalized type-II hybrid-ARQ scheme using punctured convoluntional codes," *IEEE Trans. Commun.*, **38**(11) (1990), 1938–1946.

[38] S. B. Wicker and M. J. Bartz, "Type-II hybrid-ARQ protocols using punctured MSD codes," *IEEE Trans. Commun.*, **42**(234) (1994), 1431–1440.

[39] A. Shiozaki, "Adaptive type-II hybrid broadcast ARQ system," *IEEE Trans. Commun.*, **44**(4) (1996), 420–422.

[40] R. J. Benice and A. H. Frey, Jr., "An analysis of retransmission systems," *IEEE Trans. Commun. Technol*, **12**(4) (1964), 135–145.

[41] V. I. Levenshtein, "Bounds on the probability of undetected error," *Problemy Perdachi Informatsii*, **13**(1) (1977), 3–18.

[42] T. Klove and V. I. Korzhik, *Error Detecting Codes*, Boston, MA, Kluwer Academic, 1995.

[43] M. J. Miller and S. Lin, "The analysis of some selective-repeat ARQ schems with finite receiver buffer," *IEEE Trans. Commun.*, **29**(9) (1981), 1307–1315.

[44] E. J. Weldon, "An improved selective-repeat ARQ strategy," *IEEE Trans. Commun.*, **30**(3) (1982), 480–486.

[45] S. A. Vanstone and P. C. van Oorshot, *An Introduction to Error Correcting Codes with Applications*, Boston, MA, Kluwer Academic, 1989.

[46] A. Poli and L. Huguet, *Error Correcting Codes: Theory and Applications*, Hemel Hempstead, Prentice-Hall, 1992.

[47] S. B. Wicker, *Error Control Systems for Digital Communication and Storage*, Englewood Cliffs, NJ, Prentice-Hall, 1995.

# 10

# Correction of Error-Bursts and Erasures

In the previous five chapters, we have been primarily concerned with codes and coding techniques for channels on which transmission errors occur independently, i.e., random errors. However, there are communication and data-storage systems where errors tend to be localized in nature. For example, in fading channels, noise is localized *in time*; whereas, in storage channels, physical defects are localized *in space*. Localizations of noise in time and physical defects in space result in errors that cluster into *bursts*, referred to as *bursts of errors* (or *error-bursts*). An error-burst may introduce multiple errors into a small range of the transmitted codewords which consequently requires a strong random-error-correcting capability. While the majority of the transmitted codewords are error free, the error-correcting capability is wasted. In general, codes designed for correcting random errors are not efficient for correcting error-bursts. Therefore, it is desirable to design codes and coding techniques specifically for correcting error-bursts. Codes of this kind are called *burst-error-correcting codes*.

In a communication or storage system, if the receiver is able to detect the locations where the received symbols are unreliable, it may choose to erase these received symbols rather than making hard-decisions for them which are very likely to create errors. Erasing these unreliable received symbols results in *symbol losses* at the *known locations*. Transmitted symbols that have been erased are referred to as *erasures*. There are two basic types of erasures, random erasures and bursts of erasures. Erasures occurring at random locations, each with the same probability of occurrence, are referred to as *random erasures*; whereas erasures clustered into bursts of locations are referred to as *bursts of erasures* (or *erasure-bursts*). Random-error-correcting codes, such as BCH, RS, finite geometry, and RM codes are effective in correcting (or recovering) erased symbols.

This chapter consists of two parts. The first part focuses on codes and coding techniques for correcting *single* and/or *multiple* random-error-bursts, and the second part is concerned with codes and coding techniques for recovering random erasures and/or erasure-bursts. More on correcting erasures or erasure-bursts using LDPC codes will be presented in Chapters 11, 14, and 16.

## 10.1  Definitions and Structures of Burst-Error-Correcting Codes

A *burst* of length $l$ is defined as a vector $\mathbf{v}$ whose nonzero components are confined to $l$ *consecutive locations*, the first and last of which are nonzero. If all the components confined to these $l$ consecutive locations of $\mathbf{v}$ are nonzero, the burst is referred to as a *solid burst*. For example, the vector $\mathbf{v} = (0\ 0\ 0\ 0\ 0\ 1\ 1\ 0\ 0\ 1\ 0\ 1\ 0\ 0\ 0\ 0\ 0\ 0)$ is a burst of length 7 and the vector $\mathbf{u} = (0\ 0\ 0\ 0\ 1\ 1\ 1\ 1\ 1\ 0\ 0\ 0\ 0\ 0\ 0\ 0\ 0\ 0)$ is a solid burst of length 5. A vector of a long burst can be viewed as multiple short bursts. For example, the vector $\mathbf{v} = (0\ 0\ 0\ 0\ 0\ 1\ 1\ 0\ 0\ 1\ 0\ 1\ 0\ 0\ 0\ 0\ 0\ 0)$ of burst length 7 can be viewed as a vector with two bursts, one of length 2 and the other with length 3. If an error vector is a burst of length $l$, it is called an *error-burst* of length $l$. A linear block code which is capable of correcting all error-bursts of length $l$ or less but not all the error-bursts of length $l + 1$ is called an *l-burst-error-correcting code*, or the code is said to have a *burst-error-correcting capability* of $l$.

For a given code length $n$ and a burst-error-correcting capability $l$, it is desirable to construct an $(n, k)$ code with a redundancy $n - k$ as small as possible. In the following, we shall first develop a basic structure that a burst-error-correcting $(n, k)$ linear block code must possess; and then establish certain restrictions on the redundancy $n - k$ for a given $l$, or restrictions on $l$ for a given redundancy $n - k$.

**Theorem 10.1** *A necessary condition for an $(n, k)$ linear block code $C$ to be able to correct all error-bursts of length $l$ or less is that no burst of length $2l$ or less can be a nonzero codeword in $C$.*

*Proof* Suppose there exists a burst $\mathbf{v}$ of length $2l$ or less as a codeword in $C$. The codeword $\mathbf{v}$ can be expressed as a vector sum of two bursts $\mathbf{u}$ and $\mathbf{w}$ of lengths $l$ or less (except the degenerate case in which $\mathbf{v}$ is a burst of length 1), i.e., $\mathbf{v} = \mathbf{u} + \mathbf{w}$. By the construction of a standard array presented in Section 3.9, $\mathbf{u}$ and $\mathbf{w}$ must be in the same coset of a standard array for $C$. If one of these two bursts is used as a coset leader (a correctable error pattern), the other will be an uncorrectable error pattern. As a result, the code would not be able to correct all error-bursts of length $l$ or less. Therefore, in order to correct all burst-errors of length $l$ or less, no codeword in $C$ can be a burst of length $2l$ or less. ▲▲

Theorem 10.1 says that an $(n, k)$ linear block code $C$ with burst-error-correcting capability $l$ does not have codewords which are bursts of length $2l$ or less, except the zero codeword.

**Theorem 10.2** *Consider an $(n,k)$ linear block code $C$ which has no burst of length $b$ or less as a codeword. Then, the number $n - k$ of parity-check symbols in $C$ is at least $b$, i.e.,*

$$n - k \geq b. \tag{10.1}$$

*Proof* Consider the vectors whose nonzero components are confined to the first $b$ positions. There are a total of $2^b$ such vectors. No two such vectors can be in the same coset of a standard array for the $(n,k)$ linear block code $C$; otherwise, their vector sum, which is a burst of length $b$ or less, would be a codeword in $C$. Therefore, these $2^b$ vectors must be in $2^b$ distinct cosets. Because there are a total of $2^{n-k}$ distinct cosets in a standard array of the code $C$, we must have $n - k \geq b$. This proves the theorem. ▲▲

It follows from Theorems 10.1 and 10.2 that we obtain the following restriction on the number of parity-check symbols of an $l$-burst-error-correcting code.

**Theorem 10.3** *For an $(n,k)$ linear block code to have a burst-error-correcting capability $l$, its number $n - k$ of parity-check symbols must satisfy the following lower bound:*

$$n - k \geq 2l. \tag{10.2}$$

The lower bound on $n - k$ given by (10.2) in Theorem 10.3 implies that the burst-error-correcting capability $l$ of an $(n,k)$ linear block code is *upper bounded* as follows:

$$l \leq \begin{cases} (n-k)/2, & \text{for even } n - k; \\ (n-k-1)/2, & \text{for odd } n - k. \end{cases} \tag{10.3}$$

The upper bound on the burst-error-correcting capability $l$ of an $(n,k)$ linear block code given by (10.3) is called the *Reiger bound* [1]. A code whose burst-error-correcting capability meets the Reiger bound is said to be *optimal*. A measure of closeness of the burst-error-correcting capability $l$ of an $(n,k)$ linear block code to its Reiger bound is the difference between $n - k$ and $2l$, i.e.,

$$\delta_{\text{discrep}} = n - k - 2l, \tag{10.4}$$

called the *optimal-discrepancy*. Then, an $l$-burst-error-correcting $(n,k)$ code is optimal if its optimal-discrepancy $\delta_{\text{discrep}}$ is zero for even $n - k$ and 1 for odd $n - k$, i.e., either $n - k - 2l = 0$ or $n - k - 2l = 1$.

Another measure of the efficiency of an $(n,k)$ linear block code with burst-error-correcting capability $l$ is the following ratio:

$$\eta_{\text{burst}} = \frac{2l}{n - k}, \tag{10.5}$$

which is called the *burst-error-correcting efficiency* of the code. An optimal burst-error-correcting code has a burst-error-correcting efficiency $\eta_{\text{burst}} = 1$ or $1 - 1/(n - k)$.

It can be proved that if an $(n, k)$ code is designed to correct all error-bursts of length $l$ or less and simultaneously detect all error-bursts of length $\delta$ where $\delta \geq l$, then the number $n - k$ of parity-check symbols of the code must be at least $l + \delta$, i.e., satisfy the following bound [1] (see Problem 10.1):

$$n - k \geq l + \delta. \tag{10.6}$$

Consider an $(n, k)$ cyclic code $C$ over GF(2). It is uniquely specified by a generator polynomial $\mathbf{g}(X) = 1 + g_1 X + \cdots + g_{n-k-1} X^{n-k-1} + X^{n-k}$ of degree $n - k$. Because each code polynomial $\mathbf{v}(X)$ in $C$ is a polynomial of degree $n - 1$ or less and is a multiple of (or divisible by) $\mathbf{g}(X)$, no nonzero polynomial of the following form can be a code polynomial in $C$:

$$X^j (a_0 + a_1 X + \cdots + a_{n-k-1} X^{n-k-1}) \bmod X^n + 1, \tag{10.7}$$

with $0 \leq j < n$. This implies that an $(n, k)$ cyclic code has *no codeword* which is a burst of length less than $n - k + 1$, and it has $n$ bursts of length $n - k + 1$ (i.e., its generator polynomial $\mathbf{g}(X)$ and its $n - 1$ cyclic-shifts). Hence, if an error-burst of length $n - k$ or less occurs during the transmission of a codeword in $C$, the receiver will be able to detect its existence. It is clear that an $l$-burst-error-correcting cyclic code is capable of correcting any *end-around* error-burst of length $l$ or less. An end-around error-burst is a burst which consists of part of the errors in one end of the error vector $\mathbf{e}$ and part of the errors in the other end of $\mathbf{e}$. For example, the vector $(1\ 0\ 1\ 0\ 0\ 0\ 0\ 0\ 0\ 0\ 0\ 0\ 0\ 0\ 1\ 1\ 0\ 1)$ is an end-around burst of length 7.

Cyclic codes are effective in correcting error-bursts. The first class of cyclic codes for correcting error-bursts was introduced by Fire in 1959 [2], and these are called *Fire codes*. Fire codes can be decoded easily with the simple error-trapping technique presented in Chapter 4. Besides Fire codes, other cyclic codes, optimal or nearly optimal, for correcting short single error-bursts have been found [2–25] since the late 1950s.

In the next six sections, we first demonstrate a simple decoding method for single burst-error-correcting cyclic codes, and then present various cyclic codes and coding techniques for correcting single and multiple error-bursts. Mostly, we will consider only binary cyclic codes. All the following developments for binary cyclic codes can be generalized to nonbinary cyclic codes.

## 10.2 Decoding of Single Burst-Error-Correcting Cyclic Codes

An $l$-burst-error-correcting cyclic code can be easily decoded by the error-trapping technique presented in Chapter 4. This decoding process is based on the following facts. Let $C$ be an $(n, k)$ cyclic code in systematic form with generator polynomial $\mathbf{g}(X) = 1 + g_1 X + \cdots + g_{n-k-1}X^{n-k-1} + X^{n-k}$. Suppose a code polynomial $\mathbf{v}(X) = v_0 + v_1 X + \cdots + v_{n-1}X^{n-1}$ is transmitted. Let $\mathbf{r}(X) = r_0 + r_1 X + \cdots + r_{n-1}X^{n-1}$ be the received polynomial and $\mathbf{e}(X) = e_0 + e_1 X + \cdots + e_{n-1}X^{n-1}$ be the error polynomial contained in the received polynomial $\mathbf{r}(X)$, i.e., $\mathbf{r}(X) = \mathbf{v}(X) + \mathbf{e}(X)$. The syndrome of $\mathbf{r}(X)$ is the remainder obtained by dividing $\mathbf{r}(X)$ with $\mathbf{g}(X)$ and is a polynomial of degree $n - k - 1$ or less,

$$
\begin{aligned}
\mathbf{s}(X) &= \mathbf{r}(X) \bmod \mathbf{g}(X) \\
&= \mathbf{e}(X) \bmod \mathbf{g}(X) \\
&= s_0 + s_1 X + \cdots + s_{n-k-1}X^{n-k-1}.
\end{aligned}
\tag{10.8}
$$

Suppose the errors are confined to the first $l$ high-order parity-check positions $X^{n-k-l}, \ldots, X^{n-k-2}, X^{n-k-1}$ of $\mathbf{r}(X)$, i.e., $\mathbf{e}(X) = X^{n-k-l} + \cdots + X^{n-k-2} + X^{n-k-1}$. It follows from (10.8) that the $l$ high-order coefficients $s_{n-k-l}, \ldots, s_{n-k-2}, s_{n-k-1}$ of the syndrome $\mathbf{s}(X)$ are identical to the errors confined to the $l$ positions $X^{n-k-l}, \ldots, X^{n-k-2}, X^{n-k-1}$ of $\mathbf{r}(X)$, i.e., $e_{n-k-l} = s_{n-k-l}, \ldots,$ $e_{n-k-2} = s_{n-k-2}, e_{n-k-1} = s_{n-k-1}$, and the $n - k - l$ low-order coefficients $s_0,$ $s_1, \ldots, s_{n-k-l-1}$ of $\mathbf{s}(X)$ are zeros, i.e., $s_0 = s_1 = \cdots = s_{n-k-l-1} = 0$.

Suppose the errors in $\mathbf{e}(X)$ are not confined to the $l$ high-order locations $X^{n-k-l}, \ldots, X^{n-k-2}, X^{n-k-1}$ of $\mathbf{r}(X)$, but confined to certain $l$ consecutive locations of $\mathbf{r}(X)$ (including the end-around case). Then, after a certain number of cyclic-shifts of $\mathbf{r}(X)$, say $i$ cyclic-shifts, $1 \le i < n$, the errors will be shifted to the positions $X^{n-k-l}, \ldots, X^{n-k-2}, X^{n-k-1}$ of $\mathbf{r}^{(i)}(X)$, the $i$th cyclic-shift of $\mathbf{r}(X)$. Let

$$
\mathbf{s}^{(i)}(X) = s_0^{(i)} + s_1^{(i)} X + \cdots + s_{n-k-1}^{(i)}X^{n-k-1}
\tag{10.9}
$$

be the syndrome of $\mathbf{r}^{(i)}(X)$. Then, the first $l$ high-order syndrome bits of $\mathbf{s}^{(i)}(X)$ are identical to the errors in the locations $X^{n-k-l}, \ldots, X^{n-k-2}, X^{n-k-1}$ of $\mathbf{r}^{(i)}(X)$ and the $n-k-l$ low-order bits of $\mathbf{s}^{(i)}(X)$ are zeros. In this case, errors in the burst are trapped in the locations $X^{n-k-l}, \ldots, X^{n-k-2}, X^{n-k-1}$ of $\mathbf{r}^{(i)}(X)$.

Based on the above error-trapping process, an error-trapping decoder for an $l$-burst-error-correcting cyclic code can be implemented with a simple feedback shift-register, a buffer, and a multi-input OR gate as shown in Fig. 10.1. The decoding procedure is described in the following steps.

**Step 1**. Gate 1 and Gate 2 are turned on and Gate 3 is turned off. The syndrome $\mathbf{s}(X)$ is formed by shifting the entire received vector $\mathbf{r}$ into the syndrome register. At the same time, the $k$ received information symbols, $r_{n-k}, \ldots, r_{n-2}, r_{n-1}$ are stored in the buffer register.

**Step 2**. Gate 2 is turned on and Gates 1 and 3 are turned off. The syndrome register starts to shift. A soon as its $n-k-l$ leftmost stages contain only zeros, its $l$ rightmost stages contain the error-burst pattern. In this case, errors are trapped in the first $l$ stages of the syndrome register and error correction can be made. Three cases must be considered.

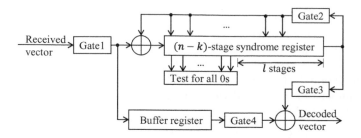

Figure 10.1 An error-trapping decoder for $l$-burst-error-correcting cyclic codes.

**Step 3.** If the $n - k - l$ leftmost stages of the syndrome register contain zeros after the $i$th cyclic-shift for $0 \leq i \leq n - k - l$, then the errors of the burst are confined to the parity-check section of $\mathbf{r}(X)$. Thus, the $k$ received information symbols in the buffer are error free. In this case, Gate 4 is then turned on and the $k$ error-free information symbols in the buffer are shifted out to the user (or data sink). If the $n - k - l$ leftmost stages of the syndrome register never contain $n - k - l$ zeros during the first $n - k - l$ shifts of the syndrome register, then the error-burst is not confined to the $n - k$ parity-check locations of $\mathbf{r}(X)$.

**Step 4.** If the $n - k - l$ leftmost stages of the syndrome register contain all zeros after the $(n - k - l + i)$th cyclic-shift of the syndrome register for $1 \leq i \leq l$, then the error-burst is confined to the locations $X^{n-i}, \ldots,$ $X^{n-1}, X^0, \ldots, X^{l-i-1}$ of $\mathbf{r}(X)$. (This is an end-around burst.) In this case, the $l - i$ bits contained in the $l - i$ rightmost stages of the syndrome shift-register are identical to errors at the parity-check positions $X^0, \ldots, X^{l-i-1}$ of $\mathbf{r}(X)$, and the $i$ bits contained in the next $i$ stages of the syndrome register match the errors at the locations $X^{n-i}, \ldots, X^{n-1}$ of $\mathbf{r}(X)$. At this instant, a clock starts to count from $(n - k - l + i + 1)$. The syndrome register is then shifted (in step with the clock) with Gate 2 turned off. As soon as the clock has counted up to $n - k$, the $i$ rightmost bits in the syndrome register match the errors at the locations $X^{n-i}, \ldots, X^{n-1}$ of $\mathbf{r}(X)$. Then, Gates 3 and 4 are turned on and the received information symbols are read out of the buffer register and corrected by the error bits shifted out from the syndrome register.

**Step 5.** If the $n - k - l$ leftmost stages of the syndrome register never contain all zeros by the time that the syndrome register has been shifted $n - k$ times, then the received information symbols are read out of the buffer register one at a time with Gate 4 turned on. At the same time, the syndrome register is shifted with Gate 2 turned on. As soon as the $n - k - l$ leftmost stages of the syndrome register contain all zeros, the symbols in the $l$ rightmost stages of the syndrome register match the errors in the next $l$ received information symbols to come out of the buffer register. Gate 3 is then turned on, and the erroneous information symbols are corrected by the error symbols coming out from the syndrome register with Gate 2 turned off.

If the leftmost $n - k - l$ stages of the syndrome register never contain zeros by the time the $k$ information symbols have been read out of the buffer register, then an error-burst of length longer than $l$ or an uncorrectable error-burst has been detected.

The error-trapping decoder described above corrects only error-bursts of length $l$ or less. The number of these error-bursts is $n2^{l-1}$ which, for large $n$, is only a small fraction of the $2^{n-k}$ correctable error-burst patterns (coset leaders). It is possible to modify the above decoding in such a way that it can correct all the correctable error-bursts of length $n - k$ or less. That is, in addition to correcting all the error-bursts of length $l$ or less, the decoder also corrects those correctable error-bursts of lengths $l + 1$ to $n - k$. This modified decoder operates as follows. The entire received vector is first shifted into the syndrome register. Before performing any error correction, the syndrome register is cyclically shifted $n$ times with the feedback gate (Gate 2) turned on. During these cyclic-shifts, the length $b$ of the shortest error-burst which appears in the $b$ rightmost stages of the syndrome register is recorded by a counter. This burst of the shortest length is assumed to be the error-burst that contaminates the transmitted codeword. Having completed the above precorrection shifts, the decoder begins its correction process. The syndrome register starts to shift again. As soon as the shortest error-burst reappears in the $b$ rightmost stages of the syndrome register, the decoder starts to make correction as described above. The above decoding is an optimal decoding for burst-error-correcting codes, proposed by Gallager [26].

## 10.3 Fire Codes

Let $\mathbf{p}(X)$ be an irreducible polynomial over GF(2) of degree $m$. Let $n_0$ be the smallest integer for which $\mathbf{p}(X)$ divides $X^{n_0} + 1$. The integer $n_0$ is called the *period* of $\mathbf{p}(X)$ and divides $2^m - 1$. Let

$$k_0 = (2^m - 1)/n_0. \tag{10.10}$$

Let $b_0$ be a positive integer such that $1 \leq b_0 \leq m$ and $2b_0 - 1$ is not divisible by $n_0$, i.e., $\mathbf{p}(X)$ does not divide $X^{2b_0-1} + 1$. In fact, $\mathbf{p}(X)$ and $X^{2b_0-1} + 1$ are relatively prime. Let $n$ be the least common multiple (LCM) of $2b_0 - 1$ and $n_0$, i.e.,

$$n = \text{LCM}\{2b_0 - 1, n_0\}. \tag{10.11}$$

Let $c_0$ be the greatest common divisor (GCD) of $2b_0 - 1$ and $n_0$, i.e.,

$$c_0 = \text{GCD}\{2b_0 - 1, n_0\}. \tag{10.12}$$

It follows from (10.11) and (10.12) that

$$n = \frac{2b_0 - 1}{c_0} n_0. \tag{10.13}$$

Form the following polynomial of degree $m + 2b_0 - 1$ over GF(2):

$$\mathbf{g}(X) = (X^{2b_0 - 1} + 1)\mathbf{p}(X), \tag{10.14}$$

which is a factor of $X^n + 1$. The cyclic code over GF(2) of length $n$ generated by the polynomial $\mathbf{g}(X) = (X^{2b_0-1}+1)\mathbf{p}(X)$ given by (10.14) is a *Fire code* [2], named after its inventor and denoted by $C_{\text{Fire}}$, with parameters $b_0$, $c_0$, and $k_0$. Such a Fire code is called a $(b_0, c_0, k_0)$-*Fire code*. It is an $(n, n - m - 2b_0 + 1)$ cyclic code. If $\mathbf{p}(X)$ is a primitive polynomial of degree $m$, then the period of $\mathbf{p}(X)$ is $n_0 = 2^m - 1$. In this case, $k_0 = 1$ and the $(b_0, c_0, 1)$-Fire code is called a *primitive Fire code*.

The $(b_0, c_0, k_0)$-Fire code $C_{\text{Fire}}$ is capable of correcting any single error-burst of length $b_0$ or less and the parameter $b_0$ is called the designed burst-error-correcting capability of $C_{\text{Fire}}$. To prove this, it is sufficient to show that all the error-bursts of length $b_0$ or less are in different cosets of a standard array of $C_{\text{Fire}}$. Hence, they can be used as coset leaders and form correctable error-burst patterns. The proof is left as an exercise (see Problem 10.3). We call the $(b_0, c_0, k_0)$-Fire code $C_{\text{Fire}}$ a $b_0$-*burst-error-correcting Fire code*. Fire codes can be decoded with a simple error-trapping decoder as shown in Fig. 10.1.

**Example 10.1** Consider the primitive polynomial $\mathbf{p}(X) = 1 + X^2 + X^5$ over GF(2) of degree 5 whose period is $n_0 = 2^5 - 1 = 31$. Hence, $k_0 = 1$. Let $b_0 = 5$. Because $2b_0 - 1 = 9$ and $n_0 = 31$ are relatively prime, $n = \text{LCM}\{9, 31\} = 279$ and $c_0 = \text{GCD}\{9, 31\} = 1$. The $(5, 1, 1)$-Fire code generated by

$$\mathbf{g}(X) = (1 + X^9)(1 + X^2 + X^5) = 1 + X^2 + X^5 + X^9 + X^{11} + X^{14}$$

is a $(279, 265)$ cyclic code with rate 0.9498 and it is capable of correcting any single error-burst of length 5 or less. The burst-error-correcting efficiency of the code is $\eta_{\text{burst}} = 10/14 = 0.714$. The error-trapping decoder for the $(5, 1, 1)$-Fire code is shown in Fig. 10.2. ▲▲

In 2016, Zhou, Lin, and Abdel-Ghaffar [27] showed that if the irreducible polynomial $\mathbf{p}(X)$ is chosen properly, the true burst-error-correcting capability, denoted by $b_t$, of the $(b_0, c_0, k_0)$-Fire code $C_{\text{Fire}}$ generated by $\mathbf{g}(X) = (X^{2b_0-1} + 1)\mathbf{p}(X)$, may be larger than its designed burst-error-correcting capability $b_0$, i.e., $b_t \geq b_0$.

**Example 10.2** Let $b_0 = 6$. The polynomial $\mathbf{p}(X) = 1 + X + X^2 + X^3 + X^{10}$ is an irreducible polynomial over GF(2) of degree 10 whose period is $n_0 = 341$. Hence, $k_0 = (2^{10} - 1)/341 = 3$. The LCM and GCD of $2b_0 - 1 = 11$ and $n_0 = 341$ are $n = 341$ and $c_0 = 11$, respectively. The $(6, 11, 3)$-Fire code generated by the polynomial

$$\begin{aligned}
\mathbf{g}(X) &= (1 + X^{11})(1 + X + X^2 + X^3 + X^{10}) \\
&= 1 + X + X^2 + X^3 + X^{10} + X^{11} + X^{12} + X^{13} + X^{14} + X^{21}
\end{aligned}$$

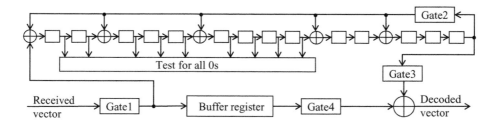

Figure 10.2 An error-trapping decoder for the $(5, 1, 1)$-Fire code in Example 10.1.

is a $(341, 320)$ cyclic code with rate $0.9384$ whose designed burst-error-correcting capability is $b_0 = 6$. However, its true burst-error-correcting capability is $b_t = 8$, two larger than its designed value. Hence, the code is capable of correcting any single error-burst of length 8 or less. The optimal-discrepancy of the code is $\delta_{\mathrm{discrep}} = 5$ and its burst-error-correcting efficiency is $\eta_{\mathrm{burst}} = 0.7619$. ▲▲

Even though Fire codes are not nearly optimal, they are constructed systematically and form a large class of cyclic codes. They can be decoded with the simple error-trapping decoding method. Moreover, their true burst-error-correcting capabilities may be *far larger than* their designed values. Table 10.1 gives a list of irreducible polynomials which produce Fire codes with true burst-error-correcting capabilities larger than their designed values. For each code, the polynomial $\mathbf{p}(X)$ is represented in an octal notation. For example, $\mathbf{p}(X) = 1 + X + X^2 + X^4 + X^6$ is represented as 127 in octal (the corresponding binary notation is $\mathbf{b} = 001010111$ with the coeffient of the highest order $X^6$ as the leftmost bit of $\mathbf{b}$ and the coefficient of the least order $X^0$ as the rightmost bit of $\mathbf{b}$). For more on the structural properties and the true burst-error-correcting capabilities of Fire codes, readers are referred to [27].

Table 10.1 A list of Fire codes which have true burst-error-correcting capabilities larger than their designed values [27].

| $b_0$ | $c_0$ | $k_0$ | $m$ | $b_t$ | $\mathbf{p}(X)$ | $n$ | rate | $b_0$ | $c_0$ | $k_0$ | $m$ | $b_t$ | $\mathbf{p}(X)$ | $n$ | rate |
|---|---|---|---|---|---|---|---|---|---|---|---|---|---|---|---|
| 2 | 3 | 1 | 10 | 4 | 2745 | 1023 | 0.9873 | 7 | 13 | 7 | 12 | 7 | 17703 | 585 | 0.9538 |
| 6 | 11 | 1 | 10 | 8 | 2415 | 1023 | 0.9795 | 8 | 15 | 7 | 12 | 12 | 14315 | 585 | 0.9538 |
| 6 | 11 | 3 | 10 | 8 | 2017 | 341 | 0.9384 | 2 | 3 | 13 | 12 | 5 | 12111 | 315 | 0.9524 |
| 2 | 3 | 11 | 10 | 5 | 2065 | 93 | 0.8602 | 3 | 5 | 13 | 12 | 6 | 17513 | 315 | 0.946 |
| 2 | 3 | 31 | 10 | 5 | 3043 | 93 | 0.6061 | 4 | 7 | 13 | 12 | 7 | 17513 | 315 | 0.9397 |
| 6 | 11 | 31 | 10 | 9 | 3043 | 33 | 0.3636 | 5 | 9 | 13 | 12 | 8 | 14373 | 315 | 0.9333 |
| 2 | 3 | 1 | 12 | 4 | 15647 | 4095 | 0.9963 | 8 | 15 | 13 | 12 | 12 | 16401 | 315 | 0.9143 |
| 3 | 5 | 1 | 12 | 5 | 15437 | 4095 | 0.9985 | 2 | 3 | 15 | 12 | 5 | 13077 | 273 | 0.9451 |
| 4 | 7 | 1 | 12 | 6 | 14433 | 4.95 | 0.9958 | 4 | 7 | 15 | 12 | 7 | 10621 | 273 | 0.9304 |

Table 10.2 (*continued*)

| 5 | 9 | 1 | 12 | 7 | 15621 | 4.95 | 0.9958 | 7 | 13 | 15 | 12 | 10 | 13077 | 273 | 0.9084 |
|---|---|---|----|---|-------|------|--------|---|----|----|----|----|-------|-----|--------|
| 7 | 13 | 1 | 12 | 10 | 12255 | 4095 | 0.9939 | 2 | 3 | 21 | 12 | 5 | 10065 | 195 | 0.9231 |
| 8 | 15 | 1 | 12 | 10 | 16047 | 4095 | 0.9934 | 3 | 5 | 21 | 12 | 6 | 17657 | 195 | 0.9128 |
| 2 | 3 | 7 | 12 | 5 | 17703 | 585 | 0.9744 | 7 | 13 | 21 | 12 | 10 | 10065 | 195 | 0.8718 |
| 3 | 5 | 7 | 12 | 6 | 13157 | 585 | 0.9709 | 8 | 15 | 21 | 12 | 11 | 10065 | 195 | 0.8615 |
| 5 | 9 | 7 | 12 | 7 | 12153 | 585 | 0.9573 | | | | | | | | |

## 10.4 Short Optimal and Nearly Optimal Single Burst-Error-Correcting Cyclic Codes

Besides Fire codes, other cyclic or shortened cyclic codes, optimal or nearly optimal, for correcting short single error-burst have been constructed [5, 7, 8] either analytically or by computer search since the discovery of Fire codes. A list of these codes with discrepancies (optimal or nearly optimal) $n - k - 2l = 0$, 1, and 2 is given in Table 10.2 (the generator polynomial $\mathbf{g}(X)$ represented in an octal form). Although their lengths are short, by using these codes as base codes, long powerful optimal or nearly optimal burst-error-correcting codes can be constructed using the interleaving and product coding techniques presented in Chapter 9. This will be discussed in two later sections.

Table 10.2 A list of optimal and nearly optimal cyclic and shortened cyclic $l$-burst-error-correcting codes [25].

| $n-k-2l$ | $n$ | $k$ | $l$ | $\mathbf{g}(X)$ | $n-k-2l$ | $n$ | $k$ | $l$ | $\mathbf{g}(X)$ | $n-k-2l$ | $n$ | $k$ | $l$ | $\mathbf{g}(X)$ |
|---|---|---|---|---|---|---|---|---|---|---|---|---|---|---|
| 0 | 7 | 3 | 2 | 35 | 1 | 21 | 10 | 5 | 7 707 | 2 | 39 | 27 | 5 | 13 617 |
| 0 | 15 | 9 | 3 | 171 | 1 | 23 | 12 | 5 | 5 343 | 2 | 41 | 21 | 9 | 6 647 133 |
| 0 | 15 | 7 | 4 | 721 | 1 | 27 | 20 | 3 | 311 | 2 | 51 | 41 | 4 | 3 501 |
| 0 | 15 | 5 | 5 | 2 467 | 1 | 31 | 20 | 5 | 4 673 | 2 | 51 | 35 | 7 | 304 251 |
| 0 | 19 | 11 | 4 | 1 151 | 1 | 38 | 29 | 4 | 1 151 | 2 | 55 | 35 | 9 | 7 164 555 |
| 0 | 21 | 9 | 6 | 14 515 | 1 | 48 | 37 | 5 | 4 501 | 2 | 57 | 39 | 8 | 1 341 035 |
| 0 | 21 | 7 | 7 | 47 343 | 1 | 63 | 50 | 6 | 22 377 | 2 | 63 | 55 | 3 | 711 |
| 0 | 21 | 5 | 8 | 214 537 | 1 | 63 | 48 | 7 | 105 437 | 2 | 63 | 53 | 4 | 2 263 |
| 0 | 21 | 3 | 9 | 1 647 235 | 1 | 63 | 46 | 8 | 730 535 | 2 | 63 | 51 | 5 | 16 447 |
| 0 | 27 | 17 | 5 | 2 671 | 1 | 63 | 44 | 9 | 2 002 353 | 2 | 63 | 49 | 6 | 61 303 |
| 0 | 34 | 22 | 6 | 15 173 | 1 | 67 | 54 | 6 | 36 365 | 2 | 73 | 63 | 4 | 2 343 |
| 0 | 38 | 24 | 7 | 114 361 | 1 | 96 | 79 | 7 | 114 361 | 2 | 85 | 75 | 4 | 2 651 |
| 0 | 50 | 34 | 8 | 224 531 | 1 | 103 | 88 | 8 | 501 001 | 2 | 85 | 73 | 5 | 10 131 |
| 0 | 56 | 38 | 9 | 1 505 773 | 2 | 17 | 9 | 3 | 471 | 2 | 105 | 91 | 6 | 70 521 |
| 0 | 59 | 39 | 10 | 4 003 351 | 2 | 21 | 15 | 2 | 123 | 2 | 131 | 119 | 5 | 15 163 |
| 1 | 15 | 10 | 2 | 65 | 2 | 31 | 25 | 2 | 161 | 2 | 169 | 155 | 6 | 55 725 |
| 1 | 21 | 14 | 3 | 171 | 2 | 31 | 21 | 4 | 3 551 | | | | | |
| 1 | 21 | 12 | 4 | 11 663 | 2 | 35 | 23 | 5 | 13 627 | | | | | |

## 10.5 Interleaved Codes for Correcting Long Error-Bursts

The interleaving method presented in Section 9.1 is an effective technique for constructing long powerful burst-error-correcting codes from short efficient burst-error-correcting codes. Interleaving is simply an attempt to distribute the errors in a long burst over codewords in the transmitted sequence and thereby make the number of errors in each individual deinterleaved vector small enough to allow the decoder of the base code to correct the errors.

Let $C$ be an $(n, k)$ linear block code with burst-error-correcting capability $l$. Let $\lambda$ be the designed interleaving degree. Let $\mathbf{v}_i = (v_{i,0}, v_{i,1}, \ldots, v_{i,n-1})$ be a codeword in $C$. Take $\lambda$ codewords $\mathbf{v}_0, \mathbf{v}_1, \ldots, \mathbf{v}_{\lambda-1}$ in $C$ and arrange them as rows of a $\lambda \times n$ *codeword array* $\mathbf{W}$ as shown in (10.15). Then, we transmit the codeword array $\mathbf{W}$ column by column. The code symbols in the $\lambda$ codewords are interleaved. Interleaving $\lambda$ codewords in $C$ results in a $(\lambda n, \lambda k)$ interleaved code $C_{\text{int}}^{(\lambda)}$. The interleaved code $C_{\text{int}}^{(\lambda)}$ is capable of correcting any single error-burst of length $\lambda l$ or less.

$$\mathbf{W} = \begin{bmatrix} v_{0,0} & v_{0,1} & \cdots & v_{0,j} & \cdots & v_{0,n-1} \\ v_{1,0} & v_{1,1} & \cdots & v_{1,j} & \cdots & v_{1,n-1} \\ \vdots & \vdots & \vdots & \vdots & \vdots & \vdots \\ v_{\lambda-1,0} & v_{\lambda-1,1} & \cdots & v_{\lambda-1,j} & \cdots & v_{\lambda-1,n-1} \end{bmatrix}. \tag{10.15}$$

In the decoding process, we first deinterleave a received vector $\mathbf{r}$ into $\lambda$ constituent vectors $\mathbf{r}_0, \mathbf{r}_1, \ldots, \mathbf{r}_{\lambda-1}$ and then arrange them back as rows of a $\lambda \times n$ array, called a *received array*. Next, we decode each constituent vector in the received array. If during the transmission of an interleaved codeword, a single error-burst of length $\lambda l$ or less occurs, regardless of its starting location, deinterleaving the received vector $\mathbf{r}$ distributes the errors in the burst into the $\lambda$ received constituent vectors $\mathbf{r}_0, \mathbf{r}_1, \ldots, \mathbf{r}_{\lambda-1}$. Each received constituent vector $\mathbf{r}_i$, $0 \leq i < \lambda$, contains an error-burst of length $l$ or less, which is within the burst-error-correcting capability $l$ of the $(n, k)$ base code $C$. Decoding each of the $\lambda$ received constituent vectors, $\mathbf{r}_0, \mathbf{r}_1, \ldots, \mathbf{r}_{\lambda-1}$, separately, we can recover the $\lambda$ transmitted codewords $\mathbf{v}_0, \mathbf{v}_1, \ldots, \mathbf{v}_{\lambda-1}$ in $C$.

Note that the interleaved code $C_{\text{int}}^{(\lambda)}$ has the same burst-error-correcting efficiency. By interleaving short codes with maximum possible burst-error-correcting capability, it is possible to construct codes of any practical length with maximum burst-error-correcting capabilities. Therefore, the interleaving technique reduces the problem of searching for long efficient burst-error-correcting codes to the search of optimal or nearly optimal short burst-error-correcting codes.

If the base code $C$ is cyclic, the interleaved code $C_{\text{int}}^{(\lambda)}$ is also cyclic. In this case, an error-trapping decoder designed for $C$ can be used to decode all the $\lambda$ received constituent vectors, one at a time. In fact, an error-trapping decoder for the interleaved code $C_{\text{int}}^{(\lambda)}$ can be derived from the one for the base code $C$. This is simply achieved by replacing each register-stage of the decoder for $C$

by $\lambda$ stages without changing the other connections. This essentially allows the decoder circuitry to look at successive rows of the received array in successive decoding cycles. Decoding the entire received vector in this way may reduce the decoding delay for large interleaving degrees.

**Example 10.3** Consider the $(34, 22)$ cyclic code $C_1$ of rate 0.6470 given in Table 10.2 whose generator polynomial is $\mathbf{g}_1(X) = 1 + X + X^3 + X^4 + X^5 + X^6 + X^9 + X^{11} + X^{12}$. This code is capable of correcting any single error-burst of length 6 or less, i.e., $l = 6$. Because $n - k - 2l = 0$, it is an optimal single burst-error-correcting code. Interleaving this code with degree $\lambda = 1000$, we obtain a $(34\,000, 22\,000)$ interleaved cyclic code $C_{\text{int},1}^{(1000)}$ whose generator polynomial is $\mathbf{g}_1^{(1000)}(X) = 1 + X^{1000} + X^{3000} + X^{4000} + X^{5000} + X^{6000} + X^{9000} + X^{11000} + X^{12000}$. This interleaved code is capable of correcting any single error-burst of length 6000 or less. It is an optimal burst-error-correcting code.

Next, we consider the $(63, 50)$ cyclic code $C_2$ with rate 0.7936 given in Table 10.2 whose generator polynomial is $\mathbf{g}_2(X) = 1 + X + X^2 + X^3 + X^4 + X^5 + X^6 + X^7 + X^{10} + X^{13}$. This code is capable of correcting any single error-burst of length 6 or less, i.e., $l = 6$. Because $n - k$ is odd and $n - k - 2l = 13 - 12 = 1$, $C_2$ is an optimal burst-error-correcting code. If we interleave this code with degree $\lambda = 1000$, we obtain a $(63\,000, 50\,000)$ interleaved cyclic code $C_{\text{int},2}^{(1000)}$ whose generator polynomial is $\mathbf{g}_2^{(1000)}(X) = 1 + X^{1000} + X^{2000} + X^{3000} + X^{4000} + X^{5000} + X^{6000} + X^{7000} + X^{10000} + X^{13000}$. This interleaved code is capable of correcting any single error-burst of length 6000 or less. ▲▲

In many communication channels, errors occur neither independently at random nor in a single burst, but in mixed types. In this case, random-error-correcting codes or single burst-error-correcting codes will be either inefficient or inadequate in combating these mixed types of errors. Hence, it is desirable to design codes or coding techniques that are capable of correcting random errors and/or multiple randomly scattered error-bursts. There are several methods of constructing such codes. Again, interleaving is one such method.

Let $C$ be an $(n, k)$ code which is capable of correcting $t$ or fewer random errors over a span of $n$ code symbols. Interleaving $C$ with degree $\lambda$, we obtain a $(\lambda n, \lambda k)$ interleaved code $C_{\text{int}}^{(\lambda)}$. This interleaved code is capable of correcting any combination of random error-bursts which corrupt no more than $t$ transmitted columns of a codeword array. Deinterleaving the received vector $\mathbf{r}$, the errors in these random error-bursts in $\mathbf{r}$ will be distributed into $\lambda$ received constituent vectors $\mathbf{r}_0, \mathbf{r}_1, \ldots, \mathbf{r}_{\lambda-1}$, each containing no more than $t$ random errors. Decoding each of these constituent received vectors with a random-error-correcting decoder for $C$ will correct all the errors in $\mathbf{r}$. Clearly, the interleaved code is also capable of correcting $t$ random-error-bursts, each of length $\lambda$ or less and each confined to a transmitted column of a codeword array (a maximum of $\lambda t$ errors).

**Example 10.4** Consider the $(1023, 781)$ Euclidean geometry code $C_{\text{EG}}$ given in Table 7.6. This code is capable of correcting 16 or fewer random errors with a simple one-step majority-logic decoder. Suppose we interleave this code with degree $\lambda = 30$. We obtain a $(30\,690, 23\,430)$ cyclic interleaved code. This code is capable of correcting any combination of random error-bursts which corrupt no more than 16 transmitted columns of a $30 \times 1023$ codeword array. It is also capable of correcting 16 random error-bursts of lengths 30 or less, each confined to a transmitted column of a codeword array (a maximum of 480 errors). ▲▲

BCH codes can also be interleaved to correct multiple random error-bursts. In decoding, each row of the received array is decoded with the Berlekamp–Massey algorithm.

## 10.6 Product Codes for Correcting Error-Bursts

The two-dimensional direct-product technique presented in Section 9.2 may be regarded as a two-dimensional interleaving. Let $C_1$ and $C_2$ be $(n_1, k_1)$ and $(n_2, k_2)$ linear block codes, respectively. In taking the direct product of $C_1$ and $C_2$, suppose we use $C_1$ as the row code and $C_2$ as the column code. If we transmit the codeword array as shown in Fig. 9.1 column by column, then the row code $C_1$ is interleaved by a degree $n_2$. However, if we transmit the codeword array row by row, the column code $C_2$ is interleaved by a degree $n_1$. This two-dimensional interleaving structure of the direct product allows us to construct product codes which are very effective in correcting error-bursts, single or multiple.

Assume that $C_1$ and $C_2$ have single-burst-error-correcting capabilities $l_1$ and $l_2$, respectively. If we transmit a codeword array in the product code $C_1 \times C_2$ column by column and decode its corresponding received array row by row, we can correct a single error-burst of length $n_2 l_1$ or less. On the other hand, if we transmit a codeword array in the product code $C_1 \times C_2$ row by row and decode its corresponding received array column by column, we can correct a single error-burst of length $n_1 l_2$ or less. Therefore, the single-burst-error-correcting capability of the product code $C_1 \times C_2$ is

$$l = \max\{n_2 l_1, n_1 l_2\}. \tag{10.16}$$

Suppose the two component codes $C_1$ and $C_2$ are two random-error-correcting codes with error-correcting capabilities $t_1$ and $t_2$, respectively. If we transmit a product codeword array in the product code $C_1 \times C_2$ column by column and decode a received array row by row, $C_1 \times C_2$ is capable of correcting $t_1$ or few random error-bursts, each of length $n_2$ or less. In this case, the product code $C_1 \times C_2$ can correct up to $n_2 t_1$ errors. However, if we transmit the codeword array in $C_1 \times C_2$ row by row and decode the received array column by column, then $C_1 \times C_2$ is capable of correcting $t_2$ or few random error-bursts, each of length $n_1$ or less. In this case, the product code $C_1 \times C_2$ can correct up to $n_1 t_2$ errors.

**Example 10.5** Let $C_1$ and $C_2$ be the $(127, 92)$ and $(63, 45)$ BCH codes with random error-correcting capabilities 5 and 3, respectively. The product code

$C_1 \times C_2$ of $C_1$ and $C_2$ can correct either 5 or less random error-bursts, each of length 63 or less, or 3 or less random error-bursts, each of length 127 of less. ▲▲

## 10.7 Phased-Burst-Error-Correcting Codes

Let $C$ be an $(n, k)$ linear block code whose length $n$ is a multiple of $m$, say $n = cm$ for some integer $c$. Each codeword $\mathbf{v} = (v_0, v_1, \ldots, v_{cm-1})$ in $C$ can be divided into $c$ *sections*, $\mathbf{v}_0, \mathbf{v}_1, \ldots, \mathbf{v}_{c-1}$, each consisting of $m$ consecutive code symbols where the $i$th section $\mathbf{v}_i$ is given as follows:

$$\mathbf{v}_i = (v_{im}, v_{im+1}, \ldots, v_{(i+1)m-1}),$$

with $0 \le i < c$. Then, $\mathbf{v} = (\mathbf{v}_0, \mathbf{v}_1, \ldots, \mathbf{v}_{c-1})$.

An error-burst of length $lm$, $1 \le l < c$, is called an $lm$-*phased-error-burst* if all the errors in the burst are confined to $l$ consecutive sections of a received vector $\mathbf{r} = (\mathbf{r}_0, \mathbf{r}_1, \ldots, \mathbf{r}_{c-1})$. For $1 \le b < c$, an $(n, k)$ linear block code $C$ of length $n = cm$ is referred to as a $bm$-*phased-burst-error-correcting code* if it is capable of correcting all the phased-error-bursts confined to $b$ or few consecutive sections of a received vector $\mathbf{r} = (\mathbf{r}_0, \mathbf{r}_1, \ldots, \mathbf{r}_{c-1})$. Because a single error-burst of length $(b-1)m + 1$, no matter where it starts, can affect at most $b$ consecutive sections of a received vector $\mathbf{r}$, a $bm$-phased-burst-error-correcting code is capable of correcting any single error-burst of length $(b-1)m+1$ or less. Hence, a $bm$-phased-burst-error-correcting code is also a single $((b-1)m + 1)$-burst-error-correcting code.

### 10.7.1 Interleaved and Product Codes

If we regard each column of a codeword array in a $(\lambda n, \lambda k)$ interleaved code $C_{\text{int}}^{(\lambda)}$ of a single $l$-burst-error-correcting code $(n, k)$ code $C$ as a section, then $C_{\text{int}}^{(\lambda)}$ is an $l\lambda$-phased-burst-error-correcting code. Similarly, the product code $C_1 \times C_2$ of two linear block codes $C_1$ and $C_2$ with single-burst-error-correcting capabilities $l_1$ and $l_2$, respectively, can be regarded either as an $n_2 l_1$-phased-burst-error-correcting code (with each column of a codeword array as a section) or as an $n_1 l_2$-phased-burst-error-correcting code (with each row of a codeword array as a section).

### 10.7.2 Codes Derived from RS Codes

Let $\alpha$ be a primitive element of the field $\text{GF}(2^m)$. Then, any element $\beta$ in $\text{GF}(2^m)$ can be expressed as a linear sum of $\alpha^0 = 1, \alpha, \alpha^2, \ldots, \alpha^{m-1}$ as follows:

$$\beta = b_0 + b_1\alpha + b_2\alpha^2 + \cdots + b_{m-1}\alpha^{m-1},$$

with $b_i \in \text{GF}(2)$, $0 \le i < m$. The $m$-tuple $(b_0, b_1, \ldots, b_{m-1})$ is the binary representation of $\beta \in \text{GF}(2^m)$, which is also called an $m$-*bit byte*.

Consider a $2^m$-ary $(2^m - 1, 2^m - 1 - 2t)$ RS code $C_{\text{RS}}$ with minimum distance $2t + 1$, which is capable of correcting $t$ or fewer random symbol errors over

$GF(2^m)$. If each code symbol in a codeword is represented by its corresponding $m$-bit byte, we obtain a binary $((2^m - 1)m, (2^m - 1 - 2t)m)$ linear block code with $2mt$ parity-check bits. This binary code is called the *binary image*, denoted by $C_{RS,b}$, of the RS code $C_{RS}$. Each codeword in $C_{RS,b}$ comprises $2^m - 1$ sections (or bytes), each consisting of $m$ bits. Suppose we use this binary code for error control over a binary input channel. Transmission of a codeword in $C_{RS,b}$ is equivalent to the transmission of its corresponding codeword in $C_{RS}$. At the channel output, the binary received vector is divided into $2^m - 1$ $m$-bit bytes and each $m$-bit byte is transformed back into a symbol in $GF(2^m)$. Thus, if an error pattern affects $t$ or fewer bytes, it affects $t$ or fewer code symbols in a codeword of $C_{RS}$. Using the Berlekamp–Massey decoding algorithm, these symbol errors will be corrected. This implies that, in binary transmission, the binary image of a $(2^m - 1, 2^m - 1 - 2t)$ RS code $C_{RS}$ over $GF(2^m)$ is capable of correcting any combination of $t$ or fewer random phased-burst-errors, each confined to an $m$-bit byte. Hence, it is a multiple-phased-burst-error-correcting code, called a *t-byte-error-correcting code*.

A single bit error in a byte is a byte error and an $m$-bit error in a byte is also a byte error. Binary images derived from RS codes are more effective against burst-errors than random errors, because burst-errors usually involve several bit errors per byte and thus relatively few byte errors. A $t$-byte-error-correcting code is capable of correcting a single error-burst of length $(t-1)m+1$ or less. In general, the binary image $C_{RS,b}$ of an RS code $C_{RS}$ is capable of correcting any combination of multiple random error-bursts of various lengths as long as the error-bursts together do not affect more than $t$ bytes of a transmitted codeword in $C_{RS}$.

**Example 10.6** Consider the $(255, 239)$ RS code over $GF(2^8)$ that is capable of correcting eight or fewer symbol errors over $GF(2^8)$. The binary image of this RS code is a $(2040, 1912)$ code $C_{RS,b}$ which is capable of correcting eight or fewer 8-bit byte errors. It can correct a single error-burst of length 57 or less with burst-error-correcting efficiency $\eta_{burst} = 2 \times 57/128 = 0.8906$. The binary image $C_{RS,b}$ is capable of correcting two random error-bursts, one of length 17 and the other one of length 33, and other combinations of random error-bursts as long as they do not affect more than 8 bytes.                    ▲▲

### 10.7.3  Burton Codes

Burton codes [11] form a class of single phased-burst-error-correcting cyclic codes. Let $\mathbf{p}(X)$ be an irreducible polynomial over $GF(2)$ of degree $m$ and period $n_0$. Let $n = \text{LCM}\{m, n_0\}$. Then, $n = cm$ for some integer $c$. For any positive integer $m$, there exists a $(cm, (c-2)m)$ cyclic code, called *Burton code* [11] and denoted by $C_{BT}$, whose generator polynomial is given as follows:

$$\mathbf{g}(X) = (X^m + 1)\mathbf{p}(X). \tag{10.17}$$

Each codeword in $C_{BT}$ can be divided into $c$ sections, each consisting of $m$ bits. The code $C_{BT}$ is capable of correcting any single error-burst of length

$m$ or less confined to a single section of a transmitted codeword. To show the burst-error-correcting characteristic of a Burton code, we only need to prove that no two error-bursts of lengths $m$ or less confined in two different sections can exist in the same coset of a standard array for the code. The proof is left as an exercise (see Problem 10.9). Note that Burton codes may be regarded as a class of *generalized Fire codes*.

A Burton code can be decoded with an error-trapping decoder as described in Section 10.2, except that the contents of the $m$ leftmost stages of the syndrome register are tested for zeros at every $m$th shift. Interleaving a Burton code with degree $\lambda$, we obtain an interleaved Burton code $C_{\text{int,BT}}^{(\lambda)}$ that is capable of correcting any single $m\lambda$-phased-error-burst with errors confined to $m$ transmitted columns corresponding to one section of a codeword in $C_{\text{BT}}$ of a codeword array.

**Example 10.7** Suppose we use the primitive polynomial $\mathbf{p}(X) = 1 + X^3 + X^7$ of degree 7 over $\text{GF}(2)$ for constructing a Burton code. The period of $\mathbf{p}(X)$ is $2^7 - 1 = 127$. Because 7 and 127 are relatively prime, their LCM is $n = 7 \times 127 = 889$. The constructed Burton code is an $(889, 875)$ cyclic code $C_{\text{BT}}$ with generator polynomial $\mathbf{g}(X) = (1 + X^7)(1 + X^3 + X^7)$. A codeword in $C_{\text{BT}}$ consists of 127 sections, each of 7 bits. This code is capable of correcting any single error-burst of length 7 or less confined to a section of a received vector. ▲▲

Note that the phased-burst-error-correcting characteristic of a Burton code is very similar to that of the binary image of a single-symbol-error-correcting RS code over $\text{GF}(2^m)$.

## 10.8  Characterization and Correction of Erasures

As pointed out in Chapter 1 and the introduction of this chapter, in a communication or storage system, if the receiver is able to detect the locations where the received symbols are unreliable, the receiver may choose to erase these received symbols rather than to make hard-decisions of these unreliable received symbols which are very likely to create errors. Erasing these unreliable received symbols results in *symbol losses* at the *known locations*. Transmitted symbols that have been erased are referred to as *erasures*. There are two basic types of erasures, *random* and *burst*. Erasures that occur at random locations, each with the same probability of occurrence, are referred to as *random erasures*. A binary-input channel with this pure random-erasure characteristic is called a *binary erasure channel* (BEC) which is modeled in Fig. 10.3(a) (a reproduction of Fig. 1.7(a)). The channel output alphabet consists of three symbols, namely, "0," "1," and "?," where the symbol "?" denotes a transmitted symbol that has been erased, i.e., an erasure. A binary-input channel with both random error and random erasure characteristics is called a *binary symmetric erasure channel* (BSEC), which is modeled in Fig. 10.3(b) (a reproduction of Fig. 1.7(b)). A binary channel over which erasures are clustered into bursts of locations is called a *binary burst erasure channel* (BBEC). In this and the next two sections, we will present codes and coding techniques for correcting random erasures over BECs, random errors and erasures over BSECs, and erasure-bursts over BBECs.

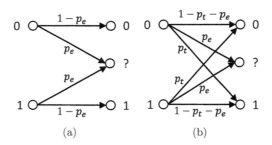

Figure 10.3 Mathematical models: (a) BEC and (b) BSEC.

## 10.8.1  Correction of Errors and Erasures over BSECs

Consider an $(n, k)$ linear block code $C$ with minimum distance $d_{\min}$. As shown in Chapter 3, this code is capable of correcting $t = \lfloor (d_{\min} - 1)/2 \rfloor$ or fewer random errors over a BSC. The parameter $t$ is called the *random-error-correcting capability* of $C$. In the following, we show that this code can be used to correct both random errors and erasures over BSECs.

Let $\nu$ and $e$ be two nonnegative integers such that

$$2\nu + e + 1 \leq d_{\min}. \tag{10.18}$$

Then, $C$ is capable of correcting all combinations of $\nu$ random errors and $e$ random erasures. A simple way to achieve this is to decode a received vector $\mathbf{r}$ in two steps. In the first step, we replace the $e$ erased positions in $\mathbf{r}$ with 0s which gives a modified received vector, denoted by $\mathbf{r}_{\text{zero}}$ and called the *0-replacement* of $\mathbf{r}$ (or 0-replaced received vector). Then, we decode the 0-replacement vector $\mathbf{r}_{\text{zero}}$ using a decoder designed for $C$. In the second step, we replace the $e$ erased positions in $\mathbf{r}$ with 1s which gives a modified received vector, denoted by $\mathbf{r}_{\text{one}}$ and called the *1-replacement* of $\mathbf{r}$ (or 1-replaced received vector). Then, we decode the 1-replacement vector $\mathbf{r}_{\text{one}}$ using the designed decoder for $C$.

Decoding $\mathbf{r}_{\text{zero}}$ and $\mathbf{r}_{\text{one}}$ separately results in two decoded vectors. If $\nu$ and $e$ satisfy the condition given by (10.18), at least one of the two decoded vectors is the transmitted codeword. To show this, we express (10.18) in terms of the random-error-correcting capability $t$ of $C$ as follows:

$$\nu + e/2 \leq t. \tag{10.19}$$

Assume that when the $e$ erasures in the received vector $\mathbf{r}$ are replaced with 0's, $e^* \leq e/2$ errors are introduced in those $e$ erased positions in $\mathbf{r}$. Then, the 0-replacement vector $\mathbf{r}_{\text{zero}}$ contains a total of $\nu + e^* \leq t$ errors that are guaranteed to be correctable. In this case, decoding of $\mathbf{r}_{\text{zero}}$ gives the transmitted codeword. However, if $e^* > e/2$, then only $e - e^* < e/2$ errors are introduced in $\mathbf{r}_{\text{one}}$ when we replace the $e$ erased positions in $\mathbf{r}$ with 1s. In this case, the 1-replacement vector $\mathbf{r}_{\text{one}}$ contains $\nu + e - e^* < t$ errors which are correctable. Then, decoding of $\mathbf{r}_{\text{one}}$ gives the transmitted codeword. Therefore, with the above two-step decoding process, at least one of the two decoded vectors is the transmitted codeword. In the case for which both $\mathbf{r}_{\text{zero}}$ and $\mathbf{r}_{\text{one}}$ are decoded into the same codeword in $C$, we choose either one as the decoded codeword. If $\mathbf{r}_{\text{zero}}$ and $\mathbf{r}_{\text{one}}$ are decoded into two different codewords in $C$, we choose the one which is closer in Hamming distance to the received vector $\mathbf{r}$ as the decoded codeword.

BCH codes are effective in correcting combinations of random errors and erasures over BSECs using the Berlekamp–Massey decoding algorithm in two steps, one step for decoding the 0-replacement of the received vector $\mathbf{r}$ and the other step for decoding the 1-replacement of $\mathbf{r}$.

**Example 10.8** Suppose we choose the primitive $(1023, 848)$ BCH code $C_{\text{BCH}}$ (see Table 5.1) as the code for correcting random errors and erasures over a BSEC. This code has a guaranteed random-error-correcting capability $t = 18$. In the transmission of a codeword in $C_{\text{BCH}}$, if $\nu$ random errors and $e$ random erasures occur and $\nu + e/2 \leq 18$, the two-step decoding process using the Berlekamp–Massey algorithm will correct both the random errors and erasures. For example, during the transmission of a codeword, if 14 random errors and 8 erasures occur, the two-step decoding process using the Berlekamp–Massey algorithm will correct those random errors and erasures. ▲▲

Finite-geometry codes are less efficient than BCH codes; however, they can be decoded with the OSML decoding, which is much simpler than the Berlekamp–Massey algorithm. In decoding, both the 0-replaced and 1-replaced received vectors, $\mathbf{r}_{\text{zero}}$ and $\mathbf{r}_{\text{one}}$, are decoded with the OSML decoding and then the decoded codeword which is closer to the received vector is chosen.

**Example 10.9** Consider the cyclic $(4095, 3367)$ EG code constructed based on the lines of the two-dimensional Euclidean geometry $\text{EG}(2, 2^6)$ over $\text{GF}(2^6)$ not passing through the origin (given in Table 7.6). Its parity-check matrix is a $4095 \times 4095$ circulant with weight 64. The code has rate 0.822 and minimum distance 65. At each code symbol position, we can form 64 orthogonal parity-check sums. Using the OSML decoding, this code is capable of correcting any combination of $\nu$ random errors and $e$ random erasures such that $2\nu + e \leq 64$. For example, if the received vector contains 20 random errors and 24 random erasures, it will be correctly decoded.

A BCH code of length 4095 based on $\text{GF}(2^{12})$ with minimum distance 65 has a rate higher than the $(4095, 3367)$ EG code; however, decoding such a long BCH code in two steps using the Berlekamp–Massey algorithm will be very complex. ▲▲

## 10.8.2 Correction of Erasures over BECs

If we use an $(n, k)$ linear block code $C$ with minimum distance $d_{\min}$ for correcting erasures only over a BEC, then it follows from the two-step decoding presented in Section 10.8.1 that $C$ is capable of correcting $d_{\min} - 1$ or fewer random erasures. That is, the erasure-correcting capability of $C$ is $d_{\min} - 1$. For example, the $(1023, 848)$ BCH code given in Example 10.8 is capable of correcting 36 or fewer random erasures over a BEC. In this section, we describe a simple successive decoding algorithm to correct random erasures over a BEC.

Consider an $(n, k)$ linear block code $C$ of length $n$ given by the null space of a $J \times n$ matrix $\mathbf{H}$ with $J$ rows $\mathbf{h}_0, \mathbf{h}_1, \ldots, \mathbf{h}_{J-1}$,

$$
\mathbf{H} = \begin{bmatrix} \mathbf{h}_0 \\ \mathbf{h}_1 \\ \vdots \\ \mathbf{h}_{J-1} \end{bmatrix}, \tag{10.20}
$$

where $J \geq n - k$ (i.e., $\mathbf{H}$ may contain redundant rows) and for $0 \leq i < J$, the $i$th row $\mathbf{h}_i$ is

$$
\mathbf{h}_i = (h_{i,0}, h_{i,1}, \ldots, h_{i,n-1}). \tag{10.21}
$$

A binary $n$-tuple $\mathbf{v} = (v_0, v_1, \ldots, v_{n-1})$ is a codeword in $C$ if and only if the following constraint is satisfied: for $0 \leq i < J$,

$$
s_i = \langle \mathbf{v}, \mathbf{h}_i \rangle = v_0 h_{i,0} + v_1 h_{i,1} + \cdots + v_{n-1} h_{i,n-1} = 0, \tag{10.22}
$$

i.e., $\mathbf{v}$ is orthogonal to all the $J$ rows of $\mathbf{H}$. The sum $s_i$ is called a *parity-check sum*. The code symbol $v_j$ is said to be checked by the parity-check sum $s_i$ (or by the row $\mathbf{h}_i$) if $v_j$ is involved in the calculation of the parity-check sum $s_i$, i.e., $h_{i,j} = 1$. Then, $v_j$ is the linear sum of the other code symbols checked by $s_i$ as follows:

$$
v_j = \sum_{l=0, l \neq j}^{n-1} v_l h_{i,l}. \tag{10.23}
$$

Suppose a codeword $\mathbf{v} = (v_0, v_1, \ldots, v_{n-1})$ in $C$ is transmitted over a BEC. Let $\mathbf{r} = (r_0, r_1, \ldots, r_{n-1})$ be the received vector and

$$
\mathcal{E} = \{j_1, j_2, \ldots, j_\tau\} \tag{10.24}
$$

be the set of $l$ locations in $\mathbf{r}$, $0 \leq j_1 < j_2 < \cdots < j_\tau < n$, where the transmitted code symbols are erased, i.e., $r_{j_1} = r_{j_2} = \cdots = r_{j_\tau} = ?$. The set $\mathcal{E}$ displays the pattern of erased symbols in $\mathbf{r}$ and is called an *erasure pattern*. Define the following index set:

$$
\mathcal{E}^c = \{0, 1, \ldots, n-1\} \setminus \{j_1, j_2, \ldots, j_\tau\}. \tag{10.25}
$$

Then, $\mathcal{E}^c$ is the set of locations in $\mathbf{r}$ where the transmitted code symbols are correctly received, i.e., $r_l = v_l$ for $l \in \mathcal{E}^c$. Decoding $\mathbf{r}$ involves in determining the value of $r_{j_l}$ for each $j_l \in \mathcal{E}$. An erasure pattern $\mathcal{E}$ is said to be recoverable (or correctable) if the value of each erased transmitted code symbol $r_{j_l}$ with $j_l \in \mathcal{E}$ can be uniquely determined from the correctly received code symbols indexed by $\mathcal{E}^c$.

In the following, we present a simple successive decoding method for correcting erasures. For each erased position $j_l$ in $\mathcal{E} = \{j_1, j_2, \ldots, j_\tau\}$, find a row $\mathbf{h}_{i_l}$ in $\mathbf{H}$ that checks only the erased code symbol $r_{j_l}$ and none of the other $\tau - 1$ erased code symbols with locations in $\mathcal{E}$; then, it follows from (10.23) that the value of the erased code symbol $r_{j_l}$ can be determined from the correctly received code symbols that are checked by $\mathbf{h}_{i_l}$ (or by the parity-check sum $s_{i_l}$). Once the erased code symbol $r_{j_l}$ is recovered, its location index $j_l$ is removed from the erasure pattern $\mathcal{E}$. The removal of $j_l$ from the erasure pattern $\mathcal{E}$ results in a new erasure pattern $\mathcal{E}_1$ with $\tau - 1$ erasures. Next, we find another row $h_{i_s}$ in $\mathbf{H}$ that checks only the erased code symbol $r_{j_s}$ and none of the other remaining $\tau - 2$ erased code symbols with locations in $\mathcal{E}_1$; then, it follows from (10.23) that the value of the erased code symbol $r_{j_s}$ with $j_s \in \mathcal{E}_1$ can be correctly determined from the correctly received code symbols (may include the recovered code symbol $r_{j_l}$) that are checked by $\mathbf{h}_{i_s}$. The above erasure-by-erasure recovering procedure continues until either all the erasures in $\mathcal{E}$ are recovered, or at the certain point no rows in $\mathbf{H}$ can be found to check only one erasure of the remaining erasures in $\mathcal{E}$. In either case, we stop the erasure recovering process. In the second case, we declare a decoding failure. We refer to the above decoding of erasures as a *successive peeling* (SPL) decoding.

Let $t_\mathcal{E}$ be the largest positive integer for which any erasure pattern $\mathcal{E}$ with $t_\mathcal{E}$ or fewer erasures is correctable by an $(n, k)$ linear block code $C$ with the SPL-decoding. We call $t_\mathcal{E}$ the *SPL-erasure-correcting capability* of the code $C$ and the code $C$ is said to be $t_\mathcal{E}$-*SPL-decodable*. The parity-check matrix $\mathbf{H}$ used for the SPL-decoding is said to have an *orthogonal peeling structure*. If the minimum distance of $C$ is $d_{\min}$ and $t_\mathcal{E} = d_{\min} - 1$, we say that $C$ is *maximal SPL-decodable* and $\mathbf{H}$ possesses a maximal orthogonal peeling structure. SPL-decoding of a binary linear block code requires only modulo-2 computations that can be easily implemented.

A class of cyclic codes which have parity-check matrices with a maximal orthogonal peeling structure is the class of two-dimensional finite-geometry codes, namely, the Euclidean and projective-geometry codes presented in Chapter 7. Consider an Euclidean geometry code $C_{\mathrm{EG}}$ of length $2^{2s} - 1$ constructed based on the two-dimensional Euclidean geometry $\mathrm{EG}(2, 2^s)$ over $\mathrm{GF}(2^s)$. A maximal orthogonal parity-check matrix $\mathbf{H}_{\mathrm{EG}}$ for such a code is formed by using the incidence vectors of all the $2^{2s} - 1$ lines in $\mathrm{EG}(2, 2^s)$ not passing through the origin as rows. As shown in Chapter 7, for each code symbol in a codeword of $C_{\mathrm{EG}}$, there are $2^s$ rows orthogonal on it. During the transmission of a codeword in $C_{\mathrm{EG}}$ over a BEC, if there are $2^s$ or fewer code symbols being erased, there is at least one row in $\mathbf{H}_{\mathrm{EG}}$ which checks one and only one erasure and no others.

Consequently, all the erased symbols in the received vector can be recovered with the SPL-decoding. Besides being able to correct all the erasure patterns with $2^s$ or fewer erasures, the EG code $C_{\mathrm{EG}}$ can also correct many erasure patterns with more than $2^s$ erasures depending on their distributions. However, there is at least one erasure pattern $\mathcal{E}$ with $2^s + 1$ erasures that is uncorrectable. Consider the $2^s$ rows in $\mathbf{H}_{\mathrm{EG}}$ which check a specific erased symbol but each of these rows also checks another erased symbol (e.g., the incidence vectors of an intersecting bundle in $\mathrm{EG}(2, 2^s)$). In this case, the erasure pattern consists of $2^s + 1$ erasures. Because each row checks two erasures, the erased symbol checked by all the $2^s$ rows cannot be recovered from the check-sum formed by any of the $2^s$ rows orthogonal on it with the SPL-decoding. Hence, the maximum number of random erasures that are guaranteed to be correctable is $2^s$. Because the EG code $C_{\mathrm{EG}}$ has minimum distance $2^s + 1$, it is maximal SPL-decodable. Similarly, a two-dimensional projective-geometry code is also maximal SPL-decodable.

In Section 10.8.1, we showed that an $(n, k)$ linear block code $C$ with minimum distance $d_{\min}$ is capable of correcting $d_{\min} - 1$ or fewer erasures over a BEC with the two-step decoding using a designed decoding algorithm for $C$. We also showed that a BCH code with a designed minimum distance $d_{\min}$ is capable of correcting $d_{\min} - 1$ or fewer erasures with a two-step decoding using the Berlekamp–Massey algorithm in each step. However, so far, it is not known whether a parity-check matrix for a BCH code with maximal orthogonal peeling structure exists. Any OSML decodable code is SPL-decodable. An extensive coverage of OSML decodable codes can be found in the text by Lin and Costello [25].

**Example 10.10** Consider the cyclic $(1023, 781)$ EG code $C_{\mathrm{EG}}$ constructed based on the two-dimensional Euclidean geometry $\mathrm{EG}(2, 2^5)$ over $\mathrm{GF}(2^5)$ given in Table 7.6. Based on the incidence vectors of the 1023 lines in $\mathrm{EG}(2, 2^5)$ not passing through the origin, a $1023 \times 1023$ matrix $\mathbf{H}_{\mathrm{EG}}$ with an orthogonal peeling structure can be constructed. For each code symbol in a codeword, there are 32 rows in $\mathbf{H}_{\mathrm{EG}}$ orthogonal on it. Hence, if 32 or fewer code symbols are erased during the transmission of a codeword over a BEC, there is at least one row in $\mathbf{H}_{\mathrm{EG}}$ that checks one and only one erasure and no others. Consequently, all the erasures in the received vector can be recovered. Because the minimum distance of the code is 33, this code is maximal SPL-decodable.

There is a primitive $(1023, 778)$ BCH code $C_{\mathrm{BCH}}$ with a designed minimum distance 51 and about the same rate as the $(1023, 781)$ EG code $C_{\mathrm{EG}}$ given above. This BCH code is capable of correcting 50 or fewer random erasures with the two-step decoding using the Berlekamp–Massey algorithm. Obviously, in correcting random erasures, $C_{\mathrm{BCH}}$ is more powerful than $C_{\mathrm{EG}}$. However, decoding the BCH code $C_{\mathrm{BCH}}$ with the two-step decoding using the Berlekamp–Massey algorithm for correcting 50 or fewer (or even 32 or fewer) erasures is much more complex than decoding the EG code $C_{\mathrm{EG}}$ with the SPL-decoding. Hence, there is a tradeoff between erasure-correcting capability and decoding complexity. ▲▲

The $(4095, 3367)$ EG code constructed based on the two-dimensional Euclidean geometry over $GF(2^6)$ given in Example 10.9 is capable of correcting 64 or fewer random erasures over a BEC.

## 10.8.3 RM Codes for Correcting Random Erasures

An RM code $C_{RM}$ is capable of correcting random erasures over a BEC using the SPL-decoding in *conjunction* with the SCIR (successive-cancellation information retrieval) process presented in Chapter 8, called an *SPL/SCIR-decoding*. Erasure correction is based on the layer and orthogonal structures of the generator matrix of $C_{RM}$. In this section, we show that an RM code $C_{RM}$ with minimum distance $d_{min}$ is capable of correcting $d_{min} - 1$ or fewer random erasures with the SPL/SCIR-decoding.

To present the SPL/SCIR-decoding, we give a brief review of the encoding and some basic structures of an RM code. Consider the $r$th-order RM code $C_{RM}(r, m)$ of length $2^m$ with minimum distance $d_{min} = 2^{m-r}$ whose generator matrix $\mathbf{G}_{RM}(r, m)$ with a multiple layer structure is given by (8.7). Let $\mathbf{a} = (\mathbf{a}_0, \mathbf{a}_1, \dots, \mathbf{a}_r)$ be the message to be encoded. The message $\mathbf{a}$ consists of $r + 1$ sections. For $0 \le l \le r$, the $l$th section $\mathbf{a}_l$ consists of $\binom{m}{l}$ information symbols of $\mathbf{a}$ in the set $\mathbf{A}_l$. The set $\mathbf{A}_l$ of information symbols is given by (8.26) and is repeated below:

$$\mathbf{A}_l = \{a_{i_1, i_2, \dots, i_l} : i_t \in \{0, 1, 2, \dots, m\}, 0 \le t \le l\}. \tag{10.26}$$

In encoding, the message $\mathbf{a}$ is encoded into a codeword $\mathbf{b}$ in $C_{RM}(r, m)$ of the form:

$$\mathbf{b} = \mathbf{a}\mathbf{G}_{RM}(r, m) = \mathbf{b}_0 + \mathbf{b}_1 + \cdots + \mathbf{b}_r,$$

given by (8.30), where $\mathbf{b}_l = \mathbf{a}_l \mathbf{G}_l$, $0 \le l \le r$, is the $l$th constituent codeword (or the $l$th layer) of $\mathbf{b}$.

Suppose the codeword $\mathbf{b}$ in $C_{RM}(r, m)$ is transmitted. Let $\mathbf{r} = (r_0, r_1, r_2, \dots, r_{2^m-1})$ be the received vector. In SCIR-decoding, we recover the transmitted message $\mathbf{a} = (\mathbf{a}_0, \mathbf{a}_1, \dots, \mathbf{a}_r)$, one section at a time in the order, $\mathbf{a}_r, \mathbf{a}_{r-1}, \dots, \mathbf{a}_0$, by successive cancellations of the estimates of $\mathbf{b}_r, \mathbf{b}_{r-1}, \dots, \mathbf{b}_1$ from $\mathbf{r}$.

In SPL/SCIR-decoding of $\mathbf{r}$ to correct erasures, we first form $2^{m-r}$ *independent estimates* of each information symbol $a_{i_1, i_2, \dots, i_r}$ in the set $\mathbf{A}_r = \{a_{i_1, i_2, \dots, i_r} : i_t \in \{1, 2, \dots, m\}, 1 \le t \le r\}$ which contains the information symbols in the last section $\mathbf{a}_r$ of $\mathbf{a}$, using the received code symbols (erased and unerased) in $\mathbf{r}$ as given by (8.65) which is repeated as follows:

$$E_e(a_{i_1, i_2, \dots, i_r}) = \sum_{k \in \mathbf{B}_{r,e}(t_e)} r_k \tag{10.27}$$

with $1 \le e \le 2^{m-r}$, where the index set $\mathbf{B}_{r,e}(t_e)$ is defined by (8.38). As shown in Chapter 8, the $2^{m-r}$ estimates given by (10.27) are *mutually disjoint*. If, during the transmission of the codeword $\mathbf{b}$, $2^{m-r} - 1$ or fewer code symbols are erased, there is at least one of the estimates given by (10.27) which contains no erased code symbol. Then, this estimate is identical to the transmitted information symbol $a_{i_1,i_2,\ldots,i_r}$. Hence, all the information symbols in $\mathbf{A}_r$ can be recovered.

Once the information symbols in $\mathbf{A}_r$ have been recovered, we try to recover the other erased code symbols in $\mathbf{r}$ from the estimates given by (10.27). If all the erased code symbols are recovered, the decoded vector is identical to the transmitted codeword $\mathbf{b}$. In this case, based on the decoded vector, we can recover the rest of the information symbols using the SCIR process. In the case when not all the erased code symbols in $\mathbf{r}$ can be recovered, we form the $r$th constituent codeword $\mathbf{b}_r$ based on the decoded message section $\mathbf{a}_r$, i.e., $\mathbf{b}_r = \mathbf{a}_r \mathbf{G}_r$. Canceling $\mathbf{b}_r$ from $\mathbf{r}$ in the known position, we obtain a modified received vector $\mathbf{r} - \mathbf{b}_r$ that contains at most the same number of erasures as in $\mathbf{r}$, which is less than $2^{m-r} - 1$. Next, we use the known code symbols in $\mathbf{r} - \mathbf{b}_r$ to recover the information symbols in the $(r-1)$th section $\mathbf{a}_{r-1}$ of the message $\mathbf{a}$, i.e., the information symbols in

$$\mathbf{A}_{r-1} = \{a_{i_1,i_2,\ldots,i_{r-1}} : i_t \in \{1, 2, \ldots, m\}, 1 \le t \le r - 1\}. \tag{10.28}$$

For each information symbol $a_{i_1,i_2,\ldots,i_{r-1}}$ in $\mathbf{A}_{r-1}$, we form $2^{m-r+1}$ independent estimates. Because the number of erasures in $\mathbf{r} - \mathbf{b}_r$ is less than $2^{m-r}$, there are more than half of the estimates of the transmitted information symbol $a_{i_1,i_2,\ldots,i_{r-1}}$ which are identical to $a_{i_1,i_2,\ldots,i_{r-1}}$. Hence, all the information symbols in the $(r-1)$th section $\mathbf{a}_{r-1}$ of $\mathbf{a}$ can be recovered. Then, based on the recovered information symbols in $\mathbf{A}_{r-1}$, we try to recover the erased code symbols in $\mathbf{b} - \mathbf{b}_r$. If all the erased code symbols in $\mathbf{b} - \mathbf{b}_r$ are recovered, the decoded vector is identical to $\mathbf{r} - \mathbf{b}_r$. In this case, we can recover the rest of the information symbols by the SCIR process. However, when not all the erased code symbols in $\mathbf{r} - \mathbf{b}_r$ can be recovered, we repeat the SPL/SCIR process to recover the information symbols in the $(r-2)$th section $\mathbf{a}_{r-2}$ of $\mathbf{a}$. The SPL/SCIR decoding process continues until all the information symbols in $\mathbf{a}$ are recovered. Then, the codeword $\mathbf{b}$ can be reconstructed from the decoded information symbols if needed.

**Example 10.11** Consider the $(16, 11)$ second-order RM code $C_{\mathrm{RM}}(2, 4)$ with minimum distance $d_{\min} = 4$ constructed in Example 8.2 whose generator matrix $\mathbf{G}_{\mathrm{RM}}(2, 4)$ is given by (8.24). Each message

$$\mathbf{a} = (a_0, a_4, a_3, a_2, a_1, a_{3,4}, a_{2,4}, a_{1,4}, a_{2,3}, a_{1,3}, a_{1,2})$$
$$= (\mathbf{a}_0, \mathbf{a}_1, \mathbf{a}_2)$$

for this code consists of 11 information symbols which are divided into 3 sections, $\mathbf{a}_0 = (a_0)$, $\mathbf{a}_1 = (a_4, a_3, a_2, a_1)$, and $\mathbf{a}_2 = (a_{3,4}, a_{2,4}, a_{1,4}, a_{2,3}, a_{1,3}, a_{1,2})$. The codeword for the message $\mathbf{a}$ is given by

$$\mathbf{b} = (b_0, b_1, b_2, b_3, b_4, b_5, b_6, b_7, b_8, b_9, b_{10}, b_{11}, b_{12}, b_{13}, b_{14}, b_{15})$$
$$= \mathbf{b}_0 + \mathbf{b}_1 + \mathbf{b}_2$$
$$= a_0 \mathbf{v}_0 + a_4 \mathbf{v}_4 + a_3 \mathbf{v}_3 + a_2 \mathbf{v}_2 + a_1 \mathbf{v}_1 + a_{3,4} \mathbf{v}_3 \mathbf{v}_4$$
$$\quad + a_{2,4} \mathbf{v}_2 \mathbf{v}_4 + a_{1,4} \mathbf{v}_1 \mathbf{v}_4 + a_{2,3} \mathbf{v}_2 \mathbf{v}_3 + a_{1,3} \mathbf{v}_1 \mathbf{v}_3 + a_{1,2} \mathbf{v}_1 \mathbf{v}_2.$$

The 16 code symbols in $\mathbf{b}$ are given by (8.33). Suppose the codeword $\mathbf{b}$ is transmitted over a BEC. Let

$$\mathbf{r} = (b_0, b_1, ?, b_3, b_4, ?, b_6, b_7, b_8, b_9, b_{10}, b_{11}, ?, b_{13}, b_{14}, b_{15})$$

be the received vector in which the code symbols $b_2$, $b_5$, $b_{12}$ are erased.

The zeroth level of decoding of $\mathbf{r}$ is to recover the information symbols $a_{3,4}$, $a_{2,4}$, $a_{1,4}$, $a_{2,3}$, $a_{1,3}$, $a_{1,2}$ in the second section $\mathbf{a}_2$ of $\mathbf{a}$. To achieve this, we form four independent estimates of each information symbol in $\mathbf{a}_2$ using the received code symbols as follows:

$$
\begin{aligned}
&a_{1,2} = b_0 + b_1 + ? + b_3, \quad && a_{1,2} = b_4 + ? + b_6 + b_7, \\
&a_{1,2} = b_8 + b_9 + b_{10} + b_{11}, \quad && a_{1,2} = ? + b_{13} + b_{14} + b_{15}, \\
&a_{1,3} = b_0 + b_1 + b_4 + ?, \quad && a_{1,3} = ? + b_3 + b_6 + b_7, \\
&a_{1,3} = b_8 + b_9 + ? + b_{13}, \quad && a_{1,3} = b_{10} + b_{11} + b_{14} + b_{15}, \\
&a_{1,4} = b_0 + b_1 + b_8 + b_9, \quad && a_{1,4} = ? + b_3 + b_{10} + b_{11}, \\
&a_{1,4} = b_4 + ? + ? + b_{13}, \quad && a_{1,4} = b_6 + b_7 + b_{14} + b_{15}, \\
&a_{2,3} = b_0 + ? + b_4 + b_6, \quad && a_{2,3} = b_1 + b_3 + ? + b_7, \\
&a_{2,3} = b_8 + b_{10} + ? + b_{14}, \quad && a_{2,3} = b_9 + b_{11} + b_{13} + b_{15}, \\
&a_{2,4} = b_0 + ? + b_8 + b_{10}, \quad && a_{2,4} = b_1 + b_3 + b_9 + b_{11}, \\
&a_{2,4} = b_4 + b_6 + ? + b_{14}, \quad && a_{2,4} = ? + b_7 + b_{13} + b_{15}, \\
&a_{3,4} = b_0 + b_4 + b_8 + ?, \quad && a_{3,4} = b_1 + ? + b_9 + b_{13}, \\
&a_{3,4} = ? + b_6 + b_{10} + b_{14}, \quad && a_{3,4} = b_3 + b_7 + b_{11} + b_{15}.
\end{aligned}
$$

$$(10.29)$$

From (10.29), we see that among the four independent estimates of each information symbol in $\mathbf{a}_2$, there is at least one estimate which contains no erasure. Hence, all six information symbols in $\mathbf{a}_2$ can be recovered. With these recovered information symbols, we can recover all three erased code symbols $b_2$, $b_5$, $b_{12}$. With the three erased code symbols in $\mathbf{r}$ replaced by the recovered code symbols, we obtain the transmitted codeword $\mathbf{b}$. Then, in the next two steps of SCIR decoding, we can recover all five information symbols in the sections $\mathbf{a}_1$ and $\mathbf{a}_0$ of $\mathbf{a}$ by using $\mathbf{b} - \mathbf{b}_2$ and $\mathbf{b} - \mathbf{b}_2 - \mathbf{b}_1$, respectively. ▲▲

**Example 10.12** For $m = 10$ and $r = 5$, the fifth-order RM code $C_{\mathrm{RM}}(5, 10)$ is a $(1024, 637)$ code with minimum distance $d_{\min} = 32$. With the SPL/SCIR-decoding, this code can correct up to 31 random erasures over a BEC. ▲▲

## 10.9  Correcting Erasure-Bursts over BBECs

Over BBECs, erasures often cluster into bursts, called *bursts of erasures* (or *erasure-bursts*). In this case, codes that are capable of correcting erasure-bursts are needed. An erasure-burst is said to have length $b$ if the erasures are confined to $b$ consecutive locations, the first and last of which are erasures. This section is concerned with codes, especially cyclic codes, which are capable of correcting erasure-bursts, single and multiple. First, we show that cyclic codes are effective for correcting a single erasure-burst with the SPL-decoding.

## 10.9.1 Cyclic Codes for Correcting Single Erasure-Burst

Let $\mathbf{v} = (v_0, v_1, \ldots, v_{n-1})$ be a nonzero $n$-tuple over $\mathrm{GF}(2)$. The first (the leftmost) 1-component is called the *leading* 1 of $\mathbf{v}$ and the last (rightmost) 1-component is called the *tailing* 1 of $\mathbf{v}$. If $\mathbf{v}$ has only one 1-component, the leading 1 and the tailing 1 of $\mathbf{v}$ are the same. A *zero-span* of $\mathbf{v}$ is defined as a sequence of *consecutive zeros* between two 1-components of $\mathbf{v}$. The zeros to the right of the tailing 1 of $\mathbf{v}$ together with the zeros to the left of the leading 1 of $\mathbf{v}$ also form a zero-span, called the *end-around zero-span*. The number of zeros in a zero-span is called the *length* of the zero-span. A zero-span of length zero is called a *null zero-span*. A nonzero $n$-tuple $\mathbf{v}$ may have multiple nonnull zero-spans. A zero-span of maximum length is called a *maximal zero-span* of $\mathbf{v}$. For example, the 24-tuple

$$\mathbf{v} = (0\ 0\ 0\ 1\ 0\ 0\ 0\ 1\ 0\ 0\ 1\ 1\ 0\ 0\ 0\ 0\ 0\ 0\ 0\ 0\ 1\ 0\ 0)$$

has four nonnull zero-spans of lengths 3, 2, 9, and 5. The length of the end-around zero-span of $\mathbf{v}$ is 5. The maximal zero-span of $\mathbf{v}$ has length 9.

Let $C$ be a binary $(n, k)$ cyclic code with generator polynomial

$$\mathbf{g}(X) = 1 + g_1 X + g_2 X^2 + \cdots + g_{n-k-1} X^{n-k-1} + X^{n-k}. \tag{10.30}$$

Recall that $X^n + 1 = \mathbf{g}(X)\mathbf{h}(X)$, where

$$\mathbf{h}(X) = \frac{X^n + 1}{\mathbf{g}(X)} = 1 + h_1 X + h_2 X^2 + \cdots + h_{k-1} X^{k-1} + X^k, \tag{10.31}$$

is called the parity-check polynomial of the code. The reciprocal of $\mathbf{h}(X)$ is given as follows (as shown in (4.19)):

$$\mathbf{h}^*(X) = 1 + h_{k-1} X + h_{k-2} X^2 + \cdots + h_1 X^{k-1} + X^k, \tag{10.32}$$

which is the generator polynomial of the dual code $C_d$ of $C$ that is also cyclic. The $n$-tuple corresponding to the polynomial $\mathbf{h}^*(X)$ is

$$\mathbf{h}^* = (1\ h_{k-1}\ h_{k-2}\ \ldots\ h_1\ 1\ 0\ 0\ 0\ \ldots\ 0\ 0\ 0), \tag{10.33}$$

which is called the *parity-check vector* of the cyclic code $C$. As shown in (4.20), $\mathbf{h}^*$ and its $n - k - 1$ cyclic-shifts form the parity-check matrix $\mathbf{H}$ of the code $C$, an $(n - k) \times n$ matrix of rank $n - k$.

From (10.33), we see that the parity-check vector $\mathbf{h}^*$ of $C$ has a zero-span of length $n - k - 1$, which is the *longest zero-span* in $\mathbf{h}^*$ as shown in the following theorem.

**Theorem 10.4** *The maximum length of a zero-span in the parity-check vector* $\mathbf{h}^*$ *of an* $(n, k)$ *cyclic code over* $\text{GF}(2)$ *is* $n - k - 1$.

*Proof* First, from (10.33), we know that $\mathbf{h}^*$ has a zero-span of length $n - k - 1$ which consists of the $n - k - 1$ zeros in the rightmost $n - k - 1$ positions of $\mathbf{h}^*$. For $k < n - k$, the theorem is obviously true. Hence, we only need to prove the theorem for $k \geq n-k$. Let $\mathbf{z}_0$ denote the zero-span that consists of the rightmost $n - k - 1$ zeros of $\mathbf{h}^*$. Suppose there is a zero-span $\mathbf{z}$ of length $l \geq n - k$ in $\mathbf{h}^*$. Because $h_0 = 1$ and $h_k = 1$, this zero-span $\mathbf{z}$ starts at position $i$ for some $i$ and ends at the position $i + l - 1$ with $1 \leq i \leq k - l$. We shift $\mathbf{h}^*$ cyclically to the right until the zero-span $\mathbf{z}$ is shifted to the rightmost $l$ positions of a new $n$-tuple $\mathbf{h}^{**} = (h_0^{**}, h_1^{**}, \ldots, h_{n-1-l}^{**}, 0, 0, \ldots, 0)$, where $h_0^{**} = h_{n-1-l}^{**} = 1$. The $n$-tuple $\mathbf{h}^{**}$ is a codeword in the dual code $C_d$ of $C$ because $C_d$ is a cyclic code. This new $n$-tuple $\mathbf{h}^{**}$ and its $l$ cyclic-shifts form a set of $l + 1 > n - k$ linearly independent vectors that are codewords in $C_d$ of $C$. This contradicts the fact that $C_d$, of dimension $n - k$, has at most $n - k$ linearly independent codewords. This proves the theorem. ▲▲

Next, we form an $n \times n$ matrix with the parity-check vector $\mathbf{h}^*$ and its $n - 1$ cyclic-shifts as rows as follows:

$$
\mathbf{H}_n =
\left[
\begin{array}{ccccccccccc}
1 & h_{k-1} & h_{k-2} & \cdots & & h_1 & 1 & 0 & 0 & \cdots \cdots & 0 \\
0 & 1 & h_{k-1} & h_{k-2} & \cdots & & h_1 & 1 & 0 & \cdots \cdots & 0 \\
0 & 0 & 1 & h_{k-1} & h_{k-2} & \cdots & & h_1 & 1 & \cdots \cdots & 0 \\
\cdots & \cdots & \cdots & \cdots & \cdots & \cdots & \cdots & \cdots & \cdots & \cdots & \cdots\cdots \\
\cdots & \cdots & \cdots & \cdots & \cdots & \cdots & \cdots & \cdots & \cdots & \cdots & \cdots\cdots \\
\cdots & \cdots & \cdots & \cdots & \cdots & \cdots & \cdots & \cdots & \cdots & \cdots & \cdots\cdots \\
0 & 0 & \cdots & & 0 & 1 & h_{k-1} & h_{k-2} & \cdots & \cdots & h_1 \quad 1 \\
\hline
1 & 0 & 0 & \cdots & & 0 & 1 & h_{k-1} & h_{k-2} & \cdots & \cdots h_1 \\
h_1 & 1 & 0 & 0 & \cdots & & 0 & 1 & h_{k-1} & h_{k-2} & \cdots h_2 \\
\cdots & \cdots & \cdots & \cdots & \cdots & \cdots & \cdots & \cdots & \cdots & \cdots & \cdots\cdots \\
\cdots & \cdots & \cdots & \cdots & \cdots & \cdots & \cdots & \cdots & \cdots & \cdots & \cdots\cdots \\
\cdots & \cdots & \cdots & \cdots & \cdots & \cdots & \cdots & \cdots & \cdots & \cdots & \cdots\cdots \\
h_{k-1} & h_{k-2} & \cdots & & h_1 & 1 & 0 & 0 & \cdots & \cdots & 0 \quad 1
\end{array}
\right].
\tag{10.34}
$$

The $n \times n$ matrix $\mathbf{H}_n$ forms a parity-check matrix of the cyclic code $C$, in which the first $n - k$ rows are linearly independent and they form a full-rank parity-check matrix $\mathbf{H}_{n-k}$ of $C$ as shown in (4.20). The lower $k$ rows of $\mathbf{H}_n$ are redundant rows. Because the top row $\mathbf{h}^*$ has a zero-span of length $n - k - 1$ and every row below it is a cyclic-shift of $\mathbf{h}^*$, every row of $\mathbf{H}_n$ has a zero-span of maximal length $n - k - 1$. This zero-span structure is called a *zero-covering span* of $\mathbf{H}_n$. From (10.34), we see that at every column position $j$, $0 \leq j < n$, there is a row in $\mathbf{H}_n$ with a 1-component at the position $j$ followed by a *span* of $n - k - 1$ zeros, including the end-around case.

Based on the zero-covering span structure of the parity-check matrix $\mathbf{H}_n$, the $(n, k)$ cyclic code $C$ is capable of correcting any single erasure-burst of length $n - k$ or less using the SPL-decoding. Suppose a codeword $\mathbf{v}$ in $C$ is transmitted over a BBEC. Let $\mathbf{r} = (r_0, r_1, \ldots, r_{n-1})$ be the received vector with an erasure-burst pattern $\mathcal{E}$ of length $n - k$ or less. Suppose the starting position of the erasure-burst $\mathcal{E}$ is $j$ with $0 \leq j < n$. Then, there exists a row $\mathbf{h}_i = (h_{i,0}, h_{i,1}, \ldots, h_{i,n-1})$ in $\mathbf{H}_n$ whose $j$th component $h_{i,j}$ is equal to 1 and followed by a zero-span of length $n - k - 1$. Therefore, the row $\mathbf{h}_i$ checks only the erasure at the $j$th position in $\mathbf{r}$ but no other erasure in $\mathbf{r}$. Setting the inner product $\langle \mathbf{r}, \mathbf{h}_i \rangle = 0$, we have the following equation:

$$r_0 h_{i,0} + r_1 h_{i,1} + \cdots + r_{n-1} h_{i,n-1} = 0$$

which contains only one unknown $r_j$, the erased symbol at the $j$th position. From the above equation, we can determine the value of the $j$th transmitted symbol $v_j$ as follows:

$$v_j = \sum_{l=0, l \neq j}^{n-1} r_l h_{i,l}. \tag{10.35}$$

Once the code symbol $v_j$ is recovered, the location index $j$ is removed from the erasure-burst pattern $\mathcal{E}$ and we obtain a new erasure-burst pattern $\mathcal{E}_1$ of shorter length. Then, we perform the SPL-decoding on $\mathcal{E}_1$ to recover another erasure. The SPL-decoding continues until all erasures in $\mathcal{E}$ are recovered.

Note that the maximum length of a single erasure-burst that an $(n, k)$ cyclic code $C$ can correct is equal to the number $n - k$ of parity-check symbols of the code. Hence, the code $C$ is optimal in terms of correcting a single erasure-burst.

**Example 10.13** Consider the primitive $(1023, 848)$ BCH code $C_{\mathrm{BCH}}$ given in Example 10.8. This code can correct any single erasure-burst of length 175 or less with the SPL-decoding. As shown in Example 10.8, this code can also be used to correct 36 or fewer random erasures with the two-step decoding method using the Berlekamp–Massey algorithm. Consequently, we can use this code to correct both random erasures and erasure-bursts as follows. When a vector $\mathbf{r}$ is received, we check whether the erasures are confined to 175 consecutive positions. If they are confined to 175 consecutive positions, regardless of the number of erasures in these positions, we can correct the erasures as a single erasure-burst using the SPL-decoding. If the erasures in $\mathbf{r}$ are not confined to 175 consecutive positions in $\mathbf{r}$ but the number of erasures is 36 or less, we can correct the erasures with the two-step decoding method using the Berlekamp–Massey algorithm. This two-phase decoding process enhances the erasure-correcting performance with an additional SPL-decoding circuitry. ▲▲

From the above example, we readily see that any BCH code can be used to correct both random erasures and a single erasure-burst. Even more, we can use a BCH code to correct random errors, random erasures, and a single erasure-burst (see Problem 10.11).

## 10.9.2    Correction of Multiple Random Erasure-Bursts

To correct multiple random erasure-bursts, we can interleave a random erasure-correcting code. Let $C$ be an $(n, k)$ code which is capable of correcting $t_\mathcal{E}$ or fewer random erasures. Interleaving $C$ with degree $\lambda$, we obtain a $(\lambda n, \lambda k)$ linear block code $C_{\text{int}}^{(\lambda)}$ which is capable of correcting $t_\mathcal{E}$ or fewer random erasure-bursts of lengths $\lambda$ or less as long as the erasures do not affect more than $t_\mathcal{E}$ transmitted columns of the transmitted codeword array in $C_{\text{int}}^{(\lambda)}$. In decoding, we first deinterleave the received vector $\mathbf{r}$ and arrange it into a $\lambda \times n$ received array. The deinterleaving and array arrangement process distributes the erasures in the $t_\mathcal{E}$ or fewer random erasure-bursts in the received vector $\mathbf{r}$ into $\lambda$ rows of the received array, each containing no more than $t_\mathcal{E}$ erasures. Decode each row of the received array with the two-step decoding algorithm with the decoder designed for the base code $C$ (or by using the SPL-decoding algorithm). Then, all the erasures contained in the $t_\mathcal{E}$ or fewer random erasure-bursts of lengths $\lambda$ or less will be recovered. If $C$ is a BCH code, we can use the Berlekamp–Massey algorithm for the two-step decoding of $C$. If $C$ is a finite-geometry code, we can decode $C$ with the SPL-decoding method.

**Example 10.14** As was shown in Example 10.8, the primitive $(1023, 848)$ BCH code $C_{\text{BCH}}$ is capable of correcting 36 or fewer random erasures over a BEC with the two-step decoding method using the Berlekamp–Massey algorithm. If we interleave this code by degree $\lambda = 100$, the interleaved code $C_{\text{int,BCH}}^{(100)}$ is capable of correcting 36 or fewer random erasure-bursts of various lengths as long as the erasures in the bursts do not affect more than 36 transmitted columns of the transmitted codeword array in $C_{\text{int,BCH}}^{(100)}$. It is capable of correcting 36 random erasure-bursts of lengths 100 or less, each confined to a transmitted column of a codeword array. In this case, the maximum number of erasures that can be corrected in the bursts is 3600.

If we use the $(1023, 781)$ Euclidean-geometry code $C_{\text{EG}}$ given in Example 10.10 and interleave it with degree $\lambda = 100$, we obtain an interleaved code $C_{\text{int,EG}}^{(100)}$. Using the simple SPL-decoding algorithm, $C_{\text{int,EG}}^{(100)}$ is capable of correcting 32 or fewer random erasure-bursts of various lengths as long as the erasures in the bursts do not affect more than 32 transmitted columns of the transmitted codeword array in $C_{\text{int,EG}}^{(100)}$.                      ▲▲

The direct-product technique presented in Section 9.2 can be used to construct product codes for correcting multiple random erasure-bursts from short random-erasure-correcting codes (see Problem 10.14).

# 10.10    RS Codes for Correcting Random Errors and Erasures

Thus far, we have considered only binary codes and coding techniques for correcting binary random errors and/or erasures. In this section, we show that

RS codes are very effective for correcting combinations of random symbol errors and erasures, or purely random symbol erasures over *nonbinary erasure channels* (NBECs).

Consider a $(2^m - 1, 2^m - 2t - 1)$ RS code $C_{\text{RS}}$ over GF$(2^m)$ with minimum distance $2t + 1$, which is capable of correcting $t$ or fewer random symbol errors over GF$(2^m)$ using the Berlekamp–Massey (or Euclidean) decoding algorithm presented in Chapter 6. In this section, we show that such an RS code is capable of correcting $\nu$ random symbol errors and $e$ random symbol erasures over an NBEC provided that the following inequality holds:

$$2\nu + e \leq 2t. \tag{10.36}$$

Let $\mathbf{v}(X)$ and $\mathbf{r}(X)$ be the transmitted code polynomial and the received polynomial, respectively. Suppose the received polynomial $\mathbf{r}(X)$ contains $\nu$ symbol errors at locations $X^{i_1}$, $X^{i_2}, \ldots,$ $X^{i_\nu}$, and $e$ symbol erasures at locations $X^{j_1}$, $X^{j_2}, \ldots,$ $X^{j_e}$. Because the locations of erased symbols are known, we need only to find the locations and values of the symbol errors and the values of the erased symbols in the decoding process. The erasure-location numbers corresponding to the erased locations $X^{j_1}$, $X^{j_2}, \ldots,$ $X^{j_e}$ are $\alpha^{j_1}, \alpha^{j_2}, \ldots, \alpha^{j_e}$, where $\alpha$ is a primitive element of GF$(2^m)$. Form the erasure-location polynomial as follows:

$$\beta(X) = \prod_{l=1}^{e}(1 - \alpha^{j_l}X). \tag{10.37}$$

Next, we fill the $e$ erased locations in $\mathbf{r}(X)$ with zeros (or any arbitrary symbol from GF$(2^m)$). The replacement of $e$ erased symbols in $\mathbf{r}(X)$ by zeros can introduce up to $e$ additional symbol errors. Let $\mathbf{r}^*(X)$ denote the modified received polynomial, i.e., the 0-replacement of $\mathbf{r}(X)$. Let

$$\sigma(X) = \prod_{k=1}^{\nu}(1 - \alpha^{i_k}X) \tag{10.38}$$

be the error-location polynomial of the errors in $\mathbf{r}(X)$ at locations $X^{i_1}$, $X^{i_2}, \ldots,$ $X^{i_\nu}$. Then, the error/erasure-location polynomial for the 0-replacement polynomial $\mathbf{r}^*(X)$ is

$$\begin{aligned}
\gamma(X) &= \beta(X)\sigma(X) \\
&= \sum_{i=1}^{e+\nu}(1 - \alpha^{\gamma_i}X),
\end{aligned} \tag{10.39}$$

in which $\beta(X)$ is known.

Now, the decoding objective is to find $\sigma(X)$ and the error-value evaluator $\mathbf{Z}_0(X)$ for $\mathbf{r}^*(X)$. To find $\sigma(X)$ and $\mathbf{Z}_0(X)$, we first compute the syndrome polynomial

$$\mathbf{S}(X) = S_1 + S_2X + S_3X^2 + \cdots + S_{2t}X^{2t-1} \tag{10.40}$$

from the 0-replacement received polynomial $\mathbf{r}^*(X)$. Then, the key-equation for solving $\boldsymbol{\sigma}(X)$ and $\mathbf{Z}_0(X)$ becomes

$$\beta(X)\boldsymbol{\sigma}(X)\mathbf{S}(X) = \mathbf{Z}_0(X) \mod X^{2t}. \tag{10.41}$$

The decoding problem is to find the solution $(\boldsymbol{\sigma}(X),\ \mathbf{Z}_0(X))$ of the key-equation such that $\boldsymbol{\sigma}(X)$ has a minimum degree $\nu \le t$, and $\deg(\mathbf{Z}_0(X)) < \nu + e$.

Because $\beta(X)$ and $\mathbf{S}(X)$ are known, we can combine them to simplify the key-equation. Let

$$\mathbf{T}(X) = [\beta(X)\mathbf{S}(X)]_{2t} = T_1 + T_2 X + T_3 X^2 + \cdots + T_{2t-1}X^{2t-1} \tag{10.42}$$

denote the polynomial that consists of the first $2t$ terms in the product $\beta(X)\mathbf{S}(X)$ from $X^0$ to $X^{2t-1}$, which is called the *modified syndrome polynomial*. Then, the key-equation given by (10.41) is reduced to the following form:

$$\boldsymbol{\sigma}(X)\mathbf{T}(X) = \mathbf{Z}_0(X) \mod X^{2t}. \tag{10.43}$$

This key-equation can be solved by using either the Berlekamp–Massey or the Euclidean algorithm presented in Chapter 6. In the following, we outline the steps of the Euclidean algorithm for correcting random symbol errors and erasures. The steps are given below.

(1) Compute the erasure-location polynomial $\beta(X)$ using the erasure locations in the received polynomial $\mathbf{r}(X)$.

(2) Form the 0-replacement received polynomial $\mathbf{r}^*(X)$ by replacing the erased symbols in $\mathbf{r}(X)$ with zeros.

(3) Compute the syndrome polynomial $\mathbf{S}(X)$ of $\mathbf{r}^*(X)$ and the modified syndrome polynomial $\mathbf{T}(X) = [\beta(X)\mathbf{S}(X)]_{2t}$.

(4) Set the following initial conditions:

$$\mathbf{Z}_0^{(-1)}(X) = X^{2t},\ \mathbf{Z}_0^{(0)}(X) = \mathbf{T}(X),$$
$$\boldsymbol{\sigma}^{(-1)}(X) = 0,\quad \boldsymbol{\sigma}^{(0)}(X) = 1. \tag{10.44}$$

(5) Execute the Euclidean algorithm iteratively as described in Section 6.11 until a step $i$ is reached where the pair $(\boldsymbol{\sigma}^{(i)}(X),\ \mathbf{Z}_0^{(i)}(X))$ is the solution of the key-equation given by (10.43) such that $\boldsymbol{\sigma}(X)$ has a minimum degree $\nu \le t$, and $\deg(\mathbf{Z}_0(X)) < \nu + e$. Then, we set

$$\boldsymbol{\sigma}(X) = \boldsymbol{\sigma}^{(i)}(X), \mathbf{Z}_0(X) = \mathbf{Z}_0^{(i)}(X). \tag{10.45}$$

(6) Find the roots of $\boldsymbol{\sigma}(X)$ and determine the error locations in $\mathbf{r}^*(X)$.

(7) Determine the values of errors and erasures from $\mathbf{Z}_0(X)$ and $\gamma(X) = \beta(X)\boldsymbol{\sigma}(X)$. Take the derivative $\gamma'(X)$ of the error/erasure-location polynomial $\gamma(X)$ which is given by:

$$\gamma'(X) = \frac{d\gamma(X)}{dX} = -a \sum_{l=1}^{\nu+e} \alpha^{\gamma_l} \prod_{i=1, i\neq l}^{\nu+e} (1 - \alpha^{\gamma_i} X), \tag{10.46}$$

where $a \neq 1$ is a constant that may appear in $\boldsymbol{\sigma}(X)$ when the Euclidean algorithm is used. For $1 \leq k \leq \nu$, compute the error values with the following formula:

$$e_{i_k} = \frac{-\mathbf{Z}_0(\alpha^{-i_k})}{\gamma'(\alpha^{-i_k})}. \tag{10.47}$$

For $1 \leq l \leq e$, compute the values of the erased symbols using the following formula:

$$f_{j_l} = \frac{-\mathbf{Z}_0(\alpha^{-j_l})}{\gamma'(\alpha^{-j_l})}. \tag{10.48}$$

(8) Based on the error locations, the erasure locations, the error and erasure values, we form the error/erasure polynomial $\mathbf{e}(X)$. Subtracting $\mathbf{e}(X)$ from $\mathbf{r}^*(X)$, we obtain the decoded polynomial $\mathbf{v}^*(X) = \mathbf{r}^*(X) - \mathbf{e}(X)$. If the condition $2\nu + e \leq 2t$ holds, the decoded polynomial $\mathbf{v}^*(X)$ is identical to the transmitted code polynomial $\mathbf{v}(X)$.

If we want to use the $(2^m - 1, 2^m - 2t - 1)$ RS code $C_{\mathrm{RS}}$ for correcting symbol erasures only, we set $\nu = 0$. Then, the RS code $C_{\mathrm{RS}}$ is capable of correcting $2t$ symbol erasures. In this case, we set $\boldsymbol{\beta}(X) = 1$ with initial condition $\boldsymbol{\sigma}^{(-1)}(X) = 0$ and $\boldsymbol{\sigma}^{(0)}(X) = 1$. The decoding process is exactly the same as described above. Because the RS code $C_{\mathrm{RS}}$ is capable of correcting up to $2t$ random symbol erasures, which is equal to the number of its parity-check symbols, the code is optimal in correcting random symbol erasures.

In the above explanation of correcting random errors and erasures, we used a primitive RS code over a field $\mathrm{GF}(2^m)$ of characteristic 2. However, primitive or nonprimitive RS codes over any field $\mathrm{GF}(q^m)$, where $q$ is a power of a prime, can be used for correcting random symbol errors and erasure using either the Euclidean or the Berlekamp–Massey algorithms. Moreover, they are optimal.

**Example 10.15** For $t = 3$, consider the triple-symbol-error-correcting $(15, 9)$ RS code $C_{\mathrm{RS}}$ over $\mathrm{GF}(2^4)$ generated by $\mathbf{g}(X) = (X + \alpha)(X + \alpha^2)(X + \alpha^3)(X + \alpha^4)(X + \alpha^5)(X + \alpha^6)$, where $\alpha$ is a primitive element of $\mathrm{GF}(2^4)$. Over the 16-ary symmetric erasure channel, this code is capable of correcting any combination of $\nu$ random symbol errors and $e$ erasures provided that $2\nu + e \leq 2t = 6$. Suppose the all-zero codeword is transmitted and the received vector is

$$\mathbf{r} = (0\ 0\ 0\ ?\ 0\ 0\ ?\ 0\ 0\ \alpha\ 0\ 0\ \alpha^4\ 0\ 0),$$

where "?" denotes an erasure. The received polynomial corresponding to this received vector is

$$\mathbf{r}(X) = (?)X^3 + (?)X^6 + \alpha X^9 + \alpha^4 X^{12}.$$

There are two erased symbols in $\mathbf{r}(X)$ at the locations $X^3$ and $X^6$ and two errors at the locations $X^9$ and $X^{12}$. The location numbers of the two erased symbols are $\alpha^3$ and $\alpha^6$. Using Table 2.9 for computation, the erasure-location polynomial is

$$\beta(X) = (1 - \alpha^3 X)(1 - \alpha^6 X) = (1 + \alpha^3 X)(1 + \alpha^6 X) = 1 + \alpha^2 X + \alpha^9 X^2.$$

Replacing the erased symbols with zeros, we obtain the following 0-replacement received polynomial:

$$\mathbf{r}^*(X) = \alpha X^9 + \alpha^4 X^{12}.$$

The syndrome components computed from $\mathbf{r}^*(X)$ are:

$$S_1 = \mathbf{r}^*(\alpha) = \alpha^8, \ S_2 = \mathbf{r}^*(\alpha^2) = \alpha^{11}, \ S_3 = \mathbf{r}^*(\alpha^3) = \alpha^9,$$
$$S_4 = \mathbf{r}^*(\alpha^4) = 0, \quad S_5 = \mathbf{r}^*(\alpha^5) = 1, \quad S_6 = \mathbf{r}^*(\alpha^6) = \alpha^8.$$

The syndrome polynomial is then

$$\mathbf{S}(X) = \alpha^8 + \alpha^{11} X + \alpha^9 X^2 + X^4 + \alpha^8 X^5,$$

and the modified syndrome polynomial is

$$\mathbf{T}(X) = [\beta(X)\mathbf{S}(X)]_6 = \alpha^8 + \alpha^{14} X + \alpha^4 X^2 + \alpha^3 X^3 + \alpha^{14} X^4 + X^5.$$

Using the Euclidean algorithm to decode $\mathbf{r}^*(X)$, we set the initial conditions as follows:

$$\mathbf{Z}_0^{(-1)}(X) = X^6, \mathbf{Z}_0^{(0)}(X) = \mathbf{T}(X), \boldsymbol{\sigma}^{(-1)}(X) = 0, \boldsymbol{\sigma}^{(0)}(X) = 1.$$

Executing the algorithm, we obtain Table 10.3. At step $i = 2$, we find that the pair $(\boldsymbol{\sigma}^{(2)}(X), \mathbf{Z}_0^{(2)}(X))$ is the solution of the key-equation $\beta(X)\boldsymbol{\sigma}(X)\mathbf{S}(X) = \mathbf{Z}_0(X) \bmod X^6$. The degree $\nu$ of $\boldsymbol{\sigma}^{(2)}(X)$ is 2 and the degree of $\mathbf{Z}_0^{(2)}(X)$ is 3, which is less than $\nu + e = 4$, i.e., $\deg(\mathbf{Z}_0^{(2)}(X)) < 4$. Then, we set $\boldsymbol{\sigma}(X) = \boldsymbol{\sigma}^{(2)}(X)$ and $\mathbf{Z}_0(X) = \mathbf{Z}_0^{(2)}(X)$. This gives the error-location and the error-value evaluation polynomials as follows:

$$\boldsymbol{\sigma}(X) = \alpha(1 + \alpha^8 X + \alpha^6 X^2) = \alpha(1 + \alpha^9 X)(1 + \alpha^{12} X),$$
$$\mathbf{Z}_0(X) = \alpha^9 + \alpha^8 X + \alpha X^2 + \alpha X^3.$$

The two roots of $\boldsymbol{\sigma}(X)$ are $\alpha^{-9}$ and $\alpha^{-12}$. The inverses of these two roots give the error-location numbers $\alpha^9$ and $\alpha^{12}$.

Using $\boldsymbol{\sigma}(X)$ and $\beta(X)$ and following (10.39), we compute the error/erasure-location polynomial as follows:

$$\gamma(X) = \beta(X)\boldsymbol{\sigma}(X) = \alpha(1 + \alpha^3 X)(1 + \alpha^6 X)(1 + \alpha^9 X)(1 + \alpha^{12} X).$$

The derivative of $\gamma(X)$ is given by:

$$\gamma'(X) = \alpha^4(1 + \alpha^6 X)(1 + \alpha^9 X)(1 + \alpha^{12} X) + \alpha^7(1 + \alpha^3 X)(1 + \alpha^9 X)(1 + \alpha^{12} X)$$
$$+ \alpha^{10}(1 + \alpha^3 X)(1 + \alpha^6 X)(1 + \alpha^{12} X) + \alpha^{13}(1 + \alpha^3 X)(1 + \alpha^6 X)(1 + \alpha^9 X).$$

It follows from (10.47) that the values of errors at locations $X^9$ and $X^{12}$ are

$$e_9 = \frac{-\mathbf{Z}_0(\alpha^{-9})}{\gamma'(\alpha^{-9})} = \frac{\alpha^{13}}{\alpha^{12}} = \alpha,$$

$$e_{12} = \frac{-\mathbf{Z}_0(\alpha^{-12})}{\gamma'(\alpha^{-12})} = \frac{\alpha^3}{\alpha^{14}} = \alpha^4.$$

Following (10.48), we find that the values of the two erased symbols are:

$$f_3 = \frac{-\mathbf{Z}_0(\alpha^{-3})}{\gamma'(\alpha^{-3})} = \frac{0}{\alpha^8} = 0,$$

$$f_6 = \frac{-\mathbf{Z}_0(\alpha^{-6})}{\gamma'(\alpha^{-6})} = \frac{0}{\alpha^0} = 0.$$

Then, the decoded error/erasure polynomial is

$$\mathbf{e}(X) = \alpha X^9 + \alpha^4 X^{12}.$$

Subtracting $\mathbf{e}(X)$ from $\mathbf{r}^*(X)$, we obtain the decoded polynomial $\mathbf{v}^*(X) = \mathbf{r}^*(X) - \mathbf{e}(X) = 0$ which is the transmitted code polynomial, the zero code polynomial. The decoding is successful.

Correction of random symbol errors and erasures with an RS code can be carried out using the Berlekamp–Massey algorithm in a similar manner (see Problem 10.15). ▲▲

Table 10.3 Euclidean steps to find the solution $(\boldsymbol{\sigma}(X), \mathbf{Z}_0(X))$ of the key-equation for decoding the $(15, 9)$ RS code in Example 10.15.

| $i$ | $\mathbf{Z}_0^{(i)}(X)$ | $\mathbf{q}_i(X)$ | $\boldsymbol{\sigma}^{(i)}(X)$ |
|---|---|---|---|
| $-1$ | $X^6$ | $-$ | $0$ |
| $0$ | $\mathbf{T}(X)$ | $-$ | $1$ |
| $1$ | $\alpha^7 + \alpha^3 X + X^2 + \alpha^{10} X^3 + \alpha^8 X^4$ | $\alpha^{14} + X$ | $\alpha^{14} + X$ |
| $2$ | $\alpha^9 + \alpha^8 X + \alpha X^2 + \alpha X^3$ | $\alpha^5 + \alpha^7 X$ | $\alpha + \alpha^9 X + \alpha^7 X^2$ |

**Example 10.16** Suppose the $(15, 9)$ RS code given in Example 10.15 is used for correcting symbol erasures only. It is capable of correcting six or fewer symbol erasures. Suppose we transmit the all-one codeword in the RS code and receive the following vector:

$$\mathbf{r} = (1\ ?\ 1\ 1\ ?\ ?\ 1\ 1\ ?\ 1\ ?\ 1\ 1\ ?\ 1),$$

which consists of six erasures. The received polynomial corresponding to this received vector is

$$\mathbf{r}(X) = 1 + (?)X + X^2 + X^3 + (?)X^4 + (?)X^5 + X^6 + X^7$$
$$+ (?)X^8 + X^9 + (?)X^{10} + X^{11} + X^{12} + (?)X^{13} + X^{14}.$$

The locations of the six erased symbols in $\mathbf{r}(X)$ are $X$, $X^4$, $X^5$, $X^8$, $X^{10}$, and $X^{13}$. The location numbers of these six erased symbols are $\alpha$, $\alpha^4$, $\alpha^5$, $\alpha^8$, $\alpha^{10}$, and $\alpha^{13}$, respectively. The erasure-location polynomial is

$$\boldsymbol{\beta}(X) = (1 + \alpha X)(1 + \alpha^4 X)(1 + \alpha^5 X)(1 + \alpha^8 X)(1 + \alpha^{10} X)(1 + \alpha^{13} X)$$
$$= 1 + \alpha^3 X + \alpha^9 X^2 + \alpha X^3 + \alpha^8 X^4 + \alpha^{10} X^5 + \alpha^{11} X^6.$$

Replacing the erased symbols in $\mathbf{r}(X)$ with zeros, we obtain the following 0-replacement received polynomial:

$$\mathbf{r}^*(X) = 1 + X^2 + X^3 + X^6 + X^7 + X^9 + X^{11} + X^{12} + X^{14}.$$

The syndrome components computed from $\mathbf{r}^*(X)$ are:

$$S_1 = \mathbf{r}^*(\alpha) = \alpha^3, \quad S_2 = \mathbf{r}^*(\alpha^2) = \alpha^6, \quad S_3 = \mathbf{r}^*(\alpha^3) = \alpha^{10},$$
$$S_4 = \mathbf{r}^*(\alpha^4) = \alpha^{12}, \quad S_5 = \mathbf{r}^*(\alpha^5) = 0, \quad S_6 = \mathbf{r}^*(\alpha^6) = \alpha^5.$$

The syndrome polynomial is then

$$\mathbf{S}(X) = \alpha^3 + \alpha^6 X + \alpha^{10} X^2 + \alpha^{12} X^3 + \alpha^5 X^5,$$

and the modified syndrome polynomial is

$$\mathbf{T}(X) = [\boldsymbol{\beta}(X)\mathbf{S}(X)]_6 = \alpha^3 + \alpha X^2 + \alpha^{10} X^4.$$

Using the Euclidean decoding algorithm, we set the initial conditions as follows:

$$\mathbf{Z}_0^{(-1)}(X) = X^6, \mathbf{Z}_0^{(0)}(X) = \mathbf{T}(X), \boldsymbol{\sigma}^{(-1)}(X) = 0, \boldsymbol{\sigma}^{(0)}(X) = 1.$$

Because $\deg(\boldsymbol{\sigma}^{(0)}(X)) = 0$ and $\deg(\mathbf{Z}_0^{(0)}(X)) = \deg(\mathbf{T}(X)) = 4 < \nu + e = 0 + 6 = 6$, we have

$$\mathbf{Z}_0(X) = \mathbf{T}(X) = \alpha^3 + \alpha X^2 + \alpha^{10} X^4,$$

and

$$\boldsymbol{\sigma}(X) = 1,$$

i.e., the result $(\boldsymbol{\sigma}^{(0)}(X), \mathbf{Z}_0^{(0)}(X))$ at step 0 is the solution for the key-equation.

Using $\boldsymbol{\sigma}(X)$ and $\boldsymbol{\beta}(X)$ and following (10.39), we compute the error/erasure-location polynomial as follows:

$$\boldsymbol{\gamma}(X) = \boldsymbol{\beta}(X)\boldsymbol{\sigma}(X) = 1 + \alpha^3 X + \alpha^9 X^2 + \alpha X^3 + \alpha^8 X^4 + \alpha^{10} X^5 + \alpha^{11} X^6.$$

The derivative of $\boldsymbol{\gamma}(X)$ is given below:

$$\boldsymbol{\gamma}'(X) = \alpha^3 + \alpha X^2 + \alpha^{10} X^4.$$

Following (10.48), the value of the erased symbol at location $X^{j_l}$ with $j_l = 1$, 4, 5, 8, 10, 13 is calculated as

$$f_{j_l} = \frac{-\mathbf{Z}_0(\alpha^{-j_l})}{\gamma'(\alpha^{-j_l})} = \frac{\alpha^3 + \alpha(\alpha^{-j_l})^2 + \alpha^{10}(\alpha^{-j_l})^4}{\alpha^3 + \alpha(\alpha^{-j_l})^2 + \alpha^{10}(\alpha^{-j_l})^4} = 1.$$

Thus, the values of the erased symbols at locations $X$, $X^4$, $X^5$, $X^8$, $X^{10}$, and $X^{13}$ are all equal to 1. The erasure/error polynomial is

$$\mathbf{e}(X) = X + X^4 + X^5 + X^8 + X^{10} + X^{13},$$

and the decoded polynomial is

$$\begin{aligned} \mathbf{v}^*(X) &= \mathbf{r}^*(X) - \mathbf{e}(X) \\ &= 1 + X^2 + X^3 + X^6 + X^7 + X^9 + X^{11} + X^{12} + X^{14} \\ &\quad + X + X^4 + X^5 + X^8 + X^{10} + X^{13} \\ &= 1 + X + X^2 + X^3 + X^4 + X^5 + X^6 + X^7 + X^8 + X^9 + X^{10} + X^{11} \\ &\quad + X^{12} + X^{13} + X^{14}. \end{aligned}$$

The decoded polynomial $\mathbf{v}^*(X)$ corresponds to a all-one 15-tuple which is the transmitted codeword. Therefore, the code is able to correct six random erasures. ▲▲

If we interleave a $t$-symbol-error-correcting $(n, n - 2t)$ RS code $C_{\mathrm{RS}}$ of length $n$ over $\mathrm{GF}(2^m)$ with degree $\lambda$, the interleaved code $C_{\mathrm{int,RS}}^{(\lambda)}$ is a $(\lambda n, \lambda(n - 2t))$ code which is capable of correcting $2t$ random phased-erasure-bursts, each confined to an interleaved section of length $\lambda$. Decoding is performed on each row of the received array using either the Euclidean or the Berlekamp–Massey decoding algorithms. The maximum number of symbol erasures that can be corrected is $2\lambda t$, which is equal to the number of parity-check symbols of the interleaved code $C_{\mathrm{int,RS}}^{(\lambda)}$. The direct-product technique can be also used to construct product codes to correct random symbol erasures or multiple phased-burst-erasures with RS codes over the same field as component codes (see Problem 10.17).

For application to a BEC, we take the binary image of an $(n, n - 2t)$ RS code $C_{\mathrm{RS}}$ over $\mathrm{GF}(2^m)$ which is an $(mn, m(n - 2t))$ binary code $C_{\mathrm{RS,b}}$. Each codeword in $C_{\mathrm{RS,b}}$ consists of $n$ sections, each as an $m$-bit byte. This binary image $C_{\mathrm{RS,b}}$ is capable of correcting $2t$ random phased-erasure-bursts, each confined to an $m$-bit byte. In decoding, we first convert each received $m$-bit byte into a symbol in $\mathrm{GF}(2^m)$. If a received $m$-bit byte contains erasures, it is replaced by an erased symbol. Then, the decoding is performed on the received vector over $\mathrm{GF}(2^m)$ either using the Berlekamp–Massey algorithm or the Euclidean algorithm.

## Problems

**10.1** Prove (10.6).

**10.2** Consider the $(7,3)$ cyclic code $C$ generated by $\mathbf{g}(X) = 1 + X^2 + X^3 + X^4$ with burst-error-correcting capability $l = 2$. Build an error-trapping decoder for $C$ and use it to decode the following received vectors (show the contents of the syndrome register in each decoding step): $\mathbf{r}_0 = (0\,0\,0\,0\,1\,1\,0)$, $\mathbf{r}_1 = (0\,0\,1\,1\,0\,0\,0)$, and $\mathbf{r}_2 = (0\,0\,1\,1\,1\,0\,0)$.

**10.3** Prove that the $(b_0, c_0, k_0)$-Fire code $C_{\text{Fire}}$ with a generator polynomial in the form of (10.14) is capable of correcting any single error-burst of length $b_0$ or less.

**10.4** Consider the polynomial $\mathbf{p}(X) = 1 + X + X^4$ which is primitive over GF(2). Find the generator polynomial of a Fire code which is capable of correcting any single error-burst of length 4 or less. Build an error-trapping decoder for the constructed code.

**10.5** Choose a code from Table 10.2 to construct a new code with burst-error-correcting capability $l = 51$, length $n = 255$, and burst-error-correcting efficiency $\eta_{\text{burst}} = 1$. Devise a decoder for the constructed code.

**10.6** Consider an $(n, k)$ cyclic code $C$ with generator polynomial $\mathbf{g}(X)$. Interleave this code with a degree of $\lambda$ to obtain an interleaved $(\lambda n, \lambda k)$ code $C_{\text{int}}^{(\lambda)}$. Show that $C_{\text{int}}^{(\lambda)}$ is cyclic and has generator polynomial $\mathbf{g}(X^\lambda)$ which is obtained by replacing $X$ in $\mathbf{g}(X)$ by $X^\lambda$.

**10.7** Consider the $(15, 7)$ BCH code $C_{\text{BCH}}$. Construct an interleaved code $C_{\text{int,BCH}}^{(\lambda)}$ with an interleaving degree $\lambda = 7$. Discuss the error-correcting capability of $C_{\text{int,BCH}}^{(\lambda)}$ and build a decoder for $C_{\text{int,BCH}}^{(\lambda)}$.

**10.8** Consider the $(31, 15)$ RS code over GF($2^5$). Find the binary image of this RS code and discuss the error-correcting capability of the binary image.

**10.9** Prove that a $(cm, (c-2)m)$ cyclic Burton code $C_{\text{BT}}$ is capable of correcting any single error-burst of length $m$ or less confined to a single section of a transmitted codeword in $C_{\text{BT}}$.

**10.10** Consider the irreducible polynomial $\mathbf{p}(X) = 1 + X + X^2 + X^3 + X^{10}$ over GF(2) of degree 10 whose period is $n_0 = 341$ given in Example 10.2. Construct a Burton code with $\mathbf{p}(X)$ and discuss the error-correcting capability of the constructed code.

**10.11** Consider the $(31, 16)$ BCH code over GF(2) with error-correcting capability $t = 3$ generated by $\mathbf{g}(X) = 1 + X + X^2 + X^3 + X^5 + X^7 + X^8 + X^9 + X^{10} + X^{11} + X^{15}$. Assume that the all-zero codeword is transmitted over a BSEC. Decode the following received vectors.

(a) $\mathbf{r}_0 = (1\ 1\ ?\ 0\ 0\ 0\ 0\ 0\ 0\ 0\ 0\ 0\ 0\ 0\ 0\ 0\ 0\ 0\ 0\ 0\ 0\ 0\ 0\ 0\ 0\ 0\ 0\ 0\ 0\ 0\ 0)$.
(b) $\mathbf{r}_1 = (1\ 0\ 0\ 0\ ?\ ?\ ?\ 0\ 0\ 0\ 0\ 0\ 0\ 0\ 0\ 0\ 0\ 0\ 0\ 0\ 0\ 0\ 0\ 0\ 0\ 0\ 0\ 0\ 0\ 0\ 0\ 0)$.
(c) $\mathbf{r}_2 = (?\ 0\ 0\ 0\ ?\ ?\ ?\ 0\ ?\ ?\ 0\ 0\ 0\ 0\ 0\ 0\ 0\ 0\ 0\ 0\ 0\ 0\ 0\ 0\ 0\ 0\ 0\ 0\ 0\ 0\ 0\ 0)$.
(d) $\mathbf{r}_3 = (1\ 1\ ?\ ?\ ?\ ?\ 0\ 0\ 0\ 0\ 0\ 0\ 0\ 0\ 0.0\ 0\ 0\ 0\ 0\ 0\ 0\ 0\ 0\ 0\ 0\ 0\ 0\ 0\ 0\ 0\ 0)$.
(e) $\mathbf{r}_4 = (?\ ?\ ?\ ?\ ?\ ?\ 0\ 0\ 0\ 0\ 0\ 0\ 0\ 0\ 0\ 0\ 0\ 0\ 0\ 0\ 0\ 0\ 0\ 0\ 0\ 0\ 0\ 0\ 0\ 0\ 0\ 0)$.
(f) $\mathbf{r}_5 = (?\ ?\ ?\ ?\ ?\ ?\ 0\ 0\ 0\ 0\ 0\ 0\ 0\ 0\ 0\ 0\ 0\ 0\ 0\ 0\ 0\ 0\ 0\ 0\ 0\ ?\ ?\ ?\ ?\ ?)$.

**10.12** Consider the $(16, 11)$ second-order RM code $C_{\mathrm{RM}}(2, 4)$ given in Example 10.11. Decode the following received vectors.
(a) $\mathbf{r}_0 = (1\ 0\ ?\ 0\ 1\ ?\ 1\ 0\ 0\ 0\ 0\ 0\ ?\ 1\ 1\ 1)$.
(b) $\mathbf{r}_1 = (1\ 0\ ?\ 0\ 1\ ?\ 1\ 0\ 0\ 0\ ?\ 0\ ?\ 1\ 1\ 1)$.

**10.13** Consider the $(7, 3)$ cyclic code $C$ generated by $\mathbf{g}(X) = 1 + X^2 + X^3 + X^4$ over GF(2). Suppose $C$ is used for error correcting over a BBSE. Decode the following received vectors: $\mathbf{r}_0 = (?\ ?\ ?\ 0\ 0\ 0\ 0)$, $\mathbf{r}_1 = (?\ ?\ ?\ 0\ 0\ 0\ ?)$, and $\mathbf{r}_2 = (?\ ?\ ?\ 0\ 0\ ?\ ?)$.

**10.14** Consider two linear block codes, an $(n_1, k_1)$ code $C_1$ with erasure-correcting capability $t_{\mathcal{E},1}$ and an $(n_2, k_2)$ code $C_2$ with erasure-correcting capability $t_{\mathcal{E},2}$. Use these two codes as the component codes to construct a product code $C_1 \times C_2$. Discuss the erasure-correcting capabilities of the product code $C_1 \times C_2$.

**10.15** Use the Berlekamp–Massey algorithm to decode the received vector $\mathbf{r}$ in Example 10.15.

**10.16** Continue Example 10.16. Decode the following received vectors.
(a) $\mathbf{r}_0 = (?\ ?\ 0\ 0\ 1\ 1\ 1\ 1\ 1\ 1\ 1\ 1\ 1\ 1\ 1)$.
(b) $\mathbf{r}_1 = (?\ ?\ 1\ 1\ ?\ ?\ 1\ 1\ 1\ 1\ 1\ 1\ 1\ 1\ 1)$.
(c) $\mathbf{r}_2 = (?\ ?\ ?\ 0\ 0\ 1\ 1\ 1\ 1\ 1\ 1\ 1\ 1\ 1\ 1)$.
(d) $\mathbf{r}_3 = (?\ ?\ ?\ ?\ ?\ ?\ ?\ 1\ 1\ 1\ 1\ 1\ 1\ 1\ 1)$.

**10.17** Consider two RS codes, an $(n_1, n_1 - 2t_1)$ RS code $C_{\mathrm{RS},1}$ over GF($2^m$) and an $(n_2, n_2 - 2t_2)$ RS code $C_{\mathrm{RS},2}$ over GF($2^m$). Use these two RS codes as the component codes to construct a product code $C_{\mathrm{RS},1} \times C_{\mathrm{RS},2}$. Discuss the error- and erasure-correcting capabilities of the product code $C_{\mathrm{RS},1} \times C_{\mathrm{RS},2}$.

# References

[1] S. H. Reiger, "Codes for the correction of 'clustered' errors," *IRE Trans. Inf. Theory*, **IT-6**(1) (1960), 16–21.

[2] P. Fire, *A Class of Multiple-Error-Correcting Binary Codes for Non-Independent Errors*, Stanford Electronics Laboratories, March 1959.

[3] N. M. Abramson, "A class of systematic codes for non-independent errors," *IRE Trans. Inf. Theory*, **5**(4) (1959), 150–157.

[4] J. E. Meggitt, "Error correcting codes for correcting bursts of errors," *IBM J. Res. Dev.*, **4**(3) (1960), 708–722.

[5] B. Elspas and R. A. Short, "A note on optimum burst-error-correcting codes," *IRE Trans. Inf. Theory*, **8**(1) (1962), 39–42.

[6] A. J. Gross, "Augmented Bose–Chaudhuri codes which correct single bursts of errors," *IEEE Trans. Inf. Theory*, **9**(2) (1963), 121.

[7] E. Gorog, "Some new classes of cyclic codes used for burst error correction," *IBM J. Res. Dev.*, **7** (1963), 102–111.

[8] T. Kasami and S. Matoba, "Some efficient shortened cyclic codes for burst-error correction," *IEEE Trans. Inf. Theory*, **10**(3) (1964), 252–253.

[9] R. T. Chien, "Burst-correcting codes with high-speed decoding," *IEEE Trans. Inf. Theory*, **15**(1) (1969), 109–113.

[10] W. Wagner, "Best Fire codes with length up to 1200 bits (Corresp.)," *IEEE Trans. Inf. Theory*, **16**(5) (1970), 649–650.

[11] H. O. Burton, "Some asymptotically optimal burst-correcting codes and their relation to single-error-correcting Reed–Solomon codes," *IEEE Trans. Inf. Theory*, **17**(1) (1971), 92–95.

[12] H. J. Matt and J. L. Massey, "Determining the burst-correcting limit of cyclic codes," *IEEE Trans. Inf. Theory*, **26**(3) (1980), 289–297.

[13] K. Abdel-Ghaffar, R. J. McEliece, A. M. Odlyzko, and H. C. A. van Tilborg, "On the existence of optimum cyclic burst-correcting codes," *IEEE Trans. Inf. Theory*, **32**(6) (1986), 768–775.

[14] H. C. A. van Tilborg, "Fire codes revisited," *Discrete Math.*, **106** (1992), 479–482.

[15] K. Abdel-Ghaffar, "Achieving the Reiger bound for burst errors using two-dimensional interleaving schemes," *Proc. IEEE Int. Symp. Inf. Theory*, Ulm, Germany, 1997, p. 425.

[16] T. Etzion, "Constructions for perfect 2-burst-correcting codes," *IEEE Trans. Inf. Theory*, **47**(6) (2001), 2553–2555.

[17] M. G. Luby, M. Mitzenmacher, M. A. Sokrollahi, and D. A. Spilman, "Efficient erasure correcting codes," *IEEE Trans. Inf. Theory*, **47**(2) (2001), 569–584.

[18] P. Oswald and A. Shokrollahi, "Capacity-achieving sequences for erasure channels," *IEEE Trans. Inf. Theory*, **48**(12) (2002), 3017–3028.

[19] H. D. Pfister, I. Sason, and R. L. Urbanke, "Capacity-approaching ensembles for the binary erasure channel with bounded complexity," *IEEE Trans. Inf. Theory*, **51**(7) (2005), 2352–2379.

[20] S. Song, S. Lin, and K. Addel-Ghaffar, "Burst-correction decoding of cyclic LDPC codes," *Proc. IEEE Int. Symp. Inf. Theory*, Seattle, WA, July 9–14, 2006, pp. 1718–1722.

[21] S. Song, S. Lin, K. Abdel-Ghaffar, and W. Fong, "Erasure-burst and error-burst decoding of linear codes," *Proc. IEEE Inf. Theory Workshop*, Lake Tahoe, CA, September 2–6, 2007, pp. 132–137.

[22] W. W. Peterson and E. J. Weldon, Jr., *Error-Correcting Codes*, 2nd ed., Cambridge, MA, The MIT Press, 1972.

[23] R. J. McEliece, *The Theory of Information Theory and Coding*, Cambridge, Cambridge University Press, 2002.

[24] R. E. Blahut, *Algebraic Codes for Data Transmission*, Cambridge, Cambridge University Press, 2003.

[25] S. Lin and D. J. Costello, Jr., *Error Control Coding: Fundamentals and Applications*, 2nd ed., Upper Saddle River, NJ, Prentice-Hall, 2004.

[26] R. G. Gallager, *Information Theory and Reliable Communication*, Hoboken, NJ, John Wiley & Sons, Inc., USA, 1968.

[27] W. Zhou, S. Lin, and K. Abdel-Ghaffar, "On the maximum true burst-correcting capability of Fire codes," *IEEE Trans. Inf. Theory*, **62**(10) (2016), 5323–5342.

# 11

# Introduction to Low-Density Parity-Check Codes

Low-density parity-check (LDPC) codes form a class of linear block codes whose parity-check matrices are low-density (or sparse). This class of codes can achieve *near-capacity* (or *near Shannon limit*) [1] performance on various communication and data-storage channels. LDPC codes were discovered by Gallager in 1962 [2, 3]. Unfortunately, Gallager's outstanding work was overlooked by coding theorists for almost 35 years with one exception, which is the work of Tanner published in 1982 [4]. In his work, Tanner gave a generalization of LDPC codes and introduced a *graphical representation* of LDPC codes, now called a *Tanner graph*. Tanner's work was again overlooked for another 13 years. The significance of LDPC codes was eventually fully recognized in the middle of the 1990s with the work of MacKay, Luby, and others [5–8], who noticed, apparently independently of Gallager's work, the advantages of linear block codes with sparse (low-density) parity-check matrices.

Since then a great deal of research effort has been expended in design, construction, encoding, decoding algorithms, structural analysis, performance analysis, generalizations, and applications of these remarkable codes. Many LDPC codes have been adopted as standard codes for various current and next-generation communication systems, such as wireless (4G, 5G, etc.), optical, satellite, space, digital video broadcast (DVB), network communications, and others. Applications to high-density data-storage systems, such as flash memories and hard-disk drives, are found in data-storage products. This rapid dominance of LDPC codes in applications is the result of their capacity-approaching performance which can be achieved with *practically implementable iterative decoding algorithms based on belief-propagation* even for very long codes. Finite-geometry codes constructed in Chapter 7 form an important class of LDPC codes.

This chapter introduces LDPC codes and creates a foundation for further study of LDPC codes in later chapters. We start with the fundamental representations of LDPC codes via parity-check matrices and Tanner graphs. We then learn about the decoding advantages of linear codes that possess sparse parity-check matrices. We will see that this sparseness characteristic makes the code amenable to various iterative decoding algorithms, which in many instances provide near-optimal performance. Gallager [2, 3] of course recognized the decoding advantages of such low-density parity-check codes and he proposed a decoding algorithm for the BI-AWGN channel and a few others for the BSC. These algorithms have received much scrutiny in the past decade and are still being studied. In this chapter, we present these decoding algorithms together with several others, most of which are related to Gallager's original algorithms. We point out that some of the algorithms were independently devised by other coding researchers (e.g., MacKay and Luby [6, 7]), who were unaware of Gallager's work at the time, as well as by researchers working on graph-based problems unrelated to coding [9].

We mainly focus on binary LDPC codes in this chapter. Properties and characteristics of nonbinary LDPC codes can be obtained in a similar way. See more details of decoding nonbinary LDPC codes in [10–15].

## 11.1 Definitions and Basic Concepts

Let $\mathbf{H}$ be a matrix over GF(2). Define the *density* of $\mathbf{H}$ as the ratio of the number of 1-entries in $\mathbf{H}$ to the total number of entries in $\mathbf{H}$. The matrix $\mathbf{H}$ is said to be a *low-density matrix* (or a *sparse matrix*) if its density is *much smaller than 1/2*.

**Definition 11.1** An LDPC code is a linear block code given by the null space over GF(2) of a sparse (or low-density) parity-check $\mathbf{H}$ over GF(2).

There are two types of LDPC codes, *regular* and *irregular*, as defined in the following two definitions.

**Definition 11.2** A *regular $(n, k)$ LDPC code* is defined as the null space of a *sparse* parity-check matrix $\mathbf{H}$ of size $m \times n$ over GF(2) with the following structural properties:
(1) each row of $\mathbf{H}$ has constant weight $\rho$,
(2) each column of $\mathbf{H}$ has constant weight $\gamma$,
(3) no two rows (or two columns) of $\mathbf{H}$ have more than one 1-component in common,
(4) both $\rho$ and $\gamma$ are small compared to $n$ and $m$, respectively, i.e., $\rho << n$ and $\gamma << m$,

where $m \geq (n - k)$, i.e., $\mathbf{H}$ may not be a full-rank matrix.

The matrix $\mathbf{H}$ given in Definition 11.2 is said to be $(\gamma, \rho)$-*regular* and the code given by the null space of $\mathbf{H}$ is called a $(\gamma, \rho)$-*regular $(n, k)$ LDPC code*.

Property (3) of the parity-check matrix $\mathbf{H}$ is referred to as the *row and column (RC) constraint*. The RC-constraint on the rows and columns of $\mathbf{H}$ ensures that no $\gamma$ or fewer columns of $\mathbf{H}$ can be added (vector sum) to zero. As a result, the minimum distance $d_{\min}$ of the LDPC code is at least $\gamma + 1$ (see Section 3.7). Property (4) simply ensures that the number of 1-entries in the parity-check matrix $\mathbf{H}$ is small compared to the total number of entries of $\mathbf{H}$. Therefore, $\mathbf{H}$ has a low density of 1-entries. It is noted that Property (3), i.e., the RC-constraint, is not required in the general definition of LDPC codes. The reason for imposing this constraint on the parity-check matrix $\mathbf{H}$ is that the decoding of the LDPC code based on such a parity-check matrix will achieve good performance as will be shown in later sections of this chapter and Chapters 12–15.

**Definition 11.3** An LDPC code is said to be *irregular* if its parity-check matrix $\mathbf{H}$ has *multiple column weights* and/or *multiple row weights*.

Regular and irregular LDPC codes perform differently in two different ranges of signal to noise ratios (SNRs). In the range of low to medium SNRs, well-designed irregular LDPC codes may perform better than regular LDPC codes [16, 17]. However, in the range of medium to high SNRs, well-designed regular LDPC codes may perform better than irregular LDPC codes, especially if they can achieve much lower error rates than irregular LDPC codes.

As defined in Chapter 2, a square matrix is called a circulant if each row is a *cyclic-shift* (one place to the right or to the left) of the row above it and the first row is the cyclic-shift of the last row. A linear block code given by the null space of a circulant or a column of circulants is a cyclic code.

**Definition 11.4** An LDPC code is said to be *cyclic* if its parity-check matrix is a column of *circulants* of the same size.

**Definition 11.5** An LDPC code is said to be *quasi-cyclic (QC)* if its parity-check matrix is an *array* of circulants of the same size.

LDPC codes without cyclic or QC structures are called *general LDPC codes*. As shown in Chapter 4, encoding of a cyclic or a QC-LDPC code can be implemented with simple shift-register(s). A general LDPC code can be encoded through finding its generator matrix and performing matrix product (see Section 3.2). The complexities of decoders for LDPC codes can be greatly reduced by employing the cyclic or QC structures, as will be shown in Chapter 12.

**Example 11.1** Consider the following $5 \times 10$ matrix $\mathbf{H}$:

$$\mathbf{H} = \begin{bmatrix} 1 & 1 & 1 & 1 & 0 & 0 & 0 & 0 & 0 & 0 \\ 1 & 0 & 0 & 0 & 1 & 1 & 1 & 0 & 0 & 0 \\ 0 & 1 & 0 & 0 & 1 & 0 & 0 & 1 & 1 & 0 \\ 0 & 0 & 1 & 0 & 0 & 1 & 0 & 1 & 0 & 1 \\ 0 & 0 & 0 & 1 & 0 & 0 & 1 & 0 & 1 & 1 \end{bmatrix}. \tag{11.1}$$

Each column of $\mathbf{H}$ has weight 2 and each row of $\mathbf{H}$ has weight 4. Thus, $\mathbf{H}$ is a $(2, 4)$-regular matrix. No two rows (or two columns) in $\mathbf{H}$ have more than one 1-entry in common. Hence, it satisfies the RC-constraint. It is easy to compute the rank of **H** which is 4. Hence, **H** is not full-rank and has one redundant row. The null space over GF(2) of **H** gives a $(2, 4)$-regular $(10, 6)$ LDPC code $C$. ▲▲

**Example 11.2** In this example, we consider a matrix with multiple column weights:

$$\mathbf{H} = \begin{bmatrix} 0 & 0 & 0 & 0 & 0 & 0 & 0 & 1 & 1 & 0 & 1 & 0 & 0 & 0 & 1 \\ 1 & 0 & 0 & 0 & 0 & 0 & 0 & 0 & 1 & 1 & 0 & 1 & 0 & 0 & 0 \\ 0 & 1 & 0 & 0 & 0 & 0 & 0 & 0 & 0 & 1 & 1 & 0 & 1 & 0 & 0 \\ 0 & 0 & 1 & 0 & 0 & 0 & 0 & 0 & 0 & 0 & 1 & 1 & 0 & 1 & 0 \\ 0 & 0 & 0 & 1 & 0 & 0 & 0 & 0 & 0 & 0 & 0 & 1 & 1 & 0 & 1 \\ 1 & 0 & 0 & 0 & 1 & 0 & 0 & 0 & 0 & 0 & 0 & 0 & 1 & 1 & 0 \\ 0 & 1 & 0 & 0 & 0 & 1 & 0 & 0 & 0 & 0 & 0 & 0 & 0 & 1 & 1 \\ 1 & 0 & 1 & 0 & 0 & 0 & 1 & 0 & 0 & 0 & 0 & 0 & 0 & 0 & 1 \end{bmatrix}, \tag{11.2}$$

which is an $8 \times 15$ matrix. The matrix **H** has constant row weight 4 and multiple column weights, five columns with weight 1, four columns with weight 2, five columns with weight 3, and one column with weight 4. **H** satisfies the RC-constraint and has rank 8. The null space of **H** gives an irregular $(15, 7)$ LDPC code. ▲▲

**Example 11.3** Consider the matrix given below:

$$\mathbf{H} = \begin{bmatrix} 1 & 0 & 1 & 1 & 0 & 0 & 0 \\ 0 & 1 & 0 & 1 & 1 & 0 & 0 \\ 0 & 0 & 1 & 0 & 1 & 1 & 0 \\ 0 & 0 & 0 & 1 & 0 & 1 & 1 \\ 1 & 0 & 0 & 0 & 1 & 0 & 1 \\ 1 & 1 & 0 & 0 & 0 & 1 & 0 \\ 0 & 1 & 1 & 0 & 0 & 0 & 1 \end{bmatrix}.$$

This matrix has both column and row weights 3. We can easily check that there are no two rows (or two columns) of **H** with more than one 1-entry in common. Therefore, the matrix satisfies the RC-constraint. It is a $(3, 3)$-regular matrix. We also note that **H** has a cyclic structure, i.e., **H** is a circulant. The rank of **H** is 4. Hence, **H** has three redundant rows. The null space of **H** gives a $(7, 3)$ cyclic LDPC code with minimum distance 4 (column weight $+ 1$). ▲▲

**Example 11.4** Consider the following $15 \times 15$ matrix:

$$\mathbf{H} = \begin{bmatrix} 1 & 0 & 0 & 0 & 1 & 0 & 1 & 0 & 0 & 0 & 0 & 0 & 0 & 0 & 1 \\ 0 & 1 & 0 & 0 & 0 & 1 & 0 & 1 & 0 & 0 & 0 & 0 & 1 & 0 & 0 \\ 0 & 0 & 1 & 1 & 0 & 0 & 0 & 0 & 1 & 0 & 0 & 0 & 0 & 1 & 0 \\ 1 & 0 & 0 & 1 & 0 & 0 & 0 & 1 & 0 & 1 & 0 & 0 & 0 & 0 & 0 \\ 0 & 1 & 0 & 0 & 1 & 0 & 0 & 0 & 1 & 0 & 1 & 0 & 0 & 0 & 0 \\ 0 & 0 & 1 & 0 & 0 & 1 & 1 & 0 & 0 & 0 & 0 & 1 & 0 & 0 & 0 \\ 0 & 0 & 0 & 1 & 0 & 0 & 1 & 0 & 0 & 0 & 1 & 0 & 1 & 0 & 0 \\ 0 & 0 & 0 & 0 & 1 & 0 & 0 & 1 & 0 & 0 & 0 & 1 & 0 & 1 & 0 \\ 0 & 0 & 0 & 0 & 0 & 1 & 0 & 0 & 1 & 1 & 0 & 0 & 0 & 0 & 1 \\ 0 & 1 & 0 & 0 & 0 & 0 & 1 & 0 & 0 & 1 & 0 & 0 & 0 & 1 & 0 \\ 0 & 0 & 1 & 0 & 0 & 0 & 0 & 1 & 0 & 0 & 1 & 0 & 0 & 0 & 1 \\ 1 & 0 & 0 & 0 & 0 & 0 & 0 & 0 & 1 & 0 & 0 & 1 & 1 & 0 & 0 \\ 0 & 0 & 1 & 0 & 1 & 0 & 0 & 0 & 0 & 1 & 0 & 0 & 1 & 0 & 0 \\ 1 & 0 & 0 & 0 & 0 & 1 & 0 & 0 & 0 & 0 & 1 & 0 & 0 & 1 & 0 \\ 0 & 1 & 0 & 1 & 0 & 0 & 0 & 0 & 0 & 0 & 0 & 1 & 0 & 0 & 1 \end{bmatrix},$$

which is a $5 \times 5$ array of circulant permutation matrices and zeros matrices of size $3 \times 3$. The matrix $\mathbf{H}$ has column weight 4 and row weight 4 and rank 8. The null space of $\mathbf{H}$ gives a $(4,4)$-regular $(15,7)$ QC-LDPC code. ▲▲

Consider a binary code $C$ given by the null space of an $m \times n$ parity-check matrix $\mathbf{H} = [h_{i,j}]_{0 \leq i < m, 0 \leq j < n}$ over GF(2). Let $\mathbf{h}_0, \mathbf{h}_1, \ldots, \mathbf{h}_{m-1}$ be the $m$ rows of $\mathbf{H}$ where for $0 \leq i < m$, the $i$th row $\mathbf{h}_i$ is

$$\mathbf{h}_i = (h_{i,0}, h_{i,1}, \ldots, h_{i,n-1}).$$

An $n$-tuple $\mathbf{v}$ over GF(2), $\mathbf{v} = (v_0, v_1, \ldots, v_{n-1})$, is a codeword of $C$ if and only if $\mathbf{v}\mathbf{H}^T = \mathbf{0}$ (see Section 3.2). Let $\mathbf{v}$ be a codeword in $C$. Then for $0 \leq i < m$,

$$s_i = \langle \mathbf{v}, \mathbf{h}_i \rangle = v_0 h_{i,0} + v_1 h_{i,1} + \cdots + v_{n-1} h_{i,n-1} = 0. \qquad (11.3)$$

The $m$ sums, $s_0, s_1, \ldots, s_{m-1}$, are called *parity-check sums* that give $m$ *constraints* on the $n$ code bits (or symbols), $v_0, v_1, \ldots, v_{n-1}$, of a codeword in $C$. A code bit $v_j$ is said to be *checked* by the parity-check sum $s_i$ (or the row $\mathbf{h}_i$) if $h_{i,j} = 1$, i.e., the code bit $v_j$ is contained in the parity-check sum $s_i$.

**Example 11.5** Consider the $(10,6)$ LDPC code $C$ in Example 11.1 with a low-density parity-check matrix $\mathbf{H}$ given by (11.1). Let $\mathbf{v} = (v_0, v_1, \ldots, v_9)$ be a codeword in $C$. Then, we have $\mathbf{v}\mathbf{H}^T = \mathbf{0}$ which results in the following five parity-check sums (constraints):

$$\begin{aligned}
s_0 &= v_0 + v_1 + v_2 + v_3 = 0, \\
s_1 &= v_0 + v_4 + v_5 + v_6 = 0, \\
s_2 &= v_1 + v_4 + v_7 + v_8 = 0, \\
s_3 &= v_2 + v_5 + v_7 + v_9 = 0, \\
s_4 &= v_3 + v_6 + v_8 + v_9 = 0.
\end{aligned}$$

Each parity-check sum checks on four code bits and each code bit is checked by two different parity-check sums. Note that the code bit $v_2$ is checked by two parity-check sums $s_0$ and $s_3$. ▲▲

## 11.2 Graphical Representation of LDPC Codes

As shown in Section 2.10, a bipartite graph can be represented by its adjacency matrix, and vice versa. Recall that a bipartite graph $\mathcal{G} = (\mathcal{W}, \mathcal{E})$ consists of a set of nodes, $\mathcal{W}$, and a set of edges, $\mathcal{E}$. The node set $\mathcal{W}$ of $\mathcal{G}$ consists of two disjoint subsets $\mathcal{V}$ and $\mathcal{C}$ such that every edge in $\mathcal{E}$ connects a node in $\mathcal{V}$ with a node in $\mathcal{C}$ and does not connect two nodes that are both either in $\mathcal{V}$ or in $\mathcal{C}$. The adjacency matrix $\mathbf{A}(\mathcal{G}) = [a_{i,j}]_{0 \leq i < m, 0 \leq j < n}$ for $\mathcal{G}$ is an $m \times n$ matrix in which the columns correspond to the $n$ nodes in $\mathcal{V}$, the rows correspond to the $m$ nodes in $\mathcal{C}$, and $a_{i,j} = \ell$ if there are $\ell$ parallel edges connecting nodes $c_i$ and $v_j$; otherwise, $a_{i,j} = 0$. If $\mathbf{A}(\mathcal{G})$ is a binary matrix, there are no parallel edges in $\mathcal{G}$. A bipartite graph without parallel edges is called a *simple bipartite graph*. In this chapter, we consider only simple bipartite graphs.

An $(n, k)$ binary linear block code is completely specified by its parity-check matrix of size $m \times n$ ($m \geq n - k$) which can be represented by a bipartite graph. Thus, a binary linear block code $C$ given by the null space of a parity-check matrix $\mathbf{H}$ can be represented by a bipartite graph. The construction of such a bipartite graph is given below.

(1) The first subset $\mathcal{V}$ of nodes consists of $n$ nodes that represent the $n$ code bits of a codeword, i.e., correspond to the $n$ columns in the parity-check matrix $\mathbf{H}$. The $n$ nodes in $\mathcal{V}$, denoted by $v_0, v_1, \ldots, v_{n-1}$, are called *variable nodes* (VNs).[1]

(2) The second subset $\mathcal{C}$ of nodes consists of $m$ nodes that represent the $m$ parity-check constraints given by (11.3) that the code bits must satisfy, i.e., correspond to the $m$ rows in $\mathbf{H}$. The $m$ nodes in $\mathcal{C}$, denoted by $c_0, c_1, \ldots, c_{m-1}$, are called *check nodes* (CNs).

(3) A VN $v_j$ is connected to a CN $c_i$ by an edge if and only if the code bit $v_j$ is checked by the parity-check sum $s_i$, i.e., the entry $h_{i,j}$ in $\mathbf{H}$ is nonzero.

(4) No two VNs in $\mathcal{V}$ (or two CNs in $\mathcal{C}$) are directly connected.

This graphical representation of a binary linear block code is called the *Tanner graph* $\mathcal{G}$ [4] of the code which was first introduced by Tanner to study iterative decoding. The parity-check matrix $\mathbf{H}$ is the adjacency matrix of $\mathcal{G}$. Because $\mathbf{H}$ contains only 0s and 1s, there is no parallel edges in the Tanner graph $\mathcal{G}$. Based on the construction of Tanner graph for the code, the degree of the VN $v_j$ in $\mathcal{V}$ equals the weight of the $j$th column of $\mathbf{H}$, the degree of the CN $c_i$ in $\mathcal{C}$ equals the weight of the $i$th row of $\mathbf{H}$, and the number of edges in $\mathcal{G}$ equals the number of 1s in the parity-check matrix $\mathbf{H}$.

If the parity-check matrix $\mathbf{H}$ of a binary linear block code satisfies the RC-constraint, then no two code bits can be checked *simultaneously* by two different parity-check sums. As a result, no two VNs are simultaneously connected to two CNs in the Tanner graph of the code. Therefore, the Tanner graph of the code is free of cycles of length 4 and has a girth (the length of shortest cycle) of at least 6 (see Section 2.10.2).

A linear block code may be specified by different parity-check matrices leading to different Tanner graphs which may have different cycle distributions. Because we require that the parity-check matrix of an LDPC code, regular or irregular, satisfies the RC-constraint, the Tanner graph of such an LDPC code has girth at least 6.

The Tanner graph gives a good insight into iterative decoding of LDPC codes [4]. For an LDPC code to perform well with iterative decoding, its Tanner graph must not contain *short cycles* [14, 18], in other words, its girth must be relatively large. Cycles that affect the code performance the most are those of length 4. Therefore, in construction of LDPC codes, cycles of length 4 should be avoided. This is the reason that the RC-constraint on the parity-check matrix of an LDPC code is imposed.

In the literature, the VN and CN degree distributions of the Tanner graph of an LDPC code, especially an irregular LDPC code, are expressed by two

---

[1]Throughout this book, we use the same notation for both code bits and VNs.

polynomials, $\lambda(X)$ and $\rho(X)$, called *VN* and *CN degree distribution polynomials*, respectively. The VN degree distribution polynomial is specified as

$$\lambda(X) = \sum_{d=1}^{d_v} \lambda_d X^{d-1}, \tag{11.4}$$

where $\lambda_d$ denotes the fraction of all edges connected to VNs of degree $d$, and $d_v$ denotes the maximum VN degree. Similarly, the CN degree distribution polynomial is specified as

$$\rho(X) = \sum_{d=1}^{d_c} \rho_d X^{d-1}, \tag{11.5}$$

where $\rho_d$ denotes the fraction of all edges connected to CNs of degree $d$, and $d_c$ denotes the maximum CN degree. Note that, for a $(d_v, d_c)$-regular LDPC code, there is only one term in $\lambda(X)$ and $\rho(X)$, i.e., $\lambda(X) = X^{d_v-1}$ and $\rho(X) = X^{d_c-1}$.

Let us denote the number of VNs of degree $d$ by $N_v(d)$, the number of CNs of degree $d$ by $N_c(d)$, and the number of edges in the graph by $N_\mathcal{E}$. Then, it can be shown that

$$N_\mathcal{E} = \frac{n}{\int_0^1 \lambda(X)dX} = \frac{m}{\int_0^1 \rho(X)dX}, \tag{11.6}$$

$$N_v(d) = N_\mathcal{E}\lambda_d/d = \frac{n\lambda_d/d}{\int_0^1 \lambda(X)dX}, \tag{11.7}$$

$$N_c(d) = N_\mathcal{E}\rho_d/d = \frac{m\rho_d/d}{\int_0^1 \rho(X)dX}. \tag{11.8}$$

From the two expressions for $N_\mathcal{E}$, we can readily see that the code rate is lower bounded as follows:

$$R \geq 1 - \frac{m}{n} = 1 - \frac{\int_0^1 \rho(X)dX}{\int_0^1 \lambda(X)dX}. \tag{11.9}$$

The polynomials $\lambda(X)$ and $\rho(X)$ represent a Tanner graph's degree distributions from an "edge perspective" point of view, called *edge-perspective degree distribution polynomials*. The degree distributions may also be represented from a "node perspective" point of view using the notation $\tilde{\lambda}(X) = \sum_{d=1}^{d_v} \tilde{\lambda}_d X^{d-1}$ and $\tilde{\rho}(X) = \sum_{d=1}^{d_c} \tilde{\rho}_d X^{d-1}$, called *node-perspective VN and CN degree distribution polynomials*, to represent the VN and CN degree distributions, respectively. The coefficient $\tilde{\lambda}_d$ of $\tilde{\lambda}(X)$ is the fraction of all VNs that have degree $d$. The coefficient $\tilde{\rho}_d$ of $\tilde{\rho}(X)$ is the fraction of CNs that have degree $d$. Then, we have

$$\tilde{\lambda}_d = \frac{\lambda_d/d}{\int_0^1 \lambda(X)dX}, \tag{11.10}$$

$$\tilde{\rho}_d = \frac{\rho_d/d}{\int_0^1 \rho(X)dX}. \tag{11.11}$$

Note that, for a regular code, its edge-perspective VN and CN degree distribution polynomials are the same as its node-perspective ones.

**Example 11.6** Consider the $(10, 6)$ LDPC code $C$ given in Example 11.1 with a parity-check matrix $\mathbf{H}$ given by (11.1). The Tanner graph $\mathcal{G}$ of the code is depicted in Fig. 11.1. The graph $\mathcal{G}$ has 10 VNs which correspond to the 10 code bits of a codeword in $C$ (or the 10 columns of its parity-check matrix $\mathbf{H}$), and five CNs which correspond to the five parity-check sums of the code (or the five rows of its parity-check matrix $\mathbf{H}$).

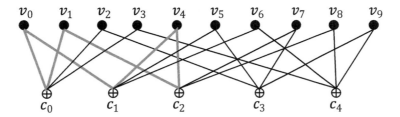

Figure 11.1 The Tanner graph of the $(10, 6)$ LDPC code given in Example 11.1.

From Fig. 11.1, we observe that VNs $v_0$, $v_1$, $v_2$, and $v_3$ are connected to CN $c_0$ in accordance with the fact that, in the zeroth row of $\mathbf{H}$, $h_{0,0} = h_{0,1} = h_{0,2} = h_{0,3} = 1$ (all other entries are zero). We also observe that analogous situation holds for CNs $c_1$, $c_2$, $c_3$, and $c_4$, which correspond to the first, second, third, and fourth rows of $\mathbf{H}$, respectively.

We can easily see that there is no cycle of length 4 in the Tanner graph in accordance with the fact that $\mathbf{H}$ satisfies the RC-constraint. A cycle of length 6, $(v_0 - c_0 - v_1 - c_2 - v_4 - c_1 - v_0)$, is marked by boldface lines in Fig. 11.1. Thus, the girth of the Tanner graph is 6. Moreover, this LDPC code has the edge/node perspective degree distribution polynomials: $\lambda(X) = \tilde{\lambda}(X) = X$, $\rho(X) = \tilde{\rho}(X) = X^3$. Because the code is regular, there is only one term in both of the degree distribution polynomials. ▲▲

**Example 11.7** In this example, we consider the irregular $(15, 7)$ LDPC code given in Example 11.2 with a parity-check matrix specified by (11.2). The Tanner graph $\mathcal{G}$ of this LDPC code is shown in Fig. 11.2.

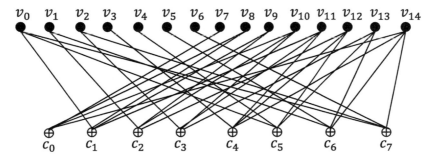

Figure 11.2 The Tanner graph of the $(15, 7)$ LDPC code given in Example 11.2.

The Tanner graph $\mathcal{G}$ has 15 VNs, 8 CNs, and 32 edges. Each CN has degree 4. There are 5 VNs with degree 1, 4 VNs with degree 2, 5 VNs with degree 3, and 1 VN with degree 4. For the 5 VNs with degree 1, there are 5 edges connected to them; then, $\lambda_1 = 5/32 = 0.156\,25$. For the 4 VNs with degree 2, there are 8 edges connected to them; then, $\lambda_2 = 8/32 = 0.25$. Similarly, we have $\lambda_3 = 15/32 = 0.46875$ and $\lambda_4 = 4/32 = 0.125$. Then, the edge perspective degree distribution polynomial for VNs is $\lambda(X) = 0.156\,25 + 0.25X + 0.468\,75X^2 + 0.125X^3$. Because every CN has degree 4, we have $\rho(X) = X^3$.

For node perspective, there are 15 VNs and 5 of them with degree 1, i.e., $\tilde{\lambda}_1 = 5/15 = 0.3333$; there are 4 VNs with degree 2, i.e., $\tilde{\lambda}_2 = 4/15 = 0.2667$. Similarly, we have $\tilde{\lambda}_3 = 5/15 = 0.3333$ and $\tilde{\lambda}_4 = 1/15 = 0.0667$. Then, the node perspective degree distribution polynomial for VNs is $\tilde{\lambda}(X) = 0.3333 + 0.2667X + 0.3333X^2 + 0.0667X^3$. Because every CN has degree 4, we have $\tilde{\rho}_4 = 8/8 = 1$ and $\tilde{\rho}(X) = X^3$.

Because the parity-check matrix $\mathbf{H}$ satisfies the RC-constraint, there is no cycle of length 4 in $\mathcal{G}$. A cycle of length 6, $(v_8 - c_0 - v_{14} - c_4 - v_{11} - c_1 - v_8)$ exists in $\mathcal{G}$. Therefore, the girth of the Tanner graph is 6. ▲▲

Because the parity-check matrix $\mathbf{H}$ of an LDPC code is the adjacency matrix of the Tanner graph $\mathcal{G}$ of the code whose columns and rows correspond to the VNs and CNs in $\mathcal{G}$, the VN and CN degree distributions (node perspective) of $\mathcal{G}$ are the column and row weight distributions of $\mathbf{H}$, respectively.

## 11.3 Original Construction of LDPC Codes

Some linear block codes may have low-density parity-check matrices; thus, they can be viewed as LDPC codes, e.g., the $(7, 3)$ cyclic code in Example 11.3. However, not all of linear block codes have a low-density parity-check matrix or it is hard to find such a matrix for them. Therefore, it is desirable to have methods to construct directly parity-check matrices which are low-density in nature.

Unlike BCH and RS codes, for which there exists essentially a single code design procedure, there are many code design approaches for LDPC codes. Many of these design approaches, including the original one proposed by Gallager, are computer-based search algorithms. Many others rely on finite mathematics, such as finite geometries [19–22], finite fields, and experimental designs. LDPC codes constructed based on finite mathematics are either cyclic or QC-LDPC codes and in general they have distinctive algebraic and/or geometric structures which facilitate the hardware implementations of their encoder and decoder.

In this section, we briefly review the original constructions of LDPC codes devised by Gallager [2, 3] and Mackay [5, 6], which are computer-based algorithms. Other constructions of LDPC codes will be discussed in Chapters 12–15.

### 11.3.1 Gallager Codes

Let $b$ be a positive integer greater than 1. To construct the parity-check matrix of a Gallager LDPC code, we first construct a $b \times b\rho$ matrix $\mathbf{H}_0$ over GF(2) with column and row weights 1 and $\rho$, respectively, where $\rho > 1$. Label the rows and columns of $\mathbf{H}_0$ from 0 to $b-1$ and 0 to $b\rho - 1$, respectively. The matrix $\mathbf{H}_0$ has the following specific form: for $i = 0, 1, \ldots, b-1$, the $i$th row contains all of its $\rho$ 1-entries in the $(i\rho)$th to $((i+1)\rho - 1)$th *consecutive* positions and 0-entries elsewhere. The matrix $\mathbf{H}_0$ is called the *base matrix*. Let $\gamma$ be a positive integer such that $1 \leq \gamma \leq \rho$. Next, we find $\gamma - 1$ permutations to permute the columns of $\mathbf{H}_0$ to form $\gamma - 1$ matrices, $\mathbf{H}_1, \mathbf{H}_2, \ldots, \mathbf{H}_{\gamma-1}$, each of size $b \times b\rho$. Then, using the $\gamma$ matrices $\mathbf{H}_0, \mathbf{H}_1, \mathbf{H}_2, \ldots, \mathbf{H}_{\gamma-1}$, we form the following $b\gamma \times b\rho$ matrix:

$$
\mathbf{H} = \begin{bmatrix} \mathbf{H}_0 \\ \mathbf{H}_1 \\ \vdots \\ \mathbf{H}_{\gamma-1} \end{bmatrix}. \tag{11.12}
$$

The matrix $\mathbf{H}$ has constant column and row weights $\gamma$ and $\rho$, respectively. The number of 1-entries in $\mathbf{H}$ is $b\gamma\rho$. The density of $\mathbf{H}$ is $1/b$. If $b$ is large, $\mathbf{H}$ is a low-density matrix. The null space over GF(2) of $\mathbf{H}$ gives a Gallager LDPC code of length $n = b\rho$ and rate at least $(\rho - \gamma)/\rho$. It is a $(\gamma, \rho)$-regular LDPC code.

Gallager showed that LDPC codes with parity-check matrices in the form of (11.12) form a class of *capacity-* (or *Shannon-limit*) approaching codes. However, he did not provide a specific way of finding permutations to permute the columns of the base matrix $\mathbf{H}_0$ to obtain a parity-check matrix $\mathbf{H}$ such that the LDPC code specified by $\mathbf{H}$ has a good distance structure and its Tanner graph does not contain short cycles, especially cycles of length 4 (i.e., to satisfy the RC-constraint). Hence, computer search is needed to find column permutations for the base matrix $\mathbf{H}_0$ to produce an LDPC code with good error performance using iterative decoding based on belief-propagation.

**Example 11.8** The following parity-check matrix $\mathbf{H}$ is the first example given by Gallager [2, 3]

$$
\mathbf{H} = \begin{bmatrix} \mathbf{H}_0 \\ \mathbf{H}_1 \\ \mathbf{H}_2 \end{bmatrix} = \left[ \begin{array}{cccccccccccccccccccc}
1&1&1&1&0&0&0&0&0&0&0&0&0&0&0&0&0&0&0&0\\
0&0&0&0&1&1&1&1&0&0&0&0&0&0&0&0&0&0&0&0\\
0&0&0&0&0&0&0&0&1&1&1&1&0&0&0&0&0&0&0&0\\
0&0&0&0&0&0&0&0&0&0&0&0&1&1&1&1&0&0&0&0\\
0&0&0&0&0&0&0&0&0&0&0&0&0&0&0&0&1&1&1&1\\
1&0&0&0&1&0&0&0&1&0&0&0&1&0&0&0&0&0&0&0\\
0&1&0&0&0&1&0&0&0&1&0&0&0&0&0&1&0&0&0\\
0&0&1&0&0&0&1&0&0&0&0&0&1&0&0&0&1&0&0\\
0&0&0&1&0&0&0&0&0&1&0&0&0&1&0&0&0&1&0\\
0&0&0&0&0&0&0&1&0&0&0&1&0&0&0&1&0&0&0&1\\
1&0&0&0&0&1&0&0&0&0&1&0&0&0&0&0&1&0&0\\
0&1&0&0&0&0&1&0&0&0&1&0&0&0&0&1&0&0&0\\
0&0&1&0&0&0&0&1&0&0&0&0&1&0&0&0&0&0&1&0\\
0&0&0&1&0&0&0&0&1&0&0&0&0&1&0&0&1&0&0&0\\
0&0&0&0&1&0&0&0&0&1&0&0&0&0&1&0&0&0&0&1
\end{array} \right],
$$

which is a $15 \times 20$ matrix with column weight 3 and row weight 4, where $b = 5$, $\rho = 4$, and $\gamma = 3$. **H** has rank 15, i.e., it is a full-rank matrix and satisfies the RC-constraint. The null space over GF(2) of **H** is a $(3, 4)$-regular $(20, 5)$ LDPC code. ▲▲

### 11.3.2 MacKay Codes

Thirty-five years after Gallager's work on LDPC codes, MacKay, unaware of Gallager's work, independently discovered the benefits of designing binary codes with sparse parity-check matrices and was the first one to show by computer simulation that these codes perform close to capacity limits on BSCs and AWGN channels. MacKay archived on a web page [23] a large number of LDPC codes designed for applications to data communication and storage systems. A few of the computer-based design algorithms proposed by MacKay are listed below in the order of increasing algorithm complexity (but not necessarily improved performance).

(1) **H** is constructed by randomly generating columns of weight $\gamma$ and uniform row weight as nearly as possible.
(2) **H** is created by randomly generating columns of weight $\gamma$, while ensuring rows of weight $\rho$, and the ones-overlap of any two columns is at most one (i.e., ensuring the RC-constraint to avoid cycles of length 4).
(3) **H** is generated as method two with additional effort to avoid short cycles.
(4) **H** is generated as method three with one more condition: there is a sub-matrix $\mathbf{H}_2$ in $\mathbf{H} = [\mathbf{H}_1 \ \mathbf{H}_2]$ which is invertible for encoding purpose.

One drawback of the MacKay codes is that they lack sufficient structure to enable low-complexity encoding. Encoding is performed by transforming **H** into the form of $\begin{bmatrix} \mathbf{P}^T \ \mathbf{I} \end{bmatrix}$ via Gauss–Jordan elimination (or by multiplying **H** by $\mathbf{H}_2^{-1}$ for method (4)), from which the generator matrix can be obtained in systematic form $\mathbf{G} = [\mathbf{I} \ \mathbf{P}]$.[2] The problem with encoding via **G** is that the submatrix **P** is generally not sparse; so for codes of practical interest, the encoding complexity is high. An efficient encoding technique based on the parity-check matrix (in an arbitrary form) was proposed in [24]. In Chapters 12–15, we give various approaches for constructing parity-check matrices with structural properties so that encoding can be facilitated with essentially no performance loss relative to less-structured parity-check matrices.

## 11.4 Decoding of LDPC Codes

There are various methods for decoding LDPC codes, ranging from low to high decoding complexities and from reasonably good to very good error performance. The major decoding methods include: (1) *one-step majority-logic* (OSML) *decoding*; (2) *bit-flipping* (BF) *decoding*; (3) *weighted BF decoding*; (4)

---

[2]We note that Chapter 3 introduces the systematic generator matrix form $\mathbf{G} = [\mathbf{P} \ \mathbf{I}]$, whereas here we use the form $\mathbf{G} = [\mathbf{I} \ \mathbf{P}]$. Both are commonly used in the literature and we use both in this book.

weighted OSML decoding; and (5) *iterative decoding* based on *belief-propagation* (IDBP). The OSML and BF decodings are hard-decision decodings. The IDBP is a soft-input and soft-output (soft-in-soft-out; SISO) decoding method (soft-decision decoding). The weighted BF and OSML algorithms are in between, commonly referred to as reliability-based decoding algorithm. The OSML decoding (presented in Chapter 7) is simplest in terms of decoding complexity. The BF decoding requires slightly higher complexity but provides better error performance than the OSML decoding. The IDBP algorithm gives the best error performance among the five types of decoding algorithms but requires much higher decoding complexity than the others. The weighted BF and OSML algorithms offer a tradeoff between error performance and decoding complexity. This section will discuss the first four decoding algorithms for LDPC codes, starting from the one with the lowest decoding complexity. The IDBP decoding will be covered in the next section.

Suppose an $(n, k)$ LDPC code $C$ is applied for error control in an AWGN channel with zero mean and one-sided power spectral density (PSD) $N_0$. Assume BPSK signaling with unit energy. A codeword $\mathbf{v} = (v_0, v_1, \ldots, v_{n-1})$ in $C$ is sent out from the channel encoder. First, the codeword $\mathbf{v}$ is mapped into a bipolar sequence $\mathbf{x} = (x_0, x_1, \ldots, x_{n-1})$, where $x_i = 1 - 2v_i = +1$ for $v_i = 0$ and $x_i = 1 - 2v_i = -1$ for $v_i = 1$ with $0 \leq i < n$. Then, $\mathbf{x}$ is transmitted over the channel. Let $\mathbf{y} = (y_0, y_1, \ldots, y_{n-1})$ be the soft-decision received vector at the output of the channel detector. For $0 \leq i < n$, the $i$th received symbol is $y_i = x_i + n_i$, where $n_i$ is a Gaussian random variable with zero mean and variance $\sigma^2 = N_0/2$ (i.e., Gaussian noise). Let $\mathbf{r} = (r_0, r_1, \ldots, r_{n-1})$ be the binary hard-decision received vector obtained as follows:

$$r_i = \begin{cases} 0, & \text{for } y_i > 0, \\ 1, & \text{for } y_i \leq 0. \end{cases} \tag{11.13}$$

Let $\mathbf{H}$ be the parity-check matrix of the LDPC code $C$ with size $m \times n$. Let $\mathbf{h}_i = (h_{i,0}, h_{i,1}, \ldots, h_{i,n-1})$ be the $i$th row of $\mathbf{H}$, $0 \leq i < m$. To check whether the hard-decision received vector $\mathbf{r}$ is a valid codeword, we compute its syndrome

$$\mathbf{s} = \mathbf{r} \cdot \mathbf{H}^T = (s_0, s_1, \ldots, s_{m-1}).$$

If $\mathbf{s} = \mathbf{0}$, $\mathbf{r}$ is a codeword; thus, we can assume that $\mathbf{r}$ is error free and $\mathbf{v} = \mathbf{r}$. If $\mathbf{s} \neq \mathbf{0}$, errors in $\mathbf{r}$ are detected. Let $\mathbf{e} = (e_0, e_1, \ldots, e_{n-1})$ be the error vector, i.e.,

$$\mathbf{r} = \mathbf{v} + \mathbf{e}.$$

Then, the syndrome of the hard-decision received vector $\mathbf{r}$ is

$$\mathbf{s} = \mathbf{r} \cdot \mathbf{H}^T = (\mathbf{v} + \mathbf{e}) \cdot \mathbf{H} = \mathbf{e} \cdot \mathbf{H}^T,$$

where the $i$th syndrome component $s_i$ of $\mathbf{s}$ is given by

$$s_i = \langle \mathbf{e}, \mathbf{h}_i \rangle = \sum_{j=0}^{n-1} e_j h_{i,j},$$

which is the inner product of the error vector $\mathbf{e}$ and the $i$th row $\mathbf{h}_i$ of the parity-check matrix $\mathbf{H}$.

A hard-decision decoding is a method to find the most probable error vector $\mathbf{e}$ based on the computed syndrome $\mathbf{s}$.

## 11.4.1 One-Step Majority-Logic Decoding

The one-step majority-logic (OSML) decoding was presented in Chapter 7 to decode finite-geometry codes. Here, we briefly discuss how to apply this decoding algorithm to LDPC codes.

Consider a $(\gamma, \rho)$-regular LDPC code $C$ of length $n$ given by the null space of an $m \times n$ parity-check matrix $\mathbf{H}$ with column and row weights $\gamma$ and $\rho$, respectively. Let $\mathbf{h}_0, \mathbf{h}_1, \ldots, \mathbf{h}_{m-1}$ be the $m$ rows of $\mathbf{H}$. For each code bit position $l$, $0 \leq l < n$, due to the RC-constraint on the rows and columns of $\mathbf{H}$, there is a set

$$A_l = \{\mathbf{h}_1^{(l)}, \mathbf{h}_2^{(l)}, \ldots, \mathbf{h}_\gamma^{(l)}\}$$

of $\gamma$ rows in $\mathbf{H}$ such that: (1) each row in $A_l$ has a 1-component at position $l$, i.e., $h_{i,l}^{(l)} = 1$ for $1 \leq i \leq \gamma$; and (2) no two rows in $A_l$ have a common 1-component at any position other than $l$. The rows in $A_l$ are said to be *orthogonal* on the bit position $l$.

Suppose a codeword $\mathbf{v} = (v_0, v_1, \ldots, v_{n-1})$ in $C$ is transmitted and $\mathbf{r} = (r_0, r_1, \ldots, r_{n-1})$ is the hard-decision received vector. Then, $\mathbf{r} = \mathbf{v} + \mathbf{e}$ where $\mathbf{e} = (e_0, e_1, \ldots, e_{n-1})$ is the error pattern introduced by the noisy channel. The $i$th syndrome component of the syndrome $\mathbf{s}$ of $\mathbf{r}$ is related to the unknown error bits as follows:

$$s_i = \langle \mathbf{e}, \mathbf{h}_i \rangle = e_0 h_{i,0} + e_1 h_{i,1} + \cdots + e_{n-1} h_{i,n-1}, \tag{11.14}$$

for $0 \leq i < m$.

Equation (11.14) gives a set of $m$ equations that relate the $n$ error bits in $\mathbf{e}$ to the $m$ computed syndrome bits. Consider the following set of $\gamma$ syndrome bits computed based on the rows of the set $A_l$:

$$S_l = \{s_i^{(l)} = \langle \mathbf{e}, \mathbf{h}_i^{(l)} \rangle : \mathbf{h}_i^{(l)} \in A_l\}, \tag{11.15}$$

where

$$
\begin{aligned}
s_i^{(l)} &= \langle \mathbf{e}, \mathbf{h}_i^{(l)} \rangle \\
&= e_0 h_{i,0}^{(l)} + e_1 h_{i,1}^{(l)} + \cdots + e_{n-1} h_{i,n-1}^{(l)}, \\
&= e_l + \sum_{j=0, j \neq l}^{n-1} e_j h_{i,j}^{(l)}
\end{aligned}
\tag{11.16}
$$

for $1 \leq i \leq \gamma$. The set $S_l$ consists of $\gamma$ parity-check sums which are orthogonal on the error bit $e_l$.

Based on the analysis in Section 7.2, with the OSML decoding of a $(\gamma, \rho)$-regular LDPC code, correct decoding is guaranteed if there are $\lfloor \gamma/2 \rfloor$ or fewer errors in the received vector $\mathbf{r}$. The OSML decoding rule can be mathematically formulated as

$$\sum_{s_j^{(l)} \in S_l} s_j^{(l)} \stackrel{?}{>} \left\lfloor \frac{\gamma}{2} \right\rfloor, \tag{11.17}$$

i.e., to check whether the sum $\sum_{s_j^{(l)} \in S_l} s_j^{(l)}$ is larger than $\lfloor \frac{\gamma}{2} \rfloor$. If it does, decode $e_l$ as "1"; otherwise, decode $e_l$ as "0." The inequality in (11.17) can be rearranged in the following form

$$\sum_{s_j^{(l)} \in S_l} (2s_j^{(l)} - 1) \stackrel{?}{>} 0. \tag{11.18}$$

The sum $\sum_{s_j^{(l)} \in S_l} (2s_j^{(l)} - 1)$ is called the *OSML decision function*.

The above OSML decoding algorithm can be summarized as Algorithm 11.1. The loop in Algorithm 11.1 can be formulated as a matrix product over integers. Let $\mathbf{f} = (f_0, f_1, \ldots, f_{n-1})$ be an $n$-tuple over integers, where $f_l = \sum_{s_j^{(l)} \in S_l} (2s_j^{(l)} - 1)$ for $0 \le l < n$. Then, we have $\mathbf{f} = (2\mathbf{s} - \mathbf{1}) \cdot \mathbf{H}$ (where $\mathbf{1}$ is an all-one $m$-tuple and the operation is carried out over integers). Accordingly, the decision rule for the OSML decoding is: if $f_l > 0$, $e_l = 1$; otherwise, $e_l = 0$.

---

**Algorithm 11.1** OSML decoding algorithm for LDPC codes

---

Inputs: $\mathbf{r} = (r_0, r_1, \ldots, r_{n-1})$, $\mathbf{H}$.
Compute the syndrome: $\mathbf{s} = \mathbf{r} \cdot \mathbf{H}^T$.
**if $\mathbf{s} = \mathbf{0}$ then**
  Output $\mathbf{r}$ as the decoded codeword and declare a decoding success.
**else**
  **for** $l = 0, 1, 2, \ldots, n - 1$ **do**
    Compute the parity-check sums in the set $S_l$.
    Decode the error bit $e_l$ as follows:

$$e_l = \begin{cases} 1, & \text{if } \sum_{s_j^{(l)} \in S_l} (2s_j^{(l)} - 1) > 0, \\ 0, & \text{if } \sum_{s_j^{(l)} \in S_l} (2s_j^{(l)} - 1) \le 0. \end{cases}$$

  **end for**
  Compute the decoded vector $\mathbf{v}^* = \mathbf{r} + \mathbf{e}$. Declare a decoding success if $\mathbf{v}^* \cdot \mathbf{H}^T = \mathbf{0}$; otherwise, declare a decoding failure.
**end if**

---

**Example 11.9** Consider the following sparse $15 \times 15$ matrix:

$$\mathbf{H} = \begin{bmatrix} 0\,0\,0\,0\,0\,0\,0\,1\,1\,0\,1\,0\,0\,0\,1 \\ 1\,0\,0\,0\,0\,0\,0\,0\,1\,1\,0\,1\,0\,0\,0 \\ 0\,1\,0\,0\,0\,0\,0\,0\,0\,1\,1\,0\,1\,0\,0 \\ 0\,0\,1\,0\,0\,0\,0\,0\,0\,0\,1\,1\,0\,1\,0 \\ 0\,0\,0\,1\,0\,0\,0\,0\,0\,0\,0\,1\,1\,0\,1 \\ 1\,0\,0\,0\,1\,0\,0\,0\,0\,0\,0\,0\,1\,1\,0 \\ 0\,1\,0\,0\,0\,1\,0\,0\,0\,0\,0\,0\,0\,1\,1 \\ 1\,0\,1\,0\,0\,0\,1\,0\,0\,0\,0\,0\,0\,0\,1 \\ 1\,1\,0\,1\,0\,0\,0\,1\,0\,0\,0\,0\,0\,0\,0 \\ 0\,1\,1\,0\,1\,0\,0\,0\,1\,0\,0\,0\,0\,0\,0 \\ 0\,0\,1\,1\,0\,1\,0\,0\,0\,1\,0\,0\,0\,0\,0 \\ 0\,0\,0\,1\,1\,0\,1\,0\,0\,0\,1\,0\,0\,0\,0 \\ 0\,0\,0\,0\,1\,1\,0\,1\,0\,0\,0\,1\,0\,0\,0 \\ 0\,0\,0\,0\,0\,1\,1\,0\,1\,0\,0\,0\,1\,0\,0 \\ 0\,0\,0\,0\,0\,0\,1\,1\,0\,1\,0\,0\,0\,1\,0 \end{bmatrix}. \tag{11.19}$$

This matrix is a circulant of weight 4 and satisfies the RC-constraint. The rank of $\mathbf{H}$ is 8 and hence there are seven redundant rows in $\mathbf{H}$. The null space of this matrix gives a $(4, 4)$-regular $(15, 7)$ cyclic LDPC code $C$ whose Tanner graph has girth 6. It has minimum distance 5.

Let $\mathbf{v} = (v_0, v_1, v_2, \ldots, v_{14})$ be a codeword in $C$. Consider the code bit $v_4$ of $\mathbf{v}$. The rows of $\mathbf{H}$ orthogonal on the code bit $v_4$ are:

$$
\begin{aligned}
&\qquad\qquad\qquad\quad \overset{\displaystyle v_4}{\downarrow} \\
\mathbf{h}_5 &= (1\,0\,0\,0\ 1\ 0\,0\,0\,0\,0\,0\,0\,1\,1\,0), \\
\mathbf{h}_9 &= (0\,1\,1\,0\ 1\ 0\,0\,0\,1\,0\,0\,0\,0\,0\,0), \\
\mathbf{h}_{11} &= (0\,0\,0\,1\ 1\ 0\,1\,0\,0\,0\,1\,0\,0\,0\,0), \\
\mathbf{h}_{12} &= (0\,0\,0\,0\ 1\ 1\,0\,1\,0\,0\,0\,1\,0\,0\,0),
\end{aligned}
$$

which contain a 1-component at the bit position 4. We can check readily that no other bit position with a 1-component is checked by more than one of these four rows. Hence, $A_4 = \{\mathbf{h}_5, \mathbf{h}_9, \mathbf{h}_{11}, \mathbf{h}_{12}\}$.

Suppose the all-zero codeword $\mathbf{0}$ is transmitted and

$$\mathbf{r} = (0\,0\,0\,0\,1\,0\,0\,0\,0\,0\,0\,0\,0\,1\,0)$$

is the hard-decision received vector. There are two errors in $\mathbf{r}$ which are at the bit positions 4 and 13, i.e., $e_4 = e_{13} = 1$. The syndrome of the received vector $\mathbf{r}$ is

$$
\begin{aligned}
\mathbf{s} = (s_0, s_1, \ldots, s_{14}) &= \mathbf{r} \cdot \mathbf{H}^T \\
&= (0\,0\,0\,1\,0\,0\,1\,0\,0\,1\,0\,1\,1\,0\,1).
\end{aligned}
$$

Because $\mathbf{s} \neq \mathbf{0}$, the existence of errors is detected in the received vector $\mathbf{r}$. In the following, we apply the OSML decoding to decode $\mathbf{r}$.

The syndrome bits computed from the four rows orthogonal on $v_4$ in $A_4$ are:

$$
\begin{aligned}
s_5 &= 0 = e_0 + e_4 + e_{12} + e_{13} = e_4 + e_0 + e_{12} + e_{13}, \\
s_9 &= 1 = e_1 + e_2 + e_4 + e_8 \quad = e_4 + e_1 + e_2 + e_8, \\
s_{11} &= 1 = e_3 + e_4 + e_6 + e_{10} \quad = e_4 + e_3 + e_6 + e_{10}, \\
s_{12} &= 1 = e_4 + e_5 + e_7 + e_{11} \quad = e_4 + e_5 + e_7 + e_{11},
\end{aligned}
$$

i.e., $S_4 = \{s_5 = 0, s_9 = 1, s_{11} = 1, s_{12} = 1\}$. The OSML decision function for $v_4$ is

$$
\sum_{s_j^{(4)} \in S_4} (2s_j^{(4)} - 1) = 2s_5 - 1 + 2s_9 - 1 + 2s_{11} - 1 + 2s_{12} - 1 = 2 > 0.
$$

Based on the OSML decision rule, the error bit $e_4$ is decoded as "1."

To determine the value of $e_5$, we find the rows of $\mathbf{H}$ that are orthogonal on the bit position 5,

$$
\begin{aligned}
&\qquad\qquad\qquad\quad\ \overset{v_5}{\downarrow} \\
\mathbf{h}_6 &= (0\,1\,0\,0\,0\,\ 1\ 0\,0\,0\,0\,0\,0\,1\,1), \\
\mathbf{h}_{10} &= (0\,0\,1\,1\,0\,\ 1\ 0\,0\,0\,1\,0\,0\,0\,0\,0), \\
\mathbf{h}_{12} &= (0\,0\,0\,0\,1\,\ 1\ 0\,1\,0\,0\,0\,1\,0\,0\,0), \\
\mathbf{h}_{13} &= (0\,0\,0\,0\,0\,\ 1\ 1\,0\,1\,0\,0\,0\,1\,0\,0),
\end{aligned}
$$

i.e., $A_5 = \{\mathbf{h}_6, \mathbf{h}_{10}, \mathbf{h}_{12}, \mathbf{h}_{13}\}$. The syndrome bits computed from $\mathbf{r}$ and the four rows in $A_5$ that are orthogonal on $e_5$ are:

$$
\begin{aligned}
s_6 &= 1 = e_1 + e_5 + e_{13} + e_{14} = e_5 + e_1 + e_{13} + e_{14}, \\
s_{10} &= 0 = e_2 + e_3 + e_5 + e_9 \quad = e_5 + e_2 + e_3 + e_9, \\
s_{12} &= 1 = e_4 + e_5 + e_7 + e_{11} \quad = e_5 + e_4 + e_7 + e_{11}, \\
s_{13} &= 0 = e_5 + e_6 + e_8 + e_{12} \quad = e_5 + e_6 + e_8 + e_{12},
\end{aligned}
$$

i.e., $S_5 = \{s_6 = 1, s_{10} = 0, s_{12} = 1, s_{13} = 0\}$. The OSML decision function for $v_5$ is

$$
\sum_{s_j^{(5)} \in S_5} (2s_j^{(5)} - 1) = 2s_6 - 1 + 2s_{10} - 1 + 2s_{12} - 1 + 2s_{13} - 1 = 0 \le 0.
$$

Based on the OSML decision rule, the error bit $e_5$ is decoded as "0."

The values of all the other error bits can be determined in the same manner. The error vector computed is $\mathbf{e} = (0\,0\,0\,0\,1\,0\,0\,0\,0\,0\,0\,0\,0\,1\,0)$. Thus, the decoded vector is

$$
\mathbf{v}^* = \mathbf{r} + \mathbf{e} = \mathbf{0}.
$$

Because $\mathbf{v}^* \cdot \mathbf{H}^T = \mathbf{0}$, $\mathbf{v}^*$ is a codeword in $C$ and the OSML decoding is successful.

It is noted that the parity-check matrix of the code $C$ has column weight $\gamma = 4$. The OSML decoding based on $\mathbf{H}$ is guaranteed to correct two or fewer errors.

Consider another hard-decision received vector $\mathbf{r}' = (1\ 0\ 0\ 0\ 1\ 0\ 0\ 0\ 0\ 0\ 0\ 0\ 1\ 0)$. The syndrome of this received vector is $\mathbf{s} = (0\ 1\ 0\ 1\ 0\ 1\ 1\ 1\ 1\ 0\ 1\ 1\ 0)$. The vector of the OSML decision function is computed as

$$\mathbf{f} = (2\mathbf{s} - \mathbf{1}) \cdot \mathbf{H} = (4, 2, 2, 0, 4, 0, 2, 2, 0, 0, 0, 2, -2, 4, 0).$$

Based on the OSML decision rule, the error vector computed based on $\mathbf{f}$ is $\mathbf{e} = (1\ 1\ 1\ 0\ 1\ 0\ 1\ 1\ 0\ 0\ 0\ 1\ 0\ 1\ 0)$. The decoded vector $\mathbf{v}^*$ is then

$$\mathbf{v}^* = \mathbf{r}' + \mathbf{e} = (0\ 1\ 1\ 0\ 0\ 0\ 1\ 1\ 0\ 0\ 0\ 1\ 0\ 0\ 0).$$

However, $\mathbf{v}^*$ is not a codeword because $\mathbf{v}^* \cdot \mathbf{H}^T = (1\ 1\ 1\ 0\ 1\ 0\ 1\ 0\ 0\ 0\ 1\ 1\ 0\ 1\ 0) \neq \mathbf{0}$. The decoder declares a decoding failure. ▲▲

As shown in Chapter 7, the OSML decoding is capable of correcting many error patterns with more than $\lfloor \gamma/2 \rfloor$ errors, depending on the error bit distributions in these error patterns. For an LDPC code to be effective with OSML decoding, its parity-check matrix must have a reasonably large column weight.

## 11.4.2 Bit-Flipping Decoding

The OSML decoding discussed in the previous section is simply a one-time calculation decoding, i.e., no iterative operation involved. In this section, we introduce a simple *iterative* hard-decision decoding algorithm to decode LDPC codes, which is devised by Gallager [2, 3] and is called the bit-flipping (BF) algorithm. In the BF decoding of an LDPC code $C$, the decoder first computes the syndrome $\mathbf{s} = (s_0, s_1, \ldots, s_{m-1})$ of the hard-decision received vector $\mathbf{r}$ based on the parity-check matrix $\mathbf{H}$ of $C$ using (11.14). If $\mathbf{s} = \mathbf{0}$, then $\mathbf{r}$ is a codeword and the decoding stops. If $\mathbf{s} \neq \mathbf{0}$, then there are *parity-check sum failures* (i.e., some syndrome bits in $\mathbf{s}$ are not equal to zero). The number of parity-check sum failures is equal to the number of syndrome bits in $\mathbf{s}$ that are not equal to 0. Then, the decoder flips (or changes) those received bits in $\mathbf{r}$ that are contained in *more than some fixed number $\delta$ of the failed parity-check sums*. This results in a modified received vector $\mathbf{r}^*$. A new syndrome $\mathbf{s}^* = \mathbf{r}^* \cdot \mathbf{H}^T$ is computed based on the modified vector $\mathbf{r}^*$. If $\mathbf{s}^* = \mathbf{0}$, the decoding stops; otherwise, the bits in $\mathbf{r}^*$ which involve in more than $\delta$ of the failed parity-check sums are flipped to generate another modified received vector. The decoder repeats this flipping process until either there is no failed parity-check sum, i.e., a codeword is found and the decoding succeeds, or no bit can be flipped, i.e., there are still failed parity-check sums but no received bit appears in more than $\delta$ failed parity-check sums after a preset number of flipping iterations.

The parameter $\delta$ is called the *threshold* of the BF decoding. We should design such a threshold $\delta$ that optimizes the decoding performance while minimizing the decoding complexity. The threshold depends on the code parameters (e.g., the column weight, the minimum distance $d_{\min}$, and the number of redundant rows in its parity-check matrix) and channel condition. Gallager derived the optimum thresholds for regular LDPC codes, but this is not included here (see details in [2, 3]).

If decoding fails for a given value of $\delta$, then the value of $\delta$ should be reduced to allow further decoding iterations. For error patterns with numbers of errors less than the error-correcting capability $\lfloor d_{\min}/2 \rfloor$ of the code, the decoding can be completed in one or few iterations; and for those with more errors, more iterations are required. To reduce excessive computations, a limit may be set on the number of maximum iterations.

Let $\mathbf{q} = (q_0, q_1, \ldots, q_{n-1})$ be an $n$-tuple over nonnegative integers, where $q_l$ is the number of parity-check sums in $S_l = \{s_i^{(l)} = \langle \mathbf{r} \cdot \mathbf{h}_i^{(l)} \rangle : \mathbf{h}_i^{(l)} \in A_l\}$ that are equal to 1 (parity-check failures), for $0 \le l < n$. Then, $\mathbf{q} = \mathbf{s} \cdot \mathbf{H}$ (the matrix operation is over integers). We call $\mathbf{q}$ the *failed parity-check sum vector*. Let $I_{\max}$ be the number of maximum iterations to be performed. With these definitions, the BF decoding is formulated as Algorithm 11.2.

---

**Algorithm 11.2** BF decoding algorithm for LDPC codes

---

Inputs: $\mathbf{r} = (r_0, r_1, \ldots, r_{n-1})$, $\mathbf{H}$, $I_{\max}$, $\delta$.
Compute the syndrome: $\mathbf{s} = \mathbf{r} \cdot \mathbf{H}^T$.
**if $\mathbf{s} = \mathbf{0}$ then**
    Output $\mathbf{r}$ as the decoded codeword and declare a decoding success.
**else**
    Set $i = 0$ and $\mathbf{r}^{(0)} = \mathbf{r}$.
    **while $\mathbf{s} \ne \mathbf{0}$ and $i < I_{\max}$ do**
        Compute $\mathbf{q} = \mathbf{s} \cdot \mathbf{H}$.
        Flip the bit $r_l$ (may be more than 1 bit) in $\mathbf{r}^{(i)}$ whose $q_l$ in $\mathbf{q}$ is greater than $\delta$ to obtain a flipped version $\mathbf{r}^{(i+1)}$ of $\mathbf{r}^{(i)}$.
        Compute $\mathbf{s} = \mathbf{r}^{(i+1)} \cdot \mathbf{H}^T$.
        Set $i \leftarrow i + 1$.
    **end while**
    If $\mathbf{s} = \mathbf{0}$, declare a decoding success and output $\mathbf{r}^{(i)}$ as the decoded codeword; otherwise, declare a decoding failure.
**end if**

---

The simplest version of BF decoding is the one without threshold $\delta$, i.e., the received bits that are flipped in each iteration are those that involve in the largest number of failed parity-check sums. The decoding is carried out as shown in Algorithm 11.3.

The BF decoding algorithms specified in Algorithms 11.2 and 11.3 can be applied to decode hard-decision received vectors for LDPC codes.

**Example 11.10** Consider the (4, 4)-regular (15, 7) LDPC code given in Example 11.9 with the parity-check matrix $\mathbf{H}$ given by (11.19). Again, we assume that the all-zero codeword $\mathbf{0}$ is transmitted and

---

**Algorithm 11.3** Simplified BF decoding algorithm for LDPC codes

---

Inputs: $\mathbf{r} = (r_0, r_1, \ldots, r_{n-1})$, $\mathbf{H}$, $I_{\max}$.
Compute the syndrome: $\mathbf{s} = \mathbf{r} \cdot \mathbf{H}^T$.
**if** $\mathbf{s} = \mathbf{0}$ **then**
    Output $\mathbf{r}$ as the decoded codeword and declare a decoding success.
**else**
    Set $i = 0$ and $\mathbf{r}^{(0)} = \mathbf{r}$.
    **while** $\mathbf{s} \neq \mathbf{0}$ and $i < I_{\max}$ **do**
        Compute $\mathbf{q} = \mathbf{s} \cdot \mathbf{H}$. Identify the maximum component in $\mathbf{q}$: $q_{\max} = \max_{q_j \in \mathbf{q}} \{q_j\}$.
        Flip the bit(s) in $\mathbf{r}^{(i)}$ with $q_l = q_{\max}$ to obtain a flipped version $\mathbf{r}^{(i+1)}$ of $\mathbf{r}^{(i)}$.
        Compute $\mathbf{s} = \mathbf{r}^{(i+1)} \cdot \mathbf{H}^T$.
        Set $i \leftarrow i + 1$.
    **end while**
    If $\mathbf{s} = \mathbf{0}$, declare a decoding success and output $\mathbf{r}^{(i)}$ as the decoded codeword; otherwise, declare a decoding failure.
**end if**

---

$$\mathbf{r} = (0\ 0\ 0\ 0\ 1\ 0\ 0\ 0\ 0\ 0\ 0\ 0\ 0\ 1\ 0)$$

is the hard-decision received vector. In the following, we use the simplified BF decoding in Algorithm 11.3 to decode the received vector. The bits in the syndrome of $\mathbf{r}$ computed based on $\mathbf{H}$ are given below:

$$s_0 = 0 = r_7 + r_8 + r_{10} + r_{14}, \quad s_1 = 0 = r_0 + r_8 + r_9 + r_{11},$$
$$s_2 = 0 = r_1 + r_9 + r_{10} + r_{12}, \quad s_3 = 1 = r_2 + r_{10} + r_{11} + r_{13},$$
$$s_4 = 0 = r_3 + r_{11} + r_{12} + r_{14}, \quad s_5 = 0 = r_0 + r_4 + r_{12} + r_{13},$$
$$s_6 = 1 = r_1 + r_5 + r_{13} + r_{14}, \quad s_7 = 0 = r_0 + r_2 + r_6 + r_{14},$$
$$s_8 = 0 = r_0 + r_1 + r_3 + r_7, \quad s_9 = 1 = r_1 + r_2 + r_4 + r_8,$$
$$s_{10} = 0 = r_2 + r_3 + r_5 + r_9, \quad s_{11} = 1 = r_3 + r_4 + r_6 + r_{10},$$
$$s_{12} = 1 = r_4 + r_5 + r_7 + r_{11}, \quad s_{13} = 0 = r_5 + r_6 + r_8 + r_{12},$$
$$s_{14} = 1 = r_6 + r_7 + r_9 + r_{13}.$$

Hence, $\mathbf{s} = (0\ 0\ 0\ 1\ 0\ 0\ 1\ 0\ 0\ 1\ 0\ 1\ 1\ 0\ 1)$, Then, the failed parity-check sum vector is

$$\mathbf{q} = \mathbf{s} \cdot \mathbf{H} = (0, 2, 2, 1, 3, 2, 2, 2, 1, 1, 2, 2, 0, 3, 1).$$

We see that the received bits $r_4$ and $r_{13}$ involve in the most of the failed parity-check sums, each in three of the six failed parity-check sums. Then, we flip the received bits $r_4$ and $r_{13}$. This results in a flipped version of the received vector $\mathbf{r}$

$$\mathbf{r}^* = (0\ 0\ 0\ 0\ 0\ 0\ 0\ 0\ 0\ 0\ 0\ 0\ 0\ 0\ 0).$$

Because $\mathbf{r}^* \cdot \mathbf{H}^T = \mathbf{0}$, the decoder stops, outputs $\mathbf{r}^*$ as the decoded codeword, and declares a decoding success.

Consider another received vector $\mathbf{r}' = (1\ 0\ 0\ 0\ 1\ 0\ 0\ 0\ 0\ 0\ 0\ 0\ 0\ 1\ 0)$ which contains three errors. The syndrome of $\mathbf{r}'$ is $\mathbf{s} = (0\ 1\ 0\ 1\ 0\ 1\ 1\ 1\ 1\ 1\ 0\ 1\ 1\ 0)$. The failed parity-check sum vector is

$$\mathbf{q} = \mathbf{s} \cdot \mathbf{H} = (4, 3, 3, 2, 4, 2, 3, 3, 2, 2, 2, 3, 1, 4, 2),$$

whose maximum entry value is $q_{\max} = 4$. The received bits $r_0$, $r_4$, and $r_{13}$ are in $q_{\max} = 4$ failed parity-check sums. Flipping these bits in the received vector $\mathbf{r}'$ results in a modified received vector $\mathbf{r}^* = \mathbf{0}$ which is the transmitted codeword. Thus, the BF decoding is able to correct the errors in $\mathbf{r}'$. Note that, in Example 11.9, the OSML decoding fails to decode the error pattern in $\mathbf{r}'$. ▲▲

## 11.4.3 Weighted One-Step Majority-Logic and Bit-Flipping Decodings

The error performance of the simple OSML and BF decodings given in Sections 11.4.1 and 11.4.2, respectively, can be improved by including *reliability information* of the received symbols to their decoding decisions. Consider the soft-decision received vector $\mathbf{y} = (y_0, y_1, \ldots, y_{n-1})$ at the output of the channel detector. Let $\mathbf{r} = (r_0, r_1, \ldots, r_{n-1})$ be the corresponding hard-decision vector of $\mathbf{y}$. For BPSK transmission over the AWGN channel, a simple measure of the reliability of a received symbol $y_l$ is its magnitude $|y_l|$: *the larger the magnitude* $|y_l|$, *the larger the reliability* of the hard-decision bit $r_l$.

For $0 \leq l < n$, consider the set

$$A_l = \{\mathbf{h}_1^{(l)}, \mathbf{h}_2^{(l)}, \ldots, \mathbf{h}_\gamma^{(l)}\}$$

of $\gamma$ rows of $\mathbf{H}$ that are orthogonal on the bit position $l$. For $1 \leq j \leq \gamma$, consider the $j$th row in $A_l$,

$$\mathbf{h}_j^{(l)} = (h_{j,0}^{(l)}, h_{j,1}^{(l)}, \ldots, h_{j,n-1}^{(l)}).$$

Let $0 \leq i_1 < i_2 < \cdots < i_\rho < n$ be the $\rho$ locations where $h_{j,i_1}^{(l)} = h_{j,i_2}^{(l)} = \cdots = h_{j,i_\rho}^{(l)} = 1$ in the row $\mathbf{h}_j^{(l)}$. Then, $\mathbf{h}_j^{(l)}$ checks the $\rho$ code bits, $v_{i_1}, v_{i_2}, \ldots, v_{i_\rho}$. The syndrome bit computed from $\mathbf{r}$ and $\mathbf{h}_j^{(l)}$ is

$$s_j^{(l)} = \langle \mathbf{r}, \mathbf{h}_j^{(l)} \rangle = r_{i_1} h_{j,i_1}^{(l)} + r_{i_2} h_{j,i_2}^{(l)} + \cdots + r_{i_\rho} h_{j,i_\rho}^{(l)}.$$

The calculation of $s_j^{(l)}$ involves in $\rho$ hard-decision received bits, $r_{i_1}$, $r_{i_2}, \ldots,$ $r_{i_\rho}$. The accuracy of $s_j^{(l)}$ depends on the accuracies of the hard-decisions on these $\rho$ bits. The reliability of the received symbol $r_{i_k}$ is measured by $|y_{i_k}|$. For $0 \leq l < n$ and $1 \leq j \leq \gamma$, define

$$|y_j|_{\min}^{(l)} = \min\{|y_{i_k}| : 1 \leq k \leq \rho\}. \tag{11.20}$$

The value $|y_j|_{\min}^{(l)}$ is used as a measure of the reliability of the syndrome bit $s_j^{(l)}$, which is called the *weight* of $s_j^{(l)}$. Define the following sum:

$$E_l \triangleq \sum_{s_j^{(l)} \in S_l} (2s_j^{(l)} - 1)|y_j|_{\min}^{(l)}, \tag{11.21}$$

which is simply a *weighted parity-check sum* that is orthogonal on the bit position $l$. A decision rule can be derived based on the weighted parity-check sum $E_l$ as follows:

$$e_l = \begin{cases} 1, & \text{if } E_l > 0, \\ 0, & \text{if } E_l \leq 0. \end{cases} \tag{11.22}$$

Incorporating the above decision rule to the OSML decoding presented in Section 11.4.1, we obtained a *weighted OSML decoding* [14, 18].

Similarly, the weighted parity-check sums can be incorporated in the BF decoding algorithm to improve the decoding performance. Such a BF decoding is called *weighted BF* decoding [14, 18–22] formulated as Algorithm 11.4.

---

**Algorithm 11.4** Weighted BF decoding algorithm for LDPC codes

---

Inputs: $\mathbf{r} = (r_0, r_1, \ldots, r_{n-1})$, $\mathbf{y} = (y_0, y_1, \ldots, y_{n-1})$, $\mathbf{H}$, $I_{\max}$.
Compute the syndrome: $\mathbf{s} = \mathbf{r} \cdot \mathbf{H}^T$.
**if** $\mathbf{s} = \mathbf{0}$ **then**
    Output $\mathbf{r}$ as the decoded codeword and declare a decoding success.
**else**
    Set $i = 0$ and $\mathbf{r}^{(0)} = \mathbf{r}$.
    **while** $\mathbf{s} \neq \mathbf{0}$ and $i < I_{\max}$ **do**
        Compute the weighted parity-check sums, $E_l$, based on (11.21) for $0 \leq l < n$.
        Flip the bit $r_l$ in $\mathbf{r}^{(i)}$ whose $E_l$ is the largest. A flipped version $\mathbf{r}^{(i+1)}$ of $\mathbf{r}^{(i)}$ is obtained.
        Compute $\mathbf{s} = \mathbf{r}^{(i+1)} \cdot \mathbf{H}^T$.
        Set $i \leftarrow i + 1$.
    **end while**
    If $\mathbf{s} = \mathbf{0}$, declare a decoding success and output $\mathbf{r}^{(i)}$ as the decoded codeword; otherwise, declare a decoding failure.
**end if**

---

Weighted OSML and BF decodings improve error performance of a given LDPC code but require some real-number computations. There are other variations of weighted BF algorithms [14, 25–34].

**Example 11.11** Consider the (4, 4)-regular (15, 7) LDPC code given in Example 11.9 with the parity-check matrix given by (11.19). Again, we assume that the all-zero codeword $\mathbf{0}$ is transmitted over an AWGN channel. Consider the following soft-decision received vector: $\mathbf{y} = (-0.2562, 1.4198, 0.2170,$

0.6911, 1.510, 1.9586, $-0.0311$, 2.163, 1.609, 0.9374, 0.8199, 0.7993, 0.7204, 1.0213, 1.0473). The corresponding hard-decision received vector is

$$\mathbf{r} = (1\ 0\ 0\ 0\ 0\ 0\ 1\ 0\ 0\ 0\ 0\ 0\ 0\ 0\ 0).$$

Because $\mathbf{s} = \mathbf{r} \cdot \mathbf{H}^T = (0\ 1\ 0\ 0\ 0\ 1\ 0\ 0\ 1\ 0\ 0\ 1\ 0\ 1\ 1) \neq \mathbf{0}$, errors are detected in the received vector $\mathbf{r}$. Here, we apply the weighted BF algorithm to decode the received vector. Consider the zeroth received bit $r_0$. The bit $r_0$ is involved in four parity-check sums, $s_1$, $s_5$, $s_7$, and $s_8$, among which $s_1$, $s_5$, and $s_8$ fail, i.e., $s_1 = 1$, $s_5 = 1$, $s_7 = 0$, and $s_8 = 1$. The received bits $r_0$, $r_8$, $r_9$, and $r_{11}$ are checked by the parity-check sum $s_1$. The weight for $s_1$ is $|y_1|_{\min}^{(0)} = \min\{|y_0|, |y_8|, |y_9|, |y_{11}|\} = \min\{0.2562, 1.609, 0.9374, 0.7993\} = |y_0| = 0.2562$. Similarly, we can find $|y_5|_{\min}^{(0)} = 0.2562$, $|y_7|_{\min}^{(0)} = 0.0311$, and $|y_8|_{\min}^{(0)} = 0.2562$. Then, the weighted parity-check sum for $r_0$ is $E_0 = (2s_1 - 1)|y_1|_{\min}^{(0)} + (2s_5 - 1)|y_5|_{\min}^{(0)} + (2s_7 - 1)|y_7|_{\min}^{(0)} + (2s_8 - 1)|y_8|_{\min}^{(0)} = 0.7376$.

Following the same calculation method, we can compute the weighted parity-check sums for all the received bits, $E_0, E_1, \ldots, E_{13}, E_{14}$, which are equal to 0.7376, $-1.7025$, $-0.6822$, $-0.6208$, $-0.7290$, $-2.0065$, 0.06210, $-1.3319$, $-0.7497$, $-0.6501$, $-1.7263$, $-1.4511$, $-1.1242$, $-0.9510$, $-2.5633$, respectively. Because $E_0$ is the largest component among the $E_l$s, $0 \leq l < 15$, we flip the received bit $r_0$ in the received vector and obtain a modified received vector $\mathbf{r}^* = (0, 0, 0, 0, 0, 0, 1, 0, 0, 0, 0, 0, 0, 0, 0)$.

Because $\mathbf{r}^* \cdot \mathbf{H}^T = (0\ 0\ 0\ 0\ 0\ 0\ 0\ 1\ 0\ 0\ 0\ 1\ 0\ 1\ 1) \neq \mathbf{0}$, there are still errors in the modified received vector $\mathbf{r}^*$. The decoder needs to perform another iteration. Following the same steps, we compute the weighted parity-check sums based on the calculated syndrome $\mathbf{s}$ of $\mathbf{r}^*$ and the soft-decision vector $\mathbf{y}$, $E_0$, $E_1, \ldots, E_{13}, E_{14}$, which are equal to $-0.7376$, $-2.2149$, $-0.6201$, $-1.1332$, $-1.2414$, $-2.0065$, 0.1242, $-1.8443$, $-1.2621$, $-1.1626$, $-1.7263$, $-1.9636$, $-1.6366$, $-1.4634$, $-2.5012$, respectively. The largest component among the $E_l$s, $0 \leq l < 15$, is $E_6$. Thus, we flip the sixth bit in $\mathbf{r}^*$ which results in the following decoded vector:

$$\mathbf{v}^* = (0\ 0\ 0\ 0\ 0\ 0\ 0\ 0\ 0\ 0\ 0\ 0\ 0\ 0\ 0).$$

Because $\mathbf{s} = \mathbf{v}^* \cdot \mathbf{H}^T = \mathbf{0}$, the decoder stops the decoding process, outputs $\mathbf{v}^*$ as the decoded vector, and declares a decoding success. ▲▲

## 11.5 Iterative Decoding Based on Belief-Propagation

An iterative soft-decision decoding algorithm for LDPC codes based on belief-propagation is a *symbol-by-symbol soft-in–soft-out* (SISO) decoding algorithm. It processes the received symbols *iteratively* to improve the *reliability* of each decoded code symbol based on the channel information, the parity-check sums computed from the hard-decisions of the received symbols, and the sparse parity-check matrix $\mathbf{H}$ of the code.

Reliability of a decoded symbol can be measured by its *marginal a posteriori probability* or its *log-likelihood ratio*. The computed reliability measures of code symbols at the end of each decoding iteration are used as inputs for the next decoding iteration. The decoding process continues until a certain *stopping criterion* is satisfied. Then, based on the computed reliability measures of code symbols, hard-decisions are made.

Iterative decoding based on belief-propagation (IDBP) is extremely efficient for decoding LDPC codes which do not contain short cycles in their Tanner graphs. The IDBP consists of a sequence of iterations. Each decoding iteration consists of two *message passings* between the VNs and CNs of the code's Tanner graph (or between code bits and their parity-check sums). First, messages are passed from the VNs to the CNs and then messages are passed from the CNs back to the VNs.

## 11.5.1   Message Passing

Consider a $(\gamma, \rho)$-regular LDPC code $C$ of length $n$ given by the null space of a sparse $m \times n$ parity-check matrix $\mathbf{H}$:

$$
\mathbf{H} = \begin{bmatrix}
h_{0,0} & h_{0,1} & \cdots & h_{0,n-2} & h_{0,n-1} \\
h_{1,0} & h_{1,1} & \cdots & h_{1,n-2} & h_{1,n-1} \\
\vdots & \vdots & \ddots & \vdots & \vdots \\
h_{m-1,0} & h_{m-1,1} & \cdots & h_{m-1,n-2} & h_{m-1,n-1}
\end{bmatrix}.
$$

Define the following two sets of indices, for $0 \leq i < m$ and $0 \leq j < n$,

$$
\begin{aligned}
N(i) &= \{j : h_{i,j} = 1, 0 \leq j < n\}, \\
M(j) &= \{i : h_{i,j} = 1, 0 \leq i < m\},
\end{aligned}
\tag{11.23}
$$

i.e., $N(i)$ contains the locations of 1s in the $i$th row of $\mathbf{H}$ and $M(j)$ contains the locations of 1s in the $j$th column of $\mathbf{H}$. For the convenience of explanation of the iterative message-passing decoding of an LDPC code, we relabel the locations of 1-entries in $N(i)$ and $M(j)$ as follows:

$$
\begin{aligned}
N(i) &= \{n_0, n_1, \ldots, n_{\rho-1}\}, \\
M(j) &= \{m_0, m_1, \ldots, m_{\gamma-1}\}.
\end{aligned}
$$

With the above two sets of indices, we can see that $N(i)$ contains the indices of the neighbor VNs of the CN $c_i$ and $M(j)$ contains the indices of the neighbor CNs of the VN $v_j$ in the Tanner graph of the LDPC code $C$.

In IDBP, each VN $v_j$ sends messages to its $\gamma$ neighbor CNs, $c_{m_0}$, $c_{m_1}$, $\ldots$, $c_{m_{\gamma-1}}$, i.e., CNs with indices in $M(j)$. The message $L_{v_j \to c_{m_l}}$ sent from VN $v_j$ to CN $c_{m_l}$ is computed based on the messages previously received from the CNs *other than* $c_{m_l}$, i.e., $L_{c_{m_0} \to v_j}, \ldots, L_{c_{m_{l-1}} \to v_j}, L_{c_{m_{l+1}} \to v_j}, \ldots, L_{c_{m_{\gamma-1}} \to v_j}$ (see Fig. 11.3). This prevents the information that was sent from $c_{m_l}$ to $v_j$ from being reused by $c_{m_l}$ itself. The message $L_{v_j \to c_{m_l}}$ sent from $v_j$ to each CN $c_{m_l}$ adjacent to it contains a reliability measure of the code bit $v_j$ (e.g., *a posteriori probability*).

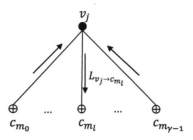

Figure 11.3 Message passing from VN $v_j$ to its neighbor (or adjacent) CNs.

In IDBP, each CN $c_i$ sends messages to its $\rho$ neighbor VNs, $v_{n_0}$, $v_{n_1}, \ldots,$ $v_{n_{\rho-1}}$, i.e., VNs with indices in $N(i)$. The message $L_{c_i \to v_{n_l}}$ sent from CN $c_i$ to VN $v_{n_l}$ is computed based on the reliability of information previously received from the VNs *other than* $v_{n_l}$, i.e., $L_{v_{n_0} \to c_i}, \ldots, L_{v_{n_{l-1}} \to c_i}, L_{v_{n_{l+1}} \to c_i}, \ldots,$ $L_{v_{n_{\rho-1}} \to c_i}$ (see Fig. 11.4). This prevents the information that was passed from $v_{n_l}$ to $c_i$ from being reused by $v_{n_l}$ itself. The message $L_{c_i \to v_{n_l}}$ sent from $c_i$ to each VN $v_{n_l}$ adjacent to it contains a reliability measure of the parity-check sum $s_i$ corresponding to CN $c_i$.

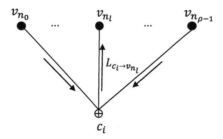

Figure 11.4 Message passing from CN $c_i$ to its neighbor (or adjacent) VNs.

The process of message passing from VNs to CNs and then from CNs back to VNs is defined as *one decoding iteration*. The way that the messages are computed at VNs and CNs and passed between these two types of nodes is intended to prevent *correlation* between successive decoding iterations.

## 11.5.2 Sum-Product Algorithm

The sum-product algorithm (SPA) is a special version of IDBP which is a near-optimal decoding algorithm proposed by Gallager in his seminal work [2, 3] in the 1960s. In this section, we will present the algorithm without any proof. More detailed developments and proofs of the SPA can be found in [2, 3, 9, 14].

The optimality criterion underlying the development of the SPA decoder is *symbol-wise maximum a posteriori* (MAP). We are interested in computing the *a posteriori probability* (APP) that a specific bit in the transmitted codeword $\mathbf{v} = (v_0, v_1, \ldots, v_{n-1})$ equals 0, given the soft-decision received vector $\mathbf{y} = (y_0, y_1, \ldots, y_{n-1})$.

Without loss of generality, we focus on the decoding of bit $v_j$. The APP ratio (also called the *likelihood ratio* (LR)) of bit $v_j$ is defined as

$$l(v_j|\mathbf{y}) \triangleq \frac{\Pr(v_j = 0|\mathbf{y})}{\Pr(v_j = 1|\mathbf{y})},$$

where $\Pr(v_j = 0|\mathbf{y})$ and $\Pr(v_j = 1|\mathbf{y})$ denote the probabilities that the transmitted bit $v_j$ equals "0" and "1" given the soft-decision received vector $\mathbf{y}$, respectively. A more numerically stable log-APP ratio, also called the *log-likelihood ratio* (LLR), is defined as

$$L(v_j|\mathbf{y}) \triangleq \log\left(\frac{\Pr(v_j = 0|\mathbf{y})}{\Pr(v_j = 1|\mathbf{y})}\right). \tag{11.24}$$

Hereafter, the natural logarithm is assumed for LLRs.

In the SPA, two decoders, VN decoder (also called VN processing unit (VNPU)) and CN decoder (also called CN processing unit (CNPU)), work cooperatively and iteratively to estimate $L(v_j|\mathbf{y})$, for $j = 0, 1, \ldots, n - 1$. The VN decoder and CN decoder are shown in Fig. 11.5 and Fig. 11.6, respectively.

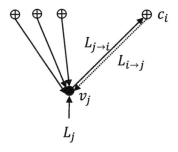

Figure 11.5 A VN decoder in an SPA decoder.

In the VN decoder, VN $v_j$ receives LLR information $L_j$ from the channel and from all of its neighbor CNs. Based on this information, excluding the message $L_{i \to j}$ previously received from CN $c_i$ (as described in Section 11.5.1), VN $v_j$ computes the message $L_{j \to i}$ to be sent to CN $c_i$. Similarly, the VN decoder computes all the messages to be sent to its neighbor CNs with indices in $M(j)$.

In the CN decoder, CN $c_i$ receives LLR information from all of its neighbor VNs. Based on this information, excluding the message $L_{j \to i}$ previously received from VN $v_j$ (as described in Section 11.5.1), CN $c_i$ computes the message $L_{i \to j}$ to be sent to VN $v_j$. Similarly, the CN decoder computes all the messages to be sent to its neighbor VNs with indices in $N(i)$.

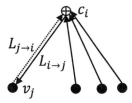

Figure 11.6 A CN decoder in an SPA decoder.

In the context of decoding of LDPC codes, as shown in Fig. 11.5, the information that VN $v_j$ sends to its neighbor CN $c_i$ is calculated as follows:

$$L_{j\to i} = L_j + \sum_{i'\in M(j)\setminus\{i\}} L_{i'\to j}, \tag{11.25}$$

where the LLR $L_j$ is computed based on the $j$th component of the soft-decision received vector $\mathbf{y} = (y_0, y_1, \ldots, y_{n-1})$ from the channel,[3] i.e.,

$$L_j = L(v_j|y_j) = \log\left(\frac{\Pr(v_j = 0|y_j)}{\Pr(v_j = 1|y_j)}\right). \tag{11.26}$$

The information $L_{j\to i}$ (or $L_{i\to j}$) is called the *extrinsic information* and $L_j$ is called the *intrinsic information*. The decision made on $v_j$ depends on the LLR obtained at the end of the iteration which is given as follows:

$$L_j^{\text{total}} = L_j + \sum_{i\in M(j)} L_{i\to j}. \tag{11.27}$$

Based on the definition of the intrinsic information $L_j$, if $L_j > 0$, it indicates that the conditional probability $\Pr(v_j = 0|y_j)$ given $y_j$ with the transmitted bit $v_j = 0$ is higher than the probability $\Pr(v_j = 1|y_j)$ with the transmitted bit $v_j = 1$, i.e., $\Pr(v_j = 0|y_j) > \Pr(v_j = 1|y_j)$. In this case, we have a higher confidence to declare that the transmitted bit is 0. Similarly, if $L_j < 0$, we have a higher confidence to declare that the transmitted bit $v_j$ is 1. However, if $L_j = 0$, it indicates $\Pr(v_j = 0|y_j) = \Pr(v_j = 1|y_j) = 1/2$. In this case, we have no confidence to declare the value of the transmitted bit $v_j$. The hard-decision on $v_j$ is made as follows (where $\hat{v}_j$ is the estimate of $v_j$):

$$\hat{v}_j = \begin{cases} 1, & \text{if } L_j^{\text{total}} < 0 \\ 0, & \text{otherwise.} \end{cases} \tag{11.28}$$

Throughout our development, we shall assume that the *flooding schedule* is employed, where all VNs process their inputs and pass extrinsic information to their neighbor CNs, and all CNs then process their inputs and pass extrinsic information back to their neighbor VNs. This message-passing procedure repeats

---

[3]Here, a discrete memoryless channel (DMC) is assumed.

in every decoding iteration, starting with VNs. After a preset maximum number of iterations of this VN/CN decoding round, or after some stopping criterion has been met, the VN decoder computes (estimates) the total LLR $L_j^{\text{total}}$ following (11.27) and makes a decision on the bit $v_j$ based on $L_j^{\text{total}}$ following (11.28).

The development of the SPA relies on the following *independence assumption*: the LLR quantities received at each node from its neighbors are independent. Recall that we impose the RC-constraint on the parity-check matrix $\mathbf{H}$ of an LDPC code in Definition 11.2. Suppose we have a parity-check matrix $\mathbf{H}$ which does not satisfy the RC-constraint, i.e., there is a $2 \times 2$ submatrix in $\mathbf{H}$ with four 1s. This $2 \times 2$ submatrix corresponds to a cycle of length 4 in the Tanner graph associated with $\mathbf{H}$, as shown in Fig. 11.7. In the first iteration of the iterative decoding, VN $v_1$ receives information from CNs $c_0$ and $c_1$ which is calculated based on the information obtained from VNs $v_0$ and $v_1$. The information passing is shown by the dashed arrow in Fig. 11.7(a). In the second iteration of the iterative decoding, VN $v_0$ receives information from CNs $c_0$ and $c_1$ which is calculated based on the information obtained from VNs $v_0$ and $v_1$. The information passing in this iteration is shown by the dotted arrow in Fig. 11.7(b). In this iteration, it can be seen that the information which VN $v_1$ sends to CNs $c_0$ and $c_1$ is calculated based on the information which was sent out by VN $v_0$ in the first iteration, i.e., in the second iteration, $v_0$ receives information which is related to the information that it sent out in the first iteration. That is, information passing in the graph gets correlated after two iterations. This violates the independence assumption. This is the major reason that we impose the RC-constraint on the parity-check matrices of LDPC codes.

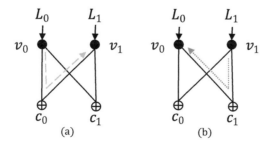

Figure 11.7 A Tanner graph with a cycle of length 4 and message passing.

As we can see, for a Tanner graph with girth $g$, information of the iterative decoding gets correlated after $g/2$ iterations. Therefore, in general, for an LDPC code to achieve good error performance with iterative decoding, it is required that the girth of its Tanner graph is relatively large. When the girth is large, the information estimates will be very accurate and the decoder will have near-optimal (MAP) performance.

A standard stopping criterion for an SPA decoder is that the decoded vector $\hat{\mathbf{v}}$ is a codeword in the LDPC code, i.e., $\hat{\mathbf{v}} \cdot \mathbf{H}^T = \mathbf{0}$. There are cases in which the decoder is trapped and not able to find a codeword. In this case, we have to stop the decoding process to avoid long decoding delay and power consumption.

Thus, we set a limit $I_{\max}$ on the maximum number of iterations. After the decoder has performed $I_{\max}$ iterations, it stops regardless of whether the decoding is successful or not.

The SPA decoder is initialized by setting all the VN to CN messages, $L_{j\to i}$, to the LLRs, $L_j$, calculated based on the soft-decision received vector $\mathbf{y}$, i.e., $L_j = L(v_j|y_j) = \log(\Pr(v_j = 0|y_j)/\Pr(v_j = 1|y_j))$. The operations inside the VN and CN decoders in the SPA are specified in Algorithm 11.5.

There are two functions used in the calculation of extrinsic information in each decoding iteration, $\tanh(x)$ and $\tanh^{-1}(x)$, which are

$$\tanh(x) = \frac{e^{2x} - 1}{e^{2x} + 1}, \quad \tanh^{-1}(x) = 1/2 \ln\left(\frac{1 + x}{1 - x}\right), \tag{11.29}$$

called *hyperbolic tangent* and *inverse hyperbolic tangent*, respectively. It can be shown that

$$\tanh(x) = \begin{cases} \frac{e^{2|x|} - 1}{e^{2|x|} + 1}, & \text{if } x \ge 0 \\ -\frac{e^{2|x|} - 1}{e^{2|x|} + 1}, & \text{if } x < 0 \end{cases}$$

$$\tanh^{-1}(x) = \begin{cases} 1/2 \ln(\frac{1+|x|}{1-|x|}), & \text{if } x \ge 0 \\ -1/2 \ln(\frac{1+|x|}{1-|x|}), & \text{if } x < 0, \end{cases} \tag{11.30}$$

i.e., $\tanh(x) = \text{sign}(x) \cdot (e^{2|x|} - 1)/(e^{2|x|} + 1)$ and $\tanh^{-1}(x) = \text{sign}(x) \cdot 1/2 \ln((1 + |x|)/(1 - |x|))$, where $\text{sign}(x)$ denotes the sign of $x$.

---

**Algorithm 11.5** The Gallager sum-product algorithm (SPA)

---

(1) **Initialization** For $0 \le j < n$, initialize $L_j$ according to (11.24) for the appropriate channel model. Then, for $0 \le i < m$ and $0 \le j < n$ with $h_{i,j} = 1$, set $L_{j\to i} = L_j$; otherwise, set $L_{j\to i} = 0$. Set $n_{\text{itr}} = 0$ which denotes the number of decoding iterations.

(2) **CN update** For $0 \le i < m$ and $0 \le j < n$, compute outgoing CN to VN message $L_{i\to j}$ for each CN using

$$L_{i\to j} = 2 \tanh^{-1}\left(\prod_{j' \in N(i) \backslash \{j\}} \tanh\left(\frac{1}{2}L_{j'\to i}\right)\right), \tag{11.31}$$

and then send it to VN $v_j$. (This step is shown diagrammatically in Fig. 11.6.)

(3) **VN update** For $0 \le i < m$ and $0 \le j < n$, compute the outgoing VN to CN message $L_{j\to i}$ for each VN using

$$L_{j\to i} = L_j + \sum_{i' \in M(j) \backslash \{i\}} L_{i'\to j}, \tag{11.32}$$

and then send it to CN $c_i$. (This step is shown diagrammatically in Fig. 11.5.)

---

---

**Algorithm 11.5** Continued

(4) **LLR update** For $j = 0, 1, \ldots, n-1$, compute

$$L_j^{\text{total}} = L_j + \sum_{i \in M(j)} L_{i \to j}. \tag{11.33}$$

(5) **Stopping criteria** For $j = 0, 1, \ldots, n-1$, set

$$\hat{v}_j = \begin{cases} 1, & \text{if } L_j^{\text{total}} < 0 \\ 0, & \text{otherwise}, \end{cases} \tag{11.34}$$

to obtain a decoded vector $\hat{\mathbf{v}} = (\hat{v}_0, \hat{v}_0, \ldots, \hat{v}_{n-1})$. If $\hat{\mathbf{v}} \cdot \mathbf{H}^T = \mathbf{0}$ or $n_{\text{itr}} = I_{\max}$, stop; else, set $n_{\text{itr}} \leftarrow n_{\text{itr}} + 1$ and go to Step (2) to perform another iteration.

---

The updated Equation (11.31) in Step (2) of Algorithm 11.5 is numerically challenging due to the presence of the product and the $\tanh(x)$ and $\tanh^{-1}(x)$ functions. We can improve the situation as follows. First, factor $L_{j \to i}$ into its sign and magnitude (or bit value and bit reliability):

$$\begin{aligned} L_{j \to i} &= \alpha_{ji} \beta_{ji}, \\ \alpha_{ji} &= \text{sign}(L_{j \to i}), \\ \beta_{ji} &= |L_{j \to i}|. \end{aligned}$$

Following (11.30) and the above definition, (11.31) can be rewritten as

$$\begin{aligned} L_{i \to j} &= 2 \tanh^{-1} \left( \prod_{j' \in N(i) \backslash \{j\}} \alpha_{j'i} \prod_{j' \in N(i) \backslash \{j\}} \tanh(1/2 \beta_{j'i}) \right) \\ &= \prod_{j' \in N(i) \backslash \{j\}} \alpha_{j'i} \cdot 2 \cdot \tanh^{-1} \left( \prod_{j' \in N(i) \backslash \{j\}} \tanh \left( \frac{1}{2} \beta_{j'i} \right) \right). \end{aligned} \tag{11.35}$$

Then, we have

$$\begin{aligned} L_{i \to j} &= \prod_{j' \in N(i) \backslash \{j\}} \alpha_{j'i} \cdot 2 \cdot \tanh^{-1} \left( \prod_{j' \in N(i) \backslash \{j\}} \tanh \left( \frac{1}{2} \beta_{j'i} \right) \right) \\ &= \prod_{j' \in N(i) \backslash \{j\}} \alpha_{j'i} \cdot 2 \cdot \tanh^{-1} \log^{-1} \log \left( \prod_{j' \in N(i) \backslash \{j\}} \tanh \left( \frac{1}{2} \beta_{j'i} \right) \right) \\ &= \prod_{j' \in N(i) \backslash \{j\}} \alpha_{j'i} \cdot 2 \cdot \tanh^{-1} \log^{-1} \sum_{j' \in N(i) \backslash \{j\}} \log \left( \tanh \left( \frac{1}{2} \beta_{j'i} \right) \right) \end{aligned}$$

which yields a new form for (11.31) in Step (2) of Algorithm 11.5:

$$\textbf{CN update } L_{i \to j} = \prod_{j' \in N(i) \setminus \{j\}} \alpha_{j'i} \cdot \phi \left( \sum_{j' \in N(i) \setminus \{j\}} \phi(\beta_{j'i}) \right), \qquad (11.36)$$

where the function $\phi(x)$ is defined as follows:

$$\phi(x) = -\log[\tanh(x/2)] = \log \frac{e^x + 1}{e^x - 1}.$$

Note that $\phi^{-1}(x) = \phi(x)$ for $x > 0$. Thus, (11.36) may be used instead of (11.31) in Step (2) of the SPA presented in Algorithm 11.5. The function $\phi(x)$ shown in Fig. 11.8 may be implemented using a look-up table to reduce computation complexity.

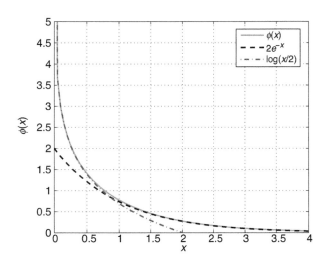

Figure 11.8 A plot of the $\phi(x)$ function together with its approximates $2e^{-x}$ and $\log(x/2)$.

Note that the SPA decoding requires real-number computations, including multiplication, addition, and comparison. The performances of SPA decoders for LDPC codes will be explored more in examples of later chapters.

### 11.5.2.1    Applications of the SPA to BI-AWGN Channel, BSC, and BEC

The LLRs, $L_j$, used in the initialization of the SPA, are calculated based on the soft-decision received vector $\mathbf{y}$. In the following, we consider the calculations of these initial LLRs for three different channels, namely, BI-AWGN channel, BSC, and BEC.

**BI-AWGN** Let $v_j$ be the $j$th transmitted bit and $x_j$ be the corresponding transmitted symbol after BPSK mapping, i.e., $x_j = +1$ if $v_j = 0$ and $x_j = -1$ if $v_j = 1$. Let $y_j$ be the soft-decision received symbol. Then, $y_j = x_j + n_j$, where $n_j$ is the channel noise (disturbance) which is independently and normally distributed random variable following $\mathcal{N}(0, \sigma^2)$ with zero mean and variance $\sigma^2$. (In practice, an estimate of $\sigma^2$ is necessary.) In this case, we have

$$\Pr(x_j = x|y_j) = \frac{1}{1 + e^{-2y_j x/\sigma^2}},$$

where $x \in \{-1, +1\}$. Then, it follows from (11.26) that the LLR $L_j$ is

$$L_j = \log\left(\frac{\Pr(v_j = 0|y_j)}{\Pr(v_j = 1|y_j)}\right) = \frac{2y_j}{\sigma^2}. \tag{11.37}$$

**BSC** In this case, we have $y_j \in \{0, 1\}$. Let $p = \Pr(y_j = b^c|v_j = b)$ be the error (transition) probability, where $b^c$ represents the complement of $b$, i.e., if $b = 0$, $b^c = 1$ and if $b = 1$, $b^c = 0$. Then,

$$\Pr(v_j = b|y_j) = \begin{cases} 1 - p, & \text{if } y_j = b, \\ p, & \text{if } y_j = b^c, \end{cases}$$

and it follows that

$$L_j = \log\left(\frac{\Pr(v_j = 0|y_j)}{\Pr(v_j = 1|y_j)}\right) = \begin{cases} \log(\frac{1-p}{p}), & \text{if } y_j = 0, \\ \log(\frac{p}{1-p}), & \text{if } y_j = 1. \end{cases} \tag{11.38}$$

**BEC** The received symbol $y_j$ is in the set $\{0, 1, \epsilon\}$ where $\epsilon$ represents an erasure. Let $p_e = \Pr(y_j = \epsilon|v_j = b)$ define the erasure probability, where $b \in \{0, 1\}$. Then,

$$\Pr(v_j = b|y_j) = \begin{cases} 1 - p_e, & \text{if } y_j = b, \\ 0, & \text{if } y_j = b^c, \\ p_e, & \text{if } y_j = \epsilon. \end{cases}$$

It follows that

$$L_j = \log\left(\frac{\Pr(v_j = 0|y_j)}{\Pr(v_j = 1|y_j)}\right) = \begin{cases} +\infty, & \text{if } y_j = 0, \\ -\infty, & \text{if } y_j = 1, \\ 0, & \text{if } y_j = \epsilon. \end{cases} \tag{11.39}$$

### 11.5.3 Min-Sum Algorithm

As mentioned in the last section, the SPA decoder for a long LDPC code has a very high decoding complexity, which may exclude it from practical applications. There are many reduced-complexity approximate-SPA decoders [14, 26, 35–45] devised in the literature which mainly focus on the complex CN update given by (11.36). Among these reduced-complexity approximate-SPA decoders, the most practical one is the decoder designed based on the *min-sum algorithm* (MSA) [38, 44]. Besides reduction in decoding complexity, an MSA decoder can provide a performance close to an SPA decoder with maybe a small or ignorable degradation. In this section, we present the MSA to decode LDPC codes.

Consider the CN update Equation (11.36) for calculating $L_{i \to j}$ in the SPA decoder. Note from the shape of $\phi(x)$ (see Fig. 11.8) that the largest term in the sum $\sum_{j' \in N(i) \backslash \{j\}} \phi(\beta_{j'i})$ corresponds to the smallest $\beta_{ji}$. Assuming that this term dominates the sum given by (11.36), we have the following approximation:

$$\phi \left( \sum_{j' \in N(i) \backslash \{j\}} \phi(\beta_{j'i}) \right) \approx \phi \left( \phi(\min_{j' \in N(i) \backslash \{j\}} \{\beta_{j'i}\}) \right)$$

$$= \min_{j' \in N(i) \backslash \{j\}} \{\beta_{j'i}\}.$$

Thus, the MSA is simply the SPA with Step (2) replaced by

$$\textbf{CN update } L_{i \to j} = \prod_{j' \in N(i) \backslash \{j\}} \alpha_{j'i} \cdot \min_{j' \in N(i) \backslash \{j\}} \{\beta_{j'i}\}. \tag{11.40}$$

To minimize the performance degradation compared to the SPA, a *scaled or attenuated MSA* is proposed [38, 44]. The resulting update equation of (11.36) for Step (2) in the SPA is modified as

$$\textbf{CN update } L_{i \to j} = \prod_{j' \in N(i) \backslash \{j\}} \alpha_{j'i} \cdot c_{\text{atten}} \cdot \min_{j' \in N(i) \backslash \{j\}} \{\beta_{j'i}\}, \tag{11.41}$$

where $0 < c_{\text{atten}} < 1$ is called the *scaled* or *attenuated factor*. A particularly convenient scaling (or attenuating) factor is $c_{\text{atten}} = 0.5$, because it is implementable by a register shift.

As shown above, decoding an LDPC code with a scaled MSA only requires real-number additions, comparisons, and one multiplication in each decoding iteration. While decoding with the SPA requires much more multiplications and extra computations of the functions $\tanh(x)$ and $\tanh^{-1}(x)$. Even though there may be a small (or negligible) performance degradation using a scaled MSA compared to the SPA, the larger reduction in computation complexity overcomes the small loss in performance.

**Example 11.12** In this example, we apply the decoding algorithms presented so far to decode a $(4095, 3367)$ LDPC code $C$ with rate 0.822. This code is constructed based on the lines in the two-dimensional Euclidean geometry $\text{EG}(2, 2^6)$ not passing through the origin (see Chapter 12). It is a cyclic LDPC code with minimum distance 65. The parity-check matrix $\mathbf{H}$ of the code is a single circulant of size $4095 \times 4095$ with weight 64 which satisfies the RC-constraint and hence has an orthogonal structure. With the OSML decoding, the code is capable of correcting all error patterns with 32 or fewer errors.

Suppose we apply this code to the BI-AWGN channel with BPSK signaling and decode it with the SPA, MSA (scaled by a factor of 0.25), BF, weighted BF, and OSML decodings. In decoding with the SPA and the MSA, we set the maximum number of iterations to 50. Figure 11.9 shows the BER (bit-error rate) performances of all the decoding algorithms. Also included in the figure is the performance of an uncoded BPSK system.

From the figure, we see that the BF decoding performs better than the OSML decoding with a 0.4dB coding gain. For this code, the weighted BF decoding performs the same as the BF decoding. The two soft-decision decodings of the code, namely, the SPA and MSA, have the best performance among all the decoding algorithms, of course at the expense of decoding complexity. From the figure, we see that the MSA performs very close to the SPA with a performance gap (in terms of SNR) less than 0.1 dB, i.e., the MSA requires no more than 0.1 dB extra SNR to achieve the same BER as the SPA. We also see from the figure that at the BER of $10^{-8}$, the OSML decoding achieves 5.6 dB coding gain over the uncoded BPSK system. The SPA and MSA have around 7.8 dB coding gain over the uncoded BPSK system. ▲▲

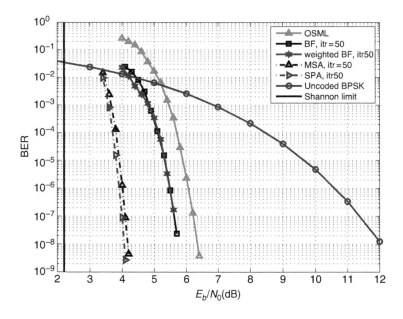

Figure 11.9 The BER performances of the $(4095, 3367)$ LDPC code given in Example 11.12 decoded with the OSML, BF, weighted BF, MSA, and SPA decodings.

More on the performance of the MSA decoding will be presented in the next four chapters. Cyclic LDPC codes constructed based on finite geometries will be presented in Chapter 12.

## 11.6 Error Performance of LDPC Codes with Iterative Decoding

### 11.6.1 Error-Floor

LDPC codes perform amazingly well with IDBP. However, with iterative decoding, *most* LDPC codes have a *common severe weakness*, known as *error-floor* [46, 47]. The error-floor of an LDPC code is characterized by the phenomenon that as the SNR continues to increase, the error probability (error performance) curve *suddenly drops* at a rate *much slower* than that in the region of low to moderate SNRs as illustrated in Fig. 11.10. In the region of low to moderate SNRs, the error performance curve drops rapidly like a *waterfall*. At a certain point of SNR, the error performance starts to drop slowly and eventually flattens out as SNR continues to increase. The region of SNRs in which the error performance of an LDPC code drops like a waterfall is called the *waterfall region*. The region of SNRs in which the dropping rate of the error performance curve starts to slow down and eventually flatten out is called the *error-floor region*.

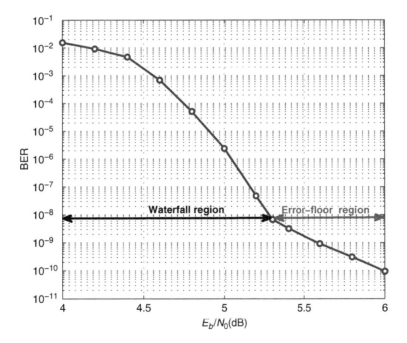

Figure 11.10 The error-floor phenomenon.

Error-floor may preclude many LDPC codes which perform very well in the waterfall region from applications where very low error rates are required, such as high-speed satellite and optical communication systems and high-density data-storage systems.

For an AWGN channel, the error-floor of an LDPC code is mostly caused by an undesirable structure, known as a *trapping set* [46, 47], in the Tanner graph of the code based on which the iterative decoding is carried out. A trapping set simply corresponds to an error pattern which prevents the iterative LDPC decoder to converge. Trapping set structure of the Tanner graph of an LDPC code will be discussed in a later part of this section.

**Example 11.13** In this example, we use an LDPC code to demonstrate the waterfall and error-floor phenomena. The code used is a $(5, 70)$-regular $(3934, 3653)$ QC-LDPC code with rate $0.9285$ whose parity-check matrix $\mathbf{H}$ is a $281 \times 3934$ matrix with constant column weight 5 and row weight 70. (The construction of this code will be given in Chapter 13.) Suppose we apply this code to a BI-AWGN channel using BPSK signaling. In decoding, we use both SPA and MSA and set the maximum number of iterations to 50. The MSA is scaled by a factor of 0.75.

The BER (bit-error rate) and BLER (block-error rate) performances of the code are shown in Fig. 11.11. We see that the code performs very well with both decoding algorithms for SNRs below 5.2 dB. In the region between 4 dB and 5.2 dB of SNRs, the performance curves of the code with both decoding algorithms drop like a waterfall. From the figure, we also see that the scaled MSA performs very close to the SPA. At the BER of $10^{-7}$, the code decoded with either SPA or the MSA performs about 1.3 dB away from the Shannon limit. However, error-floor starts to appear at the SNR of 5.2 dB and flattens out as SNR continues to increase. ▲▲

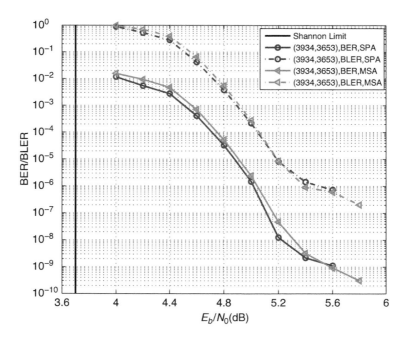

Figure 11.11 The BER and BLER performances of the $(3934, 3653)$ LDPC code given in Example 11.13 decoded with SPA and MSA.

## 11.6.2 Decoding Threshold

The error performance of LDPC codes with iterative decoding based on belief-propagation (IDBP) displays a *threshold* [16, 17] effect in the waterfall region. The data transmission is reliable beyond the threshold; otherwise, the transmission is unreliable. The Shannon limit presented in Chapter 1 is a lower bound for an AWGN channel for a code ensemble with a fixed code rate. The decoding threshold (commonly known as *convergence threshold*) here is a function of code ensemble properties (i.e., VN/CN degree distributions in the Tanner graph of the code) and the decoding algorithm (e.g., IDBP). For an AWGN channel, the decoding threshold is defined as the minimum SNR at which the decoding error probability converges to zero as the code length goes to infinity for a given LDPC code ensemble under IDBP [16, 17]. For a BSC, the decoding threshold is the maximum transition probability at which the decoding error probability converges to zero as the code length goes to infinity.

The calculation of decoding thresholds for LDPC codes is based on the estimations of bit-error rates when the code length and the number of decoding iterations go to infinity. There are many tools/algorithms to compute the decoding thresholds for LDPC codes [16, 17, 48–51]. A commonly used numerical algorithm, referred to as *density evolution* (DE), estimates the bit-error rates through tracking the average probability density function (PDF) of the messages exchanged between VNs and CNs at each decoding iteration in the limit of infinite codeword length. DE operates under the assumption that all the messages in the iterative decoding are independent, that is, the corresponding Tanner graph is cycle free. As the codeword length goes to infinity, the codes in the code ensemble will be more and more likely to be cycle free. The code behavior under IDBP with length $>10^4$ is well predicted by DE. Thus, the optimal degree distribution for an LDPC code with given code rate can be obtained by DE [16, 17]. Table 11.1 gives the decoding thresholds for several LDPC code ensembles under IDBP over AWGN channels. Also included in this table are the Shannon limits.

Besides DE, there are many other algorithms proposed to compute decoding thresholds for LDPC codes, e.g., *Gaussian approximation* (GA) [48] and

Table 11.1 Decoding thresholds under IDBP for LDPC codes over AWGN channels.

| $d_v$ | $d_c$ | Code rate $R$ | Threshold $E_b/N_0$(dB) | Shannon limit $E_b/N_0$(dB) |
|---|---|---|---|---|
| 3 | 6 | 1/2 | 1.11 | 0.185 |
| 4 | 8 | 1/2 | 1.54 | 0.185 |
| 5 | 10 | 1/2 | 2.01 | 0.185 |
| 3 | 5 | 2/5 | −0.078 | −0.24 |
| 4 | 6 | 1/3 | −0.094 | −0.51 |
| 3 | 4 | 1/4 | −2.05 | −0.796 |

*extrinsic-information-transfer* (EXIT) *chart* technique [49–51]. For the details of calculating decoding thresholds for LDPC codes using these algorithms, readers are referred to [14, 16, 17, 48–51].

### 11.6.3  Overall Performance and Its Determinating Factors

The overall performance of an LDPC code decoded with an IDBP algorithm is in general measured by:

(1) decoded error probability (bit, symbol, and block),
(2) rate of *decoding convergence*, and
(3) error-floor,

where the rate of decoding convergence is measured by *how fast* the decoding of an LDPC code converges to its performance limit, i.e., how many decoding iterations are required. In the code design, it is desired to achieve a good *balance* among these three performance measurements.

Before discussing further the error performance of LDPC codes based on an iterative decoding algorithm, we first introduce a few concepts.

**Definition 11.6** CN-redundancy (or row-redundancy) of an LDPC code is defined as the number of redundant rows in its parity-check matrix.

**Definition 11.7** In the Tanner graph $\mathcal{G}$ of an LDPC code, a VN may be connected to other VNs by paths of length 2. Given a VN $v_j$ in $\mathcal{G}$, the number of other VNs that connect to VN $v_j$ by paths of length 2 is defined as the *connection number* of VN $v_j$. The connection numbers of all VNs in the Tanner graph $\mathcal{G}$ are called *VN-connectivity* of $\mathcal{G}$.

Consider the $(10, 6)$ LDPC code in Example 11.1 with a $5 \times 10$ parity-check matrix $\mathbf{H}$ given by (11.1). The rank of $\mathbf{H}$ is 4. Thus, $\mathbf{H}$ has one redundant row, i.e., it has a CN-redundancy of 1. The Tanner graph $\mathcal{G}$ of the code is shown in Fig. 11.1. Take VN $v_0$ for example. It has a connection number of 6, i.e., VNs $v_1$, $v_2$, $v_3$, $v_4$, $v_5$, and $v_6$ connect to $v_0$ by paths of length 2. In iterative decoding, this indicates that VN $v_0$ receives messages from the other six VNs in each decoding iteration.

**Definition 11.8** Let $\mathcal{G}$ be a Tanner graph of an LDPC code given by the null space of an $m \times n$ parity-check matrix $\mathbf{H}$. For $1 \leq \kappa \leq n$, $0 \leq \tau \leq m$, a $(\kappa, \tau)$ *trapping set* [47] is a set $\mathcal{T}(\kappa, \tau)$ of $\kappa$ VNs in $\mathcal{G}$ which induces a subgraph of $\mathcal{G}$ with exactly $\tau$ odd-degree CNs (and an arbitrary number of even-degree CNs). An *elementary* $(\kappa, \tau)$ trapping set is a trapping set for which all CNs in the induced subgraph of the Tanner graph have either degree 1 or degree 2, and there are exactly $\tau$ CNs of degree 1.

**Definition 11.9** A $(\kappa, \tau)$ trapping set in the Tanner graph of an LDPC code of length $n$ is said to be *small* [52] if $\kappa \leq \sqrt{n}$ and $\tau \leq 4\kappa$, i.e., $\tau/\kappa \leq 4$.

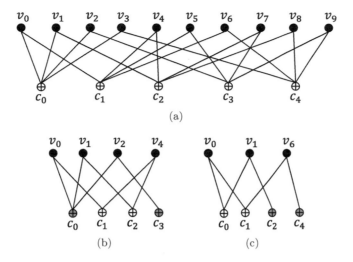

Figure 11.12 (a) The Tanner graph of the $(10,6)$ LDPC code given in Example 11.1, (b) a $(4,2)$ trapping set, and (c) a $(3,2)$ elementary trapping set.

**Definition 11.10** Let $\mathcal{G}$ be the Tanner graph of an LDPC code given by the null space of an $m \times n$ parity-check matrix $\mathbf{H}$. For $1 \le \kappa \le n$, $0 \le \tau \le m$, a $(\kappa, \tau)$ *absorbing set* [53] is a set $\mathcal{T}(\kappa, \tau)$ of $\kappa$ VNs in $\mathcal{G}$ that induces a subgraph of $\mathcal{G}$ in which each VN connects to strictly more CNs of even degrees than CNs with odd degrees. An *elementary* $(\kappa, \tau)$ absorbing set is an absorbing set for which all CNs in the induced subgraph of the Tanner graph have either degree 1 or degree 2, and there are exactly $\tau$ CNs of degree 1.

Figure 11.12(a) (a repetition of Fig. 11.1) shows the Tanner graph $\mathcal{G}$ of the $(10,6)$ LDPC code given in Example 11.1. $\mathcal{G}$ has 10 VNs and 5 CNs. Figures 11.12(b) and 11.12(c) show two subgraphs of $\mathcal{G}$ which are induced by a $(4,2)$ trapping set and a $(3,2)$ trapping set, respectively. The $(4,2)$ trapping set consists of four VNs, $v_0$, $v_1$, $v_2$ and $v_4$, and four CNs, $c_0$, $c_1$, $c_2$, and $c_3$, of which there are two CNs ($c_1$ and $c_2$) of degree 2, one CN ($c_0$) of degree 3, and one CN ($c_3$) of degree 1. The $(3,2)$ trapping set consists of three VNs, $v_0$, $v_1$, and $v_6$, and four CNs, $c_0$, $c_1$, $c_2$, and $c_4$, of which there are two CNs ($c_0$ and $c_1$) of degree 2 and two CNs ($c_2$ and $c_4$) of degree 1. According to Definitions 11.8 and 11.9, the $(4,2)$ trapping set is a small trapping set and the $(3,2)$ trapping set is an elementary trapping set.

Numerous studies of LDPC codes since their rediscovery in the late 1990s show that the error performance of an LDPC code over an AWGN channel decoded with an IDBP algorithm very much depends on the structural properties of its Tanner graph collectively, namely:

(1) girth;
(2) cycle distribution;
(3) degree distributions of VNs and CNs;

(4) VN-connectivity;

(5) CN-redundancy;

(6) trapping set structure; and

(7) minimum distance of the code. (Maybe other unknown properties.)

No one of these structural characteristics dominates the performance of all LDPC codes, although one characteristic might be responsible for the performance of a single code. For example, for an LDPC code to perform well over the AWGN channel decoded with an IDBP algorithm, the girth of its Tanner graph must be at least 6.

With IDBP, the decoded error performance of an LDPC code is in general judged by how close the code performs to the *Shannon limit* [1], the *decoding threshold*, or the *sphere packing bound* (SPB) [54]. The error performance, especially in the region of low to moderate SNRs, i.e., in the waterfall region, is mainly determined by the girth, the cycle distribution, and the VN and CN degree distributions of the Tanner graph of the code. The girth $g$ of the Tanner graph needs to be reasonably large, at least 6, and the numbers of cycles of lengths $g$, $g + 2$, $g + 4$ (called *short cycles*) should be small. The VN and CN degree distributions should be designed to push the error performance of the code close to the decoding threshold or the Shannon limit, especially for an irregular LDPC code.

Fast decoding convergence is required for high-speed communication and high-density storage systems because it can reduce power consumption and decoding delay. The rate of decoding convergence of an LDPC code very much depends on the VN-connectivity of its Tanner graph and the CN-redundancy of its parity-check matrix. For an LDPC code with high degree of VN-connectivity, each VN is connected to a large number of other VNs with paths of length 2. Hence, in each decoding iteration, each VN receives extrinsic information from many other VNs. As a result, in a few iterations, each VN collects enough extrinsic information to update its intrinsic information to a level sufficient to make a correct hard-decision with high probability.

The error-floor of an LDPC code on the AWGN channel is a measure of how low the error probability of the code can achieve. It is mostly determined by the trapping set structure of its Tanner graph and its minimum distance. Suppose, in the transmission of a codeword, an error pattern $\mathbf{e}$ with $\kappa$ errors at the locations of the $\kappa$ VNs of a $(\kappa, \tau)$ trapping set occurs. This error pattern $\mathbf{e}$ will cause $\tau$ parity-check failures (i.e., the parity-check sums are not equal to zeros, because each of these $\tau$ parity-check sums contains an odd number of errors in $\mathbf{e}$). In this case, for iterative decoding, another decoding iteration must be carried out to correct these failed parity-check sums. Iterative decoding, such as the SPA and MSA, is very susceptible to trapping sets of a code because it works locally in a distributed-processing manner. Each CN has a local processor unit to process the messages received from the VNs connected to it and each VN has a local processor unit to process the messages received from the CNs connected to it. Hopefully, these local processor units through iterations and message exchanges collect enough information to make a global optimum decision on the transmitted code bits.

In each decoding iteration, we call a CN a *satisfied* CN if it satisfies its corresponding parity-check sum constraint (i.e., its corresponding parity-check sum is equal to zero); otherwise, we call it an *unsatisfied* CN. During the decoding process, the decoder undergoes state transitions from one state to another until all the CNs satisfy their corresponding parity-check sum constraints or a predetermined maximum number of iterations is reached. The $i$th state of an iterative decoder is represented by the hard-decision sequence obtained at the end of the $i$th iteration. In the process of a decoding iteration, the messages from the satisfied CNs try to reinforce the current decoder state, while the messages from the unsatisfied CNs try to change some of the bit decisions to satisfy their own parity-check sum constraints, i.e., try to force the decoder to go to next state. If errors affect the $\kappa$ code bits (or the $\kappa$ VNs) of a $(\kappa, \tau)$ trapping set $\mathcal{T}(\kappa, \tau)$, the $\tau$ odd-degree CNs, each connected to an odd number of VNs in $\mathcal{T}(\kappa, \tau)$, will not be satisfied, while all other CNs will be satisfied. The decoder will succeed in correcting the errors in $\mathcal{T}(\kappa, \tau)$ if the messages coming from the unsatisfied CNs connected to the VNs in $\mathcal{T}(\kappa, \tau)$ are strong enough to overcome the (false or inaccurate) messages coming from the satisfied CNs. However, this may not be the case if $\tau$ is small. As a result, the decoder may not converge to a valid codeword even if more decoding iterations are performed and this nonconvergence of decoding results in an error-floor. In this case, we say that the decoder is trapped.

For the BI-AWGN channel, error patterns with a small number of errors (or low-weight error patterns) are more probable to occur than those with a larger number of errors. Consequently, in message-passing decoding algorithms, the most harmful $(\kappa, \tau)$ trapping sets are usually the small trapping sets. Extensive study and simulation results [52, 53, 55–68] show that small trapping sets result in high decoding failure rates and contribute significantly to high error-floors.

Besides small trapping sets and their distributions, undetected errors caused by small minimum weight codewords of a code also contribute considerably to the error-floor of the code. If there is no trapping set with size $\kappa$ smaller than the minimum weight of an LDPC code, then the error-floor of the code decoded with iterative decoding is dominated by the minimum weight of the code. For $\tau = 0$, $\mathcal{T}(\kappa, 0)$ is a special trapping set with no odd-degree CNs. Such a trapping set is induced by an error pattern which is identical to a codeword of weight $\kappa$ in the code. When such a trapping set exists, the decoder converges into an incorrect codeword and commits an undetected error. In this case, we say that the decoder is trapped into a *fixed point* [69].

Designing codes to avoid harmful trapping sets to mitigate the error-floor is a challenging combinatorial problem. High error-floors most commonly occur for random or pseudorandom LDPC codes constructed by computer search. LDPC codes constructed using algebraic methods (one category of the LDPC code construction methods) in general have much lower error-floors. It has been proved that Tanner graphs of several subclasses of algebraic-based LDPC codes do not contain harmful trapping sets with sizes smaller than their minimum distances (which are in general reasonably large) [69–72]. The error-floor of an LDPC code can also be lowered by taking a decoder-based strategy [73–81] to remove or reduce the effect of harmful trapping sets. See more on trapping sets in [53, 55–68].

How to design LDPC codes with good waterfall error performances, very low error-floor, and fast rate of decoding convergence without excessive decoding complexity is a challenging problem.

The Tanner graphs of cyclic LDPC codes constructed based on finite geometries (to be presented in Chapter 12) have at least four of the seven structural properties required to achieve a good balance among the three performance measurements, namely, decoded error probability, rate of decoding convergence, and error-floor. This will be discussed in the next chapter. Here, we use the $(64, 64)$-regular $(4095, 3367)$ cyclic finite-geometry LDPC code $C$ given in Example 11.12 to demonstrate its fast decoding convergence. This code has minimum distance 65. The Tanner graph $\mathcal{G}$ of $C$ has girth 6. Each VN in the Tanner graph of the code is connected to 4032 other VNs, a very large VN connection number. Thus, $\mathcal{G}$ has a very large VN-connectivity. There are no harmful trapping sets with sizes smaller than its minimum distance 65. The parity-check matrix of $C$ has 728 redundant rows. With all these structural properties, the decoding of the code should converge very quickly and have very low error-floor. Suppose we decode the code $C$ with 5, 10, 50, and 100 iterations of the MSA scaled by a factor of 0.25. The BER and BLER performances of this finite-geometry LDPC code are shown in Fig. 11.13 where the performances decoded with 5, 10, 50, and 100 iterations are labeled with itr5, itr10, itr50, and itr100, respectively. We see that the decoding of this code converges quickly. The performance with 50 decoding iterations is very close to the performance with 100 decoding iterations. At the BER of $10^{-8}$, the performance gap between 5 and 10 iterations is around 0.1 dB and the performance gap between 10 and 50 iterations about 0.08 dB.

## 11.7 Iterative Decoding of LDPC Codes over BECs

For transmitting information at a rate $R$ (information bit per channel usage) over BEC, the Shannon limit is $1 - R$. The implication of this Shannon limit is that, for erasure probability $p_e$ smaller than $1 - R$, information can be transmitted reliably by using a significantly long code with rate $R$ and conversely reliable transmission is not possible if the erasure probability $p_e$ is larger than the Shannon limit $1 - R$. Theoretical study and experimental results have shown that LDPC codes decoded with IDBP perform well not only on an AWGN channel but also on a BEC. If the SPA presented in Section 11.5.2 is applied to decode an LDPC code over a BEC, the LLRs defined by (11.39) are used as the initial values.

To demonstrate how an LDPC code performs on a BEC, we use the (4095, 3367) cyclic finite-geometry LDPC code with rate 0.8222 given in Example 11.12 as an example. Because the rate of this code is 0.8222, its Shannon limit $1 - R$

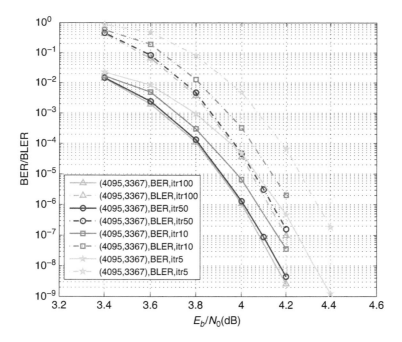

Figure 11.13 The BER and BLER performances of the $(4095, 3367)$ LDPC code given in Example 11.12 decoded with 5, 10, 50, and 100 iterations of the MSA.

is $1 - 0.8222 = 0.1778$. With the SPA decoding of maximum 50 iterations, the error performance of this code over the BEC with erasure probability $p_e$ is shown in Fig. 11.14. We see that at the unresolved (or unrecovered) erasure bit rate (UEBR) of $10^{-6}$ (or unresolved erasure block rate (UEBLR) of $5 \times 10^{-6}$), the code performs 0.095 from the Shannon limit. Hence, from Figs. 11.9, 11.13, and 11.14, we see that this code performs well on both the AWGN channel and BEC.

The hard-decision successive-peeling (SPL) algorithm presented in Section 10.8.2 can be applied to decode LDPC codes over BECs. The decoding algorithm simply works to recover the erased bits in the received vector successively until either all of the parity-check equations (or sums) formed based on the parity-check matrix **H** of the code are satisfied (i.e., all the erased code bits are recovered from the parity-check sums), or an erasure pattern $\xi$ at certain decoding stage is obtained such that no erased bit in $\xi$ can be recovered, i.e., no parity-check equation can be found which checks only one erased bit in $\xi$. This set of erasure locations in $\xi$ is called a *stopping set* which prevents the erasure recovering process to continue. In the following, we describe stopping sets from a graphical point of view.

Let $\mathcal{V}$ be a set of VNs in the Tanner graph $\mathcal{G}$ of an LDPC code and $\mathcal{C}$ be a set of CNs in $\mathcal{G}$ that are adjacent to the VNs in $\mathcal{V}$, i.e., each CN in $\mathcal{C}$ is connected to at least one VN in $\mathcal{V}$. The CNs in $\mathcal{C}$ are called the *neighbors* of the VNs in $\mathcal{V}$ and $\mathcal{C}$ is called the *neighbor set* of $\mathcal{V}$. A set $\mathcal{V}$ of VNs is called a *stopping set* of $\mathcal{G}$ if each CN in the neighbor set $\mathcal{C}$ of $\mathcal{V}$ is connected to *at least*

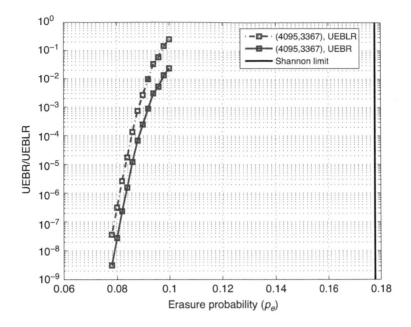

Figure 11.14 The UEBR and UEBLR performances of the $(4095, 3367)$ cyclic finite-geometry LDPC code given in Example 11.12 over a BEC.

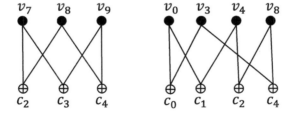

Figure 11.15 Two stopping sets of the $(10, 6)$ LDPC code given in Example 11.1.

*two* VNs in $\mathcal{V}$. The number of VNs in $\mathcal{V}$ is called the size of the stopping set. If the locations of the erased symbols in an erasure pattern $\xi$ correspond to a stopping set in the Tanner graph $\mathcal{G}$ of an LDPC code, then a parity-check sum that checks an erasure in $\xi$ also checks *at least one* other erasure in $\xi$, i.e., every parity-check sum contains at least two unknowns. As a result, no erasure in $\xi$ can be determined from any parity-check sum that checks it. Hence, the error pattern $\xi$ is unrecoverable. Figure 11.15 shows the subgraphs induced by two stopping sets, one of size 3 and the other of size 4, of the Tanner graph of the $(10, 6)$ LDPC code given in Example 11.1.

A set $\mathcal{Q}$ of VNs in $\mathcal{G}$ may contain many stopping sets. The union of any two stopping sets in $\mathcal{Q}$ is also a stopping set in $\mathcal{Q}$. The union of all the stopping sets in $\mathcal{Q}$ gives the *maximum stopping set* of $\mathcal{Q}$. A set $\mathcal{V}_{\text{sfs}}$ of VNs in $\mathcal{G}$ is called a *stopping-free set* (SFS) if it does not contain any stopping set. It is clear that any erasure pattern $\xi$ with erased symbols whose locations correspond to an SFS $\mathcal{V}_{\mathrm{sfs}}$ in $\mathcal{G}$ is recoverable.

A stopping set $\mathcal{V}_{\min}$ of *minimum size* in the Tanner graph $\mathcal{G}$ of an LDPC code is called a *minimum stopping set*. If the code symbols corresponding to the VNs in $\mathcal{V}_{\min}$ during transmission are erased, then the erased symbols form an erased pattern $\xi$ of minimum size which is not recoverable. Therefore, for correcting random erasures with iterative decoding (including the SPL), it is desirable to construct codes with *largest possible minimum stopping sets* in their Tanner graphs. A good LDPC code for random erasure correcting must have very few stopping sets. A stopping set always contains cycles. In [82], it is proved that the size of a minimum stopping set of a Tanner graph with girth 4 is two. The size of a minimum stopping set of a Tanner graph with girth 6 is $\gamma + 1$ and is $2\gamma$ for girth 8 where $\gamma$ is the column weight of the parity-check matrix of a VN-regular LDPC code, i.e., all the VNs in the Tanner graph of the code have the same degree $\gamma$. Hence, for iterative decoding of an LDPC code, the most critical cycles in the code's Tanner graph are cycles of length 4. Therefore, in code construction for BECs, cycles of length 4 must be avoided in the Tanner graph of a code. Cycles of length 4 can be avoided by constructing parity-check matrix of an LDPC code which satisfies the RC-constraint as defined in Section 11.1. It is proved in [82] that for a code with minimum distance $d_{\min}$, its Tanner graph contains at least one stopping set of size $d_{\min}$. Hence, the performance of an LDPC code over a BEC is determined by its minimum distance and the stopping set distribution in its Tanner graph $\mathcal{G}$.

In the next four chapters, methods will be presented for constructing several classes of LDPC codes whose parity-check matrices satisfy the RC-constraint. Among these classes, the class of codes constructed based on finite geometries are shown to have large minimum distances and no stopping sets with sizes smaller than their minimum distances. For example, consider the $(4095, 3367)$ cyclic finite-geometry code given in Example 11.12. The minimum distance of this code is 65. Because its parity-check matrix satisfies the RC-constraint and has column weight 64, the size of a minimum stopping set in its Tanner graph is 65. Hence, any erasure pattern with 64 or fewer erasures is recoverable with the simple SPL algorithm presented in Section 10.8.2. See more on stopping sets in [58, 82–86].

## 11.8 Categories of LDPC Code Constructions

There has been a tremendous amount of work on LDPC codes after the works of Gallager, MacKay, and Luby, including code construction, decoding algorithms, structural analysis, performance analysis, generalizations, applications, etc. In this section, we briefly introduce the construction methods for LDPC codes, most of which or their variances will be discussed in detail in the following chapters.

Most methods of constructing LDPC codes can be classified into two general categories:

(1) *graph-theoretic-based* methods, and

(2) *algebraic-based* (or *matrix-theoretic-based*) methods.

The best known graph-theoretic-based construction methods are

(1) *progressive edge-growth* (PEG), and

(2) *protograph* (PTG)-based methods,

which will be covered in Chapter 15. LDPC codes constructed based on graphs are in general *pseudorandom* or *random* codes and they lack structural properties. This class of codes is constructed with the aid of computer search. In general, it is hard to construct graph-based LDPC codes with very low error-floor.

Algebraic-based constructions of LDPC codes can be classified into three categories:

(1) construction based on *finite fields*;

(2) construction based on *finite* or *partial geometries*; and

(3) construction based on *combinatorial* (or *experimental*) *designs*.

The LDPC codes constructed based on finite geometries have been partially covered in Chapter 7 and will be discussed more in Chapters 12 and 13. The constructions based on combinatorial designs and finite fields will be covered in Chapters 13 and 14, respectively. Algebraic LDPC codes are abundant of structural properties and they are mostly either *cyclic* or *quasi-cyclic* (QC). Cyclic and QC-structures facilitate encoder and decoder implementation with reduced hardware complexity. LDPC codes constructed based on finite fields are in general quasi-cyclic. High-rate (above 0.9) LDPC codes constructed based on finite fields which achieve a very low BER, say $10^{-15}$, without visible error-floors have been constructed [87].

## 11.9 Nonbinary LDPC Codes

As mentioned at the beginning of this chapter, the fundamental concepts, structural properties, and methods of construction, encoding, and decoding developed for binary LDPC codes in the previous sections can be generalized to LDPC codes with symbols from nonbinary fields. In this section, we briefly present definitions and some structural properties of nonbinary LDPC codes.

Let $\mathrm{GF}(q)$ be a Galois field with $q$ elements, where $q$ is a power of a prime. A *q-ary $(\gamma, \rho)$-regular* LDPC code $C_q$ of length $n$ is given by the null space over $\mathrm{GF}(q)$ of a sparse parity-check matrix $\mathbf{H}_q$ of size $m \times n$ over $\mathrm{GF}(q)$ that has the following structural properties: (1) each row has weight $\rho$; (2) each column has weight $\gamma$; (3) no two rows (or two columns) of $\mathbf{H}_q$ have more than one nonzero component in common; and (4) both $\rho$ and $\gamma$ are small compared with $n$ and $m$, respectively, where $m \geq (n - k)$, i.e., $\mathbf{H}_q$ may not be a full-rank matrix. If the columns and/or rows of the parity-check matrix $\mathbf{H}_q$ have *varying* (*multiple*) weights, then the null space over $\mathrm{GF}(q)$ of $\mathbf{H}_q$ gives a *q-ary irregular* LDPC code. The subscript "$q$" in $C_q$ and $\mathbf{H}_q$ stands for "q-ary" (or "nonbinary").

If the parity-check matrix $\mathbf{H}_q$ is a single sparse circulant or a column of sparse circulants of the same size over $\mathrm{GF}(q)$, its null space over $\mathrm{GF}(q)$ gives a *q-ary cyclic LDPC code*. If $\mathbf{H}_q$ is an array of sparse circulants and/or zero matrices

(ZMs) of the same size over $GF(q)$, then the null space over $GF(q)$ of $\mathbf{H}_q$ gives a *q-ary quasi-cyclic* (QC) *LDPC code*. Encoding of $q$-ary cyclic and QC-LDPC codes can be implemented with shift-registers similar to the encoding of binary cyclic and QC codes as presented in Sections 4.5 and 4.13 with some modifications, i.e., converting the computation over $GF(2)$ to computations defined over the nonbinary field $GF(q)$.

Consider a $q$-ary code $C_q$ given by the null space over $GF(q)$ of an $m \times n$ parity-check matrix $\mathbf{H}_q = [h_{i,j}]_{0 \leq i < m, 0 \leq j < n}$ over $GF(q)$. Let $\mathbf{h}_0, \mathbf{h}_1, \ldots, \mathbf{h}_{m-1}$ be the $m$ rows of $\mathbf{H}_q$ where, for $0 \leq i < m$, the $i$th row $\mathbf{h}_i$ is

$$\mathbf{h}_i = (h_{i,0}, h_{i,1}, \ldots, h_{i,n-1}).$$

An $n$-tuple $\mathbf{v}$ over $GF(q)$, $\mathbf{v} = (v_0, v_1, \ldots, v_{n-1})$, is a codeword of $C_q$ if and only if $\mathbf{v}\mathbf{H}_q^T = \mathbf{0}$. Let $\mathbf{v}$ be a codeword in $C_q$. Then for $0 \leq i < m$,

$$s_i = \langle \mathbf{v}, \mathbf{h}_i \rangle = v_0 h_{i,0} + v_1 h_{i,1} + \cdots + v_{n-1} h_{i,n-1} = 0, \tag{11.42}$$

where the addition and multiplication are carried out with the operations defined on $GF(q)$. The $m$ sums, $s_0, s_1, \ldots, s_{m-1}$, are called *parity-check sums* that give $m$ *constraints* on the $n$ code symbols, $v_0, v_1, \ldots, v_{n-1}$, of a codeword in $C$. A code symbol $v_j$ is said to be *checked* by the parity-check sum $s_i$ (or the row $\mathbf{h}_i$) if $h_{i,j} \neq 0$, i.e., the code symbol $v_j$ is contained in the parity-check sum $s_i$.

The Tanner graph $\mathcal{G}_q$ of a $q$-ary LDPC code $C_q$ given by the null space over $GF(q)$ of a sparse $m \times n$ parity-check matrix $\mathbf{H}_q = [h_{i,j}]_{0 \leq i < m, 0 \leq j < n}$ over $GF(q)$ can be constructed in the same way as that for a binary LDPC code given in Section 11.2. The graph $\mathcal{G}_q$ has $n$ VNs that correspond to the $n$ code symbols of a codeword in $C_q$ (or the $n$ columns of $\mathbf{H}_q$) and $m$ CNs that correspond to $m$ parity-check sums on the code symbols (or the $m$ rows of $\mathbf{H}_q$). The $j$th VN $v_j$ is connected to the $i$th CN $c_i$ with an edge if and only if the $j$th code symbol $v_j$ is contained in the $i$th parity-check sum $c_i$, i.e., if and only if the entry $h_{i,j}$ of $\mathbf{H}_q$ is a nonzero element of $GF(q)$. In some literature, the edge connecting the $j$th VN $v_j$ and the $i$th CN $c_i$ is labeled by the nonzero entry $h_{i,j}$ of $\mathbf{H}_q$. (In binary case, we always have the nonzero entries in the parity-check matrices as "1" that are just omitted in their Tanner graphs.) The VN and CN degree distribution polynomials of $\mathcal{G}_q$ can be obtained following the same way as that for binary LDPC codes. If the parity-check matrix $\mathbf{H}_q$ of the $q$-ary LDPC code $C_q$ satisfies the RC-constraint, then its corresponding Tanner graph $\mathcal{G}_q$ has girth at least 6.

**Example 11.14** Consider the Galois field $GF(2^3)$ generated by the primitive polynomial $\mathbf{p}(X) = 1 + X + X^3$ over $GF(2)$ (see Table 2.10). Let $\alpha$ be a primitive element in $GF(2^3)$. Consider the following $5 \times 10$ matrix $\mathbf{H}_q$ with entries in $GF(2^3)$:

$$\mathbf{H}_q = \begin{bmatrix} \alpha^5 & \alpha^2 & \alpha & \alpha^2 & 0 & 0 & 0 & 0 & 0 & 0 \\ 1 & 0 & 0 & 0 & 1 & \alpha^6 & \alpha^6 & 0 & 0 & 0 \\ 0 & \alpha^4 & 0 & 0 & 1 & 0 & 0 & 1 & \alpha^2 & 0 \\ 0 & 0 & \alpha^5 & 0 & 0 & 1 & 0 & 1 & 0 & \alpha \\ 0 & 0 & 0 & \alpha^4 & 0 & 0 & \alpha^5 & 0 & \alpha^4 & \alpha^3 \end{bmatrix}. \tag{11.43}$$

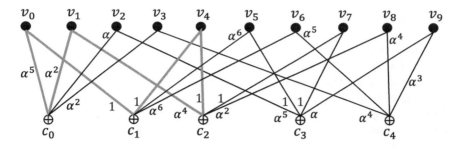

Figure 11.16 The Tanner graph of the 8-ary $(10, 5)$ LDPC code given in Example 11.14.

Each column of $\mathbf{H}_q$ has weight 2, i.e., there are two nonzero entries in each column of $\mathbf{H}_q$, and each row of $\mathbf{H}_q$ has weight 4. $\mathbf{H}_q$ is an 8-ary $(2, 4)$-regular matrix. It is clear that $\mathbf{H}_q$ satisfies the RC-constraint. The rank of $\mathbf{H}_q$ is 5, i.e., $\mathbf{H}_q$ is a full-rank matrix over $\mathrm{GF}(2^3)$. The null space over $\mathrm{GF}(2^3)$ of $\mathbf{H}_q$ gives an 8-ary $(2, 4)$-regular $(10, 5)$ LDPC code $C_q$.

The Tanner graph $\mathcal{G}_q$ of $C_q$ is shown in Fig. 11.16. The graph $\mathcal{G}_q$ has 10 VNs and 5 CNs, where each VN has degree 2 and each CN has degree 4. Because $\mathbf{H}_q$ satisfies the RC-constraint, there is no cycle of length 4 in $\mathcal{G}_q$. A cycle of length 6 marked by bold lines exists in the graph $\mathcal{G}_q$. Thus, $\mathcal{G}_q$ has girth 6. ▲▲

**Example 11.15** Consider the following matrix $\mathbf{H}_q$ over $\mathrm{GF}(2^2)$ of size $6 \times 9$

$$\mathbf{H}_q = \left[\begin{array}{ccc|ccc|ccc} \underline{1} & \alpha & 0 & 0 & \alpha & 0 & 0 & 0 & 0 \\ 0 & \underline{1} & \alpha & 0 & 0 & \alpha & 0 & 0 & 0 \\ \alpha & 0 & \underline{1} & \alpha & 0 & 0 & 0 & 0 & 0 \\ \hline 0 & 0 & \alpha^2 & 0 & 0 & 0 & \alpha & 0 & 0 \\ \alpha^2 & 0 & 0 & 0 & 0 & 0 & 0 & \alpha & 0 \\ 0 & \alpha^2 & 0 & 0 & 0 & 0 & 0 & 0 & \alpha \end{array}\right], \tag{11.44}$$

where $\alpha$ is a primitive element in $\mathrm{GF}(2^2)$. The matrix $\mathbf{H}_q$ is a $2 \times 3$ array of circulants and ZMs over $\mathrm{GF}(2^2)$ of size $3 \times 3$. $\mathbf{H}_q$ has two row weights, 2 and 3, and two column weights, 1 and 2. Hence, $\mathbf{H}_q$ is an irregular matrix. It is easy to compute the rank of $\mathbf{H}_q$, which is 6, i.e., $\mathbf{H}_q$ is a full-rank matrix over $\mathrm{GF}(2^2)$. The null space over $\mathrm{GF}(2^2)$ of $\mathbf{H}_q$ gives a 4-ary irregular $(9, 3)$ QC-LDPC code. The Tanner graph $\mathcal{G}_q$ of the code has girth at least 6 because its corresponding parity-check matrix $\mathbf{H}_q$ satisfies the RC-constraint. There is a cycle of length 6 in $\mathcal{G}_q$ whose edges (corresponding nonzero entries in $\mathbf{H}_q$) are marked by underline in $\mathbf{H}_q$. Hence, $\mathcal{G}_q$ has girth 6. ▲▲

Akin to binary LDPC codes, nonbinary LDPC codes can be classified into two major categories: (1) random-like nonbinary codes constructed by computer under certain design criteria or rules; and (2) structured nonbinary codes constructed on the basis of algebraic or combinatorial tools, such as finite fields and finite geometries. The construction methods of the two categories for binary LDPC codes given in Section 11.8 can be adopted to construct nonbinary LDPC codes. References for the constructions of nonbinary LDPC codes will be provided in the following chapters which present construction methods for binary LDPC codes.

The first study of nonbinary LDPC codes was conducted by Davey and Mackay in 1998 [10]. In their work, they generalized the SPA (see Section 11.5.2) for decoding binary LDPC codes to decode $q$-ary LDPC codes, called *QSPA*. Later, in 2000, MacKay and Davey introduced a *fast-Fourier-transform* (FFT)-based QSPA to reduce the decoding computational complexity of QSPA [88], referred to as FFT-QSPA. MacKay and Davey's work on FFT-QSPA was further improved by Barnault and Declercq in 2003 [11] and Declercq and Fossorier in 2007 [13]. A detailed derivation of the iterative decoding algorithm for nonbinary LDPC codes, namely, FFT-QSPA, is included in Appendix E [89].

# Problems

**11.1** Consider the following matrix:

$$\mathbf{H} = \begin{bmatrix} 1 & 1 & 0 & 0 & 0 \\ 0 & 1 & 1 & 0 & 0 \\ 0 & 0 & 1 & 1 & 0 \\ 0 & 0 & 0 & 1 & 1 \\ 1 & 0 & 1 & 0 & 0 \\ 0 & 1 & 0 & 1 & 0 \\ 0 & 0 & 1 & 0 & 1 \\ 1 & 0 & 0 & 1 & 0 \\ 0 & 1 & 0 & 0 & 1 \\ 1 & 0 & 0 & 0 & 1 \end{bmatrix}.$$

Check whether the above matrix $\mathbf{H}$ can be used as a parity-check matrix for an LDPC code, i.e., satisfies the conditions in Definition 11.2. If it does, compute the rank of $\mathbf{H}$ and the length and dimension of the LDPC code $C$ given by its null space. Find all the codewords in $C$ either in systematic or nonsystematic form.

**11.2** Consider the transpose $\mathbf{H}^T$ of the matrix $\mathbf{H}$ in Problem 11.1. Check whether $\mathbf{H}^T$ can be used as a parity-check matrix for an LDPC code, i.e., satisfies the conditions in Definition 11.2. If it does, compute the rank of $\mathbf{H}^T$ and the length and dimension of the LDPC code $C$ given by its null space. Find all the codewords in $C$ either in systemic or nonsystematic form.

**11.3** Consider the $(2,4)$-regular $(10,6)$ LDPC code $C$ given in Example 11.1. Find a systematic parity-check matrix and a generator matrix of the code $C$.

**11.4** Prove that the $(n,1)$ repetition code is an LDPC code and find a low-density parity-check matrix for this code.

**11.5** Consider a matrix **H** which consists of all the distinct $m$-tuples of weight 2 as its columns. Check whether **H** can be used as a parity-check matrix for an LDPC code. If it can, compute the rank of **H** and the length and dimension of the LDPC code $C$ given by its null space.

**11.6** Consider the following low-density matrix **H**:

$$\mathbf{H} = \begin{bmatrix} 1 & 1 & 0 & 1 & 0 & 0 & 0 \\ 0 & 1 & 1 & 0 & 1 & 0 & 0 \\ 0 & 0 & 1 & 1 & 0 & 1 & 0 \\ 0 & 0 & 0 & 1 & 1 & 0 & 1 \\ 1 & 0 & 0 & 0 & 1 & 1 & 0 \\ 0 & 1 & 0 & 0 & 0 & 1 & 1 \\ 1 & 0 & 1 & 0 & 0 & 0 & 1 \end{bmatrix}.$$

Find the code given by the null space of the above matrix **H** and the minimum distance of the constructed LDPC code.

**11.7** Prove (11.10) and (11.11).

**11.8** Construct the Tanner graphs associated with the LDPC codes constructed in Problems 11.1 and 11.2. Find the girths of the two Tanner graphs, a cycle of shortest length, and the VN and CN degree distribution polynomials from both edge and node perspectives.

**11.9** Construct the Tanner graph of the LDPC code constructed in Problem 11.6. Find the girth of the Tanner graph, a cycle of shortest length, and the VN and CN degree distribution polynomials from both edge and node perspectives. Derive all the parity-check sums of the code and find those parity-check sums which are orthogonal on the zeroth code bit $v_0$.

**11.10** Consider the LDPC code $C$ given by the null space of the following parity-check matrix:

$$\mathbf{H} = \begin{bmatrix} 1 & 0 & 0 & 1 & 0 & 0 & 0 & 1 & 0 & 0 & 0 & 1 & 0 & 0 & 0 \\ 1 & 0 & 0 & 0 & 1 & 0 & 0 & 0 & 1 & 0 & 0 & 0 & 1 & 0 & 0 \\ 1 & 0 & 0 & 0 & 0 & 1 & 0 & 0 & 0 & 1 & 0 & 0 & 0 & 1 & 0 \\ 1 & 0 & 0 & 0 & 0 & 0 & 1 & 0 & 0 & 0 & 1 & 0 & 0 & 0 & 1 \\ 0 & 1 & 0 & 1 & 0 & 0 & 1 & 0 & 0 & 1 & 0 & 0 & 0 & 0 & 0 \\ 0 & 1 & 0 & 0 & 1 & 0 & 0 & 1 & 0 & 0 & 1 & 0 & 0 & 0 & 0 \\ 0 & 1 & 0 & 0 & 0 & 1 & 0 & 0 & 1 & 0 & 0 & 1 & 0 & 0 & 1 \\ 0 & 0 & 1 & 1 & 0 & 0 & 0 & 0 & 1 & 0 & 0 & 0 & 0 & 1 & 0 \\ 0 & 0 & 1 & 0 & 1 & 0 & 1 & 0 & 0 & 0 & 0 & 0 & 0 & 0 & 0 \\ 0 & 0 & 1 & 0 & 0 & 1 & 0 & 1 & 0 & 0 & 0 & 0 & 1 & 0 & 0 \end{bmatrix}.$$

Find the code length and dimension. Construct the Tanner graph of $C$ and find its girth and a cycle of shortest length. Compute the VN and CN degree distribution polynomials from both edge and node perspectives. Derive all the parity-check sums of the code and find those parity-check sums which are orthogonal on the zeroth code bit $v_0$.

**11.11** Use Gallager's proposed method to construct an LDPC code with length 24 whose parity-check matrix has column weight 3 and row weight 4.

**11.12** Consider the (15, 7) cyclic LDPC code $C$ constructed in Example 11.9. Assume that the all-zero codeword $\mathbf{0}$ is transmitted and the following four hard-decision vectors are received. Use the OSML algorithm to decode them and analyze the decoding results:

$$\mathbf{r}_0 = (0\ 0\ 0\ 0\ 1\ 0\ 0\ 0\ 0\ 0\ 0\ 0\ 0\ 0\ 0),$$
$$\mathbf{r}_1 = (0\ 0\ 0\ 0\ 1\ 1\ 0\ 0\ 0\ 0\ 0\ 0\ 0\ 0\ 0),$$
$$\mathbf{r}_2 = (1\ 0\ 0\ 0\ 0\ 1\ 1\ 0\ 0\ 0\ 0\ 0\ 0\ 0\ 0),$$
$$\mathbf{r}_3 = (0\ 0\ 0\ 0\ 1\ 1\ 1\ 0\ 0\ 0\ 0\ 0\ 0\ 0\ 0).$$

**11.13** Consider the (15, 7) cyclic LDPC code $C$ constructed in Example 11.9. Assume that the all-zero codeword $\mathbf{0}$ is transmitted and the following five hard-decision vectors are received. Use the BF algorithm to decode them and analyze the decoding results (assuming $I_{\max} = 50$):

$$\mathbf{r}_0 = (0\ 0\ 0\ 0\ 1\ 0\ 0\ 0\ 0\ 0\ 0\ 0\ 0\ 0\ 0),$$
$$\mathbf{r}_1 = (0\ 0\ 0\ 0\ 1\ 1\ 0\ 0\ 0\ 0\ 0\ 0\ 0\ 0\ 0),$$
$$\mathbf{r}_2 = (1\ 0\ 0\ 0\ 0\ 1\ 1\ 0\ 0\ 0\ 0\ 0\ 0\ 0\ 0),$$
$$\mathbf{r}_3 = (0\ 0\ 0\ 0\ 1\ 1\ 1\ 0\ 0\ 0\ 0\ 0\ 0\ 0\ 0),$$
$$\mathbf{r}_4 = (1\ 0\ 0\ 0\ 0\ 1\ 0\ 0\ 0\ 0\ 0\ 0\ 0\ 0\ 1).$$

**11.14** Consider the (4, 4)-regular (15, 7) code in Example 11.9 with a parity-check matrix given by (11.19). Assume that the all-zero codeword $\mathbf{0}$ is transmitted over an AWGN channel. The following four soft-decision vectors are received. Apply the weighted BF algorithm (Algorithm 11.3) to decode them and analyze the decoding results (assuming $I_{\max} = 50$):

$\mathbf{y}_0 = (0.806\,50,\ 0.882\,80,\ 2.4063,\ 2.3302,\ 2.3379,\ 1.6339,\ -0.13990,\ 1.6771,$ $2.5390,\ 1.4615,\ 1.9768,\ 1.6862,\ 0.713\,50,\ 1.2774,\ 0.256\,80),$

$\mathbf{y}_1 = (0.970\,60,\ 0.838\,90,\ 1.6134,\ 2.0683,\ 2.0840,\ 0.156\,00,\ 1.0756,\ -0.186\,40,$ $-0.088\,100,\ 0.993\,30,\ 2.4977,\ 0.247\,90,\ 1.3629,\ 0.779\,60,\ 2.0919),$

$\mathbf{y}_2 = (1.4960,\ 2.6918,\ -1.0839,\ 1.7954,\ 1.2941,\ -0.206\,40,\ 0.600\,00,\ 1.3161,$ $4.3012,\ 3.5549,\ -0.245\,30,\ 3.7998,\ 1.6692,\ 0.941\,80,\ 1.6594),$

$\mathbf{y}_3 = (1.8681,\ -0.120\,90,\ -0.044\,500,\ 0.209\,00,\ -1.8771,\ 2.4056,\ 1.3178,$ $0.262\,30,\ 2.3391,\ -0.672\,50,\ 0.900\,10,\ 0.764\,10,\ 1.3119,\ 1.3057,\ 0.1548).$

**11.15** Prove (11.35) and (11.36).

**11.16** Prove (11.37), (11.38), and (11.39).

**11.17** Consider the $(2640, 1320)$ Margulis code [90] whose parity-check matrix is available in Mackay's website [23] listed as "Margulis2640.1320.3 ($N = 2640$, $K = 1320, M = 1320, R = 0.5$)." Assume the code is applied to a BI-AWGN channel with BPSK signaling. Compute the performance of the code decoded with 5, 10, and 50 maximum iterations of the SPA and MSA.

**11.18** Find two trapping sets of the $(10, 6)$ LDPC code in Example 11.1 which are different from those shown in Fig. 11.12.

**11.19** Find another two stopping sets of the $(10, 6)$ LDPC code in Example 11.1 which are different from those given in Fig. 11.15, one of size 4 and the other of size 3.

**11.20** Find an absorbing set of the $(10, 6)$ LDPC code in Example 11.1.

# References

[1] C. Shannon, "A mathematical theory of communication," *Bell System Tech. J.*, **27**(3) (1948), 379–423 (Part I), 623–656 (Part II).

[2] R. G. Gallager, *Low-Density Parity-Check Codes*, Cambridge, MA, The MIT Press, 1963.

[3] R. G. Gallager, "Low-density parity-check codes," *IRE Trans. Inf. Theory*, **IT-8**(1) (1962), 21–28.

[4] R. M. Tanner, "A recursive approach to low complexity codes," *IEEE Trans. Inf. Theory*, **27**(9) (1981), 533–547.

[5] D. MacKay and R. Neal, "Good codes based on very sparse matrices," in C. Boyd, ed., *Cryptography and Coding, 5th IMA Conf.* Berlin, Springer-Verlag, October 1995.

[6] D. MacKay, "Good error correcting codes based on very sparse matrices," *IEEE Trans. Inf. Theory*, **45**(3) (1999), 399–431.

[7] N. Alon and M. Luby, "A linear time erasure-resilient code with nearly optimal recovery," *IEEE Trans. Inf. Theory*, **42**(11) (1996), 1732–1736.

[8] J. Byers, M. Luby, M. Mitzenmacher, and A. Rege, "A digital fountain approach to reliable distribution of bulk data," *ACM SIGCOMM Computer Communication Rev.*, **28**(4) (1998), 56–67.

[9] J. Pearl, *Probabilistic Reasoning in Intelligent Systems*, San Mateo, CA, Morgan Kaufmann, 1988.

[10] M. C. Davey and D. J. C. MacKay, "Low-density parity check codes over GF($q$)," *IEEE Commun. Lett.*, **2**(6) (1998), 165–167.

[11] L. Barnault and D. Declercq, "Fast decoding algorithm for LDPC over GF($2^q$)," *Proc. IEEE Inf. Theory Workshop*, Paris, France, March 31–April 4, 2003, pp. 70–73.

[12] H. Wymeersch, H. Steendam, and M. Moeneclaey, "Log-domain decoding of LDPC codes over GF($q$)," in *Proc. IEEE Int. Conf. Commun.*, Paris, June 2004, pp. 772–776.

[13] D. Declercq and M. Fossorier, "Decoding algorithms for nonbinary LDPC codes over GF($q$)," *IEEE Trans. Commun.*, **55**(4) (2007), 633–643.

[14] W. E. Ryan and S. Lin, *Channel Codes: Classical and Modern*, New York, Cambridge University Press, 2009.

[15] V. Savin, "Min-Max decoding for non binary LDPC codes," *Proc. IEEE Int. Symp. Inf. Theory*, Toronto, Canada, July 6–11, 2008, pp. 960–964.

[16] T. Richardson and R. Urbanke, "The capacity of LDPC codes under message-passing decoding," *IEEE Trans. Inf. Theory*, **47**(2) (2001), 599–618.

[17] T. Richardson and R. Urbanke, "Design of capacity-approaching irregular LDPC codes," *IEEE Trans. Inf. Theory*, **47**(2) (2001), 619–637.

[18] S. Lin and D. J. Costello, Jr., *Error Control Coding: Fundamentals and Applications*, 2nd ed., Upper Saddle River, NJ, Prentice-Hall, 2004.

[19] Y. Kou, S. Lin, and M. Fossorier, "Low density parity check codes based on finite geometries: a rediscovery," *Proc. IEEE Int. Symp. Inf. Theory*, Sorrento, June 25–30, 2000, p. 200.

[20] Y. Kou, S. Lin, and M. Fossorier, "Construction of low density parity check codes: a geometric approach," *Proc. 2nd Int. Symp. Turbo Codes and Related Topics*, Brest, September 2000, pp. 137–140.

[21] Y. Kou, S. Lin, and M. Fossorier, "Low density parity-check codes: construction based on finite geometries," *Proc. IEEE Globecom*, San Francisco, CA, November 27–December 1, 2000, pp. 825–829.

[22] Y. Kou, S. Lin, and M. Fossorier, "Low-density parity-check codes based on finite geometries: a rediscovery and new results," *IEEE Trans. Inf. Theory*, **47**(11) (2001), 2711–2736.

[23] Encyclopedia of Sparse Graph Codes: www.inference.phy.cam.ac.uk/mackay/CodesFiles.html.

[24] T. J. Richardson and R. Urbanke, "Efficient encoding of low-density parity-check codes," *IEEE Trans. Inf. Theory*, **47**(2) (2001), 638–656.

[25] J. Zhang and M. P. C. Fossorier, "A modified weighted bit-flipping decoding for low-density parity-check codes," *IEEE Commun. Lett.*, **8**(3) (2004), 165–167.

[26] J. Chen and M. P. C. Fossorier, "Near optimum universal belief propagation based decoding of low-density parity check codes," *IEEE Trans. Commun.*, **50**(3) (2002), 406–614.

[27] M. Jiang, C. Zhao, Z. Shi, and Y. Chen, "An improvement on the modified weighted bit-flipping decoding algorithm for LDPC codes," *IEEE Commun. Lett.*, **9**(7) (2005), 814–816.

[28] Z. Liu and D. A. Pados, "A decoding algorithm for finite-geometry LDPC codes," *IEEE Trans. Commun.*, **53**(3) (2005), 415–421.

[29] M. Shan, C. Zhao, and M. Jian, "Improved weighted bit-flipping algorithm for decoding LDPC codes," *IET Proc. Commun.*, **152**(6) (2005), 919–922.

[30] T. Wadayama, K. Nakamura, M. Yagita, Y. Funahashi, S. Usami, and I. Takumi, "Gradient descent bitflipping algorithms for decoding LDPC codes," *IEEE Trans. Commun.*, **58**(6) (2010), 1610–1614.

[31] Q. Zhu and L. Wu, "Weighted candidate bit based bit-flipping decoding algorithms for LDPC codes," *2013 3rd International Conference on Consumer Electronics, Communications and Networks*, Xianning, China, 2013, pp. 731–734.

[32] Y. Liu, X. Niu, and M. Zhang, "Multi-threshold bit flipping algorithm for decoding structured LDPC codes," in *IEEE Commun. Lett.*, **19**(2) (2015), 127–130.

[33] T. C. Chang and Y. T. Su, "Dynamic weighted bit-flipping decoding algorithms for LDPC codes," *IEEE Trans. Commun.*, **63**(11) (2015), 3950–3963.

[34] J. Oh and J. Ha, "A two-bit weighted bit-flipping decoding algorithm for LDPC codes," *IEEE Commun. Lett.*, **22**(5) (2018), 874–877.

[35] M. Fossorier, "Iterative reliability-based decoding of low-density parity check codes," *IEEE J. Selected Areas Commun.*, **19**(5) (2001), 908–917.

[36] C. Jones, E. Valles, M. Smith, and J. Villasenor, "Approximate-min* constraint node updating for LDPC code decoding," *IEEE Military Commun. Conf.*, October 2003, pp. 157–162.

[37] M. Mansour and N. Shanbhag, "High-throughput LDPC decoders," *IEEE Trans. VLSI Systems*, **11**(6) (2003), 976–996.

[38] J. Chen, A. Dholakia, E. Eleftheriou, M. Fossorier, and X.-Y. Hu, "Reduced-complexity decoding of LDPC codes," *IEEE Trans. Commun.*, **53**(8) (2005), 1288–1299.

[39] J. Chen, R. M. Tanner, C. Jones, and Y. Li, "Improved min-sum decoding algorithms for irregular LDPC codes," *Proc. Int. Symp. Inf. Theory*, September 2005, pp. 449–453.

[40] J. Zhao, F. Zarkeshvari, and A. Banihashemi, "On implementation of min-sum algorithm and its modifications for decoding low-density parity-check (LDPC) codes," *IEEE Trans. Commun.*, **53**(4) (2005), 549–554.

[41] M. Viens, *A reduced-complexity iterative decoder for LDPC codes*, M. S. thesis, ECE Department, University of Arizona, December 2007.

[42] M. Viens and W. E. Ryan, "A reduced-complexity box-plus decoder for LDPC codes," *Fifth Int. Symp. on Turbo Codes and Related Topics*, September 2008, pp. 151–156.

[43] M. Xu, J. Wu, and M. Zhang, "A modified offset min-sum decoding algorithm for LDPC codes," *2010 3rd International Conference on Computer Science and Information Technology*, Chengdu, China, 2010, pp. 19–22.

[44] J. Chen, Y. Zhang, and R. Sun, "An improved normalized min-sum algorithm for LDPC codes," *2013 IEEE/ACIS 12th International Conference on Computer and Information Science* (ICIS), Niigata, 2013, pp. 509–512.

[45] A. A. Emran and M. Elsabrouty, "Simplified variable-scaled min sum LDPC decoder for irregular LDPC codes," *2014 IEEE 11th Consumer Communications and Networking Conference* (CCNC), Las Vegas, NV, 2014, pp. 518–523.

[46] D. MacKay and M. S. Postol, "Weaknesses of Margulis and Ramanujan-Margulis low-density parity-check codes," *Elect. Notes Theor. Comp. Sci.*, **74** (2003), 97–104.

[47] T. Richardson, "Error floor of LDPC codes," in *Proc. 41st Annual Allerton Conf. Commun., Control, Computing*, Menticello, IL, October 1–3, 2003, pp. 1426–1435.

[48] S.-Y. Chung, T. Richardson, and R. Urbanke, "Analysis of sum-product decoding of LDPC codes using a Gaussian approximation," *IEEE Trans. Inf. Theory*, **47**(2) (2001), 657–670.

[49] S. ten Brink, G. Kramer, and A. Ashikhmin, "Design of low-density parity-check codes for modulation and detection," *IEEE Trans. Commun.*, **52**(4) (2004), 670–678.

[50] A. Ashikhmin, G. Kramer, and S. ten Brink, "Extrinsic information transfer functions: model and erasure channel properties," *IEEE Trans. Inf. Theory*, **50**(11) (2004), 2657–2673.

[51] E. Sharon, A. Ashikhmin, and S. Litsyn, "EXIT functions for binary input memoryless symmetric channels," *IEEE Trans. Commun.*, **54**(7) (2006), 1207–1214.

[52] O. Milenkovic, E. Soljanin, and P. Whiting, "Asymptotic spectra of trapping sets in regular and irregular LDPC code ensembles," *IEEE Trans. Inf. Theory*, **53**(1) (2007), 39–55.

[53] L. Dolecek, Z. Zhang, V. Anantharam, M. J. Wainwright, and B. Nikolic, "Analysis of absorbing sets and fully absorbing sets of array-based LDPC codes," *IEEE Trans. Inf. Theory*, **56**(1) (2010), 181–201.

[54] C. E. Shannon, R. G. Gallager, and E. R. Berlekamp, "Lower bounds to error probability for coding on discrete memoryless channels," *Inform. Contr.*, **10** (1967), pt. I, pp. 65–103, pt. II, pp. 522–552.

[55] S. Lander and O. Milenkovic, "Algorithmic and combinatorial analysis of trapping sets in structured LDPC codes," *Proc. Int. Conf. Wireless Networks Commun. Mobile Comp.*, June 2005, pp. 630–635.

[56] S. Sankaranarayanan, S. K. Chilappagari, R. Radhakrishnan, and B. Vasic, "Failures of the Gallager B decoder: analysis and applications," *Proc. 2nd Inf. Theory Applications Workshop*, San Diego, CA, 2006, pp. 1–5.

[57] C. A. Cole, S. G. Wilson, E. K. Hall, and T. R. Giallorenzi, "A general method for finding low error-rates of LDPC codes," https://arxiv.org/abs/cs/0605051, June 2006.

[58] S. Laendner and O. Milenkovic, "LDPC codes based on Latin squares: cycle structure, stopping set, and trapping set analysis," *IEEE Trans. Commun.*, **55**(2) (2007), 303–312.

[59] M. Ivkovic, S. K. Chilappagari, and B. Vasic, "Eliminating trapping sets in low-density parity check codes by using Tanner graph covers," *IEEE Trans. Inf. Theory*, **54**(8) (2008), 3763–3768.

[60] E. Cavus, C. L. Haymes, and B. Daneshrad, "Low BER performance estimation of LDPC codes via application of importance sampling to trapping sets," *IEEE Trans. Commun.*, **57**(7) (2009), 1886–1888.

[61] B. Vasic, S. K. Chilappagari, D. V. Nguyen, and S. K. Planjery, "Trapping set ontology," *2009 47th Annual Allerton Conference on Communication, Control, and Computing* (Allerton), Monticello, IL, 2009, pp. 1–7.

[62] A. McGregor and O. Milenkovic, "On the hardness of approximating stopping and trapping sets," *IEEE Trans. Inf. Theory*, **56**(4) (2010), 1640–1650.

[63] S. Abu-Surra, D. Declercq, D. Divsalar, and W. E. Ryan, "Trapping set enumerators for specific LDPC codes," *2010 Information Theory and Applications Workshop (ITA)*, San Diego, CA, 2010, pp. 1–5.

[64] M. Karimi and A. H. Banihashemi, "Efficient algorithm for finding dominant trapping sets of LDPC codes," *IEEE Trans. Inf. Theory*, **58**(11) (2012), 6942–6958.

[65] D. G. M. Mitchell, A. E. Pusane, and D. J. Costello, "Minimum distance and trapping set analysis of protograph-based LDPC convolutional codes," *IEEE Trans. Inf. Theory*, **59**(1) (2013), 254–281.

[66] Y. Hashemi and A. H. Banihashemi, "Characterization of elementary trapping sets in irregular LDPC codes and the corresponding efficient exhaustive search algorithms," *IEEE Trans. Inf. Theory*, **64**(5) (2018), 3411–3430.

[67] Y. Hashemi and A. H. Banihashemi, "Characterization and efficient search of non-elementary trapping sets of LDPC codes with applications to stopping sets," *IEEE Trans. Inf. Theory*, **65**(2) (2019), 1017–1033.

[68] A. Dehghan and A. H. Banihashemi, "From cages to trapping sets and codewords: a technique to derive tight upper bounds on the minimum size of trapping sets and minimum distance of LDPC codes," *IEEE Trans. Inf. Theory*, **65**(4) (2019), 2062–2074.

[69] Q. Huang, Q. Diao, S. Lin, and K. Abdel-Ghaffar, "Cyclic and quasi-cyclic LDPC codes on constrained parity-check matrices and their trapping sets," *IEEE Trans. Inf. Theory*, **58**(5) (2012), 2648–2671.

[70] Q. Diao, Y. Y. Tai, S. Lin, and K. Abdel-Ghaffar, "Trapping set structure of LDPC codes on finite geometries," *Proc. IEEE Inf. Theory Applic. Workshop*, San Diego, CA, February 10–15, 2013, pp. 1–8.

[71] Q. Diao, Y. Y. Tai, S. Lin, and K. Abdel-Ghaffar, "LDPC codes on partial geometries: construction, trapping sets structure, and puncturing," *IEEE Trans. Inf. Theory*, **59**(12) (2013), 7898–7914.

[72] Q. Diao, J. Li, S. Lin, and I. F. Blake, "New classes of partial geometries and their associated LDPC codes," *IEEE Trans. Inf. Theory*, **62**(6) (2016), 2947–2965.

[73] H. Pishro-Nik and F. Fekri, "On decoding of low-density parity-check codes over the binary erasure channel," *IEEE Trans. Inf. Theory*, **50**(3) (2004), 439–454.

[74] H. Pishro-Nik and F. Fekri, "Results on punctured low-density parity-check codes and improved iterative decoding techniques," *IEEE Trans. Inf. Theory*, **53**(2) (2007), 599–614.

[75] Z. Zhang, L. Dolecek, B. Nikolic, V. Anantharam, and M. Wainwright, "Lowering LDPC error floors by post processing," *2008 IEEE GlobeCom Conf.*, November 30–December 4, 2008, pp. 1–6.

[76] Y. Han, *LDPC coding for magnetic storage: low-floor decoding algorithms, system design, and performance analysis*, Ph.D. dissertation, ECE Department, University of Arizona, August 2008.

[77] Y. Han and W. E. Ryan, "Pinning techniques for low-floor detection/decoding of LDPC-coded partial response channels," *2008 Symposium on Turbo Codes and Related Topics*, Lausanne, 2008, pp. 49–54.

[78] Y. Zhang and W. E. Ryan, "Toward low LDPC-code floors: a case study," *IEEE Trans. Commun.*, **57**(6) (2009), 1566–1573.

[79] Y. Han and W. E. Ryan, "Low-floor decoders for LDPC codes," *IEEE Trans. Commun.*, **57**(6) (2009), 1663–1673.

[80] J. Kang, Q. Huang, S. Lin, and K. Abdel-Ghaffar, "An iterative decoding algorithm with backtracking to lower the error-floors of LDPC codes," *IEEE Trans. Commun.*, **59**(1) (2011), 64–73.

[81] X. Chen, J. Kang, S. Lin, and V. Akella, "Hardware implementation of a backtracking-based reconfigurable decoder for lowering the error floor of quasi-cyclic LDPC codes," *IEEE Trans. Circuits and Systems I: Regular Papers*, **58**(12) (2011), 2931–2943.

[82] A. Orlitsky, R. Urbanke, K. Viswanathan, and J. Zhang, "Stopping sets and the girth of Tanner graphs," *Proceedings IEEE International Symposium on Inf. Theory*, Lausanne, June 30–July 5, 2002, p. 2.

[83] A. Orlitsky, K. Viswanathan, and J. Zhang, "Stopping set distribution of LDPC code ensembles," *IEEE International Symposium on Inf. Theory*, Yokohama, Japan, June 29–July 4, 2003, p. 123.

[84] K. Abdel-Ghaffar and J. H. Weber, "Generalized stopping sets and stopping redundancy," *IEEE 2007 Information Theory and Applications Workshop*, La Jolla, CA, January 29–February 2, 2007, pp. 400–404.

[85] J. Zhang, F. Fu, and D. Wan, "Stopping sets of algebraic geometry codes," *IEEE Transactions on Inf. Theory*, **60**(3) (2014), 1488–1495.

[86] A. Price and J. Hall, "A survey on trapping sets and stopping sets," http://arxiv.org/abs/1705.05996, 2017.

[87] J. Li, K. Liu, S. Lin, and K. Abdel-Ghaffar, "Algebraic quasi-cyclic LDPC codes: construction, low error-floor, large girth and a reduced-complexity decoding scheme," *IEEE Trans. Commun.*, **62**(8) (2014), 2626–2637.

[88] D. J. C. MacKay and M. C. Davey, "Evaluation of Gallager codes of short block length and hight rate applications," *Proc. IMA International Conference on Mathematics and Its Applications: Codes, Systems and Graphical Models,* New York, Springer-Verlag, 2000, pp. 113–130.

[89] J. Li, S. Lin, K. Abdel-Ghaffar, W. E. Ryan, and D. J. Costello, Jr., *LDPC Code Designs, Constructions, and Unification,* Cambridge, Cambridge University Press, 2017.

[90] G. A. Margulis, "Explicit constructions of graphs without short cycles and low density codes," *Combinatorics,* **2**(1) (1982), 71–78.

# 12

# Cyclic and Quasi-cyclic LDPC Codes on Finite Geometries

Chapter 7 showed that cyclic codes can be constructed based on the lines of two classes of finite geometries, namely, Euclidean and projective geometries. It was shown that these cyclic finite-geometry (FG) codes can be decoded with the simple one-step majority-logic decoding (OSMLD) based on the orthogonal structure of their parity-check matrices. In 2000 [1, 2], it was discovered that these FG codes form a class of LDPC codes and hence can be decoded with various iterative decoding algorithms based on belief-propagation (IDBP-algorithms), such as the SPA [3] or the MSA [4] (see Section 11.5). From the low-density point of view, these FG codes are referred to as *cyclic-FG-LDPC codes*. Since 2000, there has been a great deal of research on construction, decoding, and analysis of this class of codes [5–29]. Cyclic-FG-LDPC codes have very distinct structural properties which allow them to achieve very low error rates without visible error-floors with IDBP-algorithms [5, 9, 13, 14].

In this chapter, we first present cyclic-FG codes as LDPC codes and provide examples to demonstrate their distinctive error performance with iterative decoding. Next, we show that FG-LDPC codes with *quasi-cyclic* (QC) structure, called *QC-FG-LDPC codes*, can be constructed. QC-FG-LDPC codes form a much larger class of FG-LDPC codes than the class of cyclic-FG-LDPC codes. Also presented in this chapter are two *reduced-complexity iterative decoding algorithms*, one for decoding cyclic LDPC codes and the other for decoding QC-LDPC codes. These iterative decoding algorithms *significantly* reduce the hardware complexity and memory size of the decoder *without (or with a small) performance degradation*.

Our focus in this chapter is on binary FG-LDPC codes, cyclic and quasi-cyclic. References [28, 29] also have a good coverage of this class of LDPC codes.

Nonbinary LDPC codes can also be constructed based on finite geometries. Readers are referred to [17, 19, 24, 26] for details of nonbinary FG-LDPC codes.

## 12.1 Cyclic-FG-LDPC Codes

We first start with cyclic LDPC codes constructed based on the lines of the two-dimensional Euclidean geometry $EG(2,q)$ over $GF(q)$ with $q = 2^s$. Hereafter, unless otherwise specified, $q$ is a power of 2 and we will use $q$ and $2^s$ interchangeably throughout this chapter.

Let $EG^*(2,q)$ be the subgeometry obtained by removing the origin of $EG(2,q)$ and the lines passing through the origin. As it was shown in Chapter 7, the subgeometry $EG^*(2,q)$ of $EG(2,q)$ consists of $q^2 - 1$ points and $q^2 - 1$ lines. Each line in $EG^*(2,q)$ consists of $q$ points and each point is on $q$ lines in $EG^*(2,q)$. Using the incidence vectors of the $q^2 - 1$ lines in $EG^*(2,q)$, we can form a $(q^2 - 1) \times (q^2 - 1)$ circulant $\mathbf{H}_{EG,cyc}(2,q)$ over $GF(2)$ with weight $q$. Recall that the $q^2 - 1$ lines in $EG^*(2,q)$ form a single cyclic class. The incidence vector of any line in this cyclic class can be used as the generator $\mathbf{h}_0$ of the circulant $\mathbf{H}_{EG,cyc}(2,q)$. To form $\mathbf{H}_{EG,cyc}(2,q)$, we simply cyclically shift the generator $\mathbf{h}_0$ $q^2 - 2$ times, one place to the right each time. The subscript "cyc" in $\mathbf{H}_{EG,cyc}(2,q)$ stands for "cyclic."

The total number of entries in $\mathbf{H}_{EG,cyc}(2,q)$ is $(q^2-1)^2$ and the total number of 1-entries in $\mathbf{H}_{EG,cyc}(2,q)$ is $q(q^2-1)$. The density of $\mathbf{H}_{EG,cyc}(2,q)$ is defined as the ratio of the number of 1-entries and the number of total entries, i.e., $q(q^2-1)/(q^2-1)^2 = q/(q^2-1)$. For large $q$, say $q = 2^s$ with $s \geq 3$, the density of $\mathbf{H}_{EG,cyc}(2,q)$ is low and hence it is a low-density (or a sparse) matrix. Hence, the cyclic EG code $C_{EG,cyc}(2,q)$ given by the null space over $GF(2)$ of $\mathbf{H}_{EG,cyc}(2,q)$ is a $(q,q)$-regular LDPC code, called *cyclic-EG-LDPC code*. For $q = 2^s$, the rank of $\mathbf{H}_{EG,cyc}(2,2^s)$ is $3^s - 1$ [28, 30, 31]. Thus, the cyclic-EG-LDPC code $C_{EG,cyc}(2,2^s)$ has the following parameters:

$$
\begin{aligned}
\text{code length} \qquad & n & = 2^{2s} - 1, \\
\text{code dimension} \qquad & k & = 2^{2s} - 3^s, \\
\text{number of parity-check bits} \quad & n - k & = 3^s - 1, \\
\text{minimum distance} \qquad & d_{\min} & = 2^s + 1.
\end{aligned}
$$

The Tanner graph, denoted by $\mathcal{G}_{EG,cyc}(2,2^s)$, of the code $C_{EG,cyc}(2,2^s)$ has $2^{2s} - 1$ VNs and $2^{2s} - 1$ CNs. The girth of $\mathcal{G}_{EG,cyc}(2,2^s)$ is 6. Each VN in $\mathcal{G}_{EG,cyc}(2,2^s)$ is connected to other $2^s(2^s - 1)$ VNs through paths of length 2 and hence each VN has a connection number of $2^s(2^s - 1)$. It was shown in [20, 32] that the Tanner graph $\mathcal{G}_{EG,cyc}(2,2^s)$ of $C_{EG,cyc}(2,2^s)$ has good trapping set structure. It contains no trapping set of size smaller than $2^s + 1$ and no absorbing set of size smaller than $2^{s-1} + 1$. The absorbing set structure simply follows from the orthogonal structure of $\mathbf{H}_{EG,cyc}(2,2^s)$ that for every code symbol $v$, there are $2^s$ rows in of $\mathbf{H}_{EG,cyc}(2,2^s)$ orthogonal on it. For $\kappa \leq 2^{s-1}$, in the subgraph $\mathcal{G}_{EG,cyc}(\Delta_{EG}(\kappa))$ of $\mathcal{G}_{EG,cyc}(2,2^s)$ induced by a set $\Delta_{EG}(\kappa)$ of

$\kappa$ VNs, every VN in $\Delta_{\mathrm{EG}}(\kappa)$ is adjacent (or connected) to more CNs of odd degrees than CNs of even degrees, i.e., the number of CNs of odd degrees in $\mathcal{G}_{\mathrm{EG,cyc}}(\Delta_{\mathrm{EG}}(\kappa))$ is at least $2^{s-1}+1$. In this case, if $\kappa$ errors in a received vector occur at the locations corresponding to the $\kappa$ VNs in $\mathcal{G}_{\mathrm{EG,cyc}}(\Delta_{\mathrm{EG}}(\kappa))$, they can all be corrected with the OSMLD. Hence, for $\kappa \leq 2^{s-1}$, no absorbing set of size $\kappa$ exists in $\mathcal{G}_{\mathrm{EG,cyc}}(2, 2^s)$.

For $s \geq 4$, because the cyclic-EG-LDPC code $C_{\mathrm{EG,cyc}}(2, 2^s)$ has a large minimum distance and no small harmful trapping set in its Tanner graph $\mathcal{G}_{\mathrm{EG,cyc}}(2, 2^s)$, it can achieve a very low error rate without a visible error-floor decoded using either the SPA [3] or the MSA [4]. Because its Tanner graph also has a large VN-connectivity, decoding of $C_{\mathrm{EG,cyc}}(2, 2^s)$ with an IDBP-algorithm converges fast.

Furthermore, the size of a minimum stopping set in $\mathcal{G}_{\mathrm{EG,cyc}}(2, 2^s)$ of the cyclic-EG-LDPC code $C_{\mathrm{EG,cyc}}(2, 2^s)$ is $2^s + 1$. Hence, with the SPL-algorithm, the code $C_{\mathrm{EG,cyc}}(2, 2^s)$ is capable of correcting any erasure pattern with $2^s$ or fewer erasures. Decoding with the SPA, the code $C_{\mathrm{EG,cyc}}(2, 2^s)$ also performs well over the BEC. This will be demonstrated in examples given later.

**Example 12.1** Let $q = 2^5$. Consider the subgeometry $\mathrm{EG}^*(2, 2^5)$ of the two-dimensional Euclidean geometry $\mathrm{EG}(2, 2^5)$ over $\mathrm{GF}(2^5)$ which consists of 1023 points and 1023 lines. The incidence vectors of the lines in $\mathrm{EG}^*(2, 2^5)$ form a $1023 \times 1023$ circulant $\mathbf{H}_{\mathrm{EG,cyc}}(2, 2^5)$ of weight 32. The rank of $\mathbf{H}_{\mathrm{EG,cyc}}(2, 2^5)$ is $3^5 - 1 = 242$. Hence, $\mathbf{H}_{\mathrm{EG,cyc}}(2, 2^5)$ has 781 redundant rows. The null space over $\mathrm{GF}(2)$ of $\mathbf{H}_{\mathrm{EG,cyc}}(2, 2^5)$ gives a $(32, 32)$-regular $(1023, 781)$ cyclic-EG-LDPC code $C_{\mathrm{EG,cyc}}(2, 2^5)$ with rate 0.7634 and minimum distance 33. Its Tanner graph $\mathcal{G}_{\mathrm{EG,cyc}}(2, 2^5)$ contains no trapping set of size smaller than 33 and no absorbing set of size smaller than 17. The connection number of each VN in $\mathcal{G}_{\mathrm{EG,cyc}}(2, 2^5)$ is 992.

The BER (bit-error rate) and BLER (block-error rate) performances of the code $C_{\mathrm{EG,cyc}}(2, 2^5)$ decoded with 5, 10, and 50 iterations of the SPA over the AWGN channel using BPSK signaling are shown in Fig. 12.1(a). We see that the decoding of the code $C_{\mathrm{EG,cyc}}(2, 2^5)$ converges quickly. The BER and BLER performances of this code decoded with 5, 10, and 50 iterations of the MSA with scaling factor 0.25 are shown in Fig. 12.1(b). From Fig. 12.1(a) and Fig. 12.1(b), we see that the MSA performs almost the same as the SPA. The advantage of the MSA over the SPA is that the MSA enjoys much lower computation complexity, as shown in Section 11.5.3. ▲▲

**Example 12.2** Suppose we apply the $(32, 32)$-regular $(1023, 781)$ cyclic-EG-LDPC code $C_{\mathrm{EG,cyc}}(2, 2^5)$ constructed in Example 12.1 to the BEC. Because the column weight of the parity-check matrix of this code is 32, the size of a minimum stopping set in the Tanner graph of this code is 33. With the SPL-algorithm presented in Section 10.8.2, this code is capable of correcting any erasure pattern with 32 or fewer erasures. Because the rate of this code is $R = 0.7634$, the Shannon limit for this rate over the BEC is $1 - R = 0.2366$. The UEBR (unrecovered erasure bit rate) and UEBLR (unrecovered erasure block

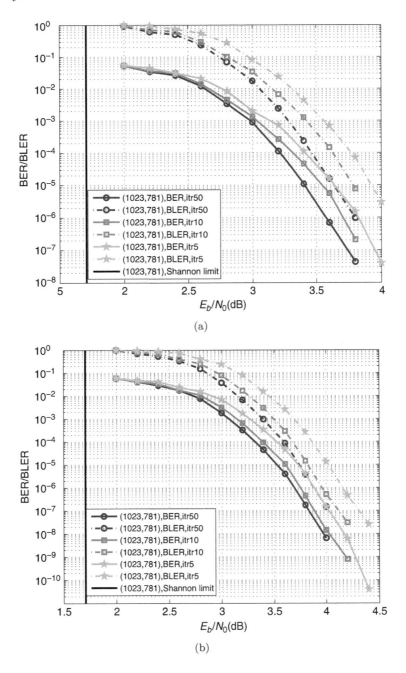

Figure 12.1 The BER and BLER performances of the $(1023, 781)$ cyclic-EG-LDPC code $C_{\mathrm{EG,cyc}}(2, 2^5)$ given in Example 12.1 decoded with 5, 10, and 50 iterations: (a) SPA and (b) MSA.

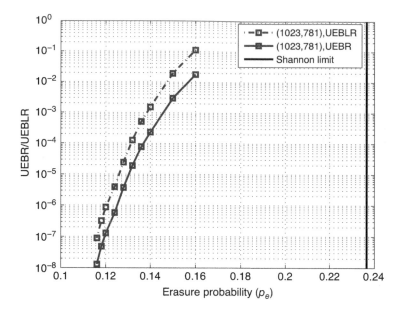

Figure 12.2 The UEBR and UEBLR performances of the $(1023, 781)$ cyclic-EG-LDPC code $C_{\mathrm{EG,cyc}}(2, 2^5)$ given in Example 12.1 over the BEC.

rate) performances of this code decoded with the SPA over the BEC are shown in Fig. 12.2. We see that the code performs well and achieves a UEBR of $10^{-8}$ without showing any error-floor. ▲▲

**Example 12.3** Consider the subgeometry $\mathrm{EG}^*(2, 2^6)$ of $\mathrm{EG}(2, 2^6)$ over $\mathrm{GF}(2^6)$ for code construction. Based on the lines in $\mathrm{EG}^*(2, 2^6)$, we construct a $4095 \times 4095$ circulant $\mathbf{H}_{\mathrm{EG,cyc}}(2, 2^6)$ with weight 64. The rank of $\mathbf{H}_{\mathrm{EG,cyc}}(2, 2^6)$ is 728. Hence, $\mathbf{H}_{\mathrm{EG,cyc}}(2, 2^6)$ has 3367 redundant rows. The null space over $\mathrm{GF}(2)$ of $\mathbf{H}_{\mathrm{EG,cyc}}(2, 2^6)$ gives a $(64, 64)$-regular $(4095, 3367)$ cyclic-EG-LDPC code $C_{\mathrm{EG,cyc}}(2, 2^6)$ with rate 0.8222 and minimum distance 65. Its Tanner graph contains no trapping set of size smaller than 65. All the 3367 redundant rows of $\mathbf{H}_{\mathrm{EG,cyc}}(2, 2^6)$ can assist the decoding to converge rapidly.

The BER and BLER performances of this code over the AWGN channel using BPSK signaling decoded with the MSA, scaled by a factor of 0.25, based on the entire parity-check matrix $\mathbf{H}_{\mathrm{EG,cyc}}(2, 2^6)$ with 5, 10, and 50 iterations, are shown in Fig. 12.3(a). At the BER of $10^{-7}$, the performance gap between 5 and 50 iterations is less than 0.2 dB and the performance gap between 10 and 50 iterations of the MSA is less than 0.1 dB.

The size of a minimum stopping set in the Tanner graph of $C_{\mathrm{EG,cyc}}(2, 2^6)$ is 65. Hence, with the SPL-algorithm, the code is capable of correcting any erasure pattern with 64 or fewer erasures over the BEC. The UEBR and UEBLR performances of this code over the BEC decoded with the SPA are shown in Fig. 12.3(b). Because the rate $R$ of the code is 0.8222, the Shannon limit for this rate is $1 - R = 0.1778$. From Fig. 12.3(b), we see that at the UEBR of $10^{-9}$, the code performs 0.1 away from the Shannon limit. ▲▲

As shown in Chapter 7, cyclic codes can also be constructed based on the lines of projective geometries over finite fields. The parity-check matrix $\mathbf{H}_{\mathrm{PG,cyc}}(2, q)$ of the cyclic PG code $C_{\mathrm{PG,cyc}}(2, q)$ constructed based on the lines of a two-dimensional projective geometry $\mathrm{PG}(2, q)$ over $\mathrm{GF}(q)$ is a $(q^2 + q + 1) \times (q^2 + q + 1)$ circulant with weight $q + 1$. The density of $\mathbf{H}_{\mathrm{PG,cyc}}(2, q)$ is $(q + 1)/(q^2 + q + 1)$ which is very small for large $q$. Hence, the cyclic PG code $C_{\mathrm{PG,cyc}}(2, q)$ is an LDPC code, referred to as a *cyclic-PG-LDPC code*.

For $q = 2^s$, the rank of $\mathbf{H}_{\mathrm{PG,cyc}}(2, 2^s)$ is $3^s + 1$. The null space over $\mathrm{GF}(2)$ of $\mathbf{H}_{\mathrm{PG,cyc}}(2, 2^s)$ gives a $(2^s+1, 2^s+1)$-regular cyclic-PG-LDPC code $C_{\mathrm{PG,cyc}}(2, 2^s)$ with the following parameters [9, 14, 28, 29]:

$$
\begin{aligned}
\text{code length} \quad & n &=& \ 2^{2s} + 2^s + 1, \\
\text{code dimension} \quad & k &=& \ 2^{2s} + 2^s - 3^s, \\
\text{number of parity-check bits} \quad & n - k &=& \ 3^s + 1, \\
\text{minimum distance} \quad & d_{\min} &=& \ 2^s + 2.
\end{aligned}
$$

The Tanner graph $\mathcal{G}_{\mathrm{PG,cyc}}(2, 2^s)$ of the cyclic-PG-LDPC code $C_{\mathrm{PG,cyc}}(2, 2^s)$ has similar structures as the Tanner graph $\mathcal{G}_{\mathrm{EG,cyc}}(2, 2^s)$ of the cyclic-EG-LDPC $C_{\mathrm{EG,cyc}}(2, 2^s)$. It has $2^{2s} + 2^s + 1$ VNs and $2^{2s} + 2^s + 1$ CNs. The girth of $\mathcal{G}_{\mathrm{PG,cyc}}(2, 2^s)$ is 6. Each VN is connected to another $2^s(2^s+1)$ VNs through paths of length 2 and hence each VN has a connection number of $2^s(2^s + 1)$. It follows from [20, 32] that $\mathcal{G}_{\mathrm{PG,cyc}}(2, 2^s)$ contains no trapping set of size smaller than $2^s + 2$ and no absorbing set of size smaller than $\lfloor (2^s + 1)/2 \rfloor + 1$. The absorbing set structure of $\mathcal{G}_{\mathrm{PG,cyc}}(2, 2^s)$ simply follows from the orthogonal structure of $\mathbf{H}_{\mathrm{PG,cyc}}(2, 2^s)$ that, for every code symbol $v$ of a codeword in $C_{\mathrm{PG,cyc}}(2, 2^s)$, there are $2^s + 1$ rows in of $\mathbf{H}_{\mathrm{PG,cyc}}(2, 2^s)$ orthogonal on $v$. For $\kappa \leq \lfloor (2^s+1)/2 \rfloor = 2^{s-1}$, in the subgraph $\mathcal{G}_{\mathrm{PG,cyc}}(\Delta_{\mathrm{PG}}(\kappa))$ of $\mathcal{G}_{\mathrm{PG,cyc}}(2, 2^s)$ induced by a set $\Delta_{\mathrm{PG}}(\kappa)$ of $\kappa$ VNs, every VN is connected to more CNs of odd degrees than CNs of even degrees. In this case, if the $\kappa$ errors occur at the locations corresponding to the $\kappa$ VNs in $\mathcal{G}_{\mathrm{PG,cyc}}(\Delta_{\mathrm{PG}}(\kappa))$, they can all be corrected with the OSMLD. Hence, for $\kappa \leq 2^{s-1}$, no absorbing set of size $\kappa$ exists in $\mathcal{G}_{\mathrm{PG,cyc}}(2, 2^s)$.

For $s \geq 3$, the cyclic-PG-LDPC code $C_{\mathrm{PG,cyc}}(2, 2^s)$ has a large minimum distance and no small harmful trapping set in its Tanner graph $\mathcal{G}_{\mathrm{PG,cyc}}(2, 2^s)$. Consequently, the cyclic-PG-LDPC code $C_{\mathrm{PG,cyc}}(2, 2^s)$ can achieve a very low error rate without a visible error-floor decoded with either the SPA or the MSA. Because its Tanner graph also has a large VN-connectivity, decoding of the code $C_{\mathrm{PG,cyc}}(2, 2^s)$ with an IDBP-algorithm converges fast.

The size of a minimum stopping set in $\mathcal{G}_{\mathrm{PG,cyc}}(2, 2^s)$ is $2^s + 2$. Hence, the code $C_{\mathrm{PG,cyc}}(2, 2^s)$ decoded with the SPL-algorithm is capable of correcting any erasure pattern with $2^s + 1$ or fewer erasures over the BEC.

**Example 12.4** Consider the two-dimensional projective geometry $\mathrm{PG}(2, 2^5)$ over $\mathrm{GF}(2^5)$ which contains 1057 points and 1057 lines. The 1057 lines of

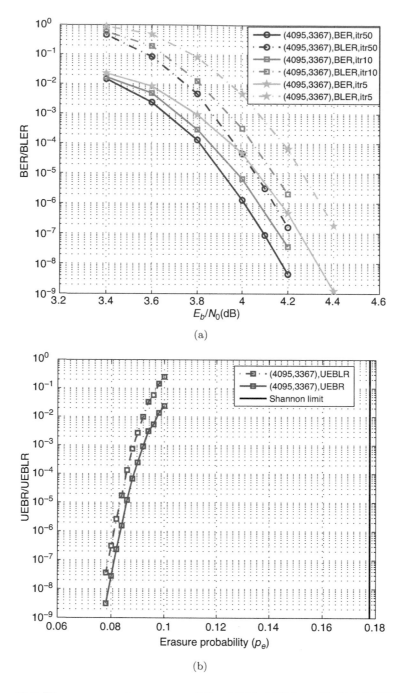

Figure 12.3 The error performances of the $(4095, 3367)$ cyclic-EG-LDPC code given in Example 12.3 over: (a) AWGN channel and (b) BEC.

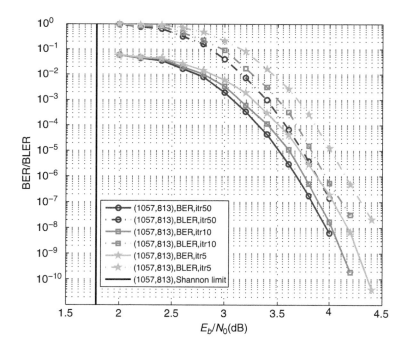

Figure 12.4 The BER and BLER performances of the $(1057, 813)$ cyclic-PG-LDPC code given in Example 12.4.

$\mathrm{PG}(2, 2^5)$ form a single cyclic class. Using the incidence vector of a line in this cyclic class as the generator, we construct a $1057 \times 1057$ circulant $\mathbf{H}_{\mathrm{PG,cyc}}(2, 2^5)$ with weight 33. The rank of $\mathbf{H}_{\mathrm{PG,cyc}}(2, 2^5)$ is $3^5 + 1 = 244$. Hence, the matrix $\mathbf{H}_{\mathrm{PG,cyc}}(2, 2^5)$ has 813 redundant rows. The null space over $\mathrm{GF}(2)$ of $\mathbf{H}_{\mathrm{PG,cyc}}(2, 2^5)$ gives a binary $(33, 33)$-regular $(1057, 813)$ cyclic-PG-LDPC code $C_{\mathrm{PG,cyc}}(2, 2^5)$ with rate 0.7691 and minimum distance 34 whose Tanner graph $\mathcal{G}_{\mathrm{PG,cyc}}(2, 2^5)$ has girth 6 and contains no harmful trapping set of size smaller than 34 and no absorbing set of size smaller than 17. Each VN in $\mathcal{G}_{\mathrm{PG,cyc}}(2, 2^5)$ is connected to other 1056 VNs in $\mathcal{G}_{\mathrm{PG,cyc}}(2, 2^5)$ by paths of length 2. Therefore, each VN in $\mathcal{G}_{\mathrm{PG,cyc}}(2, 2^5)$ has a large connection number. With the above structural properties, the cyclic-PG-LDPC code $C_{\mathrm{PG,cyc}}(2, 2^5)$ decoded with an IDBP-algorithm, such as the SPA or the MSA, can achieve a very low error rate without an error-floor and a fast rate of decoding convergence.

The BER and BLER performances of the code $C_{\mathrm{PG,cyc}}(2, 2^5)$ over the AWGN channel using BPSK signaling decoded with 5, 10, and 50 iterations of the MSA scaled by a factor 0.25 are shown in Fig. 12.4. We see that the code performs well and the decoding of the code converges quickly. At a BER of $10^{-8}$, the performance gap between 5 and 50 decoding iterations is about 0.25 dB and performance gap between 10 and 50 decoding iterations is less than 0.1 dB. ▲▲

Cyclic EG and PG codes constructed based on cyclic classes of lines in multi-dimensional Euclidean and projective geometries over finite fields in Chapter 7 are also cyclic LDPC codes with large minimum distances. The Tanner graphs of these cyclic LDPC codes contain no small trapping set of size smaller than their respective minimum distances and no absorbing set of size smaller than half of their respective minimum distances [20, 32]. Constructions of LDPC codes based on multi-dimensional Euclidean and projective geometries will be presented in Section 12.7.

## 12.2 A Complexity-Reduced Iterative Algorithm for Decoding Cyclic-FG-LDPC Codes

Cyclic-FG-LDPC codes are rich in geometric and algebraic structural properties. They have large minimum distances and good trapping set structures. Decoded with an IDBP-algorithm, such as the SPA or the MSA, they can achieve *excellent overall performance in terms of coding gain, error-floor*, and *rate of decoding convergence*. This class of codes can be decoded with various decoding algorithms from soft-decision iterative decoding algorithms to simple hard-decision one-step majority-logic decoding to provide a wide spectrum of tradeoffs between decoding complexity and error performance. However, the relatively high density and large column weight of the parity-check matrix ($\mathbf{H}_{\mathrm{EG,cyc}}(2, q)$ has column weight $q$ and $\mathbf{H}_{\mathrm{PG,cyc}}(2, q)$ has column weight $q + 1$) of a cyclic-FG-LDPC code which makes the hardware implementation complexity of a soft-decision iterative decoder quite large, and this is a critical issue in practical applications.

In this section, we present an effective *reduced-complexity* iterative decoding algorithm for a cyclic-FG-LDPC code based on the *cyclic structure* and a *cyclic grouping* of the rows of the parity-check matrix of the code. The decoding algorithm significantly reduces the decoding complexity in terms of the number of processing units, the number of connecting wires, the size of memory, and the number of real-number computations. This reduction in decoding complexity may enhance their practical applications owing to their outstanding overall performance, especially low error-floor and fast rate of decoding convergence.

Consider a cyclic-FG-LDPC code $C_{\mathrm{FG,cyc}}$ given by the null space of an $n \times n$ circulant $\mathbf{H}_{\mathrm{FG,cyc}}$ with rank $r$ and weight $\gamma$. As shown in Section 12.1, $r$ is much smaller than $n$. Let $\mathbf{h}_0, \mathbf{h}_1, \ldots, \mathbf{h}_{n-1}$ denote the $n$ rows of $\mathbf{H}_{\mathrm{FG,cyc}}$. Let $\ell$ be a positive integer with $1 \leq \ell \leq n$. For $i \geq 0$ and $1 \leq \ell \leq n$, we form a sequence of $\ell \times n$ submatrices of $\mathbf{H}_{\mathrm{FG,cyc}}$,

$$(\mathbf{H}_i)_{i=0,1,2,\ldots} = (\mathbf{H}_0, \mathbf{H}_1, \ldots, \mathbf{H}_i, \ldots), \tag{12.1}$$

each submatrix consisting of $\ell$ *consecutive rows* of $\mathbf{H}_{\mathrm{FG,cyc}}$ (including $\ell$ consecutive *end-around* rows in case $n$ is not divisible by $\ell$, i.e., $w$ rows at the bottom of the circulant and $\ell - w$ rows on the top of the circulant). The $i$th submatrix $\mathbf{H}_i$ in $(\mathbf{H}_i)_{i=0,1,2,\ldots}$ is given as follows:

$$\mathbf{H}_i = \begin{bmatrix} \mathbf{h}_{(i\ell)_n} \\ \mathbf{h}_{(i\ell+1)_n} \\ \vdots \\ \mathbf{h}_{(i\ell+\ell-1)_n} \end{bmatrix}, \tag{12.2}$$

where $(a)_n$ denotes the result of $a$ module $n$. The submatrix $\mathbf{H}_0$ consists of the top $\ell$ consecutive rows of the circulant $\mathbf{H}_{FG,cyc}$, i.e.,

$$\mathbf{H}_0 = \begin{bmatrix} \mathbf{h}_0 \\ \mathbf{h}_1 \\ \vdots \\ \mathbf{h}_{\ell-1} \end{bmatrix}. \tag{12.3}$$

All the submatrices in the sequence $(\mathbf{H}_i)_{i=0,1,2,\dots}$ have the same rank. If $\text{rank}(\mathbf{H}_0) = \text{rank}(\mathbf{H}_{FG,cyc}) = r$, then the null space of $\mathbf{H}_0$ (or $\mathbf{H}_i$ for any $i \geq 0$) gives the same code $C_{FG,cyc}$ as the entire parity-check matrix $\mathbf{H}_{FG,cyc}$. If $\ell > r$, then each submatrix $\mathbf{H}_i$ contains $\ell - r$ redundant rows; and if $\ell < r$, then the null space of $\mathbf{H}_i$ gives a *super code* of $C_{FG,cyc}$. The parameter $\ell$ is called the *grouping size*.

Based on the cyclic structure of $\mathbf{H}_{FG,cyc}$ and the construction of the sequence $(\mathbf{H}_i)_{i=0,1,2,\dots}$ of the submatrices of $\mathbf{H}_{FG,cyc}$, we readily see that, for $i > 0$, the submatrix $\mathbf{H}_i$ can be obtained from the first submatrix $\mathbf{H}_0$ by cyclically shifting all the rows of $\mathbf{H}_0$ to the right $i\ell$ positions. This cyclic structure is referred to as the *block cyclic structure*. This block cyclic structure allows us to decode the code $C_{FG,cyc}$ based on the submatrix $\mathbf{H}_0$ of $\mathbf{H}_{FG,cyc}$ *alone in a revolving manner*. Furthermore, we can also see that any consecutive $c = \lceil n/\ell \rceil$ submatrices in the sequence $(\mathbf{H}_i)_{i=0,1,2,\dots}$ cover all the rows of $\mathbf{H}_{FG,cyc}$. In the case for which $\ell$ is a factor of $n$, any consecutive $c = n/\ell$ submatrices in the sequence $(\mathbf{H}_i)_{i=0,1,2,\dots}$ form a partition of $\mathbf{H}_{FG,cyc}$ and $\mathbf{H}_c = \mathbf{H}_0$. The sequence $(\mathbf{H}_i)_{i=0,1,2,\dots}$ is a repetition of the first $c$ submatrices $\mathbf{H}_0, \mathbf{H}_1, \dots, \mathbf{H}_{c-1}$.

Iterative decoding of the cyclic-FG-LDPC code $C_{FG,cyc}$ based on its entire parity-check matrix $\mathbf{H}_{FG,cyc}$ can be carried out in terms of the consecutive submatrices in the sequence $(\mathbf{H}_i)_{i=0,1,2,\dots}$. For $i \geq 0$, the $i$th decoding iteration is defined as the reliability information updating process based on the matrix $\mathbf{H}_i$ and passing the updated reliability information to the decoder based on the next matrix $\mathbf{H}_{i+1}$ as input for decoding. Let $I_{max}$ be the preset maximum number of iterations such that $I_{max} \geq c = \lceil n/\ell \rceil$. Label the decoding iterations from 0 to $I_{max} - 1$. For $0 \leq i < I_{max}$, at the completion of the $i$th decoding iteration based on $\mathbf{H}_i$, we pass the updated reliability information of the received symbols to the decoder based on $\mathbf{H}_{i+1}$ and then we carry out the $(i+1)$th decoding iteration. At the end of each decoding iteration, a hard-decision vector $\mathbf{z}$ is formed based on the reliabilities of the decoded symbols. Using the hard-decision vector $\mathbf{z}$, we compute the syndrome $\mathbf{s} = \mathbf{z} \cdot \mathbf{H}_{FG,cyc}^T$. If $\mathbf{s} = \mathbf{0}$, $\mathbf{z}$ is a codeword in $C_{FG,cyc}$ and we stop the decoding process; otherwise, the decoding process continues until either a codeword is found or the preset number $I_{max}$ of decoding iterations is reached.

Using the block cyclic structure of the sequence $(\mathbf{H}_i)_{i=0,1,2,\dots}$, the decoding of $C_{FG,cyc}$ as described above can be carried out by *using only* the first submatrix

$\mathbf{H}_0$ in the sequence. The decoding process begins with initial symbol reliability information as input. For $i > 0$, at the completion of the $(i-1)$th decoding iteration, the reliability vector of the decoded vector is cyclically shifted to the *left* $\ell$ positions. This left-shifted reliability vector and the channel information are used as input to carry out the $i$th decoding iteration based on $\mathbf{H}_0$. Note that using the left-shifted reliability vector at the completion of the $(i-1)$th decoding iteration as the input to the $i$th decoding iteration based on $\mathbf{H}_0$ amounts to carrying out the $i$th decoding iteration based on the submatrix $\mathbf{H}_i$ of $\mathbf{H}_{\mathrm{FG,cyc}}$ using the unshifted output reliability vector of the $(i-1)$th decoding iteration based on the submatrix $\mathbf{H}_{i-1}$ as the symbol reliability information input. At the completion of each iteration, a hard-decision vector $\mathbf{z}$ is formed. Compute the syndrome $\mathbf{s}$ of $\mathbf{z}$ based on $\mathbf{H}_{\mathrm{FG,cyc}}$. If $\mathbf{s} = \mathbf{0}$, $\mathbf{z}$ is a codeword in $C_{\mathrm{FG,cyc}}$ and decoding stops; otherwise, the decoding process continues until either a codeword in $C_{\mathrm{FG,cyc}}$ is found at the end of a certain decoding iteration or the preset maximum number $I_{\max}$ of decoding iterations is reached. The entire decoding process simply revolves around $\mathbf{H}_0$ in conjunction with left-shift operations, which is referred to as a *revolving iterative decoding* (RID) [33, 34] and the submatrix $\mathbf{H}_0$ is referred to as the *decoding matrix*. In the case for which rank$(\mathbf{H}_0)$ = rank$(\mathbf{H}_{\mathrm{FG,cyc}})$, we can compute the syndrome of the decoded vector $\mathbf{z}$ at the end of each decoding iteration based on $\mathbf{H}_0$ rather than $\mathbf{H}_{\mathrm{FG,cyc}}$ and check whether $\mathbf{z} \cdot \mathbf{H}_0$ is equal to $\mathbf{0}$. Recall that any consecutive $c = \lceil n/\ell \rceil$ submatrices cover all the rows of $\mathbf{H}_{\mathrm{FG,cyc}}$. This implies that $c$ decoding iterations based on $\mathbf{H}_0$ cover all the message exchanges between the CN processing units (CNPUs) and VN processing units (VNPUs).

Consider the special case for which $\ell$ is a factor of $n$ and $c = n/\ell$. In this case, $\mathbf{H}_c = \mathbf{H}_0$ and $c$ RID iterations are equivalent to one iteration of the conventional iterative decoding based on the entire parity-check matrix $\mathbf{H}_{\mathrm{FG,cyc}}$ using either the SPA or MSA. An extreme case is that $\ell = 1$. In this case, the decoding matrix $\mathbf{H}_0$ consists of only the top row $\mathbf{h}_0$ of $\mathbf{H}_{\mathrm{FG,cyc}}$. The decoder consists of only a single CNPU and $\gamma$ (the weight of $\mathbf{h}_0$) VNPUs.

In each decoding iteration of the RID, we could use either the SPA or the MSA to update the reliabilities of the received code symbols. The MSA is the most commonly used reliability updating algorithm in practice owing to its low computational complexity. As shown in Example 12.1, if the MSA is scaled properly, it can perform as well as (or close to) the SPA. In the following, we present an RID-based algorithm in which a scaled MSA is used for reliability updating. This algorithm is referred to as an RID-MSA (or simply RMSA).

Let $\mathbf{y} = (y_0, y_1, \ldots, y_{n-1})$ be the soft-decision received vector at the output of the receiver detector. We assume that the symbols of a codeword in $C_{\mathrm{FG,cyc}}$ are transmitted over a binary AWGN channel with BPSK signaling. Let $\mathbf{H}_0 = [h_{t,j}]_{0 \leq t < \ell, 0 \leq j < n}$. For $0 \leq j < n$ and $0 \leq t < \ell$, let $N(j) = \{t : 0 \leq t < \ell, h_{t,j} = 1\}$ and $M(t) = \{j : 0 \leq j < n, h_{t,j} = 1\}$. Use the notation $(i)$ to denote that the decoder is in the $i$th decoding iteration. Let $I_{\max}$ be the maximum number of decoding iterations to be performed such that $I_{\max} \geq c = \lceil n/\ell \rceil$. For $0 \leq i < I_{\max}$, let $\mathbf{R}^{(i)} = (R_0^{(i)}, R_1^{(i)}, \ldots, R_{n-1}^{(i)})$ denote the reliability information vector at the beginning of the $i$th iteration, $\mathbf{F}^{(i)} = (F_0^{(i)}, F_1^{(i)}, \ldots, F_{n-1}^{(i)})$ denote the updated reliability information vector at the end of the $i$th iteration, and $\mathbf{y}^{(i)} = (y_0^{(i)}, y_1^{(i)}, \ldots, y_{n-1}^{(i)})$ denote the shifted version (to the left by $i\ell$ positions) of the received vector $\mathbf{y}$ at the beginning of the $i$th iteration. Let $\lambda_{\text{atten}}$ denote the scaling factor for a scaled MSA-based reliability updating. We label the iterations from 0 to $I_{\max} - 1$. Let $f$ be a positive integer such that we include the channel information for reliability update every $f$ RMSA iterations. The integer $f$ is chosen such that $f$ iterations of the RMSA cover all the rows of the parity-check matrix $\mathbf{H}_{\text{FG,cyc}}$ of $C_{\text{FG,cyc}}$ in the decoding process. We call $f$ the *channel information inclusion* (CII) *frequency*.

With the above definitions and notations, an RMSA is formulated in Algorithm 12.1.

---

**Algorithm 12.1 RMSA**

---

**Step 1. (Initialization)** Set $i = 0$. For $0 \leq j < n$, set $y_j^{(0)} = y_j$ and $R_j^{(0)} = y_j$.

**Step 2. (Iteration)** Carry out the $i$th iteration:
(1) If $i \bmod f \neq 0$, compute the updated reliability information $\mathbf{F}^{(i)}$ based on $\mathbf{R}^{(i)}$ as follows: for $0 \leq j < n$,

$$F_j^{(i)} = R_j^{(i)} + \lambda_{atten} \sum_{t \in N(j)} \left\{ \left[ \prod_{j' \in M(t) \backslash j} \text{sign}\left(R_{j'}^{(i)}\right) \right] \times \left[ \min_{j' \in M(t) \backslash j} \{|R_{j'}^{(i)}|\} \right] \right\};$$

otherwise, (i.e., for $i \bmod f = 0$), compute the updated reliability information $\mathbf{F}^{(i)}$ based on $\mathbf{y}^{(i)}$ and $\mathbf{R}^{(i)}$ as follows: for $0 \leq j < n$,

$$F_j^{(i)} = y_j^{(i)} + \lambda_{atten} \sum_{t \in N(j)} \left\{ \left[ \prod_{j' \in M(t) \backslash j} \text{sign}\left(R_{j'}^{(i)}\right) \right] \times \left[ \min_{j' \in M(t) \backslash j} \{|R_{j'}^{(i)}|\} \right] \right\}.$$

(2) Form the hard-decision vector $\mathbf{z}$ based on $\mathbf{F}^{(i)}$ and compute its syndrome $\mathbf{s} = \mathbf{z} \cdot \mathbf{H}_{\text{FG,cyc}}^T$. If $\mathbf{s} = \mathbf{0}$, go to **Step 3**; otherwise, go to **Step 2** (3).
(3) If $i = I_{\max} - 1$, stop the decoding and declare a decoding failure; otherwise, cyclically shift $\mathbf{F}^{(i)}$ and $\mathbf{y}^{(i)}$ by $\ell$ positions to the left to obtain $\mathbf{R}^{(i+1)}$ and $\mathbf{y}^{(i+1)}$, respectively, set $i \leftarrow i+1$, and return to **Step 2** (1) to perform another iteration.

**Step 3.** Cyclically shift $\mathbf{z}$ to the right by $i\ell$ positions to make it corresponding to the transmitted codeword and output the shifted version of $\mathbf{z}$ as the decoded vector.

---

To prevent (or minimize) performance loss of the RMSA based on $\mathbf{H}_0$ compared with the MSA based on the entire parity-check matrix $\mathbf{H}_{\text{FG,cyc}}$, the CII-frequency $f$ should be chosen based on the following guidelines [33, 34]. If the grouping size $\ell$ is relatively small compared with the total number of rows of $\mathbf{H}_{\text{FG,cyc}}$, $f$ should be chosen such that, for every $f$ iterations, each row of $\mathbf{H}_{\text{FG,cyc}}$ should be involved in information update at least once. However, if $\ell$ is a relatively large number, we should choose $f$ such that, for every $f$ iterations, each row of $\mathbf{H}_{\mathrm{FG,cyc}}$ should not be involved in an information update more than once. Using these guidelines for choosing $f$, we can minimize the performance loss of the RMSA based on $\mathbf{H}_0$ compared with the MSA based on the entire parity-check matrix $\mathbf{H}_{\mathrm{FG,cyc}}$.

Because the column (or row) weight of the circulant parity-check matrix $\mathbf{H}_{\mathrm{FG,cyc}}$ is $\gamma$, the number of 1-entries in $\mathbf{H}_{\mathrm{FG,cyc}}$ is $\gamma n$ and the number of 1-entries in $\mathbf{H}_0$ is $\gamma \ell$. The average column weight of $\mathbf{H}_0$ is $\gamma \ell / n$. In hardware implementation of an RID-decoder, the number of CNPUs and the number of wires required to connect the CNPUs and VNPUs are $\ell/n$ of those needed for an iterative decoder implemented based on the entire parity-check matrix $\mathbf{H}_{\mathrm{FG,cyc}}$. Even though the number of VNPUs required for an RID-decoder is still the same as the one based on the entire parity-check matrix $\mathbf{H}_{\mathrm{FG,cyc}}$, the number of wires connected to each VNPU is $\ell/n$ (on average) of that connected to a VNPU of a decoder based on the entire parity-check matrix $\mathbf{H}_{\mathrm{FG,cyc}}$. If the chosen parameter $\ell$ is much smaller than $n$, the hardware complexity of an RID-decoder will be greatly reduced compared with the one (using the same reliability updating algorithm) based on the entire parity-check matrix $\mathbf{H}_{\mathrm{FG,cyc}}$.

The RID-algorithm not only significantly reduces the hardware implementation complexity of the decoder of a cyclic-FG-LDPC code but also reduces the size of memory required to store information for message passing between VNPUs and CNPUs. To show this, we compare the size of memory required for an RMSA decoder implemented based on the decoding matrix $\mathbf{H}_0$ to that required for an MSA decoder implemented based on the entire parity-check matrix $\mathbf{H}_{\mathrm{FG,cyc}}$.

Suppose each real-number (RN) message is quantized and represented by $b$ bits ($b$-bit quantization) for both the RMSA decoder and the MSA decoder. For both decoders, $nb$ binary memory units (BMUs) are needed to store the received vector $\mathbf{y}$ and $nb$ BMUs to store the updated reliability information vector at the end of each decoding iteration. The reliability information vector at the beginning of each decoding iteration shares the same memory units with the updated reliability information vector at the end of each decoding iteration. In addition, each CNPU requires $2(b-1)$ BMUs to store the two smallest magnitudes of the messages of its adjacent VNPUs, $\lceil \log_2 \gamma \rceil$ BMUs to store the index of the smallest magnitude of the messages of its adjacent VNPUs, and $\gamma$ BMUs for the signs of the messages of its adjacent VNPUs. Because the Tanner graph of $\mathbf{H}_0$ has $\ell$ CNs, the total number of BMUs required for the RMSA decoder to store CN messages is $\ell(2(b-1) + \lceil \log_2 \gamma \rceil + \gamma)$. Because the Tanner graph of $\mathbf{H}_{\mathrm{FG,cyc}}$ has $n$ CNs, the total number of BMUs required for the MSA decoder to store CN messages is $n(2(b-1) + \lceil \log_2 \gamma \rceil + \gamma)$. Hence, the RMSA decoder only needs $\ell/n$ as many memory units as the MSA decoder, which results in a great reduction in memory size.

In [33, 34], it is shown that the computational complexity (in terms of real addition and comparison operations) of the RMSA is slightly less than the computational complexity of the MSA for the same total number of message-passing operations between the VNPUs and CNPUs.

**Example 12.5** Consider the $(64, 64)$-regular $(4095, 3367)$ cyclic-EG-LDPC code $C_{\mathrm{EG,cyc}}(2, 2^6)$ constructed in Example 12.3 based on the lines in $\mathrm{EG}^*(2, 2^6)$. Its parity-check matrix $\mathbf{H}_{\mathrm{EG,cyc}}(2, 2^6)$ is a $4095 \times 4095$ circulant with weight 64.

To decode this code with the RMSA, we first choose a grouping size $\ell$, $1 \le \ell \le 4095$. Suppose we choose $\ell = 819$ which is a factor of 4095. Let $c = 4095/819 = 5$. Partition the matrix $\mathbf{H}_{\mathrm{EG,cyc}}(2, 2^6)$ into five $819 \times 4095$ submatrices $\mathbf{H}_0, \mathbf{H}_1, \mathbf{H}_2, \mathbf{H}_3, \mathbf{H}_4$, each consisting of 819 consecutive rows of $\mathbf{H}_{\mathrm{EG,cyc}}(2, 2^6)$ and with rank 728. Hence, each of these five submatrices has 91 redundant rows. The null space over $\mathrm{GF}(2)$ of each of these five submatrices gives the same $(4095, 3367)$ cyclic-EG-LDPC code $C_{\mathrm{EG,cyc}}(2, 2^6)$. Next, we choose the CII-frequency $f = 5$ and decode the code $C_{\mathrm{EG,cyc}}(2, 2^6)$ using the RMSA based on the submatrix $\mathbf{H}_0$ of $\mathbf{H}_{\mathrm{EG,cyc}}(2, 2^6)$. The RMSA is also scaled by a factor of 0.25. With $f = 5$, we include the channel information for reliability update every five RID-iterations based on $\mathbf{H}_0$ which correspond to one MSA decoding iteration based on the entire parity-check matrix $\mathbf{H}_{\mathrm{EG,cyc}}(2, 2^6)$ of the code $C_{\mathrm{EG,cyc}}(2, 2^6)$.

The BER and BLER performances of the code over the AWGN channel using BPSK signaling decoded with 25, 50, and 250 iterations of the RMSA based on the submatrix $\mathbf{H}_0$ of $\mathbf{H}_{\mathrm{EG,cyc}}(2, 2^6)$ are shown in Fig. 12.5. Note that 25, 50, and 250 RMSA iterations are equivalent to 5, 10, and 50 MSA iterations based on the entire parity-check matrix $\mathbf{H}_{\mathrm{EG,cyc}}(2, 2^6)$ of the code. With the RMSA, every five consecutive iterations cover all 4095 rows of $\mathbf{H}_{\mathrm{EG,cyc}}(2, 2^6)$, i.e., cover 4095 CNs' message processing operations. From Fig. 12.5, we see that the performance curves cluster together and the decoding converges fast. At a BER of $10^{-6}$, the BER performance gap between 50 and 250 RMSA iterations is very small, about 0.05 dB, and the BER performance gap between 25 and 250 RMSA iterations is less than 0.2 dB. From Figs. 12.3(a) and 12.5, we see that the MSA with 50 iterations based on the entire parity-check matrix $\mathbf{H}_{\mathrm{EG,cyc}}(2, 2^6)$ performs slightly better than the RMSA with 250 iterations based on the submatrix $\mathbf{H}_0$ of $\mathbf{H}_{\mathrm{EG,cyc}}(2, 2^6)$.

The parity-check matrix $\mathbf{H}_{\mathrm{EG,cyc}}(2, 2^6)$ of the code $C_{\mathrm{EG,cyc}}(2, 2^6)$ has 262 080 1-entries. To implement an MSA decoder for the code $C_{\mathrm{EG,cyc}}(2, 2^6)$ based on the entire parity-check matrix $\mathbf{H}_{\mathrm{EG,cyc}}(2, 2^6)$, we need 4095 VNPUs, 4095 CNPUs, and 262 080 wires connecting them. The submatrix $\mathbf{H}_0$ of $\mathbf{H}_{\mathrm{EG,cyc}}(2, 2^6)$ has 52 416 1-entries which is $1/5$ of the total 1-entries of $\mathbf{H}_{\mathrm{EG,cyc}}(2, 2^6)$. Hardware implementation of an RMSA decoder for the code $C_{\mathrm{EG,cyc}}(2, 2^6)$ based on $\mathbf{H}_0$ requires 4095 VNPUs, 819 CNPUs, and 52 416 wires connecting them. The numbers of CNPUs and the numbers of wires needed to connect the CNPUs to the VNPUs are reduced by a factor of 5. In decoding, if each real-number message is quantized with six bits, the RMSA decoder has about 70 percent memory saving compared with the MSA decoder. ▲▲

**Example 12.6** Continue Example 12.5. Suppose we choose a grouping size $\ell = 1$. With this choice of $\ell$, we decode the $(4095, 3367)$ cyclic-EG-LDPC code $C_{\mathrm{EG,cyc}}(2, 2^6)$ only based on the top row of the parity-check matrix $\mathbf{H}_{\mathrm{EG,cyc}}(2, 2^6)$ of the code. In this extreme case, the RMSA decoder consists

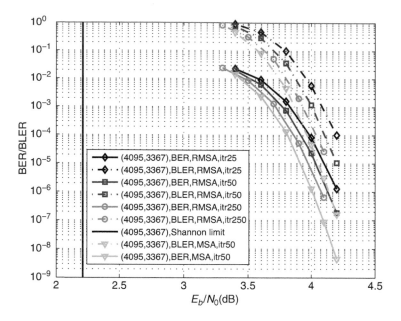

Figure 12.5 The BER and BLER performances of the $(4095, 3367)$ cyclic-EG-LDPC code in Example 12.5 using the RMSA with $\ell = 819$ and $f = 5$.

of one CNPU and 64 VNPUs. There is a huge reduction in hardware implementation complexity compared with the MSA decoder based on the entire parity-check matrix $\mathbf{H}_{\mathrm{EG,cyc}}(2, 2^6)$.

Suppose a 6-bit quantization of a real-number message is used for the RMSA decoding. Choose the CII-frequency $f = 16\,380$. With this choice of $f$, we include the channel information for reliability update every $16\,380$ RMSA iterations based on the single row matrix $\mathbf{H}_0$ which correspond to four iterations based on the entire parity-check matrix $\mathbf{H}_{\mathrm{EG,cyc}}(2, 2^6)$. The BER and BLER performances of the code $C_{\mathrm{EG,cyc}}(2, 2^6)$ over the AWGN channel using BPSK signaling decoded using the RMSA with $20\,475$ and $204\,750$ iterations (corresponding to 5 and 50 MSA iterations based on the entire parity-check matrix $\mathbf{H}_{\mathrm{EG,cyc}}(2, 2^6)$, respectively) scaled by a factor of 0.25 are shown in Fig. 12.6. Again, we see that the decoding using the RMSA converges fast, and the performance gap between $20\,475$ and $204\,750$ iterations is negligible. Also included in Fig. 12.6 are the BER and BLER performances of the code decoded using the 6-bit quantized MSA with 50 iterations based on $\mathbf{H}_{\mathrm{EG,cyc}}(2, 2^6)$, using the same scaling factor, 0.25. We see that the RMSA with $204\,750$ iterations (corresponding to 50 MSA iterations based on $\mathbf{H}_{\mathrm{EG,cyc}}(2, 2^6)$) performs about 0.2 dB away from the MSA with 50 iterations based on $\mathbf{H}_{\mathrm{EG,cyc}}(2, 2^6)$ at a BER of $10^{-6}$.

The RMSA decoder has an up to 87 percent memory saving compared with the MSA decoder. For this special case, the small error performance loss is more than compensated by the huge reduction in decoding complexity. ▲▲

**Example 12.7** In this example, we demonstrate the performances of the $(1057, 813)$ cyclic-PG-LDPC code $C_{\mathrm{PG,cyc}}(2, 2^5)$ given in Example 12.4 with the RMSA

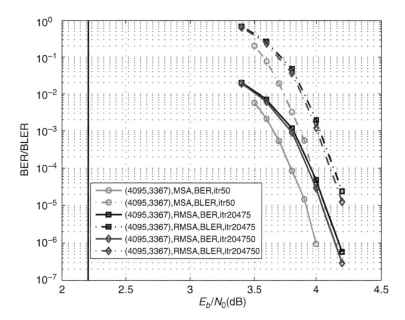

Figure 12.6 The BER and BLER performances of the $(4095, 3367)$ cyclic-EG-LDPC code decoded using the RMSA with $\ell = 1$ and $f = 16\,380$ given in Example 12.6.

using two different grouping sizes. The code $C_{\text{PG,cyc}}(2, 2^5)$ is constructed based on the two-dimensional projective geometry $\text{PG}(2, 2^5)$ over $\text{GF}(2^5)$. Its parity-check matrix $\mathbf{H}_{\text{PG,cyc}}(2, 2^5)$ is a $1057 \times 1057$ circulant with weight 33. The chosen grouping sizes are $\ell = 151$ and $\ell = 244$, where 151 is a factor of 1057 $(7 \times 151 = 1057)$ and 244 is the rank of $\mathbf{H}_{\text{PG,cyc}}(2, 2^5)$. With these choices of grouping sizes, the sizes of decoding matrices $\mathbf{H}_0$s are $151 \times 1057$ and $244 \times 1057$, respectively. Associated with the two chosen grouping sizes, the CII-frequencies $f$s are chosen to be 16 and 8, respectively.

With the choices of grouping sizes and their associated CII-frequencies, the BER and BLER performances of the code $C_{\text{PG,cyc}}(2, 2^5)$ over the AWGN channel using BPSK signaling decoded with the RMSA are shown in Fig. 12.7(a) and Fig. 12.7(b), respectively. For $\ell = 151$ and $f = 16$, the code is decoded with 35, 70, and 350 RMSA iterations based on the $151 \times 1057$ decoding matrix $\mathbf{H}_0$ which correspond to 5, 10, and 50 MSA iterations based on the entire parity-check $\mathbf{H}_{\text{PG,cyc}}(2, 2^5)$ of the code, respectively. For $\ell = 244$ and $f = 8$, the code is decoded with 25, 50, and 250 RMSA iterations based on the $244 \times 1057$ decoding matrix $\mathbf{H}_0$ which correspond to 5, 10, and 50 MSA iterations based on $\mathbf{H}_{\text{PG,cyc}}(2, 2^5)$. Also included in Figs. 12.7(a) and 12.7(b) are the BER and BLER performances of the code $C_{\text{PG,cyc}}(2, 2^5)$ decoded with 50 iterations of the MSA based on the entire parity-check matrix $\mathbf{H}_{\text{PG,cyc}}(2, 2^5)$. From the figures, we see that for the two grouping sizes, the RMSA decodings of the code converge quickly and with 350 RMSA iterations ($\ell = 151$) and 250 RMSA iterations ($\ell = 244$) perform very close to the MSA based on $\mathbf{H}_{\text{PG,cyc}}(2, 2^5)$ with 50 iterations.

For the two chosen grouping sizes, the RMSA reduces the hardware implementation complexities by factors 7 and 5, respectively. In this example, we show again that the RMSA based on a smaller decoding matrix gives almost the same error performance as the MSA based on the entire parity-check matrix of a cyclic-FG-LDPC code. ▲▲

## 12.3   QC-EG-LDPC Codes

In this section, we present two approaches to construct EG-LDPC codes with quasi-cyclic (QC) structure (see Section 4.13 for QC structure) based on a two-dimensional Euclidean geometry $EG(2, q)$ over $GF(q)$. These codes are referred to as *QC-EG-LDPC codes* or *QC-FG-LDPC codes*.

Again, we begin with the construction of QC-EG-LDPC codes based on the subgeometry $EG^*(2, q)$ of $EG(2, q)$ over $GF(q)$ with $q = 2^s$. Consider the $(q^2 - 1) \times (q^2 - 1)$ circulant $\mathbf{H}_{\text{EG,cyc}}(2, 2^s)$ over $GF(2)$ constructed based on the lines in $EG^*(2, q)$, each consisting of $q$ points.

Label the rows and columns of $\mathbf{H}_{\text{EG,cyc}}(2, 2^s)$ from 0 to $q^2 - 2$. Factor $q^2 - 1$ as the product of $q - 1$ and $q + 1$, i.e., $q^2 - 1 = (q-1)(q+1)$. Define the following index sequences: for $0 \le i, j \le q$,

$$\pi_{\text{row},i} = [i, (q+1) + i, 2(q+1) + i, \ldots, (q-2)(q+1) + i], \tag{12.4}$$

$$\pi_{\text{col},j} = [j, (q+1) + j, 2(q+1) + j, \ldots, (q-2)(q+1) + j]. \tag{12.5}$$

Let

$$\pi_{\text{row}} = [\pi_{\text{row},0}, \pi_{\text{row},1}, \ldots, \pi_{\text{row},q}], \tag{12.6}$$

and

$$\pi_{\text{col}} = [\pi_{\text{col},0}, \pi_{\text{col},1}, \ldots, \pi_{\text{col},q}]. \tag{12.7}$$

Then, $\pi_{\text{row}}$ and $\pi_{\text{col}}$ define reorderings of the row and column labels of the matrix $\mathbf{H}_{\text{EG,cyc}}(2, q)$, respectively.

Permuting the rows and columns of $\mathbf{H}_{\text{EG,cyc}}(2, q)$ based on $\pi_{\text{row}}$ and $\pi_{\text{col}}$, respectively, we obtain the following $(q + 1) \times (q + 1)$ array of $(q - 1) \times (q - 1)$ matrices over $GF(2)$:

$$\mathbf{H}_{\text{EG,qc}}(2, q) = [\mathbf{A}_{i,j}]_{0 \le i, j \le q} = \begin{bmatrix} \mathbf{A}_{0,0} & \mathbf{A}_{0,1} & \cdots & \mathbf{A}_{0,q} \\ \mathbf{A}_{1,0} & \mathbf{A}_{1,1} & \cdots & \mathbf{A}_{1,q} \\ \vdots & \vdots & \vdots & \vdots \\ \mathbf{A}_{q-1,0} & \mathbf{A}_{q-1,1} & \cdots & \mathbf{A}_{q-1,q} \\ \mathbf{A}_{q,0} & \mathbf{A}_{q,1} & \cdots & \mathbf{A}_{q,q} \end{bmatrix}$$

$$= \begin{bmatrix} \mathbf{A}_0 \\ \mathbf{A}_1 \\ \vdots \\ \mathbf{A}_{q-1} \\ \mathbf{A}_q \end{bmatrix}, \tag{12.8}$$

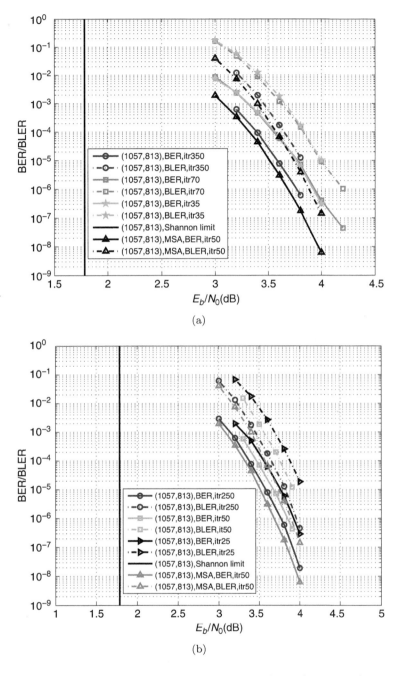

Figure 12.7 The BER and BLER performances of the $(1057, 813)$ cyclic-PG-LDPC code given in Example 12.7 decoded with the RMSA using two grouping sizes: (a) $\ell = 151$, $f = 16$, and (b) $\ell = 244$, $f = 8$.

where the subscript "qc" stands for "quasi-cyclic." The array $\mathbf{H}_{\mathrm{EG,qc}}(2,q)$ consists of $q+1$ row-blocks $(\mathbf{A}_0, \mathbf{A}_1, \ldots, \mathbf{A}_{q-1}, \mathbf{A}_q)$ and $q+1$ column-blocks of matrices of size $(q-1) \times (q-1)$. It follows from the cyclic structure and row and column weights of $\mathbf{H}_{\mathrm{EG,cyc}}(2,q)$ that among the $q+1$ submatrices in each row-block (or each column-block), $q$ of them are *circulant permutation matrices* (CPMs) of size $(q-1) \times (q-1)$ and one is a *zero matrix* (ZM) of size $(q-1) \times (q-1)$ (see proof in [35]). The row-blocks and column-blocks are called *CPM-row-blocks* and *CPM-column-blocks* of $\mathbf{H}_{\mathrm{EG,qc}}(2,q)$, respectively. Each ZM in $\mathbf{H}_{\mathrm{EG,qc}}(2,q)$ resides in a *separate CPM-row-block* and a *separate CPM-column-block*. The array $\mathbf{H}_{\mathrm{EG,qc}}(2,q)$ has a *row-block cyclic structure*. Each CPM-row-block of $\mathbf{H}_{\mathrm{EG,qc}}(2,q)$ is the cyclic-shift of the CPM-row-block above it, but when the rightmost $\mathrm{CPM}_r$ is shifted around to the leftmost position, all the rows of the rightmost $\mathrm{CPM}_r$ are *cyclically shifted one position to the right within the $q-1$ positions* of the $\mathrm{CPM}_r$, which gives another $\mathrm{CPM}_l$, i.e., the generator of $\mathrm{CPM}_l$ is the cyclic-shift of the generator of $\mathrm{CPM}_r$ one place to the right. The top CPM-row-block of $\mathbf{H}_{\mathrm{EG,qc}}(2,q)$ is the block cyclic-shift of its last CPM-row-block.

With the above CPM-row-block cyclic structure, the array $\mathbf{H}_{\mathrm{EG,qc}}(2,q)$ can be formed by using its top CPM-row-bock $\mathbf{A}_0 = [\mathbf{A}_{0,0}, \mathbf{A}_{0,1}, \ldots, \mathbf{A}_{0,q}]$, a $(q-1) \times (q^2-1)$ matrix. The $q-1$ rows in a CPM-row-block of $\mathbf{H}_{\mathrm{EG,qc}}(2,q)$ correspond to $q-1$ lines in a parallel bundle of the subgeometry $\mathrm{EG}^*(2,q)$.

The top CPM-row-block $\mathbf{A}_0$ of $\mathbf{H}_{\mathrm{EG,qc}}(2,q)$ can be formed by taking the $q-1$ rows, labeled by $0, q+1, 2(q+1), \ldots, (q-2)(q+1)$ from $\mathbf{H}_{\mathrm{EG,cyc}}(2,q)$ to form a $(q-1) \times (q^2-1)$ matrix $\mathbf{H}_0$. Then, permute the $(q^2-1)$ columns of $\mathbf{H}_0$ with the permutation $\pi_{\mathrm{col}} = [\pi_{\mathrm{col},0}, \pi_{\mathrm{col},1}, \ldots, \pi_{\mathrm{col},q}]$ defined by (12.7) where $\pi_{\mathrm{col},j} = [j, (q+1)+j, 2(q+1)+j, \ldots, (q-2)(q+1)+j]$, for $0 \le j \le q$, defined by (12.5). The column permutation of $\mathbf{H}_0$ results in the top CPM-row-block $\mathbf{A}_0 = [\mathbf{A}_{0,0}, \mathbf{A}_{0,1}, \ldots, \mathbf{A}_{0,q}]$ of $\mathbf{H}_{\mathrm{EG,qc}}(2,q)$. In fact, the $(q-1) \times (q^2-1)$ matrix $\mathbf{H}_0$ can be formed by taking the top row $\mathbf{h}_0$ of the circulant $\mathbf{H}_{\mathrm{EG,cyc}}(2,q)$ and cyclically shifting it to the right $q+1, 2(q+1), \ldots, (q-2)(q+1)$ times to obtain the other $q-2$ rows of $\mathbf{H}_0$.

**Example 12.8** Consider the field $\mathrm{GF}(2^4)$. Let $\alpha$ be a primitive element of $\mathrm{GF}(2^4)$. The 16 powers of $\alpha$, namely, $\alpha^{-\infty}, \alpha^0, \alpha, \alpha^2, \ldots, \alpha^{14}$, give the 16 elements of $\mathrm{GF}(2^4)$. Let $\beta = \alpha^5$. The four elements $0, 1, \beta, \beta^2$ form a subfield $\mathrm{GF}(2^2)$ of $\mathrm{GF}(2^4)$, i.e., $\mathrm{GF}(2^2) = \{0, 1, \beta = \alpha^5, \beta^2 = \alpha^{10}\}$. If we regard $\mathrm{GF}(2^4)$ as an extension field of $\mathrm{GF}(2^2)$, the 16 elements of $\mathrm{GF}(2^4)$ are linear sums of $\alpha^0$ and $\alpha$ in the following form: $\eta_0 \alpha^0 + \eta_1 \alpha$ with $\eta_0, \eta_1 \in \mathrm{GF}(2^2)$. The 16 elements of $\mathrm{GF}(2^4)$ as the extension field of $\mathrm{GF}(2^2)$ are listed in Table 12.1 (a reproduction of Table 7.2). The field $\mathrm{GF}(2^4)$ is a realization of the two-dimensional Euclidean $\mathrm{EG}(2,2^2)$ over $\mathrm{GF}(2^2)$ which has 16 points and 20 lines. Each line in $\mathrm{EG}(2,2^2)$ consists of four points. The 16 points of $\mathrm{EG}(2,2^2)$ are represented by the 16 elements $\alpha^{-\infty}, \alpha^0, \alpha, \alpha^2, \ldots, \alpha^{14}$ of $\mathrm{GF}(2^4)$.

Consider the subgeometry $\mathrm{EG}^*(2,2^2)$ of $\mathrm{EG}(2,2^2)$ which consists of 15 nonorigin points of $\mathrm{EG}(2,2^2)$ and 15 lines not passing through the origin. For $0 \le i, j \le 14$, let $\alpha^i$ and $\alpha^j$ be two linearly independent elements in $\mathrm{GF}(2^4)$ over

$\mathrm{GF}(2^2) = \{0, 1, \beta, \beta^2\}$. Then, the four elements in the set $\{\alpha^i + \eta\alpha^j : \eta \in \mathrm{GF}(2^2)\}$ represent four points in $\mathrm{EG}^*(2, 2^2)$ which form a line in $\mathrm{EG}^*(2, 2^2)$. For example, the elements $\alpha^3$ and $\alpha^{14}$ are two linearly independent elements in $\mathrm{GF}(2^4)$. Then, the four elements in the set

$$\{\alpha^3 + \eta\alpha^{14} : \eta \in \mathrm{GF}(2^2)\} = \{\alpha^0, \alpha^2, \alpha^6, \alpha^{14}\}$$

represent the four points which form a line in $\mathrm{EG}^*(2, 2^2)$. The incidence vector of this line is (1 0 1 0 0 0 1 0 0 0 0 0 0 0 1). Using this incidence vector as the generator, we form the following $15 \times 15$ circulant:

$$\mathbf{H}_{\mathrm{EG,cyc}}(2, 2^2) = \begin{bmatrix} 1&0&1&0&0&0&1&0&0&0&0&0&0&0&1 \\ 1&1&0&1&0&0&0&1&0&0&0&0&0&0&0 \\ 0&1&1&0&1&0&0&0&1&0&0&0&0&0&0 \\ 0&0&1&1&0&1&0&0&0&1&0&0&0&0&0 \\ 0&0&0&1&1&0&1&0&0&0&1&0&0&0&0 \\ 0&0&0&0&1&1&0&1&0&0&0&1&0&0&0 \\ 0&0&0&0&0&1&1&0&1&0&0&0&1&0&0 \\ 0&0&0&0&0&0&1&1&0&1&0&0&0&1&0 \\ 0&0&0&0&0&0&0&1&1&0&1&0&0&0&1 \\ 1&0&0&0&0&0&0&0&1&1&0&1&0&0&0 \\ 0&1&0&0&0&0&0&0&0&1&1&0&1&0&0 \\ 0&0&1&0&0&0&0&0&0&0&1&1&0&1&0 \\ 0&0&0&1&0&0&0&0&0&0&0&1&1&0&1 \\ 1&0&0&0&1&0&0&0&0&0&0&0&1&1&0 \\ 0&1&0&0&0&1&0&0&0&0&0&0&0&1&1 \end{bmatrix}.$$

Label the column and rows of $\mathbf{H}_{\mathrm{EG,cyc}}(2, 2^2)$ from 0 to 14. Factor $2^4 - 1$ as the product of $(2^2 - 1) = 3$ and $(2^2 + 1) = 5$. Following (12.4)–(12.7), we form the following row and column permutations:

$$\pi_{\mathrm{row}} = (0, 5, 10, 1, 6, 11, 2, 7, 12, 3, 8, 13, 4, 9, 14),$$
$$\pi_{\mathrm{col}} = (0, 5, 10, 1, 6, 11, 2, 7, 12, 3, 8, 13, 4, 9, 14).$$

Permuting the rows and columns of $\mathbf{H}_{\mathrm{EG,cyc}}(2, 2^2)$ with $\pi_{\mathrm{row}}$ and $\pi_{\mathrm{col}}$, respectively, we obtain the following $5 \times 5$ array of CPMs and ZMs of size $3 \times 3$:

$$\mathbf{H}_{\mathrm{EG,qc}}(2, 2^2) = \left[\begin{array}{ccc|ccc|ccc|ccc|ccc} 1&0&0&0&1&0&1&0&0&0&0&0&0&0&1 \\ 0&1&0&0&0&1&0&1&0&0&0&0&1&0&0 \\ 0&0&1&1&0&0&0&0&1&0&0&0&0&1&0 \\ \hline 1&0&0&1&0&0&0&1&0&1&0&0&0&0&0 \\ 0&1&0&0&1&0&0&0&1&0&1&0&0&0&0 \\ 0&0&1&0&0&1&1&0&0&0&0&1&0&0&0 \\ \hline 0&0&0&1&0&0&1&0&0&0&1&0&1&0&0 \\ 0&0&0&0&1&0&0&1&0&0&0&1&0&1&0 \\ 0&0&0&0&0&1&0&0&1&1&0&0&0&0&1 \\ \hline 0&1&0&0&0&0&1&0&0&1&0&0&0&1&0 \\ 0&0&1&0&0&0&0&1&0&0&1&0&0&0&1 \\ 1&0&0&0&0&0&0&0&1&0&0&1&1&0&0 \\ \hline 0&0&1&0&1&0&0&0&0&1&0&0&1&0&0 \\ 1&0&0&0&0&1&1&0&0&0&0&0&0&1&0 \\ 0&1&0&1&0&0&0&0&0&0&0&1&0&0&1 \end{array}\right] = \begin{bmatrix} \mathbf{A}_0 \\ \mathbf{A}_1 \\ \mathbf{A}_2 \\ \mathbf{A}_3 \\ \mathbf{A}_4 \end{bmatrix}.$$

We readily see that the array $\mathbf{H}_{\mathrm{EG,qc}}(2, 2^2)$ has the row-block cyclic structure. Hence, the array can be generated by cyclically shifting its top CPM-row-block $\mathbf{A}_0$. To form the top CPM-row-block $\mathbf{A}_0$, we first take the zeroth, fifth, and tenth rows of $\mathbf{H}_{\mathrm{EG,cyc}}(2, 2^2)$ to form the following $3 \times 15$ matrix:

$$\mathbf{H}_0 = \begin{bmatrix} 1\,0\,1\,0\,0\,0\,1\,0\,0\,0\,0\,0\,0\,0\,1 \\ 0\,0\,0\,0\,1\,1\,0\,1\,0\,0\,0\,1\,0\,0\,0 \\ 0\,1\,0\,0\,0\,0\,0\,0\,1\,1\,0\,1\,0\,0 \end{bmatrix}.$$

Then, we permute the columns of $\mathbf{H}_0$ with the permutation $\pi_{\mathrm{col}} = (0,\,5,\,10,\,1,\,6,\,11,\,2,\,7,\,12,\,3,\,8,\,13,\,4,\,9,\,14)$. The permutation results in the top CPM-row-block $\mathbf{A}_0$ of the array $\mathbf{H}_{\mathrm{EG,qc}}(2, 2^2)$:

$$\mathbf{A}_0 = \begin{bmatrix} 1\,0\,0 & 0\,1\,0 & 1\,0\,0 & 0\,0\,0 & 0\,0\,1 \\ 0\,1\,0 & 0\,0\,1 & 0\,1\,0 & 0\,0\,0 & 1\,0\,0 \\ 0\,0\,1 & 1\,0\,0 & 0\,0\,1 & 0\,0\,0 & 0\,1\,0 \end{bmatrix}.$$

The $3 \times 15$ matrix $\mathbf{H}_0$ can be formed by taking the top row $\mathbf{h}_0$ of the circulant $\mathbf{H}_{\mathrm{EG,cyc}}(2, 2^2)$ and cyclically shifting it to the right 5 times and 10 times to obtain the other two rows. ▲▲

**Table 12.1** $\mathrm{GF}(2^4)$ as an extension field of $\mathrm{GF}(2^2) = \{0, 1, \beta, \beta^2\}$ with $\beta = \alpha^5$.

| Power form | Polynomial form | | Vector form | Power form | Polynomial form | | Vector form |
|---|---|---|---|---|---|---|---|
| 0 | 0 | | $(0,0)$ | $\alpha^7$ | $\beta^2 +$ | $\beta\alpha$ | $(\beta^2, \beta)$ |
| 1 | 1 | | $(1,0)$ | $\alpha^8$ | $\beta^2 +$ | $\alpha$ | $(\beta^2, 1)$ |
| $\alpha$ | | $\alpha$ | $(0,1)$ | $\alpha^9$ | $\beta +$ | $\beta\alpha$ | $(\beta, \beta)$ |
| $\alpha^2$ | $\beta +$ | $\alpha$ | $(\beta, 1)$ | $\alpha^{10}$ | $\beta^2$ | | $(\beta^2, 0)$ |
| $\alpha^3$ | $\beta +$ | $\beta^2\alpha$ | $(\beta, \beta^2)$ | $\alpha^{11}$ | | $\beta^2\alpha$ | $(0, \beta^2)$ |
| $\alpha^4$ | $1 +$ | $\alpha$ | $(1,1)$ | $\alpha^{12}$ | $1 +$ | $\beta^2\alpha$ | $(1, \beta^2)$ |
| $\alpha^5$ | $\beta$ | | $(\beta, 0)$ | $\alpha^{13}$ | $1 +$ | $\beta\alpha$ | $(1, \beta)$ |
| $\alpha^6$ | | $\beta\alpha$ | $(0, \beta)$ | $\alpha^{14}$ | $\beta^2 +$ | $\beta^2\alpha$ | $(\beta^2, \beta^2)$ |

Performing column and row permutations of $\mathbf{H}_{\mathrm{EG,cyc}}(2, q)$ does not change its RC-constrained (or orthogonal) structure. Hence, $\mathbf{H}_{\mathrm{EG,qc}}(2, q)$ is an RC-constrained $(q + 1) \times (q + 1)$ array of CPMs and ZMs of size $(q - 1) \times (q - 1)$. The null space over $\mathrm{GF}(2)$ of $\mathbf{H}_{\mathrm{EG,qc}}(2, q)$ gives a QC-EG-LDPC code $C_{\mathrm{EG,qc}}(2, q)$ which is *combinatorially equivalent* to the cyclic-EG-LDPC code $C_{\mathrm{EG,cyc}}(2, q)$. The cyclic-EG-LDPC code $C_{\mathrm{EG,cyc}}(2, q)$ and the QC-EG-LDPC code $C_{\mathrm{EG,qc}}(2, q)$ have the same rate and the same minimum distance. Their Tanner graphs have the same girth, same number of cycles, and same trapping set structures. Decoding with the same IDBP algorithm, the two codes perform the same.

Let $\gamma$ and $\rho$ be two positive integers such that $1 \leq \gamma \leq \rho \leq q + 1$. Let $\mathbf{M}_{\mathrm{EG,qc}}(\gamma, \rho)$ be a $\gamma \times \rho$ subarray of $\mathbf{H}_{\mathrm{EG,qc}}(2, q)$ which is a $\gamma(q - 1) \times \rho(q - 1)$ matrix over $\mathrm{GF}(2)$ with column weight $\gamma - 1$ (or $\gamma$) and row weight $\rho - 1$ (or $\rho$) (depending on whether the ZMs in $\mathbf{H}_{\mathrm{EG,qc}}(2, q)$ are included in the subarray

$\mathbf{M}_{\mathrm{EG,qc}}(\gamma, \rho))$. The null space over $\mathrm{GF}(2)$ of $\mathbf{M}_{\mathrm{EG,qc}}(\gamma, \rho)$ gives a QC-EG-LDPC code $C_{\mathrm{EG,qc}}(\gamma, \rho)$ of length $\rho(q-1)$ with rate at least $(\rho - \gamma)/\rho$. With various choices of $\gamma$ and $\rho$, we can construct a *large family* of QC-EG-LDPC codes based on the two-dimensional Euclidean geometry $\mathrm{EG}(2,q)$ for large $q$.

A special case is that $\rho = q + 1$. In this case, $\mathbf{M}_{\mathrm{EG,qc}}(\gamma, q+1)$ consists of $\gamma$ CPM-row-blocks of $\mathbf{H}_{\mathrm{EG,qc}}(2,q)$, each corresponding to $q-1$ lines in a parallel bundle of the subgeometry $\mathrm{EG}^*(2,q)$. Hence, $\mathbf{M}_{\mathrm{EG,qc}}(\gamma, q+1)$ is the incidence matrix of a subgeometry of $\mathrm{EG}(2,q)$ which consists of $q^2 - 1$ nonorigin points of $\mathrm{EG}(2,q)$ and $\gamma$ bundles of parallel lines in $\mathrm{EG}(2,q)$ not passing the origin.

**Example 12.9** Consider the $1023 \times 1023$ circulant $\mathbf{H}_{\mathrm{EG,cyc}}(2,2^5)$ constructed based on the lines in $\mathrm{EG}(2,2^5)$ over $\mathrm{GF}(2^5)$ not passing through the origin of $\mathrm{EG}(2,2^5)$ given in Example 12.1. Label the rows and columns from 0 to 1022. Define the following row and column permutations for $\mathbf{H}_{\mathrm{EG,cyc}}(2,2^5)$:

$$\pi_{\mathrm{row}} = [\pi_{\mathrm{row},0}, \pi_{\mathrm{row},1}, \ldots, \pi_{\mathrm{row},32}],$$

$$\pi_{\mathrm{col}} = [\pi_{\mathrm{col},0}, \pi_{\mathrm{col},1}, \ldots, \pi_{\mathrm{col},32}],$$

where, for $0 \leq i, j \leq 32$,

$$\pi_{\mathrm{row},i} = [i, 33 + i, 2 \times 33 + i, \ldots, 30 \times 33 + i],$$

$$\pi_{\mathrm{col},j} = [j, 33 + j, 2 \times 33 + j, \ldots, 30 \times 33 + j].$$

Permuting the rows and columns of $\mathbf{H}_{\mathrm{EG,cyc}}(2,2^5)$ with $\pi_{\mathrm{row}}$ and $\pi_{\mathrm{col}}$, respectively, we obtain a $33 \times 33$ array $\mathbf{H}_{\mathrm{EG,qc}}(2,2^5)$ of CPMs and ZMs of size $31 \times 31$,

$$\mathbf{H}_{\mathrm{EG,qc}}(2,2^5) = [\mathbf{A}_{i,j}]_{0 \leq i,j \leq 32} = \begin{bmatrix} \mathbf{A}_{0,0} & \mathbf{A}_{0,1} & \cdots & \mathbf{A}_{0,32} \\ \mathbf{A}_{1,0} & \mathbf{A}_{1,1} & \cdots & \mathbf{A}_{1,32} \\ \vdots & \vdots & \vdots & \vdots \\ \mathbf{A}_{31,0} & \mathbf{A}_{31,1} & \cdots & \mathbf{A}_{31,32} \\ \mathbf{A}_{32,0} & \mathbf{A}_{32,1} & \cdots & \mathbf{A}_{32,32} \end{bmatrix}.$$

The null space of $\mathbf{H}_{\mathrm{EG,qc}}(2,2^5)$ gives a $(32,32)$-regular $(1023,781)$ QC-EG-LDPC code $C_{\mathrm{EG,qc}}(2,2^5)$ that is equivalent to the $(32,32)$-regular $(1023,781)$ cyclic-EG-LDPC code $C_{\mathrm{EG,cyc}}(2,2^5)$ given in Example 12.1. With the same IDBP-algorithm, the code $C_{\mathrm{EG,qc}}(2,2^5)$ performs the same as its cyclic equivalent $C_{\mathrm{EG,cyc}}(2,2^5)$ shown in Fig. 12.1.

The null space of any subarray of $\mathbf{H}_{\mathrm{EG,qc}}(2,2^5)$ gives a QC-EG-LDPC code. Suppose we choose the top four CPM-row-blocks of $\mathbf{H}_{\mathrm{EG,qc}}(2,2^5)$ to form a $4 \times 33$ subarray $\mathbf{M}_{\mathrm{EG,qc}}(4,33)$ of $\mathbf{H}_{\mathrm{EG,qc}}(2,2^5)$,

$$\mathbf{M}_{\mathrm{EG,qc}}(4,33) = [\mathbf{A}_{i,j}]_{0 \leq i < 4, 0 \leq j \leq 32} = \begin{bmatrix} \mathbf{A}_{0,0} & \mathbf{A}_{0,1} & \cdots & \mathbf{A}_{0,32} \\ \mathbf{A}_{1,0} & \mathbf{A}_{1,1} & \cdots & \mathbf{A}_{1,32} \\ \mathbf{A}_{2,0} & \mathbf{A}_{2,1} & \cdots & \mathbf{A}_{2,32} \\ \mathbf{A}_{3,0} & \mathbf{A}_{3,1} & \cdots & \mathbf{A}_{3,32} \end{bmatrix},$$

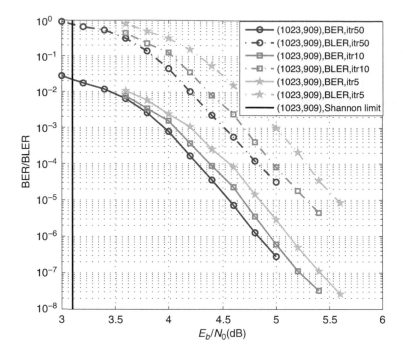

Figure 12.8 The BER and BLER performances of the $(1023, 909)$ QC-EG-LDPC code given in Example 12.9.

which is a $124 \times 1023$ matrix over $\mathrm{GF}(2)$ with constant row weight 32 and two column weights 3 and 4. The null space over $\mathrm{GF}(2)$ of $\mathbf{M}_{\mathrm{EG,qc}}(4, 33)$ gives a $(1023, 909)$ QC-EG-LDPC code $C_{\mathrm{EG,qc}}(4, 33)$ with rate 0.8886. The BER and BLER performances of the code over the AWGN channel using BPSK signaling decoded with 5, 10, and 50 iterations of the MSA with scaling factor 0.75 are shown in Fig. 12.8. ▲▲

**Example 12.10** Let $q = 2^6$. Consider the subgeometry $\mathrm{EG}^*(2, 2^6)$ of the two-dimensional Euclidean geometry $\mathrm{EG}(2, 2^6)$ over $\mathrm{GF}(2^6)$. Based on the 4095 lines of $\mathrm{EG}^*(2, 2^6)$, we construct a $4095 \times 4095$ circulant $\mathbf{H}_{\mathrm{EG,cyc}}(2, 2^6)$ over $\mathrm{GF}(2)$ with weight 64. Note that $(2^6)^2 - 1$ can be factored as the product of $2^6 - 1 = 63$ and $2^6 + 1 = 65$. Label the rows and columns of $\mathbf{H}_{\mathrm{EG,cyc}}(2, 2^6)$ from 0 to 4094. Permuting the rows and columns with the permutations defined by (12.4) to (12.7) with $q = 2^6$, we obtain a $65 \times 65$ array $\mathbf{H}_{\mathrm{EG,qc}}(2, 2^6)$ of CPMs and ZMs of size $63 \times 63$.

Let $\gamma = 4$ and $\rho = 60$. Take the top four CPM-row-blocks of $\mathbf{H}_{\mathrm{EG,qc}}(2, 2^6)$ and remove five CPM-column-blocks to form a $4 \times 60$ subarray $\mathbf{M}_{\mathrm{EG,qc}}(4, 60)$ of CPMs without ZMs. The array $\mathbf{M}_{\mathrm{EG,qc}}(4, 60)$ is a $252 \times 3780$ matrix with column and row weights 4 and 60, respectively. The null space over $\mathrm{GF}(2)$ of $\mathbf{M}_{\mathrm{EG,qc}}(4, 60)$ gives a $(4, 60)$-regular $(3780, 3543)$ QC-EG-LDPC code $C_{\mathrm{EG,qc}}(4, 60)$ with rate 0.9373. The BER and BLER performances of the code $C_{\mathrm{EG,qc}}(4, 60)$ over the AWGN channel using BPSK signaling decoded with 5, 10, and 50 iterations of the MSA are shown in Fig. 12.9. ▲▲

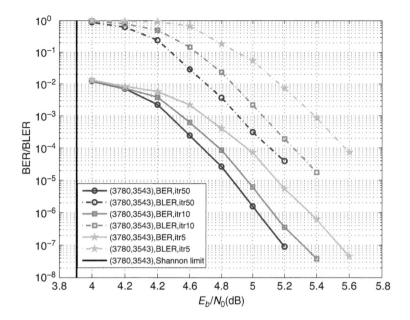

Figure 12.9 The BER and BLER performances of the $(3780, 3543)$ QC-EG-LDPC code given in Example 12.10.

## 12.4 QC-PG-LDPC Codes

Because $q^2 + q + 1$ is not divisible by $q - 1$, the $(q^2 + q + 1) \times (q^2 + q + 1)$ circulant $\mathbf{H}_{\mathrm{PG,cyc}}(2, q)$ constructed by using the incidence vectors of the lines in PG$(2, q)$ over GF$(q)$ cannot be put into an array of CPMs and ZMs of size $(q - 1) \times (q - 1)$. However, if $q^2 + q + 1$ can be factored as a product of two integers $c$ and $n$, then $\mathbf{H}_{\mathrm{PG,cyc}}(2, q)$ can be put into a $c \times c$ array $\mathbf{H}_{\mathrm{PG,qc}}(2, q)$ of circulants over GF$(2)$ of size $n \times n$ by properly permuting its rows and columns. The null space of each circulant gives a cyclic-PG-LDPC code of length $n$. Any $\gamma \times \rho$ subarray of $\mathbf{H}_{\mathrm{PG,qc}}(2, q)$ can be used as a parity-check matrix to construct a QC-PG-LDPC code.

**Example 12.11** Consider the $1057 \times 1057$ circulant $\mathbf{H}_{\mathrm{PG,cyc}}(2, 2^5)$ constructed in Example 12.4 based on PG$(2, 2^5)$ over GF$(2^5)$. The circulant $\mathbf{H}_{\mathrm{PG,cyc}}(2, 2^5)$ has weight 33. Let $c = 7$ and $n = 151$ such that $cn = 7 \times 151 = q^2 + q + 1 = 1057$. Define the following sequences:

$$
\begin{aligned}
\pi_{\mathrm{row},i} &= [i, c+i, 2c+i, \ldots, (n-1)c+i], \\
\pi_{\mathrm{col},j} &= [j, c+j, 2c+j, \ldots, (n-1)c+j], \\
\pi_{\mathrm{row}} &= [\pi_{\mathrm{row},0}, \pi_{\mathrm{row},1}, \ldots, \pi_{\mathrm{row},c-1}], \\
\pi_{\mathrm{col}} &= [\pi_{\mathrm{col},0}, \pi_{\mathrm{col},1}, \ldots, \pi_{\mathrm{col},c-1}].
\end{aligned}
$$

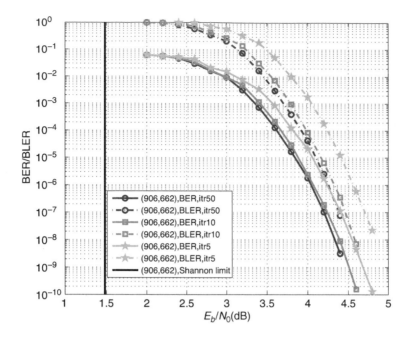

Figure 12.10 The BER and BLER performances of the $(906, 662)$ QC-PG-LDPC code given in Example 12.11.

Permuting the rows and columns of $\mathbf{H}_{\mathrm{PG,cyc}}(2, 2^5)$ based on $\pi_{\mathrm{row}}$ and $\pi_{\mathrm{col}}$ defined above, respectively, we obtain a $7 \times 7$ array $\mathbf{H}_{\mathrm{PG,qc}}(2, 2^5)$ of circulants of size $151 \times 151$. $\mathbf{H}_{\mathrm{PG,qc}}(2, 2^5)$ consists of seven circulant-row-blocks, each composed of seven circulants, and seven circulant-column-blocks, each composed of seven circulants. For each circulant-row-block of $\mathbf{H}_{\mathrm{PG,qc}}(2, 2^5)$, there are three circulants with weight 6, 3 circulants with weight 5, and one circulant with weight 0 (i.e., a ZM).

Let $\gamma = 4$ and $\rho = 6$. Take the top four circulant-row-blocks of $\mathbf{H}_{\mathrm{PG,qc}}(2, 2^5)$ and remove one circulant-column-block to form a $4 \times 6$ array $\mathbf{M}_{\mathrm{PG,qc}}(4, 6)$ of circulants and ZM of size $151 \times 151$. The array $\mathbf{M}_{\mathrm{PG,qc}}(4, 6)$ is a $604 \times 906$ matrix with five column weights, 16, 17, 21, 22, 23 (average column weight $= 19.17$), and three row weights, 27, 28, 33 (average row weight $= 28.75$). The rank of $\mathbf{M}_{\mathrm{PG,qc}}(4, 6)$ is 244 and thus it contains 360 redundant rows. The null space over GF(2) of $\mathbf{M}_{\mathrm{PG,qc}}(4, 6)$ gives a $(906, 662)$ QC-PG-LDPC code $C_{\mathrm{PG,qc}}(4, 6)$ with rate 0.73. The BER and BLER performances of the code $C_{\mathrm{PG,qc}}(4, 6)$ over the AWGN channel using BPSK signaling decoded with 5, 10, and 50 iterations of the MSA with scaling factor 0.25 are shown in Fig. 12.10. The decoding converges quickly and the code can achieve a BER of $10^{-10}$ without a visible error-floor. ▲▲

## 12.5 Construction of QC-EG-LDPC Codes by CPM-Dispersion

The $(q+1) \times (q+1)$ array $\mathbf{H}_{\mathrm{EG,qc}}(2,q) = [\mathbf{A}_{i,j}]_{0 \leq i,j \leq q}$ given by (12.8) can be represented by a $(q+1) \times (q+1)$ matrix with entries from the field $\mathrm{GF}(q)$. With this representation, QC-EG-LDPC codes of various lengths and rates can be constructed much more easily.

First, we label the CPM-row-blocks and CPM-column-blocks of the array $\mathbf{H}_{\mathrm{EG,qc}}(2,q) = [\mathbf{A}_{i,j}]_{0 \leq i,j \leq q}$ from 0 to $q$. Next, we label the columns and rows of each $(q-1) \times (q-1)$ submatrix $\mathbf{A}_{i,j}$, a CPM or a ZM, of $\mathbf{H}_{\mathrm{EG,qc}}(2,q)$ from 0 to $q-2$. Let $\beta$ be a primitive element of $\mathrm{GF}(q)$. Then, $\beta^{-\infty} = 0$, $\beta^0 = 1$, $\beta$, $\beta^2, \ldots, \beta^{q-2}$ form all the elements of $\mathrm{GF}(q)$. Let $\{-\infty, 0, 1, \ldots, q-2\}$ be the set of powers of $\beta^{-\infty} = 0$, $\beta^0 = 1$, $\beta$, $\beta^2, \ldots, \beta^{q-2}$. For $0 \leq j \leq q$, let $l_j$ be an element in the power set $\{-\infty, 0, 1, \ldots, q-2\}$. The element $l_j$ is either an integer in the range between 0 to $q-2$ or is the element $-\infty$.

Consider the top CPM-row-block of $\mathbf{H}_{\mathrm{EG,qc}}(2,q)$:

$$\mathbf{A}_0 = [\mathbf{A}_{0,0}, \mathbf{A}_{0,1}, \ldots, \mathbf{A}_{0,q}]. \tag{12.9}$$

If the $j$th submatrix $\mathbf{A}_{0,j}$ of $\mathbf{A}_0$ is a CPM and the single 1-component in the top row of $\mathbf{A}_{0,j}$ is at the location $l_j$, we represent the CPM $\mathbf{A}_{0,j}$ by the nonzero element $\beta^{l_j}$ in $\mathrm{GF}(q)$. If $\mathbf{A}_{0,j}$ is a ZM in $\mathbf{A}_0$, we represent $\mathbf{A}_{0,j}$ by the zero element $\beta^{-\infty} = 0$ of $\mathrm{GF}(q)$. The representation of the $q$ CPMs and the single ZM in the top CPM-row-block $\mathbf{A}_0 = [\mathbf{A}_{0,0}, \mathbf{A}_{0,1}, \ldots, \mathbf{A}_{0,q}]$ of the array $\mathbf{H}_{\mathrm{EG,qc}}(2,q)$ by the elements of $\mathrm{GF}(q)$ results in a $(q+1)$-tuple over $\mathrm{GF}(q)$,

$$\mathbf{b}_0 = (\beta^{l_0}, \beta^{l_1}, \ldots, \beta^{l_q}). \tag{12.10}$$

The representation of $\mathbf{A}_{0,j}$ by $\beta^{l_j}$ is *one-to-one*. Because there is only one ZM in $\mathbf{A}_0 = [\mathbf{A}_{0,0}, \mathbf{A}_{0,1}, \ldots, \mathbf{A}_{0,q}]$, there is a single 0 in the $(q+1)$-tuple $\mathbf{b}_0 = (\beta^{l_0}, \beta^{l_1}, \ldots, \beta^{l_q})$. The matrix $\mathbf{A}_{0,j}$ is called the $(q-1) \times (q-1)$ *CPM-dispersion* of $\beta^{l_j}$ (or $(q-1)$-fold CPM-dispersion). If $\beta^{l_j} = 0$, the CPM-dispersion of the zero element of $\mathrm{GF}(q)$ is a $(q-1) \times (q-1)$ ZM.

It follows from the block-cyclic structure of $\mathbf{H}_{\mathrm{EG,qc}}(2,q)$ that $\mathbf{H}_{\mathrm{EG,qc}}(2,q)$ can be represented by the following $(q+1) \times (q+1)$ matrix over $\mathrm{GF}(q)$:

$$\mathbf{B}_{\mathrm{EG,qc}}(q+1,q+1) = \begin{bmatrix} \beta^{l_0} & \beta^{l_1} & \beta^{l_2} & \ldots & \beta^{l_{q-1}} & \beta^{l_q} \\ \beta^{l_q+1} & \beta^{l_0} & \beta^{l_1} & \ldots & \beta^{l_{q-2}} & \beta^{l_{q-1}} \\ \beta^{l_{q-1}+1} & \beta^{l_q+1} & \beta^{l_0} & \ldots & \beta^{l_{q-1}} & \beta^{l_{q-2}} \\ \vdots & \vdots & \vdots & \vdots & \vdots & \vdots \\ \beta^{l_2+1} & \beta^{l_3+1} & \beta^{l_4+1} & \ldots & \beta^{l_0} & \beta^{l_1} \\ \beta^{l_1+1} & \beta^{l_2+1} & \beta^{l_3+1} & \ldots & \beta^{l_q+1} & \beta^{l_0} \end{bmatrix}. \tag{12.11}$$

From the structure of $\mathbf{B}_{\mathrm{EG,qc}}(q+1,q+1)$, we see that each row is a cyclic-shift of the row above it one position to the right, but when the rightmost component, say $\beta^{l_j}$, is shifted around to the leftmost position, it is multiplied by $\beta$, giving the element $\beta^{l_j+1}$. (Here, it is noted that $\beta \cdot \beta^{-\infty} = 0$ and $l_j + 1$ is calculated under modulo $q-1$.) Hence, the CPM-dispersion of $\beta^{l_j+1}$ is obtained by cyclically shifting all the rows of the CPM-dispersion of $\beta^{l_j}$ one place to the right. From this, we readily see that the dispersions of the nonzero entries in $\mathbf{B}_{\mathrm{EG,qc}}(q+1,q+1)$ into $(q-1) \times (q-1)$ CPMs and 0-entries into ZMs of size $(q-1) \times (q-1)$ yield the array $\mathbf{H}_{\mathrm{EG,qc}}(2,q)$. In this case, we refer to the array $\mathbf{H}_{\mathrm{EG,qc}}(2,q)$ as the *CPM-dispersion of the matrix* $\mathbf{B}_{\mathrm{EG,qc}}(q+1,q+1)$ over $\mathrm{GF}(q)$, denoted by $\mathrm{CPM}(\mathbf{B}_{\mathrm{EG,qc}}(q+1,q+1))$. The matrix $\mathbf{B}_{\mathrm{EG,qc}}(q+1,q+1)$ is called the *base matrix* of the array $\mathbf{H}_{\mathrm{EG,qc}}(2,q)$. The cyclic row-shift structure of $\mathbf{B}_{\mathrm{EG,qc}}(q+1,q+1)$ is referred to as *$\beta$-multiply cyclic-shift*. Because the $(q-1) \times (q-1)$ CPM-dispersion of $\mathbf{B}_{\mathrm{EG,qc}}(q+1,q+1)$ gives the incidence matrix $\mathbf{H}_{\mathrm{EG,qc}}(2,q)$ (in QC form) of the subgeometry $\mathrm{EG}^*(2,q)$, we also call the matrix $\mathbf{B}_{\mathrm{EG,qc}}(2,q)$ the *base matrix* of the subgeometry $\mathrm{EG}^*(2,q)$.

To construct the base matrix $\mathbf{B}_{\mathrm{EG,qc}}(q+1,q+1)$, we first construct its top row $\mathbf{b}_0 = (\beta^{l_0}, \beta^{l_1}, \ldots, \beta^{l_q})$ and then perform the $\beta$-multiply cyclic-shift of $\mathbf{b}_0$ $q$ times. The top row $\mathbf{b}_0$ of $\mathbf{B}_{\mathrm{EG,qc}}(q+1,q+1)$ can be constructed directly from a line (any line) in the subgeometry $\mathrm{EG}^*(2,q)$ of $\mathrm{EG}(2,q)$. Recall that the field $\mathrm{GF}(q^2)$ is a realization of $\mathrm{EG}(2,q)$. Let $\alpha$ be a primitive element of $\mathrm{GF}(q^2)$. The powers of $\alpha^{-\infty} = 0$, $\alpha^0 = 1$, $\alpha$, $\alpha^2, \ldots, \alpha^{q^2-2}$ represent the $q^2$ points of $\mathrm{EG}(2,q)$. Let $\beta = \alpha^{q+1}$. Then, the $q$ elements $0$, $1$, $\beta$, $\beta^2, \ldots, \beta^{q-2}$ in $\mathrm{GF}(q^2)$ form a subfield $\mathrm{GF}(q)$ of $\mathrm{GF}(q^2)$. Let $\alpha^k$ and $\alpha^l$ be two linearly independent elements in $\mathrm{GF}(q^2)$ over $\mathrm{GF}(q)$. Then, the set $\{\alpha^k + \eta\alpha^l : \eta \in \mathrm{GF}(q)\}$ of $q$ elements in $\mathrm{GF}(q^2)$ represents $q$ points on a line $L$ in $\mathrm{EG}^*(2,q)$. Let $\mathbf{v} = (v_0, v_1, \ldots, v_{q^2-2})$ be the incidence vector of the line $L$. Permute the $q^2 - 1$ components of $\mathbf{v}$ with the permutation $\pi_{\mathrm{col}}$ defined by (12.7). The permutation results in a permuted incidence vector $\mathbf{w} = (\mathbf{w}_0, \mathbf{w}_1, \ldots, \mathbf{w}_q)$ of $L$ which comprises $q+1$ sections, $\mathbf{w}_0, \mathbf{w}_1, \ldots, \mathbf{w}_q$, each consisting of $q-1$ components of $\mathbf{v}$. For $0 \leq j \leq q$, the $j$th section $\mathbf{w}_j$ of $\mathbf{w}$ consists of the $q-1$ components of $\mathbf{v}$ with indices $j$, $(q+1)+j$, $2(q+1)+j, \ldots, (q-2)(q+1)+j$. One section in $\mathbf{w}$ is a zero-section, consisting of $q-1$ zeros, and each of the other $q$ sections of $\mathbf{w}$ consists of a single 1-component and $q-2$ 0-components. For $0 \leq j \leq q$, let $l_j$ be the location of the 1-component in the $j$th section $\mathbf{w}_j$ of $\mathbf{w}$ with $l_j \in \{-\infty, 0, 1, \ldots, q-2\}$. For $0 \leq j \leq q$, we represent the $j$th section $\mathbf{w}_j$ of $\mathbf{w}$ by $\beta^{l_j}$. If $l_j = -\infty$, the $j$th section $\mathbf{w}_j$ is represented by $\beta^{-\infty} = 0$. The representation of the $q+1$ sections of $\mathbf{w}$ by elements in $\mathrm{GF}(q)$ results in a $(q+1)$-tuple over $\mathrm{GF}(q)$, $\mathbf{b}_0 = (\beta^{l_0}, \beta^{l_1}, \ldots, \beta^{l_q})$, which is the top row of the base matrix $\mathbf{B}_{\mathrm{EG,qc}}(q+1,q+1)$ given by (12.11). Then, $\mathbf{B}_{\mathrm{EG,qc}}(q+1,q+1)$ is obtained by performing the $\beta$-multiply cyclic-shifts of $\mathbf{b}_0$ $q$ times.

**Example 12.12** Consider the two-dimensional Euclidean geometry $\mathrm{EG}(2,2^2)$ over $\mathrm{GF}(2^2)$ which is realized by the field $\mathrm{GF}(2^4)$ given by Table 12.1. The field

$GF(2^4)$ is the extension field of $GF(2^2) = \{0, 1, \beta, \beta^2\}$ where $\beta = \alpha^5$ and $\alpha$ is a primitive element of $GF(2^4)$. It was shown in Example 12.8 that the four points $\alpha^0 = 1$, $\alpha^2$, $\alpha^6$, and $\alpha^{14}$ form a line $L$ in $EG^*(2, 2^2)$. The incidence vector of $L$ is $\mathbf{v} = (1\ 0\ 1\ 0\ 0\ 0\ 1\ 0\ 0\ 0\ 0\ 0\ 0\ 0\ 1)$. Label the 15 components of $\mathbf{v}$ from 0 to 14. Permuting the components of $\mathbf{v}$ with the permutation $\pi_{\mathrm{col}} = (0, 5, 10, 1, 6, 11, 2, 7, 12, 3, 8, 13, 4, 9, 14)$, we obtain a permuted incidence vector $\mathbf{w} = (\mathbf{w}_0, \mathbf{w}_1, \mathbf{w}_2, \mathbf{w}_3, \mathbf{w}_4) = (1\ 0\ 0, 0\ 1\ 0, 1\ 0\ 0, 0\ 0\ 0, 0\ 0\ 1)$ of $L$ which consists of five sections, each comprising three consecutive components of $\mathbf{w}$. Representing the five sections of $\mathbf{w}$ by $\beta^0$, $\beta^1$, $\beta^0$, $\beta^{-\infty}$, and $\beta^2$, respectively, we obtain a 5-tuple $\mathbf{b}_0 = (\beta^0, \beta^1, \beta^0, \beta^{-\infty}, \beta^2)$ over $GF(2^2)$. Using $\mathbf{b}_0$ and its four consecutive $\beta$-multiply cyclic-shifts, we form the following $5 \times 5$ matrix over $GF(2^2)$:

$$\mathbf{B}_{\mathrm{EG,qc}}(5,5) = \begin{bmatrix} \beta^0 & \beta^1 & \beta^0 & \beta^{-\infty} & \beta^2 \\ \beta^0 & \beta^0 & \beta^1 & \beta^0 & \beta^{-\infty} \\ \beta^{-\infty} & \beta^0 & \beta^0 & \beta^1 & \beta^0 \\ \beta^1 & \beta^{-\infty} & \beta^0 & \beta^0 & \beta^1 \\ \beta^2 & \beta^1 & \beta^{-\infty} & \beta^0 & \beta^0 \end{bmatrix}.$$

The $3 \times 3$ CPM-dispersions of the entries in $\mathbf{B}_{\mathrm{EG,qc}}(5,5)$ result in the $5 \times 5$ array $\mathbf{H}_{\mathrm{EG,qc}}(2, 2^2)$ of CPMs and ZMs with size $3 \times 3$ given in Example 12.8. Hence, the matrix $\mathbf{B}_{\mathrm{EG,qc}}(5,5)$ is the base matrix of the array $\mathbf{H}_{\mathrm{EG,qc}}(2, 2^2)$ (or the base matrix of the subgeometry $EG^*(2, 2^2)$ of $EG(2, 2^2)$).          ▲▲

Let $\gamma$ and $\rho$ be two positive integers with $1 \leq \gamma < \rho \leq q + 1$. Take a $\gamma \times \rho$ submatrix $\mathbf{B}_{\mathrm{EG,qc}}(\gamma, \rho)$ of $\mathbf{B}_{\mathrm{EG,qc}}(q+1, q+1)$. The $(q-1)$-fold CPM-dispersion of $\mathbf{B}_{\mathrm{EG,qc}}(\gamma, \rho)$ results in a $\gamma \times \rho$ array $\mathbf{H}_{\mathrm{EG,qc}}(\gamma, \rho)$ of CPMs and/or ZMs of size $(q-1) \times (q-1)$ that is a subarray of $\mathbf{H}_{\mathrm{EG,qc}}(2, q)$. Because $\mathbf{H}_{\mathrm{EG,qc}}(2, q)$ satisfies the RC-constraint, $\mathbf{H}_{\mathrm{EG,qc}}(\gamma, \rho)$ also satisfies the RC-constraint. The null space over $GF(2)$ of $\mathbf{H}_{\mathrm{EG,qc}}(\gamma, \rho)$ gives a QC-EG-LDPC code $C_{\mathrm{EG,qc}}(\gamma, \rho)$ of length $\rho(q-1)$ with rate at least $(\rho - \gamma)/\rho$. The girth of the Tanner graph $\mathcal{G}_{\mathrm{EG,qc}}(\gamma, \rho)$ of $C_{\mathrm{EG,qc}}(\gamma, \rho)$ is at least 6.

For various of choices of $\gamma$ and $\rho$ with $1 \leq \gamma < \rho \leq q + 1$, we can construct submatrices of $\mathbf{B}_{\mathrm{EG,qc}}(q+1, q+1)$ with various sizes. Using these submatrices as base matrices, we can construct a large family of QC-EG-LDPC codes of various lengths and rates using the CPM-dispersion construction. Using two-dimensional Euclidean geometries over various fields, we can construct a large class of QC-EG-LDPC codes.

**Example 12.13** Consider the two-dimensional Euclidean geometry $EG(2, 2^6)$ over the field $GF(2^6)$. Based on a line in $EG(2, 2^6)$ not passing through the origin of $EG(2, 2^6)$, we construct a $65 \times 65$ matrix $\mathbf{B}_{\mathrm{EG,qc}}(65, 65)$ over $GF(2^6)$ in the form of (12.11) which is the base matrix of the subgeometry $EG^*(2, 2^6)$ of $EG(2, 2^6)$. The 63-fold CPM-dispersion of $\mathbf{B}_{\mathrm{EG,qc}}(65, 65)$ gives a $65 \times 65$ array $\mathbf{H}_{\mathrm{EG,qc}}(2, 2^6)$ of CPMs and ZMs of size $63 \times 63$, a $4095 \times 4095$ matrix over $GF(2)$, which is the incidence matrix (in QC form) of the subgeometry $EG^*(2, 2^6)$.

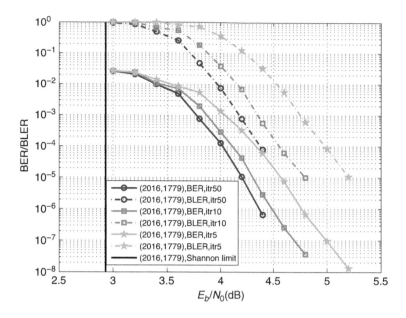

Figure 12.11 The BER and BLER performances of the $(2016, 1779)$ QC-EG-LDPC code given in Example 12.13.

Let $\gamma = 4$ and $\rho = 32$. Take a $4 \times 32$ submatrix $\mathbf{B}_{EG,qc}(4, 32)$ from $\mathbf{B}_{EG,qc}(65, 65)$ (avoiding 0-entries). The 63-fold CPM-dispersion of $\mathbf{B}_{EG,qc}(4, 32)$ gives a $4 \times 32$ array $\mathbf{H}_{EG,qc}(4, 32)$ of CPMs of size $63 \times 63$ which is a $252 \times 2016$ matrix over GF$(2)$ with column and row weights 4 and 32, respectively. The null space over GF$(2)$ of $\mathbf{H}_{EG,qc}(4, 32)$ gives a $(4, 32)$-regular $(2016, 1779)$ QC-EG-LDPC code $C_{EG,qc}(4, 32)$ with rate 0.8824. The BER and BLER performances of the code $C_{EG,qc}(4, 32)$ over the AWGN channel using BPSK signaling decoded with 5, 10, and 50 iterations of the MSA are shown in Fig. 12.11. ▲▲

Before we leave this section, we construct a QC-EG-LDPC code based on a two-dimensional Euclidean geometry EG$(2, q)$ over GF$(q)$ where $q$ is not a power of 2 but a prime number.

**Example 12.14** Set $q = 257$, which is a prime. There is a prime field GF$(257)$ with 257 elements. Let EG$(2, 257)$ be the two-dimensional Euclidean geometry over GF$(257)$. Based on a single line not passing through the origin of EG$(2, 257)$, we can construct a $258 \times 258$ base matrix $\mathbf{B}_{EG,qc}(258, 258)$ over GF$(257)$ for the subgeometry EG$^*(2, 257)$ of EG$(2, 257)$.

Set $\gamma = 4$ and $\rho = 64$. Take a $4 \times 64$ submatrix $\mathbf{B}_{EG,qc}(4, 64)$ from the base matrix $\mathbf{B}_{EG,qc}(258, 258)$, e.g., 64 columns of the top four rows of $\mathbf{B}_{EG,qc}(258, 258)$ (avoiding the 0-entries). The column and row weights of $\mathbf{B}_{EG,qc}(4, 64)$ are 4 and 64, respectively. Dispersing each entry in $\mathbf{B}_{EG,qc}(4, 64)$ into a CPM of size $256 \times 256$, we obtain a $4 \times 64$ array $\mathbf{H}_{EG,qc}(4, 64)$ of CPMs of size $256 \times 256$ which is a $1024 \times 16\,384$ matrix over GF(2) with column and row weights 4 and 64, respectively.

The null space over GF(2) of $\mathbf{H}_{\mathrm{EG,qc}}(4,64)$ gives a $(4,64)$-regular $(16\,384, 15\,363)$ QC-EG-LDPC code with rate 0.9377. The BER and BLER performances of the constructed code over the AWGN channel decoded with 5, 10, and 50 iterations of the MSA are shown in Fig. 12.12(a). From this figure, we see that at the BER of $10^{-8}$ with 50 decoding iterations, the code performs about 0.9 dB from the Shannon limit.

The UEBR and UEBLR performances of the code over the BEC are shown in Fig. 12.12(b). Because the rate of the code is 0.9377, the Shannon limit of this rate over the BEC is $1 - 0.9377 = 0.0623$. From Fig. 12.12(b), we see that the code performs very close to the Shannon limit over the BEC. ▲▲

## 12.6 Masking Techniques

Let $\mathbf{B}_{\mathrm{EG,qc}}(\gamma, \rho) = [b_{i,j}]_{0 \leq i < \gamma, 0 \leq j < \rho}$ be a $\gamma \times \rho$ submatrix of the base matrix $\mathbf{B}_{\mathrm{EG,qc}}(q+1, q+1)$ of the incidence matrix $\mathbf{H}_{\mathrm{EG,qc}}(2, q)$ of the subgeometry $\mathrm{EG}^*(2, q)$. Suppose we replace a nonzero entry in $\mathbf{B}_{\mathrm{EG,qc}}(\gamma, \rho)$ by the 0-element of GF($q$). In the CPM-dispersion $\mathbf{H}_{\mathrm{EG,qc}}(\gamma, \rho)$ of $\mathbf{B}_{\mathrm{EG,qc}}(\gamma, \rho)$, this replacement results in replacing a $(q-1) \times (q-1)$ CPM by a $(q-1) \times (q-1)$ ZM and is referred to as *masking* [16, 28, 29, 36].

Let $k$ be a nonnegative integer less than the number of nonzero entries in $\mathbf{B}_{\mathrm{EG,qc}}(\gamma, \rho)$. The replacement of $k$ nonzero entries in $\mathbf{B}_{\mathrm{EG,qc}}(\gamma, \rho)$ by $k$ zeros amounts to replacing $k$ CPMs by $k$ ZMs at the locations in $\mathbf{H}_{\mathrm{EG,qc}}(\gamma, \rho)$ which correspond to the $k$ locations of the nonzero entries in $\mathbf{B}_{\mathrm{EG,qc}}(\gamma, \rho)$ being masked. Masking $k$ CPMs in $\mathbf{H}_{\mathrm{EG,qc}}(\gamma, \rho)$ amounts to removing $k(q-1)$ edges from the Tanner graph $\mathcal{G}_{\mathrm{EG,qc}}(\gamma, \rho)$ associated with the array $\mathbf{H}_{\mathrm{EG,qc}}(\gamma, \rho)$. Removing these edges from $\mathcal{G}_{\mathrm{EG,qc}}(\gamma, \rho)$ may break many short cycles. As a result, the resultant Tanner graph, denoted by $\mathcal{G}_{\mathrm{EG,qc,mask}}(\gamma, \rho)$, may have a smaller number of short cycles, or a larger girth (larger than 6), or both. The subscript "mask" in $\mathcal{G}_{\mathrm{EG,qc,mask}}(\gamma, \rho)$ stands for "masking."

Masking the base matrix $\mathbf{B}_{\mathrm{EG,qc}}(\gamma, \rho) = [b_{i,j}]_{0 \leq i < \gamma, 0 \leq j < \rho}$ can be modeled mathematically as a *matrix product*. Let $\mathbf{Z}(\gamma, \rho) = [z_{i,j}]_{0 \leq i < \gamma, 0 \leq j < \rho}$ be a $\gamma \times \rho$ matrix with the zero and unit elements of GF($q$) as entries. Define the following product of $\mathbf{Z}(\gamma, \rho)$ and $\mathbf{B}_{\mathrm{EG,qc}}(\gamma, \rho)$:

$$
\begin{aligned}
\mathbf{B}_{\mathrm{EG,qc,mask}}(\gamma, \rho) &= \mathbf{Z}(\gamma, \rho) \otimes \mathbf{B}_{\mathrm{EG,qc}}(\gamma, \rho) \\
&= [z_{i,j} b_{i,j}]_{0 \leq i < \gamma, 0 \leq j < \rho},
\end{aligned}
\tag{12.12}
$$

where $z_{i,j} b_{i,j} = b_{i,j}$ if $z_{i,j} = 1$ and $z_{i,j} b_{i,j} = 0$ if $z_{i,j} = 0$. In this matrix product, the nonzero entries in $\mathbf{B}_{\mathrm{EG,qc}}(\gamma, \rho)$ at the locations corresponding to the zero entries in $\mathbf{Z}(\gamma, \rho)$ are replaced (or masked) by zeros. The matrix product defined by (12.12) is called the *Hadamard matrix product*.

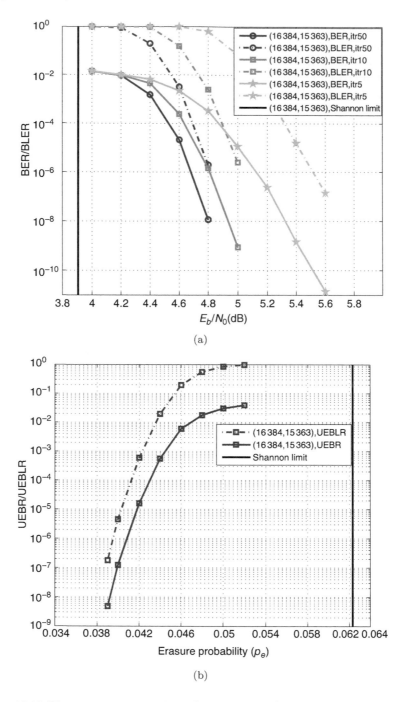

Figure 12.12 The performances of the $(16\,384, 15\,363)$ QC-EG-LDPC code given in Example 12.14 over: (a) AWGN channel and (b) BEC.

The CPM-dispersion of $\mathbf{B}_{\mathrm{EG,qc,mask}}(\gamma, \rho)$ gives a $\gamma \times \rho$ masked array, denoted by $\mathbf{H}_{\mathrm{EG,qc,mask}}(\gamma, \rho)$, of CPMs and ZMs of size $(q-1) \times (q-1)$. We call $\mathbf{Z}(\gamma, \rho)$ and $\mathbf{B}_{\mathrm{EG,qc,mask}}(\gamma, \rho)$ the *masking matrix* and the *masked base matrix*, respectively. Because $\mathbf{H}_{\mathrm{EG,qc}}(\gamma, \rho)$ satisfies the RC-constraint, $\mathbf{H}_{\mathrm{EG,qc,mask}}(\gamma, \rho)$ also satisfies the RC-constraint. The null space of $\mathbf{H}_{\mathrm{EG,qc,mask}}(\gamma, \rho)$ gives a QC-LDPC code, called the *masked QC-EG-LDPC code*, denoted by $C_{\mathrm{EG,qc,mask}}(\gamma, \rho)$.

CPM-dispersions of properly chosen base matrices in conjunction with masking can result in masked QC-EG-LDPC codes with good error performance in the waterfall region as well as low error-floors. Masking will be discussed further in Chapter 14.

**Example 12.15** Consider the two-dimensional Euclidean geometry $\mathrm{EG}(2, 257)$ over $\mathrm{GF}(257)$ in Example 12.14, where a $258 \times 258$ base matrix $\mathbf{B}_{\mathrm{EG,qc}}(258, 258)$ over $\mathrm{GF}(257)$ is constructed based on the subgeometry $\mathrm{EG}^*(2, 257)$ of $\mathrm{EG}(2, 257)$. Let $\alpha$ be a primitive element of $\mathrm{GF}(257)$. Set $\gamma = 4$ and $\rho = 8$. Take the following $4 \times 8$ submatrix $\mathbf{B}_{\mathrm{EG,qc}}(4, 8)$ from $\mathbf{B}_{\mathrm{EG,qc}}(258, 258)$ avoiding zeros:

$$\mathbf{B}_{\mathrm{EG,qc}}(4, 8) = \begin{bmatrix} \alpha^{35} & \alpha^{60} & \alpha^{127} & \alpha^{211} & \alpha^{106} & \alpha^{22} & \alpha^{146} & \alpha^{215} \\ \alpha^{113} & \alpha^{35} & \alpha^{60} & \alpha^{127} & \alpha^{211} & \alpha^{106} & \alpha^{22} & \alpha^{146} \\ \alpha^{95} & \alpha^{113} & \alpha^{35} & \alpha^{60} & \alpha^{127} & \alpha^{211} & \alpha^{106} & \alpha^{22} \\ \alpha^{255} & \alpha^{95} & \alpha^{113} & \alpha^{35} & \alpha^{60} & \alpha^{127} & \alpha^{211} & \alpha^{106} \end{bmatrix}$$

whose column and row weights are 4 and 8, respectively. The $256 \times 256$ CPM-dispersion of $\mathbf{B}_{\mathrm{EG,qc}}(4, 8)$ gives a $4 \times 8$ array $\mathbf{H}_{\mathrm{EG,qc}}(4, 8)$ of CPMs of size $256 \times 256$, a $1024 \times 2048$ matrix with column and row weights 4 and 8, respectively. The null space over $\mathrm{GF}(2)$ of $\mathbf{H}_{\mathrm{EG,qc}}(4, 8)$ gives a $(4, 8)$-regular $(2048, 1027)$ QC-EG-LDPC code $C_{\mathrm{EG,qc}}(4, 8)$ with rate 0.501. Using the cycle counting algorithm in [37, 38], it is found that the Tanner graph $\mathcal{G}_{\mathrm{EG,qc}}(4, 8)$ of $C_{\mathrm{EG,qc}}(4, 8)$ has girth 6 and contains 5632 cycles of length 6, 37 120 cycles of length 8, and 501 504 cycles of length 10. The BER and BLER performances of $C_{\mathrm{EG,qc}}(4, 8)$ over the AWGN channel decoded with 50 iterations of the MSA are shown in Fig. 12.13.

Suppose we mask the base matrix $\mathbf{B}_{\mathrm{EG,qc}}(4, 8)$ with the following $4 \times 8$ masking matrix:

$$\mathbf{Z}(4, 8) = \begin{bmatrix} 1 & 0 & 1 & 0 & 1 & 1 & 1 & 1 \\ 0 & 1 & 0 & 1 & 1 & 1 & 1 & 1 \\ 1 & 1 & 1 & 1 & 1 & 0 & 1 & 0 \\ 1 & 1 & 1 & 1 & 0 & 1 & 0 & 1 \end{bmatrix}. \tag{12.13}$$

Masking $\mathbf{B}_{\mathrm{EG,qc}}(4, 8)$ with $\mathbf{Z}(4, 8)$ results in the following $4 \times 8$ masked matrix with column and row weights 3 and 6, respectively:

$$\mathbf{B}_{\mathrm{EG,qc,mask}}(4, 8) = \begin{bmatrix} \alpha^{35} & 0 & \alpha^{127} & 0 & \alpha^{106} & \alpha^{22} & \alpha^{146} & \alpha^{215} \\ 0 & \alpha^{35} & 0 & \alpha^{127} & \alpha^{211} & \alpha^{106} & \alpha^{22} & \alpha^{146} \\ \alpha^{95} & \alpha^{113} & \alpha^{35} & \alpha^{60} & \alpha^{127} & 0 & \alpha^{106} & 0 \\ \alpha^{255} & \alpha^{95} & \alpha^{113} & \alpha^{35} & 0 & \alpha^{127} & 0 & \alpha^{106} \end{bmatrix}.$$

The 256-fold CPM-dispersion of $\mathbf{B}_{\mathrm{EG,qc,mask}}(4, 8)$ gives a $4 \times 8$ masked array $\mathbf{H}_{\mathrm{EG,qc,mask}}(4, 8)$ of CPMs and ZMs of size $256 \times 256$, which is a $1024 \times 2048$ matrix with column and row weights 3 and 6, respectively.

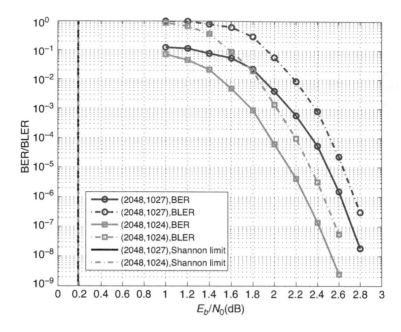

Figure 12.13 The BER and BLER performances of the unmasked $(2048, 1027)$ and the masked $(2048, 1024)$ QC-EG-LDPC codes given in Example 12.15.

The null space over GF(2) of $\mathbf{H}_{\text{EG,qc,mask}}(4, 8)$ gives a $(3, 6)$-regular $(2048, 1024)$ QC-EG-LDPC code $C_{\text{EG,qc,mask}}(4, 8)$ with rate 1/2. The Tanner graph $\mathcal{G}_{\text{EG,qc,mask}}(4, 8)$ of $C_{\text{EG,qc,mask}}(4, 8)$ has girth 8 and contains 256 cycles of length 8, and 15 872 cycles of length 10. We see that masking reduces the number of short cycles of the Tanner graph $\mathcal{G}_{\text{EG,qc}}(4, 8)$ of the unmasked code $C_{\text{EG,qc}}(4, 8)$ significantly. It also enlarges the girth of $\mathcal{G}_{\text{EG,qc}}(4, 8)$ from 6 to 8.

The BER and BLER performances of $C_{\text{EG,qc,mask}}(4, 8)$ over the AWGN channel decoded with 50 iterations of the MSA are also shown in Fig. 12.13. From the figure, we see that the masked code $C_{\text{EG,qc,mask}}(4, 8)$ outperforms the unmasked code $C_{\text{EG,qc}}(4, 8)$ by 0.2 dB in the simulation region. ▲▲

# 12.7 Construction of QC-FG-LDPC Codes by Circulant-Decomposition

Consider an $n \times n$ RC-constrained circulant $\mathbf{G}$ over GF(2) with weight $w$. Label both the rows and columns of $\mathbf{G}$ from 0 to $n - 1$. For $1 \le t \le w$, let $w_0, w_1, \dots, w_{t-1}$ be a set of positive integers such that $w_0 + w_1 + \dots + w_{t-1} = w$. Let $\mathbf{g}$ be the zeroth (*leftmost*) column of $\mathbf{G}$. Then, $\mathbf{g}$ and its $n - 1$ *downward cyclic-shifts* form the circulant $\mathbf{G}$. We call $\mathbf{g}$ the *column-generator* of $\mathbf{G}$. Partition the locations of the $w$ 1-components in $\mathbf{g}$ into $t$ disjoint sets, $\Lambda_0, \Lambda_1, \dots, \Lambda_{t-1}$. For $0 \le j < t$, $\Lambda_j$ consists of $w_j$ locations of $\mathbf{g}$ where the components are 1s.

Split $\mathbf{g}$ into $t$ columns of the same length $n$, $\mathbf{g}_0$, $\mathbf{g}_1, \ldots, \mathbf{g}_{t-1}$, with weights $w_0$, $w_1, \ldots, w_{t-1}$, respectively. For $0 \le j < t$, the $w_j$ 1-components of $\mathbf{g}_j$ are at the locations in $\Lambda_j$. For each new column $\mathbf{g}_j$, $0 \le j < t$, we form an $n \times n$ circulant $\mathbf{G}_j$ with $\mathbf{g}_j$ as its column-generator. This results in $t$ circulants, $\mathbf{G}_0$, $\mathbf{G}_1, \ldots,$ $\mathbf{G}_{t-1}$, each of size $n \times n$ and satisfying the RC-constraint. These circulants are called the *column-descendants* of $\mathbf{G}$ and they form a decomposition of $\mathbf{G}$, called a *column-decomposition* of $\mathbf{G}$. Because the 1-component location sets $\Lambda_0$, $\Lambda_1, \ldots, \Lambda_{t-1}$ of $\mathbf{g}_0$, $\mathbf{g}_1, \ldots, \mathbf{g}_{t-1}$ are disjoint, no two column-descendants of $\mathbf{G}$ have 1-entries at the same location, i.e., the locations of 1-entries in one column-descendant of $\mathbf{G}$ and the locations of 1-entries in another column-descendant of $\mathbf{G}$ are *disjoint*. Because the column-descendants $\mathbf{G}_0$, $\mathbf{G}_1, \ldots, \mathbf{G}_{t-1}$ of $\mathbf{G}$ satisfy the RC-constraint, the 1-entry location disjoint structure of these column-descendants ensures that they satisfy the *pair-wise RC-constraint, i.e., no two rows (or two columns) from either the same column-descendant $\mathbf{G}_i$ or from two different column-descendants, $\mathbf{G}_i$ and $\mathbf{G}_j$, have more than one 1-component in common.* The column-decomposition of $\mathbf{G}$ results in a row of $t$ circulants of size $n \times n$,

$$\mathbf{G}_{\text{col,decom}} = [\mathbf{G}_0, \mathbf{G}_1, \ldots, \mathbf{G}_{t-1}], \tag{12.14}$$

which is an $n \times tn$ matrix with constant row weight $w$ and satisfies the RC-constraint. The parameter $t$ is called the *column-splitting factor*. The weights $w_0$, $w_1, \ldots, w_{t-1}$ of $\mathbf{g}_0$, $\mathbf{g}_1, \ldots, \mathbf{g}_{t-1}$ are called the *column-splitting weights* of $\mathbf{G}$. We denote this set of weights by $W_{\text{col}}$, i.e., $W_{\text{col}} = \{w_0, w_1, \ldots, w_{t-1}\}$. The subscript "col, decom" in $\mathbf{G}_{\text{col,decom}}$ stands for "column-decomposition."

Let $c$ be a positive integer such that $1 \le c \le \max\{w_j : 0 \le j < t\}$. For $0 \le j < t$, let $w_{0,j}$, $w_{1,j}, \ldots, w_{c-1,j}$ be a set of nonnegative integers such that $w_{0,j} + w_{1,j} + \cdots + w_{c-1,j} = w_j$. Let $\mathbf{h}_j$ be the top row (called *row-generator*) of $\mathbf{G}_j$. For $0 \le j < t$, partition the locations of $w_j$ 1-components of $\mathbf{h}_j$ (the locations in $\Lambda_j$) into $c$ disjoints sets, $\Lambda_{0,j}$, $\Lambda_{1,j}, \ldots, \Lambda_{c-1,j}$. Split $\mathbf{h}_j$ into $c$ new rows of the same length $n$, $\mathbf{h}_{0,j}$, $\mathbf{h}_{1,j}, \ldots, \mathbf{h}_{c-1,j}$, where for $0 \le k < c$, $\mathbf{h}_{k,j}$ contains the $w_{k,j}$ 1-components of $\mathbf{h}_j$ at the locations in $\Lambda_{k,j}$. For $0 \le k < c$, we form an $n \times n$ circulant $\mathbf{G}_{k,j}$ with $\mathbf{h}_{k,j}$ as the row-generator. The $c$ circulants, $\mathbf{G}_{0,j}$, $\mathbf{G}_{1,j}, \ldots, \mathbf{G}_{c-1,j}$, are referred to as the *row-descendants* of $\mathbf{G}_j$ and they form a *row-decomposition* of $\mathbf{G}_j$. Because the 1-component location sets $\Lambda_{0,j}$, $\Lambda_{1,j}, \ldots, \Lambda_{c-1,j}$ of $\mathbf{h}_{0,j}$, $\mathbf{h}_{1,j}, \ldots, \mathbf{h}_{c-1,j}$ are disjoint, no two row-descendants of $\mathbf{G}_j$ have 1-entries at the same location. This implies that the $c$ row-descendants, $\mathbf{G}_{0,j}$, $\mathbf{G}_{1,j}, \ldots, \mathbf{G}_{c-1,j}$, of $\mathbf{G}_j$, satisfy the pair-wise RC-constraint. The above row-splitting and circulant forming result in a column of $cn \times n$ circulants,

$$\mathbf{G}_{\text{row,decom},j} = \begin{bmatrix} \mathbf{G}_{0,j} \\ \mathbf{G}_{1,j} \\ \vdots \\ \mathbf{G}_{c-1,j} \end{bmatrix} \tag{12.15}$$

which is a $cn \times n$ matrix with constant column weight $w_j$. The column array $\mathbf{G}_{\text{row,decom},j}$ is called a *row-decomposition* of $\mathbf{G}_j$ and $c$ is called the *row-splitting factor*. The weights $w_{0,j}$, $w_{1,j}, \ldots, w_{c-1,j}$ of $\mathbf{h}_{0,j}$, $\mathbf{h}_{1,j}, \ldots, \mathbf{h}_{c-1,j}$ are called *row-splitting weights* of $\mathbf{G}_j$. Because $\mathbf{G}_j$ satisfies the RC-constraint and its row-descendants $\mathbf{G}_{0,j}$, $\mathbf{G}_{1,j}, \ldots, \mathbf{G}_{c-1,j}$ satisfy the pair-wise RC-constraint, the array $\mathbf{G}_{\text{row,decom},j}$ given by (12.15) satisfies the RC-constraint. We denote the set of row-splitting weights as $W_{\text{row},j}$, i.e., $W_{\text{row},j} = \{w_{0,j}, w_{1,j}, \ldots, w_{c-1,j}\}$. Note that some of the weights in $W_{\text{row},j}$ may be zeros, i.e., some of the row-descendants of $\mathbf{G}_j$ may be ZMs.

If each circulant $\mathbf{G}_j$ in the array $\mathbf{G}_{\text{col,decom}}$ given by (12.14) is replaced by its row-decomposition $\mathbf{G}_{\text{row,decom},j}$, we obtain the following $c \times t$ array of circulants of size $n \times n$:

$$\mathbf{H}_{\text{array,decom}}(c,t) = \begin{bmatrix} \mathbf{G}_{0,0} & \mathbf{G}_{0,1} & \cdot & \mathbf{G}_{0,t-1} \\ \mathbf{G}_{1,0} & \mathbf{G}_{1,1} & \cdot & \mathbf{G}_{1,t-1} \\ \vdots & \vdots & \vdots & \vdots \\ \mathbf{G}_{c-1,0} & \mathbf{G}_{c-1,1} & \cdot & \mathbf{G}_{c-1,t-1} \end{bmatrix}. \tag{12.16}$$

The array $\mathbf{H}_{\text{array,decom}}(c,t)$ is called the *array-decomposition* [39] of the circulant $\mathbf{G}$ and is a $cn \times tn$ matrix.

Following from the pair-wise RC-constrained structure of the column-descendants $\mathbf{G}_0$, $\mathbf{G}_1, \ldots, \mathbf{G}_{t-1}$ of $\mathbf{G}$ and the row-descendants $\mathbf{G}_{0,j}$, $\mathbf{G}_{1,j}, \ldots$, $\mathbf{G}_{c-1,j}$ of $\mathbf{G}_j$, $0 \leq j < t$, the array $\mathbf{H}_{\text{array,decom}}(c,t)$ satisfies the RC-constraint. If $ct = w$, $w_0 = w_1 = \cdots = w_{t-1} = c$, and $w_{0,j} = w_{1,j} = \cdots = w_{c-1,j} = 1$, the matrix $\mathbf{H}_{\text{array,decom}}(c,t)$ is a $c \times t$ array of CPMs of size $n \times n$ which is a $cn \times tn$ matrix with constant column and row weights $c$ and $t$, respectively. The density of $\mathbf{H}_{\text{array,decom}}(c,t)$ is $1/ct$ of the density of $\mathbf{G}$. The null space over GF(2) of $\mathbf{H}_{\text{array,decom}}(c,t)$ gives a QC-LDPC code of length $tn$ whose Tanner graph has girth at least 6. The above array-decomposition of an RC-constrained circulant provides a method for constructing QC-LDPC codes of various lengths and rates.

Circulants that satisfy the RC-constraint or the pair-wise RC-constraint can be constructed from Euclidean or projective geometries over finite fields as shown in Sections 7.3.3 and 7.4.1. Based on these circulants, a family of QC-FG-LDPC codes can be constructed with array-decompositions using various column- or row-splitting factors and column- or row-splitting weight sets.

Consider the $m$-dimensional Euclidean geometry $\text{EG}(m,q)$ over GF($q$) with $q = 2^s$. Let $\text{EG}^*(m,q)$ be the subgeometry of $\text{EG}(m,q)$ obtained by removing the origin of $\text{EG}(m,q)$ and the lines passing through the origin. As shown in Section 7.3.3, the lines in $\text{EG}^*(m,q)$ can be partitioned into $K_{\text{EG}} = (q^{m-1} - 1)/(q - 1)$ cyclic classes. Based on these $K_{\text{EG}}$ cyclic classes of lines in $\text{EG}^*(m,q)$, we can form $K_{\text{EG}}$ circulants, denoted by $\mathbf{G}^{(0)}$, $\mathbf{G}^{(1)}, \ldots, \mathbf{G}^{(K_{\text{EG}}-1)}$ of size $(q^m - 1) \times (q^m - 1)$, each with weight $q$. These EG circulants satisfy the pair-wise RC-constraint. Suppose we take the first $\lambda$ of these EG circulants and form the following RC-constrained matrix:

$$\mathbf{H}_{\text{EG}}(1,\lambda) = \begin{bmatrix} \mathbf{G}^{(0)} & \mathbf{G}^{(1)} & \cdots & \mathbf{G}^{(\lambda-1)} \end{bmatrix}. \tag{12.17}$$

Let $\ell$ and $k$ by two positive integers such that $1 \leq \ell, k \leq q$. For $0 \leq d < \lambda$, we decompose the $d$th EG circulant $\mathbf{G}^{(d)}$ into the following $\ell \times k$ array of circulants of size $(q^m - 1) \times (q^m - 1)$:

$$\mathbf{H}_{\text{array,decom}}^{(d)}(\ell, k) = \begin{bmatrix} \mathbf{G}_{0,0}^{(d)} & \mathbf{G}_{0,1}^{(d)} & \cdot & \mathbf{G}_{0,k-1}^{(d)} \\ \mathbf{G}_{1,0}^{(d)} & \mathbf{G}_{1,1}^{(d)} & \cdot & \mathbf{G}_{1,k-1}^{(d)} \\ \vdots & \vdots & \vdots & \vdots \\ \mathbf{G}_{\ell-1,0}^{(d)} & \mathbf{G}_{\ell-1,1}^{(d)} & \cdot & \mathbf{G}_{\ell-1,k-1}^{(d)} \end{bmatrix}. \tag{12.18}$$

Replacing each circulant in $\mathbf{H}_{\text{EG}}(1, \lambda)$ by its $\ell \times k$ array-decomposition, we obtain the following $\ell \times \lambda k$ array of circulants of size $(q^m - 1) \times (q^m - 1)$:

$$\mathbf{H}_{\text{array,decom}}(\ell, \lambda k) = \left[ \mathbf{H}_{\text{array,decom}}^{(0)}(\ell, k) \ \mathbf{H}_{\text{array,decom}}^{(1)}(\ell, k) \cdots \mathbf{H}_{\text{array,decom}}^{(\lambda-1)}(\ell, k) \right], \tag{12.19}$$

which is an $\ell(q^m - 1) \times \lambda k(q^m - 1)$ matrix and satisfies the RC-constraint. In array-decomposition, all the $\lambda$ EG circulants in $\mathbf{H}_{\text{EG}}(1, \lambda)$ can be decomposed using the same designed set of column- or row-splitting weights or different designed sets of column- or row-splitting weights.

Let $c$ and $t$ be two factors of $q$. For $0 \leq d < \lambda$, suppose we choose the $t$ column-splitting weights $w_0, w_1, \ldots, w_{t-1}$, each equaling $c$ and the $c$ row-splitting weights $w_{0,j}, w_{1,j}, \ldots, w_{c-1,j}$, each equaling 1, i.e., $W_{\text{col}} = \{c, c, \ldots, c\}$ and $W_{\text{row},j} = \{1, 1, \ldots, 1\}$ for $0 \leq j < t$. Decompose each circulant $\mathbf{G}^{(j)}$ in $\mathbf{H}_{\text{EG}}(1, \lambda)$ with the above choices of $t$, $c$, $W_{\text{col}}$, and $W_{\text{row},j}$. The array-decomposition of $\mathbf{H}_{\text{EG}}(1, \lambda)$ is a $c \times \lambda t$ array $\mathbf{H}_{\text{array,decom}}(c, \lambda t)$ of CPMs of size $(q^m - 1) \times (q^m - 1)$, which is called the $c \times t$ *CPM-decomposition* of $\mathbf{H}_{\text{EG}}(1, \lambda)$. A special case is $t = 1$ and $c = q$. In this case, $\mathbf{H}_{\text{array,decom}}(q, \lambda)$ is a $q \times \lambda$ array of CPMs of size $(q^m - 1) \times (q^m - 1)$. The null space over GF(2) of a sub-array of $\mathbf{H}_{\text{array,decom}}(c, \lambda t)$ (or $\mathbf{H}_{\text{array,decom}}(q, \lambda)$) gives a QC-EG-LDPC code $C_{\text{EG,qc,decom}}$ whose Tanner graph has girth at least 6.

**Example 12.16** Consider the subgeometry $\text{EG}^*(3, 2^3)$ of the three-dimensional Euclidean geometry $\text{EG}(3, 2^3)$ over $\text{GF}(2^3)$. The subgeometry $\text{EG}^*(3, 2^3)$ consists of 511 nonorigin points and 4599 lines not passing through the origin of $\text{EG}(3, 2^3)$. The 4599 lines in $\text{EG}^*(3, 2^3)$ can be partitioned into nine cyclic classes. Based on these cyclic classes of lines, we can construct nine circulants of size $511 \times 511$, denoted by $\mathbf{G}^{(0)}, \mathbf{G}^{(1)}, \ldots, \mathbf{G}^{(8)}$, each with weight 8. These nine circulants satisfy the pair-wise RC-constraint. Take eight of these nine circulants and form the following $511 \times 4088$ matrix,

$$\mathbf{H}_{\text{EG}}(1, 8) = \left[ \mathbf{G}^{(0)} \ \mathbf{G}^{(1)} \ \cdots \ \mathbf{G}^{(7)} \right].$$

Choose column-splitting and row-splitting factors $t = c = 2$. Suppose we use $W_{\text{col}} = \{4, 4\}$ and $W_{\text{row},j} = \{2, 2\}$ for $0 \leq j < 2$ as the column-splitting and row-splitting weight sets, respectively. For $0 \leq d \leq 7$, decomposing each circulant

$\mathbf{G}^{(d)}$ in $\mathbf{H}_{\mathrm{EG}}(1, 8)$ based on $t = c = 2$, $W_{\mathrm{col}} = \{4, 4\}$, and $W_{\mathrm{row},j} = \{2, 2\}$ for $0 \leq j < 2$, we obtain the following $2 \times 2$ array of circulants of size $511 \times 511$:

$$\mathbf{H}^{(d)}_{\mathrm{array,decom}}(2, 2) = \begin{bmatrix} \mathbf{G}^{(d)}_{0,0} & \mathbf{G}^{(d)}_{0,1} \\ \mathbf{G}^{(d)}_{1,0} & \mathbf{G}^{(d)}_{1,1} \end{bmatrix}.$$

The weight of each circulant in $\mathbf{H}^{(d)}_{\mathrm{array,decom}}(2, 2)$ is 2. Replacing each circulant in $\mathbf{H}_{\mathrm{EG}}(1, 8)$ by its $2 \times 2$ array-decomposition, we obtain a $2 \times 16$ array of circulants of size $511 \times 511$:

$$\mathbf{H}_{\mathrm{array,decom}}(2, 16) = \left[ \mathbf{H}^{(0)}_{\mathrm{array,decom}}(2, 2) \; \mathbf{H}^{(1)}_{\mathrm{array,decom}}(2, 2) \cdots \mathbf{H}^{(7)}_{\mathrm{array,decom}}(2, 2) \right].$$

The array $\mathbf{H}_{\mathrm{array,decom}}(2, 16)$ is a $1022 \times 8176$ matrix with column and row weights 4 and 32, respectively. It has two redundant rows. The null space over $\mathrm{GF}(2)$ of $\mathbf{H}_{\mathrm{array,decom}}(2, 16)$ gives a $(8176, 7156)$ QC-EG-LDPC code $C_{\mathrm{EG,qc,decom}}$ with rate 0.8753. The BER and BLER performances of this code over the AWGN channel with 5, 10, and 50 iterations of the MSA scaled with a factor of 0.75 are shown in Fig. 12.14(a). This code is used by NASA-USA in the Landsat Data Continuity and the Interface Region Image Spectrograph missions launched in 2013 and 2014, respectively. An MSA decoder for this code was implemented in hardware. Using this decoder with 15 iterations of the MSA, the BER performance was computed down to $10^{-14}$ without a visible error-floor as shown in Fig. 12.14(b) and at the BER of $10^{-14}$, the code performs 1.5 dB from the Shannon limit. ▲▲

**Example 12.17** Continue Example 12.16. Suppose we set $t = 2$ and $c = 4$ as the column-splitting and row-splitting factors and choose $W_{\mathrm{col}} = \{4, 4\}$ and $W_{\mathrm{row},j} = \{1, 1, 1, 1\}$ for $0 \leq j < 2$ as the set of column-splitting weights and the sets of row-splitting weights, respectively. Take the first four circulants $\mathbf{G}^{(0)}$, $\mathbf{G}^{(1)}$, $\mathbf{G}^{(2)}$, $\mathbf{G}^{(3)}$ and form the following matrix:

$$\mathbf{H}_{\mathrm{EG}}(1, 4) = \left[ \mathbf{G}^{(0)} \; \mathbf{G}^{(1)} \; \mathbf{G}^{(2)} \; \mathbf{G}^{(3)} \right].$$

Decomposing each circulant in $\mathbf{H}_{\mathrm{EG}}(1, 4)$ into a $4 \times 2$ array of circulants based on $t = 2$, $c = 4$, $W_{\mathrm{col}} = \{4, 4\}$, and $W_{\mathrm{row},j} = \{1, 1, 1, 1\}$ for $0 \leq j < 2$, we obtain the following $4 \times 8$ array of CPMs of size $511 \times 511$:

$\mathbf{H}_{\mathrm{array,decom}}(4, 8)$

$$= \left[ \mathbf{H}^{(0)}_{\mathrm{array,decom}}(4, 2) \; \mathbf{H}^{(1)}_{\mathrm{array,decom}}(4, 2) \; \mathbf{H}^{(2)}_{\mathrm{array,decom}}(4, 2) \; \mathbf{H}^{(3)}_{\mathrm{array,decom}}(4, 2) \right],$$

which is a $2044 \times 4088$ matrix with column and row weights 4 and 8, respectively. Suppose we mask the array $\mathbf{H}_{\mathrm{array,decom}}(4, 8)$ with the masking matrix given by (12.13) (masking in terms of CPMs). We obtain a $4 \times 8$ masked array $\mathbf{H}_{\mathrm{array,decom,mask}}(4, 8)$ which is a $2044 \times 4088$ matrix with column and row weights 3 and 6, respectively. The null space over $\mathrm{GF}(2)$ of $\mathbf{H}_{\mathrm{array,decom,mask}}(4, 8)$

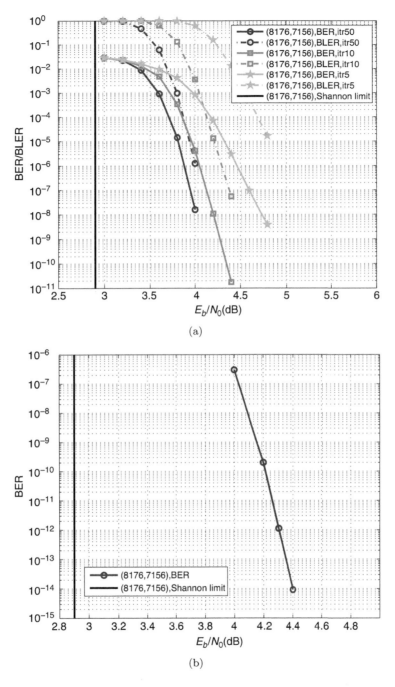

Figure 12.14 The BER and BLER performances of the $(8176, 7156)$ QC-EG-LDPC code given in Example 12.16: (a) MSA and (b) hardware MSA decoder.

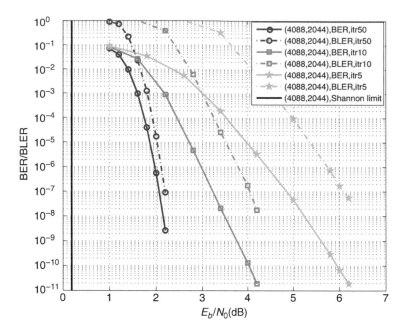

Figure 12.15 The BER and BLER performances of the $(4088, 2044)$ QC-EG-LDPC code given in Example 12.17.

gives a $(3, 6)$-regular $(4088, 2044)$ QC-EG-LDPC code $C_{\mathrm{EG,qc,decom,mask}}$ with rate 0.5. The BER and BLER performances of this code decoded with 5, 10, and 50 iterations of the MSA scaled by a factor of 0.75 are shown in Fig. 12.15. ▲▲

**Example 12.18** Continue Example 12.16. Suppose we take all nine circulants constructed in Example 12.16 and form the following matrix:

$$\mathbf{H}_{\mathrm{EG}}(1, 9) = \begin{bmatrix} \mathbf{G}^{(0)} & \mathbf{G}^{(1)} & \cdots & \mathbf{G}^{(8)} \end{bmatrix}.$$

Choose $t = 1$ and $c = 8$, $W_{\mathrm{col}} = \{8\}$, and $W_{\mathrm{row},j} = \{1, 1, 1, 1, 1, 1, 1, 1\}$ for $0 \leq j < 1$. The array-decomposition of $\mathbf{H}_{\mathrm{EG}}(1, 9)$ based on the chosen $t$, $c$, $W_{\mathrm{col}}$, and $W_{\mathrm{row},j}$ results in an $8 \times 9$ array $\mathbf{H}_{\mathrm{array,decom}}(8, 9)$ of CPMs of size $511 \times 511$.

Suppose we choose the first three CPM-row-blocks of $\mathbf{H}_{\mathrm{array,decom}}(8, 9)$ and form a $3 \times 9$ array $\mathbf{H}_{\mathrm{array,decom}}(3, 9)$ of CPMs. The array $\mathbf{H}_{\mathrm{array,decom}}(3, 9)$ is a $1533 \times 4599$ matrix with column and row weights 3 and 9, respectively. The null space over GF(2) of $\mathbf{H}_{\mathrm{array,decom}}(3, 9)$ is a $(3, 9)$-regular $(4599, 3068)$ QC-EG-LDPC code with rate 0.6671. The BER and BLER performances of this code over the AWGN channel decoded with 5, 10, and 50 iterations of the MSA scaled by a factor of 0.75 are shown in Fig. 12.16. With 50 iterations of the MSA, the code achieves a BER of $5 \times 10^{-10}$ without a visible error-floor. ▲▲

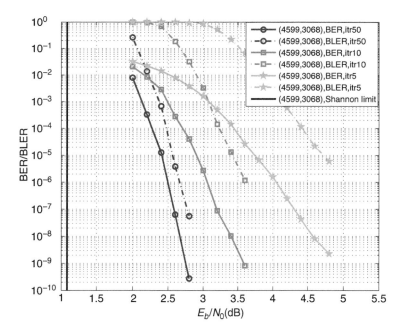

Figure 12.16 The BER and BLER performances of the $(4599, 3068)$ QC-EG-LDPC code given in Example 12.18.

Pair-wise RC-constrained circulants can also be constructed based on the lines in an $m$-dimensional projective geometry $\mathrm{PG}(m, q)$ over $\mathrm{GF}(q)$ as shown in Section 7.6. For even $m$, the lines in $\mathrm{PG}(m, q)$ can be partitioned into $K_{\mathrm{PG}}^{(1)} = (q^m - 1)/(q^2 - 1)$ cyclic classes. Based on these cyclic classes of lines, $K_{\mathrm{PG}}^{(1)}$ pair-wise RC-constrained circulants of size $(q^{m+1} - 1)/(q - 1)$ can be formed. For odd $m$, $\mathrm{PG}(m, q)$ consists of $K_{\mathrm{PG}}^{(2)} = q(q^{m-1} - 1)/(q^2 - 1)$ cyclic classes of lines. Hence, $K_{\mathrm{PG}}^{(2)}$ pair-wise RC-constrained circulants of size $(q^{m+1} - 1)/(q - 1)$ can be formed. With array-decomposition of these circulants, we can construct QC-PG-LDPC codes of various lengths and rates. Circulant decomposition is very fruitful in constructing finite-geometry-based QC-LDPC codes.

## 12.8  A Complexity-Reduced Iterative Algorithm for Decoding QC-FG-LDPC Codes

Section 12.2 presented a reduced-complexity iterative decoding scheme, called RID, to decode cyclic-FG-LDPC codes. The RID scheme cannot be applied to QC-LDPC codes because their parity-check matrices do not have row-block cyclic structure. QC-LDPC codes have parity-check arrays with CPM-structure, i.e., their parity-check matrices, $\mathbf{H}_{qc}$, are arrays of CPMs and/or ZMs. To acquire the row-block cyclic structure, *specific column and row permutations* must be performed to transform the array $\mathbf{H}_{qc}$ into another array $\mathbf{H}_{perm}$ of submatrices of the same size [34]. The permuted array $\mathbf{H}_{perm}$ possesses row-block cyclic structure but its nonzero submatrices are *no longer* CPMs. Consequently, in the hardware implementation of an RID-decoder based on the row-block cyclic structure of $\mathbf{H}_{perm}$, the *advantage of simple wire routing* due to the CPM-structure of the parity-check array $\mathbf{H}_{qc}$ [40] is lost.

In the following, we present a variance of the RID scheme, called *CPM-RID scheme* [41–43], for QC-LDPC codes whose parity-check arrays possess the CPM-structure. The CPM-RID scheme is simply devised based on the CPM-structure of the parity-check array $\mathbf{H}_{qc}$ of a QC-LDPC code. This CPM-RID scheme not only has the *same advantage* in reduction of decoding hardware complexity as the RID scheme given in Section 12.2, but also maintains the *advantage of simple wire routing* due to the CPM-structure. Because QC-FG-LDPC codes are QC-LDPC codes, the CPM-RID scheme can be applied to decode this class of QC-LDPC codes to reduce the decoding complexity. In the following, we present this decoding scheme in general for QC-LDPC codes.

Let $\mathbf{H}_{qc}$ be an $m \times n$ array of CPMs and/or ZMs of size $r \times r$. Let $\mathbf{H}_0$, $\mathbf{H}_1, \ldots, \mathbf{H}_{m-1}$ denote the $m$ CPM-row-blocks of $\mathbf{H}_{qc}$, each consisting of $n$ CPMs and/or ZMs of size $r \times r$. For $0 \leq k < m$ and $0 \leq s < r$, let $\mathbf{h}_{k,s} = [\mathbf{h}_{k,s,0}, \mathbf{h}_{k,s,1}, \ldots, \mathbf{h}_{k,s,n-1}]$ be the $s$th row in the $k$th row-block $\mathbf{H}_k$ which consists of $n$ sections, $\mathbf{h}_{k,s,0}, \mathbf{h}_{k,s,1}, \ldots, \mathbf{h}_{k,s,n-1}$, each containing $r$ components. Each section is the $s$th row of either a CPM or a ZM in the $k$th row-block $\mathbf{H}_k$. If we cyclically shift all the $n$ sections of $\mathbf{h}_{k,s}$ simultaneously one place to the right *within the sections*, we obtain the $(s+1)$th row $\mathbf{h}_{k,s+1}$ of $\mathbf{H}_k$ which also consists of $n$ sections. Each section of $\mathbf{h}_{k,s+1}$ is the $(s+1)$th row of a CPM or a ZM in the $k$th row-block $\mathbf{H}_k$. The above cyclic-shift within each section of $\mathbf{h}_{k,s}$ is referred to as *the section-wise cyclic-shift* (see Section 4.13) of the row $\mathbf{h}_{k,s}$. For $s = r - 1$, the section-wise cyclic-shift of $\mathbf{h}_{k,r-1}$ (the last row of $\mathbf{H}_k$) results in the top row $\mathbf{h}_{k,0}$ of $\mathbf{H}_k$. Consequently, all the rows of the $k$th row-block $\mathbf{H}_k$ can be obtained by section-wise cyclically shifting the top row $\mathbf{h}_{k,0}$ $r-1$ times. This section-wise cyclic-shift of the rows of $\mathbf{H}_{qc}$ *maintains the CPM-structure* of each row-block of $\mathbf{H}_{qc}$. The top row $\mathbf{h}_{k,0}$ of $\mathbf{H}_k$ is called the *generator* of $\mathbf{H}_k$.

Let $\mathbf{H}(1)$ be an $m \times nr$ matrix which consists of the top rows $\mathbf{h}_{0,0}$, $\mathbf{h}_{1,0}, \ldots$, $\mathbf{h}_{m-1,0}$ of the $m$ CPM-row-blocks $\mathbf{H}_0, \mathbf{H}_1, \ldots, \mathbf{H}_{m-1}$ of the parity-check array $\mathbf{H}_{qc}$. Then, it follows from the section-wise cyclic-shift structure of $\mathbf{H}_{qc}$ that the entire array $\mathbf{H}_{qc}$ can be obtained by section-wise cyclically shifting the rows of $\mathbf{H}(1)$ simultaneously $r - 1$ times, one place to the right at each time.

**Example 12.19** Consider the $5 \times 5$ array $\mathbf{H}_{EG,qc}(2, 2^2)$ given in Example 12.8. We denote this matrix as $\mathbf{H}_{qc}$ which is shown as follows:

$$\mathbf{H}_{qc} = \begin{bmatrix} 1 & 0 & 0 & 0 & 1 & 0 & 1 & 0 & 0 & 0 & 0 & 0 & 0 & 0 & 1 \\ 0 & 1 & 0 & 0 & 0 & 1 & 0 & 1 & 0 & 0 & 0 & 0 & 1 & 0 & 0 \\ 0 & 0 & 1 & 1 & 0 & 0 & 0 & 0 & 1 & 0 & 0 & 0 & 0 & 1 & 0 \\ \hline 1 & 0 & 0 & 1 & 0 & 0 & 0 & 1 & 0 & 1 & 0 & 0 & 0 & 0 & 0 \\ 0 & 1 & 0 & 0 & 1 & 0 & 0 & 0 & 1 & 0 & 1 & 0 & 0 & 0 & 0 \\ 0 & 0 & 1 & 0 & 0 & 1 & 1 & 0 & 0 & 0 & 0 & 1 & 0 & 0 & 0 \\ \hline 0 & 0 & 0 & 1 & 0 & 0 & 1 & 0 & 0 & 0 & 1 & 0 & 1 & 0 & 0 \\ 0 & 0 & 0 & 0 & 1 & 0 & 0 & 1 & 0 & 0 & 0 & 1 & 0 & 1 & 0 \\ 0 & 0 & 0 & 0 & 0 & 1 & 0 & 0 & 1 & 1 & 0 & 0 & 0 & 0 & 1 \\ \hline 0 & 1 & 0 & 0 & 0 & 0 & 1 & 0 & 0 & 1 & 0 & 0 & 0 & 1 & 0 \\ 0 & 0 & 1 & 0 & 0 & 0 & 0 & 1 & 0 & 0 & 1 & 0 & 0 & 0 & 1 \\ 1 & 0 & 0 & 0 & 0 & 0 & 0 & 0 & 1 & 0 & 0 & 1 & 1 & 0 & 0 \\ \hline 0 & 0 & 1 & 0 & 1 & 0 & 0 & 0 & 0 & 1 & 0 & 0 & 1 & 0 & 0 \\ 1 & 0 & 0 & 0 & 0 & 1 & 0 & 0 & 0 & 0 & 1 & 0 & 0 & 1 & 0 \\ 0 & 1 & 0 & 1 & 0 & 0 & 0 & 0 & 0 & 0 & 0 & 1 & 0 & 0 & 1 \end{bmatrix} = \begin{bmatrix} \mathbf{H}_0 \\ \mathbf{H}_1 \\ \mathbf{H}_2 \\ \mathbf{H}_3 \\ \mathbf{H}_4 \end{bmatrix}.$$

The matrix $\mathbf{H}_{qc}$ is a $5 \times 5$ array of CPMs and ZMs of size $3 \times 3$, which consists of five CPM-row-blocks, $\mathbf{H}_0$, $\mathbf{H}_1$, $\mathbf{H}_2$, $\mathbf{H}_3$, $\mathbf{H}_4$, as shown above. Taking the first rows from the five CPM-row-blocks of $\mathbf{H}_{qc}$, we obtain a $5 \times 15$ matrix $\mathbf{H}(1)$:

$$\mathbf{H}(1) = \begin{bmatrix} 1 & 0 & 0 & 0 & 1 & 0 & 1 & 0 & 0 & 0 & 0 & 0 & 0 & 0 & 1 \\ 1 & 0 & 0 & 1 & 0 & 0 & 0 & 1 & 0 & 1 & 0 & 0 & 0 & 0 & 0 \\ 0 & 0 & 0 & 1 & 0 & 0 & 1 & 0 & 0 & 0 & 1 & 0 & 1 & 0 & 0 \\ 0 & 1 & 0 & 0 & 0 & 0 & 1 & 0 & 0 & 1 & 0 & 0 & 0 & 1 & 0 \\ 0 & 0 & 1 & 0 & 1 & 0 & 0 & 0 & 0 & 1 & 0 & 0 & 1 & 0 & 0 \end{bmatrix}.$$

Section-wise cyclically shifting $\mathbf{H}(1)$ one position to the right results in the following $5 \times 15$ matrix:

$$\mathbf{H}^{(1)}(1) = \begin{bmatrix} 0 & 1 & 0 & 0 & 0 & 1 & 0 & 1 & 0 & 0 & 0 & 0 & 1 & 0 & 0 \\ 0 & 1 & 0 & 0 & 1 & 0 & 0 & 0 & 1 & 0 & 1 & 0 & 0 & 0 & 0 \\ 0 & 0 & 0 & 1 & 0 & 0 & 1 & 0 & 0 & 0 & 1 & 0 & 1 & 0 \\ 0 & 0 & 1 & 0 & 0 & 0 & 0 & 1 & 0 & 0 & 1 & 0 & 0 & 0 & 1 \\ 1 & 0 & 0 & 0 & 0 & 1 & 0 & 0 & 0 & 0 & 1 & 0 & 0 & 1 & 0 \end{bmatrix},$$

and one more position to the right, we obtain the following $5 \times 15$ matrix:

$$\mathbf{H}^{(2)}(1) = \begin{bmatrix} 0 & 0 & 1 & 1 & 0 & 0 & 0 & 0 & 1 & 0 & 0 & 0 & 0 & 1 & 0 \\ 0 & 0 & 1 & 0 & 0 & 1 & 1 & 0 & 0 & 0 & 0 & 1 & 0 & 0 & 0 \\ 0 & 0 & 0 & 0 & 1 & 0 & 0 & 1 & 1 & 0 & 0 & 0 & 0 & 1 \\ 1 & 0 & 0 & 0 & 0 & 0 & 0 & 1 & 0 & 0 & 1 & 1 & 0 & 0 \\ 0 & 1 & 0 & 1 & 0 & 0 & 0 & 0 & 0 & 0 & 1 & 0 & 0 & 1 \end{bmatrix}.$$

Obviously, the matrix $\mathbf{H}(1)$ and its two section-wise cyclic-shifts $\mathbf{H}^{(1)}(1)$ and $\mathbf{H}^{(2)}(1)$ cover all the rows in the array $\mathbf{H}_{qc}$, i.e., $\mathbf{H}_{qc}$ can be obtained by section-wise cyclically shifting $\mathbf{H}(1)$ 2 times. Section-wise cyclically shifting $\mathbf{H}^{(2)}(1)$ one position to the right, we obtain $\mathbf{H}(1)$. ▲▲

This section-wise cyclic-shift structure allows us to decode the QC-LDPC code given by the null space of the parity-check array $\mathbf{H}_{qc}$ based on the sub-matrix $\mathbf{H}(1)$ *alone* in a revolving manner similar to the RID scheme given in Section 12.2.

Each decoding iteration based on $\mathbf{H}(1)$ is called a *decoding subiteration*. At the end of each decoding subiteration, the reliabilities of the received symbols are updated with a chosen reliability updating algorithm, such as the SPA [3] or the MSA [4]. Then, the reliability vector ($nr$ components in $n$ sections) and the soft-decision received vector ($nr$ symbols in $n$ sections) are *section-wise cyclically shifted to the left by one position* and used as the input information to carry out the next decoding subiteration based on $\mathbf{H}(1)$. It is clear that any $r$ consecutive decoding subiterations performed based on $\mathbf{H}(1)$ are *equivalent to one decoding iteration based on the entire parity-check array* $\mathbf{H}_{qc}$. At the end of each decoding subiteration, the syndrome $\mathbf{s}$ of the hard-decision of the received vector (after section-wise cyclically shifting one position to the right) is computed based on the entire parity-check array $\mathbf{H}_{qc}$. If $\mathbf{s} = \mathbf{0}$, we stop the decoding process; otherwise, we continue the decoding process until a preset maximum number of decoding subiterations is reached. The decoding process simply revolves around $\mathbf{H}(1)$ iteratively similar to the RID scheme. Because the scheme is devised based on the CPM-structure (or section-wise cyclic-shift structure) of the parity-check array $\mathbf{H}_{qc}$, we call it the *CPM-RID scheme* [41–45]. The submatrix $\mathbf{H}(1)$ is called the *decoding matrix* of the CPM-RID scheme.

The advantage of the CPM-RID scheme over the RID scheme devised in Section 12.2 is that *no column and row permutation* of the parity-check array $\mathbf{H}_{qc}$ are required. Consequently, the CPM-structure of the parity-check matrix array $\mathbf{H}_{qc}$ is preserved and the simple wire routing advantage due to the CPM-structure of the parity-check array [40] is maintained.

Based on the structure of $\mathbf{H}(1)$, we can easily see that the number of rows and the number of 1-entries in $\mathbf{H}(1)$ are $1/r$th of those in $\mathbf{H}_{qc}$. Therefore, implementing a CPM-RID decoder of a QC-LDPC code, the number of CNPUs and the number of wires required to connect the CNPUs and VNPUs are reduced by *a factor of* $r$ of those required in implementing a decoder based on the entire parity-check array $\mathbf{H}_{qc}$ of the code. For large $r$, there is a tremendous reduction in hardware implementation complexity using the CPM-RID scheme. Furthermore, the decoder still enjoys the simple wire routing advantage due to the CPM-structure.

Any known reliability-updating algorithm can be incorporated with the CPM-RID scheme to decode a QC-LDPC code. In this section, we use the MSA for updating the reliabilities of the received symbols in each decoding subiteration. The incorporation of the MSA into the CPM-RID scheme can be achieved using exactly the same approach as in the RID scheme described in Section 12.2. This combination of the CPM-RID and the MSA is called CPM-RID-MSA (CPM-RMSA). In updating the reliabilities of the received symbols, there is a major difference between the CPM-RMSA and the RMSA. This will be explained later.

Let $\mathbf{y} = (y_0, y_1, \ldots, y_{nr-1})$ be the soft-decision received vector at the output of the receiver detector. We assume that the symbols of a codeword are transmitted over a binary AWGN channel with BPSK signaling. Let $I_{\max}$ be the maximum number of iterations to be performed based on the entire parity-check array $\mathbf{H}_{qc}$, each iteration consisting of $r$ decoding subiterations based on the $m \times nr$ submatrix $\mathbf{H}(1) = [h_{t,j}]_{0 \le t < m, 0 \le j < nr}$. Therefore, performing $I_{\max}$ iterations based on $\mathbf{H}_{qc}$ is equivalent to performing $I_{\max}r$ decoding subiterations based on $\mathbf{H}(1)$. We label the iterations from 0 to $I_{\max} - 1$. For each iteration, we label its subiterations from 0 to $r - 1$. We use the symbols $k$ and $i$ to denote the $k$th subiteration of the $i$th iteration, respectively. For $0 \le j < nr$, $0 \le t < m$ and $0 \le k < r$, let $L_{t \to j}^{(k)}$ denote the message sent from the $t$th CN to the $j$th VN in the Tanner graph of $\mathbf{H}(1)$ in the $k$th subiteration of each iteration. For $0 \le j < nr$ and $0 \le t < m$, let $N(j) = \{t : 0 \le t < m, h_{t,j} = 1\}$ and $M(t) = \{j : 0 \le j < nr, h_{t,j} = 1\}$. Let $\mathbf{R} = (R_0, R_1, \ldots, R_{nr-1})$ denote the reliability information vector. Let $\lambda_{\text{atten}}$ denote the attenuation factor for a scaled MSA-based reliability information updating.

With the above definitions and notations, the CPM-RMSA is formulated in Algorithm 12.2.

Here, we point out the major difference between the above CPM-RMSA and the RMSA in Section 12.2. Using the RMSA, each VN sends the same message to all its adjacent CNs in the Tanner graph of $\mathbf{H}(1)$ *without subtracting the CN message received at the end of the previous iteration*. This may result in noticeable performance degradation compared with the decoding based on the entire parity-check array $\mathbf{H}_{qc}$ using the conventional MSA if the *row-redundancy* of $\mathbf{H}_{qc}$ is small. In the CPM-RMSA, each VN still sends the same message to all its adjacent CNs. However, the message sent from a VN is the difference between the updated reliability message at the VN and the sum of the messages previously received from all its adjacent CNs (see (12.20)). Because the previous CN messages are removed from the updated reliability messages, the CPM-RMSA gives a better error performance than RMSA for decoding QC-LDPC codes whose parity-check arrays have small row redundancy.

Although the CPM-RID scheme based on $\mathbf{H}(1)$ greatly reduces the hardware complexity of the decoder of a QC-LDPC code, it increases the decoding latency. To reduce the decoding latency, we can increase the size of the decoding matrix. Let $\ell$ be a factor of $r$ and $r = c\ell$. (In the case where $\ell$ is not a factor of $r$, we choose $c = \lceil r/\ell \rceil$.) Suppose we take the top $\ell$ rows of each CPM-row-block of $\mathbf{H}_{qc}$ and form an $m\ell \times nr$ submatrix, denoted by $\mathbf{H}(\ell)$, of $\mathbf{H}_{qc}$. Then, it follows from the section-wise cyclic-shift structure that if we section-wise cyclically shift all the rows of $\mathbf{H}(\ell)$ simultaneously $c$ times, each time to the right $\ell$ positions, we obtain the entire parity-check array $\mathbf{H}_{qc}$. (In the case where $\ell$ is not a factor of $r$, section-wise cyclically shifting all the rows of $\mathbf{H}(\ell)$ simultaneously $c$ times will cover all the rows in $\mathbf{H}_{qc}$ and some rows in $\mathbf{H}_{qc}$ will be covered more than once.) Based on this structure, we can use $\mathbf{H}(\ell)$ as the matrix for decoding the QC-LDPC code given by the null space of $\mathbf{H}_{qc}$ in a revolving manner similar to the decoding based on $\mathbf{H}(1)$ described above. After each subiteration decoding based on $\mathbf{H}(\ell)$, the reliability vector must be section-wise cyclically shifted to the left $\ell$ positions for the next subiteration decoding. The CPM-RID decoding scheme based on $\mathbf{H}(\ell)$ reduces the decoder hardware implementation complexity by a factor of $c$ which is less than $r$, but it reduces the decoding latency. Decoding based on $\mathbf{H}(\ell)$ gives a tradeoff between the decoding complexity and latency.

---

**Algorithm 12.2** CPM-RMSA

---

**Step 1.** (**Initialization**) For $0 \leq j < nr$, set $R_j = y_j$. For $0 \leq j < nr$, $t \in N(j)$ and $0 \leq k < r$, set $L_{t \to j}^{(k)} = 0$. Set $i = 0$.

**Step 2.** (**Iteration**) Set $k = 0$. Carry out the $k$th subiteration in the $i$th iteration as follows.
(1) Update the reliability information vector $\mathbf{R}$ as follows: for $0 \leq j < nr$,

$$R_j \leftarrow R_j - \sum_{t \in N(j)} L_{t \to j}^{(k)}, \tag{12.20}$$

and use the updated $R_j$ as the outgoing VN to CN messages.
(2) For $0 \leq j < nr$ and $t \in N(j)$, update the outgoing CN to VN messages $L_{t \to j}^{(k)}$ as follows:

$$L_{t \to j}^{(k)} = \lambda_{\text{atten}} \left[ \prod_{j' \in M(t) \backslash j} \text{sign}\,(R_{j'}) \right] \times \left[ \min_{j' \in M(t) \backslash j} \{|R_{j'}|\} \right]. \tag{12.21}$$

(3) Update the reliability information $\mathbf{R}$ as follows: for $0 \leq j < nr$,

$$R_j \leftarrow R_j + \sum_{t \in N(j)} L_{t \to j}^{(k)}. \tag{12.22}$$

(4) Form the hard-decision vector $\mathbf{z}$ based on $\mathbf{R}$; section-wise cyclically shift $\mathbf{z}$ by $k$ positions to the right to obtain a shifted decision vector $\mathbf{z}^{(k)}$ and compute its syndrome $\mathbf{s} = \mathbf{z}^{(k)} \cdot \mathbf{H}^T$. If $\mathbf{s} = \mathbf{0}$, go to **Step 4**; otherwise, go to **Step 2**(5).
(5) Section-wise cyclically shift $\mathbf{R}$ one position to the left. If $k < r - 1$, set $k \leftarrow k + 1$ and return to **Step 2**(1) to begin another subiteration; otherwise, go to **Step 3**.

**Step 3.** If $i = I_{\max} - 1$, stop the decoding and declare a decoding failure; otherwise, set $i \leftarrow i + 1$ and return to **Step 2**.

**Step 4.** Output $\mathbf{z}^{(k)}$ as the decoded vector.

---

The CPM-RID scheme can also be applied to decode nonbinary QC-LDPC codes [44, 45].

**Example 12.20** Consider the $(4, 32)$-regular $(2016, 1779)$ QC-EG-LDPC code $C_{\text{EG,qc}}(4, 32)$ constructed in Example 12.13. The BER and BLER performances of the code over the AWGN channel decoded with 50 iterations of the MSA are show in Fig. 12.11. Its parity-check matrix $\mathbf{H}_{\text{qc}} = \mathbf{H}_{\text{EG,qc}}(4, 32)$ is a $4 \times 32$ array of CPMs of size $63 \times 63$ which is a $252 \times 2016$ matrix over GF(2). $\mathbf{H}_{\text{qc}}$ consists of four CPM-row-blocks and 32 CPM-column-blocks.

In this example, we apply the proposed CPM-RMSA to decode the code $C_{\text{EG,qc}}(4, 32)$. Set $\ell = 1$. Taking the first $\ell = 1$ row from each CPM-row-block of

$\mathbf{H}_{qc}$, we obtain a $4 \times 2016$ matrix $\mathbf{H}(1)$. Section-wise cyclically shifting $\mathbf{H}(1)$ 62 times, each time one place to the right, we cover all the rows of $\mathbf{H}_{qc}$. Using $\mathbf{H}(1)$ as the decoding matrix to decode the code $C_{EG,qc}(4, 32)$, the number of CNPUs is only four which is $1/63$ of those required by using the entire parity-check matrix $\mathbf{H}_{qc}$. One iteration of the MSA based on $\mathbf{H}_{qc}$ is equivalent to 63 subiterations based on the decoding matrix $\mathbf{H}(1)$. The BER and BLER performances of the code over the AWGN channel decoded based on $\mathbf{H}(1)$ with 315, 630, and 3150 subiterations of the CPM-RMSA (equivalent to 5, 10, and 50 iterations of MSA based on $\mathbf{H}_{qc}$) are shown in Fig. 12.17(a). Also included in this figure is the performance of the code decoded with 50 iterations of the MSA based on the entire matrix $\mathbf{H}_{qc}$. From Fig. 12.17(a), we see that the performance curves of the CPM-RMSA based on the submatrix $\mathbf{H}(1)$ with 3150 subiterations overlap with those decoded with the MSA based on the entire parity-check matrix $\mathbf{H}_{qc}$.

Suppose we choose $\ell = 3$. Take three rows (the top three rows) from the four CPM-row-block of $\mathbf{H}_{qc}$ to form a $12 \times 2016$ matrix $\mathbf{H}(3)$. Using $\mathbf{H}(3)$ as the decoding matrix, the number of CNPUs is $1/21$ of those required based on the entire parity-check matrix $\mathbf{H}_{qc}$. The BER and BLER performances of the code decoded based on $\mathbf{H}(3)$ with 105, 210, and 1050 subiterations of the CPM-RMSA are shown in Fig. 12.17(b). Also included in this figure is the performance of the code decoded with 50 iterations of the MSA based on the entire parity-check matrix $\mathbf{H}_{qc}$. From Fig. 12.17(b), we see that the performance curves of the CPM-RMSA based on the submatrix $\mathbf{H}(3)$ with 1050 subiterations overlap with those decoded with the MSA based on the entire parity-check matrix $\mathbf{H}_{qc}$. ▲▲

## 12.9 Remarks

An $(n, k)$ cyclic finite-geometry LDPC code $C_{FG}$ constructed based on a line-point incidence matrix of a finite geometry can be used to correct a single burst of erasures. To achieve this, we first determine its parity-check polynomial $\mathbf{h}(X)$ and then form a circulant parity-check matrix in the form of (10.34). Based on this circulant parity-check matrix, the code is capable of correcting any single random burst of erasures of length $n - k$ or less using the SPL-algorithm as described in Section 10.9.1. Hence, a cyclic finite-geometry code can be used to correct three types of errors: (1) random errors using an IDBP-algorithm; (2) random erasures based on its orthogonal structure; and (3) single burst of erasures based on the zero-span of its circulant parity-check matrix.

Finite-geometry LDPC codes presented in this chapter are constructed based on the lines and points in finite geometries. In [9], it is shown that finite-geometry LDPC codes can be constructed based on flats of two consecutive dimensions [9] in finite geometries with dimensions greater than 2. With this construction, a large class of finite-geometry LDPC codes can be constructed. The codes constructed are in general not cyclic but can be put in quasi-cyclic form.

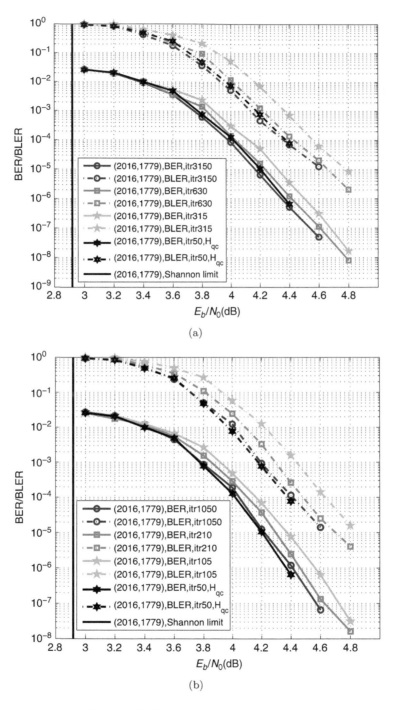

Figure 12.17 The BER and BLER performances of the $(2016, 1779)$ QC-EG-LDPC code $C_{\text{EG,qc}}(4, 32)$ given in Example 12.20 decoded with the CPM-RMSA of different sizes of decoding matrices: (a) $\ell = 1$ and (b) $\ell = 3$.

# Problems

**12.1** Consider the subgeometry $EG^*(2, 2^4)$ of the two-dimensional Euclidean geometry $EG(2, 2^4)$ over $GF(2^4)$. Construct a cyclic-EG-LDPC code based on this subgeometry by using all the lines.

(a) Calculate the length, dimension, rate, and the minimum distance of the constructed code.

(b) Compute the BER and BLER performances of the constructed code over the AWGN channel by using the SPA with 5, 10, and 50 iterations.

(c) Compute the BER and BLER performances of the constructed code over the AWGN channel by using the scaled MSA with 5, 10, and 50 iterations.

(d) Compute the performance of the constructed code over the BEC with 50 iterations of the SPA.

**12.2** Consider the projective geometry $PG(2, 2^6)$ over $GF(2^6)$. Construct a cyclic-EG-LDPC code based on this geometry by using all the lines.

(a) Calculate the length, dimension, rate, and the minimum distance of the constructed code.

(b) Compute the BER and BLER performances of the constructed code over the AWGN channel by using the SPA with 5, 10, and 50 iterations.

(c) Compute the BER and BLER performances of the constructed code over the AWGN channel by using the scaled MSA with 5, 10, and 50 iterations.

(d) Compute the performance of the constructed code over the BEC with 50 iterations of the SPA.

**12.3** Decode the cyclic-EG-LDPC code constructed in Problem 12.1 over the AWGN channel using the RMSA with different grouping sizes and number of iterations corresponding to 5, 10, and 50 of iterations of the scaled MSA based on the entire parity-check matrix of the code:

$$\text{(a) } \ell = 1, \quad \text{(b) } \ell = 5, \quad \text{(c) } \ell = 85.$$

Analyze the decoding hardware complexity of each chosen grouping size $\ell$ and compare the performances with those obtained in Problem 12.1.

**12.4** Decode the cyclic-PG-LDPC code constructed in Problem 12.2 over the AWGN channel using the RMSA with different grouping sizes and number of iterations corresponding to 5, 10, and 50 iterations of the MSA based on the entire parity-check matrix of the code:

$$\text{(a) } \ell = 1, \quad \text{(b) } \ell = 19, \quad \text{(c) } \ell = 1387.$$

Analyze the decoding hardware complexity of each chosen grouping size $\ell$ and compare the performances with those obtained in Problem 12.2.

**12.5** Consider the subgeometry $EG^*(2, 2^3)$ of $EG(2, 2^3)$ which consists of 63 nonorigin points of $EG(2, 2^3)$ and 63 lines not passing through the origin.

(a) Find a line $L$ in $\text{EG}^*(2, 2^3)$ and the eight points on this line.
(b) Find the incidence vector $\mathbf{g}$ of the line $L$.
(c) Find the circulant $\mathbf{H}_{\text{EG,cyc}}(2, 2^3)$ using the incidence vector $\mathbf{g}$ as its generator.
(d) Find the two permutations $\pi_{\text{row}}$ and $\pi_{\text{col}}$ with $q = 2^3$ in the forms of (12.6) and (12.7), respectively.
(e) Find the array $\mathbf{H}_{\text{EG,qc}}(2, 2^3)$ of CPMs and ZMs by permuting $\mathbf{H}_{\text{EG,cyc}}(2, 2^3)$ based on $\pi_{\text{row}}$ and $\pi_{\text{col}}$.

**12.6** Consider the parity-check matrix $\mathbf{H}_{\text{EG,cyc}}(2, 2^4)$ of the cyclic-EG-LDPC code constructed in Problem 12.1 which is a circulant.
(a) Find the two permutations $\pi_{\text{row}}$ and $\pi_{\text{col}}$ with $q = 2^4$ in the forms of (12.6) and (12.7), respectively.
(b) Find the array $\mathbf{H}_{\text{EG,qc}}(2, 2^4)$ of CPMs and ZMs by permuting $\mathbf{H}_{\text{EG,cyc}}(2, 2^4)$ based on $\pi_{\text{row}}$ and $\pi_{\text{col}}$.
(c) Construct a QC-LDPC code by taking a $4 \times 17$ subarray from $\mathbf{H}_{\text{EG,qc}}(2, 2^4)$ and using it as the parity-check matrix: find the code length, dimension, and rate.
(d) Calculate the BER and BLER performance over the AWGN channel of the constructed QC-LDPC code by using 5, 10, and 50 iterations of the scaled MSA.

**12.7** Following the technique described in Section 12.5, find the base matrix $\mathbf{B}_{\text{EG,qc}}(9, 9)$ of the subgeometry $\text{EG}^*(2, 2^3)$ in Problem 12.5.

**12.8** Consider the $65 \times 65$ matrix $\mathbf{B}_{\text{EG,qc}}(65, 65)$ over $\text{GF}(2^6)$ constructed on Example 12.13. Let $\gamma = 4$ and $\rho = 8$. Take a $4 \times 8$ submatrix $\mathbf{B}_{\text{EG,qc}}(4, 8)$ from $\mathbf{B}_{\text{EG,qc}}(65, 65)$ (avoiding the 0-entries). Construct a regular QC-EG-LDPC code by using $\mathbf{B}_{\text{EG,qc}}(4, 8)$ as a base matrix and its CPM-dispersion. Compute the BER and BLER performances of the constructed code over the AWGN channel by 5, 10, and 50 iterations of the scaled MSA.

**12.9** Consider the $258 \times 258$ base matrix $\mathbf{B}_{\text{EG,qc}}(258, 258)$ over $\text{GF}(257)$ for the subgeometry $\text{EG}^*(2, 257)$ of $\text{EG}(2, 257)$ constructed in Example 12.14. Take a $4 \times 8$ subamtrix $\mathbf{B}_{\text{EG,qc}}(4, 8)$ from $\mathbf{B}_{\text{EG,qc}}(258, 258)$ (avoiding the 0-entries). Use $\mathbf{B}_{\text{EG,qc}}(4, 8)$ as a base matrix and its CPM-dispersion to construct a $(4, 8)$-regular QC-EG-LDPC code.
(a) Find the code length, dimension, and rate.
(b) Compute the BER and BLER performances of the constructed code over the AWGN channel by 5, 10, and 50 iterations of the scaled MSA.

**12.10** Consider the $4 \times 8$ base matrix $\mathbf{B}_{\text{EG,qc}}(4, 8)$ constructed in Problem 12.8. Mask $\mathbf{B}_{\text{EG,qc}}(4, 8)$ with the masking matrix given by (12.13) to obtain a masked matrix $\mathbf{B}_{\text{EG,qc,mask}}(4, 8)$ over $\text{GF}(2^6)$. Construct a QC-EG-LDPC code by using

$\mathbf{B}_{\text{EG,qc,mask}}(4, 8)$ as a base matrix and its CPM-dispersion. Compute the BER and BLER performances of the constructed code over the AWGN channel by 5, 10, and 50 iterations of the scaled MSA.

**12.11** Consider the $2 \times 16$ array $\mathbf{H}_{\text{array,decom}}(2, 16)$ of circulants of size $511 \times 511$ constructed in Example 12.16. Suppose we take the first five $2 \times 2$ array-decomposition, $\mathbf{H}_{\text{array,decom}}^{(0)}(2, 2)$, $\mathbf{H}_{\text{array,decom}}^{(1)}(2, 2), \ldots, \mathbf{H}_{\text{array,decom}}^{(4)}(2, 2)$ and form the following matrix

$$\mathbf{H}_{\text{array,decom}}(2, 10) = \left[ \mathbf{H}_{\text{array,decom}}^{(0)}(2, 2)\ \mathbf{H}_{\text{array,decom}}^{(1)}(2, 2)\ \cdots\ \mathbf{H}_{\text{array,decom}}^{(4)}(2, 2) \right].$$

Use $\mathbf{H}_{\text{array,decom}}(2, 10)$ as a parity-check matrix to construct a QC-EG-LDPC code.

(a) Compute the code length, dimension, and rate.
(b) Compute the BER and BLER performances of the constructed code over the AWGN channel with 50 iterations of the scaled MSA.

**12.12** Show the section-wise cyclic-shift structure of the parity-check array $\mathbf{H}_{\text{EG,qc}}(2, 2^3)$ constructed in Problem 12.5.

**12.13** Decode the $(16\,384, 15\,363)$ QC-EG-LDPC code constructed in Example 12.14 using the CPM-RMSA with number of maximum subiterations corresponding to 5, 10, and 50 iterations based on the entire parity-check matrix of the code and with the following chosen $\ell$s.

$$\text{(a) } \ell = 1, \quad \text{(b) } \ell = 8, \quad \text{(c) } \ell = 128.$$

Analyze the decoding hardware complexity of each chosen grouping size $\ell$ and compare the performances with those obtained in Example 12.14.

# References

[1] Y. Kou, S. Lin, and M. Fossorier, "Low-density parity-check codes based on finite geometries: a rediscovery," *Proc. IEEE Int. Symp. Inf. Theory*, Sorrento, June 25–30, 2000, p. 200.

[2] Y. Kou, S. Lin, and M. Fossorier, "Construction of low-density parity-check codes: a geometric approach," *Proc. 2nd Int. Symp. Turbo Codes and Related Topics*, Brest, September 2000, pp. 137–140.

[3] D. J. C. MacKay, "Good error-correcting codes based on very sparse matrices," *IEEE Trans. Inf. Theory*, **45**(2) (1999), 399–432.

[4] J. Chen and M. P. C. Fossorier, "Near optimum universal belief propagation based decoding of low-density parity-check codes," *IEEE Trans. Commun.*, **50**(3) (2002), 406–614.

[5] Y. Kou, S. Lin, and M. Fossorier, "Low-density parity-check codes based on finite geometries: a rediscovery and new results," *IEEE Trans. Inf. Theory*, **47**(7) (2001), 2711–2736.

[6] Y. Kou, J. Xu, H. Tang, S. Lin, and K. Abdel-Ghaffar, "On circulant low-density parity-check codes," *Proc. IEEE Int. Symp. Inf. Theory*, Lausanne, June 30–July 5, 2002, p. 200.

[7] S. Lin, H. Tang, and Y. Kou, "On a class of finite geometry low-density parity-check codes," *Proc. IEEE Int. Symp. Inf. Theory*, Washington, DC, June 2001, p. 24.

[8] S. Lin, H. Tang, Y. Kou, and K. Abdel-Ghaffar, "Codes on finite geometries," *Proc. IEEE Inf. Theory Workshop*, Cairns, Australia, September 2–7, 2001, pp. 14–16.

[9] H. Tang, *Codes on finite geometries*, Ph.D. dissertation, University of California, Davis, CA, 2002.

[10] H. Tang, Y. Kou, J. Xu, S. Lin, and K. Abdel-Ghaffar, "Codes on finite geometries: old, new, majority-logic and iterative decoding," *Proc. 6th Int. Symp. Commun. Theory and Applications*, Ambleside, UK, July 2001, pp. 381–386.

[11] H. Tang, J. Xu, Y. Kou, S. Lin, and K. Abdel-Ghaffar, "On algebraic construction of Gallager low-density parity-check codes," *Proc. IEEE Int. Symp. Inf. Theory*, Lausanne, June 30–July 5 2002, p. 482.

[12] N. Kamiya and M. P. C. Fossorier, "Quasi cyclic codes from a finite affine plane," *Proc. IEEE Int. Symp. Inf. Theory*, Yokohama, Japan, June 29–July 4, 2003, p. 57.

[13] H. Tang, J. Xu, Y. Kou, S. Lin, and K. Abdel-Ghaffar, "On algebraic construction of Gallager and circulant low-density parity-check codes," *IEEE Trans. Inf. Theory*, **50**(6) (2004), 1269–1279.

[14] H. Tang, J. Xu, S. Lin, and K. Abdel-Ghaffar, "Codes on finite geometries," *IEEE Trans. Inf. Theory*, **51**(2) (2005), 572–596.

[15] Y. Y. Tai, L. Lan, L.-Q. Zeng, S. Lin, and K. Abdel-Ghaffar, "Algebraic construction of quasi-cyclic LDPC codes for AWGN and erasure channels," *IEEE Trans. Commun.*, **54**(10) (2006), 1765–1774.

[16] J. Xu, L. Chen, I. Djurdjevic, S. Lin, and K. Abdel-Ghaffar, "Construction of regular and irregular LDPC codes: geometry decomposition and masking," *IEEE Trans. Inf. Theory*, **53**(1) (2007), 121–134.

[17] L.-Q. Zeng, L. Lan, Y. Y. Tai, *et al.*, "Construction of nonbinary quasi-cyclic LDPC codes: a finite geometry approach," *IEEE Trans. Commun.*, **56**(3) (2008), 378–387.

[18] J. Zhang and M. P. C. Fossorier, "A modified weighted bit-flipping decoding for low-density parity-check codes," *IEEE Commun. Lett.*, **8**(3) (2004), 165–167.

[19] B. Zhou, J.-Y. Kang, Y. Y. Tai, S. Lin, and Z. Ding, "High performance non-binary quasi-cyclic LDPC codes on Euclidean geometries," *IEEE Trans. Commun.*, **57**(5) (2009), 1298–1311.

[20] Q. Diao, Y. Y. Tai, S. Lin, and K. Abdel-Ghaffar, "LDPC codes on partial geometries: construction, trapping sets structure, and puncturing," *IEEE Trans. Inf. Theory*, **59**(12) (2013), 7898–7914.

[21] Z. Liu and D. A. Pados, "A decoding algorithm for finite-geometry LDPC codes," *IEEE Trans. Commun.*, **53**(3) (2005), 415–421.

[22] S.-T. Xia and F.-W. Fu, "On the stopping distance of finite geometry LDPC codes," *IEEE Commun. Lett.*, **10**(5) (2006), 381–383.

[23] X. Wu, M. Jiang, C. Zhao, and X. You, "Fast weighted bit-flipping decoding of finite-geometry LDPC codes," *2006 IEEE Inf. Theory Workshop*, Chengdu, Punta del Este, 2006, pp. 132–134.

[24] X. Jiang and M. H. Lee, "Large girth non-binary LDPC codes based on finite fields and Euclidean geometries," *IEEE Signal Process. Lett.*, **16**(6) (2009), 521–524.

[25] T. M. N. Ngatched, F. Takawira, and M. Bossert, "An improved decoding algorithm for finite-geometry LDPC codes," *IEEE Trans. Commun.*, **57**(2) (2009), 302–306.

[26] Z. Mrabet, F. Ayoub, M. Belkasmi, A. Yatribi, and A. I. Z. El Abidine, "Non-binary Euclidean geometry codes: majority logic decoding," *2016 International Conference on Advanced Communication Systems and Information Security* (ACOSIS), Marrakesh, October 17–19, 2016, pp. 1–7.

[27] M. Nasseri, X. Xiao, S. Zhang, T. Wang, and S. Lin, "Concatenated finite geometry and finite field LDPC codes," *2017 11th International Conference on Signal Processing and Communication Systems* (ICSPCS), Gold Coast, QLD, December 13–15, 2017, pp. 1–8.

[28] S. Lin and D. J. Costello, Jr., *Error Control Coding: Fundamentals and Applications*, 2nd ed., Upper Saddle River, NJ, Prentice-Hall, 2004.

[29] W. E. Ryan and S. Lin, *Channel Codes: Classical and Modern*, New York: Cambridge University Press, 2009.

[30] T. Kasami, S. Lin, and W. W. Peterson, "Polynomial codes," *IEEE Trans. Inf. Theory*, **14**(6) (1968), 807–814.

[31] S. Lin, "On the number of information symbols in polynomial codes," *IEEE Trans. Inf. Theory*, **18**(6) (1972), 785–794.

[32] Q. Diao, Y. Y. Tai, S. Lin, and K. Abdel-Ghaffar, "Trapping set structure of LDPC codes on finite geometries," *Proc. IEEE Inf. Theory Applic. Workshop*, San Diego, CA, February 10–15, 2013, pp. 1–8.

[33] K. Liu, S. Lin, and K. Abdel-Ghaffar, "A revolving iterative algorithm for decoding algebraic quasi-cyclic LDPC codes," *Proc. IEEE Int. Symp. Inform. Theory (ISIT)*, Istanbul, Turkey, July 7–12, 2013, pp. 2656–2660.

[34] K. Liu, S. Lin, and K. Abdel-Ghaffar, "A revolving iterative algorithm for decoding algebraic cyclic and quasi-cyclic LDPC codes," *IEEE Trans. Commun.*, **61**(12) (2013), 4816–4827.

[35] Q. Huang, Q. Diao, S. Lin, and K. Abdel-Ghaffar, "Cyclic and quasi-cyclic LDPC codes on constrained parity-check matrices and their trapping sets," *IEEE Trans. Inf. Theory*, **58**(5) (2012), 2648–2671.

[36] Y. Liu and Y. Li, "Design of masking matrix for QC-LDPC codes," *Proc. IEEE Inf. Theory Workshop*, Seville, Spain, September 9–13, 2013, pp. 1–5.

[37] J. Li, S. Lin, and K. Abdel-Ghaffar, "Improved message-passing algorithm for counting short cycles in bipartite graphs," *Proc. Int. Symp. Info. Theory (ISIT)*, June 2015, pp. 416–420.

[38] J. Li, S. Lin, and K. Abdel-Ghaffar, "Improved trellis-based algorithm for locating and breaking cycles in bipartite graphs with applications to LDPC codes," *Proc. Inf. Theory Applic. Workshop*, San Diego, CA, February 11–16, 2018, pp. 1–8.

[39] L. Chen, J. Xu, I. Djurdjevic, and S. Lin, "Near Shannon limit quasi-cyclic low-density parity-check codes," *IEEE Trans. Commun.*, **52**(7) (2004), 1038–1042.

[40] Y. Chen and K. Parhi, "Overlapped message passing for quasi-cyclic low-density parity-check codes," *IEEE Trans. Circuits and Systems I*, **51**(6) (2004), 1106–1113.

[41] J. Li, K. Liu, S. Lin, and K. Abdel-Ghaffar, "Decoding of quasi-cyclic LDPC codes with section-wise cyclic structure," *Proc. Inf. Theory Applic. Workshop*, San Diego, CA, February 9–14, 2014, pp. 1–10.

[42] S. Lin, K. Liu, J. Li, and K. Abdel-Ghaffar, "A reduced-complexity iterative scheme for decoding quasi-cyclic low-density parity-check codes," *2014 48th Asilomar Conference on Signals, Systems and Computers*, Pacific Grove, CA, November 2–5, 2014, pp. 119–125.

[43] J. Li, K. Liu, S. Lin, and K. Abdel-Ghaffar, "Algebraic quasi-cyclic LDPC codes: construction, low error-floor, large girth and a reduced-complexity decoding scheme," *IEEE Trans. Commun.*, **62**(8) (2014), 2626–2637.

[44] K. Liu, J. Li, S. Lin, and K. Abdel-Ghaffar, "A merry-go-round decoding scheme for non-binary quasi-cyclic LDPC codes," *2014 IEEE Global Commun. Conference*, Austin, TX, December 8–12, 2014, pp. 1497–1503.

[45] J. Li, K. Liu, S. Lin, and K. Abdel-Ghaffar, "A matrix-theoretic approach to the construction of non-binary quasi-cyclic LDPC codes," *IEEE Trans. Commun.*, **63**(4) (2015), 1057–1068.

# 13

# Partial Geometries and Their Associated QC-LDPC Codes

Besides Euclidean and projective geometries, there are other types of finite geometries. One such type is known as *partial geometries*. Akin to Euclidean and projective geometries, partial geometries can be used to construct LDPC codes whose Tanner graphs have similar structural properties as those of Euclidean- and projective-geometry LDPC codes. In this chapter, we present three classes of partial geometries. In the first class, partial geometries are constructed based on *cyclic subgroups of prime orders of finite fields*. In the second class, partial geometries are constructed based on *prime fields*. The last class of partial geometries are derived from a specific class of *balanced incomplete block designs*. Based on these three classes of partial geometries, three classes of structured QC-LDPC codes can be constructed and they perform well with iterative decoding algorithms based on belief-propagation.

## 13.1   CPM-Dispersions of Finite-Field Elements

Let $\mathrm{GF}(q)$ be a finite field with $q$ elements and let $\alpha$ be a primitive element of $\mathrm{GF}(q)$. The powers of $\alpha$, namely, $\alpha^0 = 1, \alpha, \alpha^2, \ldots, \alpha^{q-2}$, give all the nonzero elements of $\mathrm{GF}(q)$. Let $\beta$ be a nonzero element in $\mathrm{GF}(q)$ of order $n$, i.e., $\beta^n = 1$, where $n$ is a factor of $q-1$ greater than 1. Suppose $q-1 = cn$ for some positive integer $c$. Then, $\beta = \alpha^c$. The $n$ elements $\beta^0 = 1, \beta, \beta^2, \ldots, \beta^{n-1}$ form a cyclic subgroup $\mathbf{S}_n$ of the multiplicative group of $\mathrm{GF}(q)$. For $0 \le j < n$, we represent the element $\beta^j$ by a *circulant permutation matrix* (CPM) over GF(2) of size $n \times n$ (with rows and columns labeled from 0 to $n - 1$) whose generator (the top row) has the unit-element "1" of GF($q$) as its single 1-component at position $j$. We denote this CPM by $\text{CPM}_n(\beta^j)$. The representation of the element $\beta^j$ by $\text{CPM}_n(\beta^j)$ is *unique* and the mapping between $\beta^j$ and $\text{CPM}_n(\beta^j)$ is *one-to-one*. This matrix representation of $\beta^j$ is referred to as the $n \times n$ *CPM-dispersion of $\beta^j$ with respect to the cyclic subgroup $\mathbf{S}_n$* (or *n-fold CPM-dispersion*) [1, 2]. We represent the 0-element of GF($q$) by a *zero matrix* (ZM) of size $n \times n$.

Let $l$ be another factor of $q - 1$ which is relatively prime to $n$. Then, $ln$ is a factor of $q - 1$. Let $\delta$ be an element in GF($q$) of order $ln$, i.e., $\delta^{ln} = 1$. Then, $\beta$ can be expressed as the $l$th power of $\delta$, i.e., $\beta = \delta^l$. Let $\mathbf{S}_{ln}$ be the cyclic subgroup of GF($q$) with order $ln$ generated by $\delta$. Then, $\mathbf{S}_{ln}$ contains $\mathbf{S}_n$ as a subgroup. Or, we say that $\mathbf{S}_{ln}$ is a super group of $\mathbf{S}_n$. If we disperse each element in $\mathbf{S}_{ln}$ by a CPM of size $ln \times ln$ as described above, then the element $\beta^j = \delta^{jl}$, $0 \le j < n$, as an element in $\mathbf{S}_{ln}$, is dispersed into a CPM of size $ln \times ln$, $\text{CPM}_{ln}(\beta^j) = \text{CPM}_{ln}(\delta^{jl})$, whose generator has its single 1-component at position $jl$. In this case, every element $\beta^j$ in $\mathbf{S}_n$ is uniquely dispersed into a CPM of size $ln \times ln$ which is referred to as the $ln \times ln$ *CPM-dispersion* (or *ln-fold CPM-dispersion*) *of $\beta^j$ with respect to the super group $\mathbf{S}_{ln}$ of $\mathbf{S}_n$*. Clearly, this CPM-dispersion of each element $\beta^j$, $0 \le j < n$, in $\mathbf{S}_n$ with respect to the super group $\mathbf{S}_{ln}$ is *unique* and the mapping between of $\beta^j$ and $\text{CPM}_{ln}(\beta^j)$ is *one-to-one with respect to $\mathbf{S}_{ln}$*. The integer $ln$ is called the *CPM-dispersion factor* with respect to $\mathbf{S}_{ln}$. Therefore, each element in $\mathbf{S}_n$ can be one-to-one dispersed into a CPM of a size equal to the order of a cyclic subgroup of GF($q$) which contains $\mathbf{S}_n$ as a subgroup (including $\mathbf{S}_n$ itself). The largest CPM-dispersion factor of an element in the cyclic subgroup $\mathbf{S}_n$ is $q - 1$, which is the order of the multiplicative group of GF($q$).

**Example 13.1** Consider the field GF($2^4$) given in Table 2.9, where $\alpha$ is a primitive element of order $2^4 - 1 = 15$. The integer 15 can be factored as the product of 3 and 5. Let $\beta = \alpha^3$. Then, the order of $\beta$ is 5 and the set $\mathbf{S}_5 = \{\beta^0 = 1, \beta, \beta^2, \beta^3, \beta^4\}$ forms a cyclic subgroup of the field GF($2^4$). The CPM-dispersions of the elements in $\mathbf{S}_5$ with respect to $\mathbf{S}_5$ itself are (with columns and rows labeled from 0 to 4):

$$\text{CPM}_5(\beta^0) = \begin{bmatrix} 1&0&0&0&0 \\ 0&1&0&0&0 \\ 0&0&1&0&0 \\ 0&0&0&1&0 \\ 0&0&0&0&1 \end{bmatrix}, \ \text{CPM}_5(\beta) = \begin{bmatrix} 0&1&0&0&0 \\ 0&0&1&0&0 \\ 0&0&0&1&0 \\ 0&0&0&0&1 \\ 1&0&0&0&0 \end{bmatrix}, \ \text{CPM}_5(\beta^2) = \begin{bmatrix} 0&0&1&0&0 \\ 0&0&0&1&0 \\ 0&0&0&0&1 \\ 1&0&0&0&0 \\ 0&1&0&0&0 \end{bmatrix},$$

$$\text{CPM}_5(\beta^3) = \begin{bmatrix} 0&0&0&1&0 \\ 0&0&0&0&1 \\ 1&0&0&0&0 \\ 0&1&0&0&0 \\ 0&0&1&0&0 \end{bmatrix}, \text{CPM}_5(\beta^4) = \begin{bmatrix} 0&0&0&0&1 \\ 1&0&0&0&0 \\ 0&1&0&0&0 \\ 0&0&1&0&0 \\ 0&0&0&1&0 \end{bmatrix}.$$

Note that the cyclic group $\mathbf{S}_5 = \{\beta^0 = 1, \beta, \beta^2, \beta^3, \beta^4\} = \{\alpha^0 = 1, \alpha^3, \alpha^6, \alpha^9, \alpha^{12}\}$ is a subgroup of the multiplicative group $\mathbf{S}_{15} = \{\alpha^0, \alpha, \alpha^2, \alpha^3, \alpha^4, \alpha^5, \alpha^6, \alpha^7, \alpha^8, \alpha^9, \alpha^{10}, \alpha^{11}, \alpha^{12}, \alpha^{13}, \alpha^{14}\}$ of $\mathrm{GF}(2^4)$. For $0 \leq j < 5$, the CPM-dispersion of the element $\beta^j = \alpha^{3j}$ in $\mathbf{S}_5$ with respect to $\mathbf{S}_{15}$ is a CPM of size $15 \times 15$ whose generator has its single 1-component at position $3j$. For example, the CPM-dispersion of $\beta^2 = \alpha^6$ with respect to $\mathbf{S}_{15}$ gives the following $15 \times 15$ CPM:

$$
\mathrm{CPM}_{15}(\beta^2) =
\begin{bmatrix}
0 & 0 & 0 & 0 & 0 & 0 & 1 & 0 & 0 & 0 & 0 & 0 & 0 & 0 & 0 \\
0 & 0 & 0 & 0 & 0 & 0 & 0 & 1 & 0 & 0 & 0 & 0 & 0 & 0 & 0 \\
0 & 0 & 0 & 0 & 0 & 0 & 0 & 0 & 1 & 0 & 0 & 0 & 0 & 0 & 0 \\
0 & 0 & 0 & 0 & 0 & 0 & 0 & 0 & 0 & 1 & 0 & 0 & 0 & 0 & 0 \\
0 & 0 & 0 & 0 & 0 & 0 & 0 & 0 & 0 & 0 & 1 & 0 & 0 & 0 & 0 \\
0 & 0 & 0 & 0 & 0 & 0 & 0 & 0 & 0 & 0 & 0 & 1 & 0 & 0 & 0 \\
0 & 0 & 0 & 0 & 0 & 0 & 0 & 0 & 0 & 0 & 0 & 0 & 1 & 0 & 0 \\
0 & 0 & 0 & 0 & 0 & 0 & 0 & 0 & 0 & 0 & 0 & 0 & 0 & 1 & 0 \\
0 & 0 & 0 & 0 & 0 & 0 & 0 & 0 & 0 & 0 & 0 & 0 & 0 & 0 & 1 \\
1 & 0 & 0 & 0 & 0 & 0 & 0 & 0 & 0 & 0 & 0 & 0 & 0 & 0 & 0 \\
0 & 1 & 0 & 0 & 0 & 0 & 0 & 0 & 0 & 0 & 0 & 0 & 0 & 0 & 0 \\
0 & 0 & 1 & 0 & 0 & 0 & 0 & 0 & 0 & 0 & 0 & 0 & 0 & 0 & 0 \\
0 & 0 & 0 & 1 & 0 & 0 & 0 & 0 & 0 & 0 & 0 & 0 & 0 & 0 & 0 \\
0 & 0 & 0 & 0 & 1 & 0 & 0 & 0 & 0 & 0 & 0 & 0 & 0 & 0 & 0 \\
0 & 0 & 0 & 0 & 0 & 1 & 0 & 0 & 0 & 0 & 0 & 0 & 0 & 0 & 0
\end{bmatrix}.
$$

Similarly, we can obtain the CPM-dispersions of all the elements in $\mathbf{S}_5$ with respect to its super group $\mathbf{S}_{15}$. ▲▲

## 13.2 Matrices with RC-Constrained Structure

As defined in Chapter 11, a low-density matrix $\mathbf{H}$ over $\mathrm{GF}(2)$ for which *no two rows (two columns) have more than one location where they both have 1-entries* is said to satisfy the *row–column* (RC) *constraint*. Such a matrix is called an *RC-constrained matrix* [1, 2]. The RC-constraint on $\mathbf{H}$ ensures that the girth of the Tanner graph of the LDPC code given by the null space over $\mathrm{GF}(2)$ of the matrix $\mathbf{H}$ is at least 6 [1, 2].

Let $\mathbf{B} = [b_{i,j}]_{0 \leq i < k, 0 \leq j < m}$ be a $k \times m$ matrix with nonzero entries from a cyclic subgroup $\mathbf{S}_n$ of order $n$ of the field $\mathrm{GF}(q)$. If we disperse each nonzero entry in $\mathbf{B}$ into a CPM of size $ln \times ln$ and a zero entry (if any) into a ZM of size $ln \times ln$ with respect to the super group $\mathbf{S}_{ln}$ of $\mathbf{S}_n$, where $l$ is a factor of $q - 1$ and relatively prime to $n$, we obtain a $k \times m$ array $\mathbf{H} = [\mathrm{CPM}_{ln}(b_{i,j})]_{0 \leq i < k, 0 \leq j < m}$ of CPMs and/or ZMs of size $ln \times ln$ which is a $kln \times mln$ matrix over $\mathrm{GF}(2)$. The array $\mathbf{H}$ is referred to as the *CPM-dispersion of* $\mathbf{B}$, denoted by $\mathrm{CPM}_{ln}(\mathbf{B})$, with a dispersion factor of $ln$ and $\mathbf{B}$ is called the *base matrix of* $\mathbf{H}$.

A *necessary and sufficient condition* on a base matrix $\mathbf{B}$ with nonzero entries from a cyclic subgroup $\mathbf{S}_n$ whose $ln$-fold CPM-dispersion $\mathrm{CPM}_{ln}(\mathbf{B})$ satisfies the RC-constraint was proved in [3–5] and is rephrased in the following theorem (proof omitted here).

**Theorem 13.1** *Let* **B** *be a matrix with nonzero entries from a cyclic subgroup* $\mathbf{S}_n$ *of order* $n$ *of the field* $\mathrm{GF}(q)$. *A necessary and sufficient condition for the CPM-dispersion of* **B**, $\mathrm{CPM}_{ln}(\mathbf{B})$, *with dispersion factor* $ln$ *where* $l$ *is factor of* $q-1$ *and relatively prime to* $n$, *to satisfy the RC-constraint is that every* $2 \times 2$ *submatrix of* **B** *either contains* at least *one zero entry* or is *nonsingular.*

For convenience, the necessary and sufficient condition on the base matrix **B** given in Theorem 13.1 is called the $2 \times 2$ *submatrix constraint* (or $2 \times 2$ *SM-constraint*).

Let **H** be the $ln \times ln$ CPM-dispersion of a $2 \times 2$ SM-constrained $k \times m$ base matrix $\mathbf{B} = [b_{i,j}]_{0 \leq i < k, 0 \leq j < m}$ with nonzero entries from a cyclic subgroup $\mathbf{S}_n$ of order $n$ of $\mathrm{GF}(q)$. Then, **H** is a $k \times m$ array of CPMs and/or ZMs of size $ln \times ln$, a $kln \times mln$ matrix over $\mathrm{GF}(2)$. It follows from Theorem 13.1 that the matrix **H** satisfies the RC-constraint. The largest density of **H** is $1/ln$, i.e., there is no ZMs in **H** or no 0-entries in the base matrix **B**. For a large dispersion factor, the density of the CPM-dispersion $\mathbf{H} = \mathrm{CPM}_{ln}(\mathbf{B})$ is quite low.

**Example 13.2** Consider the field $\mathrm{GF}(2^4)$ used in Example 13.1. Let $\beta = \alpha^5$. Then, the order of $\beta$ is 3. The set $\mathbf{S}_3 = \{\beta^0 = 1, \beta, \beta^2\}$ is a cyclic subgroup of order 3 of $\mathrm{GF}(2^4)$. Let **B** be a $2 \times 3$ matrix over $\mathbf{S}_3$ given as follows:

$$\mathbf{B} = \begin{bmatrix} 1 & \beta & \beta^2 \\ 1 & \beta^2 & \beta \end{bmatrix}.$$

It can be verified that any $2 \times 2$ submatrix of **B** is nonsingular. Hence, **B** satisfies the $2 \times 2$ SM-constraint. The CPM-dispersion $\mathrm{CPM}_3(\mathbf{B})$ of **B** with respect to the cyclic subgroup $\mathbf{S}_3$ of $\mathrm{GF}(2^4)$ is a $2 \times 3$ array of CPMs of size $3 \times 3$ given as follows:

$$\mathbf{H} = \mathrm{CPM}_3(\mathbf{B}) = \begin{bmatrix} 1 & 0 & 0 & 0 & 1 & 0 & 0 & 0 & 1 \\ 0 & 1 & 0 & 0 & 0 & 1 & 1 & 0 & 0 \\ 0 & 0 & 1 & 1 & 0 & 0 & 0 & 1 & 0 \\ 1 & 0 & 0 & 0 & 0 & 1 & 0 & 1 & 0 \\ 0 & 1 & 0 & 1 & 0 & 0 & 0 & 0 & 1 \\ 0 & 0 & 1 & 0 & 1 & 0 & 1 & 0 & 0 \end{bmatrix}.$$

The array **H** is a $6 \times 9$ matrix over $\mathrm{GF}(2)$. Because the base matrix **B** satisfies the $2 \times 2$ SM-constraint, its CPM-dispersion **H** satisfies the RC-constraint. This fact can be easily verified by examining the 1-entries in any two rows (or two columns) of **H**. The Tanner graph associated with **H** has girth 8.

Dispersing **B** with respective to the supergroup $\mathbf{S}_{15}$ of $\mathbf{S}_3$, we obtain the following $30 \times 45$ matrix $\mathbf{H} = \mathrm{CPM}_{15}(\mathbf{B})$:

$$
\left[\begin{array}{c}
100000000000000010000000000000001000000000000\\
010000000000000001000000000000000100000000000\\
001000000000000000100000000000000010000000000\\
000100000000000000010000000000000001000000000\\
000010000000000000001000000000000000100000000\\
000001000000000000000100000000000000010000000\\
000000100000000000000010000000000000001000000\\
000000010000000000000001000000000000000100000\\
000000001000000000000000100000000000000010000\\
000000000100000000000000010000000000000001000\\
000000000010000000000000001000000000000000100\\
000000000001000000000000000100000000000000010\\
000000000000100000000000000010000000000000001\\
000000000000010000000000000001100000000000000\\
000000000000001100000000000000010000000000000\\
100000000000000001000000000000010000000000000\\
010000000000000000100000000000001000000000000\\
001000000000000000010000000000000100000000000\\
000100000000000000001000000000000010000000000\\
000010000000000000000100000000000001000000000\\
000001000000000000000010000000000000100000000\\
000000100000000000000001000000000000010000000\\
000000010000000000000000100000000000001000000\\
000000001000000000000000010000000000000100000\\
000000000100000000000000001000000000000010000\\
000000000010000000000000000100000000000001000\\
000000000001000000000000000010000000000000100\\
000000000000100000000000000001000000000000010\\
000000000000010100000000000000000000000000001\\
000000000000001010000000000000100000000000000
\end{array}\right]
$$

which is a $2 \times 3$ array of CPMs of size $15 \times 15$ and satisfies the RC-constraint. ▲▲

## 13.3 Definitions and Structural Properties of Partial Geometries

Consider a system composed of a set $\mathbf{P}$ of $N$ points and a set $\mathbf{L}$ of $M$ lines where each line is a set of points. If a line $L$ contains a point $\mathbf{a}$, we say that $\mathbf{a}$ is on $L$ and $L$ passes through $\mathbf{a}$. If two points are on a line, we say that the two points are *adjacent* or *connected* by a line. If two lines pass through the same point $\mathbf{a}$, we say that the two lines *intersect* at the point $\mathbf{a}$; otherwise, they are *parallel*. The system composed of the sets $\mathbf{P}$ and $\mathbf{L}$ is a *partial geometry* if the following conditions are satisfied for some fixed integers $r \geq 2, t \geq 2$, and $c \geq 1$ [6–10].

(1) Any two points are on *at most one* line.
(2) Each point is on $r$ lines.
(3) Each line passes through $t$ points.
(4) If a point $\mathbf{a}$ is not on a line $L$, then there are exactly $c$ lines, each passing through $\mathbf{a}$ and a point on $L$.

We denote such a partial geometry (PaG) by $\mathrm{PaG}(r, t, c)$ where $r$, $t$, and $c$ are called the *parameters* of the partial geometry $\mathrm{PaG}(r, t, c)$. The parameter $c$ is called the *connection number*. Partial geometries were first introduced by Bose in 1963 [6]. A simple counting argument [9] shows that the partial geometry PaG$(r, t, c)$ has

$$N = t((t-1)(r-1) + c)/c \qquad (13.1)$$

points and

$$M = r((t-1)(r-1) + c)/c \qquad (13.2)$$

lines. Two lines in PaG$(r, t, c)$ either are parallel or intersect at *one and only one* point.

If $c = r - 1$, then the partial geometry PaG$(r, t, r-1)$ consists of $N = t^2$ points and $M = rt$ lines. In this case, a point **a** not on a line $L$ in PaG$(r, t, r-1)$ is on a *unique* line which is parallel to $L$. It can be shown that the $rt$ lines in PaG$(r, t, r-1)$ can be partitioned into $r$ parallel bundles, each consisting of $t$ parallel lines. A partial geometry PaG$(r, t, r-1)$ with parameters $r$, $t$, $r-1$ is called a *net* [11, 12]. A special type of net is $t = r$ and $c = r - 1$, i.e., a partial geometry PaG$(r, r, r-1)$ with parameters $r$, $r$, $r-1$. In this case, $N = M = t^2 = r^2$.

Examples of partial geometries are two-dimensional Euclidean and projective geometries over GF$(q)$ [11, 12]. The two-dimensional Euclidean geometry EG$(2, q)$ over GF$(q)$ is a partial geometry PaG$(q+1, q, q)$ with parameters $r = q + 1$, $t = q$, and $c = q$ which has $q^2$ points and $q(q+1)$ lines, each line consisting of $q$ points, as shown in Chapter 7. Because $c = r - 1$, EG$(2, q)$ is a net. The two-dimensional projective geometry PG$(2, q)$ over GF$(q)$ is a partial geometry PaG$(q+1, q+1, q)$ with parameters $r = t = q + 1$ and $c = q$, which consists of $q^2 + q + 1$ points and $q^2 + q + 1$ lines, each line comprising $q + 1$ points, and also is a net.

Let $\mathbf{H}_{\text{PaG}}(r, t, c)$ be the line-point incidence matrix of a partial geometry PaG$(r, t, c)$ in which the rows and columns of $\mathbf{H}_{\text{PaG}}(r, t, c)$ correspond to the lines and points in PaG$(r, t, c)$, respectively. Then, $\mathbf{H}_{\text{PaG}}(r, t, c)$ is an $M \times N$ matrix over GF$(2)$. Following the fundamental structural properties of PaG$(r, t, c)$, we find that $\mathbf{H}_{\text{PaG}}(r, t, c)$ has constant column weight $r$ and constant row weight $t$ and satisfies the RC-constraint.

In Sections 13.4 and 13.5, we will present two classes of partial geometries. These two classes of partial geometries are constructed based on Theorem 1 in [13] which is rephrased in the following theorem.

**Theorem 13.2** *Let $\mathbf{H}_{PaG}(r, t, t-1)$ be an $M \times N$ RC-constrained matrix over GF$(2)$ which is an $r \times t$ array of CPMs of size $r \times r$ where $M = r^2$ and $N = rt$. Then, $\mathbf{H}_{\text{PaG}}(r, t, t-1)$ is the line-point incidence matrix of a partial geometry PaG$(r, t, t-1)$ which has $N$ points and $M$ lines corresponding to the columns and rows of $\mathbf{H}_{\text{PaG}}(r, t, t-1)$, respectively.*

The array $\mathbf{H}_{\text{PaG}}(r, t, t-1)$ given in Theorem 13.2 is composed of $r$ CPM-row-blocks, each consisting of $t$ CPMs of size $r \times r$, and $t$ CPM-column-blocks, each consisting of $r$ CPMs of size $r \times r$. Each CPM-row-block has $r$ rows. Because no two rows in the same CPM-row-block have any location where they both have 1-components, the $r$ rows in a CPM-row-block of $\mathbf{H}_{\mathrm{PaG}}(r, t, t-1)$ correspond to $r$ parallel lines in $\mathrm{PaG}(r, t, t-1)$ which form a parallel bundle. Hence, the $r^2$ lines in $\mathrm{PaG}(r, t, t-1)$ can be partitioned into $r$ parallel bundles. Because $\mathbf{H}_{\mathrm{PaG}}(r, t, t-1)$ satisfies the RC-constraint, any two points in $\mathrm{PaG}(r, t, t-1)$ are on at most one line. Because each row of $\mathbf{H}_{\mathrm{PaG}}(r, t, t-1)$ has weight $t$, each line in $\mathrm{PaG}(r, t, t-1)$ passes through $t$ points. Because each column of $\mathbf{H}_{\mathrm{PaG}}(r, t, t-1)$ has weight $r$, each point in $\mathrm{PaG}(r, t, t-1)$ is intersected by $r$ lines. For a point $\mathbf{a}$ not on a line $L$, there are $t-1$ lines in $\mathrm{PaG}(r, t, t-1)$, each passing through the point $\mathbf{a}$ and one point on the line $L$.

Let $d$ and $k$ be two positive integers such that $1 \leq d < r$ and $d \leq k \leq t$. Take a $d \times k$ subarray $\mathbf{H}(d, k)$ from $\mathbf{H}_{\mathrm{PaG}}(r, t, t-1)$. The subarray $\mathbf{H}(d, k)$ is a $dr \times kr$ matrix over $\mathrm{GF}(2)$ with column and row weights $d$ and $k$, respectively. Because $\mathbf{H}_{\mathrm{PaG}}(r, t, t-1)$ satisfies the RC-constraint, $\mathbf{H}(d, k)$ also satisfies the RC-constraint. Then, the null space over $\mathrm{GF}(2)$ of $\mathbf{H}(d, k)$ gives a $(d, k)$-regular QC-LDPC code $C_{\mathrm{PaG}}(d, k)$ of length $kr$ and rate at least $(k-d)/k$. We call such an LDPC code $C_{\mathrm{PaG}}(d, k)$ a *quasi-cyclic partial-geometry LDPC* (QC-PaG-LDPC) *code*. Because $\mathbf{H}(d, k)$ satisfies the RC-constraint, it has orthogonal structure. Hence, the QC-PaG-LDPC code $C_{\mathrm{PaG}}(d, k)$ is OSML-decodable (see Section 7.2).

In Sections 13.4 and 13.5, two classes of partial geometries whose line-point incidence matrices have the structure given in Theorem 13.2 will be presented. Based on these classes of partial geometries, two classes of QC-PaG-LDPC codes can be constructed.

## 13.4 Partial Geometries Based on Prime-Order Cyclic Subgroups of Finite Fields and Their Associated QC-LDPC Codes

In this section, we present a class of partial geometries, $\mathrm{PaG}(p, p, p-1)$, with parameters $p$, $p$, and $p-1$ which are constructed based on *cyclic subgroups of prime orders* of finite fields [14], where $p$ is a prime. Based on these partial geometries, a class of QC-PaG-LDPC codes can be constructed.

Consider the field $\mathrm{GF}(q)$ with a primitive element $\alpha$. Let $p$ be a prime factor of $q-1$, $c = (q-1)/p$, and $\beta = \alpha^c$. Then, the order of $\beta$ is $p$ and the set $\mathbf{S}_c = \{1, \beta, \beta^2, \ldots, \beta^{p-1}\}$ forms a cyclic subgroup of $\mathrm{GF}(q)$ of order $p$. Using the $p$ elements of $\mathbf{S}_c$, we form the following $p \times p$ matrix $\mathbf{B}_{\mathrm{PaG},c}(p, p)$ over $\mathbf{S}_c$ (over $\mathrm{GF}(q)$):

$$\mathbf{B}_{\mathrm{PaG},c}(p, p) = \begin{bmatrix} 1 & 1 & 1 & \cdots & 1 \\ 1 & \beta & \beta^2 & \cdots & \beta^{p-1} \\ 1 & \beta^2 & (\beta^2)^2 & \cdots & (\beta^2)^{p-1} \\ \vdots & \vdots & \vdots & \cdots & \vdots \\ 1 & \beta^{p-1} & (\beta^{p-1})^2 & \cdots & (\beta^{p-1})^{p-1} \end{bmatrix}. \tag{13.3}$$

The subscript "$c$" in $\mathbf{B}_{\mathrm{PaG},c}(p,p)$ and $\mathbf{S}_c$ stands for "cyclic subgroup." Label the rows and columns of $\mathbf{B}_{\mathrm{PaG},c}(p,p)$ from 0 to $p-1$. Except for the zeroth row (top row), every row of $\mathbf{B}_{\mathrm{PaG},c}(p,p)$ consists of all the elements of $\mathbf{S}_c$, and except for the zeroth column (leftmost column), every column of $\mathbf{B}_{\mathrm{PaG},c}(p,p)$ also consists of all the elements of $\mathbf{S}_c$. The matrix $\mathbf{B}_{\mathrm{PaG},c}(p,p)$ has a property which is specified by the following theorem.

**Theorem 13.3** *The matrix* $\mathbf{B}_{\mathrm{PaG},c}(p,p)$ *given by* (13.3) *satisfies the* $2 \times 2$ *SM-constraint.*

*Proof* First, notice that all the entries of $\mathbf{B}_{\mathrm{PaG},c}(p,p)$ are nonzero. Let $i$, $j$, $k$, and $l$ be any four nonnegative integers such that $0 \leq i < j < p$ and $0 \leq k < l < p$. Consider the following $2 \times 2$ submatrix of $\mathbf{B}_{\mathrm{PaG},c}(p,p)$:

$$\mathbf{B}(2,2) = \begin{bmatrix} (\beta^i)^k & (\beta^i)^l \\ (\beta^j)^k & (\beta^j)^l \end{bmatrix}.$$

The determinant of $\mathbf{B}(2,2)$ is $\beta^{jl+ik} - \beta^{jk+il}$. Next, we show that $\beta^{jl+ik} - \beta^{jk+il} \neq 0$, i.e., $\mathbf{B}(2,2)$ is nonsingular. Suppose $\beta^{jl+ik} - \beta^{jk+il} = 0$. Then, we must have $\beta^{(j-i)(l-k)} = 1$. Because $j-i$ and $l-k$ are less than $p$ and $p$ is a prime, the product $(j-i)(l-k)$ cannot be divisible by $p$. The fact that the order of $\beta$ is $p$ implies that $\beta^{(j-i)(l-k)} \neq 1$ and thus $\beta^{jl+ik} - \beta^{jk+il} \neq 0$. Hence, $\mathbf{B}(2,2)$ must be nonsingular and $\mathbf{B}_{\mathrm{PaG},c}(p,p)$ satisfies the $2 \times 2$ SM-constraint. ▲▲

Dispersing each entry of $\mathbf{B}_{\mathrm{PaG},c}(p,p)$ into a CPM of size $p \times p$, we obtain a $p \times p$ array $\mathbf{H}_{\mathrm{PaG},c}(p,p)$ of CPMs of size $p \times p$. The array $\mathbf{H}_{\mathrm{PaG},c}(p,p)$ is a $p^2 \times p^2$ matrix over GF(2) with both column and row weights $p$. It follows from Theorems 13.1 and 13.3 that $\mathbf{H}_{\mathrm{PaG},c}(p,p)$ satisfies the RC-constraint. Then, it follows from Theorem 13.2 that $\mathbf{H}_{\mathrm{PaG},c}(p,p)$, as a $p \times p$ RC-constrained array of CPMs of size $p \times p$, is the line-point incidence matrix of a partial geometry $\mathrm{PaG}_c(p,p,p-1)$ which is a net. The $p^2$ columns of $\mathbf{H}_{\mathrm{PaG},c}(p,p)$ correspond to the $p^2$ points of $\mathrm{PaG}_c(p,p,p-1)$ and the $p^2$ rows of $\mathbf{H}_{\mathrm{PaG},c}(p,p)$ correspond to the $p^2$ lines of $\mathrm{PaG}_c(p,p,p-1)$. The rows of $\mathbf{H}_{\mathrm{PaG},c}(p,p)$ are the incidence vectors of the lines in $\mathrm{PaG}_c(p,p,p-1)$. The $p^2$ lines in $\mathrm{PaG}_c(p,p,p-1)$ can be partitioned into $p$ parallel bundles, each consisting of $p$ parallel lines.

The matrix $\mathbf{H}_{\mathrm{PaG},c}(p,p)$ consists of $p$ CPM-row-blocks of size $p \times p^2$ and $p$ CPM-column-blocks of size $p^2 \times p$. Each CPM-row-block (column-block) is composed of $p$ CPMs of size $p \times p$. The $p$ rows in each CPM-row-block of $\mathbf{H}_{\mathrm{PaG},c}(p,p)$ are incidence vectors of the $p$ lines in a parallel bundle of $\mathrm{PaG}_c(p,p,p-1)$. Label the rows and columns of $\mathbf{H}_{\mathrm{PaG},c}(p,p)$ from 0 to $p^2 - 1$. For $0 \leq j < p^2$, there are $p$ rows in $\mathbf{H}_{\mathrm{PaG},c}(p,p)$, each residing in a separate CPM-row-block of $\mathbf{H}_{\mathrm{PaG},c}(p,p)$, which are orthogonal on the $j$th column position. That is, every point in $\mathrm{PaG}_c(p,p,p-1)$ is intersected by $p$ lines. Hence, $\mathbf{H}_{\mathrm{PaG},c}(p,p)$ has an *orthogonal structure*. Furthermore, no two points in $\mathrm{PaG}_c(p,p,p-1)$ corresponding to two columns in a CPM-column-block of $\mathbf{H}_{\mathrm{PaG},c}(p,p)$ are on the same lines.

From the construction of the line-point incidence matrix $\mathbf{H}_{\mathrm{PaG},c}(p,p)$ of $\mathrm{PaG}_c(p,p,p-1)$, we may regard the matrix $\mathbf{B}_{\mathrm{PaG},c}(p,p)$ given by (13.3) as the *base matrix* for constructing the partial geometry $\mathrm{PaG}_c(p, p, p-1)$. Note that the base matrix $\mathbf{B}_{\mathrm{PaG},c}(p, p)$ given by (13.3) is a square Vandermonde matrix.

**Example 13.3** Consider the field $\mathrm{GF}(2^{10})$. Note that $2^{10} - 1 = 1023$ can be factored as the product of 3, 11, and 31, i.e., $1023 = 3 \times 11 \times 31$. Based on these three prime factors of $2^{10} - 1$, we can form three cyclic subgroups $\mathbf{S}_3$, $\mathbf{S}_{11}$, and $\mathbf{S}_{31}$ of $\mathrm{GF}(2^{10})$ with orders 3, 11, and 31, respectively. Hence, three PaG base matrices $\mathbf{B}_{\mathrm{PaG},c}(3, 3)$, $\mathbf{B}_{\mathrm{PaG},c}(11, 11)$, and $\mathbf{B}_{\mathrm{PaG},c}(31, 31)$ over $\mathrm{GF}(2^{10})$ in the form of (13.3) can be constructed. Using $\mathbf{B}_{\mathrm{PaG},c}(3, 3)$, $\mathbf{B}_{\mathrm{PaG},c}(11, 11)$, and $\mathbf{B}_{\mathrm{PaG},c}(31, 31)$ as base matrices in conjunction with CPM-dispersion, we can construct three partial geometries, $\mathrm{PaG}_c(3, 3, 2)$, $\mathrm{PaG}_c(11, 11, 10)$, and $\mathrm{PaG}_c(31, 31, 30)$.

For example, let $\alpha$ be a primitive element of $\mathrm{GF}(2^{10})$ and $\beta = \alpha^{341}$. The set $\mathbf{S}_3 = \{1, \beta, \beta^2\}$ forms a cyclic subgroup of $\mathrm{GF}(2^{10})$ with order 3. Using the three elements in $\mathbf{S}_3$, we form the following base matrix in the form of (13.3):

$$\mathbf{B}_{\mathrm{PaG},c}(3, 3) = \begin{bmatrix} 1 & 1 & 1 \\ 1 & \beta & \beta^2 \\ 1 & \beta^2 & \beta \end{bmatrix}.$$

The CPM-dispersion of $\mathbf{B}_{\mathrm{PaG},c}(3, 3)$ with respect to $\mathbf{S}_3 = \{1, \beta, \beta^2\}$ gives the following $3 \times 3$ array of CPMs of size $3 \times 3$:

$$\mathbf{H}_{\mathrm{PaG},c}(3, 3) = \left[\begin{array}{ccc|ccc|ccc} 1 & 0 & 0 & 1 & 0 & 0 & 1 & 0 & 0 \\ 0 & 1 & 0 & 0 & 1 & 0 & 0 & 1 & 0 \\ 0 & 0 & 1 & 0 & 0 & 1 & 0 & 0 & 1 \\ \hline 1 & 0 & 0 & 0 & 1 & 0 & 0 & 0 & 1 \\ 0 & 1 & 0 & 0 & 0 & 1 & 1 & 0 & 0 \\ 0 & 0 & 1 & 1 & 0 & 0 & 0 & 1 & 0 \\ \hline 1 & 0 & 0 & 0 & 0 & 1 & 0 & 1 & 0 \\ 0 & 1 & 0 & 1 & 0 & 0 & 0 & 0 & 1 \\ 0 & 0 & 1 & 0 & 1 & 0 & 1 & 0 & 0 \end{array}\right],$$

which is a $9 \times 9$ matrix with column and row weights 3 and rank 7. The matrix $\mathbf{H}_{\mathrm{PaG},c}(3, 3)$ is the line-point incidence matrix of the partial geometry $\mathrm{PaG}_c(3, 3, 2)$ which consists of nine points corresponding to the nine columns of $\mathbf{H}_{\mathrm{PaG},c}(3, 3)$ and nine lines corresponding to the nine rows of $\mathbf{H}_{\mathrm{PaG},c}(3, 3)$. Each line in $\mathrm{PaG}_c(3, 3, 2)$ consists of three points. Each point in $\mathrm{PaG}_c(3, 3, 2)$ is on three lines. Any two points are on at most one line because there is no $2 \times 2$ rectangular in $\mathbf{H}_{\mathrm{PaG},c}(3, 3)$ with all 1-entries. If a point $\mathbf{a}$ is not a line $L$, there are exactly two lines, each passing through the point $\mathbf{a}$ and one point on the line $L$.

The array $\mathbf{H}_{\mathrm{PaG},c}(3, 3)$ consists of three CPM-row-blocks and three CPM-column-blocks. The three rows in a CPM-row-block of $\mathbf{H}_{\mathrm{PaG},c}(3, 3)$ are the incidence vectors of three parallel lines in $\mathrm{PaG}_c(3, 3, 2)$. Label the rows and columns of $\mathbf{H}_{\mathrm{PaG},c}(3, 3)$ from 0 to 8. Consider the first, third, and eighth rows of $\mathbf{H}_{\mathrm{PaG},c}(3, 3)$ which are:

$$\begin{bmatrix} 0 & 1 & 0 & 0 & 1 & 0 & 0 & 1 & 0 \\ 1 & 0 & 0 & 0 & 1 & 0 & 0 & 0 & 1 \\ 0 & 0 & 1 & 0 & 1 & 0 & 1 & 0 & 0 \end{bmatrix}.$$

These three rows are the incidence vectors of three lines that intersect at the point corresponding to the fourth column of $\mathbf{H}_{\mathrm{PaG},c}(3,3)$, i.e., they are orthogonal on the fourth column position. ▲▲

Let $d$ and $k$ be two positive integers such that $1 \leq d < k \leq p$. Take a $d \times k$ submatrix $\mathbf{B}_{\mathrm{PaG},c}(d,k)$ from the base matrix $\mathbf{B}_{\mathrm{PaG},c}(p,p)$ of the partial geometry $\mathrm{PaG}_c(p,p,p-1)$. The $p$-fold CPM-dispersion of $\mathbf{B}_{\mathrm{PaG},c}(d,k)$ gives a $d \times k$ array $\mathbf{H}_{\mathrm{PaG},c}(d,k)$ of CPMs of size $p \times p$ which is a $dp \times kp$ matrix over $\mathrm{GF}(2)$ with column and row weights $d$ and $k$, respectively. The array $\mathbf{H}_{\mathrm{PaG},c}(d,k)$ is a subarray of $\mathbf{H}_{\mathrm{PaG},c}(p,p)$ which also satisfies the RC-constraint. The null space over $\mathrm{GF}(2)$ of $\mathbf{H}_{\mathrm{PaG},c}(d,k)$ gives a $(d,k)$-regular QC-PaG-LDPC code $C_{\mathrm{PaG},c}(d,k)$ of length $kp$ with rate at least $(k-d)/k$. If $p$ is large, we can construct a family of QC-PaG-LDPC codes of various lengths and rates based on subarrays of $\mathbf{H}_{\mathrm{PaG},c}(p,p)$.

For $1 \leq d \leq p$, suppose we choose $d$ consecutive rows, labeled by $1, 2, \ldots,$ $d$, from the base matrix $\mathbf{B}_{\mathrm{PaG},c}(p,p)$ given by (13.3) to form the following $d \times p$ submatrix of $\mathbf{B}_{\mathrm{PaG},c}(p,p)$:

$$
\mathbf{B}_{\mathrm{PaG},c}(d,p) = \begin{bmatrix} 1 & \beta & \beta^2 & \cdots & \beta^{(p-1)} \\ 1 & \beta^2 & (\beta^2)^2 & \cdots & (\beta^2)^{(p-1)} \\ \vdots & \vdots & \vdots & \ddots & \vdots \\ 1 & \beta^d & (\beta^d)^2 & \cdots & (\beta^d)^{(p-1)} \end{bmatrix}. \tag{13.4}
$$

The matrix $\mathbf{B}_{\mathrm{PaG},c}(d,p)$ is simply the conventional parity-check matrix of a $(p, p-d)$ nonprimitive RS code over $\mathrm{GF}(q)$ of length $p$, rate $(p-d)/p$, and minimum distance $d+1$ as presented in Section 6.8.

The $p$-fold CPM-dispersion $\mathbf{H}_{\mathrm{PaG},c}(d,p)$ of $\mathbf{B}_{\mathrm{PaG},c}(d,p)$ is a $dp \times p^2$ matrix over $\mathrm{GF}(2)$ with column and row weights $d$ and $p$, respectively. The matrix $\mathbf{H}_{\mathrm{PaG},c}(d,p)$ is the line-point incidence matrix of a subgeometry of $\mathrm{PaG}_c(p,p,p-1)$ which has $p^2$ points and $dp$ lines in $d$ parallel bundles of $\mathrm{PaG}_c(p,p,p-1)$, i.e., choosing $d$ parallel bundles of lines to construct $\mathbf{H}_{\mathrm{PaG},c}(d,p)$.

The null space over $\mathrm{GF}(2)$ of $\mathbf{H}_{\mathrm{PaG},c}(d,p)$ gives a $(d,p)$-regular RS-based QC-PaG-LDPC code, denoted by $C_{\mathrm{RS,PaG},c}(d,p)$ and referred to as a *QC-RS-PaG-LDPC code*. In fact, any $d$ consecutive rows of the base matrix $\mathbf{B}_{\mathrm{PaG},c}(p,p)$ given by (13.3) form a parity-check matrix of a $(p, p-d)$ RS code over $\mathrm{GF}(q)$ with minimum distance $d+1$. From this point of view, the partial geometry $\mathrm{PaG}_c(p,p,p-1)$ constructed based on the matrix $\mathbf{B}_{\mathrm{PaG},c}(p,p)$ over $\mathrm{GF}(q)$ given by (13.3) may be regarded as an *RS-based partial geometry*. The Tanner graphs of QC-PaG-LDPC codes were proved in [13, 15, 16] to have good trapping set structures. Hence, the Tanner graphs of QC-RS-PaG-LDPC codes also have good trapping set structures which indicates that these codes can achieve very low error rates without error-floors.

**Example 13.4** Let $q = 2^5$. Consider the field $\mathrm{GF}(2^5)$. Let $\alpha$ be a primitive element of $\mathrm{GF}(2^5)$. Because $2^5 - 1 = 31$ is a prime, the only prime factor of $2^5 - 1$ greater than 1 is 31. Based on the primitive element $\alpha$ and all its powers, we can construct a $31 \times 31$ matrix $\mathbf{B}_{\mathrm{PaG},c}(31,31)$ over $\mathrm{GF}(2^5)$ in the form of (13.3). The $31 \times 31$ CPM-dispersion of $\mathbf{B}_{\mathrm{PaG},c}(31, 31)$ gives a $31 \times 31$ array $\mathbf{H}_{\mathrm{PaG},c}(31, 31)$ of CPMs of size $31 \times 31$. The array $\mathbf{H}_{\mathrm{PaG},c}(31, 31)$ is a $961 \times 961$ matrix over GF(2) with both column and row weights 31 and it is the line-point incidence matrix of a partial geometry $\mathrm{PaG}_c(31, 31, 30)$. The partial geometry $\mathrm{PaG}_c(31, 31, 30)$ consists of 961 points and 961 lines in which each point is intersected by 31 lines and each line contains 31 points. The lines in $\mathrm{PaG}_c(31, 31, 30)$ can be partitioned into 31 parallel bundles, each consisting of 31 parallel lines.

Label the rows and columns of $\mathbf{B}_{\mathrm{PaG},c}(31, 31)$ from 0 to 30. Suppose we take four consecutive rows, labeled with 1, 2, 3, and 4, from $\mathbf{B}_{\mathrm{PaG},c}(31, 31)$ to form a $4 \times 31$ matrix $\mathbf{B}_{\mathrm{PaG},c}(4, 31)$ over GF($2^5$). The matrix $\mathbf{B}_{\mathrm{PaG},c}(4, 31)$ is the parity-check matrix of the $(31, 27)$ primitive RS code over GF($2^5$) with minimum distance 5 whose generator polynomial has $\alpha$, $\alpha^2$, $\alpha^3$, and $\alpha^4$ as roots. The $31 \times 31$ CPM-dispersion of $\mathbf{B}_{\mathrm{PaG},c}(4, 31)$ gives a $4 \times 31$ array $\mathbf{H}_{\mathrm{PaG},c}(4, 31)$ of CPMs of size $31 \times 31$ which is a subarray of $\mathbf{H}_{\mathrm{PaG},c}(31, 31)$. The array $\mathbf{H}_{\mathrm{PaG},c}(4, 31)$ is a $124 \times 961$ matrix over GF(2) with column and row weights 4 and 31, respectively. The 124 rows in the four CPM-row-blocks of $\mathbf{H}_{\mathrm{PaG},c}(4, 31)$ correspond to the incidence vectors of 124 lines in four parallel bundles of $\mathrm{PaG}_c(31, 31, 30)$. Hence, $\mathbf{H}_{\mathrm{PaG},c}(4, 31)$ is the line-point incidence matrix of a subgeometry of $\mathrm{PaG}_c(31, 31, 30)$ which consists of 124 lines and 961 points.

The null space over GF(2) of $\mathbf{H}_{\mathrm{PaG},c}(4, 31)$ gives a $(4, 31)$-regular $(961, 840)$ QC-RS-PaG-LDPC code $C_{\mathrm{RS,PaG},c}(4, 31)$ with rate 0.874 whose Tanner graph has girth 6. The BER and BLER performances of the code $C_{\mathrm{RS,PaG},c}(4, 31)$ over the AWGN channel decoded with 5, 10, and 50 iterations of the MSA with scaling factor 0.75 are shown in Fig. 13.1(a). We see that the code achieves a BER of $10^{-9}$ without a visible error-floor and performs 2.6 dB from the Shannon limit. At a BLER of $10^{-7}$, the code performs 1.25 dB from the sphere packing bound (SPB) [17] with 50 iterations.

For comparison, we also include in Fig. 13.1(a) the BER and BLER performances of a binary $(961, 838)$ PEG code $C_{\mathrm{peg}}$ [18]. (The construction of PEG code will be covered in Chapter 15.) From the figure, we see that the performance curves of the two codes overlap with each other down to the BER of $10^{-7}$. After that, the PEG code starts to show error-floor.

If we apply this code to the BEC, its UEBR and UEBLR performances are shown in Fig. 13.1(b). Because the rate of the code is $R = 0.875$, the Shannon limit for this rate over the BEC is $1 - 0.875 = 0.125$. From Fig. 13.1(b), we see that the code performs also well over the BEC. ▲▲

**Example 13.5** Suppose we use the field GF($2^{11}$) for code construction. Note that $2^{11} - 1 = 2047$ can be factored as the product of two primes 23 and 89, i.e., $2047 = 23 \times 89$. Based on this field and the two prime factors 23 and 89, we can construct two partial geometries $\mathrm{PaG}_c(23, 23, 22)$ and $\mathrm{PaG}_c(89, 89, 88)$.

Let $\alpha$ be a primitive element of GF($2^{11}$) and $\beta = \alpha^{23}$. Then, the order of $\beta$ is 89. Using $\beta$ as the generator, we construct a cyclic subgroup $\mathbf{S}_c = \{\beta^0 = 1, \beta, \beta^2, \ldots, \beta^{88}\}$ of GF($2^{11}$). The base matrix $\mathbf{B}_{\mathrm{PaG},c}(89, 89)$ for constructing the partial geometry $\mathrm{PaG}_c(89, 89, 88)$ is an $89 \times 89$ matrix over $\mathbf{S}_c$ in the form of

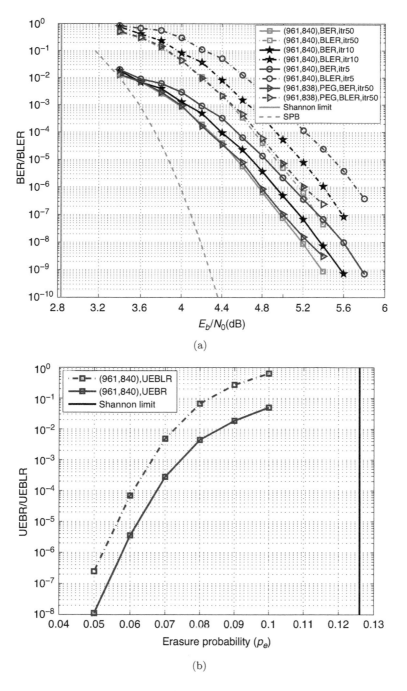

Figure 13.1 The performances of the $(961, 840)$ QC-RS-PaG-LDPC code given in Example 13.4 over: (a) AWGN channel and (b) BEC.

(13.3). The line-point incidence matrix $\mathbf{H}_{\mathrm{PaG},c}(89, 89)$ of $\mathrm{PaG}_c(89, 89, 88)$ is the $89 \times 89$ CPM-dispersion of $\mathbf{B}_{\mathrm{PaG},c}(89, 89)$. The partial geometry $\mathrm{PaG}_c(89, 89, 88)$ consists of 7921 points and 7921 lines. Each point in $\mathrm{PaG}_c(89, 89, 88)$ is intersected by 89 lines. Each line in $\mathrm{PaG}_c(89, 89, 88)$ has 89 points. The 7921 lines in $\mathrm{PaG}_c(89, 89, 88)$ can be partitioned into 89 parallel bundles, each consisting of 89 parallel lines. A point $\mathbf{a}$ not on a line $L$ is connected to the 88 points on $L$ by the other 88 lines, i.e., the connection number of $\mathrm{PaG}_c(89, 89, 88)$ is 88.

Label the rows and columns of $\mathbf{B}_{\mathrm{PaG},c}(89, 89)$ from 0 to 88. Suppose we take four consecutive rows, labeled by 1, 2, 3, and 4, from $\mathbf{B}_{\mathrm{PaG},c}(89, 89)$ to form a $4 \times 89$ matrix $\mathbf{B}_{\mathrm{PaG},c}(4, 89)$. The null space over $\mathrm{GF}(2^{11})$ of $\mathbf{B}_{\mathrm{PaG},c}(4, 89)$ gives an $(89, 85)$ RS code over $\mathrm{GF}(2^{11})$ of length 89, minimum distance 5, and rate 0.955. The $89 \times 89$ CPM-dispersion of $\mathbf{B}_{\mathrm{PaG},c}(4, 89)$ gives a $4 \times 89$ array $\mathbf{H}_{\mathrm{PaG},c}(4, 89)$ of CPMs of size $89 \times 89$ which is a $356 \times 7921$ matrix with column and row weights 4 and 89, respectively. The matrix $\mathbf{H}_{\mathrm{PaG},c}(4, 89)$ is the line-point incidence matrix of a subgeometry of $\mathrm{PaG}_c(89, 89, 88)$ which is composed of all the 7921 points of $\mathrm{PaG}_c(89, 89, 88)$ and 356 lines in four parallel bundles of $\mathrm{PaG}_c(89, 89, 88)$.

The null space over $\mathrm{GF}(2)$ of $\mathbf{H}_{\mathrm{PaG},c}(4, 89)$ gives a $(4, 89)$-regular $(7921, 7568)$ QC-RS-PaG-LDPC code $C_{\mathrm{RS,PaG},c}(4, 89)$ with rate 0.950. The BER and BLER performances of $C_{\mathrm{RS,PaG},c}(4, 89)$ over the AWGN channel decoded with 5, 10, and 50 iterations of the MSA scaled by 0.75 are shown in Fig. 13.2. With 10 iterations of the MSA, we see that the code achieves a BER of $10^{-10}$ without a visible error-floor and performs 1.38 dB from the Shannon limit. For comparison, we also include in Fig. 13.2 the BER and BLER performances of a $(7921, 7566)$ binary PEG code $C_{\mathrm{peg}}$. The performance curves of the two codes decoded with 50 iterations of the MSA overlap with each other in the simulation region.

Based on the parallel bundles of lines in $\mathrm{PaG}_c(89, 89, 88)$, we can construct a family of QC-RS-PaG-LDPC codes. ▲▲

**Example 13.6** Continue on Example 13.5. Suppose we delete the last 25 columns of the base matrix $\mathbf{B}_{\mathrm{PaG},c}(4, 89)$. We obtain a $4 \times 64$ matrix $\mathbf{B}_{\mathrm{PaG},c}(4, 64)$ with column and row weights 4 and 64, respectively. The null space over $\mathrm{GF}(2^{11})$ of $\mathbf{B}_{\mathrm{PaG},c}(4, 64)$ gives a $(64, 60)$ shortened RS code with minimum distance 5 (a shortened code of the $(89, 85)$ RS code over $\mathrm{GF}(2^{11})$ given in Example 13.5). The $89 \times 89$ CPM-dispersion of $\mathbf{B}_{\mathrm{PaG},c}(4, 64)$ gives a $4 \times 64$ array $\mathbf{H}_{\mathrm{PaG},c}(4, 64)$ of CPMs of size $89 \times 89$, a $356 \times 5696$ matrix with column and row weights 4 and 64, respectively. The null space over $\mathrm{GF}(2)$ of $\mathbf{H}_{\mathrm{PaG},c}(4, 64)$ gives a $(4, 64)$-regular $(5696, 5343)$ QC-RS-PaG-LDPC code $C_{\mathrm{RS,PaG},c}(4, 64)$ with rate 0.938.

The BER and BLER performances of $C_{\mathrm{RS,PaG},c}(4, 64)$ over the AWGN channel decoded with 5, 10, and 50 iterations of the MSA scaled by 0.75 are shown in Fig. 13.3(a). We see that the code achieves a BER of $10^{-9}$ without a visible error-floor and performs 1.3 dB from the Shannon limit with 50 iterations of the MSA.

If we remove the last 57 columns from the base matrix $\mathbf{B}_{\mathrm{PaG},c}(4, 89)$, we can construct a $4 \times 32$ array $\mathbf{H}_{\mathrm{PaG},c}(4, 32)$ of CPMs of size $89 \times 89$, a $356 \times 2848$ matrix with column and row weights 4 and 32, respectively. The null space over

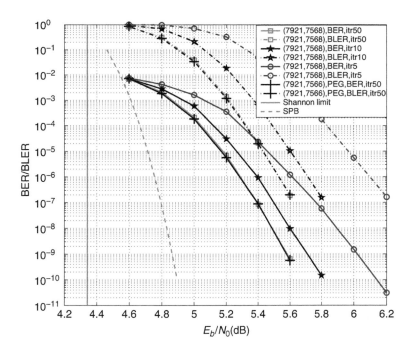

Figure 13.2 The BER and BLER performances of the $(7921, 7568)$ QC-RS-PaG-LDPC code $C_{\mathrm{RS,PaG},c}(4, 89)$ and the $(7921, 7566)$ PEG code $C_{\mathrm{peg}}$ given in Example 13.5.

GF(2) of $\mathbf{H}_{\mathrm{PaG},c}(4, 32)$ gives a $(4, 32)$-regular $(2848, 2495)$ QC-RS-PaG-LDPC code $C_{\mathrm{RS,PaG},c}(4, 32)$ with rate 0.875. The BER and BLER performances of the code $C_{\mathrm{RS,PaG},c}(4, 32)$ over the AWGN channel decoded with 5, 10, and 50 iterations of the MSA scaled by 0.75 are shown in Fig. 13.3(b).                    ▲▲

## 13.5  Partial Geometries Based on Prime Fields and Their Associated QC-LDPC Codes

In this section, we present another class of partial geometries, $\mathrm{PaG}(p, p, p - 1)$, with parameters $p$, $p$, and $p - 1$ constructed based on *prime fields* [13]. Based on this class of partial geometries, we can construct another class of QC-PaG-LDPC codes.

Let $p$ be a prime and let $\mathrm{GF}(p)$ be a prime field which consists of the following $p$ elements: $0, 1, 2, \ldots, p-1$, i.e., $\mathrm{GF}(p) = \{0, 1, 2, \ldots, p-1\}$. Form the following $p \times p$ matrix over $\mathrm{GF}(p)$ by using all the elements in $\mathrm{GF}(p)$:

$$\mathbf{B}_{\mathrm{PaG},p}(p, p) = \begin{bmatrix} 0 \cdot 0 & 0 \cdot 1 & \cdots & 0 \cdot (p-1) \\ 1 \cdot 0 & 1 \cdot 1 & \cdots & 1 \cdot (p-1) \\ 2 \cdot 0 & 2 \cdot 1 & \cdots & 2 \cdot (p-1) \\ \vdots & \vdots & \cdots & \vdots \\ (p-1) \cdot 0 & (p-1) \cdot 1 & \cdots & (p-1) \cdot (p-1) \end{bmatrix}, \quad (13.5)$$

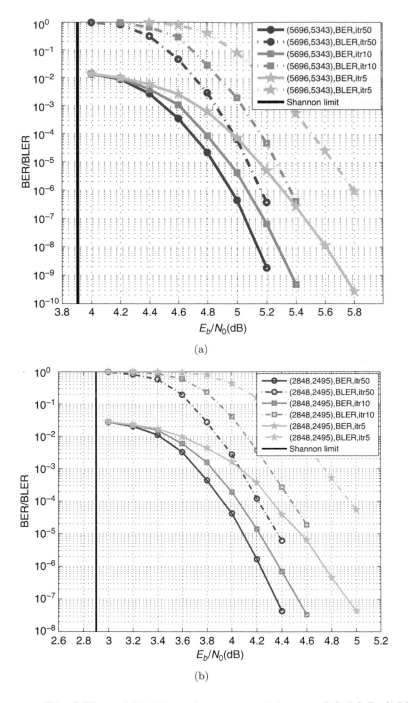

Figure 13.3 The BER and BLER performances of the two QC-RS-PaG-LDPC codes given in Example 13.6: (a) $(5696, 5343)$ and (b) $(2848, 2495)$.

where the multiplication is carried out with modulo-$p$ multiplication. The subscript "$p$" in $\mathbf{B}_{\mathrm{PaG},p}(p,p)$ stands for "prime field." Label the rows and columns of $\mathbf{B}_{\mathrm{PaG},p}(p,p)$ from 0 to $p-1$. The matrix $\mathbf{B}_{\mathrm{PaG},p}(p,p)$ has the following structural properties.

(1) All the entries in the zeroth row and the zeroth column are zeros.
(2) All the $p$ entries in any row (or column) other than the zeroth row (or the zeroth column) are different elements in $\mathrm{GF}(p)$.
(3) Two different rows (or columns) have the 0-element of $\mathrm{GF}(p)$ in common at the zeroth column (or the zeroth row) and differ in all the other $p-1$ positions.
(4) The $i$th column of $\mathbf{B}_{\mathrm{PaG},p}(p,p)$ is identical to the transpose of the $i$th row of $\mathbf{B}_{\mathrm{PaG},p}(p,p)$ for $0 \leq i < p$.

The last property implies that the transpose $\mathbf{B}_{\mathrm{PaG},p}^T(p,p)$ of $\mathbf{B}_{\mathrm{PaG},p}(p,p)$ is identical to $\mathbf{B}_{\mathrm{PaG},p}(p,p)$, i.e., $\mathbf{B}_{\mathrm{PaG},p}^T(p,p) = \mathbf{B}_{\mathrm{PaG},p}(p,p)$.

We disperse each element $j$, $0 \leq j < p$, of $\mathrm{GF}(p) = \{0,1,2,\ldots,p-1\}$, into a CPM (with columns and rows labeled from 0 to $p-1$) of size $p \times p$ whose zeroth row (or generator) has its single 1-component at the $j$th position. The 0-element of $\mathrm{GF}(p)$ is dispersed into a $p \times p$ identity matrix. (It is noted here that the CPM-dispersion is slightly different from the one used in Section 13.4.) Then, the CPM-dispersions of the entries in $\mathbf{B}_{\mathrm{PaG},p}(p,p)$ result in a $p \times p$ array $\mathbf{H}_{\mathrm{PaG},p}(p,p)$ of CPMs of size $p \times p$. The array $\mathbf{H}_{\mathrm{PaG},p}(p,p)$ consists of $p$ CPM-row-blocks and $p$ CPM-column-blocks. Each CPM-row-block has $p$ rows and each CPM-column-block has $p$ columns. The array $\mathbf{H}_{\mathrm{PaG},p}(p,p)$ is a $p^2 \times p^2$ matrix with both column and row weights $p$.

It was proved in [2] that $\mathbf{H}_{\mathrm{PaG},p}(p,p)$ satisfies the RC-constraint. Then, it follows from Theorem 13.2 that $\mathbf{H}_{\mathrm{PaG},p}(p,p)$ is the line-point incidence matrix of a partial geometry, denoted by $\mathrm{PaG}_p(p,p,p-1)$, with columns and rows corresponding to the points and lines in $\mathrm{PaG}_p(p,p,p-1)$, respectively. The partial geometry $\mathrm{PaG}_p(p,p,p-1)$ consists of $N = p^2$ points and $M = p^2$ lines. Each line in $\mathrm{PaG}_p(p,p,p-1)$ consists of $p$ points and each point is on $p$ lines. A point $\mathbf{a}$ that is not on a line $L$ is connected to the $p-1$ points on $L$ by $p-1$ lines, i.e., the connection number of a point in $\mathrm{PaG}_p(p,p,p-1)$ is $p-1$. The $p^2$ lines in $\mathrm{PaG}_p(p,p,p-1)$ can be partitioned into $p$ parallel bundles, each with $p$ parallel lines. The $p$ rows of $\mathbf{H}_{\mathrm{PaG},p}(p,p)$ in a CPM-row-block correspond to the $p$ lines in a parallel bundle of $\mathrm{PaG}_p(p,p,p-1)$. The structure of $\mathrm{PaG}_p(p,p,p-1)$ constructed based on a prime field $\mathrm{GF}(p)$ is similar to the structure of the partial geometry $\mathrm{PaG}_c(p,p,p-1)$ constructed based on a cyclic subgroup of prime order $p$ of a field $\mathrm{GF}(q)$ presented in Section 13.4. The matrix $\mathbf{H}_{\mathrm{PaG},p}(p,p)$ has also an orthogonal structure. For every column position, there are $p$ rows orthogonal on it.

More on structural properties of partial geometries based on finite fields can be found in [19].

**Example 13.7** Consider the prime field $\mathrm{GF}(5) = \{0,1,2,3,4\}$. Using the elements in $\mathrm{GF}(5)$, we form the following $5 \times 5$ matrix over $\mathrm{GF}(5)$:

$$\mathbf{B}_{\mathrm{PaG},p}(5,5) = \begin{bmatrix} 0\cdot 0 & 0\cdot 1 & 0\cdot 2 & 0\cdot 3 & 0\cdot 4 \\ 1\cdot 0 & 1\cdot 1 & 1\cdot 2 & 1\cdot 3 & 1\cdot 4 \\ 2\cdot 0 & 2\cdot 1 & 2\cdot 2 & 2\cdot 3 & 2\cdot 4 \\ 3\cdot 0 & 3\cdot 1 & 3\cdot 2 & 3\cdot 3 & 3\cdot 4 \\ 4\cdot 0 & 4\cdot 1 & 4\cdot 2 & 4\cdot 3 & 4\cdot 4 \end{bmatrix} = \begin{bmatrix} 0 & 0 & 0 & 0 & 0 \\ 0 & 1 & 2 & 3 & 4 \\ 0 & 2 & 4 & 1 & 3 \\ 0 & 3 & 1 & 4 & 2 \\ 0 & 4 & 3 & 2 & 1 \end{bmatrix}.$$

Dispersing each entry of $\mathbf{B}_{\mathrm{PaG},p}(5,5)$ into a $5 \times 5$ CPM, we obtain the following $5 \times 5$ array of CPMs of size $5 \times 5$:

$$\mathbf{H}_{\mathrm{PaG},p}(5,5) = \begin{bmatrix}
1\,0\,0\,0\,0 & 1\,0\,0\,0\,0 & 1\,0\,0\,0\,0 & 1\,0\,0\,0\,0 & 1\,0\,0\,0\,0 \\
0\,1\,0\,0\,0 & 0\,1\,0\,0\,0 & 0\,1\,0\,0\,0 & 0\,1\,0\,0\,0 & 0\,1\,0\,0\,0 \\
0\,0\,1\,0\,0 & 0\,0\,1\,0\,0 & 0\,0\,1\,0\,0 & 0\,0\,1\,0\,0 & 0\,0\,1\,0\,0 \\
0\,0\,0\,1\,0 & 0\,0\,0\,1\,0 & 0\,0\,0\,1\,0 & 0\,0\,0\,1\,0 & 0\,0\,0\,1\,0 \\
0\,0\,0\,0\,1 & 0\,0\,0\,0\,1 & 0\,0\,0\,0\,1 & 0\,0\,0\,0\,1 & 0\,0\,0\,0\,1 \\
1\,0\,0\,0\,0 & 0\,1\,0\,0\,0 & 0\,0\,1\,0\,0 & 0\,0\,0\,1\,0 & 0\,0\,0\,0\,1 \\
0\,1\,0\,0\,0 & 0\,0\,1\,0\,0 & 0\,0\,0\,1\,0 & 0\,0\,0\,0\,1 & 1\,0\,0\,0\,0 \\
0\,0\,1\,0\,0 & 0\,0\,0\,1\,0 & 0\,0\,0\,0\,1 & 1\,0\,0\,0\,0 & 0\,1\,0\,0\,0 \\
0\,0\,0\,1\,0 & 0\,0\,0\,0\,1 & 1\,0\,0\,0\,0 & 0\,1\,0\,0\,0 & 0\,0\,1\,0\,0 \\
0\,0\,0\,0\,1 & 1\,0\,0\,0\,0 & 0\,1\,0\,0\,0 & 0\,0\,1\,0\,0 & 0\,0\,0\,1\,0 \\
1\,0\,0\,0\,0 & 0\,0\,1\,0\,0 & 0\,0\,0\,0\,1 & 0\,1\,0\,0\,0 & 0\,0\,0\,1\,0 \\
0\,1\,0\,0\,0 & 0\,0\,0\,1\,0 & 1\,0\,0\,0\,0 & 0\,0\,1\,0\,0 & 0\,0\,0\,0\,1 \\
0\,0\,1\,0\,0 & 0\,0\,0\,0\,1 & 0\,1\,0\,0\,0 & 0\,0\,0\,1\,0 & 1\,0\,0\,0\,0 \\
0\,0\,0\,1\,0 & 1\,0\,0\,0\,0 & 0\,0\,1\,0\,0 & 0\,0\,0\,0\,1 & 0\,1\,0\,0\,0 \\
0\,0\,0\,0\,1 & 0\,1\,0\,0\,0 & 0\,0\,0\,1\,0 & 1\,0\,0\,0\,0 & 0\,0\,1\,0\,0 \\
1\,0\,0\,0\,0 & 0\,0\,0\,1\,0 & 0\,1\,0\,0\,0 & 0\,0\,0\,0\,1 & 0\,0\,1\,0\,0 \\
0\,1\,0\,0\,0 & 0\,0\,0\,0\,1 & 0\,0\,1\,0\,0 & 1\,0\,0\,0\,0 & 0\,0\,0\,1\,0 \\
0\,0\,1\,0\,0 & 1\,0\,0\,0\,0 & 0\,0\,0\,1\,0 & 0\,1\,0\,0\,0 & 0\,0\,0\,0\,1 \\
0\,0\,0\,1\,0 & 0\,1\,0\,0\,0 & 0\,0\,0\,0\,1 & 0\,0\,1\,0\,0 & 1\,0\,0\,0\,0 \\
0\,0\,0\,0\,1 & 0\,0\,1\,0\,0 & 1\,0\,0\,0\,0 & 0\,0\,0\,1\,0 & 0\,1\,0\,0\,0 \\
1\,0\,0\,0\,0 & 0\,0\,0\,0\,1 & 0\,0\,0\,1\,0 & 0\,0\,1\,0\,0 & 0\,1\,0\,0\,0 \\
0\,1\,0\,0\,0 & 1\,0\,0\,0\,0 & 0\,0\,0\,0\,1 & 0\,0\,0\,1\,0 & 0\,0\,1\,0\,0 \\
0\,0\,1\,0\,0 & 0\,1\,0\,0\,0 & 1\,0\,0\,0\,0 & 0\,0\,0\,0\,1 & 0\,0\,0\,1\,0 \\
0\,0\,0\,1\,0 & 0\,0\,1\,0\,0 & 0\,1\,0\,0\,0 & 1\,0\,0\,0\,0 & 0\,0\,0\,0\,1 \\
0\,0\,0\,0\,1 & 0\,0\,0\,1\,0 & 0\,0\,1\,0\,0 & 0\,1\,0\,0\,0 & 1\,0\,0\,0\,0
\end{bmatrix}.$$

The array $\mathbf{H}_{\mathrm{PaG},p}(5,5)$ is a $25 \times 25$ matrix which is the line-point incidence matrix of a partial geometry $\mathrm{PaG}_p(5,5,4)$ that consists of 25 points and 25 lines, each line with five points. The 25 lines of $\mathrm{PaG}_p(5,5,4)$ can be partitioned into five parallel bundles, each with five parallel lines. Each point in $\mathrm{PaG}_p(5,5,4)$ is intersected by five lines. The five rows in each CPM-row-block of $\mathbf{H}_{\mathrm{PaG},p}(5,5)$ are the incidence vectors of five lines in a parallel bundle of $\mathrm{PaG}_p(5,5,4)$.

Label the rows and columns of $\mathbf{H}_{\mathrm{PaG},p}(5,5)$ from 0 to 24. The fourth, eighth, twelfth, sixteenth, and twentieth rows of $\mathbf{H}_{\mathrm{PaG},p}(5,5)$ are:

$$\begin{bmatrix}
0\,0\,0\,0\,1\,0\,0\,0\,0\,1\,0\,0\,0\,0\,1\,0\,0\,0\,0\,1\,0\,0\,0\,0\,1 \\
0\,0\,0\,1\,0\,0\,0\,0\,0\,1\,1\,0\,0\,0\,0\,0\,1\,0\,0\,0\,0\,0\,1\,0\,0 \\
0\,0\,1\,0\,0\,0\,0\,0\,0\,1\,0\,1\,0\,0\,0\,0\,0\,0\,1\,0\,1\,0\,0\,0\,0 \\
0\,1\,0\,0\,0\,0\,0\,0\,0\,1\,0\,0\,1\,0\,0\,1\,0\,0\,0\,0\,0\,0\,0\,1\,0 \\
1\,0\,0\,0\,0\,0\,0\,0\,0\,1\,0\,0\,0\,1\,0\,0\,0\,1\,0\,0\,0\,1\,0\,0\,0
\end{bmatrix},$$

which are orthogonal on the ninth column of $\mathbf{H}_{\mathrm{PaG},p}(5,5)$. These five rows correspond to the five lines in $\mathrm{PaG}_p(5,5,4)$ that intersect at the point corresponding to the ninth column of $\mathbf{H}_{\mathrm{PaG},p}(5,5)$. ▲▲

Let $d$ be a positive integer less than $p$. Take $d$ rows from $\mathbf{B}_{\mathrm{PaG},p}(p,p)$ given by (13.5) and form a $d \times p$ submatrix $\mathbf{B}_{\mathrm{PaG},p}(d,p)$ of $\mathbf{B}_{\mathrm{PaG},p}(p,p)$. The CPM-dispersion of $\mathbf{B}_{\mathrm{PaG},p}(d,p)$ results in a $d \times p$ subarray $\mathbf{H}_{\mathrm{PaG},p}(d,p)$ of $\mathbf{H}_{\mathrm{PaG},p}(p,p)$. The subarray $\mathbf{H}_{\mathrm{PaG},p}(d,p)$ is a $dp \times p^2$ matrix with column and row weights $d$ and $p$, respectively. The matrix $\mathbf{H}_{\mathrm{PaG},p}(d,p)$ is the line-point incidence matrix of a subgeometry of $\mathrm{PaG}_p(p,p,p-1)$. The null space over GF(2) of $\mathbf{H}_{\mathrm{PaG},p}(d,p)$ gives a $(d,p)$-regular QC-PaG-LDPC code, denoted by $C_{\mathrm{PaG},p}(d,p)$, whose Tanner graph has girth at least 6.

Let $d$ and $k$ be two positive integers such that $1 \le d < k \le p$. Take a $d \times k$ submatrix $\mathbf{B}_{\mathrm{PaG},p}(d,k)$ from $\mathbf{B}_{\mathrm{PaG},p}(p,p)$ given by (13.5). The CPM-dispersion of $\mathbf{B}_{\mathrm{PaG},p}(d,k)$ results in a $d \times k$ subarray $\mathbf{H}_{\mathrm{PaG},p}(d,k)$ of $\mathbf{H}_{\mathrm{PaG},p}(p,p)$. The subarray $\mathbf{H}_{\mathrm{PaG},p}(d,k)$ is a $dp \times kp$ matrix with column and row weights $d$ and $k$, respectively. The null space over GF(2) of $\mathbf{H}_{\mathrm{PaG},p}(d,k)$ gives a $(d,k)$-regular QC-PaG-LDPC code, denoted by $C_{\mathrm{PaG},p}(d,k)$, whose Tanner graph has girth at least 6.

**Example 13.8** Consider the prime field GF(181). Based on this field, we can construct a $181 \times 181$ base matrix $\mathbf{B}_{\mathrm{PaG},p}(181,181)$ in the form of (13.5). The $181 \times 181$ CPM-dispersion of $\mathbf{B}_{\mathrm{PaG},p}(181,181)$ gives a $181 \times 181$ array $\mathbf{H}_{\mathrm{PaG},p}(181,181)$ of CPMs of size $181 \times 181$ which is a $32\,761 \times 32\,761$ matrix. The matrix $\mathbf{H}_{\mathrm{PaG},p}(181,181)$ is the line-point incidence matrix of a partial geometry $\mathrm{PaG}_p(181,181,180)$ with $32\,761$ points and $32\,761$ lines.

Label the rows and columns of $\mathbf{B}_{\mathrm{PaG},p}(181,181)$ from 0 to 180. Set $d = 4$ and $k = 64$. Take the 4 rows, labeled by 1, 2, 3, and 4, from $\mathbf{B}_{\mathrm{PaG},p}(181,181)$ and then take the 64 columns, labeled from 1 to 64, from the four rows. This gives a $4 \times 64$ submatrix $\mathbf{B}_{\mathrm{PaG},p}(4,64)$ of $\mathbf{B}_{\mathrm{PaG},p}(181,181)$. The CPM-dispersion of $\mathbf{B}_{\mathrm{PaG},p}(4,64)$ produces a $4 \times 64$ array $\mathbf{H}_{\mathrm{PaG},p}(4,64)$ of CPMs of size $181 \times 181$ which is a $724 \times 11\,584$ matrix with column and row weights 4 and 64, respectively. The rank of $\mathbf{H}_{\mathrm{PaG},p}(4,64)$ is 721.

The null space over GF(2) of $\mathbf{H}_{\mathrm{PaG},p}(4,64)$ gives a $(4,64)$-regular $(11\,584, 10\,863)$ QC-PaG-LDPC code $C_{\mathrm{PaG},p}(4,64)$ with rate 0.9378. The Tanner graph of the code has girth 6. Each VN is connected to other 252 VNs of the graph by paths of length 2. This high degree of connectivity allows rapid exchange of information among the VNs. In each iteration of an IDBP-algorithm, each VN receives extrinsic information from other 252 VNs. As a result, the decoding process can converge into a codeword with a small number of decoding iterations.

The BER and BLER performances of the code $C_{\mathrm{PaG},p}(4,64)$ decoded with 5, 10, 50, and 100 iterations of the MSA scaled by a factor of 0.75 over the AWGN channel using BPSK signaling are shown in Fig. 13.4(a). From the figure, we see that the code $C_{\mathrm{PaG},p}(4,64)$ performs well. With 50 iterations of the MSA, the code achieves a BER of $10^{-9}$ without a visible error-floor. At a BER of $10^{-9}$, the code performs about 1.1 dB away from the Shannon limit. Furthermore, decoding of the code converges quickly. At a BER of $10^{-9}$, the performance gap between 10 and 50 iterations is about 0.1 dB and there is no performance gap between 50 and 100 iterations.

Various choices of $d$ and $k$ can be used to construct a large family of QC-PaG-LDPC codes. Here, we construct another three QC-PaG-LDPC codes, $C_{\text{PaG},p}(4, 40)$, $C_{\text{PaG},p}(4, 80)$, and $C_{\text{PaG},p}(4, 160)$, by choosing $d = 4$ and $k = 40, 80, 160$, respectively. The code $C_{\text{PaG},p}(4, 40)$ is a $(4, 40)$-regular $(7240, 6519)$ code with rate 0.9004, $C_{\text{PaG},p}(4, 80)$ is a $(4, 80)$-regular $(14\,480, 13\,759)$ code with rate 0.9502, and $C_{\text{PaG},p}(4, 160)$ is a $(4, 160)$-regular $(28\,960, 28\,239)$ code with rate 0.9751. The Tanner graphs of these codes have girth 6. The BER performances of the three codes and the code $C_{\text{PaG},p}(4, 64)$ over the AWGN channel decoded with 50 iterations of the MSA are shown in Fig. 13.4(b). From this figure, we see that all the constructed QC-PaG-LDPC codes perform well and have no visible error-floor down to the BER of $10^{-8}$. ▲▲

**Example 13.9** Consider the $127 \times 127$ base matrix $\mathbf{B}_{\text{PaG},p}(127, 127)$ over the prime field GF(127) in the form given by (13.5). The CPM-dispersion of $\mathbf{B}_{\text{PaG},p}(127, 127)$ gives a $127 \times 127$ array $\mathbf{H}_{\text{PaG},p}(127, 127)$ of CPMs of size $127 \times 127$, a $16\,129 \times 16\,129$ matrix, which is the line-point incidence matrix of a partial geometry $\text{PaG}_p(127, 127, 126)$ with columns corresponding to the $16\,129$ points and rows corresponding to the $16\,129$ lines of $\text{PaG}_p(127, 127, 126)$.

Consider the following $4 \times 8$ submatrix $\mathbf{B}_{\text{PaG},p}(4, 8)$ of $\mathbf{B}_{\text{PaG},p}(127, 127)$ (with rows and columns chosen at random):

$$\mathbf{B}_{\text{PaG},p}(4, 8) = \begin{bmatrix} 2 & 83 & 33 & 46 & 36 & 94 & 42 & 86 \\ 109 & 15 & 84 & 94 & 57 & 43 & 3 & 115 \\ 112 & 76 & 70 & 36 & 111 & 57 & 66 & 117 \\ 31 & 80 & 67 & 78 & 50 & 60 & 16 & 63 \end{bmatrix}.$$

Masking $\mathbf{B}_{\text{PaG},p}(4, 8)$ with the following masking matrix $\mathbf{Z}(4, 8)$ (see (12.13)):

$$\mathbf{Z}(4, 8) = \begin{bmatrix} 1 & 0 & 1 & 0 & 1 & 1 & 1 & 1 \\ 1 & 1 & 1 & 1 & 1 & 0 & 1 & 0 \\ 0 & 1 & 0 & 1 & 1 & 1 & 1 & 1 \\ 1 & 1 & 1 & 1 & 0 & 1 & 0 & 1 \end{bmatrix}, \tag{13.6}$$

we obtain the following masked base matrix over GF(127):

$$\mathbf{B}_{\text{PaG},p,\text{mask}}(4, 8) = \begin{bmatrix} 2 & -\infty & 33 & -\infty & 36 & 94 & 42 & 86 \\ -\infty & 15 & -\infty & 94 & 57 & 43 & 3 & 115 \\ 112 & 76 & 70 & 36 & 111 & -\infty & 66 & -\infty \\ 31 & 80 & 67 & 78 & -\infty & 60 & -\infty & 63 \end{bmatrix},$$

where $-\infty$ is used to represent the masked entries which will be replaced by ZMs in the CPM-dispersion. The $127 \times 127$ CPM-dispersion of $\mathbf{B}_{\text{PaG},p,\text{mask}}(4, 8)$ results in a $4 \times 8$ masked array $\mathbf{H}_{\text{PaG},p,\text{mask}}(4, 8)$ of CPMs and ZMs of size $127 \times 127$. The array $\mathbf{H}_{\text{PaG},p,\text{mask}}(4, 8)$ is a masked subarray of $\mathbf{H}_{\text{PaG},p}(127, 127)$ which is a $508 \times 1016$ matrix with column and row weights 3 and 6, respectively. The null space over GF(2) of $\mathbf{H}_{\text{PaG},p,\text{mask}}(4, 8)$ gives a $(3, 6)$-regular $(1016, 508)$ QC-PaG-LDPC code with rate 0.5. The Tanner graph of this code has girth 8 and contains 889 cycles of length 8.

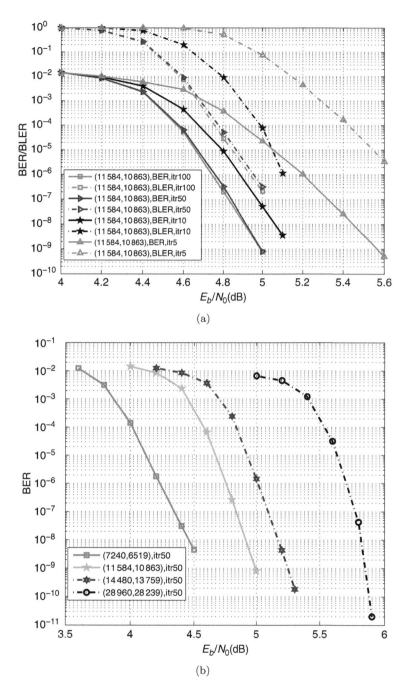

Figure 13.4 (a) The BER and BLER performances of the $(11\,584, 10\,863)$ QC-PaG-LDPC code $C_{\mathrm{PaG},p}(4, 64)$ in Example 13.8 and (b) the BER performances of the four codes in Example 13.8.

The BER and BLER performances of the code over the AWGN channel decoded with 5, 10, and 50 iterations of the MSA scaled by a factor of 0.75 are shown in Fig. 13.5(a). We see that the code achieves a BER of $10^{-9}$ without error-floor. With 50 iterations and at the BER of $10^{-9}$, the code performs 2.2 dB from the threshold (1.11 dB). Also included in the figure are the BER and BLER performances of a (1016, 508) PEG-LDPC code $C_{\text{peg}}$ decoded with 50 iterations of the MSA scaled by the same scaling factor. From Fig. 13.5(a), we can see that the performance curves of the constructed QC-PaG-LDPC code and the PEG-LDPC code overlap with each other.

If we apply this code to the BEC, its UEBR and UEBLR performances are shown in Fig. 13.5(b). Because the code has rate $R = 0.5$, the Shannon limit for this rate over the BEC is $1 - 0.5 = 0.5$. From Fig. 13.5(b), we see that the code performs also well over the BEC. ▲▲

## 13.6 Partial Geometries Based on Balanced Incomplete Block Designs and Their Associated QC-LDPC Codes

*Balanced incomplete block designs* (BIBDs) form a class of *combinatorial* (or *experimental*) designs [6, 8, 9, 11, 12, 19–24]. Most recently, combinatorial designs have been utilized to construct LDPC codes [2, 25–28]. In this section, we first show that BIBDs of a special case are partial geometries, called *BIBD associated partial geometries* (BIBD-PaGs). Next, we present two classes of BIBD-PaGs which are constructed based on prime fields under certain constraints. Based on these two classes of BIBD-PaGs, QC-PaG-LDPC codes can be constructed.

### 13.6.1 BIBDs and Partial Geometries

A BIBD is an arrangement of $N$ objects in $M$ sets that satisfies the following conditions.
(1) Each set contains $t$ *different* objects.
(2) Each object appears in $r$ *different* sets.
(3) Any pair of objects appears in $\lambda$ *different* sets.

The sets are called *blocks*. The integer parameters $N$, $M$, $r$, $t$, $\lambda$ satisfy the following two equalities:

$$tM = rN, \tag{13.7}$$

$$\lambda(N - 1) = r(t - 1). \tag{13.8}$$

Because a BIBD is characterized by these five parameters, $N$, $M$, $r$, $t$, $\lambda$, it is denoted as an $(N, M, t, r, \lambda)$-BIBD.

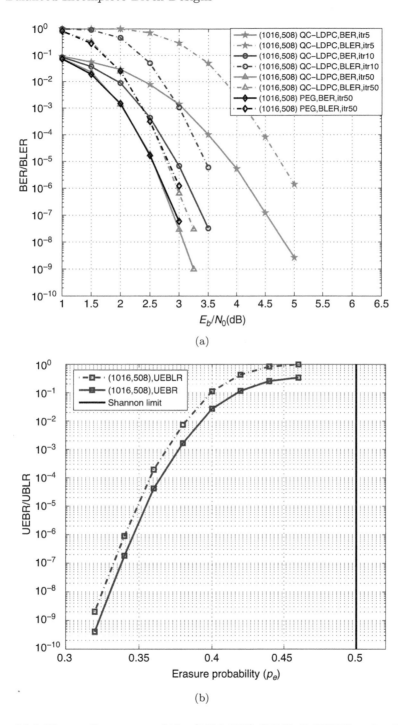

Figure 13.5 The performances of the $(1016, 508)$ QC-PaG-LDPC code given in Example 13.9 over: (a) AWGN channel and (b) BEC.

For a BIBD with $\lambda = 1$, each pair of objects appears in one and only one block. This implies that two blocks in an $(N, M, t, r, 1)$-BIBD have *at most one* object in common. By viewing the objects and blocks as points and lines, respectively, an $(N, M, t, r, 1)$-BIBD is a partial geometry, denoted by $\mathrm{PaG}_{\mathrm{BIBD}}(r, t, t)$, with $N$ points and $M$ lines in which each line consists of $t$ points and each point is intersected by $r$ lines. For a point **a** not on a line $L$, there are $t$ lines, each passing through the point **a** and one point on $L$, because any two points in $\mathrm{PaG}_{\mathrm{BIBD}}(r, t, t)$ are on one and only one line.

In the case of a BIBD with $\lambda = 1$, it follows from (13.7) and (13.8) that

$$N = r(t - 1) + 1, \tag{13.9}$$

$$M = r(r(t - 1) + 1)/t. \tag{13.10}$$

The line-point incidence matrix $\mathbf{H}_{\mathrm{PaG,BIBD}}(M, N)$ of the partial geometry $\mathrm{PaG}_{\mathrm{BIBD}}(r, t, t)$ (or the block-object incidence matrix of the $(N, M, t, r, 1)$-BIBD) is an $M \times N$ matrix over $\mathrm{GF}(2)$ with column and row weights $r$ and $t$, respectively. The rows of $\mathbf{H}_{\mathrm{PaG,BIBD}}(M, N)$ are the incidence vectors of the lines in $\mathrm{PaG}_{\mathrm{BIBD}}(r, t, t)$ (or the blocks of the $(N, M, t, r, 1)$-BIBD). The columns of $\mathbf{H}_{\mathrm{PaG,BIBD}}(M, N)$ correspond to the point of $\mathrm{PaG}_{\mathrm{BIBD}}(r, t, t)$ (or objects of the $(N, M, t, r, 1)$-BIBD). Because any two points (objects) appear on one and only one line (block), the matrix $\mathbf{H}_{\mathrm{PaG,BIBD}}(M, N)$ satisfies the RC-constraint. We call the partial geometry $\mathrm{PaG}_{\mathrm{BIBD}}(r, t, t)$ a *BIBD associated partial geometry*, BIBD-PaG in short.

**Example 13.10** Consider a set $\mathbf{S} = \{\mathbf{a}_0, \mathbf{a}_1, \mathbf{a}_2, \mathbf{a}_3, \mathbf{a}_4, \mathbf{a}_5, \mathbf{a}_6\}$ of seven objects. The blocks

$$\mathcal{B}_0 = \{\mathbf{a}_0, \mathbf{a}_1, \mathbf{a}_3\}, \quad \mathcal{B}_1 = \{\mathbf{a}_1, \mathbf{a}_2, \mathbf{a}_4\}, \quad \mathcal{B}_2 = \{\mathbf{a}_2, \mathbf{a}_3, \mathbf{a}_5\},$$
$$\mathcal{B}_3 = \{\mathbf{a}_3, \mathbf{a}_4, \mathbf{a}_6\}, \quad \mathcal{B}_4 = \{\mathbf{a}_0, \mathbf{a}_4, \mathbf{a}_5\}, \quad \mathcal{B}_5 = \{\mathbf{a}_1, \mathbf{a}_5, \mathbf{a}_6\},$$
$$\mathcal{B}_6 = \{\mathbf{a}_0, \mathbf{a}_2, \mathbf{a}_6\},$$

form a $(7, 7, 3, 3, 1)$-BIBD. Each block consists of three objects. Each object appears in three blocks and any pair of objects appears in one and only one block.

The $(7, 7, 3, 3, 1)$-BIBD forms a partial geometry $\mathrm{PaG}_{\mathrm{BIBD}}(3, 3, 3)$ which consists of seven points and seven lines. Each line consists of three points and each point is on three lines. Any two points in $\mathrm{PaG}_{\mathrm{BIBD}}(3, 3, 3)$ are on one and only one line. The line-point incidence matrix of the partial geometry $\mathrm{PaG}_{\mathrm{BIBD}}(3, 3, 3)$ is

$$\mathbf{H}_{\mathrm{PaG,BIBD}}(7, 7) = \begin{bmatrix} 1 & 1 & 0 & 1 & 0 & 0 & 0 \\ 0 & 1 & 1 & 0 & 1 & 0 & 0 \\ 0 & 0 & 1 & 1 & 0 & 1 & 0 \\ 0 & 0 & 0 & 1 & 1 & 0 & 1 \\ 1 & 0 & 0 & 0 & 1 & 1 & 0 \\ 0 & 1 & 0 & 0 & 0 & 1 & 1 \\ 1 & 0 & 1 & 0 & 0 & 0 & 1 \end{bmatrix}.$$

The matrix $\mathbf{H}_{\text{PaG,BIBD}}(7,7)$ is a circulant with weight 3 and satisfies the RC-constraint. The rank of $\mathbf{H}_{\text{PaG,BIBD}}(7,7)$ is 4. ▲▲

In the following, we present two classes of $(N, M, t, r, 1)$-BIBDs and their associated partial geometries. Both classes of $(N, M, t, r, 1)$-BIBDs are constructed based on prime fields under certain constraints. These two classes of BIBDs were discovered by Bose [20], called *Bose BIBDs*. Here, we construct the designs directly without any proof. For detailed proofs, readers are referred to [20].

## 13.6.2 Class-1 Bose $(N, M, t, r, 1)$-BIBDs

Let $b$ be a positive integer such that $12b + 1$ is a prime. If the prime field $\text{GF}(12b + 1) = \{0, 1, 2, \ldots, 12b\}$ has a primitive element $\alpha$ which satisfies the following constraint:

$$\alpha^{4b} - 1 = \alpha^c \tag{13.11}$$

where $c$ is an *odd nonnegative integer* less than $12b + 1$, then there exists a $(12b+1, b(12b+1), 4, 4b, 1)$-BIBD, denoted by $B_1(12b+1)$, in which the $12b+1$ elements of $\text{GF}(12b + 1)$ are treated as objects in the BIBD [20]. Associated to this BIBD is a partial geometry with $12b+1$ points and $b(12b+1)$ lines. In this partial geometry, each line consists of four points, each point is intersected by $4b$ lines, and for a point $\mathbf{a}$ not on a line $L$, there are four lines, each passing through the point $\mathbf{a}$ and one point on the line $L$.

To construct such a BIBD, we first form $b$ *base blocks*. Label these $b$ base blocks from 0 to $b - 1$. For $0 \leq i < b$, the $i$th base block is given as follows:

$$\mathcal{B}_{i,0} = \{\alpha^{-\infty}, \alpha^{2i}, \alpha^{2i+4b}, \alpha^{2i+8b}\}, \tag{13.12}$$

which consists of four elements in $\text{GF}(12b + 1)$. For each base block $\mathcal{B}_{i,0}$, we form $12b+1$ blocks, $\mathcal{B}_{i,0}, \mathcal{B}_{i,1}, \ldots, \mathcal{B}_{i,12b}$, by adding each element of $\text{GF}(12b+1)$ in turn to the elements in $\mathcal{B}_{i,0}$. Then, for $0 \leq j \leq 12b$, we have

$$\mathcal{B}_{i,j} = \{j + \alpha^{-\infty}, j + \alpha^{2i}, j + \alpha^{2i+4b}, j + \alpha^{2i+8b}\}, \tag{13.13}$$

where the addition is modulo-$(12b+1)$ addition. The $12b+1$ blocks, $\mathcal{B}_{i,0}, \mathcal{B}_{i,1}, \ldots, \mathcal{B}_{i,12b}$, are called *coblocks* of the base block $\mathcal{B}_{i,0}$ and they form a *translate class*, denoted by $\mathcal{T}_i$. Based on the $b$ base blocks given by (13.12), we construct $b$ translate classes $\mathcal{T}_0, \mathcal{T}_1, \ldots, \mathcal{T}_{b-1}$, each consisting of $12b + 1$ coblocks. The $b(12b+1)$ blocks in the $b$ translate classes form a $(12b+1, b(12b+1), 4, 4b, 1)$-BIBD, called *class-1 Bose-BIBD*.

Table 13.1 gives a list of $b$s such that $12b + 1$ is a prime and the prime field $\text{GF}(12b+1)$ has a primitive element $\alpha$ satisfying the condition given by (13.11).

Viewing each block in $B_1(12b + 1)$ as a *line* and each element in a block as a *point*, $B_1(12b + 1)$ is a partial geometry, denoted by $\text{PaG}_{\text{BIBD},1}(4b, 4, 4)$ and

Table 13.1 A list of $b$s for which $12b + 1$ is a prime and the field $\mathrm{GF}(12b + 1)$ satisfies the condition given by (13.11).

| $b$ | Field | $(\alpha, c)$ | $b$ | Field | $(\alpha, c)$ |
|---|---|---|---|---|---|
| 1 | GF(13) | (2, 1) | 28 | GF(337) | (10, 129) |
| 6 | GF(73) | (5, 33) | 34 | GF(409) | (21, 9) |
| 8 | GF(97) | (5, 27) | 35 | GF(421) | (2, 167) |
| 9 | GF(109) | (6, 71) | 38 | GF(457) | (13, 387) |
| 15 | GF(181) | (2, 13) | 45 | GF(541) | (2, 7) |
| 19 | GF(229) | (6, 199) | 59 | GF(709) | (2, 381) |
| 20 | GF(241) | (7, 191) | 61 | GF(733) | (6, 145) |
| 23 | GF(277) | (5, 209) | | | |

called *class-1 Bose-BIBD-PaG*, which consists of $12b + 1$ points and $b(12b + 1)$ lines. Each line contains four points and each point is on (or intersected by) $4b$ lines. Hereafter, we call a block $\mathcal{B}_{i,j}$ in $B_1(12b + 1)$ a line and each object in $B_1(12b + 1)$ a point. We label the points and lines in $\mathrm{PaG}_{\mathrm{BIBD},1}(4b, 4, 4)$ from 0 to $12b$ and 0 to $b(12b + 1) - 1$, respectively.

For $0 \leq i < b$ and $0 \leq j < 12b$, let $\mathbf{v}_{i,j}$ be the incidence vector of the line $\mathcal{B}_{i,j}$ in the translate class $\mathcal{T}_i$ which is a $(12b+1)$-tuple over $\mathrm{GF}(2)$ with 1-components at the locations labeled by the four objects (integers) in $\mathcal{B}_{i,j}$. From (13.12) and (13.13), we see that $\mathbf{v}_{i,j}$ is the *cyclic-shift* of the incidence vector $\mathbf{v}_{i,0}$ of the base line $\mathcal{B}_{i,0}$ to the right $j$ positions. Hence, the $(12b + 1) \times (12b + 1)$ matrix formed by the incidence vectors of lines in the $i$th translate class $\mathcal{T}_i$ is a circulant $\mathbf{G}_i$ with weight 4. The generator (the top row) of $\mathbf{G}_i$ is the incidence vector $\mathbf{v}_{i,0}$ of the base line $\mathcal{B}_{i,0}$ in $\mathrm{PaG}_{\mathrm{BIBD},1}(4b, 4, 4)$. Because $\mathbf{G}_i$ is a circulant, the columns of $\mathbf{G}_i$ are also the incidence vectors of the lines in the translate class $\mathcal{T}_i$ but in *reverse order* (by exchanging the roles of points and lines).

The line-point incidence matrix $\mathbf{H}_{\mathrm{PaG,BIBD},1}(b(12b+1), 12b+1)$ of the partial geometry $\mathrm{PaG}_{\mathrm{BIBD},1}(4b, 4, 4)$ is a $b(12b+1) \times (12b+1)$ matrix over $\mathrm{GF}(2)$ which consists of $b$ circulants, $\mathbf{G}_0, \mathbf{G}_1, \ldots, \mathbf{G}_{b-1}$, in a column as follows:

$$\mathbf{H}_{\mathrm{PaG,BIBD},1}(b(12b + 1), 12b + 1) = \begin{bmatrix} \mathbf{G}_0 \\ \mathbf{G}_1 \\ \vdots \\ \mathbf{G}_{b-1} \end{bmatrix}. \tag{13.14}$$

The column and row weights of $\mathbf{H}_{\mathrm{PaG,BIBD},1}(b(12b + 1), 12b + 1)$ are $4b$ and 4, respectively. The $12b + 1$ lines in a translate class $\mathcal{T}_i$ actually form a cyclic class of lines as defined in Chapter 7. By the properties of $B_1(12b + 1)$, it is straightforward to verify that the matrix $\mathbf{H}_{\mathrm{PaG,BIBD},1}(b(12b + 1), 12b + 1)$ satisfies the RC-constraint.

The number of rows in $\mathbf{H}_{\mathrm{PaG,BIBD},1}(b(12b + 1), 12b + 1)$ is $b$ times the number of its columns. For $b > 1$, the LDPC code given by the null space of $\mathbf{H}_{\mathrm{PaG,BIBD},1}(b(12b + 1), 12b + 1)$ or a subset of its constituent circulants will be a very low rate code, even though it is cyclic. For LDPC code constructions, rather than directly using $\mathbf{H}_{\mathrm{PaG,BIBD},1}(b(12b+1), 12b+1)$ or its submatrix as the parity-check matrix, we use the following matrix:·

$$\mathbf{H}^*_{\mathrm{PaG,BIBD},1}(12b+1, b(12b+1)) = \begin{bmatrix} \mathbf{G}_0 & \mathbf{G}_1 & \cdots & \mathbf{G}_{b-1} \end{bmatrix}, \qquad (13.15)$$

which is a $(12b+1) \times b(12b+1)$ matrix over GF(2) with the $b$ circulants $\mathbf{G}_0, \mathbf{G}_1, \ldots, \mathbf{G}_{b-1}$ arranged in a row. In the form of (13.15), the columns of $\mathbf{H}^*_{\mathrm{PaG,BIBD},1}(12b+1, b(12b+1))$ are the incidence vectors of the lines in $\mathrm{PaG}_{\mathrm{BIBD},1}(4b, 4, 4)$ but in reverse order. The matrix $\mathbf{H}^*_{\mathrm{PaG,BIBD},1}(12b+1, b(12b+1))$ has column and row weights 4 and $4b$, respectively, and it also satisfies the RC-constraint.

The null space over GF(2) of $\mathbf{H}^*_{\mathrm{PaG,BIBD},1}(12b+1, b(12b+1))$ gives a $(4, 4b)$-regular QC-BIBD-PaG-LDPC code $C_{\mathrm{PaG,BIBD},1}(4, 4b)$ of length $b(12b+1)$ with rate at least $(b-1)/b$ whose Tanner graph has girth at least 6. For $1 \le k < b$, we can use any $k$ circulants in the set $\{\mathbf{G}_0, \mathbf{G}_1, \ldots, \mathbf{G}_{b-1}\}$, say $\mathbf{G}_0, \mathbf{G}_1, \ldots, \mathbf{G}_{k-1}$, to form a $(12b+1) \times k(12b+1)$ matrix as follows:

$$\mathbf{H}^*_{\mathrm{PaG,BIBD},1}(12b+1, k(12b+1)) = \begin{bmatrix} \mathbf{G}_0 & \mathbf{G}_1 & \cdots & \mathbf{G}_{k-1} \end{bmatrix}. \qquad (13.16)$$

The null space over GF(2) of $\mathbf{H}^*_{\mathrm{PaG,BIBD},1}(12b+1, k(12b+1))$ gives a $(4, 4k)$-regular QC-BIBD-PaG-LDPC code $C_{\mathrm{PaG,BIBD},1}(4, 4k)$ of length $k(12b+1)$ with rate at least $(k-1)/k$.

**Example 13.11** Let $b = 1$. The integer $12 \times 1 + 1 = 13$ is a prime. Based on this prime number, we construct a prime field GF(13) = $\{0, 1, 2, 3, 4, 5, 6, 7, 8, 9, 10, 11, 12\}$ under modulo-13 addition and multiplication. The element 2 is a primitive element in GF(13) which satisfies the condition given by (13.11) with $c = 1$. The field GF(13) is the first field listed in Table 13.1. Based on this field, we can construct a $(13, 13, 4, 4, 1)$-BIBD $B_1(13)$. Because $b = 1$, there is only one translate class $\mathcal{T}_0$ which consists of all 13 blocks of $B_1(13)$. The base block of this translate class is

$$\mathcal{B}_{0,0} = \{0, 1, 2^4, 2^8\} = \{0, 1, 3, 9\}.$$

The base block $\mathcal{B}_{0,0}$ and its 12 coblocks are:

$$\{0, 1, 3, 9\}, \{1, 2, 4, 10\}, \{2, 3, 5, 11\}, \{3, 4, 6, 12\}, \{4, 5, 7, 0\}$$
$$\{5, 6, 8, 1\}, \{6, 7, 9, 2\}, \{7, 8, 10, 3\}, \{8, 9, 11, 4\}, \{9, 10, 12, 5\},$$
$$\{10, 11, 0, 6\}, \{11, 12, 1, 7\}, \{12, 0, 2, 8\},$$

which form the translate class $\mathcal{T}_0$ of $B_1(13)$. Associated to $B_1(13)$ is a partial geometry $\mathrm{PaG}_{\mathrm{BIBD},1}(4, 4, 4)$ which consists of 13 points and 13 lines. Each line consists of four points and each point is intersected by four lines. The line-point incidence matrix of $\mathrm{PaG}_{\mathrm{BIBD},1}(4, 4, 4)$ is:

$$\mathbf{H}_{\mathrm{PaG,BIBD},1}(13,13) = \begin{bmatrix} 1 & 1 & 0 & 1 & 0 & 0 & 0 & 0 & 0 & 1 & 0 & 0 & 0 \\ 0 & 1 & 1 & 0 & 1 & 0 & 0 & 0 & 0 & 0 & 1 & 0 & 0 \\ 0 & 0 & 1 & 1 & 0 & 1 & 0 & 0 & 0 & 0 & 0 & 1 & 0 \\ 0 & 0 & 0 & 1 & 1 & 0 & 1 & 0 & 0 & 0 & 0 & 0 & 1 \\ 1 & 0 & 0 & 0 & 1 & 1 & 0 & 1 & 0 & 0 & 0 & 0 & 0 \\ 0 & 1 & 0 & 0 & 0 & 1 & 1 & 0 & 1 & 0 & 0 & 0 & 0 \\ 0 & 0 & 1 & 0 & 0 & 0 & 1 & 1 & 0 & 1 & 0 & 0 & 0 \\ 0 & 0 & 0 & 1 & 0 & 0 & 0 & 1 & 1 & 0 & 1 & 0 & 0 \\ 0 & 0 & 0 & 0 & 1 & 0 & 0 & 0 & 1 & 1 & 0 & 1 & 0 \\ 0 & 0 & 0 & 0 & 0 & 1 & 0 & 0 & 0 & 1 & 1 & 0 & 1 \\ 1 & 0 & 0 & 0 & 0 & 0 & 1 & 0 & 0 & 0 & 1 & 1 & 0 \\ 0 & 1 & 0 & 0 & 0 & 0 & 0 & 1 & 0 & 0 & 0 & 1 & 1 \\ 1 & 0 & 1 & 0 & 0 & 0 & 0 & 0 & 1 & 0 & 0 & 0 & 1 \end{bmatrix}.$$

The matrix $\mathbf{H}_{\mathrm{PaG,BIBD},1}(13,13)$ consists of a single $13 \times 13$ circulant formed by the incidence vectors of the 13 lines in the translate class $\mathcal{T}_0$. It satisfies the RC-constraint. ▲▲

**Example 13.12** For $b = 8$, $12b + 1 = 97$ is a prime. As shown in Table 13.1, the prime field GF(97) has a primitive element $\alpha = 5$, which satisfies the condition given by (13.11) with $c = 27$. Based on GF(97), we can construct a $(97, 776, 4, 32, 1)$-BIBD $B_1(97)$ by constructing the following eight base blocks:

$$\mathcal{B}_{0,0} = \{0, 1, 5^{32}, 5^{64}\}, \quad \mathcal{B}_{1,0} = \{0, 5^2, 5^{34}, 5^{66}\}, \quad \mathcal{B}_{2,0} = \{0, 5^4, 5^{36}, 5^{68}\},$$
$$\mathcal{B}_{3,0} = \{0, 5^6, 5^{38}, 5^{70}\}, \quad \mathcal{B}_{4,0} = \{0, 5^8, 5^{40}, 5^{72}\}, \quad \mathcal{B}_{5,0} = \{0, 5^{10}, 5^{42}, 5^{74}\},$$
$$\mathcal{B}_{6,0} = \{0, 5^{12}, 5^{44}, 5^{76}\}, \quad B_{7,0} = \{0, 5^{14}, 5^{46}, 5^{78}\}.$$

The 776 blocks constructed from the above eight base blocks form the BIBD $B_1(97)$. Associated to $B_1(97)$ is a partial geometry $\mathrm{PaG}_{\mathrm{BIBD},1}(32,4,4)$ with 97 points and 776 lines. Each line consists of four points and each point is intersected by 32 lines. The lines in $\mathrm{PaG}_{\mathrm{BIBD},1}(32,4,4)$ can be partitioned into eight cyclic (translate) classes, $\mathcal{T}_0$, $\mathcal{T}_1$, $\mathcal{T}_2$, $\mathcal{T}_3$, $\mathcal{T}_4$, $\mathcal{T}_5$, $\mathcal{T}_6$, $\mathcal{T}_7$, each consisting of 97 lines. Using the incidence vectors of lines in these eight cyclic classes, we can form eight circulants $\mathbf{G}_0$, $\mathbf{G}_1$, $\mathbf{G}_2$, $\mathbf{G}_3$, $\mathbf{G}_4$, $\mathbf{G}_5$, $\mathbf{G}_6$, $\mathbf{G}_7$, each of size $97 \times 97$ and weight 4.

Suppose we use the eight circulants $\mathbf{G}_0$, $\mathbf{G}_1$, $\mathbf{G}_2$, $\mathbf{G}_3$, $\mathbf{G}_4$, $\mathbf{G}_5$, $\mathbf{G}_6$, $\mathbf{G}_7$ to form the following $97 \times 776$ matrix:

$$\mathbf{H}^*_{\mathrm{PaG,BIBD},1}(97,776) = \begin{bmatrix} \mathbf{G}_0 & \mathbf{G}_1 & \mathbf{G}_2 & \mathbf{G}_3 & \mathbf{G}_4 & \mathbf{G}_5 & \mathbf{G}_6 & \mathbf{G}_7 \end{bmatrix}.$$

The null space over GF(2) of $\mathbf{H}^*_{\mathrm{PaG,BIBD},1}(97,776)$ gives a $(4,32)$-regular $(776, 680)$ QC-BIBD-PaG-LDPC code $C_{\mathrm{PaG,BIBD},1}(4,32)$ with rate 0.8763. The BER and BLER performances of $C_{\mathrm{PaG,BIBD},1}(4,32)$ over the AWGN channel using BPSK signaling decoded with 5, 10, and 50 iterations of the MSA scaled by a factor of 0.75 are shown in Fig. 13.6.

For any positive integer $k \leq 8$, we can take $k$ circulants from the set $\{\mathbf{G}_0, \mathbf{G}_1, \mathbf{G}_2, \mathbf{G}_3, \mathbf{G}_4, \mathbf{G}_5, \mathbf{G}_6, \mathbf{G}_7\}$ and form a matrix with the chosen circulants arranged in a row. The null space over GF(2) of this matrix gives a QC-BIBD-PaG-LDPC code. ▲▲

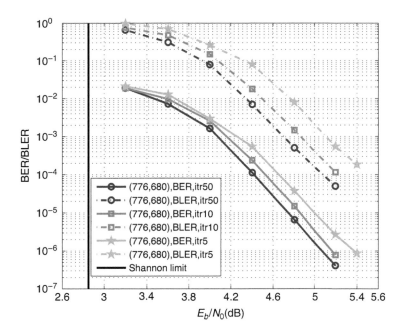

Figure 13.6 The BER and BLER performances of the $(776, 680)$ QC-BIBD-PaG-LDPC code $C_{\mathrm{PaG,BIBD},1}(4, 32)$ given in Example 13.12.

Using the $b$ base lines of $\mathrm{PaG}_{\mathrm{BIBD},1}(4b, 4, 4)$, we can construct a $4 \times b$ array of CPMs of size $(12b+1) \times (12b+1)$ which can be used to construct QC-LDPC codes. To achieve this, we first form a $4 \times b$ matrix over $\mathrm{GF}(12b+1)$ as follows (with rows and columns labeled from 0 to 3 and 0 to $b-1$, respectively):

$$
\mathbf{B}_{\mathrm{PaG,BIBD},1}(4, b) = \begin{bmatrix} \alpha^{-\infty} & \alpha^{-\infty} & \alpha^{-\infty} & \cdots & \alpha^{-\infty} \\ \alpha^0 & \alpha^2 & \alpha^4 & \cdots & \alpha^{2(b-1)} \\ \alpha^{4b} & \alpha^{2+4b} & \alpha^{4+4b} & \cdots & \alpha^{2(b-1)+4b} \\ \alpha^{8b} & \alpha^{2+8b} & \alpha^{4+8b} & \cdots & \alpha^{2(b-1)+8b} \end{bmatrix}, \quad (13.17)
$$

where, for $0 \leq i < b$, the entries of the $i$th column are the four points on the $i$th base line $\mathcal{B}_{i,0}$ given by (13.12).

Next, we disperse each entry of $\mathbf{B}_{\mathrm{PaG,BIBD},1}(4, b)$ into a CPM of size $(12b+1) \times (12b+1)$. Note that the entries of $\mathbf{B}_{\mathrm{PaG,BIBD},1}(4, b)$ are the integers in $\mathrm{GF}(12b+1) = \{0, 1, 2, \ldots, 12b\}$. If an entry in $\mathbf{B}_{\mathrm{PaG,BIBD},1}(4, b)$ is $\ell$, $0 \leq \ell \leq 12b$, the CPM-dispersion of $\ell$ is a CPM of size $(12b+1) \times (12b+1)$ whose generator has its single 1-component at the position $\ell$. The CPM-dispersion of the $0 = \alpha^{-\infty}$ element of $\mathrm{GF}(12b+1)$ is a $(12b+1) \times (12b+1)$ identity matrix. The CPM-dispersion of the entries in $\mathbf{B}_{\mathrm{PaG,BIBD},1}(4, b)$ results in the following $4 \times b$ array $\mathbf{H}_{\mathrm{PaG,BIBD},1}(4, b)$ of CPMs of size $(12b+1) \times (12b+1)$:

$$\mathbf{H}_{\text{PaG,BIBD},1}(4,b) = \begin{bmatrix} \mathbf{A}_{0,0} & \mathbf{A}_{0,1} & \mathbf{A}_{0,2} & \cdots & \mathbf{A}_{0,b-1} \\ \mathbf{A}_{1,0} & \mathbf{A}_{1,1} & \mathbf{A}_{1,2} & \cdots & \mathbf{A}_{1,b-1} \\ \mathbf{A}_{2,0} & \mathbf{A}_{2,1} & \mathbf{A}_{2,2} & \cdots & \mathbf{A}_{2,b-1} \\ \mathbf{A}_{3,0} & \mathbf{A}_{3,1} & \mathbf{A}_{3,2} & \cdots & \mathbf{A}_{3,b-1} \end{bmatrix}. \tag{13.18}$$

The array $\mathbf{H}_{\text{PaG,BIBD},1}(4,b)$ is a $4(12b+1) \times b(12b+1)$ matrix with column and row weights 4 and $b$, respectively. The matrix $\mathbf{H}_{\text{PaG,BIBD},1}(4,b)$ also satisfies the RC-constraint. Because the four points in each base line $\mathcal{B}_{i,0}$ are distinct, the four CPMs in each CPM-column-block are distinct. The modulo-2 sum of the four CPMs $\mathbf{A}_{0,i}$, $\mathbf{A}_{1,i}$, $\mathbf{A}_{2,i}$, $\mathbf{A}_{3,i}$ in the $i$th CPM-column-block of $\mathbf{H}_{\text{PaG,BIBD},1}(4,b)$ is the circulant $\mathbf{G}_i$ formed by the incidence vectors of the lines in the $i$th translate class $\mathcal{T}_i$ of $\text{PaG}_{\text{BIBD},1}(4b,4,4)$. The 4 CPMs $\mathbf{A}_{0,i}$, $\mathbf{A}_{1,i}$, $\mathbf{A}_{2,i}$, $\mathbf{A}_{3,i}$ form a *decomposition* of $\mathbf{G}_i$ (row-decomposition of $\mathbf{G}_i$). The matrix $\mathbf{B}_{\text{PaG,BIBD},1}(4,b)$ is called the *base matrix* of the array $\mathbf{H}_{\text{PaG,BIBD},1}(4,b)$.

The null space over $\text{GF}(2)$ of $\mathbf{H}_{\text{PaG,BIBD},1}(4,b)$ or any of its subarrays gives a QC-BIBD-PaG-LDPC code whose Tanner graph has girth at least 6. Masking can be performed on the base matrix $\mathbf{B}_{\text{PaG,BIBD},1}(4,b)$ or its submatrices to construct QC-BIBD-PaG-LDPC codes of various lengths and rates.

**Example 13.13** For $b = 61$, $12b + 1 = 733$ is a prime. From Table 13.1, we see that the prime field $\text{GF}(733)$ has a primitive element 6 which satisfies the condition given by (13.11). Using this field, we can construct a $(733, 44\,713, 4, 244, 1)$-BIBD $B_1(733)$. The partial geometry $\text{PaG}_{\text{BIBD},1}(244, 4, 4)$ associated to this BIBD consists of 733 points and 44 713 lines. Each line consist of four points and each point is intersected by 244 lines.

Using the 61 base lines of $\text{PaG}_{\text{BIBD},1}(244, 4, 4)$, we can form a $4 \times 61$ base matrix $\mathbf{B}_{\text{PaG,BIBD},1}(4, 61)$ over $\text{GF}(773)$ in the form of (13.17). The CPM-dispersion of $\mathbf{B}_{\text{PaG,BIBD},1}(4, 61)$ gives a $4 \times 61$ array $\mathbf{H}_{\text{PaG,BIBD},1}(4, 61)$ of CPMs of size $733 \times 733$ in the form of (13.18) which is a $2932 \times 44\,713$ matrix with column and row weights 4 and 61, respectively.

The null space over $\text{GF}(2)$ of $\mathbf{H}_{\text{PaG,BIBD},1}(4, 61)$ gives a $(4, 61)$-regular $(44\,713, 41\,784)$ QC-BIBD-PaG-LDPC code $C_{\text{PaG,BIBD},1}(4, 61)$ with rate 0.9345. The BER performances of this code over the AWGN channel using BPSK signaling decoded with 5, 10, and 50 iterations of the MSA are shown in Fig. 13.7(a). From the figure, we see that with 50 iterations of the MSA, the code achieves a BER of $5 \times 10^{-10}$ without a visible error-floor and performs less than 0.8 dB from the Shannon limit.

For $4 \le k \le b$, deleting $61 - k$ CPM-column-blocks from $\mathbf{H}_{\text{PaG,BIBD},1}(4, 61)$, we obtain a $4 \times k$ array $\mathbf{H}_{\text{PaG,BIBD},1}(4, k)$ of CPMs of size $733 \times 733$ which is a $2932 \times 733k$ matrix with column and row weights 4 and $k$, respectively. The null space over $\text{GF}(2)$ of $\mathbf{H}_{\text{PaG,BIBD},1}(4, k)$ gives a QC-BIBD-PaG-LDPC code of length $733k$ with rate at least $(k-4)/k$. For example, we set $k = 32$ and delete the last 29 CPM-column-blocks from $\mathbf{H}_{\text{PaG,BIBD},1}(4, 61)$. We obtain a $4 \times 32$ array $\mathbf{H}_{\text{PaG,BIBD},1}(4, 32)$ of CPMs of size $733 \times 733$. The null space over $\text{GF}(2)$ of $\mathbf{H}_{\text{PaG,BIBD},1}(4, 32)$ gives a $(4, 32)$-regular $(23\,456, 20\,527)$ QC-BIBD-PaG-LDPC code $C_{\text{PaG,BIBD},1}(4, 32)$ with rate 0.8748. The BER performances of this code decoded with 5, 10, and 50 iterations of the MSA with a scaling factor of 0.75 are also shown in Fig. 13.7(a).

If we apply the two codes constructed above over the BEC, the UEBR and UEBLR performances of the two codes over the BEC are shown in Fig. 13.7(b). We see that both codes perform well. ▲▲

**Example 13.14** For $b = 15$, $12b + 1 = 181$ is a prime. From Table 13.1, we see that the prime field GF(181) has a primitive element 2 which satisfies the condition given by (13.11). Using this field, we can construct a $(181, 2715, 4, 60, 1)$-BIBD $B_1(181)$. The partial geometry $\text{PaG}_{\text{BIBD},1}(60, 4, 4)$ associated to this BIBD consists of 181 points and 2715 lines. Each line consists of four points and each point is intersected by 60 lines.

Using the 15 base lines of $\text{PaG}_{\text{BIBD},1}(60, 4, 4)$, we can form a $4 \times 15$ base matrix $\mathbf{B}_{\text{PaG},\text{BIBD},1}(4, 15)$ over GF(181) in the form of (13.17). Suppose we take two submatrices, $\mathbf{B}_{\text{PaG},\text{BIBD},1}(4, 8)$ and $\mathbf{B}_{\text{PaG},\text{BIBD},1}(4, 12)$, from $\mathbf{B}_{\text{PaG},\text{BIBD},1}(4, 15)$, the first one with size $4 \times 8$ and the other one with size $4 \times 12$. The 181-fold CPM-dispersions of $\mathbf{B}_{\text{PaG},\text{BIBD},1}(4, 8)$ and $\mathbf{B}_{\text{PaG},\text{BIBD},1}(4, 12)$ give two arrays, $\mathbf{H}_{\text{PaG},\text{BIBD},1}(4, 8)$ and $\mathbf{H}_{\text{PaG},\text{BIBD},1}(4, 12)$, respectively. The array $\mathbf{H}_{\text{PaG},\text{BIBD},1}(4, 8)$ is a $4 \times 8$ array of CPMs of size $181 \times 181$ which is a $724 \times 1448$ matrix over GF(2). The array $\mathbf{H}_{\text{PaG},\text{BIBD},1}(4, 12)$ is a $4 \times 12$ array of CPMs of size $181 \times 181$ which is a $724 \times 2172$ matrix over GF(2). The null spaces over GF(2) of $\mathbf{H}_{\text{PaG},\text{BIBD},1}(4, 8)$ and $\mathbf{H}_{\text{PaG},\text{BIBD},1}(4, 12)$ give two QC-BIBD-PaG-LDPC codes, a $(4, 8)$-regular $(1448, 727)$ code $C_{\text{PaG},\text{BIBD},1}(4, 8)$ and a $(4, 12)$-regular $(2172, 1451)$ code $C_{\text{PaG},\text{BIBD},1}(4, 12)$, respectively. The BER and BLER performances of these two codes over the AWGN channel decoded 50 iterations of the MSA with a scaling factor of 0.75 are shown in Fig. 13.8.

Masking $\mathbf{B}_{\text{PaG},\text{BIBD},1}(4, 8)$ with the matrix $\mathbf{Z}(4, 8)$ given by (13.6), we obtain a $4 \times 8$ masked matrix $\mathbf{B}_{\text{PaG},\text{BIBD},1,\text{mask}}(4, 8)$ over GF(181). The 181-fold CPM-dispersion of $\mathbf{B}_{\text{PaG},\text{BIBD},1,\text{mask}}(4, 8)$ gives a $4 \times 8$ array $\mathbf{H}_{\text{PaG},\text{BIBD},1,\text{mask}}(4, 8)$ of CPMs and ZMs of size $181 \times 181$ which is a $724 \times 1448$ matrix over GF(2) with column and row weights 3 and 6, respectively. The null space over GF(2) of $\mathbf{H}_{\text{PaG},\text{BIBD},1,\text{mask}}(4, 8)$ gives a $(3, 6)$-regular $(1448, 724)$ masked QC-BIBD-PaG-LDPC code $C_{\text{PaG},\text{BIBD},1,\text{mask}}(4, 8)$ of rate 0.5. The BER and BLER performances of the masked code $C_{\text{PaG},\text{BIBD},1,\text{mask}}(4, 8)$ decoded 50 iterations of the MSA with a scaling factor of 0.75 are also shown in Fig. 13.8. From the figure, we see that the masked code $C_{\text{PaG},\text{BIBD},1,\text{mask}}(4, 8)$ outperforms the unmasked code $C_{\text{PaG},\text{BIBD},1}(4, 8)$ by 0.2 dB at a BER of $10^{-9}$.

If we mask the $4 \times 12$ base matrix $\mathbf{B}_{\text{PaG},\text{BIBD},1}(4, 12)$ with the following masking matrix:

$$\mathbf{Z}(4, 12) = \begin{bmatrix} 1 & 0 & 1 & 0 & 1 & 0 & 1 & 1 & 1 & 1 & 1 & 1 \\ 0 & 1 & 0 & 1 & 0 & 1 & 1 & 1 & 1 & 1 & 1 & 1 \\ 1 & 1 & 1 & 1 & 1 & 1 & 0 & 1 & 0 & 1 & 0 \\ 1 & 1 & 1 & 1 & 1 & 1 & 0 & 1 & 0 & 1 & 0 & 1 \end{bmatrix},$$

we obtain a $4 \times 12$ masked matrix $\mathbf{B}_{\text{PaG},\text{BIBD},1,\text{mask}}(4, 12)$ over GF(181). The 181-fold CPM-dispersion of the masked base matrix $\mathbf{B}_{\text{PaG},\text{BIBD},1,\text{mask}}(4, 12)$ gives

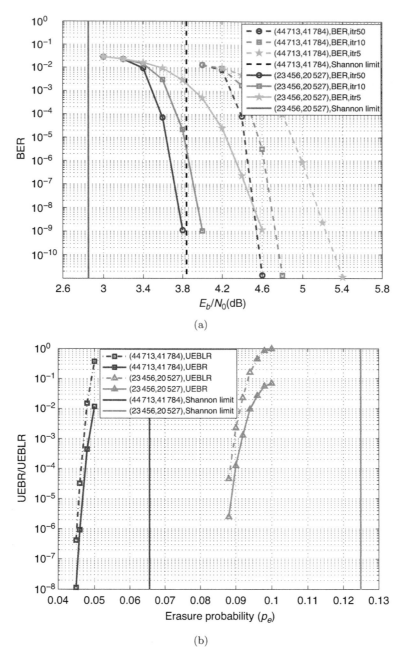

Figure 13.7 The performances of the $(44\,713, 41\,781)$ and $(23\,456, 20\,524)$ QC-BIBD-PaG-LDPC codes given in Example 13.13 over: (a) AWGN channel and (b) BEC.

a $4 \times 12$ array $\mathbf{H}_{\text{PaG,BIBD,1,mask}}(4, 12)$ of CPMs and ZMs of size $181 \times 181$ which is a $724 \times 2172$ matrix over $GF(2)$ with column and row weights 3 and 9, respectively. The null space over $GF(2)$ of $\mathbf{H}_{\text{PaG,BIBD,1,mask}}(4, 12)$ gives a $(3, 9)$-regular $(2172, 1448)$ masked QC-BIBD-PaG-LDPC code $C_{\text{PaG,BIBD,1,mask}}(4, 12)$ of rate $2/3$. The performances of the masked code $C_{\text{PaG,BIBD,1,mask}}(4, 12)$ over the AWGN channel decoded 50 iterations of the MSA with a scaling factor of 0.75 are also shown in Fig. 13.8. ▲▲

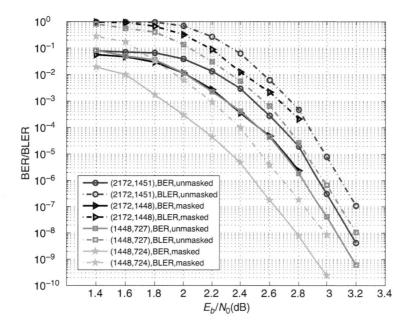

Figure 13.8 The BER and BLER performances of the four QC-BIBD-PaG-LDPC codes given in Example 13.14.

### 13.6.3 Class-2 Bose $(N, M, t, r, 1)$-BIBDs

In the following, we present another class of Bose-BIBDs for which each pair of objects appears also in one and only one block, i.e., $\lambda = 1$. The construction of this class of BIBDs is similar to the construction of BIBDs given in Section 13.6.2. Associated to this class of BIBDs is another class of partial geometries, called *class-2 BIBD-PaGs*.

Let $b$ be a positive integer such that $20b + 1$ is a prime. If the prime field $GF(20b + 1) = \{0, 1, 2, \ldots, 20b\}$ under modulo-$(20b + 1)$ addition and multiplication has a primitive element $\alpha$ which satisfies the following constraint:

$$\alpha^{4b} + 1 = \alpha^c \tag{13.19}$$

where $c$ is a *positive odd* integer less than $20b + 1$, then there exists a $(20b + 1, b(20b + 1), 5, 5b, 1)$-BIBD [20], denoted by $B_2(20b + 1)$ with $20b + 1$ objects distributed in $b(20b + 1)$ blocks. Each block consists of five objects, each object appears in $5b$ blocks, and each pair of objects appears in one and only one block. The $20b + 1$ elements in $GF(20b + 1)$ are treated as objects in $B_2(20b + 1)$.

To construct $B_2(20b + 1)$, we first form $b$ *base blocks*. Label these $b$ base blocks from 0 to $b - 1$. For $0 \leq i < b$, the $i$th base block is given as follows:

$$\mathcal{B}_{i,0} = \{\alpha^{2i}, \alpha^{2i+4b}, \alpha^{2i+8b}, \alpha^{2i+12b}, \alpha^{2i+16b}\}, \tag{13.20}$$

which consists of five elements in $GF(20b + 1)$. For each base block $\mathcal{B}_{i,0}$, we form $20b + 1$ *coblocks* $\mathcal{B}_{i,0}$, $\mathcal{B}_{i,1}, \ldots, \mathcal{B}_{i,20b}$ of $\mathcal{B}_{i,0}$ by adding each element of $GF(20b + 1)$ in turn to the elements in the base block $\mathcal{B}_{i,0}$. For $0 \leq j \leq 20b$, the $j$th coblock of $\mathcal{B}_{i,0}$ is given by

$$\mathcal{B}_{i,j} = \{j + \alpha^{2i}, j + \alpha^{2i+4b}, j + \alpha^{2i+8b}, j + \alpha^{2i+12b}, j + \alpha^{2i+16b}\}. \tag{13.21}$$

Each base block $\mathcal{B}_{i,0}$ and its $20b$ coblocks form a *translate class* $\mathcal{T}_i$ which consists of $20b + 1$ blocks. There are $b$ translate classes $\mathcal{T}_0, \mathcal{T}_1, \ldots, \mathcal{T}_{b-1}$. The $b(20b+1)$ blocks in the $b$ translate classes $\mathcal{T}_0, \mathcal{T}_1, \ldots, \mathcal{T}_{b-1}$ form a $(20b+1, b(20b+1), 5, 5b, 1)$-BIBD $B_2(20b + 1)$, called *class-2 Bose-BIBD*.

Table 13.2 gives a list of $b$s for which $20b + 1$ is a prime and the prime field $GF(20b + 1)$ has a primitive element $\alpha$ that satisfies the constraint given by (13.19).

Table 13.2 A list of $b$s for which $20b + 1$ is a prime and the field $GF(20b + 1)$ satisfies the condition given by (13.19).

| $b$ | Field | $(\alpha, c)$ | $b$ | Field | $(\alpha, c)$ |
|---|---|---|---|---|---|
| 2 | GF(41) | (6, 3) | 30 | GF(601) | (7, 79) |
| 3 | GF(61) | (2, 23) | 32 | GF(641) | (3, 631) |
| 12 | GF(241) | (7, 197) | 33 | GF(661) | (2, 657) |
| 14 | GF(281) | (3, 173) | 35 | GF(701) | (2, 533) |
| 21 | GF(421) | (2, 227) | 41 | GF(821) | (2, 713) |

Viewing each block in $B_2(20b + 1)$ as a line and each object in a block as a point, $B_2(20b + 1)$ gives a partial geometry, denoted by $PaG_{BIBD,2}(5b, 5, 5)$ and called *class-2 Bose-BIBD-PaG*, which consists of $20b + 1$ points and $b(20b + 1)$ lines. Each line consists of five points, each point is on $5b$ lines, and two points are on one and only one line. Based on the lines in the $b$ translate classes $\mathcal{T}_0, \mathcal{T}_1, \ldots, \mathcal{T}_{b-1}$, we can construct $b$ circulants of size $(20b + 1) \times (20b + 1)$, $\mathbf{G}_0, \mathbf{G}_1, \ldots, \mathbf{G}_{b-1}$. For $0 \leq i < b$, the generator of $\mathbf{G}_i$ is the incidence vector of the base line $\mathcal{B}_{i,0}$. The line-point incidence matrix of the partial geometry $PaG_{BIBD,2}(5b, 5, 5)$ is given as follows:

$$\mathbf{H}_{\mathrm{PaG,BIBD},2}(b(20b + 1), 20b + 1) = \begin{bmatrix} \mathbf{G}_0 \\ \mathbf{G}_1 \\ \vdots \\ \mathbf{G}_{b-1} \end{bmatrix}. \tag{13.22}$$

The column and row weights of $\mathbf{H}_{\text{PaG,BIBD},2}(b(20b+1), 20b+1)$ are $5b$ and 5, respectively. For $1 \leq k \leq b$, we can take any $k$ circulants from the set $\{\mathbf{G}_0, \mathbf{G}_1, \ldots, \mathbf{G}_{b-1}\}$ and arrange them in a row to form a $(20b+1) \times k(20b+1)$ matrix $\mathbf{H}^*_{\text{PaG,BIBD},2}(20b+1, k(20b+1))$. The null space over GF(2) of $\mathbf{H}^*_{\text{PaG,BIBD},2}(20b+1, k(20b+1))$ gives a $(5, 5k)$-regular QC-BIBD-PaG-LDPC $C_{\text{PaG,BIBD},2}(5, 5k)$ of length $k(20b+1)$ with rate at least $(k-1)/k$.

**Example 13.15** Let $b = 14$. The integer $20b+1 = 281$ is a prime. The element 3 in the prime field GF(281) given in Table 13.2 is a primitive element satisfying the condition given by (13.19) with $c = 173$. Using this prime field, we can construct a $(281, 3934, 5, 70, 1)$-BIBD $B_2(281)$. Associated to $B_2(281)$ is a partial geometry $\text{PaG}_{\text{BIBD},2}(70, 5, 5)$ which consists of 281 points and 3934 lines. Each line consists of five points, each point is on 70 lines, and two points are on one and only one line. There are 14 base lines in $\text{PaG}_{\text{BIBD},2}(70, 5, 5)$ which are

$$\mathcal{B}_{0,0} = \{3^0, 3^{0+56}, 3^{0+112}, 3^{0+168}, 3^{0+224}\} = \{1, 86, 90, 153, 232\},$$

$$\mathcal{B}_{1,0} = \{3^2, 3^{2+56}, 3^{2+112}, 3^{2+168}, 3^{2+224}\} = \{9, 121, 212, 248, 253\},$$

$$\mathcal{B}_{2,0} = \{3^4, 3^{4+56}, 3^{4+112}, 3^{4+168}, 3^{4+224}\} = \{29, 81, 222, 246, 265\},$$

$$\mathcal{B}_{3,0} = \{3^6, 3^{6+56}, 3^{6+112}, 3^{6+168}, 3^{6+224}\} = \{31, 137, 167, 247, 261\},$$

$$\mathcal{B}_{4,0} = \{3^8, 3^{8+56}, 3^{8+112}, 3^{8+168}, 3^{8+224}\} = \{98, 101, 109, 256, 279\},$$

$$\mathcal{B}_{5,0} = \{3^{10}, 3^{10+56}, 3^{10+112}, 3^{10+168}, 3^{10+224}\} = \{39, 56, 66, 138, 263\},$$

$$\mathcal{B}_{6,0} = \{3^{12}, 3^{12+56}, 3^{12+112}, 3^{12+168}, 3^{12+224}\} = \{32, 70, 118, 119, 223\},$$

$$\mathcal{B}_{7,0} = \{3^{14}, 3^{14+56}, 3^{14+112}, 3^{14+168}, 3^{14+224}\} = \{7, 40, 68, 219, 228\},$$

$$\mathcal{B}_{8,0} = \{3^{16}, 3^{16+56}, 3^{16+112}, 3^{16+168}, 3^{16+224}\} = \{4, 50, 63, 79, 85\},$$

$$\mathcal{B}_{9,0} = \{3^{18}, 3^{18+56}, 3^{18+112}, 3^{18+168}, 3^{18+224}\} = \{5, 36, 149, 169, 203\},$$

$$\mathcal{B}_{10,0} = \{3^{20}, 3^{20+56}, 3^{20+112}, 3^{20+168}, 3^{20+224}\} = \{43, 45, 116, 141, 217\},$$

$$\mathcal{B}_{11,0} = \{3^{22}, 3^{22+56}, 3^{22+112}, 3^{22+168}, 3^{22+224}\} = \{106, 124, 145, 201, 267\},$$

$$\mathcal{B}_{12,0} = \{3^{24}, 3^{24+56}, 3^{24+112}, 3^{24+168}, 3^{24+224}\} = \{111, 123, 155, 181, 273\},$$

$$\mathcal{B}_{13,0} = \{3^{26}, 3^{26+56}, 3^{26+112}, 3^{26+168}, 3^{26+224}\} = \{156, 209, 224, 264, 271\}.$$

Using the incidence vectors of the above 14 base lines, we can construct 14 circulants $\mathbf{G}_0, \mathbf{G}_1, \ldots, \mathbf{G}_{13}$, each of size $281 \times 281$ and weight 5. The matrix $\mathbf{H}^*_{\text{PaG,BIBD},2}(281, 3934) = [\mathbf{G}_0 \ \mathbf{G}_1 \ \ldots \ \mathbf{G}_{13}]$ is a $281 \times 3934$ matrix with column and row weights 5 and 70, respectively. The null space over GF(2) of $\mathbf{H}^*_{\text{PaG,BIBD},2}(281, 3934)$ gives a $(5, 70)$-regular $(3934, 3653)$ QC-BIBD-PaG-LDPC code $C_{\text{PaG,BIBD},2}(5, 70)$ with rate 0.9285. The BER and BLER performances of the code $C_{\text{PaG,BIBD},2}(5, 70)$ over the AWGN channel using BPSK signaling decoded with 5, 10, and 50 iterations of the MSA scaled by a factor of 0.75 are shown in Fig. 13.9. ▲▲

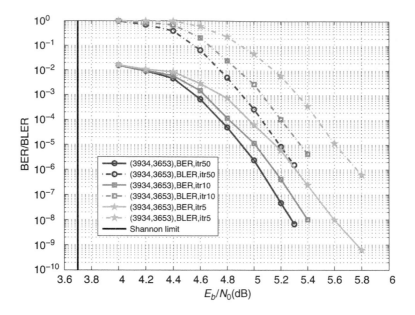

Figure 13.9 The BER and BLER performances of the $(3934, 3653)$ QC-BIBD-PaG-LDPC code given in Example 13.15.

Similar to the construction of $4 \times b$ CPM array $\mathbf{H}_{\mathrm{PaG,BIBD},1}(4, b)$ using the $b$ base lines of the class-1 partial geometry $\mathrm{PaG}_{\mathrm{BIBD},1}(4b, 4, 4)$, we can construct a $5 \times b$ array $\mathbf{H}_{\mathrm{PaG,BIBD},2}(5, b)$ of CPMs of size $(20b + 1) \times (20b + 1)$ using the $b$ base lines of the class-2 partial geometry $\mathrm{PaG}_{\mathrm{BIBD},2}(5b, 5, 5)$. To achieve this, we first form a $5 \times b$ base matrix over $\mathrm{GF}(20b + 1)$ as follows:

$$\mathbf{B}_{\mathrm{PaG,BIBD},2}(5, b) = \begin{bmatrix} \alpha^0 & \alpha^2 & \alpha^4 & \cdots & \alpha^{2(b-1)} \\ \alpha^{4b} & \alpha^{2+4b} & \alpha^{4+4b} & \cdots & \alpha^{2(b-1)+4b} \\ \alpha^{8b} & \alpha^{2+8b} & \alpha^{4+8b} & \cdots & \alpha^{2(b-1)+8b} \\ \alpha^{12b} & \alpha^{2+12b} & \alpha^{4+12b} & \cdots & \alpha^{2(b-1)+12b} \\ \alpha^{16b} & \alpha^{2+16b} & \alpha^{4+16b} & \cdots & \alpha^{2(b-1)+16b} \end{bmatrix}, \quad (13.23)$$

where, for $0 \le i < b$, the five entries in the $i$th column are the five points on the $i$th base lines $\mathcal{B}_{i,0}$ given by (13.20) of $\mathrm{PaG}_{\mathrm{BIBD},2}(5b, 5, 5)$.

Dispersing each entry of $\mathbf{B}_{\mathrm{PaG,BIBD},2}(5, b)$ into a CPM of size $(20b + 1) \times (20b + 1)$, we obtain a $5 \times b$ array of CPMs of size $(20b + 1) \times (20b + 1)$ as follows:

$$\mathbf{H}_{\mathrm{PaG,BIBD},2}(5, b) = \begin{bmatrix} \mathbf{A}_{0,0} & \mathbf{A}_{0,1} & \mathbf{A}_{0,2} & \cdots & \mathbf{A}_{0,b-1} \\ \mathbf{A}_{1,0} & \mathbf{A}_{1,1} & \mathbf{A}_{1,2} & \cdots & \mathbf{A}_{1,b-1} \\ \mathbf{A}_{2,0} & \mathbf{A}_{2,1} & \mathbf{A}_{2,2} & \cdots & \mathbf{A}_{2,b-1} \\ \mathbf{A}_{3,0} & \mathbf{A}_{3,1} & \mathbf{A}_{3,2} & \cdots & \mathbf{A}_{3,b-1} \\ \mathbf{A}_{4,0} & \mathbf{A}_{4,1} & \mathbf{A}_{4,2} & \cdots & \mathbf{A}_{4,b-1} \end{bmatrix}. \quad (13.24)$$

The array $\mathbf{H}_{\text{PaG,BIBD},2}(5,b)$ is a $5(20b+1) \times b(20b+1)$ matrix over $\text{GF}(2)$ with column and row weights 5 and $5b$, respectively. The null space over $\text{GF}(2)$ of $\mathbf{H}_{\text{PaG,BIBD},2}(5,b)$ gives a QC-BIBD-PaG-LDPC code whose Tanner graph has girth at least 6.

For any two positive integers $d$ and $m$ such that $1 \leq d \leq 5$ and $d \leq m \leq b$, let $\mathbf{B}_{\text{PaG,BIBD},2}(d,m)$ be a $d \times m$ submatrix of $\mathbf{B}_{\text{PaG,BIBD},2}(5,b)$. The null space of the CPM-dispersion of $\mathbf{B}_{\text{PaG,BIBD},2}(d,m)$ gives a $(d,m)$-regular QC-BIBD-PaG-LDPC code of length $m(20b+1)$ with rate at least $(m-d)/m$.

**Example 13.16** For $b = 32$, $20b+1 = 641$ is a prime. The prime field $\text{GF}(641)$ listed in Table 13.2 has a primitive element $\alpha = 3$ which satisfies the condition given by (13.19). Based on this field, we can construct a $(641, 20\,512, 5, 160, 1)$-BIBD $B_2(641)$. The partial geometry $\text{PaG}_{\text{BIBD},2}(160, 5, 5)$ associated to this BIBD contains $20\,512$ lines and 641 points. Each line in $\text{PaG}_{\text{BIBD},2}(160, 5, 5)$ consists of five points. Each point is intersected by 160 lines. Based on the 32 base blocks in $B_2(641)$, we construct a $5 \times 32$ base matrix $\mathbf{B}_{\text{PaG,BIBD},2}(5, 32)$ in the form of (13.23).

Using the top four rows of $\mathbf{B}_{\text{PaG,BIBD},2}(5, 32)$, we obtain a $4 \times 32$ submatrix $\mathbf{B}_{\text{PaG,BIBD},2}(4, 32)$. The CPM-dispersion of $\mathbf{B}_{\text{PaG,BIBD},2}(4, 32)$ gives a $4 \times 32$ array $\mathbf{H}_{\text{PaG,BIBD},2}(4, 32)$ of CPMs of size $641 \times 641$ which is a $2564 \times 20\,512$ matrix with column and row weights 4 and 32, respectively. The null space over $\text{GF}(2)$ of $\mathbf{H}_{\text{PaG,BIBD},2}(4, 32)$ gives a $(4, 32)$-regular $(20\,512, 17\,951)$ QC-BIBD-PaG-LDPC code $C_{\text{PaG,BIBD},2}(4, 32)$ with rate 0.8751. The BER and BLER performances of the code over the AWGN channel decoded with 5, 10, and 50 iterations of the MSA scaled by a factor of 0.75 are shown in Fig. 13.10(a).

Suppose we take the first eight columns from $\mathbf{B}_{\text{PaG,BIBD},2}(4, 32)$ to obtain a $4 \times 8$ base matrix $\mathbf{B}_{\text{PaG,BIBD},2}(4, 8)$. Masking $\mathbf{B}_{\text{PaG,BIBD},2}(4, 8)$ with the masking matrix in (13.6) gives a $4 \times 8$ masked matrix $\mathbf{B}_{\text{PaG,BIBD},2,\text{mask}}(4,8)$ with column and row weights 3 and 6, respectively. The CPM-dispersion of $\mathbf{B}_{\text{PaG,BIBD},2,\text{mask}}(4, 8)$ gives a $4 \times 8$ masked array $\mathbf{H}_{\text{PaG,BIBD},2,\text{mask}}(4, 8)$ which is a $2564 \times 5128$ matrix with column and row weights 3 and 6, respectively. The null space over $\text{GF}(2)$ of $\mathbf{H}_{\text{PaG,BIBD},2,\text{mask}}(4, 8)$ gives a $(3, 6)$-regular $(5128, 2564)$ QC-BIBD-PaG-LDPC code $C_{\text{PaG,BIBD},2,\text{mask}}(4, 8)$ with rate 0.5. The BER and BLER performances of the code over the AWGN channel decoded with 5, 10, and 50 iterations of the MSA scaled by a factor of 0.75 are shown in Fig. 13.10(b). At the BER of $10^{-9}$ with 50 iterations of the MSA, the code performs 0.9 dB from the threshold.

If we apply the two QC-BIBD-PaG-LDPC codes constructed above over the BEC, the UEBR and UEBLR performances of the two codes are shown in Fig. 13.11(a) and Fig. 13.11(b), respectively. We see that the two codes perform well over the BEC. ▲▲

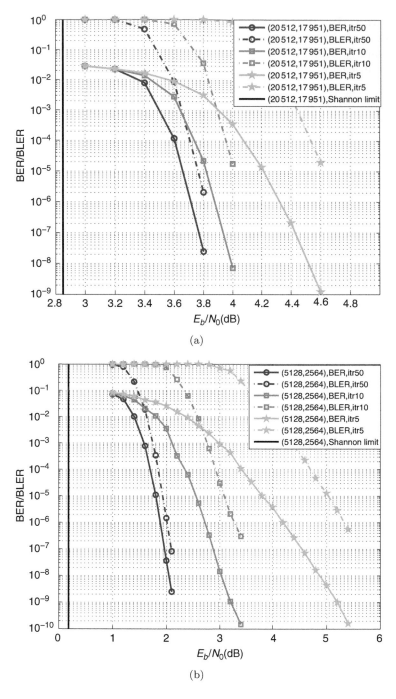

Figure 13.10 The BER and BLER performances of the two QC-BIBD-PaG-LDPC codes given in Example 13.16 over the AWGN channel: (a) $(20\,512, 17\,951)$ and (b) $(5128, 2564)$.

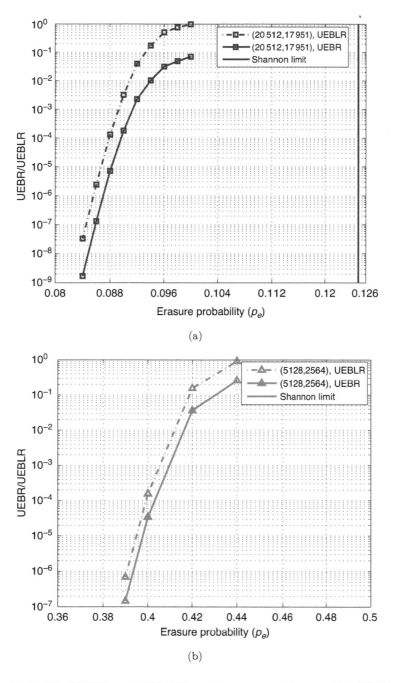

Figure 13.11 The UEBR and UEBLR performances of the two QC-BIBD-PaG-LDPC codes given in Example 13.16 over the BEC: (a) $(20\,512, 17\,951)$ and (b) $(5128, 2564)$.

## 13.7 Remarks

In this chapter, we presented three classes of partial geometries and showed that these partial geometries are very effective in constructing QC-LDPC codes. Through examples, we demonstrated that LDPC codes constructed based on these partial geometries perform well and can achieve very low error rates without visible error-floors. All partial-geometry-based LDPC codes are OSML decodable, similar to cyclic EG- and PG-LDPC codes. They can also be decoded with the CPM-RID decoding scheme presented in Section 12.8 to reduce decoding complexity.

In Section 13.6, we showed that partial geometries can be derived from BIBDs with $\lambda = 1$ and presented two small classes of BIBDs with $\lambda = 1$ which were constructed by Bose [6, 20]. Besides these two classes of Bose BIBDs with $\lambda = 1$, there are many other BIBDs with $\lambda = 1$ and these can be found in [21]. Based on these BIBDs, partial geometries and their associated LDPC codes can be constructed.

The partial-geometry QC-LDPC codes constructed in this chapter are also capable of correcting random phased bursts of erasures. This subject will be presented in the following chapter.

## Problems

**13.1** Consider the two groups $\mathbf{S}_5$ and $\mathbf{S}_{15}$ constructed in Example 13.1. Find the CPM-dispersion of $\beta^3 \in \mathbf{S}_5$ with respective to $\mathbf{S}_{15}$.

**13.2** Prove that any $2 \times 2$ submatrix of the $2 \times 3$ matrix $\mathbf{B}$ in Example 13.2 is nonsingular.

**13.3** Consider the field $\mathrm{GF}(2^3)$.
 (a) Find the cyclic subgroup of $\mathrm{GF}(2^3)$.
 (b) Find the base matrix $\mathbf{B}_{\mathrm{PaG},c}(p,p)$ by using the elements in the cyclic subgroup.
 (c) Find the CPM-dispersion $\mathbf{H}_{\mathrm{PaG},c}(p,p)$ of the base matrix $\mathbf{B}_{\mathrm{PaG},c}(p,p)$ with respective to the cyclic subgroup.
 (d) Compute the partial geometry $\mathrm{PaG}_c(p,p,p-1)$ using $\mathbf{H}_{\mathrm{PaG},c}(p,p)$ as a line-point incidence matrix.
 (e) Find the numbers of lines, points, parallel bundles, the number of lines in each parallel bundle, and the number of lines passing each point in $\mathrm{PaG}_c(p,p,p-1)$.

**13.4** Consider the field $\mathrm{GF}(2^7)$.
 (a) Find the cyclic subgroup of $\mathrm{GF}(2^7)$.
 (b) Find the base matrix $\mathbf{B}_{\mathrm{PaG},c}(p,p)$ by using the elements in the cyclic subgroup.

(c) Find the CPM-dispersion $\mathbf{H}_{\mathrm{PaG},c}(p,p)$ of the base matrix $\mathbf{B}_{\mathrm{PaG},c}(p,p)$ with respect to the cyclic subgroup.

(d) Compute the partial geometry $\mathrm{PaG}_c(p,p,p-1)$ using $\mathbf{H}_{\mathrm{PaG},c}(p,p)$ as a line-point incidence matrix.

(e) Find the numbers of lines, points, parallel bundles, the number of lines in each parallel bundle, and the number of lines passing each point in $\mathrm{PaG}_c(p,p,p-1)$.

(f) Use the line-point incidence matrix of four parallel bundles of lines of $\mathrm{PaG}_c(p,p,p-1)$ as a parity-check matrix to construct a QC-LDPC code. Compute the code length, dimension, rate, and the error performance over the AWGN channel decoded with 50 iterations of the scaled MSA.

(g) Compute the performance of the constructed code over the BEC with 50 iterations of the SPA.

**13.5** Consider the partial geometry $\mathrm{PaG}_c(89,89,88)$ constructed in Example 13.5. Construct three QC-RS-PaG-LDPC codes by using three, five, and six parallel bundles of lines in $\mathrm{PaG}_c(89,89,88)$ and compute their error performances over the AWGN channel decoded with 50 iterations of the scaled MSA.

**13.6** Prove that the CPM-dispersion $\mathbf{H}_{\mathrm{PaG},p}(p,p)$ of the base matrix $\mathbf{B}_{\mathrm{PaG},p}(p,p)$ given by (13.5) over the prime field $\mathrm{GF}(p)$ satisfies the RC-constraint.

**13.7** Consider the prime field $\mathrm{GF}(3)$.

(a) Construct a partial geometry based on this prime field; find the number of points, lines, and parallel bundles in this partial geometry.

(b) Compute the line-point incidence matrix $\mathbf{H}_{\mathrm{PaG},p}(3,3)$ of the constructed partial geometry.

(c) Find the Tanner graph $\mathcal{G}_{\mathrm{PaG},p}(3,3)$ corresponding to the line-point incidence matrix $\mathbf{H}_{\mathrm{PaG},p}(3,3)$; find the girth of $\mathcal{G}_{\mathrm{PaG},p}(3,3)$ and a cycle with the shortest length.

**13.8** Consider the partial geometry $\mathrm{PaG}_p(181,181,180)$ constructed in Example 13.8 and its line-point incidence matrix $\mathbf{H}_{\mathrm{PaG},p}(181,181)$.

(a) Construct a QC-PaG-LDPC code by using four parallel bundles of lines and removing $32\,037$ points from these lines, i.e., taking a $4 \times 8$ subarray $\mathbf{H}_{\mathrm{PaG},p}(4,8)$ from $\mathbf{H}_{\mathrm{PaG},p}(181,181)$ and using the subarray as a parity-check matrix; compute the error performance of the constructed code under AWGN channel with 50 iterations of the scaled MSA.

(b) Mask the subarray $\mathbf{H}_{\mathrm{PaG},p}(4,8)$ with the masking matrix given by (13.6) and use the masked matrix as parity-check matrix to construct a QC-LDPC code; compute the error performance of the masked code under AWGN channel with 50 iterations of the scaled MSA.

(c) Compute the performances of the constructed codes over the BEC with 50 iterations of the SPA.

**13.9** Consider the partial geometry constructed in Example 13.12 based on the class-1 $(97, 776, 4, 32, 1)$-BIBD $B_1(97)$. Eight circulants $\mathbf{G}_0$, $\mathbf{G}_1$, $\mathbf{G}_2$, $\mathbf{G}_3$, $\mathbf{G}_4$, $\mathbf{G}_5$, $\mathbf{G}_6$, $\mathbf{G}_7$ are obtained based on the eight base blocks in $B_1(97)$. Take four circulants, say $\mathbf{G}_0$, $\mathbf{G}_1$, $\mathbf{G}_2$, $\mathbf{G}_3$, to form a matrix $\mathbf{H} = [\mathbf{G}_0 \ \mathbf{G}_1 \ \mathbf{G}_2 \ \mathbf{G}_3]$. Construct a QC-BIBD-PaG-LDPC code by using $\mathbf{H}$ as a parity-check matrix and compute its error performance over the AWGN channel by using 50 iterations of the scaled MSA.

**13.10** Use the eight base lines of the partial geometry $\text{PaG}_{\text{BIBD},1}(32, 4, 4)$ constructed in Example 13.12 to construct a $4 \times 8$ array $\mathbf{H}_{\text{PaG,BIBD},1}(4, 8)$ of CPMs of size $97 \times 97$.

(a) Find the base matrix for the array $\mathbf{H}_{\text{PaG,BIBD},1}(4, 8)$.

(b) Construct a QC-BIBD-PaG-LDPC code $C_{\text{PaG,BIBD},1}(4, 8)$ by using the array $\mathbf{H}_{\text{PaG,BIBD},1}(4, 8)$ as a parity-check matrix.

(c) Compute the BER and BLER performance of $C_{\text{PaG,BIBD},1}(4, 8)$ over the AWGN channel based on 50 iterations of the scaled MSA.

(d) Mask the array $\mathbf{H}_{\text{PaG,BIBD},1}(4, 8)$ by the masking matrix given by (13.6). Use the masked array as a parity-check matrix to construct a QC-LDPC code $C_{\text{PaG,BIBD},1,\text{mask}}(4, 8)$. Compute the BER and BLER performances of $C_{\text{PaG,BIBD},1,\text{mask}}(4, 8)$ over the AWGN channel based on 50 iterations of the scaled MSA.

(e) Compute the performances of the constructed codes over the BEC with 50 iterations of the SPA.

**13.11** In Example 13.13, a $4 \times 61$ base matrix $\mathbf{B}_{\text{PaG,BIBD},1}(4, 61)$ over $\text{GF}(773)$ is constructed based on the partial geometry $\text{PaG}_{\text{BIBD},1}(244, 4, 4)$. The CPM-dispersion of $\mathbf{B}_{\text{PaG,BIBD},1}(4, 61)$ gives a $4 \times 61$ array $\mathbf{H}_{\text{PaG,BIBD},1}(4, 61)$. A QC-BIBD-PaG-LDPC code $C_{\text{PaG,BIBD},1}(4, 32)$ is constructed by taking a $4 \times 32$ sub-array from $\mathbf{H}_{\text{PaG,BIBD},1}(4, 61)$. Construct a sequence of QC-BIBD-PaG-LDPC codes by choosing $k = 8$, 12, 16, 20, 24, 28, 36, 40, 44, 48, and compute their error performances over the AWGN channel by using 50 iterations of the scaled MSA.

**13.12** Let $b = 2$.

(a) Find all the blocks in the class-2 $(20b + 1, b(20b + 1), 5, 5b, 1)$-BIBD $B_2(20b + 1)$.

(b) Find the line-point incidence matrix $\mathbf{G}_i$ based on the lines in the translate class $\mathcal{T}_i$, $0 \leq i < b$.

(c) Use $\mathbf{G}_0$ as a parity-check matrix to construct a cyclic LDPC code: compute the code length, dimension, and rate.

**13.13** Consider the 14 circulants $\mathbf{G}_0$, $\mathbf{G}_1$, ..., $\mathbf{G}_{13}$ of size $281 \times 281$ constructed in Example 13.15 based on the partial geometry $\text{PaG}_{\text{BIBD},2}(70, 5, 5)$. Take five circulants $\mathbf{G}_0$, $\mathbf{G}_1$, ..., $\mathbf{G}_4$ and arrange them in a row-block to form a $1 \times 5$ array $\mathbf{H} = [\mathbf{G}_0 \ \mathbf{G}_1 \ \ldots \ \mathbf{G}_4]$. Compute the QC-BIBD-PaG-LDPC code given by the null space of the array **H** and the BER and BLER performances of the constructed code over the AWGN channel by 50 iterations of the scaled MSA.

**13.14** Let $b = 12$.

(a) Construct a class-2 BIBD $B_2(20b + 1)$.

(b) Construct a partial geometry $\text{PaG}_{\text{BIBD},2}(5b, 5, 5)$ based on $B_2(20b + 1)$.

(c) Compute the base matrix $\mathbf{B}_{\text{PaG},\text{BIBD},2}(5, b)$ in the form of (13.24) based on $\text{PaG}_{\text{BIBD},2}(5b, 5, 5)$.

(d) Let $\mathbf{H}_{\text{PaG},\text{BIBD},2}(5, b)$ be the CPM-dispersion of $\mathbf{B}_{\text{PaG},\text{BIBD},2}(5, b)$. Take a $5 \times 10$ subarray $\mathbf{H}_{\text{PaG},\text{BIBD},2}(5, 10)$ from $\mathbf{H}_{\text{PaG},\text{BIBD},2}(5, b)$. Compute the BER and BLER performances of the code given by the null space over GF(2) of $\mathbf{H}_{\text{PaG},\text{BIBD},2}(5, 10)$ over the AWGN channel by using 50 iterations of the scaled MSA.

(e) Compute the performance of the constructed code over the BEC with 50 iterations of the SPA.

# References

[1] S. Lin and D. J. Costello, Jr., *Error Control Coding: Fundamentals and Applications*, 2nd ed., Upper Saddle River, NJ, Prentice-Hall, 2004.

[2] W. E. Ryan and S. Lin, *Channel Codes: Classical and Modern*, New York, Cambridge University Press, 2009.

[3] Q. Diao, Q. Huang, S. Lin, and K. Abdel-Ghaffar, "A transform approach for analyzing and constructing quasi-cyclic low-density parity-check codes," *Proc. IEEE Inf. Theory Applic. Workshop*, San Diego, CA, February 6–11, 2011, pp. 1–8.

[4] Q. Diao, Q. Huang, S. Lin, and K. Abdel-Ghaffar, "A matrix-theoretic approach for analyzing quasi-cyclic low-density parity-check codes," *IEEE Trans. Inf. Theory*, **58**(6) (2012), 4030–4048.

[5] J. Li, K. Liu, S. Lin, and K. Abdel-Ghaffar, "Algebraic quasi-cyclic LDPC codes: construction, low error-floor, large girth and a reduced-complexity decoding scheme," *IEEE Trans. Commun.*, **62**(8) (2014), 2626–2637.

[6] R. C. Bose, "Strongly regular graphs, partial geometries and partially balanced designs," *Pacif. J. Math.*, **13**(2) (1963), 389–419.

[7] J. H. van Lint, "Partial geometries," *Proc. Int. Congress of Mathematicians*, Warszawa, August 16–24, 1983, pp. 1579–1589.

[8] P. J. Cameron and J. H. van Lint, *Designs, Graphs, Codes, and Their Links*, Cambridge, Cambridge University Press, 1991.

[9] L. M. Batten, *Combinatorics of Finite Geometries*, 2nd ed., Cambridge, Cambridge University Press, 1997.

[10] E. J. Kamischke, *Benson's Theorem for Partial Geometries*, Master's thesis, Michigan Technological University, 2013.

[11] R. D. Carmichael, *Introduction to the Theory of Groups of Finite Orders*, New York, Dover, 1956.

[12] H. Mann, *Analysis and Design of Experiments*, New York, Dover, 1949.

[13] Q. Diao, J. Li, S. Lin, and I. F. Blake, "New classes of partial geometries and their associated LDPC codes," *IEEE Trans. Inf. Theory*, **62**(6) (2016), 2947–2965.

[14] J. Li, K. Liu, S. Lin, and K. Abdel-Ghaffar, "Construction of partial geometries and LDPC codes based on Reed–Solomon codes," *Proc. Int. Symp. Inf. Theory*, Paris, France, July 7–12, 2019, pp. 61–65.

[15] Q. Diao, Y. Y. Tai, S. Lin, and K. Abdel-Ghaffar, "Trapping set structure of LDPC codes on finite geometries," *Proc. IEEE Inf. Theory Applic. Workshop*, San Diego, CA, February 10–15, 2013, pp. 1–8.

[16] Q. Diao, Y. Y. Tai, S. Lin, and K. Abdel-Ghaffar, "LDPC codes on partial geometries: construction, trapping sets structure, and puncturing," *IEEE Trans. Inf. Theory*, **59**(12) (2013), 7898–7914.

[17] C. E. Shannon, R. G. Gallager, and E. R. Berlekamp, "Lower bounds to error probability for coding on discrete memoryless channels," *Inform. Contr.*, **10** (1967), pt. I, pp. 65–103, pt. II, pp. 522–552.

[18] X.-Y Hu, E. Eleftheriou, and D.-M. Arnold, "Regular and irregular progressive edge-growth Tanner graphs," *IEEE Trans. Inf. Theory*, **51**(1) (2005), 386–398.

[19] I. F. Blake and R. C. Mullin, *The Mathematical Theory of Coding*, New York, Academic Press, 1975.

[20] R. C. Bose, "On the construction of balanced incomplete block designs," *Ann. Eugenics*, **9** (1939), 353–399.

[21] C. J. Colbourn and J. H. Dintz (eds.), *The Handbook of Combinatorial Designs*, Boca Raton, FL, CRC Press, 1996.

[22] D. J. Finney, *An Introduction to the Theory of Experimental Design*, Chicago, IL, University of Chicago Press, 1960.

[23] M. Hall, Jr., *Combinatorial Theory*, 2nd ed., New York, Wiley, 1986.

[24] H. J. Ryser, *Combinatorial Mathematics*, New York, Wiley, 1963.

[25] B. Ammar, B. Honary, Y. Kou, J. Xu, and S. Lin, "Construction of low-density parity-check codes based on balanced incomplete block designs," *IEEE Trans. Inf. Theory*, **50**(6) (2004), 1257–1268.

[26] S. Johnson and S. R. Weller, "Regular low-density parity-check codes from combinatorial designs," *Proc. 2001 IEEE Inf. Theory Workshop*, Cairns, Australia, September 2–7 2001, pp. 90–92.

[27] L. Lan, Y. Y. Tai, S. Lin, B. Memari, and B. Honary, "New constructions of quasi-cyclic LDPC codes based on special classes of BIBDs for the AWGN and binary erasure channels," *IEEE Trans. Commun.*, **56**(1) (2008), 39–48.

[28] B. Vasic and O. Milenkovic, "Combinatorial construction of low-density parity-check codes for iterative decoding," *IEEE Trans. Inf. Theory*, **50**(6) (2004), 1156–1176.

# 14

# Quasi-cyclic LDPC Codes Based on Finite Fields

Finite fields have been applied to construct error-correcting codes for reliable information transmission and data storage [1–23] since the late 1950s. These codes are commonly called *algebraic codes*, which have nice structures and large minimum distances. The most well-known *classical* algebraic codes are BCH and RS codes presented in Chapters 5 and 6 which can be decoded with the elegant hard-decision Berlekamp–Massey iterative algorithm. Over the past 60 years, BCH and RS codes have been widely applied in many communication and data-storage systems and they are still operating nowadays. Since the rediscovery of LDPC codes in the late 1990s [24–27], extensive research has been done in constructing modern LDPC codes using algebraic methods. It was found that finite fields are also very effective for constructing modern LDPC codes with distinctive structures that not only simplify encoding and decoding implementations but also provide good error performance in both waterfall and error-floor regions.

In this chapter, we present three *flexible* methods for constructing three types of QC-LDPC codes based on finite fields. The first two types of QC-LDPC codes are constructed based on *two arbitrary subsets* of elements from a given finite field. The third type of QC-LDPC codes is constructed based on the *conventional parity-check matrices of RS codes* under certain constraints. The Tanner graphs of QC-LDPC codes of the three types have girth at least 6. Techniques, algorithms, and conditions are developed to find codes among the three types of QC-LDPC codes whose Tanner graphs have girth 8 or larger and small number of short cycles. Experimental results show that the constructed codes perform well and have low error-floors over the AWGN channel decoded with either the SPA [25], the MSA [28], or the reduced-complexity CPM-RID algorithm presented in Section 12.8. These codes also perform well over the BEC. A class of optimal LDPC codes that are capable of correcting any two random CPM-phased bursts of erasures is presented. At the end of this chapter, a new class of LDPC codes with globally coupled structure is presented.

Nonbinary QC-LDPC codes can also be constructed based on finite fields. Readers are referred to [15, 22, 23, 29, 30] for details of nonbinary finite-field-based QC-LDPC codes.

## 14.1 Construction of QC-LDPC Codes Based on CPM-Dispersion

In this section, we give a brief description of an approach to construct QC-LDPC codes based on finite fields. The first step in the construction is to choose a *base matrix* $\mathbf{B}$ over a finite field $GF(q)$ that satisfies certain constraints. The second step in the construction is to disperse each nonzero entry in $\mathbf{B}$ into a binary circulant permutation matrix (CPM) of a chosen size, say $q - 1$, and a zero entry (if any) into a zero matrix (ZM) of the same size (see the definition of CPM-dispersion in Section 12.5). The $(q - 1)$-fold CPM-dispersion, denoted by $CPM(\mathbf{B})$, of the base matrix $\mathbf{B}$ gives an array $\mathbf{H}_{qc} = CPM(\mathbf{B})$ of CPMs and/or ZMs which is a low-density matrix over $GF(2)$. The last step in the code construction is to use $\mathbf{H}_{qc}$ as a parity-check matrix. The null space over $GF(2)$ of $\mathbf{H}_{qc}$ gives a QC-LDPC code, denoted by $C_{qc}$.

Let $\mathcal{G}_{qc}$ be the Tanner graph [31] of $C_{qc}$ (or associated with $\mathbf{H}_{qc}$). To ensure that $\mathcal{G}_{qc}$ is free of cycles of length 4, i.e., the girth of $\mathcal{G}_{qc}$ is at least 6, the base matrix $\mathbf{B}$ must satisfy the *necessary and sufficient condition*, referred to as the $2 \times 2$ *SM-constraint*, given in Theorem 13.1.

In [32], a necessary and sufficient condition on a base matrix $\mathbf{B}$ over $GF(q)$ is also given for which the Tanner graph associated with the CPM-dispersion of $\mathbf{B}$ has girth at least 8. We rephrase the condition in the following theorem.

**Theorem 14.1** *Let $\mathbf{B}$ be a matrix over $GF(q)$ and $C_{qc}$ be the QC-LDPC code given by the null space over $GF(2)$ of the $(q - 1)$-fold CPM-dispersion of $\mathbf{B}$. A necessary and sufficient condition for the Tanner graph of $C_{qc}$ to have girth at least 8 is that no $2 \times 2$ or $3 \times 3$ submatrix of $\mathbf{B}$ has two identical nonzero terms in its determinant expansion.*

For convenience, we call the necessary and sufficient condition given in Theorem 14.1 the $2 \times 2/3 \times 3$ *SM-constraint*. Note that Theorem 14.1 implies the $2 \times 2$ SM-constraint. A matrix $\mathbf{B}$ over $GF(q)$ that satisfies the $2 \times 2/3 \times 3$ SM-constraint is called a $2 \times 2/3 \times 3$ *SM-constrained matrix*.

**Example 14.1** Consider the finite field $GF(2^2)$ with primitive element $\alpha$. Let

$$\mathbf{B} = \begin{bmatrix} 1 & 1 & 1 \\ 1 & \alpha & \alpha^2 \\ 1 & \alpha^2 & \alpha \end{bmatrix}$$

be a $3 \times 3$ matrix over $GF(2^2)$. It is simple to verify that $\mathbf{B}$ satisfies the $2 \times 2$ SM-constraint because any of its $2 \times 2$ submatrices is nonsingular. Following Theorem 13.1, the 3-fold CPM-dispersion $\mathbf{H}_{qc} = CPM(\mathbf{B})$ of $\mathbf{B}$ satisfies the RC-constraint and the associated Tanner graph $\mathcal{G}_{qc}$ has girth at least 6.

The matrix $\mathbf{B}$ has only one $3 \times 3$ matrix. The six terms in the determinant expansion of $\mathbf{B}$ are $1 \cdot \alpha \cdot \alpha = \alpha^2$, $1 \cdot \alpha^2 \cdot 1 = \alpha^2$, $1 \cdot 1 \cdot \alpha^2 = \alpha^2$, $1 \cdot \alpha \cdot 1 = \alpha$, $1 \cdot \alpha^2 \cdot \alpha^2 = \alpha$, and $\alpha \cdot 1 \cdot 1 = \alpha$. Because there are identical terms in the determinant expansion, following Theorem 14.1, $\mathbf{B}$ does not satisfy the $2 \times 2/3 \times 3$ SM-constraint and the Tanner graph $\mathcal{G}_{\mathrm{qc}}$ of its 3-fold CPM-dispersion $\mathbf{H}_{\mathrm{qc}} = \mathrm{CPM}(\mathbf{B})$ has girth less than 8. Therefore, $\mathcal{G}_{\mathrm{qc}}$ has girth 6.

The 3-fold CPM-dispersion $\mathbf{H}_{\mathrm{qc}} = \mathrm{CPM}(\mathbf{B})$ of $\mathbf{B}$ is

$$
\mathbf{H}_{\mathrm{qc}} = \mathrm{CPM}(\mathbf{B}) = 
\begin{bmatrix}
\underline{1} & 0 & 0 & \underline{1} & 0 & 0 & 1 & 0 & 0 \\
0 & 1 & 0 & 0 & 1 & 0 & 0 & 1 & 0 \\
0 & 0 & 1 & 0 & 0 & 1 & 0 & 0 & 1 \\
\underline{1} & 0 & 0 & 0 & 1 & 0 & 0 & 0 & \underline{1} \\
0 & 1 & 0 & 0 & 0 & 1 & 1 & 0 & 0 \\
0 & 0 & 1 & 1 & 0 & 0 & 0 & 1 & 0 \\
1 & 0 & 0 & 0 & 0 & 1 & 0 & 1 & 0 \\
0 & 1 & 0 & \underline{1} & 0 & 0 & 0 & 0 & \underline{1} \\
0 & 0 & 1 & 0 & 1 & 0 & 1 & 0 & 0
\end{bmatrix}
$$

where the six 1-entries marked with underline correspond to the six edges of a cycle of length 6 in the Tanner graph $\mathcal{G}_{\mathrm{qc}}$. ▲▲

In the rest of this chapter, we first present three specific methods for constructing $2 \times 2$ SM-constrained base matrices over finite fields. Using these base matrices in conjunction with CPM-dispersion, three types of finite-field-based QC-LDPC codes are constructed. Then, techniques, conditions, and algorithms are developed to find $2 \times 2/3 \times 3$ SM-constrained base matrices from the $2 \times 2$ SM-constrained base matrices. At the end of this chapter, we present a class of optimal QC-LDPC codes that are capable of correcting any two CPM-phased bursts of erasures, and a class of globally coupled (GC) LDPC codes.

## 14.2 Construction of Type-I QC-LDPC Codes Based on Two Subsets of a Finite Field

Let $\alpha$ be a primitive element of $\mathrm{GF}(q)$. Then, the powers of $\alpha$, namely, $\alpha^{-\infty} \triangleq 0$, $\alpha^0 = 1$, $\alpha^1$, $\alpha^2, \ldots,$ $\alpha^{q-2}$, give all the $q$ elements of $\mathrm{GF}(q)$. Let $\mathbf{L} = \{-\infty, 0, 1, \ldots, q-2\}$ be the set of powers of $\alpha^{-\infty} \triangleq 0$, $\alpha^0 = 1$, $\alpha^1$, $\alpha^2, \ldots,$ $\alpha^{q-2}$. For $1 \le m, n \le q$, let $\mathbf{S}_1 = \{\alpha^{i_0}, \alpha^{i_1}, \ldots, \alpha^{i_{m-1}}\}$ and $\mathbf{S}_2 = \{\alpha^{j_0}, \alpha^{j_1}, \ldots, \alpha^{j_{n-1}}\}$ be two *arbitrary* sets of *distinct* elements in $\mathrm{GF}(q)$ with $i_k, j_l \in \mathbf{L}$, $i_0 < i_1 < \cdots < i_{m-1}$, and $j_0 < j_1 < \cdots < j_{n-1}$. Let $\eta$ be a nonzero element in $\mathrm{GF}(q)$. Form the following $m \times n$ matrix over $\mathrm{GF}(q)$:

$$
\mathbf{B}_{\mathrm{s}}(m, n) = 
\begin{bmatrix}
\eta\alpha^{i_0} + \alpha^{j_0} & \eta\alpha^{i_0} + \alpha^{j_1} & \cdots & \eta\alpha^{i_0} + \alpha^{j_{n-1}} \\
\eta\alpha^{i_1} + \alpha^{j_0} & \eta\alpha^{i_1} + \alpha^{j_1} & \cdots & \eta\alpha^{i_1} + \alpha^{j_{n-1}} \\
\vdots & \vdots & \ddots & \vdots \\
\eta\alpha^{i_{m-1}} + \alpha^{j_0} & \eta\alpha^{i_{m-1}} + \alpha^{j_1} & \cdots & \eta\alpha^{i_{m-1}} + \alpha^{j_{n-1}}
\end{bmatrix}. \tag{14.1}
$$

The entries of the matrix $\mathbf{B}_\mathrm{s}(m,n)$ are *linear sums* of elements from the two subsets $\mathbf{S}_1$ and $\mathbf{S}_2$ of $\mathrm{GF}(q)$. The addition "+" operation in (14.1) can be replaced by the subtraction "−" operation of $\mathrm{GF}(q)$. The matrix $\mathbf{B}_\mathrm{s}(m,n)$ is said to be in *sum-form*. The subscript "s" in $\mathbf{B}_\mathrm{s}(m,n)$ stands for "sum-form." From (14.1), we can easily check and prove that the matrix $\mathbf{B}_\mathrm{s}(m,n)$ has the following *fundamental structural properties*:

(1) all the entries in a row (or a column) of $\mathbf{B}_\mathrm{s}(m,n)$ are *distinct* elements in $\mathrm{GF}(q)$;

(2) each row (or each column) of $\mathbf{B}_\mathrm{s}(m,n)$ contains *at most one zero element*; and

(3) no two rows (or two columns) in $\mathbf{B}_\mathrm{s}(m,n)$ have *identical entries* at any position.

The nonzero element $\eta$ used in forming the matrix $\mathbf{B}_\mathrm{s}(m,n)$ is called the *multiplier* of $\mathbf{B}_\mathrm{s}(m,n)$. The following theorem gives another structural property of $\mathbf{B}_\mathrm{s}(m,n)$.

**Theorem 14.2** *Any* $2 \times 2$ *submatrix of the* $m \times n$ *matrix* $\mathbf{B}_s(m,n)$ *given by* (14.1) *is nonsingular.*

*Proof* Consider a $2 \times 2$ submatrix $\mathbf{B}(k,l;s,t)$ in $\mathbf{B}_\mathrm{s}(m,n)$:

$$\mathbf{B}(k,l;s,t) = \left[ \begin{array}{cc} \eta\alpha^{i_k} + \alpha^{j_s} & \eta\alpha^{i_k} + \alpha^{j_t} \\ \eta\alpha^{i_l} + \alpha^{j_s} & \eta\alpha^{i_l} + \alpha^{j_t} \end{array} \right],$$

with four entries $\eta\alpha^{i_k} + \alpha^{j_s}$, $\eta\alpha^{i_k} + \alpha^{j_t}$, $\eta\alpha^{i_l} + \alpha^{j_s}$, and $\eta\alpha^{i_l} + \alpha^{j_t}$ from the $k$th and $l$th rows and the $s$th and $t$th columns with $0 \le k < l < m$ and $0 \le s < t < n$.

It follows from the fundamental structural properties (1), (2), and (3) of $\mathbf{B}_\mathrm{s}(m,n)$ that $\mathbf{B}(k,l;s,t)$ consists of at least two nonzero entries (or at most two zero entries). To prove that $\mathbf{B}(k,l;s,t)$ is nonsingular, there are three cases to be considered. The first case is that $\mathbf{B}(k,l;s,t)$ contains a single 0-entry. In this case, the determinant of $\mathbf{B}(k,l;s,t)$ contains a single nonzero product term, either $(\eta\alpha^{i_k} + \alpha^{j_s})(\eta\alpha^{i_l} + \alpha^{j_t})$ or $(\eta\alpha^{i_k} + \alpha^{j_t})(\eta\alpha^{i_l} + \alpha^{j_s})$. Hence, the determinant of $\mathbf{B}(k,l;s,t)$ is nonzero and $\mathbf{B}(k,l;s,t)$ is nonsingular.

The second case is that $\mathbf{B}(k,l;s,t)$ contains two 0-entries. It follows from the structural properties (2) and (3) of $\mathbf{B}_\mathrm{s}(m,n)$ that the two 0-entries must lie on either the main diagonal or the antidiagonal of $\mathbf{B}(k,l;s,t)$. In this case, the determinant of $\mathbf{B}(k,l;s,t)$ also contains a single nonzero product term, either $(\eta\alpha^{i_k} + \alpha^{j_s})(\eta\alpha^{i_l} + \alpha^{j_t})$ or $(\eta\alpha^{i_k} + \alpha^{j_t})(\eta\alpha^{i_l} + \alpha^{j_s})$, and hence $\mathbf{B}(k,l;s,t)$ is nonsingular.

The third case is that all four entries in $\mathbf{B}(k,l;s,t)$ are nonzero. In this case, the determinant of $\mathbf{B}(k,l;s,t)$ is $(\eta\alpha^{i_k} + \alpha^{j_s})(\eta\alpha^{i_l} + \alpha^{j_t}) - (\eta\alpha^{i_k} + \alpha^{j_t})(\eta\alpha^{i_l} + \alpha^{j_s})$. Suppose $\mathbf{B}(k,l;s,t)$ is singular. Then, the following equality must hold:

$$(\eta\alpha^{i_k} + \alpha^{j_s})(\eta\alpha^{i_l} + \alpha^{j_t}) - (\eta\alpha^{i_k} + \alpha^{j_t})(\eta\alpha^{i_l} + \alpha^{j_s}) = 0.$$

With some algebraic manipulations of the above equality, we have either $\alpha^{i_k} = \alpha^{i_l}$ or $\alpha^{j_s} = \alpha^{j_t}$ which contradicts the fact that all the elements in $\mathbf{S}_1$ and $\mathbf{S}_2$ are distinct. Hence, $\mathbf{B}(k,l;s,t)$ must be nonsingular.

With the above results, we conclude that $\mathbf{B}(k,l;s,t)$ is nonsingular. This proves the theorem. ▲▲

It follows from Theorems 13.1 and 14.1 that the matrix $\mathbf{B}_s(m,n)$ over $GF(q)$ given by (14.1) satisfies the $2 \times 2$ SM-constraint. Let $\mathbf{H}_{s,qc}(m,n)$ be the $(q-1)$-fold CPM-dispersion of $\mathbf{B}_s(m,n)$ which is an $m(q-1) \times n(q-1)$ matrix over $GF(2)$. Because $\mathbf{B}_s(m,n)$ satisfies the $2 \times 2$ SM-constraint, $\mathbf{H}_{s,qc}(m,n)$, as an $m(q-1) \times n(q-1)$ matrix, satisfies the RC-constraint. Hence, the null space over $GF(2)$ of $\mathbf{H}_{s,qc}(m,n)$ gives a QC-LDPC code, denoted by $C_{s,qc}(m,n)$, of length $n(q-1)$ with rate at least $(n-m)/n$ whose Tanner graph $\mathcal{G}_{s,qc}(m,n)$ has girth at least 6. The above construction is referred to as *type-I construction* of QC-LDPC codes, which is also called *sum-construction*.

With different choices of $m$, $n$, $\mathbf{S}_1$, $\mathbf{S}_2$, and the multiplier $\eta$ in $GF(q)$, we can construct a family of $2 \times 2$ SM-constrained matrices over $GF(q)$ in the sum-form given by (14.1) with various sizes. Using these matrices as base matrices and their $(q-1)$-fold CPM-dispersions as parity-check matrices, we can construct a family of QC-LDPC codes with various lengths and rates whose Tanner graphs have girths at least 6.

**Example 14.2** In this example, we use a small field for illustration of the type-I construction of QC-LDPC codes. The field used is $GF(2^4)$ which is given in Table 2.9. Let $\alpha$ be a primitive element in $GF(2^4)$. Choose $m = 4$, $n = 12$, and the following two subsets of elements from $GF(2^4)$:

$$\mathbf{S}_1 = \{1, \alpha, \alpha^2, \alpha^3\}, \quad \mathbf{S}_2 = \{1, \alpha, \alpha^2, \alpha^3, \alpha^4, \alpha^5, \alpha^6, \alpha^7, \alpha^8, \alpha^9, \alpha^{10}, \alpha^{11}\}.$$

Let $\eta = 1$. Using $\mathbf{S}_1$, $\mathbf{S}_2$, and $\eta$, we construct the following $2 \times 2$ SM-constrained $4 \times 12$ matrix over $GF(2^4)$ in the form of (14.1):

$$\mathbf{B}_s(4,12) = \begin{bmatrix} 1+1 & 1+\alpha & 1+\alpha^2 & \cdots & 1+\alpha^9 & 1+\alpha^{10} & 1+\alpha^{11} \\ \alpha+1 & \alpha+\alpha & \alpha+\alpha^2 & \cdots & \alpha+\alpha^9 & \alpha+\alpha^{10} & \alpha+\alpha^{11} \\ \alpha^2+1 & \alpha^2+\alpha & \alpha^2+\alpha^2 & \cdots & \alpha^2+\alpha^9 & \alpha^2+\alpha^{10} & \alpha^2+\alpha^{11} \\ \alpha^3+1 & \alpha^3+\alpha & \alpha^3+\alpha^2 & \cdots & \alpha^3+\alpha^9 & \alpha^3+\alpha^{10} & \alpha^3+\alpha^{11} \end{bmatrix}$$

$$= \begin{bmatrix} 0 & \alpha^4 & \alpha^8 & \alpha^{14} & \alpha & \alpha^{10} & \alpha^{13} & \alpha^9 & \alpha^2 & \alpha^7 & \alpha^5 & \alpha^{12} \\ \alpha^4 & 0 & \alpha^5 & \alpha^9 & 1 & \alpha^2 & \alpha^{11} & \alpha^{14} & \alpha^{10} & \alpha^3 & \alpha^8 & \alpha^6 \\ \alpha^8 & \alpha^5 & 0 & \alpha^6 & \alpha^{10} & \alpha & \alpha^3 & \alpha^{12} & 1 & \alpha^{11} & \alpha^4 & \alpha^9 \\ \alpha^{14} & \alpha^9 & \alpha^6 & 0 & \alpha^7 & \alpha^{11} & \alpha^2 & \alpha^4 & \alpha^{13} & \alpha & \alpha^{12} & \alpha^5 \end{bmatrix}.$$

The 15-fold CPM-dispersion of $\mathbf{B}_s(4,12)$ gives a $4 \times 12$ array $\mathbf{H}_{s,qc}(4,12)$ of CPMs and ZMs of size $15 \times 15$ which is a $60 \times 180$ matrix of rank 52 with constant row weight 11 and two column weights 3 and 4. The null space over $GF(2)$ of $\mathbf{H}_{s,qc}(4,12)$ gives a $(180, 128)$ QC-LDPC code $C_{s,qc}(4,12)$ with rate 0.711 whose Tanner graph has girth at least 6. The BER and BLER performances of the code over the AWGN channel using BPSK signaling decoded with 50 iterations of the MSA are shown in Fig. 14.1. ▲▲

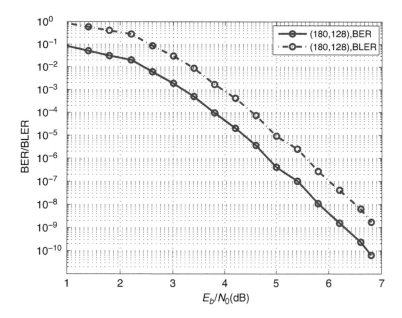

Figure 14.1 The BER and BLER performances of the $(180, 128)$ QC-LDPC code given in Example 14.2.

**Example 14.3** In this example, we construct three codes using the field $GF(2^7)$. Let $\alpha$ be a primitive element in $GF(2^7)$. Choose $m = 4$, $n = 40$, and the following two subsets of elements in $GF(2^7)$:

$$\mathbf{S}_1 = \{1, \alpha, \alpha^2, \alpha^3\}, \ \mathbf{S}_2 = \{\alpha^4, \alpha^5, \alpha^6, \dots, \alpha^{42}, \alpha^{43}\}.$$

Let $\eta = 1$. Using $\mathbf{S}_1$, $\mathbf{S}_2$, and $\eta$, we construct a $2 \times 2$ SM-constrained $4 \times 40$ matrix $\mathbf{B}_s(4, 40)$ over $GF(2^7)$ in the form of (14.1) which contains no zeros. The 127-fold CPM-dispersion of $\mathbf{B}_s(4, 40)$ gives a $4 \times 40$ array $\mathbf{H}_{s,qc}(4, 40)$ of CPMs of size $127 \times 127$, which is a $508 \times 5080$ matrix over $GF(2)$ with column and row weights 4 and 40, respectively. The rank of $\mathbf{H}_{s,qc}(4, 40)$ is 491. Hence, $\mathbf{H}_{s,qc}(4, 40)$ has 17 redundant rows. The null space of $\mathbf{H}_{s,qc}(4, 40)$ over $GF(2)$ gives a $(4, 40)$-regular $(5080, 4589)$ QC-LDPC code $C_{s,qc}(4, 40)$ of rate 0.9033 whose Tanner graph $\mathcal{G}_{s,qc}(4, 40)$ has girth 6. Each VN in $\mathcal{G}_{s,qc}(4, 40)$ is connected to other 156 VNs by paths of length 2.

The BER and BLER performances of the constructed code $C_{s,qc}(4, 40)$ over the AWGN channel using BPSK signaling decoded with 5, 10, and 50 iterations of the MSA scaled by a factor of 0.75 are shown in Fig. 14.2(a). From the figure, we see that the decoding of this code converges quickly. At a BER of $10^{-7}$, the code performs 1.1 dB away from the Shannon limit with 50 iterations.

If we apply the $(5080, 4589)$ QC-LDPC code $C_{s,qc}(4, 40)$ to the BEC, its UEBR and UEBLR performances are shown in Fig. 14.2(b). Because the code has rate $R = 0.9033$, the Shannon limit for this code rate is $1 - R = 0.0967$.

Figure 14.2(b) shows that the code performs 0.041 from the Shannon limit at a UEBR of $10^{-8}$.

Suppose we take the first eight columns from $\mathbf{B}_s(4, 40)$. We obtain a $4 \times 8$ submatrix $\mathbf{B}_s(4, 8)$ of $\mathbf{B}_s(4, 40)$. The 127-fold CPM-dispersion of $\mathbf{B}_s(4, 8)$ gives a $4 \times 8$ array $\mathbf{H}_{s,qc}(4, 8)$ of CPMs of size $127 \times 127$, which is a $508 \times 1016$ matrix over GF(2) with column and row weights 4 and 8, respectively. The rank of $\mathbf{H}_{s,qc}(4, 8)$ is 491. The null space over GF(2) of $\mathbf{H}_{s,qc}(4, 8)$ gives a $(4, 8)$-regular $(1016, 525)$ QC-LDPC code $C_{s,qc}(4, 8)$ with rate 0.5167 whose Tanner graph $\mathcal{G}_{s,qc}(4, 8)$ has girth 6. By using the cycle counting algorithm presented in [33, 34], we find that $\mathcal{G}_{s,qc}(4, 8)$ contains 3810 cycles of length 6 and 31 496 cycles of length 8.

The BER and BLER performances of the code $C_{s,qc}(4, 8)$ over the AWGN channel using BPSK signaling decoded with 50 iterations of the MSA scaled by a factor of 0.75 are shown in Fig. 14.3(a).

Suppose we mask the matrix $\mathbf{B}_s(4, 8)$ with the masking matrix $\mathbf{Z}(4, 8)$ given by (12.13). We obtain a $4 \times 8$ masked base matrix $\mathbf{B}_{s,mask}(4, 8)$. The 127-fold CPM-dispersion of $\mathbf{B}_{s,mask}(4, 8)$ gives a $4 \times 8$ masked array $\mathbf{H}_{s,qc,mask}(4, 8)$ of CPMs and ZMs of size $127 \times 127$, which is a $508 \times 1016$ matrix over GF(2) with column and row weights 3 and 6, respectively. The rank of $\mathbf{H}_{s,qc,mask}(4, 8)$ is 508. The null space over GF(2) of $\mathbf{H}_{s,qc,mask}(4, 8)$ gives a $(3, 6)$-regular $(1016, 508)$ QC-LDPC code $C_{s,qc,mask}(4, 8)$ with rate 0.5. The Tanner graph $\mathcal{G}_{s,qc,mask}(4, 8)$ of $C_{s,qc,mask}(4, 8)$ has girth 6 and contains 381 cycles of length 6 and 1270 cycles of length 8.

The BER and BLER performances of the code $C_{s,qc,mask}(4, 8)$ over the AWGN channel using BPSK signaling decoded with 50 iterations of the MSA scaled by a factor of 0.75 are also shown in Fig. 14.3(a). From the figure, we see that the masked code $C_{s,qc,mask}(4, 8)$ outperforms the unmasked code $C_{s,qc}(4, 8)$. At the BER of $10^{-7}$, it performs 1.8 dB from the threshold (1.11 dB).

The UEBR and UEBLR performances of $C_{s,qc}(4, 8)$ and $C_{s,qc,mask}(4, 8)$ over the BEC are shown in Fig. 14.3(b). From this figure, we can also see that the masked code $C_{s,qc,mask}(4, 8)$ outperforms the unmasked code $C_{s,qc}(4, 8)$ over the BEC. ▲▲

**Example 14.4** In this example, we show the flexibility of the proposed type-I code construction. To do this, we use the $4 \times 40$ base matrix $\mathbf{B}_s(4, 40)$ over GF($2^7$) constructed in Example 14.3. For $k = 4(i + 1)$, $1 \leq i \leq 9$, we take the first $k$ columns from $\mathbf{B}_s(4, 40)$ and form a $4 \times k$ base matrix $\mathbf{B}_s(4, k)$. The 127-fold CPM-dispersion of $\mathbf{B}_s(4, k)$ gives a $4 \times k$ array $\mathbf{H}_{s,qc}(4, k)$ of CPMs of size $127 \times 127$, which is a $508 \times 127k$ matrix over GF(2) with column and row weights 4 and $k$, respectively. The null space of $\mathbf{H}_{s,qc}(4, k)$ over GF(2) gives a $(4, k)$-regular QC-LDPC code of length $127k$ with rate at least $(k - 1)/k$ whose Tanner graph $\mathcal{G}_{s,qc}(4, k)$ has girth at least 6. For $i = 1, 2, \ldots, 9$, we construct nine QC-LDPC codes whose parameters are listed in Table 14.1. The BER performances of these nine codes over the AWGN channel decoded with 50 iterations of the MSA are plotted in Fig. 14.4. All the constructed codes perform well. Note that

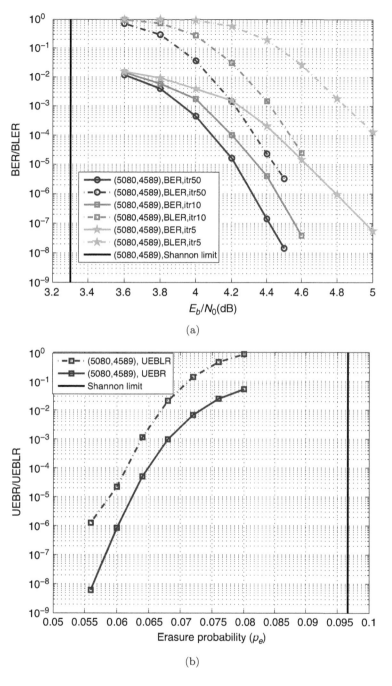

(a)

(b)

Figure 14.2 The performances of the $(5080, 4589)$ QC-LDPC code $C_{\mathrm{s,qc}}(4, 40)$ given in Example 14.3 over: (a) AWGN channel and (b) BEC.

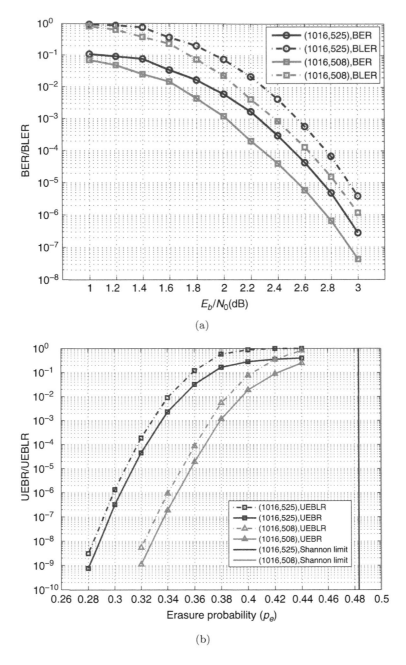

Figure 14.3 The performances of the unmasked $(1016, 525)$ and masked $(1016, 508)$ QC-LDPC codes given in Example 14.3 over: (a) AWGN channel and (b) BEC.

Table 14.1 The nine QC-LDPC codes given in Example 14.4.

| $k$ | Code | Length | Dimension | Rate |
|----|------|--------|-----------|------|
| 8  | $C_{s,qc}(4,8)$  | 1016 | 525  | 0.5167 |
| 12 | $C_{s,qc}(4,12)$ | 1524 | 1033 | 0.6778 |
| 16 | $C_{s,qc}(4,16)$ | 2032 | 1541 | 0.7584 |
| 20 | $C_{s,qc}(4,20)$ | 2540 | 2049 | 0.8067 |
| 24 | $C_{s,qc}(4,24)$ | 3048 | 2557 | 0.8389 |
| 28 | $C_{s,qc}(4,28)$ | 3556 | 3065 | 0.8619 |
| 32 | $C_{s,qc}(4,32)$ | 4064 | 3573 | 0.8792 |
| 36 | $C_{s,qc}(4,36)$ | 4572 | 4081 | 0.8926 |
| 40 | $C_{s,qc}(4,40)$ | 5080 | 4589 | 0.9033 |

the first and last codes listed in Table 14.1 are the $(1016, 525)$ and $(5080, 4589)$ QC-LDPC codes constructed in Example 14.3.                    ▲▲

Constructions of LDPC codes based on finite fields of characteristic 2 put a limitation on the code construction. To remove this limitation, we can use prime fields or fields of prime characteristics greater than 2 for code construction.

**Example 14.5** In this example, we choose the prime field $GF(131)$ for code construction. Let $\alpha = 128$ be a primitive element of this prime field. Choose the following two subsets of $GF(131)$:

$$\mathbf{S}_1 = \left\{\alpha^0 = 1, \alpha^1, \alpha^2, \alpha^3, \alpha^4, \alpha^5\right\}, \mathbf{S}_2 = \left\{\alpha^6, \alpha^7, \ldots, \alpha^{128}, \alpha^{129}\right\}$$

for the base matrix construction and set the multiplier $\eta = 1$. The two chosen sets are mutually disjoint and both consist of consecutive powers of $\alpha$. Based on these two sets and $\eta = 1$, we form the following $6 \times 124$ matrix over $GF(131)$ in the form of (14.1) which satisfies the $2 \times 2$ SM-constraint:

$$\mathbf{B}_s(6, 124) = [\alpha^i + \alpha^j]_{0 \le i \le 5, 6 \le j \le 129}.$$

All the entries in $\mathbf{B}_s(6, 124)$ are nonzero elements in $GF(131)$. The 130-fold CPM-dispersion of $\mathbf{B}_s(6, 124)$ gives a $6 \times 124$ array $\mathbf{H}_{s,qc}(6, 124)$ of CPMs of size $130 \times 130$ which is a $780 \times 16\,120$ matrix over $GF(2)$ with constant column weight 6 and constant row weight 124 and satisfies the RC-constraint.

The null space of $\mathbf{H}_{s,qc}(6, 124)$ gives a $(6, 124)$-regular $(16\,120, 15\,345)$ QC-LDPC code $C_{s,qc}(6, 124)$ of length $16\,120$ with rate 0.952. The Tanner graph $\mathcal{G}_{s,qc}(6, 124)$ of $C_{s,qc}(6, 124)$ has girth 6. Because $\mathbf{H}_{s,qc}(6, 124)$ has column weight 6, for every column position of $\mathbf{H}_{s,qc}(6, 124)$, there are six rows orthogonal on it. With this orthogonal structure, each VN in the Tanner graph $\mathcal{G}_{s,qc}(6, 124)$ of $C_{s,qc}(6, 124)$ is connected to other 738 VNs by paths of length 2. This results in a very high degree of *VN-connectivity*. This high degree of connectivity allows

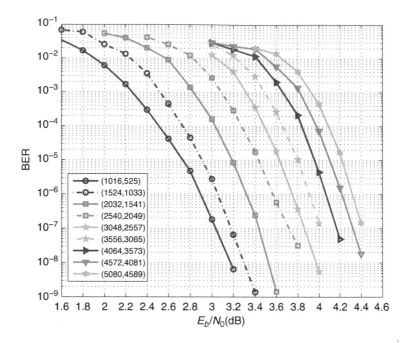

Figure 14.4 The BER performances of nine QC-LDPC codes in Example 14.4.

rapid exchange of information between all the VNs. Consequently, in a few decoding iterations, each VN processing unit (VNPU) collects enough extrinsic information from other VNPUs to update the reliability of the VN being processed to a level to make a correct hard-decision with high probability. As a result, the decoding process can converge into a codeword with a small number of decoding iterations.

The BER and BLER performances of the code over the AWGN channel decoded with 5, 10, and 50 iterations of the MSA with a scaling factor of 0.75 are shown in Fig. 14.5. The performance gap between 5 and 10 iterations is less than 0.2 dB and the performance gap between 10 and 50 iterations is less than 0.1 dB. With 50 iterations of the scaled MSA, the code achieves a BER of $10^{-14}$ without a visible error-floor. At a BER of $10^{-14}$, the code performs 1.4 dB away from the Shannon limit and achieves a 8.5 dB coding gain over the uncoded BPSK system.                                                                                ▲▲

**Example 14.6** Let $\alpha = 326$ be a primitive element of the prime field GF(331) and

$$\mathbf{S}_1 = \left\{\alpha^{112}, \alpha^{115}, \alpha^{148}, \alpha^{317}\right\}, \mathbf{S}_2 = \left\{\alpha^{24}, \alpha^{112}, \alpha^{115}, \alpha^{234}, \alpha^{236}, \alpha^{274}, \alpha^{316}, \alpha^{320}\right\}$$

be two sets of elements from GF(331). All the elements in each set are *chosen randomly* and are *distinct*. The two sets have two elements in common. We set $\eta = 1$. Based on the two chosen sets $\mathbf{S}_1$, $\mathbf{S}_2$, and $\eta = 1$, we form a $4 \times 8$ base matrix $\mathbf{B}_s(4, 8)$ over GF(331) in the form of (14.1),

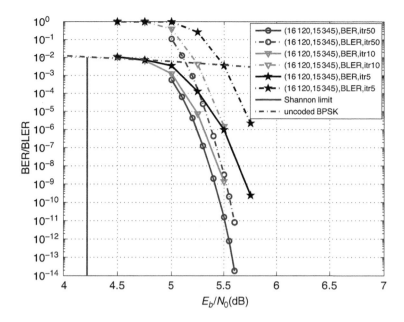

Figure 14.5 The BER and BLER performances of the $(16\,120, 15\,345)$ QC-LDPC code given in Example 14.5.

$$\mathbf{B}_{\mathrm{s}}(4,8) = \left[ \alpha^i + \alpha^j \right]_{\alpha^i \in \mathbf{S}_1, \alpha^j \in \mathbf{S}_2}$$

$$= \begin{bmatrix} \alpha^{39} & \alpha^{123} & \alpha^{79} & \alpha^{190} & \alpha^{294} & \alpha^{328} & \alpha^{297} & \alpha^{68} \\ \alpha^{139} & \alpha^{79} & \alpha^{126} & \alpha^{75} & \alpha^{158} & \alpha^{295} & \alpha^{148} & \alpha^{274} \\ \alpha^{206} & \alpha^{209} & \alpha^{287} & \alpha^{324} & \alpha^{163} & \alpha^{129} & \alpha^{202} & \alpha^{325} \\ \alpha^{181} & \alpha^{271} & \alpha^{275} & \alpha^{143} & \alpha^{208} & \alpha^{10} & \alpha^{173} & \alpha^{284} \end{bmatrix}.$$

All the entries of $\mathbf{B}_{\mathrm{s}}(4,8)$ are nonzero elements in GF(331). The 330-fold CPM-dispersion of $\mathbf{B}_{\mathrm{s}}(4,8)$ results in a $4 \times 8$ array $\mathbf{H}_{\mathrm{s,qc}}(4,8)$ of CPMs of size $330 \times 330$ which is a $1320 \times 2640$ matrix with column and row weights 4 and 8, respectively. The null space of $\mathbf{H}_{\mathrm{s,qc}}(4,8)$ gives a $(4,8)$-regular $(2640, 1323)$ QC-LDPC code $C_{\mathrm{s,qc}}(4,8)$ with rate 0.501 whose Tanner graph $\mathcal{G}_{\mathrm{s,qc}}(4,8)$ has girth at least 6. There are three $3 \times 3$ submatrices in $\mathbf{B}_{\mathrm{s}}(4,8)$ which do not satisfy the $2 \times 2/3 \times 3$ SM-constraint, each containing two identical terms in its determinant expansion. Therefore, the Tanner graph $\mathcal{G}_{\mathrm{s,qc}}(4,8)$ of the code $C_{\mathrm{s,qc}}(4,8)$ has girth 6. By using the cycle counting algorithm in [33, 34], we find that $\mathcal{G}_{\mathrm{s,qc}}(4,8)$ contains 990 cycles of length 6, 24 750 cycles of length 8, and 389 400 cycles of length 10. The BER and BLER performances of $C_{\mathrm{s,qc}}(4,8)$ over the AWGN channel decoded with 50 iterations of the MSA are shown in Fig. 14.6.

Suppose we mask the base matrix $\mathbf{B}_{\mathrm{s}}(4,8)$ with the $4 \times 8$ masking matrix $\mathbf{Z}(4,8)$ given by (12.13). We obtain a $4 \times 8$ masked matrix $\mathbf{B}_{\mathrm{s,mask}}(4,8)$ with column and row weights 3 and 6, respectively. The 330-fold CPM-dispersion of $\mathbf{B}_{\mathrm{s,mask}}(4,8)$ gives a $4 \times 8$ array $\mathbf{H}_{\mathrm{s,qc,mask}}(4,8)$ of CPMs and ZMs of size $330 \times 330$. The array $\mathbf{H}_{\mathrm{s,qc,mask}}(4,8)$ is a $1320 \times 2640$ matrix with constant column and row weights 3 and 6, respectively.

The null space of $\mathbf{H}_{\mathrm{s,qc,mask}}(4,8)$ gives a $(3,6)$-regular $(2640,1320)$ QC-LDPC code $C_{\mathrm{s,qc,mask}}(4,8)$ with rate 0.5. The Tanner graph $\mathcal{G}_{\mathrm{s,qc,mask}}(4,8)$ of $C_{\mathrm{s,qc,mask}}(4,8)$ has girth 8 and contains 990 cycles of length 8 and 8580 cycles of length 10. We see that masking the $4 \times 8$ base matrix $\mathbf{B}_{\mathrm{s}}(4,8)$ with $\mathbf{Z}(4,8)$ not only increases the girth of the Tanner graph of the unmasked code $C_{\mathrm{s,qc}}(4,8)$, from 6 to 8, but also greatly reduces the number of short cycles, e.g., from 24 750 cycles of length 8 to 990 cycles of length 8, and from 389 400 cycles of length 10 to 8580 cycles of length 10. This shows that the masking matrix $\mathbf{Z}(4,8)$ is very effective in reducing the number of short cycles in a Tanner graph.

The BER and BLER performances of the masked code $C_{\mathrm{s,qc,mask}}(4,8)$ over the AWGN channel decoded with 5, 10, and 50 iterations of the MSA scaled by a factor of 0.75 are shown in Fig. 14.6. With 50 iterations, we see that the masked code $C_{\mathrm{s,qc,mask}}(4,8)$ outperforms the unmasked code $C_{\mathrm{s,qc}}(4,8)$ by 0.3 dB at a BER of $10^{-9}$. We also see that $C_{\mathrm{s,qc,mask}}(4,8)$ achieves a BER of almost $10^{-10}$ without a visible error-floor and it performs only 2.3 dB away from the Shannon limit and 1.4 dB from the threshold (1.11 dB).

To demonstrate the flexibility of the proposed construction, we construct another $(3,6)$-regular $(2640,1320)$ QC-LDPC code based on the same field GF(331). This time, we arbitrarily choose another two sets of elements from GF(331),

$$\mathbf{S}_1 = \left\{\alpha^4, \alpha^{139}, \alpha^{29}, \alpha^{322}\right\}, \ \mathbf{S}_2 = \left\{\alpha^0, \alpha^{27}, \alpha^{58}, \alpha^{74}, \alpha^{213}, \alpha^{254}, \alpha^{262}, \alpha^{308}\right\}.$$

Based on these two sets and a randomly chosen multiplier $\eta = \alpha^{14} = 81$, we form another $4 \times 8$ base matrix $\mathbf{B}_{\mathrm{s}}^*(4,8)$ in the form of (14.1):

$$\mathbf{B}_{\mathrm{s}}^*(4,8) = [\eta\alpha^i + \alpha^j]_{\alpha^i \in \mathbf{S}_1, \alpha^j \in \mathbf{S}_2}$$

$$= \begin{bmatrix} \alpha^{149} & \alpha^{109} & \alpha^{301} & \alpha^{102} & \alpha^{265} & \alpha^{286} & \alpha^{108} & \alpha^{261} \\ \alpha^{220} & \alpha^8 & \alpha^{270} & \alpha^{175} & \alpha^{157} & \alpha^{103} & \alpha^{258} & \alpha^{13} \\ \alpha^{295} & \alpha^{168} & \alpha^{214} & \alpha^{282} & \alpha^{290} & \alpha^{251} & \alpha^{328} & \alpha^{272} \\ \alpha^{46} & \alpha^{170} & \alpha^{199} & \alpha^{275} & \alpha^{176} & \alpha^{284} & \alpha^{180} & \alpha^{195} \end{bmatrix}.$$

All the entries of $\mathbf{B}_{\mathrm{s}}^*(4,8)$ are again nonzero elements in GF(331). Masking this base matrix with the same masking matrix $\mathbf{Z}(4,8)$, we obtain a masked base matrix $\mathbf{B}_{\mathrm{s,mask}}^*(4,8)$. The 330-fold CPM-dispersion of $\mathbf{B}_{\mathrm{s,mask}}^*(4,8)$ gives another $4 \times 8$ array $\mathbf{H}_{\mathrm{s,qc,mask}}^*(4,8)$ of CPMs and ZMs of size $330 \times 330$.

The null space over GF(2) of $\mathbf{H}_{\mathrm{s,qc,mask}}^*(4,8)$ gives another $(3,6)$-regular $(2640,1320)^*$ QC-LDPC code $C_{\mathrm{s,qc,mask}}^*(4,8)$ with rate 0.5 whose Tanner graph also has girth 8 but contains only 330 cycles of length 8. The BER and BLER performances of this code decoded with 50 iterations of the MSA are also included in Fig. 14.6. From this figure, we see that the performance curves of the two masked codes, $C_{\mathrm{s,qc,mask}}(4,8)$ and $C_{\mathrm{s,qc,mask}}^*(4,8)$, overlap with each other. ▲▲

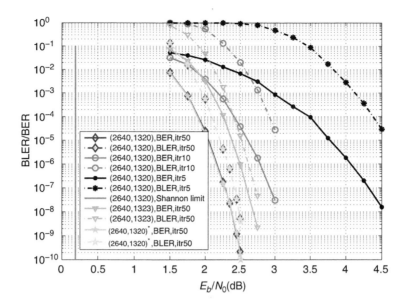

Figure 14.6 The BER and BLER performances of the three QC-LDPC codes given in Example 14.6.

A special case of the sum-construction of a base matrix in the form of (14.1) is to set $m = n = q$ and choose $\mathbf{S}_1 = \mathbf{S}_2 = \mathrm{GF}(q) = \{0, 1, \alpha, \alpha^2, \ldots, \alpha^{q-2}\}$. In this case, the base matrix is a $q \times q$ matrix over $\mathrm{GF}(q)$,

$$
\mathbf{B}_\mathrm{s}(q,q) = [\eta\alpha^i + \alpha^j]_{\alpha^i \in \mathbf{S}_1, \alpha^j \in \mathbf{S}_2}
$$

$$
= \begin{bmatrix}
0 & 1 & \alpha & \cdots & \alpha^{q-2} \\
\eta & \eta + 1 & \eta + \alpha & \cdots & \eta + \alpha^{q-2} \\
\eta\alpha & \eta\alpha + 1 & \eta\alpha + \alpha & \cdots & \eta\alpha + \alpha^{q-2} \\
\eta\alpha^2 & \eta\alpha^2 + 1 & \eta\alpha^2 + \alpha & \cdots & \eta\alpha^2 + \alpha^{q-2} \\
\vdots & \vdots & \vdots & \vdots & \vdots \\
\eta\alpha^{q-2} & \eta\alpha^{q-2} + 1 & \eta\alpha^{q-2} + \alpha & \cdots & \eta\alpha^{q-2} + \alpha^{q-2}
\end{bmatrix}. \tag{14.2}
$$

Each row and each column of $\mathbf{B}_\mathrm{s}(q,q)$ contain every element of $\mathrm{GF}(q)$ exactly once. No two rows (or two columns) in $\mathbf{B}_\mathrm{s}(q,q)$ have identical entries in any position. The matrix $\mathbf{B}_\mathrm{s}(q,q)$ is a *Latin square* over $\mathrm{GF}(q)$ of order $q$ [35].[1] Construction of QC-LDPC codes using Latin squares as base matrices was first proposed in [17].

Let $\mathbf{B}_\mathrm{s}(d,k)$ be a $d \times k$ submatrix of the Latin square $\mathbf{B}_\mathrm{s}(q,q)$ over $\mathrm{GF}(q)$ where $d$ and $k$ are two positive integers such that $1 \leq d \leq k \leq q$. The $(q-1)$-fold CPM-dispersion of $\mathbf{B}_\mathrm{s}(d,k)$ gives a $d \times k$ array $\mathbf{H}_\mathrm{s,qc}(d,k)$ of CPMs and/or

---

[1] A Latin square of order $N$ is an $N \times N$ array of elements from a set $X = \{x_0, x_1, \ldots, x_{N-1}\}$ of $N$ elements (or objects) for which each row and each column of the array contains every element in the set $X$ exactly once.

ZMs of size $(q-1) \times (q-1)$. The null space over GF(2) of $\mathbf{H}_{s,qc}(d,k)$ gives a QC-LDPC code $C_{s,qc}(d,k)$ of length $k(q-1)$ with rate at least $(k-d)/k$. The Tanner graph of $C_{s,qc}(d,k)$ has girth at least 6.

The $6 \times 124$ base matrix $\mathbf{B}_s(6,124) = [\alpha^i + \alpha^j]_{0 \le i \le 5, 6 \le j \le 129}$ over GF(131) in Example 14.5 is a submatrix of the Latin square $\mathbf{B}_s(131,131)$ over the prime field GF(131) with $\eta = 1$.

Another special case of the base matrix $\mathbf{B}_s(m,n)$ in the sum-form given by (14.1) is that the two subsets $\mathbf{S}_1$ and $\mathbf{S}_2$ are two additive subgroups of GF($q$) of orders $m$ and $n$, respectively, such that $m + n \le q$ and $\mathbf{S}_1 \cap \mathbf{S}_2 = \{0\}$. Let $\alpha^{i_0} = \alpha^{j_0} = 0$. Then, the first row of the matrix $\mathbf{B}_s(m,n)$ contains the elements of the subgroup $\mathbf{S}_2$ of GF($q$) and all the other rows of $\mathbf{B}_s(m,n)$ are cosets of $\mathbf{S}_2$ with the $m$ elements in the set $\eta\mathbf{S}_1 = \{\eta\alpha^{i_0}, \eta\alpha^{i_1}, \dots, \eta\alpha^{i_{m-1}}\}$ as coset leaders. Construction of QC-LDPC codes based on two additive subgroups of GF($q$) of orders $m$ and $n = q - m$, respectively, and $\eta = 1$ was first presented in [15] and later generalized in [16].

**Example 14.7** Suppose we use the field GF($2^7$) for code construction. Let $\alpha$ be a primitive element of GF($2^7$) and $\mathbf{S}_1$ and $\mathbf{S}_2$ be two additive subgroups of GF($2^7$) with orders 4 and 32, respectively. The two additive subgroups $\mathbf{S}_1$ and $\mathbf{S}_2$ can be constructed as follows:

$$\mathbf{S}_1 = \{c_0\alpha^5 + c_1\alpha^6 : c_0, c_1 \in \text{GF}(2)\},$$
$$\mathbf{S}_2 = \{c_0\alpha^0 + c_1\alpha + c_2\alpha^2 + c_3\alpha^3 + c_4\alpha^4 : c_0, c_1, c_2, c_3, c_4 \in \text{GF}(2)\}.$$

Let $\eta = 1$. Based on $\mathbf{S}_1$ and $\mathbf{S}_2$ and $\eta$, we construct a $4 \times 32$ base matrix $\mathbf{B}_s(4,32)$ in the form of (14.1) over GF($2^7$). $\mathbf{B}_s(4,32)$ has one zero entry at the zeroth row and zeroth column. The 127-fold CPM-dispersion of $\mathbf{B}_s(4,32)$ gives a $4 \times 32$ array $\mathbf{H}_{s,qc}(4,32)$ of CPMs and a single ZM of size $127 \times 127$, which is a $508 \times 4064$ matrix over GF(2) with two column weights 3 and 4 and two row weights 31 and 32. The null space over GF(2) of $\mathbf{H}_{s,qc}(4,32)$ gives a nearly $(4,32)$-regular $(4064, 3572)$ QC-LDPC code $C_{s,qc}(4,32)$ with rate 0.8789. The BER and BLER performances of the code $C_{s,qc}(4,32)$ over the AWGN channel decoded with 5, 10, and 50 iterations of the MSA scaled by 0.75 are shown in Fig. 14.7. ▲▲

## 14.3 Construction of Type-II QC-LDPC Codes Based on Two Subsets of a Finite Field

The construction of a $2 \times 2$ SM-constrained base matrix based on the two subsets $\mathbf{S}_1 = \{\alpha^{i_0}, \alpha^{i_1}, \dots, \alpha^{i_{m-1}}\}$ and $\mathbf{S}_2 = \{\alpha^{j_0}, \alpha^{j_1}, \dots, \alpha^{j_{n-1}}\}$, each with distinct elements of GF($q$), can be put in *product-form* as follows:

$$\mathbf{B}_p(m,n) = \begin{bmatrix} \alpha^{i_0}\alpha^{j_0} - \eta & \alpha^{i_0}\alpha^{j_1} - \eta & \cdots & \alpha^{i_0}\alpha^{j_{n-1}} - \eta \\ \alpha^{i_1}\alpha^{j_0} - \eta & \alpha^{i_1}\alpha^{j_1} - \eta & \cdots & \alpha^{i_1}\alpha^{j_{n-1}} - \eta \\ \vdots & \vdots & \ddots & \vdots \\ \alpha^{i_{m-1}}\alpha^{j_0} - \eta & \alpha^{i_{m-1}}\alpha^{j_1} - \eta & \cdots & \alpha^{i_{m-1}}\alpha^{j_{n-1}} - \eta \end{bmatrix}, \quad (14.3)$$

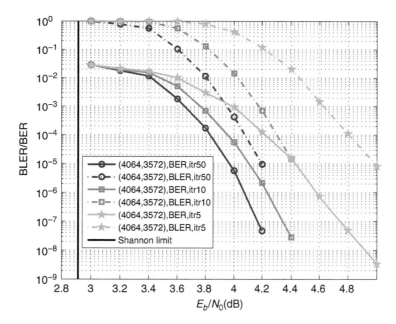

Figure 14.7 The BER and BLER performances of the $(4064, 3572)$ QC-LDPC code $C_{\mathrm{s,qc}}(4, 32)$ given in Example 14.7.

where the subtraction "$-$" can be replaced by addition "$+$." The matrix $\mathbf{B}_{\mathrm{p}}(m, n)$ has the same fundamental structural properties as the base matrix $\mathbf{B}_{\mathrm{s}}(m, n)$ in the sum-form given by (14.1). It also satisfies the $2 \times 2$ SM-constraint which can be proved in a similar way as the sum-form. The subscript "p" in $\mathbf{B}_{\mathrm{p}}(m, n)$ stands for "product-form." The $(q - 1)$-fold CPM-dispersion of $\mathbf{B}_{\mathrm{p}}(m, n)$ gives an $m \times n$ array $\mathbf{H}_{\mathrm{p,qc}}(m, n)$ of CPMs and/or ZMs of size $(q - 1) \times (q - 1)$. The null space over GF(2) of $\mathbf{H}_{\mathrm{p,qc}}(m, n)$ gives a QC-LDPC code $C_{\mathrm{p,qc}}(m, n)$ of length $n(q - 1)$ with rate at least $(n - m)/n$. The Tanner graph of $C_{\mathrm{p,qc}}(m, n)$ has girth at least 6. The above construction of QC-LDPC codes is referred to as *type-II construction*, which is also called *product-construction*.

**Example 14.8** Consider the code construction field GF(257) with a primitive element $\alpha = 254$ and choose the following two subsets:

$$\mathbf{S}_1 = \{\alpha^0, \alpha, \alpha^2, \alpha^3\}, \ \mathbf{S}_2 = \{\alpha^4, \alpha^5, \ldots, \alpha^{35}\},$$

from GF(257) with sizes $m = 4$ and $n = 32$, respectively. Let $\eta = 1$. Based on the two sets $\mathbf{S}_1$ and $\mathbf{S}_2$ and $\eta$, we can construct a $4 \times 32$ base matrix $\mathbf{B}_{\mathrm{p}}(4, 32)$ in the form of (14.3) over GF(257). $\mathbf{B}_{\mathrm{p}}(4, 32)$ has no zero entry. The 256-fold CPM-dispersion of $\mathbf{B}_{\mathrm{p}}(4, 32)$ is a $4 \times 32$ array $\mathbf{H}_{\mathrm{p,qc}}(4, 32)$ of CPMs of size $256 \times 256$, which is a $1024 \times 8192$ matrix over GF(2) with column and row weights 4 and 32, respectively. The null space over GF(2) of $\mathbf{H}_{\mathrm{p,qc}}(4, 32)$ gives a $(4, 32)$-regular $(8192, 7171)$ QC-LDPC code of rate 0.8754. The BER and BLER performances of the code over the AWGN channel decoded with 5, 10, and 50 iterations of the MSA scaled by 0.75 are shown in Fig. 14.8.                                ▲▲

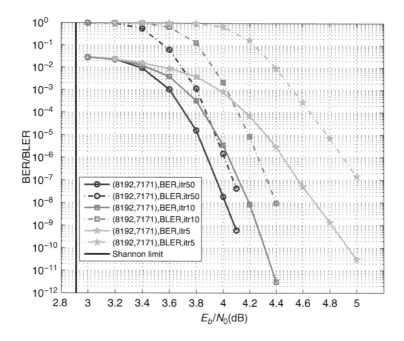

Figure 14.8 The BER and BLER performances of the $(8192, 7171)$ QC-LDPC code given in Example 14.8.

A special case of the base matrix in product-form given by (14.3) is that both $\mathbf{S}_1$ and $\mathbf{S}_2$ are chosen to be the multiplicative subgroup of $\mathrm{GF}(q)$, i.e.,

$$\mathbf{S}_1 = \mathbf{S}_2 = \mathrm{GF}(q) \setminus \{0\} = \{1, \alpha, \alpha^2, \dots, \alpha^{q-2}\}.$$

In this case, the base matrix in product-form defined by (14.3) is a $(q-1) \times (q-1)$ matrix over $\mathrm{GF}(q)$ of the following form:

$$\mathbf{B}_{\mathrm{p}}(q - 1, q - 1) = \begin{bmatrix} \alpha^0 - \eta & \alpha - \eta & \alpha^2 - \eta & \cdots & \alpha^{q-3} - \eta & \alpha^{q-2} - \eta \\ \alpha - \eta & \alpha^2 - \eta & \alpha^3 - \eta & \cdots & \alpha^{q-2} - \eta & \alpha^0 - \eta \\ \vdots & \vdots & & \ddots & & \vdots \\ \alpha^{q-2} - \eta & \alpha^0 - \eta & \alpha - \eta & \cdots & \alpha^{q-4} - \eta & \alpha^{q-3} - \eta \end{bmatrix}. \quad (14.4)$$

Note that each row of $\mathbf{B}_{\mathrm{p}}(q - 1, q - 1)$ is the cyclic-shift of the row above it one position to the left and the top row is the cyclic-shift of the last row one position to the left. Each column of $\mathbf{B}_{\mathrm{p}}(q - 1, q - 1)$ is the cyclic-shift of the column on its left one place upward and the first column is the cyclic-shift of the last column upward one place. Hence, the base matrix $\mathbf{B}_p(q-1, q-1)$ has a cyclic structure and is a $(q-1) \times (q-1)$ circulant over $\mathrm{GF}(q)$. Each row (column) of $\mathbf{B}_p(q-1, q-1)$ contains a single 0-entry and $q-2$ distinct nonzero elements in $\mathrm{GF}(q)$. Any two rows are different at every position. For $0 \leq i < q-1$, the $i$th column of $\mathbf{B}_p(q-1, q-1)$ is the transpose of its $i$th row and vice versa. Note that the matrix $\mathbf{B}_p(q-1, q-1)$ is a Latin square of order $q-1$. The construction of QC-LDPC codes by using the the base matrix (with $\eta = 1$) in the form of (14.4) was first proposed in [14]. Any submatrix of $\mathbf{B}_p(q-1, q-1)$ can be used as a base matrix for constructing a QC-LDPC code. The $4 \times 32$ base matrix $\mathbf{B}_p(4, 32)$ constructed in Example 14.8 is a submatrix of the base matrix $\mathbf{B}_p(256, 256)$ in the form of (14.4) constructed by choosing $\mathbf{S}_1 = \mathbf{S}_2 = \mathrm{GF}(257) \setminus \{0\}$ and $\eta = 1$.

The second special case of a base matrix in product-form given by (14.3) is that the two subsets $\mathbf{S}_1$ and $\mathbf{S}_2$ are cyclic subgroups of $\mathrm{GF}(q)$ with orders $m$ and $n$, respectively, such that $\mathbf{S}_1 \cap \mathbf{S}_2 = \{1\}$ and their orders $m$ and $n$ are two relatively prime factors of $q-1$. Let $\alpha^{i_0} = \alpha^{j_0} = 1$ and $\eta = 1$. In this case, all the entries in the matrix $\mathbf{B}_p(m, n)$ are distinct elements in $\mathrm{GF}(q)$. The construction of QC-LDPC codes based on two cyclic subgroups of $\mathrm{GF}(q)$ with orders $m$ and $n$, respectively, such that $mn = q-1$, was first presented in [15] and later extended in [18].

**Example 14.9** Consider the code construction field $\mathrm{GF}(173)$ with a primitive element $\alpha$. Let $m = 4$ and $n = 43$ such that $mn = q - 1 = 172$. Let $\mathbf{S}_1$ be a cyclic subgroup of $\mathrm{GF}(173)$ of order $m = 4$ and $\mathbf{S}_2$ be a cyclic subgroup of $\mathrm{GF}(173)$ of order $n = 43$. Let $\beta = \alpha^{43}$. Then, the order of $\beta$ is 4, i.e., $\beta^4 = 1$. Let $\delta = \alpha^4$. Then, the order of $\delta$ is 43, i.e., $\delta^{43} = 1$. The two cyclic subgroups $\mathbf{S}_1$ and $\mathbf{S}_2$ of $\mathrm{GF}(173)$ can be constructed as follows:

$$\mathbf{S}_1 = \{\beta^0, \beta, \beta^2, \beta^3\}, \quad \mathbf{S}_2 = \{\delta^0, \delta, \delta^2, \dots, \delta^{41}, \delta^{42}\}.$$

Based on these two sets $\mathbf{S}_1$ and $\mathbf{S}_2$ and $\eta = 1$, we construct a $4 \times 43$ matrix $\mathbf{B}_p(4, 43)$ over $\mathrm{GF}(173)$ in the form of (14.3). $\mathbf{B}_p(4, 43)$ has a zero entry at the zeroth row and zeroth column.

Take two submatrices $\mathbf{B}_p(4, 20)$ and $\mathbf{B}_p(4, 40)$ of sizes $4 \times 20$ and $4 \times 40$, respectively, from $\mathbf{B}_p(4, 43)$ (avoiding the column with 0-entry). The 172-fold CPM-dispersions of $\mathbf{B}_p(4, 20)$ and $\mathbf{B}_p(4, 40)$ give two arrays of which one is a $4 \times 20$ array $\mathbf{H}_{p,qc}(4, 20)$ of CPMs of size $172 \times 172$ and the other one is a $4 \times 40$ array $\mathbf{H}_{p,qc}(4, 40)$ of CPMs of size $172 \times 172$. $\mathbf{H}_{p,qc}(4, 20)$ is a $688 \times 3440$ matrix over $\mathrm{GF}(2)$ whose null space gives a $(4, 20)$-regular $(3440, 2755)$ QC-LDPC code $C_{p,qc}(4, 20)$ with rate 0.8009. $\mathbf{H}_{p,qc}(4, 40)$ is a $688 \times 6880$ matrix over $\mathrm{GF}(2)$ whose null space gives a $(4, 40)$-regular $(6880, 6195)$ QC-LDPC code $C_{p,qc}(4, 40)$ with rate 0.9004. The BER and BLER performances of the two codes $C_{p,qc}(4, 20)$ and $C_{p,qc}(4, 40)$ over the AWGN channel decoded with 50 iterations of the MSA scaled by 0.75 are shown in Fig. 14.9. ▲▲

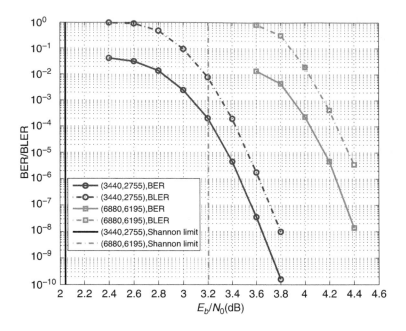

Figure 14.9 The BER and BLER performances of the $(3440, 2755)$ and $(6880, 6195)$ QC-LDPC codes given in Example 14.9.

## 14.4 Masking-Matrix Design

Masking introduced in Section 12.6 is a technique for improving the error performance of an LDPC code by removing short cycles from its Tanner graph. This was demonstrated in examples given in Chapters 12, 13, and Sections 14.2 and 14.3 of this chapter. In this section, we present 3 types of masking matrices which are quite effective in reducing short cycles from Tanner graphs of QC-LDPC codes.

### 14.4.1 Type-1 Design

In several examples given in Chapters 12, 13, and Sections 14.2 and 14.3 of this chapter, we showed that the $4 \times 8$ masking matrix defined by (12.13) is very effective for masking $4 \times 8$ base matrices to construct QC-LDPC codes of rate $1/2$ whose Tanner graphs have girths 8 or 10 and small numbers of short cycles. These codes perform very well in both waterfall and error-floor regions. We can use this masking matrix and/or its row-permuted versions as *building blocks* to construct masking matrices of larger sizes. For convenience, we reproduce the matrix as follows:

$$
\mathbf{Z}(4, 8) = \begin{bmatrix} 1 & 0 & 1 & 0 & 1 & 1 & 1 & 1 \\ 0 & 1 & 0 & 1 & 1 & 1 & 1 & 1 \\ 1 & 1 & 1 & 1 & 1 & 0 & 1 & 0 \\ 1 & 1 & 1 & 1 & 0 & 1 & 0 & 1 \end{bmatrix}. \tag{14.5}
$$

From (14.5), we see that the matrix $\mathbf{Z}(4,8)$ has a diagonal symmetrical structure. Suppose we partition this masking matrix into two $4 \times 4$ submatrices, denoted by $\mathbf{Z}_l(4,4)$ and $\mathbf{Z}_r(4,4)$. The submatrix $\mathbf{Z}_l(4,4)$ consists of the leftmost four columns of $\mathbf{Z}(4,8)$ and the submatrix $\mathbf{Z}_r(4,4)$ consists of the rightmost four columns of $\mathbf{Z}(4,8)$:

$$\mathbf{Z}_l(4,4) = \begin{bmatrix} 1 & 0 & 1 & 0 \\ 0 & 1 & 0 & 1 \\ 1 & 1 & 1 & 1 \\ 1 & 1 & 1 & 1 \end{bmatrix}, \quad \mathbf{Z}_r(4,4) = \begin{bmatrix} 1 & 1 & 1 & 1 \\ 1 & 1 & 1 & 1 \\ 1 & 0 & 1 & 0 \\ 0 & 1 & 0 & 1 \end{bmatrix}.$$

The submatrices $\mathbf{Z}_l(4,4)$ and $\mathbf{Z}_r(4,4)$ are called *left-half* and *right-half* submatrices of $\mathbf{Z}(4,8)$, respectively. The right-half submatrix $\mathbf{Z}_r(4,4)$ can be obtained by cyclically shifting the rows of the left-half submatrix $\mathbf{Z}_l(4,4)$ downward by two positions. Both $\mathbf{Z}_l(4,4)$ and $\mathbf{Z}_r(4,4)$ have constant column weight 3 and two row weights 2 and 4. Based on the pair, $\mathbf{Z}_l(4,4)$ and $\mathbf{Z}_r(4,4)$, we can construct masking matrices with alternating $\mathbf{Z}_l(4,4)$s and $\mathbf{Z}_r(4,4)$s in the form of:

$$[\mathbf{Z}_l(4,4) \; \mathbf{Z}_r(4,4) \; \mathbf{Z}_l(4,4) \; \mathbf{Z}_r(4,4) \; \cdots].$$

For an even integer $k$ with $k = 2\ell$, we form the following masking matrix:

$$\mathbf{Z}_1(4,4k) = [\mathbf{Z}_l(4,4) \; \mathbf{Z}_r(4,4) \; \mathbf{Z}_l(4,4) \; \mathbf{Z}_r(4,4) \; \cdots \; \mathbf{Z}_l(4,4) \; \mathbf{Z}_r(4,4)], \quad (14.6)$$

which consists of $\ell$ pairs of $\mathbf{Z}_l(4,4)$ and $\mathbf{Z}_r(4,4)$. In this case, $\mathbf{Z}_1(4,4k)$ has column and row weights 3 and $6\ell$, respectively. For an odd integer $k = 2\ell - 1$ with $\ell > 1$, we form the following masking matrix:

$$\mathbf{Z}_1(4,4k) = [\mathbf{Z}_l(4,4) \; \mathbf{Z}_r(4,4) \; \mathbf{Z}_l(4,4) \; \mathbf{Z}_r(4,4) \; \cdots \; \mathbf{Z}_l(4,4)], \quad (14.7)$$

which consists of $\ell - 1$ pairs of $\mathbf{Z}_l(4,4)$ and $\mathbf{Z}_r(4,4)$ and ends with the left-half submatrix $\mathbf{Z}_l(4,4)$. In this case, $\mathbf{Z}_1(4,4k)$ has column weight 3 and two different row weights $6(\ell - 1) + 2$ and $6(\ell - 1) + 4$. Note that, for $k \geq 3$ ($\ell \geq 2$), there are $3 \times 3$ submatrices in $\mathbf{Z}_1(4,4k)$ which contain all 1-entries.

We call $\mathbf{Z}_1(4,4k)$ the *type-1 masking matrix* constructed using the masking matrix $\mathbf{Z}(4,8)$ given by (14.5) as the building block. With $k = 2, 3, 4, \ldots$, we can form a sequence of masking matrices of the forms (14.6) or (14.7) for constructing QC-LDPC codes with rates $1/2, 2/3, 3/4, \ldots$.

**Example 14.10** Suppose we use the field $GF(2^7)$ to construct nine $4 \times 4k$ base matrices $\mathbf{B}_s(4,4k)$ in the form of (14.1) with $k = 2, 3, \ldots, 10$. The pairs of subsets of $GF(2^7)$ used for constructing the nine base matrices are given in Table 14.2. Next, we mask these nine base matrices $\mathbf{B}_s(4,4k)$ with nine masking matrices $\mathbf{Z}_1(4,4k)$ in the form of (14.6) for even $k$s and (14.7) for odd $k$s. Masking results in nine masked base matrices $\mathbf{B}_{s,\text{mask}}(4,4k)$.

For $2 \leq k \leq 10$, the 127-fold CPM-dispersion of the masked matrix $\mathbf{B}_{s,mask}(4, 4k)$ gives a $4 \times 4k$ masked array $\mathbf{H}_{s,qc,mask}(4, 4k)$ of CPMs and ZMs of size $127 \times 127$ which is a $508 \times 508k$ matrix with constant column weight 3. For even $k$, the row weight of $\mathbf{H}_{s,qc,mask}(4, 4k)$ is $3k$. For odd $k$, $\mathbf{H}_{s,qc,mask}(4, 4k)$ has two different row weights $3(k + 1) + 2$ and $3(k + 1) + 4$.

The null space over GF(2) of the masked array $\mathbf{H}_{s,qc,mask}(4, 4k)$ gives a QC-LDPC code $C_{s,qc,mask}(4, 4k)$ of length $508k$. For $2 \leq k \leq 10$, the nine QC-LDPC codes (listed in Table 14.2) have rates 1/2, 2/3, 3/4, 4/5, 5/6, 6/7, 7/8, 8/9, 9/10, respectively. The numbers of cycles in the Tanner graphs of these nine codes are given in Table 14.2. The BER performances of these nine masked QC-LDPC codes over the AWGN channel decoded with 50 iterations of the MSA are shown in Fig. 14.10. From the figure, we see that all the codes perform well. ▲▲

Table 14.2 The nine QC-LDPC codes constructed in Example 14.10.

| Codes | $k$ | $\mathbf{S}_1$ and $\mathbf{S}_2$ |
|---|---|---|
| $C_{s,qc,mask}(4, 8)$ | 2 | $\mathbf{S}_1 = \{\alpha^0, \alpha, \alpha^2, \alpha^3\}, \mathbf{S}_2 = \{\alpha^5, \alpha^6, \ldots, \alpha^{12}\}$ |
| $C_{s,qc,mask}(4, 12)$ | 3 | $\mathbf{S}_1 = \{\alpha^0, \alpha, \alpha^2, \alpha^3\}, \mathbf{S}_2 = \{\alpha^5, \ldots, \alpha^{12}, \alpha^{14}, \ldots, \alpha^{17}\}$ |
| $C_{s,qc,mask}(4, 16)$ | 4 | $\mathbf{S}_1 = \{\alpha^0, \alpha, \alpha^2, \alpha^3\}, \mathbf{S}_2 = \{\alpha^5, \ldots, \alpha^{12}, \alpha^{14}, \ldots, \alpha^{21}\}$ |
| $C_{s,qc,mask}(4, 20)$ | 5 | $\mathbf{S}_1 = \{\alpha^0, \alpha, \alpha^2, \alpha^3\}, \mathbf{S}_2 = \{\alpha^5, \ldots, \alpha^{12}, \alpha^{14}, \ldots, \alpha^{21}, \alpha^{23}, \ldots, \alpha^{26}\}$ |
| $C_{s,qc,mask}(4, 24)$ | 6 | $\mathbf{S}_1 = \{\alpha^0, \alpha, \alpha^2, \alpha^3\}, \mathbf{S}_2 = \{\alpha^5, \ldots, \alpha^{12}, \alpha^{14}, \ldots, \alpha^{21}, \alpha^{23}, \ldots, \alpha^{30}\}$ |
| $C_{s,qc,mask}(4, 28)$ | 7 | $\mathbf{S}_1 = \{\alpha^0, \alpha, \alpha^2, \alpha^3\}, \mathbf{S}_2 = \{\alpha^5, \ldots, \alpha^{12}, \alpha^{14}, \ldots, \alpha^{21}, \alpha^{23}, \ldots, \alpha^{30}, \alpha^{33}, \ldots, \alpha^{36}\}$ |
| $C_{s,qc,mask}(4, 32)$ | 8 | $\mathbf{S}_1 = \{\alpha^0, \alpha, \alpha^2, \alpha^3\}, \mathbf{S}_2 = \{\alpha^5, \ldots, \alpha^{12}, \alpha^{14}, \ldots, \alpha^{21}, \alpha^{23}, \ldots, \alpha^{30}, \alpha^{33}, \ldots, \alpha^{40}\}$ |
| $C_{s,qc,mask}(4, 36)$ | 9 | $\mathbf{S}_1 = \{\alpha^0, \alpha, \alpha^2, \alpha^3\}, \mathbf{S}_2 = \{\alpha^5, \ldots, \alpha^{12}, \alpha^{14}, \ldots, \alpha^{21}, \alpha^{23}, \ldots, \alpha^{30}, \alpha^{33}, \ldots, \alpha^{44}\}$ |
| $C_{s,qc,mask}(4, 40)$ | 10 | $\mathbf{S}_1 = \{\alpha^0, \alpha, \alpha^2, \alpha^3\}, \mathbf{S}_2 = \{\alpha^5, \ldots, \alpha^{12}, \alpha^{14}, \ldots, \alpha^{21}, \alpha^{23}, \ldots, \alpha^{30}, \alpha^{33}, \ldots, \alpha^{48}\}$ |

| Codes | $k$ | Length | Dimension | Rate | Cycle-6 | Cycle-8 |
|---|---|---|---|---|---|---|
| $C_{s,qc,mask}(4, 8)$ | 2 | 1016 | 508 | 1/2 | 0 | 1 143 |
| $C_{s,qc,mask}(4, 12)$ | 3 | 1524 | 1016 | 2/3 | 889 | 9 271 |
| $C_{s,qc,mask}(4, 16)$ | 4 | 2032 | 1524 | 3/4 | 1 524 | 31 115 |
| $C_{s,qc,mask}(4, 20)$ | 5 | 2540 | 2032 | 4/5 | 4 191 | 81 407 |
| $C_{s,qc,mask}(4, 24)$ | 6 | 3048 | 2540 | 5/6 | 6 604 | 178 689 |
| $C_{s,qc,mask}(4, 28)$ | 7 | 3556 | 3048 | 6/7 | 10 033 | 333 883 |
| $C_{s,qc,mask}(4, 32)$ | 8 | 4064 | 3556 | 7/8 | 15 748 | 586 105 |
| $C_{s,qc,mask}(4, 36)$ | 9 | 4572 | 4064 | 8/9 | 21 971 | 963 041 |
| $C_{s,qc,mask}(4, 40)$ | 10 | 5080 | 4572 | 9/10 | 30 480 | 1 485 773 |

## 14.4.2  Type-2 Design

Another way of constructing masking matrices using the masking matrix $\mathbf{Z}(4, 8)$ given by (14.5) as the building block is to repeat the first pair of columns and the third pair of columns in $\mathbf{Z}(4, 8)$ $k$ times to obtain a $4 \times 4k$ masking matrix, $\mathbf{Z}_2(4, 4k)$, with column weight 3 and row weight $3k$. The masking matrix $\mathbf{Z}_2(4, 4k)$ is called the *type-2 masking matrix*. For example, setting $k = 3$ gives the following masking matrix:

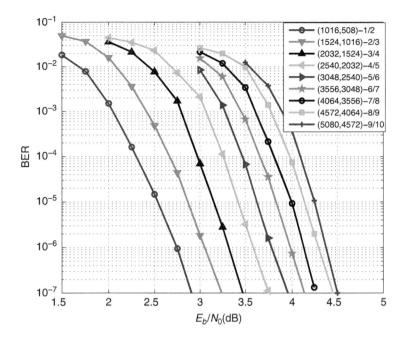

Figure 14.10 The BER performances of the nine QC-LDPC codes in Example 14.10.

$$\mathbf{Z}_2(4, 12) = \begin{bmatrix} 1\,0\,1\,0\,1\,0\,1\,1\,1\,1\,1\,1 \\ 0\,1\,0\,1\,0\,1\,1\,1\,1\,1\,1\,1 \\ 1\,1\,1\,1\,1\,1\,1\,0\,1\,0\,1\,0 \\ 1\,1\,1\,1\,1\,1\,0\,1\,0\,1\,0\,1 \end{bmatrix}. \tag{14.8}$$

**Example 14.11** Suppose we use the prime field GF(331) for code construction and let $\alpha = 326$ be a primitive element of GF(331). To construct the base matrix in sum-form given by (14.1), we use the following two subsets of elements in GF(331):

$$\mathbf{S}_1 = \{\alpha^0, \alpha^1, \alpha^2, \alpha^3\}, \ \mathbf{S}_2 = \{\alpha^{14}, \alpha^{15}, \alpha^{16}, \ldots, \alpha^{24}, \alpha^{25}\}.$$

Set the multiplier $\eta = -1$. Based on $\mathbf{S}_1$, $\mathbf{S}_2$, and $\eta$, we construct the following $4 \times 12$ base matrix $\mathbf{B}_s(4, 12)$ over GF(331) which has column and row weights 4 and 12, respectively:

$$\mathbf{B}_s(4, 12) = \begin{bmatrix} \alpha^{210} & \alpha^{88} & \alpha^{225} & \alpha^{259} & \alpha^{49} & \alpha^{310} & \alpha^{285} & \alpha^{171} & \alpha^{251} & \alpha^{287} & \alpha^{239} & \alpha^{326} \\ \alpha^{73} & \alpha^{211} & \alpha^{89} & \alpha^{226} & \alpha^{260} & \alpha^{50} & \alpha^{311} & \alpha^{286} & \alpha^{172} & \alpha^{252} & \alpha^{288} & \alpha^{240} \\ \alpha^{234} & \alpha^{74} & \alpha^{212} & \alpha^{90} & \alpha^{227} & \alpha^{261} & \alpha^{51} & \alpha^{312} & \alpha^{287} & \alpha^{173} & \alpha^{253} & \alpha^{289} \\ \alpha^{3} & \alpha^{235} & \alpha^{75} & \alpha^{213} & \alpha^{91} & \alpha^{228} & \alpha^{262} & \alpha^{52} & \alpha^{313} & \alpha^{288} & \alpha^{174} & \alpha^{254} \end{bmatrix}.$$

The code given by the null space of the 330-fold CPM-dispersion of $\mathbf{B}_s(4, 12)$ is a $(4, 12)$-regular $(3960, 2643)$ QC-LDPC code $C_{s,qc}(4, 12)$ with rate 0.6674 whose Tanner graph has girth 6 and contains 13 530 cycles of length 6, 165 660 cycles of length 8, and 3 835 590 cycles of length 10.

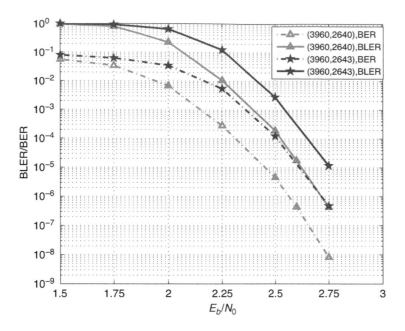

Figure 14.11 The BER and BLER performances of the unmasked $(3960, 2643)$ and the masked $(3960, 2640)$ QC-LDPC codes in Example 14.11.

Suppose we mask the base matrix $\mathbf{B}_s(4, 12)$ with $\mathbf{Z}_2(4, 12)$ given by (14.8). The masked base matrix $\mathbf{B}_{s,\mathrm{mask}}(4, 12)$ has column and row weights 3 and 9, respectively. The null space of the 330-fold CPM-dispersion of $\mathbf{B}_{s,\mathrm{mask}}(4, 12)$ is a $(3, 9)$-regular $(3960, 2640)$ QC-LDPC code $C_{s,\mathrm{qc},\mathrm{mask}}(4, 12)$ whose Tanner graph has girth 8 and contains 3960 cycles of length 8 and 117 480 cycles of length 10. We see that masking not only increases the girth from 6 to 8 but also significantly reduces the number of short cycles.

The BER and BLER performances of the unmasked $(3960, 2643)$ and the masked $(3960, 2640)$ QC-LDPC codes over the AWGN channel decoded with 50 iterations of the MSA are shown in Fig. 14.11. We see that the masked code $C_{s,\mathrm{qc},\mathrm{mask}}(4, 12)$ outperforms the unmasked code $C_{s,\mathrm{qc}}(4, 12)$. ▲▲

Note that in the construction of a type-2 masking matrix, we can repeat the first pair of columns of $\mathbf{Z}(4, 8)$ given by (14.5) $k_1$ times and the third pair of columns of $\mathbf{Z}(4, 8)$ $k_2$ times. This results in a nonuniform expansion of $\mathbf{Z}(4, 8)$. The masking matrix formed in this way is a $4 \times (2k_1 + 2k_2)$ matrix $\mathbf{Z}_2(4, 2k_1 + 2k_2)$ with column weight 3 and two row weights $k_1 + 2k_2$ and $2k_1 + k_2$. Nonuniform expansions of $\mathbf{Z}(4, 8)$ result in a larger class of masking matrices.

### 14.4.3 Type-3 Design

Let $\mathbf{A}(4, 2k)$ be a $4 \times 2k$ matrix obtained by repeating the first two columns of $\mathbf{Z}(4, 8)$ given by (14.5) $k$ times. For $\ell \geq 0$, we form a $(4 + \ell) \times 2k$ matrix

$\mathbf{A}(4 + \ell, 2k)$ by adding $\ell$ all-zero rows at the bottom of $\mathbf{A}(4, 2k)$. For $f \geq 0$, let $\mathbf{A}^{(f)}(4 + \ell, 2k)$ be a $(4 + \ell) \times 2k$ matrix obtained by *cyclically down-shifting* the rows of $\mathbf{A}(4+\ell, 2k)$ $f$ positions. For $t \geq 2$, we construct the following $(4+\ell) \times 2kt$ matrix:

$$\mathbf{Z}_3(4 + \ell, 2kt) = \left[ \mathbf{A}^{(0)}(4 + \ell, 2k) \; \mathbf{A}^{(f)}(4 + \ell, 2k) \; \cdots \; \mathbf{A}^{((t-1)f)}(4 + \ell, 2k) \right], \tag{14.9}$$

which has column weight 3 and multiple row weights. For various choices of $k$, $\ell$, $f$, and $t$, we obtain a sequence of matrices $\mathbf{Z}_3(4 + \ell, 2kt)$ of the form given by (14.9) that can be used as masking matrices. Note that if we set $k = 2$, $\ell = 0$, $f = 2$, and $t = 2$, we obtain the $4 \times 8$ masking matrix $\mathbf{Z}(4, 8)$ given by (14.5). The matrix $\mathbf{Z}_3(4 + \ell, 2kt)$ is called a *type-3 masking matrix*. Using performance simulations, we find that masking matrices of this type are also quite effective.

**Example 14.12** Again, we use the field $\mathrm{GF}(331)$ for the code construction. First, we construct a $6 \times 16$ base matrix $\mathbf{B}_s(6, 16)$ over $\mathrm{GF}(331)$ in the sum-form of (14.1) using the following two subsets of $\mathrm{GF}(331)$ and $\eta = 1$:

$$\mathbf{S}_1 = \{\alpha^0, \alpha^1, \alpha^2, \alpha^3, \alpha^4, \alpha^5\}, \; \mathbf{S}_2 = \{\alpha^{15}, \alpha^{16}, \alpha^{17}, \ldots, \alpha^{29}, \alpha^{30}\}.$$

The null space over $\mathrm{GF}(2)$ of the 330-fold CPM-dispersion of $\mathbf{B}_s(6, 16)$ gives a $(6, 16)$-regular $(5280, 3305)$ QC-LDPC code $C_{s,qc}(6, 16)$. The Tanner graph $\mathcal{G}_{s,qc}(6, 16)$ of $C_{s,qc}(6, 16)$ has girth 6 and contains $133\,980$ cycles of length 6, $4\,307\,325$ cycles of length 8, and $241\,141\,560$ cycles of length 10, a very large number of short cycles.

Setting $k = 4$, $\ell = 2$, $f = 4$, and $t = 2$, we construct the following $6 \times 16$ type-3 masking matrix in the form of (14.9):

$$\mathbf{Z}_3(6, 16) = \begin{bmatrix} 1\,0\,1\,0\,1\,0\,1\,0\,1\,1\,1\,1\,1\,1\,1\,1 \\ 0\,1\,0\,1\,0\,1\,0\,1\,1\,1\,1\,1\,1\,1\,1\,1 \\ 1\,1\,1\,1\,1\,1\,1\,1\,0\,0\,0\,0\,0\,0\,0\,0 \\ 1\,1\,1\,1\,1\,1\,1\,1\,0\,0\,0\,0\,0\,0\,0\,0 \\ 0\,0\,0\,0\,0\,0\,0\,0\,1\,0\,1\,0\,1\,0\,1\,0 \\ 0\,0\,0\,0\,0\,0\,0\,0\,1\,0\,1\,0\,1\,0\,1 \end{bmatrix}. \tag{14.10}$$

Masking $\mathbf{B}_s(6, 16)$ with $\mathbf{Z}_3(6, 16)$, we obtain a $6 \times 16$ masked base matrix $\mathbf{B}_{s,mask}(6, 16)$ with column weight 3 and three row weights 4, 8, and 12. The masked matrix $\mathbf{B}_{s,mask}(6, 16)$ satisfies the $2 \times 2/3 \times 3$ SM-constraint. The 330-fold CPM-dispersion of $\mathbf{B}_{s,mask}(6, 16)$ is a $6 \times 16$ array $\mathbf{H}_{s,qc,mask}(6, 16)$ of CPMs and ZMs of size $330 \times 330$, which is a $1980 \times 5280$ matrix over $\mathrm{GF}(2)$ with column weight 3 and three row weights 4, 8, and 12.

The null space over $\mathrm{GF}(2)$ of $\mathbf{H}_{s,qc,mask}(6, 16)$ gives an irregular $(5280, 3302)$ QC-LDPC code $C_{s,qc,mask}(6, 16)$ with rate slightly higher than $5/8$. The Tanner graph of $C_{s,qc,mask}(6, 16)$ has girth 8 and contains 4290 cycles of length 8 and $114\,840$ cycles of length 10. We see that masking not only increases the girth of

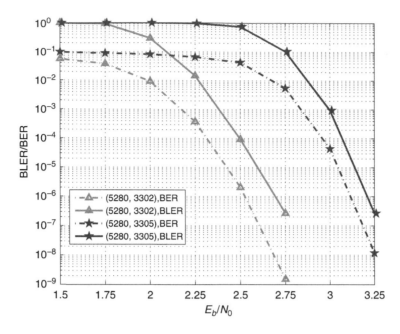

Figure 14.12 The BER and BLER performances of the unmasked $(5280, 3305)$ and the masked $(5280, 3302)$ QC-LDPC codes in Example 14.12.

the Tanner graph of the unmasked code $C_{s,qc}(6, 16)$ but also reduces the number of short cycles drastically.

The BER and BLER performances of the masked and the unmasked codes, $C_{s,qc,mask}(6, 16)$ and $C_{s,qc}(6, 16)$, over the AWGN channel decoded with 50 iterations of the MSA are shown in Fig. 14.12. We see that the masked code $C_{s,qc,mask}(6, 16)$ outperforms the unmasked code $C_{s,qc}(6, 16)$ and it reaches down to a BER of $10^{-9}$ without a visible error-floor.                    ▲▲

## 14.5   A Search Algorithm for $2 \times 2/3 \times 3$ SM-Constrained Base Matrices for Constructing Rate-$1/2$ QC-LDPC Codes

As stated in Theorem 14.1, for the Tanner graph of a QC-LDPC code constructed by CPM-dispersion of a base matrix $\mathbf{B}$ to have girth 8 or larger, the necessary and sufficient condition is that the base matrix $\mathbf{B}$ satisfies the $2 \times 2/3 \times 3$ SM-constraint. In this section, we present a search method to find $4 \times 8$ base matrices that satisfy the $2 \times 2/3 \times 3$ SM-constraint from a large $2 \times 2$ SM-constrained matrix, called *mother base matrix*.

To explain the search method, we use a $2 \times 2$ SM-constrained base matrix constructed based on two subsets of a finite field in the sum-form given by (14.1) for demonstration. First, we construct a $2 \times 2$ SM-constrained $m \times n$ matrix

$\mathbf{B}_s(m, n) = [\eta\alpha^{i_k} + \alpha^{j_l}]_{0 \le k < m, 0 \le l < n}$ over $GF(q)$. Next, we search $\mathbf{B}_s(m, n)$ to find all possible $4 \times 8$ submatrices $\mathbf{B}_s(4, 8)$ in $\mathbf{B}_s(m, n)$ from which, with proper masking, masked matrices $\mathbf{B}_{s,mask}(4, 8)$ with column weight 3 and row weight 6 can be obtained and they satisfy the $2 \times 2/3 \times 3$ SM-constraint. Then, the null space over $GF(2)$ of the $(q - 1)$-fold CPM-dispersion $\mathbf{H}_{s,qc,mask}(4, 8)$ of each of these $4 \times 8$ masked base matrices gives a $(3, 6)$-regular QC-LDPC code of length $8(q - 1)$ with girth at least 8. Using the cycle enumeration algorithm given in [33, 34], we find the number of the shortest cycles of the Tanner graph of each of these codes. Then, choose a code with the largest girth and the smallest number of shortest cycles as the code for use.

To mask a $4 \times 8$ submatrix $\mathbf{B}_s(4, 8)$ of $\mathbf{B}_s(m, n)$, we need a $4 \times 8$ masking matrix $\mathbf{Z}(4, 8)$ such that the masked $4 \times 8$ matrix $\mathbf{B}_{s,mask}(4, 8) = \mathbf{Z}(4, 8) \otimes \mathbf{B}_s(4, 8)$ has column and row weights 3 and 6, respectively. There are many such $4 \times 8$ possible masking matrices. The matrix given by (14.5) is one of them. In Example 14.6, we showed that masking a $2 \times 2$ SM-constrained $4 \times 8$ base matrix $\mathbf{B}_s(4, 8)$ with the masking matrix $\mathbf{Z}(4, 8)$ given by (14.5) not only enlarges the girth of the Tanner graph of the code constructed based on the unmasked matrix $\mathbf{B}_s(4, 8)$ but also reduces the number of short cycles drastically. In the following, we use this masking matrix to construct $4 \times 8$ masked base matrices that satisfy the $2 \times 2/3 \times 3$ SM-constraint.

Let $\mathbf{B}_{s,mask}(4, 8)$ be a $4 \times 8$ masked matrix obtained by masking a $4 \times 8$ submatrix $\mathbf{B}_s(4, 8)$ of $\mathbf{B}_s(m, n)$ with the $4 \times 8$ masking matrix $\mathbf{Z}(4, 8)$ given by (14.5). From the structure of $\mathbf{Z}(4, 8)$, we find that every $3 \times 3$ submatrix of the masked matrix $\mathbf{B}_{s,mask}(4, 8)$ contains at least one 0-entry. In fact, there are 24 $3 \times 3$ submatrices of $\mathbf{B}_{s,mask}(4, 8)$, each containing a single 0-entry, 120 $3 \times 3$ submatrices in $\mathbf{B}_{s,mask}(4, 8)$, each containing two 0-entries, and 80 $3 \times 3$ submatrices of $\mathbf{B}_{s,mask}(4, 8)$, each containing three 0-entries. Therefore, in the determinant expansion of a $3 \times 3$ submatrix of $\mathbf{B}_s(4, 8)$, there are at most four nonzero terms. Hence, to check whether $\mathbf{B}_{s,mask}(4, 8)$ satisfies the $2 \times 2/3 \times 3$ SM-constraint, the number of computations required is relatively small.

To avoid masking a 0-entry in the base matrix $\mathbf{B}_s(4, 8)$ (or masking a nonzero entry in a column of $\mathbf{B}_s(4, 8)$ which already contains a 0-entry), we restrict ourselves only to finding those $4 \times 8$ submatrices in $\mathbf{B}_s(4, 8)$ which contain no 0-entry and mask them with the masking matrix $\mathbf{Z}(4, 8)$ given by (14.5). This ensures that the masked base matrix $\mathbf{B}_{s,mask}(4, 8)$ has column and row weights 3 and 6, respectively. Any masked base matrix $\mathbf{B}_{s,mask}(4, 8)$ found that satisfies the $2 \times 2/3 \times 3$ SM-constraint gives a QC-LDPC code whose Tanner graph has girth at least 8.

The search for $4 \times 8$ masked matrices, $\mathbf{B}_{s,mask}(4, 8)$, in $\mathbf{B}_s(m, n)$ that satisfy the $2 \times 2/3 \times 3$ SM-constraint can be carried out systematically. Label the rows and columns of $\mathbf{B}_s(m, n)$ from 0 to $m - 1$ and 0 to $n - 1$, respectively. To simplify the search process, we only consider those $4 \times 8$ submatrices which are taken from four consecutive rows and eight consecutive columns in $\mathbf{B}_s(m, n)$ (no end-around rows or columns). The search begins from the first four rows of $\mathbf{B}_s(m, n)$ (the zeroth to the third rows) and moves from the leftmost edge of $\mathbf{B}_s(m, n)$ to the right, one position (or one column) at a time. Each time, we mask a

$4 \times 8$ submatrix $\mathbf{B}_s(4,8)$ with the masking matrix $\mathbf{Z}(4,8)$ given by (14.5) and check whether the masked $4 \times 8$ submatrix $\mathbf{B}_{s,\text{mask}}(4,8)$ satisfies the $2 \times 2/3 \times 3$ SM-constraint. After checking the last $4 \times 8$ submatrix of the first four rows of $\mathbf{B}_s(m,n)$, we start to search across the next four consecutive rows (the first to the fourth rows) of $\mathbf{B}_s(m,n)$ from left to right. We move one position at a time until we reach and test the last $4 \times 8$ submatrix. Then, we move back to the leftmost edge of $\mathbf{B}_s(m,n)$ and start to search across the next four consecutive rows (the second to the fifth rows) of $\mathbf{B}_s(m,n)$ from left to right. This searching process continues until we finish searching the last four consecutive rows of $\mathbf{B}_s(m,n)$.

An algorithm to perform the above search for $4 \times 8$ masked matrices in $\mathbf{B}_s(m,n)$ that satisfy the $2 \times 2/3 \times 3$ SM-constraint is given in Appendix B. We call the algorithm the $2 \times 2/3 \times 3$ *SM-constrained* $4 \times 8$ *masked matrix search algorithm* (MMSA), in short $2 \times 2/3 \times 3$ *SMC-MMSA*. Table 14.3 gives a list of numbers of $4 \times 8$ masked base matrices $\mathbf{B}_{s,\text{mask}}(4,8)$ constructed from various fields with multiplier $\eta = 1$ that produce $(3,6)$-regular QC-LDPC codes with girths 8 or 10. Also included in the table are the smallest numbers of cycles of lengths 8 and 10 for each category of $4 \times 8$ masked based matrices. The table also gives the number of cycles of length 6 for each category of unmasked $4 \times 8$ base matrices, $\mathbf{B}_s(4,8)$, to show the reduction in the numbers of short cycles.

Table 14.3 Numbers of $4 \times 8$ masked base matrices over various fields $\mathrm{GF}(q)$ that give rate-$1/2$ QC-LDPC codes with girths 8 or 10 for $\eta = 1$.

| $q$ | No. of $\mathbf{B}_{s,\text{mask}}(4,8)$ $g = 8$ | No. of *cycle*8 (smallest) | No. of *cycle*6 (unmasked) | No. of *cycle*8 (unmasked) | No. of $\mathbf{B}_{s,\text{mask}}(4,8)$ $g = 10$ | No. of *cycle*10 (smallest) |
|---|---|---|---|---|---|---|
| 53 | 220 | 806 | 1 612 | 26 286 | 0 | 0 |
| 64 | 542 | 819 | 1 890 | 25 965 | 0 | 0 |
| 89 | 1 433 | 616 | 2 640 | 26 576 | 0 | 0 |
| 128 | 6 872 | 635 | 3 810 | 28 575 | 0 | 0 |
| 181 | 15 799 | 360 | 4 140 | 27 990 | 680 | 11 880 |
| 256 | 45 977 | 255 | 6 375 | 41 820 | 5 | 9 435 |
| 257 | 36 927 | 256 | 6 400 | 46 976 | 1 479 | 11 776 |
| 331 | 65 373 | 330 | 7 920 | 34 320 | 8 970 | 11 220 |

Consider the $4 \times 8$ masked base matrices constructed from the field $\mathrm{GF}(331)$. The mother matrix used for the search is a $331 \times 331$ matrix $\mathbf{B}_s(331, 331)$ over $\mathrm{GF}(331)$ with $\eta = 1$ in the sum-form of (14.1). Using the $2 \times 2/3 \times 3$ SMC-MMSA, we find 65 373 $4 \times 8$ masked base matrices which give codes with girth 8 and 8970 masked base matrices which give codes with girth 10. Among these masked base matrices, the smallest numbers of cycles of lengths 8 and 10 in the Tanner graphs of the generated codes are 330 and 11 220, respectively. Consider the $(3,6)$-regular $(2640, 1320)$ QC-LDPC code $C_{s,qc,\text{mask}}(4,8)$ constructed in Example 14.6. The Tanner graph of this code has girth 8 and contains 330 cycles of length 8 and 8970 cycles of length 10. However, the Tanner graph of its corresponding unmasked $(4,8)$-regular $(2640, 1323)$ QC-LDPC code $C_{s,qc}(4,8)$

has girth 6 and contains 7920 cycles of length 6 and 34 320 cycles of length 8. We see that masking not only enlarges the girth of the Tanner graph of the unmasked code from 6 to 8 but also reduces the number of short cycles of lengths 6 to 10.

Table 14.4 gives a list of the numbers of $4 \times 8$ masked base matrices constructed from the field GF(331) with various choices of multipliers, $\eta$, which result in (3, 6)-regular QC-LDPC codes with girths 8 and 10. From the table, we see that, for different choices of multipliers, the numbers of cycles are almost the same for both girths 8 and 10. Furthermore, the smallest number of cycles of length 8 for the codes with girth 8 is the same, 330, and the smallest number of cycles of length 10 for codes with girth 10 is also the same, 11 220.

Table 14.4 Numbers of $4 \times 8$ masked matrices over GF(331) that give rate-1/2 QC-LDPC codes with girths 8 and 10 for different choices of $\eta$s.

| $\eta$ | No. of $\mathbf{B}_{s,mask}(4,8)$, $g = 8$ | No. of $cycle8_{smallest}$ | No. of $\mathbf{B}_{s,mask}(4,8)$, $g = 10$ | No. of $cycle10_{smallest}$ |
|---|---|---|---|---|
| $\alpha^0$ | 65 373 | 330 | 8 970 | 11 220 |
| $\alpha^7$ | 65 375 | 330 | 8 970 | 11 220 |
| $\alpha^{31}$ | 65 377 | 330 | 8 973 | 11 220 |
| $\alpha^{181}$ | 65 376 | 330 | 8 973 | 11 220 |
| $\alpha^{256}$ | 65 383 | 330 | 8 970 | 11 220 |

**Example 14.13** Table 14.3 shows that, using the masking matrix $\mathbf{Z}(4, 8)$ given by (14.5) and the $2 \times 2/3 \times 3$ SMC-MMSA, we find 542 masked base matrices of size $4 \times 8$ over GF($2^6$) with column and row weights 3 and 6, respectively, which satisfy the $2 \times 2/3 \times 3$ SM-constraint. The null space over GF(2) of the 63-fold CPM-dispersion of each of these masked base matrices gives a (3, 6)-regular (504, 252) QC-LDPC code $C_{s,qc,mask}(4, 8)$ whose Tanner graph has girth 8. One such masked base matrix $\mathbf{B}_{s,mask}(4, 8)$ is constructed based on the following two subsets of elements in GF($2^6$): $\mathbf{S}_1 = \{\alpha^7, \alpha^8, \alpha^9, \alpha^{10}\}$ and $\mathbf{S}_2 = \{\alpha^{53}, \alpha^{54}, \alpha^{55}, \alpha^{56}, \alpha^{57}, \alpha^{58}, \alpha^{59}, \alpha^{60}\}$, where $\alpha$ is a primitive element of GF($2^6$). The multiplier for the base matrix construction is $\eta = 1$. The Tanner graph of the code has girth 8 and contains 819 cycles of length 8.

The BER and BLER performances of the constructed (504, 252) QC-LDPC code $C_{s,qc,mask}(4, 8)$ over the AWGN channel decoded with 50 iterations of the MSA scaled by a factor of 0.75 are shown in Fig. 14.13. The code achieves a BER of $10^{-10}$ and a BLER of $10^{-8}$ without visible error-floors. Also included in the figure are the performances of two other (504, 252) LDPC codes, one constructed based on the PEG algorithm [36] (to be presented in Section 15.4) and the other constructed by MacKay [37]. From the figure, we see that the constructed QC-LDPC code $C_{s,qc,mask}(4, 8)$ has a lower error-floor than both the PEG and MacKay codes.                                   ▲▲

The $2 \times 2/3 \times 3$ SMC-MMSA can be used in conjunction with any of the three types of masking matrices presented in Section 14.4 to construct QC-LDPC codes with rates higher than 1/2 and girth 8 or larger using mother base matrices

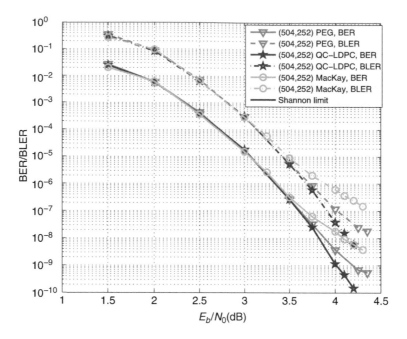

Figure 14.13 The BER and BLER performances of the $(504, 252)$ QC-LDPC code $C_{\mathrm{s,qc,mask}}(4, 8)$ and two other $(504, 252)$ LDPC codes given in Example 14.13.

in the sum-form given by (14.1) or product-form given by (14.3). The masked-based matrix over GF(331) for constructing the $(4, 12)$-regular $(3960, 2640)$ QC-LDPC code $C_{\mathrm{s,qc,mask}}(4, 12)$ given in Example 14.11 is constructed by using the $2 \times 2/3 \times 3$ SMC-MMSA in conjunction with the type-2 masking matrix given by (14.8).

It is clear that the $2 \times 2/3 \times 3$ SMC-MMSA in conjunction with a chosen masking matrix can be applied to base matrices constructed from EGs and PaGs presented in Chapters 12 and 13, respectively.

## 14.6 Designs of $2 \times 2$ SM-Constrained RS Base Matrices

Under certain constraints, the *conventional parity-check matrix* of an RS (Reed–Solomon) code can be used as a base matrix to construct a QC-LDPC code [7, 19, 23, 38–45]. Such a QC-LDPC code is called an *RS-based QC-LDPC code* (QC-RS-LDPC code) and its base matrix is called an *RS base matrix*. In this section and Sections 14.7–14.9, we first present three basic types of RS base matrices that satisfy the $2 \times 2$ SM-constraint. Next, we present a necessary and sufficient condition for a $2 \times 2$ SM-constrained RS base matrix to satisfy the $2 \times 2/3 \times 3$ SM-constraint. Based on this necessary and sufficient condition, systematic methods are developed to design $2 \times 2/3 \times 3$ SM-constrained RS matrices. Using these $2 \times 2$ SM-constrained and $2 \times 2/3 \times 3$ SM-constrained RS matrices as base matrices in conjunction with masking, QC-RS-LDPC codes with good error performance can be constructed.

Let $\mathrm{GF}(q)$ be a finite field with $q$ elements and $\beta$ be an element of order $n$ in $\mathrm{GF}(q)$ where $n$ is a factor of $q - 1$. The set $\mathbf{S}_n = \{\beta^0 = 1, \beta, \dots, \beta^{n-1}\}$ forms a cyclic subgroup of $\mathrm{GF}(q)$ with $\beta \in \mathrm{GF}(q)$ as the generator. Let $d$ be a positive integer such that $1 \le d \le n$. Form the following $d \times n$ matrix over $\mathrm{GF}(q)$:

$$
\mathbf{B}_{\mathrm{RS},n}(d,n) = [(\beta^i)^j]_{1 \le i \le d, 0 \le j < n} = \begin{bmatrix} 1 & \beta & \beta^2 & \cdots & \beta^{(n-1)} \\ 1 & \beta^2 & (\beta^2)^2 & \cdots & (\beta^2)^{(n-1)} \\ \vdots & \vdots & \vdots & \ddots & \vdots \\ 1 & \beta^d & (\beta^d)^2 & \cdots & (\beta^d)^{(n-1)} \end{bmatrix}. \quad (14.11)
$$

The null space over $\mathrm{GF}(q)$ of $\mathbf{B}_{\mathrm{RS},n}(d,n)$ gives a $q$-ary $(n, n - d)$ RS code of length $n$, dimension $n - d$, rate $(n - d)/n$, and minimum distance $d + 1$. The generator polynomial of this RS code has $\beta$, $\beta^2, \dots, \beta^d$ as roots. The matrix $\mathbf{B}_{\mathrm{RS},n}(d,n)$ is the parity-check matrix of the $(n, n - d)$ RS code in the *conventional form* as presented in Chapter 6. Let $b$ be a positive integer such that $d < b < n$. Suppose we delete $n - b$ columns from $\mathbf{B}_{\mathrm{RS},n}(d,n)$ (usually the last $n - b$ columns of $\mathbf{B}_{\mathrm{RS},n}(d,n)$). We obtain a $d \times b$ matrix $\mathbf{B}_{\mathrm{RS},n}(d,b)$ over $\mathrm{GF}(q)$ whose null space over $\mathrm{GF}(q)$ gives a $(b, b - d)$ shortened RS code with minimum distance $d + 1$. We call both $\mathbf{B}_{\mathrm{RS},n}(d,n)$ and $\mathbf{B}_{\mathrm{RS},n}(d,b)$ *RS matrices*.

In general, the RS matrix $\mathbf{B}_{\mathrm{RS},n}(d,n)$ does not satisfy the $2 \times 2$ SM-constraint. In the following, we present three cases in which the RS matrices satisfy the $2 \times 2$ SM-constraint.

In the first case, we consider a submatrix of $\mathbf{B}_{\mathrm{RS},n}(d,n)$ and the order $n$ of the cyclic subgroup $\mathbf{S}_n$ is not a prime. Suppose $n$ can be factored as a *product of two proper factors* $k$ and $m$ with $n = km$ and $k \le m$. Let $d$ be a positive integer such that $1 \le d \le k + 1$. Form the following $d \times m$ matrix over $\mathrm{GF}(q)$:

$$
\mathbf{B}_{\mathrm{RS},n}(d,m) = [(\beta^i)^j]_{1 \le i \le d, 0 \le j < m} = \begin{bmatrix} 1 & \beta & \beta^2 & \cdots & \beta^{(m-1)} \\ 1 & \beta^2 & (\beta^2)^2 & \cdots & (\beta^2)^{(m-1)} \\ \vdots & \vdots & \vdots & \ddots & \vdots \\ 1 & \beta^d & (\beta^d)^2 & \cdots & (\beta^d)^{(m-1)} \end{bmatrix}. \quad (14.12)
$$

The null space over $\mathrm{GF}(q)$ of $\mathbf{B}_{\mathrm{RS},n}(d,m)$ gives an $(m, m - d)$ shortened RS code of the $(n, n - d)$ RS code. Label the rows and columns of $\mathbf{B}_{\mathrm{RS},n}(d,m)$ from 1 to $d$ and 0 to $m - 1$, respectively. The following theorem proves that the RS matrix $\mathbf{B}_{\mathrm{RS},n}(d,m)$ given by (14.12) satisfies the $2 \times 2$ SM-constraint.

**Theorem 14.3** *The RS matrix $\mathbf{B}_{\mathrm{RS},n}(d,m)$ over $\mathrm{GF}(q)$ given by (14.12) satisfies the $2 \times 2$ SM-constraint.*

*Proof* Let $i$, $j$, $s$, and $l$ be any four nonnegative integers such that $1 \le i < j \le d$ and $0 \le s < l < m$. Consider the following $2 \times 2$ submatrix $\mathbf{B}(2,2)$ of $\mathbf{B}_{\mathrm{RS},n}(d,m)$:

$$\mathbf{B}(2,2) = \begin{bmatrix} (\beta^i)^s & (\beta^i)^l \\ (\beta^j)^s & (\beta^j)^l \end{bmatrix}. \tag{14.13}$$

Because all the entries of $\mathbf{B}_{\mathrm{RS},n}(d,m)$ are nonzero, the four entries of $\mathbf{B}(2,2)$ are nonzero. The determinant of $\mathbf{B}(2,2)$ is $\beta^{jl+is} - \beta^{js+il}$. Next, we show $\beta^{jl+is} - \beta^{js+il} \neq 0$. Suppose $\beta^{jl+is} - \beta^{js+il} = 0$. Then, we must have $\beta^{(j-i)(l-s)} = 1$. Because $0 < j - i \leq d - 1 \leq k$ and $0 < l - s < m$, the product $(j-i)(l-s)$ is less than $n = km$ and nonzero. Because the order of $\beta$ is $n$ and $(j-i)(l-s)$ is not divisible by $n$, $\beta^{(j-i)(l-s)}$ cannot be equal to 1. Hence, $\beta^{jl+is} - \beta^{js+il} \neq 0$ and $\mathbf{B}(2,2)$ must be a nonsingular matrix. Thus, the RS matrix $\mathbf{B}_{\mathrm{RS},n}(d,m)$ satisfies the $2 \times 2$ SM-constraint. ▲▲

For $1 \leq d \leq k + 1$, the RS matrix $\mathbf{B}_{\mathrm{RS},n}(d,m)$ in (14.12) is a submatrix of $\mathbf{B}_{\mathrm{RS},n}(d,n)$ in (14.11), which consists of the first $m$ columns of $\mathbf{B}_{\mathrm{RS},n}(d,n)$. In fact, any $m$ consecutive columns of $\mathbf{B}_{\mathrm{RS},n}(d,n)$ form a $2 \times 2$ SM-constrained $d \times m$ RS matrix. This can be proved in a similar manner as Theorem 14.3. We call this class of RS matrices *type-1 RS matrices*.

In the second case, we consider the entire matrix $\mathbf{B}_{\mathrm{RS},n}(d,n)$ and the order $n$ of the cyclic subgroup $\mathbf{S}_n$ is not a prime. Let $p_s$ be the smallest prime factor of $n$ and let $d$ be a positive integer such that $1 \leq d \leq p_s$. Then, the RS matrix $\mathbf{B}_{\mathrm{RS},n}(d,n)$ in the form of (14.11) satisfies the $2 \times 2$ SM-constraint. The proof of this case is similar to the proof given in Theorem 14.3. We call this class of RS matrices *type-2 RS matrices*.

The third case is that $n$ is a prime factor of $q - 1$. Let $d$ be a positive integer such that $1 \leq d < n$. Then, the RS matrix $\mathbf{B}_{\mathrm{RS},n}(d,n)$ given by (14.11) satisfies the $2 \times 2$ SM-constraint. In this case, if we set $d = n$, then the $n \times n$ matrix

$$\mathbf{B}_{\mathrm{RS},n}(n,n) = \begin{bmatrix} 1 & \beta & \cdots & \beta^{(n-1)} \\ 1 & \beta^2 & \cdots & (\beta^2)^{(n-1)} \\ \vdots & \vdots & \ddots & \vdots \\ 1 & \beta^{n-1} & \cdots & (\beta^{n-1})^{(n-1)} \\ 1 & \beta^n & \cdots & (\beta^n)^{(n-1)} \end{bmatrix} = \begin{bmatrix} 1 & \beta & \cdots & \beta^{(n-1)} \\ 1 & \beta^2 & \cdots & (\beta^2)^{(n-1)} \\ \vdots & \vdots & \ddots & \vdots \\ 1 & \beta^{n-1} & \cdots & (\beta^{n-1})^{(n-1)} \\ 1 & 1 & \cdots & 1 \end{bmatrix}$$

is the base matrix for constructing the partial geometry $\mathrm{PaG}(n,n,n-1)$ presented in Section 13.4 (by moving the last row of $\mathbf{B}_{\mathrm{RS},n}(n,n)$ to the top). We call this class of RS matrices *type-3 RS matrices*.

## 14.7 Construction of Type-III QC-LDPC Codes Based on RS Codes

The three types of $2 \times 2$ SM-constrained RS matrices presented in the last section can be used as base matrices in conjunction with CPM-dispersion to construct three types of QC-LDPC codes, named *QC-RS-LDPC codes*. We refer to this class of QC-RS-LDPC codes as the *type-III QC-LDPC codes* and call the RS matrices the *RS base matrices*.

As shown in Section 13.4, each element in a cyclic subgroup $\mathbf{S}_n = \{\beta^0 = 1, \beta, \ldots, \beta^{n-1}\}$ of GF($q$) can be uniquely dispersed into a CPM of size $n \times n$ with respect to $\mathbf{S}_n$ (i.e., $n$-fold CPM-dispersion) or into a CPM of size $ln \times ln$ with respect to a super group $\mathbf{S}_{ln}$ of $\mathbf{S}_n$ with order $ln$ (i.e., $ln$-fold CPM-dispersion).

Consider the construction of a QC-RS-LDPC code using the first type of $2 \times 2$ SM-constrained RS matrix $\mathbf{B}_{\mathrm{RS},n}(d,m)$ in the form of (14.12). All the entries of $\mathbf{B}_{\mathrm{RS},n}(d,m)$ are elements in the cyclic subgroup $\mathbf{S}_n$ of GF($q$) with order $n$. Dispersing each entry in $\mathbf{B}_{\mathrm{RS},n}(d,m)$ into a CPM of size $n \times n$ with respect to $\mathbf{S}_n$, we obtain a $d \times m$ array $\mathbf{H}_{\mathrm{RS},n}(d,m)$ of CPMs of size $n \times n$. The array $\mathbf{H}_{\mathrm{RS},n}(d,m)$ is a $dn \times mn$ matrix with column and row weights $d$ and $m$, respectively, which satisfies the RC-constraint because $\mathbf{B}_{\mathrm{RS},n}(d,m)$ satisfies the $2 \times 2$ SM-constraint. The null space over GF(2) of $\mathbf{H}_{\mathrm{RS},n}(d,m)$ gives a $(d,m)$-regular QC-RS-LDPC code, denoted by $C_{\mathrm{RS},n}(d,m)$, of length $mn$ with rate at least $(m-d)/m$, whose Tanner graph, denoted by $\mathcal{G}_{\mathrm{RS},n}(d,m)$, has girth at least 6.

If we disperse each entry in $\mathbf{B}_{\mathrm{RS},n}(d,m)$ into a CPM of size $ln \times ln$ with respect to a super group $\mathbf{S}_{ln}$ of $\mathbf{S}_n$, we obtain a $d \times m$ array $\mathbf{H}_{\mathrm{RS},ln}(d,m)$ of CPMs of size $ln \times ln$. The array $\mathbf{H}_{\mathrm{RS},ln}(d,m)$ is a $dln \times mln$ matrix with column and row weights $d$ and $m$, respectively, which also satisfies the RC-constraint. The null space over GF(2) of $\mathbf{H}_{\mathrm{RS},ln}(d,m)$ gives a $(d,m)$-regular QC-RS-LDPC code, denoted by $C_{\mathrm{RS},ln}(d,m)$, of length $mln$ with rate at least $(m-d)/m$ whose Tanner graph, denoted by $\mathcal{G}_{\mathrm{RS},ln}(d,m)$, has girth at least 6.

CPM-dispersions of a $2 \times 2$ SM-constrained RS base matrix $\mathbf{B}_{\mathrm{RS},n}(d,m)$ of type-1 with respect to super groups of $\mathbf{S}_n$ of various orders result in arrays of CPMs with different sizes. The null spaces over GF(2) of these arrays give QC-RS-LDPC codes of different lengths.

Construction of QC-RS-LDPC codes using $2 \times 2$ SM-constrained RS matrices of type-2 and type-3 as base matrices is exactly the same as using the type-1 RS matrices. Note that the RS base matrices are nonbinary matrix over a general field GF($q$) whereas the constructed parity-check matrices are binary (i.e., over GF(2)).

In the following, we use several examples to demonstrate the construction of QC-RS-LDPC codes with the three types of $2 \times 2$ SM-constrained RS matrices as base matrices and show their error performances.

**Example 14.14** Suppose we use the field GF($2^9$) for code construction. Note that $2^9 - 1 = 511$ can be factored as the product of two primes, 7 and 73. Set $n = 511$. Let $\alpha$ be a primitive element of GF($2^9$). Then, the set $\mathbf{S}_{511} = \{1, \alpha, \alpha^2, \ldots, \alpha^{510}\}$ forms the multiplicative subgroup of GF($2^9$). Set $k = 7$ and $m = 73$ such that $n = km$. Using $n = 511$, $k = 7$, $m = 73$, and $d = k + 1 = 8$, we construct a $2 \times 2$ SM-constrained $8 \times 73$ RS matrix $\mathbf{B}_{\mathrm{RS},511}(8, 73)$ in the form of (14.12) using the type-1 construction. The null space over $GF(2^9)$ of $\mathbf{B}_{RS,511}(8, 73)$ gives a $2^9$-ary $(73, 65)$ shortened RS code with minimum distance 9.

Any submatrix of $\mathbf{B}_{RS,511}(8, 73)$ can be used as a base matrix for constructing a QC-RS-LDPC code. Because the entries in $\mathbf{B}_{RS,511}(8, 73)$ are elements in the multiplicative group of $GF(2^9)$, the CPM-dispersion factor must be 511. Suppose we take a $5 \times 64$ submatrix $\mathbf{B}_{RS,511}(5, 64)$ from $\mathbf{B}_{RS,511}(8, 73)$ by deleting the last nine columns and last three rows of $\mathbf{B}_{RS,511}(8, 73)$. The 511-fold CPM-dispersion of $\mathbf{B}_{RS,511}(5, 64)$ gives a $5 \times 64$ array $\mathbf{H}_{RS,511}(5, 64)$ of CPMs of size $511 \times 511$ which is a $2555 \times 32\,704$ matrix with column and row weights 5 and 64, respectively.

The null space over $GF(2)$ of $\mathbf{H}_{RS,511}(5, 64)$ gives a $(5, 64)$-regular $(32\,704, 30\,153)$ QC-RS-LDPC code $C_{RS,511}(5, 64)$ with rate 0.9220. The Tanner graph $\mathcal{G}_{RS,511}(5, 64)$ of the code $C_{RS,511}(5, 64)$ has girth 6 and each VN is connected to other 315 VNs by paths of length 2 in $C_{RS,511}(5, 64)$. The BER and BLER performances of the code $C_{RS,511}(5, 64)$ over the AWGN channel decoded with 50 iterations of the MSA scaled by a factor of 0.7 are shown in Fig. 14.14(a). At a BER of $10^{-8}$, the code performs about 0.84 dB from the Shannon limit and about 0.27 dB from its threshold.

If we apply the constructed code to the BEC, its UEBR and UEBLR performances are shown in Fig. 14.14(b). The Shannon limit for this code over BEC is $1 - 0.9220 = 0.078$. At a UEBR of $10^{-8}$, the code performs 0.03 away from the Shannon limit. ▲▲

**Example 14.15** In this example, we use the same field $GF(2^9)$ as used in Example 14.14 for code construction. Suppose we set $n = 73$. Let $\beta = \alpha^7$. Then, the order of $\beta$ is 73 and the set $\mathbf{S}_{73} = \{1, \beta, \beta^2, \ldots, \beta^{72}\}$ forms a subgroup of the multiplicative group $\mathbf{S}_{511} = \{1, \alpha, \alpha^2, \ldots, \alpha^{510}\}$ of $GF(2^9)$. Because the order 73 of the cyclic group $\mathbf{S}_{73}$ is a prime, we can choose any integer $d$, $1 \leq d \leq 73$, to construct a $d \times 73$ RS matrix $\mathbf{B}_{RS,73}(d, 73)$ over $GF(2^9)$ that satisfies the $2 \times 2$ SM-constraint (the type-3 construction). For $0 \leq b < 73 - d$, if we delete $b$ columns from $\mathbf{B}_{RS,73}(d, 73)$, we obtain a $d \times (73 - b)$ submatrix $\mathbf{B}_{RS,73}(d, 73 - b)$ of $\mathbf{B}_{RS,73}(d, 73)$. The matrix $\mathbf{B}_{RS,73}(d, 73 - b)$ can be used as the base matrix for code construction.

Because $\mathbf{S}_{73}$ is a subgroup of the multiplicative group $\mathbf{S}_{511}$ of $GF(2^9)$, each entry in $\mathbf{B}_{RS,73}(d, 73 - b)$ can be dispersed into a CPM of size either $73 \times 73$ or $511 \times 511$. Hence, the matrix $\mathbf{B}_{RS,73}(d, 73 - b)$ can be dispersed into either a $d \times (73 - b)$ array $\mathbf{H}_{RS,73}(d, 73 - b)$ of CPMs of size $73 \times 73$ or a $d \times (73 - b)$ array $\mathbf{H}_{RS,511}(d, 73 - b)$ of CPMs of size $511 \times 511$. The null spaces over $GF(2)$ of $\mathbf{H}_{RS,73}(d, 73 - b)$ and $\mathbf{H}_{RS,511}(d, 73 - b)$ give two QC-RS-LDPC codes of lengths $73(73 - b)$ and $511(73 - b)$, respectively. For $b = 0, 1, 2, \ldots, 72 - d$, we can construct a sequence of pairs of QC-RS-LDPC codes.

Suppose we choose $d = 4$ and construct a $4 \times 73$ RS matrix $\mathbf{B}_{RS,73}(4, 73)$ using the cyclic subgroup $\mathbf{S}_{73}$. Set $b = 9$. Deleting the last nine columns of

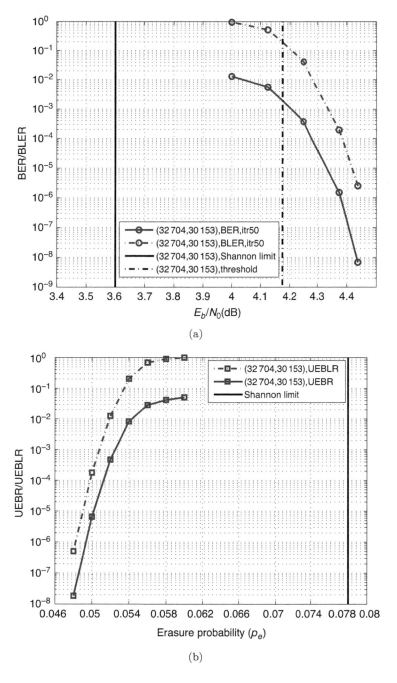

Figure 14.14 The performances of the $(32\,704, 30\,153)$ QC-RS-LDPC code given in Example 14.14 over: (a) AWGN channel and (b) BEC.

$\mathbf{B}_{RS,73}(4, 73)$, we obtain a $4 \times 64$ RS matrix $\mathbf{B}_{RS,73}(4, 64)$. The 73-fold CPM-dispersion of $\mathbf{B}_{RS,73}(4, 64)$ results in a $4 \times 64$ array $\mathbf{H}_{RS,73}(4, 64)$ of CPMs of size $73 \times 73$ which is a $292 \times 4672$ matrix with column and row weights 4 and 64, respectively.

The null space over GF(2) of $\mathbf{H}_{RS,73}(4, 64)$ gives a $(4, 64)$-regular $(4672, 4383)$ QC-RS-LDPC code $C_{RS,73}(4, 64)$ with rate 0.938. The BER and BLER performances of the code $C_{RS,73}(4, 64)$ over the AWGN channel decoded with 50 iterations of the MSA scaled by a factor 0.70 are shown in Fig. 14.15. Decoded with 50 iterations of the MSA, the code achieves a BER of $10^{-8}$ at an SNR of 5.25 dB without a visible error-floor and at a BER of $10^{-8}$, the code performs 1.35 dB from the Shannon limit and 0.911 dB from its threshold (4.3389 dB).

If we take 511-fold CPM-dispersion of $\mathbf{B}_{RS,73}(4, 64)$, we obtain a $4 \times 64$ array $\mathbf{H}_{RS,511}(4, 64)$ of CPMs of size $511 \times 511$ which is a $2044 \times 32\,704$ matrix with column and row weights 4 and 64, respectively. The null space over GF(2) of $\mathbf{H}_{RS,511}(4, 64)$ gives a $(4, 64)$-regular $(32\,704, 30\,663)$ QC-RS-LDPC code $C_{RS,511}(4, 64)$ with rate 0.9376. The BER and BLER performances of the code $C_{RS,511}(4, 64)$ over the AWGN channel decoded with 50 iterations of the MSA scaled by a factor of 0.70 are also shown in Fig. 14.15. ▲▲

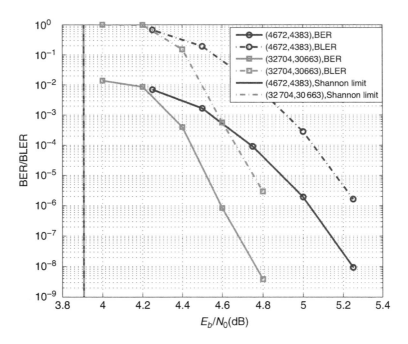

Figure 14.15 The BER and BLER performances of the two QC-RS-LDPC codes given in Example 14.15.

**Example 14.16** In this example, we construct a high-rate QC-RS-LDPC code using the field GF($2^{11}$) in conjunction with masking. Note that $2^{11} - 1$ can be factored as the product of two primes 23 and 89, i.e., $2^{11} - 1 = 2047 = 23 \times 89$.

Let $\alpha$ be a primitive element of $GF(2^{11})$ and $\beta = \alpha^{23}$ which is an element in $GF(2^{11})$ of order 89. Using $\beta$ as the generator, we construct a cyclic subgroup $\mathbf{S}_{89} = \{1, \beta, \beta^2, \ldots, \beta^{88}\}$ of order 89. Set $n = 89$ and $d = 8$. Using the cyclic group $\mathbf{S}_{89}$ and the type-3 construction, we form a $2 \times 2$ SM-constrained $8 \times 89$ RS matrix $\mathbf{B}_{RS,89}(8, 89)$.

Label the columns of $\mathbf{B}_{RS,89}(8, 89)$ from 0 to 88. Then, we choose the 64 columns in $\mathbf{B}_{RS,89}(8, 89)$ with labels 1, 2, 4, 5, 7, 8, 10, 12, 13, 14, 15, 17, 18, 19, 21, 22, 24, 25, 26, 27, 29, 30, 32, 33, 34, 35, 37, 38, 39, 40, 41, 43, 44, 46, 47, 48, 50, 51, 52, 54, 55, 57, 58, 60, 61, 62, 64, 65, 66, 68, 69, 71, 72, 74, 75, 76, 79, 80, 81, 83, 84, 86, 87, 88 to form an $8 \times 64$ submatrix $\mathbf{B}_{RS,89}(8, 64)$ of $\mathbf{B}_{RS,89}(8, 89)$. The 89-fold CPM-dispersion of $\mathbf{B}_{RS,89}(8, 64)$ gives an $8 \times 64$ array $\mathbf{H}_{RS,89}(8, 64)$ of CPMs of size $89 \times 89$ that is a $712 \times 5696$ matrix with column and row weights 8 and 64, respectively. The rank of $\mathbf{H}_{RS,89}(8, 64)$ is 705. The null space over $GF(2)$ of $\mathbf{H}_{RS,89}(8, 64)$ gives a $(8, 64)$-regular $(5696, 4991)$ QC-RS-LDPC code $C_{RS,89}(8, 64)$ of rate 0.8762. The Tanner graph of $C_{RS,89}(8, 64)$ has girth 6 and contains 14 299 986 cycles of length 6 and 4 760 166 513 cycles of length 8.

Next, we design an $8 \times 64$ masking matrix $\mathbf{Z}(8, 64)$ which consists of eight $8 \times 8$ circulants, $\mathbf{Z}_i(8, 8)$, with generators, $\mathbf{g}_i$, $0 \leq i < 8$,

$$\begin{aligned}
\mathbf{g}_0 &= [1\ 0\ 1\ 0\ 1\ 1\ 0\ 0], & \mathbf{g}_1 &= [0\ 1\ 0\ 1\ 0\ 0\ 1\ 1], \\
\mathbf{g}_2 &= [0\ 0\ 1\ 0\ 1\ 1\ 0\ 1], & \mathbf{g}_3 &= [0\ 1\ 0\ 1\ 1\ 1\ 0\ 0], \\
\mathbf{g}_4 &= [0\ 1\ 1\ 0\ 1\ 1\ 0\ 0], & \mathbf{g}_5 &= [0\ 0\ 0\ 1\ 1\ 1\ 0\ 1], \\
\mathbf{g}_6 &= [0\ 0\ 0\ 1\ 1\ 1\ 1\ 0], & \mathbf{g}_7 &= [0\ 1\ 1\ 0\ 0\ 1\ 1\ 0].
\end{aligned}$$

Arranging these eight circulants in a row, we obtain the following masking matrix:

$$\mathbf{Z}(8, 64) = [\mathbf{Z}_0(8, 8)\ \ \mathbf{Z}_1(8, 8)\ \ \ldots\ \ \mathbf{Z}_6(8, 8)\ \ \mathbf{Z}_7(8, 8)],$$

which has column and row weights 4 and 32, respectively. Masking $\mathbf{B}_{RS,89}(8, 64)$ with $\mathbf{Z}(8, 64)$, we obtain a masked matrix $\mathbf{B}_{RS,89,mask}(8, 64)$ with column and row weights 4 and 32, respectively.

The 89-fold CPM-dispersion of $\mathbf{B}_{RS,89,mask}(8, 64)$ gives an $8 \times 64$ array $\mathbf{H}_{RS,89,mask}(8, 64)$ of CPMs and ZMs of size $89 \times 89$ that is a $712 \times 5696$ matrix with column and row weights 4 and 32, respectively. The rank of $\mathbf{H}_{RS,89,mask}(8, 64)$ is 711. The null space over $GF(2)$ of $\mathbf{H}_{RS,89,mask}(8, 64)$ gives a $(4, 32)$-regular $(5696, 4985)$ QC-RS-LDPC code $C_{RS,89,mask}(8, 64)$ of rate 0.875. The Tanner graph of $C_{RS,89,mask}(8, 64)$ has girth 6 and contains 137 149 cycles of length 6 and 9 422 608 cycles of length 8. Note that the number of short cycles in the Tanner graph is greatly reduced after masking.

The BER and BLER performances of $C_{RS,89,mask}(8, 64)$ over the AWGN channel decoded with 50 iterations of the SPA are shown in Fig. 14.16. At a BER of $10^{-8}$, it performs about 1.3 dB from the Shannon limit.    ▲▲

The three examples given above show that the construction of QC-LDPC codes based on RS matrices is a flexible method for constructing LDPC codes.

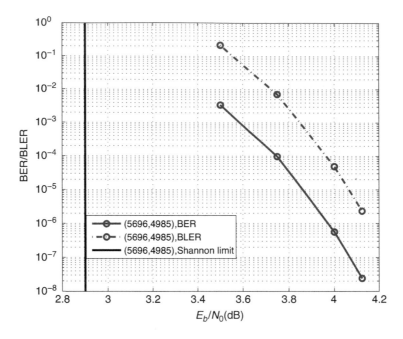

Figure 14.16 The BER and BLER performances of the $(5696, 4985)$ QC-RS-LDPC code given in Example 14.16.

RS base matrices in conjunction with masking can yield QC-LDPC codes with good error performance. Besides, RS base matrices, as $2 \times 2$ SM-constrained matrix, can be used as mother base matrices in the $2 \times 2/3 \times 3$ SMC-MMSA to construct QC-LDPC codes with large girths.

## 14.8 Construction of QC-RS-LDPC Codes with Girths at Least 8

In Section 14.7, three methods to construct $2 \times 2$ SM-constrained RS matrices were presented. Based on these matrices, QC-RS-LDPC codes whose Tanner graphs have girths at least 6 can be constructed. A $2 \times 2$ SM-constrained RS matrix unnecessarily satisfies the $2 \times 2/3 \times 3$ SM-constraint. However, if we choose columns from a $2 \times 2$ SM-constrained RS matrix that satisfy certain constraints to form a submatrix, a $2 \times 2/3 \times 3$ SM-constrained RS matrix can be constructed. Based on such an RS matrix, we can construct a QC-RS-LDPC code whose Tanner graph has girth at least 8. In this section, we present a set of constraints on choosing columns from a $2 \times 2$ SM-constrained RS matrix to form an RS matrix which also satisfies the $2 \times 2/3 \times 3$ SM-constraint.

Consider the $2 \times 2$ SM-constrained $d \times m$ RS matrix $\mathbf{B}_{\mathrm{RS},n}(d, m)$ over $\mathrm{GF}(q)$ in the form of (14.12). Label the columns of $\mathbf{B}_{\mathrm{RS},n}(d, m)$ from 0 to $m - 1$. Let

$d \leq t \leq m$. Suppose we take a set $\Lambda_t = \{l_1, l_2, \ldots, l_t\}$ of labels of $t$ columns in $\mathbf{B}_{\mathrm{RS},n}(d, m)$ with $0 \leq l_1 < l_2 < \cdots < l_t < m \leq n$. Next, we choose $t$ columns, labeled by $l_1, l_2, \ldots, l_t$, from $\mathbf{B}_{\mathrm{RS},n}(d, m)$ and form the following $d \times t$ submatrix of $\mathbf{B}_{\mathrm{RS},n}(d, m)$:

$$
\begin{aligned}
\mathbf{B}_{\mathrm{RS},n,\Lambda_t}(d, t) &= [(\beta^i)^{l_j}]_{1 \leq i \leq d, 1 \leq j \leq t} \\
&= \begin{bmatrix}
\beta^{l_1} & \beta^{l_2} & \beta^{l_3} & \cdots & \beta^{l_t} \\
(\beta^2)^{l_1} & (\beta^2)^{l_2} & (\beta^2)^{l_3} & \cdots & (\beta^2)^{l_t} \\
\vdots & \vdots & \vdots & \ddots & \vdots \\
(\beta^d)^{l_1} & (\beta^d)^{l_2} & (\beta^d)^{l_3} & \cdots & (\beta^d)^{l_t}
\end{bmatrix}.
\end{aligned} \tag{14.14}
$$

The following theorem [45] gives a set of necessary and sufficient conditions on the labels in $\Lambda_t = \{l_1, l_2, \ldots, l_t\}$ such that the $d \times t$ RS matrix $\mathbf{B}_{\mathrm{RS},n,\Lambda_t}(d, t)$ in (14.14) satisfies the $2 \times 2/3 \times 3$ SM-constraint. The development and proof of this theorem are given in Appendix C.

**Theorem 14.4 ([45])** *The $2 \times 2$ SM-constrained RS matrix $\mathbf{B}_{\mathrm{RS},n,\Lambda_t}(d, t)$ given by (14.14) satisfies the $2 \times 2/3 \times 3$ SM-constraint if and only if for any six integers $i_1$, $i_2$, $i_3$, $x_1$, $x_2$, $x_3$ with $1 \leq i_1 < i_2 < i_3 \leq t$ and $0 \leq x_1 < x_2 < x_3 < d$, the labels in $\Lambda_t = \{l_1, l_2, \ldots, l_t\}$ satisfy the following six inequalities:*

$$
\begin{aligned}
((x_2 - x_1)(l_{i_2} - l_{i_1}) + (x_3 - x_2)(l_{i_3} - l_{i_1}))_n &\neq 0, \\
((x_2 - x_1)(l_{i_2} - l_{i_1}) + (x_3 - x_2)(l_{i_2} - l_{i_3}))_n &\neq 0, \\
((x_2 - x_1)(l_{i_3} - l_{i_1}) + (x_3 - x_2)(l_{i_3} - l_{i_2}))_n &\neq 0, \\
((x_2 - x_1)(l_{i_3} - l_{i_2}) + (x_3 - x_2)(l_{i_3} - l_{i_1}))_n &\neq 0, \\
((x_2 - x_1)(l_{i_2} - l_{i_3}) + (x_3 - x_2)(l_{i_2} - l_{i_1}))_n &\neq 0, \\
((x_2 - x_1)(l_{i_3} - l_{i_1}) + (x_3 - x_2)(l_{i_2} - l_{i_1}))_n &\neq 0,
\end{aligned} \tag{14.15}
$$

*where $(a)_n$ denotes a modulo $n$.*

From the six inequalities given by (14.15), we readily see that if the labels of a set $\Lambda_t = \{l_1, l_2, \ldots, l_t\}$ satisfy the six inequalities, then the labels of the set $\Lambda_t^* = \{(r + l_1)_n, (r + l_2)_n, \ldots, (r + l_t)_n\}$ obtained by adding an integer $r$, $0 \leq r < n$, to each label in $\Lambda_t$ also satisfy the six inequalities given by (14.15). Using the label set $\Lambda_t^*$, we can construct another $d \times t$ RS matrix which also satisfies the $2 \times 2/3 \times 3$ SM-constraint. For convenience, we call $\Lambda_t$ the $2 \times 2/3 \times 3$ *SM-constrained column-label set.*

Theorem 14.4 instructs how to choose columns from the $2 \times 2$ SM-constrained $d \times m$ RS matrix $\mathbf{B}_{\mathrm{RS},n}(d, m)$ given by (14.12) to form a $d \times t$ RS matrix $\mathbf{B}_{\mathrm{RS},n,\Lambda_t}(d, t)$ which satisfies the $2 \times 2/3 \times 3$ SM-constraint. Consequently, the Tanner graph associated with the CPM-dispersion of $\mathbf{B}_{\mathrm{RS},n,\Lambda_t}(d, t)$ has girth at least 8. The matrix $\mathbf{B}_{\mathrm{RS},n,\Lambda_t}(d, t)$ is called a *type-1* $2 \times 2/3 \times 3$ *SM-constrained RS matrix.*

The six conditions in Theorem 14.4 for a submatrix of a type-1 $2 \times 2$ SM-constrained RS matrix to satisfy the $2 \times 2/3 \times 3$ SM-constraint also apply to construct $2 \times 2/3 \times 3$ constrained matrices using type-2 and type-3 $2 \times 2$ SM-constrained RS matrices. In both cases, we replace $m$ by $n$. The constructed matrices are called *type-2 and type-3 $2 \times 2/3 \times 3$ SM-constrained RS matrices.*

All the six inequalities given by (14.15) are expressed in terms of the labels of the columns chosen from $\mathbf{B}_{\mathrm{RS},n}(d,m)$. To find such a $2 \times 2/3 \times 3$ SM-constrained RS matrix, computations required are relatively simple.

**Example 14.17** In this example, we use the same field $\mathrm{GF}(2^9)$ as Example 14.14 to construct a rate-1/2 QC-RS-LDPC code whose Tanner graph has girth 8. First, we factor $2^9 - 1 = 511$ as the product of 7 and 73. Set $n = 511$, $k = 7$, and $m = 73$. Choose $d = 4$. Using the type-1 construction, we form a $2 \times 2$ SM-constrained $4 \times 73$ RS matrix $\mathbf{B}_{\mathrm{RS},73}(4,73)$ in the form of (14.12). From the 73 column labels, we find a set $\Lambda_8$ of eight column labels 2, 5, 9, 15, 26, 42, 64, and 72 which satisfy the six inequalities given by (14.15) in Theorem 14.4. Hence, the set $\Lambda_8 = \{2, 5, 9, 15, 26, 42, 64, 72\}$ forms a $2 \times 2/3 \times 3$ SM-constrained column-label set. Choose the eight columns from $\mathbf{B}_{\mathrm{RS},73}(4,73)$ with labels in $\Lambda_8$ and form a $4 \times 8$ submatrix $\mathbf{B}_{\mathrm{RS},73,\Lambda_8}(4,8)$ of $\mathbf{B}_{\mathrm{RS},73}(4,73)$. Then, the matrix $\mathbf{B}_{\mathrm{RS},73,\Lambda_8}(4,8)$ satisfies the $2 \times 2/3 \times 3$ SM-constraint.

The 511-fold CPM-dispersion of $\mathbf{B}_{\mathrm{RS},73,\Lambda_8}(4,8)$ gives a $4 \times 8$ RC-constrained array $\mathbf{H}_{\mathrm{RS},511,\Lambda_8}(4,8)$ of CPMs of size $511 \times 511$ which is a $2044 \times 4088$ matrix over $\mathrm{GF}(2)$ with column and row weights 4 and 8, respectively. The null space over $\mathrm{GF}(2)$ of $\mathbf{H}_{\mathrm{RS},511,\Lambda_8}(4,8)$ gives a $(4,8)$-regular $(4088, 2047)$ QC-RS-LDPC code $C_{\mathrm{RS},511,\Lambda_8}(4,8)$ of rate 0.501, slightly higher than $1/2$. The Tanner graph $\mathcal{G}_{\mathrm{RS},511,\Lambda_8}(4,8)$ of the code has girth 8. The numbers of short cycles of lengths 8, 10, 12, and 14 in $\mathcal{G}_{\mathrm{RS},511,\Lambda_8}(4,8)$ are 87 892, 623 420, 12 511 835, and 192 430 366, respectively. The total number of such short cycles in $\mathcal{G}_{\mathrm{RS},511,\Lambda_8}(4,8)$ is 205 653 483.

Suppose we mask the RS matrix $\mathbf{B}_{\mathrm{RS},73,\Lambda_8}(4,8)$ with the $4 \times 8$ masking matrix $\mathbf{Z}(4,8)$ given by (14.5). We obtain a $4 \times 8$ masked RS matrix $\mathbf{B}_{\mathrm{RS},73,\Lambda_8,\mathrm{mask}}(4,8)$. The 511-fold CPM-dispersion of $\mathbf{B}_{\mathrm{RS},73,\Lambda_8,\mathrm{mask}}(4,8)$ gives a $4 \times 8$ masked array $\mathbf{H}_{\mathrm{RS},511,\Lambda_8,\mathrm{mask}}(4,8)$ of CPMs and ZMs of size $511 \times 511$ which is a $2044 \times 4088$ matrix with column and row weights 3 and 6, respectively. The null space over $\mathrm{GF}(2)$ of $\mathbf{H}_{\mathrm{RS},511,\Lambda_8,\mathrm{mask}}(4,8)$ gives a $(3,6)$-regular $(4088, 2044)$ QC-RS-LDPC code $C_{\mathrm{RS},511,\Lambda_8,\mathrm{mask}}(4,8)$ of rate $1/2$. Its Tanner graph $\mathcal{G}_{\mathrm{RS},511,\Lambda_8,\mathrm{mask}}(4,8)$ also has girth 8. The numbers of short cycles of lengths 8, 10, 12, and 14 in $\mathcal{G}_{\mathrm{RS},511,\Lambda_8,\mathrm{mask}}(4,8)$ are 1022, 14 308, 141 547, and 1 016 890, respectively. The total number of short cycles in $\mathcal{G}_{\mathrm{RS},511,\Lambda_8,\mathrm{mask}}(4,8)$ is 1 173 767.

Comparing the short cycle distributions of the masked Tanner graph to those of the unmasked one, we see that masking results in an enormous reduction in short cycles. The reduction of the number of cycles of length 8 is from 87 892 to 1022, and the reduction of the total number of short cycles is from 205 653 483 to 1 173 767, a reduction by a factor of more than 175.

The BER and BLER performances of the unmasked code $C_{\mathrm{RS},511,\Lambda_8}(4,8)$ and the masked code $C_{\mathrm{RS},511,\Lambda_8,\mathrm{mask}}(4,8)$ over the AWGN channel decoded with 50 iterations of the MSA scaled by factors of 0.7 and 0.75, respectively, are shown in Fig. 14.17. We see that masking results in a significant performance improvement. The masked code achieves a BER of $10^{-9}$ at an SNR of 2.18 dB and shows no error-floor. At a BER of $10^{-9}$, it performs about 1.9 dB from the Shannon limit and 1.09 dB from the threshold. This performance improvement of the masked code is mainly due to the large reduction of short cycles and the change of degree distribution from a $(4,8)$-regular distribution to a $(3,6)$-regular distribution because of masking. ▲▲

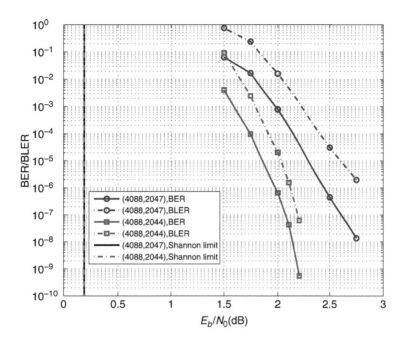

Figure 14.17 The BER and BLER performances of the unmasked $(4088, 2047)$ and the masked $(4088, 2044)$ QC-RS-LDPC codes given in Example 14.17.

A special case of Theorem 14.4 is $d = 4$. Consider a $4 \times t$ RS matrix $\mathbf{B}_{\mathrm{RS},n,\Lambda_t}(4,t)$ in the form of (14.14) constructed from a $2 \times 2$ SM-constrained RS matrix $\mathbf{B}_{\mathrm{RS},n}(4,n)$ based on a chosen set $\Lambda_t$ of column labels. Because $0 \le x_1 < x_2 < x_3 < 4$, there are four possibilities for $(x_1, x_2, x_3)$, which are $(0,1,2)$, $(0,1,3)$, $(0,2,3)$, and $(1,2,3)$. Substituting these four possibilities of $(x_1, x_2, x_3)$ into the six inequalities given by (14.15) in Theorem 14.4, we obtain the following corollary.

**Corollary 14.1** *Let* $\mathbf{B}_{\mathrm{RS},n,\Lambda_t}(4,t)$ *be a* $4 \times t$ *RS matrix formed by the columns of the RS matrix* $\mathbf{B}_{\mathrm{RS},n}(4,n)$ *labeled by the elements in a column-label set*

$\Lambda_t = \{l_1, l_2, \ldots, l_t\}$. *The RS matrix* $\mathbf{B}_{\mathrm{RS},n,\Lambda_t}(4,t)$ *satisfies the* $2 \times 2/3 \times 3$ *SM-constraint if and only if for any three labels* $l_{i_1}$, $l_{i_2}$, *and* $l_{i_3}$ *in* $\Lambda_t$ *with* $0 \le l_{i_1} < l_{i_2} < l_{i_3} < n$, *the following nine conditions hold:*

$$n \nmid l_{i_3} - 2l_{i_2} + l_{i_1}, \quad n \nmid l_{i_3} - 3l_{i_2} + 2l_{i_1}, \quad n \nmid 2l_{i_3} - 3l_{i_2} + l_{i_1},$$
$$n \nmid l_{i_3} + l_{i_2} - 2l_{i_1}, \quad n \nmid 2l_{i_3} + l_{i_2} - 3l_{i_1}, \quad n \nmid l_{i_3} + 2l_{i_2} - 3l_{i_1}, \quad (14.16)$$
$$n \nmid 2l_{i_3} - l_{i_2} - l_{i_1}, \quad n \nmid 3l_{i_3} - l_{i_2} - 2l_{i_1}, \quad n \nmid 3l_{i_3} - 2l_{i_2} - l_{i_1},$$

*where* $a \nmid b$ *denotes that* $a$ *is not divisible by* $b$.

Let $\mathbf{H}_{\mathrm{RS},n,\Lambda_t}(4,t)$ be the $n$-fold CPM-dispersion of $\mathbf{B}_{\mathrm{RS},n,\Lambda_t}(4,t)$ whose column labels satisfy the nine conditions given by (14.16) in Corollary 14.1. Then, the null space over GF(2) of $\mathbf{H}_{\mathrm{RS},n,\Lambda_t}(4,t)$ gives a QC-RS-LDPC code $C_{\mathrm{RS},n,\Lambda_t}(4,t)$ whose Tanner graph has girth at least 8.

**Example 14.18** In this example, we construct a QC-RS-LDPC code of rate 1/2 using the field GF($2^8$). First, we factor $2^8 - 1$ as the product of three primes 3, 5, and 17. Set $n = 5 \times 17 = 85$. Note that 5 is the smallest prime factor of $n = 85$. Next, we choose $d = 4$. Using the type-2 construction, we construct a $2 \times 2$ SM-constrained $4 \times 85$ RS matrix $\mathbf{B}_{\mathrm{RS},85}(4,85)$ based on a cyclic subgroup $\mathbf{S}_{85}$ of GF($2^8$) which is generated by an element $\beta \in \mathrm{GF}(2^8)$ of order 85. We can take any submatrix of $\mathbf{B}_{\mathrm{RS},85}(4,85)$ as a base matrix for constructing a QC-LDPC code whose Tanner graph has girth at least 6.

Set $t = 8$. We find that the eight column labels 2, 5, 7, 13, 20, 31, 48, and 54 satisfy the nine conditions given by (14.16) in Corollary 14.1. Hence, $\Lambda_8 = \{2, 5, 7, 13, 20, 31, 48, 54\}$ forms a $2 \times 2/3 \times 3$ SM-constrained column-label set. Form a $4 \times 8$ RS matrix $\mathbf{B}_{\mathrm{RS},85,\Lambda_8}(4,8)$ with eight columns chosen from $\mathbf{B}_{\mathrm{RS},85}(4,85)$ based on $\Lambda_8$. Then, $\mathbf{B}_{\mathrm{RS},85,\Lambda_8}(4,8)$ satisfies the $2 \times 2/3 \times 3$ SM-constraint. The 85-fold CPM-dispersion of $\mathbf{B}_{\mathrm{RS},85,\Lambda_8}(4,8)$ gives a $4 \times 8$ array $\mathbf{H}_{\mathrm{RS},85,\Lambda_8}(4,8)$ of CPMs of size $85 \times 85$ which is a $340 \times 680$ matrix over GF(2) with column and row weights 4 and 8, respectively. The rank of $\mathbf{H}_{\mathrm{RS},85,\Lambda_8}(4,8)$ is 337.

The null space over GF(2) of $\mathbf{H}_{\mathrm{RS},85,\Lambda_8}(4,8)$ gives a $(4,8)$-regular $(680,343)$ QC-RS-LDPC code $C_{\mathrm{RS},85,\Lambda_8}(4,8)$ with rate 0.5044 and minimum distance 10. Because $\mathbf{B}_{\mathrm{RS},85,\Lambda_8}(4,8)$ satisfies the $2 \times 2/3 \times 3$ SM-constraint, the Tanner graph $\mathcal{G}_{\mathrm{RS},85,\Lambda_8}(4,8)$ of $C_{\mathrm{RS},85,\Lambda_8}(4,8)$ has girth at least 8. Using the algorithm given in [33, 34] for cycle enumeration, we find that $\mathcal{G}_{\mathrm{RS},85,\Lambda_8}(4,8)$ contains 32 810 cycles of length 8, 386 240 cycles of length 10, 7 256 535 cycles of length 12, and 128 090 240 cycles of length 14. The total number of short cycles of lengths 8, 10, 12, and 14 in $\mathcal{G}_{\mathrm{RS},85,\Lambda_8}(4,8)$ is 135 765 825, a very large number of short cycles.

The BER and BLER performances of the $(680,343)$ QC-RS-LDPC code $C_{\mathrm{RS},85,\Lambda_8}(4,8)$ over the AWGN channel decoded with 50 iterations of the MSA with a scaling factor of 0.7 are shown in Fig. 14.18. The code starts to have an error-floor below the BLER of $10^{-4}$ even though its Tanner graph has girth 8. This poor error performance is caused by the large number of short cycles and a small minimum distance of 10.

Suppose we mask the RS matrix $\mathbf{B}_{\mathrm{RS},85,\Lambda_8}(4,8)$ with the $4 \times 8$ masking $\mathbf{Z}(4,8)$ given by (14.5). We obtain a $4 \times 8$ masked matrix $\mathbf{B}_{\mathrm{RS},85,\Lambda_8,\mathrm{mask}}(4,8)$ with column and row weights 3 and 6, respectively. The 85-fold CPM-dispersion of $\mathbf{B}_{\mathrm{RS},85,\Lambda_8,\mathrm{mask}}(4,8)$ gives a $4 \times 8$ masked array $\mathbf{H}_{\mathrm{RS},85,\Lambda_8,\mathrm{mask}}(4,8)$ of CPMs and ZMs of size $85 \times 85$. The array $\mathbf{H}_{\mathrm{RS},85,\Lambda_8,\mathrm{mask}}(4,8)$ is a $340 \times 680$ matrix with column and row weights 3 and 6, respectively.

The null space of $\mathbf{H}_{\mathrm{RS},85,\Lambda_8,\mathrm{mask}}(4,8)$ gives a $(3,6)$-regular $(680, 340)$ QC-RS-LDPC code $C_{\mathrm{RS},85,\Lambda_8,\mathrm{mask}}(4,8)$ with rate 0.5 and minimum distance 34 (a very large minimum distance for an LDPC code). The numbers of cycles of lengths 8, 10, 12, and 14 in the Tanner graph $\mathcal{G}_{\mathrm{RS},85,\Lambda_8,\mathrm{mask}}(4,8)$ of $C_{\mathrm{RS},85,\Lambda_8,\mathrm{mask}}(4,8)$ are 1020, 9945, 85 170, and 720 970, respectively. The masked Tanner graph $\mathcal{G}_{\mathrm{RS},85,\Lambda_8,\mathrm{mask}}(4,8)$ also has girth 8 but the number of cycles of length 8 is only 1020 which is much smaller than the 32 810 cycles of length 8 in the unmasked Tanner graph $\mathcal{G}_{\mathrm{RS},85,\Lambda_8}(4,8)$. The total number of cycles of lengths 8, 10, 12, and 14 in $\mathcal{G}_{\mathrm{RS},85,\Lambda_8,\mathrm{mask}}(4,8)$ is 817 105.

Comparing the short cycle distributions of the unmasked Tanner graph $\mathcal{G}_{\mathrm{RS},85,\Lambda_8}(4,8)$ and the masked Tanner graph $\mathcal{G}_{\mathrm{RS},85,\Lambda_8,\mathrm{mask}}(4,8)$, we find that there is a very large reduction in short cycles from 135 765 825 to 817 105, a reduction by a factor of almost 166. We see that masking not only reduces the number of short cycles of the Tanner graph of the unmasked code but also increases its minimum distance from 10 to 34. Again, it shows that the masking matrix given by (14.5) is an effective masking matrix.

The BER and BLER performances of $C_{\mathrm{RS},85,\Lambda_8,\mathrm{mask}}(4,8)$ over the AWGN channel decoded with 50 iterations of the MSA with a scaling factor of 0.75 are also shown in Fig. 14.18. We see that the masked code $C_{\mathrm{RS},85,\Lambda_8,\mathrm{mask}}(4,8)$ performs very well, much better than the unmasked code below the BER of $10^{-4}$. It achieves a BER of $10^{-9}$ without a visible error-floor.                 ▲▲

# 14.9    A Special Class of QC-RS-LDPC Codes with Girth 8

In this section, we present a special class of $2 \times 2/3 \times 3$ SM-constrained RS matrices. Consider the $d \times n$ RS matrix $\mathbf{B}_{\mathrm{RS},n}(d,n)$ given by (14.11) which may not satisfy the $2 \times 2$ SM-constraint. Set $d = 4$. Then, we obtain a $4 \times n$ RS matrix $\mathbf{B}_{\mathrm{RS},n}(4,n)$. Suppose 3 is a factor of $n$, i.e., $3|n$. Label the columns of $\mathbf{B}_{\mathrm{RS},n}(4,n)$ from 0 to $n-1$. Next, we partition the column labels of $\mathbf{B}_{\mathrm{RS},n}(4,n)$ into $n/3$ disjoint triplets,

$$(0, n/3, 2n/3), (1, 1 + n/3, 1 + 2n/3), \ldots,$$
$$\ldots, (i, i + n/3, i + 2n/3), \ldots, (n/3 - 1, 2n/3 - 1, n - 1).$$

From each triplet $(i, i + n/3, i + 2n/3)$, $0 \le i < n/3$, we take any label $j_i$ and form a label set $\Lambda_{n/3} = \{j_0, j_1, \ldots, j_{n/3-1}\}$ with $n/3$ column labels. Next, we select $n/3$ columns, labeled with $j_0, j_1, \ldots, j_{n/3-1}$, from $\mathbf{B}_{\mathrm{RS},n}(4,n)$ and form

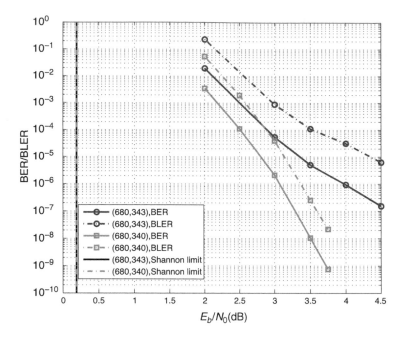

Figure 14.18 The BER and BLER performances of the unmasked $(680, 343)$ and the masked $(680, 340)$ QC-RS-LDPC codes given in Example 14.18.

a $4 \times n/3$ submatrix $\mathbf{B}_{\mathrm{RS},n,\Lambda_{n/3}}(4, n/3)$. Following the proof of Theorem 14.3, it can be shown that the $4 \times n/3$ RS matrix $\mathbf{B}_{\mathrm{RS},n,\Lambda_{n/3}}(4, n/3)$ satisfies the $2 \times 2$ SM-constraint. This gives another construction of $2 \times 2$ SM-constrained RS matrices.

Suppose we choose a subset $\Lambda_t$ of $t$ elements from the column-label set $\Lambda_{n/3}$ with $t < n/3$ such that for any three labels $l_{i_1}$, $l_{i_2}$, and $l_{i_3}$ in $\Lambda_t$ with $l_{i_1} < l_{i_2} < l_{i_3}$, all nine conditions given by (14.16) in Corollary 14.1 hold. Then, $\Lambda_t$ is a $2 \times 2/3 \times 3$ SM-constrained column-label set. The columns in $\mathbf{B}_{\mathrm{RS},n}(4, n)$ labeled by the elements in $\Lambda_t$ form a $4 \times t$ RS matrix $\mathbf{B}_{\mathrm{RS},n,\Lambda_t}(4, t)$ which satisfies the $2 \times 2/3 \times 3$ SM-constraint. Hence, the null space over $\mathrm{GF}(2)$ of the $n$-fold CPM-dispersion $\mathbf{H}_{\mathrm{RS},n,\Lambda_t}(4, t)$ of $\mathbf{B}_{\mathrm{RS},n,\Lambda_t}(4, t)$ gives a QC-RS-LDPC code whose Tanner graph has girth at least 8.

**Example 14.19** In this example, we use the largest cyclic subgroup (the multiplicative group) of $\mathrm{GF}(2^8)$ for code construction. The order $n$ of this group is 255 which is divisible by 3. Let $\beta$ be a primitive element of $\mathrm{GF}(2^8)$. We first construct a $4 \times 255$ RS matrix $\mathbf{B}_{\mathrm{RS},255}(4, 255)$ in the form of (14.11). This RS matrix does not satisfy the $2 \times 2$ SM-constraint. Because 3 is a factor of 255, we can partition the column labels of $\mathbf{B}_{\mathrm{RS},255}(4, 255)$ into 85 disjoint triplets, $(i, i + 85, i + 170)$, $0 \leq i < 85$. For $1 \leq t \leq 85$, suppose we choose a set of $t$ triplets from the 85 triplets. From each triplet in this set, we pick up one number. This gives a set $\Lambda_t = \{l_1, l_2, \ldots, l_t\}$ of labels of $t$ columns in $\mathbf{B}_{\mathrm{RS},255}(4, 255)$.

Using the columns of $\mathbf{B}_{\text{RS},255}(4,255)$ labeled by $\Lambda_t$, we form a $4 \times t$ RS matrix $\mathbf{B}_{\text{RS},255,\Lambda_t}(4,t)$. This matrix satisfies the $2 \times 2$ SM-constraint. If the labels in $\Lambda_t$ satisfy all nine conditions given in Corollary 14.1, then the $4 \times t$ matrix $\mathbf{B}_{\text{RS},255,\Lambda_t}(4,t)$ satisfies the $2 \times 2/3 \times 3$ SM-constraint. The null space over $\text{GF}(2)$ of the 255-fold CPM-dispersion of $\mathbf{B}_{\text{RS},255,\Lambda_t}(4,t)$ gives a QC-RS-LDPC code of length $255t$ whose Tanner graph has girth at least 8.

Suppose we want to construct a QC-RS-LDPC code with rate equal or close to $1/2$. We set $t = 8$. From the 85 triplets, $(i, i + 85, i + 170)$, $0 \leq i < 85$, we find the set $\Lambda_8 = \{2, 5, 7, 13, 20, 32, 54, 60\}$ of column labels which satisfy the nine conditions given by (14.16). The column labels in $\Lambda_8$ are the first numbers of the following eight triplets:

$$(2, 87, 172), \quad (5, 90, 175), \quad (7, 92, 177), \quad (13, 98, 183),$$
$$(20, 105, 190), \quad (32, 117, 202), \quad (54, 139, 224), \quad (60, 145, 230).$$

Take eight columns, labeled by $\Lambda_8$, from $\mathbf{B}_{\text{RS},255}(4,255)$ and form a $4 \times 8$ RS matrix $\mathbf{B}_{\text{RS},255,\Lambda_8}(4,8)$. Then, $\mathbf{B}_{\text{RS},255,\Lambda_8}(4,8)$ satisfies the $2 \times 2/3 \times 3$ SM-constraint. The 255-fold CPM-dispersion of $\mathbf{B}_{\text{RS},255,\Lambda_8}(4,8)$ gives a $4 \times 8$ array $\mathbf{H}_{\text{RS},255,\Lambda_8}(4,8)$ of CPMs of size $255 \times 255$. It is a $1020 \times 2040$ matrix with column and row weights 4 and 8, respectively. The rank of $\mathbf{H}_{\text{RS},255,\Lambda_8}(4,8)$ is 1015 and hence it is not a full-rank matrix.

The null space over $\text{GF}(2)$ of $\mathbf{H}_{\text{RS},255,\Lambda_8}(4,8)$ gives a $(4,8)$-regular $(2040, 1025)$ QC-RS-LDPC code $C_{\text{RS},255,\Lambda_8}(4,8)$ with rate 0.5025 which is slightly higher than $1/2$. Because $\mathbf{B}_{\text{RS},255,\Lambda_8}(4,8)$ satisfies the $2 \times 2/3 \times 3$ SM-constraint, the Tanner graph $\mathcal{G}_{\text{RS},255,\Lambda_8}(4,8)$ of $C_{\text{RS},255,\Lambda_8}(4,8)$ has girth at least 8. The distribution of cycles of lengths 8, 10, 12, and 14 in $\mathcal{G}_{\text{RS},255,\Lambda_8}(4,8)$ is $\{53\,805, 407\,490, 8\,168\,670, 133\,452\,720\}$. The Tanner graph $\mathcal{G}_{\text{RS},255,\Lambda_8}(4,8)$ contains $53\,805$ cycles of length 8. The total number of cycles of length 8, 10, 12, and 14 is $142\,082\,685$, a very large number of short cycles.

Suppose we mask the RS matrix $\mathbf{B}_{\text{RS},255,\Lambda_8}(4,8)$ with the masking matrix given by (14.5). We obtain a $4 \times 8$ masked RS matrix $\mathbf{B}_{\text{RS},255,\Lambda_8,\text{mask}}(4,8)$ whose 255-fold CPM-dispersion gives a $4 \times 8$ array $\mathbf{H}_{\text{RS},255,\Lambda_8,\text{mask}}(4,8)$ of CPMs and ZMs of size $255 \times 255$. The array $\mathbf{H}_{\text{RS},255,\Lambda_8,\text{mask}}(4,8)$ is a $1020 \times 2040$ matrix with column and row weights 3 and 6, respectively. The rank of $\mathbf{H}_{\text{RS},255,\Lambda_8,\text{mask}}(4,8)$ is 1020 and hence $\mathbf{H}_{\text{RS},255,\Lambda_8,\text{mask}}(4,8)$ is a full-rank matrix.

The null space over $\text{GF}(2)$ of $\mathbf{H}_{\text{RS},255,\Lambda_8,\text{mask}}(4,8)$ gives a $(3,6)$-regular $(2040, 1020)$ QC-RS-LDPC code $C_{\text{RS},255,\Lambda_8,\text{mask}}(4,8)$ with rate 0.5. The distribution of cycles of lengths 8, 10, 12, and 14 in its Tanner graph $\mathcal{G}_{\text{RS},255,\Lambda_8,\text{mask}}(4,8)$ is $\{765, 10\,200, 84\,405, 743\,580\}$. The masked Tanner graph $\mathcal{G}_{\text{RS},255,\Lambda_8,\text{mask}}(4,8)$ also has girth 8 but the number of cycles of length 8 is only 765 which is much smaller than the $53\,805$ cycles of length 8 in the unmasked Tanner graph $\mathcal{G}_{\text{RS},255,\Lambda_8}(4,8)$. The total number of cycles of length 8, 10, 12, and 14 in $\mathcal{G}_{\text{RS},255,\Lambda_8,\text{mask}}(4,8)$ is $838\,950$. Comparing the cycle distributions of the unmasked and masked Tanner graphs, $\mathcal{G}_{\text{RS},255,\Lambda_8}(4,8)$ and $\mathcal{G}_{\text{RS},255,\Lambda_8,\text{mask}}(4,8)$, we find that there is a very large reduction in short cycles from $142\,082\,685$ to $838\,950$, a reduction by a factor of almost 169.

The BER and BLER performances of $C_{\mathrm{RS},255,\Lambda_8}(4,8)$ and $C_{\mathrm{RS},255,\Lambda_8,\mathrm{mask}}(4,8)$ over the AWGN channel decoded with 50 iterations of the MSA scaled by factors of 0.70 and 0.75, respectively, are shown in Fig. 14.19. We see that the masking improves the error performances upon the unmasked code. This performance improvement is the result of the large reduction of short cycles and the change of degree distributions of the Tanner graph after masking, from $(4,8)$-regular to $(3,6)$-regular. The masked QC-RS-LDPC code $C_{\mathrm{RS},255,\Lambda_8,\mathrm{mask}}(4,8)$ achieves a BER of almost $10^{-10}$ without a visible error-floor, and it has a 0.3 dB coding gain over the unmasked code. ▲▲

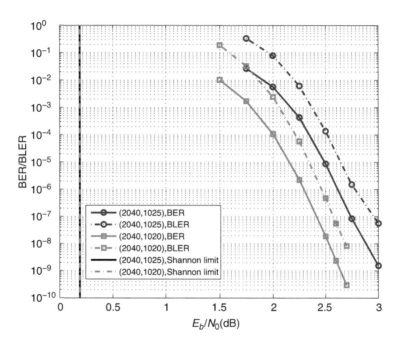

Figure 14.19 The BER and BLER performances of the unmasked $(2040, 1025)$ and masked $(2040, 1020)$ QC-RS-LDPC codes given in Example 14.19.

**Example 14.20** In this example, we continue using the field $\mathrm{GF}(2^8)$ to construct two QC-RS-LDPC codes, one with rate $2/3$ and the other with rate $3/4$. From the column-label partition given in the last example, we choose two label sets, $\Lambda_{12} = \{0, 1, 4, 9, 11, 20, 24, 35, 41, 49, 90, 225\}$ and $\Lambda_{16} = \{1, 3, 6, 13, 21, 32, 44, 59, 64, 73, 77, 83, 111, 212, 226, 239\}$. Each of these two column-label sets satisfies the nine conditions in Corollary 14.1. Using these two column-label sets, we form a $4 \times 12$ RS matrix $\mathbf{B}_{\mathrm{RS},255,\Lambda_{12}}(4, 12)$ and a $4 \times 16$ RS matrix $\mathbf{B}_{\mathrm{RS},255,\Lambda_{16}}(4, 16)$, respectively, both satisfying the $2 \times 2/3 \times 3$ SM-constraint.

The null spaces over $\mathrm{GF}(2)$ of the 255-fold CPM-dispersions of the two RS matrices give a $(4, 12)$-regular $(3060, 2049)$ QC-RS-LDPC code and a $(4, 16)$-regular $(4080, 3065)$ QC-RS-LDPC code, respectively. The Tanner graphs of both codes have girth 8, one with 259 845 cycles of length 8 and the other with 688 500 cycles of length 8. The distributions of short cycles of lengths 8, 10, and 12 in the Tanner graphs of these two codes are {259 845, 3 886 710, 116 167 800} and {688 500, 17 485, 860, 703 291 020}, respectively. We see that both Tanner graphs have very large numbers of short cycles.

Suppose we mask the RS matrices $\mathbf{B}_{\mathrm{RS},255,\Lambda_{12}}(4,12)$ and $\mathbf{B}_{\mathrm{RS},255,\Lambda_{16}}(4,16)$ with the following two masking matrices, respectively:

$$\mathbf{Z}(4,12) = \begin{bmatrix} 1 & 0 & 1 & 0 & 1 & 0 & 1 & 1 & 1 & 1 & 1 & 1 \\ 0 & 1 & 0 & 1 & 0 & 1 & 1 & 1 & 1 & 1 & 1 & 1 \\ 1 & 1 & 1 & 1 & 1 & 1 & 0 & 1 & 0 & 1 & 0 \\ 1 & 1 & 1 & 1 & 1 & 1 & 0 & 1 & 0 & 1 & 0 & 1 \end{bmatrix}, \tag{14.17}$$

$$\mathbf{Z}(4,16) = \begin{bmatrix} 1 & 1 & 1 & 0 & 1 & 1 & 1 & 0 & 1 & 1 & 1 & 0 & 1 & 1 & 1 & 0 \\ 0 & 1 & 1 & 1 & 0 & 1 & 1 & 1 & 0 & 1 & 1 & 1 & 0 & 1 & 1 & 1 \\ 1 & 0 & 1 & 1 & 1 & 0 & 1 & 1 & 1 & 0 & 1 & 1 & 1 & 0 & 1 & 1 \\ 1 & 1 & 0 & 1 & 1 & 1 & 0 & 1 & 1 & 1 & 0 & 1 & 1 & 1 & 0 & 1 \end{bmatrix}. \tag{14.18}$$

The above masking matrices are designed by using the masking-matrix design methods presented in Section 14.4. Masking $\mathbf{B}_{\mathrm{RS},255,\Lambda_{12}}(4,12)$ and $\mathbf{B}_{\mathrm{RS},255,\Lambda_{16}}(4,16)$ results in two masked RS matrices $\mathbf{B}_{\mathrm{RS},255,\Lambda_{12},\mathrm{mask}}(4,12)$ and $\mathbf{B}_{\mathrm{RS},255,\Lambda_{16},\mathrm{mask}}(4,16)$, respectively. The null spaces over $\mathrm{GF}(2)$ of the 255-fold CPM-dispersions of the two masked RS matrices give a $(3,9)$-regular $(3060, 2040)$ masked QC-RS-LDPC code and a $(3,12)$-regular $(4080, 3060)$ masked QC-RS-LDPC code with rates $2/3$ and $3/4$, respectively. The Tanner graphs of both the masked codes have girth 8. The distributions of short cycles of lengths 8, 10, and 12 in the Tanner graphs of the two masked QC-RS-LDPC codes are {7905, 105 825, 1 444 320} and {32 640, 495 210, 9 570 915}, respectively. We see that masking reduces the short cycles of the Tanner graphs of the two unmasked QC-RS-LDPC codes drastically.

Consider the first unmasked $(3060, 2049)$ QC-RS-LDPC code. Its Tanner graph has a total of 120 314 355 cycles of lengths 8, 10, and 12. However, the Tanner graph of its corresponding masked $(3060, 2040)$ QC-RS-LDPC code has a total of 1 558 050 cycles of length 8, 10, and 12. We see that masking reduces the total number of short cycles of length 8, 10, and 12 by a factor of 77, a large reduction. Consider the second unmasked $(4080, 3065)$ QC-RS-LDPC code. Its Tanner graph contains a total of 721 465 380 cycles of length 8, 10, and 12. However, the Tanner graph of the masked $(4080, 3060)$ QC-RS-LDPC code contains a total of 10 098 765 cycles of lengths 8, 10, and 12. In this case, masking reduces the total number of short cycles of lengths 8, 10, and 12 of the unmasked Tanner graph by a factor of more than 71, again a large reduction.

The BER and BLER performances of the above four codes over the AWGN channel decoded with 50 iterations of the MSA are shown in Fig. 14.20. The MSA scaled factor for both unmasked codes is 0.70 and the one for both masked codes is 0.75. We see that masking improves the error performances. ▲▲

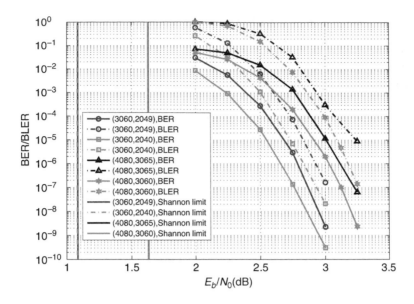

Figure 14.20 The BER and BLER performances of the four QC-RS-LDPC codes given in Example 14.20.

## 14.10 Optimal Codes for Correcting Two Random CPM-Phased Erasure-Bursts

Consider a QC-LDPC code $C_{qc}$ given by the null space of a $d \times m$ array $\mathbf{H}_{qc}$ of CPMs and/or ZMs of size $n \times n$. A codeword $\mathbf{v} = (\mathbf{v}_0, \mathbf{v}_1, \ldots, \mathbf{v}_{m-1})$ in $C_{qc}$ consists of $m$ sections, each containing $n$ code symbols, which correspond to the $n$ columns of a CPM-column-block in $\mathbf{H}_{qc}$. The sections of $\mathbf{v}$ are called *CPM-sections*. Suppose $\mathbf{v}$ is transmitted over a *binary burst erasure channel* (BBEC). An erasure-burst with erasures confined to a CPM-section of $\mathbf{v}$ is called a *CPM-phased erasure-burst*. The longest length of such an erasure-burst is $n$. If all the $n$ code symbols in a CPM-section of $\mathbf{v}$ are erased, the erasure-burst is said to be *solid*; otherwise, it is not solid, i.e., *at least one* of the code symbols in the CPM-section is not erased.

In the following, we present a class of RS-based QC-LDPC codes which are capable of correcting *two random CPM-phased erasure-bursts* over the BBEC. The erasure-burst correction of the codes in this class is characterized by the *local cycle structure* of their Tanner graphs.

Consider the field GF($q$) with $\alpha$ as a primitive element. Let $n$ be a prime factor of $q - 1$ such that $q - 1 = cn$. Let $\beta = \alpha^c$. Then, the order of $\beta$ is $n$ and the set $\mathbf{S}_n = \{1, \beta, \beta^2, \ldots, \beta^{n-1}\}$ forms a cyclic subgroup of GF($q$). Form the following $2 \times 2$ SM-constrained RS matrix of size $2 \times n$ over GF($q$):

$$\mathbf{B}_{\mathrm{RS},n}(2,n) = \left[ \begin{array}{ccccc} 1 & \beta & \beta^2 & \ldots & \beta^{n-1} \\ 1 & \beta^2 & (\beta^2)^2 & \ldots & (\beta^{n-1})^2 \end{array} \right]. \tag{14.19}$$

Label the two rows in $\mathbf{B}_{\mathrm{RS},n}(2,n)$ with 0 and 1 and its $n$ columns from 0 to $n-1$. Because $n$ is a prime, the $n$ entries of each row of $\mathbf{B}_{\mathrm{RS},n}(2,n)$ are distinct which are the $n$ elements in $\mathbf{S}_n$. Except for the zeroth column, the two entries in each column of $\mathbf{B}_{\mathrm{RS},n}(2,n)$ are different. The null space over $\mathrm{GF}(q)$ of $\mathbf{B}_{\mathrm{RS},n}(2,n)$ gives a $q$-ary $(n, n-2)$ RS code of length $n$ with minimum distance 3.

The $n$-fold CPM-dispersion of $\mathbf{B}_{\mathrm{RS},n}(2,n)$ gives a $2 \times n$ array $\mathbf{H}_{\mathrm{RS},n}(2,n)$ of CPMs of size $n \times n$ which is a $2n \times n^2$ matrix with column and row weights 2 and $n$, respectively. Because the modulo-2 sum of all the $2n$ rows of $\mathbf{H}_{\mathrm{RS},n}(2,n)$ gives a zero $n^2$-tuple and no $2n-1$ or fewer rows of $\mathbf{H}_{\mathrm{RS},n}(2,n)$ can sum up to a zero $n^2$-tuple, the rank of $\mathbf{H}_{\mathrm{RS},n}(2,n)$ is $2n-1$. The null space of $\mathbf{H}_{\mathrm{RS},n}(2,n)$ over $\mathrm{GF}(2)$ gives a $(2,n)$-regular $(n^2, n^2-2n+1)$ QC-RS-LDPC code $C_{\mathrm{RS},n}(2,n)$ of length $n^2$ and rate $R = 1 - (2n-1)/n^2$.

The Tanner graph $\mathcal{G}_{\mathrm{RS},n}(2,n)$ associated with $\mathbf{H}_{\mathrm{RS},n}(2,n)$ consists of $n^2$ VNs and $2n$ CNs. The $n$ VNs corresponding to the $n$ columns of a CPM-column-block of $\mathbf{H}_{\mathrm{RS},n}(2,n)$ are not connected to each other. The $n$ CNs corresponding to the $n$ rows of a CPM-row-block of $\mathbf{H}_{\mathrm{RS},n}(2,n)$ check $n$ *disjoint sets* of VNs, each consisting of $n$ VNs.

For $0 \le r < t < n$, consider the following $2 \times 2$ subarray of $\mathbf{H}_{\mathrm{RS},n}(2,n)$:

$$\mathbf{H}_{\mathrm{sub}}(r,t) = \left[ \begin{array}{cc} \mathrm{CPM}(\beta^r) & \mathrm{CPM}(\beta^t) \\ \mathrm{CPM}(\beta^{2r}) & \mathrm{CPM}(\beta^{2t}) \end{array} \right]. \tag{14.20}$$

Let $\mathcal{G}_{\mathrm{sub}}(r,t)$ be the subgraph of $\mathcal{G}_{\mathrm{RS},n}(2,n)$ associated with the subarray $\mathbf{H}_{\mathrm{sub}}(r,t)$ of $\mathbf{H}_{\mathrm{RS},n}(2,n)$. Then, the subgraph $\mathcal{G}_{\mathrm{sub}}(r,t)$ is composed of *a single cycle of length $4n$*, i.e., the $2n$ VNs and $2n$ CNs of $\mathcal{G}_{\mathrm{sub}}(r,t)$ are on a single cycle of length $4n$. This *local cycle structure* of the Tanner graph $\mathcal{G}_{\mathrm{RS},n}(2,n)$ associated with the array $\mathbf{H}_{\mathrm{RS},n}(2,n)$ is referred to as the $2 \times 2$ *CPM-array cycle structure*. The proof of this local cycle structure is given in Appendix D.

Each codeword $\mathbf{v} = (\mathbf{v}_0, \mathbf{v}_1, \ldots, \mathbf{v}_{n-1})$ in $C_{\mathrm{RS},n}(2,n)$ consists of $n$ CPM-sections of length $n$. Two code symbols in $\mathbf{v}$ are said to be *neighbors* if they correspond to two VNs in the Tanner graph $\mathcal{G}_{\mathrm{RS},n}(2,n)$ which are connected by a path of length 2. It follows from the $2 \times 2$ CPM-array cycle structure of $\mathcal{G}_{\mathrm{RS},n}(2,n)$ of the code $C_{\mathrm{RS},n}(2,n)$ that each code symbol in a CPM-section $\mathbf{v}_i$ has *exactly two neighbors* in a different CPM-section $\mathbf{v}_j$ of $\mathbf{v}$. This neighbor structure is referred to as *adjacency structure* of the code symbols of a codeword in $C_{\mathrm{RS},n}(2,n)$.

Based on the local cycle structure of its Tanner graph, the code $C_{\mathrm{RS},n}(2,n)$ is capable of correcting any two *random* CPM-phased erasure-bursts, one is solid and the other one is not solid (referred to as *two mutually semi-solid CPM-phased erasure-bursts*). The maximum number of erasures which can be recovered is $2n-1$ and it is equal to the number of parity-check symbols of the code $C_{\mathrm{RS},n}(2,n)$.

Suppose, during the transmission of a codeword $\mathbf{v} = (\mathbf{v}_0, \mathbf{v}_1, \ldots, \mathbf{v}_{n-1})$, two CPM-phased erasure-bursts occur, one in the $i$th section $\mathbf{v}_i$ and the other one in the $j$th section $\mathbf{v}_j$, with one CPM-phased erasure-burst not solid. It follows from the adjacency structure of $C_{\mathrm{RS},n}(2,n)$ that there exists at least one erased symbol $v$ in the $j$th (or $i$th) CPM-section which is the neighbor of an *unerased* code symbol in the $i$th (or $j$th) CPM-section of $\mathbf{v}$. In this case, there is a row $\mathbf{h}$ in $\mathbf{H}_{\mathrm{RS},n}(2,n)$ which checks $v$ as the *only* erased symbol. Take the inner product of $\mathbf{h}$ and $\mathbf{v}$ to form a check sum $\sum$. The check sum $\sum$ contains $v$ as the only *unknown*. From $\sum$, we recover $v$ by taking the modulo-2 sum of the unerased symbols in $\sum$. After recovering the erased symbol $v$, we find another erased code symbol which is the neighbor of an unerased code symbol or the newly recovered code symbol $v$ in the two CPM-phased erasure-bursts confined in the two CPM-sections, $\mathbf{v}_i$ and $\mathbf{v}_j$ of $\mathbf{v}$. We recover this erased code symbol in exactly the same manner as the recovery of the first erased code symbol $v$. Continue this recovery process until all the erased code symbols in the CPM-sections $\mathbf{v}_i$ and $\mathbf{v}_j$ of $\mathbf{v}$ are recovered. Because all the erased code symbols in $\mathbf{v}_i$ and $\mathbf{v}_j$ correspond to the $2n$ VNs on the single cycle of length $4n$ in $\mathcal{G}_{\mathrm{sub}}(i,j)$, there is always an unerased or recovered code symbol which is the neighbor of an erased code symbol. Hence, every erased code symbol in the CPM-sections $\mathbf{v}_i$ and $\mathbf{v}_j$ of $\mathbf{v}$ can be recovered. The erasure recovery process will never fail as long as there is at least one unerased code symbol in two CPM-phased erasure-bursts which are confined to two CPM-sections of a transmitted codeword $\mathbf{v}$ in $C_{\mathrm{RS},n}(2,n)$. The above erasure recovery process is simply the *successive peeling* (SPL) algorithm presented in Section 10.8.

The maximum number of recoverable erasures confined to two CPM-sections of a received vector in $C_{\mathrm{RS},n}(2,n)$ is $2n-1$, which is equal to the number of parity-check symbols of the code. Hence, the code $C_{\mathrm{RS},n}(2,n)$ is optimal, i.e., its erasure-correcting efficiency is $\eta = 1$.

Let $m$ be a positive integer such that $2 \leq m \leq n$ and $\mathbf{B}_{\mathrm{RS},n}(2,m)$ be a submatrix of $\mathbf{B}_{\mathrm{RS},n}(2,n)$. Clearly, the Tanner graph of the QC-RS-LDPC code $C_{\mathrm{RS},n}(2,m)$ generated by the $n$-fold CPM-dispersion of $\mathbf{B}_{\mathrm{RS},n}(2,m)$ also has the $2 \times 2$ CPM-array cycle structure. Hence, it is capable of correcting any two mutually semi-solid CPM-phased erasure-bursts and is also optimal.

**Example 14.21** Consider the field $\mathrm{GF}(2^9)$ with $\alpha$ as a primitive element. Note that $2^9 - 1 = 511$ can be factored as the product of two primes 7 and 73. Set $n = 73$ and let $\beta = \alpha^7$. Then, $\beta$ is an element in $\mathrm{GF}(2^9)$ of order 73. The set $\mathbf{S}_{73} = \{1, \beta, \beta^2, \ldots, \beta^{72}\}$ forms a cyclic subgroup of $\mathrm{GF}(2^9)$. Construct a $2 \times 73$ RS matrix $\mathbf{B}_{\mathrm{RS},n}(2,73)$ with elements from $\mathbf{S}_{73}$ in the form of (14.19). Dispersing each entry in $\mathbf{B}_{\mathrm{RS},73}(2,73)$ into a $73 \times 73$ CPM, we obtain a $2 \times 73$ array $\mathbf{H}_{\mathrm{RS},73}(2,73)$ of CPMs of size $73 \times 73$ which is a $146 \times 5329$ matrix with rank 145.

The null space over $\mathrm{GF}(2)$ of $\mathbf{H}_{\mathrm{RS},73}(2,73)$ gives a $(5329, 5184)$ QC-RS-LDPC code $C_{\mathrm{RS},73}(2,73)$ with rate 0.9737. Each codeword in $C_{\mathrm{RS},73}(2,73)$ consists of 73 CPM-sections, each of 73 code symbols. The code $C_{\mathrm{RS},73}(2,73)$ is capable of correcting any two random CPM-phased erasure-bursts, one not solid, with a maximum number of 145 recoverable erasures which is equal to the number of parity-check symbols of $C_{RS,73}(2,73)$. Hence, the code $C_{RS,73}(2,73)$ is optimal in correction of two random mutually semi-solid CPM-phased erasure-bursts.

Suppose we set $m = 10$. Take any 10 columns from $\mathbf{B}_{RS,73}(2,73)$ and form a $2 \times 10$ matrix $\mathbf{B}_{RS,73}(2,10)$. The 73-fold CPM-dispersion of $\mathbf{B}_{RS,73}(2,10)$ gives a $2 \times 10$ array $\mathbf{H}_{RS,73}(2,10)$ which is a $146 \times 730$ matrix with rank 145. The null space over $GF(2)$ of $\mathbf{H}_{RS,73}(2,10)$ gives a $(730, 585)$ QC-RS-LDPC code $C_{RS,73}(2,10)$ with rate 0.8013 which is capable of recovering two mutually semi-solid CPM-phased erasure-bursts with maximum 145 erasures. ▲▲

Using interleaving, we can construct long powerful codes for correcting long erasure-bursts using $C_{RS,n}(2,m)$ as the base code. For example, if we interleave the second code $C_{RS,73}(2,10)$ given in Example 14.21 with degree 100, the interleaved code is a $(73\,000, 58\,500)$ code which is capable of correcting a single erasure-burst with maximum 14\,500 erasures and many other patterns of erasure-bursts as long as, after deinterleaving, each row of the received code array contains erasures confined in two mutually semi-solid CPM-sections.

Let $d$ be a positive integer with $2 < d < n$. Suppose we form the following $2 \times 2$ SM-constrained $d \times n$ matrix $\mathbf{B}_{RS,n}(d,n)$ over $\mathbf{S}_n = \{1, \beta, \beta^2, \ldots, \beta^{n-1}\}$:

$$\mathbf{B}_{RS,n}(d,n) = \begin{bmatrix} 1 & \beta & \beta^2 & \cdots & \beta^{n-1} \\ 1 & \beta^2 & (\beta^2)^2 & \cdots & (\beta^2)^{n-1} \\ \vdots & \vdots & \ddots & \vdots & \vdots \\ 1 & \beta^d & (\beta^d)^2 & \cdots & (\beta^d)^{n-1} \end{bmatrix}. \tag{14.21}$$

The matrix $\mathbf{B}_{RS,n}(d,n)$ is the conventional parity-check matrix of an $(n, n-d)$ RS code over $GF(2^s)$ with minimum distance $d+1$. The $n$-fold CPM-dispersion of $\mathbf{B}_{RS,n}(d,n)$ is a $d \times n$ array $\mathbf{H}_{RS,n}(d,n)$ of CPMs of size $n \times n$ which is a $dn \times n^2$ matrix over $GF(2)$ with column and row weights $d$ and $n$, respectively. Following the structural properties developed in Section 14.6, any two CPM-row-blocks have the same structure as the top two CPM-row-blocks of $\mathbf{H}_{RS,n}(d,n)$. Hence, the Tanner graph associated with any two CPM-row-blocks of $\mathbf{H}_{RS,n}(d,n)$ has the same $2 \times 2$ CPM-array cycle structure as the Tanner graph associated with the top two CPM-row-blocks of $\mathbf{H}_{RS,n}(d,n)$. Consequently, the Tanner graph $\mathcal{G}_{RS,n}(d,n)$ associated with the entire array $\mathbf{H}_{RS,n}(d,n)$ has the $2 \times 2$ CPM-array cycle structure, i.e., any subgraph of $\mathcal{G}_{RS,n}(d,n)$ associated with a $2 \times 2$ subarray $\mathbf{H}_{sub}(r,t)$ of $\mathbf{H}_{RS,n}(d,n)$ is composed of a single cycle of length $4n$. Hence, the QC-RS-LDPC code $C_{RS,n}(d,n)$ specified by $\mathbf{H}_{RS,n}(d,n)$ can be applied to either an AWGN channel for correcting random errors or a BBEC for correcting random CPM-phased erasure-bursts.

**Example 14.22** Suppose we use the cyclic subgroup $\mathbf{S}_{73}$ of $GF(2^9)$ constructed in Example 14.21 for code construction. Setting $n = 73$ and $d = 4$, we construct a $4 \times 73$ RS base matrix $\mathbf{B}_{RS,73}(4,73)$ in the form of (14.21) using the 4 elements $\beta, \beta^2, \beta^3$, and $\beta^4$ from $\mathbf{S}_{73}$. Deleting the last nine columns from $\mathbf{B}_{RS,73}(4,73)$, we obtain a $4 \times 64$ RS matrix $\mathbf{B}_{\mathrm{RS},73}(4,64)$. The 73-fold CPM-dispersion of $\mathbf{B}_{\mathrm{RS},73}(4,64)$ results in a $4 \times 64$ array $\mathbf{H}_{\mathrm{RS},73}(4,64)$ of CPMs of size $73 \times 73$ which is a $292 \times 4672$ matrix of rank 289 with column and row weights 4 and 64, respectively.

The null space over GF(2) of $\mathbf{H}_{\mathrm{RS},73}(4,64)$ gives a $(4,64)$-regular $(4672, 4383)$ QC-RS-LDPC code $C_{\mathrm{RS},73}(4,64)$ of rate 0.9381. Applying this code to the BBEC, we can use any two CPM-row-blocks of $\mathbf{H}_{\mathrm{RS},73}(4,64)$ to correct up to two random mutually semi-solid CPM-phased erasure-bursts (a maximum of 145 erasures).

If we apply the constructed code to the AWGN channel, the BER and BLER performances of the code $C_{\mathrm{RS},73}(4,64)$ decoded with 5, 10, and 50 iterations of the MSA scaled by a factor of 0.75 are shown in Fig. 14.21(a). With 10 decoding iterations, the code achieves a BER of $10^{-10}$ at SNR $= 5.6$ dB without a visible error-floor. At a BER of $10^{-10}$, the code performs 1.8 dB away from the Shannon limit. The code also performs well over the BEC as shown in Fig. 14.21(b). ▲▲

## 14.11 Globally Coupled LDPC Codes

Consider the base matrix $\mathbf{B}_{\mathrm{p}}(q-1,q-1)$ given by (14.3). Suppose we set $\eta = 1$ and permute the rows of $\mathbf{B}_{\mathrm{p}}(q-1,q-1)$. We obtain a $(q-1) \times (q-1)$ circulant in the following form:

$$\mathbf{B}_{\mathrm{p}}(q-1,q-1) = \begin{bmatrix} \alpha^0 - 1 & \alpha - 1 & \alpha^2 - 1 & \cdots & \alpha^{q-3} - 1 & \alpha^{q-2} - 1 \\ \alpha^{q-2} - 1 & \alpha^0 - 1 & \alpha - 1 & \cdots & \alpha^{q-4} - 1 & \alpha^{q-3} - 1 \\ \vdots & \vdots & \vdots & \ddots & \vdots & \vdots \\ \alpha - 1 & \alpha^2 - 1 & \alpha^3 - 1 & \cdots & \alpha^{q-2} - 1 & \alpha^0 - 1 \end{bmatrix}. \quad (14.22)$$

From (14.22), we find that the matrix $\mathbf{B}_{\mathrm{p}}(q-1,q-1)$ has the following structural properties: (1) each row is the cyclic-shift of the row above it one place to the right and the top row is the cyclic-shift of the last row one place to the right; (2) each column is the cyclic-shift of the column on it left downward one place and the leftmost column is the cyclic-shift of the rightmost column downward one place; (3) it contains $q-1$ zero entries lying on its main diagonal; (4) all the entries in a row (column) are different; and (5) it satisfies the $2 \times 2$ SM-constraint.

Suppose $q-1$ can be factored as the product of two positive integers $l$ and $r$, i.e., $lr = q-1$. Then, $\mathbf{B}_{\mathrm{p}}(q-1,q-1)$ can be partitioned into as an $r \times r$ array of $l \times l$ submatrices in the following form with a block-cyclic structure:

$$\mathbf{B}_{\mathrm{p}}(q-1,q-1) = \begin{bmatrix} \mathbf{W}_{0,0} & \mathbf{W}_{0,1} & \cdots & \mathbf{W}_{0,r-1} \\ \mathbf{W}_{0,r-1} & \mathbf{W}_{0,0} & \cdots & \mathbf{W}_{0,r-2} \\ \vdots & \vdots & \ddots & \vdots \\ \mathbf{W}_{0,1} & \mathbf{W}_{0,2} & \cdots & \mathbf{W}_{0,0} \end{bmatrix}. \quad (14.23)$$

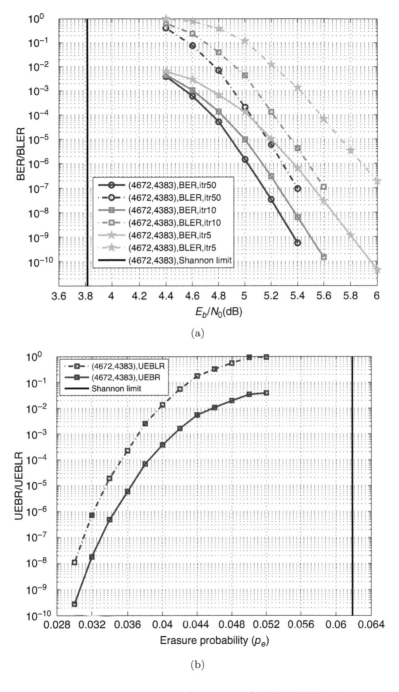

Figure 14.21 The performances of the (4672, 4383) QC-RS-LDPC code in Example 14.22 over: (a) AWGN channel and (b) BEC.

For $0 \leq j < r$, $1 \leq n \leq l$, and $1 \leq m < l$, we take an $m \times n$ submatrix $\mathbf{R}_{0,j}$ from $\mathbf{W}_{0,j}$. The submatrices $\mathbf{R}_{0,0}, \mathbf{R}_{0,1}, \ldots, \mathbf{R}_{0,r-1}$ are taken from $\mathbf{W}_{0,0}, \mathbf{W}_{0,1}, \ldots, \mathbf{W}_{0,r-1}$ under the following *location restriction*: for $j' \neq j$, the locations of the entries of $\mathbf{R}_{0,j}$ in $\mathbf{W}_{0,j}$ are *identical to* the locations of the entries of $\mathbf{R}_{0,j'}$ in $\mathbf{W}_{0,j'}$. Next, using these $r$ submatrices $\mathbf{R}_{0,0}, \mathbf{R}_{0,1}, \ldots, \mathbf{R}_{0,r-1}$, we form the following $r \times r$ array $\mathbf{R}(m, n)$ of $m \times n$ submatrices over GF($q$):

$$
\mathbf{R}(m, n) = \begin{bmatrix} \mathbf{R}_{0,0} & \mathbf{R}_{0,1} & \cdots & \mathbf{R}_{0,r-1} \\ \mathbf{R}_{0,r-1} & \mathbf{R}_{0,0} & \cdots & \mathbf{R}_{0,r-2} \\ \vdots & \vdots & \ddots & \vdots \\ \mathbf{R}_{0,1} & \mathbf{R}_{0,2} & \cdots & \mathbf{R}_{0,0} \end{bmatrix}, \tag{14.24}
$$

which is a submatrix of $\mathbf{B}_{\mathrm{p}}(q-1, q-1)$ with block-cyclic structure and satisfies the $2 \times 2$ SM-constraint.

In forming the $r \times r$ block-cyclic array $\mathbf{R}(m, n)$ from the $r \times r$ array $\mathbf{B}_{\mathrm{p}}(q-1, q-1)$ of $l \times l$ submatrices over GF($q$) given by (14.23), there are $l-m$ rows in each row-block and $l-n$ columns in each column-block of $\mathbf{B}_{\mathrm{p}}(q-1, q-1)$ which are unused. So, a total of $r(l-m)$ rows of $\mathbf{B}_{\mathrm{p}}(q-1, q-1)$ are not used in forming the array $\mathbf{R}(m, n)$. We denote the set of $r(l-m)$ unused rows of $\mathbf{B}_{\mathrm{p}}(q-1, q-1)$ by $\Pi$. For each row $\mathbf{w}$ in $\Pi$, we remove the components at the locations which correspond to the columns that are not used in forming the array $\mathbf{R}(m, n)$ from $\mathbf{B}_{\mathrm{p}}(q-1, q-1)$. This results in a shortened row $\mathbf{w}^* = (\mathbf{w}_{0,0}, \mathbf{w}_{0,1}, \ldots, \mathbf{w}_{0,r-1})$ which consists of $r$ sections, each section of $n$ components. The locations of the $n$ components of the $i$th section $\mathbf{w}_{0,i}$ of $\mathbf{w}^*$ correspond to the locations of the $n$ columns of the submatrix $\mathbf{R}_{0,i}$ of the submatrix $\mathbf{W}_{0,i}$ in the array $\mathbf{B}_{\mathrm{p}}(q-1, q-1)$.

Let $\Pi^*$ denote the set of $r(l-m)$ shortened versions of the rows in $\Pi$. The set $\Pi^*$ of rows and the sets of rows in the row-blocks of $\mathbf{R}(m, n)$ are *disjoint*. Let $s$ and $t$ be two positive integers with $1 \leq s \leq r(l-m)$ and $1 \leq t \leq r$. Take $s$ rows from $\Pi^*$ and remove the last $r-t$ sections from each of the $s$ chosen rows. With these $s$ shortened rows, we form an $s \times nt$ matrix $\mathbf{X}_{\mathrm{gc,cn}}(s, t)$ over GF($q$). Next, we form the following array over GF($q$):

$$
\mathbf{R}_{\mathrm{gc,cn}}(m, n, s, t) = \begin{bmatrix} \mathbf{R}_{0,0} & & & \\ & \mathbf{R}_{0,0} & & \\ & & \ddots & \\ & & & \mathbf{R}_{0,0} \\ \hdashline & & \mathbf{X}_{\mathrm{gc,cn}}(s, t) & \end{bmatrix}. \tag{14.25}
$$

The matrix $\mathbf{R}_{\mathrm{gc,cn}}(m, n, s, t)$ consists of two submatrices. The top submatrix of $\mathbf{R}_{\mathrm{gc,cn}}(m, n, s, t)$ is a $t \times t$ diagonal array, denoted by $\mathrm{diag}(\mathbf{R}_{0,0}, \mathbf{R}_{0,0}, \ldots, \mathbf{R}_{0,0})$, with $t$ copies of $\mathbf{R}_{0,0}$ lying on its main diagonal and zero matrices elsewhere. The lower submatrix $\mathbf{X}_{\mathrm{gc,cn}}(s, t)$ of $\mathbf{R}_{\mathrm{gc,cn}}(m, n, s, t)$ is an $s \times nt$ matrix over GF($q$). The matrix $\mathbf{R}_{\mathrm{gc,cn}}(m, n, s, t)$ is an $(mt + s) \times nt$ matrix over GF($q$) which is a submatrix of $\mathbf{B}_{\mathrm{p}}(q-1, q-1)$ and satisfies the $2 \times 2$ SM-constraint.

Disperse each nonzero entry in $\mathbf{R}_{\mathrm{gc,cn}}(m, n, s, t)$ into a CPM of size $(q-1) \times (q-1)$ and each 0-entry into a ZM of size $(q-1) \times (q-1)$. Let $\mathrm{CPM}(\mathbf{R}_{0,0})$ and $\mathrm{CPM}(\mathbf{X}_{\mathrm{gc,cn}}(s, t))$ denote the CPM-dispersions of $\mathbf{R}_{0,0}$ and $\mathbf{X}_{\mathrm{gc,cn}}(s, t)$, respectively. Then, the CPM-dispersion of $\mathbf{R}_{\mathrm{gc,cn}}(m, n, s, t)$ gives the following $(mt+s) \times nt$ array $\mathbf{H}_{\mathrm{gc,cn,qc}}(q-1, q-1)$ of CPMs and ZMs of size $(q-1) \times (q-1)$:

$$\mathbf{H}_{\mathrm{gc,cn,qc}}(q-1, q-1) =$$

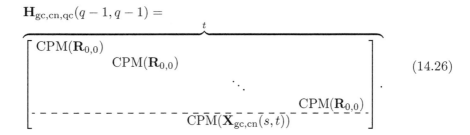

$$(14.26)$$

From (14.26), we readily see that the Tanner graph $\mathcal{G}_{\mathrm{gc,cn,qc}}(q-1, q-1)$ associated with the matrix $\mathbf{H}_{\mathrm{gc,cn,qc}}(q-1, q-1)$ is composed of $t$ *disjoint local Tanner graphs*, each consisting of $n(q-1)$ *local* VNs and $m(q-1)$ *local* CNs, which are connected together by a group of $s(q-1)$ *global* CNs which correspond to the $s(q-1)$ rows of the matrix $\mathrm{CPM}(\mathbf{X}_{\mathrm{gc,cn}}(s, t))$. The Tanner graph $\mathcal{G}_{\mathrm{gc,cn,qc}}(q-1, q-1)$ is called a *CN-based globally coupled* (GC) *Tanner graph*. The subscript "gc, cn, qc" in $\mathcal{G}_{\mathrm{gc,cn,qc}}(q-1, q-1)$ stands for "CN-based globally coupled quasi-cyclic."

The null space of $\mathbf{H}_{\mathrm{gc,cn,qc}}(q-1, q-1)$ gives a QC-LDPC code of length $nt(q-1)$, denoted by $C_{\mathrm{gc,cn,qc}}$, is called a *CN-based globally coupled QC-LDPC code*, simply CN-GC-QC-LDPC code. Each copy of $\mathrm{CPM}(\mathbf{R}_{0,0})$ in $\mathbf{H}_{\mathrm{gc,cn,qc}}(q-1, q-1)$ is called a *local parity-check matrix* and the matrix $\mathrm{CPM}(\mathbf{X}_{\mathrm{gc,cn}}(s, t))$ is called the *global coupling matrix*. The matrix $\mathbf{H}_{\mathrm{gc,cn,qc}}(q-1, q-1)$ is called the *global parity-check matrix*. The null space of each local parity-check matrix gives a QC-LDPC code of length $n(q-1)$, called a *local code*. The Tanner graph of each local code is called a *local graph*. The Tanner graph associated with the global coupling matrix $\mathrm{CPM}(\mathbf{X}_{\mathrm{gc,cn}}(s, t))$ is called the *global coupling graph*. Because $\mathbf{H}_{\mathrm{gc,cn,qc}}(q-1, q-1)$ satisfies the RC-constraint, its associated global graph $\mathcal{G}_{\mathrm{gc,cn,qc}}(q-1, q-1)$ has at least 6.

From (14.26), we see that each codeword $\mathbf{v} = (\mathbf{v}_0, \mathbf{v}_1, \ldots, \mathbf{v}_{t-1})$ in $C_{\mathrm{gc,cn,qc}}$ consists of $t$ local codewords, each from a separate local code. These $t$ local codewords $\mathbf{v}_0, \mathbf{v}_1, \ldots, \mathbf{v}_{t-1}$ in sequence must satisfy the $s(q-1)$ parity-check constraints specified by the rows of $\mathrm{CPM}(\mathbf{X}_{\mathrm{gc,cn}}(s, t))$, i.e., $\mathbf{v} \cdot \mathrm{CPM}(\mathbf{X}_{\mathrm{gc,cn}}(s, t))^T = \mathbf{0}$.

For different choices of parameters, $r$, $l$, $m$, $n$, $s$, and $t$, we can construct a family of CN-GC-QC-LDPC codes from a given field GF($q$) with various lengths and rates.

**Example 14.23** Let GF(127) be the field for code construction. Based on this field, we construct a $126 \times 126$ base matrix $\mathbf{B}_{\mathrm{p}}(126, 126)$ over GF(127) in the form of (14.22). Factor 126 as the product of 3 and 42 and set $r = 3$ and $l = 42$.

Next, we partition $\mathbf{B}_p(126, 126)$ into a $3 \times 3$ array of $42 \times 42$ submatrices in the form of (14.23), whose first row-block consists of three submatrices $\mathbf{W}_{0,0}$, $\mathbf{W}_{0,1}$, $\mathbf{W}_{0,2}$ of size $42 \times 42$.

Set $m = 2$, $n = 42$, $s = 2$, and $t = 3$. Taking three $2 \times 42$ submatrices $\mathbf{R}_{0,0}$, $\mathbf{R}_{0,1}$, $\mathbf{R}_{0,2}$ from the submatrices $\mathbf{W}_{0,0}$, $\mathbf{W}_{0,1}$, $\mathbf{W}_{0,2}$, respectively, under the location-constraint as described above, we obtain a $3 \times 3$ array $\mathbf{R}(2, 42)$ of $2 \times 42$ submatrices over GF(127) in the form of (14.24). The matrices $\mathbf{R}_{0,0}$, $\mathbf{R}_{0,1}$, $\mathbf{R}_{0,2}$ are taken arbitrarily from $\mathbf{W}_{0,0}$, $\mathbf{W}_{0,1}$, $\mathbf{W}_{0,2}$ under the location constraint, respectively. The submatrix $\mathbf{R}_{0,0}$ is a $2 \times 42$ matrix over GF(127). The set $\Pi$ consists of $r(l - m) = 3 \times (42 - 2) = 120$ rows of length 126 of the matrix $\mathbf{B}_p(126, 126)$. The set of rows in $\Pi$ are disjoint with the rows in $\mathbf{R}_{0,0}$, $\mathbf{R}_{0,1}$, $\mathbf{R}_{0,2}$. Take two rows from $\Pi$ to form a $2 \times 126$ matrix $\mathbf{X}_{gc,cn}(2, 3)$. Next, we form the matrix $\mathbf{R}_{gc,cn}(2, 42, 2, 3)$ in the form of (14.25) which is an $8 \times 126$ matrix over GF(127). The CPM-dispersion of $\mathbf{R}_{gc,cn}(2, 42, 2, 3)$ gives an $8 \times 126$ array $\mathbf{H}_{gc,cn,qc}(126, 126)$ of CPMs and ZMs of size $126 \times 126$, which is a $1008 \times 15\,876$ matrix over GF(2). The null space over GF(2) of $\mathbf{H}_{gc,cn,qc}(126, 126)$ gives a $(15\,876, 14\,871)$ CN-GC-QC-LDPC code $C_{gc,cn,qc}$ with rate 0.9367.

The BER and BLER error performances of $C_{gc,cn,qc}$ over the AWGN channel with BPSK signaling decoded with 50 iterations of the MSA scaled by a factor of 0.75 are shown in Fig. 14.22(a). From Fig. 14.22(a), we see that, decoded with 50 iterations of the scaled MSA, the bit-error probability of the code drops all the way down to a BER of $10^{-10}$ without a visible error-floor and the code performs within 1.2 dB from the Shannon limit.

The code also performs well over the BEC. The UEBR and UEBLR performances of the code are shown in Fig. 14.22(b). ▲▲

**Example 14.24** In this example, we use the same field GF(127) in Example 14.23 for code construction. Factor $126 = 6 \times 21$ and set $r = 6$ and $l = 21$. Partition the matrix $\mathbf{B}_p(126, 126)$ into a $6 \times 6$ array of $21 \times 21$ submatrices, whose first row-block consists of six $21 \times 21$ submatrices $\mathbf{W}_{0,0}$, $\mathbf{W}_{0,1}, \ldots, \mathbf{W}_{0,5}$.

Set $m = 3$, $n = 21$, $t = 6$, and $s = 1$. Taking six $3 \times 21$ submatrices, $\mathbf{R}_{0,0}$, $\mathbf{R}_{0,1}, \ldots, \mathbf{R}_{0,5}$, from the submatrices $\mathbf{W}_{0,0}$, $\mathbf{W}_{0,1}, \ldots, \mathbf{W}_{0,5}$, respectively, we obtain a $6 \times 6$ array $\mathbf{R}(3, 21)$ of $3 \times 21$ submatrices over GF(127) in the form of (14.24). The matrices $\mathbf{R}_{0,0}$, $\mathbf{R}_{0,1}, \ldots, \mathbf{R}_{0,5}$ are taken from $\mathbf{W}_{0,0}$, $\mathbf{W}_{0,1}, \ldots, \mathbf{W}_{0,5}$ under the location constraint, respectively. The submatrix $\mathbf{R}_{0,0}$ is a $3 \times 21$ matrix over GF(127). The set $\Pi$ consists of $r(l - m) = 6 \times (21 - 3) = 108$ rows of length 126 in the matrix $\mathbf{B}_p(126, 126)$. Then, the top submatrix of the base matrix $\mathbf{R}_{gc,cn}(3, 21, 1, 6)$ in the form of (14.25) is a $6 \times 6$ array with six copies of $\mathbf{R}_{0,0}$ lying on its main diagonal, and lower submatrix is a $1 \times 126$ matrix $\mathbf{X}_{gc,cn}(1, 6)$ which is a single row taken from $\Pi$. The matrix $\mathbf{R}_{gc,cn}(3, 21, 1, 6)$ is a $19 \times 126$ matrix over GF(127) and satisfies the $2 \times 2$ SM-constraint. The 126-fold CPM-dispersion of $\mathbf{R}_{gc,cn}(3, 21, 1, 6)$ gives a $19 \times 126$ array $\mathbf{H}_{gc,cn,qc}(126, 126)$ of CPMs and ZMs of size $126 \times 126$ which is a $2394 \times 15\,876$ binary matrix. The null space of $\mathbf{H}_{gc,cn,qc}(126, 126)$ gives a $(15\,876, 13\,494)$ CN-GC-QC-LDPC code $C_{gc,cn,qc}$ with rate 0.85.

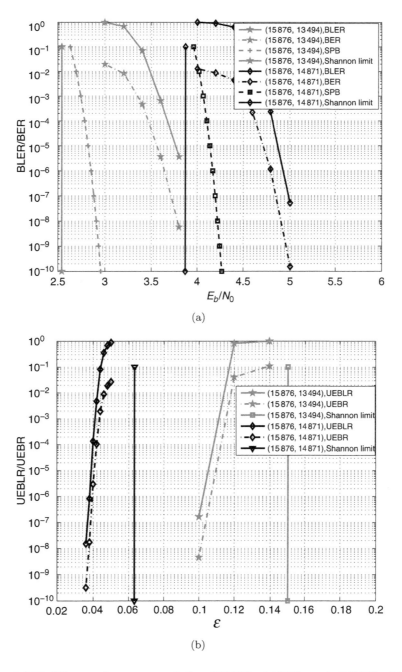

Figure 14.22 The performances of the (15 876, 14 871) and (15 876, 13 494) CN-QC-GC-LDPC codes given in Examples 14.23 and 14.24, respectively, over: (a) AWGN channel and (b) BEC.

The BER and BLER performances of $C_{gc,cn,qc}$ over the AWGN channel with BPSK signaling decoded with 50 iterations of the MSA are shown in Fig. 14.22(a). Decoded with 50 iterations of the MSA, the code performs within 1.5 dB from the Shannon limit at the BER of $10^{-8}$. The UEBLR and UEBR performances of $C_{gc,cn,qc}$ over the BEC are shown in Fig. 14.22(b).  ▲▲

A CN-GC-QC-LDPC code $C_{gc,cn,qc}$ can be decoded with a local/global two-phase BP-decoding scheme. Suppose a global codeword $\mathbf{v} = (\mathbf{v}_0, \mathbf{v}_1, \ldots, \mathbf{v}_{t-1})$ is transmitted. Let $\mathbf{r} = (\mathbf{r}_0, \mathbf{r}_1, \ldots, \mathbf{r}_{t-1})$ be the received vector. Decoding starts with a local phase of decoding, where each received local vector $\mathbf{r}_j$, $0 \le j < t$, is decoded using a local iterative decoder for the local code with a fixed number of iterations. If all the local received vectors are successfully decoded, the $t$ decoded local codewords form a decoded global vector $\mathbf{v}^* = (\mathbf{v}_0^*, \mathbf{v}_1^*, \ldots, \mathbf{v}_{t-1}^*)$. Then, we check whether this decoded global vector satisfies the $s(q-1)$ parity-check constraints imposed by the $s(q-1)$ global CNs. If it does, the decoded global vector $\mathbf{v}^*$ is a codeword in the CN-GC-QC-LDPC code $C_{gc,cn,qc}$ and it is delivered to the user(s). If the decoded vector $\mathbf{v}^*$ does not satisfy the $s(q-1)$ global CN-constraints, a global decoder based on the global parity-check matrix $\mathbf{H}_{gc,cn,qc}(q-1, q-1)$ of the code $C_{gc,cn,qc}$ is activated to decode the received vector $\mathbf{r}$, iteratively.

The local phase of decoding is designed to correct random errors scattered among $t$ received local vectors in $\mathbf{r}$. There are two possible approaches to switch the decoding from the local decoding phase to the global decoding phase. The first approach is to complete the local decoding of all the received local vectors (successfully and unsuccessfully decoded) and then switch to the global phase decoding to process the received vector $\mathbf{r}$ globally. The other approach is to switch to the global phase decoding as soon as a local decoder fails to decode a received local vector. The global phase of decoding is needed only when the distribution of errors does not allow for the error correction of each received local vector or when the number of errors is large.

The local and global decoding processes are carried on iteratively until either the received vector is successfully decoded or a preset maximum number of global decoding iterations is reached.

If we remove the first copy $\mathrm{CPM}(\mathbf{R}_{0,0})$ from the matrix $\mathbf{H}_{gc,cn,qc}(q-1, q-1)$ without changing other parts, we obtain the following $(m(t-1)+s) \times nt$ array $\mathbf{H}^*_{gc,cn,qc}(q-1, q-1)$ of CPMs and ZMs of size $(q-1) \times (q-1)$:

$$\mathbf{H}^*_{gc,cn,qc}(q-1, q-1)$$

$$= \begin{bmatrix} \mathbf{O} & & & \\ & \mathrm{CPM}(\mathbf{R}_{0,0}) & & \\ & & \ddots & \\ & & & \mathrm{CPM}(\mathbf{R}_{0,0}) \\ \hline & & \mathrm{CPM}(\mathbf{X}_{gc,cn}(s,t)) & \end{bmatrix}, \tag{14.27}$$

where $\mathbf{O}$ is an $m(q-1) \times n(q-1)$ ZM. Then, the null space of $\mathbf{H}^*_{\text{gc,cn,qc}}(q-1, q-1)$ gives a concatenated QC-LDPC code as presented in Section 9.3.1. In this case, the QC-LDPC code given by the null space of $\text{CPM}(\mathbf{R}_{0,0})$ is the outer code and the QC-LDPC code given by the null space of $\text{CPM}(\mathbf{X}_{\text{gc,cn}}(s,t))$ is the inner code.

## 14.12 Remarks

In this chapter, we showed that finite fields are very effective in constructing QC-LDPC codes. Any nonbinary field can be used to construct QC-LDPC codes in conjunction with CPM-dispersion and/or masking. All the finite-field-based LDPC codes are OSML-decodable.

In Section 14.10, we presented a class of optimal QC-LDPC codes that are capable of correcting two random CPM-phased bursts of erasures over a binary burst erasure channel (BBEC). The erasure correction characteristic is based on the $2 \times 2$ CPM-array cycle structure of their Tanner graphs. This class of codes is constructed based on cyclic groups of prime orders of finite fields. With this observation, we readily see that optimal QC-LDPC codes for correcting two random CPM-phased erasure-bursts can also be constructed from the three types of partial geometries presented in Chapter 13. For example, consider the BIBD-PaG base matrix $\mathbf{B}_{\text{PaG,BIBD},1}(4, b)$ over the prime field $\text{GF}(12b+1)$ given by (13.17). The CPM-dispersion of $\mathbf{B}_{\text{PaG,BIBD},1}(4, b)$ results in a $4 \times b$ array $\mathbf{H}_{\text{PaG,BIBD},1}(4, b)$ of CPMs of size $(12b+1) \times (12b+1)$ given by (13.18). The Tanner graph associated with $\mathbf{H}_{\text{PaG,BIBD},1}(4, b)$ has the $2 \times 2$ CPM-array cycle structure. The null space of any two CPM-row blocks of $\mathbf{H}_{\text{PaG,BIBD},1}(4, b)$ gives a $(b(12b+1), (b-2)(12b+1)+1)$ QC-LDPC code with $2(12b+1)-1$ parity-check symbols which is capable of correcting any two random mutually semi-solid CPM-phased erasure-bursts with a maximum of $2(12b+1)-1$ erasures. Consider the BIBD-PaG base matrix $\mathbf{B}_{\text{PaG,BIBD},1}(4, 15)$ (with $b = 15$) constructed based on the prime field $\text{GF}(181)$ in Example 13.14. The null space of the $181 \times 181$ CPM-dispersion of any two rows of $\mathbf{B}_{\text{PaG,BIBD},1}(4, 15)$ gives a $(2715, 2354)$ QC-LDPC code with 361 parity-check symbols and rate 0.867. This code is capable of correcting two random mutually semi-solid CPM-phased erasure-bursts over a BBEC with a maximum of 361 erasures. Hence, the code is optimal in terms of correcting two random mutually semi-solid CPM-phased erasure-bursts.

## Problems

**14.1** Consider the field $\text{GF}(2^3)$. Let $n = m = 8$ and $\mathbf{S}_1 = \mathbf{S}_2 = \text{GF}(2^3)$. Construct two matrices over $\text{GF}(2^3)$ in the form of (14.1) by choosing two multipliers $\eta_1 = \alpha^0$ and $\eta_2 = \alpha$, where $\alpha$ is a primitive element in $\text{GF}(2^3)$.

**14.2** Consider the field $\text{GF}(2^8)$. Choose the following two sets

$$\mathbf{S}_1 = \{1, \alpha, \alpha^2, \alpha^3\}, \ \mathbf{S}_2 = \{1, \alpha, \alpha^2, \alpha^3, \dots, \alpha^{18}, \alpha^{19}\},$$

and a multiplier $\eta = 1$. Use $\mathbf{S}_1$, $\mathbf{S}_2$, and $\eta$ to construct a $4 \times 20$ matrix $\mathbf{B}_s(4, 20)$ in the form of (14.1). Construct a QC-LDPC code using the 255-fold CPM-dispersion of $\mathbf{B}_s(4, 20)$ as a parity-check matrix. Compute the BER and BLER performances of the constructed QC-LDPC code over the AWGN channel by 5, 10, and 50 iterations of the scaled MSA. Compute its performance over the BEC.

**14.3** Consider the field $\mathrm{GF}(2^8)$ with a primitive element $\alpha$. Let $k$ be a positive integer. Choose the following two sets

$$\mathbf{S}_1 = \{1, \alpha, \alpha^2, \alpha^3\}, \ \mathbf{S}_2 = \{1, \alpha, \alpha^2, \alpha^3, \dots, \alpha^{k-2}, \alpha^{k-1}\}.$$

Let $\eta = \alpha$. Use $\mathbf{S}_1$, $\mathbf{S}_2$, and $\eta$ to construct a $4 \times k$ matrix $\mathbf{B}_s(4, k)$ in the form of (14.1). Construct a family of QC-LDPC codes by choosing $k = 8$, 12, 16, 24, 28, 36, 40, and compute their error performances over the AWGN channel by using 50 iterations of the scaled MSA. Compute its performance over the BEC.

**14.4** Consider the two sets $\mathbf{S}_1$ and $\mathbf{S}_2$ over $\mathrm{GF}(331)$ chosen in Example 14.6. Let $\eta = \alpha^2$. Construct a $4 \times 8$ matrix $\mathbf{B}_s(4, 8)$ over $\mathrm{GF}(331)$ in the form of (14.1) based on $\mathbf{S}_1$, $\mathbf{S}_2$, and $\eta$. Construct a QC-LDPC code given by the null space over $\mathrm{GF}(2)$ of the 330-fold CPM-dispersion of the constructed matrix $\mathbf{B}_s(4, 8)$. Calculate the BER and BLER performances of the constructed QC-LDPC code over the AWGN channel using 50 iterations of the scaled MSA. Compute its performance over the BEC.

**14.5** Consider the field $\mathrm{GF}(2^6)$. Use the sum-construction to construct a Latin square of order 64. Based on this Latin square, do the following.
  (a) Construct a QC-LDPC code with length 504 and rate around 0.5.
  (b) Construct a QC-LDPC code with length 504 and rate 0.5 (*Hint:* masking).
  (c) Calculate the BER and BLER performances of the two constructed QC-LDPC codes over the AWGN channel using 50 iterations of the scaled MSA.
  (d) Calculate the performances of the two constructed QC-LDPC codes over the BEC.

**14.6** Prove that the matrix $\mathbf{B}_p(m, n)$ over $\mathrm{GF}(q)$ in the form of (14.3) satisfies the $2 \times 2$ SM-constraint.

**14.7** Consider the base matrix $\mathbf{B}_s(4, 8)$ constructed in Problem 14.4. Mask $\mathbf{B}_s(4, 8)$ with the masking matrix given by (14.5) to obtain a masked matrix $\mathbf{B}_{s,\mathrm{mask}}(4, 8)$. Compute the QC-LDPC code given by the null space over $\mathrm{GF}(2)$ of the 330-fold CPM-dispersion of $\mathbf{B}_{s,\mathrm{mask}}(4, 8)$ and its error performances over the AWGN channel by using 50 iterations of the scaled MSA.

**14.8** Consider the eight matrices, $\mathbf{B}_s(4, k)$ with $k = 8$, 12, 16, 24, 28, 32, 36, 40, constructed in Problem 14.3. Mask these eight matrices with the type-1 masking matrices given by (14.6) or (14.7). Use the eight masked base matrices and their 255-fold CPM-dispersions to construct eight masked QC-LDPC codes. Calculate the error performances of these codes over the AWGN channel by using 50 iterations of the scaled MSA.

**14.9** Consider the $4 \times 40$ matrix $\mathbf{B}_s(4, 40)$ constructed in Problem 14.3. Take a $4 \times 12$ submatrix $\mathbf{B}_s(4, 12)$ from $\mathbf{B}_s(4, 40)$ (avoiding the 0-entries). Mask $\mathbf{B}_s(4, 12)$ with the type-2 masking matrix given by (14.8) to obtain a masked matrix $\mathbf{B}_{s,mask}(4, 12)$. Construct a QC-LDPC code given by the null space over GF(2) of the 255-fold CPM-dispersion of the constructed matrix $\mathbf{B}_{s,mask}(4, 12)$. Calculate the BER and BLER performances of the constructed QC-LDPC code over the AWGN channel using 50 iterations of the scaled MSA.

**14.10** Consider the field $GF(2^8)$. Choose two sets, $\mathbf{S}_1$ of size 6 and $\mathbf{S}_2$ of size 16, and a multiplier $\eta$.
   (a) Construct a $6 \times 16$ matrix $\mathbf{B}_s(6, 16)$ in the form of (14.1) without any 0-entries.
   (b) Calculate the QC-LDPC code given by the null space of the 255-fold CPM-dispersion of the $6 \times 16$ matrix $\mathbf{B}_s(6, 16)$ and its BER and BLER performances over the AWGN channel using 50 iterations of the scaled MSA.
   (c) Mask $\mathbf{B}_s(6, 16)$ with the type-3 masking matrix given by (14.10) to obtain a masked base matrix $\mathbf{B}_{s,mask}(6, 16)$ over $GF(2^8)$. Calculate the QC-LDPC code given by the null space of the 255-fold CPM-dispersion of the masked base matrix $\mathbf{B}_{s,mask}(6, 16)$ and its BER and BLER performances over the AWGN channel using 50 iterations of the scaled MSA.
   (d) Calculate the performances of the constructed QC-LDPC codes over the BEC.

**14.11** In Example 14.13, a $(504, 252)$ QC-LDPC code whose Tanner graph has girth 8 is constructed using the $2 \times 2/3 \times 3$ SMC-MMSA. Construct another $(504, 252)$ QC-LDPC code whose Tanner graph also has girth 8. Find the corresponding masked base matrix and the equivalent two subsets. (The two subsets must be different from those given in Example 14.13.)

**14.12** Prove that type-2 RS matrices in the form of (14.11) satisfy the $2 \times 2$ SM-constraint.

**14.13** Prove that type-3 RS matrices in the form of (14.11) satisfy the $2 \times 2$ SM-constraint.

**14.14** Consider the field $GF(2^7)$. Let $d = 4$ and $n = 127$. Construct a $d \times n$ type-3 RS matrix $\mathbf{B}_{RS,n}(d, n)$.
   (a) Find the RS code given by the null space over $GF(2^7)$ of $\mathbf{B}_{RS,n}(d, n)$.
   (b) Take a $4 \times 32$ submatrix $\mathbf{B}_{RS,n}(4, 32)$ from $\mathbf{B}_{RS,n}(d, n)$. Compute the QC-RS-LDPC code given by the null space of the 127-fold CPM-dispersion of $\mathbf{B}_{RS,n}(4, 32)$ and its BER and BLER performances over the AWGN channel decoded with 5, 10, and 50 iterations of the scaled MSA.
   (c) Take a $4 \times 8$ submatrix $\mathbf{B}_{RS,n}(4, 8)$ from $\mathbf{B}_{RS,n}(d, n)$. Compute the QC-RS-LDPC code given by the null space of the 127-fold CPM-dispersion of $\mathbf{B}_{RS,n}(4, 8)$ and its BER and BLER performances over the AWGN channel decoded with 5, 10, and 50 iterations of the scaled MSA.
   (d) Calculate the performances of the constructed QC-RS-LDPC codes over the BEC.

**14.15** Consider the field $GF(2^8)$. Let $n = 85$ which is a factor of $2^8 - 1$.
(a) Construct a type-1 RS matrix $\mathbf{B}_{RS,n}(d, n)$ of size $5 \times 85$.
(b) Compute the RS code given by the null space over $GF(2^8)$ of $\mathbf{B}_{RS,n}(d, n)$.
(c) Take a $4 \times k$ submatrix from $\mathbf{B}_{RS,n}(d, n)$. For $k = 8, 12, 16, 20, 24, 28, 32,$ 36, 40, construct a family of $(4, k)$-regular QC-RS-LDPC codes based on the submatrices, $\mathbf{B}_{RS,n}(4, k)$, and their 85-fold CPM-dispersions. Compute the BER and BLER performances of the constructed QC-RS-LDPC codes over the AWGN channel decoded with 50 iterations of the scaled MSA.
(d) Compute the performances of the constructed QC-RS-LDPC codes over the BEC.

**14.16** Consider the $5 \times 85$ type-1 RS matrix $\mathbf{B}_{RS,n}(5, 85)$ constructed in Problem 14.15. Take four rows, say the last four rows, from $\mathbf{B}_{RS,n}(5, 85)$ to form a $4 \times 85$ matrix $\mathbf{B}_{RS,n}(4, 85)$ over $GF(2^8)$. Based on the matrix $\mathbf{B}_{RS,n}(4, 85)$ and Theorem 14.4, do the following.
(a) Construct a $(4, 8)$-regular QC-RS-LDPC code of length 680 whose Tanner graph has girth at least 8. Find the short cycle distributions of the Tanner graph and the column-label set $\Lambda_t$.
(b) Construct a $(4, 8)$-regular QC-RS-LDPC code of length 2040 whose Tanner graph has girth at least 8. Find the short cycle distributions of the Tanner graph and the column-label set $\Lambda_t$.
(c) Compute the BER and BLER performances of the two constructed QC-RS-LDPC codes over the AWGN channel decoded with 50 iterations of the scaled MSA.

**14.17** Consider the $4 \times 127$ type-3 RS matrix $\mathbf{B}_{RS,n}(4, 127)$ constructed in Problem 14.14. Using this RS matrix and Corollary 14.1, do the following.
(a) Construct a $(4, 8)$-regular QC-RS-LDPC code of length 1016 whose Tanner graph has girth at least 8. Find the short cycle distributions of the Tanner graph, the column-label set $\Lambda_t$, and the base matrix $\mathbf{B}_{RS,n,\Lambda_t}(4, 8)$.
(b) Construct a $(4, 12)$-regular QC-RS-LDPC code of length 1524 whose Tanner graph has girth at least 8. Find the short cycle distributions of the Tanner graph, the column-label set $\Lambda_t$, and the base matrix $\mathbf{B}_{RS,n,\Lambda_t}(4, 12)$.
(c) Design a $4 \times 8$ masking matrix $\mathbf{Z}(4, 8)$. Mask $\mathbf{B}_{RS,n,\Lambda_t}(4, 8)$ with $\mathbf{Z}(4, 8)$ to construct a $(3, 6)$-regular QC-RS-LDPC code of length 1016.
(d) Design a $4 \times 12$ masking matrix $\mathbf{Z}(4, 12)$. Mask $\mathbf{B}_{RS,n,\Lambda_t}(4, 12)$ with $\mathbf{Z}(4, 12)$ to construct a $(3, 9)$-regular QC-RS-LDPC code of length 1524.
(e) Compute the BER and BLER performances of the four constructed QC-RS-LDPC codes over the AWGN channel decoded with 50 iterations of the scaled MSA.

**14.18** Prove that the $4 \times n/3$ submatrix $\mathbf{B}_{RS,n,\Lambda_{n/3}}(4, n/3)$ constructed in Section 14.9 following the specially chosen column-label set $\Lambda_{n/3} = \{j_0, j_1, \ldots, j_{n/3-1}\}$ satisfies the $2 \times 2$ SM-constraint.

**14.19** In Example 14.19, the column-label set $\Lambda_8 = \{2, 5, 7, 13, 20, 32, 54, 60\}$ is used to construct a QC-RS-LDPC code with girth 8. Suppose we choose the second numbers of the triplets to form another column-label set $\Lambda_8^* = \{87, 90, 92, 98, 105, 117, 139, 145\}$. Construct a $4 \times 8$ RS matrix $\mathbf{B}_{\text{RS}, 255, \Lambda_8^*}(4, 8)$ based on $\Lambda_8^*$.

  (a) Show that the matrix $\mathbf{B}_{\text{RS}, 255, \Lambda_8^*}(4, 8)$ also satisfies the $2 \times 2/3 \times 3$ SM-constraint.

  (b) Construct a QC-RS-LDPC code of length 2040 based on $\mathbf{B}_{\text{RS}, 255, \Lambda_8^*}(4, 8)$. Find the short cycle distributions of the constructed code and its error performance over the AWGN channel.

  (c) Choose another eight triplets (different from those given in Example 14.19) to construct a different RS matrix $\mathbf{B}_{\text{RS}, 255, \Lambda_8'}(4, 8)$ which satisfies the $2 \times 2/3 \times 3$ SM-constraint.

  (d) Construct a QC-RS-LDPC code of length 2040 based on $\mathbf{B}_{\text{RS}, 255, \Lambda_8'}(4, 8)$. Find the short cycle distributions of the constructed code and its error performance over the AWGN channel.

  (e) Follow the same construction method to construct a $(4, 20)$-regular QC-RS-LDPC code of length 5100 with rate around 4/5 whose Tanner graph has girth at least 8. Find the column-label set and the short cycle distributions of the constructed code, and compute its error performance over the AWGN channel.

**14.20** Prove Theorem 14.4 based on the Theorem 14.1.

# References

[1] A. Hocquenghem, "Codes correcteurs d'erreurs," *Chiffres*, **2** (1959), 147–156.

[2] R. C. Bose and D. K. Ray-Chaudhuri, "On a class of error correcting binary group codes," *Inf. Control*, **3** (1960), 68–79.

[3] I. S. Reed and G. Solomon, "Polynomial codes over certain finite fields," *J. Soc. Indust. Appl. Math.* **8** (1960), 300–304.

[4] S. Lin and D. J. Costello, Jr., *Error Control Coding: Fundamentals and Applications*, 2nd ed., Upper Saddle River, NJ, Prentice-Hall, 2004.

[5] W. E. Ryan and S. Lin, *Channel Codes: Classical and Modern*, New York, Cambridge University Press, 2009.

[6] J. L. Fan, "Array codes as low-density parity-check codes," *Proc. 2nd Int. Symp. on Turbo Codes and Related Topics*, Brest, September 2000, pp. 543–546.

[7] I. Djurdjevic, J. Xu, K. Abdel-Ghaffar, and S. Lin, "A class of low-density parity-check codes constructed based on Reed-Solomon codes with two information symbols," *IEEE Commun. Lett.*, **7**(7) (2003), 317–319.

[8] L. Chen, J. Xu, I. Djurdjevic, and S. Lin, "Near-Shannon limit quasi-cyclic low-density parity-check codes," *IEEE Trans. Commun.*, **52**(7) (2004), 1038–1042.

[9] M. P. C. Fossorier, "Quasi-cyclic low-density parity-check codes from circulant permutation matrices," *IEEE Trans. Inf. Theory*, **50**(8) (2004), 1788–1793.

[10] H. Tang, J. Xu, Y. Kou, S. Lin, and K. Abdel-Ghaffar, "On algebraic construction of Gallager and circulant low-density parity-check codes," *IEEE Trans. Inf. Theory*, **50**(6) (2004), 1269–1279.

[11] L. Chen, L. Lan, I. Djurdjevic, S. Lin, and K. Abdel-Ghaffar, "An algebraic method for constructing quasi-cyclic LDPC codes," *Proc. Int. Symp. on Inf. Theory and Its Applications (ISITA)*, Parma, October 2004, pp. 535–539.

[12] L. Lan, L.-Q. Zeng, Y. Y. Tai, S. Lin, and K. Abdel-Ghaffar, "Constructions of quasi-cyclic LDPC codes for AWGN and binary erasure channels based on finite fields and affine permutations," *Proc. IEEE Int. Symp. on Inf. Theory*, Adelaide, September 2005, pp. 2285–2289.

[13] Y. Y. Tai, L. Lan, L.-Q. Zeng, S. Lin, and K. Abdel Ghaffar, "Algebraic construction of quasi-cyclic LDPC codes for AWGN and erasure channels," *IEEE Trans. Commun.*, **54**(10) (2006), 1765–1774.

[14] L. Lan, L. Zeng, Y. Y. Tai, *et al.*, "Construction of quasi-cyclic LDPC codes for AWGN and binary erasure channels: a finite field approach," *IEEE Trans. Inf. Theory*, **53**(7) (2007), 2429–2458.

[15] S. Song, B. Zhou, S. Lin, and K. Abdel-Ghaffar, "A unified approach to the construction of binary and nonbinary quasi-cyclic LDPC codes based on finite fields," *IEEE Trans. Commun.*, **57**(1) (2009), 84–93.

[16] J. Kang, Q. Huang, L. Zhang, B. Zhou, and S. Lin, "Quasi-cyclic LDPC codes: an algebraic construction," *IEEE Trans. Commun.*, **58**(5) (2010), 1383–1396.

[17] L. Zhang, Q. Huang, S. Lin, and K. Abdel-Ghaffar, "Quasi-cyclic LDPC codes: an algebraic construction, rank analysis, and codes on Latin squares," *IEEE Trans. Commun.*, **58**(11) (2010), 3126–3139.

[18] L. Zhang, S. Lin, K. Abdel-Ghaffar, Z. Ding, and B. Zhou, "Quasi-cyclic LDPC codes on cyclic subgroups of finite fields," *IEEE Trans. Commun.*, **59**(9) (2011), 2330–2336.

[19] Q. Diao, Q. Huang, S. Lin, and K. Abdel-Ghaffar, "A matrix-theoretic approach for analyzing quasi-cyclic low-density parity-check codes," *IEEE Trans. Inf. Theory*, **58**(6) (2012), 4030–4048.

[20] J. Li, K. Liu, S. Lin, and K. Abdel-Ghaffar, "Quasi-cyclic LDPC codes on two arbitrary sets of a finite field," *Proc. Int. Symp. Info. Theory (ISIT)*, Honolulu, HI, June 29–July 4, 2014, pp. 2454–2458.

[21] J. Li, K. Liu, S. Lin, and K. Abdel-Ghaffar, "Algebraic quasi-cyclic LDPC codes: construction, low error-floor, large girth and a reduced-complexity decoding scheme," *IEEE Trans. Commun.*, **62**(8) (2014), 2626–2637.

[22] J. Li, K. Liu, S. Lin, and K. Abdel-Ghaffar, "A matrix-theoretic approach to the construction of non-binary quasi-cyclic LDPC codes," *IEEE Trans. Commun.*, **63**(4) (2015), 1057–1068.

[23] J. Li, S. Lin, K. Abdel-Ghaffar, W. E. Ryan, and D. J. Costello, Jr., *LDPC Code Designs, Constructions, and Unification*, Cambridge, Cambridge University Press, 2017.

[24] D. MacKay and R. Neal, "Good codes based on very sparse matrices," *Cryptography and Coding, 5th IMA Conf.,* in C. Boyd, ed., Berlin, Springer-Verlag, October 1995, pp. 100–111.

[25] D. MacKay, "Good error correcting codes based on very sparse matrices," *IEEE Trans. Inf. Theory*, **45**(3) (1999), 399–431.

[26] N. Alon and M. Luby, "A linear time erasure-resilient code with nearly optimal recovery," *IEEE Trans. Inf. Theory*, **42**(11) (1996), 1732–1736.

[27] J. Byers, M. Luby, M. Mitzenmacher, and A. Rege, "A digital fountain approach to reliable distribution of bulk data," *ACM SIGCOMM Computer Communication Rev.*, **28**(4) (1998), 56–67.

[28] J. Chen and M. P. C. Fossorier, "Near optimum universal belief propagation based decoding of low-density parity check codes," *IEEE Trans. Commun.*, **50**(3) (2002), 406–614.

[29] L. Zeng, L. Lan, Y. Y. Tai, S. Song, S. Lin, and K. Abdel-Ghaffar, "Constructions of nonbinary quasi-cyclic LDPC codes: a finite field approach," *IEEE Trans. Commun.*, **56**(4) (2008), 545–554.

[30] B. Zhou, J. Kang, S. Song, S. Lin, K. Abdel-Ghaffar, and M. Xu, "Construction of non-binary quasi-cyclic LDPC codes by arrays and array dispersions," *IEEE Trans. Commun.*, **57**(6) (2009), 1652–1662.

[31] R. M. Tanner, "A recursive approach to low complexity codes," *IEEE Trans. Inf. Theory*, **IT-27**(5) (1981), 533–547.

[32] Q. Diao, Q. Huang, S. Lin, and K. Abdel-Ghaffar, "A transform approach for analyzing and constructing quasi-cyclic low-density parity-check codes," *Proc. IEEE Inf. Theory Applic. Workshop*, San Diego, CA, February 6–11, 2011, pp. 1–8.

[33] J. Li, S. Lin, and K. Abdel-Ghaffar, "Improved message-passing algorithm for counting short cycles in bipartite graphs," *Proc. Int. Symp. Inf. Theory (ISIT)*, Hong Kong, China, June 2015, pp. 416–420.

[34] J. Li, S. Lin, and K. Abdel-Ghaffar, "Improved trellis-based algorithm for locating and breaking cycles in bipartite graphs with applications to LDPC codes," *Proc. Inf. Theory Applic. Workshop*, San Diego, CA, February 11–16, 2018, pp. 1–8.

[35] R. Lidl and H. Niederreiter, *Introduction to Finite Fields and Their Applications,* revised ed., Cambridge, Cambridge University Press, 1994.

[36] X.-Y Hu, E. Eleftheriou, and D.-M. Arnold, "Regular and irregular progressive edge-growth Tanner graphs," *IEEE Trans. Inf. Theory*, **51**(1) (2005), 386–398.

[37] Encyclopedia of Sparse Graph Codes: www.inference.phy.cam.ac.uk/mackay/CodesFiles.html

[38] J. Li, K. Liu, S. Lin, and K. Abdel-Ghaffar, "Reed–Solomon based nonbinary LDPC codes," *Proc. Int. Symp. Inf. Theory and Its Applications (ISITA)*, Monterey, CA, October 30–November 2, 2016, pp. 384–388.

[39] J. Li, K. Liu, S. Lin, and K. Abdel-Ghaffar, "Reed–Solomon based globally coupled quasi-cyclic LDPC codes," *Proc. Inf. Theory Applic. Workshop*, San Diego, CA, February 12–17, 2017, pp. 1–10.

[40] J. Li, K. Liu, S. Lin, and K. Abdel-Ghaffar, "Reed–Solomon based nonbinary globally coupled LDPC codes: correction of random errors and bursts of erasures," *Proc. Int. Symp. Inf. Theory*, Aachen, Germany, June 25–30, 2017, pp. 381–385.

[41] J. Li, K. Liu, S. Lin, and K. Abdel-Ghaffar, "Construction of partial geometries and LDPC codes based on Reed–Solomon codes," *Proc. Int. Symp. Inf. Theory*, Paris, France, July 7–12, 2019, pp. 61–65.

[42] X. Xiao, W. E. Ryan, B. Vasic, S. Lin, and K. Abdel-Ghaffar, "Reed–Solomon-based quasi-cyclic LDPC codes: designs, cycle structure and erasure correction," *Proc. Inf. Theory Applic. Workshop*, San Diego, CA, 2018, pp. 1–10.

[43] X. Xiao, B. Vasic, S. Lin, K. Abdel-Ghaffar, and W. E. Ryan, "Girth-eight Reed–Solomon based QC-LDPC codes," *2018 10th Int. Symp. Turbo Codes and Iterative Information Processing (ISTC)*, December 3–7, 2018, pp. 1–5.

[44] X. Xiao, B. Vasic, S. Lin, K. Abdel-Ghaffar, and W. E. Ryan, "Quasi-cyclic LDPC codes for correcting multiple phased bursts of erasures," *Proc. Int. Symp. Inf. Theory*, July 7–12, 2019, Paris, France, pp. 71–75.

[45] X. Xiao, B. Vasic, S. Lin, K. Abdel-Ghaffar, and W. E. Ryan, "Reed–Solomon based quasi-cyclic LDPC codes: designs, girth, cycle structure, and reduction of short cycles," *IEEE Trans. Commun.*, **67**(8) (2019), 5275–5286.

# 15

# Graph-Theoretic LDPC Codes

In general, LDPC codes are classified into two categories based on their construction methods: algebraic methods and graphical methods. LDPC codes constructed based on finite geometries and finite fields are classified as algebraic LDPC codes, such as the cyclic and quasi-cyclic LDPC codes presented in Chapters 12 to 14. Methods for constructing algebraic LDPC codes are, in general, systematic. In the construction of an algebraic LDPC code, its parity-check matrix is first constructed and then we take the null space of the parity-check matrix to produce the code. Algebraic LDPC codes have abundant algebraic structural properties which facilitate their encoding and decoding implementations. LDPC codes constructed based on graphical methods are put in the category of graph-theoretic LDPC codes. In the construction of a *graph-theoretic LDPC code*, the Tanner graph of the code is first constructed from which the adjacency matrix of the constructed Tanner graph is then formed. The null space of the adjacency matrix, used as the parity-check matrix, gives a graph-theoretic LDPC code. Construction of a graph-theoretic LDPC code, in general, requires computer aid. However, well-designed graph-theoretic LDPC codes give very good error performances, especially in the waterfall region (i.e., low-to-medium SNR region).

The best-known graphical methods for constructing graph-theoretic LDPC codes are the *protograph-based* method and the *progressive edge-growth* method. The protograph-based method for constructing graph-theoretic LDPC codes was first introduced by Thorpe in 2003 [1]. Using this method for code construction, the first step is to design a relatively small well-designed bipartite graph, called a *protograph* (PTG), with a near-capacity iterative decoding threshold as a base (or a building block). Then, take copies of this protograph and permute the edges among the copies according to certain rules to connect them into a Tanner graph of larger size. Next, form the adjacency matrix of the constructed Tanner graph and use it as the parity-check matrix to produce an LDPC code, called a

*PTG-based LDPC code*, or simply a PTG-LDPC code. The construction process is commonly referred to as *copy-and-permute*. LDPC codes constructed based on the ensembles of protographs are *capacity-approaching* LDPC codes [2, 3]. Even though the ensembles of PTG-LDPC codes contain good codes, there is no systematic method (so far) to construct these good codes and a computer search is then needed. In this chapter, we first present the graphical copy-and-permute method to construct LDPC codes based on protographs and then give an interpretation of this graphical construction from a *matrix-theoretic* point of view. Based on this matrix-theoretic interpretation of the copy-and-permute process, we present an algebraic construction of PTG-LDPC codes based on the adjacency matrices associated with the chosen protographs.

The second graphical method for constructing LDPC codes presented in this chapter is the progressive edge-growth (PEG) method [4, 5]. Construction with this method begins with a set of VNs (variable nodes) and a set of CNs (check nodes) without edges connecting them. Then, edges are progressively added to connect the VNs and CNs by applying a set of rules and a given VN degree profile (distribution). Edges are added to a VN *one at a time* using an edge-selection procedure until the number of edges added to a VN is equal to its specified degree. This progressive addition of edges to a VN is called the *edge-growth*. At the end of edge-growth of all the VNs, we obtain a connected Tanner graph. Then, the null space of the adjacency matrix of this Tanner graph gives an LDPC code, called a PEG-LDPC code. Besides the protograph-based and the progressive edge-growth-based constructions of graphical LDPC codes, there are other graphical constructions of LDPC codes, although these will not be presented in this chapter. Readers who are interested in more on graphical constructions of LDPC codes (including both binary and nonbinary) are referred to References [2, 3, 6–25] listed at the end of this chapter.

## 15.1 Protograph-Based LDPC Codes

A *protograph* (PTG) is defined as a bipartite graph, denoted by $\mathcal{G}_{\text{ptg}}$, of *relatively small size* which consists of a set of nodes called VNs, a set of nodes called CNs, and a set of edges which connect the VNs to the CNs (or CNs to VNs) in the graph. Construction of an LDPC code based on a designed protograph $\mathcal{G}_{\text{ptg}}$ begins with the construction of the Tanner graph $\mathcal{G}$ of the code by expanding $\mathcal{G}_{\text{ptg}}$. Then, form the adjacency matrix $\mathbf{H}$ of the Tanner graph $\mathcal{G}$. The null space over GF(2) of $\mathbf{H}$ gives a *protograph-based LDPC code* (PTG-LDPC code). The construction of the Tanner graph $\mathcal{G}$ consists of two steps. The first step is to take $k$ copies of the protograph $\mathcal{G}_{\text{ptg}}$, i.e., duplicate $\mathcal{G}_{\text{ptg}}$ $k$ times. The second step is to form a single connected bipartite graph of larger size *without parallel edges* by permuting the edges of the individual copies among the $k$ copies of $\mathcal{G}_{\text{ptg}}$. The resultant bipartite graph $\mathcal{G}$ is then used as the Tanner graph to construct a PTG-LDPC code. The parity-check matrix of the code is the adjacency matrix $\mathbf{H}$ of the graph $\mathcal{G}$.

Let $\mathcal{G}_{\mathrm{ptg}} = (\mathcal{V}, \mathcal{C}, \mathcal{E})$ be a protograph for code construction which consists of a set $\mathcal{V} = \{v_0, v_1, \ldots, v_{n-1}\}$ of $n$ VNs, a set $\mathcal{C} = \{c_0, c_1, \ldots, c_{m-1}\}$ of $m$ CNs, and a set $\mathcal{E} = \{(i,j)\}$ of edges, where $(i,j)$ denotes the edge that connects the $i$th CN $c_i$ to the $j$th VN $v_j$. If a VN $v_j$ is connected to a CN $c_i$ by more than one edge, the edges connecting the VN $v_j$ and the CN $c_i$ are referred to as *parallel edges*. A protograph without parallel edges is called a *simple graph*. Let $b_{i,j}$ denote the number of parallel edges in $\mathcal{E}$ that connect the CN $c_i$ and the VN $v_j$ in $\mathcal{G}_{\mathrm{ptg}}$. Then, the total number of edges in $\mathcal{E}$ is $\lambda = \sum_{0 \leq i < m, 0 \leq j < n} b_{i,j}$. If $b_{i,j} = 0$, $v_j$ and $c_i$ are not connected. If $b_{i,j} = 1$, $v_j$ and $c_i$ are connected by a single edge in $\mathcal{E}$.

Let $k \geq 1$ be a positive integer. Take $k$ copies of $\mathcal{G}_{\mathrm{ptg}}$ and denote these $k$ copies by $\mathcal{G}_{\mathrm{ptg},0}, \mathcal{G}_{\mathrm{ptg},1}, \ldots, \mathcal{G}_{\mathrm{ptg},k-1}$. For $0 \leq j < n$, we group the copies of the $j$th VN $v_j$ in the $k$ copies of $\mathcal{G}_{\mathrm{ptg}}$ into a set of $k$ VNs, denoted by $\Phi_j = \{v_{0,j}, v_{1,j} \ldots, v_{k-1,j}\}$, where $v_{l,j}$ denotes the $j$th VN $v_j$ of $\mathcal{G}_{\mathrm{ptg}}$ in the $l$th copy $\mathcal{G}_{\mathrm{ptg},l}$ of $\mathcal{G}_{\mathrm{ptg}}$ with $0 \leq l < k$, i.e., the $k$ VNs in $\Phi_j$ are copies of the $j$th VN $v_j$ in $\mathcal{G}_{\mathrm{ptg}}$. We refer to the $k$ VNs in $\Phi_j$ as the *type-$j$ VNs*. For $0 \leq i < m$, we group the copies of the $i$th CN $c_i$ in the $k$ copies of $\mathcal{G}_{\mathrm{ptg}}$ into a set of $k$ CNs, denoted by $\Omega_i = \{c_{0,i}, c_{1,i}, \ldots, c_{k-1,i}\}$, where $c_{l,i}$ denotes the $i$th CN $c_i$ of $\mathcal{G}_{\mathrm{ptg}}$ in the $l$th copy $\mathcal{G}_{\mathrm{ptg},l}$ of $\mathcal{G}_{\mathrm{ptg}}$ with $0 \leq l < k$, i.e., the $k$ CNs in $\Omega_i$ are copies of the $i$th CN $c_i$ in $\mathcal{G}_{\mathrm{ptg}}$. We refer to the $k$ CNs in $\Omega_i$ as the *type-$i$ CNs*. Next, we group the $k\lambda$ edges in the $k$ copies of $\mathcal{G}_{\mathrm{ptg}}$ into $\lambda$ edge-sets, denoted by $\mathcal{E}_0, \mathcal{E}_1, \ldots, \mathcal{E}_{\lambda-1}$, each consisting of $k$ copies of an edge in the set $\mathcal{E}$ of $\mathcal{G}_{\mathrm{ptg}}$. For $0 \leq l < \lambda$, if the edges in $\mathcal{E}_l$ are the $k$ copies of an edge $(i,j)$ in the edge-set $\mathcal{E}$ of $\mathcal{G}_{\mathrm{ptg}}$, then the edges in $\mathcal{E}_l$ are said to be *type-$(i,j)$ edges*. Because there are $b_{i,j}$ parallel edges connecting the VN $v_j$ and the CN $c_i$ in $\mathcal{G}_{\mathrm{ptg}}$, there are $kb_{i,j}$ type-$(i,j)$ edges in the $k$ copies of $\mathcal{G}_{\mathrm{ptg}}$, i.e., there are $b_{i,j}$ sets of type-$(i,j)$ edges among the $\lambda$ edge-sets $\mathcal{E}_0, \mathcal{E}_1, \ldots, \mathcal{E}_{\lambda-1}$. The above process is referred to as a *copy-and-grouping operation*.

The next step in the construction of a PTG-LDPC code is to connect the $k$ copies $\mathcal{G}_{\mathrm{ptg},0}, \mathcal{G}_{\mathrm{ptg},1}, \ldots, \mathcal{G}_{\mathrm{ptg},k-1}$ of $\mathcal{G}_{\mathrm{ptg}}$ to produce a single connected bipartite graph without parallel edges, denoted by $\mathcal{G}_{\mathrm{ptg}}(k,k)$. To achieve this, we permute the edges of the individual copies of $\mathcal{G}_{\mathrm{ptg}}$ among the $k$ copies of $\mathcal{G}_{\mathrm{ptg}}$. The permutation is carried out under the *restriction*: for $0 \leq j < n$ and $0 \leq i < m$, a type-$j$ VN in $\Phi_j$ can only connect to one of the type-$i$ CNs in $\Omega_i$ by a single type-$(i,j)$ edge (if any) and vice versa. The edge permutations among the $k$ copies of $\mathcal{G}_{\mathrm{ptg}}$ result in a simple connected bipartite graph $\mathcal{G}_{\mathrm{ptg}}(k,k)$, with $nk$ VNs, $mk$ CNs, and $\lambda k$ edges. The bipartite graph $\mathcal{G}_{\mathrm{ptg}}(k,k)$ has the same degree distributions as the protograph $\mathcal{G}_{\mathrm{ptg}}$. The pair $(k,k)$ in $\mathcal{G}_{\mathrm{ptg}}(k,k)$ indicates that each VN in $\mathcal{G}_{\mathrm{ptg}}$ is expanded into $k$ VNs in $\mathcal{G}_{\mathrm{ptg}}(k,k)$ and each CN in $\mathcal{G}_{\mathrm{ptg}}$ is expanded into $k$ CNs in $\mathcal{G}_{\mathrm{ptg}}(k,k)$.

The above process to construct a bipartite graph $\mathcal{G}_{\mathrm{ptg}}(k,k)$ of larger size without parallel edges by taking $k$ copies of a protograph $\mathcal{G}_{\mathrm{ptg}}$ of a smaller size and permuting the edges of the individual copies of $\mathcal{G}_{\mathrm{ptg}}$ among the copies of $\mathcal{G}_{\mathrm{ptg}}$ is referred to as the *copy-and-permute* operation. The copy-and-permute operation expands the protograph $\mathcal{G}_{\mathrm{ptg}}$ by a factor of $k$. The parameter $k$ is called the *expansion factor* (or *lifting degree*).

Let $\mathbf{H}_{\mathrm{ptg}}(k, k)$ be the adjacency matrix of $\mathcal{G}_{\mathrm{ptg}}(k, k)$ which is a $km \times kn$ matrix over GF(2) with rows corresponding to the $km$ CNs in $\mathcal{G}_{\mathrm{ptg}}(k, k)$ and columns corresponding to the $kn$ VNs in $\mathcal{G}_{\mathrm{ptg}}(k, k)$. If the expansion factor $k$ is large enough, $\mathbf{H}_{\mathrm{ptg}}(k, k)$ is a low-density matrix. Using $\mathbf{H}_{\mathrm{ptg}}(k, k)$ as a parity-check matrix, the null space over GF(2) of $\mathbf{H}_{\mathrm{ptg}}(k, k)$ gives a PTG-LDPC code, denoted by $C_{\mathrm{ptg}}$, whose Tanner graph is $\mathcal{G}_{\mathrm{ptg}}(k, k)$. The edge permutations among the copies of $\mathcal{G}_{\mathrm{ptg}}$ should be carried out in such a way that the girth of the resultant Tanner graph $\mathcal{G}_{\mathrm{ptg}}(k, k)$ of $C_{\mathrm{ptg}}$ is at least 6. To achieve this, computer aid is generally needed.

The copy-and-permute construction of a PTG-LDPC code based on a protograph $\mathcal{G}_{\mathrm{ptg}}$ is a graph-theoretic construction which can be viewed from a matrix-theoretic point of view. Let

$$\mathbf{B}_{\mathrm{ptg}} = \begin{bmatrix} b_{0,0} & b_{0,1} & \cdots & b_{0,n-1} \\ b_{1,0} & b_{1,1} & \cdots & b_{1,n-1} \\ \vdots & \vdots & \ddots & \vdots \\ b_{m-1,0} & b_{m-1,1} & \cdots & b_{m-1,n-1} \end{bmatrix} \tag{15.1}$$

be the adjacency matrix of $\mathcal{G}_{\mathrm{ptg}}$ with $n$ VNs and $m$ CNs. It is an $m \times n$ matrix over *nonnegative integers* with rows corresponding to the $m$ CNs of $\mathcal{G}_{\mathrm{ptg}}$ and columns corresponding to the $n$ VNs of $\mathcal{G}_{\mathrm{ptg}}$. The entry $b_{i,j}$ at the location $(i, j)$ of $\mathbf{B}_{\mathrm{ptg}}$ represents the number of parallel edges connecting the $i$th CN $c_i$ and the $j$th VN $v_j$ in $\mathcal{G}_{\mathrm{ptg}}$. The adjacency matrix $\mathbf{B}_{\mathrm{ptg}}$ is called the *protomatrix* associated with $\mathcal{G}_{\mathrm{ptg}}$.

Based on the restriction on the copy-and-permute operation performed on the $k$ copies of $\mathcal{G}_{\mathrm{ptg}}$ to connect them into a Tanner graph $\mathcal{G}_{\mathrm{ptg}}(k, k)$, we readily see that the connections of the $k$ type-$i$ CNs in $\Omega_i$ to the $k$ type-$j$ VNs in $\Phi_j$ via edges of type-$(i, j)$ can be specified by a $k \times k$ regular matrix $\mathbf{A}_{i,j}$ over GF(2) as follows:

$$\mathbf{A}_{i,j} = \begin{bmatrix} (a_{i,j})_{0,0} & (a_{i,j})_{0,1} & \cdots & (a_{i,j})_{0,k-1} \\ (a_{i,j})_{1,0} & (a_{i,j})_{1,1} & \cdots & (a_{i,j})_{1,k-1} \\ \vdots & \vdots & \ddots & \vdots \\ (a_{i,j})_{k-1,0} & (a_{i,j})_{k-1,1} & \cdots & (a_{i,j})_{k-1,k-1} \end{bmatrix} . \tag{15.2}$$

The rows of $\mathbf{A}_{i,j}$ correspond to the $k$ type-$i$ CNs in $\Omega_i$ and the columns of $\mathbf{A}_{i,j}$ correspond to the $k$ type-$j$ VNs in $\Phi_j$. For $0 \leq l, t < k$, the entry $(a_{i,j})_{l,t}$ in $\mathbf{A}_{i,j}$ at the location $(l, t)$ is 1 if and only if the $l$th type-$i$ CN in $\Omega_i$ and the $t$th type-$j$ VN in $\Phi_j$ are connected by a type $(i, j)$-edge in $\mathcal{G}_{\mathrm{ptg}}(k, k)$; otherwise, $(a_{i,j})_{l,t} = 0$. Because the $i$th CN $c_i$ and the $j$th VN $v_j$ are connected by $b_{i,j}$ parallel edges in $\mathcal{G}_{\mathrm{ptg}}$, it follows from the restriction on the edge permutations among the $k$ copies of $\mathcal{G}_{\mathrm{ptg}}$ that each column (row) of $\mathbf{A}_{i,j}$ contains $b_{i,j}$ 1-entries. Hence, $\mathbf{A}_{i,j}$ is a $k \times k$ regular matrix with both column and row weights $b_{i,j}$. If $b_{i,j} = 0$, $\mathbf{A}_{i,j}$ is a $k \times k$ ZM and if $b_{i,j} = 1$, $\mathbf{A}_{i,j}$ is a $k \times k$ permutation matrix (PM). The matrix $\mathbf{A}_{i,j}$ is called the *connection matrix* for the type-$i$ CNs in $\Omega_i$ and type-$j$ VNs in $\Phi_j$. Hence, the adjacency matrix $\mathbf{H}_{\mathrm{ptg}}(k, k)$ of the Tanner graph $\mathcal{G}_{\text{ptg}}(k,k)$ of the PTG-LDPC code $C_{\text{ptg}}$ constructed based on $\mathcal{G}_{\text{ptg}}$ is an $m \times n$ array of regular matrices of size $k \times k$ as follows:

$$\mathbf{H}_{\text{ptg}}(k,k) = [\mathbf{A}_{i,j}]_{0 \leq i < m, 0 \leq j < n} = \begin{bmatrix} \mathbf{A}_{0,0} & \mathbf{A}_{0,1} & \cdots & \mathbf{A}_{0,n-1} \\ \mathbf{A}_{1,0} & \mathbf{A}_{1,1} & \cdots & \mathbf{A}_{1,n-1} \\ \vdots & \vdots & \ddots & \vdots \\ \mathbf{A}_{m-1,0} & \mathbf{A}_{m-1,1} & \cdots & \mathbf{A}_{m-1,n-1} \end{bmatrix}, \quad (15.3)$$

which is the parity-check matrix of the PTG-LDPC code $C_{\text{ptg}}$.

Based on the above matrix interpretation of edge permutations to connect copies of a protograph $\mathcal{G}_{\text{ptg}}$, the copy-and-permute process to construct a PTG-LDPC code can be expressed matrix-theoretically in the following four steps: (1) design a protograph $\mathcal{G}_{\text{ptg}}$ with $n$ VNs and $m$ CNs; (2) form the protomatrix $\mathbf{B}_{\text{ptg}} = [b_{i,j}]_{0 \leq i < m, 0 \leq j < n}$ over nonnegative integers associated with $\mathcal{G}_{\text{ptg}}$; (3) replace each entry $b_{i,j}$, $0 \leq i < m, 0 \leq j < n$, in $\mathbf{B}_{\text{ptg}}$ by a $k \times k$ regular matrix $\mathbf{A}_{i,j}$ over GF(2) with both column and row weights $b_{i,j}$ to form an $m \times n$ array $\mathbf{H}_{\text{ptg}}(k,k)$ of regular matrices of size $k \times k$; and (4) take the null space over GF(2) of $\mathbf{H}_{\text{ptg}}(k,k)$ to produce a PTG-LDPC code $C_{\text{ptg}}$. In this case, permutations of edges among the copies of a protograph $\mathcal{G}_{\text{ptg}}$ are performed by replacing each entry $b_{i,j}$ in the protomatrix $\mathbf{B}_{\text{ptg}}$ of $\mathcal{G}_{\text{ptg}}$ with a connection matrix $\mathbf{A}_{i,j}$, $0 \leq i < m, 0 \leq j < n$. The connection matrix $\mathbf{A}_{i,j}$ specifies the connections between the $k$ type-$i$ CNs and the $k$ type-$j$ VNs in the $k$ copies of $\mathcal{G}_{\text{ptg}}$. Replacement of $b_{i,j}$ by $\mathbf{A}_{i,j}$ is referred to as the *matrix-dispersion* of the entry $b_{i,j}$ in $\mathbf{B}_{\text{ptg}}$. The array $\mathbf{H}_{\text{ptg}}(k,k)$ is referred to as the $k \times k$ *matrix-dispersion* of the protomatrix $\mathbf{B}_{\text{ptg}}$. Finding the connection matrices, $\mathbf{A}_{i,j}$, $0 \leq i < m, 0 \leq j < n$, to avoid short cycles in the Tanner graph $\mathcal{G}_{\text{ptg}}(k,k)$ associated with the parity-check matrix $\mathbf{H}_{\text{ptg}}(k,k)$ may require computer searches.

In fact, by using the above matrix-theoretic approach to construct a PTG-LDPC code, we can skip the first step. First, we design a protomatrix $\mathbf{B}_{\text{ptg}}$ over the nonnegative integers. Second, we carry out the matrix-dispersion of $\mathbf{B}_{\text{ptg}}$ to form the parity-check matrix $\mathbf{H}_{\text{ptg}}(k,k)$. Then, take the null space over GF(2) of $\mathbf{H}_{\text{ptg}}(k,k)$ to produce a desired PTG-LDPC code $C_{\text{ptg}}$. In this manner, the matrix-theoretic construction of a PTG-LDPC code is similar to the algebraic construction of an LDPC code based on a partial geometry or a finite field using CPM-dispersion of a base matrix as presented in Chapters 12–14.

Preferably, the copy-and-permute operation on the copies of a protograph is carried out in such a way (with computer aid) that the connection matrices are *circulants*. In this case, the parity-check matrix $\mathbf{H}_{\text{ptg}}(k,k)$ of a PTG-LDPC code $C_{\text{ptg}}$ is an $m \times n$ array of circulants and/or ZMs of size $k \times k$ and the code $C_{\text{ptg}}$ given by the null space over GF(2) of $\mathbf{H}_{\text{ptg}}(k,k)$ is a QC-PTG-LDPC code.

The error performance of a PTG-LDPC code depends on the choice of the protograph (protomatrix), its VN and CN degree distributions (column and row weight distributions), and the permutations of the edges among the copies (connection matrices) to avoid short cycles in the Tanner graph of the resultant code. Protographs (or protomatrices) are generally designed with good decoding thresholds based on a computer-aided density evolution search procedure [2, 3]. A good decoding threshold, in general, gives an LDPC code with a good waterfall performance, *but not necessarily a low error-floor or a fast rate of decoding convergence.*

**Example 15.1** Consider the protograph $\mathcal{G}_{\mathrm{ptg}}$ shown in Fig. 15.1(a), which consists of three VNs, two CNs, and six edges. The VN $v_0$ is connected to the CN $c_0$ by two parallel edges. The protomatrix $\mathbf{B}_{\mathrm{ptg}}$ associated with $\mathcal{G}_{\mathrm{ptg}}$ is a $2 \times 3$ matrix given as follows:

$$\mathbf{B}_{\mathrm{ptg}} = [b_{i,j}]_{0 \leq i < 2, 0 \leq j < 3} = \begin{bmatrix} 2 & 1 & 1 \\ 0 & 1 & 1 \end{bmatrix}. \tag{15.4}$$

Suppose we choose an expansion factor $k = 3$ and take three copies of $\mathcal{G}_{\mathrm{ptg}}$ as shown in Fig. 15.1(b). The nine VNs in the three copies of $\mathcal{G}_{\mathrm{ptg}}$ are grouped into three types, type-0 $\Phi_0 = \{v_{0,0}, v_{1,0}, v_{2,0}\}$, type-1 $\Phi_1 = \{v_{0,1}, v_{1,1}, v_{2,1}\}$, and type-2 $\Phi_2 = \{v_{0,2}, v_{1,2}, v_{2,2}\}$, each consisting of three copies of a VN in $\mathcal{G}_{\mathrm{ptg}}$, as shown in Fig. 15.1(b). The six CNs in the three copies of $\mathcal{G}_{\mathrm{ptg}}$ are grouped into two types, type-0 $\Omega_0 = \{c_{0,0}, c_{1,0}, c_{2,0}\}$ and type-1 $\Omega_1 = \{c_{0,1}, c_{1,1}, c_{2,1}\}$, each consisting of three copies of a CN in $\mathcal{G}_{\mathrm{ptg}}$, as shown in Fig. 15.1(b). Denote the two parallel edges connecting VN $v_0$ and CN $c_0$ in $\mathcal{G}_{\mathrm{ptg}}$ by $(0,0)_0$ and $(0,0)_1$. The 18 edges $(i,j)$s in the three copies of $\mathcal{G}_{\mathrm{ptg}}$ are grouped into six sets, $\mathcal{E}_0 = \{(0,0)_0, (0,0)_0, (0,0)_0\}$, $\mathcal{E}_1 = \{(0,0)_1, (0,0)_1, (0,0)_1\}$, $\mathcal{E}_2 = \{(0,1), (0,1), (0,1)\}$, $\mathcal{E}_3 = \{(0,2), (0,2), (0,2)\}$, $\mathcal{E}_4 = \{(1,1), (1,1), (1,1)\}$, and $\mathcal{E}_5 = \{(1,2), (1,2), (1,2)\}$.

Suppose we permute the edges among the three copies of $\mathcal{G}_{\mathrm{ptg}}$ and connect the three types of VNs and the two types of CNs following the restriction on edge permutation and connection. This results in a connected bipartite graph $\mathcal{G}_{\mathrm{ptg}}(3,3)$ shown in Fig. 15.1(c) which consists of nine VNs, six CNs, and 18 edges. The girth of $\mathcal{G}_{\mathrm{ptg}}(3,3)$ is 6. A cycle of length 6 is shown in Fig. 15.1(c) marked by boldface lines.

From the connections of the nine VNs and the six CNs in $\mathcal{G}_{\mathrm{ptg}}(3,3)$ shown in Fig. 15.1(c), we find the adjacency matrix $\mathbf{H}_{\mathrm{ptg}}(3,3)$ of $\mathcal{G}_{\mathrm{ptg}}(3,3)$ which is a $6 \times 9$ matrix over GF(2) given as follows:

$$\mathbf{H}_{\mathrm{ptg}}(3,3) = \begin{bmatrix} 1 & 1 & 0 & 0 & 0 & 1 & 0 & 0 & 1 \\ 0 & 1 & 1 & 1 & 0 & 0 & 1 & 0 & 0 \\ 1 & 0 & 1 & 0 & 1 & 0 & 0 & 1 & 0 \\ 0 & 0 & 0 & 0 & 1 & 0 & 1 & 0 & 0 \\ 0 & 0 & 0 & 0 & 0 & 1 & 0 & 1 & 0 \\ 0 & 0 & 0 & 1 & 0 & 0 & 0 & 0 & 1 \end{bmatrix}. \tag{15.5}$$

It is straightforward to check that $\mathbf{H}_{\mathrm{ptg}}(3,3)$ satisfies the RC-constraint. From Fig. 15.1(c), we identify the six connection matrices that specify the

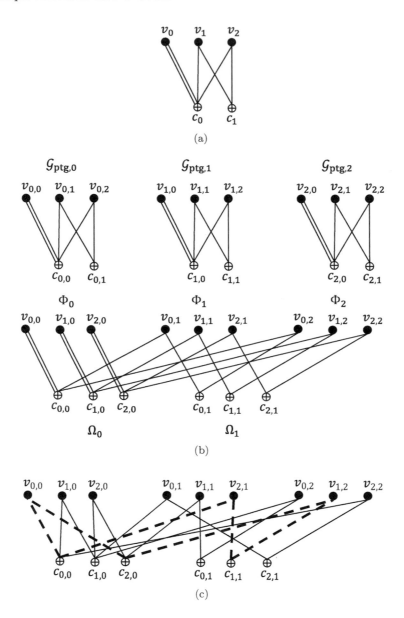

Figure 15.1 (a) The protograph $\mathcal{G}_{\text{ptg}}$, (b) three copies of the protograph $\mathcal{G}_{\text{ptg}}$ and the grouping of their VNs and CNs, and (c) the connected bipartite graph $\mathcal{G}_{\text{ptg}}(3,3)$ given in Example 15.1.

connections of the three types of VNs and two types of CNs among the three copies of the protograph $\mathcal{G}_{\mathrm{ptg}}$ as follows:

$$\mathbf{A}_{0,0} = \begin{bmatrix} 1 & 1 & 0 \\ 0 & 1 & 1 \\ 1 & 0 & 1 \end{bmatrix}, \ \mathbf{A}_{0,1} = \begin{bmatrix} 0 & 0 & 1 \\ 1 & 0 & 0 \\ 0 & 1 & 0 \end{bmatrix}, \ \mathbf{A}_{0,2} = \begin{bmatrix} 0 & 0 & 1 \\ 1 & 0 & 0 \\ 0 & 1 & 0 \end{bmatrix},$$

$$\mathbf{A}_{1,0} = \begin{bmatrix} 0 & 0 & 0 \\ 0 & 0 & 0 \\ 0 & 0 & 0 \end{bmatrix}, \ \mathbf{A}_{1,1} = \begin{bmatrix} 0 & 1 & 0 \\ 0 & 0 & 1 \\ 1 & 0 & 0 \end{bmatrix}, \ \mathbf{A}_{1,2} = \begin{bmatrix} 1 & 0 & 0 \\ 0 & 1 & 0 \\ 0 & 0 & 1 \end{bmatrix}.$$

$$(15.6)$$

Because the VN $v_0$ and the CN $c_0$ in $\mathcal{G}_{\mathrm{ptg}}$ are connected by two parallel edges, the connection matrix $\mathbf{A}_{0,0}$ which connects the three type-0 VNs and three type-0 CNs has both column and row weights 2. Because the VN $v_0$ and the CN $c_1$ in $\mathcal{G}_{\mathrm{ptg}}$ are not connected, there is no connection between a type-0 VN and a type-1 CN in $\mathcal{G}_{\mathrm{ptg}}(3,3)$ and hence the matrix $\mathbf{A}_{1,0}$ is a $3 \times 3$ ZM. Furthermore, we find that each of the five nonzero connection matrices given in (15.6) is a circulant of size $3 \times 3$. Hence, $\mathbf{H}_{\mathrm{ptg}}(3,3)$ is a $2 \times 3$ array of five circulants and one ZM of size $3 \times 3$.

Using the matrix-theoretic construction approach, $\mathbf{H}_{\mathrm{ptg}}(3,3)$ can be obtained by replacing the six entries of $\mathbf{B}_{\mathrm{ptg}}$ by their corresponding connection matrices, i.e., for $0 \leq i < 2$ and $0 \leq j < 3$, the entry $b_{i,j}$ in $\mathbf{B}_{\mathrm{ptg}}$ is replaced by the connection matrix $\mathbf{A}_{i,j}$.

The null space over $\mathrm{GF}(2)$ of $\mathbf{H}_{\mathrm{ptg}}(3,3)$ gives a QC-PTG-LDPC code $C_{\mathrm{ptg}}$ of rate $1/3$ whose Tanner graph has girth 6. ▲▲

The major step in the copy-and-permute construction of a PTG-LDPC code is to permute the edges among the copies of a chosen protograph $\mathcal{G}_{\mathrm{ptg}}$ to obtain a Tanner graph without parallel edges and with girth at least 6. If there are too many parallel edges between pairs of VNs and CNs in $\mathcal{G}_{\mathrm{ptg}}$ and the expansion factor $k$ is large, permuting parallel edges among the copies of $\mathcal{G}_{\mathrm{ptg}}$ under the restriction of edge permutations may require a large amount of computer searches to avoid short cycles in the resultant Tanner graph $\mathcal{G}_{\mathrm{ptg}}(k,k)$. To simplify this problem, the copy-and-permute process can be carried out in two steps. The first step is to remove parallel edge. The second step is to construct the PTG-LDPC code with a proper chosen expansion factor $k$.

Let $b_{\max}$ denote the largest number of parallel edges connecting a VN and a CN in a protograph $\mathcal{G}_{\mathrm{ptg}}$ with $n$ VNs and $m$ CNs, i.e.,

$$b_{\max} = \max_{0 \leq i < m, 0 \leq j < n}\{b_{i,j}\}, \qquad (15.7)$$

where $b_{i,j}$ is the number of parallel edges connecting the $j$th VN $v_j$ to the $i$th CN $c_i$ in $\mathcal{G}_{\mathrm{ptg}}$. Let $c$ be a positive integer such that $c \geq b_{\max}$. In the first step of the copy-and-permute process for constructing a PTG-LDPC code based on $\mathcal{G}_{\mathrm{ptg}}$, we take $c$ copies of $\mathcal{G}_{\mathrm{ptg}}$ and permute the edges among these $c$ copies to obtain a connected bipartite graph, denoted by $\mathcal{G}_{\mathrm{ptg},c}$, which is composed of $cn$ VNs and $cm$ CNs and contains no parallel edges. The subscript "$c$" in $\mathcal{G}_{\mathrm{ptg},c}$ stands for the number of copies of $\mathcal{G}_{\mathrm{ptg}}$ taken in the first step.

In the second step of the copy-and-permute process, we use $\mathcal{G}_{\text{ptg},c}$ as the protograph and choose an expansion factor $k$ such that $kcn$ gives the required code length. Next, we take $k$ copies of $\mathcal{G}_{\text{ptg},c}$ and connect these copies by permuting the edges among the copies. This results in the Tanner graph, denoted by $\mathcal{G}_{\text{ptg},c}(k,k)$, of a PTG-LDPC code $C_{\text{ptg}}$ of length $kcn$. Note that the Tanner graph $\mathcal{G}_{\text{ptg},c}(k,k)$ is composed of $kc$ copies of $\mathcal{G}_{\text{ptg}}$.

In the following, we give constructions of three PTG-LDPC codes based on two small protographs using the copy-and-permute construction process.

**Example 15.2** Consider the protograph $\mathcal{G}_{\text{ptg}}$ shown in Fig. 15.2(a) which consists of four VNs, each of degree 3, and two CNs, each of degree 6. The largest number of parallel edges connecting a VN and a CN in $\mathcal{G}_{\text{ptg}}$ is $b_{\max} = 2$. The protomatrix associated with $\mathcal{G}_{\text{ptg}}$ is a $2 \times 4$ matrix given by

$$\mathbf{B}_{\text{ptg}} = [b_{i,j}]_{0 \leq i < 2, 0 \leq j < 4} = \begin{bmatrix} 2 & 1 & 2 & 1 \\ 1 & 2 & 1 & 2 \end{bmatrix}. \tag{15.8}$$

Suppose we carry out the copy-and-permute process in two steps to construct a PTG-LDPC code of length 680 and rate 1/2 based on $\mathcal{G}_{\text{ptg}}$. In the first step of construction, we set the expansion factor to $c = b_{\max} = 2$. Take two copies of $\mathcal{G}_{\text{ptg}}$ and permute their edges to connect them. This results in a bipartite graph $\mathcal{G}_{\text{ptg},2}$ with eight VNs and four CNs as shown in Fig. 15.2(b) which contains no parallel edges. Each VN in $\mathcal{G}_{\text{ptg},2}$ has degree 3 and each CN in $\mathcal{G}_{\text{ptg},2}$ has degree 6. The protomatrix associated with $\mathcal{G}_{\text{ptg},2}$ is given by

$$\mathbf{B}_{\text{ptg},2} = \begin{bmatrix} 1 & 0 & 1 & 0 & 1 & 1 & 1 & 1 \\ 0 & 1 & 0 & 1 & 1 & 1 & 1 & 1 \\ 1 & 1 & 1 & 1 & 1 & 0 & 1 & 0 \\ 1 & 1 & 1 & 1 & 0 & 1 & 0 & 1 \end{bmatrix}. \tag{15.9}$$

In the second step of the copy-and-permute process, we use $\mathcal{G}_{\text{ptg},2}$ as a protograph and take $k = 85$ copies of $\mathcal{G}_{\text{ptg},2}$. Then, we permute the edges among the 85 copies of $\mathcal{G}_{\text{ptg},2}$ to connect them with computer aid. This results in a Tanner graph $\mathcal{G}_{\text{ptg},2}(85,85)$ with 680 VNs, each of degree 3, and 340 CNs, each of degree 6. The 680 VNs of $\mathcal{G}_{\text{ptg},2}(85,85)$ are grouped into eight types of VNs and the 340 CNs are grouped into four types of CNs. For $0 \leq i < 4, 0 \leq j < 8$, the 85 type-$j$ VNs are connected to the 85 type-$i$ CNs under the restriction on edge permutations and their connection matrix $\mathbf{A}_{i,j}$ is either a CPM of size $85 \times 85$ or a ZM of size $85 \times 85$. Hence, the adjacency matrix of $\mathcal{G}_{\text{ptg},2}(85,85)$ is a $4 \times 8$ array $\mathbf{H}_{\text{ptg},2}(85,85) = [\mathbf{A}_{i,j}]_{0 \leq i < 4, 0 \leq j < 8}$ of CPMs and ZMs of size $85 \times 85$ which is a $340 \times 680$ matrix with column and row weights 3 and 6, respectively.

Label the columns and rows of each CPM in $\mathbf{H}_{\text{ptg},2}(85,85)$ from 0 to 84. For $0 \leq i < 4, 0 \leq j < 8$, we specify the CPM $\mathbf{A}_{i,j}$ at the location $(i,j)$ in $\mathbf{H}_{\text{ptg},2}(85,85)$ by a nonnegative integer $l$, $0 \leq l < 85$, which is the location of the single 1-component of the generator of the CPM $\mathbf{A}_{i,j}$. There are many possible permutations of edges among the 85 copies of $\mathcal{G}_{\text{ptg},2}$, each of which results in a $4 \times 8$ array of CPMs and/or ZMs of size $85 \times 85$. One such permutation results in the following $4 \times 8$ array of CPMs and ZMs of size $85 \times 85$:

$$\mathbf{H}_{\mathrm{ptg},2}(85,85) = \begin{bmatrix} 2 & -\infty & 7 & -\infty & 20 & 31 & 48 & 54 \\ -\infty & 10 & -\infty & 26 & 40 & 62 & 11 & 23 \\ 6 & 15 & 21 & 39 & 60 & -\infty & 59 & -\infty \\ 8 & 20 & 28 & 52 & -\infty & 39 & -\infty & 46 \end{bmatrix}, \qquad (15.10)$$

where $-\infty$ represents a ZM.

The null space over GF(2) of $\mathbf{H}_{\mathrm{ptg},2}(85,85)$ gives a $(680,340)$ QC-PTG-LDPC code $C_{\mathrm{ptg}}$. The Tanner graph $\mathcal{G}_{\mathrm{ptg},2}(85,85)$ of $C_{\mathrm{ptg}}$ has girth 8 and contains 1020 cycles of 8. The total number of cycles of lengths 8, 10, 12, and 14 in $\mathcal{G}_{\mathrm{ptg},2}(85,85)$ is 817 105. The BER and BLER performances of $C_{\mathrm{ptg}}$ over the AWGN channel decoded with 5, 10, and 50 iterations of the MSA scaled by a factor of 0.75 are shown in Fig. 15.2(c). The constructed QC-PTG-LDPC code turns out to be identical to the $(680,340)$ QC-RS-LDPC code with minimum distance 34 given in Example 14.18. Furthermore, the protomatrix $\mathbf{B}_{\mathrm{ptg},2}$ given by (15.9) for the second step of the copy-and-permute construction process is identical to the masking matrix given by (14.5).

In the second step of the copy-and-permute process, suppose we set the expansion fact $k$ to 511 and take 511 copies of $\mathcal{G}_{\mathrm{ptg},2}$. Permuting the edges among the 511 copies of $\mathcal{G}_{\mathrm{ptg},2}$ under the restriction on edge permutations with computer aid, we obtain a Tanner graph $\mathcal{G}_{\mathrm{ptg},2}(511,511)$ with 4088 VNs and 2040 CNs which has girth 8 and 1022 cycles of length 8. The numbers of short cycles of lengths 8, 10, 12, and 14 in $\mathcal{G}_{\mathrm{ptg},2}(511,511)$ are 1022, 14 308, 141 547, and 1 016 890, respectively. The total number of short cycles in $\mathcal{G}_{\mathrm{ptg},2}(511,511)$ is 1 173 767. The adjacency matrix of $\mathcal{G}_{\mathrm{ptg},2}(511,511)$ is a $4 \times 8$ array $\mathbf{H}_{\mathrm{ptg},2}(511,511)$ of CPMs and ZMs of size $511 \times 511$ given below:

$$\mathbf{H}_{\mathrm{ptg},2}(511,511) = \begin{bmatrix} 2 & -\infty & 9 & -\infty & 26 & 42 & 64 & 72 \\ -\infty & 10 & -\infty & 30 & 52 & 84 & 128 & 144 \\ 6 & 15 & 27 & 45 & 78 & -\infty & 192 & -\infty \\ 8 & 20 & 36 & 60 & -\infty & 168 & -\infty & 288 \end{bmatrix}. \qquad (15.11)$$

The array $\mathbf{H}_{\mathrm{ptg},2}(511,511)$ is a $2044 \times 4088$ matrix over GF(2) with column and row weights 3 and 6, respectively. The null space over GF(2) of $\mathbf{H}_{\mathrm{ptg},2}(511,511)$ gives a $(3,6)$-regular $(4088,2044)$ QC-PTG-LDPC code $C^*_{\mathrm{ptg}}$ with rate 1/2. The BER and BLER performances of $C^*_{\mathrm{ptg}}$ over the AWGN channel decoded with 5, 10, and 50 iterations of the MSA scaled by a factor of 0.75 are shown in Fig. 15.2(d). The $(4088,2044)$ QC-PTG-LDPC code $C_{\mathrm{ptg}}$ turns out to be identical to the $(4088,2044)$ QC-RS-LDPC code given in Example 14.17.

The UEBR and UEBLR performances of the two constructed codes over the BEC are plotted in Fig. 15.3, which shows that the two codes also perform well over the BEC. ▲▲

In Example 15.2, we constructed two regular QC-PTG-LDPC codes of different lengths with the same rate based on the same protograph using two different expansion factors in the second step of the copy-and-permute process. In the next example, we construct an irregular QC-PTG-LDPC code based on a protograph which was designed in [3].

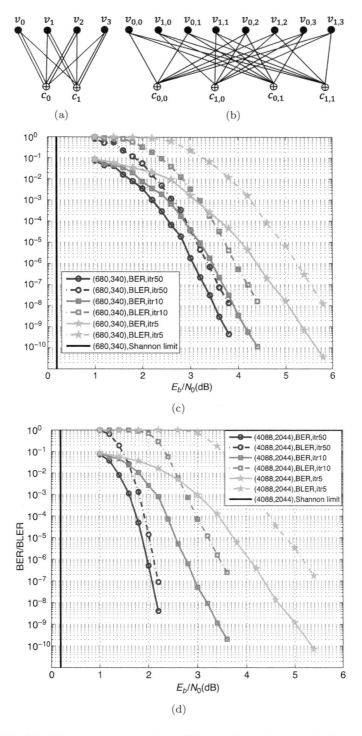

Figure 15.2 (a) The protograph $\mathcal{G}_{\mathrm{ptg}}$, (b) the bipartite graph $\mathcal{G}_{\mathrm{ptg},2}$, (c) the performances of the $(680, 340)$ QC-PTG-LDPC code, and (d) the performances of the $(4088, 2044)$ QC-PTG-LDPC code given in Example 15.2.

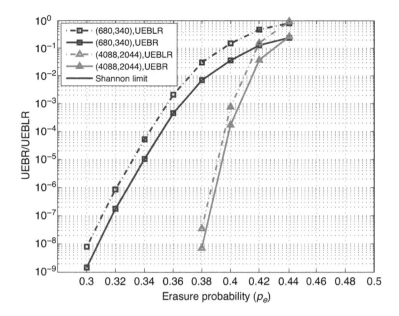

Figure 15.3 The UEBR and UEBLR performances of the $(680, 340)$ and $(4088, 2044)$ QC-PTG-LDPC codes given in Example 15.2 over BEC.

**Example 15.3** The protograph $\mathcal{G}_{\text{ptg}}$ used for code construction in this example is shown in Fig. 15.4(a) which was designed in [3]. It is composed of eight VNs and four CNs. Among the eight VNs in $\mathcal{G}_{\text{ptg}}$, there are one VN of degree 16 and seven VNs of degree 3. The four CNs in $\mathcal{G}_{\text{ptg}}$ have degrees 8, 10, 7, and 12, respectively. The VN $v_0$ is connected to each of the four CNs with four parallel edges. The largest number of parallel edges connecting a VN and a CN in $\mathcal{G}_{\text{ptg}}$ is $b_{\max} = 4$. The protomatrix associated with $\mathcal{G}_{\text{ptg}}$ is a $4 \times 8$ matrix given below:

$$\mathbf{B}_{\text{ptg}} = \begin{bmatrix} 4 & 2 & 0 & 0 & 1 & 0 & 0 & 1 \\ 4 & 0 & 1 & 1 & 1 & 1 & 1 & 1 \\ 4 & 1 & 0 & 0 & 1 & 0 & 1 & 0 \\ 4 & 0 & 2 & 2 & 0 & 2 & 1 & 1 \end{bmatrix}. \tag{15.12}$$

Suppose we want to construct a PTG-LDPC code of length 5792 with rate $1/2$ based on $\mathcal{G}_{\text{ptg}}$ using the copy-and-permute construction in two steps. In the first step, we set the expansion factor $c = b_{\max} = 4$. Take four copies of $\mathcal{G}_{\text{ptg}}$ and connect them by permuting the edges among the copies. This results in a bipartite graph $\mathcal{G}_{\text{ptg},4}$ with 32 VNs and 16 CNs which has no parallel edge. The graph $\mathcal{G}_{\text{ptg},4}$ is specified by the following $16 \times 32$ adjacency matrix:

$$\mathbf{B}_{\mathrm{ptg},4} = \begin{bmatrix}
1&1&0&0&0&0&0&1&1&1&0&0&1&0&0&0&1&0&0&0&0&0&0&0&1&0&0&0&0&0&0&0 \\
1&0&1&0&0&0&1&0&1&0&0&1&0&0&0&1&1&0&0&0&1&0&0&0&1&0&0&0&0&1&0&0 \\
1&1&0&0&0&0&0&0&1&0&0&0&1&0&0&0&1&0&0&0&0&1&0&1&0&0&0&0&0&0&0&0 \\
1&0&0&0&0&1&1&0&1&0&0&1&0&1&0&0&1&0&1&0&0&0&0&0&1&0&1&1&0&0&0&1 \\
1&0&0&0&0&0&0&0&1&1&0&0&0&0&0&0&1&1&1&0&0&1&0&0&0&1&0&0&0&0&0&0 \\
1&0&0&0&0&1&0&0&1&0&1&0&0&0&1&0&1&0&0&1&0&0&0&1&1&0&0&0&1&0&0&0 \\
1&0&0&0&0&0&0&1&1&0&0&0&0&0&0&1&0&0&0&1&0&0&0&0&1&0&0&0&0&0&1&0 \\
1&0&1&1&0&0&0&1&1&0&0&0&0&1&1&0&1&0&0&1&0&1&0&0&1&0&1&0&0&0&0&0 \\
1&0&0&0&0&0&0&0&1&0&0&0&0&0&0&0&1&1&0&0&0&0&0&1&1&1&0&0&1&0&0&0 \\
1&0&0&0&1&0&0&0&1&0&0&0&0&1&0&0&1&0&1&0&0&0&1&0&1&0&0&1&0&0&0&1 \\
1&0&0&0&0&0&1&0&1&0&0&0&0&0&0&0&1&1&0&0&0&0&0&1&0&0&0&1&0&0&0 \\
1&0&1&0&0&0&0&0&1&0&1&1&0&0&0&1&1&0&0&0&1&1&0&1&0&1&0&0&1&0&0 \\
1&1&0&0&1&0&0&0&1&0&0&0&0&0&0&0&0&1&0&0&0&0&0&0&1&1&0&0&0&0&0&1 \\
1&0&0&1&0&0&0&1&1&0&0&0&1&0&0&0&1&0&0&0&1&0&0&1&0&1&0&0&0&1&0 \\
1&0&0&0&1&0&0&0&1&0&0&0&0&0&0&1&0&1&0&0&0&0&0&0&1&1&0&0&0&0&0&0 \\
1&0&0&1&0&1&0&0&1&0&1&0&1&0&0&0&0&0&1&0&1&1&0&0&0&1&1&0&0&0&0&1&1&0
\end{bmatrix}.$$

In the second step of the copy-and-permute construction process, we use $\mathcal{G}_{\mathrm{ptg},4}$ as the protograph. Set the expansion factor $k = 181$ and take 181 copies of $\mathcal{G}_{\mathrm{ptg},4}$. Permuting the edges among the 181 copies of $\mathcal{G}_{\mathrm{ptg},4}$ under the restriction on edge permutations, we obtain a Tanner graph $\mathcal{G}_{\mathrm{ptg},4}(181,181)$ with 5792 VNs and 2896 CNs. The 5792 VNs of $\mathcal{G}_{\mathrm{ptg},4}(181,181)$ are grouped into 32 types of VNs, each consisting of 181 VNs, and the 2896 CNs are grouped into 16 types of CNs, each consisting of 181 CNs. For $0 \le i < 16$ and $0 \le j < 32$, the connection matrix $\mathbf{A}_{i,j}$ for the 181 type-$i$ CNs and the 181 type-$j$ VNs in $\mathcal{G}_{\mathrm{ptg},4}(181,181)$ is either a CPM of size $181 \times 181$ or a ZM of size $181 \times 181$. Hence, the adjacency matrix of $\mathcal{G}_{\mathrm{ptg},4}(181,181)$ is a $16 \times 32$ array $\mathbf{H}_{\mathrm{ptg},4}(181,181)$ of CPMs and ZMs of size $181 \times 181$ which is a $2896 \times 5792$ matrix with two different column weights 3 and 16, and 4 different row weights 7, 8, 10, and 12.

The null space over GF(2) of $\mathbf{H}_{\mathrm{ptg},4}(181,181)$ gives an irregular $(5792, 2896)$ QC-PTG-LDPC code with rate 1/2. The BER and BLER performances of the code over the AWGN channel decoded with 100 iterations of the SPA are shown in Fig. 15.4(b). The error performance of the $(5792, 2896)^*$ PTG-LDPC code constructed in [3, Fig. 11] are also shown in Fig. 15.4(b). The code constructed in this example performs very close to the code given in [3, Fig. 11]. ▲▲

For a given protograph with different choices of expansion factors, $c$ and $k$, in the two steps of the copy-and-permute process, we can construct a family of QC-PTG-LDPC codes of different lengths with the same rate as demonstrated in Example 15.2. Note that the copy-and-permute process can be carried out in more than two steps.

## 15.2 A Matrix-Theoretic Method for Constructing Protograph-Based LDPC Codes

In this section, we present an algebraic method for constructing PTG-LDPC codes from a *matrix-theoretic* point of view. The construction is based on *expanding the protomatrix* associated with a protograph rather than the protograph itself.

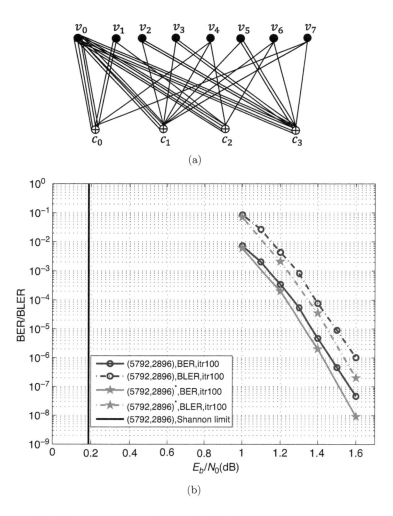

(a)

(b)

Figure 15.4 (a) The protograph $\mathcal{G}_{\mathrm{ptg}}$ and (b) the BER and BLER performances of the $(5792, 2896)$ QC-PTG-LDPC code given in Example 15.3.

Let $\mathcal{G}_{\mathrm{ptg}}$ be a protograph with $n$ VNs, $v_0$, $v_1, \ldots, v_{n-1}$, and $m$ CNs, $c_0$, $c_1, \ldots, c_{m-1}$, and let $\mathbf{B}_{\mathrm{ptg}} = [b_{i,j}]_{0 \le i < m, 0 \le j < n}$ be the protomatrix associated with $\mathcal{G}_{\mathrm{ptg}}$ which is an $m \times n$ matrix over nonnegative integers.

Let $k$ be a positive integer such that $k \ge b_{\max} = \max_{0 \le i < m, 0 \le j < n}\{b_{i,j}\}$. Decompose the protomatrix $\mathbf{B}_{\mathrm{ptg}}$ into $k$ matrices $\mathbf{D}_0, \mathbf{D}_1, \ldots, \mathbf{D}_{k-1}$, each of size $m \times n$. In decomposition, the entry $b_{i,j}$ at the location $(i, j)$, $0 \le i < m, 0 \le j < n$, in $\mathbf{B}_{\mathrm{ptg}}$ is split into $b_{i,j}$ 1-entries which are distributed in $b_{i,j}$ different matrices among the $k$ matrices $\mathbf{D}_0, \mathbf{D}_1, \ldots, \mathbf{D}_{k-1}$ at the location $(i, j)$s. In the decomposition of $\mathbf{B}_{\mathrm{ptg}}$, ZMs are allowed, i.e., there may be ZMs among the matrices $\mathbf{D}_0, \mathbf{D}_1, \ldots, \mathbf{D}_{k-1}$. Therefore, for $0 \le e < k$, each matrix $\mathbf{D}_e$ is either an $m \times n$ binary matrix or an $m \times n$ ZM. Furthermore, we require that the sum (using integer addition) of the matrices $\mathbf{D}_0, \mathbf{D}_1, \ldots, \mathbf{D}_{k-1}$ gives the protomatrix $\mathbf{B}_{\mathrm{ptg}}$, i.e.,

$$\mathbf{D}_0 + \mathbf{D}_1 + \cdots + \mathbf{D}_{k-1} = \mathbf{B}_{\mathrm{ptg}}. \tag{15.13}$$

The matrices $\mathbf{D}_0, \mathbf{D}_1, \ldots, \mathbf{D}_{k-1}$ are referred to as the *constituent matrices* of $\mathbf{B}_{\mathrm{ptg}}$ in the decomposition. The set $\Psi = \{\mathbf{D}_0, \mathbf{D}_1, \ldots, \mathbf{D}_{k-1}\}$ is called the *decomposition set* of $\mathbf{B}_{\mathrm{ptg}}$, which may contain identical matrices. Note that, if the protograph $\mathcal{G}_{\mathrm{ptg}}$ contains no parallel edges, its protomatrix $\mathbf{B}_{\mathrm{ptg}}$ is a binary matrix. In this case, the constituent matrices $\mathbf{D}_0, \mathbf{D}_1, \ldots, \mathbf{D}_{k-1}$ of $\mathbf{B}_{\mathrm{ptg}}$ are *mutually disjoint*, i.e., no two constituent matrices have 1-entries at the same location. The parameter $k$ is referred to as the *decomposition factor*.

Arrange the $k$ matrices $\mathbf{D}_0, \mathbf{D}_1, \ldots, \mathbf{D}_{k-1}$ into a *row-block*, and then cyclically shift this row-block $k-1$ times, each time by $n$ positions (or one constituent matrix $\mathbf{D}_e$) to the right. The cyclic-shift operation results in a $k \times k$ array $\mathbf{H}_{\mathrm{ptg,cyc}}(m, n)$ of $m \times n$ matrices with a *block-cyclic structure* shown below:

$$\mathbf{H}_{\mathrm{ptg,cyc}}(m, n) = \begin{bmatrix} \mathbf{D}_0 & \mathbf{D}_1 & \cdots & \mathbf{D}_{k-1} \\ \mathbf{D}_{k-1} & \mathbf{D}_0 & \cdots & \mathbf{D}_{k-2} \\ \vdots & \vdots & \ddots & \vdots \\ \mathbf{D}_1 & \mathbf{D}_2 & \cdots & \mathbf{D}_0 \end{bmatrix}. \tag{15.14}$$

The array $\mathbf{H}_{\mathrm{ptg,cyc}}(m, n)$ consists of $k$ row-blocks and $k$ column-blocks of matrices of size $m \times n$. Each row-block is the *cyclic-shift* of the row-block above it one constituent matrix to the right (or $n$ positions to the right) and the top row-block is the cyclic-shift of the last (bottom) row-block one constituent matrix to the right. Each column-block is the cyclic-shift of the column-block on its left one constituent matrix (or $m$ positions) downward and the first (leftmost) column-block is the cyclic-shift of the last (rightmost) column-block one constituent matrix downward. The subscript "cyc" in $\mathbf{H}_{\mathrm{ptg,cyc}}(m, n)$ stands for "block-cyclic structure." Because the constituent matrices of each row-block (or column-block) are the same, i.e., $\mathbf{D}_0, \mathbf{D}_1, \ldots, \mathbf{D}_{k-1}$, the integer sum of the constituent matrices in each row-block (or each column-block) of $\mathbf{H}_{\mathrm{ptg,cyc}}(m, n)$ is a copy of the protomatrix $\mathbf{B}_{\mathrm{ptg}}$. Therefore, the block-cyclic array $\mathbf{H}_{\mathrm{ptg,cyc}}(m, n)$ is composed of $k$ copies of the protomatrix $\mathbf{B}_{\mathrm{ptg}}$ associated with the protograph $\mathcal{G}_{\mathrm{ptg}}$, i.e., an expansion of $\mathbf{B}_{\mathrm{ptg}}$ by a factor of $k$.

Let $\mathcal{G}_{\mathrm{ptg,cyc}}(k, k)$ denote the Tanner graph associated with the $k \times k$ block-cyclic array $\mathbf{H}_{\mathrm{ptg,cyc}}(m, n)$. Then, $\mathcal{G}_{\mathrm{ptg,cyc}}(k, k)$ is composed of $k$ copies of the protograph $\mathcal{G}_{\mathrm{ptg}}$ interconnected by permuting the edges among the copies. Suppose we take each column-block of $\mathbf{H}_{\mathrm{ptg,cyc}}(m, n)$ as a copy of the protomatrix $\mathbf{B}_{\mathrm{ptg}}$ (in the decomposed form). Then, the interconnections of the $k$ copies of the protograph $\mathcal{G}_{\mathrm{ptg}}$ are based on (or specified by) the rows of $\mathbf{H}_{\mathrm{ptg,cyc}}(m, n)$. If we take each row-block of $\mathbf{H}_{\mathrm{ptg,cyc}}(m, n)$ as a copy of the protomatrix $\mathbf{B}_{\mathrm{ptg}}$, then the interconnections of the $k$ copies of the protograph $\mathcal{G}_{\mathrm{ptg}}$ are based on (or specified by) the columns of $\mathbf{H}_{\mathrm{ptg,cyc}}(m, n)$. Hence, the Tanner graph $\mathcal{G}_{\mathrm{ptg,cyc}}(k, k)$ associated with the $k \times k$ block-cyclic array $\mathbf{H}_{\mathrm{ptg,cyc}}(m, n)$ is an expansion (by a factor of $k$) of the protograph $\mathcal{G}_{\mathrm{ptg}}$. Hence, the expansion factor of a protograph is the decomposition factor of its associated protomatrix and vice versa.

Using the $k \times k$ block-cyclic array $\mathbf{H}_{\mathrm{ptg,cyc}}(m,n)$ as the parity-check matrix, its null space gives a QC-PTG-LDPC code, denoted by $C_{\mathrm{ptg,cyc}}$, of length $nk$ in the form for which cyclically shifting each codeword in $C_{\mathrm{ptg,cyc}}$ $n$ positions to the right also gives a codeword in $C_{\mathrm{ptg,cyc}}$. This simply follows from the block-cyclic structure of $\mathbf{H}_{\mathrm{ptg,cyc}}(m,n)$ and represents a definition of QC codes first introduced in the 1960s [26–31] (see Section 4.13).

The $k \times k$ block-cyclic array $\mathbf{H}_{\mathrm{ptg,cyc}}(m,n)$ can be put in a QC form, commonly used nowadays, as an array of circulants of the same size. Label the rows and columns of $\mathbf{H}_{\mathrm{ptg,cyc}}(m,n)$ from 0 to $mk-1$ and 0 to $nk-1$, respectively. Define the following index sequences:

$$\pi_{\mathrm{row}}^{(0)} = [0, m, 2m, \ldots, (k-1)m], \tag{15.15}$$

$$\pi_{\mathrm{row}} = [\pi_{\mathrm{row}}^{(0)}, \pi_{\mathrm{row}}^{(0)} + 1, \ldots, \pi_{\mathrm{row}}^{(0)} + m - 1], \tag{15.16}$$

$$\pi_{\mathrm{col}}^{(0)} = [0, n, 2n, \ldots, (k-1)n], \tag{15.17}$$

$$\pi_{\mathrm{col}} = [\pi_{\mathrm{col}}^{(0)}, \pi_{\mathrm{col}}^{(0)} + 1, \ldots, \pi_{\mathrm{col}}^{(0)} + n - 1]. \tag{15.18}$$

Then, $\pi_{\mathrm{row}}$ and $\pi_{\mathrm{col}}$ define a permutation of the rows and a permutation of the columns of $\mathbf{H}_{\mathrm{ptg,cyc}}(m,n)$, respectively. If we first permute the rows of $\mathbf{H}_{\mathrm{ptg,cyc}}(m,n)$ based on $\pi_{\mathrm{row}}$ and then permute the columns of $\mathbf{H}_{\mathrm{ptg,cyc}}(m,n)$ based on $\pi_{\mathrm{col}}$, we obtain an $m \times n$ array $\mathbf{H}_{\mathrm{ptg,qc}}(k,k)$ of $k \times k$ matrices as follows:

$$\mathbf{H}_{\mathrm{ptg,qc}}(k,k) = \begin{bmatrix} \mathbf{A}_{0,0} & \mathbf{A}_{0,1} & \cdots & \mathbf{A}_{0,n-1} \\ \mathbf{A}_{1,0} & \mathbf{A}_{1,1} & \cdots & \mathbf{A}_{1,n-1} \\ \vdots & \vdots & \ddots & \vdots \\ \mathbf{A}_{m-1,0} & \mathbf{A}_{m-1,1} & \cdots & \mathbf{A}_{m-1,n-1} \end{bmatrix}, \tag{15.19}$$

where the subscript "qc" in $\mathbf{H}_{\mathrm{ptg,qc}}(k,k)$ stands for the "quasi-cyclic structure" of $\mathbf{H}_{\mathrm{ptg,qc}}(k,k)$.

Let $\mathcal{G}_{\mathrm{ptg,qc}}(k,k)$ denote the Tanner graph associated with the $m \times n$ array $\mathbf{H}_{\mathrm{ptg,qc}}(k,k)$ given by (15.19). Then, $\mathcal{G}_{\mathrm{ptg,qc}}(k,k)$ and the Tanner graph $\mathcal{G}_{\mathrm{ptg,cyc}}(k,k)$ associated with the $k \times k$ block-cyclic array $\mathbf{H}_{\mathrm{ptg,cyc}}(m,n)$ are *isomorphic* (or *structurally identical*). One can be obtained from the other by permuting the locations of VNs and CNs. For $0 \leq i < m$ and $0 \leq j < n$, the $k \times k$ matrix $\mathbf{A}_{i,j}$ in $\mathbf{H}_{\mathrm{ptg,qc}}(k,k)$ is the connection matrix that specifies the edge connections between the $k$ type-$i$ CNs and the $k$ type-$j$ VNs in $\mathcal{G}_{\mathrm{ptg,qc}}(k,k)$ among the $k$ copies of the protograph $\mathcal{G}_{\mathrm{ptg}}$. Based on the block-cyclic structure of the array $\mathbf{H}_{\mathrm{ptg,cyc}}(m,n)$ and the row and column permutations, $\pi_{\mathrm{row}}$ and $\pi_{\mathrm{col}}$, performed on $\mathbf{H}_{\mathrm{ptg,cyc}}(m,n)$, we readily see that each connection matrix $\mathbf{A}_{i,j}$ in $\mathbf{H}_{\mathrm{ptg,qc}}(k,k)$ given by (15.19) is either a circulant or a ZM of size $k \times k$.

The generator of each circulant $\mathbf{A}_{i,j}$ in $\mathbf{H}_{\mathrm{ptg,qc}}(k,k)$ can be determined directly from the decomposition set $\{\mathbf{D}_0, \mathbf{D}_1, \ldots, \mathbf{D}_{k-1}\}$ of $\mathbf{B}_{\mathrm{ptg}}$. For $0 \leq i < m, 0 \leq j < n$ and $0 \leq e < k$, let $g_{i,j,e}$ be the entry at location $(i,j)$ of the $m \times n$ constituent matrix $\mathbf{D}_e$. Then, the $k$-tuple $\mathbf{g}_{i,j} = (g_{i,j,0}, g_{i,j,1}, \ldots, g_{i,j,k-1})$ is the generator of the constituent circulant $\mathbf{A}_{i,j}$ of the array $\mathbf{H}_{\mathrm{ptg,qc}}(k,k)$. (If $\mathbf{g}_{i,j}$ is a zero $k$-tuple, $\mathbf{A}_{i,j}$ is a ZM of size $k \times k$.) Hence, we can construct the array $\mathbf{H}_{\mathrm{ptg,qc}}(k,k)$ from these circulant generators directly without performing the row and column permutations on $\mathbf{H}_{\mathrm{ptg,cyc}}(m,n)$. This is achieved simply by replacing the nonzero entry $b_{i,j}$ at location $(i,j)$ of the protomatrix $\mathbf{B}_{\mathrm{ptg}} = [b_{i,j}]_{0 \leq i < m, 0 \leq j < n}$ of the protograph $\mathcal{G}_{\mathrm{ptg}}$ with the $k \times k$ connection circulant $\mathbf{A}_{i,j}$ generated by $\mathbf{g}_{i,j} = (g_{i,j,0}, g_{i,j,1}, \ldots, g_{i,j,k-1})$. If $b_{i,j} = 0$, $\mathbf{A}_{i,j}$ is a $k \times k$ ZM. Let $\delta$ be the number of nonzero entries in the protomatrix $\mathbf{B}_{\mathrm{ptg}}$. Then, the array $\mathbf{H}_{\mathrm{ptg,qc}}(k,k)$ consists of $\delta$ circulants of size $k \times k$ (they may not be all distinct). If the entries of the protomatrix $\mathbf{B}_{\mathrm{ptg}}$ are binary, then $\mathbf{H}_{\mathrm{ptg,qc}}(k,k)$ is an $m \times n$ array of CPMs and/or ZMs of size $k \times k$.

Using $\mathbf{H}_{\mathrm{ptg,qc}}(k,k)$ as the parity-check matrix, its null space over GF(2) gives a QC-PTG-LDPC code, denoted by $C_{\mathrm{ptg,qc}}$, which is isomorphic to the QC-PTG-LDPC code $C_{\mathrm{ptg,cyc}}$ given by the null space over GF(2) of $\mathbf{H}_{\mathrm{ptg,cyc}}(m,n)$. A codeword $\mathbf{v}$ in $C_{\mathrm{ptg,qc}}$ consists of $n$ sections, each containing $k$ bits. If we cyclically shift all $n$ sections of the codeword $\mathbf{v}$ simultaneously one position to the right within each section, we obtain another codeword in $C_{\mathrm{ptg,qc}}$. This QC-structure is referred to as a *section-wise cyclic structure* (see Section 4.13). Therefore, a QC-LDPC code can be put in either the block-cyclic form or the section-wise cyclic form. All the QC-LDPC codes constructed in Chapters 12–14 are section-wise cyclic codes. If a QC-LDPC code has both block and section-wise cyclic structures without transforming from one form to the other by row and column permutations, it is called a *doubly QC-LDPC code*.

With the above construction of a QC-PTG-LDPC code, the graphical copy-and-permute process is replaced by a matrix decomposition process. It is clear that there are many possible decompositions of the protomatrix $\mathbf{B}_{\mathrm{ptg}}$, each resulting in a different $k \times k$ array $\mathbf{H}_{\mathrm{ptg,cyc}}(m,n)$ (or $m \times n$ array $\mathbf{H}_{\mathrm{ptg,qc}}(k,k)$) and hence a different block-cyclic QC-PTG-LDPC code $C_{\mathrm{ptg,cyc}}$ (or a different section-wise cyclic QC-PTG-LDPC code $C_{\mathrm{ptg,qc}}$).

The girth and cycle distributions of the Tanner graph $\mathcal{G}_{\mathrm{ptg,cyc}}(k,k)$ (or $\mathcal{G}_{\mathrm{ptg,qc}}(k,k)$), and hence the performance of a QC-PTG-LDPC code $C_{\mathrm{ptg,cyc}}$ (or $C_{\mathrm{ptg,qc}}$), very much depend on the decomposition of the protomatrix $\mathbf{B}_{\mathrm{ptg}}$, i.e., the distribution of its nonzero entries, and the *arrangement* (or *order*) of the nonzero constituent matrices in the first row-block of the cyclic array $\mathbf{H}_{\mathrm{ptg,cyc}}(m,n)$ given by (15.14). The distribution of the nonzero entries in $\mathbf{B}_{\mathrm{ptg}}$ and the order of the nonzero constituent matrices in the decomposition set $\Psi = \{\mathbf{D}_0, \mathbf{D}_1, \ldots, \mathbf{D}_{k-1}\}$ should be carried out in such a way that the array $\mathbf{H}_{\mathrm{ptg,cyc}}(m,n)$ satisfies the RC-constraint. It is clear that if $\mathbf{H}_{\mathrm{ptg,cyc}}(m,n)$ satisfies the RC-constraint, the array $\mathbf{H}_{\mathrm{ptg,qc}}(k,k)$ also satisfies the RC-constraint. In this case, the Tanner graph $\mathcal{G}_{\mathrm{ptg,cyc}}(k,k)$ ($\mathcal{G}_{\mathrm{ptg,qc}}(k,k)$) has girth at least 6.

Once we choose a specific order-arrangement of the nonzero matrices in the decomposition $\{\mathbf{D}_0, \mathbf{D}_1, \ldots, \mathbf{D}_{k-1}\}$, the top row-block of the block-cyclic array $\mathbf{H}_{\mathrm{ptg,cyc}}(m,n)$ is formed with the nonzero matrices arranged in the chosen order. To specify a nonzero constituent matrix $\mathbf{D}_e$, we need only to specify the locations of its 1-entries.

The general guidelines to decompose the protomatrix $\mathbf{B}_{\mathrm{ptg}}$ of the protograph $\mathcal{G}_{\mathrm{ptg}}$ are: (1) each nonzero constituent matrix $\mathbf{D}_e$ in the decomposition set $\Psi = \{\mathbf{D}_0, \mathbf{D}_1, \ldots, \mathbf{D}_{k-1}\}$ satisfies the RC-constraint; (2) any two nonzero constituent matrices, $\mathbf{D}_e$ and $\mathbf{D}_j$, satisfy the *pair-wise* RC-constraint; and (3) the number of 1-entries in each nonzero constituent matrix is as small as possible. The composition of the 1-entries in each nonzero constituent matrix in $\{\mathbf{D}_0, \mathbf{D}_1, \ldots, \mathbf{D}_{k-1}\}$ is a major factor in making $\mathbf{H}_{\text{ptg,cyc}}(m, n)$ to satisfy the RC-constraint. It is clear that, if the protomatrix $\mathbf{B}_{\text{ptg}}$ satisfies the RC-constraint, the nonzero constituent matrices in $\{\mathbf{D}_0, \mathbf{D}_1, \ldots, \mathbf{D}_{k-1}\}$ satisfy both the RC- and pair-wise RC-constraints and thus the array $\mathbf{H}_{\text{ptg,cyc}}(m, n)$ also satisfies the RC-constraint. In decomposing a protomatrix $\mathbf{B}_{\text{ptg}}$ to obtain a PTG-LDPC code whose Tanner graph has no short cycles or a small number of short cycles, computer aid is needed.

The algebraic construction of a QC-PTG-LDPC code presented above is simply based on the decomposition of the protomatrix $\mathbf{B}_{\text{ptg}}$ associated with a chosen protograph $\mathcal{G}_{\text{ptg}}$. Because the protomatrix is simply the adjacency matrix of the protograph $\mathcal{G}_{\text{ptg}}$ which uniquely specifies the protograph, in design of a PTG-LDPC code, we only need to design the protomatrix $\mathbf{B}_{\text{ptg}}$. Then, the algebraic construction of a QC-PTG-LDPC code of length $kn$ with rate $(n - m)/n$ (or close to it) is carried out with the following steps: (1) design an appropriate $m \times n$ protomatrix $\mathbf{B}_{\text{ptg}}$ and choose the decomposition factor $k$; (2) decompose $\mathbf{B}_{\text{ptg}}$ into $k$ constituent matrices $\mathbf{D}_0, \mathbf{D}_1, \ldots, \mathbf{D}_{k-1}$; (3) form a $k \times k$ array $\mathbf{H}_{\text{ptg,cyc}}(m, n)$ of matrices of size $m \times n$ in the form of (15.14); (4) permute the rows and columns of the $k \times k$ array $\mathbf{H}_{\text{ptg,cyc}}(m, n)$ based on the column and row permutations, $\pi_{\text{col}}$ and $\pi_{\text{row}}$, defined by (15.18) and (15.16), respectively, to obtain an $m \times n$ array $\mathbf{H}_{\text{ptg,qc}}(k, k)$ of circulants and/or ZMs of size $k \times k$ in the form of (15.19); and (5) take the null space over $\text{GF}(2)$ of $\mathbf{H}_{\text{ptg,qc}}(k, k)$ to produce a QC-PTG-LDPC code with section-wise cyclic structure. If a QC-PTG-LDPC code with block-cyclic structure is preferred, we can use the block-cyclic array $\mathbf{H}_{\text{ptg,cyc}}(m, n)$ constructed in Step (3) as the parity-check matrix.

If we need only the array $\mathbf{H}_{\text{ptg,qc}}(k, k)$ as the parity-check matrix of a QC-PTG-LDPC code, we can construct it directly from the constituent matrices $\mathbf{D}_0, \mathbf{D}_1, \ldots, \mathbf{D}_{k-1}$ of the protomatrix $\mathbf{B}_{\text{ptg}}$. Find the generators of all the connection circulants in $\mathbf{H}_{\text{ptg,qc}}(k, k)$ from $\mathbf{D}_0, \mathbf{D}_1, \ldots, \mathbf{D}_{k-1}$. From these generators, we construct all the connection circulants $\mathbf{A}_{i,j}$, $0 \leq i < m, 0 \leq j < n$, in the $m \times n$ array $\mathbf{H}_{\text{ptg,qc}}(k, k)$. Then, replace each nonzero entry $b_{i,j}$ in $\mathbf{B}_{\text{ptg}} = [b_{i,j}]_{0 \leq i < m, 0 \leq j < n}$ by its corresponding connection circulant $\mathbf{A}_{i,j}$ and a zero entry in $\mathbf{B}_{\text{ptg}}$ by a $k \times k$ ZM. Hence, the code construction process is reduced to three steps: design of a protomatrix $\mathbf{B}_{\text{ptg}}$, decomposition of the protomatrix $\mathbf{B}_{\text{ptg}}$, and replacement of the entries in $\mathbf{B}_{\text{ptg}}$ by their corresponding connection circulants or ZMs.

If the protomatrix $\mathbf{B}_{\text{ptg}}$ contains too many entries greater than one, we can carry out the decomposition of $\mathbf{B}_{\text{ptg}}$ in two steps as described earlier. Let $c$ be a positive integer such that $c \geq b_{\max} = \max_{0 \leq i < m, 0 \leq j < n}\{b_{i,j}\}$. In the first step of decomposition, we decompose $\mathbf{B}_{\text{ptg}}$ into a $cm \times cn$ binary matrix $\mathbf{B}_{\text{ptg},c}$. The matrix $\mathbf{B}_{\text{ptg},c}$ is the adjacency matrix of a bipartite graph $\mathcal{G}_{\text{ptg},c}$ obtained by taking $c$ copies of $\mathcal{G}_{\text{ptg}}$ and connecting them by permuting edges among the $c$ copies. In the second step of decomposition, we use $\mathbf{B}_{\text{ptg},c}$ as a protomatrix. Decompose $\mathbf{B}_{\text{ptg},c}$ by a factor of $k$ and form a $cm \times cn$ array $\mathbf{H}_{\text{ptg},c,\text{qc}}(k, k)$ of circulants and/or ZMs of size $k \times k$ in the form of (15.19). Then, the null space over GF(2) of $\mathbf{H}_{\text{ptg,c,qc}}(k, k)$ gives a QC-PTG-LDPC code of length $kcn$. The Tanner graph $\mathcal{G}_{\text{ptg,c,qc}}(k, k)$ associated with $\mathbf{H}_{\text{ptg,c,qc}}(k, k)$ is a bipartite graph with $kcn$ VNs and $kcm$ CNs obtained by taking $kc$ copies of the protograph $\mathcal{G}_{\text{ptg}}$ and connecting them by edges permutations in two steps. Note that the two-step decomposition of the protomatrix $\mathbf{B}_{\text{ptg}}$ is equivalent to the two-step expansion of the protograph $\mathcal{G}_{\text{ptg}}$.

In the following, we use a simple example to illustrate the process of decomposing a protomatrix to construct a QC-PTG-LDPC code, in either the block-cyclic form or section-wise cyclic form.

**Example 15.4** In this example, we replace the copy-and-permute of the protograph $\mathcal{G}_{\text{ptg}}$ shown in Fig. 15.1(a) of Example 15.1 by decomposing its associated protomatrix. The protomatrix associated with $\mathcal{G}_{\text{ptg}}$ is a $2 \times 3$ matrix:

$$\mathbf{B}_{\text{ptg}} = \begin{bmatrix} 2 & 1 & 1 \\ 0 & 1 & 1 \end{bmatrix}. \tag{15.20}$$

Set the decomposition factor $k = 3$ and decompose $\mathbf{B}_{\text{ptg}}$ into the following three $2 \times 3$ constituent matrices:

$$\mathbf{D}_0 = \begin{bmatrix} 1 & 0 & 0 \\ 0 & 0 & 1 \end{bmatrix}, \mathbf{D}_1 = \begin{bmatrix} 1 & 0 & 0 \\ 0 & 1 & 0 \end{bmatrix}, \mathbf{D}_2 = \begin{bmatrix} 0 & 1 & 1 \\ 0 & 0 & 0 \end{bmatrix}. \tag{15.21}$$

Note that $\mathbf{D}_0 + \mathbf{D}_1 + \mathbf{D}_2 = \mathbf{B}_{\text{ptg}}$. Using these three constituent matrices, we form the following $3 \times 3$ array of $2 \times 3$ matrices with block-cyclic structure in the form of (15.14):

$$\mathbf{H}_{\text{ptg,cyc}}(2, 3) = \begin{bmatrix} 1 & 0 & 0 & 1 & 0 & 0 & 0 & 1 & 1 \\ 0 & 0 & 1 & 0 & 1 & 0 & 0 & 0 & 0 \\ \hline 0 & 1 & 1 & 1 & 0 & 0 & 1 & 0 & 0 \\ 0 & 0 & 0 & 0 & 0 & 1 & 0 & 1 & 0 \\ \hline 1 & 0 & 0 & 0 & 1 & 1 & 1 & 0 & 0 \\ 0 & 1 & 0 & 0 & 0 & 0 & 0 & 0 & 1 \end{bmatrix}. \tag{15.22}$$

Permute the columns and rows of $\mathbf{H}_{\text{ptg,cyc}}(2, 3)$ based on the column and row permutations, $\pi_{\text{col}}$ and $\pi_{\text{row}}$, defined by (15.18) and (15.16) with $m = 2$, $n = 3$, and $k = 3$, respectively. This results in the following $2 \times 3$ array of circulants and ZMs of size $3 \times 3$ in the form given by (15.19):

$$\mathbf{H}_{\text{ptg,qc}}(3, 3) = \begin{bmatrix} \mathbf{A}_{0,0} & \mathbf{A}_{0,1} & \mathbf{A}_{0,2} \\ \mathbf{A}_{1,0} & \mathbf{A}_{1,1} & \mathbf{A}_{1,2} \end{bmatrix}$$

$$= \begin{bmatrix} 1 & 1 & 0 & 0 & 0 & 1 & 0 & 0 & 1 \\ 0 & 1 & 1 & 1 & 0 & 0 & 1 & 0 & 0 \\ 1 & 0 & 1 & 0 & 1 & 0 & 0 & 1 & 0 \\ \hline 0 & 0 & 0 & 0 & 1 & 0 & 1 & 0 & 0 \\ 0 & 0 & 0 & 0 & 0 & 1 & 0 & 1 & 0 \\ 0 & 0 & 0 & 1 & 0 & 0 & 0 & 0 & 1 \end{bmatrix}. \tag{15.23}$$

The above array $\mathbf{H}_{\mathrm{ptg,qc}}(3,3)$ is the same as the array given by (15.5) in Example 15.1. From the structure of the array $\mathbf{H}_{\mathrm{ptg,qc}}(3,3)$, we see that the array is obtained by replacing each nonzero entry in the protomatrix $\mathbf{B}_{\mathrm{ptg}}$ by a circulant of size $3 \times 3$ whose generator can be determined directly from the three constituent matrices $\mathbf{D}_0$, $\mathbf{D}_1$, and $\mathbf{D}_2$ of $\mathbf{B}_{\mathrm{ptg}}$ given by (15.21). The generators are:

$$\mathbf{g}_{0,0} = (1\ 1\ 0),\ \mathbf{g}_{0,1} = (0\ 0\ 1),\ \mathbf{g}_{0,2} = (0\ 0\ 1),$$
$$\mathbf{g}_{1,0} = (0\ 0\ 0),\ \mathbf{g}_{1,1} = (0\ 1\ 0),\ \mathbf{g}_{1,2} = (1\ 0\ 0).$$

The Tanner graph $\mathcal{G}_{\mathrm{ptg,qc}}(3,3)$ associated with $\mathbf{H}_{\mathrm{ptg,qc}}(3,3)$ is shown in Fig. 15.1(c) and is composed of three copies of the protograph $\mathcal{G}_{\mathrm{ptg}}$ connected by permuting the edges among the copies.

If we use the array $\mathbf{H}_{\mathrm{ptg,cyc}}(2,3)$ given by (15.22) as the parity-check matrix, the null space over GF(2) of $\mathbf{H}_{\mathrm{ptg,cyc}}(2,3)$ gives a $(9,3)$ QC-PTG-LDPC code $C_{\mathrm{ptg,cyc}}$ with block-cyclic structure. If we use the array $\mathbf{H}_{\mathrm{ptg,qc}}(3,3)$ given by (15.23) as the parity-check matrix, the null space over GF(2) of $\mathbf{H}_{\mathrm{ptg,qc}}(3,3)$ gives a $(9,3)$ QC-PTG-LDPC code with section-wise cyclic structure. ▲▲

In the next example, we reconstruct the two QC-PTG-LDPC codes, namely, the $(680, 340)$ and the $(4088, 2044)$ codes given in Example 15.2, using the algebraic decomposition method in two steps.

**Example 15.5** Consider the $2 \times 4$ protomatrix given by (15.8) which is reproduced below:

$$\mathbf{B}_{\mathrm{ptg}} = \begin{bmatrix} 2 & 1 & 2 & 1 \\ 1 & 2 & 1 & 2 \end{bmatrix}. \tag{15.24}$$

The protomatrix $\mathbf{B}_{\mathrm{ptg}}$ is the adjacency matrix of the protograph $\mathcal{G}_{\mathrm{ptg}}$ shown in Fig. 15.2(a) which contains parallel edges.

To construct the $(680, 340)$ QC-PTG-LDPC code $C_{\mathrm{ptg,qc}}$ given in Example 15.2, we decompose $\mathbf{B}_{\mathrm{ptg}}$ in two steps. In the first step, we choose a decomposition factor $c = 2$ and decompose $\mathbf{B}_{\mathrm{ptg}}$ into two constituent matrices:

$$\mathbf{D}_0 = \begin{bmatrix} 1 & 0 & 1 & 0 \\ 0 & 1 & 0 & 1 \end{bmatrix}, \mathbf{D}_1 = \begin{bmatrix} 1 & 1 & 1 & 1 \\ 1 & 1 & 1 & 1 \end{bmatrix}. \tag{15.25}$$

Using the two constituent matrices $\mathbf{D}_0$ and $\mathbf{D}_1$, we form the following $2 \times 2$ block-cyclic array of $2 \times 4$ matrices in the form of (15.14):

$$\mathbf{B}_{\mathrm{ptg,2}} = \begin{bmatrix} 1 & 0 & 1 & 0 & 1 & 1 & 1 & 1 \\ 0 & 1 & 0 & 1 & 1 & 1 & 1 & 1 \\ 1 & 1 & 1 & 1 & 1 & 0 & 1 & 0 \\ 1 & 1 & 1 & 1 & 0 & 1 & 0 & 1 \end{bmatrix}, \tag{15.26}$$

which is a $4 \times 8$ matrix with column and row weights 3 and 6, respectively. The matrix $\mathbf{B}_{\mathrm{ptg,2}}$ is the same as the one given by (15.9) and is the adjacency matrix of the graph $\mathcal{G}_{\mathrm{ptg},2}$ shown in Fig. 15.2(b). The graph $\mathcal{G}_{\mathrm{ptg},2}$ contains no parallel edges and is composed of two copies of $\mathcal{G}_{\mathrm{ptg}}$. The matrix $\mathbf{B}_{\mathrm{ptg},2}$ contains 24 1-entries.

Table 15.1 The nonzero constituent matrices of the protomatrix $\mathbf{B}_{\mathrm{ptg},2}$ and the locations of their 1-entries with decomposition factor $k = 85$ given in Example 15.5.

| $\mathbf{D}_e$ | 1-entries | $\mathbf{D}_e$ | 1-entries | $\mathbf{D}_e$ | 1-entries | $\mathbf{D}_e$ | 1-entries |
|---|---|---|---|---|---|---|---|
| $\mathbf{D}_2$ | $(0,0)$ | $\mathbf{D}_9$ | $(2,2)$ | $\mathbf{D}_{31}$ | $(0,5)$ | $\mathbf{D}_{50}$ | $(2,6)$ |
| $\mathbf{D}_3$ | $(1,0)$ | $\mathbf{D}_{10}$ | $(3,2)$ | $\mathbf{D}_{32}$ | $(1,5)$ | $\mathbf{D}_{51}$ | $(3,6)$ |
| $\mathbf{D}_4$ | $(2,0)$ | $\mathbf{D}_{20}$ | $(0,4)$ | $\mathbf{D}_{33}$ | $(2,5)$ | $\mathbf{D}_{54}$ | $(0,7)$ |
| $\mathbf{D}_5$ | $(3,0)$ | $\mathbf{D}_{21}$ | $(1,4)$ | $\mathbf{D}_{34}$ | $(3,5)$ | $\mathbf{D}_{55}$ | $(1,7)$ |
| $\mathbf{D}_7$ | $(0,2)$ | $\mathbf{D}_{22}$ | $(2,4)$ | $\mathbf{D}_{48}$ | $(0,6)$ | $\mathbf{D}_{56}$ | $(2,7)$ |
| $\mathbf{D}_8$ | $(1,2)$ | $\mathbf{D}_{23}$ | $(3,4)$ | $\mathbf{D}_{49}$ | $(1,6)$ | $\mathbf{D}_{57}$ | $(3,7)$ |

In the second step of decomposition, we use $\mathbf{B}_{\mathrm{ptg},2}$ as the protomatrix for code construction. Set the decomposition factor $k = 85$ and decompose $\mathbf{B}_{\mathrm{ptg},2}$ into 85 constituent matrices, $\mathbf{D}_0, \mathbf{D}_1, \ldots, \mathbf{D}_{84}$, each of size $4 \times 8$. Among these 85 constituent matrices of $\mathbf{B}_{\mathrm{ptg},2}$, 24 of them are nonzero matrices, each containing a single 1-entry, and all the other 61 matrices are ZMs. The 24 nonzero constituent matrices of $\mathbf{B}_{\mathrm{ptg},2}$ and the locations of their 1-entries are given in Table 15.1. In the table, the $(i, j)$ entry $(0 \le i < 4, 0 \le j < 8)$ in the "1-entries" column indicates that the 1-entry in the constituent matrix $\mathbf{D}_e$ is at the $i$th row and the $j$th column. Using the 85 constituent matrices $\mathbf{D}_0, \mathbf{D}_1, \ldots, \mathbf{D}_{84}$ of $\mathbf{B}_{\mathrm{ptg},2}$, we form an $85 \times 85$ array $\mathbf{H}_{\mathrm{ptg},2,\mathrm{cyc}}(4,8)$ of $4 \times 8$ matrices with block-cyclic structure in the form of (15.14).

Permuting the columns and rows of $\mathbf{H}_{\mathrm{ptg},2,\mathrm{cyc}}(4,8)$ using the column and row permutations, $\pi_{\mathrm{col}}$ and $\pi_{\mathrm{row}}$, defined by (15.18) and (15.16) with $m = 4$, $n = 8$, and $k = 85$, respectively, we obtain the following $4 \times 8$ array of CPMs and ZMs of size $85 \times 85$ in which the CPMs are represented by positive integers less than 85 as described in Example 15.2:

$$\mathbf{H}_{\mathrm{ptg},2,\mathrm{qc}}(85,85) = \begin{bmatrix} 2 & -\infty & 7 & -\infty & 20 & 31 & 48 & 54 \\ -\infty & 10 & -\infty & 26 & 40 & 62 & 11 & 23 \\ 6 & 15 & 21 & 39 & 60 & -\infty & 59 & -\infty \\ 8 & 20 & 28 & 52 & -\infty & 39 & -\infty & 46 \end{bmatrix},$$

where $-\infty$ represents a ZM. The array $\mathbf{H}_{\mathrm{ptg},2,\mathrm{qc}}(85,85)$ given above is the same as the array $\mathbf{H}_{\mathrm{ptg},2}(85,85)$ given by (15.10). Hence, the null space over GF(2) of $\mathbf{H}_{\mathrm{ptg},2,\mathrm{qc}}(85,85)$ gives the same $(680, 340)$ QC-PTG-LDPC code $C_{\mathrm{ptg},\mathrm{qc}}$ as the one constructed in Example 15.2 based on the protograph $\mathcal{G}_{\mathrm{ptg}}$ shown in

Fig. 15.2(a) using the copy-and-permute construction in two steps. Note that the array $\mathbf{H}_{\mathrm{ptg,2,qc}}(85, 85)$ can be constructed directly from the constituent matrices $\mathbf{D}_0, \mathbf{D}_1, \ldots, \mathbf{D}_{84}$.

To construct the $(4088, 2044)$ QC-PTG-LDPC code $C_{\mathrm{ptg,qc}}$ given in Example 15.2 with the algebraic decomposition method, in the second step of decomposition of the protomatrix $\mathbf{B}_{\mathrm{ptg,2}}$, we choose the decomposition factor $k = 511$ and decompose $\mathbf{B}_{\mathrm{ptg,2}}$ into 511 constituent matrices, $\mathbf{D}_0, \mathbf{D}_1, \ldots, \mathbf{D}_{510}$, each of size $4 \times 8$. Among these 511 matrices, 24 of them are nonzero matrices, each containing a single 1-entry, and all the other 487 matrices are ZMs. The locations of the 1-entries in these 24 nonzero matrices are given in Table 15.2. Using the constituent matrices, $\mathbf{D}_0, \mathbf{D}_1, \ldots, \mathbf{D}_{510}$, of $\mathbf{B}_{\mathrm{ptg,2}}$, we form a $511 \times 511$ array $\mathbf{H}_{\mathrm{ptg,2,cyc}}(4, 8)$ of $4 \times 8$ matrices with block-cyclic structure in the form of (15.14). The array $\mathbf{H}_{\mathrm{ptg,2,cyc}}(4, 8)$ is a $2044 \times 4088$ matrix with column and row weights 3 and 6, respectively.

Table 15.2 The nonzero constituent matrices of the protomatrix $\mathbf{B}_{\mathrm{ptg,2}}$ and the locations of their 1-entries with decomposition factor $k = 511$ given in Example 15.5.

| $\mathbf{D}_e$ | 1-entries | $\mathbf{D}_e$ | 1-entries | $\mathbf{D}_e$ | 1-entries | $\mathbf{D}_e$ | 1-entries |
|---|---|---|---|---|---|---|---|
| $\mathbf{D}_2$ | $(0,0)$ | $\mathbf{D}_{11}$ | $(2,2)$ | $\mathbf{D}_{42}$ | $(0,5)$ | $\mathbf{D}_{66}$ | $(2,6)$ |
| $\mathbf{D}_3$ | $(1,0)$ | $\mathbf{D}_{12}$ | $(3,2)$ | $\mathbf{D}_{43}$ | $(1,5)$ | $\mathbf{D}_{67}$ | $(3,6)$ |
| $\mathbf{D}_4$ | $(2,0)$ | $\mathbf{D}_{26}$ | $(0,4)$ | $\mathbf{D}_{44}$ | $(2,5)$ | $\mathbf{D}_{72}$ | $(0,7)$ |
| $\mathbf{D}_5$ | $(3,0)$ | $\mathbf{D}_{27}$ | $(1,4)$ | $\mathbf{D}_{45}$ | $(3,5)$ | $\mathbf{D}_{73}$ | $(1,7)$ |
| $\mathbf{D}_9$ | $(0,2)$ | $\mathbf{D}_{28}$ | $(2,4)$ | $\mathbf{D}_{64}$ | $(0,6)$ | $\mathbf{D}_{74}$ | $(2,7)$ |
| $\mathbf{D}_{10}$ | $(1,2)$ | $\mathbf{D}_{29}$ | $(3,4)$ | $\mathbf{D}_{65}$ | $(1,6)$ | $\mathbf{D}_{75}$ | $(3,7)$ |

Permute the columns and rows of $\mathbf{H}_{\mathrm{ptg,2,cyc}}(4, 8)$ using the column and row permutations, $\pi_{\mathrm{col}}$ and $\pi_{\mathrm{row}}$, defined by (15.18) and (15.16) with $m = 4$, $n = 8$, and $k = 511$, respectively. This results in the following $4 \times 8$ array of CPMs and ZMs of size $511 \times 511$ in which the CPMs are represented by nonnegative integers less than 511:

$$\mathbf{H}_{\mathrm{ptg,2,qc}}(511, 511) = \begin{bmatrix} 2 & -\infty & 9 & -\infty & 26 & 42 & 64 & 72 \\ -\infty & 10 & -\infty & 30 & 52 & 84 & 128 & 144 \\ 6 & 15 & 27 & 45 & 78 & -\infty & 192 & -\infty \\ 8 & 20 & 36 & 60 & -\infty & 168 & -\infty & 288 \end{bmatrix}. \quad (15.27)$$

The array $\mathbf{H}_{\mathrm{ptg,2,qc}}(511, 511)$ is identical to the array $\mathbf{H}_{\mathrm{ptg,2}}(511, 511)$ given by (15.11) in Example 15.2. Hence, the null space over $\mathrm{GF}(2)$ of $\mathbf{H}_{\mathrm{ptg,2,qc}}(511, 511)$ gives the same $(4088, 2044)$ QC-PTG-LDPC code constructed in Example 15.2 using the copy-and-permute construction method in two steps. ▲▲

**Example 15.6** In this example, we use the same protomatrix given by (15.26) (or (15.9)) in Example 15.5 (or Example 15.2) to construct another QC-PTG-LDPC code with rate 0.5 using a decomposition factor $k = 330$. Following the general decomposition guidelines, we decompose $\mathbf{B}_{\text{ptg},2}$ into 330 constituent matrices, $\mathbf{D}_0, \mathbf{D}_1, \ldots, \mathbf{D}_{329}$, each of size $4 \times 8$. Among these 330 matrices, 23 of them are nonzero matrices and all the other 307 matrices are ZMs. Among the 23 nonzero matrices, one contains two 1-entries and each of the other 22 contains a single 1-entry. The locations of the 1-entries in these 23 nonzero constituent matrices are given in Table 15.3. Using the constituent matrices, $\mathbf{D}_0, \mathbf{D}_1, \ldots, \mathbf{D}_{329}$, of $\mathbf{B}_{\text{ptg},2}$, we form a $330 \times 330$ array $\mathbf{H}_{\text{ptg},2,\text{cyc}}(4,8)$ of $4 \times 8$ matrices with block-cyclic structure in the form of (15.14).

Table 15.3 The nonzero constituent matrices of the protomatrix $\mathbf{B}_{\text{ptg},2}$ and the locations of their 1-entries with decomposition factor $k = 330$ given in Example 15.6.

| $\mathbf{D}_e$ | 1-entries | $\mathbf{D}_e$ | 1-entries | $\mathbf{D}_e$ | 1-entries | $\mathbf{D}_e$ | 1-entries |
|---|---|---|---|---|---|---|---|
| $\mathbf{D}_0$ | $(2,5)$ | $\mathbf{D}_{68}$ | $(0,7)$ | $\mathbf{D}_{81}$ | $(2,2)$ | $\mathbf{D}_{298}$ | $(1,6)$ |
| $\mathbf{D}_1$ | $(3,5)$ | $\mathbf{D}_{69}$ | $(1,7)$ | $\mathbf{D}_{82}$ | $(3,2)$ | $\mathbf{D}_{299}$ | $(2,6)$ |
| $\mathbf{D}_{39}$ | $(0,0)$ | $\mathbf{D}_{70}$ | $(2,7)$ | $\mathbf{D}_{294}$ | $(0,4)$ | $\mathbf{D}_{300}$ | $(3,6)$ |
| $\mathbf{D}_{40}$ | $(1,0)$ | $\mathbf{D}_{71}$ | $(3,7)$ | $\mathbf{D}_{295}$ | $(1,4)$ | $\mathbf{D}_{328}$ | $(0,5)$ |
| $\mathbf{D}_{41}$ | $(2,0)$ | $\mathbf{D}_{79}$ | $(0,2)$ | $\mathbf{D}_{296}$ | $(2,4)$ | $\mathbf{D}_{329}$ | $(1,5)$ |
| $\mathbf{D}_{42}$ | $(3,0)$ | $\mathbf{D}_{80}$ | $(1,2)$ | $\mathbf{D}_{297}$ | $(3,4), (0,6)$ | | |

Permuting the columns and rows of $\mathbf{H}_{\text{ptg},2,\text{cyc}}(4,8)$ based on the column and row permutations, $\pi_{\text{col}}$ and $\pi_{\text{row}}$, defined by (15.18) and (15.16) with $m = 4$ and $n = 8$, respectively, we obtain the following $4 \times 8$ array of CPMs and ZMs of size $330 \times 330$ in which the CPMs are represented by positive integers less than 330:

$$\mathbf{H}_{\text{ptg},2,\text{qc}}(330,330) = \begin{bmatrix} 39 & -\infty & 79 & -\infty & 294 & 328 & 297 & 68 \\ -\infty & 79 & -\infty & 75 & 158 & 295 & 148 & 274 \\ 206 & 209 & 287 & 324 & 163 & -\infty & 202 & -\infty \\ 181 & 271 & 275 & 143 & -\infty & 10 & -\infty & 284 \end{bmatrix}. \tag{15.28}$$

The array $\mathbf{H}_{\text{ptg},2,\text{qc}}(330,330)$ is a $1320 \times 2640$ matrix with column and row weights, 3 and 6, respectively. The null space over GF(2) of $\mathbf{H}_{\text{ptg},2,\text{qc}}(330,330)$ gives a $(2640, 1320)$ QC-PTG-LDPC code $C_{\text{ptg,qc}}$ with section-wise cyclic structure. The Tanner graph $\mathcal{G}_{\text{ptg},2,\text{qc}}(330,330)$ of the code has girth 8 and contains 990 cycles of length 8 and 8580 cycles of length 10. If we use the block-cyclic array $\mathbf{H}_{\text{ptg},2,\text{cyc}}(4,8)$ as the parity-check matrix, we obtain a $(2640, 1320)$ QC-PTG-LDPC code $C_{\text{ptg,cyc}}$ with block-cyclic structure.

The BER and BLER performances of the code $C_{\text{ptg,qc}}$ (or $C_{\text{ptg,cyc}}$) over the AWGN channel decoded with 5, 10, and 50 iterations of the MSA scaled by a factor of 0.75 are shown in Fig. 15.5(a). The code achieves a BER of $10^{-10}$ without a visible error-floor. The code also performs well the BEC as shown by Fig. 15.5(b). ▲▲

In the last two examples, we showed that using the same protomatrix with different decomposition factors, we can construct QC-PTG-LDPC codes of different lengths with the same rate. We also note that the protomatix given by (15.24) consists of two copies of the following $2 \times 2$ base matrix:

$$\mathbf{B}_0 = \begin{bmatrix} 2 & 1 \\ 1 & 2 \end{bmatrix}. \tag{15.29}$$

The graph $\mathcal{G}_0$ specified by this base matrix is shown in Fig. 15.6. The protograph $\mathcal{G}_{\mathrm{ptg}}$ shown in Fig. 15.2(a) is simply obtained by taking two copies of $\mathcal{G}_0$ and connecting them by edge permutations. The base protomatrix given by (15.29) can be used as a building block to construct a larger protomatrix.

**Example 15.7** In this example, we construct a QC-PTG-LDPC code of length 3060 with rate 2/3. In the construction, we use the following $2 \times 6$ protomatrix:

$$\mathbf{B}_{\mathrm{ptg}} = \begin{bmatrix} 2 & 1 & 2 & 1 & 2 & 1 \\ 1 & 2 & 1 & 2 & 1 & 2 \end{bmatrix}, \tag{15.30}$$

which consists of three copies of the base protomatrix $\mathbf{B}_0$ given by (15.29). The protograph $\mathcal{G}_{\mathrm{ptg}}$ specified by $\mathbf{B}_{\mathrm{ptg}}$ is shown in Fig. 15.7(a). The protograph $\mathcal{G}_{\mathrm{ptg}}$ consists of six VNs, each of degree 3, and two CNs, each of degree 9. The largest number $b_{\max}$ of parallel edges connecting a VN and a CN in $\mathcal{G}_{\mathrm{ptg}}$ is 2.

To construct the desired QC-PTG-LDPC code, we decompose the protomatrix $\mathbf{B}_{\mathrm{ptg}}$ in two steps. In the first step, we choose the decomposition factor $c = b_{\max} = 2$ and decompose $\mathbf{B}_{\mathrm{ptg}}$ into following two binary $2 \times 6$ constituent matrices:

$$\mathbf{D}_0 = \begin{bmatrix} 1 & 0 & 1 & 0 & 1 & 0 \\ 0 & 1 & 0 & 1 & 0 & 1 \end{bmatrix}, \mathbf{D}_1 = \begin{bmatrix} 1 & 1 & 1 & 1 & 1 & 1 \\ 1 & 1 & 1 & 1 & 1 & 1 \end{bmatrix}. \tag{15.31}$$

Using the two constituent matrices $\mathbf{D}_0$ and $\mathbf{D}_1$, we form the following $2 \times 2$ block-cyclic array of $2 \times 6$ matrices in the form of (15.14):

$$\mathbf{B}_{\mathrm{ptg},2} = \begin{bmatrix} 1 & 0 & 1 & 0 & 1 & 0 & 1 & 1 & 1 & 1 & 1 & 1 \\ 0 & 1 & 0 & 1 & 0 & 1 & 1 & 1 & 1 & 1 & 1 & 1 \\ 1 & 1 & 1 & 1 & 1 & 1 & 1 & 0 & 1 & 0 & 1 & 0 \\ 1 & 1 & 1 & 1 & 1 & 1 & 0 & 1 & 0 & 1 & 0 & 1 \end{bmatrix}. \tag{15.32}$$

The graphical representation of $\mathbf{B}_{\mathrm{ptg},2}$ is a bipartite graph $\mathcal{G}_{\mathrm{ptg},2}$ with 12 VNs, each of degree 3, and 4 CNs, each of degree 9.

In the second step of decomposing $\mathbf{B}_{\mathrm{ptg}}$, we use $\mathbf{B}_{\mathrm{ptg},2}$ as the protomatrix and choose the decomposition factor $k = 255$. Next, we decompose $\mathbf{B}_{\mathrm{ptg},2}$ into 255 constituent matrices, $\mathbf{D}_0, \mathbf{D}_1, \ldots, \mathbf{D}_{254}$, each of size $4 \times 12$. Among the 255 constituent matrices of $\mathbf{B}_{\mathrm{ptg},2}$, 36 of them are nonzero matrices and the

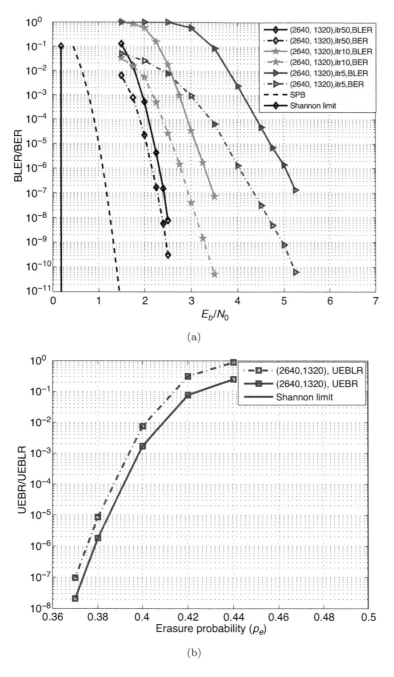

(a)

(b)

Figure 15.5 The performances of the $(2640, 1320)$ QC-PTG-LDPC code given in Example 15.6 over: (a) AWGN channel and (b) BEC.

Figure 15.6 The protograph $\mathcal{G}_0$ specified by the protomatrix $\mathbf{B}_0$ given by (15.29).

other 294 are ZMs. Each nonzero constituent matrix of $\mathbf{B}_{\mathrm{ptg},2}$ contains a single 1-entry. The 36 nonzero constituent matrices and the locations of their 1-entries are given in Table 15.4. Using the 255 constituent matrices $\mathbf{D}_0, \mathbf{D}_1, \ldots, \mathbf{D}_{254}$, we form a $255 \times 255$ block-cyclic array $\mathbf{H}_{\mathrm{ptg},2,\mathrm{cyc}}(4, 12)$ in the form (15.14) which is a $1020 \times 3060$ matrix with column and row weights 3 and 9, respectively.

Table 15.4 The nonzero constituent matrices of the protomatrix $\mathbf{B}_{\mathrm{ptg},2}$ and the locations of their 1-entries with decomposition factor $k = 255$ given in Example 15.7.

| $\mathbf{D}_e$ | 1-entries | $\mathbf{D}_e$ | 1-entries | $\mathbf{D}_e$ | 1-entries | $\mathbf{D}_e$ | 1-entries |
|---|---|---|---|---|---|---|---|
| $\mathbf{D}_0$ | $(0,0)$ | $\mathbf{D}_{12}$ | $(1,4)$ | $\mathbf{D}_{37}$ | $(2,7)$ | $\mathbf{D}_{52}$ | $(3,9)$ |
| $\mathbf{D}_1$ | $(1,0)$ | $\mathbf{D}_{13}$ | $(2,4)$ | $\mathbf{D}_{38}$ | $(3,7)$ | $\mathbf{D}_{90}$ | $(0,10)$ |
| $\mathbf{D}_2$ | $(2,0)$ | $\mathbf{D}_{14}$ | $(3,4)$ | $\mathbf{D}_{41}$ | $(0,8)$ | $\mathbf{D}_{91}$ | $(1,10)$ |
| $\mathbf{D}_3$ | $(3,0)$ | $\mathbf{D}_{24}$ | $(0,6)$ | $\mathbf{D}_{42}$ | $(1,8)$ | $\mathbf{D}_{92}$ | $(2,10)$ |
| $\mathbf{D}_4$ | $(0,2)$ | $\mathbf{D}_{25}$ | $(1,6)$ | $\mathbf{D}_{43}$ | $(2,8)$ | $\mathbf{D}_{93}$ | $(3,10)$ |
| $\mathbf{D}_5$ | $(1,2)$ | $\mathbf{D}_{26}$ | $(2,6)$ | $\mathbf{D}_{44}$ | $(3,8)$ | $\mathbf{D}_{225}$ | $(0,11)$ |
| $\mathbf{D}_6$ | $(2,2)$ | $\mathbf{D}_{27}$ | $(3,6)$ | $\mathbf{D}_{49}$ | $(0,9)$ | $\mathbf{D}_{226}$ | $(1,11)$ |
| $\mathbf{D}_7$ | $(3,2)$ | $\mathbf{D}_{35}$ | $(0,7)$ | $\mathbf{D}_{50}$ | $(1,9)$ | $\mathbf{D}_{227}$ | $(2,11)$ |
| $\mathbf{D}_{11}$ | $(0,4)$ | $\mathbf{D}_{36}$ | $(1,7)$ | $\mathbf{D}_{51}$ | $(2,9)$ | $\mathbf{D}_{228}$ | $(3,11)$ |

Permuting the columns and rows of $\mathbf{H}_{\mathrm{ptg},2,\mathrm{cyc}}(4, 12)$ based on the column and row permutations, $\pi_{\mathrm{col}}$ and $\pi_{\mathrm{row}}$, defined by (15.18) and (15.16) with $m = 4$, $n = 12$, and $k = 255$, respectively, we obtain the following $4 \times 12$ array of CPMs and ZMs of size $255 \times 255$ in which the CPMs are represented by positive integers less than 255:

$$\mathbf{H}_{\mathrm{ptg},2,\mathrm{qc}}(255, 255) = \begin{bmatrix} 0 & -\infty & 4 & -\infty & 11 & -\infty & 24 & 35 & 41 & 49 & 90 & 225 \\ -\infty & 2 & -\infty & 18 & -\infty & 40 & 48 & 70 & 82 & 98 & 180 & 195 \\ 0 & 3 & 12 & 27 & 33 & 60 & 72 & -\infty & 123 & -\infty & 15 & -\infty \\ 0 & 4 & 16 & 36 & 44 & 80 & -\infty & 140 & -\infty & 196 & -\infty & 135 \end{bmatrix}.$$

The null space of $\mathbf{H}_{\mathrm{ptg},2,\mathrm{qc}}(255, 255)$ gives a $(3060, 2640)$ QC-PTG-LDPC code $C_{\mathrm{ptg},\mathrm{qc}}$ of rate $2/3$ with section-wise cyclic structure. The Tanner graph

$\mathcal{G}_{\mathrm{ptg},2,\mathrm{qc}}(255,255)$ of the code has girth 8. If we use the block-cyclic array $\mathbf{H}_{\mathrm{ptg},2,\mathrm{cyc}}(4,12)$ as the parity-check matrix, we obtain a $(3060,2040)$ QC-PTG-LDPC code $C_{\mathrm{ptg},\mathrm{cyc}}$ with block-cyclic structure which is equivalent to the $(3060,2040)$ QC-PTG-LDPC code $C_{\mathrm{ptg},\mathrm{qc}}$ with section-wise cyclic structure.

The BER and BLER performances of the code $C_{\mathrm{ptg},\mathrm{qc}}$ (or $C_{\mathrm{ptg},\mathrm{cyc}}$) over the AWGN channel decoded with 5, 10, and 50 iterations of the MSA scaled by a factor of 0.75 are shown in Fig. 15.7(b). With 50 iterations of the MSA, at SNR $= 3$ dB, the code achieves a BER of $10^{-10}$ without a visible error-floor. ▲▲

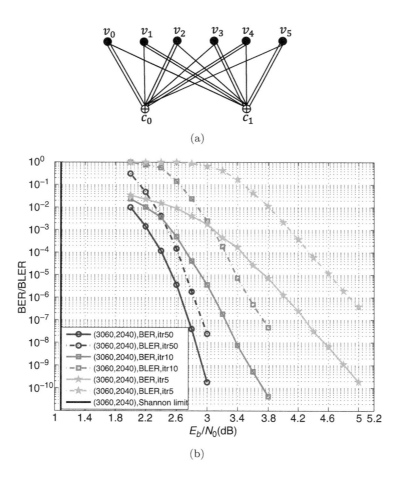

Figure 15.7 (a) The protograph $\mathcal{G}_{\mathrm{ptg}}$ and (b) the BER and BLER performances of the $(3060,2040)$ QC-PTG-LDPC code given in Example 15.7.

In the last three examples, we constructed QC-PTG-LDPC codes using the matrix decomposition method in two steps. In the next example, we construct a long QC-PTG-LDPC code with high rate by decomposing a protomatrix in one step.

**Example 15.8** In this example, the protomatrix used for code construction is a $2 \times 16$ matrix given by

$$\mathbf{B}_{\text{ptg}} = \begin{bmatrix} 2 \ 2 \ 2 \ 2 \ 2 \ 2 \ 2 \ 2 \ 2 \ 2 \ 2 \ 2 \ 2 \ 2 \ 2 \ 2 \\ 2 \ 2 \ 2 \ 2 \ 2 \ 2 \ 2 \ 2 \ 2 \ 2 \ 2 \ 2 \ 2 \ 2 \ 2 \ 2 \end{bmatrix}. \qquad (15.33)$$

The protograph $\mathcal{G}_{\text{ptg}}$ specified by the above protomatrix is a bipartite graph with 16 VNs, each of degree 4, and 2 CNs, each of degree 32. Each VN in $\mathcal{G}_{\text{ptg}}$ is connected to a CN in $\mathcal{G}_{\text{ptg}}$ by two parallel edges.

In one-step decomposition of the protomatrix $\mathbf{B}_{\text{ptg}}$, we choose the decomposition factor $k = 511$ and decompose $\mathbf{B}_{\text{ptg}}$ into 511 constituent matrices, $\mathbf{D}_0, \mathbf{D}_1, \ldots, \mathbf{D}_{510}$, each of size $2 \times 16$. In decomposing $\mathbf{B}_{\text{ptg}}$, each 2-entry is split into two 1-entries which are put into two different constituent matrices. Among the 511 constituent matrices of $\mathbf{B}_{\text{ptg}}$, there are 64 nonzero matrices. Each nonzero constituent matrix contains a single 1-entry. The 64 nonzero constituent matrices in the decomposition of $\mathbf{B}_{\text{ptg}}$ and the locations of their 1-entries are given in Table 15.5. Based on this decomposition of $\mathbf{B}_{\text{ptg}}$, we construct a $511 \times 511$ block-cyclic array $\mathbf{H}_{\text{ptg,cyc}}(2, 16)$ of matrices of size $2 \times 16$ in the form of (15.14), which is a $1022 \times 8176$ matrix, satisfies the RC-constraint, and has rank 1020.

Permuting the columns and rows of $\mathbf{H}_{\text{ptg,cyc}}(2, 16)$ based on the column and row permutations, $\pi_{\text{col}}$ and $\pi_{\text{row}}$, defined by (15.18) and (15.16) with $m = 2$, $n = 16$, and $k = 511$, respectively, we obtain a $2 \times 16$ array $\mathbf{H}_{\text{ptg,qc}}(511, 511) = [\mathbf{A}_{i,j}]_{0 \le i < 2, 0 \le j < 16}$ of circulants of size $511 \times 511$ in the form of (15.19). The weight of each circulant $\mathbf{A}_{i,j}$ in $\mathbf{H}_{\text{ptg,qc}}(511, 511)$ is 2. The generators of the 32 circulants in $\mathbf{H}_{\text{ptg,qc}}(511, 511)$ are given in Table 15.6 where the pair $(a, b)$ in the "1-entries in $\mathbf{g}_{i,j}$" column denotes the two locations of the 1-components in the generator $\mathbf{g}_{i,j}$ of the circulant $\mathbf{A}_{i,j}$.

The null space over GF(2) of $\mathbf{H}_{\text{ptg,qc}}(511, 511)$ gives a $(4, 32)$-regular $(8176, 7156)$ QC-PTG-LDPC code with rate 0.875 whose Tanner graph has girth 6 and 112 420 cycles of length 6. The BER and BLER performances of the code over the AWGN channel decoded with 5, 10, and 50 iterations of the MSA are shown in Fig. 15.8. With 50 iterations of the MSA, the code achieves a BER of $10^{-8}$ at SNR = 4 dB which is 1.1 dB away from the Shannon limit and less than 0.7 dB from SPB [32].                                    ▲▲

## 15.3  Masking Matrices as Protomatrices

The protomatrices used for constructing QC-PTG-LDPC codes in Examples 15.5, 15.6, and 15.7 at the second step of matrix decomposition are actually the masking matrices defined in (12.13) and (14.8) that were used for reducing short cycles in the Tanner graphs of QC-LDPC codes constructed based on finite geometries and finite fields in Chapters 12 to 14. From the error performances of the QC-PTG-LDPC codes constructed in these examples, these masking matrices seem to be also good protomatrices. A good question is whether the three

Table 15.5 The nonzero constituent matrices of the protomatrix $\mathbf{B}_{\mathrm{ptg}}$ and the locations of their 1-entries with decomposition factor $k = 511$ given in Example 15.8.

| $\mathbf{D}_e$ | 1-entries | $\mathbf{D}_e$ | 1-entries | $\mathbf{D}_e$ | 1-entries | $\mathbf{D}_e$ | 1-entries |
|---|---|---|---|---|---|---|---|
| $\mathbf{D}_{16}$ | $(0, 13)$ | $\mathbf{D}_{18}$ | $(1, 12)$ | $\mathbf{D}_{35}$ | $(1, 3)$ | $\mathbf{D}_{45}$ | $(0, 1)$ |
| $\mathbf{D}_{55}$ | $(0, 3)$ | $\mathbf{D}_{56}$ | $(1, 14)$ | $\mathbf{D}_{62}$ | $(1, 10)$ | $\mathbf{D}_{65}$ | $(1, 5)$ |
| $\mathbf{D}_{69}$ | $(1, 14)$ | $\mathbf{D}_{79}$ | $(1, 10)$ | $\mathbf{D}_{84}$ | $(1, 5)$ | $\mathbf{D}_{86}$ | $(0, 12)$ |
| $\mathbf{D}_{93}$ | $(0, 6)$ | $\mathbf{D}_{106}$ | $(1, 4)$ | $\mathbf{D}_{111}$ | $(0, 2)$ | $\mathbf{D}_{131}$ | $(1, 6)$ |
| $\mathbf{D}_{133}$ | $(0, 12)$ | $\mathbf{D}_{137}$ | $(1, 7)$ | $\mathbf{D}_{149}$ | $(0, 5)$ | $\mathbf{D}_{162}$ | $(0, 8)$ |
| $\mathbf{D}_{180}$ | $(1, 0)$ | $\mathbf{D}_{190}$ | $(0, 14)$ | $\mathbf{D}_{209}$ | $(0, 0)$ | $\mathbf{D}_{219}$ | $(0, 8)$ |
| $\mathbf{D}_{221}$ | $(1, 6)$ | $\mathbf{D}_{229}$ | $(0, 5)$ | $\mathbf{D}_{230}$ | $(1, 7)$ | $\mathbf{D}_{232}$ | $(1, 3)$ |
| $\mathbf{D}_{242}$ | $(1, 12)$ | $\mathbf{D}_{244}$ | $(0, 9)$ | $\mathbf{D}_{263}$ | $(1, 11)$ | $\mathbf{D}_{268}$ | $(0, 14)$ |
| $\mathbf{D}_{273}$ | $(0, 11)$ | $\mathbf{D}_{284}$ | $(0, 0)$ | $\mathbf{D}_{303}$ | $(1, 8)$ | $\mathbf{D}_{305}$ | $(1, 0)$ |
| $\mathbf{D}_{306}$ | $(1, 4)$ | $\mathbf{D}_{314}$ | $(0, 9)$ | $\mathbf{D}_{315}$ | $(1, 1)$ | $\mathbf{D}_{317}$ | $(0, 3)$ |
| $\mathbf{D}_{324}$ | $(0, 15)$ | $\mathbf{D}_{330}$ | $(0, 7)$ | $\mathbf{D}_{337}$ | $(0, 7)$ | $\mathbf{D}_{338}$ | $(0, 15)$ |
| $\mathbf{D}_{343}$ | $(1, 13)$ | $\mathbf{D}_{359}$ | $(0, 13)$ | $\mathbf{D}_{364}$ | $(0, 6)$ | $\mathbf{D}_{373}$ | $(1, 2)$ |
| $\mathbf{D}_{384}$ | $(1, 2)$ | $\mathbf{D}_{388}$ | $(1, 9)$ | $\mathbf{D}_{389}$ | $(1, 15)$ | $\mathbf{D}_{404}$ | $(1, 13)$ |
| $\mathbf{D}_{406}$ | $(0, 10)$ | $\mathbf{D}_{419}$ | $(0, 11)$ | $\mathbf{D}_{432}$ | $(1, 1)$ | $\mathbf{D}_{441}$ | $(0, 10)$ |
| $\mathbf{D}_{448}$ | $(1, 9)$ | $\mathbf{D}_{455}$ | $(1, 15)$ | $\mathbf{D}_{461}$ | $(1, 11)$ | $\mathbf{D}_{466}$ | $(0, 4)$ |
| $\mathbf{D}_{467}$ | $(0, 4)$ | $\mathbf{D}_{469}$ | $(0, 1)$ | $\mathbf{D}_{475}$ | $(1, 8)$ | $\mathbf{D}_{482}$ | $(0, 2)$ |

Table 15.6 The generators of the circulants in the array $\mathbf{H}_{\mathrm{ptg,qc}}(511, 511)$ given in Example 15.8 and the locations of their 1-entries.

| $\mathbf{A}_{i,j}$ | 1-entries in $\mathbf{g}_{i,j}$ | $\mathbf{A}_{i,j}$ | 1-entries in $\mathbf{g}_{i,j}$ | $\mathbf{A}_{i,j}$ | 1-entries in $\mathbf{g}_{i,j}$ | $\mathbf{A}_{i,j}$ | 1-entries in $\mathbf{g}_{i,j}$ |
|---|---|---|---|---|---|---|---|
| $\mathbf{A}_{0,0}$ | $(209,284)$ | $\mathbf{A}_{0,8}$ | $(162,219)$ | $\mathbf{A}_{1,0}$ | $(180,305)$ | $\mathbf{A}_{1,8}$ | $(303,475)$ |
| $\mathbf{A}_{0,1}$ | $(45,469)$ | $\mathbf{A}_{0,9}$ | $(244,314)$ | $\mathbf{A}_{1,1}$ | $(315,432)$ | $\mathbf{A}_{1,9}$ | $(388,448)$ |
| $\mathbf{A}_{0,2}$ | $(111,482)$ | $\mathbf{A}_{0,10}$ | $(406,441)$ | $\mathbf{A}_{1,2}$ | $(373,384)$ | $\mathbf{A}_{1,10}$ | $(62,79)$ |
| $\mathbf{A}_{0,3}$ | $(55,317)$ | $\mathbf{A}_{0,11}$ | $(273,419)$ | $\mathbf{A}_{1,3}$ | $(35,232)$ | $\mathbf{A}_{1,11}$ | $(263,461)$ |
| $\mathbf{A}_{0,4}$ | $(466,467)$ | $\mathbf{A}_{0,12}$ | $(86,133)$ | $\mathbf{A}_{1,4}$ | $(106,306)$ | $\mathbf{A}_{1,12}$ | $(18,242)$ |
| $\mathbf{A}_{0,5}$ | $(149,229)$ | $\mathbf{A}_{0,13}$ | $(16,359)$ | $\mathbf{A}_{1,5}$ | $(65,84)$ | $\mathbf{A}_{1,13}$ | $(343,404)$ |
| $\mathbf{A}_{0,6}$ | $(93,364)$ | $\mathbf{A}_{0,14}$ | $(190,268)$ | $\mathbf{A}_{1,6}$ | $(131,221)$ | $\mathbf{A}_{1,14}$ | $(56,69)$ |
| $\mathbf{A}_{0,7}$ | $(330,337)$ | $\mathbf{A}_{0,15}$ | $(324,338)$ | $\mathbf{A}_{1,7}$ | $(137,230)$ | $\mathbf{A}_{1,15}$ | $(389,455)$ |

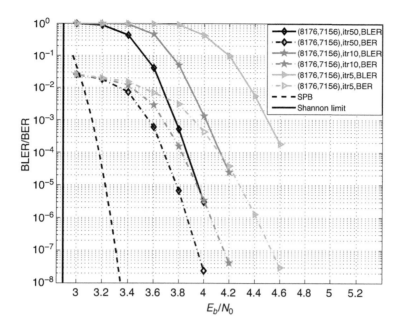

Figure 15.8 The BER and BLER performances of the (8176, 7156) QC-PTG-LDPC code given in Example 15.8.

types of masking matrices defined in Section 14.4 are also good protomatrices. So far, there is no answer to this question. But many examples, besides Examples 15.5, 15.6, and 15.7, show that some good masking matrices are indeed good protomatrices. This will be further demonstrated in this section.

**Example 15.9** Consider the following $6 \times 16$ matrix:

$$\mathbf{B}_{\mathrm{ptg}} = \begin{bmatrix} 1 & 0 & 1 & 0 & 1 & 0 & 1 & 0 & 1 & 1 & 1 & 1 & 1 & 1 & 1 & 1 \\ 0 & 1 & 0 & 1 & 0 & 1 & 0 & 1 & 1 & 1 & 1 & 1 & 1 & 1 & 1 & 1 \\ 1 & 1 & 1 & 1 & 1 & 1 & 1 & 1 & 0 & 0 & 0 & 0 & 0 & 0 & 0 & 0 \\ 1 & 1 & 1 & 1 & 1 & 1 & 1 & 1 & 0 & 0 & 0 & 0 & 0 & 0 & 0 & 0 \\ 0 & 0 & 0 & 0 & 0 & 0 & 0 & 0 & 1 & 0 & 1 & 0 & 1 & 0 & 1 & 0 \\ 0 & 0 & 0 & 0 & 0 & 0 & 0 & 0 & 1 & 0 & 1 & 0 & 1 & 0 & 1 & 1 \end{bmatrix}. \tag{15.34}$$

This matrix is the masking matrix used in Example 14.12 to construct a $(5280, 3302)$ QC-LDPC code based on the prime field GF(331). It is a type-3 masking matrix presented in Section 14.4.3. In this example, we use this masking matrix as the protomatrix $\mathbf{B}_{\mathrm{ptg}}$ to construct an irregular QC-PTG-LDPC code of the same length and the same rate as the code constructed in Example 14.12. The protograph $\mathcal{G}_{\mathrm{ptg}}$ associated with $\mathbf{B}_{\mathrm{ptg}}$ is a bipartite graph with 16 VNs and six CNs. All the VNs have the same degree 3. Among the six CNs, there are two CNs of degree 4, two CNs of degree 8, and two CNs of degree 12.

In code construction, we choose the decomposition factor $k = 330$ and decompose the protomatrix $\mathbf{B}_{\mathrm{ptg}}$ into 330 constituent matrices, $\mathbf{D}_0, \mathbf{D}_1, \ldots, \mathbf{D}_{329}$, each of size $6 \times 16$. Among these 330 constituent matrices of $\mathbf{B}_{\mathrm{ptg}}$, 48 of them are nonzero matrices and the other 207 are ZMs. Each of the nonzero constituent matrix contains a single 1-entry. The nonzero constituent matrices in the decomposition of $\mathbf{B}_{\mathrm{ptg}}$ and the locations of their 1-entries are given in Table 15.7. Using the constituent matrices of $\mathbf{B}_{\mathrm{ptg}}$, we form a $330 \times 330$ block-cyclic array $\mathbf{H}_{\mathrm{ptg,cyc}}(6,16)$ of matrices of size $6 \times 16$ in the form of (15.14) which is a $1980 \times 5280$ matrix with constant column weight 3 and three row weights 4, 8, and 12. The rank of $\mathbf{H}_{\mathrm{ptg,cyc}}(6,16)$ is 1978. The null space over GF(2) of $\mathbf{H}_{\mathrm{ptg,cyc}}(6,16)$ gives a $(5280, 3302)$ irregular QC-PTG-LDPC code $C_{\mathrm{ptg,cyc}}$ with block-cyclic structure. Each codeword in $C_{\mathrm{ptg,cyc}}$ consists of 330 sections, each composed of 16 code symbols.

Table 15.7 The nonzero constituent matrices of the protomatrix $\mathbf{B}_{\mathrm{ptg}}$ and the locations of their 1-entries with decomposition factor $k = 330$ given in Example 15.9.

| $\mathbf{D}_e$ | 1-entries | $\mathbf{D}_e$ | 1-entries | $\mathbf{D}_e$ | 1-entries | $\mathbf{D}_e$ | 1-entries |
|---|---|---|---|---|---|---|---|
| $\mathbf{D}_{10}$ | $(3,0)$ | $\mathbf{D}_{41}$ | $(2,7)$ | $\mathbf{D}_{49}$ | $(0,8)$ | $\mathbf{D}_{50}$ | $(1,9)$ |
| $\mathbf{D}_{53}$ | $(4,12)$ | $\mathbf{D}_{54}$ | $(5,13)$ | $\mathbf{D}_{66}$ | $(2,1)$ | $\mathbf{D}_{67}$ | $(3,2)$ |
| $\mathbf{D}_{99}$ | $(0,15)$ | $\mathbf{D}_{137}$ | $(0,9)$ | $\mathbf{D}_{138}$ | $(1,10)$ | $\mathbf{D}_{151}$ | $(2,5)$ |
| $\mathbf{D}_{152}$ | $(3,6)$ | $\mathbf{D}_{158}$ | $(0,11)$ | $\mathbf{D}_{159}$ | $(1,12)$ | $\mathbf{D}_{164}$ | $(0,6)$ |
| $\mathbf{D}_{165}$ | $(1,7)$ | $\mathbf{D}_{167}$ | $(1,8)$ | $\mathbf{D}_{168}$ | $(4,10)$ | $\mathbf{D}_{169}$ | $(5,11)$ |
| $\mathbf{D}_{201}$ | $(2,3)$ | $\mathbf{D}_{202}$ | $(0,14)$ | $\mathbf{D}_{202}$ | $(3,4)$ | $\mathbf{D}_{203}$ | $(1,15)$ |
| $\mathbf{D}_{217}$ | $(0,13)$ | $\mathbf{D}_{218}$ | $(1,14)$ | $\mathbf{D}_{230}$ | $(0,12)$ | $\mathbf{D}_{231}$ | $(1,13)$ |
| $\mathbf{D}_{237}$ | $(2,0)$ | $\mathbf{D}_{238}$ | $(3,1)$ | $\mathbf{D}_{289}$ | $(0,2)$ | $\mathbf{D}_{290}$ | $(1,3)$ |
| $\mathbf{D}_{291}$ | $(2,4)$ | $\mathbf{D}_{292}$ | $(3,5)$ | $\mathbf{D}_{296}$ | $(0,4)$ | $\mathbf{D}_{297}$ | $(1,5)$ |
| $\mathbf{D}_{298}$ | $(2,6)$ | $\mathbf{D}_{299}$ | $(3,7)$ | $\mathbf{D}_{300}$ | $(4,8)$ | $\mathbf{D}_{301}$ | $(5,9)$ |
| $\mathbf{D}_{320}$ | $(0,10)$ | $\mathbf{D}_{321}$ | $(1,11)$ | $\mathbf{D}_{324}$ | $(0,0)$ | $\mathbf{D}_{324}$ | $(4,14)$ |
| $\mathbf{D}_{325}$ | $(1,1)$ | $\mathbf{D}_{325}$ | $(5,15)$ | $\mathbf{D}_{326}$ | $(2,2)$ | $\mathbf{D}_{327}$ | $(3,3)$ |

If we permute the columns and rows of $\mathbf{H}_{\mathrm{ptg,cyc}}(6,16)$ based on the column and row permutations, $\pi_{\mathrm{col}}$ and $\pi_{\mathrm{row}}$, defined by (15.18) and (15.16) with $m = 6$, $n = 16$, and $k = 330$, respectively, we obtain the following $6 \times 16$ array of CPMs and ZMs of size $330 \times 330$ in which the CPMs are represented by positive integers less than 330:

$\mathbf{H}_{\mathrm{ptg,qc}}(330,330)$

$$= \begin{bmatrix} 324 & -\infty & 289 & -\infty & 296 & -\infty & 164 & -\infty & 49 & 137 & 320 & 158 & 230 & 217 & 202 & 99 \\ -\infty & 325 & -\infty & 290 & -\infty & 297 & -\infty & 165 & 167 & 50 & 138 & 321 & 159 & 231 & 218 & 203 \\ 237 & 66 & 326 & 201 & 291 & 151 & 298 & 41 & -\infty & -\infty & -\infty & -\infty & -\infty & -\infty & -\infty & -\infty \\ 10 & 238 & 67 & 327 & 202 & 292 & 152 & 299 & -\infty & -\infty & -\infty & -\infty & -\infty & -\infty & -\infty & -\infty \\ -\infty & -\infty & -\infty & -\infty & -\infty & -\infty & -\infty & -\infty & 300 & -\infty & 168 & -\infty & 53 & -\infty & 324 & -\infty \\ -\infty & -\infty & -\infty & -\infty & -\infty & -\infty & -\infty & -\infty & -\infty & 301 & -\infty & 169 & -\infty & 54 & -\infty & 325 \end{bmatrix}.$$

The null space over GF(2) of $\mathbf{H}_{\text{ptg,qc}}(330, 330)$ gives a (5280, 3302) irregular QC-PTG-LDPC code $C_{\text{ptg,qc}}$ with section-wise cyclic structure that is equivalent to the QC-PTG-LDPC code $C_{\text{ptg,cyc}}$ with block-cyclic structure. The code $C_{\text{ptg,qc}}$ is the same as the (5280, 3302) QC-LDPC code constructed in Example 14.12. As shown in Fig. 14.12, the code performs well. ▲▲

**Example 15.10** Consider the following $4 \times 16$ masking matrix:

$$\mathbf{B}_{\text{ptg}} = \begin{bmatrix} 1 & 1 & 1 & 0 & 1 & 1 & 1 & 0 & 1 & 1 & 1 & 0 & 1 & 1 & 1 & 0 \\ 0 & 1 & 1 & 1 & 0 & 1 & 1 & 1 & 0 & 1 & 1 & 1 & 0 & 1 & 1 & 1 \\ 1 & 0 & 1 & 1 & 1 & 0 & 1 & 1 & 1 & 0 & 1 & 1 & 1 & 0 & 1 & 1 \\ 1 & 1 & 0 & 1 & 1 & 1 & 0 & 1 & 1 & 1 & 0 & 1 & 1 & 1 & 0 & 1 \end{bmatrix}. \tag{15.35}$$

Use $\mathbf{B}_{\text{ptg}}$ as a protomatrix for constructing a QC-PTG-LDPC code using the matrix-decomposition method. Choose the decomposition factor $k = 255$ and decompose $\mathbf{B}_{\text{ptg}}$ into 255 binary constituent matrices, $\mathbf{D}_0, \mathbf{D}_1, \ldots, \mathbf{D}_{254}$, each of size $4 \times 16$. Among these 255 constituent matrices of $\mathbf{B}_{\text{ptg}}$, 48 of them are nonzero matrices and the other 207 are ZMs. Each of the nonzero constituent matrix contains a single 1-entry. The nonzero constituent matrices in the decomposition of $\mathbf{B}_{\text{ptg}}$ and the locations of their 1-entries are given in Table 15.8. Using the 255 constituent matrices of $\mathbf{B}_{\text{ptg}}$, we form a $255 \times 255$ block-cyclic array $\mathbf{H}_{\text{ptg,cyc}}(4, 16)$ of matrices of size $4 \times 16$ in the form of (15.14) which is a $1020 \times 4080$ matrix with column and row weights 3 and 12, respectively.

Table 15.8 The nonzero constituent matrices of the protomatrix $\mathbf{B}_{\text{ptg}}$ and the locations of their 1-entries with decomposition factor $k = 255$ given in Example 15.10.

| $\mathbf{D}_e$ | 1-entries | $\mathbf{D}_e$ | 1-entries | $\mathbf{D}_e$ | 1-entries | $\mathbf{D}_e$ | 1-entries |
|---|---|---|---|---|---|---|---|
| $\mathbf{D}_1$ | $(0,0)$ | $\mathbf{D}_1$ | $(3,8)$ | $\mathbf{D}_3$ | $(0,1)$ | $\mathbf{D}_3$ | $(2,0)$ |
| $\mathbf{D}_4$ | $(3,0)$ | $\mathbf{D}_6$ | $(0,2)$ | $\mathbf{D}_6$ | $(1,1)$ | $\mathbf{D}_{12}$ | $(1,2)$ |
| $\mathbf{D}_{12}$ | $(3,1)$ | $\mathbf{D}_{18}$ | $(2,2)$ | $\mathbf{D}_{21}$ | $(0,4)$ | $\mathbf{D}_{26}$ | $(1,3)$ |
| $\mathbf{D}_{32}$ | $(0,5)$ | $\mathbf{D}_{37}$ | $(3,9)$ | $\mathbf{D}_{39}$ | $(2,3)$ | $\mathbf{D}_{44}$ | $(0,6)$ |
| $\mathbf{D}_{52}$ | $(3,3)$ | $\mathbf{D}_{63}$ | $(2,4)$ | $\mathbf{D}_{64}$ | $(0,8)$ | $\mathbf{D}_{64}$ | $(1,5)$ |
| $\mathbf{D}_{73}$ | $(0,9)$ | $\mathbf{D}_{77}$ | $(0,10)$ | $\mathbf{D}_{77}$ | $(3,11)$ | $\mathbf{D}_{78}$ | $(2,12)$ |
| $\mathbf{D}_{83}$ | $(3,13)$ | $\mathbf{D}_{84}$ | $(3,4)$ | $\mathbf{D}_{88}$ | $(1,6)$ | $\mathbf{D}_{111}$ | $(0,12)$ |
| $\mathbf{D}_{118}$ | $(1,7)$ | $\mathbf{D}_{128}$ | $(3,5)$ | $\mathbf{D}_{132}$ | $(2,6)$ | $\mathbf{D}_{146}$ | $(1,9)$ |
| $\mathbf{D}_{154}$ | $(1,10)$ | $\mathbf{D}_{166}$ | $(1,11)$ | $\mathbf{D}_{168}$ | $(2,14)$ | $\mathbf{D}_{169}$ | $(1,13)$ |
| $\mathbf{D}_{177}$ | $(2,7)$ | $\mathbf{D}_{189}$ | $(3,12)$ | $\mathbf{D}_{191}$ | $(3,15)$ | $\mathbf{D}_{192}$ | $(2,8)$ |
| $\mathbf{D}_{197}$ | $(1,14)$ | $\mathbf{D}_{207}$ | $(2,15)$ | $\mathbf{D}_{212}$ | $(0,13)$ | $\mathbf{D}_{223}$ | $(1,15)$ |
| $\mathbf{D}_{226}$ | $(0,14)$ | $\mathbf{D}_{231}$ | $(2,10)$ | $\mathbf{D}_{236}$ | $(3,7)$ | $\mathbf{D}_{249}$ | $(2,11)$ |

Permuting the columns and rows of $\mathbf{H}_{\text{ptg,cyc}}(4, 16)$ based on the column and row permutations, $\pi_{\text{col}}$ and $\pi_{\text{row}}$, defined by (15.18) and (15.16) with $m = 4$,

$n = 16$, and $k = 255$, respectively, we obtain the following $4 \times 16$ array of CPMs and ZMs of size $255 \times 255$ in which the CPMs are represented by positive integers less than 255:

$\mathbf{H}_{\text{ptg,qc}}(255, 255)$

$$= \begin{bmatrix} 1 & 3 & 6 & -\infty & 21 & 32 & 44 & -\infty & 64 & 73 & 77 & -\infty & 111 & 212 & 226 & -\infty \\ -\infty & 6 & 12 & 26 & -\infty & 64 & 88 & 118 & -\infty & 146 & 154 & 166 & -\infty & 169 & 197 & 223 \\ 3 & -\infty & 18 & 39 & 63 & -\infty & 132 & 177 & 192 & -\infty & 231 & 249 & 78 & -\infty & 168 & 207 \\ 4 & 12 & -\infty & 52 & 84 & 128 & -\infty & 236 & 1 & 37 & -\infty & 77 & 189 & 83 & -\infty & 191 \end{bmatrix}.$$

The Tanner graph $\mathcal{G}_{\text{ptg,qc}}(255, 255)$ associated with $\mathbf{H}_{\text{ptg,qc}}(255, 255)$ has girth 8. The numbers of cycles of lengths 8, 10, and 12 in $\mathcal{G}_{\text{ptg,qc}}(255, 255)$ are $32\,640$, $495\,210$, and $9\,570\,915$, respectively. The null space over GF(2) of $\mathbf{H}_{\text{ptg,qc}}(255, 255)$ gives a $(3, 12)$-regular $(4080, 3060)$ QC-PTG-LDPC code $C_{\text{ptg,qc}}$ with rate $3/4$. The BER and BLER performances of the constructed code over the AWGN channel decoded with 5, 10, and 50 iterations of the MSA scaled by 0.75 are shown in Fig. 15.9. At the BER of $10^{-8}$ with 50 decoding iterations, the code performs 1.6 dB away from the Shannon limit. ▲▲

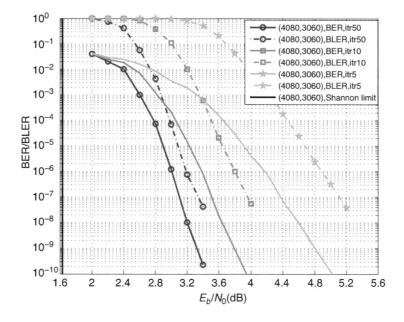

Figure 15.9 The BER and BLER performances of the $(4080, 3060)$ QC-PTG-LDPC code given in Example 15.10.

**Example 15.11** In this example, we consider a relatively large protomatrix $\mathbf{B}_{\text{ptg}}$ for code construction, which is a $12 \times 63$ binary matrix (the matrix is constructed by using the PEG algorithm which will be presented in Section 15.4) whose column and row weight distributions are given in Table 15.9.

Table 15.9 Column and row weight distributions of the $12 \times 63$ protomatrix $\mathbf{B}_{\mathrm{ptg}}$ used in Example 15.11.

| Column weight distribution | | | Row weight distribution | |
|---|---|---|---|---|
| Column weight | Number of columns | | Row weight | Number of rows |
| 3 | 26 | | 23 | 11 |
| 4 | 25 | | 24 | 1 |
| 8 | 9 | | | |
| 9 | 3 | | | |

The protograph $\mathcal{G}_{\mathrm{ptg}}$ specified by this protomatrix has 63 VNs, 12 CNs, and 277 edges whose node degree distributions are given below:

$$\begin{aligned} \rho(X) &= 0.4052X^2 + 0.3927X^3 + 0.1466X^7 + 0.0555X^8, \\ \gamma(X) &= 0.9167X^{22} + 0.0833X^{23}. \end{aligned} \quad (15.36)$$

In the following, we use this matrix as the protomatrix to construct a QC-PTG-LDPC code $C_{\mathrm{ptg,qc}}$ which has the same length and dimension as the $(3969, 3213)^*$ QC-LDPC code $C_{\mathrm{qc}}^*$ constructed in [33]. The QC-LDPC code $C_{\mathrm{qc}}^*$ in [33] is constructed through masking by using a $12 \times 63$ masking matrix which has similar VN and CN degree distributions (or column and row weight distributions) as the protomatrix $\mathbf{B}_{\mathrm{ptg}}$.

In code construction, we choose the decomposition factor $k = 63$ and decompose the protomatrix $\mathbf{B}_{\mathrm{ptg}}$ into 63 constituent matrices $\mathbf{D}_0, \mathbf{D}_1, \ldots, \mathbf{D}_{62}$, each of size $12 \times 63$. The nonzero constituent matrices in the decomposition of $\mathbf{B}_{\mathrm{ptg}}$ and the locations of their 1-entries are given in Table 15.10. Using the constituent matrices of $\mathbf{B}_{\mathrm{ptg}}$, we form a $63 \times 63$ block-cyclic array $\mathbf{H}_{\mathrm{ptg,cyc}}(12, 63)$ of matrices of size $12 \times 63$ in the form of (15.14) which is a $756 \times 3969$ matrix.

Permuting the columns and rows of $\mathbf{H}_{\mathrm{ptg,cyc}}(12, 63)$ based on the column and row permutations, $\pi_{\mathrm{col}}$ and $\pi_{\mathrm{row}}$, defined by (15.18) and (15.16) with $m = 12$, $n = 63$, and $k = 63$, respectively, we obtain a $12 \times 63$ array $\mathbf{H}_{\mathrm{ptg,qc}}(63, 63)$ of CPMs and ZMs of size $63 \times 63$ in which the CPMs are represented by positive integers less than 63 and shown in Table 15.11.

The null space over $\mathrm{GF}(2)$ of $\mathbf{H}_{\mathrm{ptg,qc}}(63, 63)$ gives an irregular $(3969, 3213)$ QC-PTG-LDPC code $C_{\mathrm{ptg,qc}}$ with rate 0.801. For comparison with the $(3969, 3213)^*$ QC-LDPC code $C_{\mathrm{qc}}^*$ constructed in [33], we decode $C_{\mathrm{ptg,qc}}$ with the SPA using 5, 10, 50, and 100 iterations. The BER and BLER performances of the two codes over the AWGN channel are shown in Fig. 15.10(a). From the figure, we see that the performance curves of the two codes overlap with each other.

If we choose the decomposition factor $k = 127$, decomposition of the protomatrix $\mathbf{B}_{\mathrm{ptg}}$ specified by Table 15.12 will give an irregular $(8001, 6477)$ QC-PTG-LDPC code with rate 0.801. The BER and BLER performances of the constructed code over the AWGN channel decoded with 5, 10, and 50 iterations of the MSA scaled by a factor of 0.75 are included in Fig. 15.10(b). Table 15.13 gives the constituent CPMs of the $12 \times 63$ parity-check array $\mathbf{H}_{\mathrm{ptg,qc}}(127, 127)$ of the code. ▲▲

Table 15.10 The nonzero constituent matrices of the protomatrix $\mathbf{B}_{\mathrm{ptg}}$ and the locations of their 1-entries with decomposition factor $k = 63$ given in Example 15.11.

| $\mathbf{D}_{i,j}$ | 1-entries |
|---|---|
| $\mathbf{D}_0$ | (1,6), (2,12), (3,32), (4,24), (5,62), (6,1), (7,26), (8,48), (9,45), (10,61), (11,25) |
| $\mathbf{D}_1$ | (0,6), (2,7), (3,13), (4,33), (5,25), (6,0), (7,2), (8,27), (9,49), (10,46), (11,62) |
| $\mathbf{D}_2$ | (0,12), (1,7), (3,8), (4,14), (5,34), (6,26), (7,1), (8,3), (9,28), (10,50), (11,47) |
| $\mathbf{D}_3$ | (0,32), (1,13), (2,8), (4,9), (5,15), (6,35), (7,27), (8,2), (9,4), (10,29), (11,51) |
| $\mathbf{D}_4$ | (0,24), (1,33), (2,14), (3,9), (5,10), (6,16), (7,36), (8,28), (9,3), (10,5), (11,30) |
| $\mathbf{D}_5$ | (0,62), (1,25), (2,34), (3,15), (4,10), (6,11), (7,17), (8,37), (9,29), (10,4), (11,6) |
| $\mathbf{D}_6$ | (0,1), (1,0), (2,26), (3,35), (4,16), (5,11), (7,12), (8,18), (9,38), (10,30), (11,5) |
| $\mathbf{D}_7$ | (0,26), (1,2), (2,1), (3,27), (4,36), (5,17), (6,12), (8,13), (9,19), (10,39), (11,31) |
| $\mathbf{D}_8$ | (0,48), (1,27), (2,3), (3,2), (4,28), (5,37), (6,18), (7,13), (9,14), (10,20), (11,40) |
| $\mathbf{D}_9$ | (0,45), (1,49), (2,28), (3,4), (4,3), (5,29), (6,38), (7,19), (8,14), (10,15), (11,21) |
| $\mathbf{D}_{10}$ | (0,61), (1,46), (2,50), (3,29), (4,5), (5,4), (6,30), (7,39), (8,20), (9,15), (11,16) |
| $\mathbf{D}_{11}$ | (0,25), (1,62), (2,47), (3,51), (4,30), (5,6), (6,5), (7,31), (8,40), (9,21), (10,16) |
| $\mathbf{D}_{12}$ | (0,2), (1,26), (2,0), (3,48), (4,52), (5,31), (6,7), (7,6), (8,32), (9,41), (10,22), (11,17) |
| $\mathbf{D}_{13}$ | (0,35), (1,3), (2,27), (3,1), (4,49), (5,53), (6,32), (7,8), (8,7), (9,33), (10,42), (11,23) |
| $\mathbf{D}_{14}$ | (0,52), (1,36), (2,4), (3,28), (4,2), (5,50), (6,54), (7,33), (8,9), (9,8), (10,34), (11,43) |
| $\mathbf{D}_{15}$ | (0,23), (1,53), (2,37), (3,5), (4,29), (5,3), (6,51), (7,55), (8,34), (9,10), (10,9), (11,35) |
| $\mathbf{D}_{16}$ | (0,33), (1,24), (2,54), (3,38), (4,6), (5,30), (6,4), (7,52), (8,56), (9,35), (10,11), (11,10) |
| $\mathbf{D}_{17}$ | (0,47), (1,34), (2,25), (3,55), (4,39), (5,7), (6,31), (7,5), (8,53), (9,57), (10,36), (11,12) |
| $\mathbf{D}_{18}$ | (0,27), (1,48), (2,35), (3,26), (4,56), (5,40), (6,8), (7,32), (8,6), (9,54), (10,58), (11,37) |
| $\mathbf{D}_{19}$ | (0,56), (1,28), (2,49), (3,36), (4,27), (5,57), (6,41), (7,9), (8,33), (9,7), (10,55), (11,59) |
| $\mathbf{D}_{20}$ | (0,59), (1,57), (2,29), (3,50), (4,37), (5,28), (6,58), (7,42), (8,10), (9,34), (10,8), (11,56) |

Table 15.10 (*continued*)

| | |
|---|---|
| $D_{21}$ | (0, 42), (1, 60), (2, 58), (3, 30), (4, 51), (5, 38), (6, 29), (7, 59), (8, 43), (9, 11), (10, 35), (11, 9) |
| $D_{22}$ | (0, 50), (1, 43), (2, 61), (3, 59), (4, 31), (5, 52), (6, 39), (7, 30), (8, 60), (9, 44), (10, 12), (11, 36) |
| $D_{23}$ | (0, 15), (1, 51), (2, 44), (3, 62), (4, 60), (5, 32), (6, 53), (7, 40), (8, 31), (9, 61), (10, 45), (11, 13) |
| $D_{24}$ | (0, 4), (1, 16), (2, 52), (3, 45), (4, 0), (5, 61), (6, 33), (7, 54), (8, 41), (9, 32), (10, 62), (11, 46) |
| $D_{25}$ | (0, 11), (1, 5), (2, 17), (3, 53), (4, 46), (5, 1), (6, 62), (7, 34), (8, 55), (9, 42), (10, 33), (11, 0) |
| $D_{26}$ | (0, 7), (1, 12), (2, 6), (3, 18), (4, 54), (5, 47), (6, 2), (7, 0), (8, 35), (9, 56), (10, 43), (11, 34) |
| $D_{27}$ | (0, 18), (1, 8), (2, 13), (3, 7), (4, 19), (5, 55), (6, 48), (7, 3), (8, 1), (9, 36), (10, 57), (11, 44) |
| $D_{28}$ | (0, 41), (1, 19), (2, 9), (3, 14), (4, 8), (5, 20), (6, 56), (7, 49), (8, 4), (9, 2), (10, 37), (11, 58) |
| $D_{29}$ | (0, 60), (1, 42), (2, 20), (3, 10), (4, 15), (5, 9), (6, 21), (7, 57), (8, 50), (9, 5), (10, 3), (11, 38) |
| $D_{30}$ | (0, 46), (1, 61), (2, 43), (3, 21), (4, 11), (5, 16), (6, 10), (7, 22), (8, 58), (9, 51), (10, 6), (11, 4) |
| $D_{31}$ | (0, 34), (1, 47), (2, 62), (3, 44), (4, 22), (5, 12), (6, 17), (7, 11), (8, 23), (9, 59), (10, 52), (11, 7) |
| $D_{32}$ | (0, 3), (1, 35), (2, 48), (3, 0), (4, 45), (5, 23), (6, 13), (7, 18), (8, 12), (9, 24), (10, 60), (11, 53) |
| $D_{33}$ | (0, 16), (1, 4), (2, 36), (3, 49), (4, 1), (5, 46), (6, 24), (7, 14), (8, 19), (9, 13), (10, 25), (11, 61) |
| $D_{34}$ | (0, 31), (1, 17), (2, 5), (3, 37), (4, 50), (5, 2), (6, 47), (7, 25), (8, 15), (9, 20), (10, 14), (11, 26) |
| $D_{35}$ | (0, 13), (1, 32), (2, 18), (3, 6), (4, 38), (5, 51), (6, 3), (7, 48), (8, 26), (9, 16), (10, 21), (11, 15) |
| $D_{36}$ | (0, 54), (1, 14), (2, 33), (3, 19), (4, 7), (5, 39), (6, 52), (7, 4), (8, 49), (9, 27), (10, 17), (11, 22) |
| $D_{37}$ | (0, 44), (1, 55), (2, 15), (3, 34), (4, 20), (5, 8), (6, 40), (7, 53), (8, 5), (9, 50), (10, 28), (11, 18) |
| $D_{38}$ | (0, 49), (1, 45), (2, 56), (3, 16), (4, 35), (5, 21), (6, 9), (7, 41), (8, 54), (9, 6), (10, 51), (11, 29) |
| $D_{39}$ | (0, 43), (1, 50), (2, 46), (3, 57), (4, 17), (5, 36), (6, 22), (7, 10), (8, 42), (9, 55), (10, 7), (11, 52) |
| $D_{40}$ | (0, 55), (1, 44), (2, 51), (3, 47), (4, 58), (5, 18), (6, 37), (7, 23), (8, 11), (9, 43), (10, 56), (11, 8) |
| $D_{41}$ | (0, 28), (1, 56), (2, 45), (3, 52), (4, 48), (5, 59), (6, 19), (7, 38), (8, 24), (9, 12), (10, 44), (11, 57) |
| $D_{42}$ | (0, 21), (1, 29), (2, 57), (3, 46), (4, 53), (5, 49), (6, 60), (7, 20), (8, 39), (9, 25), (10, 13), (11, 45) |
| $D_{43}$ | (0, 39), (1, 22), (2, 30), (3, 58), (4, 47), (5, 54), (6, 50), (7, 61), (8, 21), (9, 40), (10, 26), (11, 14) |

Table 15.10 (*continued*)

| | |
|---|---|
| $D_{44}$ | (0, 37), (1, 40), (2, 23), (3, 31), (4, 59), (5, 48), (6, 55), (7, 51), (8, 62), (9, 22), (10, 41), (11, 27) |
| $D_{45}$ | (0, 9), (1, 38), (2, 41), (3, 24), (4, 32), (5, 60), (6, 49), (7, 56), (8, 52), (9, 0), (10, 23), (11, 42) |
| $D_{46}$ | (0, 30), (1, 10), (2, 39), (3, 42), (4, 25), (5, 33), (6, 61), (7, 50), (8, 57), (9, 53), (10, 1), (11, 24) |
| $D_{47}$ | (0, 17), (1, 31), (2, 11), (3, 40), (4, 43), (5, 26), (6, 34), (7, 62), (8, 51), (9, 58), (10, 54), (11, 2) |
| $D_{48}$ | (0, 8), (1, 18), (2, 32), (3, 12), (4, 41), (5, 44), (6, 27), (7, 35), (8, 0), (9, 52), (10, 59), (11, 55) |
| $D_{49}$ | (0, 38), (1, 9), (2, 19), (3, 33), (4, 13), (5, 42), (6, 45), (7, 28), (8, 36), (9, 1), (10, 53), (11, 60) |
| $D_{50}$ | (0, 22), (1, 39), (2, 10), (3, 20), (4, 34), (5, 14), (6, 43), (7, 46), (8, 29), (9, 37), (10, 2), (11, 54) |
| $D_{51}$ | (0, 53), (1, 23), (2, 40), (3, 11), (4, 21), (5, 35), (6, 15), (7, 44), (8, 47), (9, 30), (10, 38), (11, 3) |
| $D_{52}$ | (0, 14), (1, 54), (2, 24), (3, 41), (4, 12), (5, 22), (6, 36), (7, 16), (8, 45), (9, 48), (10, 31), (11, 39) |
| $D_{53}$ | (0, 51), (1, 15), (2, 55), (3, 25), (4, 42), (5, 13), (6, 23), (7, 37), (8, 17), (9, 46), (10, 49), (11, 32) |
| $D_{54}$ | (0, 36), (1, 52), (2, 16), (3, 56), (4, 26), (5, 43), (6, 14), (7, 24), (8, 38), (9, 18), (10, 47), (11, 50) |
| $D_{55}$ | (0, 40), (1, 37), (2, 53), (3, 17), (4, 57), (5, 27), (6, 44), (7, 15), (8, 25), (9, 39), (10, 19), (11, 48) |
| $D_{56}$ | (0, 19), (1, 41), (2, 38), (3, 54), (4, 18), (5, 58), (6, 28), (7, 45), (8, 16), (9, 26), (10, 40), (11, 20) |
| $D_{57}$ | (0, 58), (1, 20), (2, 42), (3, 39), (4, 55), (5, 19), (6, 59), (7, 29), (8, 46), (9, 17), (10, 27), (11, 41) |
| $D_{58}$ | (0, 57), (1, 59), (2, 21), (3, 43), (4, 40), (5, 56), (6, 20), (7, 60), (8, 30), (9, 47), (10, 18), (11, 28) |
| $D_{59}$ | (0, 20), (1, 58), (2, 60), (3, 22), (4, 44), (5, 41), (6, 57), (7, 21), (8, 61), (9, 31), (10, 48), (11, 19) |
| $D_{60}$ | (0, 29), (1, 21), (2, 59), (3, 61), (4, 23), (5, 45), (6, 42), (7, 58), (8, 22), (9, 62), (10, 32), (11, 49) |
| $D_{61}$ | (0, 10), (1, 30), (2, 22), (3, 60), (4, 62), (5, 24), (6, 46), (7, 43), (8, 59), (9, 23), (10, 0), (11, 33) |
| $D_{62}$ | (0, 5), (1, 11), (2, 31), (3, 23), (4, 61), (5, 0), (6, 25), (7, 47), (8, 44), (9, 60), (10, 24), (11, 1) |

Table 15.11 The entries of $\mathbf{H}_{\mathrm{ptg,qc}}(63,63)$ given in Example 15.11.

| $\mathbf{A}_{i,j}$ | 0 | 1 | 2 | 3 | 4 | 5 | 6 | 7 | 8 | 9 | 10 | 11 | 12 | 13 | 14 | 15 | 16 | 17 | 18 | 19 | 20 | 21 | 22 | 23 | 24 | 25 | 26 | 27 | 28 | 29 | 30 | 31 |
|---|---|---|---|---|---|---|---|---|---|---|---|---|---|---|---|---|---|---|---|---|---|---|---|---|---|---|---|---|---|---|---|---|
| 0 | $-\infty$ | 6 | 12 | 32 | 24 | 62 | 1 | 26 | 48 | 45 | 61 | 25 | 2 | 35 | 52 | 23 | 33 | 47 | 27 | 56 | 59 | 42 | 50 | 15 | 4 | 11 | 7 | 18 | 41 | 60 | 46 | 34 |
| 1 | 6 | $-\infty$ | 7 | 13 | 33 | 25 | 0 | 2 | 27 | 49 | 46 | 62 | 26 | 3 | 36 | 53 | 24 | 34 | 48 | 28 | 57 | 60 | 43 | 51 | 16 | 5 | 12 | 8 | 19 | 42 | 61 | 47 |
| 2 | 12 | 7 | $-\infty$ | 8 | 14 | 34 | 26 | 1 | 3 | 28 | 50 | 47 | 0 | 27 | 4 | 37 | 54 | 25 | 35 | 49 | 29 | 58 | 61 | 44 | 52 | 17 | 6 | 13 | 9 | 20 | 43 | 62 |
| 3 | 32 | 13 | 8 | $-\infty$ | 9 | 15 | 35 | 27 | 2 | 4 | 29 | 51 | 48 | 1 | 28 | 5 | 38 | 55 | 26 | 36 | 50 | 30 | 59 | 62 | 45 | 53 | 18 | 7 | 14 | 10 | 21 | 44 |
| 4 | 24 | 33 | 14 | 9 | $-\infty$ | 10 | 16 | 36 | 28 | 3 | 5 | 30 | 52 | 49 | 2 | 29 | 6 | 39 | 56 | 27 | 37 | 51 | 31 | 60 | 0 | 46 | 54 | 19 | 8 | 15 | 11 | 22 |
| 5 | 62 | 25 | 34 | 15 | 10 | $-\infty$ | 11 | 17 | 37 | 29 | 4 | 6 | 31 | 53 | 50 | 3 | 30 | 7 | 40 | 57 | 28 | 38 | 52 | 32 | 61 | 1 | 47 | 55 | 20 | 9 | 16 | 12 |
| 6 | 1 | 0 | 26 | 35 | 16 | 11 | $-\infty$ | 12 | 18 | 38 | 30 | 5 | 7 | 32 | 54 | 51 | 4 | 31 | 8 | 41 | 58 | 29 | 39 | 53 | 33 | 62 | 2 | 48 | 56 | 21 | 10 | 17 |
| 7 | 26 | 2 | 1 | 27 | 36 | 17 | 12 | $-\infty$ | 13 | 19 | 39 | 31 | 6 | 8 | 33 | 55 | 52 | 5 | 32 | 9 | 42 | 59 | 30 | 40 | 54 | 34 | 0 | 3 | 49 | 57 | 22 | 11 |
| 8 | 48 | 27 | 3 | 2 | 28 | 37 | 18 | 13 | $-\infty$ | 14 | 20 | 40 | 32 | 7 | 9 | 34 | 56 | 53 | 6 | 33 | 10 | 43 | 60 | 31 | 41 | 55 | 35 | 1 | 4 | 50 | 58 | 23 |
| 9 | 45 | 49 | 28 | 4 | 3 | 29 | 38 | 19 | 14 | $-\infty$ | 15 | 21 | 41 | 33 | 8 | 10 | 35 | 57 | 54 | 7 | 34 | 11 | 44 | 61 | 32 | 42 | 56 | 36 | 2 | 5 | 51 | 59 |
| 10 | 61 | 46 | 50 | 29 | 5 | 4 | 30 | 39 | 20 | 15 | $-\infty$ | 16 | 22 | 42 | 34 | 9 | 11 | 36 | 58 | 55 | 8 | 35 | 12 | 45 | 62 | 33 | 43 | 57 | 37 | 3 | 6 | 52 |
| 11 | 25 | 62 | 47 | 51 | 30 | 6 | 5 | 31 | 40 | 21 | 16 | $-\infty$ | 17 | 23 | 43 | 35 | 10 | 12 | 37 | 59 | 56 | 9 | 36 | 13 | 46 | 0 | 34 | 44 | 58 | 38 | 4 | 7 |

| $\mathbf{A}_{i,j}$ | 0 | 1 | 2 | 3 | 4 | 5 | 6 | 7 | 8 | 9 | 10 | 11 | 12 | 13 | 14 | 15 | 16 | 17 | 18 | 19 | 20 | 21 | 22 | 23 | 24 | 25 | 26 | 27 | 28 | 29 | 30 | 31 |
|---|---|---|---|---|---|---|---|---|---|---|---|---|---|---|---|---|---|---|---|---|---|---|---|---|---|---|---|---|---|---|---|---|
| 0 | 32 | 33 | 34 | 35 | 36 | 37 | 38 | 39 | 40 | 41 | 42 | 43 | 44 | 45 | 46 | 47 | 48 | 49 | 50 | 51 | 52 | 53 | 54 | 55 | 56 | 57 | 58 | 59 | 60 | 61 | 62 | |
| 1 | 35 | 16 | 31 | 13 | 54 | 44 | 49 | 43 | 55 | 28 | 21 | 39 | 37 | 9 | 30 | 17 | 8 | 38 | 22 | 53 | 14 | 51 | 36 | 40 | 19 | 58 | 57 | 20 | 29 | 10 | 5 | |
| 2 | 48 | 5 | 32 | 18 | 14 | 55 | 45 | 50 | 56 | 44 | 55 | 43 | 23 | 38 | 10 | 31 | 18 | 9 | 39 | 23 | 54 | 15 | 53 | 37 | 41 | 20 | 59 | 58 | 21 | 30 | 11 | |
| 3 | 0 | 36 | 5 | 48 | 33 | 15 | 56 | 46 | 51 | 4 | 36 | 55 | 24 | 41 | 39 | 11 | 32 | 19 | 44 | 40 | 24 | 52 | 14 | 53 | 16 | 46 | 30 | 44 | 59 | 22 | 31 | |
| 4 | 45 | 49 | 28 | 6 | 19 | 34 | 16 | 57 | 37 | 38 | 30 | 26 | 42 | 25 | 42 | 40 | 41 | 33 | 35 | 40 | 16 | 55 | 56 | 17 | 59 | 60 | 61 | 21 | 45 | 60 | 23 | |
| 5 | 23 | 1 | 50 | 38 | 7 | 20 | 35 | 17 | 11 | 19 | 60 | 21 | 31 | 56 | 50 | 26 | 44 | 28 | 43 | 44 | 36 | 13 | 16 | 27 | 54 | 17 | 56 | 57 | 23 | 24 | 61 | |
| 6 | 13 | 46 | 2 | 51 | 39 | 8 | 21 | 36 | 43 | 9 | 20 | 54 | 55 | 60 | 34 | 35 | 27 | 42 | 36 | 15 | 45 | 42 | 26 | 57 | 28 | 29 | 47 | 41 | 47 | 58 | 0 | |
| 7 | 18 | 24 | 37 | 17 | 52 | 40 | 9 | 22 | 36 | 57 | 26 | 42 | 26 | 52 | 61 | 51 | 52 | 13 | 34 | 21 | 36 | 55 | 14 | 43 | 58 | 59 | 21 | 22 | 43 | 42 | 25 | |
| 8 | 12 | 14 | 11 | 43 | 4 | 53 | 41 | 10 | 23 | 11 | 20 | 14 | 23 | 49 | 57 | 62 | 35 | 34 | 43 | 35 | 24 | 13 | 44 | 15 | 16 | 19 | 40 | 57 | 20 | 45 | 47 | |
| 9 | 24 | 19 | 16 | 26 | 49 | 5 | 54 | 43 | 55 | 28 | 21 | 49 | 41 | 47 | 26 | 30 | 47 | 29 | 46 | 38 | 31 | 43 | 24 | 28 | 45 | 46 | 30 | 40 | 25 | 28 | 44 | |
| 10 | 60 | 13 | 21 | 16 | 27 | 50 | 6 | 31 | 47 | 60 | 23 | 26 | 27 | 56 | 50 | 46 | 32 | 28 | 29 | 47 | 48 | 38 | 18 | 25 | 16 | 60 | 57 | 18 | 58 | 59 | 60 | |
| 11 | 53 | 25 | 26 | 15 | 22 | 18 | 29 | 52 | 8 | 33 | 19 | 41 | 20 | 48 | 24 | 2 | 55 | 60 | 54 | 3 | 32 | 49 | 50 | 15 | 20 | 41 | 19 | 28 | 49 | 33 | 1 | |

Table 15.12 The nonzero constituent matrices of the protomatrix $\mathbf{B}_{\mathrm{ptg}}$ and the locations of their 1-entries with decomposition factor $k = 127$ given in Example 15.11.

| $\mathbf{D}_{i,j}$ | 1-entries |
|---|---|
| $\mathbf{D}_0$ | $(0, 30), (1, 61), (2, 6), (5, 13), (6, 2), (8, 43), (9, 36), (10, 33), (11, 27)$ |
| $\mathbf{D}_1$ | $(1, 31), (2, 62), (3, 7), (6, 14), (7, 3), (9, 44), (10, 37), (11, 34)$ |
| $\mathbf{D}_2$ | $(0, 31), (2, 32), (4, 8), (7, 15), (8, 4), (10, 45), (11, 38)$ |
| $\mathbf{D}_3$ | $(0, 62), (1, 32), (3, 33), (5, 9), (8, 16), (9, 5), (11, 46)$ |
| $\mathbf{D}_4$ | $(0, 7), (2, 33), (4, 34), (6, 10), (7, 0), (9, 17), (10, 6)$ |
| $\mathbf{D}_5$ | $(1, 8), (3, 34), (5, 35), (7, 11), (8, 1), (10, 18), (11, 7)$ |
| $\mathbf{D}_6$ | $(2, 9), (4, 35), (6, 36), (8, 12), (9, 2), (11, 19)$ |
| $\mathbf{D}_7$ | $(0, 14), (3, 10), (5, 36), (7, 37), (9, 13), (10, 3)$ |
| $\mathbf{D}_8$ | $(0, 3), (1, 15), (3, 0), (4, 11), (6, 37), (8, 38), (10, 14), (11, 4)$ |
| $\mathbf{D}_9$ | $(1, 4), (2, 16), (4, 1), (5, 12), (7, 38), (9, 39), (11, 15)$ |
| $\mathbf{D}_{10}$ | $(0, 44), (2, 5), (3, 17), (5, 2), (6, 13), (8, 39), (10, 40)$ |
| $\mathbf{D}_{11}$ | $(0, 37), (1, 45), (3, 6), (4, 18), (6, 3), (7, 14), (9, 40), (11, 41)$ |
| $\mathbf{D}_{12}$ | $(0, 34), (1, 38), (2, 46), (4, 7), (5, 19), (7, 4), (8, 15), (10, 41)$ |
| $\mathbf{D}_{13}$ | $(0, 28), (1, 35), (2, 39), (3, 47), (5, 8), (6, 20), (8, 5), (9, 16), (11, 42)$ |
| $\mathbf{D}_{14}$ | $(1, 29), (2, 36), (3, 40), (4, 48), (6, 9), (7, 21), (9, 6), (10, 17)$ |
| $\mathbf{D}_{15}$ | $(0, 6), (2, 30), (3, 37), (4, 41), (5, 49), (6, 0), (7, 10), (8, 22), (10, 7), (11, 18)$ |
| $\mathbf{D}_{16}$ | $(1, 7), (3, 31), (4, 38), (5, 42), (6, 50), (7, 1), (8, 11), (9, 23), (11, 8)$ |
| $\mathbf{D}_{17}$ | $(2, 8), (4, 32), (5, 39), (6, 43), (7, 51), (8, 2), (9, 12), (10, 24)$ |
| $\mathbf{D}_{18}$ | $(3, 9), (5, 33), (6, 40), (7, 44), (8, 52), (9, 3), (10, 13), (11, 25)$ |
| $\mathbf{D}_{19}$ | $(4, 10), (6, 34), (7, 41), (8, 45), (9, 53), (10, 4), (11, 14)$ |
| $\mathbf{D}_{20}$ | $(5, 11), (7, 35), (8, 42), (9, 46), (10, 54), (11, 5)$ |
| $\mathbf{D}_{21}$ | $(6, 12), (8, 36), (9, 43), (10, 47), (11, 55)$ |
| $\mathbf{D}_{22}$ | $(0, 40), (7, 13), (9, 37), (10, 44), (11, 48)$ |
| $\mathbf{D}_{23}$ | $(1, 41), (8, 14), (10, 38), (11, 45)$ |
| $\mathbf{D}_{24}$ | $(2, 42), (9, 15), (11, 39)$ |
| $\mathbf{D}_{25}$ | $(0, 56), (3, 43), (10, 16)$ |
| $\mathbf{D}_{26}$ | $(1, 57), (4, 44), (11, 17)$ |
| $\mathbf{D}_{27}$ | $(0, 55), (2, 58), (5, 45)$ |
| $\mathbf{D}_{28}$ | $(1, 56), (3, 59), (6, 46)$ |
| $\mathbf{D}_{29}$ | $(0, 12), (2, 57), (4, 60), (7, 47)$ |
| $\mathbf{D}_{30}$ | $(1, 13), (3, 58), (5, 61), (8, 48)$ |
| $\mathbf{D}_{31}$ | $(2, 14), (4, 59), (6, 62), (9, 49)$ |
| $\mathbf{D}_{32}$ | $(0, 1), (1, 0), (3, 15), (5, 60), (10, 50)$ |
| $\mathbf{D}_{33}$ | $(1, 2), (2, 1), (4, 16), (6, 61), (11, 51)$ |
| $\mathbf{D}_{34}$ | $(2, 3), (3, 2), (5, 17), (7, 62)$ |
| $\mathbf{D}_{35}$ | $(0, 11), (3, 4), (4, 3), (6, 18), (11, 0)$ |
| $\mathbf{D}_{36}$ | $(1, 12), (4, 5), (5, 4), (7, 19)$ |
| $\mathbf{D}_{37}$ | $(0, 49), (2, 13), (5, 6), (6, 5), (8, 20)$ |
| $\mathbf{D}_{38}$ | $(0, 10), (1, 50), (3, 14), (6, 7), (7, 6), (9, 21), (10, 0)$ |
| $\mathbf{D}_{39}$ | $(1, 11), (2, 51), (4, 15), (7, 8), (8, 7), (10, 22), (11, 1)$ |
| $\mathbf{D}_{40}$ | $(0, 57), (2, 12), (3, 52), (5, 16), (8, 9), (9, 8), (11, 23)$ |

Table 15.12 (*continued*)

| | |
|---|---|
| $D_{41}$ | $(0, 21), (1, 58), (3, 13), (4, 53), (6, 17), (9, 10), (10, 9)$ |
| $D_{42}$ | $(1, 22), (2, 59), (4, 14), (5, 54), (7, 18), (10, 11), (11, 10)$ |
| $D_{43}$ | $(2, 23), (3, 60), (5, 15), (6, 55), (8, 19), (11, 12)$ |
| $D_{44}$ | $(3, 24), (4, 61), (6, 16), (7, 56), (9, 20)$ |
| $D_{45}$ | $(0, 9), (4, 25), (5, 62), (7, 17), (8, 57), (9, 0), (10, 21)$ |
| $D_{46}$ | $(0, 50), (1, 10), (5, 26), (8, 18), (9, 58), (10, 1), (11, 22)$ |
| $D_{47}$ | $(1, 51), (2, 11), (6, 27), (9, 19), (10, 59), (11, 2)$ |
| $D_{48}$ | $(2, 52), (3, 12), (7, 28), (10, 20), (11, 60)$ |
| $D_{49}$ | $(3, 53), (4, 13), (8, 29), (11, 21)$ |
| $D_{50}$ | $(0, 36), (4, 54), (5, 14), (9, 30)$ |
| $D_{51}$ | $(0, 45), (1, 37), (5, 55), (6, 15), (10, 31)$ |
| $D_{52}$ | $(1, 46), (2, 38), (6, 56), (7, 16), (11, 32)$ |
| $D_{53}$ | $(2, 47), (3, 39), (7, 57), (8, 17)$ |
| $D_{54}$ | $(3, 48), (4, 40), (8, 58), (9, 18)$ |
| $D_{55}$ | $(4, 49), (5, 41), (9, 59), (10, 19)$ |
| $D_{56}$ | $(0, 26), (5, 50), (6, 42), (10, 60), (11, 20)$ |
| $D_{57}$ | $(0, 24), (1, 27), (6, 51), (7, 43), (11, 61)$ |
| $D_{58}$ | $(0, 39), (1, 25), (2, 28), (7, 52), (8, 44)$ |
| $D_{59}$ | $(1, 40), (2, 26), (3, 29), (8, 53), (9, 45)$ |
| $D_{60}$ | $(2, 41), (3, 27), (4, 30), (9, 54), (10, 46)$ |
| $D_{61}$ | $(3, 42), (4, 28), (5, 31), (10, 55), (11, 47)$ |
| $D_{62}$ | $(4, 43), (5, 29), (6, 32), (11, 56)$ |
| $D_{63}$ | $(0, 2), (2, 0), (5, 44), (6, 30), (7, 33)$ |
| $D_{64}$ | $(0, 15), (1, 3), (3, 1), (6, 45), (7, 31), (8, 34)$ |
| $D_{65}$ | $(1, 16), (2, 4), (4, 2), (7, 46), (8, 32), (9, 35)$ |
| $D_{66}$ | $(2, 17), (3, 5), (5, 3), (8, 47), (9, 33), (10, 36)$ |
| $D_{67}$ | $(0, 41), (3, 18), (4, 6), (6, 4), (9, 48), (10, 34), (11, 37)$ |
| $D_{68}$ | $(1, 42), (4, 19), (5, 7), (7, 5), (10, 49), (11, 35)$ |
| $D_{69}$ | $(0, 22), (2, 43), (5, 20), (6, 8), (8, 6), (11, 50)$ |
| $D_{70}$ | $(0, 17), (1, 23), (3, 44), (6, 21), (7, 9), (9, 7)$ |
| $D_{71}$ | $(1, 18), (2, 24), (4, 45), (7, 22), (8, 10), (10, 8)$ |
| $D_{72}$ | $(2, 19), (3, 25), (5, 46), (8, 23), (9, 11), (11, 9)$ |
| $D_{73}$ | $(3, 20), (4, 26), (6, 47), (9, 24), (10, 12)$ |
| $D_{74}$ | $(0, 53), (4, 21), (5, 27), (7, 48), (10, 25), (11, 13)$ |
| $D_{75}$ | $(0, 20), (1, 54), (5, 22), (6, 28), (8, 49), (11, 26)$ |
| $D_{76}$ | $(0, 58), (1, 21), (2, 55), (6, 23), (7, 29), (9, 50)$ |
| $D_{77}$ | $(0, 43), (1, 59), (2, 22), (3, 56), (7, 24), (8, 30), (10, 51)$ |
| $D_{78}$ | $(1, 44), (2, 60), (3, 23), (4, 57), (8, 25), (9, 31), (11, 52)$ |
| $D_{79}$ | $(2, 45), (3, 61), (4, 24), (5, 58), (9, 26), (10, 32)$ |
| $D_{80}$ | $(3, 46), (4, 62), (5, 25), (6, 59), (10, 27), (11, 33)$ |
| $D_{81}$ | $(0, 42), (4, 47), (6, 26), (7, 60), (11, 28)$ |
| $D_{82}$ | $(0, 59), (1, 43), (5, 48), (7, 27), (8, 61)$ |
| $D_{83}$ | $(0, 5), (1, 60), (2, 44), (5, 0), (6, 49), (8, 28), (9, 62)$ |
| $D_{84}$ | $(1, 6), (2, 61), (3, 45), (6, 1), (7, 50), (9, 29)$ |

Table 15.12 (*continued*)

| | |
|---|---|
| $D_{85}$ | $(0, 33), (2, 7), (3, 62), (4, 46), (7, 2), (8, 51), (10, 30)$ |
| $D_{86}$ | $(0, 38), (1, 34), (3, 8), (5, 47), (8, 3), (9, 52), (11, 31)$ |
| $D_{87}$ | $(0, 25), (1, 39), (2, 35), (4, 9), (6, 48), (9, 4), (10, 53)$ |
| $D_{88}$ | $(1, 26), (2, 40), (3, 36), (5, 10), (7, 49), (10, 5), (11, 54)$ |
| $D_{89}$ | $(0, 18), (2, 27), (3, 41), (4, 37), (6, 11), (8, 50), (11, 6)$ |
| $D_{90}$ | $(0, 47), (1, 19), (3, 28), (4, 42), (5, 38), (7, 12), (9, 51)$ |
| $D_{91}$ | $(1, 48), (2, 20), (4, 29), (5, 43), (6, 39), (8, 13), (10, 52)$ |
| $D_{92}$ | $(0, 13), (2, 49), (3, 21), (5, 30), (6, 44), (7, 40), (9, 14), (11, 53)$ |
| $D_{93}$ | $(1, 14), (3, 50), (4, 22), (6, 31), (7, 45), (8, 41), (10, 15)$ |
| $D_{94}$ | $(2, 15), (4, 51), (5, 23), (7, 32), (8, 46), (9, 42), (11, 16)$ |
| $D_{95}$ | $(0, 51), (3, 16), (5, 52), (6, 24), (8, 33), (9, 47), (10, 43)$ |
| $D_{96}$ | $(1, 52), (4, 17), (6, 53), (7, 25), (9, 34), (10, 48), (11, 44)$ |
| $D_{97}$ | $(2, 53), (5, 18), (7, 54), (8, 26), (10, 35), (11, 49)$ |
| $D_{98}$ | $(3, 54), (6, 19), (8, 55), (9, 27), (11, 36)$ |
| $D_{99}$ | $(4, 55), (7, 20), (9, 56), (10, 28)$ |
| $D_{100}$ | $(5, 56), (8, 21), (10, 57), (11, 29)$ |
| $D_{101}$ | $(6, 57), (9, 22), (11, 58)$ |
| $D_{102}$ | $(0, 29), (7, 58), (10, 23)$ |
| $D_{103}$ | $(0, 61), (1, 30), (8, 59), (11, 24)$ |
| $D_{104}$ | $(0, 32), (1, 62), (2, 31), (9, 60)$ |
| $D_{105}$ | $(1, 33), (3, 32), (10, 61)$ |
| $D_{106}$ | $(0, 46), (2, 34), (4, 33), (11, 62)$ |
| $D_{107}$ | $(0, 19), (1, 47), (3, 35), (5, 34)$ |
| $D_{108}$ | $(0, 54), (1, 20), (2, 48), (4, 36), (6, 35)$ |
| $D_{109}$ | $(1, 55), (2, 21), (3, 49), (5, 37), (7, 36)$ |
| $D_{110}$ | $(2, 56), (3, 22), (4, 50), (6, 38), (8, 37)$ |
| $D_{111}$ | $(0, 52), (3, 57), (4, 23), (5, 51), (7, 39), (9, 38)$ |
| $D_{112}$ | $(1, 53), (4, 58), (5, 24), (6, 52), (8, 40), (10, 39)$ |
| $D_{113}$ | $(0, 48), (2, 54), (5, 59), (6, 25), (7, 53), (9, 41), (11, 40)$ |
| $D_{114}$ | $(1, 49), (3, 55), (6, 60), (7, 26), (8, 54), (10, 42)$ |
| $D_{115}$ | $(2, 50), (4, 56), (7, 61), (8, 27), (9, 55), (11, 43)$ |
| $D_{116}$ | $(0, 16), (3, 51), (5, 57), (8, 62), (9, 28), (10, 56)$ |
| $D_{117}$ | $(0, 23), (1, 17), (4, 52), (6, 58), (10, 29), (11, 57)$ |
| $D_{118}$ | $(0, 27), (1, 24), (2, 18), (5, 53), (7, 59), (11, 30)$ |
| $D_{119}$ | $(0, 35), (1, 28), (2, 25), (3, 19), (6, 54), (8, 60)$ |
| $D_{120}$ | $(1, 36), (2, 29), (3, 26), (4, 20), (7, 55), (9, 61)$ |
| $D_{121}$ | $(2, 37), (3, 30), (4, 27), (5, 21), (8, 56), (10, 62)$ |
| $D_{122}$ | $(0, 8), (3, 38), (4, 31), (5, 28), (6, 22), (8, 0), (9, 57)$ |
| $D_{123}$ | $(1, 9), (4, 39), (5, 32), (6, 29), (7, 23), (9, 1), (10, 58)$ |
| $D_{124}$ | $(2, 10), (5, 40), (6, 33), (7, 30), (8, 24), (10, 2), (11, 59)$ |
| $D_{125}$ | $(0, 4), (3, 11), (4, 0), (6, 41), (7, 34), (8, 31), (9, 25), (11, 3)$ |
| $D_{126}$ | $(0, 60), (1, 5), (4, 12), (5, 1), (7, 42), (8, 35), (9, 32), (10, 26)$ |

Table 15.13 The entries of $\mathbf{H}_{\mathrm{ptg,qc}}(127,127)$ given in Example 15.11.

| $A_{i,j}$ | 0 | 1 | 2 | 3 | 4 | 5 | 6 | 7 | 8 | 9 | 10 | 11 | 12 | 13 | 14 | 15 | 16 | 17 | 18 | 19 | 20 | 21 | 22 | 23 | 24 | 25 | 26 | 27 | 28 | 29 | 30 | 31 |
|---|---|---|---|---|---|---|---|---|---|---|---|---|---|---|---|---|---|---|---|---|---|---|---|---|---|---|---|---|---|---|---|---|
| 0 | $-\infty$ | 32 | 63 | 8 | 125 | 83 | 15 | 4 | 122 | 45 | 38 | 35 | 29 | 92 | 7 | 64 | 116 | 70 | 89 | 107 | 75 | 41 | 69 | 117 | 57 | 87 | 56 | 118 | 13 | 102 | 0 | 2 |
| 1 | 32 | $-\infty$ | 33 | 64 | 9 | 126 | 84 | 16 | 5 | 123 | 46 | 39 | 36 | 30 | 93 | 8 | 65 | 117 | 71 | 90 | 108 | 76 | 42 | 70 | 118 | 58 | 88 | 57 | 119 | 14 | 103 | 1 |
| 2 | 63 | 33 | $-\infty$ | 34 | 65 | 10 | 0 | 85 | 17 | 6 | 124 | 47 | 40 | 37 | 31 | 94 | 9 | 66 | 118 | 72 | 91 | 109 | 77 | 43 | 71 | 119 | 59 | 89 | 58 | 120 | 15 | 104 |
| 3 | 8 | 64 | 34 | $-\infty$ | 35 | 66 | 11 | 1 | 86 | 18 | 7 | 125 | 48 | 41 | 38 | 32 | 95 | 10 | 67 | 119 | 73 | 92 | 110 | 78 | 44 | 72 | 120 | 60 | 90 | 59 | 121 | 16 |
| 4 | 125 | 9 | 65 | 35 | $-\infty$ | 36 | 67 | 12 | 2 | 87 | 19 | 8 | 126 | 49 | 42 | 39 | 33 | 96 | 11 | 68 | 120 | 74 | 93 | 111 | 79 | 45 | 73 | 121 | 61 | 91 | 60 | 122 |
| 5 | 83 | 126 | 10 | 66 | 36 | $-\infty$ | 37 | 68 | 13 | 3 | 88 | 20 | 9 | 0 | 50 | 43 | 40 | 34 | 97 | 12 | 69 | 121 | 75 | 94 | 112 | 80 | 46 | 74 | 122 | 62 | 92 | 61 |
| 6 | 15 | 84 | 66 | 11 | 67 | 37 | $-\infty$ | 38 | 69 | 14 | 4 | 89 | 21 | 10 | 1 | 51 | 44 | 41 | 35 | 98 | 13 | 70 | 122 | 76 | 95 | 113 | 81 | 47 | 75 | 123 | 63 | 93 |
| 7 | 4 | 16 | 85 | 1 | 12 | 68 | 38 | $-\infty$ | 39 | 70 | 15 | 5 | 90 | 22 | 11 | 2 | 52 | 45 | 42 | 36 | 99 | 14 | 71 | 123 | 77 | 96 | 114 | 82 | 48 | 76 | 124 | 64 |
| 8 | 122 | 5 | 17 | 86 | 2 | 13 | 69 | 39 | $-\infty$ | 40 | 71 | 16 | 6 | 91 | 23 | 12 | 3 | 53 | 46 | 43 | 37 | 100 | 15 | 72 | 124 | 78 | 97 | 115 | 83 | 49 | 77 | 125 |
| 9 | 45 | 123 | 6 | 18 | 87 | 3 | 14 | 70 | 40 | $-\infty$ | 41 | 72 | 17 | 7 | 92 | 24 | 13 | 4 | 54 | 47 | 44 | 38 | 101 | 16 | 73 | 125 | 79 | 98 | 116 | 84 | 50 | 78 |
| 10 | 38 | 46 | 124 | 7 | 19 | 88 | 4 | 15 | 71 | 41 | $-\infty$ | 42 | 73 | 18 | 8 | 93 | 25 | 14 | 5 | 55 | 48 | 45 | 39 | 102 | 17 | 74 | 126 | 80 | 99 | 117 | 85 | 51 |
| 11 | 35 | 39 | 47 | 125 | 8 | 20 | 89 | 5 | 16 | 72 | 42 | $-\infty$ | 43 | 74 | 19 | 9 | 94 | 26 | 15 | 6 | 56 | 49 | 46 | 40 | 103 | 18 | 75 | 0 | 81 | 100 | 118 | 86 |

| $A_{i,j}$ | 32 | 33 | 34 | 35 | 36 | 37 | 38 | 39 | 40 | 41 | 42 | 43 | 44 | 45 | 46 | 47 | 48 | 49 | 50 | 51 | 52 | 53 | 54 | 55 | 56 | 57 | 58 | 59 | 60 | 61 | 62 |
|---|---|---|---|---|---|---|---|---|---|---|---|---|---|---|---|---|---|---|---|---|---|---|---|---|---|---|---|---|---|---|---|
| 0 | 104 | 85 | 12 | 119 | 50 | 11 | 86 | 58 | 22 | 67 | 81 | 77 | 10 | 51 | 106 | 90 | 113 | 37 | 46 | 95 | 111 | 74 | 108 | 27 | 25 | 40 | 76 | 82 | 126 | 103 | 62 |
| 1 | 3 | 12 | 86 | 4 | 120 | 51 | 12 | 87 | 59 | 23 | 68 | 82 | 78 | 107 | 52 | 113 | 91 | 114 | 38 | 47 | 96 | 112 | 75 | 109 | 28 | 26 | 41 | 77 | 83 | 0 | 104 |
| 2 | 2 | 119 | 13 | 120 | 87 | 107 | 4 | 110 | 24 | 69 | 24 | 109 | 28 | 114 | 41 | 108 | 92 | 38 | 78 | 99 | 98 | 51 | 112 | 75 | 77 | 83 | 104 | 0 | 116 | 77 | 1 |
| 3 | 105 | 4 | 107 | 14 | 88 | 15 | 122 | 13 | 88 | 60 | 83 | 76 | 77 | 115 | 27 | 42 | 115 | 93 | 93 | 84 | 84 | 99 | 84 | 113 | 76 | 42 | 43 | 78 | 84 | 78 | 85 |
| 4 | 17 | 3 | 50 | 7 | 108 | 121 | 80 | 85 | 61 | 25 | 70 | 84 | 80 | 78 | 30 | 77 | 54 | 116 | 80 | 81 | 71 | 85 | 77 | 13 | 28 | 78 | 29 | 79 | 79 | 85 | 80 |
| 5 | 123 | 106 | 6 | 89 | 15 | 122 | 16 | 15 | 90 | 62 | 26 | 71 | 28 | 8 | 74 | 88 | 14 | 55 | 85 | 110 | 27 | 72 | 109 | 31 | 112 | 113 | 44 | 28 | 45 | 44 | 45 |
| 6 | 62 | 18 | 108 | 109 | 110 | 9 | 8 | 90 | 17 | 91 | 63 | 27 | 28 | 82 | 65 | 30 | 82 | 56 | 95 | 42 | 26 | 64 | 28 | 73 | 29 | 113 | 16 | 17 | 45 | 116 | 31 |
| 7 | 94 | 124 | 107 | 6 | 85 | 17 | 123 | 19 | 92 | 41 | 50 | 99 | 100 | 73 | 87 | 15 | 87 | 83 | 112 | 95 | 18 | 28 | 44 | 29 | 74 | 80 | 84 | 85 | 84 | 1 | 34 |
| 8 | 65 | 95 | 19 | 66 | 126 | 31 | 76 | 113 | 93 | 19 | 95 | 115 | 99 | 45 | 30 | 89 | 67 | 16 | 18 | 19 | 96 | 93 | 121 | 66 | 30 | 75 | 89 | 90 | 78 | 82 | 116 |
| 9 | 126 | 66 | 64 | 0 | 83 | 68 | 32 | 12 | 61 | 87 | 22 | 61 | 116 | 21 | 96 | 31 | 96 | 68 | 86 | 87 | 22 | 2 | 99 | 122 | 55 | 31 | 104 | 33 | 43 | 105 | 83 |
| 10 | 79 | 0 | 97 | 66 | 78 | 91 | 2 | 114 | 94 | 92 | 2 | 21 | 100 | 45 | 101 | 96 | 32 | 77 | 91 | 20 | 88 | 60 | 62 | 117 | 123 | 47 | 56 | 105 | 120 | 121 | 106 |
| 11 | 52 | 80 | 1 | 68 | 98 | 67 | 2 | 24 | 113 | 11 | 115 | 23 | 96 | 3 | 61 | 22 | 97 | 69 | 33 | 78 | 88 | 21 | 62 | 117 | 101 | 124 | 48 | 57 | 106 | 57 | 106 |

In the above examples, we demonstrated again that good masking matrices for constructing QC-LDPC codes are also good protomatrices for constructing PTG-LDPC codes.

Before we leave this section, we show that a QC-LDPC code constructed by CPM-dispersion of a base matrix over a finite field can be viewed as a QC-PTG-LDPC code. Consider a QC-LDPC code $C_{qc}$ whose parity-check matrix is an $m \times n$ array $\mathbf{H}_{qc}(k, k)$ of CPMs and/or ZMs of size $k \times k$ constructed by CPM-dispersion of an $m \times n$ base matrix over a finite field. Let $\pi_{col}^{-1}$ and $\pi_{row}^{-1}$ be the inverses of the column and row permutations, $\pi_{col}$ and $\pi_{row}$, defined by (15.18) and (15.16), respectively. If we permute the columns and then the rows of $\mathbf{H}_{qc}(k, k)$ with $\pi_{col}^{-1}$ and $\pi_{row}^{-1}$, respectively, we obtain a $k \times k$ array $\mathbf{H}_{cyc}(m, n)$ of matrices of size $m \times n$ with block-cyclic structure in the form of (15.14). The sum of the $km \times n$ matrices in the top row-block of $\mathbf{H}_{cyc}(m, n)$ is an $m \times n$ matrix over nonnegative integer. This matrix may be regarded as a protomatrix $\mathbf{B}_{ptg}$ and the $k$ matrices in the top row-block of $\mathbf{H}_{cyc}(m, n)$ may be regarded as the constituent matrices in the decomposition of $\mathbf{B}_{ptg}$. From this point of view, the QC-LDPC code $C_{qc}$ is a QC-PTG-LDPC code. Hence, QC-LDPC codes based on finite geometries and finite fields and QC-PTG-LDPC codes are closely related.

## 15.4 LDPC Codes on Progressive Edge-Growth Algorithms

LDPC codes whose Tanner graphs are cycle free [34] have optimum performance under iterative decoding based on belief-propagation, such as the SPA and MSA. As shown in the previous chapters, short cycles in the Tanner graph of an LDPC code impair the error performance of iterative decoding. Thus, LDPC codes with large girth are desirable. In Chapter 14, masking is proposed to reduce the number of short cycles in Tanner graphs by removing a certain number of edges to enlarge their girths. In this section, an effective graph-based method for constructing LDPC codes with large girth is presented. This construction method is called the *progressive-edge-growth* (PEG) algorithm proposed in [4, 5] which is widely used in computer-aided code design. The PEG algorithm is the opposite of the masking technique. In the PEG algorithm, the construction begins with a set of $n$ VNs and a set of $m$ CNs with no edges connecting VNs and CNs, i.e., a *bipartite graph without edges*. Then, edges are progressively added to connect VNs and CNs by applying a set of rules and a given VN degree distribution. Because the VNs of low degrees are most susceptible to error (they receive the least amount of information from their neighbor VNs), edge placement begins with the VNs of the lowest degree and progresses to VNs of increasing (or nondecreasing) degrees. Edges are added to a VN one at a time using an *edge-selection procedure* until the number of edges added to a VN is equal to its degree. This progressive adding of edges to a VN is called *edge-growth*. Edge-growth is performed on one VN at a time. After the completion

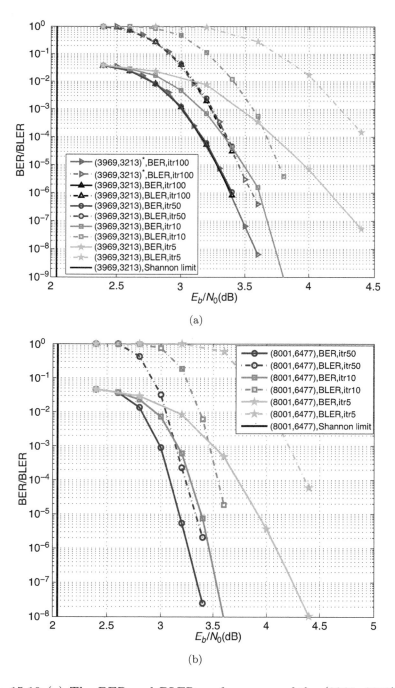

(a)

(b)

Figure 15.10 (a) The BER and BLER performances of the (3969, 3213) QC-PTG-LDPC code $C_{\mathrm{ptg,qc}}$ given in Example 15.11 and the (3969, 3213)* QC-LDPC code $C_{\mathrm{qc}}^*$ in [33] and (b) the BER and BLER performances of the $(8001, 6477)$ QC-PTG-LDPC code given in Example 15.11.

of edge-growth of one VN, we move to the next VN. Edge-growth moves from VNs of the lowest degree to VNs of the highest degree. When all the VNs have completed their edge-growth, we obtain a Tanner graph whose VNs have the specified degrees. The edge-selection procedure is devised to maximize the girth of the constructed Tanner graph.

Before presenting the edge-growth procedure, we give some definitions. Consider a bipartite graph $\mathcal{G} = (\mathcal{V}, \mathcal{C}, \mathcal{E})$ with a set $\mathcal{V} = \{v_0, v_1, \ldots, v_{n-1}\}$ of $n$ VNs, a set $\mathcal{C} = \{c_0, c_1, \ldots, c_{m-1}\}$ of $m$ CNs, and a set $\mathcal{E} = \{(v_i, c_j)\}$ of edges where $(v_i, c_j)$ represents the edge connecting the $i$th VN $v_i$ in $\mathcal{V}$ and the $j$th CN $c_j$ in $\mathcal{C}$. Consider the VN $v_i$. Let $g_i$ be the length of the shortest cycle passing through $v_i$, called the *local girth* of $v_i$. Then, the girth $g$ of $\mathcal{G}$ is given by $g = \min_{0 \leq i < n}\{g_i\}$, called the *global girth*. The edge-growth procedure for constructing a Tanner graph is devised to maximize the local girth of every VN (a best-effort algorithm) and thus the global girth $g$.

For a given VN $v_i$ in a Tanner graph $\mathcal{G}$, let $N_{v_i}^{(l)}$ denote the set of CNs in $\mathcal{G}$ connecting to $v_i$ within a distance of $2l + 1$. The distance between two nodes in a graph is defined as the length of a *shortest path* between them. The shortest paths that connect $v_i$ to the CNs in $N_{v_i}^{(l)}$ can be represented by a tree $\mathcal{T}_{v_i}$ with $v_i$ as the root as shown in Fig. 15.11. The tree $\mathcal{T}_{v_i}$ consists of $l + 1$ levels and each level consists of two layers of nodes, a layer of VNs and a layer of CNs. We label the levels from 0 to $l$ and the level with the root $v_i$ is labeled as level 0. Each path in $\mathcal{T}_{v_i}$ consists of a sequence of alternate VNs and CNs and each path in $\mathcal{T}_{v_i}$ terminates at a CN in $N_{v_i}^{(l)}$. Every node on a path in $\mathcal{T}_{v_i}$ appears *once and only once*. The path tree $\mathcal{T}_{v_i}$ with VN $v_i$ as its root can be constructed progressively. Starting from the VN $v_i$, we transverse all the edges, $(v_i, c_{j_0}), (v_i, c_{j_1}), \ldots, (v_i, c_{j_{d_{v_i}-1}})$, that connect $v_i$ to its $d_{v_i}$ *nearest-neighbor* CNs, $c_{j_0}, c_{j_1}, \ldots, c_{j_{d_{v_i}-1}}$. This results in the zeroth level of the path tree $\mathcal{T}_{v_i}$. Next, we transverse all the edges that connect the CNs in the zeroth level to their respective nearest-neighbor VNs in the first level. Then, we transverse all the edges that connect the VNs at the first layer of the first level of the tree $\mathcal{T}_{v_i}$ to their respective nearest-neighbor CNs in the second layer of the first level. This completes the first level of the tree $\mathcal{T}_{v_i}$. The transversing process continues level by level until the $l$th level, or a level from which the tree cannot grow further, has been reached. It is clear that the distance between $v_i$ and a VN in the $l$th level is $2l$ and the distance between $v_i$ and a CN in the $l$th level is $2l + 1$. If there is an edge in $\mathcal{G}$ that connects a CN in the $l$th level of the tree $\mathcal{T}_{v_i}$ to the root VN $v_i$, then adding this edge between the CN and $v_i$ will create a cycle of length $2(l + 1)$ in $\mathcal{G}$ passing through $v_i$. The set $N_{v_i}^{(l)}$ is referred to as the *neighborhood within depth $l$* of $v_i$. Let $\bar{N}_{v_i}^{(l)}$ be the *complementary set* of $N_{v_i}^{(l)}$, i.e., $\bar{N}_{v_i}^{(l)} = \mathcal{C} \backslash N_{v_i}^{(l)}$.

Using the concepts of the local girth of a VN $v_i$, the neighborhood $N_{v_i}^{(l)}$ within depth $l$ of $v_i$, and the tree representation of the paths that connect $v_i$ to the CNs in $N_{v_i}^{(l)}$, a progressive edge-growth (PEG) procedure was devised by

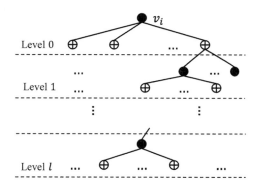

Figure 15.11 Tree representation of the neighborhood $N_{v_i}^{(l)}$ within depth $l$ of a VN $v_i$.

Hu *et al.* [4, 5] to construct Tanner graphs with relatively large girth. Suppose for the given number $n$ of VNs, the given number $m$ of CNs, and the degree sequence $D_v = (d_{v_0}, d_{v_1}, \ldots, d_{v_{n-1}})$ of VNs, we have completed the edge-growth of the first $i$ VNs, $v_0, v_1, \ldots, v_{i-1}$ with $1 \le i \le n$, using the PEG procedure. At this point, we have a *partially connected* Tanner graph. Next, we grow the edges incident from the $i$th VN $v_i$ to $d_{v_i}$ CNs. The growth is done one edge at a time. Suppose $k$ edges have been added to $v_i$ with $1 \le k \le d_{v_i}$. Before the $k$th edge is added to $v_i$, we construct a path tree $\mathcal{T}_{v_i}$ with $v_i$ as the root as shown in Fig. 15.11 based on the current partially connected Tanner graph, denoted $\mathcal{G}_{v_i,k}$. We keep growing the tree $\mathcal{T}_{v_i}$ until it reaches a level, say the $l$th level, such that one of the following two situations occurs: (1) the tree $\mathcal{T}_{v_i}$ cannot grow further but the cardinality $|N_{v_i}^{(l)}|$ of $N_{v_i}^{(l)}$ is smaller than $m$; or (2) $\bar{N}_{v_i}^{(l)} \ne \emptyset$ but $\bar{N}_{v_i}^{(l+1)} = \emptyset$ (where $\emptyset$ denotes an empty set). The first situation implies that not all CNs can be reached from $v_i$ under the current partially connected graph $\mathcal{G}_{v_i,k}$. In this case, we choose a CN $c_j$ in $\bar{N}_{v_i}^{(l)}$ with the lowest degree and connect $v_i$ and $c_j$ with a new edge. This prevents creating an additional cycle passing through $v_i$. The second situation implies that all the CNs are reachable from $v_i$. In this case, we choose the CN at the $(l+1)$th level that has lowest degree. Then, add an edge between this chosen CN and $v_i$. Adding such an edge creates a cycle of length $2(l+2)$ passing through $v_i$. By doing this, we maximize the local girth $g_i$ of $v_i$.

The PEG procedure for constructing a Tanner graph with maximized local girths can be put into an algorithm, called the *PEG algorithm* [4, 5], as shown in Algorithm 15.1.

By using the adjacency matrix of the constructed Tanner graph $\mathcal{G}$ as a parity-check matrix, an LDPC code with length $n$ can be constructed, called a *PEG-LDPC code*. The $(961, 838)$ PEG code $C_{\mathrm{peg}}$ in Example 13.4 as a comparable code to the $(961, 840)$ QC-RS-PaG-LDPC code $C_{\mathrm{PaG,c}}(4, 31)$ is constructed

---

**Algorithm 15.1 PEG algorithm** [4, 5]

---

**for** $i = 0$ to $n - 1$ **do**

    **for** $k = 0$ to $d_{v_i} - 1$ **do**

        **if** $k = 0$ **then**

            $\mathcal{E}_i^{(0)} \leftarrow \mathrm{edge}(v_i, c_j)$, where $\mathcal{E}_i^{(0)}$ is the first edge incident to $v_i$ and $c_j$ is a CN that has the lowest degree in the current partially connected Tanner graph.

        **else**

            grow a path tree $\mathcal{T}_{v_i}$ with $v_i$ as the root to a level, say the $l$th level, based on the current partially connected Tanner graph such that either the cardinality of $N_{v_i}^{(l)}$ stops increasing (i.e., the path tree cannot grow further) but is less than $m$, or $\bar{N}_{v_i}^{(l)} \neq \emptyset$ but $\bar{N}_{v_i}^{(l+1)} = \emptyset$, then $\mathcal{E}_i^{(k)} \leftarrow \mathrm{edge}(v_i, c_j)$, where $\mathcal{E}_i^{(k)}$ is the $k$th edge incident to $v_i$ and $c_j$ is a CN chosen from $\bar{N}_{v_i}^{(l)}$ that has the lowest degree.

        **end if**

    **end for**

**end for**

---

using the PEG algorithm. The $(7921, 7566)$ binary PEG code $C_{\mathrm{peg}}$ in Example 13.5 is also constructed by PEG algorithm. Besides using the PEG algorithm to construct LDPC codes directly, the algorithm can be applied to construct relatively small matrices following certain VN/CN degree distributions, such as masking matrices or protomatrices. In Example 15.11, the $12 \times 63$ protomatrix $\mathbf{B}_{\mathrm{ptg}}$ used for constructing QC-PTG-LDPC code is constructed by PEG algorithm following the degree distribution given in Table 15.9.

The PEG algorithm given above can be improved to construct irregular LDPC codes with better performance in the high-SNR region without performance degradation in the low-SNR region. An improved PEG algorithm was presented in [35, 36]. To present this improved PEG algorithm, we first introduce a new concept. A cycle $\mathfrak{C}$ of length $2t$ in a Tanner graph $\mathcal{G}$ consists of $t$ VNs and $t$ CNs. Let

$$\epsilon = \sum_{i=0}^{t-1}(d_{v_i} - 2), \tag{15.37}$$

where $d_{v_i}$ is the degree of the VN $v_i$ on $\mathfrak{C}$. The sum $\epsilon$ is simply the number of edges that connect the cycle $\mathfrak{C}$ to the rest of the Tanner graph, i.e., the edges outside the cycle. The sum $\epsilon$ is a measure of the *connectivity* of the cycle $\mathfrak{C}$ to the rest of the Tanner graph. The larger this connectivity, the more messages from outside of the cycle $\mathfrak{C}$ are available for the VNs on $\mathfrak{C}$. For this reason, $\epsilon$ is called the *approximate cycle extrinsic* (ACE) *message degree* [37, 38].

The improved PEG algorithm presented in [35, 36] is identical to the PEG algorithm except for the case in which $k \geq 1$, $\bar{N}_{v_i}^{(l)} \neq \emptyset$, and $\bar{N}_{v_i}^{(l+1)} = \emptyset$, when building the path tree $\mathcal{T}_{v_i}$ with $v_i$ as the root. In this case, there may be more than one candidate CN with the lowest degree. Let $\mathcal{C}_{v_i}^{(l,k)}$ denote the set of candidate CNs in $\bar{N}_{v_i}^{(l)}$ with the lowest degree. For any CN $c_j \in \mathcal{C}_{v_i}^{(l,k)}$, there is at least one path of length $2(l+1)+1$ between $v_i$ and $c_j$, but no shorter path between them. Hence, the placement of edge $\mathcal{E}_{v_i}^{(k)}$ between $v_i$ and $c_j$ will create at least one new cycle of length $2(l+2)$, but no shorter cycles. In the PEG algorithm, a CN is chosen randomly from $\mathcal{C}_{v_i}^{(l,k)}$. However, in the improved PEG algorithm presented in [37, 38], a CN $c_{\max}$ is chosen from $\mathcal{C}_{v_i}^{(l,k)}$ such that the new cycle created by adding an edge between $v_i$ and $c_{\max}$ has the largest possible ACE message degree. It was shown in [35, 36] that this modified PEG algorithm results in LDPC codes with better error-floor performance than the irregular LDPC codes constructed with the original PEG algorithm presented above. Clearly, the improvement is at the expense of additional computational complexity, but this applies once only, during the code design. This improved PEG algorithm is called the ACE-based PEG (ACE-PEG) algorithm.

**Example 15.12** In this example, we use the PEG and ACE-PEG algorithms to construct an irregular $(4088, 2044)$ LDPC code of rate $1/2$. The VN and CN degree distributions of the Tanner graph for the code are

$$\rho(X) = 0.451X + 0.3931X^2 + 0.1558X^9,$$
$$\gamma(X) = 0.7206X^6 + 0.2994X^7.$$

The Tanner graphs of the constructed codes based on the two algorithms have girth 8. The Tanner graph of the code based on the PEG algorithm contains 70 591 cycles of length 8 and 1 553 803 cycles of length 10. The Tanner graph of the code based on the ACE-PEG algorithm contains 79 299 cycles of length 8 and 1 628 364 cycles of length 10. The BER and BLER performances of the two constructed codes over the AWGN channel decoded with 50 iterations of the SPA are shown in Fig. 15.12. From this figure, we see that the code constructed based on the ACE-PEG algorithm outperforms the one based on the PEG algorithm at the error-floor region, because the code based on the ACE-PEG algorithm has a higher connectivity. High connectivity can lower the error-floor as pointed out in Chapter 11. ▲▲

The parity-check matrices of LDPC codes constructed by the PEG algorithm or its improved version lack hardware-friendly structures. To obtain LDPC codes which are hardware-friendly, Li and Kumar [39] proposed an algorithm to construct QC-LDPC codes by modifying the PEG algorithm. They put a quasi-cyclic constraint on the PEG algorithm (see details in [39]). Multiple algorithms have been proposed to improve the PEG algorithm to construct well-performing QC-LDPC codes, see [40–42].

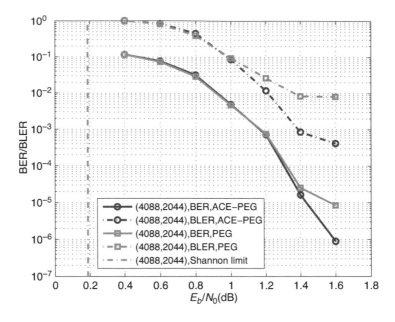

Figure 15.12 The BER and BLER performances of the $(4088, 2044)$ LDPC codes constructed by the PEG and ACE-PEG algorithms in Example 15.12.

## 15.5 Remarks

Besides the class of protograph-based LDPC codes, there are two other classes of *photograph-like* LDPC codes. The codes in the first class are known as *spatially coupled* (SC) LDPC codes [17–22]. The Tanner graph of an SC-LDPC code is a *semi-infinite chain of small local graphs* in which each local graph is connected to its *adjacent* local graphs on both sides of it, except the first and last ones, as shown in Fig. 15.13. Every VN is only connected to CNs that are confined to a *small span* of consecutive locations and every CN is only connected to VNs that are confined to a *small span* of consecutive locations.

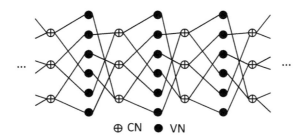

Figure 15.13 The Tanner graph of an SC-LDPC code.

The parity-check matrix $\mathbf{H}_{sc}$ of an SC-LDPC code is a semi-infinite array of small sparse local matrices with *diagonal structure* (*stair-case structure*) whose nonzero entries are confined to a *diagonal band* as given below:

$$\mathbf{H}_{sc} = \begin{bmatrix} \mathbf{H}_{0,0} & & & & & \\ \mathbf{H}_{0,1} & \mathbf{H}_{0,0} & & & & \\ \mathbf{H}_{0,2} & \mathbf{H}_{0,1} & \mathbf{H}_{0,0} & & & \\ \vdots & \vdots & \vdots & \vdots & \vdots & \\ \mathbf{H}_{0,e-1} & \mathbf{H}_{0,e-2} & \mathbf{H}_{0,e-3} & & & \\ & \mathbf{H}_{0,e-1} & \mathbf{H}_{0,e-2} & \mathbf{H}_{0,e-3} & & \\ & & \mathbf{H}_{0,e-1} & \mathbf{H}_{0,e-2} & & \\ & & & \ddots & \ddots & \end{bmatrix}. \tag{15.38}$$

It has been proved in [17, 18] that the ensemble of SC-LDPC codes contains channel-capacity-approaching codes as the length of the chain of local graphs approaching infinity. Similar to protograph-based LDPC codes, matrix-theoretic methods can be devised to construct quasi-cyclic SC-LDPC (QC-SC-LDPC) codes using finite fields in conjunction with CPM-dispersion [43].

In decoding of an SC-LDPC code, we can either terminate the semi-infinite parity-check array $\mathbf{H}_{sc}$ and decode the code as a block LDPC code or use the *sliding window decoding* as proposed in [19]. Note that the matrix $\mathbf{H}_{sc}$ given by (15.38) is a conventional parity-check matrix of a convolutional code [31].

The second class of protograph-like LDPC codes is the globally coupled (GC) LDPC codes presented in Section 14.11.

The PEG algorithm presented in Section 15.4 (or its enhanced version) is commonly used to construct irregular LDPC codes based on designed VN and CN degree distributions. PEG-LDPC codes are often used as references for performance comparison with LDPC codes constructed using methods other than PEG, especially algebraic methods. The PEG algorithm is also used to construct irregular/regular masking matrices or protomatrices for constructing finite-geometry and finite-field-based QC-LDPC codes or protograph-based LDPC codes.

# Problems

**15.1** Consider the protograph $\mathcal{G}_{ptg}$ shown in Fig. 15.1(a) in Example 15.1. Choose an expansion factor $k = 4$ and take $k = 4$ copies of $\mathcal{G}_{ptg}$. Follow the copy-and-permute operation presented in Section 15.1 to construct a connected bipartite graph $\mathcal{G}_{ptg}(4, 4)$ which has girth at least 6 based on the four copies of $\mathcal{G}_{ptg}$. Compute the adjacency matrix $\mathbf{H}_{ptg}(4, 4)$ of the Tanner graph $\mathcal{G}_{ptg}(4, 4)$ in the form of (15.3). Consider the following two cases.

(a) The connection matrix $\mathbf{A}_{i,j}$ is just a regular matrix or ZM.

(b) All the connection matrices, $\mathbf{A}_{i,j}$, are binary circulants or ZMs.

**15.2** Consider the protograph $\mathcal{G}_{\mathrm{ptg}}$ shown in Fig. 15.1(a) in Example 15.1. In Problem 15.1, $\mathcal{G}_{\mathrm{ptg}}$ is copied four times to construct a connected Tanner graph of large size. Here, construct a large Tanner graph based on $\mathcal{G}_{\mathrm{ptg}}$ in two steps.

(a) Step 1. Choose $c = 2$, take $c = 2$ copies of $\mathcal{G}_{\mathrm{ptg}}$, and perform the copy-and-permute operation to obtain a connected Tanner graph $\mathcal{G}_{\mathrm{ptg},2}$. Find the adjacency matrix of $\mathcal{G}_{\mathrm{ptg},2}$.

(b) Step 2. Take $k = 2$ copies of the protograph $\mathcal{G}_{\mathrm{ptg},2}$. Perform the copy-and-permute operation on the two copies to obtain a connected Tanner graph $\mathcal{G}_{\mathrm{ptg},2}(2,2)$. Find the adjacency matrix of $\mathcal{G}_{\mathrm{ptg},2}(2,2)$.

**15.3** Consider the $4 \times 8$ matrix $\mathbf{B}_{\mathrm{ptg},2}$ given by (15.9) in Example 15.2. Use the Tanner graph $\mathcal{G}_{\mathrm{ptg},2}$ of $\mathbf{B}_{\mathrm{ptg},2}$ as a protograph. Take $k = 170$ copies of $\mathcal{G}_{\mathrm{ptg},2}$. Perform the copy-and-permute operation on these $k = 170$ copies to obtain a connected Tanner graph $\mathcal{G}_{\mathrm{ptg},2}(170,170)$ with girth at least 6 and its adjacency matrix $\mathbf{H}_{\mathrm{ptg},2}(170,170)$ is an array of CPMs and ZMs. Compute the QC-PTG-LDPC code $C_{\mathrm{ptg}}$ given by the null space of $\mathbf{H}_{\mathrm{ptg},2}(170,170)$ and its BER and BLER performances over the AWGN channel with 5, 10, and 50 iterations of the scaled MSA. Compute the error performance of $C_{\mathrm{ptg}}$ over the BEC.

**15.4** Consider the $4 \times 8$ matrix $\mathbf{B}_{\mathrm{ptg},2}$ given by (15.9) in Example 15.2. Use the Tanner graph $\mathcal{G}_{\mathrm{ptg},2}$ of $\mathbf{B}_{\mathrm{ptg},2}$ as a protograph. Take $k = 63$ copies of $\mathcal{G}_{\mathrm{ptg},2}$. Perform the copy-and-permute operation on these $k = 63$ copies to obtain a connected Tanner graph $\mathcal{G}_{\mathrm{ptg},2}(63,63)$ with girth at least 6 and its adjacency matrix $\mathbf{H}_{\mathrm{ptg},2}(63,63)$ is an array of CPMs and ZMs. Compute the QC-PTG-LDPC code $C_{\mathrm{ptg}}$ given by the null space of $\mathbf{H}_{\mathrm{ptg},2}(63,63)$ and its BER and BLER performances over the AWGN channel with 5, 10, and 50 iterations of the scaled MSA. Compute the error performance of $C_{\mathrm{ptg}}$ over the BEC.

**15.5** Replace the copy-and-permute operation in Problem 15.1 by decomposing its associated protomatrix $\mathbf{B}_{\mathrm{ptg}}$. Set the decomposition factor to $k = 4$.

(a) Find the decomposition set $\Psi = \{\mathbf{D}_0, \mathbf{D}_1, \mathbf{D}_2, \mathbf{D}_3\}$.

(b) Verify $\mathbf{D}_0 + \mathbf{D}_1 + \mathbf{D}_2 + \mathbf{D}_3 = \mathbf{B}_{\mathrm{ptg}}$.

(c) Compute the block-cyclic matrix $\mathbf{H}_{\mathrm{ptg,cyc}}(2,3)$ in the form of (15.14).

(d) Calculate the column and row permutations, $\pi_{\mathrm{col}}$ and $\pi_{\mathrm{row}}$, defined by (15.18) and (15.16) where $n = 3$, $m = 2$, and $k = 4$, respectively.

(e) Permute $\mathbf{H}_{\mathrm{ptg,cyc}}(2,3)$ based on $\pi_{\mathrm{col}}$ and $\pi_{\mathrm{row}}$ to obtain an array $\mathbf{H}_{\mathrm{ptg,qc}}(4,4)$ of circulants or ZMs.

**15.6** Find another decomposition set $\Psi^* = \{\mathbf{D}_0^*, \mathbf{D}_1^*, \mathbf{D}_2^*, \mathbf{D}_3^*\}$ and redo the five items in Problem 15.5.

**15.7** Use the $4 \times 8$ matrix $\mathbf{B}_{\mathrm{ptg},2}$ in Problem 15.3 as a protomatrix. Choose a decomposition factor $k = 170$. Use the algebraic decomposition method to construct a QC-PTG-LDPC code $C_{\mathrm{ptg,qc}}$ of the same length and dimension as the one constructed in Problem 15.3. Find the decomposition set and the section-wise cyclic-shift parity-check matrix $\mathbf{H}_{\mathrm{ptg,qc}}(170,170)$. Compute the error performance of the constructed code over the AWGN channel by 50 iterations of the scaled MSA.

**15.8** Use the $4 \times 8$ matrix $\mathbf{B}_{\mathrm{ptg},2}$ in Problem 15.4 as a protomatrix. Choose a decomposition factor $k = 63$. Use the algebraic decomposition method to construct a QC-PTG-LDPC code $C_{\mathrm{ptg,qc}}$ of the same length and dimension as the one constructed in Problem 15.4. Find the decomposition set and the section-wise cyclic-shift parity-check matrix $\mathbf{H}_{\mathrm{ptg,qc}}(63,63)$. Compute the error performance of the constructed code over the AWGN channel by 50 iterations of the scaled MSA.

**15.9** Consider the $2 \times 6$ protomatrix $\mathbf{B}_{\mathrm{ptg}}$ given in Example 15.7. Suppose we decompose $\mathbf{B}_{\mathrm{ptg}}$ into the following two $2 \times 6$ constituent matrices in the first step of the algebraic decomposition method:

$$\mathbf{D}_0 = \begin{bmatrix} 1 & 1 & 1 & 0 & 1 & 0 \\ 0 & 1 & 0 & 1 & 1 & 1 \end{bmatrix}, \mathbf{D}_1 = \begin{bmatrix} 1 & 0 & 1 & 1 & 1 & 1 \\ 1 & 1 & 1 & 1 & 0 & 1 \end{bmatrix}.$$

Use the two constituent matrices $\mathbf{D}_0$ and $\mathbf{D}_1$ to form a $2 \times 2$ block-cyclic array $\mathbf{B}_{\mathrm{ptg},2}$ of $2 \times 6$ matrices in the form of (15.14). Use $\mathbf{B}_{\mathrm{ptg},2}$ as a protomatrix and a decomposition factor $k = 255$ to construct a QC-PTG-LDPC code of the same length and dimension as the $(3060, 2640)$ QC-PTG-LDPC code $C_{\mathrm{ptg,qc}}$ of section-wise cyclic-shift constructed in Example 15.7. Find the short cycle distribution of the Tanner graph of the constructed code. Compute the error performances of the constructed code over the AWGN channel with 50 iterations of the MSA.

**15.10** Construct the $12 \times 63$ protomatrix $\mathbf{B}_{\mathrm{ptg}}$ used in Example 15.11 following the degree distribution given by (15.36) by the PEG algorithm.

**15.11** Construct a $(961, 838)$ PEG-LDPC code $C_{\mathrm{peg}}$ whose parity-check matrix has constant column weight 4. Compute its error performance over the AWGN channel by 50 iterations of the scaled MSA.

**15.12** Consider the following node degree distributions:

$$\rho(X) = 0.451X + 0.3931X^2 + 0.1558X^9, \ \gamma(X) = 0.7206X^6 + 0.2994X^7.$$

Based on the above degree distribution, construct an irregular LDPC code of length 2044 and rate $1/2$ using the PEG and ACE-PEG algorithms. Compute the error performances of the constructed codes over the AWGN channel with 50 iterations of the MSA or SPA.

# References

[1] J. Thorpe, "Low-density parity-check (LDPC) codes constructed from protographs," Jet Propulsion Laboratory, Pasadena, CA, IPN Progress Report 42–154, August 2003.

[2] G. Liva and M. Chiani, "Protograph LDPC codes design based on EXIT analysis," *Proc. IEEE GlobeCom*, Washington, DC, November 26–30, 2007, pp. 3250–3254.

[3] D. Divsalar, S. Dolinar, C. R. Jones, and K. Andrews, "Capacity-approaching protograph codes," *IEEE J. Sel. Areas Commun.*, **27**(6) (2009), 876–888.

[4] X.-Y. Hu, E. Eleftheriou, and D.-M. Arnold, "Progressive edge-growth Tanner graphs," *Proc. IEEE GlobeCom*, San Antonio, TX, November 25–29, 2001, pp. 995–1001.

[5] X.-Y. Hu, E. Eleftheriou, and D.-M. Arnold, "Regular and irregular progressive edge-growth Tanner graphs," *IEEE Trans. Inf. Theory*, **51**(1) (2005), 386–398.

[6] D. Divsalar, S. Dolinar, and C. R. Jones, "Low-rate LDPC codes with simple protograph structure," *Proc. IEEE Int. Symp. Inf. Theory*, Adelaide, SA, September 4–9, 2005, pp. 1622–1626.

[7] D. Divsalar, C. R. Jones, S. Dolinar, and J. Thorpe, "Protograph based LDPC codes with minimum distance linearly growing with block size," *Proc. IEEE GlobeCom*, St. Louis, MO, November 28–December 2, 2005, p. 5.

[8] D. Divsalar, S. Dolinar, and C. R. Jones, "Construction of protograph LDPC codes with linear minimum distance," *Proc. IEEE Int. Symp. Inf. Theory*, Seattle, WA, July 9–14, 2006, pp. 664–668.

[9] D. Divsalar, S. Dolinar, and C. R. Jones, "Short protograph-based LDPC codes," *Proc. IEEE MILCOM*, Orlando, FL, October 29–31, 2007, vol. 3, pp. 1–6.

[10] S. Abu-Surra, D. Divsalar, and W. E. Ryan, "On the existence of typical minimum distance for protograph-based LDPC codes," *Proc. IEEE Inf. Theory Workshop*, San Diego, CA, January 31–February 5, 2010, pp. 1–7.

[11] B. K. Butler and P. H. Siegel, "On distance properties of quasi-cyclic protograph-based LDPC codes," *Proc. IEEE Int. Symp. Inf. Theory*, Austin, TX, June 13–18, 2010, pp. 809–813.

[12] S. Abu-Surra, D. Divsalar, and W. E. Ryan, "Enumerators for protograph-based ensembles of LDPC and generalized LDPC codes," *IEEE Trans. Inf. Theory*, **57**(2) (2011), 858–886.

[13] T. V. Nguyen, A. Nosratinia, and D. Divsalar, "The design of rate-compatible protograph LDPC codes," *IEEE Trans. Commun.*, **60**(10) (2012), 2841–2850.

[14] D. G. M. Mitchell, R. Smarandache, and D. J. Costello, Jr., "Quasi-cyclic LDPC codes based on pre-lifted protographs," *IEEE Trans. Inf. Theory*, **60**(10) (2014), 5856–5874.

[15] T. Chen, K. Vakilinia, D. Divsalar, and R. D. Wesel, "Protograph-based raptor-like LDPC codes," *IEEE Trans. Commun.*, **63**(5) (2015), 1522–1532.

[16] S. V. S. Ranganathan, D. Divsalar, and R. D. Wesel, "Quasi-cyclic protograph-based raptor-like LDPC codes for short block-lengths," *IEEE Trans. Inf. Theory*, **65**(6) (2019), 3758-3777.

[17] S. Kudekar, T. J. Richardson, and R. L. Urbanke, "Threshold saturation via spatial coupling: why convolutional LDPC ensembles perform so well over the BEC," *IEEE Trans. Inf. Theory*, **57**(2) (2011), 803–834.

[18] S. Kudekar, T. Richardson, and R. Urbanke, "Spatially coupled ensembles universally achieve capacity under belief propagation," *IEEE Trans. Inf. Theory*, **59**(12) (2013), 7761–7813.

[19] D. J. Costello, Jr., L. Dolecek, T. E. Fuja, *et al.*, "Spatially coupled sparse codes on graphs: theory and practice," *IEEE Commun. Mag.*, **52**(7) (2014), 168–176.

[20] D. G. M. Mitchell, M. Lentmaier, and D. J. Costello, Jr., "Spatially coupled LDPC codes constructed from protographs," *IEEE Trans. Inf. Theory*, **61**(9) (2015), 4866–4889.

[21] S. V. S. Ranganathan, K. Vakilinia, L. Dolecek, D. Divsalar, and R. D. Wesel, "Some results on spatially coupled protograph LDPC codes," *2016 Information Theory and Applications Workshop (ITA)*, La Jolla, CA, 2016, pp. 1–6.

[22] H. Esfahanizadeh, A. Hareedy, and L. Dolecek, "Finite-length construction of high performance spatially-coupled codes via optimized partitioning and lifting," *IEEE Trans. Commun.*, **67**(1) (2019), 3–16.

[23] B.-Y. Chang, L. Dolecek, and D. Divsalar, "EXIT chart analysis and design of non-binary protograph-based LDPC codes," *Proc. IEEE MILCOM 2011*, Baltimore, MD, November 7–10, 2011, pp. 566–571.

[24] L. Dolecek, D. Divsalar, Y. Sun, and B. Amiri, "Non-binary protograph-based LDPC codes: enumerators, analysis, and designs," *IEEE Trans. Inf. Theory*, **60**(7) (2014), 3913–3941.

[25] A. Hareedy, C. Lanka, N. Guo, and L. Dolecek, "A combinatorial methodology for optimizing non-binary graph-based codes: theoretical analysis and applications in data storage," *IEEE Trans. Inf. Theory*, **65**(4) (2019), 2128–2154.

[26] R. L. Townsend and E. J. Weldon, Jr., "Self-orthogonal quasi-cyclic codes," *IEEE Trans. Inf. Theory*, **13**(2) (1967), 183–195.

[27] C. L. Chen, W. W. Peterson, and E. J. Weldon, Jr., "Some results on quasi-cyclic codes," *Inf. and Control*, **5**(5) (1969), 407–423.

[28] M. Karlin, "New binary coding results by circulants," *IEEE Trans. Inf. Theory*, **15**(1) (1969), 81–92.

[29] W. W. Peterson and E. J. Weldon, Jr., *Error-Correcting Codes*, 2nd ed., Cambridge, MA, The MIT Press, 1972.

[30] T. Kasami, "A Gilbert–Varshamov bound for quasi-cyclic codes of rate 1/2," *IEEE Trans. Inf. Theory*, **20**(5) (1974), 679.

[31] S. Lin and D. J. Costello, Jr., *Error Control Coding: Fundamentals and Applications*, 2nd ed., Upper Saddle River, NJ, Prentice-Hall, 2004.

[32] C. E. Shannon, R. G. Gallager, and E. R. Berlekamp, "Lower bounds to error probability for coding on discrete memoryless channels," *Inform. Contr.*, **10** (1967), pt. I, pp. 65–103, pt. II, pp. 522–552.

[33] L. Lan, L. Zeng, Y. Y. Tai, L. Chen, S. Lin, and K. Abdel-Ghaffar, "Construction of quasi-cyclic LDPC codes for AWGN and binary erasure channels: a finite field approach," *IEEE Trans. Inf. Theory*, **53**(7) (2007), 2429–2458.

[34] R. Tanner, "A recursive approach to low complexity codes," *IEEE Trans. Inf. Theory*, **27**(5) (1981), 533–547.

[35] H. Xiao and A. Banihashemi, "Improved progressive-edge-growth (PEG) construction of irregular LDPC codes," *2004 IEEE Global Telecommunications Conf.*, Dallas, TX, November 29–December 3, 2004, pp. 489–492.

[36] H. Xiao and A. H. Banihashemi, "Improved progressive-edge-growth (PEG) construction of irregular LDPC codes," *IEEE Commun. Lett.*, **8**(12) (2004), 715–717.

[37] T. Tian, C. Jones, J. Villasenor, and R. Wesel, "Construction of irregular LDPC codes with low error floors," *2003 IEEE Int. Conf. Commun.*, Anchorage, AK, May 2003, pp. 3125–3129.

[38] T. Tian, C. Jones, J. D. Villasenor, and R. D. Wesel, "Selective avoidance of cycles in irregular LDPC code construction," *IEEE Trans. Commun.*, **52**(8) (2004), 1242–1247.

[39] Z. Li and B. V. K. V. Kumar, "A class of good quasi-cyclic low-density parity check codes based on progressive edge growth graph," *Conference Record of the Thirty-Eighth Asilomar Conference on Signals, Systems and Computers*, Pacific Grove, CA, November 2004, pp. 1990–1994.

[40] M. Diouf, D. Declercq, M. Fossorier, S. Ouya, and B. Vasi, "Improved PEG construction of large girth QC-LDPC codes," *2016 9th International Symposium on Turbo Codes and Iterative Information Processing (ISTC)*, Brest, September 2016, pp. 146–150.

[41] X. He, L. Zhou, and J. Du, "The new multi-edge metric-constrained PEG/QC-PEG algorithms for designing the binary LDPC codes with better cycle-structures," https://arxiv.org/abs/1605.05123, May 2016.

[42] X. He, L. Zhou, and J. Du, "PEG-like design of binary QC-LDPC codes based on detecting and avoiding generating small cycles," *IEEE Trans. Commun.*, **66**(5) (2018), 1845–1858.

[43] J. Li, S. Lin, K. Abdel-Ghaffar, W. E. Ryan, and D. J. Costello, Jr., *LDPC Code Designs, Constructions, and Unification*, Cambridge, Cambridge University Press, 2017.

# 16

# Collective Encoding and Soft-Decision Decoding of Cyclic Codes of Prime Lengths in Galois Fourier Transform Domain

Using a block code for error control, each codeword is generated and transmitted *independently*. At the receiving end, each received vector is decoded independently without using the reliability information of previously decoded received vectors. Except for LDPC codes, most of the well-known classes of block codes, such as BCH, RS, cyclic RM, and QR codes, are mostly decoded with algebraic hard-decision decoding algorithms devised based on their specific algebraic structures. The parity-check matrices of these codes used for decoding are not sparse and do not satisfy the RC-constraint, and hence they cannot be used for decoding with iterative decoding algorithms based on belief-propagation such as the SPA and MSA. So far, it is unknown whether there exist spare parity-check matrices of any type for these well-known codes. There are soft-decision decoding algorithms for decoding these codes but they are suitable only for decoding relatively short codes. For decoding long codes, these decoding algorithms require enormously large decoding complexity.

In this chapter, we present a *universal coding scheme* [1–3] for encoding and iterative soft-decision decoding of cyclic codes of prime lengths, binary or nonbinary. The key features of this coding scheme are *collective encoding* and *collective decoding*. The encoding of a cyclic code is performed on a *collection* of codewords which are mapped through *Hadamard permutations*, *cascading*, *interleaving*, and *Galois Fourier transform* (GFT) into a codeword in a *compound code* with a sparse RC-constrained *binary* parity-check matrix for transmission. Using this matrix, *binary iterative decoding* is applied to jointly decode a collection of received vectors. The joint-decoding allows *reliability information sharing* among the received vectors corresponding to the codewords in the transmitted collection during the iterative decoding process. This *joint-decoding* and *information sharing* can achieve a *joint-decoding gain* over the maximum likelihood decoding (MLD) of *individual* codewords. The binary iterative decoding can be performed efficiently and reduces the decoding complexity significantly.

The organization of this chapter is as follows. Section 16.1 first gives a brief review of cyclic codes of prime lengths and then introduces their Hadamard equivalents, which are also cyclic. Section 16.2 presents a method for composing a group of codewords in a cyclic code $C$ (or a Hadamard equivalent of the code) of prime length into a codeword in a *composite* cyclic code. In Section 16.3, we first combine the composite cyclic code of $C$ and its Hadamard equivalents formed in Section 16.2 into a *compound code* in the GFT domain through cascading, interleaving, and GFT. Every codeword in the compound code consists of a collection of codewords in $C$. In Section 16.4, we present the key structures of compound code in the GFT domain and show that the compound code has a *binary sparse parity-check matrix* which satisfies the RC-constraint. Sections 16.5 and 16.6 present a collective encoding scheme and a collective iterative soft-decision decoding scheme of the compound code, respectively. In Section 16.7, measures of error performance and computation complexity of the collective decoding scheme for the compound code are discussed. In Sections 16.8, 16.9, and 16.10, we apply the collective coding scheme to RS codes, BCH codes, and QR codes of prime lengths, respectively. Section 16.11 presents two techniques to construct a family of *rate-compatible* compound codes with various lengths and rates from a given cyclic base code. These rate-compatible compound codes can be decoded using the same decoder. Section 16.12 presents the erasure correcting capability of the compound code.

## 16.1 Cyclic Codes of Prime Lengths and Their Hadamard Equivalents

Let $C$ be a cyclic code of *prime length* $n$ and dimension $n - m$ over the finite field GF($q$) of characteristic 2, say $q = 2^\kappa$, for some positive integer $\kappa$. Let $\mathbf{g}(X)$ be the generator polynomial of $C$. Then, $\mathbf{g}(X)$ has $m$ distinct roots in some extension field, GF($q^\tau$), of GF($q$), where $\tau$ is a positive integer. In terms of its roots, $\mathbf{g}(X)$ is in the following form:

$$\mathbf{g}(X) = \prod_{i=0}^{m-1} \left( X - \beta^{l_i} \right), \tag{16.1}$$

where $\beta$ is an element in GF($q^\tau$) of order $n$ which divides $q^\tau - 1$; $l_0, l_1, \ldots, l_{m-1}$ are integers satisfying $0 \le l_0 < l_1 < \cdots < l_{m-1} < n$; and $\beta^{l_0}, \beta^{l_1}, \ldots, \beta^{l_{m-1}}$ are the $m$ roots of $\mathbf{g}(X)$. The code $C$ has an $m \times n$ parity-check matrix over GF($q^\tau$) of the following form:

$$\mathbf{B} = [\beta^{jl_i}]_{0 \le i < m, 0 \le j < n} = \begin{bmatrix} 1 & \beta^{l_0} & \beta^{2l_0} & \cdots & \beta^{(n-1)l_0} \\ 1 & \beta^{l_1} & \beta^{2l_1} & \cdots & \beta^{(n-1)l_1} \\ \vdots & \vdots & \vdots & \ddots & \vdots \\ 1 & \beta^{l_{m-1}} & \beta^{2l_{m-1}} & \cdots & \beta^{(n-1)l_{m-1}} \end{bmatrix}. \tag{16.2}$$

A vector $\mathbf{v} = (v_0, v_1, \ldots, v_{n-1})$ of length $n$ over GF($q$) is a codeword in $C$ if *and only if* it is *orthogonal* to every row of $\mathbf{B}$, i.e., $\mathbf{v}\mathbf{B}^T = \mathbf{0}$, where $T$ denotes transpose. In the polynomial form, the vector $\mathbf{v}$ is represented by a polynomial $\mathbf{v}(X) = v_0 + v_1 X + v_2 X^2 + \cdots + v_{n-1} X^{n-1}$ over GF($q$) with degree $n-1$ or less. Then, $\mathbf{v}(X)$ is a *code polynomial* if and only if $\mathbf{v}(X)$ is divisible by the generator polynomial $\mathbf{g}(X)$ of $C$, i.e., if and only if $\mathbf{v}(X)$ has $\beta^{l_0}, \beta^{l_1}, \ldots, \beta^{l_{m-1}}$ as roots. From the *Singleton bound* [4] (see Lemma 3.1), the minimum Hamming distance of $C$ is at most $m + 1$.

Because $n$ is a prime, all the $n$ entries in each row of $\mathbf{B}$ are distinct. Label the columns of $\mathbf{B}$ from 0 to $n - 1$. Except for the zeroth column, all the $m$ entries in each column of $\mathbf{B}$ are distinct. With these structural properties, it can be proved that $\mathbf{B}$ satisfies the $2 \times 2$ SM-constraint. The proof is similar to the proof given in Theorem 13.3.

Let $t$ be a nonnegative integer. The $t$th *Hadamard power*, denoted by $\mathbf{B}^{\circ t}$, of $\mathbf{B}$ is defined as the matrix obtained by raising each entry $\beta^{jl_i}$ in $\mathbf{B}$ to the $t$th power [5] as follows:

$$\mathbf{B}^{\circ t} = [\beta^{tjl_i}]_{0 \le i < m, 0 \le j < n} = \begin{bmatrix} 1 & (\beta^{l_0})^t & (\beta^{2l_0})^t & \cdots & (\beta^{(n-1)l_0})^t \\ 1 & (\beta^{l_1})^t & (\beta^{2l_1})^t & \cdots & (\beta^{(n-1)l_1})^t \\ \vdots & \vdots & \vdots & \ddots & \vdots \\ 1 & (\beta^{l_{m-1}})^t & (\beta^{2l_{m-1}})^t & \cdots & (\beta^{(n-1)l_{m-1}})^t \end{bmatrix}. \tag{16.3}$$

For $1 \le t < n$, the $j$th column of $\mathbf{B}$, for $0 \le j < n$, is the $(jt^{-1})_n$th column of $\mathbf{B}^{\circ t}$, where $(u)_n$ denotes the least nonnegative integer congruent to $u$ modulo $n$, i.e., $(u)_n = u - \lfloor u/n \rfloor n$, and $t^{-1}$ is the inverse of $t$ modulo $n$, which exists for all $1 \le t < n$ as $n$ is a prime.

Let $\pi_t(j) = (jt^{-1})_n$ for $1 \le t < n$ and $0 \le j < n$. Because $n$ is a prime and $1 \le t < n$, $\pi_t(j) \ne \pi_t(j')$ for $j \ne j'$. Hence, $\pi_t$ is a permutation on $\{0, 1, \ldots, n-1\}$ which is called the $t$th *Hadamard permutation*. Therefore, for $1 \le t < n$, $\mathbf{B}^{\circ t}$ is simply a column permutation of $\mathbf{B}$. Clearly, $\mathbf{B}^{\circ t}$, $1 \le t < n$, has the same rank as $\mathbf{B}$, which is $m$. Note that $\mathbf{B}^{\circ 1} = \mathbf{B}$. For $t = 0$, $\mathbf{B}^{\circ 0}$ is the all-one matrix of size $m \times n$. Because $\mathbf{B}$ satisfies the $2 \times 2$ SM-constraint, $\mathbf{B}^{\circ t}$, $1 \le t < n$, also satisfies the $2 \times 2$ SM-constraint.

For $1 \le t < n$, let $C^{(t)}$ be the code given by the null space over GF($q$) of $\mathbf{B}^{\circ t}$. The codewords of $C^{(t)}$ can be obtained from $C$ by applying the permutation $\pi_t$ to the codewords of $C$. We call $C^{(t)}$ the $t$th *Hadamard equivalent* of $C$. Hereafter, we use the notation $\pi_t$ to denote both the $t$th Hadamard permutation of the columns of a matrix with $n$ columns and the $t$th Hadamard permutation of the coordinates of a vector of length $n$. Applying the permutation $\pi_t$ to a codeword in $C$ results in a codeword in $C^{(t)}$.

As stated earlier, an $n$-tuple $\mathbf{v} = (v_0, v_1, \ldots, v_{n-1})$ over GF($q$) is in the null space of $\mathbf{B}$ *if and only if* the polynomial $\mathbf{v}(X) = v_0 + v_1 X + v_2 X^2 + \cdots + v_{n-1}X^{n-1}$ has $\beta^{l_0}, \beta^{l_1}, \ldots, \beta^{l_{m-1}}$ as roots. Hence, an $n$-tuple $\mathbf{u} = (u_0, u_1, \ldots, u_{n-1})$ over GF($q$) is in the null space of $\mathbf{B}^{\circ t}$, i.e., $\mathbf{u}(\mathbf{B}^{\circ t})^T = 0$, if and only if the polynomial $\mathbf{u}(X) = u_0 + u_1 X + u_2 X^2 + \cdots + u_{n-1}X^{n-1}$ has $\beta^{tl_0}, \beta^{tl_1}, \ldots, \beta^{tl_{m-1}}$ as roots. We therefore conclude that, for $1 \le t < n$, $C^{(t)}$ is an $(n, n-m)$ *cyclic code* over GF($q$) with generator polynomial $\mathbf{g}^{(t)}(X) = \prod_{i=0}^{m-1}(X - \beta^{tl_i})$. For $t = 0$, the null space over GF($q$) of $\mathbf{B}^{\circ 0}$ gives a single parity-check (SPC) code $C^{(0)}$, which is an $(n, n-1)$ cyclic code over GF($q$) generated by $\mathbf{g}^{(0)}(X) = X - 1$. For $t = 1$, $\mathbf{B}^{\circ 1} = \mathbf{B}$ and $C^{(1)} = C$. The code $C^{(0)}$ has minimum distance 2 whereas each of the $n - 1$ Hadamard equivalents of $C$, $C^{(1)}, C^{(2)}, \ldots, C^{(n-1)}$, has the same minimum distance as $C$, which is at most $m + 1$.

**Example 16.1** Set $\kappa = 1$ and $\tau = 3$. The extension field GF($2^3$) of GF(2) generated by the primitive polynomial $\mathbf{p}(X) = 1 + X + X^3$ over GF(2) is shown in Table 16.1 (a repetition of Table 2.10).

Let $\alpha$ be a primitive element in GF($2^3$). Consider the $(7, 4)$ cyclic code $C$ over GF(2) (see Table 4.1) whose generator polynomial is $\mathbf{g}(X) = 1 + X + X^3$ which has $\alpha$, $\alpha^2$, and $\alpha^4$ as its three roots. Then, the parity-check matrix $\mathbf{B}$ of $C$ in the form of (16.2) is given bellow:

$$\mathbf{B} = \begin{bmatrix} 1 & \alpha & \alpha^2 & \alpha^3 & \alpha^4 & \alpha^5 & \alpha^6 \\ 1 & \alpha^2 & \alpha^4 & \alpha^6 & \alpha & \alpha^3 & \alpha^5 \\ 1 & \alpha^4 & \alpha & \alpha^5 & \alpha^2 & \alpha^6 & \alpha^3 \end{bmatrix}. \tag{16.4}$$

Note that $C$ is a Hamming code or a single-error-correcting BCH code of prime length 7.

For $t = 2$, the second Hadamard power of $\mathbf{B}$ is

$$\mathbf{B}^{\circ 2} = \begin{bmatrix} 1 & (\alpha)^2 & (\alpha^2)^2 & (\alpha^3)^2 & (\alpha^4)^2 & (\alpha^5)^2 & (\alpha^6)^2 \\ 1 & (\alpha^2)^2 & (\alpha^4)^2 & (\alpha^6)^2 & (\alpha)^2 & (\alpha^3)^2 & (\alpha^5)^2 \\ 1 & (\alpha^4)^2 & (\alpha)^2 & (\alpha^5)^2 & (\alpha^2)^2 & (\alpha^6)^2 & (\alpha^3)^2 \end{bmatrix}$$

$$= \begin{bmatrix} 1 & \alpha^2 & \alpha^4 & \alpha^6 & \alpha & \alpha^3 & \alpha^5 \\ 1 & \alpha^4 & \alpha & \alpha^5 & \alpha^2 & \alpha^6 & \alpha^3 \\ 1 & \alpha & \alpha^2 & \alpha^3 & \alpha^4 & \alpha^5 & \alpha^6 \end{bmatrix}.$$

By checking, we see that $\mathbf{B}^{\circ 2}$ is a column-permutation of $\mathbf{B}$, i.e., $\pi_2 = (0, 4, 1, 5, 2, 6, 3)$.

For $t = 3$, the third Hadamard power of $\mathbf{B}$ is

$$\mathbf{B}^{\circ 3} = \begin{bmatrix} 1 & \alpha^3 & \alpha^6 & \alpha^2 & \alpha^5 & \alpha & \alpha^4 \\ 1 & \alpha^6 & \alpha^5 & \alpha^4 & \alpha^3 & \alpha^2 & \alpha \\ 1 & \alpha^5 & \alpha^3 & \alpha & \alpha^6 & \alpha^4 & \alpha^2 \end{bmatrix}.$$

Again by checking, we see that $\mathbf{B}^{\circ 3}$ is also a column-permutation of $\mathbf{B}$, i.e., $\pi_3 = (0, 5, 3, 1, 6, 4, 2)$, and $\mathbf{B}^{\circ 3} \ne \mathbf{B}^{\circ 2} \ne \mathbf{B}$. The generator polynomial of the third Hadamard equivalent $C^{(3)}$ of $C$ is $\mathbf{g}^{(3)}(X) = 1 + X^2 + X^3$ which has $\alpha^3$, $\alpha^5$, and $\alpha^6$ as roots. ▲▲

Table 16.1 GF($2^3$) generated by $\mathbf{p}(X) = 1 + X + X^3$ over GF(2).

| Power form | Polynomial form | | | Vector form | Decimal form |
|---|---|---|---|---|---|
| $0 = \alpha^{-\infty}$ | 0 | | | (0 0 0) | 0 |
| $1 = \alpha^0$ | 1 | | | (1 0 0) | 1 |
| $\alpha$ | | $\alpha$ | | (0 1 0) | 2 |
| $\alpha^2$ | | | $\alpha^2$ | (0 0 1) | 4 |
| $\alpha^3$ | 1 | $+\ \alpha$ | | (1 1 0) | 3 |
| $\alpha^4$ | | $\alpha$ | $+\ \alpha^2$ | (0 1 1) | 6 |
| $\alpha^5$ | 1 | $+\ \alpha$ | $+\ \alpha^2$ | (1 1 1) | 7 |
| $\alpha^6$ | 1 | $+$ | $\alpha^2$ | (1 0 1) | 5 |

# 16.2 Composing, Cascading, and Interleaving a Cyclic Code of Prime Length and Its Hadamard Equivalents

## 16.2.1 Composing

For $0 < t < n$, consider $\tau$ codewords $\mathbf{c}_{0,t}, \mathbf{c}_{1,t}, \ldots, \mathbf{c}_{\tau-1,t}$ in the $t$th Hadamard equivalent $C^{(t)}$ of $C$, where $\mathbf{c}_{l,t} = (c_{l,t,0}, c_{l,t,1}, \ldots, c_{l,t,n-1})$, $0 \le l < \tau$, and $c_{l,t,j} \in \mathrm{GF}(q)$ for $0 \le j < n$. For $t = 0$, the $\tau$ codewords $\mathbf{c}_{0,0}, \mathbf{c}_{1,0}, \ldots, \mathbf{c}_{\tau-1,0}$ are codewords from $C^{(0)}$. These $\tau$ codewords of length $n$ over GF($q$) can be composed as a vector $\mathbf{c}_t$ of length $n$ over the extension field GF($q^\tau$) of GF($q$) as follows:

$$\mathbf{c}_t = \sum_{i=0}^{\tau-1} \alpha^i \mathbf{c}_{i,t} = \mathbf{c}_{0,t} + \alpha\mathbf{c}_{1,t} + \cdots + \alpha^{\tau-1}\mathbf{c}_{\tau-1,t} \tag{16.5}$$

$$= (c_{t,0}, c_{t,1}, \ldots, c_{t,n-1}),$$

where $\alpha$ is a primitive element in GF($q^\tau$). Note that the $j$th component $c_{t,j}$ of $\mathbf{c}_t$, $0 \le j < n$, is the *weighted* sum of the $j$th components $c_{0,t,j}, c_{1,t,j}, \ldots, c_{\tau-1,t,j}$ of $\mathbf{c}_{0,t}, \mathbf{c}_{1,t}, \ldots, \mathbf{c}_{\tau-1,t}$, weighted by $\alpha^0, \alpha, \ldots, \alpha^{\tau-1}$, respectively, i.e.,

$$c_{t,j} = c_{0,t,j} + \alpha c_{1,t,j} + \cdots + \alpha^{\tau-1}c_{\tau-1,t,j}. \tag{16.6}$$

The vector $\mathbf{c}_t$ is called the *composite codeword* of the $\tau$ codewords $\mathbf{c}_{0,t}$, $\mathbf{c}_{1,t}, \ldots, \mathbf{c}_{\tau-1,t}$ in $C^{(t)}$, and $\mathbf{c}_{0,t}, \mathbf{c}_{1,t}, \ldots, \mathbf{c}_{\tau-1,t}$ are referred to as the *constituent codewords* of $\mathbf{c}_t$.

For $0 < t < n$, let $C_\tau^{(t)}$ be the *collection* of all composite codewords, $\mathbf{c}_t$, corresponding to all sequences (groups) $(\mathbf{c}_{0,t}, \mathbf{c}_{1,t}, \ldots, \mathbf{c}_{\tau-1,t})$ of $\tau$ codewords in $C^{(t)}$. The collection $C_\tau^{(t)}$ has $q^{(n-m)\tau}$ composite codewords over $\mathrm{GF}(q^\tau)$ which form an $(n, n-m)$ linear code over $\mathrm{GF}(q^\tau)$. Because $C^{(t)}$ is a cyclic code, $C_\tau^{(t)}$ is also a cyclic code. The code $C_\tau^{(t)}$ is called the $q^\tau$-ary *composite code* of the $q$-ary code $C^{(t)}$ which consists of $\tau$ weighted copies of $C^{(t)}$. The composite code $C_\tau^{(t)}$ is simply the null space over $\mathrm{GF}(q^\tau)$ of the $t$th Hadamard power $\mathbf{B}^{\circ t}$ of the parity-check matrix $\mathbf{B}$ of the cyclic base code $C$, i.e., the composite code $C_\tau^{(t)}$ of $C^{(t)}$ has the same parity-check matrix $\mathbf{B}^{\circ t}$ as $C^{(t)}$. Note that $C_\tau^{(t)}$ is a *super code* of $C^{(t)}$ and $C^{(t)}$ is a *subfield subcode* of $C_\tau^{(t)}$. The codewords in $C_\tau^{(t)}$ are the $t$th Hadamard permutations of the codewords in $C_\tau^{(1)}$. The $n-1$ composite codes $C_\tau^{(1)}, C_\tau^{(2)}, \ldots, C_\tau^{(n-1)}$ are Hadamard equivalent codes over $\mathrm{GF}(q^\tau)$. For $t = 0$, $C_\tau^{(0)}$ is an $(n, n-1)$ SPC code over $\mathrm{GF}(q^\tau)$.

A composite codeword $\mathbf{c}_t$ in $C_\tau^{(t)}$ can be decomposed to retrieve the constituent codewords $\mathbf{c}_{0,t}, \mathbf{c}_{1,t}, \ldots, \mathbf{c}_{\tau-1,t}$ in $C^{(t)}$ as every element in $\mathrm{GF}(q^\tau)$ can be written in a *unique way* as a polynomial in $\alpha$ over $\mathrm{GF}(q)$ of degree less than $\tau$, as shown in (16.6).

**Example 16.2** Consider the $(7, 4)$ cyclic Hamming code $C$ given in Example 16.1. From Table 4.1, we find the following three codewords in $C$:

$$\mathbf{c}_{0,1} = (1\ 1\ 0\ 1\ 0\ 0\ 0), \mathbf{c}_{1,1} = (0\ 1\ 1\ 0\ 1\ 0\ 0), \mathbf{c}_{2,1} = (0\ 0\ 0\ 1\ 1\ 0\ 1).$$

Using (16.5), (16.6), and Table 16.1, the composite codeword of $\mathbf{c}_{0,1}$, $\mathbf{c}_{1,1}$, and $\mathbf{c}_{2,1}$ is

$$\begin{aligned}
\mathbf{c}_1 &= \mathbf{c}_{0,1} + \alpha\mathbf{c}_{1,1} + \alpha^2\mathbf{c}_{2,1} \\
&= (1\ \alpha^3\ \alpha\ \alpha^6\ \alpha^4\ 0\ \alpha^2)
\end{aligned}$$

which is a codeword in the $2^3$-ary composite code $C_3^{(1)}$ of the cyclic code $C$. ▲▲

## 16.2.2 Cascading and Interleaving

Let $C_{\mathrm{casc}}$ be the code over $\mathrm{GF}(q^\tau)$ of length $n^2$ obtained by cascading codewords of the composite codes $C_\tau^{(0)}, C_\tau^{(1)}, \ldots, C_\tau^{(n-1)}$ in this order. A codeword $\mathbf{c}_{\mathrm{casc}}$ in $C_{\mathrm{casc}}$ is in the form of $\mathbf{c}_{\mathrm{casc}} = (\mathbf{c}_0, \mathbf{c}_1, \ldots, \mathbf{c}_{n-1})$ with $\mathbf{c}_t$ in $C_\tau^{(t)}$ for $0 \le t < n$. The code $C_{\mathrm{casc}}$ has length $n^2$ and dimension $(n-1) + (n-m)(n-1) = (n-1)(n-m+1)$. We call $C_{\mathrm{casc}}$ the *cascaded composite* (CC) code of the cyclic base code $C$, its $n-2$ Hadamard equivalents $C^{(t)}$, $2 \le t < n$, and the SPC code $C^{(0)}$. A parity-check matrix of the CC code $C_{\mathrm{casc}}$ is of the following form:

$$\mathbf{H}_{\mathrm{casc}} = \mathrm{diag}(\mathbf{B}^{\circ 0}, \mathbf{B}^{\circ 1}, \ldots, \mathbf{B}^{\circ(n-1)}) = \begin{bmatrix} \mathbf{B}^{\circ 0} & & & \\ & \mathbf{B}^{\circ 1} & & \\ & & \ddots & \\ & & & \mathbf{B}^{\circ(n-1)} \end{bmatrix}, \quad (16.7)$$

which is an $n \times n$ *diagonal array* with the $m \times n$ matrices $\mathbf{B}^{\circ 0}, \mathbf{B}^{\circ 1}, \ldots, \mathbf{B}^{\circ(n-1)}$ lying on its main diagonal and $m \times n$ ZMs elsewhere. The array $\mathbf{H}_{\text{casc}}$ is an $mn \times n^2$ matrix over $\text{GF}(q^{\tau})$ with rank $1 + m(n-1)$.

Label the rows of $\mathbf{H}_{\text{casc}}$ from 0 to $mn - 1$ and columns from 0 to $n^2 - 1$. Define the following index sequences:

$$\pi_{\text{row}}^{(0)} = [0, m, 2m, \ldots, (n-1)m], \tag{16.8}$$

$$\pi_{\text{row}} = [\pi_{\text{row}}^{(0)}, \pi_{\text{row}}^{(0)} + 1, \ldots, \pi_{\text{row}}^{(0)} + m - 1], \tag{16.9}$$

$$\pi_{\text{col}}^{(0)} = [0, n, 2n, \ldots, (n-1)n], \tag{16.10}$$

$$\pi_{\text{col}} = [\pi_{\text{col}}^{(0)}, \pi_{\text{col}}^{(0)} + 1, \ldots, \pi_{\text{col}}^{(0)} + n - 1]. \tag{16.11}$$

Then, $\pi_{\text{row}}$ and $\pi_{\text{col}}$ define a permutation on $\{0, 1, \ldots, mn - 1\}$ and a permutation on $\{0, 1, \ldots, n^2 - 1\}$, respectively.

Let $\mathbf{c}_{\text{casc}} = (\mathbf{c}_0, \mathbf{c}_1, \ldots, \mathbf{c}_{n-1})$ be a codeword in the CC code $C_{\text{casc}}$. Then, $\mathbf{c}_{\text{casc}}$ is a vector of length $n^2$ over $\text{GF}(q^{\tau})$. Label the coordinates of $\mathbf{c}_{\text{casc}}$ from 0 to $n^2 - 1$. Applying the permutation $\pi_{\text{col}}$ to the coordinates of $\mathbf{c}_{\text{casc}}$, we obtain a vector $\mathbf{c}_{\text{casc}}^{\pi} = (\mathbf{c}_0^{\pi}, \mathbf{c}_1^{\pi}, \ldots, \mathbf{c}_{n-1}^{\pi})$, where $\mathbf{c}_i^{\pi}$, $0 \leq i < n$, is a vector of length $n$ whose $j$th component, $0 \leq j < n$, is the $i$th component in $\mathbf{c}_j$. Note that the vector $\mathbf{c}_{\text{casc}}^{\pi}$ is obtained by *interleaving* the $n$ codewords $\mathbf{c}_0, \mathbf{c}_1, \ldots, \mathbf{c}_{n-1}$ from $C_{\tau}^{(0)}, C_{\tau}^{(1)}, \ldots, C_{\tau}^{(n-1)}$, respectively. The interleaving can be performed by arranging the $n$ codewords $\mathbf{c}_0, \mathbf{c}_1, \ldots, \mathbf{c}_{n-1}$ as rows of an $n \times n$ array $\Lambda$ (see Section 9.1). Then, the $n$ columns of the array $\Lambda$ are $\mathbf{c}_0^{\pi}, \mathbf{c}_1^{\pi}, \ldots, \mathbf{c}_{n-1}^{\pi}$.

Let $C_{\text{casc}}^{\pi}$ be the collection of all vectors, $\mathbf{c}_{\text{casc}}^{\pi}$, corresponding to all codewords, $\mathbf{c}_{\text{casc}}$, in $C_{\text{casc}}$. Then, $C_{\text{casc}}^{\pi}$ is a linear code over $\text{GF}(q^{\tau})$ of length $n^2$ and dimension $(n-1)(n-m+1)$ which are equal to those of $C_{\text{casc}}$. The code $C_{\text{casc}}^{\pi}$ is the *interleaved code* of the CC code $C_{\text{casc}}$. We call $C_{\text{casc}}^{\pi}$ the *interleaved cascaded composite* (ICC) code of $C^{(0)}, C^{(1)}, \ldots, C^{(n-1)}$, or simply the ICC code of the base cyclic code $C$.

A parity-check matrix $\mathbf{H}_{\text{casc}}^{\pi}$ for the code $C_{\text{casc}}^{\pi}$ can be obtained by applying the permutations $\pi_{\text{col}}$ and $\pi_{\text{row}}$ to the columns and rows of the parity-check matrix $\mathbf{H}_{\text{casc}}$ of $C_{\text{casc}}$ given by (16.7), respectively. Permuting first the rows of $\mathbf{H}_{\text{casc}}$ based on $\pi_{\text{row}}$ and then columns of $\mathbf{H}_{\text{casc}}$ based on $\pi_{\text{col}}$ (or vice versa), we obtain an $mn \times n^2$ matrix $\mathbf{H}_{\text{casc}}^{\pi}$ over $\text{GF}(q^{\tau})$, where $\pi$ denotes $\pi_{\text{row}}$ and $\pi_{\text{col}}$ collectively, i.e., $\pi = (\pi_{\text{row}}, \pi_{\text{col}})$. Then, the matrix $\mathbf{H}_{\text{casc}}^{\pi}$ over $\text{GF}(q^{\tau})$ is the parity-check matrix of the ICC code $C_{\text{casc}}^{\pi}$. Based on the diagonal structure of $\mathbf{H}_{\text{casc}}$ given by (16.7) and the permutation $\pi$, we readily see that the matrix $\mathbf{H}_{\text{casc}}^{\pi}$ is an $m \times n$ array of $n \times n$ diagonal matrices over $\text{GF}(q^{\tau})$ of the following form:

$$\mathbf{H}_{\text{casc}}^{\pi} = [\mathbf{D}_{e,f}]_{0 \leq e < m, 0 \leq f < n} = \begin{bmatrix} \mathbf{D}_{0,0} & \mathbf{D}_{0,1} & \cdots & \mathbf{D}_{0,n-1} \\ \mathbf{D}_{1,0} & \mathbf{D}_{1,1} & \cdots & \mathbf{D}_{1,n-1} \\ \vdots & \vdots & \vdots & \vdots \\ \mathbf{D}_{m-1,0} & \mathbf{D}_{m-1,1} & \cdots & \mathbf{D}_{m-1,n-1} \end{bmatrix}, \tag{16.12}$$

where, for $0 \le e < m$ and $0 \le f < n$, $\mathbf{D}_{e,f}$ is an $n \times n$ diagonal matrix over $\mathrm{GF}(q^\tau)$ of the following form:

$$\mathbf{D}_{e,f} = \mathrm{diag}(1, \beta^{fl_e}, \beta^{2fl_e}, \ldots, \beta^{(n-1)fl_e}) = \begin{bmatrix} 1 & & & & \\ & \beta^{fl_e} & & & \\ & & \ddots & & \\ & & & \beta^{(n-1)fl_e} \end{bmatrix}, \quad (16.13)$$

which is formed by the root $\beta^{l_e}$ of the generator polynomial $\mathbf{g}(X)$ of the $(n, n-m)$ cyclic base code $C$. Because the column and row permutations do not change the rank of a matrix, $\mathbf{H}^\pi_{\mathrm{casc}}$ has the same rank, $1 + m(n-1)$, as $\mathbf{H}_{\mathrm{casc}}$.

**Example 16.3** Consider the $(7, 4)$ cyclic Hamming code $C$ given in Example 16.1 whose parity-check matrix over $\mathrm{GF}(2^3)$ is

$$\mathbf{B} = \begin{bmatrix} 1 & \alpha & \alpha^2 & \alpha^3 & \alpha^4 & \alpha^5 & \alpha^6 \\ 1 & \alpha^2 & \alpha^4 & \alpha^6 & \alpha & \alpha^3 & \alpha^5 \\ 1 & \alpha^4 & \alpha & \alpha^5 & \alpha^2 & \alpha^6 & \alpha^3 \end{bmatrix},$$

where $\alpha$ is a primitive element in $\mathrm{GF}(2^3)$. Then, it follows from (16.12) and (16.13) that the parity-check matrix $\mathbf{H}^\pi_{\mathrm{casc}}$ of the ICC code $C^\pi_{\mathrm{casc}}$ is given by

$$\mathbf{H}^\pi_{\mathrm{casc}} = \begin{bmatrix} \mathbf{D}_{0,0} & \mathbf{D}_{0,1} & \mathbf{D}_{0,2} & \mathbf{D}_{0,3} & \mathbf{D}_{0,4} & \mathbf{D}_{0,5} & \mathbf{D}_{0,6} \\ \mathbf{D}_{1,0} & \mathbf{D}_{1,1} & \mathbf{D}_{1,2} & \mathbf{D}_{1,3} & \mathbf{D}_{1,4} & \mathbf{D}_{1,5} & \mathbf{D}_{1,6} \\ \mathbf{D}_{2,0} & \mathbf{D}_{2,1} & \mathbf{D}_{2,2} & \mathbf{D}_{2,3} & \mathbf{D}_{2,4} & \mathbf{D}_{2,5} & \mathbf{D}_{2,6} \end{bmatrix}, \quad (16.14)$$

where for $0 \le e < 3$ and $0 \le f < 7$, $\mathbf{D}_{e,f}$ is a $7 \times 7$ matrix over $\mathrm{GF}(2^3)$ given as follows:

$$\mathbf{D}_{e,f} = \mathrm{diag}(1, \alpha^{fl_e}, \alpha^{2fl_e}, \alpha^{3fl_e}, \alpha^{4fl_e}, \alpha^{5fl_e}, \alpha^{6fl_e}). \quad (16.15)$$

Because the roots of the generator polynomial $\mathbf{g}(X)$ of $C$ are $\alpha$, $\alpha^2$, and $\alpha^4$, for $f = 1$, $e = 0, 1, 2$, we have $l_0 = 1$, $l_1 = 2$, and $l_2 = 4$. It follows from (16.15) that $\mathbf{D}_{0,1}$, $\mathbf{D}_{1,1}$, and $\mathbf{D}_{2,1}$ are three $7 \times 7$ diagonal matrices over $\mathrm{GF}(2^3)$ given by

$$\mathbf{D}_{0,1} = \begin{bmatrix} 1 & & & & & & \\ & \alpha & & & & & \\ & & \alpha^2 & & & & \\ & & & \alpha^3 & & & \\ & & & & \alpha^4 & & \\ & & & & & \alpha^5 & \\ & & & & & & \alpha^6 \end{bmatrix}, \mathbf{D}_{1,1} = \begin{bmatrix} 1 & & & & & & \\ & \alpha^2 & & & & & \\ & & \alpha^4 & & & & \\ & & & \alpha^6 & & & \\ & & & & \alpha & & \\ & & & & & \alpha^3 & \\ & & & & & & \alpha^5 \end{bmatrix}, \mathbf{D}_{2,1} = \begin{bmatrix} 1 & & & & & & \\ & \alpha^4 & & & & & \\ & & \alpha & & & & \\ & & & \alpha^5 & & & \\ & & & & \alpha^2 & & \\ & & & & & \alpha^6 & \\ & & & & & & \alpha^3 \end{bmatrix}.$$

▲▲

## 16.3 Galois Fourier Transform of ICC Codes

Let $\beta$ be an element of order $n$ in $\mathrm{GF}(q^\tau)$. Then, the set $\mathbf{S}_n = \{1, \beta, \beta^2, \ldots, \beta^{n-1}\}$ forms a cyclic subgroup of $\mathrm{GF}(q^\tau)$. Form the following $n \times n$ *Vandermonde matrix* over $\mathbf{S}_n$:

$$\mathbf{V} = [\beta^{ij}]_{0 \le i,j < n} = \begin{bmatrix} 1 & 1 & 1 & \cdots & 1 \\ 1 & \beta & \beta^2 & \cdots & \beta^{n-1} \\ 1 & \beta^2 & (\beta^2)^2 & \cdots & (\beta^2)^{n-1} \\ 1 & \beta^3 & (\beta^3)^2 & \cdots & (\beta^3)^{n-1} \\ \vdots & \vdots & \ddots & \vdots & \vdots \\ 1 & (\beta)^{n-1} & (\beta^{n-1})^2 & \cdots & (\beta^{n-1})^{n-1} \end{bmatrix}, \quad (16.16)$$

which is nonsingular (see Section 2.9.4). Its inverse is given by $\mathbf{V}^{-1} = [\beta^{-ij}]_{0 \le i,j < n}$, i.e., $\mathbf{V} \cdot \mathbf{V}^{-1} = \mathbf{I}_n$, where $\mathbf{I}_n$ is an $n \times n$ identity matrix.

Let $\mathbf{a} = (a_0, a_1, \ldots, a_{n-1})$ be an $n$-tuple over $\mathrm{GF}(q^\tau)$ and $\beta$, as before, be an element of order $n$ in $\mathrm{GF}(q^\tau)$. The $n$-tuple

$$\mathbf{b} = (b_0, b_1, \ldots, b_{n-1}) = \mathbf{a} \cdot \mathbf{V} \qquad (16.17)$$

is defined as the *Galois Fourier transform* (GFT) [6] of $\mathbf{a}$, denoted by $\mathcal{F}(\mathbf{a})$, i.e., $\mathbf{b} = \mathcal{F}(\mathbf{a})$. For $0 \le j < n$, the $j$th component of $\mathbf{b}$ is given by

$$b_j = a_0 + a_1\beta^j + a_2\beta^{2j} + \cdots + a_{n-1}\beta^{(n-1)j}, \qquad (16.18)$$

which is the inner product of $\mathbf{a} = (a_0, a_1, \ldots, a_{n-1})$ and the $j$th column $[1, \beta^j, \beta^{2j}, \ldots, \beta^{(n-1)j}]^T$ of $\mathbf{V}$. The $n$-tuple $\mathbf{a}$ can be retrieved through $\mathbf{a} = \mathbf{b} \cdot \mathbf{V}^{-1}$ which is called the *inverse GFT* of the vector $\mathbf{b}$, denoted by $\mathcal{F}^{-1}(\mathbf{b})$.

**Example 16.4** Consider the composite codeword $\mathbf{c}_1 = (1 \ \alpha^3 \ \alpha \ \alpha^6 \ \alpha^4 \ 0 \ \alpha^2)$ over $\mathrm{GF}(2^3)$ in the composite code of the $(7, 4)$ cyclic Hamming code over $\mathrm{GF}(2)$ constructed in Example 16.2. Because $n = 7$, we have $\mathbf{S}_7 = \{1, \alpha, \alpha^2, \ldots, \alpha^6\}$ where $\alpha$ is a primitive element of $\mathrm{GF}(2^3)$ (i.e., $\alpha^7 = 1$). Then, the $7 \times 7$ Vandermonde matrix and its inverse over $\mathrm{GF}(2^3)$ are given as follows:

$$\mathbf{V} = \begin{bmatrix} 1 & 1 & 1 & 1 & 1 & 1 & 1 \\ 1 & \alpha & \alpha^2 & \alpha^3 & \alpha^4 & \alpha^5 & \alpha^6 \\ 1 & \alpha^2 & \alpha^4 & \alpha^6 & \alpha & \alpha^3 & \alpha^5 \\ 1 & \alpha^3 & \alpha^6 & \alpha^2 & \alpha^5 & \alpha & \alpha^4 \\ 1 & \alpha^4 & \alpha & \alpha^5 & \alpha^2 & \alpha^6 & \alpha^3 \\ 1 & \alpha^5 & \alpha^3 & \alpha & \alpha^6 & \alpha^4 & \alpha^2 \\ 1 & \alpha^6 & \alpha^5 & \alpha^4 & \alpha^3 & \alpha^2 & \alpha \end{bmatrix}, \mathbf{V}^{-1} = \begin{bmatrix} 1 & 1 & 1 & 1 & 1 & 1 & 1 \\ 1 & \alpha^6 & \alpha^5 & \alpha^4 & \alpha^3 & \alpha^2 & \alpha \\ 1 & \alpha^5 & \alpha^3 & \alpha & \alpha^6 & \alpha^4 & \alpha^2 \\ 1 & \alpha^4 & \alpha & \alpha^5 & \alpha^2 & \alpha^6 & \alpha^3 \\ 1 & \alpha^3 & \alpha^6 & \alpha^2 & \alpha^5 & \alpha & \alpha^4 \\ 1 & \alpha^2 & \alpha^4 & \alpha^6 & \alpha & \alpha^3 & \alpha^5 \\ 1 & \alpha & \alpha^2 & \alpha^3 & \alpha^4 & \alpha^5 & \alpha^6 \end{bmatrix}.$$

The GFT of the composite codeword $\mathbf{c}_1 = (1 \ \alpha^3 \ \alpha \ \alpha^6 \ \alpha^4 \ 0 \ \alpha^2)$ is

$$\mathcal{F}(\mathbf{c}_1) = \mathbf{c}_1 \cdot \mathbf{V} = (\alpha^5 \ 0 \ 0 \ \alpha^4 \ 0 \ 1 \ 1).$$

▲▲

Let $\mathbf{c}_{\mathrm{casc}}^{\pi} = (\mathbf{c}_0^{\pi}, \mathbf{c}_1^{\pi}, \ldots, \mathbf{c}_{n-1}^{\pi})$ be a codeword in $C_{\mathrm{casc}}^{\pi}$, where each of $\mathbf{c}_0^{\pi}$, $\mathbf{c}_1^{\pi}, \ldots, \mathbf{c}_{n-1}^{\pi}$ is an $n$-tuple over $\mathrm{GF}(q^{\tau})$ and called a(n) (*interleaved*) *section* of $\mathbf{c}_{\mathrm{casc}}^{\pi}$. For $0 \le i < n$, let $\mathbf{c}_i^{\pi,\mathcal{F}} = \mathcal{F}(\mathbf{c}_i^{\pi}) = \mathbf{c}_i^{\pi}\mathbf{V}$ be the GFT of the $i$th section $\mathbf{c}_i^{\pi}$. Let $\mathbf{c}_{\mathrm{casc}}^{\pi,\mathcal{F}} = (\mathbf{c}_0^{\pi,\mathcal{F}}, \mathbf{c}_1^{\pi,\mathcal{F}}, \ldots, \mathbf{c}_{n-1}^{\pi,\mathcal{F}})$. Then,

$$\mathbf{c}_{\mathrm{casc}}^{\pi,\mathcal{F}} = \mathbf{c}_{\mathrm{casc}}^{\pi}\mathrm{diag}(\underbrace{\mathbf{V}, \mathbf{V}, \ldots, \mathbf{V}}_{n}), \tag{16.19}$$

where $\mathrm{diag}(\mathbf{V}, \mathbf{V}, \ldots, \mathbf{V})$ is an $n \times n$ diagonal array with $n$ copies of the Vandermonde matrix $\mathbf{V}$ lying on its main diagonal. The vector $\mathbf{c}_{\mathrm{casc}}^{\pi,\mathcal{F}}$ is referred to as the *GFT* of the codeword $\mathbf{c}_{\mathrm{casc}}^{\pi}$. The codeword $\mathbf{c}_{\mathrm{casc}}^{\pi}$ in the ICC code $C_{\mathrm{casc}}^{\pi}$ can be retrieved from $\mathbf{c}_{\mathrm{casc}}^{\pi,\mathcal{F}}$ as follows:

$$\mathbf{c}_{\mathrm{casc}}^{\pi} = \mathbf{c}_{\mathrm{casc}}^{\pi,\mathcal{F}}\mathrm{diag}(\underbrace{\mathbf{V}^{-1}, \mathbf{V}^{-1}, \ldots, \mathbf{V}^{-1}}_{n}), \tag{16.20}$$

which is referred to as the *inverse GFT* of $\mathbf{c}_{\mathrm{casc}}^{\pi,\mathcal{F}}$.

Let $C_{\mathrm{casc}}^{\pi,\mathcal{F}}$ be the collection of all the GFTs, $\mathbf{c}_{\mathrm{casc}}^{\pi,\mathcal{F}}$, of all the codewords in $C_{\mathrm{casc}}^{\pi}$. Then, $C_{\mathrm{casc}}^{\pi,\mathcal{F}}$ is a linear code over $\mathrm{GF}(q^{\tau})$ of length $n^2$ and dimension $(n-1)(n-m+1)$, which are equal to those of the ICC code $C_{\mathrm{casc}}^{\pi}$. We call the code $C_{\mathrm{casc}}^{\pi,\mathcal{F}}$ the GFT of the ICC code $C_{\mathrm{casc}}^{\pi}$, simply the *GFT-ICC* code. Hence, $C_{\mathrm{casc}}^{\pi,\mathcal{F}}$ is composed of $n-1$ copies of the Hadamard equivalents of the composite cyclic code $C_{\tau}^{(1)} = C_{\tau}$ (i.e., $C_{\tau}^{(1)}, C_{\tau}^{(2)}, \ldots, C_{\tau}^{(n-1)}$), and one composite SPC code $C_{\tau}^{(0)}$ in the *GFT domain*. Each codeword in $C_{\mathrm{casc}}^{\pi,\mathcal{F}}$ is composed of a collection of $(n-1)\tau$ codewords in the cyclic base code $C$ and a collection of $\tau$ codewords in the SPC code $C^{(0)}$.

From (16.19) and (16.20), we can readily verify that the following matrix:

$$\mathbf{H}_{\mathrm{casc}}^{\pi,\mathcal{F}^{-1}} = \mathrm{diag}(\underbrace{\mathbf{V}, \mathbf{V}, \ldots, \mathbf{V}}_{m})\mathbf{H}_{\mathrm{casc}}^{\pi}\mathrm{diag}(\underbrace{\mathbf{V}^{-1}, \mathbf{V}^{-1}, \ldots, \mathbf{V}^{-1}}_{n}) \tag{16.21}$$

is a parity-check matrix of the GFT-ICC code $C_{\mathrm{casc}}^{\pi,\mathcal{F}}$ (see Problem 16.7). It follows from (16.12) and (16.21) that we have

$$\begin{aligned}
\mathbf{H}_{\mathrm{casc}}^{\pi,\mathcal{F}^{-1}} &= [\mathbf{V}\mathbf{D}_{e,f}\mathbf{V}^{-1}]_{0 \le e < m, 0 \le f < n} \\
&= [\mathbf{D}_{e,f}^{\mathcal{F}^{-1}}]_{0 \le e < m, 0 \le f < n},
\end{aligned} \tag{16.22}$$

where $\mathbf{D}_{e,f}^{\mathcal{F}^{-1}} = \mathbf{V}\mathbf{D}_{e,f}\mathbf{V}^{-1}$ is an $n \times n$ matrix. Because the GFT operation does not change the rank of $\mathbf{H}_{\mathrm{casc}}^{\pi}$, the rank of $\mathbf{H}_{\mathrm{casc}}^{\pi,\mathcal{F}^{-1}}$ is the same as that of $\mathbf{H}_{\mathrm{casc}}^{\pi}$, i.e., $1 + m(n-1)$. Hence, $C_{\mathrm{casc}}^{\pi,\mathcal{F}}$ is an $(n^2, (n-m-1)(n-1))$ linear code over $\mathrm{GF}(q^{\tau})$ with rate $(n-m+1)(n-1)/n^2$.

In summary, the entire coding process developed in Sections 16.2 and 16.3 consists of the following sequence of steps.

A cyclic base code $\rightarrow$ Hadamard permutations $\rightarrow$ composing $\rightarrow$ cascading
$\rightarrow$ interleaving $\rightarrow$ GFT $\rightarrow$ a GFT-ICC code.

# 16.4   Structural Properties of GFT-ICC Codes

The parity-check matrix $\mathbf{H}_{\text{casc}}^{\pi,\mathcal{F}^{-1}}$ of the GFT-ICC code $C_{\text{casc}}^{\pi,\mathcal{F}}$ has several structural properties that are relevant to its use in iterative decoding based on belief-propagation (IDBP). In this section, we develop these structural properties.

For $0 \leq e < m$ and $0 \leq f < n$, consider the $(e,f)$-submatrix $\mathbf{D}_{e,f}^{\mathcal{F}^{-1}} = \mathbf{V}\mathbf{D}_{e,f}\mathbf{V}^{-1}$ of $\mathbf{H}_{\text{casc}}^{\pi,\mathcal{F}^{-1}}$. Label the rows and columns of $\mathbf{D}_{e,f}^{\mathcal{F}^{-1}}$ from 0 to $n-1$. As shown by (16.13), $\mathbf{D}_{e,f} = \text{diag}(1, \beta^{fl_e}, \beta^{2fl_e}, \ldots, \beta^{(n-1)fl_e})$ is an $n \times n$ diagonal matrix over $\text{GF}(q^\tau)$. Multiplying the product $\mathbf{V}\mathbf{D}_{e,f}\mathbf{V}^{-1}$ out, we find that the $(i,j)$-entry, $0 \leq i,j < n$, of $\mathbf{D}_{e,f}^{\mathcal{F}^{-1}}$ is the following sum:

$$\sum_{k=0}^{n-1} \beta^{(i-j+fl_e)k}. \tag{16.23}$$

This sum is zero unless $i - j$ assumes the *unique value* modulo $n$ satisfying $\beta^{i-j+fl_e} = 1$, i.e., $i - j \equiv -fl_e$ (module $n$), in which case the sum equals 1 (see Problem 16.8). Consequently, the single 1-component in the top row (or the zeroth row) of $\mathbf{D}_{e,f}^{\mathcal{F}^{-1}} = \mathbf{V}\mathbf{D}_{e,f}\mathbf{V}^{-1}$ appears in the column numbered $(fl_e)_n$ and each row of $\mathbf{D}_{e,f}^{\mathcal{F}^{-1}}$ is the *cyclic-shift* of the row above it one place to the right and the top row of $\mathbf{D}_{e,f}^{\mathcal{F}^{-1}}$ is the cyclic-shift of the bottom row of $\mathbf{D}_{e,f}^{\mathcal{F}^{-1}}$ one place to the right. Hence, the matrix $\mathbf{D}_{e,f}^{\mathcal{F}^{-1}}$ is a *binary circulant permutation matrix* (CPM) whose generator (top row) has its single 1-component at the location $(fl_e)_n$.

It follows from the above developments that $\mathbf{H}_{\text{casc}}^{\pi,\mathcal{F}^{-1}} = [\mathbf{D}_{e,f}^{\mathcal{F}^{-1}}]_{0 \leq e < m, 0 \leq f < n}$ is an $m \times n$ array of CPMs of size $n \times n$ which is an $mn \times n^2$ binary matrix with column and row weights $m$ and $n$, respectively.

From (16.2), we see that for $0 \leq e < m$ and $0 \leq f < n$, $\beta^{(fl_e)_n}$ is an entry in the parity-check matrix $\mathbf{B}$ of the cyclic base code $C$ and the $n$-fold CPM-dispersion of $\beta^{(fl_e)_n}$ is a CPM whose generator (top row) has a single 1-component at the location $(fl_e)_n$. With these facts, the array $\mathbf{H}_{\text{casc}}^{\pi,\mathcal{F}^{-1}}$ is simply the $n$-fold CPM-dispersion of $\mathbf{B}$, i.e., replacing all the entries of $\mathbf{B}$ by their corresponding $n$-fold CPM-dispersions.

Because $\mathbf{H}_{\text{casc}}^{\pi,\mathcal{F}^{-1}}$ is an array of CPMs, the GFT-ICC code $C_{\text{casc}}^{\pi,\mathcal{F}}$ is a quasi-cyclic (QC) code over $\text{GF}(q^\tau)$. For $n \geq 7$, $\mathbf{H}_{\text{casc}}^{\pi,\mathcal{F}^{-1}}$ is a *low-density parity-check* (LDPC) *matrix*. In this case, from the LDPC code point of view, the GFT-ICC code $C_{\text{casc}}^{\pi,\mathcal{F}}$ can be regarded as a $q^\tau$-ary QC-LDPC code of length $n^2$ and dimension $(n-1)(n+1-m)$. To display the LDPC structure, we use the following notations for $\mathbf{H}_{\text{casc}}^{\pi,\mathcal{F}^{-1}}$ and $C_{\text{casc}}^{\pi,\mathcal{F}}$:

$$\mathbf{H}_{\text{casc}}^{\pi,\mathcal{F}^{-1}} = \mathbf{H}_{\text{LDPC}}, \quad C_{\text{casc}}^{\pi,\mathcal{F}} = C_{\text{LDPC}}. \tag{16.24}$$

We call $C_{\text{LDPC}}$ the *GFT-ICC-LDPC code* which is induced by the cyclic base code $C$ of prime length with the parity-check matrix $\mathbf{B}$ given by (16.2), and we call $\mathbf{B}$ the *base matrix* of $C_{\text{LDPC}}$.

The parity-check matrix $\mathbf{H}_{\mathrm{LDPC}}$ of the $q^\tau$-ary QC-LDPC code $C_{\mathrm{LDPC}}$ has several properties which are relevant to its use in IDBP. First, because the base matrix $\mathbf{B}$ satisfies the $2 \times 2$ SM-constraint and $\mathbf{H}_{\mathrm{LDPC}}$ is the $n$-fold CPM-dispersion of $\mathbf{B}$, $\mathbf{H}_{\mathrm{LDPC}}$ satisfies the RC-constraint and has orthogonal structure. The RC-constraint ensures that the Tanner graph associated with $\mathbf{H}_{\mathrm{LDPC}}$ has girth at least 6, which is typically required for iterative decoding algorithms, such as the SPA or the MSA, to achieve good performance as shown in Chapters 11–15. Next, the orthogonal structure of $\mathbf{H}_{\mathrm{LDPC}}$ ensures that, at each code symbol position of a codeword in $C_{\mathrm{LDPC}}$, there are $m$ rows in $\mathbf{H}_{\mathrm{LDPC}}$ which are orthogonal to this position. Hence, the minimum distance of $C_{\mathrm{LDPC}}$ is at least $m + 1$.

Note that the minimum distance of $C_{\mathrm{LDPC}}$ depends only on the *number of roots* of the generator polynomial $\mathbf{g}(X)$ of the cyclic base code $C$ but *not on the choices of the roots* as long as all the conjugates of each chosen root are included. This is interesting, because the cyclic base code $C$ that we started with and, therefore, the classical cascading scheme, have minimum distance at most $m + 1$ based on the Singleton bound [4]. Furthermore, this potential increase in minimum distance does not come at the cost of decreased rate. On the contrary, the rate of the cyclic base code $C$ and its associated classical cascading scheme is $(n - m)/n$ whereas the rate of the GFT-ICC-LDPC code $C_{\mathrm{LDPC}}$ is $(n - 1)(n - m + 1)/n^2 = (n - m)/m + (m - 1)/n$, which is greater than $(n - m)/n$. Hence, there is a rate gain.

Because $\mathbf{H}_{\mathrm{LDPC}}$ is RC-constrained with constant column weight $m$, its Tanner graph has no *absorbing sets* of size less than $\lfloor m/2 \rfloor + 1$ or *trapping sets* of size less than $m - 3$ [7, 8]. This implies that the code $C_{\mathrm{LDPC}}$ given by the null space of $\mathbf{H}_{\mathrm{LDPC}}$ decoded with the SPA or the MSA does not suffer from high error-floors if $m$ is reasonably large.

Finally, and most importantly, $\mathbf{H}_{\mathrm{LDPC}}$ is a *binary matrix* even though its null space $C_{\mathrm{LDPC}}$ is a nonbinary code over $\mathrm{GF}(q^\tau)$, where $q = 2^\kappa$. This binary property considerably simplifies the decoding of $C_{\mathrm{LDPC}}$ with an IDBP algorithm. Consider a vector $\mathbf{r} = (r_0, r_1, \ldots, r_{n^2-1})$ over $\mathrm{GF}(q^\tau)$ with $q = 2^\kappa$. Because each component $r_j$, $0 \leq j < n^2$, of $\mathbf{r}$ is a symbol over $\mathrm{GF}(2^{\kappa\tau})$, it can be uniquely represented by a $\kappa\tau$-tuple $(r_{j,0}, r_{j,1}, \ldots, r_{j,\kappa\tau-1})$ over $\mathrm{GF}(2)$, whose components are the coefficients of the following polynomial over $\mathrm{GF}(2^{\kappa\tau})$ of degree less than $\kappa\tau$:

$$r_j = r_{j,0} + r_{j,1}\alpha + r_{j,2}\alpha^2 + \cdots + r_{j,\kappa\tau-1}\alpha^{\kappa\tau-1}, \qquad (16.25)$$

where $\alpha$ is a primitive element in $\mathrm{GF}(2^{\kappa\tau})$. For $0 \leq j < n^2$, representing each symbol $r_j$ in $\mathbf{r}$ by its corresponding $\kappa\tau$-tuple $(r_{j,0}, r_{j,1}, \ldots, r_{j,\kappa\tau-1})$ over $\mathrm{GF}(2)$, we can decompose $\mathbf{r}$ into $\kappa\tau$ *constituent vectors* over $\mathrm{GF}(2)$:

$$\mathbf{r}_{b,i} = (r_{0,i}, r_{1,i}, \ldots, r_{n^2-1,i}) \qquad (16.26)$$

with $0 \leq i < \kappa\tau$, where the subscript "$b$" in $\mathbf{r}_{b,i}$ stands for "binary." Clearly, we can reconstruct $\mathbf{r}$ from its $\kappa\tau$ binary constituent vectors.

With the binary decomposition of $\mathbf{r}$, we readily see that $\mathbf{r}$ is in the null space over $\mathrm{GF}(2^{\kappa\tau})$ of the binary matrix $\mathbf{H}_{\mathrm{LDPC}}$ if and only if each of its $\kappa\tau$

binary constituent vectors, $\mathbf{r}_{b,i} = (r_{0,i}, r_{1,i}, \ldots, r_{n^2-1,i})$, $0 \leq i < \kappa\tau$, is in the null space over GF(2) of $\mathbf{H}_{\mathrm{LDPC}}$, i.e., $\mathbf{r}_{b,i}\mathbf{H}_{\mathrm{LDPC}}^T = \mathbf{0}$. Consequently, decoding a received vector $\mathbf{r}$ over $\mathrm{GF}(2^{\kappa\tau})$ using an IDBP algorithm based on $\mathbf{H}_{\mathrm{LDPC}}$ can be implemented by performing $\kappa\tau$ decodings of its $\kappa\tau$ *binary constituent vectors* based on $\mathbf{H}_{\mathrm{LDPC}}$ using a *binary IDBP algorithm*. This reduces the decoding complexity from a function of $(2^{\kappa\tau})^2 = 2^{2\kappa\tau}$ for direct implementation of a $2^{\kappa\tau}$-ary IDBP algorithm [9], or $\kappa\tau 2^{\kappa\tau}$ for fast Fourier transform implementation [10–13], to a linear function of $\kappa\tau$, namely, $\kappa\tau$ times the complexity of a binary IDBP algorithm.

## 16.5 Collective Encoding of GFT-ICC-LDPC Codes

Based on the developments in Sections 16.1–16.4, we can encode and decode the constituent codes of a GFT-ICC-LDPC code $C_{\mathrm{LDPC}}$ *collectively*. In this section, we present a collective encoding scheme for $C_{\mathrm{LDPC}}$. A collective iterative soft-decision decoding scheme for $C_{\mathrm{LDPC}}$ will be presented in Section 16.6.

In encoding, a collection of messages are encoded into a collection of codewords in the base cyclic code $C$ of prime length. The codewords in this collection are then combined through composition, Hadamard permutation, cascading, interleaving, and GFT into a sequence, which is a codeword in the GFT-ICC-LDPC code $C_{\mathrm{LDPC}}$ for transmission. A collective encoding scheme for $C_{\mathrm{LDPC}}$ is shown in Fig. 16.1.

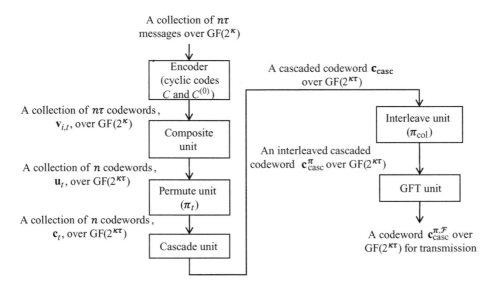

Figure 16.1 A collective encoding scheme for a GFT-ICC-LDPC code $C_{\mathrm{LDPC}}$.

The collective encoding process consists of the following steps.

**Enc-1**. For $0 \leq i < \tau$ and $t = 0$, encode a message $\mathbf{M}_{i,0}$ composed of $n - 1$ $2^{\kappa}$-ary information symbols into a codeword, $\mathbf{v}_{i,0}$, in the $(n, n - 1)$ SPC code $C^{(0)}$ over $\mathrm{GF}(2^{\kappa})$. For $0 \leq i < \tau$, $1 \leq t < n$, encode a message $\mathbf{M}_{i,t}$ composed of $n - m$ $2^{\kappa}$-ary information symbols into a codeword, $\mathbf{v}_{i,t}$, in the cyclic base code $C$ over $\mathrm{GF}(2^{\kappa})$.

**Enc-2**. Divide the $(n - 1)\tau$ codewords in $C$ generated in step **Enc-1** into $n - 1$ groups, denoted by $\mathbf{G}_t$, $1 \leq t < n$, each consisting of $\tau$ codewords. The $\tau$ codewords in $C^{(0)}$ form a group, denoted by $\mathbf{G}_0$. For $1 \leq t < n - 1$, compose the $\tau$ codewords in each group $\mathbf{G}_t$ into a $2^{\kappa\tau}$-ary codeword $\mathbf{u}_t$ in the composite code $C_{\tau}^{(1)}$. For $t = 0$, compose the $\tau$ codewords in $C^{(0)}$ into a composite codeword $\mathbf{u}_0$ in the $(n, n - 1)$ composite SPC code $C_{\tau}^{(0)}$.

**Enc-3**. For $1 \leq t < n$, apply the Hadamard permutation $\pi_t$ on the composite codeword $\mathbf{u}_t$ formed in step **Enc-2** to obtain a codeword $\mathbf{c}_t$ in $C_{\tau}^{(t)}$, the $t$th Hadamard equivalent of the composite code $C_{\tau}^{(1)}$. No permutation is performed on the codeword $\mathbf{u}_0$ and we set $\mathbf{u}_0 = \mathbf{c}_0$.

**Enc-4**. Cascade the $n$ composite codewords $\mathbf{c}_0, \mathbf{c}_1, \ldots, \mathbf{c}_{n-1}$ generated in step **Enc-3** to form a cascaded codeword $\mathbf{c}_{\mathrm{casc}} = (\mathbf{c}_0, \mathbf{c}_1, \ldots, \mathbf{c}_{n-1})$ over $\mathrm{GF}(2^{\kappa\tau})$ in the CC code $C_{\mathrm{casc}}$.

**Enc-5**. Apply the permutation $\pi_{\mathrm{col}}$ to $\mathbf{c}_{\mathrm{casc}}$, i.e., interleave $\mathbf{c}_{\mathrm{casc}}$, to obtain the interleaved cascaded composite codeword $\mathbf{c}_{\mathrm{casc}}^{\pi} = (\mathbf{c}_0^{\pi}, \mathbf{c}_1^{\pi}, \ldots, \mathbf{c}_{n-1}^{\pi})$ in the ICC code $C_{\mathrm{casc}}^{\pi}$.

**Enc-6**. Take the GFT of each component $\mathbf{c}_t^{\pi}$, $0 \leq t < n$, in $\mathbf{c}_{\mathrm{casc}}^{\pi}$ to obtain $\mathbf{c}_t^{\pi, \mathcal{F}}$ and form the vector $\mathbf{c}_{\mathrm{casc}}^{\pi, \mathcal{F}} = (\mathbf{c}_0^{\pi, \mathcal{F}}, \mathbf{c}_1^{\pi, \mathcal{F}}, \ldots, \mathbf{c}_{n-1}^{\pi, \mathcal{F}})$ of length $n^2$ over $\mathrm{GF}(2^{\kappa\tau})$ for transmission. The vector $\mathbf{c}_{\mathrm{casc}}^{\pi, \mathcal{F}}$ is a codeword in the GFT-ICC-LDPC code $C_{\mathrm{casc}}^{\pi, \mathcal{F}} = C_{\mathrm{LDPC}}$ which has $\mathbf{H}_{\mathrm{casc}}^{\pi, \mathcal{F}^{-1}} = \mathbf{H}_{\mathrm{LDPC}}$ given by (16.22) as a parity-check matrix.

Note that with the above collective encoding, encoding operations, i.e., messages to codewords, are performed only in step **Enc-1**. The operations performed in the rest of the five steps are simply compositions, Hadamard permutations, cascading, interleaving, and GFT. The implementation of the above encoding process is relatively simple. First, it needs only two encoders, one for the SPC code $C^{(0)}$, which is a cyclic code with generator polynomial $\mathbf{g}^{(0)}(X) = X - 1$, and one for the cyclic base code $C^{(1)} = C$. Because $C^{(0)}$ is an SPC code, its encoder is very simple. Because $C$ is a cyclic code, its encoder can be implemented with a simple feedback shift-register with feedback connections based on the coefficients of its generator polynomial $\mathbf{g}(X)$ (see Section 4.5). Composing $\tau$ codewords in $C$ (or $C^{(0)}$) to form a composite codeword is based on representing a sequence of $\tau$ symbols in $\mathrm{GF}(2^{\kappa})$ as a symbol in $\mathrm{GF}(2^{\kappa\tau})$. This can be easily implemented by a look-up table. Note that, for most practical applications, $\tau$ is small, e.g., equal to 8. The Hadamard permutation, cascading, and interleaving operation can be performed together using a memory unit to store the vectors as rows in an $n \times n$ matrix $\Lambda$. The GFT is performed efficiently with a fast algorithm [14, 15] on each column of the matrix $\Lambda$ as it is read out from the memory unit for transmission.

In the encoding, we can change the order of step **Enc-2** and step **Enc-3** as follows. In step **Enc-2**\*, we perform the Hadamard permutations on the codewords generated at step **Enc-1** and then, in step **Enc-3**\*, we perform the composition of the $\tau$ codewords in each permuted group.

The encoding can be performed in a way such that all the $n\tau$ constituent codewords in $C_{\text{LDPC}}$ are codewords in the cyclic base code $C$. This is done by modifying the encoding of the SPC code $C^{(0)}$ as follows. A message $\mathbf{M}_{i,0}$, $0 \le i < \tau$, of $n - m$ information symbols with the first symbol set to zero "0" is first encoded into a codeword $\mathbf{v}_{i,0}$ in $C$. In systematic form, the first code symbol in $\mathbf{v}_{i,0}$ is zero "0." Removing this zero symbol, we obtain $n - 1$ code symbols. Then, we encode these $n - 1$ code symbols into a codeword $\mathbf{c}_{i,0}$ in $C^{(0)}$, i.e., adding an overall parity-check symbol. Composing the $\tau$ codewords $\mathbf{c}_{0,0}, \mathbf{c}_{1,0}, \dots, \mathbf{c}_{\tau-1,0}$ in $C^{(0)}$, we obtain a composite codeword $\mathbf{c}_0$ in $C_{\tau}^{(0)}$ and use it in the rest of the encoding steps. This results in an output codeword $\mathbf{c}_{\text{casc}}^{\pi,\mathcal{F}}$ in the GFT-ICC-LDPC code $C_{\text{LDPC}}$ which comprises $n\tau$ codewords in the cyclic base code $C$. In this case, the code rate is $(n - m)/n - 1/n$, a little loss of rate.

## 16.6 Collective Iterative Soft-Decision Decoding of GFT-ICC-LDPC Codes

Let $\mathbf{r} = (r_0, r_1, \dots, r_{n^2-1})$, where each $r_j$, $0 \le j < n^2$, is in $\mathrm{GF}(q^\tau)$ with $q = 2^\kappa$, be the received vector (the output of the detector) corresponding to the transmitted codeword $\mathbf{c}_{\text{casc}}^{\pi,\mathcal{F}}$. The decoding process at the receiver side is illustrated in Fig. 16.2.

The process of decoding the received vector $\mathbf{r}$ consists of the following steps.

**Dec-1.** Decompose the received vector $\mathbf{r}$ into $\kappa\tau$ binary constituent vectors $\mathbf{r}_{b,i} = (r_{0,i}, r_{1,i}, \dots, r_{n^2-1,i})$ for $0 \le i < \kappa\tau$. Decode these $\kappa\tau$ binary constituent vectors using a *binary* IDBP-algorithm applied to the binary LDPC matrix $\mathbf{H}_{\text{LDPC}}$. Combine the $\kappa\tau$ decoded binary vectors to form an estimate $\tilde{\mathbf{c}}_{\text{casc}}^{\pi,\mathcal{F}} = (\tilde{\mathbf{c}}_0^{\pi,\mathcal{F}}, \tilde{\mathbf{c}}_1^{\pi,\mathcal{F}}, \dots, \tilde{\mathbf{c}}_{n-1}^{\pi,\mathcal{F}})$ of $\mathbf{c}_{\text{casc}}^{\pi,\mathcal{F}}$ formed by the encoder at step **Enc-6**.

**Dec-2.** Take the inverse GFTs of the decoded sequences $\tilde{\mathbf{c}}_0^{\pi,\mathcal{F}}, \tilde{\mathbf{c}}_1^{\pi,\mathcal{F}}, \dots, \tilde{\mathbf{c}}_{n-1}^{\pi,\mathcal{F}}$ to obtain an estimate $\tilde{\mathbf{c}}_{\text{casc}}^{\pi}$ of the ICC codeword $\mathbf{c}_{\text{casc}}^{\pi}$ formed by the encoder at step **Enc-5**.

**Dec-3.** Apply the inverse permutation $\pi_{\text{col}}^{-1}$ to $\tilde{\mathbf{c}}_{\text{casc}}^{\pi}$, i.e., deinterleave $\tilde{\mathbf{c}}_{\text{casc}}^{\pi}$, to obtain a sequence $\tilde{\mathbf{c}}_{\text{casc}} = (\tilde{\mathbf{c}}_0, \tilde{\mathbf{c}}_1, \dots, \tilde{\mathbf{c}}_{n-1})$ of $n$ vectors, each of length $n$, which is an estimate of $\mathbf{c}_{\text{casc}}$ formed by the encoder at step **Enc-4**. The $n$ vectors $\tilde{\mathbf{c}}_0, \tilde{\mathbf{c}}_1, \dots, \tilde{\mathbf{c}}_{n-1}$ in $\tilde{\mathbf{c}}_{\text{casc}}$ are estimates of the $n$ composite codewords $\mathbf{c}_0, \mathbf{c}_1, \dots, \mathbf{c}_{n-1}$ formed by the encoder in step **Enc-3**.

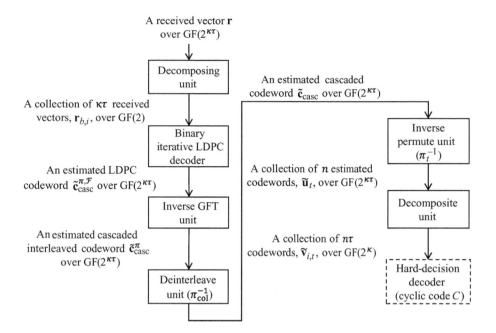

Figure 16.2 A collective iterative soft-decision decoding scheme for a GFT-ICC-LDPC code $C_{\text{LDPC}}$.

**Dec-4.** Permute $\tilde{\mathbf{c}}_t$, for $1 \leq t < n$, using the inverse Hadamard permutation $\pi_t^{-1}$ to obtain an estimate $\tilde{\mathbf{u}}_t$ of the composite codeword $\mathbf{u}_t$ formed by the encoder in step **Enc-2**.

**Dec-5.** Decompose $\tilde{\mathbf{u}}_t$, for each $t$, $0 \leq t < n$, into $\tau$ sequences $\tilde{\mathbf{v}}_{0,t}, \tilde{\mathbf{v}}_{1,t}, \ldots,$ $\tilde{\mathbf{v}}_{\tau-1,t}$ which are the estimates of the codewords $\mathbf{v}_{0,t}, \mathbf{v}_{1,t}, \ldots, \mathbf{v}_{\tau-1,t}$ in $C$ formed by the encoder in step **Enc-1**. For $0 \leq i < \tau$ and $1 \leq t < n$, if $\tilde{\mathbf{v}}_{i,t}$ is a codeword in $C$, then a message $\tilde{\mathbf{M}}_{i,t}$ is deduced from $\tilde{\mathbf{v}}_{i,t}$ which is an estimate of the message $\mathbf{M}_{i,t}$. For $t = 0$ and $0 \leq i < \tau$, if $\tilde{\mathbf{v}}_{i,0}$ is a codeword in $C^{(0)}$, then a message $\tilde{\mathbf{M}}_{i,0}$ is deduced from $\tilde{\mathbf{v}}_{i,0}$ which is an estimate of the message $\mathbf{M}_{i,0}$.

**Dec-6.** (Optional step) Apply HDD (hard-decision decoding) to decode $\tilde{\mathbf{v}}_{i,t}$, for $0 \leq i < \tau$, $1 \leq t < n$, using the base code $C$ to correct errors if detected.

With the above collective decoding scheme, decoding is only performed in the first step, step **Dec-1**. Because an iterative soft-decision decoding (ISDD) algorithm is used, we call this decoding scheme a *GFT-ISDD scheme*. Any iterative soft-decision BP-decoding algorithm can be used in conjunction with the GFT-ISDD scheme to implement step **Dec-1**. The two commonly used iterative BP-algorithms for decoding LDPC codes are the SPA [16] and the MSA [17] (see Chapter 11). As pointed out in Chapter 11, the MSA (when scaled properly) can perform very close to the SPA with significant reduction in decoding complexity.

For this reason, the MSA (or a variation) is commonly used in practice. In the following, we use the MSA in the first step of the GFT-ISDD scheme for decoding $C_{\mathrm{LDPC}} = C_{\mathrm{casc}}^{\pi,\mathcal{F}}$ based on the binary parity-check matrix $\mathbf{H}_{\mathrm{LDPC}} = \mathbf{H}_{\mathrm{casc}}^{\pi,\mathcal{F}^{-1}}$. We denote such combination as the GFT-ISDD/MSA. Let $I_{\max}$ denote the maximum number of iterations performed on each constituent received vector. For $0 \leq l < I_{\max}$, let $\mathbf{r}_{b,i}^{(l)}$ be the hard-decision of the $i$th constituent vector of $\mathbf{r}$ at the $l$th iteration and let $\mathbf{r}_{b,i}^{(-1)} = \mathbf{r}_{b,i}$ which is the received $i$th constituent vector of $\mathbf{r}$. The GFT-ISDD/MSA algorithm is formulated in Algorithm 16.1.

---

**Algorithm 16.1** GFT-ISDD/MSA

---

Inputs: $\mathbf{r} = (r_0, r_1, \ldots, r_{n^2-1})$.

**Step 1.** Initialization: set $l = 0$ and compute LLRs for the received code bits of all the $\kappa\tau$ constituent received vectors based on the detection of the channel output.

**Step 2.** Perform the $l$th GFT-ISDD/MSA iteration to update the LLRs of the bits in $\mathbf{r}_{b,i}$, $0 \leq i < \kappa\tau$, for which $\mathbf{r}_{b,i}^{(l-1)}\mathbf{H}_{\mathrm{LDPC}}^T \neq \mathbf{0}$. Compute $\mathbf{r}_{b,i}^{(l)}$.

**Step 3.** Check $\mathbf{r}_{b,i}^{(l)}\mathbf{H}_{\mathrm{LDPC}}^T$, $0 \leq i < \kappa\tau$. If $\mathbf{r}_{b,i}^{(l)}\mathbf{H}_{\mathrm{LDPC}}^T = \mathbf{0}$ for all $i = 0$, $1, \ldots, \kappa\tau - 1$, go to **Step 4**. Otherwise, save $\mathbf{r}_{b,i}^{(l)}$s for which $\mathbf{r}_{b,i}^{(l)}\mathbf{H}_{\mathrm{LDPC}}^T = \mathbf{0}$. If $l = I_{\max}$, declare a decoding failure; otherwise, set $l \leftarrow l + 1$ and go to **Step 2**.

**Step 4.** Stop GFT-ISDD/MSA decoding and group the $\kappa\tau$ decoded binary vectors into an estimate of the transmitted codeword $\mathbf{c}_{casc}^{\pi,\mathcal{F}}$.

---

## 16.7  Analysis of the GFT-ISDD Scheme

### 16.7.1  Performance Measurements

The most important feature of the GFT-ISDD coding scheme is that the decoding is performed on a *collection* of received codewords *jointly*. During the decoding process, the reliability information of each decoded codeword is shared by the others to enhance the *overall reliability* of all the decoded codewords. This *joint-decoding* and *information sharing* may result in an error performance per decoded codeword better than the error performance of a received codeword decoded individually by using MLD. This will be demonstrated through applications to three different classes of codes, namely, BCH, RS, and QR codes, given in Sections 16.8–16.10. This gain over MLD is referred to as *joint-decoding gain*. For a long code, the joint-decoding gain can be large. However, for a long code, computing its MLD performance is practically impossible. In this case, we will use the union bound (UB) on its MLD performance [18, 19], denoted by UB-MLD, for comparison. For large SNR, the UB is very tight.

In performance simulations, we assume transmission over an AWGN channel using BPSK signaling. In measuring the error performance of the GFT-ISDD/MSA, we use the *frame error rate* (FER), denoted by $P_{\mathrm{FER}}$, as the probability that a codeword in the induced GFT-ICC-LDPC code $C_{\text{LDPC}} = C_{\text{casc}}^{\pi,\mathcal{F}}$ is not decoded correctly. A frame consists of $(n-1)\tau$ codewords in $C$ and $\tau$ codewords in the SPC code $C^{(0)}$. We use the *block-error rate* (BLER), denoted by $P_{\text{BLER}}$, as the average probability that a codeword among these $(n-1)\tau$ codewords in $C$ is not decoded correctly. Simulations show that the gap between the FER and BLER is very small, less than *one-tenth* of a dB in SNR. This reflects the fact that if a frame is not decoded correctly, then most of the $(n-1)\tau$ codewords in $C$ are not decoded correctly. In the classical cascading scheme, $(n-1)\tau$ codewords from the code $C$ are *independently* transmitted and decoded *individually* using a certain decoding algorithm. The BLER of the classical cascading scheme is the probability that a codeword in $C$ is not decoded correctly. We denote this BLER with $P_{\text{BLER}}^{\star}$ and compare it with the $P_{\text{FER}}$ and $P_{\text{BLER}}$ of the GFT-ISDD/MSA.

### 16.7.2 Complexity

Recall that when using the GFT-ISDD/MSA to decode a $2^{\kappa\tau}$-ary received vector $\mathbf{r}$ of length $n^2$, we decode its $\kappa\tau$ binary constituent vectors based on the binary LDPC matrix $\mathbf{H}_{\text{LDPC}}$. In computing the complexity of decoding each binary constituent vector with the MSA, we count only the number of real-number comparisons and additions. For a scaled MSA, each check-node processing unit (CNPU) requires only two real-number multiplications, which is a very small part of the total computations and can be ignored. When we say a real-number computation, we mean either a real-number addition or a real-number comparison. For each $i$, $0 \le i < \kappa\tau$, the number of real-number computations required to update the reliabilities of the symbols of each binary constituent received vector in each decoding iteration is at most $mn(3n + \lceil \log_2 n \rceil - 2) \approx 3mn^2$ [20], which is, in general, proportional to the number $mn^2$ of 1-entries in the binary LDPC matrix $\mathbf{H}_{\text{LDPC}}$. Because every iteration updates the reliabilities of $\kappa\tau$ binary received vectors, the total number of real-number computations required for updating a $2^{\kappa\tau}$-ary received vector per iteration of the GFT-ISDD/MSA is $N_{\text{GFT-ISDD/MSA}} = \kappa\tau mn(3n + \lceil \log_2 n \rceil - 2) \approx 3\kappa\tau mn^2$. Because the GFT-ISDD/MSA decoding is performed on a collection of $n\tau$ codewords, the average number of real-number computations required to decode a single codeword using the GFT-ISDD/MSA per iteration is $N_{\text{avg}} = \kappa m(3n + \lceil \log_2 n \rceil - 2) \approx 3\kappa mn$. If we set the maximum number of iterations to be performed with the GFT-ISDD/MSA decoder to $I_{\text{max}}$, then the maximum number of real computations required to decode a single codeword in the cyclic base code with the GFT-ISDD/MSA is $N_{\text{max}} = \kappa m(3n + \lceil \log_2 n \rceil - 2)I_{\text{max}} \approx 3\kappa mnI_{\text{max}}$.

## 16.8 Joint Decoding of RS Codes with GFT-ISDD Scheme

The collective coding scheme can be applied to *any type of cyclic codes* of prime lengths, *binary* or *nonbinary*. In this section, and in Sections 16.9 and 16.10, we apply the collective coding scheme to RS codes, BCH codes, and quadratic residue (QR) codes to demonstrate its performance, respectively.

Consider the matrix $\mathbf{B} = [\beta^{jl_i}]_{0 \leq i < m, 0 \leq j < n}$ over $\mathrm{GF}(q^\tau)$, where $q = 2^\kappa$ and $0 \leq l_0, l_1, \ldots, l_{m-1} < n$. Suppose we set $\tau = 1$. Then, $\mathbf{B}$ is a matrix over $\mathrm{GF}(q)$. Let $\beta$ be an element of $\mathrm{GF}(q)$ of order $n$, which is a prime factor of $q - 1$, and let $l_0, l_1, \ldots, l_{m-1}$ be $m$ *consecutive integers modulo* $n$. Then, the null space over $\mathrm{GF}(q)$ of $\mathbf{B}$ gives an $(n, n - m)$ cyclic RS code over $\mathrm{GF}(q)$, $C_{\mathrm{RS}}$, of prime length $n$ and dimension $n - m$ with minimum distance $m + 1$ (see Chapter 6). Commonly, $l_0, l_1, \ldots, l_{m-1}$ are taken to be 1, 2, $\ldots$, $m$. The parity-check matrix $\mathbf{B}$ of $C_{\mathrm{RS}}$ is denoted by $\mathbf{B}_{\mathrm{RS}}$.

The GFT-ICC-LDPC code $C_{\mathrm{LDPC}}$ induced by $C_{\mathrm{RS}}$, called the *GFT-ICC-RS-LDPC* code and denoted by $C_{\mathrm{RS,LDPC}} = C_{\mathrm{RS,casc}}^{\pi, \mathcal{F}}$, is an $(n^2, (n - m + 1)(n - 1))$ QC-LDPC code over $\mathrm{GF}(2^\kappa)$ with minimum distance at least $m + 1$. Every codeword in $C_{\mathrm{RS,LDPC}}$ comprises $n - 1$ codewords in the RS code $C_{\mathrm{RS}}$ and one codeword in the $(n, n - 1)$ SPC code $C^{(0)}$ over $\mathrm{GF}(2^\kappa)$. Note that both the RS base code $C_{\mathrm{RS}}$ and the induced GFT-ICC-RS-LDPC code $C_{\mathrm{RS,LDPC}}$ are over the *same field*, $\mathrm{GF}(2^\kappa)$, but the parity-check matrix $\mathbf{H}_{\mathrm{RS,LDPC}}$ is *binary* which is an $m \times n$ array of CPMs of size $n \times n$.

The encoding of $C_{\mathrm{RS,LDPC}}$ follows steps **Enc-1**, **Enc-3**, **Enc-4**, **Enc-5**, and **Enc-6**. Because $\tau = 1$, no composition is needed and thus step **Enc-2** is skipped. The decoding of $C_{\mathrm{RS,LDPC}}$ follows steps **Dec-1** to **Dec-6** (step **Dec-6** is optional). In step **Dec-1**, a binary IDBP algorithm is applied to decode the $\kappa$ binary constituent vectors of a received vector $\mathbf{r}$ based on the binary LDPC matrix $\mathbf{H}_{\mathrm{RS,LDPC}}$ of $C_{\mathrm{RS,LDPC}}$. Through the decoding process, the decoder attempts to recover the $n - 1$ transmitted codewords in $C_{\mathrm{RS}}$ from the received vector $\mathbf{r}$.

**Example 16.5** Let $\beta$ be a primitive element of $\mathrm{GF}(2^5)$. The order of $\beta$ is 31, which is a prime. Let $n = 31$, $m = 6$, and $l_0 = 1$, $l_1 = 2$, $l_2 = 3$, $l_3 = 4$, $l_4 = 5$, $l_5 = 6$. Consider the $(31, 25)$ RS code $C_{\mathrm{RS}}$ over $\mathrm{GF}(2^5)$ ($q = 2^\kappa$, $\kappa = 5$) whose generator polynomial $\mathbf{g}(X)$ has $\beta, \beta^2, \beta^3, \beta^4, \beta^5, \beta^6$ as roots, i.e.,

$$\mathbf{g}_{\mathrm{RS}}(X) = (X - \beta)(X - \beta^2)(X - \beta^3)(X - \beta^4)(X - \beta^5)(X - \beta^6).$$

The code $C_{\mathrm{RS}}$ has a $6 \times 31$ parity-check matrix $\mathbf{B}_{\mathrm{RS}}$ over $\mathrm{GF}(2^5)$ in the form of (16.2).

The GFT-ICC-RS-LDPC code $C_{\mathrm{RS,LDPC}}$ induced by $C_{\mathrm{RS}}$ comprises 30 copies of $C_{\mathrm{RS}}$ ($C_{\mathrm{RS}}$ and its 29 Hadamard equivalents) and one $(31, 30)$ SPC code over $\mathrm{GF}(2^5)$ in cascade. The parity-check matrix $\mathbf{H}_{\mathrm{RS,LDPC}}$ of $C_{\mathrm{RS,LDPC}}$ is a $6 \times 31$ array of binary CPMs of size $31 \times 31$ which is obtained by the 31-fold binary CPM-dispersion of the RS base matrix $\mathbf{B}_{\mathrm{RS}}$. The matrix $\mathbf{H}_{\mathrm{RS,LDPC}}$ is a $186 \times 961$ binary matrix with column and row weights 6 and 31, respectively, and has rank $1 + m(n - 1) = 1 + 6 \times (31 - 1) = 181$. Hence, the GFT-ICC-RS-LDPC code $C_{\mathrm{RS,LDPC}}$, the null space over $\mathrm{GF}(2^5)$ of $\mathbf{H}_{\mathrm{RS,LDPC}}$, is a $(961, 780)$ QC-LDPC code over $\mathrm{GF}(2^5)$ with rate 0.8116.

Suppose we decode $C_{\mathrm{RS,LDPC}}$ with a maximum of 50 iterations of the GFT-ISDD/MSA scaled by a factor of 0.625 (without performing HDD in step

**Dec-6**). The FER and BLER performances are shown in Fig. 16.3, labeled by RS-FT-50, FER and RS-FT-50, BLER. We see that the gap between the FER and BLER curves is invisible and they basically overlap with each other. Also included in Fig. 16.3 are the block-error performances of the RS code decoded with the Berlekamp–Massey hard-decision decoding algorithm (BM-HDDA) [18, 21, 22] and MLD.

Also shown in Fig. 16.3 is the SPB [23] on the BLER, a lower bound on MLD. We see that the GFT-ISDD/MSA improves upon the performance of MLD of the $(31, 25)$ RS code $C_{\mathrm{RS}}$ below the BER of $10^{-5}$, with each codeword decoded individually. This performance improvement is the result of the fact that we do not decode the collection of 30 codewords in the RS code individually, but rather *jointly* through the powerful induced GFT-ICC-RS-LDPC code. At a BLER of $10^{-6}$, the joint-decoding gain of the GFT-ISDD/MSA over MLD of the RS code is 0.3 dB. The coding gain over the BM-HDDA is almost 3 dB at a BLER of $10^{-5}$. At the BLER of $10^{-8}$, the performance of the GFT-ISDD/MSA meets the SPB on MLD and the joint-decoding gain over MLD is about 1 dB.

If we decode the $(31, 25)$ RS code using the GFT-ISDD/MSA with 5 and 10 iterations, the performance gap between 10 and 50 iterations at a BLER of $10^{-8}$ is only about 0.2 dB, and the performance gap between 5 and 50 iterations is less than 0.5 dB, as shown in Fig. 16.3.

The number of real-number computations required in each iteration of the GFT-ISDD/MSA is 89 280. Because each iteration is performed on 30 codewords from the $(31, 25)$ RS base code $C_{\mathrm{RS}}$ and one codeword from the $(31, 30)$ SPC, the average number of real computations required for updating each codeword is 2880. At the SNR of 4.9 dB, the code achieves a BLER of $10^{-6}$ and the decoding takes an average of two iterations to converge. Therefore, at the SNR of 4.9 dB, the average number of real-number computations required to decode a codeword in $C_{\mathrm{RS}}$ is 5760. ▲▲

In the next example, we consider a longer RS code over a larger field than the one given in Example 16.5.

**Example 16.6** In this example, we consider decoding a collection of codewords from the $(127, 119)$ RS code $C_{\mathrm{RS}}$ over $\mathrm{GF}(2^7)$ of rate 0.937 and minimum distance 9 using the GFT-ISDD/MSA. Let $\beta$ be a primitive element of $\mathrm{GF}(2^7)$. The generator polynomial $\mathbf{g}_{\mathrm{RS}}(X)$ of $C_{\mathrm{RS}}$ has $\beta$, $\beta^2$, $\beta^3$, $\beta^4$, $\beta^5$, $\beta^6$, $\beta^7$, $\beta^8$ as roots. The parity-check matrix $\mathbf{B}_{\mathrm{RS}}$ of $C_{\mathrm{RS}}$ is an $8 \times 127$ matrix over $\mathrm{GF}(2^7)$ in the form of (16.2) with $l_0, l_1, \ldots, l_6, l_7 = 1, 2, \ldots, 7, 8$.

The GFT-ICC-RS-LDPC code $C_{\mathrm{RS,LDPC}}$ induced by $C_{\mathrm{RS}}$ comprises 126 copies of $C_{\mathrm{RS}}$ ($C_{\mathrm{RS}}$ and its 125 Hadamard equivalents) and one $(127, 126)$ SPC code over $\mathrm{GF}(2^7)$ in cascade. The parity-check matrix $\mathbf{H}_{\mathrm{RS,LDPC}}$ of $C_{\mathrm{RS,LDPC}}$ is an $8 \times 127$ array of binary CPMs of size $127 \times 127$ which is obtained by the 127-fold binary CPM-dispersion of the RS base matrix $\mathbf{B}_{\mathrm{RS}}$. The matrix $\mathbf{H}_{\mathrm{RS,LDPC}}$ is a $1016 \times 16\,129$ binary matrix with column and row weights 8 and 127, respectively, and has rank $1 + m(n-1) = 1 + 8 \times (127-1) = 1009$. Hence, the GFT-ICC-RS-LDPC code $C_{\mathrm{RS,LDPC}}$, the null space over $\mathrm{GF}(2^7)$ of $\mathbf{H}_{\mathrm{RS,LDPC}}$, is a $(16\,129, 15\,120)$ QC-LDPC code over $\mathrm{GF}(2^7)$ with rate 0.9374. Each codeword

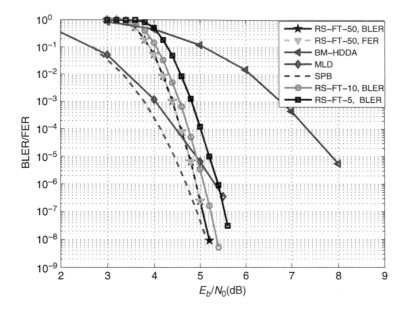

Figure 16.3 The FER and BLER performances of the $(31, 25)$ RS code given in Example 16.5 decoded by the GFT-ISDD/MSA and other decoding algorithms.

in $C_{\mathrm{RS,LDPC}}$ comprises 126 codewords in $C_{\mathrm{RS}}$ and a single codeword in the $(127, 126)$ SPC code $C^{(0)}$ over $\mathrm{GF}(2^7)$.

The BLER performances of the code $C_{\mathrm{RS,LDPC}}$ decoded with 5, 10, and 50 iterations of the GFT-ISDD/MSA are shown in Fig. 16.4(a), denoted by RS-FT-5, RS-FT-10, and RS-FT-50, respectively. We see that the GFT-ISDD/MSA converges very fast and the performance curves drop sharply. The FER with 50 iterations of the MSA is also shown in the figure. Also included in Fig. 16.4(a) is the block-error performance of the $(127, 119)$ RS code decoded with the BM-HDDA. We see that the performance of the GFT-ISDD/MSA is far superior compared with that of the BM-HDDA. MLD performance, as represented by its union bound, denoted by UB-MLD, which is quite tight for large SNR and feasible to compute for long codes, and the SPB [23] are also shown in the figure. With 50 iterations of the MSA, at a BLER of $10^{-5}$, the GFT-ISDD/MSA performs less than 0.3 dB from the SPB and achieves more than 0.2 dB gain over MLD.

The average number of decoding iterations required to decode a received vector as a function of SNR is plotted in Fig. 16.4(b), which shows the fast rate of decoding convergence using the GFT-ISDD/MSA to decode 126 codewords in the $(127, 119)$ RS code $C_{\mathrm{RS}}$ collectively. The number of real-number computations required for updating each codeword per iteration of the GFT-ISDD/MSA is 21 616. If the maximum number of iterations is set to 5, the number of real-number computations required to decode a codeword is 108 080. ▲▲

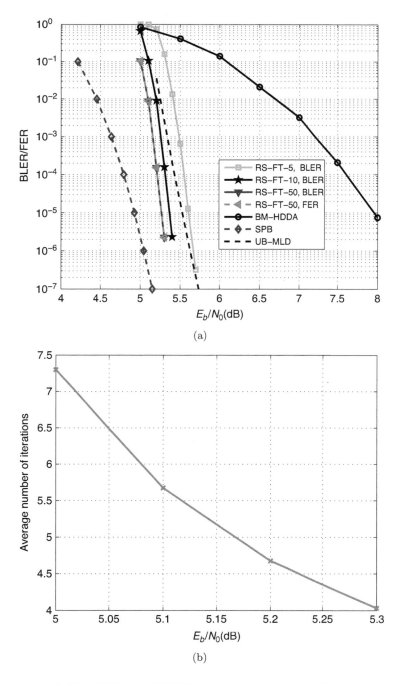

(a)

(b)

Figure 16.4 (a) The FER and BLER performances of the $(127, 119)$ RS code given in Example 16.6 decoded by the GFT-ISDD/MSA and other decoding algorithms and (b) the average number of iterations required to decode the $(127, 119)$ RS code in Example 16.6 vs. $E_b/N_0$ (dB).

# 16.9 Joint Decoding of BCH Codes with GFT-ISDD Scheme

In Section 16.8, we applied the collective GFT-ISDD coding scheme to encode and decode RS codes of prime lengths. In this section, we apply the same scheme to encode and decode binary BCH codes of prime lengths.

This time, we set $q = 2$, i.e., $\kappa = 1$. Let $n$ be a prime factor of $2^\tau - 1$ and let $\beta$ be an element of order $n$ in $\mathrm{GF}(2^\tau)$. Form the set $\mathbf{S}_{\mathrm{BCH}} = \{\beta^{l_0}, \beta^{l_1}, \ldots, \beta^{l_{m-1}}\}$ of $m$ elements which contains $2t$ consecutive powers of $\beta$ and their conjugates. The cyclic code whose generator polynomial has the $m$ elements in $\mathbf{S}_{\mathrm{BCH}}$ as roots is a $t$-error-correcting $(n, n - m)$ binary BCH code, $C_{\mathrm{BCH}}$, which has a parity-check matrix $\mathbf{B}_{\mathrm{BCH}}$ in the form (16.2). Commonly, the $2t$ consecutive elements in the set $\mathbf{S}_{\mathrm{BCH}}$ are $\beta, \beta^2, \ldots, \beta^{2t}$ with $l_0 = 1, l_1 = 2, \ldots, l_{2t-1} = 2t$. The designed minimum distance of $C_{\mathrm{BCH}}$ is $2t + 1$ based on the BCH bound as shown in Chapter 5 and at most $m + 1$ based on the Singleton bound.

With a GFT-ISDD coding scheme, the GFT-ICC-LDPC code induced by $C_{\mathrm{BCH}}$, called the GFT-ICC-BCH-LDPC code and denoted by $C_{\mathrm{BCH,LDPC}}$, is an $(n^2, (n - m + 1)(n - 1))$ code over $\mathrm{GF}(2^\tau)$ with minimum distance at least $m + 1$, which is in general greater than the designed distance $2t + 1$ of the $t$-error-correcting base BCH code $C_{\mathrm{BCH}}$. Every codeword in $C_{\mathrm{BCH,LDPC}}$ comprises $(n - 1)\tau$ codewords in $C_{\mathrm{BCH}}$ and $\tau$ codewords in the binary $(n, n - 1)$ SPC code $C^{(0)}$. The parity-check matrix $\mathbf{H}_{\mathrm{BCH,LDPC}}$ of $C_{\mathrm{BCH,LDPC}}$ is the $n$-fold binary CPM-dispersion of the parity-check matrix $\mathbf{B}_{\mathrm{BCH}}$ of the base code $C_{\mathrm{BCH}}$ which is an $m \times n$ array of CPMs of size $n \times n$. $\mathbf{H}_{\mathrm{BCH,LDPC}}$ is an $mn \times n^2$ binary matrix with column and row weights $m$ and $n$, respectively.

Encoding $C_{\mathrm{BCH,LDPC}}$ follows the six steps of collective encoding scheme presented in Section 16.5. In encoding, $(n - 1)\tau$ codewords in $C_{\mathrm{BCH}}$ and $\tau$ codewords in the binary $(n, n - 1)$ SPC code $C^{(0)}$ are combined into a codeword in $C_{\mathrm{BCH,LDPC}}$. Decoding $C_{\mathrm{BCH,LDPC}}$ is performed by using the GFT-ISDD scheme following steps **Dec-1**–**Dec-6** as presented in Section 16.6. In the first step of the decoding process, step **Dec-1**, a binary IDBP-algorithm is applied to decode the $\tau$ binary constituent vectors of the received vector $\mathbf{r}$ based on the binary LDPC matrix $\mathbf{H}_{\mathrm{BCH,LDPC}}$.

**Example 16.7** Set $\kappa = 1$ and $\tau = 7$. Let $\beta$ be a primitive element of $\mathrm{GF}(2^7)$. The order of $\beta$ is 127, which is a prime. Set $n = 127$ and $t = 2$. Based on $\beta, \beta^2, \beta^3, \beta^4$ and their 10 conjugates $\beta^6, \beta^8, \beta^{12}, \beta^{16}, \beta^{24}, \beta^{32}, \beta^{48}, \beta^{64}, \beta^{65}, \beta^{96}$, we construct a $14 \times 127$ matrix $\mathbf{B}_{\mathrm{BCH}}$ over $\mathrm{GF}(2^7)$ in the form given in (16.2). The null space over $\mathrm{GF}(2)$ of $\mathbf{B}_{\mathrm{BCH}}$ gives a double-error-correcting $(127, 113)$ binary BCH code $C_{\mathrm{BCH}}$ with rate 0.89 and minimum distance exactly 5. The generator polynomial $\mathbf{g}_{\mathrm{BCH}}(X)$ of $C_{\mathrm{BCH}}$ has $\beta, \beta^2, \beta^3, \beta^4, \beta^6, \beta^8, \beta^{12}, \beta^{16}, \beta^{24}, \beta^{32}, \beta^{48}, \beta^{64}, \beta^{65}, \beta^{96}$ as roots.

The GFT-ICC-BCH-LDPC code $C_{\mathrm{BCH,LDPC}}$ induced by $C_{\mathrm{BCH}}$ is a $(16\,129, 14\,364)$ QC-LDPC code over $\mathrm{GF}(2^7)$ of rate 0.8905 with minimum distance at least 15 which is three times larger than the distance of the base BCH code. Hence, the collective encoding enlarges the minimum distance of the base code.

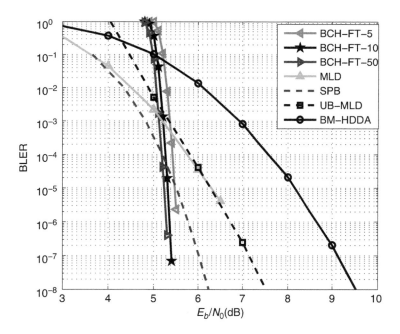

Figure 16.5 The BLER performances of the $(127, 113)$ BCH code given in Example 16.7 decoded by the GFT-ISDD/MSA, BM-HDDA, and MLD.

Each codeword in $C_{\text{BCH,LDPC}}$ consists of $7 \times 126 = 882$ codewords in $C_{\text{BCH}}$ and seven codewords in the $(127, 126)$ binary SPC code $C^{(0)}$. The parity-check matrix $\mathbf{H}_{\text{BCH,LDPC}}$ of $C_{\text{BCH,LDPC}}$ is the 127-fold binary CPM-dispersion of $\mathbf{B}_{\text{BCH}}$ which is a $14 \times 127$ array of CPMs of size $127 \times 127$, a $1778 \times 16\,129$ matrix over GF(2) with column and row weights 14 and 127, respectively.

The BLER performances of $C_{\text{BCH,LDPC}}$ decoded with the GFT-ISDD/MSA using 5, 10, and 50 iterations are shown in Fig. 16.5 and denoted by BCH-FT-5, BCH-FT-10, and BCH-FT-50, respectively. We see that the GFT-ISDD/MSA decoding converges very quickly because of the large connectivity of the Tanner graph associated with the decoding matrix $\mathbf{H}_{\text{BCH,LDPC}}$. The performance gap between 5 and 50 iterations is within 0.2 dB and the gap between 10 and 50 iterations is about 0.05 dB. From Fig. 16.5, we also see that error-rate curves drop sharply like a waterfall. This is because of the enlargement of the minimum distance resulting from collective encoding.

At the BLER of $10^{-6}$, the GFT-IDSS/MSA decoding of $C_{\text{BCH}}$ with 50 iterations achieves more than 1.5 dB and 3.4 dB coding gains over the MLD and BM-HDDA, respectively, with codewords in $C_{\text{BCH}}$ decoded independently. Furthermore, we see that for SNR greater than 5 dB, the MLD performance and its UB overlap with each other, i.e., UB on MLD is tight. ▲▲

In the next example, we show that even a simple cyclic Hamming code (a single-error-correcting BCH code) of prime length encoded and decoded collectively in the GFT domain can also achieve a very good error performance.

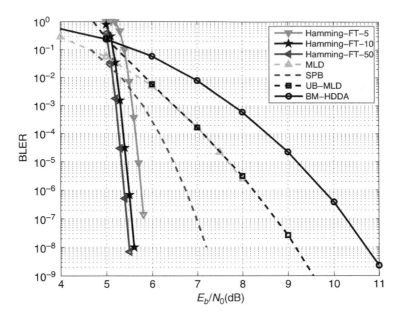

Figure 16.6 The BLER performances of the (127, 120) Hamming code given in Example 16.8 decoded by the GFT-ISDD/MSA, the BM-HDDA, and MLD.

**Example 16.8** Consider the $(127, 120)$ cyclic binary Hamming code $C_{\mathrm{Ham}}$ of rate 0.945 and minimum distance 3 generated by a primitive polynomial of degree 7 over GF(2), $\mathbf{g}_{\mathrm{Ham}}(X) = 1 + X^3 + X^7$. The polynomial $\mathbf{g}_{\mathrm{Ham}}(X)$ has a primitive element $\beta$ of GF($2^7$) and its six conjugates $\beta^2, \beta^4, \beta^8, \beta^{16}, \beta^{32}$, and $\beta^{64}$ as roots. Using these seven roots, we form a $7 \times 127$ parity-check matrix $\mathbf{B}_{\mathrm{Ham}}$ of $C_{\mathrm{Ham}}$ in the form of (16.2). The subscript "Ham" stands for "Hamming."

The 127-fold CPM-dispersion of the base matrix $\mathbf{B}_{\mathrm{Ham}}$ gives a $7 \times 127$ array $\mathbf{H}_{\mathrm{Ham,LDPC}}$ of $127 \times 127$ binary CPMs, which is an $889 \times 16\,129$ binary matrix of rank 883 with column and row weights 7 and 127, respectively. The GFT-ICC-Hamming-LDPC code $C_{\mathrm{Ham,LDPC}}$ induced by $C_{\mathrm{Ham}}$ is the null space over GF($2^7$) of $\mathbf{H}_{\mathrm{Ham,LDPC}}$, which is a $(16\,129, 15\,246)$ QC-LDPC code over GF($2^7$) with rate 0.9452 and minimum distance at least 8. Again, we see that the collective encoding enlarges the minimum distance of the base code from 3 to at least 8.

The BLER performances of the code $C_{\mathrm{Ham,LDPC}}$ decoded with 5, 10, and 50 iterations of GFT-ISDD/MSA are shown in Fig. 16.6, denoted by Hamming-FT-5, Hamming-FT-10, and Hamming-FT-50, respectively. We see that, at a BLER of $10^{-5}$, the GFT-ISDD/MSA with 50 iterations achieves 4.9 dB, 3.3 dB, and 1.6 dB gains over the BM-HDDA, MLD, and the SPB, respectively. With five iterations of the GFT-ISDD/MSA, the gains are 4.5, 2.9, and 1.2 dBs, respectively. ▲▲

# 16.10 Joint Decoding of QR Codes with GFT-ISDD Scheme

Quadratic residue (QR) codes were introduced in Section 4.12. They are binary cyclic codes of prime lengths. As pointed out in Section 4.12, some of the QR codes are optimal in the sense that they have the largest minimum distance among all codes of the same length and dimension. However, in general, they are hard to decode up to their true error-correcting capabilities with HDD algorithms, such as the BM-HDDA. Clever decoding techniques have been developed for particular QR codes [24, 25] (Table 4.7 gives a list of QR codes). The GFT-ISDD scheme can be used to decode any QR code with superior performance compared to HDD.

The most well-known QR code is the $(23, 12)$ Golay code presented in Example 4.11. This code is a perfect code with minimum distance 7. In the following, we apply the collective coding scheme to encode and decode this code and show its superior performance compared with the HDD algorithm (HDDA) and the MLD of the code with each codeword encoded and decoded individually.

**Example 16.9** As shown in Example 4.11, the generator polynomial of the $(23, 12)$ Golay code $C_{\text{Golay}}$ is

$$\mathbf{g}_{\text{Golay}}(X) = 1 + X + X^5 + X^6 + X^7 + X^9 + X^{11},$$

which has $\beta, \beta^2, \beta^3, \beta^4, \beta^6, \beta^8, \beta^9, \beta^{12}, \beta^{13}, \beta^{16}$, and $\beta^{18}$ as roots where $\beta$ is an element of order 23 in $\text{GF}(2^{11})$. The parity-check matrix $\mathbf{B}_{\text{Golay}}$ of $C_{\text{Golay}}$ in the form of (16.2) is an $11 \times 23$ matrix over $\text{GF}(2^{11})$ with $l_0, l_1, l_2, l_3, l_4, l_5, l_6, l_7, l_8, l_9, l_{10} = 1, 2, 3, 4, 6, 8, 9, 12, 13, 16, 18$.

To apply the collective coding to $C_{\text{Golay}}$, we set $n = 23$, $m = 11$, $\kappa = 1$, and $\tau = 11$. In encoding, $(n-1)\tau = 242$ codewords in $C_{\text{Golay}}$ and $\tau = 11$ codewords in the $(23, 22)$ SPC code are composed and transformed into a codeword of length $n^2 = 529$ over $\text{GF}(2^{11})$ in the induced $(529, 286)$ GFT-ICC-Golay-LDPC code, denoted by $C_{\text{Golay,LDPC}}$. The parity-check matrix $\mathbf{H}_{\text{Golay,LDPC}}$ of $C_{\text{Golay,LDPC}}$ is the 23-fold CPM-dispersion of $\mathbf{B}_{\text{Golay}}$ which is an $11 \times 23$ array of CPMs of size $23 \times 23$, a $253 \times 529$ binary matrix with column and row weights 11 and 23, respectively. Hence, the minimum distance of $C_{\text{Golay,LDPC}}$ is at least 12.

Decoding $C_{\text{Golay,LDPC}}$ is based on the binary matrix $\mathbf{H}_{\text{Golay,LDPC}}$. Suppose we decode $C_{\text{Golay,LDPC}}$ using the GFT-ISDD/MSA with a scaling factor of 0.5. Figure 16.7 shows the BLERs of the GFT-ISDD/MSA using 5, 10, and 50 iterations of the MSA in the first decoding step, denoted by Golay-FT-5, Golay-FT-10, and Golay-FT-50, respectively.

We see that the decoding converges quickly. Also included in Fig. 16.7 are the block-error performances of the HDD, MLD, and the SPB applied to decode codewords in $C_{\text{Golay}}$ independently. We see that below the BLER of $10^{-5}$, the GFT-ISDD/MSA with just 10 iterations outperforms MLD, and below the

BLER of $10^{-6}$ it outperforms the SPB and MLD. At the BLER of $10^{-8}$, the GFT-ISDD/MSA with 10 iterations achieves 3 dB and 1.1 dB joint-decoding gains over the HDDA and MLD, respectively. ▲▲

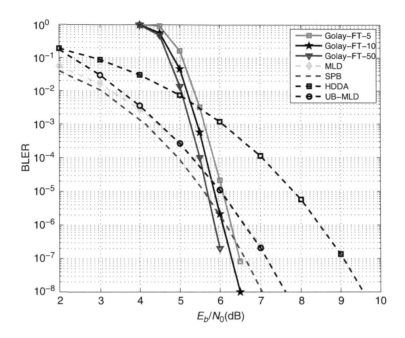

Figure 16.7 The BLER performances of the $(23, 12)$ Golay code given in Example 16.9 decoded by the GFT-ISDD/MSA, HDDA, and MLD.

## 16.11 Code Shortening and Rate Reduction

In this section, we present two techniques to construct a family of GFT-ICC codes with various lengths and rates based on a given cyclic base code. Codes in this family are commonly referred to as *rate-compatible codes* and they can be decoded essentially by using the same decoder.

### 16.11.1 Shortened GFT-ICC Codes

In many applications, it is desired to use a code of composite length rather than prime length. For these applications, we can shorten a code of prime length to a desired length. In Section 4.9, we presented a technique for shortening a *systematic* cyclic code to obtain a code with desired length and/or rate. This technique can also be used to shorten a GFT-ICC code induced by a cyclic base code of prime length.

In the following, we apply the GFT-ICC-LDPC coding scheme in conjunction with the shortening technique presented in Section 4.9 to construct a shortened GFT-ICC-LDPC code. Let $C$ be an $(n, n - m)$ cyclic code over $GF(q)$ of prime length $n$ whose parity-check matrix $\mathbf{B}$ is an $m \times n$ matrix over $\mathrm{GF}(q^\tau)$ in the form of (16.2). The GFT-ICC-LDPC code induced by $C$ is a QC-LDPC code $C_{\mathrm{LDPC}}$ over $\mathrm{GF}(q^\tau)$ of length $n^2$, dimension $(n-m+1)(n-1)$, and rate $(n-m+1)(n-1)/n^2$. Suppose we want to shorten the GFT-ICC code $C_{\mathrm{LDPC}}$ from length $n^2$ into a desired length $nn'$, where $m < n' < n$. Let $C_{\mathrm{LDPC},n'}$ denote the desired shortened code of length $nn'$. The construction and encoding process of $C_{\mathrm{LDPC},n'}$ is given as follows. There are some slight changes to the first three steps in encoding $C_{\mathrm{LDPC}}$ presented in Section 16.5.

For $0 \le i < \tau$ and $1 \le t < n$, let $\mathbf{M}'_{i,t}$ be a message composed of $n'-m$ $q$-ary symbols and for $0 \le i < \tau$ and $t = 0$, let $\mathbf{M}'_{i,0}$ be a message composed of $n'-1$ $q$-ary symbols. In the first step, denoted by $\mathbf{Enc\text{-}1}_{n'}$, of encoding $C_{\mathrm{LDPC},n'}$, we append each message $\mathbf{M}'_{i,t}$, $0 \le i < \tau$ and $0 \le t < n$, with $n-n'$ zeros on its right and encode it into a systematic codeword $\mathbf{c}_{i,t}$ in the code $C^{(t)}$ of length $n$, the $t$th Hadamard equivalent of $C$. Because each message $\mathbf{M}'_{i,t}$ is appended with $n-n'$ zeros on its right and the encoding is systematic, the rightmost $n-n'$ code symbols of each codeword $\mathbf{c}_{i,t}$ are zeros.

In the second step, denoted $\mathbf{Enc\text{-}2}_{n'}$, of encoding $C_{\mathrm{LDPC},n'}$, for each $t$, $0 \le t < n$, the $\tau$ codewords, $\mathbf{c}_{i,t}$, $0 \le i < \tau$, are composed to form a codeword $\mathbf{c}_t$ of length $n$ in the composite code $C_\tau^{(t)}$ over $\mathrm{GF}(q^\tau)$. The $n-n'$ rightmost code symbols of each composite codeword $\mathbf{c}_t$ are zeros. Note that the above two steps of encoding $C_{\mathrm{LDPC},n'}$ combine the first three steps of encoding $C_{\mathrm{LDPC}}$ to form a composite codeword $\mathbf{c}_t$. Here, the codeword $\mathbf{c}_t$ is not obtained by applying Hadamard permutation $\pi_t$ to a codeword in the composite code $C_\tau^{(1)}$ formed in step $\mathbf{Enc\text{-}2}$ of encoding $C_{\mathrm{LDPC}}$, because its permutation may not result in zeros in the rightmost $n-n'$ positions in $\mathbf{c}_t$, which is the key in shortening.

The next three steps of encoding $C_{\mathrm{LDPC},n'}$ are the same as steps $\mathbf{Enc\text{-}4}$, $\mathbf{Enc\text{-}5}$, and $\mathbf{Enc\text{-}6}$ of encoding $C_{\mathrm{LDPC}}$. In the third step, denoted by $\mathbf{Enc\text{-}3}_{n'}$, of encoding $C_{\mathrm{LDPC},n'}$, we cascade the $n$ composite codewords $\mathbf{c}_0$, $\mathbf{c}_1, \ldots, \mathbf{c}_{n-1}$ generated in step $\mathbf{Enc\text{-}2}_{n'}$ to form a cascaded composite codeword $\mathbf{c}_{\mathrm{casc}} = (\mathbf{c}_0, \mathbf{c}_1, \ldots, \mathbf{c}_{n-1})$ over $\mathrm{GF}(q^\tau)$ in the cascaded composite (CC) code $C_{\mathrm{casc}}$.

In the fourth step, denoted by $\mathbf{Enc\text{-}4}_{n'}$, of encoding $C_{\mathrm{LDPC},n'}$, we interleave the component codewords $\mathbf{c}_0$, $\mathbf{c}_1, \ldots$, $\mathbf{c}_{n-1}$ in $\mathbf{c}_{\mathrm{casc}}$ to obtain an interleaved cascaded codeword $\mathbf{c}_{\mathrm{casc}}^\pi = (\mathbf{c}_0^\pi, \mathbf{c}_1^\pi, \ldots, \mathbf{c}_{n-1}^\pi)$ in the interleaved cascaded composite (ICC) code $C_{\mathrm{casc}}^\pi$ by applying the permutation $\pi_{\mathrm{col}}$ defined by (16.11) to $\mathbf{c}_{\mathrm{casc}}$. Note that $\mathbf{c}_{\mathrm{casc}}^\pi$ can be obtained by arranging $\mathbf{c}_0, \mathbf{c}_1, \ldots$, $\mathbf{c}_{n-1}$ as rows of an $n \times n$ array $\Lambda$. Then, the columns of this array give $\mathbf{c}_0^\pi, \mathbf{c}_1^\pi, \ldots, \mathbf{c}_{n-1}^\pi$. Because each of $\mathbf{c}_0, \mathbf{c}_1, \ldots, \mathbf{c}_{n-1}$ ends with zeros in the rightmost $n-n'$ positions, $\mathbf{c}_{n'}^\pi = \mathbf{c}_{n'+1}^\pi = \cdots = \mathbf{c}_{n-1}^\pi$ are zero $n$-tuples, i.e., the rightmost $n-n'$ columns of $\Lambda$ are columns of zeros. Hence, the $n-n'$ rightmost sections of $\mathbf{c}_{\mathrm{casc}}^\pi = (\mathbf{c}_0^\pi, \mathbf{c}_1^\pi, \ldots, \mathbf{c}_{n-1}^\pi)$ are zero sections.

In the fifth step, denoted by $\mathbf{Enc\text{-}5}_{n'}$, of encoding $C_{\mathrm{LDPC},n'}$, the GFT is performed on the sequences $\mathbf{c}_0^\pi, \mathbf{c}_1^\pi, \ldots, \mathbf{c}_{n-1}^\pi$ to obtain the GFT sequences $\mathbf{c}_0^{\pi,\mathcal{F}}, \mathbf{c}_1^{\pi,\mathcal{F}}, \ldots, \mathbf{c}_{n-1}^{\pi,\mathcal{F}}$. Because the GFT of a sequence of zeros is a sequence of zeros, $\mathbf{c}_{n'}^{\pi,\mathcal{F}} = \mathbf{c}_{n'+1}^{\pi,\mathcal{F}} = \cdots = \mathbf{c}_{n-1}^{\pi,\mathcal{F}}$ are sequences of zeros. In transmission, only the sequences $\mathbf{c}_0^{\pi,\mathcal{F}}, \mathbf{c}_1^{\pi,\mathcal{F}}, \ldots, \mathbf{c}_{n'-1}^{\pi,\mathcal{F}}$, composed of $nn'$ symbols in $\mathrm{GF}(q^\tau)$, are transmitted.

The collection of all sequences, $(\mathbf{c}_0^{\pi,\mathcal{F}}, \mathbf{c}_1^{\pi,\mathcal{F}}, \ldots, \mathbf{c}_{n'-1}^{\pi,\mathcal{F}})$, gives the desired shortened code $C_{\mathrm{LDPC},n'}$ of $C_{\mathrm{LDPC}}$ with length $nn'$, dimension $(n'-1)+(n-1)(n'-m)$, and rate $((n'-1)+(n-1)(n'-m))/nn'$. The parity-check matrix $\mathbf{H}_{\mathrm{LDPC},n'}$ is obtained by deleting the last $n(n-n')$ columns of $\mathbf{H}_{\mathrm{LDPC}}$ which is the $n$-fold CPM-dispersion of the parity-check matrix $\mathbf{B}$ of the cyclic base code $C$. Because $\mathbf{H}_{\mathrm{LDPC}}$ is an $m \times n$ array of CPMs of size $n \times n$, $\mathbf{H}_{\mathrm{LDPC},n'}$ is an $m \times n'$ array of CPMs of size $n \times n$. Hence, the shortened code $C_{\mathrm{LDPC},n'}$ of $C_{\mathrm{LDPC}}$ is also a QC-LDPC code and has minimum distance at least $m + 1$. Note that $\mathbf{H}_{\mathrm{LDPC},n'}$ can be constructed by taking the $n$-fold CPM-dispersion of the $m \times n'$ matrix $\mathbf{B}_{n'}$ which consists of the first $n'$ columns of the parity-check matrix $\mathbf{B}$ of the cyclic base code $C$.

At the receiving end, the decoder first decodes the received vector of length $nn'$ based on the LDPC matrix $\mathbf{H}_{\mathrm{LDPC},n'}$ using a binary IDBP-algorithm, say GFT-ISDD/MSA, and then appends the decoded vector with $n(n-n')$ zeros. This gives an estimate $\tilde{\mathbf{c}}^{\pi,\mathcal{F}} = (\tilde{\mathbf{c}}_0^{\pi,\mathcal{F}}, \tilde{\mathbf{c}}_1^{\pi,\mathcal{F}}, \ldots, \tilde{\mathbf{c}}_{n-1}^{\pi,\mathcal{F}})$ of the codeword $\mathbf{c}^{\pi,\mathcal{F}} = (\mathbf{c}_0^{\pi,\mathcal{F}}, \mathbf{c}_1^{\pi,\mathcal{F}}, \ldots, \mathbf{c}_{n-1}^{\pi,\mathcal{F}})$ formed in the encoding step **Enc-5$_{n'}$**. Then, it follows steps **Dec-2** and **Dec-3** to obtain $\tilde{\mathbf{c}}_{0,t}, \tilde{\mathbf{c}}_{1,t}, \ldots, \tilde{\mathbf{c}}_{\tau-1,t}$, which are the estimates of the codewords $\mathbf{c}_{0,t}, \mathbf{c}_{1,t}, \ldots, \mathbf{c}_{\tau-1,t}$ in $C^{(t)}$. From these estimates, the messages $\mathbf{M}'_{i,t}$, $0 \le i < \tau$, $0 \le t < n$, can be retrieved (by removing the appended $n - n'$ zeros and the parity-check bits).

Note that, in encoding $C_{\mathrm{LDPC},n'}$, Hadamard permutation is not performed on a composite codeword. Hence, inverse Hadamard permutation is not required in decoding $C_{\mathrm{LDPC},n'}$.

**Example 16.10** In this example, we give two shortened GFT-ICC-RS-LDPC codes based on the triple-symbol-error-correcting $(127, 121)$ RS code $C_{\mathrm{RS}}$ over $\mathrm{GF}(2^7)$ whose generator polynomial has $\beta, \beta^2, \beta^3, \beta^4, \beta^5, \beta^6$ as roots, where $\beta$ is a primitive element of $\mathrm{GF}(2^7)$. The parity-check matrix $\mathbf{B}_{\mathrm{RS}}$ in the form of (16.2) is a $6 \times 127$ matrix over $\mathrm{GF}(2^7)$.

The GFT-ICC-RS-LDPC code $C_{\mathrm{RS,LDPC}}$ induced by $C_{\mathrm{RS}}$ is a $(16\,129, 15\,366)$ QC-LDPC code over $\mathrm{GF}(2^7)$ with rate 0.9527. The BLER performance of this code decoded with 50 iterations of the GFT-ISDD/MSA is shown in Fig. 16.8(a). Also included in the figure is the BLER performance of the $(127, 121)$ RS base code decoded with the BM-HDDA.

Suppose we shorten $C_{\mathrm{RS,LDPC}}$ by choosing $n' = 64$. This results in a $(64, 58)$ shortened RS code $C_{\mathrm{RS},64}$ of rate 0.9062, whose parity-check matrix $\mathbf{B}_{\mathrm{RS},64}$ is a $6 \times 64$ matrix which consists of the first 64 columns of $\mathbf{B}_{\mathrm{RS}}$. By using this shortened RS code as the base code and following the encoding process presented above, we obtain a shortened $(8128, 7371)$ GFT-ICC-RS-LDPC code $C_{\mathrm{RS,LDPC},64}$ code over $\mathrm{GF}(2^7)$ of rate 0.9068 in the GFT domain. Each transmitted codeword in $C_{\mathrm{RS,LDPC},64}$ consists of 126 codewords in $C_{\mathrm{RS},64}$ and one codeword in the $(64, 63)$ SPC code. The BLER performances of the shortened GFT-ICC-RS-LDPC code $C_{\mathrm{RS,LDPC},64}$ decoded with 5, 10, and 50 iterations of the GFT-ISDD/MSA are also shown in Fig. 16.8(a).

Choosing $n' = 32$, the shortened RS code $C_{\text{RS},32}$ is a $(32, 26)$ code over $\text{GF}(2^7)$. The GFT-ICC-RS-LDPC code induced by $C_{\text{RS},32}$ is a $(4064, 3307)$ GFT-ICC-RS-LDPC code $C_{\text{RS,LDPC},32}$ of rate 0.8137 in the GFT domain. The BLER performances of $C_{\text{RS,LDPC},32}$ decoded using 5, 10, and 50 iterations of the GFT-ISDD/MSA are shown in Fig. 16.8(b). Also included in Fig. 16.8(b) is the BLER performance of the code $C_{\text{RS}}$ decoded with 50 iterations of the GFT-ISDD/MSA. ▲▲

## 16.11.2 Reductions of Code Rate

By using the shortening technique proposed in Section 16.11.1, we not only reduce the length of the GFT-ICC-LDPC code for information transmission but also reduce the code rate. However, in some applications, we may want to reduce the code rate but maintain the code length. One approach to accomplish this goal is presented in the following.

We start with an $(n, n-m)$ cyclic code $C$ over $\text{GF}(q)$ of prime length $n$ as the base code. For $0 \le t < n$, let $\lambda_t$ be a nonnegative integer such that $0 \le \lambda_t \le \tau$. In the first step of the collective GFT-ICC encoding, we set $\tau - \lambda_t$ messages of the $\tau$ messages to *zero messages*. Encode the $\lambda_t$ nonzero messages into $\lambda_t$ nonzero codewords in $C$ and the $\tau - \lambda_t$ zero messages into $\tau - \lambda_t$ zero codewords in $C$. Then, we follow the rest of the collective GFT-ICC encoding steps. At the end of step **Enc-6**, we obtain a codeword $\mathbf{c}_{\text{casc},*}^{\pi,\mathcal{F}}$ over $\text{GF}(q^\tau)$ of length $n^2$ in the GFT domain, which contains $(n-1)\lambda_0 + (n-m)(\lambda_1 + \lambda_2 + \cdots + \lambda_{n-1})$ information symbols over $\text{GF}(q)$. The rate of the resultant GFT-ICC-LDPC code, denoted by $C_{\text{LDPC}}(\lambda_0, \lambda_1, \ldots, \lambda_{n-1})$, is $((n-1)\lambda_0 + (n-m)(\lambda_1 + \lambda_2 + \cdots + \lambda_{n-1}))/\tau n^2$. Different choices of the set $\{\lambda_0, \lambda_1, \ldots, \lambda_{n-1}\}$ of parameters result in different codes, $C_{\text{LDPC}}(\lambda_0, \lambda_1, \ldots, \lambda_{n-1})$, of the same length $n^2$ but with different rates. As an example, if we set $\lambda_0 = \lambda_1 = \cdots = \lambda_{n-1} = \tau - 1$, the code $C_{\text{LDPC}}(\tau - 1, \tau - 1, \ldots, \tau - 1)$ is then a code over $\text{GF}(q^\tau)$ with rate $(\tau - 1)(n-1)(n-m+1)/\tau n^2$. If we set $\lambda_0 = \lambda_1 = \cdots = \lambda_{n-1} = 1$, the code $C_{\text{LDPC}}(1, 1, \ldots, 1)$ is then a code over $\text{GF}(q^\tau)$ with rate $(n-1)(n-m+1)/\tau n^2$.

The choice $\lambda_0 = \lambda_1 = \cdots = \lambda_{n-1} = \tau$ gives the code $C_{\text{LDPC}}(\tau, \tau, \ldots, \tau) = C_{\text{casc}}^{\pi,\mathcal{F}}$. All these codes can be decoded with the same decoder as the code $C_{\text{casc}}^{\pi,\mathcal{F}}$.

**Example 16.11** Consider the $2^7$-ary $(16\,129, 14\,364)$ GFT-ICC-BCH-LDPC code $C_{\text{BCH,LDPC}}$ with rate 0.8906 given in Example 16.7. The base code in the construction is a $(127, 113)$ BCH code over $\text{GF}(2)$. If we set $\lambda_0 = \lambda_1 = \cdots = \lambda_{126} = 6$ in the encoding process, then the resultant code $C_{\text{BCH,LDPC}}(6, 6, \ldots, 6)$ is a $(16\,129, 11\,970)$ QC-LDPC code with rate 0.7421. If we set $\lambda_0 = \lambda_1 = \cdots = \lambda_{n-1} = 1$, the GFT-ICC encoding results in a $(16\,129, 2394)$ code $C_{\text{BCH,LDPC}}(1, 1, \ldots, 1)$ with rate 0.1484. This example shows that with collective GFT-ICC encoding based on a given base code, GFT-ICC codes with a wide spectrum of rates can be constructed. The BLER performances of the $(16\,129, 11\,970)$ code $C_{\text{BCH,LDPC}}(6, 6, \ldots, 6)$ decoded with 5, 10, and 50 iterations of the GFT-ISDD/MSA are shown in Fig. 16.9. ▲▲

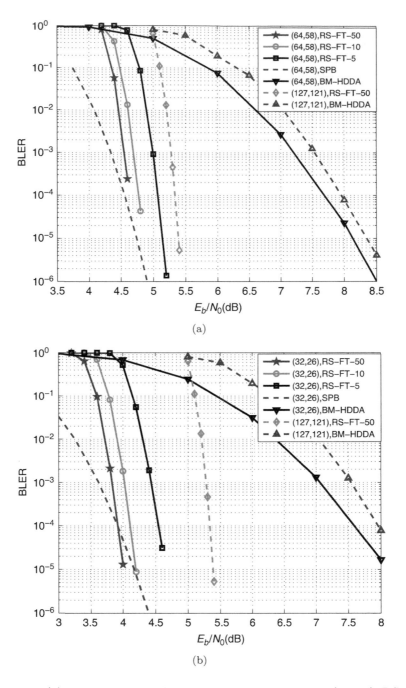

Figure 16.8 (a) The BLER performances of the shortened $(64, 58)$ RS code over $GF(2^7)$ and the $(127, 121)$ RS code over $GF(2^7)$ and (b) the BLER performances of the shortened $(32, 26)$ RS code over $GF(2^7)$ and the $(127, 121)$ RS code over $GF(2^7)$ given in Example 16.10 decoded by the GFT-ISDD/MSA and the BM-HDDA.

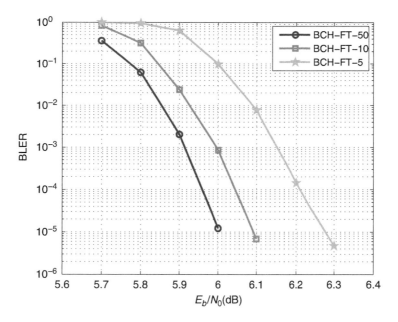

Figure 16.9 The BLER performances of the $(16\,129, 11\,970)$ QC-LDPC code $C_{\mathrm{BCH,LDPC}}(6, 6, \ldots, 6)$ in Example 16.11 decoded by the GFT-ISDD/MSA.

## 16.12 Erasure Correction of GFT-ICC-RS-LDPC Codes

In this section, we show that collective encoding and decoding of an $(n, n-2)$ RS code over $\mathrm{GF}(2^\kappa)$ is effective in correcting two random CPM-phased bursts of symbol erasures.

Let $\beta$ be an element of order $n$ in $\mathrm{GF}(2^\kappa)$, where $n$ is a prime factor of $2^\kappa - 1$. Consider the $(n, n-2)$ RS code $C_{\mathrm{RS}}$ over $\mathrm{GF}(2^\kappa)$ whose generator polynomial has $\beta$ and $\beta^2$ as roots. Then, the GFT-ICC-RS-LDPC code $C_{\mathrm{RS,LDPC}} = C_{\mathrm{casc,RS}}^{\pi, \mathcal{F}}$ induced by the base RS code $C_{\mathrm{RS}}$ is an $(n^2, n^2 - 2n + 1)$ QC-LDPC code over $\mathrm{GF}(2^\kappa)$. Every codeword in $C_{\mathrm{RS,LDPC}}$ comprises $(n-1)$ codewords in $C_{\mathrm{RS}}$ and one codeword in the $(n, n-1)$ SPC code $C_{\mathrm{RS}}^{(0)}$ over $\mathrm{GF}(2^\kappa)$. The parity-check matrix $\mathbf{H}_{\mathrm{RS,LDPC}}$ of $C_{\mathrm{RS,LDPC}}$ is a $2 \times n$ array of CPMs of size $n \times n$ which is a $2n \times n^2$ binary matrix with rank $2n - 1$ and column and row weights 2 and $n$, respectively. Hence, $C_{\mathrm{RS,LDPC}}$ is an $(n^2, n^2 - 2n + 1)$ QC-LDPC code over $\mathrm{GF}(2^\kappa)$ with $2n - 1$ parity-check symbols.

Let $\mathcal{G}_{\mathrm{RS,LDPC}}$ be the Tanner graph associated with $\mathbf{H}_{\mathrm{RS,LDPC}}$. Because $n$ is prime, the subgraph $\mathcal{G}_{\mathrm{sub}}$ of $\mathcal{G}_{\mathrm{RS,LDPC}}$ associated with any $2 \times 2$ subarray of $\mathbf{H}_{\mathrm{RS,LDPC}}$ is composed of a single cycle of length $4n$, i.e., the $2n$ VNs and $2n$ CNs of $\mathcal{G}_{\mathrm{sub}}$ are on a single cycle of length $4n$ (see Section 14.10). This local cycle structure of the Tanner graph $\mathcal{G}_{\mathrm{RS,LDPC}}$ allows $C_{\mathrm{RS,LDPC}}$ to correct any $2n - 1$ or fewer random symbol erasures confined in two CPM-sections of a codeword in $C_{\text{RS,LDPC}}$ transmitted over a $2^{\kappa}$-ary erasure channel using the SPL algorithm presented in Section 10.8.

Suppose a codeword $\mathbf{c}_{\text{casc}}^{\pi,\mathcal{F}}$ in $C_{\text{RS,LDPC}}$ is transmitted. Let $\mathbf{r} = (r_0, r_1, \ldots, r_{n^2-1})$ be the received vector, where each $r_j$, $0 \leq j < n^2$, is an element in $\text{GF}(2^{\kappa})$ or an erasure. In the erasure recovery process, for each erased symbol in $\mathbf{r}$, we find a row $\mathbf{h}$ in $\mathbf{H}_{\text{RS,LDPC}}$ that checks this erased symbol only. Form the check-sum by setting the inner product of $\mathbf{r}$ and $\mathbf{h}$ to zero, i.e., $\langle \mathbf{r}, \mathbf{h} \rangle = 0$. Then, the erased symbol in $\mathbf{r}$ checked by $\mathbf{h}$ is equal to the sum over $\text{GF}(2^{\kappa})$ of the unerased code symbols in $\mathbf{r}$ that are checked by $\mathbf{h}$. Owing to the orthogonal structure of $\mathbf{H}_{\text{RS,LDPC}}$, for each erased symbol in $\mathbf{r}$, there is at least one row $\mathbf{h}$ in $\mathbf{H}_{\text{RS,LDPC}}$ that checks this erased symbol alone as long as there are $2n - 1$ or fewer erased symbols in $\mathbf{r}$ which are confined in two CPM-sections of $\mathbf{r}$. Note that in decoding $\mathbf{r}$, we do not need to decompose it into its binary constituent vectors.

Once all the erased symbols in $\mathbf{r}$ have been recovered, we obtain the transmitted codeword $\mathbf{c}_{\text{casc}}^{\pi,\mathcal{F}}$ in $C_{\text{RS,LDPC}}$. Then, through inverse GFT, deinterleaving, and inverse Hadamard permutation, we can retrieve the $n - 1$ codewords in the base RS code $C_{\text{RS}}$ and one codeword in the $(n, n-1)$ SPC code over $\text{GF}(2^{\kappa})$ contained in $\mathbf{c}_{\text{casc}}^{\pi,\mathcal{F}}$.

The maximum number of symbol erasures that the $2^{\kappa}$-ary $(n^2, n^2 - 2n + 1)$ GFT-ICC-RS-LDPC code $C_{\text{RS,LDPC}}$ can recover is $2n - 1$, which is equal to its number of parity-check symbols. Hence, $C_{\text{RS,LDPC}}$ is optimal in correcting two random mutually semi-solid CPM-phased bursts of symbol erasures. Because $\mathbf{H}_{\text{RS,LDPC}}$ is binary, the SPL erasure recovery process requires only addition operations over $\text{GF}(2^{\kappa})$. Hence, the implementation of the erasure decoder for $C_{\text{RS,LDPC}}$ is relatively simple.

For $2 < m < n$, the GFT-ICC-RS-LDPC code $C_{\text{RS,LDPC}}$ induced by the $(n, n-m)$ RS code $C_{\text{RS}}$ over $\text{GF}(2^{\kappa})$ can be used for correcting random symbol errors or two random CPM-phased bursts of erasures. In correcting erasures, we can use any two CPM-row-blocks of the parity-check array of the GFT-ICC-RS-LDPC code $C_{\text{RS,LDPC}}$ as described in Section 14.10.

**Example 16.12** Let $\beta$ be a primitive element in $\text{GF}(2^5)$ and let $C_{\text{RS}}$ be the $(31, 29)$ RS code over $\text{GF}(2^5)$ whose generator polynomial has $\beta$ and $\beta^2$ as roots. The GFT-ICC-RS-LDPC code $C_{\text{RS,LDPC}}$ induced by $C_{\text{RS}}$ is a $(961, 900)$ QC-LDPC code whose Tanner graph has the $2 \times 2$ CPM-array cycle structure with cycle length of $4 \times 31 = 124$. The code is capable of correcting 61 random symbol erasures that are confined in two CPM-sections of a received vector.

The GFT-ICC-RS-LDPC code induced by the $(127, 119)$ RS code over $\text{GF}(2^7)$ can correct two random CPM-phased erasure-bursts (using any two CPM-row-blocks) with a maximum of 253 symbol erasures. As shown in Fig. 16.4(a), the code also performs very well over the AWGN channel.     ▲▲

Note that any cyclic code of prime length in GFT-ICC form can be used for correcting two random CPM-phased bursts of symbol erasures.

## 16.13   Remarks

The collective encoding through Hadamard permutations, composition, cascading, interleaving, and GFT presented in this chapter is basically a coding technique to construct long powerful nonbinary LDPC codes (or linear block codes) from short binary or nonbinary cyclic codes of prime lengths. Collective decoding in the GFT domain also provides a partial solution for decoding QR codes effectively, not hard-decision but soft-decision.

If we take the null space over $GF(2)$ of the binary parity-check matrix $\mathbf{H}_{casc}^{\pi,\mathcal{F}^{-1}}$ given by (16.21), we obtain a binary QC-LDPC code.

# Problems

**16.1** Prove that the parity-check matrix of a cyclic code in the form of (16.2) satisfies the $2 \times 2$ SM-constraint.

**16.2** Consider the parity-check matrix $\mathbf{B}$ given by (16.4) of the $(7, 4)$ Hamming code $C$ in Example 16.1. Find the fourth Hadamard power $\mathbf{B}^{\circ 4}$ of $\mathbf{B}$ and the generator polynomial of the cyclic code $C^{(4)}$ given by the null space over $GF(2)$ of $\mathbf{B}^{\circ 4}$, i.e., the fourth Hadamard equivalent of $C$.

**16.3** Consider the $(7, 4)$ cyclic Hamming code $C$ given in Example 16.1 and its three codewords

$$\mathbf{c}_{0,1} = (1\ 1\ 1\ 1\ 1\ 1\ 1), \mathbf{c}_{1,1} = (0\ 1\ 0\ 0\ 0\ 1\ 1), \mathbf{c}_{2,1} = (1\ 0\ 0\ 1\ 0\ 1\ 1).$$

Find the composite codeword $\mathbf{c}_1$ of the above three codewords in $C$ in the $2^3$-ary composite code $C_3^{(1)}$ of $C$.

**16.4** Consider the parity-check matrix $\mathbf{H}_{casc}^{\pi}$ given by (16.14) of the ICC code $C_{casc}^{\pi}$ of the $(7, 4)$ cyclic Hamming code $C$ in Example 16.3. Find all the submatrices, $\mathbf{D}_{e,f}$, $0 \le e < 3$ and $0 \le f < 7$, of $\mathbf{H}_{casc}^{\pi}$.

**16.5** In Example 16.4, the GFT $\mathcal{F}(\mathbf{c}_1)$ of a composite codeword $\mathbf{c}_1$ is calculated for the $(7, 4)$ Hamming code $C$. Use $\mathcal{F}(\mathbf{c}_1)$ to retrieve $\mathbf{c}_1$.

**16.6** Find the GFT of the composite codeword calculated in Problem 16.3.

**16.7** Show that the matrix $\mathbf{H}_{casc}^{\pi,\mathcal{F}^{-1}}$ given by (16.21) is the parity-check matrix of the GFT-ICC code $C_{casc}^{\pi,\mathcal{F}}$.

**16.8** Show that the sum given by (16.23) is equal to 1 if and only if $i - j \equiv -fl_e$ module $n$.

**16.9** Consider the $(7, 3)$ binary cyclic code $C$ generated by $\mathbf{g}(X) = 1 + X^2 + X^3 + X^4$.

(a) Find the parity-check matrix $\mathbf{B}$ in the form of (16.2) of the code $C$.

(b) Find the codewords in $C$ corresponding to the three messages: (1 0 0), (1 1 0), and (1 0 1).

(c) Find the composite codeword $\mathbf{c}_1$ in the $2^3$-ary composite code $C_3^{(1)}$ of $C$ of the above three codewords in $C$.

(d) Find the second and third Hadamard powers of $\mathbf{B}$.

(e) Find the generator polynomials of the second and third Hadamard equivalents of $C$.

(f) Find the parity-check matrix $\mathbf{H}_{\text{casc}}^{\pi}$ of the ICC code $C_{\text{casc}}^{\pi}$ of $C$.

(g) Find the parity-check matrix $\mathbf{H}_{\text{casc}}^{\pi,\mathcal{F}^{-1}}$ of the GFT-ICC code $C_{\text{casc}}^{\pi,\mathcal{F}}$ of $C$.

**16.10** Consider the double-error-correcting $(127, 123)$ RS code $C_{\text{RS}}$ over $GF(2^7)$. Using $C_{\text{RS}}$ as the base code, construct a GFT-ICC-RS-LDPC code using the GFT-ICC scheme. Compute its BLER performances by using GFT-ISDD/MSA with 5, 10, and 50 iterations and BM-HDDA. Analyze the decoding complexity of a GFT-ISDD/MSA decoder for the constructed code.

**16.11** Consider the $(89, 85)$ RS code $C_{\text{RS}}$ over $GF(2^{11})$. Use $C_{\text{RS}}$ as the base code to construct a GFT-ICC-RS-LDPC code using the GFT-ICC scheme. Compute its BLER performances by using GFT-ISDD/MSA and BM-HDDA.

**16.12** Construct a binary triple-error-correcting BCH $C_{\text{BCH}}$ code of length 127 based on the field $GF(2^7)$. Using the constructed BCH code $C_{\text{BCH}}$ as the cyclic base code and the GFT-ICC coding scheme, construct a GFT-ICC-BCH-LDPC code. Compute the BLER performances of the constructed code by using GFT-ISDD/MSA with 5, 10, and 50 iterations.

**16.13** Construct a binary $(17, 9)$ QR code $C_{QR}$ with minimum distance 5 based on the field $GF(2^8)$ generated by the primitive polynomial $p(X) = 1 + X^2 + X^3 + X^4 + X^8$. Using the constructed QR code $C_{QR}$ as the cyclic base code and the GFT-ICC coding scheme, construct a GFT-ICC-QR-LDPC code. Compute the BLER performances of the constructed code by using GFT-ISDD/MSA with 5, 10, and 50 iterations.

**16.14** Consider the $(16\,129, 15\,366)$ GFT-ICC-RS-LDPC code $C_{\text{RS,LDPC}}$ over $GF(2^7)$ induced by the $(127, 121)$ RS code $C_{\text{RS}}$ over $GF(2^7)$ in Example 16.10. Based on $C_{\text{RS,LDPC}}$ and $n' = 96$, construct a shortened GFT-ICC-RS-LDPC code $C_{\text{RS,LDPC,96}}$ and compute its BLER performance.

**16.15** Consider the $(16\,129, 14\,364)$ GFT-ICC-BCH-LDPC code $C_{\text{BCH,LDPC}}$ over $GF(2^7)$ with rate 0.8906 given in Example 16.7. Construct a GFT-ICC-BCH-LDPC code based on $C_{\text{BCH,LDPC}}$ by choosing $\lambda_0 = \lambda_1 = \cdots = \lambda_{n-1} = 3$. Compute its BLER performances using GFT-ISDD/MSA with 5, 10, and 50 iterations.

# References

[1] S. Lin, K. Abdel-Ghaffar, J. Li, and K. Liu, "A novel coding scheme for encoding and iterative soft-decision decoding of binary BCH codes of prime lengths," *Proc. Inf. Theory and Applications (ITA)*, San Diego, CA, February 11–16, 2018, pp. 1–10.

[2] S. Lin, K. Abdel-Ghaffar, J. Li, and K. Liu, "Collective encoding and iterative soft-decision decoding of cyclic codes of prime lengths in Galois-Fourier transform domain," *2018 10th International Symposium on Turbo Codes and Iterative Information Processing (ISTC)*, December 3–7, 2018, pp. 1–8.

[3] S. Lin, K. Abdel-Ghaffar, J. Li, and K. Liu, "A scheme for collective encoding and iterative soft-decision decoding of cyclic codes of prime lengths: applications to Reed–Solomon, BCH, and quadratic residue codes," *IEEE Trans. Inf. Theory*, **66**(9) (2020), 5358–5378.

[4] R. C. Singleton, "Maximum distance $q$-ary codes," *IEEE Trans. Inf. Theory*, **10**(2) (1964), 116–118.

[5] R. A. Horn and C. R. Johnson, *Matrix Analysis*, Cambridge, Cambridge University Press, 1985.

[6] R. E. Blahut, *Theory and Practice of Error Control Codes*, Reading, MA, Addison-Wesley, 1983.

[7] Q. Huang, Q. Diao, S. Lin, and K. Abdel-Ghaffar, "Trapping sets of structured LDPC codes," *2011 IEEE International Symposium on Information Theory Proceedings*, St. Petersburg, 2011, pp. 1086–1090,

[8] Q. Diao, Y. Y. Tai, S. Lin, and K. Abdel-Ghaffar, "LDPC codes on partial geometries: construction, trapping set structure, and puncturing," *IEEE Trans. Inf. Theory*, **59**(12) (2013), 7898–7914.

[9] M. Davey and D. J. C. MacKay, "Low density parity check codes over GF($q$)," *IEEE Commun. Lett.*, **2**(6) (1998), 165–167.

[10] D. J. C. MacKay and M. C. Davey, "Evaluation of Gallager codes of short block length and high rate applications," *Proc. IMA Int. Conf. Math. Applic.: Codes, Systems and Graphical Models*, New York, Springer-Verlag, 2000, pp. 113–130.

[11] L. Barnault and D. Declercq, "Fast decoding algorithm for LDPC over GF($2^q$)," *Proc. IEEE Inform. Theory Workshop,* Paris, France, March 31–April 4, 2003, pp. 70–73.

[12] D. Declercq and M. Fossorier, "Decoding algorithms for nonbinary LDPC codes over GF($q$)," *IEEE Trans. Commun.*, **55**(4) (2007), 633–643.

[13] W. E. Ryan and S. Lin, *Channel Codes: Classical and Modern*, New York, Cambridge University Press, 2009.

[14] E. Brigham, *Fast Fourier Transform and Its Applications*, Upper Saddle River, NJ, Prentice-Hall, 1988.

[15] S. Balaji Girisankar, M. Nasseri, J. Priscilla, S. Lin, and V. Akella, "Multiplier-free implementation of Galois field Fourier transform on a FPGA," *IEEE Trans. Circuits and Systems II: Express Briefs*, **66**(11) (2019), 1815–1819.

[16] D. MacKay, "Good error correcting codes based on very sparse matrices," *IEEE Trans. Inf. Theory*, **45**(3) (1999), 399–431.

[17] J. Chen and M. P. C. Fossorier, "Near optimum universal belief propagation based decoding of low-density parity-check codes," *IEEE Trans. Inf. Theory*, **50**(3) (2002), 406–414.

[18] S. Lin and D. J. Costello, Jr., *Error Control Coding: Fundamentals and Applications*, 2nd ed., Upper Saddle River, NJ, Prentice-Hall, 2004.

[19] J. Proakis and P. Massoud Salehi, *Digital Communications*, 5th ed., New York, McGraw-Hill, 2007.

[20] M. Fossorier, M. Mihaljevic, and H. Imai, "Reduced complexity iterative decoding of low-density parity check codes based on belief propagation," *IEEE Trans. Commun.*, **47**(5) (1999), 673–680.

[21] E. R. Berlekamp, *Algebraic Coding Theory*, New York, McGraw-Hill, 1968.

[22] J. L. Massey, "Shift-register synthesis and BCH decoding," *IEEE Trans. Inf. Theory*, **15**(1) (1969), 122–127.

[23] C. E. Shannon, R. G. Gallager, and E. R. Berlekamp, "Lower bounds to error probability for coding on discrete memoryless channels," *Inform. Contr.*, **10** (1967), pt. I, pp. 65–103, pt. II, pp. 522–552.

[24] T.-K. Truong, Y. Chang, Y.-H. Chen, and C. D. Lee, "Algebraic decoding of (103, 52, 19) and (113, 57, 15) quadratic residue codes," *IEEE Trans. Commun.*, **53**(5) (2005), 749–754.

[25] Y. Li, Q. Chen, H. Liu, and T.-K. Truong, "Performance and analysis of quadratic residue codes of lengths less than 100," 2014. [Online]. Available: arXiv:1408.5674v1 [cs.IT]

# 17

# Polar Codes

Polar codes, discovered by Arikan [1] in 2009, form a class of codes which provably achieve the capacity for a wide range of channels. Construction, encoding, and decoding of these codes are based on the phenomenon of channel polarization. In this chapter, we introduce polar codes from an algebraic point of view. First, we present a class of linear codes, called *Kronecker codes*, which are constructed by taking Kronecker (tensor) products of a *kernel* as presented in Chapter 9. Then, we show that polar codes form a special subclass of Kronecker codes through polarized encoding in which code bits in positions of a selected set are *frozen*. From the Kronecker product point of view, some algebraic structural properties of polar codes are developed. Based on these structural properties, we show that polar codes can be encoded *recursively in multilevels* using the channel polarization information and decoded with *successive cancellations*.

Coverage of this chapter includes: (1) Kronecker matrices and their structural properties; (2) Kronecker mappings and their logical implementations; (3) Kronecker vector spaces and codes; (4) definition, properties, and encoding of polar codes; (5) retrieval of polarized coded information; (6) channel polarization; (7) construction of polar codes; and (8) successive cancellation decoding of polar codes.

## 17.1  Kronecker Matrices and Their Structural Properties

In Section 9.5, we presented a special class of matrices which are formed by taking multifold Kronecker products of a $2 \times 2$ binary base matrix either in the form given by (9.25) or (9.32). These matrices are the base matrices for constructing polar codes. We call them *Kronecker matrices*. In this section, we present some essential structural properties of these Kronecker matrices.

Consistent with most works on polar codes, we use the $2 \times 2$ matrix given in the form of (9.32) as the base matrix which is reproduced as follows:

$$\mathbf{G}_{2\times2} = \begin{bmatrix} 1 & 0 \\ 1 & 1 \end{bmatrix}. \tag{17.1}$$

The matrix $\mathbf{G}_{2\times2}$ is known as a *kernel* in the construction of polar codes.

To assist in understanding the flow of the construction of polar codes and their encoding and decoding methods, we repeat the construction of Kronecker products presented in Section 9.5 using the matrix $\mathbf{G}_{2\times2}$ given by (17.1) as the base matrix. To simplify the notation in the developments of the rest of sections in this chapter, we use the notation $\mathbf{G}^{(\ell)}$ to denote the $\ell$-fold Kronecker matrix (or array) $\mathbf{G}_{2^\ell\times2^\ell}$ which we defined in (9.31) (see Section 9.5), i.e.,

$$\mathbf{G}^{(\ell)} \triangleq \mathbf{G}_{2^\ell\times2^\ell}. \tag{17.2}$$

The 2-fold Kronecker product of the kernel $\mathbf{G}^{(1)} = \mathbf{G}_{2\times2}$, denoted by $\mathbf{G}^{(2)}$, is a $2 \times 2$ lower triangular array of three copies of the kernel $\mathbf{G}^{(1)}$ given as follows:

$$\mathbf{G}^{(2)} = \mathbf{G}^{(1)} \otimes \mathbf{G}^{(1)} = \begin{bmatrix} \mathbf{G}^{(1)} & \mathbf{O} \\ \mathbf{G}^{(1)} & \mathbf{G}^{(1)} \end{bmatrix}. \tag{17.3}$$

Replacing each copy of the kernel in $\mathbf{G}^{(2)}$ by the $2\times2$ matrix given by (17.1), we obtain the following $4 \times 4$ matrix over GF(2):

$$\mathbf{G}^{(2)} = \begin{bmatrix} 1 & 0 & 0 & 0 \\ 1 & 1 & 0 & 0 \\ 1 & 0 & 1 & 0 \\ 1 & 1 & 1 & 1 \end{bmatrix}, \tag{17.4}$$

which is a lower triangular matrix with $3^2 = 9$ 1-entries lying on and below its main diagonal. We call this matrix the *2-fold Kronecker matrix*.

The 3-fold Kronecker matrix formed by the kernel $\mathbf{G}^{(1)}$ is the Kronecker product of the kernel $\mathbf{G}^{(1)}$ and the 2-fold Kronecker matrix $\mathbf{G}^{(2)}$ given as follows:

$$\begin{aligned} \mathbf{G}^{(3)} &= \mathbf{G}^{(1)} \otimes \mathbf{G}^{(2)} \\ &= \begin{bmatrix} \mathbf{G}^{(2)} & \mathbf{O} \\ \mathbf{G}^{(2)} & \mathbf{G}^{(2)} \end{bmatrix} \\ &= \begin{bmatrix} \mathbf{G}^{(1)} & \mathbf{O} & \mathbf{O} & \mathbf{O} \\ \mathbf{G}^{(1)} & \mathbf{G}^{(1)} & \mathbf{O} & \mathbf{O} \\ \mathbf{G}^{(1)} & \mathbf{O} & \mathbf{G}^{(1)} & \mathbf{O} \\ \mathbf{G}^{(1)} & \mathbf{G}^{(1)} & \mathbf{G}^{(1)} & \mathbf{G}^{(1)} \end{bmatrix}, \end{aligned} \tag{17.5}$$

which is a $4 \times 4$ triangular array of $3^2 = 9$ copies of the kernel $\mathbf{G}^{(1)}$. Replacing each copy of the kernel $\mathbf{G}^{(1)}$ in the array $\mathbf{G}^{(3)}$ with the $2 \times 2$ matrix given by (17.1), we obtain the following Kronecker matrix:

$$\mathbf{G}^{(3)} = \begin{bmatrix} 1 & 0 & 0 & 0 & 0 & 0 & 0 & 0 \\ 1 & 1 & 0 & 0 & 0 & 0 & 0 & 0 \\ 1 & 0 & 1 & 0 & 0 & 0 & 0 & 0 \\ 1 & 1 & 1 & 1 & 0 & 0 & 0 & 0 \\ 1 & 0 & 0 & 0 & 1 & 0 & 0 & 0 \\ 1 & 1 & 0 & 0 & 1 & 1 & 0 & 0 \\ 1 & 0 & 1 & 0 & 1 & 0 & 1 & 0 \\ 1 & 1 & 1 & 1 & 1 & 1 & 1 & 1 \end{bmatrix}, \tag{17.6}$$

which is an $8 \times 8$ lower triangular matrix with $3^3 = 27$ 1-entries lying on and below its main diagonal.

The 4-fold Kronecker matrix formed by the kernel $\mathbf{G}^{(1)}$ is the Kronecker product of the kernel $\mathbf{G}^{(1)}$ and the 3-fold Kronecker matrix $\mathbf{G}^{(3)}$ given as follows:

$$\mathbf{G}^{(4)} = \mathbf{G}^{(1)} \otimes \mathbf{G}^{(3)}$$

$$= \begin{bmatrix} \mathbf{G}^{(3)} & \mathbf{O} \\ \mathbf{G}^{(3)} & \mathbf{G}^{(3)} \end{bmatrix}$$

$$= \begin{bmatrix} \mathbf{G}^{(1)} & \mathbf{O} & \mathbf{O} & \mathbf{O} & \mathbf{O} & \mathbf{O} & \mathbf{O} & \mathbf{O} \\ \mathbf{G}^{(1)} & \mathbf{G}^{(1)} & \mathbf{O} & \mathbf{O} & \mathbf{O} & \mathbf{O} & \mathbf{O} & \mathbf{O} \\ \mathbf{G}^{(1)} & \mathbf{O} & \mathbf{G}^{(1)} & \mathbf{O} & \mathbf{O} & \mathbf{O} & \mathbf{O} & \mathbf{O} \\ \mathbf{G}^{(1)} & \mathbf{G}^{(1)} & \mathbf{G}^{(1)} & \mathbf{G}^{(1)} & \mathbf{O} & \mathbf{O} & \mathbf{O} & \mathbf{O} \\ \mathbf{G}^{(1)} & \mathbf{O} & \mathbf{O} & \mathbf{O} & \mathbf{G}^{(1)} & \mathbf{O} & \mathbf{O} & \mathbf{O} \\ \mathbf{G}^{(1)} & \mathbf{G}^{(1)} & \mathbf{O} & \mathbf{O} & \mathbf{G}^{(1)} & \mathbf{G}^{(1)} & \mathbf{O} & \mathbf{O} \\ \mathbf{G}^{(1)} & \mathbf{O} & \mathbf{G}^{(1)} & \mathbf{O} & \mathbf{G}^{(1)} & \mathbf{O} & \mathbf{G}^{(1)} & \mathbf{O} \\ \mathbf{G}^{(1)} & \mathbf{G}^{(1)} & \mathbf{G}^{(1)} & \mathbf{G}^{(1)} & \mathbf{G}^{(1)} & \mathbf{G}^{(1)} & \mathbf{G}^{(1)} & \mathbf{G}^{(1)} \end{bmatrix}, \quad (17.7)$$

which is an $8 \times 8$ array of $3^3 = 27$ copies of the kernel $\mathbf{G}^{(1)}$. Replacing each copy of the kernel with the $2 \times 2$ matrix in the form of (17.1), we obtain the following $16 \times 16$ lower triangular matrix with $3^4 = 81$ 1-entries:

$$\mathbf{G}^{(4)} = \begin{bmatrix} 1&0&0&0&0&0&0&0&0&0&0&0&0&0&0&0 \\ 1&1&0&0&0&0&0&0&0&0&0&0&0&0&0&0 \\ 1&0&1&0&0&0&0&0&0&0&0&0&0&0&0&0 \\ 1&1&1&1&0&0&0&0&0&0&0&0&0&0&0&0 \\ 1&0&0&0&1&0&0&0&0&0&0&0&0&0&0&0 \\ 1&1&0&0&1&1&0&0&0&0&0&0&0&0&0&0 \\ 1&0&1&0&1&0&1&0&0&0&0&0&0&0&0&0 \\ 1&1&1&1&1&1&1&1&0&0&0&0&0&0&0&0 \\ 1&0&0&0&0&0&0&0&1&0&0&0&0&0&0&0 \\ 1&1&0&0&0&0&0&0&1&1&0&0&0&0&0&0 \\ 1&0&1&0&0&0&0&0&1&0&1&0&0&0&0&0 \\ 1&1&1&1&0&0&0&0&1&1&1&1&0&0&0&0 \\ 1&0&0&0&1&0&0&0&1&0&0&0&1&0&0&0 \\ 1&1&0&0&1&1&0&0&1&1&0&0&1&1&0&0 \\ 1&0&1&0&1&0&1&0&1&0&1&0&1&0&1&0 \\ 1&1&1&1&1&1&1&1&1&1&1&1&1&1&1&1 \end{bmatrix}. \quad (17.8)$$

The above constructions of 2-, 3-, and 4-fold Kronecker matrices demonstrate that Kronecker matrices with $\mathbf{G}^{(1)}$ as a kernel can be constructed *recursively*. For $\ell \geq 1$, the $\ell$-fold Kronecker matrix $\mathbf{G}^{(\ell)}$ constructed based on the kernel $\mathbf{G}^{(1)}$ is the Kronecker product of the kernel $\mathbf{G}^{(1)}$ and the $(\ell-1)$-fold Kronecker matrix $\mathbf{G}^{(\ell-1)}$ given as follows:[1]

$$\mathbf{G}^{(\ell)} = \mathbf{G}^{(1)} \otimes \mathbf{G}^{(\ell-1)} = \begin{bmatrix} \mathbf{G}^{(\ell-1)} & \mathbf{O} \\ \mathbf{G}^{(\ell-1)} & \mathbf{G}^{(\ell-1)} \end{bmatrix}. \quad (17.9)$$

For $\ell = 1$, the 1-fold (or single-fold) Kronecker matrix is the kernel $\mathbf{G}^{(1)}$.

It follows from the recursive construction of multifold Kronecker matrices with $\mathbf{G}^{(1)}$ as a kernel that the $\ell$-fold Kronecker matrix $\mathbf{G}^{(\ell)}$ is a $2^{\ell-1} \times 2^{\ell-1}$ lower

---

[1] For $\ell = 1$, we define $\mathbf{G}^{(\ell-1)} = \mathbf{G}^{(0)} = 1$.

triangular array with $3^{\ell-1}$ copies of $\mathbf{G}^{(1)}$ lying on and below its main diagonal, which is a $2^{\ell} \times 2^{\ell}$ lower triangular matrix over GF(2) with $3^{\ell}$ 1-entries lying on and below its main diagonal. The Kronecker matrix $\mathbf{G}^{(\ell)}$ has $\ell + 1$ different row (column) weights which are $2^0 = 1$, $2^1 = 2$, $2^2 = 4, \ldots, 2^{\ell-1}$ and $2^{\ell}$. The number of rows in $\mathbf{G}^{(\ell)}$ with weight $2^{\ell-i}$, $0 \leq i \leq \ell$, is $\binom{\ell}{i}$ (binomial coefficient). Except for one row (one column), i.e., the top row (rightmost column), of $\mathbf{G}^{(\ell)}$, every row (column) of $\mathbf{G}^{(\ell)}$ has even weight. The entries of the bottom row and the leftmost column of $\mathbf{G}^{(\ell)}$ are all 1-entries. For example, consider the 4-fold Kronecker matrix given by (17.8). By checking the 16 rows of the matrix, we find that there are

$$\binom{4}{0} = 1, \binom{4}{1} = 4, \binom{4}{2} = 6, \binom{4}{3} = 4, \binom{4}{4} = 1$$

rows of weights 16, 8, 4, 2 and 1, respectively.

It follows from the triangular structure of the Kronecker matrix $\mathbf{G}^{(\ell)}$ that the rows of $\mathbf{G}^{(\ell)}$ are *linearly independent* and hence $\mathbf{G}^{(\ell)}$ is a full-rank matrix with rank $2^{\ell}$. Furthermore, $\mathbf{G}^{(\ell)}$ is its own inverse (see Problem 17.1), i.e., $(\mathbf{G}^{(\ell)})^{-1} = \mathbf{G}^{(\ell)}$ and

$$\mathbf{G}^{(\ell)} \cdot \mathbf{G}^{(\ell)} = \mathbf{I}_{2^{\ell}}, \tag{17.10}$$

where $\mathbf{I}_{2^{\ell}}$ is the $2^{\ell} \times 2^{\ell}$ identity matrix. As an example, consider the 2-fold Kronecker matrix $\mathbf{G}^{(2)}$ given by (17.4). The matrix product of $\mathbf{G}^{(2)}$ with itself is

$$\mathbf{G}^{(2)} \cdot \mathbf{G}^{(2)} = \begin{bmatrix} 1 & 0 & 0 & 0 \\ 1 & 1 & 0 & 0 \\ 1 & 0 & 1 & 0 \\ 1 & 1 & 1 & 1 \end{bmatrix} \cdot \begin{bmatrix} 1 & 0 & 0 & 0 \\ 1 & 1 & 0 & 0 \\ 1 & 0 & 1 & 0 \\ 1 & 1 & 1 & 1 \end{bmatrix} = \begin{bmatrix} 1 & 0 & 0 & 0 \\ 0 & 1 & 0 & 0 \\ 0 & 0 & 1 & 0 \\ 0 & 0 & 0 & 1 \end{bmatrix} = \mathbf{I}_4, \tag{17.11}$$

i.e., $(\mathbf{G}^{(2)})^{-1} = \mathbf{G}^{(2)}$.

## 17.2 Kronecker Mappings and Their Logical Implementations

For $\ell \geq 1$, let $\mathbf{u} = (u_0, u_1, \ldots, u_{2^{\ell}-1})$ be a $2^{\ell}$-tuple over GF(2). Consider the following mapping $f_{\text{Kron}}^{(\ell)}$ which maps $\mathbf{u}$ to another $2^{\ell}$-tuple $\mathbf{v} = (v_0, v_1, \ldots, v_{2^{\ell}-1})$ over GF(2):

$$f_{\text{Kron}}^{(\ell)} : \mathbf{u} \longrightarrow \mathbf{v}, \tag{17.12}$$

where the mapping function $f_{\text{Kron}}^{(\ell)}$ is defined as

$$\mathbf{v} = f_{\text{Kron}}^{(\ell)}(\mathbf{u}) = \mathbf{u} \cdot \mathbf{G}^{(\ell)}. \tag{17.13}$$

The mapping $f_{\text{Kron}}^{(\ell)}$ is called the $\ell$-*fold Kronecker mapping* and $\mathbf{v}$ is called the Kronecker mapping of $\mathbf{u}$. Because the rows of $\mathbf{G}^{(\ell)}$ are linearly independent, the mapping $f_{\text{Kron}}^{(\ell)}$ is a *one-to-one mapping*. Define the following mapping from **v** back to **u**:

$$f_{\text{Kron}}^{-(\ell)} : \mathbf{v} \longrightarrow \mathbf{u}. \tag{17.14}$$

The mapping $f_{\text{Kron}}^{-(\ell)}$ is called the *$\ell$-fold inverse Kronecker mapping* and **u** is called the inverse Kronecker mapping of **v**. Because $\mathbf{G}^{(\ell)}$ is its own inverse, $f_{\text{Kron}}^{(\ell)}$ and $f_{\text{Kron}}^{-(\ell)}$ are identical. Hence,

$$\mathbf{u} = f_{\text{Kron}}^{-(\ell)}(\mathbf{v}) = \mathbf{v} \cdot (\mathbf{G}^{(\ell)})^{-1} = \mathbf{v} \cdot \mathbf{G}^{(\ell)}. \tag{17.15}$$

**Example 17.1** Let $\mathbf{u} = (u_0, u_1)$ be a 2-tuple over $GF(2)$ and $f_{\text{Kron}}^{(1)}$ be the mapping function defined by the single-fold Kronecker matrix $\mathbf{G}^{(1)}$, i.e., the kernel given by (17.1). Then, the Kronecker mapping of $\mathbf{u} = (u_0, u_1)$ gives the following 2-tuple over $GF(2)$:

$$\begin{aligned}
\mathbf{v} &= (v_0, v_1) \\
&= f_{\text{Kron}}^{(1)}(\mathbf{u}) = \mathbf{u} \cdot \mathbf{G}^{(1)} \\
&= (u_0, u_1) \cdot \begin{bmatrix} 1 & 0 \\ 1 & 1 \end{bmatrix} \\
&= (u_0 + u_1, u_1),
\end{aligned}$$

i.e., $v_0 = u_0 + u_1$ and $v_1 = u_1$.

The 1-fold Kronecker mapping can be implemented with a single XOR gate (i.e., modulo-2 addition) which combines the input $u_0$ and $u_1$ to the output $v_0$ as shown in Fig. 17.1.                                         ▲▲

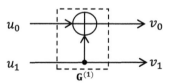

Figure 17.1 The 1-fold Kronecker mapping circuit.

**Example 17.2** Let $\mathbf{u} = (u_0, u_1, u_2, u_3)$ be a 4-tuple over $GF(2)$ and $f_{\text{Kron}}^{(2)}$ be the mapping function defined by the 2-fold Kronecker matrix $\mathbf{G}^{(2)}$ given by (17.4). Then, the Kronecker mapping of $\mathbf{u} = (u_0, u_1, u_2, u_3)$ gives the following 4-tuple over $GF(2)$:

$$\begin{aligned}
\mathbf{v} &= (v_0, v_1, v_2, v_3) \\
&= f_{\text{Kron}}^{(2)}(\mathbf{u}) = \mathbf{u} \cdot \mathbf{G}^{(2)} \\
&= (u_0, u_1, u_2, u_3) \cdot \begin{bmatrix} \mathbf{G}^{(1)} & \mathbf{O} \\ \mathbf{G}^{(1)} & \mathbf{G}^{(1)} \end{bmatrix}.
\end{aligned} \tag{17.16}$$

It follows (17.1) and (17.16) that we have

$$\mathbf{v} = (v_0, v_1, v_2, v_3)$$

$$= (u_0, u_1, u_2, u_3) \cdot \begin{bmatrix} 1 & 0 & 0 & 0 \\ 1 & 1 & 0 & 0 \\ 1 & 0 & 1 & 0 \\ 1 & 1 & 1 & 1 \end{bmatrix}. \tag{17.17}$$

Multiplying out the product of (17.17), we find that the components of $\mathbf{v} = (v_0, v_1, v_2, v_3)$ are:

$$\begin{aligned} v_0 &= u_0 + u_1 + u_2 + u_3, \\ v_1 &= \phantom{u_0 + {}} u_1 \phantom{{}+ u_2} + u_3, \\ v_2 &= \phantom{u_0 + u_1 + {}} u_2 + u_3, \\ v_3 &= \phantom{u_0 + u_1 + u_2 + {}} u_3. \end{aligned} \tag{17.18}$$

The inverse Kronecker mapping of $\mathbf{v} = (v_0, v_1, v_2, v_3)$ is

$$\begin{aligned} \mathbf{u} &= (u_0, u_1, u_2, u_3) \\ &= f_{\mathrm{Kron}}^{-(2)}(\mathbf{v}) = \mathbf{v} \cdot (\mathbf{G}^{(2)})^{-1} \\ &= \mathbf{v} \cdot \mathbf{G}^{(2)} \\ &= (v_0, v_1, v_2, v_3) \cdot \begin{bmatrix} \mathbf{G}^{(1)} & \mathbf{O} \\ \mathbf{G}^{(1)} & \mathbf{G}^{(1)} \end{bmatrix} \\ &= (v_0, v_1, v_2, v_3) \cdot \begin{bmatrix} 1 & 0 & 0 & 0 \\ 1 & 1 & 0 & 0 \\ 1 & 0 & 1 & 0 \\ 1 & 1 & 1 & 1 \end{bmatrix}. \end{aligned} \tag{17.19}$$

Multiplying out the product given by (17.19), we find $\mathbf{u} = (u_0, u_1, u_2, u_3)$ whose components in terms of the components of $\mathbf{v}$ are given below:

$$\begin{aligned} u_0 &= v_0 + v_1 + v_2 + v_3, \\ u_1 &= \phantom{v_0 + {}} v_1 \phantom{{}+ v_2} + v_3, \\ u_2 &= \phantom{v_0 + v_1 + {}} v_2 + v_3, \\ u_3 &= \phantom{v_0 + v_1 + v_2 + {}} v_3. \end{aligned} \tag{17.20}$$

Suppose $\mathbf{u} = (1\ 0\ 0\ 1)$. Using (17.18), we find that the 2-fold Kronecker mapping of $\mathbf{u} = (1\ 0\ 0\ 1)$ is $\mathbf{v} = (0\ 1\ 1\ 1)$. Using (17.20), we find that the 2-fold inverse Kronecker mapping of $\mathbf{v} = (0\ 1\ 1\ 1)$ is $\mathbf{u} = (1\ 0\ 0\ 1)$. From (17.18) and (17.20), we find that the two sets of equalities are identical in form and one can be obtained from the other by interchanging $u_i$ and $v_i$ for $0 \le i < 4$. Table 17.1 displays the 2-fold Kronecker mappings of the 16 4-tuples over $\mathrm{GF}(2)$. ▲▲

**Example 17.3** This example is a continuation of Example 17.2. In this example, we show that the 2-fold Kronecker mapping can be carried out in two levels. First, we divide $\mathbf{u} = (u_0, u_1, u_2, u_3)$ into two sections of the same length, $\mathbf{u}_0 = (u_0, u_1)$ and $\mathbf{u}_1 = (u_2, u_3)$, and $\mathbf{v} = (v_0, v_1, v_2, v_3)$ into two sections of the same length, $\mathbf{v}_0 = (v_0, v_1)$ and $\mathbf{v}_1 = (v_2, v_3)$. Then, $\mathbf{u} = (\mathbf{u}_0, \mathbf{u}_1)$ and $\mathbf{v} = (\mathbf{v}_0, \mathbf{v}_1)$. From (17.16), we find that

$$\mathbf{v}_0 = \mathbf{u}_0 \cdot \mathbf{G}^{(1)} + \mathbf{u}_1 \cdot \mathbf{G}^{(1)}, \tag{17.21}$$

$$\mathbf{v}_1 = \mathbf{u}_1 \cdot \mathbf{G}^{(1)}. \tag{17.22}$$

Table 17.1 The 2-fold Kronecker mappings of the 16 4-tuples over GF(2).

| $\mathbf{u} = (u_0, u_1, u_2, u_3)$ | $\mathbf{v} = (v_0, v_1, v_2, v_3)$ | $\mathbf{u} = (u_0, u_1, u_2, u_3)$ | $\mathbf{v} = (v_0, v_1, v_2, v_3)$ |
|---|---|---|---|
| (0 0 0 0) | (0 0 0 0) | (1 0 0 0) | (1 0 0 0) |
| (0 0 0 1) | (1 1 1 1) | (1 0 0 1) | (0 1 1 1) |
| (0 0 1 0) | (1 0 1 0) | (1 0 1 0) | (0 0 1 0) |
| (0 0 1 1) | (0 1 0 1) | (1 0 1 1) | (1 1 0 1) |
| (0 1 0 0) | (1 1 0 0) | (1 1 0 0) | (0 1 0 0) |
| (0 1 0 1) | (0 0 1 1) | (1 1 0 1) | (1 0 1 1) |
| (0 1 1 0) | (0 1 1 0) | (1 1 1 0) | (1 1 1 0) |
| (0 1 1 1) | (1 0 0 1) | (1 1 1 1) | (0 0 0 1) |

From (17.21) and (17.22), we see that $\mathbf{v}$ can be formed in two levels. In the first-level mapping, we form $\mathbf{u}_0 \cdot \mathbf{G}^{(1)}$ and $\mathbf{u}_1 \cdot \mathbf{G}^{(1)}$. In the second-level mapping, we first take the sum of $\mathbf{u}_0 \cdot \mathbf{G}^{(1)}$ and $\mathbf{u}_1 \cdot \mathbf{G}^{(1)}$ and then connect the sum $\mathbf{u}_0 \cdot \mathbf{G}^{(1)} + \mathbf{u}_1 \cdot \mathbf{G}^{(1)}$ and $\mathbf{u}_1 \cdot \mathbf{G}^{(1)}$ to form the Kronecker mapping $\mathbf{v} = (\mathbf{v}_0, \mathbf{v}_1)$ of $\mathbf{u} = (\mathbf{u}_0, \mathbf{u}_1)$.

Using (17.16), (17.21), and (17.22), we can express $\mathbf{v} = (\mathbf{v}_0, \mathbf{v}_1)$ in terms of $\mathbf{u}_0$, $\mathbf{u}_1$, and $\mathbf{G}^{(1)}$ as follows:

$$
\begin{aligned}
\mathbf{v} &= (\mathbf{v}_0, \mathbf{v}_1) \\
&= (\mathbf{u}_0, \mathbf{u}_1) \cdot \begin{bmatrix} \mathbf{G}^{(1)} & \mathbf{O} \\ \mathbf{G}^{(1)} & \mathbf{G}^{(1)} \end{bmatrix} \\
&= (\mathbf{u}_0 \cdot \mathbf{G}^{(1)}, \mathbf{u}_1 \cdot \mathbf{G}^{(1)}) \cdot \begin{bmatrix} \mathbf{I}_2 & \mathbf{O} \\ \mathbf{I}_2 & \mathbf{I}_2 \end{bmatrix} \\
&= (\mathbf{u}_0 \cdot \mathbf{G}^{(1)}, \mathbf{u}_1 \cdot \mathbf{G}^{(1)}) \cdot \mathbf{I}_{\text{con},2},
\end{aligned} \tag{17.23}
$$

where the matrix $\mathbf{I}_{\text{con},2}$ is

$$
\mathbf{I}_{\text{con},2} = \begin{bmatrix} \mathbf{I}_2 & \mathbf{O} \\ \mathbf{I}_2 & \mathbf{I}_2 \end{bmatrix} = \begin{bmatrix} 1 & 0 & 0 & 0 \\ 0 & 1 & 0 & 0 \\ 1 & 0 & 1 & 0 \\ 0 & 1 & 0 & 1 \end{bmatrix} \tag{17.24}
$$

with $\mathbf{I}_2$ as the $2 \times 2$ identity matrix. Then,

$$
\begin{aligned}
\mathbf{v} &= (\mathbf{v}_0, \mathbf{v}_1) \\
&= (\mathbf{u}_0 \cdot \mathbf{G}^{(1)}, \mathbf{u}_1 \cdot \mathbf{G}^{(1)}) \cdot \mathbf{I}_{\text{con},2} \\
&= (\mathbf{u}_0 \cdot \mathbf{G}^{(1)} + \mathbf{u}_1 \cdot \mathbf{G}^{(1)}, \mathbf{u}_1 \cdot \mathbf{G}^{(1)}) \\
&= (u_0 + u_1 + u_2 + u_3, u_1 + u_3, u_2 + u_3, u_3),
\end{aligned} \tag{17.25}
$$

which is exactly the same Kronecker mapping $\mathbf{v}$ of $\mathbf{u}$ given by (17.18).

From (17.25), we see that the matrix $\mathbf{I}_{\text{con},2}$ tells how to connect $\mathbf{u}_0 \cdot \mathbf{G}^{(1)}$ and $\mathbf{u}_1 \cdot \mathbf{G}^{(1)}$ formed in the first-level mapping to produce the mapping $\mathbf{v}$ of $\mathbf{u}$ in the second level. We call the matrix $\mathbf{I}_{\text{con},2}$ the *2-fold connection matrix*, where the subscript "con" in $\mathbf{I}_{\text{con},2}$ stands for "connection."

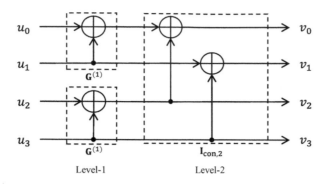

Figure 17.2 The 2-fold Kronecker mapping circuit.

Based on (17.21), (17.22), and (17.25), we can implement the 2-fold Kronecker mapping in two levels with $2^2 = 4$ XOR gates, two in the first-level mapping and two in the second-level mapping, as shown in Fig. 17.2. ▲▲

**Example 17.4** In this example, we demonstrate the process of the 3-fold Kronecker mapping of a $2^3$-tuple $\mathbf{u} = (u_0, u_1, u_2, u_3, u_4, u_5, u_6, u_7)$ over GF(2) into another $2^3$-tuple $\mathbf{v} = (v_0, v_1, v_2, v_3, v_4, v_5, v_6, v_7)$. The mapping function $f_{\text{Kron}}^{(3)}$ is defined by the 3-fold Kronecker matrix $\mathbf{G}^{(3)}$, which is

$$\mathbf{G}^{(3)} = \begin{bmatrix} \mathbf{G}^{(2)} & \mathbf{O} \\ \mathbf{G}^{(2)} & \mathbf{G}^{(2)} \end{bmatrix}. \tag{17.26}$$

If we divide the 8-tuple $\mathbf{u}$ into two sections of the same length, $\mathbf{u}_0 = (u_0, u_1, u_2, u_3)$ and $\mathbf{u}_1 = (u_4, u_5, u_6, u_7)$, and the 8-tuple $\mathbf{v}$ into two sections of the same length, $\mathbf{v}_0 = (v_0, v_1, v_2, v_3)$ and $\mathbf{v}_1 = (v_4, v_5, v_6, v_7)$, then, the 3-fold Kronecker mapping of $\mathbf{u}$ is given as follows:

$$\begin{aligned} \mathbf{v} &= (\mathbf{v}_0, \mathbf{v}_1) \\ &= (\mathbf{u}_0, \mathbf{u}_1) \cdot \begin{bmatrix} \mathbf{G}^{(2)} & \mathbf{O} \\ \mathbf{G}^{(2)} & \mathbf{G}^{(2)} \end{bmatrix} \\ &= (\mathbf{u}_0 \cdot \mathbf{G}^{(2)}, \mathbf{u}_1 \cdot \mathbf{G}^{(2)}) \cdot \begin{bmatrix} \mathbf{I}_{2^2} & \mathbf{O} \\ \mathbf{I}_{2^2} & \mathbf{I}_{2^2} \end{bmatrix} \\ &= (\mathbf{u}_0 \cdot \mathbf{G}^{(2)}, \mathbf{u}_1 \cdot \mathbf{G}^{(2)}) \cdot \mathbf{I}_{\text{con},3}, \end{aligned} \tag{17.27}$$

where

$$\mathbf{I}_{\text{con},3} = \begin{bmatrix} \mathbf{I}_{2^2} & \mathbf{O} \\ \mathbf{I}_{2^2} & \mathbf{I}_{2^2} \end{bmatrix}, \tag{17.28}$$

and $\mathbf{I}_{2^2}$ is the $2^2 \times 2^2$ $(4 \times 4)$ identity matrix. It follows from (17.27) that

$$\begin{aligned} \mathbf{v} &= (\mathbf{v}_0, \mathbf{v}_1) \\ &= (\mathbf{u}_0 \cdot \mathbf{G}^{(2)} + \mathbf{u}_1 \cdot \mathbf{G}^{(2)}, \mathbf{u}_1 \cdot \mathbf{G}^{(2)}), \end{aligned} \tag{17.29}$$

with $\mathbf{v}_0 = \mathbf{u}_0 \cdot \mathbf{G}^{(2)} + \mathbf{u}_1 \cdot \mathbf{G}^{(2)}$ and $\mathbf{v}_1 = \mathbf{u}_1 \cdot \mathbf{G}^{(2)}$. Because the matrix $\mathbf{I}_{con,3}$ tells how to connect $\mathbf{u}_0 \cdot \mathbf{G}^{(2)}$ and $\mathbf{u}_1 \cdot \mathbf{G}^{(2)}$ to form the 8-tuple $\mathbf{v}$, the 3-fold Kronecker mapping of $\mathbf{u}$, we call $\mathbf{I}_{con,3}$ the *3-fold connection matrix*.

In Example 17.3, we showed that $\mathbf{u}_0 \cdot \mathbf{G}^{(2)}$ and $\mathbf{u}_1 \cdot \mathbf{G}^{(2)}$ can be formed in two levels of mapping with four pairs $(u_0, u_1)$, $(u_2, u_3)$, $(u_4, u_5)$, and $(u_6, u_7)$ as inputs, respectively. Hence, the 3-fold Kronecker mapping $\mathbf{v}$ of $\mathbf{u}$ can be carried out recursively in three levels. The outputs of the first-level mapping are four pairs $(u_0 + u_1, u_1)$, $(u_2 + u_3, u_3)$, $(u_4 + u_5, u_5)$, and $(u_6 + u_7, u_7)$. Then, the four pairs of outputs of the first-level mapping are used as the inputs of the second level of the mapping which produce two 4-tuples:

$$\mathbf{u}_0 \cdot \mathbf{G}^{(2)} = ((u_0, u_1) \cdot \mathbf{G}^{(1)}, (u_2, u_3) \cdot \mathbf{G}^{(1)}) \cdot \mathbf{I}_{con,2}$$
$$= (u_0 + u_1 + u_2 + u_3, u_1 + u_3, u_2 + u_3, u_3), \tag{17.30}$$

$$\mathbf{u}_1 \cdot \mathbf{G}^{(2)} = ((u_4, u_5) \cdot \mathbf{G}^{(1)}, (u_6, u_7) \cdot \mathbf{G}^{(1)}) \cdot \mathbf{I}_{con,2}$$
$$= (u_4 + u_5 + u_6 + u_7, u_5 + u_7, u_6 + u_7, u_7), \tag{17.31}$$

at the outputs of the second-level mapping. These two 4-tuples are then used as the inputs of the third-level mapping and connected by using the connection matrix $\mathbf{I}_{con,3}$. The outputs of the third-level mapping give the 8-tuple $\mathbf{v} = (v_0, v_1, v_2, v_3, v_4, v_5, v_6, v_7) = (\mathbf{u}_0 \cdot \mathbf{G}^{(2)} + \mathbf{u}_1 \cdot \mathbf{G}^{(2)}, \mathbf{u}_1 \cdot \mathbf{G}^{(2)})$, the Kronecker mapping of $\mathbf{u}$, with

$$\begin{aligned}
v_0 &= u_0 + u_1 + u_2 + u_3 + u_4 + u_5 + u_6 + u_7, \\
v_1 &= \phantom{u_0 + } u_1 \phantom{+ u_2} + u_3 \phantom{+ u_4} + u_5 \phantom{+ u_6} + u_7, \\
v_2 &= \phantom{u_0 + u_1 + } u_2 + u_3 \phantom{+ u_4 + u_5} + u_6 + u_7, \\
v_3 &= \phantom{u_0 + u_1 + u_2 + } u_3 \phantom{+ u_4 + u_5 + u_6} + u_7, \\
v_4 &= \phantom{u_0 + u_1 + u_2 + u_3 + } u_4 + u_5 + u_6 + u_7, \\
v_5 &= \phantom{u_0 + u_1 + u_2 + u_3 + u_4 + } u_5 \phantom{+ u_6} + u_7, \\
v_6 &= \phantom{u_0 + u_1 + u_2 + u_3 + u_4 + u_5 + } u_6 + u_7, \\
v_7 &= \phantom{u_0 + u_1 + u_2 + u_3 + u_4 + u_5 + u_6 + } u_7.
\end{aligned} \tag{17.32}$$

The logical circuit for the 3-fold Kronecker mapping $f_{Kron}^{(3)}$ is shown in Fig. 17.3 and it consists of three levels of XOR gates, i.e., a network of $3 \times 2^2 = 12$ XOR gates.

Interchanging $v_i$ and $u_i$ in the equalities of (17.32), we obtain the components of the inverse Kronecker mapping $\mathbf{u}$ of $\mathbf{v}$ with

$$\begin{aligned}
u_0 &= v_0 + v_1 + v_2 + v_3 + v_4 + v_5 + v_6 + v_7, \\
u_1 &= \phantom{v_0 + } v_1 \phantom{+ v_2} + v_3 \phantom{+ v_4} + v_5 \phantom{+ v_6} + v_7, \\
u_2 &= \phantom{v_0 + v_1 + } v_2 + v_3 \phantom{+ v_4 + v_5} + v_6 + v_7, \\
u_3 &= \phantom{v_0 + v_1 + v_2 + } v_3 \phantom{+ v_4 + v_5 + v_6} + v_7, \\
u_4 &= \phantom{v_0 + v_1 + v_2 + v_3 + } v_4 + v_5 + v_6 + v_7, \\
u_5 &= \phantom{v_0 + v_1 + v_2 + v_3 + v_4 + } v_5 \phantom{+ v_6} + v_7, \\
u_6 &= \phantom{v_0 + v_1 + v_2 + v_3 + v_4 + v_5 + } v_6 + v_7, \\
u_7 &= \phantom{v_0 + v_1 + v_2 + v_3 + v_4 + v_5 + v_6 + } v_7.
\end{aligned} \tag{17.33}$$

Note that the circuit shown in Fig. 17.3 can be used to implement the inverse Kronecker mapping of $\mathbf{v}$ with $v_0, v_1, \ldots, v_7$ as inputs. ▲▲

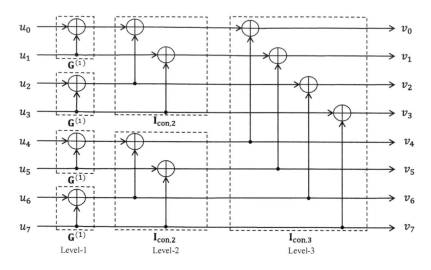

Figure 17.3 The 3-fold Kronecker mapping circuit.

In the above, we used four examples to illustrate the basic concepts of Kronecker mapping and its inverse mapping. The last three examples demonstrate that the Kronecker mapping can be carried out level by level and implemented with a network of XOR gates. In the following, we consider a general case in which the Kronecker mapping is carried out in multilevels recursively.

Consider the $\ell$-fold Kronecker mapping which is specified by the $\ell$-fold Kronecker matrix $\mathbf{G}^{(\ell)}$. This one-to-one mapping maps a $2^\ell$-tuple $\mathbf{u} = (u_0, u_1, \ldots, u_{2^\ell-1})$ into a unique $2^\ell$-tuple $\mathbf{v} = (v_0, v_1, \ldots, v_{2^\ell-1})$ as follows:

$$\mathbf{v} = f_{\text{Kron}}^{(\ell)}(\mathbf{u}) = \mathbf{u} \cdot \mathbf{G}^{(\ell)}. \tag{17.34}$$

For recursive mapping process, we express $\mathbf{G}^{(\ell)}$ as follows:

$$\mathbf{G}^{(\ell)} = \begin{bmatrix} \mathbf{G}^{(\ell-1)} & \mathbf{O} \\ \mathbf{G}^{(\ell-1)} & \mathbf{G}^{(\ell-1)} \end{bmatrix}. \tag{17.35}$$

First, we divide the $2^\ell$-tuple $\mathbf{u} = (u_0, u_1, \ldots, u_{2^\ell-1})$ into two sections of the same length, denoted by $\mathbf{u}_{\text{left},\ell-1} = (u_0, u_1, \ldots, u_{2^{\ell-1}-1})$ and $\mathbf{u}_{\text{right},\ell-1} = (u_{2^{\ell-1}}, u_{2^{\ell-1}+1}, \ldots, u_{2^\ell-1})$, where $\mathbf{u}_{\text{left},\ell-1}$ consists of the leftmost $2^{\ell-1}$ bits of $\mathbf{u}$ and $\mathbf{u}_{\text{right},\ell-1}$ consists of the rightmost $2^{\ell-1}$ bits of $\mathbf{u}$. Hence,

$$\mathbf{u} = (\mathbf{u}_{\text{left},\ell-1}, \mathbf{u}_{\text{right},\ell-1}), \tag{17.36}$$

where $\mathbf{u}_{\text{left},\ell-1}$ and $\mathbf{u}_{\text{right},\ell-1}$ are called the *left* and *right* sections of $\mathbf{u}$, respectively. Next, we divide $\mathbf{v} = (v_0, v_1, \ldots, v_{2^\ell-1})$ into two sections of the same length, denoted by $\mathbf{v}_{\text{left},\ell-1} = (v_0, v_1, \ldots, v_{2^{\ell-1}-1})$ and $\mathbf{v}_{\text{right},\ell-1} = (v_{2^{\ell-1}}, v_{2^{\ell-1}+1}, \ldots, v_{2^\ell-1})$, where $\mathbf{v}_{\text{left},\ell-1}$ consists of the leftmost $2^{\ell-1}$ bits of $\mathbf{v}$ and $\mathbf{v}_{\text{right},\ell-1}$ consists of the rightmost $2^{\ell-1}$ bits of $\mathbf{v}$. Hence,

$$\mathbf{v} = (\mathbf{v}_{\text{left},\ell-1}, \mathbf{v}_{\text{right},\ell-1}), \tag{17.37}$$

where $\mathbf{v}_{\text{left},\ell-1}$ and $\mathbf{v}_{\text{right},\ell-1}$ are called the *left* and *right* sections of $\mathbf{v}$, respectively.

It follows from (17.34) to (17.37) that the mapping given by (17.34) can be expressed as follows:

$$
\begin{aligned}
\mathbf{v} &= \left(\mathbf{v}_{\text{left},\ell-1}, \mathbf{v}_{\text{right},\ell-1}\right) \\
&= \left(\mathbf{u}_{\text{left},\ell-1}, \mathbf{u}_{\text{right},\ell-1}\right) \cdot \begin{bmatrix} \mathbf{G}^{(\ell-1)} & \mathbf{O} \\ \mathbf{G}^{(\ell-1)} & \mathbf{G}^{(\ell-1)} \end{bmatrix} \\
&= \left(\mathbf{u}_{\text{left},\ell-1} \cdot \mathbf{G}^{(\ell-1)}, \mathbf{u}_{\text{right},\ell-1} \cdot \mathbf{G}^{(\ell-1)}\right) \cdot \begin{bmatrix} \mathbf{I}_{2^{\ell-1}} & \mathbf{O} \\ \mathbf{I}_{2^{\ell-1}} & \mathbf{I}_{2^{\ell-1}} \end{bmatrix} \\
&= \left(\mathbf{u}_{\text{left},\ell-1} \cdot \mathbf{G}^{(\ell-1)}, \mathbf{u}_{\text{right},\ell-1} \cdot \mathbf{G}^{(\ell-1)}\right) \cdot \mathbf{I}_{\text{con},\ell},
\end{aligned}
\tag{17.38}
$$

where

$$
\mathbf{I}_{\text{con},\ell} = \begin{bmatrix} \mathbf{I}_{2^{\ell-1}} & \mathbf{O} \\ \mathbf{I}_{2^{\ell-1}} & \mathbf{I}_{2^{\ell-1}} \end{bmatrix}
\tag{17.39}
$$

is a $2 \times 2$ lower triangular array of three copies of the identity matrix $\mathbf{I}_{2^{\ell-1}}$ of size $2^{\ell-1} \times 2^{\ell-1}$ that is a $2^\ell \times 2^\ell$ lower triangular matrix.

From (17.38) and (17.39), we find that

$$
\mathbf{v}_{\text{left},\ell-1} = \mathbf{u}_{\text{left},\ell-1} \cdot \mathbf{G}^{(\ell-1)} + \mathbf{u}_{\text{right},\ell-1} \cdot \mathbf{G}^{(\ell-1)},
\tag{17.40}
$$

$$
\mathbf{v}_{\text{right},\ell-1} = \mathbf{u}_{\text{right},\ell-1} \cdot \mathbf{G}^{(\ell-1)},
\tag{17.41}
$$

and

$$
\begin{aligned}
\mathbf{v} &= \left(\mathbf{v}_{\text{left},\ell-1}, \mathbf{v}_{\text{right},\ell-1}\right) \\
&= \left(\mathbf{u}_{\text{left},\ell-1} \cdot \mathbf{G}^{(\ell-1)} + \mathbf{u}_{\text{right},\ell-1} \cdot \mathbf{G}^{(\ell-1)}, \mathbf{u}_{\text{right},\ell-1} \cdot \mathbf{G}^{(\ell-1)}\right).
\end{aligned}
\tag{17.42}
$$

Equations (17.40), (17.41), and (17.42) show that the calculation of the $\ell$-fold Kronecker mapping $\mathbf{v}$ of the $2^\ell$-tuple $\mathbf{u}$ is reduced to the calculation of the $(\ell-1)$-fold mappings $\mathbf{u}_{\text{left},\ell-1} \cdot \mathbf{G}^{(\ell-1)}$ and $\mathbf{u}_{\text{right},\ell-1} \cdot \mathbf{G}^{(\ell-1)}$ of the left and right sections, $\mathbf{u}_{\text{left},\ell-1}$ and $\mathbf{u}_{\text{right},\ell-1}$, of $\mathbf{u}$, respectively, and connecting them based on the matrix $\mathbf{I}_{\text{con},\ell}$. Because the function of $\mathbf{I}_{\text{con},\ell}$ is to connect the two $(\ell-1)$-fold Kronecker mappings $\mathbf{u}_{\text{left},\ell-1} \cdot \mathbf{G}^{(\ell-1)}$ and $\mathbf{u}_{\text{right},\ell-1} \cdot \mathbf{G}^{(\ell-1)}$ to form the $\ell$-fold Kronecker mapping $\mathbf{v}$ of $\mathbf{u}$, we call it the *$\ell$-fold connection matrix*.

The left part $\mathbf{I}_{\text{con},\ell,\text{left}}$ of the connection matrix $\mathbf{I}_{\text{con},\ell}$ has $2^{\ell-1}$ columns (the leftmost $2^{\ell-1}$ columns), each of weight 2, and the right part $\mathbf{I}_{\text{con},\ell,\text{right}}$ of $\mathbf{I}_{\text{con},\ell}$ has $2^{\ell-1}$ columns (the rightmost $2^{\ell-1}$ columns), each of weight 1. Label the $2^{\ell-1}$ columns of $\mathbf{I}_{\text{con},\ell,\text{left}}$ from 0 to $2^{\ell-1} - 1$. For $0 \le j < 2^{\ell-1}$, consider the $j$th column of $\mathbf{I}_{\text{con},\ell,\text{left}}$ whose two 1-entries are at the $j$th and $(2^{\ell-1} + j)$th locations. In the matrix product of (17.38), the locations of these two 1-entries tell us that the $j$th component of $\mathbf{v}$ is the sum of the $j$th entry in the $(\ell-1)$-fold Kronecker mapping $\mathbf{u}_{\text{left},\ell-1} \cdot \mathbf{G}^{(\ell-1)}$ of $\mathbf{u}_{\text{left},\ell-1}$ and the $j$th entry in the $(\ell-1)$-fold Kronecker mapping $\mathbf{u}_{\text{right},\ell-1} \cdot \mathbf{G}^{(\ell-1)}$ of $\mathbf{u}_{\text{right},\ell-1}$. The right part $\mathbf{I}_{\text{con},\ell,\text{right}}$ of the connection matrix $\mathbf{I}_{\text{con},\ell}$ indicates that the rightmost $2^{\ell-1}$ bits in the right section $\mathbf{v}_{\text{right},\ell-1}$ of the $\ell$-fold Kronecker mapping $\mathbf{v}$ of $\mathbf{u}$ are identical to the $2^{\ell-1}$

bits of the $(\ell-1)$-fold Kronecker mapping $\mathbf{u}_{\text{right},\ell-1} \cdot \mathbf{G}^{(\ell-1)}$ of $\mathbf{u}_{\text{right},\ell-1}$. Thus, the connection matrix $\mathbf{I}_{\text{con},\ell}$ tells how to connect the two $(\ell-1)$-fold Kronecker mappings of the sections $\mathbf{u}_{\text{left},\ell-1}$ and $\mathbf{u}_{\text{right},\ell-1}$ of $\mathbf{u}$ to form the $\ell$-fold Kronecker mapping $\mathbf{v}$ of $\mathbf{u}$.

The above construction of the $\ell$-fold Kronecker mapping is carried out by: (1) dividing $\mathbf{u}$ into two sections; (2) forming their Kronecker mappings; and (3) connecting them. This process is referred to as the *splitting* (or *decomposition*) process.

Following the same idea of splitting, the $(\ell-1)$-fold Kronecker mappings in (17.38) can be reduced to the calculation of the $(\ell-2)$-fold Kronecker mappings. Consider the $(\ell-1)$-fold Kronecker mapping of the $2^{\ell-1}$-tuple $\mathbf{u}_{\text{left},\ell-1}$. Let $\mathbf{u}_{\text{left},\ell-1} = [\mathbf{u}_{\text{left},\ell-2}, \mathbf{u}_{\text{right},\ell-2}]$, where $\mathbf{u}_{\text{left},\ell-2}$ is a $2^{\ell-2}$-tuple consisting of the leftmost $2^{\ell-2}$ bits of $\mathbf{u}_{\text{left},\ell-1}$ and $\mathbf{u}_{\text{right},\ell-2}$ is a $2^{\ell-2}$-tuple consisting of the rightmost $2^{\ell-2}$ bits of $\mathbf{u}_{\text{left},\ell-1}$. Then,

$$
\begin{aligned}
\mathbf{u}_{\text{left},\ell-1} \cdot \mathbf{G}^{(\ell-1)} &= (\mathbf{u}_{\text{left},\ell-2}, \mathbf{u}_{\text{right},\ell-2}) \cdot \begin{bmatrix} \mathbf{G}^{(\ell-2)} & \mathbf{O} \\ \mathbf{G}^{(\ell-2)} & \mathbf{G}^{(\ell-2)} \end{bmatrix} \\
&= (\mathbf{u}_{\text{left},\ell-2} \cdot \mathbf{G}^{(\ell-2)}, \mathbf{u}_{\text{right},\ell-2} \cdot \mathbf{G}^{(\ell-2)}) \cdot \begin{bmatrix} \mathbf{I}_{2^{\ell-2}} & \mathbf{O} \\ \mathbf{I}_{2^{\ell-2}} & \mathbf{I}_{2^{\ell-2}} \end{bmatrix} \\
&= (\mathbf{u}_{\text{left},\ell-2} \cdot \mathbf{G}^{(\ell-2)}, \mathbf{u}_{\text{right},\ell-2} \cdot \mathbf{G}^{(\ell-2)}) \cdot \mathbf{I}_{\text{con},\ell-1},
\end{aligned}
\tag{17.43}
$$

where $\mathbf{G}^{(\ell-2)}$ is the $(\ell-2)$-fold Kronecker product matrix and $\mathbf{I}_{\text{con},\ell-1}$ is the $(\ell-1)$-fold connection matrix.

Similarly, we can calculate the $(\ell-1)$-fold Kronecker mapping $\mathbf{u}_{\text{right},\ell-1} \cdot \mathbf{G}^{(\ell-1)}$ of the right section $\mathbf{u}_{\text{right},\ell-1}$ of $\mathbf{u}$ as follows:

$$
\begin{aligned}
\mathbf{u}_{\text{right},\ell-1} \cdot \mathbf{G}^{(\ell-1)} &= [\mathbf{u}^*_{\text{left},\ell-2}, \mathbf{u}^*_{\text{right},\ell-2}] \cdot \begin{bmatrix} \mathbf{G}^{(\ell-2)} & \mathbf{O} \\ \mathbf{G}^{(\ell-2)} & \mathbf{G}^{(\ell-2)} \end{bmatrix} \\
&= [\mathbf{u}^*_{\text{left},\ell-2} \cdot \mathbf{G}^{(\ell-2)}, \mathbf{u}^*_{\text{right},\ell-2} \cdot \mathbf{G}^{(\ell-2)}] \cdot \begin{bmatrix} \mathbf{I}_{2^{\ell-2}} & \mathbf{O} \\ \mathbf{I}_{2^{\ell-2}} & \mathbf{I}_{2^{\ell-2}} \end{bmatrix} \\
&= [\mathbf{u}^*_{\text{left},\ell-2} \cdot \mathbf{G}^{(\ell-2)}, \mathbf{u}^*_{\text{right},\ell-2} \cdot \mathbf{G}^{(\ell-2)}] \cdot \mathbf{I}_{\text{con},\ell-1},
\end{aligned}
\tag{17.44}
$$

where $\mathbf{u}^*_{\text{left},\ell-2}$ is a $2^{\ell-2}$-tuple consisting of the leftmost $2^{\ell-2}$ bits of $\mathbf{u}_{\text{right},\ell-1}$ and $\mathbf{u}^*_{\text{right},\ell-2}$ is a $2^{\ell-2}$-tuple consisting of the rightmost $2^{\ell-2}$ bits of $\mathbf{u}_{\text{right},\ell-1}$.

Continuing the above splitting (or decomposition) process, the calculation of the $\ell$-fold Kronecker mapping will be reduced to the calculations of the single-fold Kronecker mappings based on the kernel $\mathbf{G}^{(1)}$ and the connection operations using the series of the connection matrices, $\mathbf{I}_{\text{con},2}, \mathbf{I}_{\text{con},3}, \ldots, \mathbf{I}_{\text{con},\ell-1}$, $\mathbf{I}_{\text{con},\ell}$. Hence, the $\ell$-fold Kronecker mapping can be achieved *level by level*. The calculations of the single-fold Kronecker mappings based on $\mathbf{G}^{(1)}$ are called the *level-1 mapping* operation. The connection operation of the outputs from the level-1 mapping based on the 2-fold connection matrix $\mathbf{I}_{\text{con},2}$ is called the *level-2 mapping* operation. For $2 \le i \le \ell$, the connection operation of the outputs from level-$(i-1)$ mapping based on $\mathbf{I}_{con,i}$ is called the *level-i mapping* operation. Therefore, the $\ell$-fold Kronecker mapping can be achieved by $\ell$ levels of mapping operations.

The logical implementation of the $\ell$-fold Kronecker mapping of a $2^\ell$-tuple $\mathbf{u}$ can be achieved by a network of $\ell$ levels of XOR gates as shown in Fig. 17.4. Each level of the network requires $2^{\ell-1}$ XOR gates and hence, a total of $\ell 2^{\ell-1}$ XOR gates is required to implement the $\ell$-fold Kronecker mapping. For example, the logical circuit for the 3-fold Kronecker mapping of a $2^3$-tuple over GF(2) shown in Fig. 17.3 is a network of three levels of XOR gates which comprises $3 \times 2^2 = 12$ XOR gates.

Note that the network shown in Fig. 17.4 can also be used for the inverse Kronecker mapping of $\mathbf{v}$ to retrieve $\mathbf{u}$.

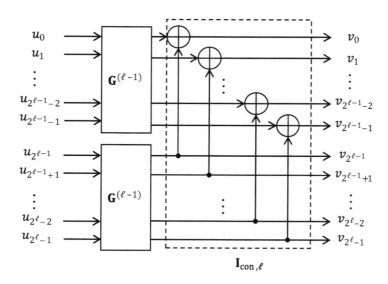

Figure 17.4 The $\ell$-fold Kronecker mapping circuit.

Kronecker matrices and mappings form the *base* for constructing polar codes, their encoding and decoding which will be shown in later sections.

For $\ell \geq 1$, the $\ell$-fold Kronecker mapping of a $2^\ell$-tuple $\mathbf{u} = (u_0, u_1, \ldots, u_{2^\ell-1})$ over GF(2) into a $2^\ell$-tuple $\mathbf{v} = (v_0, v_1, \ldots, v_{2^\ell-1})$ over GF(2) given by (17.34) results in $2^\ell$ sums of bits in $\mathbf{u}$, each giving a bit in $\mathbf{v}$. We call these sums *mapping sums*. Following the lower triangular structure of the mapping matrix $\mathbf{G}^{(\ell)}$, we readily see that some bits in $\mathbf{u}$ appear in more mapping sums than the other bits in $\mathbf{u}$. For example, consider the eight mapping sums given in (17.32) resulting from mapping an 8-tuple $\mathbf{u} = (u_0, u_1, u_2, u_3, u_4, u_5, u_6, u_7)$ into an 8-tuple $\mathbf{v} = (v_0, v_1, v_2, v_3, v_4, v_5, v_6, v_7)$. We see that the bit $u_7$ in $\mathbf{u}$ appears in all the eight mapping sums in (17.32), each of the three bits $u_3, u_5, u_6$ in $\mathbf{u}$ appears in four mapping sums, each of the three bits $u_1, u_2, u_4$ in $\mathbf{u}$ appears in two mapping sums, and the bit $u_0$ in $\mathbf{u}$ appears in only one mapping sum.

In the following, we present the distribution of bits in a $2^\ell$-tuple $\mathbf{u} = (u_0, u_1, \ldots, u_{2^\ell-1})$ in the $2^\ell$ mapping sums formed by the $\ell$-fold Kronecker mapping $\mathbf{v} = \mathbf{u} \cdot \mathbf{G}^{(\ell)}$. Let $t$ be a nonnegative integer such that $0 \leq t < 2^\ell$. The radix-2 expansion (representation) of $t$ is the real sum given as follows:

$$t = t_0 2^0 + t_1 2^1 + t_2 2^2 + \cdots + t_{\ell-1} 2^{\ell-1}, \tag{17.45}$$

where the coefficient $t_j$ is either 0 or 1, $0 \leq j < \ell$. Let

$$w(t) = t_0 + t_1 + t_2 + \cdots + t_{\ell-1}. \tag{17.46}$$

The sum $w(t)$ given by (17.46) is defined as the *radix-2 weight* of $t$. For $0 \leq t < 2^\ell$, it can be proved that the bit $u_t$ in $\mathbf{u}$ appears in $2^{w(t)}$ mapping sums formed by the $\ell$-fold Kronecker mapping $\mathbf{u} \cdot \mathbf{G}^{(\ell)}$ of $\mathbf{u}$ (see Problem 17.3). For $0 < t < 2^\ell - 1$, there are $\binom{\ell}{w(t)}$ bits in $\mathbf{u}$, each appearing in $2^{w(t)}$ mapping sums formed by the mapping $\mathbf{u} \cdot \mathbf{G}^{(\ell)}$. For $t = 0$, the bit $u_0$ of $\mathbf{u}$ appears only in one mapping sum, and for $t = 2^\ell - 1$, the bit $u_{2^\ell-1}$ in $\mathbf{u}$ appears in all the $2^\ell$ mapping sums.

**Example 17.5** Consider the eight mapping sums formed by the 3-fold Kronecker mapping of a $2^3$-tuple $\mathbf{u} = (u_0, u_1, u_2, u_3, u_4, u_5, u_6, u_7)$ given by (17.32) in Example 17.4. The radix-2 expansion of 5 is

$$5 = 1 \times 2^0 + 0 \times 2^1 + 1 \times 2^2.$$

The radix-2 weight of 5 is $w(5) = 1 + 0 + 1 = 2$. From (17.32), we find that the bit $u_5$ in $\mathbf{u}$ appears in $2^{w(5)} = 2^2 = 4$ of the eight mapping sums, which are the zeroth, first, fourth, and fifth mapping sums. The radix-2 expansion of 4 is

$$4 = 0 \times 2^0 + 0 \times 2^1 + 1 \times 2^2.$$

The radix-2 weight of 4 is $w(4) = 0 + 0 + 1 = 1$. Hence, the bit $u_4$ of $\mathbf{u}$ appears in $2^{w(t)} = 2^1 = 2$ of the eight mapping sums which are the zeroth and fourth sums.

Among the indices in the label set $\mathcal{I} = \{0, 1, 2, 3, 4, 5, 6, 7\}$ of $\mathbf{u}$, there are one label with radix-2 weight 0, three labels with radix-2 weight 1, three labels with radix-2 weight 2, and one label with radix-2 weight 3. Hence, there is $\binom{3}{0} = 1$ bit, $u_0$ in $\mathbf{u}$, which appears in only one mapping sum; there are $\binom{3}{1} = 3$ bits, $u_1$, $u_2$ and $u_4$ in $\mathbf{u}$, each appearing in $2^1 = 2$ mapping sums; there are $\binom{3}{2} = 3$ bits, $u_3$, $u_5$ and $u_6$ in $\mathbf{u}$, each appearing in $2^2 = 4$ mapping sums; and there is $\binom{3}{3} = 1$ bit, $u_7$ in $\mathbf{u}$, which appears in all the $2^3 = 8$ mapping sums. ▲▲

**Example 17.6** Consider the 4-fold Kronecker mapping defined by the 4-fold Kronecker matrix $\mathbf{G}^{(4)}$ given by (17.8). This mapping maps a $2^4$-tuple $\mathbf{u} = (u_0, u_1, \ldots, u_{14}, u_{15})$ over GF(2) uniquely into a $2^4$-tuple $\mathbf{v} = (v_0, v_1, \ldots, v_{14}, v_{15})$ over GF(2), i.e., $\mathbf{v} = \mathbf{u} \cdot \mathbf{G}^{(4)}$. The mapping is specified by the following 16 sums, each giving a bit in $\mathbf{v}$:

$$
\begin{aligned}
v_0 &= u_0+u_1+u_2+u_3+u_4+u_5+u_6+u_7+u_8+u_9+u_{10}+u_{11}+u_{12}+u_{13}+u_{14}+u_{15},\\
v_1 &= \phantom{u_0+}u_1\phantom{+u_2}+u_3\phantom{+u_4}+u_5\phantom{+u_6}+u_7\phantom{+u_8}+u_9\phantom{+u_{10}}+u_{11}\phantom{+u_{12}}+u_{13}\phantom{+u_{14}}+u_{15},\\
v_2 &= \phantom{u_0+u_1+}u_2+u_3\phantom{+u_4+u_5}+u_6+u_7\phantom{+u_8+u_9}+u_{10}+u_{11}\phantom{+u_{12}+u_{13}}+u_{14}+u_{15},\\
v_3 &= \phantom{u_0+u_1+u_2+}u_3\phantom{+u_4+u_5+u_6}+u_7\phantom{+u_8+u_9+u_{10}}+u_{11}\phantom{+u_{12}+u_{13}+u_{14}}+u_{15},\\
v_4 &= \phantom{u_0+u_1+u_2+u_3+}u_4+u_5+u_6+u_7\phantom{+u_8+u_9+u_{10}+u_{11}}+u_{12}+u_{13}+u_{14}+u_{15},\\
v_5 &= \phantom{u_0+u_1+u_2+u_3+u_4+}u_5\phantom{+u_6}+u_7\phantom{+u_8+u_9+u_{10}+u_{11}+u_{12}}+u_{13}\phantom{+u_{14}}+u_{15},\\
v_6 &= \phantom{u_0+u_1+u_2+u_3+u_4+u_5+}u_6+u_7\phantom{+u_8+u_9+u_{10}+u_{11}+u_{12}+u_{13}}+u_{14}+u_{15},\\
v_7 &= \phantom{u_0+u_1+u_2+u_3+u_4+u_5+u_6+}u_7\phantom{+u_8+u_9+u_{10}+u_{11}+u_{12}+u_{13}+u_{14}}+u_{15},\\
v_8 &= \phantom{u_0+u_1+u_2+u_3+u_4+u_5+u_6+u_7+}u_8+u_9+u_{10}+u_{11}+u_{12}+u_{13}+u_{14}+u_{15},\\
v_9 &= \phantom{u_0+u_1+u_2+u_3+u_4+u_5+u_6+u_7+u_8+}u_9\phantom{+u_{10}}+u_{11}\phantom{+u_{12}}+u_{13}\phantom{+u_{14}}+u_{15},\\
v_{10} &= \phantom{u_0+u_1+u_2+u_3+u_4+u_5+u_6+u_7+u_8+u_9+}u_{10}+u_{11}\phantom{+u_{12}+u_{13}}+u_{14}+u_{15},\\
v_{11} &= \phantom{u_0+u_1+u_2+u_3+u_4+u_5+u_6+u_7+u_8+u_9+u_{10}+}u_{11}\phantom{+u_{12}+u_{13}+u_{14}}+u_{15},\\
v_{12} &= \phantom{u_0+u_1+u_2+u_3+u_4+u_5+u_6+u_7+u_8+u_9+u_{10}+u_{11}+}u_{12}+u_{13}+u_{14}+u_{15},\\
v_{13} &= \phantom{u_0+u_1+u_2+u_3+u_4+u_5+u_6+u_7+u_8+u_9+u_{10}+u_{11}+u_{12}+}u_{13}\phantom{+u_{14}}+u_{15},\\
v_{14} &= \phantom{u_0+u_1+u_2+u_3+u_4+u_5+u_6+u_7+u_8+u_9+u_{10}+u_{11}+u_{12}+u_{13}+}u_{14}+u_{15},\\
v_{15} &= \phantom{u_0+u_1+u_2+u_3+u_4+u_5+u_6+u_7+u_8+u_9+u_{10}+u_{11}+u_{12}+u_{13}+u_{14}+}u_{15}.
\end{aligned}
\tag{17.47}
$$

Consider the bit $u_9$ in $\mathbf{u}$. The radix-2 expansion of 9 is

$$
9 = 1 \times 2^0 + 0 \times 2^1 + 0 \times 2^2 + 1 \times 2^3.
$$

The radix-2 weight of 9 is $w(9) = 1 + 0 + 0 + 1 = 2$. Then, the bit $u_9$ in $\mathbf{u}$ appears in $2^{w(9)} = 2^2 = 4$ of the 16 mapping sums given by (17.47). There are $\binom{4}{2} = 6$ labels, 3, 5, 6, 9, 10, and 12, with radix-2 weight equal to 2; hence, each of the six bits, $u_3$, $u_5$, $u_6$, $u_9$, $u_{10}$, and $u_{12}$ in $\mathbf{u}$, appears in four mapping sums in (17.47).

Consider the bit $u_{13}$ in $\mathbf{u}$. The radix-2 expansion of 13 is

$$
13 = 1 \times 2^0 + 0 \times 2^1 + 1 \times 2^2 + 1 \times 2^3.
$$

The radix-2 weight of 13 is $w(13) = 1 + 0 + 1 + 1 = 3$. Then, the bit $u_{13}$ in $\mathbf{u}$ appears in $2^{w(13)} = 2^3 = 8$ mapping sums in (17.47). There are $\binom{4}{3} = 4$ bits, $u_7$, $u_{11}$, $u_{13}$, and $u_{14}$ in $\mathbf{u}$ with radix-2 weight equal to 3 and each of these four bits in $\mathbf{u}$ appears in $2^3 = 8$ mapping sums in (17.47). $\blacktriangle\blacktriangle$

A bit in $\mathbf{u}$ that appears in more mapping sums is more likely to be retrieved when some of the bits in $\mathbf{v}$ are erased during transmission over a BEC. Consider the fifth bit $u_5$ in the $2^3$-tuple $\mathbf{u} = (u_0, u_1, u_2, u_3, u_4, u_5, u_6, u_7)$. This bit appears in the zeroth, first, fourth, and fifth mapping sums given in (17.32). If, during transmission of its Kronecker mapping $\mathbf{v} = (v_0, v_1, v_2, v_3, v_4, v_5, v_6, v_7)$, the bits $v_0$, $v_1$, $v_2$, $v_4$, and $v_6$ in $\mathbf{v}$ are erased, the bit $u_5$ in $\mathbf{u}$ can be retrieved from $\mathbf{v}$ by using the fifth equality in (17.33).

## 17.3 Kronecker Vector Spaces and Codes

It was shown in Section 17.1 that a Kronecker matrix is a full-rank matrix and all its rows are linearly independent. If we use these rows as vectors of a basis, then all the linear combinations over GF(2) of the vectors in this basis form a vector space over GF(2). We call this vector space a *Kronecker vector space*. In this section, we present a class of linear block codes which are subspaces of the Kronecker vector spaces. The codes in this class are called *Kronecker codes*. This class of codes contains RM (Reed–Muller) codes as a subclass. In Section 17.4, we will show that polar codes form another subclass of Kronecker codes whose encoding and decoding are based on *channel polarization*.

Let $\ell$ be a positive integer. Consider the $\ell$-fold Kronecker matrix $\mathbf{G}^{(\ell)}$ in the form of (17.9). $\mathbf{G}^{(\ell)}$ is a $2^\ell \times 2^\ell$ matrix over GF(2) and each row of $\mathbf{G}^{(\ell)}$ is a $2^\ell$-tuple over GF(2). If we use the rows of $\mathbf{G}^{(\ell)}$ to form a basis, denoted by $\mathbf{B}_{\mathrm{Kron}}(\ell)$, then the vector space spanned by the vectors in $\mathbf{B}_{\mathrm{Kron}}(\ell)$ is the $2^\ell$-dimensional vector space, denoted by $\mathbf{V}_{\mathrm{Kron}}(\ell)$, of all the $2^{2^\ell}$ $2^\ell$-tuples over GF(2). We call this vector space the *$\ell$-fold Kronecker vector space*. It was shown in Chapter 9 that $\mathbf{V}_{\mathrm{Kron}}(\ell)$ is a realization of the $\ell$-dimensional Euclidean geometry EG$(\ell, 2)$ over GF(2). The vectors in $\mathbf{V}_{\mathrm{Kron}}(\ell)$ are the incidence vectors of the flats in EG$(\ell, 2)$. From this geometric point of view, the rows of $\mathbf{B}_{\mathrm{Kron}}(\ell)$ are the incidence vectors of flats in EG$(\ell, 2)$ that contain the origin (or any fixed point) of EG$(\ell, 2)$.

**Definition 17.1** Let $k$ be a positive integer less than or equal to $2^\ell$, i.e., $1 \le k \le 2^\ell$. Let $\mathbf{B}_{\mathrm{Kron}}(k, \ell)$ be a submatrix of $\mathbf{B}_{\mathrm{Kron}}(\ell)$ that consists of any $k$ rows of $\mathbf{B}_{\mathrm{Kron}}(\ell)$. The subspace of $\mathbf{V}_{\mathrm{Kron}}(\ell)$ spanned by $\mathbf{B}_{\mathrm{Kron}}(k, \ell)$ gives a $(2^\ell, k)$ linear block code over GF(2) which is a *$k$-dimensional Kronecker code* of length $2^\ell$, denoted by $C_{\mathrm{Kron}}(k, \ell)$. The generator matrix of $C_{\mathrm{Kron}}(k, \ell)$, denoted by $\mathbf{G}_{\mathrm{Kron}}(k, \ell)$, is $\mathbf{B}_{\mathrm{Kron}}(k, \ell)$, i.e., $\mathbf{G}_{\mathrm{Kron}}(k, \ell) = \mathbf{B}_{\mathrm{Kron}}(k, \ell)$.

Let $r$ be an nonnegative integer such that $0 \le r \le \ell$. If we select all the rows with weights $2^{\ell-r}, 2^{\ell-r+1}, \ldots, 2^\ell$ from $\mathbf{B}_{\mathrm{Kron}}(\ell)$ to form a sub-basis $\mathbf{B}_{\mathrm{Kron}}(k, \ell)$, then $\mathbf{B}_{\mathrm{Kron}}(k, \ell)$ consists of

$$k = 1 + \binom{2^\ell}{1} + \binom{2^\ell}{2} + \cdots + \binom{2^\ell}{r} \tag{17.48}$$

rows. The linear combinations over GF(2) of the $k$ rows in $\mathbf{B}_{\mathrm{Kron}}(k, \ell)$ give a $(2^\ell, k)$ Kronecker code $C_{\mathrm{Kron}}(k, \ell)$ which is the $r$th order RM code RM$(r, \ell)$ of length $2^\ell$ with minimum distance $2^{\ell-r}$ as shown in Chapter 9. Hence, RM codes form a subclass of Kronecker codes, called *Kronecker–RM codes*. The rows in $\mathbf{B}_{\mathrm{Kron}}(k, \ell)$ are the incidence vectors of the $(\ell - r)$-, $(\ell - r + 1)$-, $\ldots$, $\ell$-flats in the $\ell$-dimensional Euclidean geometry EG$(\ell, 2)$ over GF(2) that contain a fixed point of EG$(\ell, 2)$. If we arrange the rows in $\mathbf{B}_{\mathrm{Kron}}(k, \ell)$ in the form of (8.13), we can decode Kronecker–RM codes with majority-logic decoding using successive cancellations in $r + 1$ levels, as presented in Chapters 8 and 10 for correcting random errors and erasures.

Because the rows of the generator matrix $\mathbf{G}_{\mathrm{Kron}}(k, \ell)$ of a $k$-dimensional Kronecker code $C_{\mathrm{Kron}}(k, \ell)$ are incidence vectors of a selected set of flats of various dimensions in the $\ell$-dimensional Euclidean geometry EG$(\ell, 2)$ over GF(2), $C_{\mathrm{Kron}}(k, \ell)$ may be regarded as a finite-geometry code. From this geometric point of view, Kronecker codes form a class of finite-geometry codes and they can be considered as a class of generalized RM codes. If $\mathbf{G}_{\mathrm{Kron}}(k, \ell)$ does not contain the top row, the row with weight 1, of $\mathbf{G}^{(\ell)}$, any two codewords in $C_{\mathrm{Kron}}(k, \ell)$ are *orthogonal* to each other, i.e., their inner product is equal to zero. The dual code of such a Kronecker code is also a Kronecker code. Furthermore, for $k = 2^{\ell-1}$, the rate-1/2 Kronecker code $C_{\mathrm{Kron}}(k, \ell)$ is *self-dual*, i.e., the dual code of $C_{\mathrm{Kron}}(k, \ell)$ is identical to $C_{\mathrm{Kron}}(k, \ell)$.

A Kronecker code also has the orthogonal structural property similar to an RM code. Let $\mathbf{c} = (c_0, c_1, \ldots, c_{k-1})$ be a message of $k$ information bits and let $\mathbf{v} = (v_0, v_1, \ldots, v_{2^\ell-1})$ be the codeword for $\mathbf{c}$ in a $k$-dimensional Kronecker code $C_{\mathrm{Kron}}(k, \ell)$, i.e., $\mathbf{v} = \mathbf{c} \cdot \mathbf{G}_{\mathrm{Kron}}(k, \ell)$. For $0 \leq i < k$, let $c_i$ be the $i$th information bit in $\mathbf{c}$ and let $\mathbf{z}_i$ be the $i$th row of the generator matrix $\mathbf{G}_{\mathrm{Kron}}(k, \ell)$ of $C_{\mathrm{Kron}}(k, \ell)$. If the weight of $\mathbf{z}_i$ is $2^{w_i}$, then $2^{w_i}$ independent determinations of the information bit $c_i$ can be formed by taking the sums of bits in $2^{w_i}$ disjoint subsets of the bits in $\mathbf{v}$ or its descendant (see Section 8.4), each consisting of $2^{\ell-w_i}$ bits in $\mathbf{v}$ or its descendant. Based on this orthogonal structure, the code $C_{\mathrm{Kron}}(k, \ell)$ can be decoded with the multilevel majority-logic decoding for correcting random errors as described in Section 8.5 or the SPL/SCIR decoding for correcting random erasures presented in Section 10.8.2.

**Example 17.7** Consider the 4-fold Kronecker matrix $\mathbf{G}^{(4)}$ given by (17.8). Form the following $8 \times 16$ matrix using the third, fifth, seventh, ninth, eleventh, thirteenth, fourteenth, and fifteenth rows of $\mathbf{G}^{(4)}$:

$$
\mathbf{G}_{\mathrm{Kron}}(8, 4) = \begin{bmatrix} \mathbf{z}_0 \\ \mathbf{z}_1 \\ \mathbf{z}_2 \\ \mathbf{z}_3 \\ \mathbf{z}_4 \\ \mathbf{z}_5 \\ \mathbf{z}_6 \\ \mathbf{z}_7 \end{bmatrix} = \begin{bmatrix} 1\,1\,1\,1\,0\,0\,0\,0\,0\,0\,0\,0\,0\,0\,0\,0 \\ 1\,1\,0\,0\,1\,1\,0\,0\,0\,0\,0\,0\,0\,0\,0\,0 \\ 1\,1\,1\,1\,1\,1\,1\,0\,0\,0\,0\,0\,0\,0\,0 \\ 1\,1\,0\,0\,0\,0\,0\,0\,1\,1\,0\,0\,0\,0\,0\,0 \\ 1\,1\,1\,1\,0\,0\,0\,0\,1\,1\,1\,1\,0\,0\,0\,0 \\ 1\,1\,0\,0\,1\,1\,0\,0\,1\,1\,0\,0\,1\,1\,0\,0 \\ 1\,0\,1\,0\,1\,0\,1\,0\,1\,0\,1\,0\,1\,0\,1\,0 \\ 1\,1\,1\,1\,1\,1\,1\,1\,1\,1\,1\,1\,1\,1\,1\,1 \end{bmatrix}. \tag{17.49}
$$

The matrix $\mathbf{G}_{\mathrm{Kron}}(8, 4)$ has three rows of weight 4, four rows of weight 8, and one row of weight 16. The code generated by $\mathbf{G}_{\mathrm{Kron}}(8, 4)$ is a $(16, 8)$ Kronecker code $C_{\mathrm{Kron}}(8, 4)$ of rate 1/2 with minimum distance 4. It is a subcode of the $(16, 11)$ second-order RM code $C_{\mathrm{RM}}(2, 4)$.

Let $\mathbf{c} = (c_0, c_1, c_2, c_3, c_4, c_5, c_6, c_7)$ be a message of eight information bits to be encoded and let $\mathbf{v} = (v_0, v_1, \ldots, v_{14}, v_{15})$ be the codeword of $\mathbf{c}$ in $C_{\mathrm{Kron}}(8, 4)$. Using the approach presented in Section 8.4, we can retrieve the information bits in $\mathbf{c}$ from $\mathbf{v}$ or its descendants by successive cancellation. For example, to retrieve $c_0$ from $\mathbf{v}$, we can use any of the following four independent determinations of $c_0$:

$$
\begin{aligned}
c_0 &= v_0 + v_4 + v_8 + v_{12}, \\
c_0 &= v_1 + v_5 + v_9 + v_{13}, \\
c_0 &= v_2 + v_6 + v_{10} + v_{14}, \\
c_0 &= v_3 + v_7 + v_{11} + v_{15}.
\end{aligned}
$$

To retrieve the information bit $c_6$, we can use any of the following eight independent determinations for $c_6$ after the first level of cancellation:

$$c_6 = v_0 + v_1,$$
$$c_6 = v_2 + v_3,$$
$$c_6 = v_4 + v_5,$$
$$c_6 = v_6 + v_7,$$
$$c_6 = v_8 + v_9,$$
$$c_6 = v_{10} + v_{11},$$
$$c_6 = v_{12} + v_{13},$$
$$c_6 = v_{14} + v_{15}.$$

The code is capable of correcting three random erasures with the SPL/SCIR decoding. ▲▲

## 17.4 Definition and Polarized Encoding of Polar Codes

In this section, we give a mathematical formulation of a polar code and its polarized encoding.

With $k \leq 2^\ell$, let $\mathbf{c} = (c_0, c_1, \ldots, c_{k-1})$ be a message of $k$ information bits and let $\mathbf{u} = (u_0, u_1, \ldots, u_{2^\ell - 1})$ be a $2^\ell$-tuple over GF(2). Let $\mathcal{I} = \{0, 1, 2, \ldots, 2^\ell - 1\}$ be the set of indices of the components of the $2^\ell$-tuple $\mathbf{u} = (u_0, u_1, \ldots, u_{2^\ell - 1})$. Divide the index set $\mathcal{I}$ into two disjoint subsets, $\mathcal{I}_{\text{free}}$ and $\mathcal{I}_{\text{frozen}}$, which consist of $k$ and $2^\ell - k$ indices in $\mathcal{I}$, respectively. Let

$$\mathcal{I}_{\text{free}} = \{i_0, i_1, \ldots, i_{k-1} : i_s \in \mathcal{I}, 0 \leq s < k\}, \tag{17.50}$$

$$\mathcal{I}_{\text{frozen}} = \mathcal{I} \setminus \mathcal{I}_{\text{free}} = \{j_0, j_1, \ldots, j_{2^\ell - k - 1} : j_t \in \mathcal{I}, 0 \leq t < 2^\ell - k\}. \tag{17.51}$$

Next, we construct a $2^\ell$-tuple $\mathbf{u} = \mu(\mathbf{c}) = (u_0, u_1, \ldots, u_{2^\ell - 1})$ over GF(2) by assigning the components of $\mathbf{u}$ at the locations in $\mathcal{I}_{\text{free}}$ to the information bits in $\mathbf{c}$ and the components of $\mathbf{u}$ at the locations in $\mathcal{I}_{\text{frozen}}$ to zeros. Then, we map (or encode) the $2^\ell$-tuple $\mathbf{u} = \mu(\mathbf{c})$ into a $2^\ell$-tuple $\mathbf{v} = (v_0, v_1, \ldots, v_{2^\ell - 1})$ over GF(2) using the $\ell$-fold Kronecker mapping $f_{\text{Kron}}^{(\ell)}$ as follows:

$$\mathbf{v} = f_{\text{Kron}}^{(\ell)}(\mathbf{u}) = \mathbf{u} \cdot \mathbf{G}^{(\ell)} = \mu(\mathbf{c}) \cdot \mathbf{G}^{(\ell)}. \tag{17.52}$$

The $2^\ell$-tuple $\mathbf{v}$ is a vector in the Kronecker vector space $\mathbf{V}_{\text{Kron}}(\ell)$. The vectors $\mathbf{u} = \mu(\mathbf{c})$ and $\mathbf{v} = \mu(\mathbf{c}) \cdot \mathbf{G}^{(\ell)}$ are called the *encoding vector* and *codeword* of the message $\mathbf{c}$, respectively. Such encoding of a message is referred to as *polarized encoding*. It is clear that $\mathbf{c}$ can be uniquely retrieved from $\mathbf{v}$ by performing inverse Kronecker mapping and removing the bits at the locations labeled by the indices in $\mathcal{I}_{\text{frozen}}$.

Because there are $2^k$ messages of $k$ information bits, there are $2^k$ encoding vectors associated with these messages, each constructed based on the partition of the location index set $\mathcal{I}$ into two disjoint location index sets, $\mathcal{I}_{\text{free}}$ and $\mathcal{I}_{\text{frozen}}$. Because the Kronecker mapping defined by $\mathbf{G}^{(\ell)}$ is one-to-one, the message $\mathbf{c}$ is uniquely encoded into a vector $\mathbf{v}$ in the Kronecker vector space $\mathbf{V}_{\text{Kron}}(\ell)$. The $k$ rows in $\mathbf{G}^{(\ell)}$ selected for encoding each message $\mathbf{c}$ or its corresponding encoding vector $\mathbf{u} = \mu(\mathbf{c})$ are the rows labeled with indices in $\mathcal{I}_{\text{free}}$. The bits in $\mathbf{u}$ at the locations in $\mathcal{I}_{\text{free}}$ are called *free bits* and the bits in $\mathbf{u}$ at the locations in $\mathcal{I}_{\text{frozen}}$ are called *frozen bits*. The locations given in $\mathcal{I}_{\text{free}}$ and $\mathcal{I}_{\text{frozen}}$ are called *free* and *frozen bit locations*, respectively. The rows in the Kronecker matrix $\mathbf{G}^{(\ell)}$ labeled with indices in $\mathcal{I}_{\text{free}}$ and $\mathcal{I}_{\text{frozen}}$ are called the *free* and *frozen rows*, respectively. The partition $\mathcal{I}_{\text{free}}/\mathcal{I}_{\text{frozen}}$ of $\mathcal{I}$ is called a *polarized partition*. The encoding $\mathbf{c}$ into $\mathbf{v} = \mu(\mathbf{c}) \cdot \mathbf{G}^{(\ell)}$ is referred to as *polarized encoding* and $\mu(\mathbf{c})$ is called the *polarized mapping* of $\mathbf{c}$. Clearly, the $2^k$ polarized codewords $\mathbf{v} = \mu(\mathbf{c}) \cdot \mathbf{G}^{(\ell)}$ for the $2^k$ messages, $\mathbf{c}$, form a $k$-dimensional subspace of the Kronecker vector space $\mathbf{V}_{\text{Kron}}(\ell)$ which gives a $(2^\ell, k)$ linear block code.

Polarized partition of the code bit positions for information transmission is carried out by using an elegant technique called *channel polarization*. The $2^\ell$ copies of a given binary-input discrete memoryless channel (BI-DMC) are polarized into $2^\ell$ virtual bit channels, some of which are more reliable for transmitting information, called *good channels*, than the others, which are much less reliable (or useless) for information transmission and are called *bad channels*. Each of these $2^\ell$ virtual bit channels is associated with a code bit coordinate (or position) in the encoding vector $\mathbf{u}$ of a message $\mathbf{c}$. Hence, we call these virtual bit channels the *bit-coordinate channels*. In polar-code construction, we choose a set of good virtual bit-coordinate channels for transmitting the information bits in $\mathbf{c}$. The coordinates of these chosen good channels specify the locations (in $\mathcal{I}_{\text{free}}$) in the encoding vector $\mathbf{u}$ for carrying the information bits of $\mathbf{c}$. The coordinates of the other poor bit-coordinate channels specify the locations (in $\mathcal{I}_{\text{frozen}}$) in the encoding vector $\mathbf{u}$ which are frozen by setting to zeros (or any fixed value), i.e., carry no information bit in $\mathbf{c}$. Each chosen set of good bit-coordinate channels results in a polar code. The channel polarization technique that instructs how to choose good bit-coordinate channels will be presented in Section 17.6. In the following, we give a mathematical definition of a polar code.

**Definition 17.2** Let $k$ be a positive integer such that $1 \leq k \leq 2^\ell$ and let $\mathbf{M} = \{\mathbf{c} = (c_0, c_1, \ldots, c_{k-1}) : c_i \in \text{GF}(2), 0 \leq i < k\}$ be the $k$-dimensional message space with $2^k$ messages, each consisting of $k$ information bits. The following set of $2^k$ vectors in the $\ell$-fold Kronecker vector space $\mathbf{V}_{\text{Kron}}(\ell)$,

$$C_{\text{p}}(k, \ell) = \{\mathbf{v} = \mu(\mathbf{c}) \cdot \mathbf{G}^{(\ell)} : \mathbf{c} \in \mathbf{M}\}, \tag{17.53}$$

is defined as a $(2^\ell, k)$ *polar code*, denoted by $C_{\text{p}}(k, \ell)$, where $\mu(\mathbf{c})$ is a polarized mapping of a message $\mathbf{c}$.

For each choice of $k$ free bit positions (or $2^\ell - k$ frozen bit positions), there exists a $(2^\ell, k)$ polar code $C_{\text{p}}(k, \ell)$ whose generator matrix consists of the $k$ rows in the $\ell$-fold Kronecker matrix $\mathbf{G}^{(\ell)}$ specified by the indices in the free index set $\mathcal{I}_{\text{free}}$. Hence, we regard $\mathbf{G}^{(\ell)}$ as the *universal generator matrix* of a $(2^\ell, k)$ polar code. How to choose $k$ rows (i.e., $\mathcal{I}_{\text{free}}$) from $\mathbf{G}^{(\ell)}$ to generate a $(2^\ell, k)$ polar code that performs well over a channel will be presented in Sections 17.6 and 17.7. Note that polar codes simply form a subclass of Kronecker codes.

The circuit shown in Fig. 17.4 for the $\ell$-fold Kronecker mapping can be used for encoding any $(2^\ell, k)$ polar code $C_{\text{p}}(k, \ell)$ for any $k \leq 2^\ell$. In encoding a message $\mathbf{c} = (c_0, c_1, \ldots, c_{k-1})$, we apply the $k$ information bits in $\mathbf{c}$ to the inputs labeled by the indices in $\mathcal{I}_{\text{free}}$ and set the other $2^\ell - k$ inputs to zeros. This universal encoding feature is quite useful for an adaptive error control system for different rates of information transmission.

**Example 17.8** Set $\ell = 3$ and $k = 4$. Consider an $(8, 4)$ polar code $C_{\text{p}}(4, 3)$ with the 3-fold Kronecker matrix $\mathbf{G}^{(3)}$ given by (17.6) as its universal generator matrix. For convenience, we reproduce $\mathbf{G}^{(3)}$ as follows:

$$\mathbf{G}^{(3)} = \begin{bmatrix} \mathbf{g}_0 \\ \mathbf{g}_1 \\ \mathbf{g}_2 \\ \mathbf{g}_3 \\ \mathbf{g}_4 \\ \mathbf{g}_5 \\ \mathbf{g}_6 \\ \mathbf{g}_7 \end{bmatrix} = \begin{bmatrix} 1\,0\,0\,0\,0\,0\,0\,0 \\ 1\,1\,0\,0\,0\,0\,0\,0 \\ 1\,0\,1\,0\,0\,0\,0\,0 \\ 1\,1\,1\,1\,0\,0\,0\,0 \\ 1\,0\,0\,0\,1\,0\,0\,0 \\ 1\,1\,0\,0\,1\,1\,0\,0 \\ 1\,0\,1\,0\,1\,0\,1\,0 \\ 1\,1\,1\,1\,1\,1\,1\,1 \end{bmatrix}, \tag{17.54}$$

where $\mathbf{g}_0, \mathbf{g}_1, \ldots, \mathbf{g}_7$ denotes the eight rows of $\mathbf{G}^{(3)}$.

Let $\mathbf{c} = (c_0, c_1, c_2, c_3)$ be the message to be encoded and let $\mathbf{u} = \mu(\mathbf{c}) = (u_0, u_1, u_2, u_3, u_4, u_5, u_6, u_7)$ be the encoding vector associated with $\mathbf{c}$. Suppose the indices in $\mathbf{u}$ chosen to be frozen are 0, 1, 2, and 4 and the free indices are 3, 5, 6, and 7, i.e., $\mathcal{I}_{\text{frozen}} = \{0, 1, 2, 4\}$ and $\mathcal{I}_{\text{free}} = \{3, 5, 6, 7\}$. Then, the encoding vector for $\mathbf{c}$ is $\mathbf{u} = \mu(\mathbf{c}) = (0, 0, 0, c_0, 0, c_1, c_2, c_3)$, i.e., $u_0 = 0$, $u_1 = 0$, $u_2 = 0$, $u_3 = c_0$, $u_4 = 0$, $u_5 = c_1$, $u_6 = c_2$, and $u_7 = c_3$. In encoding, the rows $\mathbf{g}_3$, $\mathbf{g}_5$, $\mathbf{g}_6$, and $\mathbf{g}_7$ of $\mathbf{G}^{(3)}$ are selected and the rows $\mathbf{g}_0$, $\mathbf{g}_1$, $\mathbf{g}_2$, and $\mathbf{g}_4$ are omitted. Hence, the codeword for the message $\mathbf{c}$ is given below:

$$\mathbf{v} = \mu(\mathbf{c}) \cdot \mathbf{G}^{(3)}$$
$$= c_0 \mathbf{g}_3 + c_1 \mathbf{g}_5 + c_2 \mathbf{g}_6 + c_3 \mathbf{g}_7.$$

The mapping circuit shown in Fig. 17.3 is used for encoding and the encoding is performed in three levels. The eight bits in $\mathbf{u}$ are applied to the inputs of the circuit. The outputs of the third level of the circuit give the codeword $\mathbf{v}$ of the message $\mathbf{c}$ for transmission.

Suppose $c = (1\ 0\ 1\ 1)$ is the message to be encoded. Then, the encoding vector for $\mathbf{c}$ is $\mathbf{u} = \mu(\mathbf{c}) = (0\ 0\ 0\ 1\ 0\ 0\ 1\ 1)$. Applying $\mathbf{u} = (0\ 0\ 0\ 1\ 0\ 0\ 1\ 1)$ to the inputs of the circuit, the outputs of the first level of the circuit give the vector $(0\ 0\ 1\ 1\ 0\ 0\ 0\ 1)$, the outputs of the second level of the circuit give the vector $(1\ 1\ 1\ 1\ 0\ 1\ 0\ 1)$, and the outputs of the third level of the circuit give the polarized codeword $\mathbf{v} = (1\ 0\ 1\ 0\ 0\ 1\ 0\ 1)$ of the message $\mathbf{c}$.

If we apply $\mathbf{v} = (1\ 0\ 1\ 0\ 0\ 1\ 0\ 1)$ to the inputs of the circuit, the output of the circuit will be $\mathbf{u} = \mu(\mathbf{c}) = (0\ 0\ 0\ 1\ 0\ 0\ 1\ 1)$ (see Problem 17.2). From $\mathbf{u}$ and $\mathcal{I}_{\text{free}} = \{3, 5, 6, 7\}$, we can retrieve the message $\mathbf{c} = (1\ 0\ 1\ 1)$. Note that the $(8, 4)$ polar code $C_{\text{p}}(4, 3)$ given above is the first-order RM code $\text{RM}(1, 3)$ of length 8. ▲▲

Polar codes perform well with successive cancellation decoding [1] and its improved algorithms [2–7]. However, it is still unknown how to determine their minimum distances. A lower bound is given as follows. Let $r_{\min}$ be the smallest positive integer less than or equal to $\ell$, such that the generator matrix of the $r_{\min}$th-order RM code $C_{\text{RM}}(r_{\min}, \ell)$ contains the $k$ rows of the generator matrix of the $(2^{\ell}, k)$ polar code $C_{\text{p}}(k, \ell)$. Then, $C_{\text{p}}(k, \ell)$ is a subcode of the $r_{\min}$th-order RM code $C_{\text{RM}}(r_{\min}, \ell)$. Hence, the minimum distance of $C_{\text{p}}(k, \ell)$ is at least $2^{\ell - r_{\min}}$. In fact, the minimum distance of the code is the smallest weight of the rows in its generator matrix.

**Example 17.9** The four rows chosen from the universal generator matrix $\mathbf{G}^{(3)}$ for encoding the $(8, 4)$ polar code $C_{\text{p}}(4, 3)$ given in Example 17.8 are contained in the generator matrix of the first-order RM code $C_{\text{RM}}(1, 3)$. Hence, the minimum distance of $C_{\text{p}}(4, 3)$ is at least 4. Because the row $\mathbf{g}_3$ of weight 4 is a codeword in $C_{\text{p}}(4, 3)$, the minimum distance of $C_{\text{p}}(4, 3)$ is exactly 4. ▲▲

If we put RM codes of length $2^{\ell}$ in Kronecker product form, they can be encoded with a single universal encoder, i.e., the circuit for $\ell$-fold Kronecker mapping shown in Fig. 17.4.

# 17.5 Successive Information Retrieval from a Polarized Codeword

Because the mapping defined by the $\ell$-fold Kronecker matrix $\mathbf{G}^{(\ell)}$ is one-to-one and $\mathbf{G}^{(\ell)}$ is self-inverse, the message $\mathbf{c} = (c_0, c_1, \ldots, c_{k-1})$ contained in its corresponding polarized codeword $\mathbf{v} = (v_0, v_1, \ldots, v_{k-1}) = \mu(\mathbf{c}) \cdot \mathbf{G}^{(\ell)}$ can be uniquely retrieved from $\mathbf{v}$ in two steps opposite to its encoding operation. In the first step, we perform the inverse $\ell$-fold Kronecker mapping of $\mathbf{v}$ to retrieve the encoding vector $\mu(\mathbf{c})$ of $\mathbf{c}$ as follows:

$$\mu(\mathbf{c}) = (u_0, u_1, \ldots, u_{2^{\ell}-1}) = \mathbf{v} \cdot \mathbf{G}^{(\ell)}. \tag{17.55}$$

In the second step, we retrieve $\mathbf{c}$ from $\mu(\mathbf{c})$ by removing the bits at the frozen locations labeled by the indices in $\mathcal{I}_{\text{frozen}}$.

The two steps can be carried out at the same time to retrieve the information bits in the message $\mathbf{c} = (c_0, c_1, \ldots, c_{k-1})$ *successively one bit at a time*. The retrieval process starts from the rightmost information bit $c_{k-1}$ and then the rest of the information bits, $c_{k-2}, \ldots, c_1, c_0$, in this order, based on the retrieved information bits and the frozen bit positions in $\mathcal{I}_{\text{frozen}}$. We illustrate this successive information retrieval process with an example.

**Example 17.10** Consider the $(8, 4)$ polar code $C_\mathrm{p}(4, 3)$ given in Example 17.8 with frozen bit positions in $\mathcal{I}_\mathrm{frozen} = \{0, 1, 2, 4\}$ and free bit positions in $\mathcal{I}_\mathrm{free} = \{3, 5, 6, 7\}$. Let $\mathbf{v} = (v_0, v_1, v_2, v_3, v_4, v_5, v_6, v_7) = \mu(\mathbf{c}) \cdot \mathbf{G}^{(3)}$ be a codeword in $C_\mathrm{p}(4, 3)$ from which we intend to retrieve the four information bits in the message $\mathbf{c} = (c_0, c_1, c_2, c_3)$. First, we apply the inverse Kronecker mapping on $\mathbf{v}$ to obtain the encoding vector $\mu(\mathbf{c})$ of $\mathbf{c}$, which is given as follows:

$$\mu(\mathbf{c}) = \mathbf{u} = (u_0, u_1, u_2, u_3, u_4, u_5, u_6, u_7) = \mathbf{v} \cdot \mathbf{G}^{(3)}. \tag{17.56}$$

Multiplying out the above equation, we obtain the following eight equalities:

$$
\begin{aligned}
0 &= u_0 = v_0 + v_1 + v_2 + v_3 + v_4 + v_5 + v_6 + v_7, \\
0 &= u_1 = \quad\;\; v_1 \quad\;\; + v_3 \quad\;\; + v_5 \quad\;\; + v_7, \\
0 &= u_2 = \quad\;\; v_2 + v_3 \quad\;\; + v_6 + v_7, \\
c_0 &= u_3 = \quad\;\; v_3 \quad\;\; + v_7, \\
0 &= u_4 = \quad\;\; v_4 + v_5 + v_6 + v_7, \\
c_1 &= u_5 = \quad\;\; v_5 \quad\;\; + v_7, \\
c_2 &= u_6 = \quad\;\; v_6 + v_7, \\
c_3 &= u_7 = \quad\;\; v_7.
\end{aligned} \tag{17.57}
$$

Label the above eight equalities from 0 to 7. From the seventh equality of (17.57), we find $u_7 = v_7$. Because the index 7 is a free bit position, then we retrieve $c_3$ by setting $c_3 = u_7 = v_7$. Replacing $v_7$ by $c_3$ in the sixth equality of (17.57), we find $u_6 = v_6 + c_3$. Because the index 6 is a free bit position, then we retrieve $c_2$ by setting $c_2 = u_6 = c_3 + v_6$. Replacing $v_7$ with $c_3$ in the fifth equality of (17.57), we find $u_5 = v_5 + c_3$. Because the index 5 is a free bit position, then we retrieve $c_1$ by setting $c_1 = u_5 = v_5 + c_3$. Because the index 4 is a frozen bit position, we set $u_4 = 0$. Replacing $v_7$ with $c_3$ in the third equality of (17.57), we find $u_3 = v_3 + c_3$. Because the index 3 is a free bit position, then we retrieve $c_0$ by setting $c_0 = u_3 = v_3 + c_3$. Because the indices 0, 1, and 2 are frozen bit positions, we set $u_0 = u_1 = u_2 = 0$. Summarizing the above results, the four information bits in the message $\mathbf{c}$ corresponding to the codeword $\mathbf{v}$ in $C_\mathrm{p}(4, 3)$ are given below:

$$
\begin{aligned}
c_3 &= v_7, \\
c_2 &= c_3 + v_6, \\
c_1 &= c_3 + v_5, \\
c_0 &= c_3 + v_3.
\end{aligned} \tag{17.58}
$$

Suppose $\mathbf{v} = (v_0, v_1, v_2, v_3, v_4, v_5, v_6, v_7)$ is transmitted over a noiseless channel (transmit the code symbols from right to left). The four information bits in the message $\mathbf{c}$ contained in $\mathbf{v}$ can be retrieved successively as follows. When $v_7$ is received, we set $c_3 = v_7$. When $v_6$ is received, we compute $c_2$ by taking the modulo-2 sum of $c_3$ and $v_6$. When $v_5$ is received, we compute $c_1$ by taking the sum of $c_3$ and $v_5$. When $v_4$ is received, we set $u_4$ to zero. When $v_3$ is received, we compute $c_0$ by taking the sum of $c_3$ and $v_3$. Because the next three received bits are at the frozen positions, we set $u_2 = u_1 = u_0 = 0$.

With the above successive retrieving process, as soon as the transmitted codeword $\mathbf{v}$ is completely received, its corresponding message $\mathbf{c}$ is completely retrieved. ▲▲

**Example 17.11** Set $\ell = 4$ and $k = 8$. Consider a $(16, 8)$ polar code $C_\mathrm{p}(8, 4)$ with the following 4-fold Kronecker matrix $\mathbf{G}^{(4)}$ as its universal generator matrix:

$$
\mathbf{G}^{(4)} = \begin{bmatrix} \mathbf{g}_0 \\ \mathbf{g}_1 \\ \mathbf{g}_2 \\ \mathbf{g}_3 \\ \mathbf{g}_4 \\ \mathbf{g}_5 \\ \mathbf{g}_6 \\ \mathbf{g}_7 \\ \mathbf{g}_8 \\ \mathbf{g}_9 \\ \mathbf{g}_{10} \\ \mathbf{g}_{11} \\ \mathbf{g}_{12} \\ \mathbf{g}_{13} \\ \mathbf{g}_{14} \\ \mathbf{g}_{15} \end{bmatrix} = \begin{bmatrix} 1&0&0&0&0&0&0&0&0&0&0&0&0&0&0&0 \\ 1&1&0&0&0&0&0&0&0&0&0&0&0&0&0&0 \\ 1&0&1&0&0&0&0&0&0&0&0&0&0&0&0&0 \\ 1&1&1&1&0&0&0&0&0&0&0&0&0&0&0&0 \\ 1&0&0&0&1&0&0&0&0&0&0&0&0&0&0&0 \\ 1&1&0&0&1&1&0&0&0&0&0&0&0&0&0&0 \\ 1&0&1&0&1&0&1&0&0&0&0&0&0&0&0&0 \\ 1&1&1&1&1&1&1&1&0&0&0&0&0&0&0&0 \\ 1&0&0&0&0&0&0&0&1&0&0&0&0&0&0&0 \\ 1&1&0&0&0&0&0&0&1&1&0&0&0&0&0&0 \\ 1&0&1&0&0&0&0&0&1&0&1&0&0&0&0&0 \\ 1&1&1&1&0&0&0&0&1&1&1&1&0&0&0&0 \\ 1&0&0&0&1&0&0&0&1&0&0&0&1&0&0&0 \\ 1&1&0&0&1&1&0&0&1&1&0&0&1&1&0&0 \\ 1&0&1&0&1&0&1&0&1&0&1&0&1&0&1&0 \\ 1&1&1&1&1&1&1&1&1&1&1&1&1&1&1&1 \end{bmatrix}. \tag{17.59}
$$

Let $\mathbf{c} = (c_0, c_1, c_2, c_3, c_4, c_5, c_6, c_7)$ be a message of eight information bits to be encoded and let $\mathbf{u} = \mu(\mathbf{c}) = (u_0, u_1, \ldots, u_{14}, u_{15})$ be the encoding 16-tuple associated with $\mathbf{c}$. Suppose we choose the following two index sets $\mathcal{I}_\mathrm{frozen} = \{0, 1, 2, 3, 4, 5, 6, 8\}$ and $\mathcal{I}_\mathrm{free} = \{7, 9, 10, 11, 12, 13, 14, 15\}$. Then, the encoding vector for $\mathbf{c}$ is $\mathbf{u} = \mu(\mathbf{c}) = (0, 0, 0, 0, 0, 0, 0, c_0, 0, c_1, c_2, c_3, c_4, c_5, c_6, c_7)$, i.e., $u_0 = u_1 = u_2 = u_3 = u_4 = u_5 = u_6 = u_8 = 0$, $u_7 = c_0$, $u_9 = c_1$, $u_{10} = c_2$, $u_{11} = c_3$, $u_{12} = c_4$, $u_{13} = c_5$, $u_{14} = c_6$, and $u_{15} = c_7$. The polarization specified by the partition $\mathcal{I}_\mathrm{free}/\mathcal{I}_\mathrm{frozen}$ of the index set $\mathcal{I} = \{0, 1, \ldots, 15\}$ gives the $(16, 8)$ polar code $C_\mathrm{p}(8, 4)$.

In encoding, the eight rows $\mathbf{g}_7$, $\mathbf{g}_9$, $\mathbf{g}_{10}$, $\mathbf{g}_{11}$, $\mathbf{g}_{12}$, $\mathbf{g}_{13}$, $\mathbf{g}_{14}$, and $\mathbf{g}_{15}$ of $\mathbf{G}^{(4)}$ are used for encoding information bits and the other eight rows $\mathbf{g}_0$, $\mathbf{g}_1$, $\mathbf{g}_2$, $\mathbf{g}_3$, $\mathbf{g}_4$, $\mathbf{g}_5$, $\mathbf{g}_6$, and $\mathbf{g}_8$ are omitted. Hence, the codeword for the message $\mathbf{c}$ is given below:

$$
\begin{aligned}
\mathbf{v} &= (v_0, v_1, \ldots, v_{14}, v_{15}) \\
&= \mu(\mathbf{c}) \cdot \mathbf{G}^{(4)} \\
&= c_0 \mathbf{g}_7 + c_1 \mathbf{g}_9 + c_2 \mathbf{g}_{10} + c_3 \mathbf{g}_{11} + c_4 \mathbf{g}_{12} + c_5 \mathbf{g}_{13} + c_6 \mathbf{g}_{14} + c_7 \mathbf{g}_{15}.
\end{aligned}
$$

Encoding of $\mu(\mathbf{c})$ (or $\mathbf{c}$) can be achieved in four levels using a 4-level network of 32 $(4 \times 2^3)$ XOR gates.

Suppose $\mathbf{v}$ is transmitted over a noiseless channel. The eight information bits in the message $\mathbf{c}$ contained in $\mathbf{v}$ can be retrieved successively in the following order:

$$
\begin{aligned}
c_7 &= v_{15}, \\
c_6 &= c_7 + v_{14}, \\
c_5 &= c_7 + v_{13}, \\
c_4 &= c_7 + v_{12} + v_{13} + v_{14}, \\
c_3 &= c_7 + v_{11}, \\
c_2 &= c_7 + v_{10} + v_{11} + v_{14}, \\
c_1 &= c_7 + v_9 + v_{11} + v_{13}, \\
c_0 &= c_7 + v_7.
\end{aligned}
$$

From the above order of information retrieval, we see that all eight information bits in the message $\mathbf{c}$ are retrieved as soon as the entire polarized codeword $\mathbf{v}$ is completely received.

Note that the $(16, 8)$ polar code $C_{\mathrm{p}}(8, 4)$ constructed based on the polarized partition $\mathcal{I}_{\mathrm{free}}/\mathcal{I}_{\mathrm{frozen}}$ of the index set $\mathcal{I}$ given above is a subcode of the second-order $\mathrm{RM}(2, 4)$ code, a $(16, 13)$ code with minimum distance 4. Because the generator matrix of $C_{\mathrm{p}}(8, 4)$ contains three rows of weight 4, the minimum distance of $C_{\mathrm{p}}(8, 4)$ is exactly 4. ▲▲

The process of retrieving information from a polarized codeword illustrated in Examples 17.10 and 17.11 applies to any polar code. The information bits in the message $\mathbf{c} = (c_0, c_1, \ldots, c_{k-1})$ contained in a polarized codeword $\mathbf{v}$ of a polar code are retrieved *successively one bit at a time*. The retrieval process starts from the rightmost information bit $c_{k-1}$ and then the rest of the information bits, $c_{k-2}, \ldots, c_1, c_0$, in this order. Each information bit is retrieved based on the information bits that have already been successively retrieved and the information of frozen locations. Over a noiseless channel, all the information bits will be completely retrieved as soon as the transmitted polarized codeword is completely received by the channel decoder.

The above information retrieval process is referred to as a *successive bit-level information retrieval* (SBLIR) *process*. This SBLIR process is the basis for the successive cancellation decoding of a polar code in which reliability information of the received bits in a received vector is used to decode each bit in the encoding vector $\mu(\mathbf{c})$ of a message $\mathbf{c}$ and retrieve the information bits in $\mathbf{c}$ one bit at a time.

## 17.6   Channel Polarization

As defined in Section 17.4, a $(2^\ell, k)$ polar code $C_{\mathrm{p}}(k, \ell)$ is a $k$-dimensional Kronecker code of length $2^\ell$ characterized by a polarized partition of the $2^\ell$ bit positions, labeled by $\mathcal{I} = \{0, 1, \ldots, 2^\ell - 2, 2^\ell - 1\}$, of an encoding vector $\mathbf{u} = (u_0, u_1, \ldots, u_{2^\ell-1})$ into two disjoint subsets $\mathcal{I}_{\mathrm{free}}$ and $\mathcal{I}_{\mathrm{frozen}}$ of sizes $k$ and $2^\ell - k$, respectively. The $k$ information bits of a message $\mathbf{c} = (c_0, c_1, \ldots, c_{k-1})$ are assigned to the $k$ free positions in $\mathbf{u}$ with labels in $\mathcal{I}_{\mathrm{free}}$ and zeros (or ones) are assigned to the $2^\ell - k$ frozen positions in $\mathbf{u}$ with labels in $\mathcal{I}_{\mathrm{frozen}}$.

The performance of a polar code very much depends on the polarized partition $\mathcal{I}_{\mathrm{free}}/\mathcal{I}_{\mathrm{frozen}}$ of the label set, $\mathcal{I} = \{0, 1, \ldots, 2^\ell - 2, 2^\ell - 1\}$, of the bit positions of the encoding vector $\mathbf{u}$. The partition should be done in such a way that, after Kronecker encoding, the code bits at the free bit positions can be reliably recovered (or error free). The location partition is based on a technique known as *channel polarization* which was discovered by Arikan in 2009 [1].

Before presenting the technique of channel polarization, we need to introduce some basic concepts and essential elements in information theory that are relevant in the development of channel polarization. For more on information theory, readers are referred to [8].

### 17.6.1 Some Elements of Information Theory

Let $X$ be a discrete random variable with alphabet $\mathcal{X}$ and probability mass function $p_X(x) = P(X = x)$, $x \in \mathcal{X}$, denoted by $X \sim p_X(x)$. The entropy of $X$, denoted by $H(X)$, is defined as

$$H(X) \triangleq \sum_{x \in \mathcal{X}} p_X(x) \log_2 \left( \frac{1}{p_X(x)} \right), \tag{17.60}$$

which represents the amount of *uncertainty* in the outcome of a realization of the random variable $X$. The entropy $H(X)$ of a random variable $X$ is nonnegative, i.e., $H(X) \geq 0$.

The joint entropy $H(X, Y)$ of a pair of discrete random variables $(X, Y)$ with a joint probability distribution $p_{X,Y}(x, y)$, $x \in \mathcal{X}$ and $y \in \mathcal{Y}$, is defined as

$$H(X, Y) \triangleq \sum_{(x,y) \in \mathcal{X} \times \mathcal{Y}} p_{X,Y}(x, y) \log_2 \left( \frac{1}{p_{X,Y}(x, y)} \right), \tag{17.61}$$

and the conditional entropy $H(X|Y)$ is defined as

$$H(X|Y) \triangleq \sum_{(x,y) \in \mathcal{X} \times \mathcal{Y}} p_{X,Y}(x, y) \log_2 \left( \frac{1}{p_{X|Y}(x|y)} \right), \tag{17.62}$$

which is the amount of uncertainty *remaining* in $X$ after observing the realization of the random variable $Y$.

Following the chain rule of joint entropy developed in information theory [8], the joint entropy $H(X, Y)$ can be expressed as the following sums:

$$\begin{aligned} H(X, Y) &= H(X) + H(Y|X) \\ &= H(Y) + H(X|Y). \end{aligned} \tag{17.63}$$

The mutual information between two random variables, $X$ and $Y$, is defined as follows:

$$\begin{aligned} I(X; Y) &\triangleq \sum_{(x,y) \in \mathcal{X} \times \mathcal{Y}} p_{X,Y}(x, y) \log_2 \left( \frac{p_{X,Y}(x, y)}{p_X(x) p_Y(y)} \right) \\ &= H(X) - H(X|Y), \end{aligned} \tag{17.64}$$

which is the *reduction* in the uncertainty of $X$ with the knowledge of $Y$ and is nonnegative, i.e., $I(X; Y) \geq 0$. Because $I(X; Y)$ is mutual information, it is clear that

$$I(X; Y) = I(Y; X) = H(Y) - H(Y|X). \tag{17.65}$$

It follows from (17.63), (17.64), and (17.65) that we have

$$I(X; Y) = H(X) + H(Y) - H(X, Y). \tag{17.66}$$

The conditional mutual information of two random variables $X$ and $Y$ given the third random variable $Z$ with alphabet $\mathcal{Z}$ is defined as

$$
\begin{aligned}
I(X;Y|Z) &= H(X|Z) - H(X|Y,Z) \\
&= H(X|Z) + H(Y|Z) - H(X,Y|Z).
\end{aligned}
\tag{17.67}
$$

The mutual information follows the chain rule, i.e.,

$$
I(X_0, X_1, \ldots, X_{n-1}; Y) = \sum_{i=0}^{n-1} I(X_i; Y|X_{i-1}, X_{i-2}, \ldots, X_0).
\tag{17.68}
$$

Consider a DMC $W$ with input $X$ and output $Y$, denoted by $W : X \to Y$, with transition probability $W(Y|X)$ (i.e., $p_{Y|X}(y|x)$ with $x \in \mathcal{X}$ and $y \in \mathcal{Y}$) (see Fig. 1.4 for $\mathcal{X} = \{0, 1\}$). For such a channel, at each instance of time, it accepts an input $x \in \mathcal{X}$ and outputs $y \in \mathcal{Y}$ with probability $p_{Y|X}(y|x)$. The capacity of $W$, called *channel capacity*, is defined as follows:

$$
C = I(W) \triangleq \max_{p_X(x)} I(X;Y).
\tag{17.69}
$$

As stated in Chapter 1, Shannon in his landmark paper [9] published in 1948 proved that, over a DMC with capacity $C$, if the rate $R$ of information transmission is less that $C$, i.e., $R < C$, there exist codes with rate $R$ which can achieve arbitrarily small decoding error probability.

For a binary symmetric channel (BSC) $W_{\mathrm{BSC}}$ with transition probability $p$ as shown in Fig. 1.4 and the input $X$ following a Bernoulli distribution with probability $1/2$, its channel capacity is given by

$$
I(W_{\mathrm{BSC}}) = 1 - H(p)
\tag{17.70}
$$

where

$$
H(p) = -p \log_2(p) - (1-p) \log_2(1-p).
\tag{17.71}
$$

For a binary erasure channel (BEC) $W_{\mathrm{BEC}}$ with erasure probability $\epsilon$ as shown in Fig. 1.7(a) and the input $X$ following a Bernoulli distribution with probability $1/2$, its channel capacity is given by

$$
I(W_{\mathrm{BEC}}) = 1 - \epsilon.
\tag{17.72}
$$

A BI-DMC $W$ is said to be *perfect* (or *noiseless*) if its capacity is 1, i.e., $I(W) = 1$. Over a perfect channel, information transmitted is error free and hence no error-control coding is needed. A BI-DMC with channel capacity $I(W) = 0$ is a *useless* channel. Over such a channel, reliable information cannot be achieved with or without coding. The channel capacity $I(W)$ of an ordinary (physical) BI-DMC is in the range of $(0, 1)$, i.e., $0 < I(W) < 1$.

## 17.6.2  Polarization Process

We now present the technique of channel polarization for BI-DMCs only. Channel polarization, introduced by Arikan [1, 10], is a channel manipulation technique to transform a group of identical and independent BI-DMCs, $W$s, with the same input alphabet $\mathcal{X} = \{0, 1\}$, the same output alphabet $\mathcal{Y}$, and the same transition probability distribution $W(Y|X)$ (i.e., copies of $W$) into *two extreme subgroups*, one containing good (perfect or nearly perfect) channels and the other containing bad (useless or nearly useless) channels. Then, we can use the good channels for information transmission. This technique of polarizing a group of identical and independent channels into two groups with opposite channel capacities, one close or equal to 1 and the other close or equal to 0, is referred to as *channel polarization* (or polarization of a BI-DMC $W$). We call the polarized channels *bit-coordinate channels*. The transformation used for channel polarization is referred to as *polarization transformation*. The Kronecker mapping defined in Section 17.2 is a special type of polarization transformation used by Arikan and we will use it in the rest of this chapter. Polarization transformations based on kernels other than the kernel given by (17.1) can be found in [11–21].

In the following, we present the polarization process of a BI-DMC $W$ without any proof. For the simplicity of illustration, polarization of a BEC($\epsilon$) with erasure probability $\epsilon$ is used in three examples by taking two, four, and eight copies of BEC($\epsilon$).

Consider the two copies of a BI-DMC $W$ as shown in Fig. 17.5(a), $W : X \to Y$ with transition probability $W(Y|X)$, one with input $X_0$ and output $Y_0$ and the other one with input $X_1$ and output $Y_1$. The total channel capacity of the two BI-DMCs shown in Fig. 17.5(a) is

$$2I(W) = 2I(X_0; Y_0) = 2I(X_1; Y_1). \tag{17.73}$$

Suppose we combine the two BI-DMCs using the $2 \times 2$ Kronecker kernel $\mathbf{G}^{(1)}$ given by (17.1) as shown in Fig. 17.5(b), where $U_0$ and $U_1$ are two independently and identically distributed random variables which are uniform in $\{0, 1\}$. Then, $X_0 = U_0 + U_1$ and $X_1 = U_1$. The combined channel, called a *1-fold vector channel* and denoted by $W^2$, has the 2-tuple $(U_0, U_1)$ as the input and the 2-tuple $(Y_0, Y_1)$ as the output with channel capacity

$$I(U_0, U_1; Y_0, Y_1) = 2I(W). \tag{17.74}$$

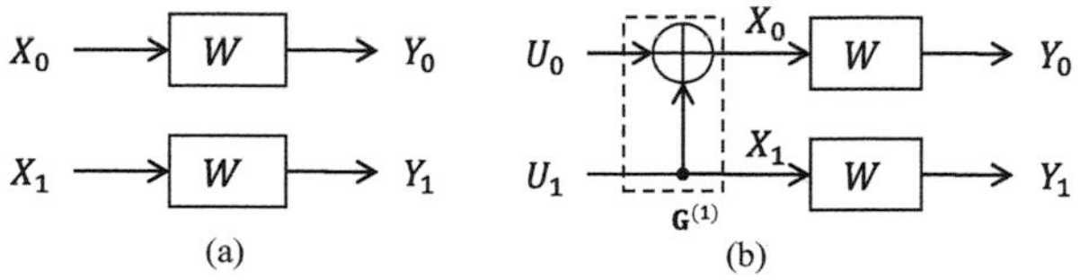

Figure 17.5 (a) Two identical and independent channels $W$ and (b) a combined 1-fold vector channel $W^2$.

The BI-DMC $W$ is called the *base channel* for constructing the 1-fold vector channel $W^2$. The transition probability of the vector channel $W^2$ is

$$W^2(Y_0, Y_1 | U_0, U_1) = W(Y_0 | U_0 + U_1) W(Y_1 | U_1). \tag{17.75}$$

Because the Kronecker mapping between $(U_0, U_1)$ and $(X_0, X_1)$ is one-to-one, we have

$$
\begin{aligned}
I(U_0, U_1; Y_0, Y_1) &= I(X_0, X_1; Y_0, Y_1) \\
&= I(X_0; Y_0) + I(X_1; Y_1) \\
&= 2I(W).
\end{aligned} \tag{17.76}
$$

Following the chain rule of mutual information, the left-hand side of the above equation can be written as

$$
\begin{aligned}
I(U_0, U_1; Y_0, Y_1) &= I(U_0; Y_0, Y_1) + I(U_1; Y_0, Y_1 | U_0) \\
&= I(U_0; Y_0, Y_1) + I(U_1; Y_0, Y_1, U_0).
\end{aligned} \tag{17.77}
$$

Equation (17.77) implies that we can achieve the channel capacity of the vector channel $W^2$ in two stages (or steps).[2] In the first stage, we transmit the information through a *virtual channel* $W^- : U_0 \to (Y_0, Y_1)$, which has capacity $I(U_0; Y_0, Y_1)$. If $U_0$ is known at the first stage, we can decode the information transmitted through the *second virtual channel* $W^+ : U_1 \to (Y_0, Y_1, U_0)$, which has capacity $I(U_1; Y_0, Y_1, U_0)$. From this point of view, the combined vector channel $W^2$ can be split into *two virtual channels*:

$$
\begin{aligned}
W^- &: U_0 \to (Y_0, Y_1), \\
W^+ &: U_1 \to (Y_0, Y_1, U_0),
\end{aligned} \tag{17.78}
$$

which are called *bit-coordinate channels*. The transition probabilities of these two bit-coordinate channels are

$$
W^-(Y_0, Y_1 | U_0) = \frac{1}{2} \sum_{u_1 \in \{0,1\}} W(y_0 | u_0 + u_1) W(y_1 | u_1), \tag{17.79}
$$

$$
W^+(Y_0, Y_1, U_0 | U_1) = \frac{1}{2} W(y_0 | u_0 + u_1) W(y_1 | u_1).
$$

These two bit-coordinate channels $W^+$ and $W^-$ have the following two properties (see Problem 17.4):

$$
\begin{aligned}
I(W^+) + I(W^-) &= 2I(W), \\
I(W^-) \le I(W) &\le I(W^+).
\end{aligned} \tag{17.80}
$$

---

[2] We have $I(U_1; Y_0, Y_1 | U_0) = I(U_1; Y_0, Y_1, U_0)$ because $U_0$ and $U_1$ are assumed to be independently and identically distributed random variables.

The first equality in (17.80) shows that the channel capacity is conserved. The second inequality shows that the 1-fold Kronecker mapping from $(U_0, U_1)$ to $(X_0, X_1)$ splits the vector channel $W^2$ into two bit-coordinate channels, $W^+$ and $W^-$, in which the bit-coordinate channel, $W^+$, has a larger channel capacity than the bit-coordinate channel, $W^-$.

The above *combine-and-split* process polarizes two identical and independent BI-DMCs into two channels in opposite directions in terms of their channel capacities and hence it is referred to as *channel polarization*, which is the key to construction of polar codes.

At this point, we clarify the signs "+" and "−" used in $W^+$ and $W^-$. These two signs stand for *increment* and *decrement* of channel capacities of the polarized bit-coordinate channels $W^+$ and $W^-$, respectively.

**Example 17.12** In this example, we use a BEC to illustrate the combine-and-split polarization process and compute the capacities and erasure probabilities of the polarized BEC bit-coordinate channels.

Consider a BEC($\epsilon$) $W$ with erasure probability $\epsilon$ (see also Fig. 1.7(a)). The channel capacity of $W$ is $I(W) = 1 - \epsilon$ as given by (17.72). Take two copies of $W$ and combine them into a vector channel $W^2$ as shown in Fig. 17.5(b). The total capacity and erasure probability of $W^2$ are $2(1 - \epsilon)$ and $2\epsilon$, respectively. Polarizing the vector channel $W^2$ with the 1-fold Kronecker mapping matrix $\mathbf{G}^{(1)}$, we obtain two bit-coordinate BECs, denoted by $W^- : U_0 \to (Y_0, Y_1)$ and $W^+ : U_1 \to (Y_0, Y_1, U_0)$, respectively. From Fig. 17.5(b), we see that the input $U_0$ of the bit-coordinate channel $W^-$ can be recovered only if both $Y_0$ and $Y_1$ are not erased. The probability that both $Y_0$ and $Y_1$ are not erased is $(1 - \epsilon)^2$. Hence, $W^-$ is a BEC with erasure probability

$$\begin{aligned} \epsilon_{W^-} &= 1 - (1 - \epsilon)^2 = 2\epsilon - \epsilon^2 \\ &\triangleq f^-(\epsilon), \end{aligned} \qquad (17.81)$$

where the function $f^-(\epsilon) = 2\epsilon - \epsilon^2$ is called a *negatively polarized erasure function*.

Because the total erasure probability of the vector channel $W^2$ is $2\epsilon$, the erasure probability of the bit-coordinate channel $W^+$ is

$$\begin{aligned} \epsilon_{W^+} &= 2\epsilon - \epsilon_{W^-} = \epsilon^2 \\ &\triangleq f^+(\epsilon), \end{aligned} \qquad (17.82)$$

where the function $f^+(\epsilon) = \epsilon^2$ is called a *positively polarized erasure function*. In this case, the input $U_1$ of the bit-coordinate channel $W^+$ can be recovered as long as one of two outputs $Y_0$ and $Y_1$ (with $U_0$ predetermined) is not erased.

From (17.81) and (17.82), we find that the channel capacities of the two bit-coordinate channels $W^-$ and $W^+$ are

$$\begin{aligned} I(W^-) &= 1 - \epsilon_{W^-} = 1 - (2\epsilon - \epsilon^2) = 1 - 2\epsilon + \epsilon^2, \\ I(W^+) &= 1 - \epsilon_{W^+} = 1 - \epsilon^2. \end{aligned} \qquad (17.83)$$

The difference of the capacities of the two bit-coordinate channels $W^+$ and $W^-$ is

$$I(W^+) - I(W^-) = 1 - \epsilon^2 - (1 - 2\epsilon + \epsilon^2) = 2\epsilon(1 - \epsilon). \qquad (17.84)$$

Because $0 \le \epsilon \le 1$, the difference $I(W^+) - I(W^-)$ is nonnegative, i.e., $I(W^+) - I(W^-) \ge 0$. The difference $I(W^+) - I(W^-)$ is zero *if and only if* either $\epsilon = 1$ or $\epsilon = 0$, i.e., the base channel $W$ is either perfect or useless. In this case, polarization is not needed. However, an ordinary (physical) BEC is neither perfect nor completely useless. Hence, $I(W^+) - I(W^-) > 0$ and the bit-coordinate channel $W^+$ has a larger capacity than the bit-coordinate channel $W^-$.

Assume the base channel BEC($\epsilon$) $W$ chosen for polarization has erasure probability $\epsilon = 0.5$ and $I(W) = 0.5$. From (17.83), we find $I(W^-) = 0.25$ and $I(W^+) = 0.75$. The capacity $I(W^+)$ of the bit-coordinate channel $W^+$ is *three times larger* than that of the bit-coordinate channel $W^-$. This shows the polarization effect of the combine-and-split process presented above. ▲▲

Next, we consider the bit-coordinate channel $W^+$ as the base channel for polarization. If we apply the above channel combine-and-split operation to a vector channel $(W^+)^2$ formed by two copies of the base channel $W^+$, we obtain another two bit-coordinate channels, denoted by $W^{++}$ and $W^{+-}$, as shown in Fig. 17.6(a). Following the same analysis and argument given above, we have the following properties for these two bit-coordinate channels:

$$I(W^{++}) + I(W^{+-}) = 2I(W^+),$$
$$I(W^{+-}) \le I(W^+) \le I(W^{++}). \qquad (17.85)$$

Similarly, if we apply the above channel combine-and-split operation to the vector channel $(W^-)^2$ formed by two copies of the base channel $W^-$, we obtain another two polarized bit-coordinate channels, denoted by $W^{-+}$ and $W^{--}$, as shown in Fig. 17.6(b). Following the same analysis and reasoning, we have the following properties for these two polarized bit-coordinate channels:

$$I(W^{-+}) + I(W^{--}) = 2I(W^-),$$
$$I(W^{--}) \le I(W^-) \le I(W^{-+}). \qquad (17.86)$$

The four polarized bit-coordinate channels, $W^{--}$, $W^{-+}$, $W^{+-}$, and $W^{++}$, are obtained by combining four copies of the base channel $W$ into a vector channel $W^4$, and then polarizing and splitting the vector channel $W^4$. Polarization is performed in two levels, as shown in Fig. 17.7(a). In the first level, we divide the four copies of $W$ into two pairs. Then, we combine the two copies of $W$ in each pair and split them in two bit-coordinate channels, $W^-$ and $W^+$. The first-level combine-and-split results in two copies of the bit-coordinate channel $W^-$ and two copies of the bit-coordinate channel $W^+$. In the second level, we perform combine-and-split operation to $W^-$ and $W^+$, respectively, as described above, to produce the four bit-coordinate channels, $W^{--}$, $W^{-+}$ and $W^{+-}$,

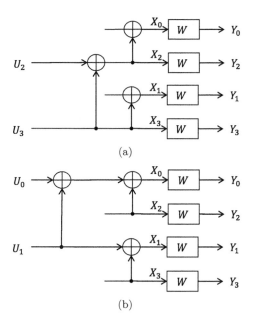

Figure 17.6 (a) A combined vector channel with $W^+$ as the base channel and (b) a combined vector channel with $W^-$ as the base channel.

$W^{++}$, respectively. We call the two copies of $W^-$ and two copies of $W^+$ the *first-level bit-coordinate channels* and $W^{--}, W^{-+}, W^{+-}$, and $W^{++}$ the *second-level bit-coordinate channels*.

The four bit-coordinate channels are:

$$
\begin{aligned}
W^{--} &: U_0 \to ((Y_0, Y_2), (Y_1, Y_3)), \\
W^{-+} &: U_1 \to ((Y_0, Y_2), (Y_1, Y_3), U_0), \\
W^{+-} &: U_2 \to ((Y_0, Y_2, U_0 + U_1), (Y_1, Y_3, U_1)), \\
W^{++} &: U_3 \to ((Y_0, Y_2, U_0 + U_1), (Y_1, Y_3, U_1), U_2).
\end{aligned}
\tag{17.87}
$$

The transition probability of the vector channel $W^4$ is:

$$
\begin{aligned}
W^4(Y_0, Y_1, Y_2, Y_3 | U_0, U_1, U_2, U_3) &= W^2(Y_0, Y_2 | U_0, U_2) W^2(Y_1, Y_3 | U_1, U_3) \\
&= W(Y_0 | U_0 + U_1 + U_2 + U_3) W(Y_1 | U_1 + U_3) W(Y_2 | U_2 + U_3) W(Y_3 | U_3).
\end{aligned}
\tag{17.88}
$$

To simplify expressions to be developed in the rest of this section, we introduce some notation as follows. Let $\mathbf{Q}^n$ denote a set with $n+1$ components, $Q_0, Q_1, \ldots, Q_n$, i.e., $\mathbf{Q}^n = (Q_0, Q_1, \ldots, Q_n)$. For $0 \le k \le n$, let $\mathbf{Q}^k$ denote a subset of $\mathbf{Q}^n$ which contains the first $(k+1)$ components of $\mathbf{Q}^n$, i.e., $\mathbf{Q}^k = (Q_0, Q_1, \ldots, Q_k)$. Let $Q^{-1}$ denotes an empty set, i.e., $\mathbf{Q}^{-1} = \emptyset$. Let $\mathbf{Q}^n_e$ denote

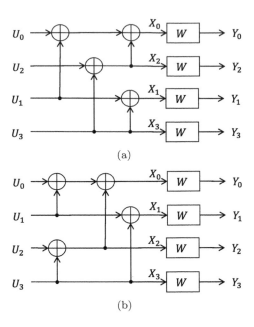

Figure 17.7 (a) A combined vector channel $W^4$ and (b) the combined vector channel $W^4$ after rewire.

the subset of $\mathbf{Q}^n$ with components of even indices, i.e., $Q_j$, where $0 \leq j \leq n$ is an even integer. Let $\mathbf{Q}^n_o$ denote the subset of $\mathbf{Q}^n$ with components of odd indices, i.e., $Q_i$, where $0 \leq i \leq n$ is an odd integer. With this notation, the four bit-coordinate channels given by (17.87) can be simplified as follows:

$$
\begin{aligned}
W^{--} &: U_0 \to (\mathbf{Y}^3, \mathbf{U}^{-1}), \\
W^{-+} &: U_1 \to (\mathbf{Y}^3, \mathbf{U}^0), \\
W^{+-} &: U_2 \to (\mathbf{Y}^3, \mathbf{U}^1), \\
W^{++} &: U_3 \to (\mathbf{Y}^3, \mathbf{U}^2),
\end{aligned}
\tag{17.89}
$$

where $\mathbf{Y}^3 = (Y_0, Y_1, Y_2, Y_3)$, $\mathbf{U}^0 = \emptyset$, $\mathbf{U}^0 = (U_0)$, $\mathbf{U}^1 = (U_0, U_1)$, and $\mathbf{U}^2 = (U_0, U_1, U_2)$. The transition probability of the vector channel $W^4$ is simplified as follows:

$$
W^4(\mathbf{Y}^3 | \mathbf{U}^3) = W^2(\mathbf{Y}^3_e | \mathbf{U}^3_e) W^2(\mathbf{Y}^3_o | \mathbf{U}^3_o).
\tag{17.90}
$$

Figure 17.7(b) shows the combined vector channel $W^4$ after rewire. From this figure, we can see that the channel combination is achieved by the 2-fold Kronecker mapping based on the Kronecker matrix $\mathbf{G}^{(2)}$ in two levels.

From (17.80), (17.85), and (17.86), we have

$$
I(W^{--}) \leq I(W^-) \leq I(W) \leq I(W^+) \leq I(W^{++}).
\tag{17.91}
$$

From the above inequality, we see that, after the second-level combine-and-split polarization of four BI-DMCs, the worse bit-coordinate channel $(W^-)$ obtained

at the first level is split into an even worse bit-coordinate channel $W^{--}$, and
the better channel $(W^+)$ obtained at the first level is split into an even better
bit-coordinate channel $W^{++}$, i.e., the Kronecker mapping in the second level
further polarizes the capacities of bit-coordinate channels.

The above polarization of a vector channel formed by four copies of a base
BI-DMC $W$ using the 2-fold Kronecker mapping matrix $\mathbf{G}^{(2)}$ is called the 2-*fold
channel polarization*. The polarization process is carried out in two levels.

**Example 17.13** Continue Example 17.12. Two bit-coordinate channels, $W^+$
and $W^-$, are obtained after combining and splitting two BECs with erasure
probability $\epsilon$, where $W^+$ is a BEC with erasure probability $\epsilon_{W^+} = \epsilon^2$ and $W^-$
is a BEC with erasure probability $\epsilon_{W^-} = 2\epsilon - \epsilon^2$. Following the same calculation
with $\epsilon$ replaced by $\epsilon_{W^+}$ and $\epsilon_{W^-}$ in (17.82) and (17.81), we compute the erasure
probabilities of the four bit-coordinate channels $W^{--}$, $W^{-+}$, $W^{+-}$, and $W^{++}$
specified by (17.87) or (17.89) as follows:

$$
\begin{aligned}
\epsilon_{W^{--}} &= f^-(\epsilon_{W^-}) = f^-(f^-(\epsilon)) \\
&= 2(2\epsilon - \epsilon^2) - (2\epsilon - \epsilon^2)^2 = 4\epsilon - 6\epsilon^2 + 4\epsilon^3 - \epsilon^4 \\
\epsilon_{W^{-+}} &= f^+(\epsilon_{W^-}) = f^+(f^-(\epsilon)) \\
&= (2\epsilon - \epsilon^2)^2 = 4\epsilon^2 - 4\epsilon^3 + \epsilon^4 \\
\epsilon_{W^{+-}} &= f^-(\epsilon_{W^+}) = f^-(f^+(\epsilon)) \\
&= 2\epsilon^2 - (\epsilon^2)^2 = 2\epsilon^2 - \epsilon^4 \\
\epsilon_{W^{++}} &= f^+(\epsilon_{W^+}) = f^+(f^+(\epsilon)) \\
&= (\epsilon^2)^2 = \epsilon^4.
\end{aligned}
\tag{17.92}
$$

The sum of the erasure probabilities of the four polarized bit-coordinate
channels is $4\epsilon$, which is equal to the total erasure probability of the 2-fold vector
channel $W^4$.

From the erasure probabilities given by (17.92), we find the capacities of the
four polarized second-level bit-coordinate channels as follows:

$$
\begin{aligned}
I(W^{--}) &= 1 - \epsilon_{W^{--}} = 1 - 4\epsilon + 6\epsilon^2 - 4\epsilon^3 + \epsilon^4, \\
I(W^{-+}) &= 1 - \epsilon_{W^{-+}} = 1 - 4\epsilon^2 + 4\epsilon^3 - \epsilon^4, \\
I(W^{+-}) &= 1 - \epsilon_{W^{+-}} = 1 - 2\epsilon^2 + \epsilon^4, \\
I(W^{++}) &= 1 - \epsilon_{W^{++}} = 1 - \epsilon^4.
\end{aligned}
\tag{17.93}
$$

The sum of the channel capacities of the four polarized bit-coordinate channels is $4I(W) = 4(1 - \epsilon)$. We can also verify that for $0 < \epsilon < 1$, the following
strict inequalities hold

$$
I(W^{--}) < I(W^{-+}) < I(W^{+-}) < I(W^{++}).
$$

Assume the channel erasure probability of the BEC $W$ is $\epsilon = 0.5$. The
channel capacities of the four polarized second-level bit-coordinate channels
are calculated as: $I(W^{--}) = 0.0625$, $I(W^{-+}) = 0.4375$, $I(W^{+-}) = 0.5625$,

$I(W^{++}) = 0.9375$. We see that doubling the number of copies of a base channel and adding another level of polarization, the capacities of good and bad bit-coordinate channel are polarized further apart. For example, the bit-coordinate channel $W^{--}$ has capacity 0.0625, close to 0, while the bit-coordinate channel $W^{++}$ has capacity 0.9375, close to 1. ▲▲

Suppose we combine eight copies of a base BI-DMC $W$ into a vector channel $W^8$ and polarize it in three levels by using the 3-fold Kronecker mapping matrix $\mathbf{G}^{(3)}$. In the first level of polarization, we group the eight copies of base channel $W$ in $W^8$ into four pairs of $W$s. Then, we polarize each pair of identical base channels into two first-level bit-coordinate channels, $W^-$ and $W^+$. Each of the first-level bit-coordinate channels, $W^-$ and $W^+$, appears four times. Next, we group the eight first-level bit-coordinate channels into two groups, $\{W^-, W^-, W^-, W^-\}$ and $\{W^+, W^+, W^+, W^+\}$. Then, we use $W^-$ and $W^+$ as base channels, respectively, and perform the combine-and-split operation in two levels to each group as described above. This results in eight 3-level bit-coordinate channels as given by

$$
\begin{aligned}
W^{---} &: U_0 \to (\mathbf{Y}^7, \mathbf{U}^{-1}) \\
W^{--+} &: U_1 \to (\mathbf{Y}^7, \mathbf{U}^0) \\
W^{-+-} &: U_2 \to (\mathbf{Y}^7, \mathbf{U}^1) \\
W^{-++} &: U_3 \to (\mathbf{Y}^7, \mathbf{U}^2) \\
W^{+--} &: U_4 \to (\mathbf{Y}^7, \mathbf{U}^3) \\
W^{+-+} &: U_5 \to (\mathbf{Y}^7, \mathbf{U}^4) \\
W^{++-} &: U_6 \to (\mathbf{Y}^7, \mathbf{U}^5) \\
W^{+++} &: U_7 \to (\mathbf{Y}^7, \mathbf{U}^6),
\end{aligned}
\tag{17.94}
$$

which are shown in Fig. 17.8 (with rewiring). The transition probability of the vector channel $W^8$ is

$$
W^8(\mathbf{Y}^7|\mathbf{U}^7) = W^4(\mathbf{Y}_e^7|\mathbf{U}_e^7)W^4(\mathbf{Y}_o^7|\mathbf{U}_o^7). \tag{17.95}
$$

From Fig. 17.8, we can see that the combine-and-split operation of the vector channel $W^8$ formed by eight copies of the BI-DMC $W$ is achieved by the 3-fold Kronecker mapping with the Kronecker matrix $\mathbf{G}^{(3)}$. The capacities of the eight bit-coordinate channels have the following properties:

$$
\begin{aligned}
&I(W^{---}) + I(W^{--+}) + I(W^{-+-}) + I(W^{-++}) \\
&+ I(W^{+--}) + I(W^{+-+}) + I(W^{++-}) + I(W^{+++}) = 8I(W),
\end{aligned}
\tag{17.96}
$$

and

$$
\begin{aligned}
I(W^{---}) &\leq I(W^{--}) \leq I(W^{--+}), \\
I(W^{-+-}) &\leq I(W^{-+}) \leq I(W^{-++}), \\
I(W^{+--}) &\leq I(W^{+-}) \leq I(W^{+-+}), \\
I(W^{++-}) &\leq I(W^{++}) \leq I(W^{+++}).
\end{aligned}
\tag{17.97}
$$

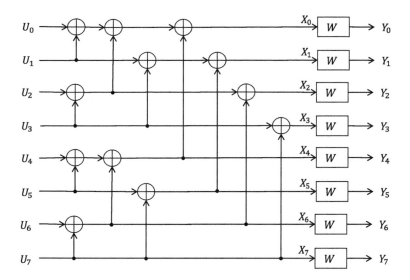

Figure 17.8 A combined vector channel $W^8$.

The above combine-and-split polarization process is carried out in three levels recursively based on the 3-fold Kronecker mapping matrix $\mathbf{G}^{(3)}$ and is called *3-fold Kronecker polarization*.

**Example 17.14** In this example, we polarize a vector channel $W^8$ formed by taking eight copies of a BEC($\epsilon$), $W$, with erasure probability $\epsilon$. Polarizing $W^8$ in three levels using the 3-fold Kronecker mapping matrix $\mathbf{G}^{(3)}$, we obtain eight bit-coordinate BECs, $W^{---}$, $W^{--+}$, $W^{-+-}$, $W^{-++}$, $W^{+--}$, $W^{+-+}$, $W^{++-}$, and $W^{+++}$, with the following erasure probabilities:

$$\begin{aligned}
\epsilon_{W^{---}} &= f^-(\epsilon_{W^{--}}) = f^-(f^-(f^-(\epsilon))) \\
&= 8\epsilon - 28\epsilon^2 + 56\epsilon^3 - 70\epsilon^4 + 56\epsilon^5 - 28\epsilon^6 + 8\epsilon^7 - \epsilon^8, \\
\epsilon_{W^{--+}} &= f^+(\epsilon_{W^{--}}) = f^+(f^-(f^-(\epsilon))) \\
&= 16\epsilon^2 - 48\epsilon^3 + 68\epsilon^4 - 56\epsilon^5 + 28\epsilon^6 - 8\epsilon^7 + \epsilon^8, \\
\epsilon_{W^{-+-}} &= f^-(\epsilon_{W^{-+}}) = f^-(f^+(f^-(\epsilon))) \\
&= 8\epsilon^2 - 8\epsilon^3 - 14\epsilon^4 + 32\epsilon^5 - 24\epsilon^6 + 8\epsilon^7 - \epsilon^8, \\
\epsilon_{W^{-++}} &= f^+(\epsilon_{W^{-+}}) = f^+(f^+(f^-(\epsilon))) \\
&= 16\epsilon^4 - 32\epsilon^5 + 24\epsilon^6 - 8\epsilon^7 + \epsilon^8, \\
\epsilon_{W^{+--}} &= f^-(\epsilon_{W^{+-}}) = f^-(f^-(f^+(\epsilon))) \\
&= 4\epsilon^2 - 6\epsilon^4 + 4\epsilon^6 - \epsilon^8, \\
\epsilon_{W^{+-+}} &= f^+(\epsilon_{W^{+-}}) = f^+(f^-(f^+(\epsilon))) \\
&= 4\epsilon^4 - 4\epsilon^6 + \epsilon^8, \\
\epsilon_{W^{++-}} &= f^-(\epsilon_{W^{++}}) = f^-(f^+(f^+(\epsilon))) \\
&= 2\epsilon^4 - \epsilon^8, \\
\epsilon_{W^{+++}} &= f^+(\epsilon_{W^{++}}) = f^+(f^+(f^+(\epsilon))) \\
&= \epsilon^8.
\end{aligned} \tag{17.98}$$

Hence, the channel capacities for the eight bit-coordinate channels are:

$$I(W^{---}) = 1 - \epsilon_{W^{---}} = 1 - 8\epsilon + 28\epsilon^2 - 56\epsilon^3 + 70\epsilon^4 - 56\epsilon^5 + 28\epsilon^6 - 8\epsilon^7 + \epsilon^8,$$
$$I(W^{--+}) = 1 - \epsilon_{W^{--+}} = 1 - 16\epsilon^2 + 48\epsilon^3 - 68\epsilon^4 + 56\epsilon^5 - 28\epsilon^6 + 8\epsilon^7 - \epsilon^8,$$
$$I(W^{-+-}) = 1 - \epsilon_{W^{-+-}} = 1 - 8\epsilon^2 + 8\epsilon^3 + 14\epsilon^4 - 32\epsilon^5 + 24\epsilon^6 - 8\epsilon^7 + \epsilon^8,$$
$$I(W^{-++}) = 1 - \epsilon_{W^{-++}} = 1 - 16\epsilon^4 + 32\epsilon^5 - 24\epsilon^6 + 8\epsilon^7 - \epsilon^8,$$
$$I(W^{+--}) = 1 - \epsilon_{W^{+--}} = 1 - 4\epsilon^2 + 6\epsilon^4 - 4\epsilon^6 + \epsilon^8,$$
$$I(W^{+-+}) = 1 - \epsilon_{W^{+-+}} = 1 - 4\epsilon^4 + 4\epsilon^6 - \epsilon^8,$$
$$I(W^{++-}) = 1 - \epsilon_{W^{++-}} = 1 - 2\epsilon^4 + \epsilon^8,$$
$$I(W^{+++}) = 1 - \epsilon_{W^{+++}} = 1 - \epsilon^8.$$

$$(17.99)$$

The sum of the channel capacities of these eight bit-coordinate channels is

$$I(W^{---}) + I(W^{--+}) + I(W^{-+-}) + I(W^{-++})$$
$$+ I(W^{+--}) + I(W^{+-+}) + I(W^{++-}) + I(W^{+++}) = 8(1 - \epsilon) = 8I(W).$$

We can also verify that the following inequality holds:

$$I(W^{---}) \leq I(W^{--}) \leq I(W^{++}) \leq I(W^{+++}).$$

Assume the channel erasure probability of the base BEC $W$ is $\epsilon = 0.5$. The channel capacities of the eight bit-coordinate channels are calculated as follows:

$$I(W^{---}) = 0.0039,$$
$$I(W^{--+}) = 0.1211,$$
$$I(W^{-+-}) = 0.1914,$$
$$I(W^{+--}) = 0.3164,$$
$$I(W^{-++}) = 0.6836,$$
$$I(W^{+-+}) = 0.8086,$$
$$I(W^{++-}) = 0.8789,$$
$$I(W^{+++}) = 0.9961,$$

$$(17.100)$$

which are arranged in the order of increasing channel capacities. The bit-coordinate channel $I(W^{---})$ has the smallest channel capacity 0.0039 (close to 0) which is much smaller than the channel capacity of the base channel ($I(W) = 1 - \epsilon = 0.5$), and the bit-coordinate channel $I(W^{+++})$ has the largest channel capacity 0.9961 (close to 1) which is much larger than the channel capacity of the base channel. There are four bit-coordinate channels with capacities larger than the capacity of the base channel.

Suppose the base BEC $W$ has $\epsilon = 0.1$. In this case, the channel capacities of the eight bit-coordinate channels are:

$$I(W^{---}) = 0.4305,$$
$$I(W^{--+}) = 0.8817,$$
$$I(W^{-+-}) = 0.9291,$$
$$I(W^{+--}) = 0.9606,$$
$$I(W^{-++}) = 0.9987,$$
$$I(W^{+-+}) = 0.9996,$$
$$I(W^{++-}) = 0.9998,$$
$$I(W^{+++}) = 1.0000.$$

In this case, among the eight bit-coordinate channels, four of them have capacities close to 1, and one is perfect.

It can be checked that the capacities of the eight polarized bit-coordinate channels are related to the row-weight distribution of the Kronecker mapping matrix $\mathbf{G}^{(3)}$. The capacity of the bit-coordinate channel with input $U_i$, $0 \leq i \leq 7$, is proportional to the weight and the location of its corresponding row in $\mathbf{G}^{(3)}$. In general, the larger weight of a row in $\mathbf{G}^{(3)}$ the larger capacity of its corresponding bit-coordinate channel. ▲▲

We can continue the above combine-and-split polarization process to $\ell$ levels with $\ell \geq 1$. In this case, the vector channel $W^N$ to be polarized is composed of $N = 2^\ell$ copies of a base BI-DMC $W$. The vector channel $W^N$ has transition probability:

$$W^N(\mathbf{Y}^{N-1}|\mathbf{U}^{N-1}) = W^{N/2}(\mathbf{Y}_{\mathrm{e}}^{N-1}|\mathbf{U}_{\mathrm{e}}^{N-1})W^{N/2}(\mathbf{Y}_{\mathrm{o}}^{N-1}|\mathbf{U}_{\mathrm{o}}^{N-1}), \quad (17.101)$$

i.e., the $\ell$-fold vector channel $W^N$ is constructed recursively from two $(\ell - 1)$-fold vector channels. Splitting the vector channel $W^N$ produces $N$ bit-coordinate channels. For $0 \leq i < N$, let $W_N^{(i)}$ denote the $i$th bit-coordinate channel:

$$W_N^{(i)} : U_i \to (\mathbf{Y}^{N-1}, \mathbf{U}^{i-1}), \quad (17.102)$$

in which $i$ is called the *bit-coordinate channel index*.

With the increment and decrement sign representation, i.e., "+" and "−," the $i$th bit-coordinate channel $W_N^{(i)}$ corresponds to the bit-coordinate channel $W^{s_0 s_1 \cdots s_{\ell-1}}$ with $s_j \in \{+, -\}$ for $0 \leq j < \ell$. In Examples 17.13 and 17.14, we use only $W^{s_0 s_1 \cdots s_{\ell-1}}$ to represent each bit-coordinate channel and have not provided the bit-coordinate channel index, $i$, i.e., how to obtain the channel index based on the increment and decrement sign sequence $s_0 s_1 \cdots s_{\ell-1}$ and vice versa. There is a one-to-one mapping between the bit-coordinate channel index and the sequence $s_0 s_1 \cdots s_{\ell-1}$. As seen from these two examples, the sequence $s_0 s_1 \cdots s_{\ell-1}$ determines the calculation of capacities of the bit-coordinate channels for BECs. If we set $s_j = 1$ when $s_j$ is the increment sign "+" and set $s_j = 0$ when $s_j$ is the decrement sign "$-$," then the sequence $s_0 s_1 \cdots s_{\ell-1}$ is a binary sequence of $\ell$ bits. In the following, we will explore the mapping between the bit-coordinate channel index $i$ and the binary sequence, $s_0 s_1 \cdots s_{\ell-1}$.

Before presenting more on the polarized bit-coordinate channels, we define some new notation. Let $i$ be a nonnegative integer such that $0 \le i < N = 2^\ell$. Express $i$ in radix-2 form,

$$i = i_0 + i_1 2 + i_2 2^2 + \cdots i_{\ell-1} 2^{\ell-1},$$

where $i_j$ is equal to either 0 or 1 for $0 \le j < \ell$. The radix-2 representation of $i$ is $(i_0, i_1, \ldots, i_{\ell-1})$. For $0 \le j < \ell$, let $s_j = i_{\ell-1-j}$. Then,

$$(s_0, s_1, s_2, \ldots, s_{\ell-2}, s_{\ell-1}) = (i_{\ell-1}, i_{\ell-2}, i_{\ell-3}, \ldots, i_1, i_0), \qquad (17.103)$$

i.e., the sequence $s_0, s_1, \ldots, s_{\ell-2}, s_{\ell-1}$ is the coefficients of the radix-2 representation of the integer $i$ in reversed order.

Consider the bit-coordinate channel $W_N^{(i)}$ with index $i$, $0 \le i < N$. Let $(i_0, i_1, \ldots, i_{\ell-1})$ be the radix-2 representation of $i$. Then, the $\ell$-bit sequence $s_0 s_1 \ldots s_{\ell-2} s_{\ell-1}$ corresponding to the bit-coordinate channel $W_N^{(i)}$ is given by the reversed order of the radix-2 representation of $i$ as shown in (17.103). For example, for $\ell = 3$ and $i = 4$, the radix-2 representation of 4 is $i = 4 = 0 + 0 \cdot 2 + 1 \cdot 2^2$ and thus the coefficient sequence of its radix-2 representation is $(i_0, i_1, i_2) = (0\ 0\ 1)$. Then, $(s_0, s_1, s_2) = (1\ 0\ 0)$. Converting the 1 and 0 in the sequence $(s_0, s_1, s_2)$ to the increment and decrement signs "+" and "$-$," respectively, we have the notation $(+\ -\ -)$. Hence, the fourth polarized bit-coordinate channel $W_8^{(4)}$ of the vector channel $W^8$ is $W^{+--} : U_4 \to (Y^7, U^3)$ as shown in (17.94). With the one-to-one mapping between the bit-coordinate channel index $i$ and the sequence $s_0 s_1 \cdots s_{\ell-2} s_{\ell-1}$, the equations given by (17.100) can be rephrased as following:

$$
\begin{aligned}
I(W_8^{(0)}) &= 0.0039, \\
I(W_8^{(1)}) &= 0.1211, \\
I(W_8^{(2)}) &= 0.1914, \\
I(W_8^{(4)}) &= 0.3164, \\
I(W_8^{(3)}) &= 0.6836, \\
I(W_8^{(5)}) &= 0.8086, \\
I(W_8^{(6)}) &= 0.8789, \\
I(W_8^{(7)}) &= 0.9961.
\end{aligned}
\qquad (17.104)
$$

The equations in (17.104) show that the zeroth bit-coordinate channel has the smallest channel capacity and the seventh bit-coordinate channel has the largest channel capacity.

Based on the above developments, we summarize the recursive process of the $\ell$-level polarization of an $\ell$-fold vector channel $W^{2^\ell}$ that comprises $2^\ell$ copies of a base BI-DMC $W$ as follows. In the first level, the $\ell$-fold vector channel $W^N$ is polarized into two $(\ell-1)$-fold vector channels, denoted by $W^{2^{\ell-1},-}$ and $W^{2^{\ell-1},+}$, respectively. In the second-level polarization, the $(\ell-1)$-fold vector channel $W^{2^{\ell-1},-}$ is polarized into two $(\ell-2)$-fold vector channels, denoted by $W^{2^{\ell-2},--}$ and $W^{2^{\ell-2},-+}$, respectively; and the $(\ell-1)$-fold vector channel $W^{2^{\ell-1},+}$ is polarized into two $(\ell-2)$-fold vector channels, denoted by $W^{2^{\ell-2},+-}$ and $W^{2^{\ell-2},++}$, respectively. The second-level polarization results in four $(\ell-2)$-fold vector channels, $W^{2^{\ell-2},--}$, $W^{2^{\ell-2},-+}$, $W^{2^{\ell-2},+-}$, and $W^{2^{\ell-2},++}$. In the third-level polarization, we polarize each of the four $(\ell-2)$-fold vector channels into two $(\ell-3)$-fold vector channels. The split-in-two polarization continues until we complete the $\ell$th-level polarization. For $1 \le k \le \ell$, there are $2^k$ $(\ell-k)$-fold vector channels in the $k$th-level polarization. Each of these $(\ell-k)$-fold vector channels comprises $2^{\ell-k}$ bit-coordinate channels with different transition probability distribution and different channel capacity, and is split into two $(\ell-k-1)$-fold vector channels at the next level of polarization. At the end of the $\ell$th-level polarization, we obtain $2^\ell$ polarized bit-coordinate channels.

The channel polarization process in $\ell$ levels can be viewed as an $\ell$-level tree as shown in Fig. 17.9.

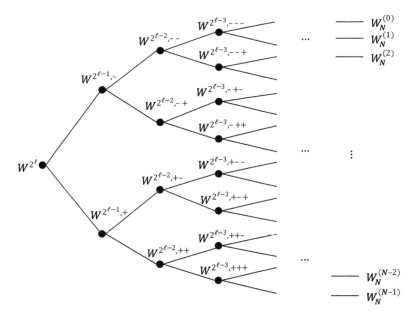

Figure 17.9 An $\ell$-level channel polarization tree.

### 17.6.3 Channel Polarization Theorem

The major theorem based on which polar codes are constructed was proved by Arikan [1]. In the following, we present the theorem without a proof.

**Theorem 17.1 (Channel Polarization Theorem, [1])** *Let $N = 2^\ell$ and $W^N$ be a vector channel formed by taking $N$ copies of any BI-DMC $W$ with channel capacity $I(W)$. Let $\{W_N^{(i)} : 0 \le i < N\}$ be the set of $N$ bit-coordinate channels obtained by polarizing the vector channel $W^N$. Then, for any fixed $\delta \in (0,1)$, as $\ell$ goes to infinity, the fraction of bit-coordinate channels in $\{W_N^{(i)} : 0 \le i < N\}$ for which $I(W_N^{(i)}) \in (1-\delta,1]$ goes to $I(W)$, and the fraction of bit-coordinate channels in $\{W_N^{(i)}; 0 \le i < N\}$ for which $I(W_N^{(i)}) \in [0,\delta)$ goes to $1 - I(W)$.*

Theorem 17.1 can be interpreted as follows: when $N$ is sufficiently large, there are about $NI(W)$ of the $N = 2^\ell$ bit-coordinate channels which are noiseless (i.e., with capacity approaching 1) and there are about $N(1 - I(W))$ bit-coordinate channels which are useless (i.e., with capacity approaching 0). Thus, we can use the $NI(W)$ channels to transmit information, calling them *free channels*, while keeping the other $N(1 - I(W))$ channels *frozen*, calling them *frozen channels*, i.e., we do not use these $N(1 - I(W))$ channels to transmit any information. For these frozen channels, we can transmit a fixed value of bit, e.g., 0, which is known to the decoder.

Consider the eight BEC bit-coordinate channels obtained in Example 17.14 with channel capacities given by (17.100) or (17.104) with $\epsilon = 0.5$. Suppose we transmit four information bits, $c_0$, $c_1$, $c_2$, and $c_3$, through the BEC vector channel $W^8$. Based on the channel capacities of the eight bit-coordinate channels, we can choose four bit-coordinate channels which have the largest channel capacities, i.e., $W_8^{(3)}$, $W_8^{(5)}$, $W_8^{(6)}$, and $W_8^{(7)}$, to transmit the four information bits. The other four bit-coordinate channels are frozen. The choice of free and frozen channels specifies the free and frozen code bit positions of a codeword of eight bits, and hence specifies the specific $(8,4)$ polar code $C_p(4,3)$ of rate $1/2$ given in Example 17.8. The information transmission is shown in Fig. 17.10, where we transmit fixed bit values of zeros on these frozen channels.

**Example 17.15** In Examples 17.12, 17.13, and 17.14, we considered the channel polarization for two, four, and eight copies of a BEC with erasure probability $\epsilon$, and calculated the erasure probabilities and capabilities of their polarized bit-coordinate channels. In this example, we consider the polarization of a vector channel which is formed by taking $N = 2^\ell$ copies of a BEC with erasure probability $\epsilon$.

For $0 \le i < N$, consider the $i$th bit-coordinate channel $W_N^{(i)} : U_i \to (\mathbf{Y}^{N-1}, \mathbf{U}^{i-1})$. The radix-2 representation of the channel index $i$ is $i = i_0 + i_1 2 + i_2 2^2 + \cdots + i_{\ell-1} 2^{\ell-1}$. Based on the recursive calculations of erasure probabilities of bit-coordinate channels as shown in these three examples, the erasure probability of the $i$th bit-coordinate channel $W_N^{(i)}$ is given as follows:

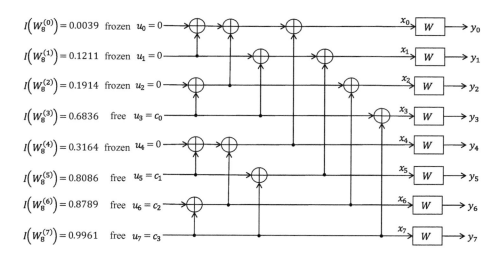

Figure 17.10 The information transmission using the $(8,4)$ polar code $C_p(4,3)$ given in Example 17.8 over the BEC vector channel $W^8$ with the base BEC channel BEC(0.5).

$$\epsilon_{W_N^{(i)}} = f^{i_0}(f^{i_1}(\cdots f^{i_{\ell-2}}(f^{i_{\ell-1}}(\epsilon)))), \qquad (17.105)$$

where $f^{i_j}(x)$ is the positively polarized erasure function $f^+(x) = x^2$ (see (17.82)) for $i_j = 1$ and the negatively polarized erasure function $f^-(x) = 2x - x^2$ (see (17.81)) for $i_j = 0$. The channel capacity of the bit-coordinate channel $W_N^{(i)}$ is given as follows:

$$I(W_N^{(i)}) = 1 - \epsilon_{W_N^{(i)}}. \qquad (17.106)$$

Based on Theorem 17.1, as $\ell$ (or $N$) goes to infinity, the fraction of good bit-coordinate channels (with capacity approaching 1) goes to $I(W) = 1 - \epsilon$ and the fraction of bad bit-coordinate channels (with capacity approaching 0) goes to $1 - I(W) = 1 - (1 - \epsilon) = \epsilon$.

Figure 17.11 shows the bit-coordinate channel capacities for $N = 16$ and $N = 64$, and Fig. 17.12 shows these for $N = 256$ and $N = 1024$ with BEC(0.5) as the base channel whose erasure probability is $\epsilon = 0.5$, where the $x$-axis represents the bit-coordinate channel index and $y$-axis is the corresponding bit-coordinate channel capacity. To show better the effect of channel polarization, we sort the bit-coordinate channel capacities from low to high and plot the sorted bit-coordinate channel capacities in Figs. 17.13 and 17.14. From these figures, we can see that, as $N$ increases through the power $2^\ell$ of 2, the bit-coordinate channels are separated into two groups, one group with capacity approaching 1 (perfect channels) and the other group with capacity approaching 0 (useless channels). ▲▲

Polarization can be applied to various types of channels. In this section, we just presented the channel polarization of the simplest binary input channel, i.e.,

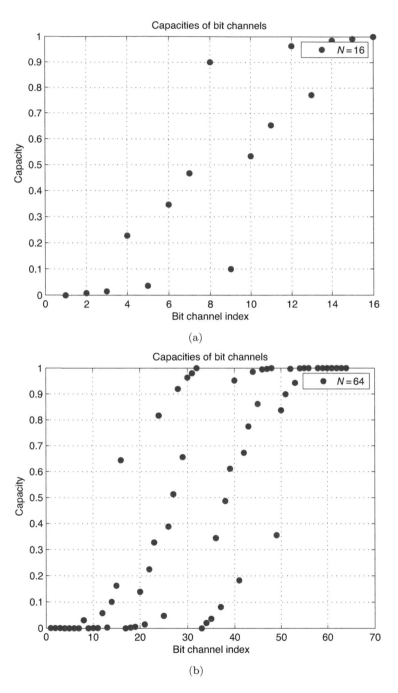

Figure 17.11 The bit-coordinate channel capacities for BEC channel polarization with (a) $N = 16$ and (b) $N = 64$ for BEC(0.5).

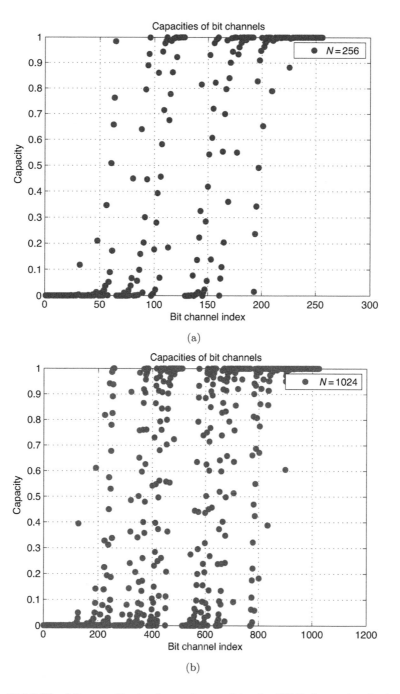

Figure 17.12 The bit-coordinate channel capacities for BEC channel polarization with (a) $N = 256$ and (b) $N = 1024$ for BEC(0.5).

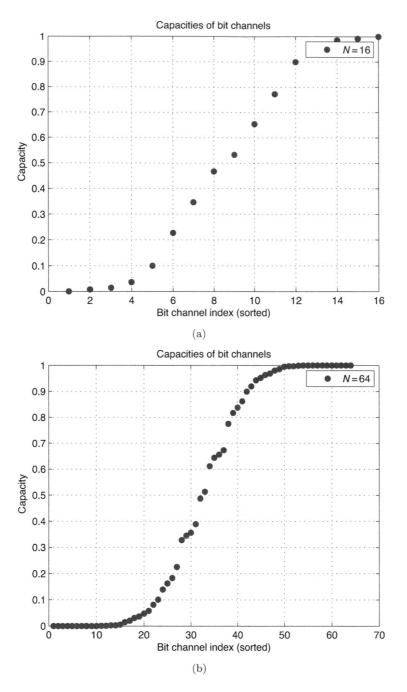

Figure 17.13 The bit-coordinate channel capacities for BEC channel polarization after sorting with (a) $N = 16$ and (b) $N = 64$ for BEC(0.5).

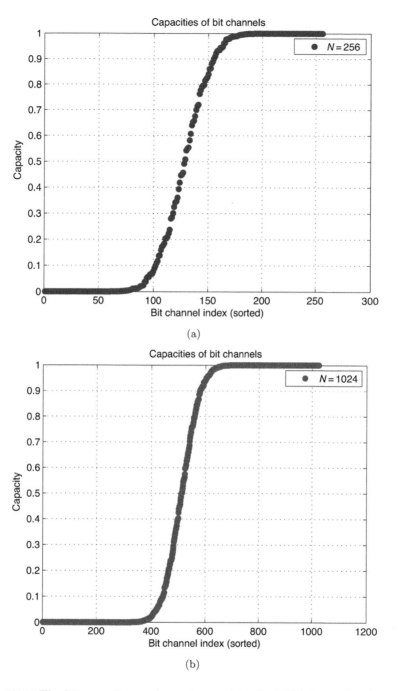

Figure 17.14 The bit-coordinate channel capacities for BEC channel polarization after sorting with (a) $N = 256$ and (b) $N = 1024$ for BEC(0.5).

the binary erasure channel (BEC), and the calculation of the channel capacities of its associated bit-coordinate channels. Polarization of other types of channels, e.g., BSC and AWGN channels, can be found in references [1, 22–30]. In fact, it has been proved that all the BI-DMCs have the same polarization effect. Additionally, channel polarization can be applied to nonbinary channels [31–34]. The channel polarization for BI-DMCs presented in this section is based on the binary $2 \times 2$ kernel, $\mathbf{G}^{(1)}$, given by (17.1). Nonbinary kernels or binary kernels of larger sizes can also be used to achieve channel polarization [11–21].

## 17.7  Construction of Polar Codes

Channel polarization based on the $\ell$-fold Kronecker mapping with the $\ell$-fold Kronecker matrix $\mathbf{G}^{(\ell)}$ developed in Section 17.6 shows that a fraction of good bit-coordinate channels (i.e., free channels), $I(W)$, have larger capacities than other bit-coordinate channels (i.e., frozen channels). Thus, we can use these good bit-coordinate channels to transmit information reliably. Therefore, we can use the indices of these good bit-coordinate channels to form a free index set $\mathcal{I}_{\text{free}}$ for constructing polar codes (as presented in Section 17.4) to achieve large channel capacity. The encoding of polar codes is achieved by using the universal generator matrix $\mathbf{G}^{(\ell)}$ or using a $k \times 2^\ell$ generator matrix whose $k$ rows are those from $\mathbf{G}^{(\ell)}$ labeled by indices in the free index set $\mathcal{I}_{\text{free}}$, as described in Section 17.4.

The reason that we call the polarized bit channels the *bit-coordinate channels* is because we use the indices of the polarized bit-coordinate channels to specify the *coordinates* of the code bits in the encoding vector $\mathbf{u}$ for transmitting information bits of a message $\mathbf{c}$.

**Example 17.16** In Example 17.14, we constructed a BEC vector channel $W^8$ by combining $N = 2^3 = 8$ copies of BEC(0.5) with erasure probability $\epsilon = 0.5$ and calculated the capacities of the eight bit-coordinate channels:

$$I(W_8^{(0)}) = 0.0039,$$
$$I(W_8^{(1)}) = 0.1211,$$
$$I(W_8^{(2)}) = 0.1914,$$
$$I(W_8^{(4)}) = 0.3164,$$
$$I(W_8^{(3)}) = 0.6836,$$
$$I(W_8^{(5)}) = 0.8086,$$
$$I(W_8^{(6)}) = 0.8789,$$
$$I(W_8^{(7)}) = 0.9961.$$

The rank of these bit-coordinate channel capacities (from high to low) is 7, 6, 5, 3, 4, 2, 1, 0, i.e., the seventh bit-coordinate channel $W_8^{(7)}$ has the largest channel capacity and the zeroth bit-coordinate channel $W_8^{(0)}$ has the smallest capacity.

An $(8, 4)$ polar code $C_p(4, 3)$ was constructed in Example 17.8 by using $\mathcal{I}_{\text{free}} = \{3, 5, 6, 7\}$ and $\mathcal{I}_{\text{frozen}} = \{0, 1, 2, 4\}$. The chosen indices in $\mathcal{I}_{\text{free}}$ are the code bit indices whose corresponding bit-coordinate channels, $W_8^{(3)}$, $W_8^{(5)}$, $W_8^{(6)}$ and $W_8^{(7)}$, have the largest capacities among the eight bit-coordinate channels. The capacities of these bit-coordinate channels are larger than the unpolarized $\text{BEC}(0.5)$, whose capacity is 0.5.

The generator matrix $\mathbf{G}$ of the polar code $C_p(4, 3)$ is a submatrix of the universal generator matrix $\mathbf{G}^{(3)}$ given by (17.54). The rows of $\mathbf{G}$ are those in $\mathbf{G}^{(3)}$ with labels in $\mathcal{I}_{\text{free}} = \{3, 5, 6, 7\}$:

$$\mathbf{G} = \begin{bmatrix} \mathbf{g}_3 \\ \mathbf{g}_5 \\ \mathbf{g}_6 \\ \mathbf{g}_7 \end{bmatrix} = \begin{bmatrix} 1 & 1 & 1 & 1 & 0 & 0 & 0 & 0 \\ 1 & 1 & 0 & 0 & 1 & 1 & 0 & 0 \\ 1 & 0 & 1 & 0 & 1 & 0 & 1 & 0 \\ 1 & 1 & 1 & 1 & 1 & 1 & 1 & 1 \end{bmatrix}. \tag{17.107}$$

Let $\mathbf{c} = (c_0, c_1, c_2, c_3)$ be a 4-tuple message to be encoded, $\mathbf{u} = \mu(\mathbf{c}) = (u_0, u_1, \ldots, u_7)$ be the encoding vector of $\mathbf{c}$, and $\mathbf{v} = (v_0, v_1, \ldots, v_7)$ be the corresponding codeword of $\mathbf{c}$.[3] With $\mathbf{G}$ as the generator matrix, we have $\mathbf{v} = \mathbf{c}\mathbf{G}$. If the universal generator matrix $\mathbf{G}^{(3)}$ is used for encoding, we have $\mathbf{v} = \mathbf{u}\mathbf{G}^{(3)} = \mu(\mathbf{c})\mathbf{G}^{(3)}$. ▲▲

**Example 17.17** Suppose we set $\ell = 4$ and construct a BEC vector channel $W^{16}$ by combining $N = 2^4 = 16$ copies of BECs with erasure probability $\epsilon = 0.5$. Next, we polarize the vector channel $W^{16}$ with the 4-fold Kronecker mapping matrix $\mathbf{G}^{(4)}$. Polarization of the vector channel $W^{16}$ results in 16 polarized bit-coordinate channels with capacities:

$$I(W_{16}^{(0)}) = 0.000\,01, \; I(W_{16}^{(9)}) = 0.532\,69,$$
$$I(W_{16}^{(1)}) = 0.007\,79, \; I(W_{16}^{(10)}) = 0.653\,82,$$
$$I(W_{16}^{(2)}) = 0.014\,66, \; I(W_{16}^{(12)}) = 0.772\,47,$$
$$I(W_{16}^{(4)}) = 0.036\,63, \; I(W_{16}^{(7)}) = 0.899\,88,$$
$$I(W_{16}^{(8)}) = 0.100\,11, \; I(W_{16}^{(11)}) = 0.963\,36,$$
$$I(W_{16}^{(3)}) = 0.227\,52, \; I(W_{16}^{(13)}) = 0.985\,33,$$
$$I(W_{16}^{(5)}) = 0.346\,17, \; I(W_{16}^{(14)}) = 0.992\,20,$$
$$I(W_{16}^{(6)}) = 0.467\,30, \; I(W_{16}^{(15)}) = 0.999\,98.$$

The rank of these bit-coordinate channel capacities is 15, 14, 13, 11, 7, 12, 10, 9, 6, 5, 3, 8, 4, 2, 1, 0, i.e., the fifteenth bit-coordinate channel $W_{16}^{(15)}$ has the largest capacity and the zeroth bit-coordinate channel $W_{16}^{(0)}$ has the smallest capacity.

We can use the eight bit-coordinate channels with the largest channel capacities to specify the locations of code bits for eight information bits, i.e., the bit

---

[3]The codeword $\mathbf{v}$ is equivalent to the vector $\mathbf{x} = (x_0, x_1, \ldots, x_7)$ shown in Fig. 17.9.

positions labeled with 15, 14, 13, 11, 7, 12, 10, and 9, as the positions for the eight information bits, and set the other eight code bits at positions 6, 5, 3, 8, 4, 2, 1, 0, to zeros. This partition of the 16 code bit positions results in the $(16, 8)$ polar code $C_p(8, 4)$ with rate $1/2$ given in Example 17.11 with the free index set $\mathcal{I}_{\text{free}} = \{7, 9, 10, 11, 12, 13, 14, 15\}$ and the frozen index set $\mathcal{I}_{\text{frozen}} = \{0, 1, 2, 3, 4, 5, 6, 8\}$.

Suppose we want to construct a $(16, 12)$ polar code $C_p(12, 4)$ of rate $3/4$. First, we choose a free index set $\mathcal{I}_{\text{free}}$ of size $k = 12$. Based on the rank of the 16 bit-coordinate channel capacities, we pick the 12 bit-coordinate channels with the largest capacities, i.e., bit-coordinate channels with indices 15, 14, 13, 11, 7, 12, 10, 9, 6, 5, 3, 8. Then, set $\mathcal{I}_{\text{free}} = \{3, 5, 6, 7, 8, 9, 10, 11, 12, 13, 14, 15\}$ and $\mathcal{I}_{\text{frozen}} = \{0, 1, 2, 4\}$. Let $\mathbf{c} = (c_0, c_1, \ldots, c_{10}, c_{11})$ be a message of 12 information bits to be encoded. Based on the chosen free index set $\mathcal{I}_{\text{free}}$, we map the 12 information bits in $\mathbf{c}$ to an encoding vector $\mathbf{u} = \mu(\mathbf{c})$, i.e., $u_0 = 0$, $u_1 = 0$, $u_2 = 0$, $u_3 = c_0$, $u_4 = 0$, $u_5 = c_1$, $u_6 = c_2$, $u_7 = c_3$, $u_8 = c_4$, $u_9 = c_5$, $u_{10} = c_6$, $u_{11} = c_7$, $u_{12} = c_8$, $u_{13} = c_9$, $u_{14} = c_{10}$, and $u_{15} = c_{11}$. The codeword $\mathbf{v}$ for $\mathbf{c}$ is formed based on the universal generator matrix $\mathbf{G}^{(4)}$ given by (17.59) as follows

$$\mathbf{v} = \mathbf{u} \cdot \mathbf{G}^{(4)}.$$

Alternately, the encoding can be achieved by using the generator matrix $\mathbf{G}$ for the $(16, 12)$ polar code $C_p(12, 4)$. The rows in $\mathbf{G}$ labeled with indices in $\mathcal{I}_{\text{free}}$ from $\mathbf{G}^{(4)}$ are chosen to form the generator matrix:

$$\mathbf{G} = \begin{bmatrix} \mathbf{g}_3 \\ \mathbf{g}_5 \\ \mathbf{g}_6 \\ \mathbf{g}_7 \\ \mathbf{g}_8 \\ \mathbf{g}_9 \\ \mathbf{g}_{10} \\ \mathbf{g}_{11} \\ \mathbf{g}_{12} \\ \mathbf{g}_{13} \\ \mathbf{g}_{14} \\ \mathbf{g}_{15} \end{bmatrix} = \begin{bmatrix} 1\,1\,1\,1\,0\,0\,0\,0\,0\,0\,0\,0\,0\,0\,0\,0 \\ 1\,1\,0\,0\,1\,1\,0\,0\,0\,0\,0\,0\,0\,0\,0\,0 \\ 1\,0\,1\,0\,1\,0\,1\,0\,0\,0\,0\,0\,0\,0\,0\,0 \\ 1\,1\,1\,1\,1\,1\,1\,1\,0\,0\,0\,0\,0\,0\,0\,0 \\ 1\,0\,0\,0\,0\,0\,0\,0\,1\,0\,0\,0\,0\,0\,0\,0 \\ 1\,1\,0\,0\,0\,0\,0\,0\,1\,1\,0\,0\,0\,0\,0\,0 \\ 1\,0\,1\,0\,0\,0\,0\,0\,1\,0\,1\,0\,0\,0\,0\,0 \\ 1\,1\,1\,1\,0\,0\,0\,0\,1\,1\,1\,1\,0\,0\,0\,0 \\ 1\,0\,0\,0\,1\,0\,0\,0\,1\,0\,0\,0\,1\,0\,0\,0 \\ 1\,1\,0\,0\,1\,1\,0\,0\,1\,1\,0\,0\,1\,1\,0\,0 \\ 1\,0\,1\,0\,1\,0\,1\,0\,1\,0\,1\,0\,1\,0\,1\,0 \\ 1\,1\,1\,1\,1\,1\,1\,1\,1\,1\,1\,1\,1\,1\,1\,1 \end{bmatrix}.$$

Then, $\mathbf{v} = \mathbf{c} \cdot \mathbf{G}$. ▲▲

The computation complexity required to polarize a BI-DMC in $\ell$-fold is in the order of $\ell 2^\ell$. For large $\ell$, computation required to construct a long polar code can be expensive.

## 17.8 Successive Cancellation Decoding

The channel polarization presented in Section 17.6 is the foundation on which polar code construction is based. In the encoding of an $(N, k)$ polar code $C_p(k, \ell)$ with $N = 2^\ell$, $1 \leq k \leq N$, and $\ell \geq 1$, a message $\mathbf{c} = (c_0, c_1, \ldots, c_{k-1})$ of $k$ information bits is mapped into an encoding vector $\mathbf{u} = (u_0, u_1, \ldots, u_{N-1})$ of $N$ bits in which the $i$th bit $u_i$, $0 \leq i < N$, is transmitted *virtually* over the $i$th polarized bit-coordinate channel $W_N^{(i)} : U_i \to (\mathbf{Y}^{N-1}, \mathbf{U}^{i-1})$ of a vector channel $W^N$ formed by $N$ copies of a base BI-DMC $W$. Some of the polarized bit-coordinate channels, called good channels, have larger capacities than the others, called bad channels. The good channels are more reliable for information transmission (or close to error-free information transmission for codes long enough) than the bad channels. Once the encoding vector $\mathbf{u}$ is formed, it is mapped into a codeword $\mathbf{v} = (v_0, v_1, \ldots, v_{N-1})$ by the $\ell$-fold Kronecker matrix $\mathbf{G}^{(\ell)}$ for transmission.[4] The encoding $\mathbf{u}$ into $\mathbf{v}$ is carried out in $\ell$ levels successively in a recursive manner as presented in Sections 17.2 and 17.4.

In this section, we present the decoding algorithm devised by Arikan [1], called *successive cancellation* (SC) decoding. For convenience, we repeat some notation used earlier and also introduce some new notation. The concepts of the SC decoding are relatively simple, but the notation in describing the decoding algorithm is burdensome.

To present the essence of the SC decoding, we use the block diagram shown in Fig. 17.15 for a general polar-coded system to display the transmission of code bits of an encoding vector over their associated polarized bit-coordinate channels. For $0 \le i < N$, $U_i$ is the input and $\mathbf{U}^{i-1}$ and $\mathbf{Y}^{N-1}$ are the outputs of the $i$th polarized bit-coordinate channel, i.e., $W_N^{(i)} : U_i \to (\mathbf{Y}^{N-1}, \mathbf{U}^{i-1})$. Let $\mathbf{u} = (u_0, u_1, \ldots, u_{N-1})$, $\mathbf{x} = (x_0, x_1, \ldots, x_{N-1})$, and $\mathbf{y} = (y_0, y_1, \ldots, y_{N-1})$ be the encoding vector, transmitted codeword, and the received vector, respectively. The $i$th bit $u_i$ is transmitted virtually over the $i$th polarized bit-coordinate channel $W_N^{(i)}$.

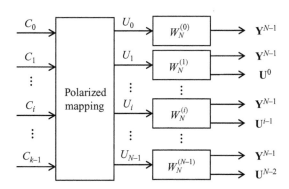

Figure 17.15 Block diagram of a polar coded system.

At the receiving end, when $\mathbf{y} = (y_0, y_1, \ldots, y_{N-1})$ is received, the decoder decodes $\mathbf{y}$ into an estimate, denoted by $\hat{\mathbf{u}} = (\hat{u}_0, \hat{u}_1, \ldots, \hat{u}_{N-1})$, of the encoding vector $\mathbf{u} = (u_0, u_1, \ldots, u_{N-1})$, where $\hat{u}_i$ denotes the decoded estimate of the $i$th bit $u_i$ of $\mathbf{u}$. With the SC decoding, the bits in $\mathbf{u} = (u_0, u_1, \ldots, u_{N-1})$ are decoded *one at a time successively (recursively)*. For $0 \le i < N$, decoding the $i$th code bit $u_i$ is based on the received vector $\mathbf{y}$ and the $i$ decoded estimates, $\hat{u}_0, \hat{u}_1, \ldots, \hat{u}_{i-1}$ of the $i$ bits $u_0, u_1, \ldots, u_{i-1}$ that precede $u_i$. Therefore, to decode $u_i$, we need to decode its preceding $i$ bits $u_0, u_1, \ldots, u_{i-1}$.

---

[4]In this section, we use $\mathbf{v}$ and $\mathbf{x}$ exchangeably, i.e., $\mathbf{x} = \mathbf{v}$, $x_i = v_i$ for $0 \le i < N$.

After decoding the $i$th code bit $u_i$, we proceed to decode the $(i+1)$th bit $u_{i+1}$, which was transmitted over the $(i+1)$th bit-coordinate channel, $W_N^{(i+1)}$ : $U_{i+1} \to (\mathbf{Y}^{N-1}, \mathbf{U}^i)$, based on the received vector $\mathbf{y}$ and the $i+1$ decoded estimates $\hat{u}_0, \hat{u}_1, \ldots, \hat{u}_i$ of $u_0, u_1, \ldots, u_i$. This *bit-by-bit decoding* continues until the bit $u_{N-1}$ is decoded.

Hence, the SC decoding algorithm is a *bit-by-bit peeling algorithm*. The reliability of peeling of the $i$th bit $u_i$ depends on reliabilities provided by the $i$th bit-coordinate channel $W_N^{(i)} : U_i \to (\mathbf{Y}^{N-1}, \mathbf{U}^{i-1})$ and the $i$ preceding decoding estimates $\hat{u}_0, \hat{u}_1, \ldots, \hat{u}_{i-1}$ of $u_0, u_1, \ldots, u_{i-1}$. If $\hat{u}_0, \hat{u}_1, \ldots, \hat{u}_{i-1}$ are error free and $W_N^{(i)}$ is perfect, then the decoded estimate $\hat{u}_i$ of the transmitted code $u_i$ will be error free. The SC decoding process is shown in Fig. 17.16. Note that, in decoding, the frozen bits (with values preset to zeros at the transmitter) are utilized to decode the information bits located in the free positions of the encoding vector $\mathbf{u}$. Once an estimate $\hat{\mathbf{u}}$ of $\mathbf{u}$ is obtained, the estimated information bits in $\mathbf{c}$ are obtained by removing the bits of $\hat{\mathbf{u}}$ in the frozen positions.

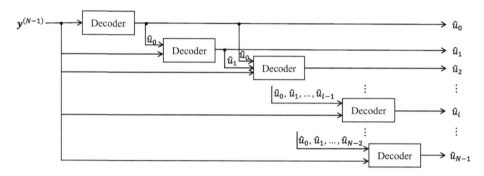

Figure 17.16 A block diagram of the SC decoding process for a polar code of length $N = 2^\ell$.

Another key feature of SC decoding of an encoding vector $\mathbf{u}$ is the *message passing between levels, forward* and *backward*, to compute the reliability of each bit in $\mathbf{u}$. Decoding starts from the highest level, level $(\ell - 1)$, and decoding decision of each bit in $\mathbf{u}$ is made at the zeroth level.

In Sections 17.8.1 and 17.8.2, we illustrate the SC decoding process using two short codes of lengths $N = 2$ and 4 (i.e., $\ell = 1$ and 2). Once the key steps of the SC decoding are presented and understood, we present the general SC decoding process with $N = 2^\ell$ in Section 17.8.3. In presenting the decoding process of each of these two short polar codes, we also present the basic architecture of an SC decoder.

## 17.8.1 SC Decoding of Polar Codes of Length $N = 2$

Here, we consider the SC decoding of a polar code of length $N = 2$ and its implementation. The decoding process and the decoder architecture are shown in Fig. 17.17, where the symbol $\boxed{\text{LLR}}$ denotes the LLR (log-likelihood ratio) calculation unit, $x_0$ and $x_1$ are two transmitted bits, and $y_0$ and $y_1$ are two received symbols.[5] The LLRs of the two transmitted bits $x_0$ and $x_1$, denoted by $\mathrm{LLR}(x_0|y_0)$ and $\mathrm{LLR}(x_1|y_1)$, respectively, can be calculated based on the received symbols $y_0$ and $y_1$ and the characteristics of the base channel $W$ (see Section 11.5.2.1 for the LLR calculation for different BI-DMCs). The SC decoder finds the estimates, $\hat{u}_0$ and $\hat{u}_1$, of the two bits $u_0$ and $u_1$ in the encoding vector $\mathbf{u}$ based on $\mathrm{LLR}(x_0|y_0)$ and $\mathrm{LLR}(x_1|y_1)$, i.e., calculate the reliabilities of $u_0$ and $u_1$ based on $\mathrm{LLR}(x_0|y_0)$ and $\mathrm{LLR}(x_1|y_1)$. Then, decoding $y_0$ and $y_1$ into $\hat{u}_0$ and $\hat{u}_1$ can be achieved in one level. Such an SC decoder is called the *SC kernel decoder*, which is the smallest decoding unit. As will be shown in Section 17.8.3, an SC decoder for a polar code of length $N = 2^\ell$ is composed of $\ell 2^{\ell-1}$ SC kernel decoders.

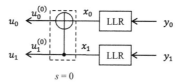

Figure 17.17 An SC decoder for a polar code of length $N = 2$.

To present the SC decoding process conveniently, we introduce some extra notations. Let $\mathbf{u}^{(s)} = (u_0^{(s)}, u_1^{(s)}, \ldots, u_{N-1}^{(s)})$ be the estimated encoding vector at level-$s$ with $0 \le s < \ell$. Let $\mathrm{LLR}_0^{(s)}, \mathrm{LLR}_1^{(s)}, \ldots, \mathrm{LLR}_{N-1}^{(s)}$ be the LLRs of the $N$ bits in $\mathbf{u}^{(s)}$ at the decoding level-$s$. For $s = 0$, we have $\mathbf{u}^{(0)} = \mathbf{u}$ and $\mathrm{LLR}_0^{(0)}, \mathrm{LLR}_1^{(0)}, \ldots, \mathrm{LLR}_{N-1}^{(0)}$ are called *reliability measures* of the bits in the encoding vector $\mathbf{u}$. The decoding process starts from level-$(\ell-1)$ to level-0 and with message passing forward and backward to compute the reliabilities (LLRs) of each bit in the encoding vector $\mathbf{u}$. Based on the computed reliabilities, the bits in $\mathbf{u}$ are estimated. The decoding decision of each bit is made at the zeroth level of decoding process.

Now, we present the SC decoding for a polar code of length $N = 2$. The SC decoder block diagram is shown in Fig. 17.17. Based on the single-fold inverse Kronecker mapping as shown in Section 17.2, we have $u_0 = x_0 + x_1$ which simply specifies a single-parity-check (SPC) code. Then, the reliability $\mathrm{LLR}(u_0)$ of $u_0$ is calculated as follows (see more details in Section 11.5 in the development of iterative decoding based on belief-propagation for LDPC codes):

$$\mathrm{LLR}(u_0) = 2\tanh^{-1}\left(\tanh\left(\frac{\mathrm{LLR}(x_0|y_0)}{2}\right)\tanh\left(\frac{\mathrm{LLR}(x_1|y_1)}{2}\right)\right). \quad (17.108)$$

---

[5]The log-likelihood ratio (LLR) is also known as the log-APP (a posteriori probability) ratio, which is defined as $\mathrm{LLR}(x_j|y_j) \triangleq \log\left(\frac{\Pr(x_j=0|y_j)}{\Pr(x_j=1|y_j)}\right)$, where $x_j$ is the transmitted bit and $y_j$ is the received symbol from a certain channel. LLR represents the reliability of the transmitted bit $x_j$, i.e., if $\mathrm{LLR}(x_j|y_j) > 0$, the probability of $x_j = 0$ is larger than that of $x_j = 1$ and the larger $|\mathrm{LLR}(x_j|y_j)|$ the more likely $x_j = 0$, and vice versa. See Section 11.5.2 for more information on LLR.

To simplify the calculation of (17.108), we can use the min-sum algorithm (MSA) as applied to an LDPC code in Section 11.5.3 to reduce the complexity of the sum-product algorithm (SPA):

$$
\begin{aligned}
\text{LLR}(u_0) &= 2\tanh^{-1}\left(\tanh\left(\frac{\text{LLR}(x_0|y_0)}{2}\right)\tanh\left(\frac{\text{LLR}(x_1|y_1)}{2}\right)\right) \\
&\approx \text{sign}(\text{LLR}(x_0|y_0))\text{sign}(\text{LLR}(x_1|y_1))\min\{|\text{LLR}(x_0|y_0)|, |\text{LLR}(x_1|y_1)|\} \\
&\triangleq f(\text{LLR}(x_0|y_0), \text{LLR}(x_1|y_1)).
\end{aligned}
$$

(17.109)

Based on $\text{LLR}(u_0)$, the estimate of $u_0$ is given by the following decision rule:

$$
\hat{u}_0 = \begin{cases} 0 \text{ if } u_0 \text{ is a frozen bit;} \\ 0 \text{ else if } \text{LLR}(u_0) > 0; \\ 1 \text{ otherwise.} \end{cases}
$$

(17.110)

Given $\hat{u}_0$, we have $u_1 = x_1$ and $u_1 = \hat{u}_0 + x_0$. If $\hat{u}_0 = 0$, we have $u_1 = x_1$ and $u_1 = x_0$ which can be interpreted as a repetition code. Then, the reliability of $u_1$, $\text{LLR}(u_1)$, is calculated as follows:

$$
\text{LLR}(u_1) = \text{LLR}(x_0|y_0) + \text{LLR}(x_1|y_1).
$$

(17.111)

If $\hat{u}_0 = 1$, we have $u_1 = x_1$ and $u_1 = \bar{x}_0$ (i.e., the complement of $x_0$) which can be interpreted as a repetition code with one repetition bit inverted. In this case, the reliability of $u_1$, $\text{LLR}(u_1)$, is calculated as follows:

$$
\text{LLR}(u_1) = \text{LLR}(x_1|y_1) - \text{LLR}(x_0|y_0).
$$

(17.112)

Combining (17.111) and (17.112), the reliability of $u_1$ is calculated as follows:

$$
\begin{aligned}
\text{LLR}(u_1) &= \text{LLR}(x_1|y_1) + (-1)^{\hat{u}_0}\text{LLR}(x_0|y_0) \\
&\triangleq g(\text{LLR}(x_0|y_0), \text{LLR}(x_1|y_1), \hat{u}_0).
\end{aligned}
$$

(17.113)

Based on $\text{LLR}(u_1)$ computed above, we make the following estimate of $u_1$ using the same decision rule of $u_0$ given by (17.110):

$$
\hat{u}_1 = \begin{cases} 0 \text{ if } u_1 \text{ is a frozen bit;} \\ 0 \text{ else if } \text{LLR}(u_1) > 0; \\ 1 \text{ otherwise.} \end{cases}
$$

(17.114)

Equation (17.113) shows that the LLR of $u_1$ depends on the estimate $\hat{u}_0$ of the first bit $u_0$ in $\mathbf{u}$. Thus, the decision of estimate of $u_1$ depends on the estimate $\hat{u}_0$ of $u_0$.

From (17.110) and (17.114), we obtain the two estimates $\hat{u}_0$ and $\hat{u}_1$ of the two bits $u_0$ and $u_1$ in the encoding vector $\mathbf{u}$. The SC decoding for $N = 2$ completes.

The above SC decoding process of a polar code of length $N = 2$ is displayed in Fig. 17.18 where the figures labeled with (a), (b), and (c) show the message-passing directions and the final decisions of the bits $u_0$ and $u_1$ at the end of the decoding process.

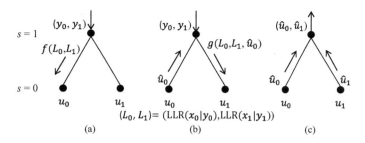

Figure 17.18 A tree structure of an SC decoder of size $N = 2$.

**Example 17.18** Assume a polar codeword of length $N = 2$ is transmitted over a BEC with erasure probability $\epsilon = 0.5$ (a very high erasure probability). The received vector is $\mathbf{y} = (y_0, y_1) = (-1, 0)$, where $-1$ represents an erased bit. The LLRs of the two transmitted bits $x_0$ and $x_1$ are calculated as follows (see Section 11.5.2.1):

$$\text{LLR}(x_0|y_0) = 0,$$
$$\text{LLR}(x_1|y_1) = +\infty.$$

In practical application, $\text{LLR}(x_1|y_1) = +\infty$ can be taken as any relatively large positive number, e.g., $\text{LLR}(x_1|y_1) = +10$. (If $\text{LLR}(x_i|y_i) = -\infty$, it can be taken as any negative number, e.g., $\text{LLR}(x_i|y_i) = -10$.)

Based on (17.109), the LLR of the first bit $u_0$ in the encoding vector $\mathbf{u} = (u_0, u_1)$ is computed as follows:

$$\text{LLR}(u_0) = f(\text{LLR}(x_0|y_0), \text{LLR}(x_1|y_1))$$
$$= \text{sign}(\text{LLR}(x_0|y_0))\text{sign}(\text{LLR}(x_1|y_1))\min\{|\text{LLR}(x_0|y_0)|, |\text{LLR}(x_1|y_1)|\}$$
$$= 0.$$

Based on (17.110), the estimate of $u_0$ is $\hat{u}_0 = 0$. Based on $\hat{u}_0$ and (17.113), the LLR of the second bit $u_1$ of $\mathbf{u}$ is

$$\text{LLR}(u_1) = g(\text{LLR}(x_0|y_0), \text{LLR}(x_1|y_1), \hat{u}_0)$$
$$= \text{LLR}(x_1|y_1) + (-1)^{\hat{u}_0}\text{LLR}(x_0|y_0) = +10.$$

Using the decision rule given by (17.114), the estimate of $u_1$ is $\hat{u}_1 = 0$. Hence, the decoded encoding vector is $\hat{\mathbf{u}} = (\hat{u}_0, \hat{u}_1) = (0, 0)$. ▲▲

### 17.8.2  SC Decoding of Polar Codes of Length $N = 4$

Now, we consider a polar code with length $N = 4$. The SC decoding process for this polar code is displayed in Fig. 17.19 which can be achieved through *two levels* of decoding. To simplify the notation for presentation, we denote the LLR, $\text{LLR}(x_i|y_i)$, of a transmitted bit $x_i$ as $\text{LLR}(y_i)$. The channel LLRs are $\text{LLR}_0^{(2)} = \text{LLR}(y_0)$, $\text{LLR}_1^{(2)} = \text{LLR}(y_1)$, $\text{LLR}_2^{(2)} = \text{LLR}(y_2)$, and $\text{LLR}_3^{(2)} = \text{LLR}(y_3)$.

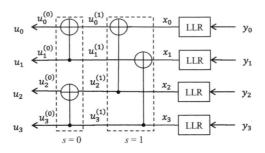

Figure 17.19 An SC decoder with size $N = 4$.

The SC decoder of the code consists of four SC kernel decoders (and four LLR computation units) as shown in Fig. 17.20 in which each SC kernel decoder is marked by boldface. The SC decoding starts at level-1 ($s = 1$). From Fig. 17.20(a), we see that to compute the LLR of bit $u_0^{(1)}$, we need to compute $\mathrm{LLR}_0^{(2)}$ and $\mathrm{LLR}_2^{(2)}$ first and then use (17.109) to compute $\mathrm{LLR}(u_0^{(1)}) = f(\mathrm{LLR}_0^{(2)}, \mathrm{LLR}_2^{(2)})$.

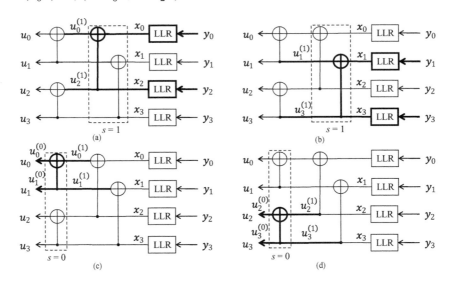

Figure 17.20 The SC decoding process of a polar code of length $N = 4$.

From Fig. 17.20(b), we see that we can compute the LLR of the bit $u_1^{(1)}$ based on (17.109) using $\mathrm{LLR}(u_1^{(1)}) = f(\mathrm{LLR}_1^{(2)}, \mathrm{LLR}_3^{(2)})$. Once $\mathrm{LLR}(u_0^{(1)})$ and $\mathrm{LLR}(u_1^{(1)})$ have been computed, decoding moves to level-0 ($s = 0$). Because the LLRs of $u_0^{(1)}$ and $u_1^{(1)}$ are known, then the estimates $\hat{u}_0$ and $\hat{u}_1$ of $u_0$ and $u_1$ can be obtained by using the SC kernel decoder shown in Fig. 17.20(c) as described in Section 17.8.1. Then, we have $\hat{u}_0^{(1)} = \hat{u}_0 + \hat{u}_1$ which is the estimate of $u_0^{(1)}$ and $\hat{u}_1^{(1)} = \hat{u}_1$. Feeding back $\hat{u}_0^{(1)}$ to the SC kernel decoder in Fig. 17.20(a) at level-1 ($s = 1$) and following the calculation given by (17.113), the LLR of $u_2^{(1)}$ can be computed as follows:

$$\text{LLR}_2^{(1)} = \text{LLR}(u_2^{(1)}) = g(\text{LLR}_0^{(2)}, \text{LLR}_2^{(2)}, \hat{u}_0^{(1)}).$$

Feeding back $\hat{u}_1^{(1)}$ to the SC kernel decoder in Fig. 17.20(b) at level-1 ($s = 1$) and following the calculation given by (17.113), the LLR of $u_3^{(1)}$ can be obtained:

$$\text{LLR}_3^{(1)} = \text{LLR}(u_3^{(1)}) = g(\text{LLR}_1^{(2)}, \text{LLR}_3^{(2)}, \hat{u}_1^{(1)}).$$

Once the LLRs of $u_2^{(1)}$ and $u_3^{(1)}$ have been computed, the decoding at level-0 starts again. Because the LLRs of $u_2^{(1)}$ and $u_3^{(1)}$ are known, then the estimates $\hat{u}_2$ and $\hat{u}_3$ of $u_2$ and $u_3$ can be obtained by the SC kernel decoder shown in Fig. 17.20(d) as described in Section 17.8.1. Thus, the four bits in the encoding vector $\mathbf{u} = (u_0, u_1, u_2, u_3)$ are decoded which results in an estimate $\hat{\mathbf{u}} = (\hat{u}_0, \hat{u}_1, \hat{u}_2, \hat{u}_3)$ of $\mathbf{u}$. The SC decoding is complete.

The above decoding of the four bits in an encoding vector $\mathbf{u}$ of a polar code is carried out in two levels with *forward-and-backward message passing* between level-1 and level-0 as displayed by the message-passing and decoding trees as shown in Fig. 17.21 labeled with (a), (b), (c), (d), (e), and (f).

**Example 17.19** Assume that the base channel $W$ is an AWGN channel and the LLR-vector associated with the received vector $\mathbf{y} = (y_0, y_1, y_2, y_3)$ is

$$(\text{LLR}(y_0), \text{LLR}(y_1), \text{LLR}(y_2), \text{LLR}(y_3)) = (\text{LLR}_0^{(2)}, \text{LLR}_1^{(2)}, \text{LLR}_2^{(2)}, \text{LLR}_3^{(2)})$$

$$= (+2.0, +1.0, +1.5, -0.5). \tag{17.115}$$

Suppose there is no frozen bit. Following the SC decoding steps described above, based on $\text{LLR}(y_0) = \text{LLR}_0^{(2)} = +2.0$, $\text{LLR}(y_2) = \text{LLR}_2^{(2)} = +1.5$, and (17.109), the LLR of the bit $u_0^{(1)}$ at level-1 can be computed using the SC kernel decoder shown in Fig. 17.20(a):

$$\text{LLR}(u_0^{(1)}) = f(\text{LLR}_0^{(2)}, \text{LLR}_2^{(2)}) = +1.5. \tag{17.116}$$

Similarly, the LLR of the bit $u_1^{(1)}$ at level-1 is (using the SC kernel decoder shown in Fig. 17.20(b))

$$\text{LLR}(u_1^{(1)}) = f(\text{LLR}_1^{(2)}, \text{LLR}_3^{(2)}) = -1.0. \tag{17.117}$$

Based on the two LLRs, $\text{LLR}(u_0^{(1)})$ and $\text{LLR}(u_1^{(1)})$, the LLRs of the bits $u_0$ and $u_1$ and their estimates at level-0 can be computed using the SC kernel decoder as shown in Fig. 17.20(c):

$$\text{LLR}(u_0) = f(\text{LLR}(u_0^{(1)}), \text{LLR}(u_1^{(1)})) = -1.0, \Longrightarrow \hat{u}_0^{(0)} = \hat{u}_0 = 1,$$

$$\text{LLR}(u_1) = g(\text{LLR}(u_0^{(1)}), \text{LLR}(u_1^{(1)}), \hat{u}_0) = -2.5, \Longrightarrow \hat{u}_1^{(0)} = \hat{u}_1 = 1. \tag{17.118}$$

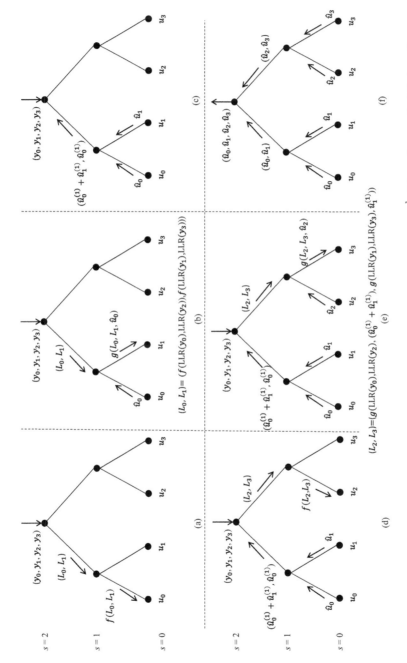

Figure 17.21 Message-passing and decision trees in decoding a polar code of length 4 with SC decoding.

Feeding back the estimate $\hat{u}_0^{(1)} = \hat{u}_0^{(0)} + \hat{u}_1^{(0)} = 0$ to the SC kernel decoder shown in Fig. 17.20(a), we compute the LLR of the bit $u_2^{(1)}$:

$$\text{LLR}(u_2^{(1)}) = g(\text{LLR}_0^{(2)}, \text{LLR}_2^{(2)}, \hat{u}_0^{(1)}) = +3.5. \tag{17.119}$$

Feeding back the estimate $\hat{u}_1^{(1)} = \hat{u}_1^{(0)} = 1$ to the SC kernel decoder shown in Fig. 17.20(b), we compute the LLR of the bit $u_3^{(1)}$:

$$\text{LLR}(u_3^{(1)}) = g(\text{LLR}_1^{(2)}, \text{LLR}_3^{(2)}, \hat{u}_1^{(1)}) = -1.5. \tag{17.120}$$

Based on the two LLRs, $\text{LLR}(u_2^{(1)})$ and $\text{LLR}(u_3^{(1)})$, the LLRs of the bits $u_2$ and $u_3$ and their estimates can be computed using the SC kernel decoder as shown in Fig. 17.20(d):

$$\text{LLR}(u_2) = f(\text{LLR}(u_2^{(1)}), \text{LLR}(u_3^{(1)})) = -1.5, \Longrightarrow \hat{u}_2^{(0)} = \hat{u}_2 = 1,$$
$$\text{LLR}(u_3) = g(\text{LLR}(u_2^{(1)}), \text{LLR}(u_3^{(1)}), \hat{u}_2) = -5.0, \Longrightarrow \hat{u}_3^{(0)} = \hat{u}_3 = 1. \tag{17.121}$$

It follows from the above bit-by-bit SC decoding process that the decoded estimate of the encoding vector $\mathbf{u} = (u_0, u_1, u_2, u_3)$ is $\hat{\mathbf{u}} = (\hat{u}_0, \hat{u}_1, \hat{u}_2, \hat{u}_3) = (1, 1, 1, 1)$. This completes the decoding.

Assume $u_0$ in $\mathbf{u}$ is a frozen bit, i.e., $u_0 = 0$ is known to the SC decoder. Here, we redecode the received vector $\mathbf{y}$ whose LLRs are given by (17.115). Following (17.116) and (17.117), we compute $\text{LLR}(u_0^{(1)}) = +1.5$ and $\text{LLR}(u_1^{(1)}) = -1.0$. With these two LLRs as the inputs to the SC kernel decoder shown in Fig. 17.20(c), we calculate the LLR of $u_0$:

$$\text{LLR}(u_0) = f(\text{LLR}(u_0^{(1)}), \text{LLR}(u_1^{(1)})) = -1.0.$$

Because $u_0$ is a frozen bit, we have $\hat{u}_0^{(0)} = \hat{u}_0 = 0$. Then,

$$\text{LLR}(u_1) = g(\text{LLR}(u_0^{(1)}), \text{LLR}(u_1^{(1)}), \hat{u}_0) = +0.5, \Longrightarrow \hat{u}_1^{(0)} = \hat{u}_1 = 0.$$

Feeding back the estimate $\hat{u}_0^{(1)} = \hat{u}_0^{(0)} + \hat{u}_1^{(0)} = 0$ to the SC kernel decoder shown in Fig. 17.20(a), we compute the LLR of the bit $u_2^{(1)}$:

$$\text{LLR}(u_2^{(1)}) = g(\text{LLR}_0^{(2)}, \text{LLR}_2^{(2)}, \hat{u}_0^{(1)}) = +3.5.$$

Feeding back the estimate $\hat{u}_1^{(1)} = \hat{u}_1^{(0)} = 0$ to the SC kernel decoder shown in Fig. 17.20(b), we compute the LLR of the bit $u_3^{(1)}$:

$$\text{LLR}(u_3^{(1)}) = g(\text{LLR}_1^{(2)}, \text{LLR}_3^{(2)}, \hat{u}_1^{(1)}) = +0.5.$$

Based on the two LLRs, $\text{LLR}(u_2^{(1)})$ and $\text{LLR}(u_3^{(1)})$, the LLRs of the bits $u_2$ and $u_3$ and their estimates can be computed using the SC kernel decoder as shown in Fig. 17.20(d):

$$\mathrm{LLR}(u_2) = f(\mathrm{LLR}(u_2^{(1)}), \mathrm{LLR}(u_3^{(1)})) = +0.5, \Longrightarrow \hat{u}_2^{(0)} = \hat{u}_2 = 0,$$

$$\mathrm{LLR}(u_3) = g(\mathrm{LLR}(u_2^{(1)}), \mathrm{LLR}(u_3^{(1)}), \hat{u}_2) = +4, \Longrightarrow \hat{u}_3^{(0)} = \hat{u}_3 = 0.$$

It follows from the above bit-by-bit SC decoding process, the decoded estimate of the encoding vector $\mathbf{u} = (u_0, u_1, u_2, u_3)$ is $\hat{\mathbf{u}} = (\hat{u}_0, \hat{u}_1, \hat{u}_2, \hat{u}_3) = (0,0,0,0)$. This completes the decoding. ▲▲

## 17.8.3    SC Decoding of Polar Codes of Length $N = 2^\ell$

In Sections 17.8.1 and 17.8.2, we used two polar codes of lengths 2 and 4, respectively, to illustrate the key steps of the SC decoding process and basic architecture of an SC decoder. In this subsection, we present the SC decoding process for a polar code of length $N = 2^\ell$ with $\ell \geq 1$.

In decoding, the message passing and estimated bit feedback process is the same as decoding a polar code of length 4 presented in Section 17.8.2. The LLRs of the bits in the encoding vector $\mathbf{u}$ can be computed via recursions.

Let $\mathrm{LLR}_0^{(\ell)}$, $\mathrm{LLR}_1^{(\ell)}, \ldots, \mathrm{LLR}_{N-1}^{(\ell)}$ be the channel LLRs where $\mathrm{LLR}_i^{(\ell)} = \mathrm{LLR}(y_i) = \mathrm{LLR}(x_i|y_i)$, $0 \leq i < N$. For $s = \ell - 1, \ell - 2, \ldots, 1, 0$, consider the calculations of the LLRs of the bits at level-$s$. For $t = 1, 3, 5, \ldots, 2^{\ell-s} - 1$,

$$\begin{aligned} \mathrm{LLR}_i^{(s)} &= f(\mathrm{LLR}_i^{(s+1)}, \mathrm{LLR}_{i+2^s}^{(s+1)}), & (t-1)2^s \leq i < t2^s, \\ \mathrm{LLR}_i^{(s)} &= g(\mathrm{LLR}_i^{(s+1)}, \mathrm{LLR}_{i+2^s}^{(s+1)}, \hat{u}_i^{(s+1)}), & t2^s \leq i < (t+1)2^s. \end{aligned} \tag{17.122}$$

Then, the estimates of the bits at level-$s$ are based on the estimates made in the $(s+1)$-level and the connection matrix of the $\ell$-fold Kronecker mapping presented in Section 17.2. The estimates are given as follows: for $t = 1, 3, 5, \ldots, 2^{\ell-s} - 1$,

$$\begin{aligned} \hat{u}_i^{(s)} &= \hat{u}_i^{(s+1)} + \hat{u}_{i+2^s}^{(s+1)}, & (t-1)2^s \leq i < t2^s, \\ \hat{u}_i^{(s)} &= \hat{u}_{i+2^s}^{(s+1)}, & t2^s \leq i < (t+1)2^s. \end{aligned} \tag{17.123}$$

The final estimates of the bits in the encoding vector $\mathbf{u} = (u_0, u_1, \ldots, u_{N-1})$ are made at level-0 and given as follows:

$$\hat{u}_i = \hat{u}_i^{(0)}, 0 \leq i < N.$$

The SC decoder of a polar code of length $N = 2^l$ consists of $\ell 2^{\ell-1}$ SC kernel decoders.

**Example 17.20** Consider the $(256, 128)$ polar code transmitted over a BEC $W$. The unrecoverable erasure probabilities, i.e., UEBR (unrecoverable erasure bit rate) and UEBLR (unrecoverable erasure block rate), of the code decoded with the SC decoding are included in Fig. 17.22. ▲▲

Computation complexity of SC decoding of a polar code of length $N = 2^\ell$ is in the order of $N \log(N)$, $\mathcal{O}(N \log(N))$. For large $N$ (or $\ell$), $\mathcal{O}(N \log(N))$ can be very large.

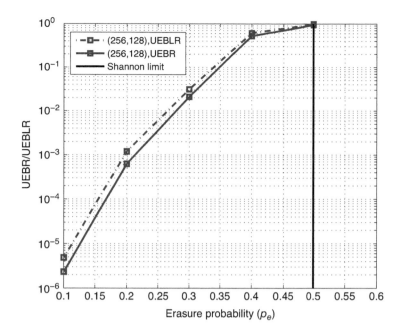

Figure 17.22 The UEBR and UEBLR of the $(256, 128)$ polar code over BEC decoded with the SC decoding in Example 17.20.

## 17.9 Remarks

Even though polar codes approach the capacity of a BI-DMC with SC decoding as their lengths grow to infinity through powers of 2, polar codes of practical lengths do not perform as well as some other well-known codes such as LDPC codes. The main reason for this weakness is that SC decoding is not strong enough for decoding polar codes of short to medium lengths. To improve the performance of polar codes of these lengths, several enhanced SC decoding algorithms have been devised [2–7], such as the successive cancellation list (SCL) [4, 5] and CRC-aided SCL decodings [6, 7]. These enhanced SC decodings improve the performance of polar codes at the expense of decoding complexity and/or code rates. Because the objective of this chapter is to introduce the basic concepts, structural properties, construction, and the base decoding algorithm (SC decoding) of polar codes, the enhanced SC decoding algorithms will not be included in this fundamental book. For these enhanced SC decoding, readers are referred to [2–7]. In fact, enhancing the base SC decoding to decode long polar codes to achieve high performance with practical decoding complexity and no (small) rate loss is a good research problem.

As pointed out in Section 17.4, polar codes and RM codes are two subclasses of Kronecker codes and they have the same universal generator matrices, Kronecker matrices. A polar code is constructed based on polarization of a BI-DMC in terms of its capacity. However, an RM code is constructed based on choosing the rows of its universal generator matrix with weights equal to and above a set minimum. From this point of view, an RM code may be regarded as a polarized code in terms of row weight polarization. In Chapter 8, we showed that multilevel majority-logic decoding is basically a successive cancellation decoding in one direction (no backward process). In decoding, the information bits corresponding to largest degree of vector products, $\mathbf{v}_{i_1}\mathbf{v}_{i_2}\ldots\mathbf{v}_{i_l}$, are decoded first (highest-level decoding) and the decoded information bits are then peeled off from the received vector which results in a modified received vector. In the next-highest level of decoding, the information bits corresponding to the second-largest degree of vector products are decoded and the decoded information bits are peeled off from the modified received vector in the previous level of decoding. The peeling-off process continues until the information bit corresponding to the single row of zero vector product degree (the zeroth level of decoding) is reached. Furthermore, we also showed that a polar code is a subcode of an RM code with some rows frozen, and an RM code has a much larger minimum distance than a polar code with same or even higher rate. A good question is whether this close relationship between these two types of Kronecker codes can be used to enhance the performance of each other.

## Problems

**17.1** Prove (17.10).

**17.2** Use the circuit in Fig. 17.3 to calculate the inverse 3-fold Kronecker mapping $\mathbf{u}$ of the codeword $\mathbf{v} = (1\ 0\ 1\ 0\ 0\ 1\ 0\ 1)$ in the $(8,4)$ polar code $C_{\mathrm{p}}(4,3)$ constructed in Example 17.8.

**17.3** For $0 \leq t < 2^\ell$, prove that the bit $u_t$ in $\mathbf{u} = (u_0, u_1, \ldots, u_{2^\ell-1})$ labeled by $t$ appears in $2^{w(t)}$ mapping sums formed by the $\ell$-fold Kronecker mapping $\mathbf{u} \cdot \mathbf{G}^{(\ell)}$ of $\mathbf{u}$, where $w(t)$ is the radix-2 weight of $t$ given by (17.46).

**17.4** Prove the two properties of the two bit-coordinate channels given by (17.80).

**17.5** Let $W$ be a BSC with transition probability $p$, $0 \leq p \leq 1$. Compute the channel capacities of the two bit-coordinate channels shown in Fig. 17.5(b).

**17.6** Consider a BEC $W$ with erasure probability $\epsilon = 0.2$. Set $\ell = 3$ and form a channel polarization by taking $N = 2^\ell = 8$ copies of the BEC $W$. Compute the channel capacities of all eight bit-coordinate channels.

**17.7** Use the channel polarization formed in Problem 17.6 to construct a $(8,3)$ polar code $C_{\mathrm{p}}(3,3)$. Compute the codeword for the message $\mathbf{c} = (1\ 1\ 0)$.

**17.8** RM codes select rows with the largest Hamming weights from the Kronecker matrix $\mathbf{G}^{(\ell)}$ (see details in Chapter 9). Construct an $(8,4)$ RM code and encode the message $\mathbf{c} = (1\ 1\ 0\ 0)$ with the $(8,4)$ RM code.

**17.9** Compare the minimum Hamming distances of the polar code in Problem 17.7 and the RM code in Problem 17.8.

**17.10** Consider a BEC $W$ with erasure probability $\epsilon = 0.5$. Set $\ell = 5$ and form a channel polarization by taking $N = 2^\ell = 32$ copies of the BEC $W$. Compute the channel capacities of all 32 bit-coordinate channels.

**17.11** Use the channel polarization formed in Problem 17.10 to construct a $(32, 16)$ polar code $C_{\mathrm{p}}(16, 5)$. Compute the codeword for the message $\mathbf{c} = (1\ 1\ 0\ 0\ 0\ 1\ 0\ 1\ 0\ 1\ 0\ 0\ 1\ 0\ 0\ 0)$.

**17.12** Redo Problems 17.10 and 17.11 by considering a BEC $W$ with $\epsilon = 0.3$.

**17.13** Suppose the bit $u_1$ in the encoding vector $\mathbf{u} = (u_0, u_1, u_2, u_3)$ is a frozen bit in Example 17.19. Use the SC decoder to redecode the received vector $\mathbf{y}$ in Example 17.19.

**17.14** Assume a $(4, 3)$ polar code of length $N = 4$ is transmitted over a BEC with erasure probability $\epsilon = 0.5$ with the first bit $u_0$ in the encoding vector $\mathbf{u} = (u_0, u_1, u_2, u_3)$ as the frozen bit. Suppose $\mathbf{y} = (0, 1, -1, 0)$ is the received vector (where $-1$ denotes an erased bit). Use the SC decoder to decode the received vector $\mathbf{y}$.

**17.15** Construct a $(128, 64)$ polar code which is transmitted over the BEC. Compute the UEBR and UEBLR performances of the constructed code over BEC decoded with SC decoding.

**17.16** Construct a $(512, 256)$ polar code which is transmitted over the BEC. Compute the UEBR and UEBLR performances of the constructed code over BEC decoded with SC decoding.

**17.17** Construct a $(128, 64)$ polar code which is transmitted over the AWGN channel. Compute the BER and BLER performances of the constructed code over the AWGN channel decoded with SC decoding.

# References

[1] E. Arikan, "Channel polarization: a method for constructing capacity-achieving codes for symmetric binary-input memoryless channels," *IEEE Trans. Inf. Theory*, **55**(7) (2009), 3051–3073.

[2] A. Alamdar-Yazdi and F. R. Kschischang, "A simplified successive-cancellation decoder for polar codes," *IEEE Commun. Lett.*, **15**(12) (2011), 1378–1380.

[3] K. Chen, K. Niu, and J. Lin, "Improved successive cancellation decoding of polar codes," *IEEE Trans. Commun.*, **61**(8) (2013), 3100–3107.

[4] I. Tal and A. Vardy, "List decoding of polar codes," *IEEE Trans. Inf. Theory*, **61**(5) (2015), 2213–2226.

[5] A. Balatsoukas-Stimming, M. B. Parizi, and A. Burg, "LLR-based successive cancellation list decoding of polar codes," *IEEE Trans. Signal Processing*, **63**(19) (2015), 5165–5179.

[6] K. Niu and K. Chen, "CRC-aided decoding of polar codes," *IEEE Commun. Lett.*, **16**(10) (2012), 1668–1671.

[7] B. Li, H. Shen, and D. Tse, "An adaptive successive cancellation list decoder for polar codes with cyclic redundancy check," *IEEE Commun. Lett.*, **16**(12) (2012), 2044–2047.

[8] T. M. Cover and J. A. Thomas, *Elements of Information Theory*, 2nd ed., Wiley Series in Telecommunications, Hoboken, NJ, Wiley, 2006.

[9] C. Shannon, "A mathematical theory of communication," *Bell System Tech. J.* (1948), pp. 379–423 (Part I), pp. 623–656 (Part II).

[10] E. Arikan, "Channel combining and splitting for cutoff rate improvement," *IEEE Trans. Inf. Theory*, **52**(2) (2006), 628–639.

[11] N. Presman, O. Shapira, and S. Litsyn, "Polar codes with mixed kernels," *Proc. IEEE Int. Symp. Inf. Theory (ISIT)*, St. Petersburg, July 31–August 5, 2011, pp. 6–10.

[12] N. Presman, O. Shapira, and S. Litsyn, "Binary polar code kernels from code decompositions," *Proc. IEEE Int. Symp. Inf. Theory (ISIT)*, St. Petersburg, July 31–August 5, 2011, pp. 179–183.

[13] V. Miloslavskaya and P. Trifonov, "Design of binary polar codes with arbitrary kernel," *Proc. IEEE Inf. Theory Workshop (ITW)*, Lausanne, Switzerland, September 3–7, 2012, pp. 119–123.

[14] S. E. Anderson and G. L. Matthews, "Exponents of polar codes using algebraic geometric code kernels," *Designs Codes Cryptogr.*, **73**(2) (2014), 699–717.

[15] M. -K. Lee and K. Yang, "The exponent of a polarizing matrix constructed from the Kronecker product," *Designs Codes Cryptogr.*, **70**(3) (2014), 313–322.

[16] H. Lin, S. Lin, and K. Abdel-Ghaffar, "Binary kernel matrices of maximum exponents of polar codes of dimensions up to sixteen," *2015 Information Theory and Applications Workshop (ITA)*, San Diego, CA, 2015, pp. 84–89.

[17] H. Lin, S. Lin, and K. A. S. Abdel-Ghaffar, "Linear and nonlinear binary kernels of polar codes of small dimensions with maximum exponents," *IEEE Trans. Inf. Theory*, **61**(10) (2015), 5253–5270.

[18] M. Ye and A. Barg, "Polar codes using dynamic kernels," *Proc. IEEE Int. Symp. Inf. Theory (ISIT)*, Hong Kong, June 14–19, 2015, pp. 231–235.

[19] M. Benammar, V. Bioglio, F. Gabry, and I. Land, "Multi-kernel polar codes: proof of polarization and error exponents," *IEEE Information Theory Workshop (ITW)*, Kaohsiung, November 6–10, 2017, pp. 101–105.

[20] L. Cheng, L. Zhang, and Q. Sun, "Exponents of hybrid multi-kernel polar codes," *IEEE 10th International Symposium on Turbo Codes & Iterative Information Processing (ISTC)*, Hong Kong, 3–7 December 2018, pp. 1–5.

[21] C. Y. Xia, C. Tsui, and Y. Z. Fan, "Construction of multi-kernel polar Codes with kernel substitution," *IEEE Wireless Commun. Lett.*, **9**(11) (2020), 1879–1883.

[22] E. Arikan and E. Telatar, "On the rate of channel polarization," *Proc. IEEE Int. Symp. Inf. Theory* (ISIT), Seoul, June 28–July 3, 2009, pp. 1493–1495.

[23] E. Arikan and E. Telatar, "On the rate of channel polarization," https://arxiv.org/abs/0807.3806, accessed 2020.

[24] S. B. Korada, E. Sasoglu, and R. Urbanke, "Polar codes: characterization of exponent bounds and constructions," *IEEE Trans. Inf. Theory*, **56**(12) (2010), 6253–6264.

[25] A. Eslami and H. Pishro-Nik, "A practical approach to polar codes," *Proc. IEEE Int. Symp. Inf. Theory (ISIT)*, St. Petersburg, July 31–August 5, 2011, pp. 16–20.

[26] R. Pedarsani, S. H. Hassani, I. Tal, and E. Telatar, "On the construction of polar codes," *Proc. IEEE Int. Symp. Inf. Theory (ISIT)*, St. Petersburg, July 31–August 5, 2011, pp. 11–15.

[27] Eren Sasoglu, "Polarization and polar codes," *Trends Commun. Info. Theory*, **8**(4) (2012), 259–381.

[28] P. Trifonov, "Efficient design and decoding of polar codes," *IEEE Trans. Commun.*, **60**(11) (2012), 3221–3227.

[29] I. Tal and A. Vardy, "How to construct polar codes," *IEEE Trans. Inf. Theory*, **59**(10) (2013), 6562–6582.

[30] H. Li and J. Yuan, "A practical construction method for polar codes in AWGN channels," *IEEE 2013 Tencon–Spring*, Sydney, NSW, April 17–19, 2013, pp. 223–226.

[31] E. Sasoglu, E. Telatar, and E. Arikan, "Polarization for arbitrary discrete memoryless channels," *IEEE Information Theory Workshop*, Taormina, October 11–16, 2009, pp. 144–148.

[32] R. Mori and T. Tanaka, "Channel polarization on $q$-ary discrete memoryless channels by arbitrary kernels," *Proc. IEEE Int. Symp. Inf. Theory (ISIT)*, Austin, TX, June 2010, pp. 894–898.

[33] A. G. Sahebi and S. S. Pradhan, "Multilevel polarization of polar codes over arbitrary discrete memoryless channels," *49th Annual Allerton Conference on Communication, Control, and Computing* (Allerton), Monticello, IL, September 28–30, 2011, pp. 1718–1725.

[34] W. Park and A. Barg, "Polar codes for $q$-ary channels, $q = 2^r$," *IEEE Trans. Inf. Theory*, **59**(2) (2013), 955–969.

# Appendix A

# Factorization of $X^n + 1$ over GF(2)

Table A.1 Factorization of $X^n + 1$ over GF(2) with $1 \leq n \leq 31$.

| $n$ | $X^n + 1 =$ |
|---|---|
| 1 | $(1+X)$ |
| 2 | $(1+X)^2$ |
| 3 | $(1+X)(1+X+X^2)$ |
| 4 | $(1+X)^4$ |
| 5 | $(1+X)(1+X+X^2+X^3+X^4)$ |
| 6 | $(1+X)^2(1+X+X^2)^2$ |
| 7 | $(1+X)(1+X+X^3)(1+X^2+X^3)$ |
| 8 | $(1+X)^8$ |
| 9 | $(1+X)(1+X+X^2)(1+X^3+X^6)$ |
| 10 | $(1+X)^2(1+X+X^2+X^3+X^4)^2$ |
| 11 | $(1+X)(1+X+X^2+X^3+\cdots+X^8+X^9+X^{10})$ |
| 12 | $(1+X)^4(1+X+X^2)^4$ |
| 13 | $(1+X)(1+X+X^2+X^3+\cdots+X^{10}+X^{11}+X^{12})$ |
| 14 | $(1+X)^2(1+X+X^3)^2(1+X^2+X^3)^2$ |
| 15 | $(1+X)(1+X+X^2)(1+X+X^4)(1+X^3+X^4)(1+X+X^2+X^3+X^4)$ |
| 16 | $(1+X)^{16}$ |
| 17 | $(1+X)(1+X+X^2+X^4+X^6+X^7+X^8)(1+X^3+X^4+X^5+X^8)$ |
| 18 | $(1+X)^2(1+X+X^2)^2(1+X^3+X^6)^2$ |
| 19 | $(1+X)(1+X+X^2+X^3+\cdots+X^{16}+X^{17}+X^{18})$ |
| 20 | $(1+X)^4(1+X+X^2+X^3+X^4)^4$ |
| 21 | $(1+X)(1+X+X^2)(1+X^2+X^3)(1+X+X^3)(1+X^2+X^4+X^5+X^6)(1+X+$ $X^2+X^4+X^6)$ |
| 22 | $(1+X)^2(1+X+X^2+X^3+\cdots+X^8+X^9+X^{10})^2$ |
| 23 | $(1+X)(1+X+X^5+X^6+X^7+X^9+X^{11})(1+X^2+X^4+X^5+X^6+X^{10}+X^{11})$ |
| 24 | $(1+X)^8(1+X+X^2)^8$ |
| 25 | $(1+X)(1+X+X^2+X^3+X^4)(1+X+X^5+X^{10}+X^{15}+X^{20})$ |
| 26 | $(1+X)^2(1+X+X^2+X^3+\cdots+X^{10}+X^{11}+X^{12})^2$ |
| 27 | $(1+X)(1+X+X^2)(1+X^3+X^6)(1+X^9+X^{18})$ |
| 28 | $(1+X)^4(1+X+X^3)^4(1+X^2+X^3)^4$ |
| 29 | $(1+X)(1+X+X^2+X^3+\cdots+X^{26}+X^{27}+X^{28})$ |
| 30 | $(1+X)^2(1+X+X^2)^2(1+X+X^4)^2(1+X^3+X^4)^2(1+X+X^2+X^3+X^4)^2$ |
| 31 | $(1+X)(1+X^2+X^5)(1+X+X^2+X^3+X^5)(1+X+X^2+X^4+X^5)(1+X+X^3+$ $X^4+X^5)(1+X^2+X^3+X^4+X^5)$ |

# Appendix B

# A $2 \times 2/3 \times 3$ SM-Constrained Masked Matrix Search Algorithm

---

**Algorithm B.1 The $3 \times 3$ SM-MMSA**

---

for $s = 0 : n - 4$ do
    for $k = 0 : m - 8$ do
        // Take a $4 \times 8$ submatrix from $\mathbf{B}_s(m, n)$
        $\mathbf{B}_{s,\mathrm{mask}}(4, 8) = \mathbf{B}_s(s : s + 3, k : k + 7)$
        // Masking
        for $ss = 1 : 4$ do
            for $kk = 1 : 8$ do
                if $\mathbf{Z}(ss, kk) == 0$ then
                    $\mathbf{B}_{s,\mathrm{mask}}(4, 8)(ss, kk) = 0$
                end if
            end for
        end for

        // Two methods to check the girth of the Tanner graph corresponding to the base matrix $\mathbf{B}_{s,mask}(4, 8)$:
        // Method-1
        (1) Check whether every $3 \times 3$ submatrix of $\mathbf{B}_{s,mask}(4, 8)$ satisfies the $3 \times 3$ SM-constraint
        // Method-2
        (2) Disperse $\mathbf{B}_{s,mask}(4, 8)$ into an array $\mathbf{H}_{s,qc,mask}(4, 8)$, and use a cycle counting algorithm to find the girth of the Tanner graph corresponding to $\mathbf{H}_{s,qc,mask}(4, 8)$

        if (1) holds or girth in (2) $\geq 8$ then
            Save $s$ and $k$
        end if

    end for
end for

---

# Appendix C

# Proof of Theorem 14.4

Consider the $d \times t$ RS matrix:

$$\mathbf{B}_{\mathrm{RS},n,\Lambda_t}(d,t) = [(\beta^i)^{l_j}]_{1 \leq i \leq d, 1 \leq j \leq t} \tag{C.1}$$

defined by (14.14) which consists of the $t$ columns of the RS matrix $\mathbf{B}_{\mathrm{RS},n}(d,m)$ over $\mathrm{GF}(q)$ labeled by the column-label set $\Lambda_t = \{l_1, l_2, \ldots, l_t\}$ with $0 \leq l_1 < l_2 < \cdots < l_t < m \leq n$, which is a submatrix of $\mathbf{B}_{\mathrm{RS},n}(d,m)$. Because $\mathbf{B}_{\mathrm{RS},n}(d,m)$ satisfies the $2 \times 2$ SM-constraint, $\mathbf{B}_{\mathrm{RS},n,\Lambda_t}(d,t)$ also satisfies the $2 \times 2$ SM-constraint.

The $n$-fold CPM-dispersion of $\mathbf{B}_{\mathrm{RS},n,\Lambda_t}(d,t)$ gives a $d \times t$ array $\mathbf{H}_{\mathrm{RS},n,\Lambda_t}(d,t)$ of CPMs of size $n \times n$ which is a $dn \times tn$ matrix over $\mathrm{GF}(2)$ with column and row weights $d$ and $t$, respectively. The null space of $\mathbf{H}_{\mathrm{RS},n,\Lambda_t}(d,t)$ gives a $(d,t)$-regular QC-RS-LDPC code $C_{\mathrm{RS},n,\Lambda_t}(d,t)$ of length $tn$. Because $\mathbf{B}_{\mathrm{RS},n,\Lambda_t}(d,t)$ satisfies the $2 \times 2$ SM-constraint, the array $\mathbf{H}_{\mathrm{RS},n,\Lambda_t}(d,t)$, as a matrix, satisfies the RC-constraint and its associated Tanner graph has girth at least 6.

The array $\mathbf{H}_{\mathrm{RS},n,\Lambda_t}(d,t)$ is composed of $t$ CPM-column-blocks, each consisting of $d$ CPMs of size $n \times n$. Each CPM-column-block consists of $n$ columns, each containing $d$ 1-entries residing in $d$ *separate* CPMs. For a column-label $l_j$, $1 \leq j \leq t$, in $\Lambda_t$, let $Col(l_j)$ denote the CPM-column-block of $\mathbf{H}_{\mathrm{RS},n,\Lambda_t}(d,t)$ which is the CPM-dispersion of the column in $\mathbf{B}_{\mathrm{RS},n,\Lambda_t}(d,t)$ labeled by $l_j$. Then,

$$Col(l_j) = \left[ \mathrm{CPM}_n(\beta^{l_j})^T, \mathrm{CPM}_n(\beta^{2l_j})^T, \ldots, \mathrm{CPM}_n(\beta^{dl_j})^T \right]^T \tag{C.2}$$

and

$$\mathbf{H}_{\mathrm{RS},n,\Lambda_t}(d,t) = [Col(l_1), Col(l_2), \ldots, Col(l_t)]. \tag{C.3}$$

Label the columns in a CPM-column-block from 0 to $n-1$. Consider the $k$th column $\mathbf{c}_k$ of $Col(l_j)$ with $0 \leq k < n$. For $1 \leq i \leq d$, the $i$th 1-entry in $\mathbf{c}_k$ is located at the position $(i-1)_n + (k - il_j)_n$ where $(x)_n$ denotes the least nonnegative integer equal to $x$ modulo $n$. The $d$-tuple

$$\mathbf{Loc}(l_j, k) = ((k-l_j)_n, n + (k-2l_j)_n, 2n + (k-3l_j)_n, \ldots, (d-1)n + (k-dl_j)_n) \tag{C.4}$$

specifies the locations of the $d$ 1-entries in the $k$th column $\mathbf{c}_k$ of $Col(l_j)$. We call this $d$-tuple, $\mathbf{Loc}(l_j, k)$, the 1-entry *location-vector* of the $k$th column $\mathbf{c}_k$ of $Col(l_j)$ and their components the *1-entry location numbers*.

Let $\mathbf{c}_{k'}$ be another column in $Col(l_j)$ with $k \neq k'$, i.e., $\mathbf{c}_k$ and $\mathbf{c}_{k'}$ are two different columns in $Col(l_j)$. It follows from the structure of a CPM-column-block of $\mathbf{H}_{\mathrm{RS},n,\Lambda_t}(d,t)$ that the location-vectors $\mathbf{Loc}(l_j, k)$ and $\mathbf{Loc}(l_j, k')$ of columns $\mathbf{c}_k$ and $\mathbf{c}_{k'}$ have no common component. If the location vectors of two columns $\mathbf{c}_k$ and $\mathbf{c}_{k'}$ in $\mathbf{H}_{\mathrm{RS},n,\Lambda_t}(d,t)$ have a *common* 1-entry location number $x$, these two columns must reside in two *different* CPM-column-blocks of $\mathbf{H}_{\mathrm{RS},n,\Lambda_t}(d,t)$ and there is *one and only one* row in a CPM-row-block that has a 1-entry at the location $x$ in column $\mathbf{c}_k$ and a 1-entry at the location $x$ in column $\mathbf{c}_{k'}$.

Let $\mathcal{G}_{\mathrm{RS},n,\Lambda_t}(d,t)$ be the Tanner graph associated with the parity-check matrix $\mathbf{H}_{\mathrm{RS},n,\Lambda_t}(d,t)$. Divide the $nt$ VNs of $\mathcal{G}_{\mathrm{RS},n,\Lambda_t}(d,t)$ into $t$ disjoint sets, $\Lambda(l_{i_1}), \Lambda(l_{i_2}), \ldots, \Lambda(l_{i_t})$, each set $\Lambda(l_{i_j})$ consisting of $n$ VNs which correspond to the $n$ columns of a CPM-column-block $Col(l_{i_j})$. Such a set of VNs is called a *CPM-VN-set*. It follows from the CPM structure of $\mathbf{H}_{\mathrm{RS},n,\Lambda_t}(d,t)$ that the $n$ VNs in each CPM-VN-set $\Lambda(l_{i_j})$ are connected to $n$ distinct CNs which correspond to $n$ rows of a CPM-row-block of $\mathbf{H}_{\mathrm{RS},n,\Lambda_t}(d,t)$. No two VNs in the same CPM-VN-set $\Lambda(l_{i_j})$ are connected to the same CN.

Because $\mathbf{H}_{\mathrm{RS},n,\Lambda_t}(d,t)$ satisfies the RC-constraint, $\mathcal{G}_{\mathrm{RS},n,\Lambda_t}(d,t)$ has girth at least 6. In the following, we first investigate the scenarios under which cycles of length 6, called 6-cycles, exist in $\mathcal{G}_{\mathrm{RS},n,\Lambda_t}(d,t)$ and then develop the conditions on the labels of $\Lambda_t$ under which $\mathcal{G}_{\mathrm{RS},n,\Lambda_t}(d,t)$ does not contain 6-cycles so that the girth of $\mathcal{G}_{\mathrm{RS},n,\Lambda_t}(d,t)$ is at least 8. Hence, the RS matrix $\mathbf{B}_{\mathrm{RS},n,\Lambda_t}(d,t)$ satisfies the $2 \times 2/3 \times 3$ SM-constraint.

Suppose $\mathcal{G}_{\mathrm{RS},n,\Lambda_t}(d,t)$ contains a 6-cycle, denoted by $C_6$. This cycle consists of three VNs, denoted by $v_1, v_2, v_3$, and three CNs, denoted by $c_1, c_2, c_3$. Then, $v_1, v_2, v_3$ correspond to three columns in three distinct CPM-column-blocks of $\mathbf{H}_{\mathrm{RS},n,\Lambda_t}(d,t)$, say, $Col(l_{i_1}), Col(l_{i_2}), Col(l_{i_3})$, respectively, where without loss of generality, we assume $1 \leq i_1 < i_2 < i_3 \leq t$, which implies that $l_{i_1} < l_{i_1} < l_{i_3}$. Suppose $v_1, v_2, v_3$ correspond to columns $k_1, k_2, k_3$ in the three CPM-column-blocks $Col(l_{i_1}), Col(l_{i_2}), Col(l_{i_3})$, respectively, where $0 \leq k_1, k_2, k_3 < n$. For $j = 1, 2, 3$, the 1-entry location-vector of column $k_j$ is

$$\mathbf{Loc}(l_{i_j}, k_j) = ((k_j - l_{i_j})_n, n + (k_j - 2l_{i_j})_n, 2n + (k_j - 3l_{i_j})_n, \ldots, (d-1)n + (k_j - dl_{i_j})_n).$$
$$(\text{C.5})$$

Suppose the three CNs $c_1, c_2, c_3$ on $C_6$ correspond to the three rows, numbered by $r_1, r_2, r_3$, in $\mathbf{H}_{\mathrm{RS},n,\Lambda_t}(d,t)$. Assume that $0 \leq r_1 < r_2 < r_3 < dn$. For $1 \leq i \leq 3$, let $r_i = x_i n + (r_i)_n$ where $x_i = \lfloor r_i/n \rfloor < d$ and $0 \leq (r_i)_n < n$. For $1 \leq i \leq 3$, $(r_i)_n$ and $x_i$ simply indicate that the row $r_i$ is the row labeled by $(r_i)_n$ in the $x_i$th CPM-row-block of $\mathbf{H}_{\mathrm{RS},n,\Lambda_t}(d,t)$. Because the rows $r_1, r_2, r_3$ are in different CPM-row-blocks and $r_1 < r_2 < r_3$, it follows that $x_1 < x_2 < x_3$.

The 6-cycle $C_6$ has six possible configurations, denoted by $C_{6,\tau}, 1 \leq \tau \leq 6$, which are in the following form:

$$C_{6,1} = (v_1 \to c_1 \to v_2 \to c_2 \to v_3 \to c_3 \to v_1),$$
$$C_{6,2} = (v_1 \to c_1 \to v_2 \to c_3 \to v_3 \to c_2 \to v_1),$$
$$C_{6,3} = (v_1 \to c_1 \to v_3 \to c_3 \to v_2 \to c_2 \to v_1),$$
$$C_{6,4} = (v_1 \to c_2 \to v_2 \to c_1 \to v_3 \to c_3 \to v_1),$$
$$C_{6,5} = (v_1 \to c_2 \to v_3 \to c_1 \to v_2 \to c_3 \to v_1),$$
$$C_{6,6} = (v_1 \to c_1 \to v_3 \to c_2 \to v_2 \to c_3 \to v_1),$$

(C.6)

where "$\to$" represents an edge that connects a VN (or CN) to a CN (or VN).

For $1 \leq \tau \leq 6$, each possible configuration $C_{6,\tau}$ of $C_6$ corresponds to a *unique sequence* $L_{6,\tau}$ of six different locations of 1-entries that reside in six separate CPMs in $\mathbf{H}_{\mathrm{RS},n,\Lambda_t}(d,t)$ with the *ending* location the same as the *starting* location. This sequence $L_{6,\tau}$ is called the *location-sequence* (LS) of 1-entries of the configuration $C_{6,\tau}$. The six 1-entry location-sequences of the six possible configurations of $C_6$ are given below:

$$L_{6,1} = (((k_1 - x_1 l_{i_1})_n, k_1), ((k_2 - x_1 l_{i_2})_n, k_2), ((k_2 - x_2 l_{i_2})_n, k_2), ((k_3 - x_2 l_{i_3})_n, k_3),$$
$$((k_3 - x_3 l_{i_3})_n, k_3), ((k_1 - x_3 l_{i_1})_n, k_1), ((k_1 - x_1 l_{i_1})_n, k_1)),$$

$$L_{6,2} = (((k_1 - x_1 l_{i_1})_n, k_1), ((k_2 - x_1 l_{i_2})_n, k_2), ((k_2 - x_3 l_{i_2})_n, k_2), ((k_3 - x_3 l_{i_3})_n, k_3),$$
$$((k_3 - x_2 l_{i_3})_n, k_3), ((k_1 - x_2 l_{i_1})_n, k_1), ((k_1 - x_1 l_{i_1})_n, k_1)),$$

$$L_{6,3} = (((k_1 - x_1 l_{i_1})_n, k_1), ((k_3 - x_1 l_{i_3})_n, k_3), ((k_3 - x_3 l_{i_3})_n, k_3), ((k_2 - x_3 l_{i_2})_n, k_2),$$
$$((k_2 - x_2 l_{i_2})_n, k_2), ((k_1 - x_2 l_{i_1})_n, k_1), ((k_1 - x_1 l_{i_1})_n, k_1)),$$

$$L_{6,4} = (((k_1 - x_2 l_{i_1})_n, k_1), ((k_2 - x_2 l_{i_2})_n, k_2), ((k_2 - x_1 l_{i_2})_n, k_2), ((k_3 - x_1 l_{i_3})_n, k_3),$$
$$((k_3 - x_3 l_{i_3})_n, k_3), ((k_1 - x_3 l_{i_1})_n, k_1), ((k_1 - x_2 l_{i_1})_n, k_1)),$$

$$L_{6,5} = (((k_1 - x_2 l_{i_1})_n, k_1), ((k_3 - x_2 l_{i_3})_n, k_3), ((k_3 - x_1 l_{i_3})_n, k_3), ((k_2 - x_1 l_{i_2})_n, k_2),$$
$$((k_2 - x_3 l_{i_2})_n, k_2), ((k_1 - x_3 l_{i_1})_n, k_1), ((k_1 - x_2 l_{i_1})_n, k_1)),$$

$$L_{6,6} = (((k_1 - x_1 l_{i_1})_n, k_1), ((k_3 - x_1 l_{i_3})_n, k_3), ((k_3 - x_2 l_{i_3})_n, k_3), ((k_2 - x_2 l_{i_2})_n, k_2),$$
$$((k_2 - x_3 l_{i_2})_n, k_2), ((k_1 - x_3 l_{i_1})_n, k_1), ((k_1 - x_1 l_{i_1})_n, k_1)).$$

In $\mathbf{H}_{\mathrm{RS},n,\Lambda_t}(d,t)$, the location patterns specified by the six location-sequences for the six configurations of $C_6$ are displayed in Fig. C.1.

Consider the 1-LS $L_{6,\tau}, 1 \leq \tau \leq 6$, for the possible configuration $C_{6,\tau}$ of $C_6$. This 1-LS consists of three pairs of 1-entry locations: $(((k_e - x_f l_{i_e})_n, k_e), ((k_{e'} - x_f l_{i_{e'}})_n, k_{e'}))$, $1 \leq e, e' \leq 3, e \neq e'$, and $1 \leq f \leq 3$. The two 1-entry locations, $((k_e - x_f l_{i_e})_n, k_e)$ and $((k_{e'} - x_f l_{i_{e'}})_n, k_{e'})$, in each pair are in the same row (the $(r_f)_n$th row) of the $x_f$th CPM-row-block of $\mathbf{H}_{\mathrm{RS},n,\Lambda_t}(d,t)$ and hence their first coordinates must be equal. This results in three equalities: $(k_e - x_f l_{i_e})_n = (k_{e'} - x_f l_{i_{e'}})_n$ for $f = 1, 2, 3$.

For example, for the configuration $C_{6,1}$, the three equalities are:

$$(k_1 - x_1 l_{i_1})_n = (k_2 - x_1 l_{i_2})_n,$$
$$(k_2 - x_2 l_{i_2})_n = (k_3 - x_2 l_{i_3})_n,$$
$$(k_3 - x_3 l_{i_3})_n = (k_1 - x_3 l_{i_1})_n.$$

(C.7)

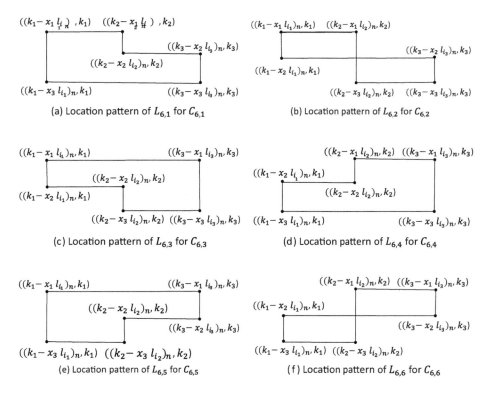

(a) Location pattern of $L_{6,1}$ for $C_{6,1}$

(b) Location pattern of $L_{6,2}$ for $C_{6,2}$

(c) Location pattern of $L_{6,3}$ for $C_{6,3}$

(d) Location pattern of $L_{6,4}$ for $C_{6,4}$

(e) Location pattern of $L_{6,5}$ for $C_{6,5}$

(f) Location pattern of $L_{6,6}$ for $C_{6,6}$

Figure C.1 Location patterns of six configurations of a cycle-6 $C_6$.

Adding these three equalities and with some algebraic manipulations, we obtain:

$$((x_2 - x_1)(l_{i_2} - l_{i_1}) + (x_3 - x_2)(l_{i_3} - l_{i_1}))_n = 0. \qquad (C.8)$$

Then, the equality given by (C.8) is the necessary and sufficient condition for the existence of $C_6$ in configuration $C_{6,1}$.

Similarly, for the other five possible configurations of $C_6$, we can derive the necessary and sufficient conditions for their existence, which are:

$$
\begin{aligned}
C_{6,2} &: ((x_2 - x_1)(l_{i_2} - l_{i_1}) + (x_3 - x_2)(l_{i_2} - l_{i_3}))_n = 0, \\
C_{6,3} &: ((x_2 - x_1)(l_{i_3} - l_{i_1}) + (x_3 - x_2)(l_{i_3} - l_{i_2}))_n = 0, \\
C_{6,4} &: ((x_2 - x_1)(l_{i_3} - l_{i_2}) + (x_3 - x_2)(l_{i_3} - l_{i_1}))_n = 0, \\
C_{6,5} &: ((x_2 - x_1)(l_{i_2} - l_{i_3}) + (x_3 - x_2)(l_{i_2} - l_{i_1}))_n = 0, \\
C_{6,6} &: ((x_2 - x_1)(l_{i_3} - l_{i_1}) + (x_3 - x_2)(l_{i_2} - l_{i_1}))_n = 0.
\end{aligned}
\qquad (C.9)
$$

Hence, the Tanner graph associated with $\mathbf{H}_{\mathrm{RS},n,\Lambda_t}(d,t)$ given by the $n$-fold CPM-dispersion of the RS matrix $\mathbf{B}_{\mathrm{RS},n,\Lambda_t}(d,t)$ in (C.1) contains cycles of length

6 if and only if there are six integers $i_1$, $i_2$, $i_3$, $x_1$, $x_2$, $x_3$, where $1 \leq i_1 < i_2 < i_3 \leq t$ and $0 \leq x_1 < x_2 < x_3 < d$, such that at least one of the six equalities given by (C.9) holds. From this we deduce the following theorem.

**Theorem C.1** *The Tanner graph associated with the parity-check matrix* $\mathbf{H}_{\mathrm{RS},n,\Lambda_t}(d,\ t)$ *has girth at least 8 if and only if for any six integers* $i_1$, $i_2$, $i_3$, $x_1$, $x_2$, $x_3$, *where* $1 \leq i_1 < i_2 < i_3 \leq t$ *and* $0 \leq x_1 < x_2 < x_3 < d$, *such that all the following six inequalities hold:*

$$((x_2 - x_1)(l_{i_2} - l_{i_1}) + (x_3 - x_2)(l_{i_3} - l_{i_1}))_n \neq 0,$$
$$((x_2 - x_1)(l_{i_2} - l_{i_1}) + (x_3 - x_2)(l_{i_2} - l_{i_3}))_n \neq 0,$$
$$((x_2 - x_1)(l_{i_3} - l_{i_1}) + (x_3 - x_2)(l_{i_3} - l_{i_2}))_n \neq 0,$$
$$((x_2 - x_1)(l_{i_3} - l_{i_2}) + (x_3 - x_2)(l_{i_3} - l_{i_1}))_n \neq 0,$$
$$((x_2 - x_1)(l_{i_2} - l_{i_3}) + (x_3 - x_2)(l_{i_2} - l_{i_1}))_n \neq 0,$$
$$((x_2 - x_1)(l_{i_3} - l_{i_1}) + (x_3 - x_2)(l_{i_2} - l_{i_1}))_n \neq 0.$$

$$(\text{C.10})$$

It follows from Theorem 14.3 and Theorem C.1 that we have Theorem 14.4.

# Appendix D

# The $2 \times 2$ CPM-Array Cycle Structure of the Tanner Graph of $C_{\mathrm{RS},n}(2, n)$

The Tanner graph $\mathcal{G}_{\mathrm{RS},n}(2, n)$ of the $(2, n)$-regular QC-RS-LDPC code $C_{\mathrm{RS},n}$ $(2, n)$ consists of $n^2$ VNs and $2n$ CNs. Each VN $v$ is connected to two CNs and each CN $c$ is connected to $n$ VNs. Each VN $v$ is connected to $n - 1$ other VNs through the same CN $c$ by paths of length 2 which are called *neighbor* VNs of $v$. Hence, each VN in $\mathcal{G}_{\mathrm{RS},n}(2, n)$ has $2(n-1)$ neighbor VNs. Because $\mathbf{H}_{\mathrm{RS},n}(2, n)$ satisfies the RC-constraint, $\mathcal{G}_{\mathrm{RS},n}(2, n)$ contains no cycle of length 4.

Let $\mathrm{CPM}(\ell)$ denote an $n \times n$ CPM in which the generator has its single 1 in position $\ell$, $0 \le \ell < n$. Then, $\mathrm{CPM}(\ell)$ has $n$ 1-entries in position $(i, (i + \ell)_n)$, $0 \le i < n$, where $(x)_n$ denotes the least nonnegative integer equaling to $x$ modulo $n$. For $0 \le i, j < n$, a 1-entry resides in position $(i, j)$ of $\mathrm{CPM}(\ell)$ if and only if $j - i \equiv \ell \pmod{n}$.

For $0 \le r < t < n$, consider the following $2 \times 2$ subarray of $\mathbf{H}_{\mathrm{RS},n}(2, n)$:

$$\mathbf{H}_{\mathrm{sub}}(r, t) = \begin{bmatrix} \mathrm{CPM}(\beta^r) & \mathrm{CPM}(\beta^t) \\ \mathrm{CPM}(\beta^{2r}) & \mathrm{CPM}(\beta^{2t}) \end{bmatrix}, \tag{D.1}$$

which consists of the $r$th and $t$th CPM-column-blocks of $\mathbf{H}_{\mathrm{RS},n}(2, n)$. Let $a = (r)_n$, $b = (2r)_n$, $c = (2t)_n$, and $d = (t)_n$. Then, $0 \le a, b, c, d < n$. It is clear that $a = r$ and $d = t$. The numbers, $a$, $b$, $c$, and $d$ specify the locations of the single 1-entries of the generators of the four CPMs in $\mathbf{H}_{\mathrm{sub}}(r, t)$. Replace the notations of the four CPMs in $\mathbf{H}_{\mathrm{sub}}(r, t)$ by $\mathrm{CPM}(a)$, $\mathrm{CPM}(b)$, $\mathrm{CPM}(c)$, and $\mathrm{CPM}(d)$, respectively. Then, the $2 \times 2$ array $\mathbf{H}_{\mathrm{sub}}(r, t)$ given by (D.1) is put in the following form:

$$\mathbf{H}_{\text{sub}}(r,t) = \begin{bmatrix} \text{CPM}(a) & \text{CPM}(d) \\ \text{CPM}(b) & \text{CPM}(c) \end{bmatrix}. \tag{D.2}$$

Let $\mathcal{G}_{\text{sub}}(r,t)$ be the Tanner graph associated with the $2 \times 2$ array $\mathbf{H}_{\text{sub}}(r,t)$ of CPMs. Then, the Tanner graph $\mathcal{G}_{\text{sub}}(r,t)$ has $2n$ VNs and $2n$ CNs. Consider a cycle in $\mathcal{G}_{\text{sub}}(r,t)$. Then, its length should be a *multiple* of 4, say $4\ell$, $2 \leq \ell \leq n$. Such a cycle corresponds to a *sequence* of 1-entries in $4\ell$ positions in $\mathbf{H}_{\text{sub}}(r,t)$ of the following form (*traced in counter-clock-wise column-row order*):

$$(i_0,j_0),(i_1,j_0),(i_1,j_1),(i_2,j_1),(i_2,j_2),(i_3,j_2),(i_3,j_3),(i_4,j_3),\ldots$$
$$\ldots,(i_{2\ell-2},j_{2\ell-2}),(i_{2\ell-1},j_{2\ell-2}),(i_{2\ell-1},j_{2\ell-1}),(i_0,j_{2\ell-1}),(i_0,j_0), \tag{D.3}$$

where (1) $i_0, i_1, \ldots, i_{2\ell-1}$ are distinct and $j_0, j_1, \ldots, j_{2\ell-1}$ are distinct; (2) $(i_0,j_0),(i_2,j_2),\ldots,(i_{2\ell-2},j_{2\ell-2})$ are the position of $\ell$ 1-entries in $\text{CPM}(a)$; (3) $(i_1,j_0),(i_3,j_2),\ldots,(i_{2\ell-1},j_{2\ell-2})$ are the position of $\ell$ 1-entries in $\text{CPM}(b)$; (4) $(i_1,j_1),(i_3,j_3),\ldots,(i_{2\ell-1},j_{2\ell-1})$ are the position of $\ell$ 1-entries in $\text{CPM}(c)$; and (5) $(i_0,j_1),(i_2,j_3),\ldots,(i_{2\ell-2},j_{2\ell-1})$ are the position of $\ell$ 1-entries in $\text{CPM}(d)$.

The positions of 1-entries in the four CPMs $\text{CPM}(a)$, $\text{CPM}(b)$, $\text{CPM}(c)$, and $\text{CPM}(d)$ must satisfy the following conditions, respectively: (1) $j_s - i_s \equiv a \pmod{n}$ for $s = 0, 2, \ldots, 2\ell - 2$; (2) $j_{s-1} - i_s \equiv b \pmod{n}$ for $s = 1, 3, \ldots, 2\ell - 1$; (3) $j_s - i_s \equiv c \pmod{n}$ for $s = 1, 3, \ldots, 2\ell - 1$; and (4) $j_{s+1} - i_s \equiv d \pmod{n}$ for $s = 0, 2, \ldots, 2\ell - 2$. Following the constraints on the positions of 1-entries in $\text{CPM}(a)$, $\text{CPM}(b)$, $\text{CPM}(c)$, and $\text{CPM}(d)$, we have

$$\sum_{s=0}^{2\ell-1} j_s - \sum_{s=0}^{2\ell-1} i_s \equiv a\ell + c\ell \quad \text{modulo } n \tag{D.4}$$

$$\sum_{s=0}^{2\ell-1} j_s - \sum_{s=0}^{2\ell-1} i_s \equiv b\ell + d\ell \quad \text{modulo } n. \tag{D.5}$$

It follows from (D.4) and (D.5) that we have $a\ell + c\ell \equiv b\ell + d\ell$ modulo $n$. Hence, the product $\ell(a - b + c - d)$ is divisible by $n$, i.e., $(\ell(a-b+c-d))_n \equiv 0$. From this, we readily see that the *shortest cycle* in $\mathbf{H}_{\text{sub}}(r,t)$ has length $4\lambda$, where $\lambda = n/\text{GCD}\{n, (a-b+c-d)_n\}$ and $\text{GCD}\{n, (a-b+c-d)_n\}$ denotes the greatest common divisor of $n$ and $(a-b+c-d)_n$. Because $n$ is a prime, $\text{GCD}\{n, (a-b+c-d)_n\} = 1$. Hence, $\lambda = n$ and $\mathcal{G}_{\text{sub}}(r,t)$ is composed of a *single cycle* of length $4n$ which includes the $2n$ VNs, $2n$ CNs, and $4n$ edges of $\mathcal{G}_{\text{sub}}(r,t)$.

# Appendix E

# Iterative Decoding Algorithm for Nonbinary LDPC Codes

## E.1   Introduction

This appendix will present the most popular decoding algorithm for LDPC codes over $\mathrm{GF}(q)$, where $q = 2^s$, $s > 1$. This algorithm relies on the fast Hadamard transform (FHT).

We will assume that each $q$-ary symbol is transmitted as a group of $s$ bits over the binary-input AWGN channel, although the results can be extended to nonbinary-input channels. We will let $n$ and $m$ denote the number of columns and rows, respectively, of the code's (low-density) parity-check matrix over $\mathrm{GF}(q)$, $\mathbf{H} = [h_{i,j}]_{0 \le i < m, 0 \le j < n}$ with $h_{i,j} \in \mathrm{GF}(q)$.

Recall that, for binary LDPC codes, the decoder iteratively updates estimates on $P_j[0]$ and $P_j[1]$, the probabilities that code bit $v_j$ is 0 and 1, respectively, for $j = 0, 1, \ldots, n - 1$. For mathematical convenience and reduced implementation complexity, the decoder estimates these probabilities indirectly by instead iteratively estimating the log-likelihood ratio (LLR) for $v_j$, given by

$$L_j = \log \left[ \frac{P_j[0 \mid \mathbf{y}]}{P_j[1 \mid \mathbf{y}]} \right], \tag{E.1}$$

where the dependency of these probabilities on the channel output vector $\mathbf{y}$ is made explicit in the a posteriori probabilities $P_j[0 \mid \mathbf{y}]$ and $P_j[1 \mid \mathbf{y}]$ for code bit $v_j$. Additional condition is implied, namely, the code constraints as encapsulated in the code's parity-check matrix $\mathbf{H}$.

For $q$-ary LDPC codes, a pair of probabilities or a single LLR for each code symbol $v_j$ is not appropriate because $v_j$ can take on one of $q$ values and so $q$ probabilities must be estimated. These $q$ probabilities form a probability mass function (pmf). Thus, the $q$-ary LDPC code decoder must propagate pmfs instead of LLRs along the edges of the code's Tanner graph representation, which consists of VNs and CNs connected by edges labeled with elements from GF($q$). Once the decoder discontinues its iterative computations, its decision for $v_j$ is that value of $v \in$ GF($q$) which maximizes the a posteriori probability $P_j[v \mid \mathbf{y}]$.

## E.2    Algorithm Derivation

As with the binary LDPC code decoding algorithm, the algorithm for nonbinary LDPC codes receives information from the channel and sends that information up to the CN processors. The CN processors then send their processed outputs to the VN processors which take these inputs along with the channel information to produce information that is sent back to the CN processors. The VN/CN processing iterations continue until a codeword is found or the preset maximum number of iterations is reached.

For the AWGN channel with two-sided power spectral density $N_0/2$, the appropriate bitwise information computed from the channel output $y$ for candidate binary input $b \in \{\pm 1\}$ is

$$\Pr(b \mid y) = p(y \mid b) \Pr(b)/p(y) = \frac{1}{1 + \exp(-4yb/N_0)}. \qquad \text{(E.2)}$$

These bit-wise probabilities are converted by a preprocessor to symbol-wise probabilities by computing appropriate products of the former. For example, for $s = 4$, suppose $\alpha^2 \in$ GF($2^4$) has the binary representation $[0\ 0\ 1\ 0]$, where $\alpha$ is a primitive element of GF($q$). Then $\Pr(\alpha^2 \mid \overline{y}) = \Pr(0 \mid y_3) \Pr(0 \mid y_2) \Pr(1 \mid y_1) \Pr(0 \mid y_0)$, where $\overline{y} = [y_3\ y_2\ y_1\ y_0]$ is the group of channel outputs corresponding to $s = 4$ consecutive binary inputs. Each VN processor receives $q$ such symbol-wise probabilities from the decoder preprocessor; that is, each VN processor receives a (conditional) pmf on the elements of GF($q$). The preprocessor pmf for VN $j$ will be denoted by $\boldsymbol{P}_j$, for $j = 0, 1, \ldots, n-1$. The $q$ elements of $\boldsymbol{P}_j$ will be denoted by $P_j(0), P_j(1), P_j(\alpha), P_j(\alpha^2), \ldots, P_j(\alpha^{q-2})$, where

$$P_j(\beta) = \Pr(v_j = \beta \mid \overline{y}) \qquad \text{(E.3)}$$

for all $\beta \in$ GF($q$), with $P_j(\beta)$ given by a product of $s$ probabilities $\Pr(b \mid y)$.

Given the channel outputs and computed pmfs, we now need to develop the VN and CN update equations involved in the iterative decoding. To aid the discussion, we define the *CN neighborhood* of VN $j$, for $0 \le j < n$, to be

$$M_j = \{i : h_{i,j} \ne 0\}. \qquad \text{(E.4)}$$

Similarly, we define the *VN neighborhood* of CN $i$, for $0 \le i < m$, to be

$$N_i = \{j : h_{i,j} \ne 0\}. \qquad \text{(E.5)}$$

## E.2.1 VN Update

As is customary, we regard each VN to be a repetition code so that all of the edges leaving a VN carry the same value in $\mathrm{GF}(q)$. (These values are later scaled by the edge labels on their way up to the CNs.) Moreover, we assume that the messages from the CNs to the VNs are independent. Because the optimum repetition code decoder adds LLRs, or multiplies probabilities, it follows that, for each $\beta \in \mathrm{GF}(q)$, the (extrinsic) message $m_{j \to i}(\beta)$ to be sent from VN $j$ to CN $i$ is given by

$$m_{j \to i}(\beta) = P_j(\beta) \prod_{k \in M_j \setminus i} m_{k \to j}(\beta), \qquad (\text{E.6})$$

where $m_{k \to j}(\beta)$ is the message sent from CN $k$ to VN $j$ about the probability that code symbol $v_j$ is equal to $\beta$. Note that the independence assumption on the incoming messages at VN $j$ actually holds only for the first $g/2$ iterations, where $g$ is the girth of the code's Tanner graph. Note also, because of the exclusion $M_j \setminus i$ in the product, that the message $m_{j \to i}(\beta)$ does not send CN $i$ information that it already has.

We emphasize that, whereas a single computation of the form (E.6) is performed in the binary case, $q$ such computations are performed for nonbinary LDPC codes, one for each value of $\beta \in \mathrm{GF}(q)$.

The computation of the CN-to-VN messages is much more involved, as is its development, which we will do in stages.

## E.2.2 CN Update: Complex Version

The CN update equation takes more time to develop, but the concepts are well known to students of coding. First, note that a CN is considered to be a nonbinary single-parity check (SPC) code. Thus, the $i$th CN, representing the $i$th row, $[h_{i,0} \ h_{i,1} \ \dots \ h_{i,n-1}]$, of the parity-check matrix $\mathbf{H}$, corresponds to the parity-check equation

$$\sum_{j \in N_i} v_j h_{i,j} = 0. \qquad (\text{E.7})$$

In this equation, $v_j \in \mathrm{GF}(q)$ is the code symbol corresponding to VN $j$, $h_{i,j}$s with $j = 0, 1, \dots, n-1$ are the nonzero elements of the $i$th row of $\mathbf{H}$, and addition and multiplication are over $\mathrm{GF}(q)$.

Letting $v'_j = v_j h_{i,j}$, the above sum can be rewritten as

$$\sum_{j \in N_i} v'_j = 0. \qquad (\text{E.8})$$

The above equation leads to the CN update equation given CN inputs from neighboring VNs. In particular, consider the $i$th CN, its neighborhood of VNs, $N_i$, and their pmfs. From these pmfs (messages), we compute the pmf to be sent from CN $i$ to VN $j$, for all $j$s in $N_i$. The probability density function (pdf) to be sent to VN $j$ is the pmf of

$$v'_j = \sum_{\ell \in N_i \setminus j} v'_\ell. \tag{E.9}$$

When a discrete-valued random variable, $v'_j$, is the sum of independent discrete-valued random variables, $v'_\ell$, the pmf of $v'_j$ is given by the *cyclic* convolution of the pmfs of the $v'_\ell$s.

It is important to highlight here an important detail regarding the convolution of pmfs of elements of GF($q$). Consider two random variables, $X$ and $Y$, taking values in GF($q$). Let their corresponding pmfs be denoted by $\mathbf{p}_X$ and $\mathbf{p}_Y$. We are interested in determining the pmf of their sum, $Z = X + Y$, noting that the sum is performed via the modulo-2 addition of the binary $s$-tuple representations of $X$ and $Y$. Thus, it is convenient to represent the elements of GF($q$) by their corresponding binary $s$-tuples or their decimal equivalents. Then, we have that

$$\mathbf{p}_Z(z) = \sum_{x,y:\, x+y=z} \mathbf{p}_X(x)\mathbf{p}_Y(y) = \sum_x \mathbf{p}_X(x)\mathbf{p}_Y(z-x), \tag{E.10}$$

or, using the cyclic convolution operator shorthand,

$$\mathbf{p}_Z = \mathbf{p}_X \circledast \mathbf{p}_X. \tag{E.11}$$

This generalizes to a sum of more than two random variables. To apply this result to (E.9), we let $\mathbf{m}_{i \to j} = [m_{i \to j}(0)\ m_{i \to j}(1)\ \cdots\ m_{i \to j}(\alpha^{q-2})]$ be the conditional pmf of $v'_j$ and $\mathbf{m}_{\ell \to i} = [m_{\ell \to i}(0)\ m_{\ell \to i}(1)\ \cdots\ m_{\ell \to i}(\alpha^{q-2})]$ be the conditional pmf of $v'_\ell$. It follows from the developments given above that the conditional pdf of $v'_j$ is

$$\mathbf{m}_{i \to j} = \underset{\ell \in N_i \setminus j}{\circledast}\ \mathbf{m}_{\ell \to i}. \tag{E.12}$$

### E.2.3 CN Update: Fast Hadamard Transform Version

The convolution of multiple pmfs to obtain the pmf of $v'_j$ is obviously quite complex. The computational complexity may be vastly reduced by using fast Hadamard transform (FHT) techniques to perform the convolutions. To introduce the FHT-based algorithm, some preliminary information on the Hadamard transform itself is necessary. We will present the fast Hadamard transform algorithm later.

The Hadamard transform of a length-$q$, real-valued (row) vector $\mathbf{p} = [\, p_0\ p_1\ \cdots\ p_{q-1}\,]$, considered to be a pmf here, is given by

$$\mathbf{P} = \mathcal{H}(\mathbf{p}) = \mathbf{p}\mathbf{H}_q, \tag{E.13}$$

where $\mathbf{H}_q$ is recursively defined as

$$\mathbf{H}_q = \frac{1}{\sqrt{2}} \begin{bmatrix} \mathbf{H}_{q/2} & \mathbf{H}_{q/2} \\ \mathbf{H}_{q/2} & -\mathbf{H}_{q/2} \end{bmatrix}, \tag{E.14}$$

with the initial condition

$$\mathbf{H}_2 = \frac{1}{\sqrt{2}} \begin{bmatrix} 1 & 1 \\ 1 & -1 \end{bmatrix}. \tag{E.15}$$

Letting $\boldsymbol{\phi}_x$ represent the $x$th row of $\mathbf{H}_q$, the Hadamard transform of $\mathbf{p}$ may be written as

$$\mathbf{P} = \sum_{x=0}^{q-1} p_x \boldsymbol{\phi}_x. \tag{E.16}$$

Because $\mathbf{H}_q$ is recursively defined, the following properties of the Hadamard transform can be proved by induction.

(1) $\mathbf{H}_q^T = \mathbf{H}_q$ (where the superscript $T$ denotes matrix transpose).

(2) $\mathbf{H}_q \mathbf{H}_q^T = \mathbf{H}_q^T \mathbf{H}_q = \mathbf{I}_q$ ($\mathbf{I}_q$ is the $q \times q$ identity matrix). In particular, the inverse Hadamard transform is performed by $\mathbf{H}_q^T = \mathbf{H}_q^{-1}$, that is, $\mathcal{H}^{-1}(\mathbf{P}) = \mathbf{P}\mathbf{H}_q^T = \mathbf{p}\mathbf{H}_q\mathbf{H}_q^T = \mathbf{p}$. A corollary to this property is that the rows of $\mathbf{H}_q$ are orthonormal $\boldsymbol{\phi}_x \boldsymbol{\phi}_y^T = \boldsymbol{\delta}_{x-y}$, where $\boldsymbol{\delta}_n$ is the Kronecker delta function.[1]

(3) The component-wise multiplication, denoted by $\odot$, of the $x$th and $y$th rows of $\mathbf{H}_q$ is $\boldsymbol{\phi}_x \odot \boldsymbol{\phi}_y = \frac{1}{\sqrt{2}} \boldsymbol{\phi}_{x \oplus y}$, where $x \oplus y$ is component-wise mod-2 addition of the binary representations of $x$ and $y$. For example, $13 \oplus 10 = 7$.

Observe that the component-wise mod-2 addition mentioned in Property (3) is identical to addition of two elements in $\mathrm{GF}(2^s)$, although, following convention, we have used $\oplus$ for the former and $+$ for the latter. Property (3) is critical to the proof of the following cyclic convolution theorem for Hadamard transforms that is central to the low-complexity computation of the cyclic convolution of pmfs of random variables that take values in $\mathrm{GF}(q)$. It will be seen in the proof that the indices to the rows of $\mathbf{H}_q$ are put into a one-to-one correspondence with the elements ($s$-tuples) of the additive group in $\mathrm{GF}(q)$.

**Theorem E.1** *Consider two independent random variables, $X$ and $Y$, defined on $\mathrm{GF}(q)$ with pmfs $\mathbf{p}_X$ and $\mathbf{p}_Y$. Then, the pmf $\mathbf{p}_Z$ of their sum in $\mathrm{GF}(q)$, $Z = X + Y$, can be computed from $\sqrt{2}\, \mathcal{H}^{-1}[\mathcal{H}(\mathbf{p}_X) \odot \mathcal{H}(\mathbf{p}_Y)]$.*

*Proof* For simplicity of notation, instead of $p_X(x)$ and $p_Y(y)$, $p(x)$ and $p(y)$ will be written for individual components of $\mathbf{p}_X$ and $\mathbf{p}_Y$, which are usually written

---

[1] The Kronecker delta function $\boldsymbol{\delta}_n$ is defined as: $\boldsymbol{\delta}_n = 1$ if $n = 0$; otherwise, $\boldsymbol{\delta}_n = 0$.

as $p_X(x)$ and $p_Y(y)$. All summations below are from 0 to $q-1$, or their binary $s$-tuple equivalents.

$$
\begin{aligned}
\sqrt{2}\,\mathcal{H}^{-1}\left[\mathcal{H}(\mathbf{p}_X)\odot\mathcal{H}(\mathbf{p}_Y)\right] &= \sqrt{2}\,\mathcal{H}^{-1}\left[\sum_x p(x)\boldsymbol{\phi}_x\odot\sum_y p(y)\boldsymbol{\phi}_y\right] \\
&= \sqrt{2}\,\mathcal{H}^{-1}\left[\sum_x\sum_y p(x)\boldsymbol{\phi}_x\odot p(y)\boldsymbol{\phi}_y\right] \\
&= \sqrt{2}\,\mathcal{H}^{-1}\left[\sum_x\sum_y p(x)p(y)\frac{1}{\sqrt{2}}\boldsymbol{\phi}_{x\oplus y}\right] \\
&= \sum_x\sum_y\sum_z p(x)p(y)\boldsymbol{\phi}_{x\oplus y}\boldsymbol{\phi}_z^T \qquad\text{(E.17)} \\
&= \sum_x\sum_y\sum_z p(x)p(y)\boldsymbol{\delta}_{(x\oplus y)-z} \\
&= \sum_{x,y:x\oplus y=z} p(x)p(y) \\
&= \sum_x p(x)p(z-x) \\
&= \mathbf{p}_X\circledast\mathbf{p}_Y\,.
\end{aligned}
$$

The first two lines are obvious, the third follows from Property (3), the fourth from Properties (1) and (2), the fifth from the corollary to Property (2), and the last three are obvious. ▲▲

**Example E.1** Although the developments so far have focused on nonbinary LDPC codes, the results are applicable to binary LDPC codes as well, that is, the case $q=2$. Substitution of $\mathbf{p}_X = [p_X(0)\ \ p_X(1)]$ and $\mathbf{p}_Y = [p_Y(0)\ \ p_Y(1)]$ into $\sqrt{2}\,\mathcal{H}^{-1}[\mathcal{H}(\mathbf{p}_X)\odot\mathcal{H}(\mathbf{p}_Y)]$ yields $\mathbf{p}_X\circledast\mathbf{p}_Y = [\,p_X(0)p_Y(0)+p_X(1)p_Y(1)\ \ p_X(0)p_Y(1)+p_X(1)p_Y(0)\,]$. Clearly, the first component of this vector is equal to $\Pr[X+Y=0]$ and the second component is equal to $\Pr[X+Y=1]$. ▲▲

**The Fast Hadamard Transform**

Efficient computation of a Hadamard transform, called the fast Hadamard transform (FHT) (which is also called fast Fourier transform (FFT)), is possible because of the recursive nature of Hadamard matrices. Further, from Properties (1) and (2), the Hadamard transform and the inverse Hadamard transform are identical, so we need only discuss one fast algorithm.

Presentation of the FHT is easiest by way of example. Let us consider the 16-ary case so that $q=16$, $\mathbf{p}=[p_0\ p_1\ \cdots\ p_{15}]$, and $\mathbf{P}=\mathcal{H}(\mathbf{p})=\mathbf{p}\mathbf{H}_{16}$. Observe that

$$
\mathbf{H}_{16} = \frac{1}{\sqrt{2}}\begin{bmatrix}\mathbf{H}_8 & \mathbf{H}_8 \\ \mathbf{H}_8 & -\mathbf{H}_8\end{bmatrix} = \frac{1}{\sqrt{2}}\begin{bmatrix}\mathbf{H}_8 & \mathbf{0} \\ \mathbf{0} & \mathbf{H}_8\end{bmatrix}\begin{bmatrix}\mathbf{I}_8 & \mathbf{I}_8 \\ \mathbf{I}_8 & -\mathbf{I}_8\end{bmatrix}. \qquad\text{(E.18)}
$$

Figure E.1 illustrates the implementation of this equation, where $\mathbf{p}_0^7$ is the subvector $[p_0 \ p_1 \ \cdots \ p_7]$ of $\mathbf{p}$ and similarly for the other subvectors. In the figure, quantities along edges that diverge are identical and quantities along edges that merge are added.

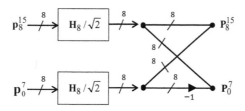

Figure E.1 Diagram of implementation of $\mathbf{P} = \mathbf{pH}_{16}$.

This implementation does not look advantageous until we observe that $\mathbf{H}_8$ decomposes as

$$\mathbf{H}_8 = \frac{1}{\sqrt{2}} \begin{bmatrix} \mathbf{H}_4 & \mathbf{0} \\ \mathbf{0} & \mathbf{H}_4 \end{bmatrix} \begin{bmatrix} \mathbf{I}_4 & \mathbf{I}_4 \\ \mathbf{I}_4 & -\mathbf{I}_4 \end{bmatrix}, \tag{E.19}$$

which is implemented as in Fig. E.2 for the input $\mathbf{p}_0^7$. The implementation for the input $\mathbf{p}_8^{15}$ is essentially identical. The setup in Fig. E.2 and the analogous one for $\mathbf{p}_8^{15}$ replace the two $\mathbf{H}_8$ blocks in Fig. E.1. Next, each of the $\mathbf{H}_4$ blocks in the implementation of $\mathbf{H}_8$ are replaced by the implementation of

$$\mathbf{H}_4 = \frac{1}{\sqrt{2}} \begin{bmatrix} \mathbf{H}_2 & \mathbf{0} \\ \mathbf{0} & \mathbf{H}_2 \end{bmatrix} \begin{bmatrix} \mathbf{I}_2 & \mathbf{I}_2 \\ \mathbf{I}_2 & -\mathbf{I}_2 \end{bmatrix}, \tag{E.20}$$

which results in eight $\mathbf{H}_2$ blocks.

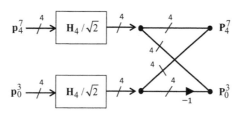

Figure E.2 Diagram of implementation of $\mathbf{P} = \mathbf{p}_0^7\mathbf{H}_8$.

The culmination of all of these steps is shown in Fig. E.3 which is a concatenation of $\log_2(q) = 4$ levels of "butterflies," with eight butterflies in the first level, four in the second, two in the third, and one in the last. The scale factor of $1/4$ at the output accounts for the four factors of $1/\sqrt{2}$ that would occur in each of the four stages, but combined near the output.

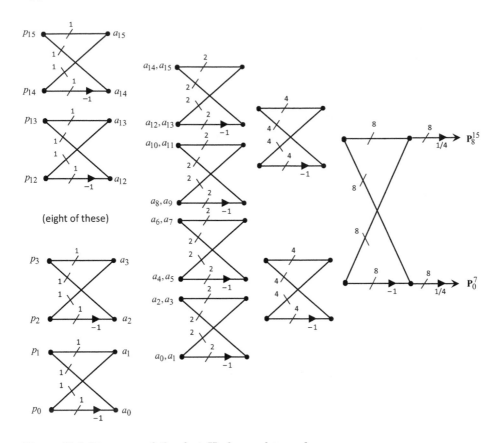

Figure E.3 Diagram of the fast Hadamard transform.

## E.3 The Nonbinary LDPC Decoding Algorithm

Below we summarize the nonbinary LDPC code decoding algorithm mentioned in Chapter 11, called the fast Fourier transform $q$-ary sum-product algorithm (FFT-QSPA). The messages $m_{j\to i}(\beta)$ computed in step (2) correspond to $v_j$, whereas the convolutions in (E.12) correspond to $v_j'$, so the translations in steps (3) and (5) are necessary. Each of the steps is to be applied to all $\beta \in \mathrm{GF}(q)$ and all $i$ and $j$ for which $h_{i,j} \neq 0$, with $0 \leq i < m$ and $0 \leq j < n$.

(1) Initialize the probabilities $P_j(\beta)$ according to (E.2) and (E.3) and the discussion in between those equations. Initialize also the CN messages $m_{i\to j}(\beta) = 1$.

(2) Compute the VN messages $m_{j\to i}(\beta)$ according to (E.6).

(3) Translate the messages $m_{j\to i}(\beta)$ to the messages $m_{j\to i}(\beta')$, according to $m_{j\to i}(\beta') = m_{j\to i}(\beta)$, where $\beta' = h_{i,j}\beta$.

(4) Using the Hadamard transform convolution theorem and the FHT algorithm, compute the CN messages:

$$\mathbf{m}_{i \to j} = \mathcal{H}^{-1} \left\{ \prod_{\ell \in N_i \setminus j} \mathcal{H}\{\mathbf{m}_{\ell \to i}\} \right\} . \qquad (\text{E.21})$$

(5) Translate the messages $m_{i \to j}(\beta')$ to the messages $m_{i \to j}(\beta)$, according to $m_{i \to j}(\beta) = m_{i \to j}(\beta')$, where $\beta' = h_{i,j}\beta$.

(6) Make symbol decisions using

$$\hat{v}_j = \text{argmax}_\beta \, m_j(\beta) , \qquad (\text{E.22})$$

where

$$m_j(\beta) = P_j(\beta) \prod_{k \in M_j} m_{k \to j}(\beta) . \qquad (\text{E.23})$$

If a codeword (if $\hat{\mathbf{v}}\mathbf{H}^T = \mathbf{0}$) is found, or if the maximum number of iterations is reached, then stop; otherwise, go to step (2).

# Index

$2 \times 2$ SM-constraint, 521, 563
$2 \times 2/3 \times 3$ SM-constrained matrix, 563
$2 \times 2/3 \times 3$ SM-constraint, 563
0-replacement, 383
1-replacement, 383

absorbing set, 443, 465, 695
  elementary absorbing set, 443
ACE (approximate cycle extrinsic), 674
additive white Gaussian noise, 6
affine geometries, 266
algebraic codes, 562
APP (a posteriori probability), 430
ARQ (automatic repeat request), 15
  ACK (positive acknowledgment), 15, 353
  continuous ARQ, 15
  GBN-ARQ, 353
  go-back-$N$ ARQ, 15
  NAK (negative acknowledgment), 15, 353
  SR-ARQ, 15
  ST-ARQ, 353
  stop-and-wait ARQ, 15
  SW-ARQ, 353
array-decomposition, 498
AWGN, *see* additive white Gaussian noise

balanced incomplete block designs, 518

BCH (Bose–Chaudhuri–Hocquenghem) codes, 185, 685
  BCH bound, 194
  Berlekamp–Massey iterative algorithm, 201
  binary BCH codes, 185
  binary parity-check matrix, 191
  code construction field, 187
  consecutive powers, 186
  designed error-correcting capability, 185
  designed minimum distance, 185
  elementary-symmetric functions, 199
  error-location numbers, 198
  error-location polynomial, 199
  even-weight BCH code, 195
  Newton's identities, 200
  nonbinary BCH codes, 220
  nonbinary parity-check matrix, 191
  nonprimitive·BCH codes, 213
  primitive BCH codes, 187
  root field, 187
  syndrome, 197
Berlekamp–Massey iterative algorithm, 201
  error-location numbers, 198
  error-location polynomial, 199
  error-value evaluator, 231
  error values, 226

generalized Newton's
identities, 228
key-equation, 235
Newton's identities, 200
BF (bit-flipping) decoding, 416,
422
threshold, 422
BIBDs, *see* balanced incomplete
block designs
base blocks, 541, 550
BIBD-PaG, 540
blocks, 538
class-1 Bose-BIBD, 541
class-1 Bose-BIBD-PaG, 542
class-2 Bose-BIBD-PaG, 549
class-2 Bose-BIBD, 550
objects, 538
sets, 538
binary burst erasure channel, 14
binary erasure channel, 14
binary image, 381
binary symmetric channel, 7
binary symmetric erasure channel,
14
bipartite graph, 82
adjacency matrix, 82
bit-error rate, 12
block-cyclic structure, 642
block-error rate, 11
BM-HDDA, 703
bound
BCH bound, 194, 222, 236
Hamming bound, 115
Reiger bound, 369
Singleton bound, 108, 686
sphere packing bound, 444,
528
UB (union bound), 700
UB-MLD, 700
bounded distance decoding, 118,
263
burst-error-correcting capability,
368
bursts
$m$-bit byte, 380
burst-error-correcting
efficiency, 370

erasure-bursts, 367
error-bursts, 367
optimal-discrepancy, 369
phased-error-burst, 380
solid burst, 368, 608
Burton codes, 381

cascading, 684
Cayley table, 27
CC (cascaded composite), 689
ceiling function, 114
channel
AWGN channel, 6
BBEC, *see also* binary burst
erasure channel, 383, 608
BEC, *see also* binary erasure
channel, 382, 435, 746
BI-AWGN channel, 6, 435
BI-DMC, 7, 747
BSC, *see also* binary
symmetric channel, 435,
746
BSEC, *see also* binary
symmetric erasure
channel, 383
compound channels, 14
DMC, *see* discrete memoryless
channel
NBEC, *see* nonbinary erasure
channel
noiseless channel, 746
perfect channel, 746
useless channel, 746
channel capacity, 12
channel information inclusion,
475
channel polarization, 744, 747
1-fold vector channel, 747
2-fold channel polarization,
753
3-fold Kronecker polarization,
755
base channel, 748
bit-coordinate channels, 747,
748, 766
channel polarization theorem,
760

channel polarization (cont.)
    combine-and-split process, 749
    free channels, 760
    frozen channels, 757, 760
    polarization transformation, 747
    virtual channel, 748
Chien search algorithm, 229
Chien search procedure, 211
circulant-decomposition, 496
    column-decomposition, 497
    column-descendants, 497
    column-generator, 496
    row-decomposition, 497
    row-descendants, 497
    row-splitting factor, 498
    row-splitting weights, 498
CN-redundancy, 442
CNPU (CN processing unit), 474
coding gain, 16
collective decoding, 684
collective encoding, 684
combinatorial designs, 538
complex-number field, 33
component-wise mod-2 addition, 797
component-wise multiplication, 797
composite codeword, 688
compound code, 684
concatenation, 341
    global errors, 341, 343
    inner code, 342
    inner encoding, 342
    interleaver, 346
    local errors, 341, 343
    outer code, 342
    outer encoding, 342
    parallel concatenated coding system, 341
    serial concatenated coding system, 341
    turbo concatenated code, 347
    turbo decoding, 347
    type-1 serial concatenated code, 342

    type-2 serial concatenated code, 345
connection number, 442
constituent codewords, 688
constituent vectors, 695
convergence threshold, 441
copy-and-grouping operation, 630
coset
    coset decomposition, 31
    coset leader, 30
    left coset, 30
    partition, 31
    right coset, 30
CPM, *see also* matrix, circulant permutation matrix (CPM), 482, 519, 563
CPM-column-blocks, 482
CPM-decomposition, 499
CPM-dispersion, 489
    base matrix, 490, 520
    CPM-dispersion factor, 519
    $ln$-fold CPM-dispersion, 519
    $n$-fold CPM-dispersion, 519
    $(q-1)$-fold CPM-dispersion, 489
CPM-RID scheme, 504
    CPM-RID-MSA, 506
    CPM-RMSA, 506
    decoding matrix, 506
    decoding subiteration, 506
    section-wise cyclic-shift, 504
CPM-row-blocks, 482
CRC-aided SCL decoding, 779
cyclic codes, 125
    BCH codes, 185
    binary cyclic codes, 125
    code polynomial, 126
    cyclic-shift, 126
    error polynomial, 142
    error-trapping decoding, 150
    generator codeword, 133
    generator matrix, 134
    generator polynomial, 130
    Meggitt decoding process, 146
    minimum-degree code polynomial, 127

nonbinary cyclic codes, 176
parity-check matrix, 135
parity-check polynomial, 135
QC (quasi-cyclic) codes, 161
QR code, 159
received polynomial, 142
reciprocal polynomial, 135
syndrome polynomial, 142
cyclic convolution, 796
cyclic grouping, 472
cyclic redundancy check code, 158
cyclic RM code, 324
cyclic shift-register-adder-
   accumulator circuit,
   165

DE (density evolution), 441
decoding
   MAP, *see* maximum a
      posteriori
   MLD, *see* maximum likelihood
      decoding
decoding convergence, 442
decoding regions, 109
decoding threshold, 441, 633
decomposition factor, 642
decomposition set, 642
demodulator, 7
direct product code, 335
   codeword array, 335
   received array, 338
discrete memoryless channel, 7

EG, *see* Euclidean geometries
entropy, 745
   channel capacity, 746
   conditional entropy, 745
   joint entropy, 745
   mutual information, 745
   probability mass function, 745
   uncertainty, 745
erasures, 367
   CPM-phased erasure-burst,
      608
   erasure-bursts, 367
   random erasures, 367

error detection
   decoding error, 101
   error pattern, 100
   error-detecting capability, 109
   guaranteed detectable error
      patterns, 109
   Hamming weight, 103
   minimum weight, 103
   syndrome, 101
   undetectable error pattern,
      101
   undetected error rate, 104
   undetected error rate upper
      bound, 105
   weight distribution, 103
   weight spectrum, 103
error rate
   BER, *see also* bit-error rate,
      16
   BLER, *see also* block-error
      rate, 16, 243
   FER, *see* frame error rate
   SER, *see also* symbol-error
      rate, 243
   UEBLR, *see* unresolved
      erasure block rate
   UEBR, *see* unresolved erasure
      bit rate
error-control strategies
   ARQ, 331
   ARQ (automatic repeat
      request), 15
   FEC, 15, 331
   HARQ (hybrid-ARQ), 16, 332
error-correcting codes, 4
   BCH codes, 185
   block codes, 4
      code dimension, 4, 89
      code length, 4, 89
      code rate, 4, 89
      codeword, 89
      message, 89
      parity-check bits, 90
      redundant bits, 90

error-correcting codes (cont.)
 burst-error-correcting codes,
  14, 367
 convolutional codes, 4, 5
  memory order, 5
 cyclic codes, 125
 error-correcting capability, 115
 finite-geometry codes, 258
 LDPC codes, 406
 linear block codes, 89
 perfect code, 115
 QR codes, *see* quadratic
  residue
 RM (Reed–Muller) codes, 303
 SPC code, *see* single
  parity-check
error-detection codes, 105
 CRC code, *see* cyclic
  redundancy check code
error-floor region, 439
error-location numbers, 198
error-location polynomial, 199
error-trapping decoder, 373
error-trapping decoding, 150
Euclidean algorithm, 220
Euclidean geometries, 266
 $\mu$-flat (hyperplane), 268
 affine geometries, 266
 cyclic class of lines, 276
 intersecting bundle, 268
 one-dimensional subspace, 267
 origin, 267
 parallel bundle, 267
 subgeometry, 275
 two-dimensional Euclidean
  geometries, 278
Euclidean geometry codes
 cyclic-EG-LDPC code, 465
 masked QC-EG-LDPC codes,
  495
 QC-EG-LDPC code, 480
 two-dimensional cyclic EG
  code, 279
Euler $\phi$ function, 38
Euler totient function, 38

EXIT (extrinsic-information-
  transfer chart),
  442
experimental designs, 538
extended code, 119

FG, *see also* finite geometries, 464
FG codes, *see also* finite-geometry
  codes, 464
field, 31
 cardinality, 33
 characteristic, 34
 extension field, 33, 221
 finite field, 33
 Galois field, 35
 ground field, 50
 infinite field, 33
 order, 33
 prime field, 33, 35
 subfield, 33, 222
finite field, 35, 562
 binary field, 35
 conjugate, 51, 222
 conjugate roots, 49
 conjugate set, 51
 conjugates with respective to
  GF($q$), 226
 decimal form, 48
 exponent, 51
 nonbinary field, 67
 polynomial basis, 46
 polynomial form, 48
 polynomial representation, 46
 power form, 48
 primitive element, 38
 vector form, 48
finite geometries, 258
 Euclidean geometries, 258
 incidence matrix, 259
 incidence vector, 259
 intersecting bundle, 259
 line-point adjacency matrix,
  259
 lines, 259
 orthogonal structure, 259
 partial geometries, 518, 522

points, 259
projective geometries, 258
finite group, 26
additive group, 27
cyclic group, 29
generator, 29
modulo-$m$ addition, 26
modulo-$m$ multiplication, 26
multiplicative group, 28
finite-geometry codes, 258, 261
cyclic-EG-LDPC code, 465
cyclic-PG-LDPC code, 469
FG-LDPC codes, 464
QC-EG-LDPC code, 480
QC-PaG-LDPC codes, 524
QC-PG-LDPC code, 487
Fire codes, 370, 374
designed burst-error-correcting
capability, 374
flooding schedule, 431
floor function, 114
forward error correction, 15
frame error rate, 700

GA (Gaussian approximation), 441
Galois Fourier transform, 74
GCD (greatest common divisor),
243
generalized Fire codes, 382
generalized Newton's identities,
228
GFT, *see also* Galois Fourier
transform, 692
GFT domain, 685
GFT-ICC codes, 693
GFT-ICC-BCH-LDPC codes, 706
GFT-ICC-Golay-LDPC code, 709
GFT-ICC-QR-LDPC codes, 709
GFT-ICC-RS-LDPC codes, 702
GFT-ISDD scheme, 699
GFT-ISDD/MSA, 700
Golay code, 115, 159
graph, 78
acyclic graph, 81
adjacency matrix, 81

bipartite graph, 82
branches, 78
CNs (check nodes), 411
cycle, 80
degree distribution, 78
edge set, 78
edges, 78
end-node, 78
finite graph, 78
girth, 80
incidence matrix, 81
infinite graph, 78
irregular graph, 79
node set, 78
nodes, 78
parallel edge, 78
path, 79
regular graph, 79
simple graph, 79
subgraph, 79
Tanner graphs, 411
tree, 81
vertices, 78
VNs (variable nodes), 411
graph-theoretic LDPC codes, 628
connection matrix, 631
copy-and-permute, 629, 630
edge-growth, 629, 670
lifting degree, 630
protomatrix, 631
PTG-based LDPC code, 629
PTG-LDPC codes, 629
QC-PTG-LDPC codes, 632
group, 25
abelian group, 26
cardinality, 26
commutative group, 26
finite group, 26
identity element, 26
infinite group, 26
inverse, 26
order, 26
subgroup, 29

Hadamard equivalent, 686
Hadamard permutation, 684, 686

Hadamard power, 686, 689
Hamming codes, 95, 125, 154
    cyclic Hamming codes, 188
    distance-4 Hamming code,
        158, 195
Hamming distance, 11, 105
hard-decision decoding, 9
    BF decoding, 416
    multilevel majority-logic
        decoding, 303
    OSMLD, 263
    SPL (successive peeling), 386
HARQ (hybrid-ARQ), 16
HDD, 709
HDD algorithm, 709
hyperbolic tangent, 433

ICC (interleaved cascaded
        composite), 690
IDBP, *see also* iterative decoding
        based on belief-
        propagation, 694
information sharing, 685
interleaving, 332, 684
    base code, 332
    constituent codewords, 332
    interleaved code, 332
    interleaved matrix, 333
    interleaved vector, 332
    interleaving degree, 332
inverse GFT, 692
inverse hyperbolic tangent, 433
ISDD (iterative soft-decision
        decoding), 699
iterative decoding based on
        belief-propagation, 417

joint-decoding, 685
joint-decoding gain, 685, 700

key-equation, 235, 247
Kronecker codes, 721, 736
    Kronecker–RM codes, 736
Kronecker mapping
    $\ell$-fold Kronecker mapping, 724
    $\ell$-fold connection matrix, 731

2-fold connection matrix, 727
3-fold Kronecker mapping, 728
decomposition process, 732
inverse Kronecker mapping,
    725
splitting process, 732
Kronecker matrices, 721
    $\ell$-fold Kronecker matrix, 723
    2-fold Kronecker matrix, 722
    3-fold Kronecker matrix, 722
    4-fold Kronecker matrix, 723
    kernel, 722
Kronecker product, 348
Kronecker vector space, 736

Latin square, 575
LCM (least common multiple), 186
LDPC codes, *see* low-density
        parity-check codes
    cyclic LDPC codes, 408
    doubly QC-LDPC codes, 644
    finite-field-based QC-LDPC
        codes, 564
    Gallager LDPC codes, 415
    GFT-ICC-LDPC code, 694
    graph-theoretic LDPC code,
        628
    irregular LDPC codes, 408
    low-density matrix, 407
    MacKay codes, 416
    PEG-LDPC codes, 528, 629
    QC-LDPC codes, 408
    QC-RS-LDPC codes, 590
    RC-constraint, 408, 520
    regular LDPC codes, 407
    RS-based QC-LDPC codes,
        590
    sparse matrix, 407
LFSR, *see also* linear feedback
        shift-register, 139
linear block codes, 89
    **P**-submatrix, 96
    $k$-dimensional subspace, 90
    $q$-ary linear block code, 120
    binary linear block code, 90,
        120

cyclic codes, 125
dual code, 93
dual space, 93
elementary row operations, 98
extended code, 119
generator matrix, 92
minimum distance, 107
minimum weight, 107
mother code, 119
nonbinary linear block code, 120
null space, 93
parity submatrix, 97
parity-check equations, 97
parity-check matrix, 93
parity-check sums, 97
row space, 92
shortened code, 119
standard array, 110
systematic form, 96
linear feedback shift-register, 125, 139
LLR, *see* log-likelihood ratio
local cycle structure, 608, 609
local girth, 672
log-likelihood ratio, 428
low-density parity-check codes, 406
LR (likelihood ratio), 430

MacWilliams identity, 104
masking, 493, 580
building blocks, 580
Hadamard matrix product, 493
masked base matrix, 495
masking matrix, 495
mother base matrix, 586
SMC-MMSA, *see*
SM-constrained $4 \times 8$
masked matrix search
algorithm
type-1 masking matrix, 581
type-2 masking matrix, 582
type-3 masking matrix, 585
matrix
binary matrix, 67

circulant, 70
circulant permutation matrix (CPM), 70
column weight, 68
dual space, 72
full column rank matrix, 68
full row rank matrix, 68
generator, 70
Hadamard matrix product, 493
identity matrix, 70
inverse, 70
irregular matrix, 68
nonbinary matrix, 73
nonsingular matrix, 68
null space, 72
permutation matrix, 70
rank, 68
redundant columns, 68
redundant rows, 68
regular matrix, 68
row space, 71
row weight, 68
singular matrix, 68
square matrix, 67
transpose, 68
Vandermonde determinant, 78
Vandermonde matrix, 73
ZM (zero matrix), 164
matrix-dispersion, 632
maximum a posteriori, 10
maximum-distance separable, 108
maximum likelihood decoding, 11
MDS, *see* maximum-distance separable
MDS code, 108, 236
Meggitt decoding process, 146
minimum-distance decoding, 12, 109, 113
modulator, 6
mother code, 119
MSA (min-sum algorithm), 437
attenuating factor, 437
scaling factor, 437
multilevel majority-logic decoding, 303
multiplicative inverse, 28

nonbinary BCH codes, 220
  $q$-ary BCH codes, 222
nonbinary erasure channel, 395

one-step majority-logic decoding,
    258
orthogonal syndrome check-sums,
    262
OSML decoding, *see also* one-step
    majority-logic decoding,
    419
  OSML decision function, 419
OSMLD, *see also* one-step
    majority-logic decoding,
    466
OSMLD decodable, 263

PaG (partial geometry) codes
  QC-BIBD-PaG-LDPC codes,
    543
  QC-PaG-LDPC codes, 524
  QC-RS-PaG-LDPC codes,
    527
PaGs (partial geometries), 522
  BIBD-PaGs, 538
  class-1 Bose-BIBD-PaG, 542
  class-2 Bose-BIBD-PaG, 549
  connection number, 523
  net, 523
  parameters, 522
  RS-based partial geometry,
    527
pair-wise RC-constraint, 497, 498,
    645
parity-check sums, 261, 410
parity-check vector, 391
PEG (progressive edge-growth),
    628, 670, 673
  ACE, 674
  ACE-PEG, 675
  PEG-LDPC code, 673
perfect code, 115
polar codes, 721
  encoding vector, 738
  free bit locations, 739

free bits, 739
frozen bit locations, 739
frozen bits, 739
polarized encoding, 738
polarized partition, 739
polynomial
  irreducible polynomial, 41
  minimal polynomial, 52
  primitive polynomial, 42
  reciprocal polynomial, 135
primitive Fire codes, 374
product-form, 576
projective geometries, 288
projective geometry codes, 292
  cyclic-PG-LDPC code, 469
  QC-PG-LDPC code, 487
PSD (power spectral density), 417
PTG (protograph), 628, 629
puncturing, 119

QC (quasi-cyclic) codes, 161
  CSRAA, *see* cyclic shift-
    register-adder-
    accumulator circuit
  QC structure, 161
  section size, 162
  section-wise cyclic-shift, 162
  semi-systematic circulant
    form, 169
  shifting constraint, 161
  systematic circulant form, 164
QR (quadratic residue) codes, 685
quadratic residue, 159

rate compatible, 685
rational-number field, 33
RC-constrained matrix, 520
RC-constraint, 408, 520
real-number field, 33
RID (revolving iterative decoding),
    474
  CII, *see* channel information
    inclusion
  RID-MSA, 474
  RMSA, 474

RM (Reed–Muller) codes, 303
  dual code, 309
  inclusion chain, 309
  multilevel encoding, 313
  mutually orthogonal structure,
      309
  self-dual, 310
  successive cancellation
      information retrieval, 317
row-redundancy, 442, 507
RS (Reed–Solomon) codes, 220,
      685
  nonprimitive RS code, 238
  primitive RS code, 236
  probability of an undetected
      error, 252
  weight distribution, 251
RS matrices, 591
  RS base matrices, 593
  type-1 $2 \times 2/3 \times 3$
      SM-constrained RS
      matrix, 599
  type-1 RS matrices, 592
  type-2 $2 \times 2/3 \times 3$
      SM-constrained RS
      matrix, 600
  type-2 RS matrices, 592
  type-3 $2 \times 2/3 \times 3$
      SM-constrained RS
      matrix, 600
  type-3 RS matrices, 592

SC (successive cancellation)
      decoding, 769
  SC kernel decoder, 771
SCIR, *see also* successive
      cancellation information
      retrieval, 388
SCL (successive cancellation list)
      decoding, 779
section-wise cyclic structure, 644
self-dual, 66, 310
Shannon limit, 18
shortened code, 119

simplex code, 157
single parity-check, 120
SM-constrained $4 \times 8$ masked
      matrix search algorithm,
      588
SNR (signal-to-noise ratio), 16
soft-decision decoding, 9
  IDBP, 417
  MSA, 437
  SPA, 429
  weighted BF decoding, 416
  weighted OSML decoding, 417
SPA (sum-product algorithm), 429,
      433
SPB, *see* bound, sphere packing
      bound
SPL (successive peeling), 386, 610
SPL/SCIR-decoding, 388
standard array, 110
  coset, 110
  coset leader, 110
  optimal standard array, 113
  syndrome decoding, 117
  table look-up decoding, 117
stopping set, 447
  maximum stopping set, 448
  minimum stopping set, 449
  SFS (stopping-free set), 448
successive cancellation information
      retrieval, 317
sum-form, 565
symbol-error rate, 16

Tanner graphs, 411
  degree distribution
      polynomials, 412
trapping set, 439, 442, 695
  elementary trapping set, 442
  small trapping set, 442
two-sided power spectral density,
      794

uncoded BPSK system, 16
universal coding scheme, 684
unresolved erasure bit rate, 447
unresolved erasure block rate, 447

Vandermonde matrices, 73, 692
vector space, 60
    basis, 64
    component-wise addition, 62
    component-wise product, 304
    dull space, 65
    inner product, 65
    linear combination, 64
    linearly dependent, 64
    linearly independent, 64
    null space, 65
    orthogonal, 65
    scalar multiplication, 61, 62
    subspace, 64
    vector, 61
    vector addition, 61
    vector product, 304

    zero vector, 61
VNPU (VN processing unit), 474

waterfall region, 439
weight distribution, 103
weight enumerator, 104, 156
weight spectrum, 103
weighted BF decoding, 416, 426
weighted OSML decoding, 417, 426

zero divisor, 83
zero-span, 135, 391
    end-around zero-span, 391
    longest zero-span, 391
    maximal zero-span, 391
    null zero-span, 391
    zero-covering span, 392
ZM, *see* matrix, ZM (zero matrix)